COMPREHENSIVE SERIES IN PHOTOSCIENCES

Series Editors

Donat-P. Häder

Professor of Botany

and

Giulio Jori

Professor of Chemistry

European Society for Photobiology

COMPREHENSIVE SERIES IN PHOTOSCIENCES

Series Editors: Donat-P. Häder and Giulio Jori

Titles in this Series

Volume 1 Photomovement
Edited by: D.-P. Häder and M. Lebert

Volume 2 Photodynamic Therapy and Fluorescence Diagnosis in Dermatology
Edited by: P.-G. Calzavara-Pinton, R.-M. Szeimies and B. Ortel

Volume 3 Sun Protection in Man
Edited by: P.U. Giacomoni

COMPREHENSIVE SERIES IN PHOTOSCIENCES – VOLUME 1

PHOTOMOVEMENT

Editors
Donat-P. Häder, Dr. rer. nat.
Professor of Botany

and

Michael Lebert, Dr. rer. nat.
Department of Botany and Pharmaceutical Biology
Friedrich-Alexander University
Erlangen, Germany

2001
ELSEVIER
AMSTERDAM – LONDON – NEW YORK – OXFORD – PARIS – SHANNON – TOKYO

ELSEVIER SCIENCE B.V.
Sara Burgerhartstraat 25
P.O. Box 211, 1000 AE Amsterdam, The Netherlands

First edition 2001

Library of Congress Cataloging in Publication Data
A catalog record from the Library of Congress has been applied for.

ISBN: 0-444-50706-X
ISSN: 1568-461X

Printed and bound by Antony Rowe Ltd, Eastbourne
Transferred to digital printing 2005

SERIES EDITORS' PREFACE

"Its not the substance, it's the dose which makes something poisonous!" When Paracelsius, a German physician of the 14th century made this statement he probably did not think about light as one of the most obvious environmental stress factors. But his statement applies as well to light. While we need light for example for vitamin D production too much light might cause skin cancer. The dose makes the difference. These diverse findings of light effects attracted the attention of scientists for centuries. The photosciences represent a dynamic multidisciplinary field which includes such diverse subjects as behavioral responses of single cells, cures for certain types of cancer and protective potential of tanning lotions. It includes photobiology and photochemistry, photomedicine as well as the technology for light production, filtering and measurement. Light is a common theme in all these areas. In the last decades a more molecular centered approach changed both, the depth and the quality of the theoretical as well as the experimental foundation of photosciences.

An example for the relationship between global environment and the biosphere is the recent discovery of ozone depletion and the resulting increase in high energy ultraviolet radiation. The hazardous effects of high energy ultraviolet radiation on all living systems is now well established. This discovery of the result of ozone depletion put photosciences in the center of public interest with the result that in an unparalleled effort scientists and politicians worked closely together to come to international agreements to stop the pollution of the atmosphere.

The changed recreational behavior and the correlation with several diseases in which sunlight or artificial light sources play a major role in the causation of clinical conditions (e.g. porphyrias, polymorphic photodermatoses, Xeroderma pigmentosum and skin cancers) have been well documented. As a result in some countries (i.e. Australia) public services inform people about the potential risk of extended periods of sun exposure for every day. The problems are often aggravated by the phototoxic or photoallergic reactions produced by a variety of environmental pollutants, food additives or therapeutic and cosmetic drugs. On the other hand, if properly used, light-stimulated processes can induce important beneficial effects in biological systems, such as the elucidation of several aspects of cell structure and function. Novel developments are centered around photodiagnostic and phototherapeutic modalities for the treatment of cancer, artherosclerosis, several autoimmune diseases, neonatal jaundice and others. In addition, classic research areas like vision and photosynthesis are still very active. Some out of these developments are unique to photobiology, since the peculiar physico-chemical properties of electronically excited biomolecules often lead to the promotion of reactions which are characterized by high levels of selectivity in space and time.

Besides the biologically centered areas, technical developments have paved the way for the harnessing of solar energy to produce warm water and electricity or the development of environmentally friendly techniques for addressing problems of large social impact (e.g. the decontamination of polluted waters). While also in use in Western countries, these techniques are of great interest for developing countries.

The European Society for Photobiology (ESP) is an organization for developing and coordinating the very different fields of photosciences in terms of public knowledge and scientific interests. Due to the ever increasing demand for a comprehensive overview over the photosciences the ESP decided to initiate an encyclopedic series, the 'Comprehensive Series of Photosciences'. This series is intended to give an in-depth coverage over all the very different fields related to light effects. It will allow investigators, physicians, students, industry and laypersons to obtain an updated record of the state-of-the-art in specific fields, including a ready access to the recent literature. Most importantly, such reviews give a critical evaluation of the directions that the field is taking, outline hotly debated or innovative topics and even suggest a redirection if appropriate. It is our intention to produce the monographs at a sufficiently high rate to generate a timely coverage of both well established and emerging topics. As a rule, the individual volumes are commissioned; however, comments, suggestions or proposals for new subjects are welcome. We are proud to present this first volume of the series which covers the field of 'Photomovement'.

Donat-P. Häder and Giulio Jori
Summer 2000

VOLUME PREFACE

The last comprehensive volume on the physiology of movement was published 1979, more than 20 years ago in the context of the 'Encyclopedia of Plant Physiology'. In the preface of that volume the editors stated "... against the background of the rapidly evolving field of molecular biology, plant movements were considered, by some scientists, as 'classical' (= old-fashioned) topics which might not contribute much to 'modern' biology." When the original decision was taken to try to assemble a new volume centered around photomovement of plants, the editors were confronted with the same type of hesitations. We consider the volume in your hands as the best argument that light-controlled signal transduction chains are by no means old-fashioned but on the contrary are in the center of modern biology.

This volume emphasizes the involvement of all facets of biology in the analysis of environmentally controlled movement responses. This includes biophysics, biochemistry, molecular biology and as an integral part of any approach to a closer understanding, physiology. The initial euphoria about molecular biology as the final solution for any problem has dwindled and the field agrees now that only the combined efforts of all facets of biology will at some day answer the question posed more than hundred years ago: "How can plants see?" One conclusion can be drawn from the current knowledge as summarized in this volume: The answer will most likely not be the same for all systems.

However, progress in the molecular understanding of photoresponses is naturally not on the same level for all systems. While in some systems the photoreceptor and the main components of the signal transduction chains were identified, in many other systems the knowledge is far from complete. The editors strongly believe that this volume will intensify and stimulate further research based on the comprehensive summary of results and findings in every article and the potential application of methods, hypotheses and ideas to other systems.

Finally the editors would like to thank all the authors for their work, their cooperation and their gracious acceptance of editorial comments. Specifically, we would like to thank all authors who agreed to write chapters on very short notice which resulted from last minute changes in the outlining. U. Trenz is acknowledged for skillful and very patient final preparation of the manuscripts. Last but not least we would like to thank Elsevier for the efficient production and excellent layout.

Donat-P. Häder
Michael Lebert
Erlangen, Germany

THE EDITORS

Donat-P. Häder, Dr. rer. nat., is a Professor of Botany, Department of Botany and Pharmaceutical Biology at the Friedrich-Alexander University at Erlangen, Germany. He received his doctoral degree and his habilitation from the University of Marburg. He had a research associate position at MSU, DOE, East Lansing, U.S.A. and was visiting scientist at the Chemistry Department, Lubbock, TX, U.S.A., CNR Pisa, Italy and the National Research Lab, Okazaki, Japan.

Professor Häder has worked on the photomovement of microorganisms, the effect of solar ultraviolet radiation on phytoplankton and is involved in space biology studying the effect of microgravity on motility in flagellates. He is a member of a Committee on Ecology for the German ministry for science and technology, expert for an Enquete commission of the German Parliament and a member of a UNEP commission on the effects of the ozone destruction.

One of the tools for his research activities is a real time image analysis system developed over the last fifteen years. He has published over 380 original papers and has been involved in eleven books as author, translator or editor.

Michael Lebert, Dr. rer. nat., is a senior scientist at the Department of Botany and Pharmaceutical Biology at the Friedrich-Alexander University at Erlangen, Germany. He received his doctoral degree from the University of Munich. He was a postdoctoral fellow at the WSU, Pullman, U.S.A.

Dr. Lebert has worked on environmentally controlled signal transduction chains in microorganisms for 15 years. This includes the relevance of light and gravity on the behavioral reactions of motile protists and bacteria. In addition, he is interested in the effect of ultraviolet radiation on aquatic ecosystems. His special interest is in the interface between biology, electronics and computers.

CONTRIBUTORS

Judith P. Armitage, Prof. Dr.
Microbiology Unit
Department of Biochemistry
University of Oxford
Oxford OX1 3QU
United Kingdom

Silvia E. Braslavsky, Prof. Dr.
Max-Planck-Institut für Strahlenchemie
P.O.B. 101365
D-45413 Mülheim an der Ruhr
Germany

Richard W. Castenholz, Prof. Dr.
Department of Biology
University of Oregon
Eugene
Oregon 97403
USA

Enrique Cerdá-Olmedo, Prof. Dr.
Departamento de Genética
Facultad de Biología
Universidad de Sevilla
E-41012 Sevilla
Spain

Stanley Cohn, Prof. Dr.
DePaul University
Department of Biological Sciences
2325 N. Clifton Ave.
Chicago, IL 60614
U.S.A.

Luis María Corrochano, Dr.
Departamento de Genética
Facultad de Biología
Universidad de Sevilla
E-41012 Sevilla
Spain

Wim Crielaard
Laboratory for Microbiology
E.C. Slater Institute
University of Amsterdam
Nieuwe Achtergracht 127
1018 WS Amsterdam
The Netherlands

Werner Deininger
Institut für Biochemie I
Universitätsstr. 31
93040 Regensburg
Germany

Mayumi Erata, Dr.
Global Environmental Forum
24–18 Inari-mae
Tsukuba, Ibaraki 305–0061
Japan

Paul R. Fisher, Dr.
La Trobe University
Department Microbiol.
Bundoova Vic. 3083
1 Joynt Street
Macleod VIC 3085
Australia

Ken Foster, Prof. Dr.
Dept. of Physics
Syracuse University
Syracuse NY 13210
USA

Paul Galland, Prof. Dr.
Fachbereich Biologie – Botanik
Lahnberge
35032 Marburg
Germany

Ferran Garcia-Pichel, Prof. Dr.
Department of Microbiology
Arizona State University
Tempe, AZ 85287–2701
USA

Francesco Ghetti, Dr.
Istituto di Biofisica
Consiglio Nazionale delle Ricerche
Area della Ricerca di Pisa
Via Alfieri 1
Localita' San Cataldo
56010 GHEZZANO – PISA
Italy

Elena G. Govorunova
Biology Department
Moscow State University
119899 Moscow
Russia

Paolo Gualtieri, Dr.
Istituto di Biofisica
Consiglio Nazionale delle Ricerche
Area della Ricerca di Pisa
Via Alfieri 1
Localita' San Cataldo
56010 GHEZZANO – PISA
Italy

J. Woodland Hastings, Prof. Dr.
Biological Laboratories
Harvard Univ.
Cambridge, MA 02138–2020
USA

Wolfgang Haupt, Prof. emer., Dr.
Erlenstr. 28
91341 Röttenbach
Germany

Peter Hegemann, Prof. Dr.
Institut für Biochemie I
Universitätsstr. 31
93040 Regensburg
Germany

Klaas J. Hellingwerf
Laboratory for Microbiology
E.C. Slater Institute
University of Amsterdam
Nieuwe Achtergracht 127
1018 WS Amsterdam
The Netherlands

Moritoshi Iino, Prof. Dr.
Botanical Gardens
Graduate School of Science
Osaka City University
Kisaichi, Katano-shi
Osaka, 576-0004
Japan

Takatoshi Kagawa
National Institute for Basic Biology,
Myodaijicho
Okazaki 444–8585
Japan

Dov Koller, Prof. Dr.
Institute of Life Sciences
The Hebrew University
Jerusalem 91904
Israel

Remco Kort
Laboratory for Microbiology
E.C. Slater Institute
University of Amsterdam
Nieuwe Achtergracht 127
1018 WS Amsterdam
The Netherlands

Georg Kreimer, Dr.
Friedrich-Alexander-Universität
Institut für Botanik und
Pharmazeutische Biologie
Staudtstr. 5
D-91058 Erlangen
Germany

Michael Lebert, Dr.
Institut für Botanik und
Pharmazeutische Biologie
Staudtstr. 5
91058 Erlangen
Germany

Francesco Lenci, Dr.
Istituto di Biofisica
Consiglio Nazionale delle Ricerche
Area della Ricerca di Pisa
Via Alfieri 1
Localita' San Cataldo
56010 GHEZZANO – PISA
Italy

Wolfgang Marwan, Prof. Dr.
Biologisches Institut II
Universität Freiburg
Schänzlestr. 1
79104 Freiburg
Germany

Peter Nick, Prof. Dr.
Institut für Biologie II
Schänzlestr. 1
79104 Freiburg
Germany

Oleg A. Sineshchekov, Dr.
Biology Department
Moscow State University
119899 Moscow
Russia

Pill-Soon Song, Prof. Dr.
Department of Chemistry
Univ. of Nebraska
Lincoln NE 68588–0376
USA

John L. Spudich, Prof. Dr.
Department of Microbiology
and Molecular Genetics
University of Texas
Medical School, Houston
Texas, 77030–1501
USA

Masamitsu Wada, Prof. Dr.
Tokyo Metropolitan University
Minami-osawa 1–1
Hachioji
Tokyo 192–0397
Japan

Gottfried Wagner, Prof. Dr.
Inst. f. Allg. Botanik
und Pflanzenphysiologie
Senckenbergstr. 17–21
35390 Gießen
Germany

Masakatsu Watanabe, Dr.
National Inst. for Basic Biology
Okazaki National Research Inst.
Okazaki Aichi 444
Japan

René M. Williams
Max-Planck-Institut für Strahlenchemie
P.O.B. 101365
D-45413 Mülheim an der Ruhr
Germany

David C. Wood, Prof. Dr.
Dept. of Neuroscience
Univ. Pittsburgh
446 Crawford Hall
Pittsburgh, PA 15260
USA

TABLE OF CONTENTS

Photomovement
D.-P. Häder and M. Lebert, editors.

Chapter 1

Photomovement: past and future

Wolfgang Haupt

Table of contents

The present volume refers to photomovement in a broad sense, comprising:

1. several types of light responses of motile organisms, i.e. modulation of movement in time and/or space (bacteria, lower algae, slime molds, ciliates),
2. light-controlled reversible or irreversible redistribution of cell organelles, particularly chloroplasts (green plants at all levels of organization), and
3. phototropism and photonasty, i.e. bending response with and without respect to the light direction, the former concerning unicellular and multicellular organisms (including higher plants), the latter mainly regulation of stomatal aperture in higher plants, but also flower movements etc.

To understand the term "photomovement" in this broad sense, a short view of the history is necessary, starting with the responses of motile organisms to light. The terminological development can be traced back at least to the turn of the century, when Rothert [1] and Pfeffer [2] tried to distinguish the various types of light responses by proper terms. Remaining inconsistencies repeatedly stimulated authors to improve the terminology (cf. [3]). Today, the most widely accepted terminology was proposed by Diehn et al. [4]. According to these authors, there are three types of responses of motile organisms to light, as will be shown below, viz., photokinesis, photophobic response and phototaxis. Although for each of these responses different parameters of the light signal are important, the final result might occasionally appear similar, e.g. accumulation in or dispersal from particular regions of the environment, and it is sometimes difficult to attribute unequivocally such an observation to one of the three types of response or to a coaction of them. This complex situation calls for a common term for all light-controlled responses of those motile organisms. It is certainly misleading to use, for this purpose, the term "phototaxis" in a broad sense (as can still be found, e.g. in [5]), rather than to restrict it to its present-day definition (see below). To circumvent this difficulty, the term "photomotion" had been used at several international meetings in the seventies, but soon it was replaced by "photomovement" (e.g. [6]). It is hard to discover who was the first to propose this latter term.

Thereafter, because of the superficial similarity with the behavior of motile organisms, light-oriented intracellular movement of chloroplasts was included in chapters on photomovement (e.g. [7]). Finally, if one realizes that light-controlled bending or curving responses of cells or plant organs, i.e. phototropism and photonasty, are being considered as movements, it is consequent to apply to these responses, too, the term photomovement (e.g. [8]).

Photomovement responses can bear a relation to the light direction, thus being vectorial with respect to it, or their direction can be determined exclusively by the morphology and physiology of the organism or organ in question, thus being scalar with respect to the light signal. In the former case, the organism is required to recognize the vectorial property of the light signal, to process it accordingly and to have the respective degrees of freedom for the activity of the motor apparatus (e.g. [9]). Besides this vectorial parameter of the light signal, there is also the possibility that the time pattern of light is important.

A widely observed feature in photomovement is the inversion of the sign of response, e.g. orientation away from instead of toward the light source (negative vs. positive response). Such inversion primarily is found with increasing strength of the signal, but can depend also on the environment, e.g. on chemical signals or even on an additional, independent light signal (see examples in the respective chapters).

To each of the various photoresponses, a generalized scheme can be applied (e.g. [10]), viz. **perception** of the signal – **transduction** (signal processing) with amplification – terminal **response** (modulation of movement); occasionally, the more indifferent terms **input**, **black box** and **output** are preferred. This frame is a valuable basis for structuring scientific questions that are under investigation or that are worth being investigated. Some general topics will be listed accordingly.

a. **Perception**. There is a huge variety of photoreceptor (photosensory) pigments, specific for taxonomic groups and for the various responses (cf. [8], and references therein). Their chemical nature is a basis for sensing fluence rate (intensity) and/or fluence (or dose) of the light signal as well as its wavelength range. As an additional requirement for sensing the vectorial properties of light (direction of propagation, polarization), localization or compartmentation of the sensory pigment(s) and their association with oriented cell structures can be important, again varying between taxonomical groups and resulting in fundamentally different principles of directional sensing (cf. [8,11]).

   From the point of photobiology, perception is in the center of interest, and progress is now being made in analyzing biochemistry and biophysics of perception, including transfer of the information to the transduction chain. Accordingly, most chapters of the present volume are preferentially concerned with perception and its molecular background, irrespective of the particular responses. It is a challenge to find principles that can be generalized in spite of the diversity in detail.

b. **Transduction**. Only little knowledge is available so far about the sequence of steps in a transduction chain that are started by photoperception and that finally result in the control of movement. The current research is concerned with the respective contribution of biochemical and biophysical processes (for the latter cf., e.g. membrane properties or ion transport). Moreover, for vectorial responses the transduction processes are required to remain strictly localized. Last but not least, a particular challenge is the nature of the amplification processes, which are thought to be an integral part of transduction. Certainly, the transduction chains are the most complex part of photomovement, and in no case has understanding already gone beyond isolated pieces of the mosaic (hence the term "black box").

c. **Response**. A fascinating topic in photomovement is the diversity of molecular structures that underly the motor apparatus, from actin-myosin or tubulin-dynein interaction to mechanisms for differential growth in cells or tissues and for turgor regulation (cf. respective chapters in [12]). Since each movement requires energy, the energy-providing system and the energy transfer to the motor apparatus is a central topic as well as the checkpoints, where the controlling internal signals from the transduction chain are channeled into the response system. The most complicated task for the organism is the spatially different control of vectorial movement and the respective coordinations.

d. Beyond this perception-transduction-response scheme, it is important to elucidate the efficiency of the respective responses for survival, as this is an indispensable basis for understanding their evolution.

Recent progress at all levels of the perception-transduction-response system has seriously benefited from molecular and genetic approaches. They were (and still are) particularly successful in a few favored model systems, as, e.g. photophobic response in *Halobacterium*, phototaxis in *Chlamydomonas*, phototropism in *Phycomyces* and *Arabidopsis* (cf. references in [13–16]). However, even in cases of most advanced research, genetic and molecular approaches always require a sound fundament of so-called "classical" research, be it physiological, biochemical, biophysical or cell structural, which is still the main approach in some other systems.

As a basis for understanding the specific chapters of this book, a short survey will be given for the three groups of photomovement, looking back to well-known facts and forward to scientific questions that are specific for the respective response type.

1.1 Light responses of motile organisms

Photomovement of motile organisms can belong to either of three fundamentally different types, viz., photokinesis, photophobic response and phototaxis (cf., e.g. [6]):

a. **Photokinesis**: The steady-state velocity depends on the intensity (fluence rate) of light, ideally in a well-defined function without hysteresis. This requires that no adaptation occurs, the response is independent of the time pattern of fluence rate. In some well-investigated organisms, photokinesis has been found as a trivial dependence on the current photosynthetic energy [8]. This makes those examples particularly interesting in which photokinesis is a true response to a light signal, including amplification processes in transduction [17].
b. **Photophobic response**: Upon a change in fluence rate (light-on, light-off, step-up, step-down), a transient change in velocity is observed, frequently starting with a stop response and comprising reversal of movement. Afterwards the velocity returns to its former level even in a constant new fluence rate. The direction of light, however, is unimportant for the response [8]. Notice that velocity is a vectorial term; thus, transient changes in velocity can comprise transient changes in speed as well as in direction. As a result, after full recovery the direction of movement may have changed with respect to the environment, although it is identical to that before with respect to the morphology of the organism. The transience of the response requires an effective adaptation, with a time constant longer than that of the change in signal intensity. This is a particular challenge for research, as well as comparison of steps in the transduction chain. As in taxonomically distant organisms particular ions and/or transmembrane potentials appear to play a central role in signal transduction [18], the question can be raised whether part of those steps may be common to organisms as different as, e.g. *Halobacterium*, gliding cyanobacteria, *Chlamydomonas*.
c. **Phototaxis**: The direction of movement is controlled by the light direction, but the time pattern of light has no influence. The final result, viz., accumulation as oriented with respect to the light source, can be based on fundamentally different mechanisms. This diversity concerns perception as well as response.

   Perception of the light direction always requires comparison of light signals, either simultaneously at different photoreceptor sites in the organism, establishing a **spatial gradient**, or at different points in time (**temporal gradient**), if the signal intensity reaching the photoreceptor is modulated by the movement [19]. In both cases, diversity is found how the gradient is established, viz., by shading, reflection, refraction or dichroism [9].

Among detailed questions of directional sensing, filamentous cyanobacteria of the *Phormidium* type may be mentioned which have only to discriminate between "light from front" and "light from rear"; it is still open whether this decision is made in individual cells (in all or in privileged cells?) or by integration over the whole filament. A similar question holds for the "steering" of the pseudoplasmodium of *Dictyostelium*. Still more enigmatic might be the perception of light direction in the flat cell of *Micrasterias*, which can reorient if the light direction deviates from normal to the cell surface (cf. [9]). In this latter case, perception of light direction in some sun-tracking leaves might serve as a model for future approach (cf. chapter by Koller, this volume).

For the **response**, no general statement is possible. This concerns primarily the motor apparatus (see respective chapters in [12]). In cilia (eukaryotic flagella), the tubulin-dynein system is almost certainly active in all examples; but details of coordination, which is necessary to ensure the proper direction of force generation, are not yet fully understood, although realistic models are available (cf. [20]). Much less knowledge is available for the various types of gliding in respective prokaryotic and eukaryotic organisms and their control by internal signals from the transduction chain. Moreover, fundamental diversity is found for the **types of response**, which range from a statistical trial-and-error or biased-random mechanism to precise steering of the individual into the desired direction [8]. Accordingly, the common term phototaxis describes only the final result (viz., accumulation, see above) rather than an immediate response to the signal.

This raises **terminological problems** and might suggest that in future the term "phototaxis" be replaced by several more specific terms. Superficially, such a terminological "evolution" would appear to be in line with the history (cf. [5]): In earlier time, phototaxis and photophobic response were jointly called "phototaxis", considering the fact that both types of response can result in patterns of accumulation in the medium. Whether or not this accumulation is related to the light direction was indicated by prefixes (topo- and phobo-phototaxis, respectively; cf. [2]). Still more complicated, additional prefixes (eu-, pseudo-) were proposed to distinguish between particular mechanisms of phototactic behavior. Although such a sophisticated terminology appeared to be logical, it has the serious disadvantage that a given behavior could get its name not before it had been thoroughly analyzed. Thus, the present-day more general terminology appears adequate for practical use, and this holds also for phototaxis.

As a more serious terminological question, one may ask whether really all present-day observations fit into the actual terminological scheme. Two examples deserve attention.

1. The photophobic response of *Halobacterium* is not simply a change in movement upon light-on (step-up) or light-off (step-down), but the autonomous random reversals are speeded up to premature reversals or are delayed, by the respective photophobic stimulus (e.g. [21]). This is not a real problem, but requires a more sophisticated definition of "photophobic response".

2. Premature or delayed autonomous reversal of the filamentous cyanobacterium *Phormidium* as controlled by the light direction is not by itself a "phototactic response", but results statistically in phototactic accumulation [8]; however, none of the other terms can be applied for this immediate light effect either.

For comparison, in graviresponses it has been found that the steady-state velocity can depend on the gravity vector with respect to the direction of movement [22] and might therefore be described as a "vectorial gravikinesis", thus broadening the definition of kinesis. A vectorial photokinesis, however, is not yet known. This may suggest to consider the response of *Phormidium* as a "vectorial photokinesis", which then would require new definitions.

1.2 Light-controlled movement of cell organelles

There are three types of light-controlled movement of cell organelles:

1. **Photodinesis** is light control (induction, acceleration, retardation) of rotational movement of cytoplasm (cyclosis), usually including the chloroplasts (cf. [23]). The overall distribution of chloroplasts in the cell remains unchanged. The light direction is unimportant; thus, photodinesis is a strictly scalar response.
2. **Orientational movement of chloroplasts** results in rearrangement or repositioning of these organelles with respect to the light direction. Usually this is interpreted as a temporal adaptation to the light environment so as to ensure optimal light harvesting and to minimize photodamage.

   In early times, chloroplast orientation has been considered as "phototaxis" of chloroplasts (e.g. [24]). However, it is now generally accepted that the light signal and its direction is not perceived by the organelle proper, but by the surrounding cytoplasm, and this intracellular "environment" controls the organelles' movement (cf. [25]). The same holds true for photodinesis and for polarity induction. Thus, similarity with phototaxis and photokinesis is only superficial, and indeed, the term "phototaxis of chloroplasts" is now disappearing.

   Based on an early extraordinary monograph by [26], recent research has successfully been extended to modern physiology, including promising molecular and genetic approaches (cf. reviews by [25,27]; cf. also [28] for historical aspects).

   Open questions of **perception** concern, e.g. multiple photoreceptor pigments, i.e. those examples in which two or even three photoreceptor pigments can act independently of each other. This raises the question as to the convergence of separate transduction chains and to the master reaction that collects the different flows of information and finally controls the same response (cf. the respective discussion in [28]).

   As to the **transduction chains**, a number of likely steps has been proposed, as concluded from recent experiments in a few model systems. But all these steps (e.g. calcium/calmodulin, changes in membrane properties, reorganisation and anchoring of cytoskeleton elements) are still under "pro-and-con" discussion, as respective observations cannot yet prove that these factors are integral parts of the flow of information and what are their causal connections [25,28,29].

The **motor apparatus** is almost certainly the actin-myosin system, at least in most of the examples investigated so far. In detail, however, its particular way of action for these responses is not yet fully understood, and some diversity may be expected, comprising also the possible points of attack for the controlling transduction chain. These questions are complicated because of the directionality and possibly also because of multiple transduction chains. Moreover, a contribution of additional cytoskeletal elements to the movement cannot be excluded yet (cf. [25]).

3. **Induction of cell polarity** is sometimes also included here, because cell organelles are redistributed with respect to the light direction. In the spore of *Equisetum* or in the zygote of *Fucus*, e.g. chloroplasts (or phaeoplasts, respectively) move toward the light, nucleus and other organelles away from it (cf. [30,31]). Superficially, this looks comparable to the orientational movements in the preceding paragraphs. However, polarity induction is part of cell differentiation and usually becomes irreversible very soon, whereas the typical chloroplast movements occur in differentiated cells, are reversible and almost infinitely repeatable.

As a particular question in this developmental process, the fast stabilization of the originally labile light-induced polarity is a particular challenge for cell biologists, but might go far beyond photobiological research.

1.3 Phototropic and photonastic curvatures

Light-induced or light-controlled curvatures of plant cells or organs can be oriented with respect to the light source (phototropism), or they are not related to the direction of the light signal. In the latter case, the direction of response is determined by the anatomical or physiological polarity of the respective cells, tissues or organs (photonasty).

Phototropism is most spectacular in etiolated seedlings and in some model systems of lower plants. As in nature the effective light source is usually the sun, the response was called, in the early time of plant physiology, **helio**tropism (e.g. [32–35]). However, repeatable and reproducible quantitative experiments require artificial light sources, and therefore the more general term **photo**tropism has replaced very soon the earlier term (cf. [2,36]); similarly, the earlier **geo**tropism has recently changed to **gravi**tropism, replacing the particular signal by the more general one [37]. However, terminological relation to the sun is still found in "sun tracking", which denotes a particular type of bending toward light. Whereas the "classical" phototropism usually is a single bending in one plane toward (or away from) a stationary light signal, a sun-tracking organ follows the sun, i.e. a moving light signal, in several dimensions, i.e. with bending in azimuth (horizontally), in elevation (vertically), and even with torsion [see chapter by Koller, this volume].

Phototropism is usually the result of unequal growth of opposite flanks, i.e. of opposite cell-wall regions in unicellular organs or of opposite tissues in multicellular organs. In many **sun-tracking** organs, however, since they have almost completed their growth, turgor changes appear to be more important than growth modulation.

After the basic work of Darwin [33], grass coleoptiles became the favored systems for **phototropism**, mainly oat (*Avena*), later on followed by corn (*Zea*). Until the fifties, a rough understanding appeared to have been reached, with a flavin as photoreceptor pigment and carotenoids as shading pigments for ensuring an absorption gradient, with unequal auxin distribution according to the repeatedly (and sometimes tacitly) modified Cholodny-Went theory, and thus with redistribution of growth.

However, not only important details remained hidden in this model, but also some of the basic steps became questioned again ([38], there also a summary of historical aspects). It is true, the nature of the photoreceptor substance as a flavin appears to be close to be disentangled, due to recent genetic and molecular approach (cf. [39]). However, among others, distribution of sensitivity along the coleoptile as well as the absorption gradient across the organ became complicated by the discovery of light piping through the coleoptile [40]. Moreover, full understanding of stimulus processing and of final response requires detailed knowledge at the cellular level rather than simple measurements of the integral responses of the whole organ [38]. This also requires to reinvestigate the role of auxin, which might only be a permissive condition rather than an essential link in the transduction chain. This would then suggest that new substances and/or processes for signal transmission be considered. The respective chapters of this volume will contribute to some of these questions.

In phototropism of single cells of lower plants, some of these problems appear to be less important, due to less complicated organization. However, in single-cell systems the response depends qualitatively on whether growth is located at the tip or at subapical regions of the cell ("bulging vs. bowing"; cf. [41]).

Phototropism of the sporangiophore of *Phycomyces* is a good example for the stepwise progress in knowledge. The first fundamental step was the discovery by Buder [42] that the sporangiophore acts as a collecting lens, thus establishing an absorption gradient opposite to the light direction. This initiated many more detailed "classical" experiments until the next important step, when Delbrück [43] applied kybernetic models to relate light-growth response and phototropism to each other. This kept again scientists busy until genetic approaches were introduced ([44], further references in [15] and see chapter by Cerdá-Olmedo, this volume) and opened new views. Among the new results is the fact that several independent responses are controlled by a single photoreceptor and that, vice versa, the bending response can be induced by various independent signals. It is a particular challenge to find out where and how the separate transduction chains converge.

More recently, there is increasing benefit from those genetic approaches also in perception and transduction of higher plants, e.g. in coleoptiles, and most promising, the *Arabidopsis* seedling is becoming a central model system for phototropism and a preferred system for genetic approach in plant physiology in general (cf. [39,45]).

In **sun tracking**, the directionality is the most interesting question at all levels of the perception-transduction-response model (cf. [46], and chapter by Koller, this volume). How is, e.g. distinction being made in perception of obliquely incident light, whether it falls on the leaf toward the tip or toward the base? How are these opposite signals transduced differentially, how is the turgor modulated respectively? How are those signals perceived and transduced that have a component from the side? Moreover, what mechanism is behind those cases where, during the night, the leaf appears to anticipate the direction of sunrise, as it orients endogenously to it? In the particular response of sun tracking, substantial "classical" experimentation is still needed before extensive genetic approaches appear promising.

Photonastic movements of leaves are operated by turgor changes in particular motor cells of pulvini. However, in addition to the direct photonastic light effects, the circadian rhythm is a main controlling factor of the well-investigated "sleeping movements", and usually the direct photonastic signal plays a minor role. Accordingly, this kind of photomovement is treated only marginally in the present volume, particularly as recent experiments on the "photonastic component" of leaf movements are limited. However, cf. chapter by Koller, this volume.

Photonastic movement of **guard cells** in higher plants controls stomatal aperture. The motor system for these movements are turgor changes in guard cells and subsidiary cells. As photonastic responses are scalar with respect to the signal direction, they might be expected to be less complicated than vectorial photoresponses. However, there are at least two separate photoreceptor systems, viz. cryptochrome and photosynthetic pigments, which may start two completely different and interacting or competing transduction chains. Moreover, stomatal aperture is under the control of multiple factors, i.e. of water potential, of intracellular CO_2 concentration and of phytohormones in addition to light, and these signals interact in a complicated way. The interaction of the signals can be competition with each other or mutual support, and light effects operate partly via the feedback systems of water potential and/or CO_2 (cf. [47,48]). It is thus a main (but difficult) task to analyze this most complicated multifactorial network.

Among all kinds of photomovement, stomatal control is the most challenging response with respect to disentangle its ecological significance. It has to ensure the optimal compromise between most effective light harvesting for photosynthesis on the one hand, and surviving in adverse conditions on the other hand, and this optimum strongly depends on the environmental conditions.

1.4 Concluding remarks

The present volume is not exclusively structured according to response types or to taxonomic groups. Rather, a combination of both principles can be found, and this appears particularly adequate as the level of knowledge is very different for the various response types and also for various taxonomic groups. As mentioned above, this level ranges from description of basic phenomena via successful analyses at the organismal and cellular level until most advanced and promising molecular and genetic approaches. The reader will be aware of this diversity as a challenge to become stimulated to ask new and promising questions.

References

1. W. Rothert (1901). Beobachtungen und Betrachtungen über taktische Reizerscheinungen. *Flora*, **88**, 371–421.
2. W. Pfeffer (1904). *Pflanzenphysiologie, 2. Auflage*. W. Engelmann, Leipzig.
3. G.S. Fraenkel, D.L. Gunn (1960). *The orientation of animals*. Dover, New York.
4. B. Diehn, M. Feinleib, W. Haupt, E. Hildebrand, F. Lenci, W. Nultsch (1977). Terminology of behavioral responses of motile microorganisms. *Photochem. Photobiol.*, **26**, 559–560.

5. W. Haupt (1959). Die Phototaxis der Algen. In: E. Bünning (Ed.), *Physiologie der Bewegungen. Band 17/1 Handbuch der Pflanzenphysiologie* (pp. 318–370). Springer, Berlin, Göttingen, Heidelberg.
6. D.-P. Häder (1979). Photomovement. In: W. Haupt, M.E. Feinleib (Eds), *Physiology of Movement, Vol. 7 of Encyclopedia of Plant Physiology* (pp. 268–309). Springer, Berlin, Heidelberg, New York.
7. W. Haupt. Photomovement (1986). In: R. Kendrick, G.H.M. Kronenberg (Eds), *Photomorphogenesis in Plants* (pp. 415–441). Nijhoff, Dordrecht, Boston, Lancaster.
8. W. Nultsch (1991). Survey of photomotile responses in microorganisms. In: F. Lenci, F. Ghetti, G. Colombetti, D.-P. Häder, P.-S. Song (Eds), *Biophysics of Photoreceptors and Photomovements in Microorganisms* (pp. 1–5). Plenum, New York, London.
9. W. Haupt (1996). Overview of photosensing in plant physiology. In: R.C. Jennings, G. Zucchelli, F. Ghetti, G. Colombetti (Eds), *Light as an Energy Source and Information Carrier in Plant Physiology* (pp. 169–183). Plenum Press, New York, London.
10. W. Haupt (1991). Introduction to photosensory transduction chains. In: F. Lenci, F. Ghetti, G. Colombetti, D.-P. Häder, P.-S. Song (Eds), *Biophysics of Photoreceptors and Photomovements in Microorganisms* (pp. 7–19). Plenum, New York, London.
11. M. Kraml (1994). Light direction and polarization. In: R. Kendrick, G.H.M. Kronenberg (Eds), *Photomorphogenesis in Plants* (2nd ed., pp. 417–445). Kluwer, Dordrecht, Boston, London.
12. W. Haupt, M.E. Feinleib (Eds) (1979). *Physiology of Movement, Vol. 7 of Encyclopedia of Plant Physiology.* Springer, Berlin, Heidelberg, New York.
13. J.L. Spudich (1991). Color discriminating pigments in *Halobacterium halobium.* In: F. Lenci, F. Ghetti, G. Colombetti, D.-P. Häder, P.-S. Song (Eds), *Biophysics of Photoreceptors and Photomovements in Microorganisms* (pp. 243–248). Plenum, New York, London.
14. P. Hegemann (1991). Photoreception in *Chlamydomonas.* In: F. Lenci, F. Ghetti, G. Colombetti, D.-P. Häder, P.-S. Song (Eds), *Biophysics of Photoreceptors and Photomovements in Microorganisms* (pp. 223–229). Plenum, New York, London.
15. E. Cerdá-Olmedo, V. Martin-Rojas (1996). Phototropism in *Phycomyces.* In: R.C. Jennings, G. Zucchelli, F. Ghetti, G. Colombetti (Eds), *Light as an Energy Source and Information Carrier in Plant Physiology* (pp. 293–299). Plenum Press, New York, London.
16. W.R. Briggs, E. Liscum, P.W. Oeller, J.M. Palmer (1996). Photomorphogenic systems. In: R.C. Jennings, G. Zucchelli, F. Ghetti, G. Colombetti (Eds), *Light as an Energy Source and Information Carrier in Plant Physiology* (pp. 159–167). Plenum Press, New York, London.
17. D.-P. Häder, M.A. Häder (1989). Effects of solar UV-B irradiation on photomovement and motility in photosynthetic and colorless flagellates. *Environ. Exp. Bot.*, **29**, 273–282.
18. W. Haupt, D.-P. Häder (1994). Photomovement. In: R. Kendrick, G.H.M. Kronenberg (Eds), *Photomorphogenesis in Plants* (2nd ed., pp. 707–732). Kluwer, Dordrecht, Boston, London.
19. M.E. Feinleib (1980). Photomotile responses in flagellates. In: F. Lenci, G. Colombetti (Eds), *Sensory Transduction in Aneural Organisms* (pp. 45–68). Plenum Press, New York, London.
20. G. Colombetti, R. Marangoni (1991). Mechanisms and strategies of photomovements in flagellates. In: F. Lenci, F. Ghetti, G. Colombetti, D.-P. Häder, P.-S. Song (Eds), *Biophysics of Photoreceptors and Photomovements in Microorganisms* (pp. 53–71). Plenum, New York, London.
21. A. Schimz, E. Hildebrand (1991). Processing of photosensory signals in *Halobacterium.* Common features of the bacterial signalling chain and of information processing in higher developed organisms. In: F. Lenci, F. Ghetti, G. Colombetti, D.-P. Häder, P.-S. Song (Eds),

Biophysics of Photoreceptors and Photomovements in Microorganisms (pp. 231–241). Plenum, New York, London.

22. H. Machemer, S. Machemer-Röhnisch, R. Bräucker, K. Takahashi (1991). Gravikinesis in *Paramecium*: Theory and isolation of a physiological response to the natural gravity vector. *J. Comp. Physiol. A*, **168**, 1–12.
23. K. Seitz (1979). Cytoplasmic streaming and cyclosis of chloroplasts. In: W. Haupt, M.E. Feinleib (Eds), *Physiology of Movement, Vol. 7 of Encyclopedia of Plant Physiology* (pp. 150–169). Springer, Berlin, Heidelberg, New York.
24. W. Haupt (1959). Chloroplastenbewegung. In: E. Bünning (Ed.), *Physiologie der Bewegungen. Band 17/1 Handbuch der Pflanzenphysiologie* (pp. 278–317). Springer, Berlin, Göttingen, Heidelberg.
25. M. Wada, F. Grolig, W. Haupt (1993). Light-oriented chloroplast positioning. Contribution to progress in photobiology. *J. Photochem. Photobiol. B*, **17**, 3–25.
26. G. Senn (1908). *Die Gestalts- und Lageveränderung der Pflanzen-Chromatophoren*. W. Engelmann, Leipzig.
27. G. Wagner (1995). Intracellular movement. *Progr. Bot.*, **57**, 68–80.
28. W. Haupt (1998). Chloroplast movement: from phenomenology to molecular biology. *Progr. Bot.*, **60**, 3–36.
29. G. Wagner, U. Russ, H. Quader (1992). Calcium, a regulator of cytoskeletal activity and cellular competence. In: D. Menzel (Ed.), *The cytoskeleton of the algae* (pp. 411–424). CRC Press, Boca Raton.
30. M.H. Weisenseel (1979). Induction of Polarity. In: W. Haupt, M.E. Feinleib (Eds), *Physiology of Movement, Vol. 7 of Encyclopedia of Plant Physiology* (pp. 485–505). Springer, Berlin, Heidelberg, New York.
31. D.L. Kropf (1992). Establishment of cellular polarity in fucoid zygotes. *Microbiol. Rev.*, **56**, 316–339.
32. A.P. de Candolle (1832). *Physiologie végétale ou exposition des forces et des fonctions vitales des végétaux*. Béchet Jeune, Paris.
33. C. Darwin (1880). *The Power of Movement in Plants*. Appleton Comp, New York.
34. W. Rothert (1894). Über Heliotropismus. *Beitr. Biol. Pflanzen*, **7**, 1–212.
35. G. Rina, R. Ardumo-Jolande (1927). Contributo studio dell'eliotropismo nelle piante. L'azione di diverse sostanze eccitanti sopra di esso. *Natura (Milano)*, **18**, 1–27.
36. P. Boysen-Jensen, N. Nielsen (1911). La transmission d'irritation phototropique dans l'*Avena*. *Bull. Acad. Roy. Danmark*, **1**, 3–24.
37. A. Sievers, D. Volkmann (1979). Gravitropism in single cells. In: W. Haupt, M.E. Feinleib (Eds), *Physiology of Movement, Vol. 7 of Encyclopedia of Plant Physiology* (pp. 567–572). Springer, Berlin, Heidelberg, New York.

38 R.D. Firn (1994). Phototropism. In: R. Kendrick, G.H.M. Kronenberg (Eds), *Photomorphogenesis in Plants* (2n ed., pp. 659–681). Kluwer, Dordrecht, Boston, London.

39 W.R. Briggs, E. Liscum (1997). The role of mutants in the search for the photoreceptor for phototropism in higher plants. *Plant Cell Environ.*, **20**, 768–772.

40 D.F. Mandoli, W.R. Briggs (1984). Fiber optics in plants. *Sci. Amer.*, **251**, 90–98.

41 R. Hertel (1980). Phototropism of Lower Plants. In: F. Lenci, G. Colombetti (Eds), *Sensory Transduction in Aneural Organisms* (pp. 89–105). Plenum Press, New York, London.

42 J. Buder (1918). Die Inversion des Phototropismus bei *Phycomyces*. *Ber. Deutsch. Bot. Ges.*, **36**, 104–105.

43 M. Delbrück, W. Reichardt (1956). System analysis for the light growth reaction in *Phycomyces*. In: D. Rudnick (Ed.), *Cell Mechanism in Differentiation and Growth* (pp. 3–44). Princeton Univ. Press.

44 T. Ootaki, A.C. Lighty, M. Delbrück (1973). Complementation between mutants of *Phycomyces* deficient with respect to carotenogenesis. *Mol. Gen. Genet.*, 57–70.
45 A.R. Cashmore (1997). The cryptochrome family of photoreceptors. *Plant Cell Environ.*, **20**, 764–767.
46 D. Koller, S. Ritter, W.R. Briggs, E. Schäfer (1990). Action dichroism in perception of vectorial photo-excitation in the solar-tracking leaf of *Lavatera cretica. L., Planta*, **181**, 184–190.
47 K. Raschke (1979). Movements of stomata. In: W. Haupt, M.E. Feinleib (Eds), *Physiology of Movement. Vol. 7 of Encyclopedia of Plant Physiology* (pp. 383–441). Springer, Berlin, Heidelberg, New York.
48 E. Zeiger (1994). The photobiology of stomatal movements. In: R. Kendrick, G.H.M. Kronenberg (Eds), *Photomorphogenesis in Plants* (2nd ed., pp. 683–706). Kluwer, Dordrecht, Boston, London.

Photomovement
D.-P. Häder and M. Lebert, editors.

Chapter 2

Triggering of photomovement – molecular basis

René M. Williams and Silvia E. Braslavsky

Table of contents

Abstract

This introductory chapter describes, from a molecular perspective, the events that occur upon excitation of the chromophores of photoreceptors triggering photomovement (comprising phototaxis, photophobic responses and photokinesis in lower organisms as well as phototropism and photonasty in higher plants). A general description is given of the processes of *cis-trans* isomerization and photoinduced electron transfer, and excursions are made into fundamental biophysical themes. Special attention is given to the influence of the medium (the protein moiety in the photoreceptors) on both the ground and excited state properties of the chromophores, and examples are offered of special cases. Furthermore, some photophysical properties of several photoreceptor chromophore models are compiled. The chromophores undergoing *cis-trans* isomerization are open-chain tetrapyrroles (present in phytochromes), *para*-hydroxycinnamoyl anion (in photoactive yellow protein) and retinal (in sensory rhodopsins in halobacteria and in some flagellate algae). Blue light photoreceptor chromophores such as flavins and the perylenquinone-type (e.g. hypericin-derivatives: stentorin and blepharismin) which most likely undergo electron transfer as the primary photochemical process are also discussed.

Abbreviations: cry1 and cry2, cryptochrome 1 and 2, respectively; FTIR, Fourier-transform infrared; phyA and phyB, phytochromes A and B, respectively; P_r and P_{fr}, red and far-red absorbing forms of phytochrome, respectively; PYP, photoactive yellow protein.

2.1 General considerations

The term photomovement comprises phototaxis, photophobic responses, and photokinesis in lower organisms [1] as well as phototropism and photonasty in higher plants. Light absorption by photoreceptors is the first step in the sequence of molecular events leading to photomovement. This trivial assessment means that there should be a light-absorbing molecule for every light-triggered process. Excitation of the photoreceptor part absorbing solar radiation (the chromophore) promotes in femtoseconds one electron to an electronically excited state with a different orbital configuration than that of the ground state. The nuclei of the photoreceptor molecule do not move during this ultrashort process. They start moving immediately after excitation to accommodate to the new situation. This nuclear relaxation and movement can be viewed as a travel along a potential energy surface, in which every position on the surface corresponds to a different nuclear configuration. In this way a situation with the lowest energy is attained.

The potential energy surfaces of both the ground and the excited state are determined by the nuclear geometry and by the interactions with the environment of the photoreceptor molecule. Thus, the relaxation contains, in addition to nuclear reorganization, medium rearrangements and energy dissipation into the environment. In the same way as the nuclei of the excited chromophore react to the change in its electronic distribution, the environment reacts to the changes in nuclear and electronic configuration of the photoreceptor on its way to the lowest energy point [2,3].

2.1.1 Nature of the bond

In all photoreceptors for photomovement the environment of the chromophore is the apoprotein into which it is embedded forming different types of bonds. In most cases one of the bonds is covalent such as the thioether bond of the open-chain tetrapyrrole to a cystein residue in phytochromes [4,5], the thioester bond of the *para*-hydroxycinnamoyl anion to a cystein residue in photoactive yellow protein (PYP) from bacteria [6], and the protonated Schiff base formed by the aldehyde retinal chromophore with a lysine residue in all retinal proteins [7]. In other cases only non-covalent bonds may be responsible for the chromophore-protein association, as reported for *Arabidopsis* NPH1, a flavoprotein with the properties of a photoreceptor for phototropism [8] as well as for cryptochrome 1, another flavoprotein blue light receptor in *Arabidopsis* [9]. Non-covalent is also the association between bacteriochlorophyll and the proteins in the photosynthetic reaction centers which are considered to give the signal for light-intensity induced changes in swimming speed in non-sulfur bacteria such as *Rhodobacter spaeroides* [10].

2.1.2 Primary photochemical process

A primary photochemical process is an elementary chemical process undergone by an electronically excited molecular entity and yielding a primary photoproduct [11]. From the point of view of the primary photochemical process, biological photoreceptors recognizing the quantity and the quality of light in the environment of the organism (photosensors) can be classified in two groups: those having a chromophore undergoing a *cis-trans* isomerization, and those having a chromophore producing an electron transfer. The former group, which absorbs from the blue to the red region of the spectrum (depending of the particular system), has been well characterized and consists of retinal proteins such as e.g. sensory rhodopsins in halobacteria [7,12–14] open-chain tetrapyrrole proteins such as e.g. phytochromes in higher plants, in cyanobacteria [15,16], in mosses and ferns and in some algae (for general references see [17,18]), and the xanthopsins, e.g. the photoactive yellow protein, PYP, found in eubacteria such as *Ecothiorhodospira halophila*, which contain a deprotonated *para*-hydroxycinnamoyl ester [19–21].

The second group, which consists mostly of blue-light-sensitive chromoproteins, is less well characterized in terms of the primary process *in vivo* and its mode of action; prominent members of this group are those containing a perylenequinone-type chromophore such as, e.g. stentorin [22] and blepharismin [23], a flavin-type photoreceptor implied in phototropism in *Arabidopsis thaliana* (*Arabidopsis* NPH1; [8]), and the cry1 and cry2 cryptochromes. The latter mediate many blue-light responses, including phototropism [24], and contain dual light-harvesting chromophores (like photolyases), i.e. flavin adenine dinucleotide (FAD), and either a deazaflavin [25] or a pterin [26], but they lack photolyase activity.

In the case of photosynthetic bacteria it is very difficult to discriminate between the photosynthetic electron transfer process providing chemical energy and the sensing process. In fact, any inhibitor or stimulus altering electron transport also alters many cellular functions, including flagellar activity governing phototaxis. The induction of the phototactic reaction chain in cyanobacteria (in particular in the cyanobacterium *Anabaena variabilis*) has been attributed to a combined action of chlorophyll *a*, phycobiliproteins and carotenoids [1]. Based on the action spectra of the photoresponses, chlorophyll *a* has been implied in photokinesis, phototaxis, and step-up and step-down photophobic responses of *Phormidium uncinatum* [1]. Should chlorophyll be involved, the primary process falls into the classification given above, in view of the fact that light absorption by chlorophyll produces an ion pair by electron transfer. It remains to be seen whether the recently discovered prokaryotic phytochrome [15,16] is involved in controlling photomovement in cyanobacteria, including *A. variabilis* and *P. uncinatum*.

In the case of *A. variabilis*, at high fluences, singlet molecular oxygen was postulated to be produced by energy transfer from the triplet chlorophyll *a* (in turn resulting from the recombination of the initially produced ion pair) to ground state molecular oxygen (a photodynamic mechanism) and to somehow induce a photophobic response of the cells in order to avoid photobleaching [27]. The photodynamic control of the photophobic response has not been confirmed for other microorganisms, such as the ciliates *Blepharisma* and *Stentor* whose photomotile responses, even though triggered by pigments which may act as photosensitizers, are not mediated by photodynamic reactions [28].

The role of active electron transport in the photoresponses of bacteria in general, and in particular of the non-sulfur bacteria *Rhodobacter sphaeroides* and *Rhodospirillum centenum* has been discussed by Armitage and co-workers [29,30]. This research group has recently reported that the response of these bacteria to increases and decreases in the intensity of light of different wavelengths is indeed regulated by photosynthetic electron transfer [10]. Also recently, a gene encoding a protein with 48% homology to the known PYP proteins has been isolated from *Rhodobacter sphaeroides* and the authors speculated about the possible involvement of PYP in the blue-light responses of these bacteria [31]. The presence of the PYP chromophore, i.e. *trans*–4-hydroxycinnamic acid, in cultures of phototropically cultured *Rhodobacter sphaeroides* cells was also reported [31]. However, mutants of *Rhodobacter sphaeroides* lacking the *pyp* gene were not affected with respect to their blue-light photophobic response, casting serious doubts on the PYP involvement in these responses [32]. As indicated by Armitage [29], it remains to be demonstrated that the primary photoreceptor for photomovement in these photosynthetic bacteria is effectively bacteriochlorophyll.

In principle, it is possible to think of photoreceptors based on other primary photoreactions, such as, e.g. a proton photodetachment. Such reactions are common in several molecule types which drastically change their p*K* value upon excitation, as is the case with aromatic amines and alcohols in their first excited singlet state [33,34]. However, so far no photoreceptor controlling photomovement has been found in which such a primary process has been demonstrated. A proton transfer as the primary process of the hypericin-related photoreceptor in ciliates was proposed [35] but the validity of this hypothesis was later questioned (vide infra, [36]). Proton transfers from and to the chromophore do take place in photoinduced cycles in photosensors at later cycle steps such as is the case in the sensory rhodopsins [37] and in PYP [38]. The transferred proton may, in turn, trigger the signal transduction chain. However, proton translocation through the membrane in these cases is not the primary photochemical process of the chromophore absorbing light.

Identity of blue-light photoreceptors. Several controversies exist regarding the identity of the blue-light absorbing photoreceptors. One of these controversies concerns the photoreceptor in the flagellate *Euglena gracilis* with an absorption maximum around 450 nm. Theoretical considerations as well as experimental evidence have been offered supporting a retinal nature of the chromophore [39,40]. Experimental data have also been offered for the involvement of pterins and flavins in photoperception by *Euglena* [41]. There have been some attempts to reconcile the rhodopsin and the flavin hypotheses by postulating the combined participation of both types of chromophores. These attempts, however, have been the object of criticism in view of the lack of experimental findings supporting the combination hypothesis [42]. In the case of the flagellate green alga *Chlamydomonas reinhardtii,* a new type of retinal-binding protein named chlamyopsin was found, which is not a typical seven helix receptor [43]. The chromophore in chlamyrhodopsin is the all-*trans* retinal which isomerizes to the 13-*cis* polyene [44].

Another long standing controversy is related to the nature of the ubiquitous cryptochrome, the blue-light photoreceptor for phototropism in higher plants and for other blue and UV-mediated effects in the plant world. Horwitz and Berrocal [45] give a spectroscopic view regarding the various possible blue-light receptors such as flavins, carotenoids and pterins. In particular, zeaxanthin has been postulated to play a role in phototropic bending in maize coleoptiles [46]. The study of mutants and the biochemical characterization of the mutated genes have led to the firm identification of cry1 and cry2 [24] as well as of NPH1 [8] as proteins implicated in the phototropic response in higher plants. In addition, genetic evidence indicate that phytochromes A and B are also required for normal phototropism in *Arabidopsis* [47].

It is of fundamental importance to recognize that the overall similarity of many blue-light action spectra with the absorption spectra of flavoproteins, carotenoids, and retinal proteins makes very difficult, if not impossible, the identification of the photoreceptors by optical methods only. This is also the case for red-absorbing chromophores such as present in phytochrome and the chlorophylls.

2.2 Chromophore-protein interactions

The nature of the link alone does not describe the chromophore-protein interactions, which are further influenced by the secondary, tertiary and even the quaternary structures of the protein. In fact, in all photoreceptors the nature of the chromophore-protein interactions is strongly linked to the particular function of each photoreceptor. Such interactions are responsible for the properties of these complex systems, which often are fundamentally different from those of the separate chromophore and apoprotein entities. E.g. the fact that bacteriorhodopsin acts as an energy converter in bacteria whereas rhodopsin acts as a photosensor in animals underscores the specificity of these interactions, since both pigments possess the same chromophore (albeit in different configurations, i.e. retinal is all-*trans* in light-adapted bacteriorhodopsin and 11-*cis* in rhodopsin) embedded in a different protein. Moreover, the same chromophore configuration (all-*trans*) linked to a slightly different protein acts as an energy converter in bacteriorhodopsin, and as light perceptor in the sensory rhodopsins I and II, all three found in the same organism [37].

The opsin shift in animal eyes represents a typical example of protein influence on absorption spectra of the chromophore [48]. Furthermore, it has been clearly shown that just a slight modification of the chromophore results in changes in the photophysical properties of the chromoprotein unit [49,50].

In addition, it is known that the photophysical properties and photochemical behavior of retinals in solution [51] are fundamentally different to those in the chromoproteins. In fact, the protein environment shortens so much the lifetime of the first singlet excited state in the chromoprotein retinal as to make intersystem crossing (a very important deactivation process in solution) unable to compete with fast isomerization.

The conformation about the single bond attaching the β-ionone ring to the polyene chain of the chromophore retinal is also tuned by the protein. In retinal proteins in animals the ring is in general twisted ca. 50 degrees with respect to the polyene chain in a highly conserved 6-s-*cis* conformation whereas in chlamyrhodopsin and in sensory rhodopsin II from archaeobacteria the conformation has been shown to be 6-s-*trans* [52], similar to the case in bacteriorhodopsin, in halorhodopsin, and in sensory rhodopsin I. It has been shown that the conformation of the β-ionone ring required for proton pumping in retinal assembled with synthetic retinal analogs is 6-s-*trans* coplanar with the polyene chain [53–55].

Protein tuning of the chromoprotein function is encountered in other photoreceptors, such as in phytochromes and phycocyanins, both containing open-chain tetrapyrrole chromophores [56,57,60]. The chromophore model compounds biliverdin and phycocyanobilin dimethyl ester have been reported to be mainly in a helical conformation in organic solvents, with a relatively small concentration of stretched conformations (the amount depending on the solvent) [58,59]. In all biliproteins (in phytochrome and phycocyanin as well as in phycoerythrin), however, the protein stabilizes a stretched conformation of the tetrapyrrole chromophore as is evident from analyses of absorption spectra [56,57,60] and from the X-ray structure when available (see e.g. [61]).

Another example of large protein-determined shift in the absorption and change in the photophysical properties is that occurring with *trans-p*-hydroxycinnamic acid, with an absorption maximum at ca. 300 nm in toluene and methanol, whereas when bonded to cystein in PYP, the absorption maximum in water shifts to 446 nm in the deprotonated ground state [62] and to 350 and 510 [63] nm in the various intermediates formed in the photocycle [64]. The processes leading to these photoinduced shifts are isomerization of the double bond concomitant with changes in the specific chromophore-amino acids interactions and most probably changes in the protonation state of the phenol group.

The above arguments imply that the absorption spectra and all other photophysical and photochemical properties of the bare chromophores in solution have little or no significance with respect to the properties in the chromoprotein. The absorption spectra as a key feature of the photoreceptors will strongly reflect the chromophore-protein interactions inasmuch as the absorption properties of molecules are very sensitive to the environment. Some rather simple compounds (e.g. pyridinium-*N*-phenoxide betaine dyes, [65,66]) are so sensitive towards their environment that they are used as solvent polarity probes.

In turn, the chromophore conformation determines the protein conformation as it is evidenced, e.g. by the changes in the α-helix content observed upon $P_r \rightarrow P_{fr}$ phototransformation in phytochrome A [67].

2.3 From molecular properties to signal transduction

There are several possibilities regarding the mechanism of signal transduction after excitation. Excitation might induce nuclear movements of the chromophore that generate in turn movement in the protein, provoking a cascade of steps leading to signal transduction [42]. In fact, an identical photoinduced reaction may activate different processes depending on the reaction partners. E.g. photoinduced proton release from the Schiff base may induce energy storage through ATP synthesis in bacteriorhodopsin (upon proton transport through the membrane) or may activate movement through the coupling of the sensor and a transducer membrane protein, such as is the case in the sensory rhodopsins I and II [37,68,69].

Activation of a Ca^{2+} channel has been found to be the light triggered action after excitation of chlamyrhodopsin [44].

In phytochrome, a series of intermediates species are produced upon excitation of the red-adapted P_r [70,71], involving double bond isomerization as well as chromophore and protein conformational changes. The question remains about the mechanism of molecular interaction between the conformationally different P_{fr} and the signal transducing partners. A further unanswered question is the possible interaction of long-lived intermediates in the photocycle with specific partners.

Using recombinant *Avena* and *Mesotaenium* phytochromes it has been recently shown that eukaryotic phytochromes autophosphorylate, and are serine/threonine protein kinases. The authors speculate that phytochrome most probably catalyzes intra-molecular phosphotransfer between the subunits of the phytochrome homodimer upon P_r to P_{fr} phototransformation [72]. It has been further speculated that this autophos-phorylation serves a regulatory role of phytochrome activity or modulates phytochrome association with other molecules [18]. Indeed, a novel phytochrome-binding protein (PKS1, phytochrome kinase substrate 1) has been identified and shown to be a substrate for light-regulated phytochrome kinase activity in vitro [73].

Recently, a novel helix-loop-helix protein called PIF3, localized in the cell nucleus, has been shown to interact with phytochromes A and B [74]. In addition, in several laboratories nuclear translocation of phytochrome is being demonstrated. Since the report by Sakamoto and Nagatani [75] that phytochrome B from *Arabidopsis* localizes in the nucleus in a light-dependent manner, several laboratories have shown that there is a nuclear translocation of phytochrome and that there are nuclear proteins which associate with phytochrome. Such studies can be expected to rapidly unravel the mode of action of phytochrome [76,77].

The protein nph1 (called phototropin, [78]) autophosphorylates in vivo and in vitro and has all motifs found in serine/threonine kinases. Therefore, the function of this particular flavoprotein should be related also to its light-activated kinase activity [8,78].

In the case of the cryptochromes the mechanism of signal transduction linked to the modulation of phototropism should be related to the presence of two chromophores, which confers the protein the ability of harvesting a relatively broad band of light and transfer an electron to an appropriate electron acceptor, similar to the case of the photolyases [26]. Interactions between red and blue photoreceptors already observed in the past during physiological studies have been now confirmed using mutants, in particular interactions between phyA, phyB, and cry1 during *Arabidopsis* development [79].

Furthermore, it is conceivable that the process inducing movement is a direct result of the change of the excited molecule itself. In fact, it is possible to imagine a moving system as a direct result of electrostriction [80]. Electrostriction is a volume contraction of the medium upon creation of a dipole in this medium. As a result of excitation a large increase in the dipole moment of the chromophore takes place and large electrostrictive effects may be expected [81,82]. However, this will obviously depend on the specific chromophore-protein interactions. If strong hydrogen and/or salt bridges determine the interaction, more specific movements might be expected upon chromophore excitation than those solely described by the electrostrictive effect [83].

In Table 1 we have compiled the organisms, the photoreceptors and their chromophores and primary processes (when known), as well as the authors of the corresponding chapters treating the respective organisms in this book. As already outlined, in some cases more than one type of chromophore has been claimed to be responsible for photomovement in one and the same species. In other cases, no postulations have been made. In the following sections we describe the events that occur in the chromophores that are embedded in the proteins of the various receptors either found or postulated to be responsible for photomovement.

2.4 *Cis-trans* isomerization

2.4.1 Principles

The simplest case of *cis-trans* isomerization is exemplified by 2-butene. UV excitation of the π-π^* transition promotes an electron to an anti-bonding molecular orbital [3,84,85].

According to Mulliken the energy of the excited singlet state is minimized when the two p-orbitals are orthogonal to each other [86]. Figure 1 represents the energy of the various states of 2-butene as a function of the torsional angle with the different orbital diagrams for each position. Relaxation of the twisted excited state leads to generation of the ground states of both *cis*- and *trans*-2-butene. The crossing to the ground state potential energy surface can, in principle, proceed via the singlet or via the triplet state (Figure 1).

The symmetric curves in Figure 1 are influenced in their form and separation by the medium around the chromophore and by the substituents on the chromophore itself. Figure 2 visualizes the induction of asymmetry on the potential energy surfaces by a specific medium interaction. Such an interaction will clearly change the isomerization quantum yields.

During the thorough investigations of the isomerization of the relatively simple compound *trans*-stilbene (1,2-diphenylethylene) it has been established that there is a

Table 1. Organism or phenomenon described by authors in this volume, photoreceptor found or postulated, chromophore and primary process involved (in some cases postulated primary process, see text)

Organism or phenomenon	Authors	Photoreceptor	Chromophore	Primary process
Purple bacteria	Armitage	photosynthetic apparatus	special pair bacteriochlorophyll	electron transfer
Halobacteria	Spudich	sensory rhodopsins	retinal	*cis-trans*
Ectothiorhodospira halophila	Crielaard et al.	photoactive yellow protein	*para*-hydroxy cinnamoyl anion	*cis-trans*
Flagellate green algae	Kreimer Sineshchekov & Govorunova	rhodopsin	retinal	*cis-trans*
Chlamydomonas	Hegemann & Deininger	chlamyopsin	retina	*cis-trans*
Euglena gracilis	Gualtieri Lebert	rhodopsin blue light receptor	retinal flavins, pterins	*cis-trans* electron transfer
Cryptophyceae	Watanabe	phycobilin	tetrapyrrole	
Diatoms	Cohn	not known		
Algal mats (cyanobacteria)	García-Pichel & Castenholz	photosynthetic apparatus	chlorophyll	electron transfer
Algal plastid movement	Wagner	phytochrome	tetrapyrrole	*cis-trans*
Circadian rhythms	Woodland-Hastings	phototropin cryptochromes	flavin	electron transfer
Ciliates	Lenci et al. Wood	stentorin and blepharismin	perylenequinones	electron transfer
Dictyostelium discoideum	Fisher		protoporphyrin IX	
Physarum polycephalum	Marwan	phytochrome	tetrapyrrole	*cis-trans*
Phycomyces	Cerdá-Olmedo & Corrochano Galland	blue-light far UV	flavin	electron transfer
Higher plants phototropism	Iino	cryptochrome	flavin	electron transfer
Higher plants cytoskeleton	Nick	phytochrome	tetrapyrrole	*cis-trans*
Higher plants solar tracking	Koller	phytochrome and blue light	tetrapyrrole and flavin	*cis-trans* electron transfer
Fern chloroplasts movement	Wada & Kagawa	phytochrome and blue light	tetrapyrrole flavin	*cis-trans* electron transfer

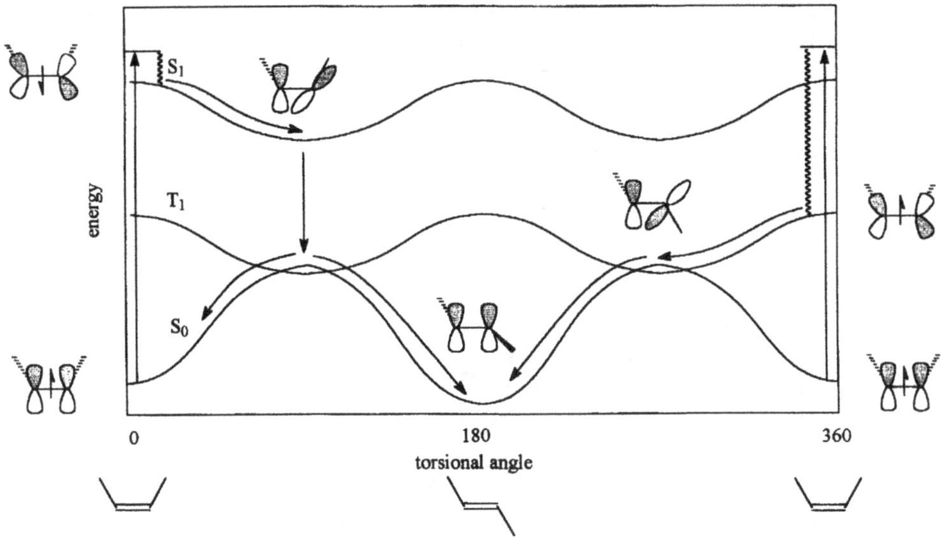

Figure 1. Potential energy surfaces of the singlet and the triplet state as a function of the torsional angle, with the orbitals involved in the *cis-trans* isomerization of *cis*–2-butene. The orbital description outside the Figure represents the S_1 state (left) and the triplet state (right). In the ground state the π^* orbitals are virtual because they are empty.

small activation barrier in the excited state between the *trans* and the perpendicular state [87]. This was attributed to mixing of various higher states depending on the twisting angle. The S_1 state has a minimum at 0°, and its energy increases as the angle increases. However, this state crosses with a doubly excited state configuration (P*), which decreases in energy as the angle evolves to 90° (Figure 3).

This example already shows the complexity of simple isomerizations. Moreover, the conformation of the ground and excited states plays an important role in the photochemistry of the olefin. This becomes even more important for olefins and polyenes bearing relatively large substituents which lower the energy of some of the rotamers with respect to the others, favoring thus the reaction from the lowest energy rotamers [88]. In particular, e.g. conformation-specific double bond isomerization from singlet states has been demonstrated for styrylanthracenes [89].

As already mentioned, the selection of particular rotamers is part of the protein influence in the case of the open-chain tetrapyrroles [58].

The movement of the atoms of large chromophores after excitation is more complex than in the case of 2-butene. Instead of rotation around the double bond, an ensemble of cooperative atom movements rearranges the structure from *cis* to *trans*. For anchored polyenes confined in a rigid medium (such as would be the protein-chromophore cavity) the so-called "Hula Twist (HT)" and "Bicycle Pedal (BP)" mechanisms have been proposed [90,91]. In the HT mechanism the lowest-energy movement involves the concerted rotation of two adjacent bonds (Figure 4). It is a "one photon-one bond" isomerization mechanism. In the BP mechanism two double bonds isomerize simultaneously. It is a "one photon-two bond" isomerization process. In view of the fact

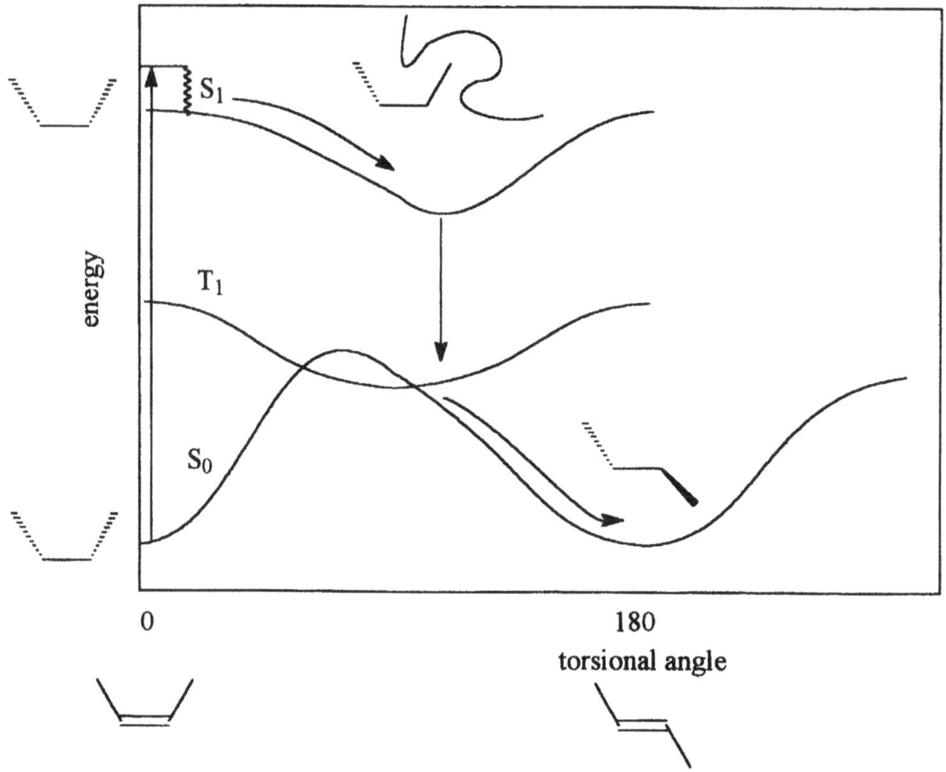

Figure 2. Induction of asymmetry in the potential energy surface due to a specific medium interaction for an isomerization reaction.

that in polyenes only one bond isomerizes with one photon, the BP mechanism was proposed to play a role in the thermal processes. The HT mechanism has been proposed as the mode of operation in a low temperature study of previtamin D [92]. The authors discuss the implications of their mechanistic conclusions for the retinal Schiff base photoisomerizations.

The HT mechanism operates in PYP isomerization leading to the first intermediate observed during time resolved X-ray studies [93]. The so-called "flipping of its thioester linkage with the protein" appears to be exactly what would be expected to happen when the HT process occurs. Flipping the thioester bond instead of moving the aromatic ring minimizes the number of atoms that move and reduces the distance they must travel, thus avoiding collisions during the initial photochemical reaction [93].

2.5 Photosensors with isomerizable chromophores

As already stated, so far three types of photoreceptors have been reported to be based on a *cis-trans* isomerization: retinal proteins, open-chain tetrapyrrole proteins, and proteins containing the *para*-hydroxycinnamoyl anion (Figures 5 to 7).

After excitation, three energy wasting processes may compete with the key isomerization reaction. These processes are light emission, heat release (vibrational relaxation) and a possible additional photochemical process changing the molecule by bond formation or breaking. Formation of the triplet state is not observed in these chromoproteins. A summary is given in Table 2 of the quantum yields of phototransformation, lifetimes, and energies of the first intermediates involved in the photoinduced process for three examples of this class, i.e. PYP [62, 94–96], phy-A [97–102], and SR-I [14, 103]. The photoinduced processes in the latter bear similarities to those in bacteriorhodopsin [104].

In all photoreceptors with isomerizable chromophores (photosensors and bacteriorhodopsin) fluorescence is extremely low (if detectable at all) and the isomerization yield is far from 100%, ca. 65% in retinal proteins and even as low as 16% in the case of phyA [105].

Table 2. Properties of some photosensors undergoing photoinduced *cis-trans* isomerization of the chromophore

	para-hydroxycinnamoyl anion based- PYP	retinal based- SRI	open chain tetrapyrrole based- phyA
λ_{max} (nm) lowest energy band	446[c]	587[g]	666[i]
Isomerized bond	C2=C3 *trans*→*cis*	C13=C14 *trans*→*cis*	C15=C16 *cis*→*trans*
Φ_{iso}	0.35[d]	0.4±0.05[g]	0.16[i]
Φ_{ic} $(1-\Phi_{iso}-\Phi_{fl})$	0.65	0.6±0.05	0.84
Φ_{fl}	3.5×10^{-3} (rt)[e] 0.07 (77 K)[c] 2.5×10^{-3}[f]	<0.02[g]	10^{-3}[j]
$^1E_{00}$ [kJ/mol][a] [nm]	255 kJ/mol[e] 469	187 kJ/mol[g] 640	176.4[k] 678
τ_{s1}	12 ps[e]	<1 ns[g] <3 ps[h]	5–16 ps (85%)[l] 40–60 ps (13%) 150–300 ps (2%)
$E_{first\ transient}$ [kJ/mol] (relative to ground state)	$E_{pr}=120\pm30$[d]	$E_{610}=142\pm12$[g]	$E_{I700}=150\pm7$[k]
$\Delta V_{gs\text{-}first\ transient}$ [ml/mol][b]	−14±2[d]	+5.5±0.3[g]	+7±2[k]

[a] From the crossing of fluorescence and absorption spectra. [b] Interpreted as reaction structural volume change from the ground state to the first transient, as derived from laser-induced optoacoustic spectroscopy. [c] [62]. [d] [96]. [e] [95]. [f] [94]. [g] [103]. [h] [14]. [i] [97]. [j] [98]. [k] [102]. [l] [99–101].

A complete review on the studies performed on the photophysics and photoinduced cycle in phytochrome has been written by Sineshchekov [70]. Using femtosecond time-resolved absorption spectroscopy of phyA, Andel III et al. [106] have concluded that the primary photoisomerization believed to be at C-15 = C-16 [107] is an ultrafast process, similar to that of rhodopsin, occurring on a femtosecond time-scale.

Excited open-chain tetrapyrroles in solution promptly relax to the ground state, showing no permanent photochemistry [58], in contrast to the ultrafast photoisomerization and subsequent thermal steps in the micro- to milliseconds time ranges induced in phytochrome [70,71]. The difference is in part due to the protein enforcement of an extended conformation in the otherwise helical chromophore as well as to chromophore-protein interactions of various types.

The strong environmental influence on the potential energy surfaces and multiplicity of the excited states of the chromophore in photosensors with isomerizable chromophore is already evidenced by the isomerization quantum yields which are always larger for the chromoproteins than for the free chromophore in solution, as well as by the fact that in the chromoproteins isomerization occurs exclusively from the singlet state whereas, e.g. in the case of retinals in solution the triplet state plays an important role [51]. These chromophore-protein interactions determine the lifetimes and spectral properties of the various intermediates in the photoinduced cycles.

Quantum chemical calculations have shown that the ground-state properties of a retinal Schiff base depend on its protonation state and charge environment. The

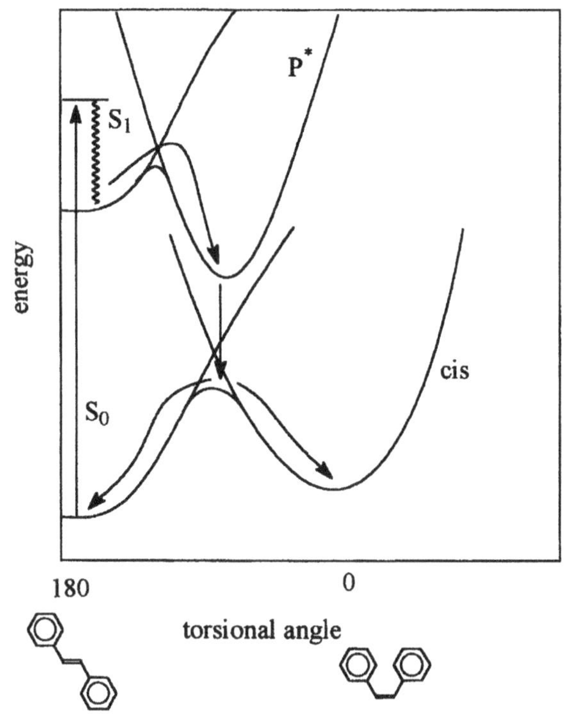

Figure 3. Potential energy surfaces postulated for the isomerization of *trans*-stilbene [87].

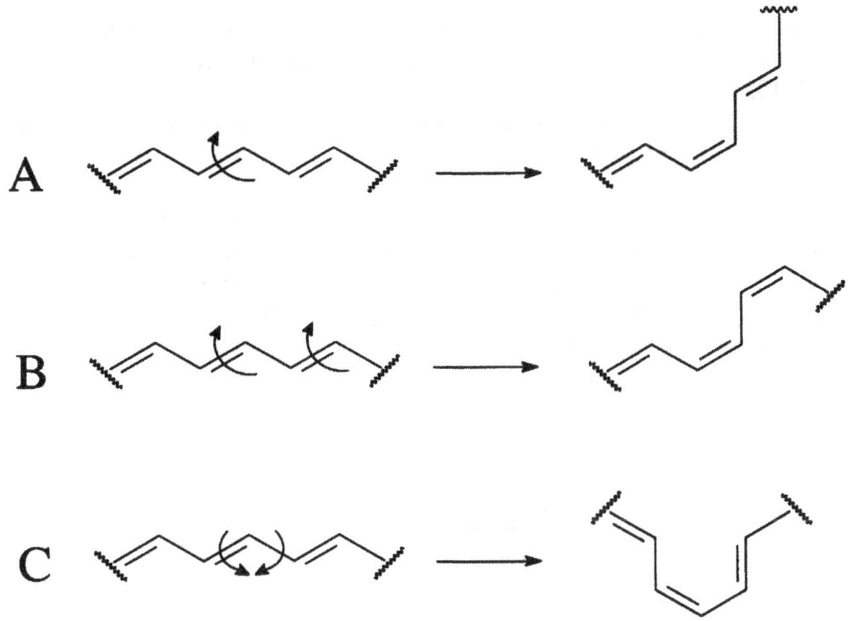

Figure 4. Three possible ways of double-bond isomerization as exemplified by a five-centers array. (**A**) One-bond rotation, (**B**) 'Bicycle Pedal (BP)', and (**C**) 'Hula Twist (HT)' mechanisms [91].

protonation of the Schiff base by the surrounding amino acids and counterions can reduce the double bond isomerization barriers and increase the single bond rotation barriers. This means that protonation (as one of the environmental factors provided by

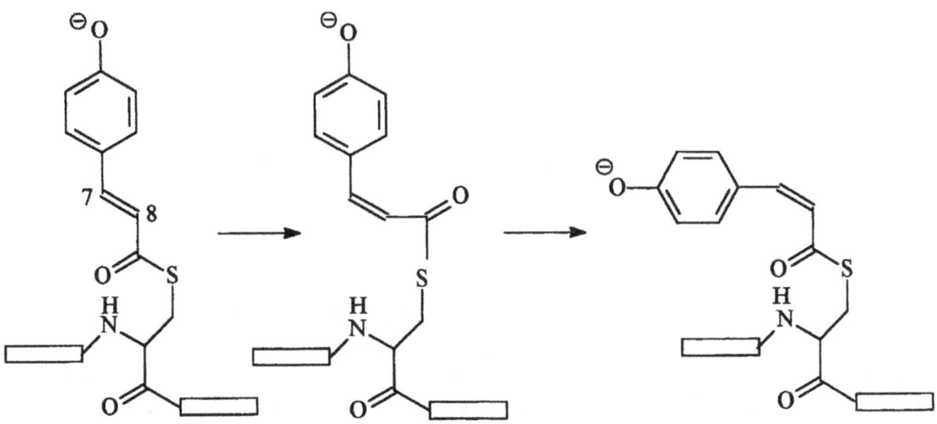

Figure 5. The reaction of 2-trans to 2-cis-para-hydroxycinnamoyl anion linked by a thioester bond to a cysteine residue.

the protein) may favor isomerization of double bonds while impairing rotations around formal single bonds. The influence of the protein dielectric constant has also been discussed [108].

A recent study with a bacteriorhodopsin mutant has offered further spectroscopic evidence for the charge control of the conformational changes in the photocycle of this retinal protein [109].

Similar to the case of retinal proteins, molecular dynamics calculations of the ground and excited states of the open-chain tetrapyrrole phycoviolobilin chromophore in the α-subunit of C-phycoerythrocyanin revealed that the chromophore conformation determines the active-site dynamics. Strong coupling of the excited states localized in the chromophore and charge transfer states from the surrounding polar residues (from the protein) provides favorable prerequisites for fast excited-state surface crossing in

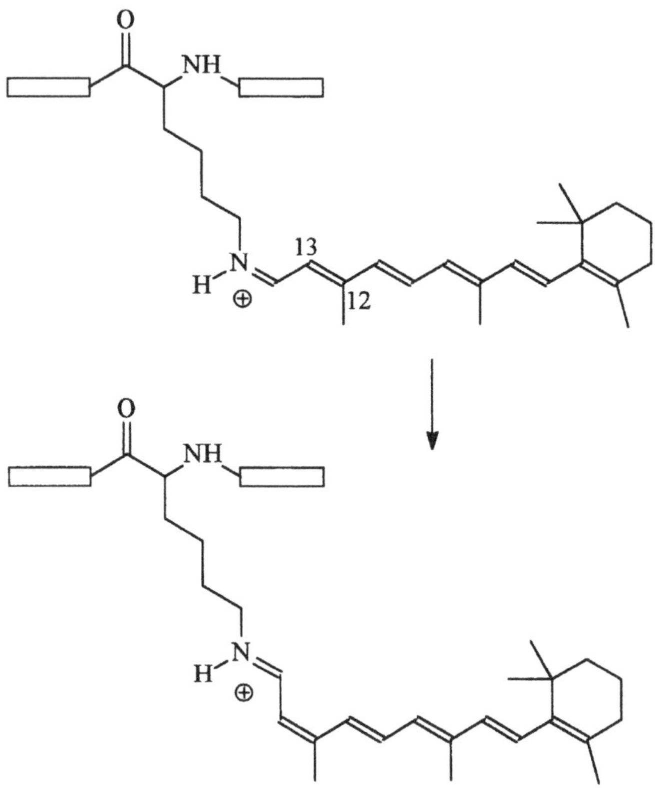

Figure 6. The photoisomerization of the lysine-linked all-trans protonated retinal Schiff base to the 13,14-cis isomer as taking place in halobacterial pigments. The conformations of the lysine chain in the parent all-trans isomer as well as in the 13–14 isomerized chromophore are those proposed by Liu et al. [175] for bacteriorhodopsin in order to keep constant the longitudinal distance between the β-ionone ring and the lysine anchoring end upon double bond isomerization. The isomerized structure is an approximation of the possible structure of the M (II) transient species upon excitation of bacteriorhodopsin, including single bond rotation [176].

competition with other deactivation processes [110]. These calculations have implications for the understanding of the fast isomerization in phytochrome. Modern quantum-dynamic methods make possible the calculation of the form of the potential energy surfaces for the isomerization of complex systems. These calculations indicate that these surfaces can intersect in a conical way, thus opening an easy path for the crossing from the excited surface to the ground state surface. This concept of conical intersections turns out to be very important for all isomerizations [111,112].

It is within the current paradigm that *cis-trans* isomerization is the primary event for the systems described above. In the case of phytochrome this paradigm has been confirmed already using chemical methods [108]. In the case of SRI this has also been supported by experimental evidence [113]. Kort et al. [20] already proved the isomerization of the PYP chromophore *in vivo* and a time-resolved X-ray study of PYP [91] has confirmed this paradigm inasmuch as it clearly showed the isomerized chromophore. However, the prerequisite of isomerization has been questioned. For example, by exchanging the chromophore in PYP by a non-isomerizable, locked compound no loss of photocycle induction was observed [114]. Based on similar

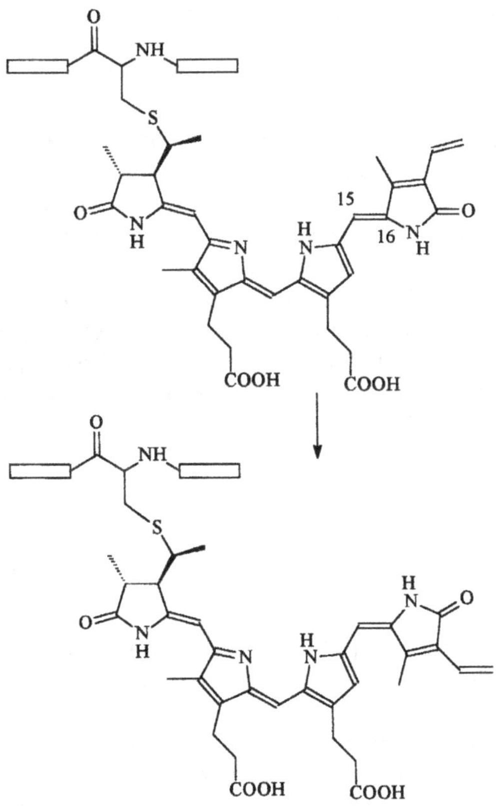

Figure 7. The photoisomerization of (15-Z)-phytochromobilin, which is linked to a cysteine via a thioether bond, to the (15-E)-phytochromobilin. The nitrogen in ring B is shown protonated [120,121].

observations with bacteriorhodopsin, it has been proposed that the mere formation of the excited state can trigger protein movements [115a]. Recent photothermal studies with recombinant opsin assembled with a retinal unable to isomerize have demonstrated, however, that retinal isomerization is a pre-requisite for protein activation [115b].

Powerful tools for establishing the molecular changes in photoreceptors are vibrational spectroscopies (such as infrared and Raman spectroscopy) in combination with quantum mechanical calculations. Novel insights have been gained into mechanistic aspects of various photoreceptors by using resonance Raman spectroscopy. With this technique the vibrational bands directly affected by the excitation and isomerization of the chromophore can be identified. For the retinal proteins and for phytochrome the *cis-trans* isomerization reaction was confirmed to be the primary event [48,116,117]. In combination with isotope labeling of the chromophore, resonance Raman and Fourier transform infrared (FTIR) spectroscopy can identify the bonds involved in the different steps of the photocycles. Using FTIR it has been possible to determine which groups become protonated upon deprotonation of the Schiff base in rhodopsin, in bacteriorhodopsin, and in sensory rhodopsin II from *Natronobacterium pharaonis* [118,119], thus offering a detailed picture of the chromophore-protein interactions during the photocycle. The application of FTIR to phytochrome has allowed to confirm that the chromophore is protonated in the red-absorbing form, Pr, in the far-red absorbing form, Pfr, as well as in several of the intermediates produced in the photocycle [120,121].

Using this technique it has been established that protein regions very near the chromophore respond to chromophore isomerization as early as the K intermediate in bacteriorhodopsin [122] and the K-like intermediate in sensory rhodopsin II [118]. A promising technique for a detailed study of photocycles is time-resolved step-scan spectroscopy, which allows time-resolved IR spectra after excitation [119].

FTIR has been successfully applied to the study of the interactions in PYP between a buried carboxylic group in the protein and the chromophore, in the chromoprotein parent state as well as in the photoinduced transients. These experiments showed that upon light excitation the residue glutamate 46 becomes deprotonated concomitant with chromophore protonation [123].

2.6 Electron transfer

2.6.1 Basic principles

In light-induced electron transfer [124–128] an electron is transferred from a donor (D) to an acceptor (A). Prior to electron transfer one of the components is excited by light absorption (Figure 8).

By this simple sequence, light energy is converted into electrochemical energy. Subsequent to the electron transfer a charge transfer state is created (a dipolar species) consisting of the radical cation of the donor ($D^{+\bullet}$) and the radical anion of the acceptor ($A^{-\bullet}$). Charge separation can also occur between two identical molecules, provided that they are easily reduced as well as oxidized. Electron transfer should be regarded as an extra deactivation path of the locally excited (singlet) state parallel to internal

$$D + A \xrightarrow[\text{excitation}]{h\nu} D^{*} + A$$

$$D^{*} + A \xrightarrow[\text{electron transfer}]{} D^{+\bullet} + A^{-\bullet}$$

Figure 8. Excitation of a donor followed by electron transfer.

conversion (ic), intersystem crossing (isc) to the triplet manifold, emission (f), and photochemical reaction (Figure 9).

Thus, electron transfer should decrease both the emission quantum yield and lifetime. Obviously, other mechanisms may be responsible for fluorescence quenching such as energy transfer, proton transfer, hydrogen bonding, external heavy atom effect and photoreaction. Aggregation may also affect the photophysics and reduce the emission yields.

Electron transfer depends on the redox properties of the donor-acceptor couple and on the excitation energy. In polar solvents the driving force for electron transfer is given by the Gibbs free energy change for charge separation, ΔG_{cs} (eq. 1, the Rehm-Weller equation).

$$\Delta G_{cs}(\mathrm{eV}) = [E_{ox}(\mathrm{D}) - E_{red}(\mathrm{A})] - z^2\, e^2/(\varepsilon_S d) - E_{00} \quad (1)$$

In other words, in order to have a spontaneous electron transfer, the energy put into the system by excitation (E_{00}, the singlet or triplet state energy) should be larger than the energy it costs to oxidize the donor [E_{ox}(D), the oxidation potential of the donor] and

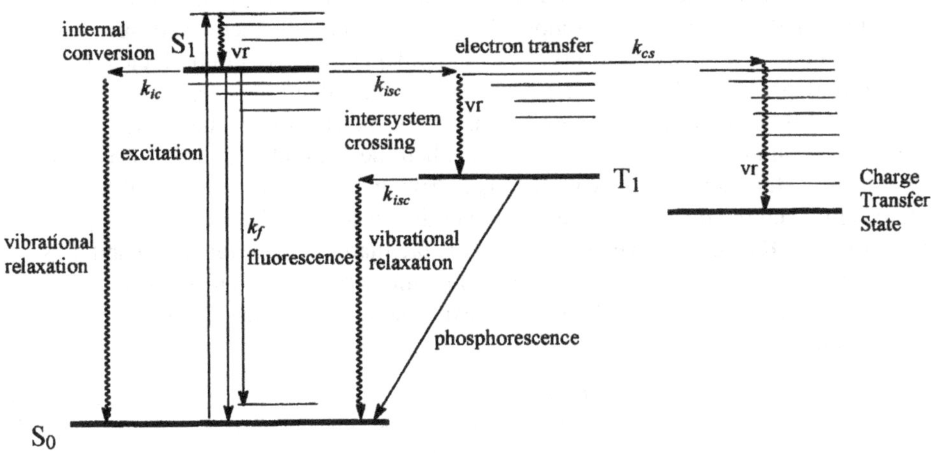

Figure 9. Term diagram including electron transfer as one of the decay pathways of the first excited singlet state.

reduce the acceptor [E_{red}(A), the reduction potential of the acceptor] to form an ion pair. The third term on the right (usually negligible in polar solvents) is the free energy for bringing the two radical ions into an encounter distance d in a solvent with dielectric constant ε_S. Equation 1 may not be valid if significant structural changes are accompanying the electron transfer (which may be the case in biological systems) since it assumes that entropy changes are negligible.

The most commonly studied property in order to determine electron transfer rates and efficiencies is the fluorescence of the excited species. In the absence of electron transfer, the fluorescence quantum yield (Φ_f) and the fluorescence lifetime (τ_f) of a given species are described by using the rate constants of the processes (k_f, k_{ic}, and k_{isc}, Figure 9). In the presence of competing processes, such as the electron transfer to an electron acceptor with a rate constant k_{cs}, the resulting expressions for τ_f and Φ_f are as follows:

$$\Phi_f = \frac{k_f}{k_f + k_{ic} + k_{isc} + k_{cs}}$$

$$\tau_f = \frac{1}{k_f + k_{ic} + k_{isc} + k_{cs}} \qquad (2)$$

The lifetime and quantum yield of the excited state in the absence of electron transfer is regarded to be the reference value (τ_f and Φ_f) and the charge separation rates (k_{cs}) may be evaluated with:

$$k_{cs} = 1/\tau - 1/\tau_{ref}$$

$$k_{cs} = (\Phi_{ref}/\Phi - 1)/\tau_{ref} \qquad (3)$$

The values of rate parameters determining charge separation (k_{cs}) and recombination may be probed by using the time-resolved radical cation and radical anion absorptions with laser flash photolysis in addition to emission data (see e.g. [129]). Other techniques such as time-resolved microwave conductivity [130], laser-induced optoacoustic spectroscopy (also called photoacoustic calorimetry, [131]), resonance Raman [117], electron spin resonance [132], chemically induced nuclear polarization (CIDNP) [133–136] and infrared spectroscopy [119] have afforded kinetic and spectroscopic information on electron transfer processes in model systems and in biological units.

The Marcus theory of electron transfer [137] implies that barrierless electron transfer can occur if the reorganization energy, λ, has the same absolute value as the free energy of charge separation, i.e. $\lambda = -\Delta G_{cs}$. This is called the optimal region. For $\lambda > -\Delta G_{cs}$ the electron transfer rate increases with larger driving forces. However, in the so-called Marcus inverted region, for $\lambda < -\Delta G_{cs}$ the rate **decreases** with larger driving forces. The barrier is very **thin** in the inverted region. This facilitates nuclear tunneling and thus, for processes in the inverted region (charge recombination is often an inverted region process) the observed rates are higher than expected from the classic Marcus theory. The relative value of λ is denoted by the dashed line in Figure 10.

How is the electron transferred from D to A? One of the questions arising when electron transfer processes are studied, is how are we supposed to view the actual transfer of the electron. Colloquially, electrons are often said to jump from D to A, to hop around or to be injected into an acceptor. It has to be realized that when speaking of electrons one can only speak of the probability of finding the electron in a certain region in space.

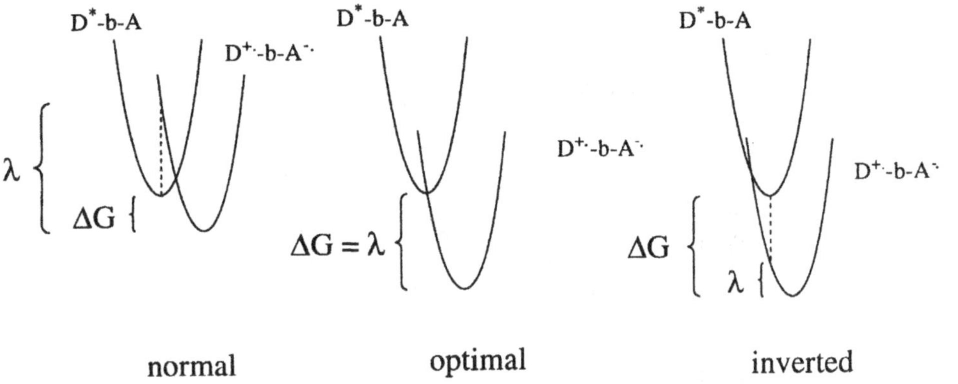

Figure 10. The different regions postulated by the Marcus theory of electron transfer, together with the representations of the Gibbs free energy change and the reorganization energy, λ.

Electron transfer in a system consisting of D and A connected by a bridge b, in which D has the lowest excitation energy, is visualized in Figures 11A and 11B. Bridge b represents any medium between donor and acceptor, not oxidized or reduced during the process.

The three positions on the potential energy curves representing the excited state ($D^* - b - A$) and the charge separated state ($D^{+\bullet} - b - A^{-\bullet}$) are denoted with numbers (1, 2, 3) in Figure 11A and described in further detail in Figure 11B.

Upon excitation of D, situation ($D^* - b - A$) is reached, described with a potential energy curve by making use of the harmonic approximation. An electron is promoted from a low lying state to a state in which most of the electron density is located on D. However, as electronic coupling is present, there is also a very small chance of finding the electron on the acceptor site. After excitation, relaxation leads to the bottom of the first potential energy well and the barrier can be reached.

Having reached the barrier, two things may happen, i.e. (a) crossing to the product state, or (b) remaining in the initial state. At this point, at the top of the barrier, there is a 50% probability of finding the electron on the acceptor side: the orbital that is inherently linked with this situation has 50% electron probability density on the donor and 50% on the acceptor side. This is often described as the situation where the electron is formally transferred as the system is now on the potential energy curve of the product.

Thus, excitation and product formation can be described as a travel across a potential energy surface involving nuclear and solvent reorganization accompanied by an evolution of electron density probabilities on the donor and acceptor sides.

During these events the energy gap ΔE (the energy difference between the initial and the final state at a certain nuclear coordinate) changes drastically. It decreases to become zero at the top of the barrier (now the mixing between the two states is the strongest, like in an orbital interaction diagram). Descending into the $D^{+\bullet} - b - A^{-\bullet}$ potential energy curve, ΔE increases again.

The electron transfer rate can thus be viewed as the time it takes for the wave function (or the time-dependent Schrödinger equation) to evolve from one zero-order state to

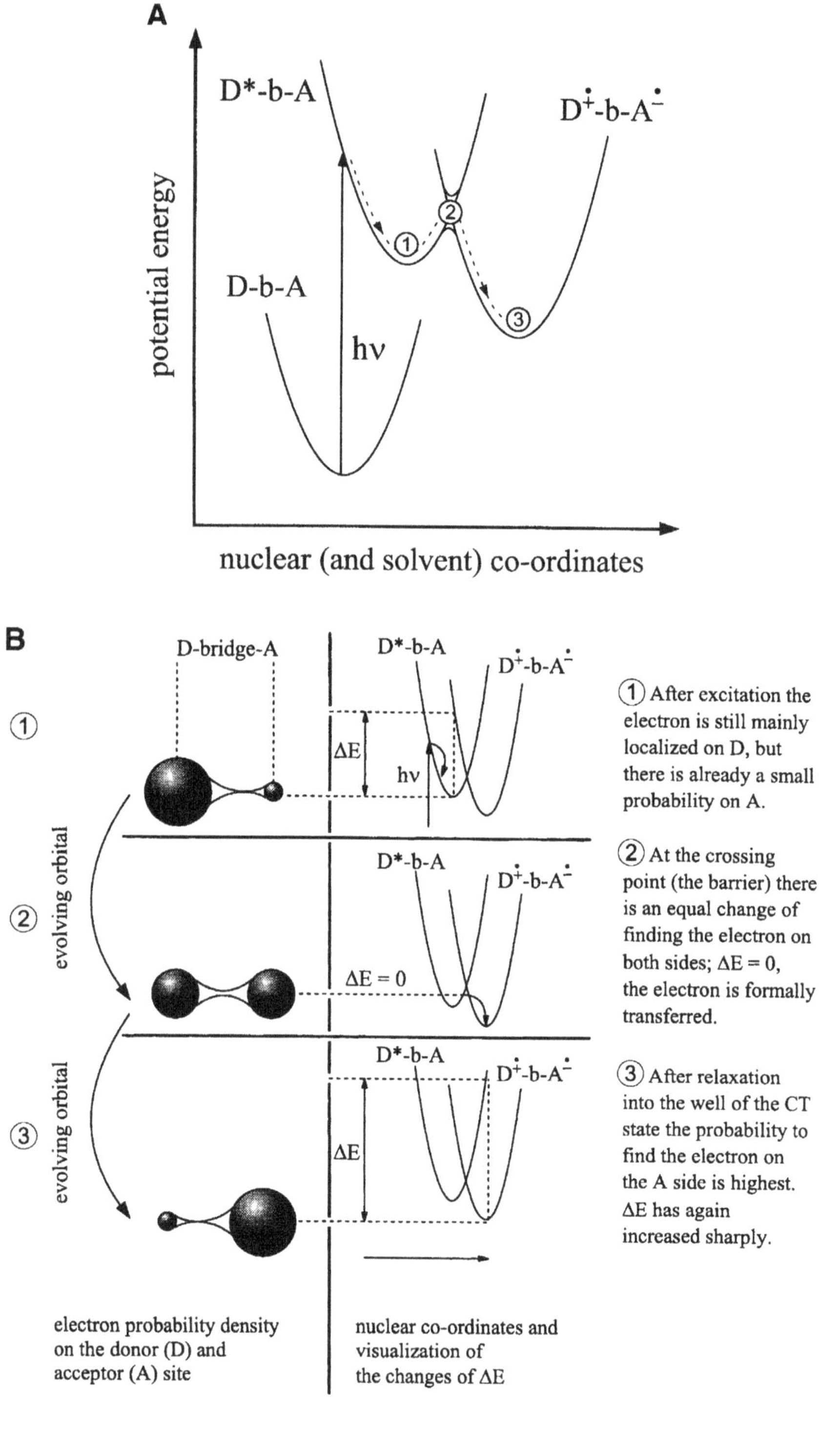
A
potential energy
D*-b-A
D-b-A
hν
nuclear (and solvent) co-ordinates
B
D-bridge-A
evolving orbital
evolving orbital
ΔE
ΔE = 0
ΔE
hν
① After excitation the electron is still mainly localized on D, but there is already a small probability on A.
② At the crossing point (the barrier) there is an equal change of finding the electron on both sides; ΔE = 0, the electron is formally transferred.
③ After relaxation into the well of the CT state the probability to find the electron on the A side is highest. ΔE has again increased sharply.
electron probability density on the donor (D) and acceptor (A) site
nuclear co-ordinates and visualization of the changes of ΔE

Figure 11. (A) Three positions on the potential energy surface representing the ground, the excited, and the charge separated state of a donor-bridge-acceptor system; 1, 2, and 3 are also represented in (B) Left: the electron density probability (also referred to as electronic position) at the donor and at the acceptor site. The evolution of the electron density during the process is displayed. Right: the two parabola represent the initial reactant state and the final product state. As the process proceeds the position on the potential energy surface changes, and thereby the energy gap between the two states becomes smaller (until the barrier is reached) to increase again in the final state (adapted from [137]).

another one. This time becomes longer if the electronic coupling between the two states is weak. The electron density is never the highest on the bridge, i.e. the electron does not localize on the medium between donor an acceptor.

The distance between donor and acceptor and the medium dielectric constant play a crucial role in electron transfer reactions. This has been demonstrated by using D – b – A model systems with various flexible and rigid bridges (see e.g. [138,139], for a review see [140]).

An interesting analysis regarding biological systems is presented by Moser et al. [141]. Although in this analysis the Marcus theory of electron transfer was applied to the bacterial photosynthetic reaction center and models thereof, the conclusions are of general validity to any other electron transfer reaction.

It is important to realize that electron transfer successfully leads to ion separation only when recombination is inhibited, as it is the case in the highly ordered, membrane-embedded photosynthetic reaction centers. Otherwise, recombination is a fast process and predominates. Thus, the charges are to be kept separated by arranging donor and acceptor in such a way that efficient electron transfer occurs. A very ingenious construction of such a system, including a synthetic D – b – A system in an artificial membrane, is described by Steinberg-Yfrach et al. [142]. The artificial unit was able to pump protons upon light excitation, driving then the action of the ATP synthase.

Therefore, the structure in which donor and acceptor are embedded in the protein plays a crucial role in electron transfer [143]. In the case of the flavoproteins it has been shown that the protein determines the redox and protonation state of the chromophore. In the case of the photolyases this is generally $FADH^-$ [144].

Besides photoinduced electron transfer the process called **optical** electron transfer may also play a role. During an optical electron transfer the charge transfer state is directly populated by light excitation in the structureless, long-wavelength charge transfer absorption band, if present.

A simple frontier molecular orbital (FMO) description of the photoinduced electron transfer in a D/A system, in which the acceptor has the lowest excited state, is given in Figure 12. The highest occupied MO (HOMO) and the lowest unoccupied MO (LUMO) of D and A are labeled ***d***, ***d****** and ***a***, ***a****** respectively. Local excitation of the acceptor ($\boldsymbol{a} \rightarrow \boldsymbol{a}^*$) is followed by electron transfer involving orbital interactions between ***d*** and ***a***, whereas charge recombination involves interactions between ***a****** and ***d***. The FMO description is undoubtedly an oversimplification, which may neglect, for example, important interactions between charge transfer configurations and locally excited configurations [145–146]. Nevertheless, the description is a useful starting point for

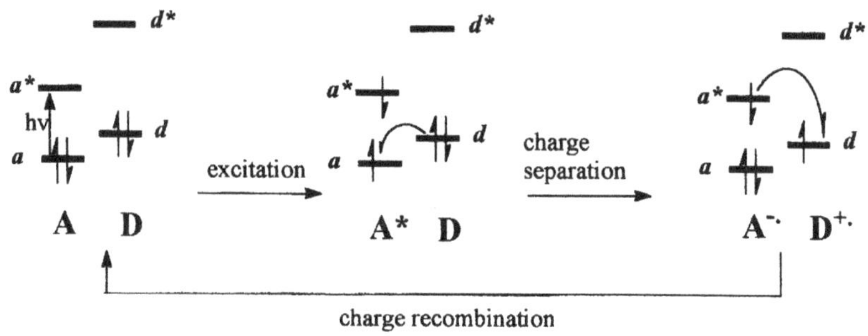

Figure 12. Frontier molecular orbital (FMO) description of photoinduced electron transfer. The highest occupied MO (HOMO) and the lowest unoccupied MO (LUMO) of D and A are labelled d, d* and a, a*, respectively.

identifying orbital interactions that contribute to the electronic coupling involved in the photoinduced charge separation (see e.g. [147]).

2.7 Photosensors with chromophores undergoing electron transfer

It has been shown that various molecules postulated as chromophores in photomovement may actually be involved in electron transfer interactions. They can function as electron donors which can be the case of the perylenequinone chromophores stentorin [36] and blepharismin [148]. It has been demonstrated that the chromophore model hypericin may also function as an electron acceptor [149].

Proton and/or electron transfer were suggested as the primary processes underlying the phototransduction chain in ciliates [35]. More recently, fluorescence quenching studies and EPR spectra analysis indicated that electron transfer from the excited singlet state to an acceptor residue in a protein (such as cystine) should be considered as the primary photoprocess for the photophobic response in *Stentor coeruleus* [36]. However, recent laser-induced optoacoustic studies with blepharismin indicate that the laser-induced structural volume changes strongly depend on the protonation state of the chromophore, suggesting that this parameter indeed could play a role in the first molecular event in signal transduction [150].

As already pointed out, it has been demonstrated that *Arabidopsis thaliana* contains a flavoprotein involved in phototropism. The chromophore was identified as flavin mono nucleotide (FMN) [8]. Cryptochromes also contain FAD and, in addition, methylenetetrahydrofolate (MTHF) [26] but they do not function as photolyases. Flavins are often involved in electron transfer in biological systems. However, the rather well documented class of special enzymes containing flavins, i.e. the photolyases, differ from other flavoproteins in the presence of the second chromophore and in the fact that the amino acids involved in flavin binding are highly conserved. This binding site is also highly conserved in the cryptochromes [151–153], which constitutes a strong argument in favor of a flavin as chromophore in cryptochrome. The photoreceptor protein itself with the chromophore attached to it has so far eluded isolation from any photomotile organism.

It is important to keep in mind the currently accepted mechanism of photolyases which are active in the repair of DNA lesions [154]. The mechanism involves energy transfer from the light-absorbing antenna, composed of MTHF or 7,8-didemethyl–8-hydroxy–5-deazaflavin (8-HDF), to $FADH^-$ (deprotonated and reduced flavin adenine dinucleotide). The latter functions as an electron donor to the thymine dimer [144]. After the generation of the neutral radical $FADH^-$ and the radical anion of one of the thymine monomers, a second charge transfer occurs which leads to the regeneration of the $FADH^-$ and the two thymine monomers [155]. The efficiency of this process is 0.7 to 0.9 [154]. The antenna enhances the overall efficiency of light absorption, as do chlorophyll antennae in plants and algae and phycocyanobilins in red algae and cyanobacteria.

Some of the structures that may play a role in electron transfer-based photomovement are given in Figure 13. The chromophores are flavins [8,156], pterins [157–159], stentorin [22], and blepharismin [160–161].

Properties relevant to a possible electron transfer process of two chromophore model compounds studied in solution or embedded in lipid vesicles, such as hypericin [149,162–169] and riboflavin [156,170–174], are collected in Table 3.

2.8 Outlook

It is clear that much more should be done to firmly establish the mechanisms of action of photoperceptors triggering photomovement and to learn about the molecular partners

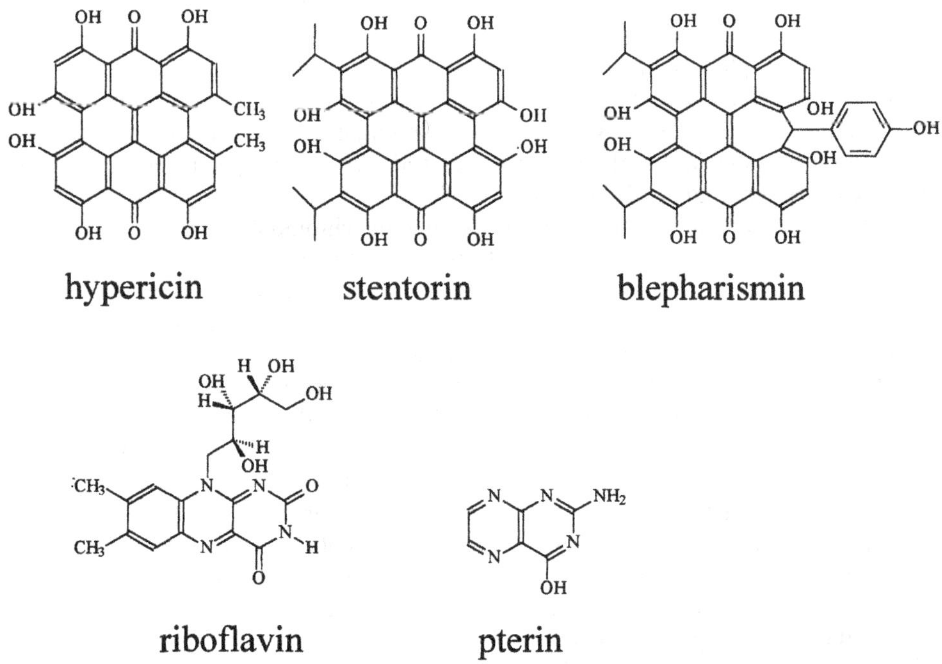

Figure 13. The structures of a flavin, pterin, hypericin, stentorin, and blepharismin.

Table 3. Some photophysical and electrochemical properties of chromophore model compounds able to participate in electron transfer reactions

	Riboflavin	Hypericin
λ_{max} [nm]	450[a]	596[g]
E_{ox} [V]	+ 1.85 (E^0)[b]	+ 0.93 (vs SCE CH_3CN)[h]
E_{red} [V]	–0.195 (E^0)[b] –1.1 (vs $AgNO_3$ CH_3CN)[c]	–0.92 (vs SCE CH_3CN)[i]
Φ_{fl}	0.32[d]	0.3[j]
Φ_{ic}	<0.18	0.27
Φ_T (Φ_Δ)	>0.5 (H_2O)[e]	0.43 (liposomes)[k]
$^1E_{00}$	490 nm[d] 2.53 eV 244.1 kJ/mol	596 nm[g] 2.08 eV 200.7 kJ/mol
$^3E_{00}$	604 nm[e] 2.05 eV 197.8 kJ/mol	758 nm[l] 1.63 eV 157.8 kJ/mol
τ_{S1}	5.2 ns[c]	5.5 ns[g]
τ_{T1}	0.17 s (77 K)[d] 1.7 s (1.8 K)[f]	2.79 ms (1.2 K)[l]

[a] [156]. [b] [174]. [c] [173]. [d] [171]. [e] [170]. [f] [172]. [g] [149]. [h] [165]. [i] [167]. [j] [162]. [k] [169]. [l] [164].

reacting with the photomodified chromophore or chromoprotein. Radical cations and anions should be detectable if electron transfer is involved. However, should the electron transfer process be very fast in vivo, the detection might be precluded. In any case, the combination of molecular biology approaches and the various spectroscopic methods have already brought great advances in this field.

In view of the fact that the protein pocket around the chromophores plays an essential role, it is of great interest to look at the influence of amino acid exchanges in the direct environment of the chromophores, keeping in mind that: (a) in electron transfer reactions the dielectric constant of the whole medium between donor and acceptor, the distance between them, and the reorganization energy determine the occurrence and the rate of electron transfer [141,143], and (b) in photoisomerizations specific interactions between the chromophore and the amino acids residues determine the chromophore conformation [56–58] and electrostatic interactions between chromophore and protein influence the rate and efficiency of the process [108–110].

With the knowledge of the redox properties of photoreceptors it should be possible to identify possible candidates as electron donating or accepting species. Thus, in the case of the perylenequinone compounds, energy considerations with hypericin as chromophore model led to the conclusion that they may undergo an electron transfer interaction with another perylenequinone molecule [149].

The relatively recent discovery of PYP and its chromophore indicates that other blue-light photoreceptors might exist. Extremophile bacteria would be interesting candidates to find such photoreceptors. Organisms reacting to bioluminescence of deep-sea fish or organisms living under other extreme-low-light conditions, such as in the Arctic waters, might possess photoreceptors different from those so far known.

Acknowledgements

We thank Anthony R. Cashmore, Wolfgang Gärtner, Donat-P. Häder, Klaas J. Hellingwerf, Francesco Lenci, Aba Losi, and Jack Saltiel for their comments and suggestions upon careful reading various versions of the manuscript, and Kurt Schaffner for his constant support and fruitful discussions. In spite of this great help, we assume of course full responsibility for the mistakes and missing references. Birgit Deckers and Willi Schlamann were very helpful with the figures. R.M.W. was a recipient of a Marie Curie Research Training Grant under the Training and Mobility of Researchers (TMR) programme of the European Commission (EC, Fourth Frame Programme).

References

1. W. Nultsch, D.-P. Häder (1988). Photomovement in motile organisms – II. *Photochem. Photobiol.*, **47**, 837–869.
2. N.J. Turro (1978). *Modern Molecular Photochemistry.*Benjamin, Menlo Park, CA.
3. T.H. Lowry, K.S. Richardson (1987). *Mechanism and Theory in Organic Chemistry.* Harper and Row, New York.
4. H.P. Koester, W. Rüdiger, D.J. Chapman (1975). About the linkage between chromophore and protein in biliproteins. Degradation studies and spectral analysis of biliproteins. *Lieb. Ann. Chem.*, 1582–1593.
5. J.C. Lagarias, H. Rapoport (1980). Chromopeptides from phytochrome. The structure and linkage of the Pr form of the phytochrome chromophore. *J. Am. Chem. Soc.*, **102**, 4821–4828.
6. W.D. Hoff, P. Duex, K. Hard, B. Devreese, I.M. Nugteren Roodzant, W. Crielaard, R. Boelens, R. Kaptein, J. Beeumen, K.J. Hellingwerf (1994). Thiol ester-linked *p*-coumaric acid as a new photoactive prostetic group in a protein with rhodopsin-like photochemistry. *Biochemistry*, **33**, 13959–13962.
7. M. Ottolenghi, M. Sheves, (Eds) (1995). Special issue on retinal proteins. *Isr. J. Chem.*, 35.
8. J.M. Christie, P. Reymond, G.K. Powell, P. Bernasconi, A.A. Raibekas, E. Liscum, W.R. Briggs (1998). *Arabidopsis* NPH1: A flavoprotein with the properties of a photoreceptor for phototropism. *Science*, **282**, 1698–1701.
9. C. Lin, D.E. Robertson, M. Ahmad, A.A. Raibekas, M. Schuman Jorns, P.L. Dutton, A.R. Cashmore (1995). Association of flavin adenine dinucleotide with the *Arabidopsis* blue light receptor CRY1. *Science*, **269**, 968–970.
10. S. Romagnoli, J.P. Armitage (1999). Roles of chemosensory pathways in transient changes in swimming speed of *Rhodobacter sphaeroides* induced by changes in photosynthetic electron transport. *J. Bacteriol.*, **181**, 34–39.
11. J.W. Verhoeven (1996). Glossary of terms used in photochemistry. *Pure Appl. Chem.*, **68**, 2223–2286.
12. F. Siebert (1990). Retinal proteins. In: H Dürr, H. Bouas-Laurent (Eds), *Photochromism: Molecules and Systems* (pp. 756–785). Elsevier, Amsterdam.

13. J.L. Spudich, D.N. Zacks, R.A. Bogomolni (1995). Microbial sensory rhodopsin: Photochemistry and function. *Isr. J. Chem.*, **35**, 495–513.
14. W.D. Hoff, K.H. Jung, J.L. Spudich (1997). Molecular mechanism of photosignaling by archaeal sensory rhodopsins. *Ann. Rev. Biophys. Biochem. Struct.*, **26**, 223–258.
15. J. Hughes, T. Lamparter, F. Mittmann, E. Hartmann, W. Gärtner, A. Wilde, T. Börner (1997). A prokaryotic phytochrome. *Nature*, **386**, 663.
16. K.-C. Yeh, S.H. Wu, J.T. Murphy, J.C. Lagarias (1997). A cyanobacterial phytochrome two-component light sensory system. *Science*, **277**, 1505–1508.
17. H. Smith (1997). Special Issue on Phytochrome and Photomorphogenesis. *Plant, Cell and Environment*, 20.
18. P.H. Quail (1997). The phytochromes: a biochemical mechanism of signaling in sight? *Bio Essays*, **19**, 571–579.
19. T.E. Meyer, E. Yakali, M.A. Cusanovich, G. Tollin (1987). Properties of a water-soluble, yellow protein isolated from a halophilic phototrophic bacterium that has photochemical activity analogous to sensory rhodopsin. *Biochemistry*, **26**, 418–423.
20. R. Kort, H. Vonk, X. Xu, W.D. Hoff, W. Crielaard, K.J. Hellingwerf (1996). Evidence for *trans-cis* isomerization of the *p*-coumaric acid chromophore as the photochemical basis of the photocycle of photoactive yellow protein. *FEBS Lett.*, **382**, 73–78.
21. G. Rubinstenn, G.W. Vuister, F.A.A. Mulder, P.E. Dux, R. Boelens, K.J. Hellingwerf, R. Kaptein (1998). Structural and dynamic changes of photoactive yellow protein during its photocycle in solution. *Nature Struct. Biol.*, **5**, 568–570.
22. N. Tao, M. Orlando, J.S. Hyon, M. Gross, P.S. Song (1993). A new photoreceptor molecule from *Stentor coeruleus*. *J. Am. Chem. Soc.*, **115**, 1526–1528.
23. G. Checcucci, F. Lenci, F. Ghetti, P.S. Song (1991). A videomicroscopic study of the effect of a singlet oxygen quencher on *Blepharisma japonicum* photobehavior. *J. Photochem. Photobiol. B: Biol.*, **11**, 49–55.
24. M. Ahmad, J.A. Jarillo, O. Smirnova, A.R. Cashmore (1998). Cryptochrome blue-light photoreceptors of *Arabidopsis* implicated in phototropism. *Nature*, **392**, 720–722.
25. M. Ahmad, A.R. Cashmore (1996). Seeing blue. The discovery of cryptochrome. *Plant. Mol. Biol.*, **30**, 851–861.
26. K. Malhotra, S.T. Kim, A. Batschauer, L. Dawut, A. Sancar (1995). Putative blue-light photoreceptors from *Arabidopsis thaliana* and *Sinapis alba* with a high-degree of sequence homology to DNA photolyase contain the two photolyase cofactors but lack DNA-repair activity. *Biochemistry*, **34**, 6892–6899.
27. W. Nultsch, H. Schuchardt, F. Koenig (1983). Effects of sodium azide on phototaxis of the blue-green alga *Anabaena variabilis* and consequences to the two-photoreceptor systems-hypothesis. *Arch Microbiol.*, **134**, 33–37.
28. F. Ghetti, G. Checcucci, F. Lenci (1992). Photosensitized reactions as primary molecular events in photomovements in microorganisms. *J. Photochem. Photobiol. B: Biol.*, **15**, 185–198.
29. J.P. Armitage (1997). Behavioural responses of bacteria to light and oxygen. *Arch. Microbiol.*, **168**, 249–261.
30. M.J. Sackett, J.P. Armitage, E.E. Sherwood, T.P. Pitta (1997). Photoresponses of the purple nonsulfur bacteria *Rhodospirillum centenum* and *Rhodobacter sphaeroides*. *J. Bacteriol.*, **179**, 6764–6768.
31. R. Kort, M.K. Phillips-Jones, D.M.F. van Aalten, A. Haker, S.M. Hoffer, K.J. Hellingwerf, W. Crielaard (1998). Sequence, chromophore extraction and 3-D model of the photoactive yellow protein from *Rhodobacter sphaeroides*. *Biochim. Biophys. Acta*, **1385**, 1–6.
32. R. Kort (1999). *Studies on a bacterial photosensor.* Dissertation, University of Amsterdam.

33. A. Weller (1952). Quantitative Untersuchungen der Fluoreszenzumwandlung bei Naphtolen. *Z. Electrochem.*, **56**, 662–668.
34. H. Shizuka (1985). Excited state proton-transfer reactions and proton-induced quenching of aromatic compounds. *Acc, Chem. Res.*, **18**, 141–147.
35. E.B. Walker, M. Yoon, P.S. Song (1981). The pH dependence of photosensory response in *Stentor coeruleus* and model systems. *Biochim. Biophys. Acta*, **634**, 289–308.
36. T.A. Wells, A. Losi, R. Dai, S.M. Park, J. Goldbeck, P.S. Song (1997). Electron transfer quenching and photoinduced EPR of hypericin and the ciliate photoreceptor stentorin. *J. Phys. Chem. A*, **101**, 366–372.
37. D. Oesterhelt (1998). The structure and mechanism of the family of retinal proteins from halophilic archaea. *Current Opinion in Structural Biology*, **8**, 489–500.
38. Y. Imamoto, K. Mihara, O. Hisatomi, F. Tokunaga, N. Bojkova, K. Yoshihara (1997). Evidence for proton transfer from Glu–46 to the chromophore during the photocycle of photoactive yellow protein. *J. Biol. Chem.*, **272**, 12905–12908.
39. P. Gualtieri (1993). A biological point of view on photoreception (no-imaging vision) in algae. *J. Photochem. Photobiol. B: Biol.*, **18**, 95–100.
40. P.J. Gualtieri (1993). *Euglena gracilis*: is the photoreception enigma solved? *J. Photochem. Photobiol. B: Biol.*, **19**, 3–14.
41. D.-P. Häder, M. Lebert (1998). The photoreceptor for phototaxis in the photosynthetic flagellate *Euglena gracilis*. *Photochem. Photobiol.*, **68**, 260–265.
42. F. Lenci, N. Angelini, A. Sgabarosa (1996). Molecular basis of photoreception. *NATO ASI Ser., Ser. A*, **287**, 147–157.
43. W. Deininger, P. Kroger, U. Hegemann, F. Lottspeich, P. Hegemann (1995). Chlamyrhodopsin represents a new-type of sensory photoreceptor. *EMBO J.*, **14**, 5849–5858.
44. P. Hegemann (1997). Vision in microalgae. *Planta*, **203**, 265–274.
45. B.A. Horwitz, G.M. Berrocal (1997). A spectroscopic view of some recent advances in the study of blue light photoreception. *Bot. Acta*, **110**, 360–368.
46. M.A. Quiñones, E. Zeiger (1994). A putative role of the xanthophyll, zeaxanthin, in blue light photoreception of corn coleoptiles. *Science*, **264**, 558–561.
47. A.K. Janoudi, R. Konjevit, G. Whitelam, W. Gordon, K.L. Poff (1997). Both phytochrome A and phytochrome B are required for the normal expression of phototropism in *Arabidopsis thaliana* seedlings. *Physiol. Plantarum*, **101**, 278–282.
48. R.A. Mathies, S.W. Lin, J.B. Ames, W.T. Pollard (1991). From femtoseconds to biology: mechanism of bacteriorhodopsin's light-driven proton pump. *Ann. Rev. Biophys. Biophys. Chem.*, **20**, 491–518.
49. D. Koch, W. Gärtner (1997). Steric hindrance between chromophore substituents as the driving force of rhodopsin photoisomerization: 10-methyl-13-demethyl retinal containing rhodopsin. *Photochem. Photobiol.*, **65**, 181–186.
50. Q. Wang, G.G. Kochendörfer, R.W. Schoenlein, P.J. Verdegem, J. Lugtenburg, R.A. Mathies, C.V. Shank (1997). Femtosecond spectroscopy of a 13-demethylrhodopsin visual pigment analogue: the role of nonbonded interactions in the isomerization process. *J. Phys. Chem.*, **100**, 17388–17394.
51. A. Feis, B. Wegewijs, W. Gärtner, S.E. Braslavsky (1997). Role of the triplet state in retinal photoisomerization as studied by laser-induced optoacoustic spectroscopy. *J. Phys. Chem. B*, **101**, 7620–7627.
52. M. Sakamoto, A. Wada, A. Akai, M. Ito, T. Goshima, T. Takahashi (1998). Evidence for the archaebacterial-type conformation about the bond between the β-ionone ring and the polyene chain of the chromophore retinal in chlamyrhodopsin. *FEBS Lett.*, **434**, 335–338.
53. G.S. Harbison, S.O. Smith, J.A. Pardoen, J.M. Courtin, J. Lugtenburg, J. Herzfeld, R.A. Mathies, R.G. Griffin (1985). Solid-state C-13 NMR detection of a perturbed 6-s-*trans*

chromophore in bacteriorhodopsin. *Biochemistry*, **24**, 6955–6962.
54. R. Vandersteen, P.L. Biesheuvel, R.A. Mathies, J. Lugtenburg (1986). Retinal analogs with locked 6–7 conformations show that bacteriorhodopsin requires the 6-*s-trans* conformation of the chromophore. *J. Am. Chem. Soc.*, **108**, 6410–6411.
55. D.R. Baselt, S.P.A. Fodor, R. van der Stehen, J. Lugtenburg, R.A. Bogomolni, R.A. Mathis (1989). Halorhodopsin and sensory rhodopsin contain a C-6-C-t-*trans* retinal chromophore. *Biophysical J.*, **55**, 193–196.
56. K. Schaffner, S.E. Braslavsky, A.R. Holzwarth (1990). Photophysics and photochemistry of phytochrome. *Adv. Photochem.*, **15**, 229–277.
57. K. Schaffner, S.E. Braslavsky, A.R. Holzwarth, Protein environment, photophysics and photochemistry of prosthetic biliprotein chromophores. In: H.-J. Schneider, H. Dürr (Eds), *Frontiers of Supramolecular Organic Chemistry and Photochemistry* (pp. 421–452). VCH, Weinheim, 1991.
58. S.E. Braslavsky, A.R. Holzwarth, K. Schaffner (1983). Solution conformations, photophysics, and photochemistry of bile pigments: bilirubin and biliverdin dimethyl esters and related linear tetrapyrroles. *Angew. Chem. Int. Edit.*, **22**, 656–674.
59. S.E. Braslavsky, D. Schneider, K. Heihoff, S. Nonell, P.F. Aramendía, K. Schaffner (1991). Phytochrome models 11. Photophysics and photochemistry of phycocyanobilin dimethyl ester. *J. Am. Chem. Soc.*, **113**, 7322–7334.
60. H. Scheer (1981). Biliproteins. *Angew. Chem. Int. Ed.*, **20**, 241–261.
61. T. Schirmer, W. Bode, R. Huber (1987). Refined three-dimensional structures of two cyanobacterial c-phycocyanins at 2.1 and 2.5 angstroem resolution. *J. Mol. Biol.*, **196**, 677–695.
62. T.E. Meyer (1985). Isolation and characterization of soluble cytochromes, ferredoxins and other chromophoric proteins from the halophilic phototrophic bacterium *Ectothiorhodospira halophila*. *Biochim. Biophys. Acta*, **806**, 175–183.
63. L. Ujj, S. Devanathan, T.E. Meyer, M.A. Cusanovich, G. Tollin, G.H. Atkinson (1999). New photocycle intermediates in the photoactive yellow protein from *Ectothiorhodospira halophila*: Picosecond transient absorption spectroscopy. *Biophys, J.*, **75**, 406–412.
64. W.D. Hoff, I.H.M. van Stokkum, H.J. van Ramesdonk, M. van Brederode, A.M. Brouwer, J.C. Fitch, T.E. Meyer, R. van Grondelle, K.J. Hellingwerf (1994). Measurement and global analysis of the absorbance changes in the photocycle of the photoactive yellow protein from *Ectothiorhodospira halophila*. *Biophys. J.*, **67**, 1691–1705.
65. A. Reichardt (1979). Empirical parameters of solvent polarity as linear free-energy relationships. *Angew. Chem. Int. Ed.*, **18**, 98–110.
66. P.M.E. Mancini, M.G. Gasparri, L.R. Vottero (1995). Determination of the empirical polarity parameter E(T)(30) for binary solvent mixtures. *J. Phys. Org. Chem.*, **8**, 617–625.
67. L. Deforce, S. Tokutomi, P.S. Song (1994). Phototransformation of pea phytochrome-A induces an increase in alpha-helical folding of the apoprotein – comparison with a monocot phytochrome-A and CD analysis by different methods. *Biochemistry*, **33**, 4918–4922.
68. D. Oesterhelt, J. Tittor, E. Bamberg (1992). A unifying concept for ion translocation by retinal proteins. *J. Bioenerg. B*, **24**, 181–191.
69. J.L. Spudich (1994). Protein-protein interaction converts a proton pump into a sensory receptor. *Cell*, **79**, 747–750.
70. V.A. Sineshchekov (1995). Photobiophysics and photochemistry of the heterogeneous phytochrome system. *Biochim. Biophys. Acta*, **1228**, 125–164.
71. S.E. Braslavsky, W. Gärtner, K. Schaffner (1997). Phytochrome photoconversion. *Plant, Cell and Environment*, **20**, 700–706.
72. K.-C. Yeh, J.C. Lagarias (1998). Eukaryotic phytochromes: Light regulated serine/threonine protein kinases with histidine kinase ancestry. *Proc. Natl. Acad. Sci. USA*, **95**,

13976–13981.
73. A. Fankhauser, K.-C. Yeh, C. Lagarias, H. Zhang, T.D. Chory, J. Elich (1999). PKS1, a substrate phosphorylated by phytochrome that modulates light signaling in *Arabidopsis*. *Science*, **284**, 1539–1541.
74. M. Ni, J.M. Tepperman, P.H. Quail (1998). PIF3, a phytochrome-interacting factor necessary for normal photoinduced signal transduction, is a novel basic helix-loop-helix protein. *Cell*, **95**, 657–667.
75. K. Sakamoto (1996). A Nagatani, Nuclear localization activity of phytochrome B. *Plant Journal*, **10**, 859–868.
76. A. Fankhauser, J. Chory (1999). Photomorphogenesis: Light receptor kinases in plants. *Curr. Biology*, **9**, R123-R126.
77. A. Batschauer (1999). Light perception in higher plants. *Cell. Molec. Life Sci.*, **55**, 153–166.
78. J.M. Christie, M. Salomon, K. Nozue, M. Wada, W.R. Briggs (1999). LOV (light, oxygen, or voltage) domains of the blue-light photoreceptor phototropin (nph1): Binding sites for the chromophore flavin mononucleotide. *Proc. Natl. Acad. Sci. USA*, **96**, 8779-8783.
79. N.M. Neff, J. Chory (1998). Genetic interactions between phytochrome A, phytochrome B, and cryptochrome 1 during *Arabidopsis* development. *Plant Phys.*, **118**, 27–36.
80. K. Uchino (1997). New applications of photostrictive ferroics. *Mat. Res. Innov.*, **1**, 163–168.
81. J. Morais, R.R. Hung, J.J. Grabowski, M.B. Zimmt (1993). Charge transfer state photophysics in a rigid molecule: Competition between electron transfer processes in the Marcus normal and inverted regions. *J. Phys. Chem.*, **97**, 13138–13144.
82. A. Wegewijs, M.N. Paddon-Row, S.E. Braslavsky (1998). Volume change associated with large photoinduced dipole formation in a rigid donor-acceptor compound: New approach to optoacoustic volume determination. *J. Phys. Chem. A*, *102*, 8812–8818.
83. C.D. Borsarelli, S.E. Braslavsky (1998). Volume changes correlate with enthalpy changes during the photoinduced formation of the ^{3}(MLCT) state of ruthenium(II) bipyridine cyano complexes in the presence of salts. A case of the entropy-enthalpy compensation effect. *J. Phys. Chem. B*, **102**, 6231–6238.
84. J. Saltiel, Y.-P. Sun (1990). Cis-trans isomerization of C=C double bonds. In: H. Dürr, H. Bouas-Laurent (Eds), *Photochromism: Molecules and Systems (pp. 64–163). Elsevier, Amsterdam.*
85. D.G. Whitten (1993). Photochemistry and photophysics of *trans*-stilbene and related alkenes in surfactant assemblies. *Acc. Chem. Res.*, **26**, 502–509.
86. R.S. Mulliken (1932). Electronic structures of polyatomic molecules and valence. II. Quantum theory of the double bond. *Phys. Rev.*, **41**, 751–758.
87. D.H. Waldeck (1991). Photoisomerization dynamics of stilbenes. *Chem. Rev.*, **91**, 415–436.
88. H.J.C. Jacobs, E. Havinga (1979). Photochemistry of vitamin D and its isomers and of simple trienes. *Adv. Photochem.*, **11**, 305–373.
89. J. Saltiel, Y.X. Zhang, D.F. Sears (1996). Highly efficient conformer-specific adiabatic *cis*→*trans* photoisomerization of *cis*-1-(2-anthryl)-2-phenylethene in S-1. *J. Am. Chem. Soc.*, **118**, 2811–2817.
90. A. Warshel (1976). Bicycle-pedal model for first step in vision process. *Nature*, **260**, 679–683.
91. R.S.H. Liu, D.T. Browne (1986). Bioorganic studies of visual pigments. 3. A bioorganic view of the chemistry of vision – HT-n and BP-m,n mechanisms for reactions of confined, anchored polyenes. *Acc. Chem. Res.*, **19**, 42–48.
92. A.M. Müller, S. Lochbrunner, W.E. Schmid, W. Fu (1998). Low temperature photo-

chemistry of provitamine D: Hula twist isomerization of a triene. *Angew. Chem.*, **110**, 520–522.
93. U.K. Genick, S.M. Soltis, P. Kuhn (1998). Structure at 0.85 angstrom resolution of an early protein photocycle intermediate. *Nature*, **392**, 206–209.
94. T.E. Meyer, G. Tollin, T.P. Causgrove, P. Cheng, R.E. Blankenship (1991). Picosecond decay kinetics and quantum yield of fluorescence of the photoactive yellow protein from the halophilic purple phototrophic bacterium, *Ectothiorhodospira halophila. Biophys.J.*, **59**, 988–991.
95. W.D. Hoff, S.L. Kwa, R. Grondelle, K.J. Hellingwerf (1992). Low temperature absorbance and fluorescence spectroscopy of the photoactive yellow protein from *Ectothiorhodospira halophila. Photochem. Photobiol.*, **56**, 529–539.
96. M.E. Van Brederode, T. Gensch, W.D. Hoff, K.J. Hellingwerf, S.E. Braslavsky (1995). Photoinduced volume change and energy storage associated with the early transformations of the photoactive yellow protein from *Ectothiorhodospira halophila. Biophys.J.*, **68**, 1101–1109.
97. J.M. Kelly, J.C. Lagarias (1985). Photochemistry of 124-kilodalton *Avena* phytochrome under constant illumination in vitro. *Biochemistry*, **24**, 6003–6010.
98. C.G. Colombano, S.E. Braslavsky, A.R. Holzwarth, K. Schaffner (1990). Fluorescence quantum yields of 124-kDa phytochrome from oat upon excitation within different absorption bands. *Photochem. Photobiol.*, **52**, 19–22.
99. A.R. Holzwarth, E. Venuti, S.E. Braslavsky, K. Schaffner (1992). The phototransformation process in phytochrome.I. Ultrafast fluorescence component and kinetics model for the initial $P_r \rightarrow P_{fr}$ transformation steps in native phytochrome. *Biochim. Biophys. Acta*, **1140**, 59–68.
100. H. Kandori, K. Yoshihara, S. Tokutomi (1992). Primary process of phytochrome:initial step of photomorphogenesis in green plants. *J. Am. Chem. Soc.*, **114**, 10958–10959.
101. S. Savikhin, T. Wells, P.S. Song, W.S. Struve (1993). Ultrafast pump – probe spectroscopy of native etiolated oat phytochrome. *Biochemistry*, **32**, 7512–7518.
102. T. Gensch, M.S. Churio, S.E. Braslavsky, K. Schaffner (1996). Primary quantum yield and volume change of phytochrome-A phototransformation determined by laser-induced optoacoustic spectroscopy. *Photochem. Photobiol.*, **63**, 719–725.
103. A. Losi, S.E. Braslavsky, W. Gärtner, J.L. Spudich (1999). Time resolved absorption and photothermal measurements with Sensory Rhodopsin I from *Halobacterium salinarum. Biophys. J.*, **76**, 2183–2191.
104. G.G. Kochendörfer, R.A. Mathies (1995). Ultrafast spectroscopy of rhodopsin. Photochemistry at its best. *Isr. J. Chem.*, **35**, 211–226.
105. T. Gensch, K.J. Hellingwerf, S.E. Braslavsky, K. Schaffner (1998). Photoequilibrium in the primary steps of the photoreceptors phytochrome A and photoactive yellow protein. *J. Phys. Chem. A*, **102**, 5398–5405.
106. A. Andel III, K.C. Hasson, F. Gai, P.A. Anfinrud, R.A. Mathies (1997). Femtosecond time-resolved spectroscopy of the primary photochemistry of phytochrome. *Biospectroscopy*, **3**, 421–433.
107. W. Rüdiger, F. Thümmler (1991). Phytochrome, the visual pigment of plants. *Angew. Chem. Int. Ed.*, **30**, 1216–1228.
108. P. Tavan, K. Schulten, D. Oesterhelt (1985). The effect of protonation and electrical interactions on the stereochemistry of retinal schiff bases. *Biophys. J.*, **47**, 415–430.
109. H.J. Sass, R. Gessenich, M.H. Koch, D. Oesterhelt, N.A. Dencher, G. Buldt, G. Rapp (1998). Evidence for charge-controlled conformational changes in the photocycle of bacteriorhodopsin. *Biophys. J.*, **75**, 399–405.
110. A. Scharnagl, S.F. Fischer (1993). Reversible photochemistry in the alpha-subunit of

phycoerythrocyanin-characterization of chromophore and protein by molecular dynamics and quantum chemical calculations. *Photochem. Photobiol.*, **57**, 63–70.

111. M. Ben-Nun, T.J. Martínez (1998). Ab initio molecular dynamics study of *cis-trans* isomerization of ethylene. *Chem. Phys. Lett.*, **298**, 57–65.
112. M. Ben-Nun, F. Molnar, H. Lu, J.C. Philips, T.J. Martínez, K. Schulten (1998). Quantum dynamics of the femtosecond photoisomerization of retinal in bacteriorhodopsin. *Faraday Discuss.*, **110**, 447–462.
113. A. Yan, T. Takahashi, R. Johnson, F. Derguini, K. Nakanishi, J.L. Spudich (1990). All *trans/* 13 *cis* isomerization of retinal is required for phototaxis signaling by sensory rhodopsin in *Halobacterium halobium. Biophys.J.*, **57**, 807–814.
114. R. Cordfunke, R. Kort, A. Pierik, B. Gobets, G.J. Koomen, J.W. Verhoeven, K.J. Hellingwerf (1998). *trans/cis* (*Z/E*) photoisomerization of the chromophore of photoactive yellow protein is not a prerequisite for the initiation of the photocycle of this photoreceptor protein. *Proc. Nat. Acad. Sci. USA*, **95**, 7396–7401.
115. a) I. Rousso, Y. Gat, A. Lewis, M. Sheves, M. Ottolenghi (1998). Effective light-induced hydroxylamine reactions occur with C-13 = C-14 nonisomerizable bacteriorhodopsin pigments. *Biophys. J.*, **75**, 413–417. b) A. Losi, I. Michler, W. Gärtner, S.E. Braslavsky (2000). Time-resolved thermodynamic changes photoinduced in 5,12-*trans*–locked bacteriorhodopsin. Evidence that retinal isomerization is required for protein activation. *Photochem. Photobiol.*, **72**, 590-597.
116. P. Hildebrandt, C. Kneip, J. Matysik, K. Nemeth, M. ldiko, F. Mark, K. Schaffner (1999). Vibrational analysis of linear tetrapyrroles. Implications for interpreting the Resonance Raman spectra of phytochrome. *Rec. Res. Dev. Phys. Chem.*, **3**, 63–77.
117. P. Hildebrandt, S. Lecomte (1999). Biochemical applications of Raman spectroscopy. In: J. Linder, G. Tranter, J. Holmes, (Eds), *Encyclopedia of Spectroscopy and Spectrometry. Academic Press, London.*
118. M. Engelhard, B. Scharf, F. Siebert (1996). Protonation changes during the photocycle of sensory rhodopsin II from *Natrobacterium pharaonis. FEBS Lett.*, **395**, 195–198.
119. A. Siebert (1997). Monitoring the mechanism of biological reactions by infrared spectroscopy. *Mikrochim. Acta [Suppl.]*, **14**, 43–50.
120. A. Foerstendorf, E. Mummert, E. Schäfer, H. Scheer, F. Siebert (1996). Fourier-transform infrared spectroscopy of phytochrome: difference spectra of the intermediates of the photoreactions. *Biochemistry*, **35**, 10793–10799.
121. A. Kneip, P. Hildebrandt, W. Schlamann, S.E. Braslavsky, K. Schaffner (1999). Protonation state and structural changes of the tetrapyrrol chromophore during the $P_r \rightarrow P_{fr}$ phototransformation of phytochrome. A resonance Raman spectroscopic study. *Biochemistry*, **38**, 15185–15192.
122. T. Kluge, J. Olejnik, L. Smilowitz, K.J. Rothschild (1998). Conformational changes in the core structure of bacteriorhodopsin. *Biochemistry*, **37**, 10279–10285.
123. A.H. Xie, W.D. Hoff, A.R. Kroon, K.J. Hellingwerf (1996). Glu46 donates a proton to the 4-hydroxycinnamate anion chromophore during the photocycle of photoactive yellow protein. *Biochemistry*, **35**, 14671–14678.
124. M.A. Fox, M. Chanon (1988). *Photoinduced Electron Transfer.* Elsevier Science Publishers, Amsterdam.
125. G.L. Closs, J.R. Miller (1988). Intramolecular long-distance electron transfer in organic molecules. *Science*, **240**, 440–447.
126. M.R. Wasielewski (1992). Photoinduced electron-transfer in supramolecular systems for artificial photosynthesis. *Chem. Rev.*, **92**, 435–461.
127. G.J. Kavarnos (1993). *Fundamentals of Photoinduced Electron Transfer.* VCH Publishers Inc.

128. R.A. Marcus (1993). Electron-transfer reactions in chemistry – theory and experiment (Nobel lecture). *Angew. Chem. Int. Ed.*, **32**, 1111–1121.
129. R.V. Bensasson, E.J. Land, T.G. Truscott (1993). *Excited States and Free Radicals in Biology and Medicine*. Contributions from Flash Photolysis and Pulse Radiolysis. Oxford Univ. Press.
130. J.M. Warman, S.A. Jonker, W. Schuddeboom, M.P. Dehaas, M.N. Paddon-Row, J.W. Verhoeven, K.A. Zachariasse (1993). Straight, bent and twisted intramolecular charge separated states as seen by time-resolved microwave conductivity (TRMC). *Pure Appl. Chem.*, **65**, 1723–1728.
131. S.E. Braslavsky, G.E. Heibel (1992). Time-resolved photothermal and photoacoustic methods applied to photoinduced processes in solution. *Chem. Rev.*, **92**, 1381–1410.
132. M.R. Symons (1997). Electron movement through proteins and DNA. *Free Radical Bio. Med.*, **22**, 1271–1276.
133. M. Goez, J. Rozwadowski, B. Marciniak (1996). Photoinduced electron transfer, decarboxylation, and radical fragmentation of cysteine derivatives: A chemically induced dynamic nuclear polarization study. *J. Am. Chem. Soc.*, **118**, 2882–2891.
134. T.V. Leshina (1996). Investigation of the one-electron-transfer stages in complex organic reactions using spin effects. *Kin. Catal.*, **37**, 620–626.
135. N.E. Polyakov, V.S. Bashurova, P.V. Schastnev, T.V. Leshina (1997). Photoinitiated electron transfer interaction of all-trans retinal with electron donors and acceptors. *J. Photochem. Photobiol. A: Chem.*, **107**, 55–62.
136. K. Schaffner, H. Fischer (1993). Electron-transfer to triplet C–60 and subsequent radical-ion reactions studied by time-resolved CIDNP. *J. Phys. Chem.*, **97**, 13149–13151.
137. R.A. Marcus, N. Sutin (1985). Electron transfers in chemistry and biology. *Biochim. Biophys. Acta*, **811**, 265–322.
138. B.P. Paulson, L.A. Curtiss, B. Bal, G.L. Closs, J.R. Miller (1996). Investigation of through-bond coupling dependence on spacer structure. *J. Am. Chem. Soc.*, **118**, 378–387.
139. A. Kusiauskas, P.A. Liddell, S.C. Hung, S. Lin, S. Stone, G.R. Seely, A.L. Moore, T.A. Moore, D. Gust (1997). Structural effects on photoinduced electron transfer in carotenoid-porphyrin-quinone triads. *J. Phys. Chem.*, **B101**, 429–440.
140. J.W. Verhoeven (1999). From close contact to long-range intramolecular electron transfer. *Adv. Chem. Phys.*, **106**, 603–644.
141. C.C. Moser, J.M. Keske, K. Warncke, R.S. Farid, P.L. Dutton (1992). Nature of biological electron transfer. *Nature*, **355**, 796–802.
142. A. Steinberg-Yfrach, J.L. Rigaud, E.N. Durantini, A.L. Moore, D. Gust, D. Gust (1998). Light-driven production of ATP catalysed by F_0F_1-ATP synthase in an artificial photosynthetic membrane. *Nature*, **392**, 479–482.
143. D.N. Beratan, S.S. Skourtis (1998). Electron transfer mechanisms. *Curr. Opinion Chem. Biol.*, **2**, 235–243.
144. A. Sancar (1994). Structure and function of DNA photolyase. *Biochemistry*, **33**, 2–9.
145. M. Bixon, J. Jortner, J. Cortes, H. Heitele, B. Michel (1994). Energy gap law for nonradiative charge transfer in isolated and in solvated supermolecules. *J. Phys. Chem.*, **98**, 7289–7299.
146. M. Bixon, J. Jortner (1999). Electron transfer – From isolated molecules to biomolecules. *Adv. Chem. Phys.*, **106**, 35–202.
147. R.M. Williams, M. Koeberg, J.M. Lawson, Y.Z. An, Y. Rubin, M.N. Paddon-Row, J.W. Verhoeven (1996). Photoinduced electron transfer to C_{60} across extended 3- and 11-bond hydrocarbon bridges: Creation of a long-lived charge-separated state. *J. Org. Chem.*, **61**, 5055–5062.
148. N. Angelini, A. Quaranta, G. Checcucci, P.S. Song, F. Lenci (1998). Electron transfer

fluorescence quenching of *Blepharisma japonicum* photoreceptor pigments. *Photochem. Photobiol.*, **68**, 864–868.

149. G.M. Xia, X.Z. He, Y.L. Zhou, M.H. Zhang, T. Shen (1998). A study of photoinduced electron transfer and redox properties of hypericin. *J. Photochem. Photobiol. A: Chem.*, **114**, 31–35.
150. A. Losi, A. Vecli, C. Viappiani (1999). Photoinduced structural volume changes in aqueous solutions of blepharismin. *Photochem. Photobiol.*, **69**, 435–442.
151. S. Kanai, R. Kikuno, H. Toh, H. Ryo, T. Todo (1997). Molecular evolution of the photolyase-blue-light photoreceptor family. *J. Mol. Evol.*, **45**, 535–548.
152. P. Emery, W.V. So, M. Kaneko, J.C. Hall, M. Rosbash (1998). CRY, a *Drosophila* clock and light-regulated cryptochrome, is a major contributor to circadian rhythm resetting and photosensitivity. *Cell*, **95**, 669–679.
153. A.R. Cashmore, J.A. Jarillo, Y.-J Wu, D. Liu (1999). Cryptochromes: blue light receptors for plants and animals. *Science*, **284**, 760–765.
154. T. Carell, R. Epple (1998). Repair of UV light induced DNA lesions: a comparative study with model compounds. *Eur.J. Org. Chem.*, **7**, 1245–1258.
155. S.T. Kim, M. Volk, G. Rousseau, P.F. Heelis, A. Sancar, M. Beyerle (1994). Laser flash photolysis on an intermediate in the repair of pyrimidine dimers by DNA photolyase. *J. Am. Chem. Soc.*, **116**, 3115–3116.
156. A. Walsh (1980). Flavin coenzymes: at the crossroads of biological redox chemistry. *Acc. Chem. Res.*, **13**, 148–155.
157. P. Galland, H. Senger (1988). The role of pterins in the photoreception and metabolism of plants. *Photochem. Photobiol.*, **48**, 811–820.
158. P. Galland, P. Keiner, D. Dörnemann, H. Senger, B. Brodhun, D.P. Häder (1990). Pterin- and flavin-like fluorescence associated with isolated flagella of *Euglena gracilis*. *Photochem. Photobiol.*, **51**, 675–680.
159. K.V. Neverov, E.A.Mironov, T.A. Lyudnikova, A.A. Krasnovsky, M.S. Kritsky (1996). Phosphorecence analysis of the triplet state of pterins in connection with their photoreceptor function in biochemical systems. *Biochemistry*, **61**, 1149–1155.
160. A. Checcucci, R.S. Shoemaker, R. Cerny, N. Tao, J.S. Hyon, D. Gioffre, F. Ghetti, F. Lenci, P.S. Song, E. Bini (1997). Chemical structure of blepharismin, the photosensor pigment for *Blepharisma japonicum*. *J. Am. Chem. Soc.*, **119**, 5762–5763.
161. A. Spitzner, G. Hofle, I. Klein, S. Pohlan, D. Ammermann, L. Jaenicke (1998). On the structure of oxyblepharismin and its formation from blepharismin. *Tetrahedron Lett.*, **39**, 4003–4006.
162. A. Racinet, P. Jardon, R. Gautron (1988). Generation of singlet oxygen 1-delta-g photosensitized by hypericin, kinetic-study in non ionic micellar medium. *J. Chim. Phys. Phys.-Chim. Biol.*, **85**, 971–977.
163. A. Bouirig, D. Eloy, P. Jardon (1992). Formation and reactivity of singlet oxygen sensitized by hypericin in dipalmitoylphosphatidylcholine liposomes. Evidence of a retarded photooxidation. *J. Chim. Phys.*, **89**, 1391–1411.
164. A. Angerhofer, H. Falk, J. Meyer, G. Schoppel (1993). Lowest excited triplet-states of hypericin and isohypericin. *J. Photochem. Photobiol. B: Biol.*, **20**, 133–137.
165. A. Redepenning, N.B. Tao (1993). Measurement of formal potentials for hypericin in dimethylsulfoxide. *Photochem. Photobiol.*, **58**, 532–535.
166. A.P, Darmanyan, L. Burel, D. Eloy, P. Jardon (1994). Singlet oxygen production by hypericin in various solvents. *J. Chim. Phys.*, **91**, 1774–1785.
167. F. Gerson, G. Gescheidt, P. Haering, Y. Mazur, D. Freeman, H. Spreitzer, J. Daub (1995). Electron-acceptor properties of hypericin and its salts: An ESR/ENDOR and electrochemical study. *J. Am. Chem. Soc.*, **117**, 11861–11866.

168. A. Eloy, A. LePellec, P. Jardon (1996). Protonation et deprotonation dans l'état fondamental et le premier état singulet de l'hypéricine en milieu micellaire nonique. *J. Chim. Phys.*, **93**, 442–457.
169. A. Ehrenberg, J.L. Anderson, C.S. Foote (1998). Kinetics and yield of singlet oxygen photosensitized by hypericin in organic and biological media. *Photochem. Photobiol.*, **68**, 135–140.
170. P.S. Song, T.A. Moore (1968). Mechanism of the photodephosphorylation of menadiol diphosphate. A model for bioquantum conversion. *J. Am. Chem. Soc.*, **90**, 6507–6514.
171. M. Sun, T.A. Moore, P.S. Song (1972). Molecular luminescence studies of flavins.I. The excited states of flavins. *J. Am. Chem. Soc.*, **94**, 1730–1740.
172. T.A. Moore, A.L. Kwiram (1974). Investigation of triplet-state of flavines and flavoproteins by optical detection of magnetic-resonance. *Biochemistry*, **13**, 5403–5407.
173. M. Yasuda, Y. Nishinaka, T. Nakazono, T. Hamasaki, N. Nakamura, T. Shiragami, C. Pac, K. Shima (1998). Photochemistry of flavins in micelles: Specific effects of anionic surfactants on the monomerization of thymine cyclobutane dimers photosensitized by tetra-O-acyl riboflavins. *Photochem. Photobiol.*, **67**, 192–197.
174. N.H. Andersen, M. Hervas, J.A. Navarro, M.A. De la Rosa, J. Ulstrup (1998). Photosensitized electron transfer reactions of cytochrome c_4 from *Pseudomonas stutzeri* with flavins and methyl viologen. *Inorg. Chim. Acta*, **272**, 109–114.
175. R.S.H. Liu, C.W. Liu, X.-Y. Li, A.E. Asato (1991). Butyl conformational reorganization as a possible explanation for the longitudinal flexibility of the binding site of bacteriorhodopsin. The azulene and C–22 retinoid analogs. *Photochem. Photobiol.*, **54**, 625–631.
176. K. Schulten, P. Tavan (1978). A mechanism for the light-driven proton pump of *Halobacterium halobium*. *Nature*, **272**, 85–86.

Photomovement
D.-P. Häder and M. Lebert, editors.

Chapter 3

Action spectroscopy of photomovement

Kenneth W. Foster

Table of contents

Abstract

Action spectroscopy was first used to suggest the nature of pigments, such as in determining the pigments directing photomovement in microorganisms. More recently, by incorporating non-native chromophores, action spectroscopy of photomovement has been used as a tool to study the mechanisms of action of pigments, chromophore synthesis, and the molecular environments of chromophores within photoreceptor pigments. The evolution of photoreceptors has been advanced by the identification of photopigments in microorganisms belonging to different branches on the evolutionary tree suggesting a branching of the crown eukaryotes. Theoretical considerations, experimental techniques, difficulties, practical advice and the applications are discussed.

3.1 Introduction

Purpose. This review presents to a biological audience how to carry out, interpret and understand what has been done with the powerful technique of action spectroscopy. The motivation was that more could be done if the principles underlying the methods were understood. Thereby, the field might advance more quickly and with greater impact.

Activation spectroscopies. A variety of spectral measurements are associated with activation of a pigment by light or other means. For example, the excitation spectrum for fluorescence is an activation spectroscopy. Some aspect of the actual result of absorption of light is measured. Action spectroscopy is a special case of activation spectroscopy in which the "effect" of light is usually a biological response. This review is restricted to this subclass of activation spectroscopy particularly as associated with photomovement.

What Is an Action Spectrum? An action spectrum is a spectrum, a measurement as a function of wavelength or photon energy, that is proportional to the activation cross-section of a pigment that causes a biological response or effect. This involves an absorption process that leads to a molecular transformation of interest (such as fluorescence or behavior). This transformation occurs in only a fraction, the quantum efficiency, of the absorptions. For example, rhodopsins under certain conditions yield a behavioral response about two-thirds of the time or fluorescence in one part in ten thousandths of the time. The activation cross-section is then the quantum efficiency for activation of the response times the absorption cross-section of the pigment. The absorption cross-section is a quantitative measure of pigment absorption as discussed further below. Frequently, the measure of action is the light intensity of a particular wavelength to give a fixed response, most typically the threshold. The reciprocal of this threshold intensity gives a measure, the sensitivity, which is proportional to the activation cross section. Only such a true spectrum can be correlated to absorption spectra.

Photomovement. Free-swimming cells swim in some pattern spatially related to a continuous source of light giving them the property of phototaxis (Figure 1). This

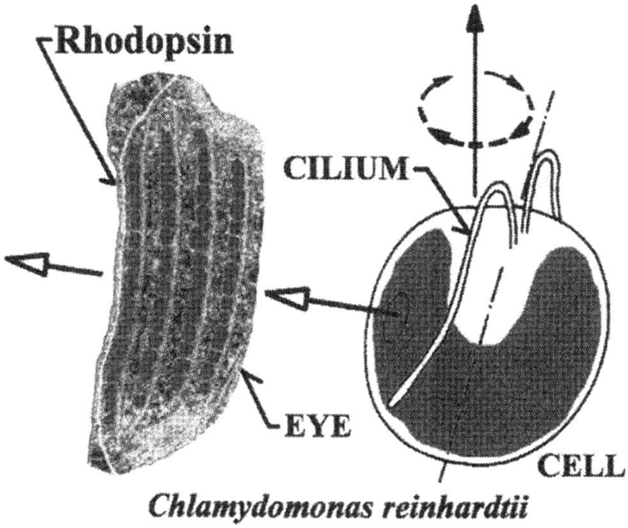

Figure 1. Drawing of *Chlamydomonas reinhardtii* with an electron micrograph of its eye, arrow points in the direction it sees with maximum sensitivity. Solid arrows show direction of swimming and simultaneous conical scanning. The open arrows show the direction that the eye or antenna is maximally sensitive. The figure is modified from Figures 7 and 23 [22].

response is graded being proportional to the amplitude of the modulated continuous stimulus. In the absence of orienting stimuli, free swimming cells rotationally diffuse. For example, the orientation and direction of swimming of *Chlamydomonas* cells has a relaxation time of 2 seconds. Cells confined to a surface may crawl with the same phototaxis property. Cells that are solitary may bend or induce growth in a light dependent manner, which is called phototropism or polarotropism if in response to polarized light. Naturally, given these behaviors there necessarily must be responses to transient stimuli. If photoreceptor molecules are oriented in one direction then they have oriented transition dipole moments. Absorption by light is proportional to the square of the cosine of the angle between the electric component of the light and the dipole moments of the receptor resulting in detection of polarized light.

Cells may also change their motion because of a large change in light intensity, which may trigger an action potential, typically called a photophobic, stop or ecclitic response [1]. This is an ungraded response. A stop of locomotion, a change of direction and a resumption of on-going motion characterize this response. The pattern is typical of the species and usually is the same regardless of the nature of the stimulus. A pause and reversal of motion may occur before the change of direction. The new direction may be due to rotational diffusion during the pause or due to a specific response with respect the orientation of the body. The turn may or may not be directional.

Subcellular organelles such as chloroplasts may also rotate, redistribute or cluster because of light stimuli. Finally, the rate of movement is dependent on the illumination in photokinesis. The direction of the light beam is immaterial. Typically, light is providing the energy for movement.

3.1.1 Activation compared with absorption spectroscopy

In absorption spectroscopy, the absorption cross section of the pigment is measured from the light transmitted through the sample. Situations when action spectroscopy are preferred are: (a) when the virtue of its selectivity in a complex environment is important, and (b) when its association with a particular response is needed. Regular light absorption spectroscopy works best on pure nonscattering material. In acoustical spectroscopy, the adverse effect of light scattering is minimized, but the result is still the absorption of all components in the sample. Action spectroscopy is selective even in a complex environment, because it can be directly associated with a particular function and whether a particular action occurred. It can be done in vivo under normal physiological conditions; however, light scattering can still adversely influence the measurement.

Historically, the data for action spectroscopy usually is different from that gathered for absorption spectral studies. This is not inherent to the spectroscopy, but a choice of the experimenter. Typically in action spectroscopy, the light intensity is varied to give a relatively constant response, whereas in absorption spectroscopy, the response is measured at a relatively constant light intensity. Consequently, in action spectroscopy, measurement of sensitivity has a fractional or relative error compared to the fixed or absolute error associated with most forms of absorption spectroscopy. For this reason alone, in action spectra, in order to insure an equal error in ordinate values, the logarithm of the receptor sensitivity should be plotted.

3.1.2 Uses of action spectroscopy

In biology, action spectroscopy is readily assayed by the change in expression of some gene [2], an observed morphological change [references in [3]], the production of oxygen as in photosynthesis [4] or some aspect of behavior such as sight [5]. The primary focus in this chapter is on the application of action spectroscopy to photomovement of small organisms or suborganelles whose shortest dimension is less than 1 mm. Typically, this restricts hydrodynamics to the low Reynolds number regime (on the order of 1 or less). The movement is confined to either creeping (phototaxis), differential growth (phototropism) or if free swimming to conical-scanning light trackers. An example of creeping is the behavior of the forminiferin, *Amphistegina radiata* [6]. Differential growth is represented by the zygomycete *Phycomyces blakesleeanus* [7]. The green alga *Chlamydomonas reinhardtii* [8], chytridiomycete zoospores such as *Allomyces reticulatus* [9], the stramenopile *Mallomonas,* the euglenoid *Euglena gracilis* and animal larvae such as that of sponges and jellyfish [10] conically scan as they swim freely. For this review, I will limit consideration to those situations in which one may continuously monitor the behavior of the organism within the time scale of its response.

The most common applications of action spectroscopy in photomovement have been to either infer the nature of an organisms light environment or to suggest the nature of a particular photoreceptor or photoreceptor system responsible for a response. In photomovement studies, the uses of action spectroscopy have been broadened. The mechanism of activation of pigments has been studied in vivo ([11–14]. Chromophore synthesis [15], induction of gene expression [16] the molecular environments of chromophores within photoreceptor pigments [13] have been well documented. The evolution of photoreceptors has received some attention.

One major advantage is its extraordinary sensitivity compared to other methods such as enzyme assays or other spectroscopies. A cell has its own amplification system for sensory stimuli. Therefore, it is possible with the threshold phototaxis assay to measure the activity of a single molecule on the average per cell and hence the response of a photoreceptor replaced by an analog whose sensitivity is four orders of magnitude less than the native one. For example, the sensitivity with *trans*-β-apo-14′-carotenal incorporated was 0.1% of that for all-*trans*-retinal in *Chlamydomonas* [15]. Due to its extra length of four carbons, it can barely fit into the binding site. Clearly, this response would not have been seen with the typical in vitro enzyme assay [17] for rhodopsin activation, which has a steady-state background level between 3 and 5%.

A second major advantage is that the conditions of testing are inherently physiological. If it happens, unlike in almost any other spectroscopy, there are no questions about whether it occurs in vivo because it is in vivo.

3.1.3 Reviews of action spectroscopy

One may read Mast [18] for the early literature and the status at that time. A must read is the excellent primer by Duysens [19]. Then for details one may read from the following partial list: [3,20–28].

Reviewer's Caveat. There has been no attempt to be exhaustive with respect to the literature and please inform me if great ideas have been missed. Examples have been taken only from the Eukarya. Häder [29] has reviewed Archaeal and Bacterial photomovement. My emphasis has been on suggestions to broaden the application of photomovement action spectroscopy and on refinements in the traditional applications. On biological terminology, I have been guided by the advice of Manton [30] to use cilia uniformly instead of eukaryotic flagella, preserving the word flagella for the rotating motile structures of bacteria and archaea, and Andersen et al. [31] for other terms.

3.2 History

According to Mast [18] the study of the response of microorganisms to different colors of light owes its early development primarily to Darwin's Origin of the Species (1859) and the desire "to find evidence of mental faculties in the lower organisms". However, Mast (1871–1947) had the modern view that the "results . . . will throw light on the nature of these chemical changes in the organisms, which are associated with the reactions to light". He emphasized the necessity of spectral purity (monochromatic light), calibration so as to give the "relative stimulating efficiency of the different wavelengths", and careful measurement. "It is also highly essential to ascertain the stimulating efficiency for all regions in the spectrum that are at all effective, not merely for those that are most effective". These guiding principles are as important today as they were over 80 years ago. His spectra were relative action spectra in which the standard was varied by precise attenuation (reduction in light intensity) to exactly match the effect of the test wavelength. In these organisms, euglenoids, green algae, or insect larvae, if two stimuli incident from 90 degrees apart were equivalent the cells would swim or move bisecting this angle. By matching the standard intensity and reading off the attenuation from his Lummer-Brodhun rotating-sector attenuator, he obtained the relative effectiveness of the two stimuli. As was true for his time, he plotted his spectra in terms of response per incident energy rather than photon flux density. Historically, we may excuse this oversight since it took Milliken until 1916 to demonstrate the truth of Einstein's 1905 photon theory, "bold, not to say reckless, hypothesis", [32]. What has come to be known as the Stark-Einstein equivalence theory [33] means that if an electronic excitation occurs, then one photon absorbed has the possibility of leading to the excitation of a single molecule. This means that the sensitivity of response is proportional to the number of potentially absorbed photons rather than the total incident energy. This is why the response per number of photons and not per energy must be used when constructing action spectra. With this modernization, a few of Mast's [18] action spectra are included in Fig. 5B for those that look rhodopsin-like and in Fig. 6A for those that look flavoprotein-like. One may note the easy identification in hindsight of rhodopsin for all the green algae and animals and flavoprotein receptors for all the euglenoids.

3.3 Action spectroscopy to infer the nature of the light environment

For an organism to optimally respond in nature there needs to be both a spatial and spectral match between a receptor and its light environment. Nature only cares about the net result of the matching, and not whether it is easy to assign the different contributing factors, like the component due to the absorption spectrum of the photopigment. However, if the question is predicting the light environment of an isolated organism, then one cares only about the spectral match and not how it is achieved. If one has reason to believe that this photoresponse is important enough to the organism to have been optimized then one may tentatively assume this spectral match to predict the light environment. In this case, one wishes to measure response at several realistic light levels in the environment away from threshold. The important caveat is that nature does not typically have optimized systems.

3.4 Action spectroscopy to suggest the nature of the pigment

In this case, all appropriate corrections should be made to remove any error due to cell screening, self-screening of pigments, the presence of other reactions, cell geometry, polarization, etc. The classical admonition was that one could do action spectroscopy when one had parallel logarithmic photon irradiance-response curves, the system is weakly absorbing and if only one primary reaction influenced the effect [19]. One must take into account that these criteria are rarely met in photomovement studies.

3.4.1 Phototaxis of free-swimming organisms

If the cells are not close to a surface, small free-swimming organisms necessarily rotate as they swim. Hence, if they have localized photoreceptors, the internal absorption and scattering modulates the light signal for phototaxis. The magnitude of response (R) will be proportional to the magnitude of this wavelength-dependent modulation, A (λ). In detail, $R(\lambda) \propto n_r \cdot \varphi_r \cdot \sigma_r(\lambda) \cdot I_r(\lambda) \cdot A(\lambda) \cdot \tau$.

n_r is the number of receptor molecules. n_r is not generally constant for more than a day or in some cases a few hours, so that an action spectrum should be completed in an equivalent short time.

φ_r is the quantum efficiency or the probability an absorbed photon leads to the response.

φ_r is in general constant over each absorption band, but not necessarily in different bands.

$\sigma_r(\lambda)$, the absorption cross section, is the probability that one pigment molecule in a thin layer of the given area will be excited by one photon passing through the layer. In other words, it is the area associated with each absorbing molecule such that if a photon hits that area it will be absorbed, otherwise it will pass through.

I_r is the photon intensity at the receptor [in general $I_r = I_i \cdot T_e(\lambda)$, where $T_e(\lambda)$ is the fraction transmitted to the receptor from the externally incident light, I_i].

$T_e(\lambda)$ is constant and near one typically only when the receptor is in the cell surface membrane.

$A(\lambda)$ is the fractional absorption of the screen that modulates the intensity on the receptor as the cells rotate.

τ is the integration time of the detector. τ varies with temperature, but is otherwise thought to be constant with wavelength.

These multiple wavelength (λ) dependent parts cannot be separated in a one parameter determination. If n_r, φ_r, $T_e(\lambda)$, and τ are fairly constant then one may determine a spectrum.

If the spectrum is constructed by measuring the intensity that gives some fixed response at different wavelengths we have the equal response action spectrum, $1/I(\lambda) \propto \sigma_r(\lambda)\ A(\lambda)$. If the stimulus light contains equal numbers of photons at each wavelength, we have the equal stimulus action spectrum. $I(\lambda)$ is constant, so that $R \propto \sigma_r(\lambda)\ A(\lambda)$. If both intensity and response are allowed to vary, we obtain the response-per-intensity action spectrum, $R/I(\lambda) \propto \sigma_r(\lambda)\ A(\lambda)$. All have the same form being proportional to the product of the receptor and screen extinctions [22]. For proper interpretation of these spectra each parameter needs to be separated or made wavelength independent [8]. Since this has been rarely done, most published spectra of free-swimming organisms are the products of the activation spectrum and screening-absorption spectrum [22].

3.4.2 Photon-irradiance response curves of absolute action spectra

An absolute action spectrum, as distinct from a relative action spectrum, is one in which response at a single wavelength, e.g. threshold for phototaxis, is measured against no stimulation. In a relative action spectrum two stimulated conditions are compared. The modulation frequency in *Chlamydomonas* is 2 Hz at 20°C compares to the orientation relaxation time of 2 s at the same temperature. Measurement of the threshold action spectrum can separate the wavelength-dependent screening from the estimate of pigment absorption. There is an additional argument for threshold action spectra. Near threshold, if a pigment is made because of excitation of the first pigment, the concentration of this second pigment will be low and therefore have little possibility of being activated by light in the experiment.

Threshold action spectra are particularly free from artifacts for free-swimming microorganisms whose photopigment is on the exterior membrane of the cell. In this case, there is no absorption between the light source and the pigment $[T(\lambda) \approx 1]$. Moreover, self-screening is negligible since there is only a single molecular layer of pigment. Most motile green algae cells and chytridiomycete zoospores are of this sort. Since the cell rotates, the light received by the receptor is modulated by the absorption and scattering of light through the cell compared to the direct unfiltered absorption. The magnitude of phototaxis response for any particular stimulus is a product of the magnitude of this modulation times the cell's photoreceptor system response (response/modulation) at that frequency of temporal modulation. One potential problem is presence of quarter-wave stacks or other light guiding devices (light antennae) which modify the incident light intensity in the region of the receptor as a function of wavelength (Figure 2). Therefore, it is valuable to use mutants that suppress these structures [8]. Note the maximum theoretical increase is a factor of four in the electrical energy density (the part of the electromagnetic wave that results in absorption) and this will not be nearly approached in practice. With the appropriate assay, the linear extrapolation of the supra-threshold responses to threshold gives an accurate relative

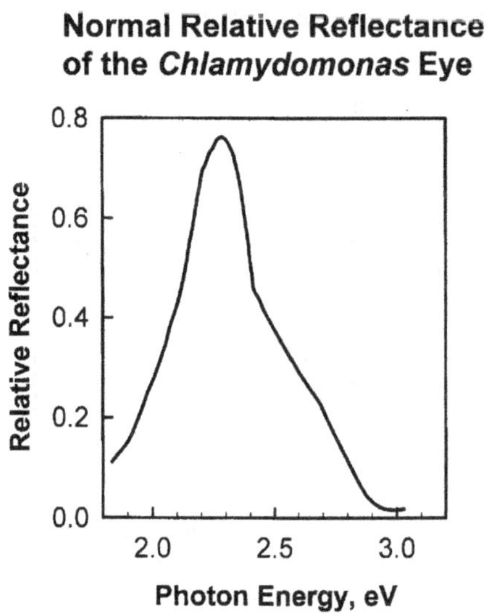

Figure 2. Measured reflection from the eye of *Chlamydomonas reinhardtii* using a Leitz microspectrophotometer (Foster, unpublished).

value of the sensitivity of the pigment at each wavelength. The threshold is not influenced by the absorption and scattering modulation. The slope of response versus the logarithm of the irradiance is proportional to the amplitude of the stimulus modulation and may strongly vary with wavelength. The slopes are proportional to the fraction of light transmitted. Figure 3A shows the effect of high and low contrast on the slope of the irradiance response curve in a cell population assay. In other words, the slope can be used to estimate the pigment absorption and the threshold the amount of pigment.

3.4.3 Assays that give straight-line fits with log irradiance

Straight-line fits can be obtained when Weber-Fechner's law [34] applies and the data is analyzed appropriately. For the free-swimming cell, response is proportional to ln $[(I_{dark}+I_r)/(I_{dark}+T_b(\lambda)I_r]$. I_r is the light intensity at the receptor with the minimum attenuation during the rotation cycle. $T_b(\lambda)$ is the fraction of light transmitted between the views with minimum and maximum attenuation in one rotation cycle. I_{dark} is the thermally activated spontaneous activations of receptor in the dark. Near threshold, the response is proportional to $[(I_{dark}+I_r)-(I_{dark}+T_b(\lambda)I_r]/[(I_{dark}+I_r)+(I_{dark}+T_b(\lambda)I_r)] = [I_rA(\lambda)]/[2I_{dark}+(1+T_b(\lambda))I_r]$, where $A(\lambda)$ is $1-T_b(\lambda)$. $A(\lambda)$ is the fractional absorption of the screen between the two compared samples of the irradiance at wavelength (λ). There is also the potential for light loss between the incident light I_i and that which reaches the receptor which in the minimally attenuated view, i.e. $I_r = I_i \cdot T_e(\lambda)$. In the

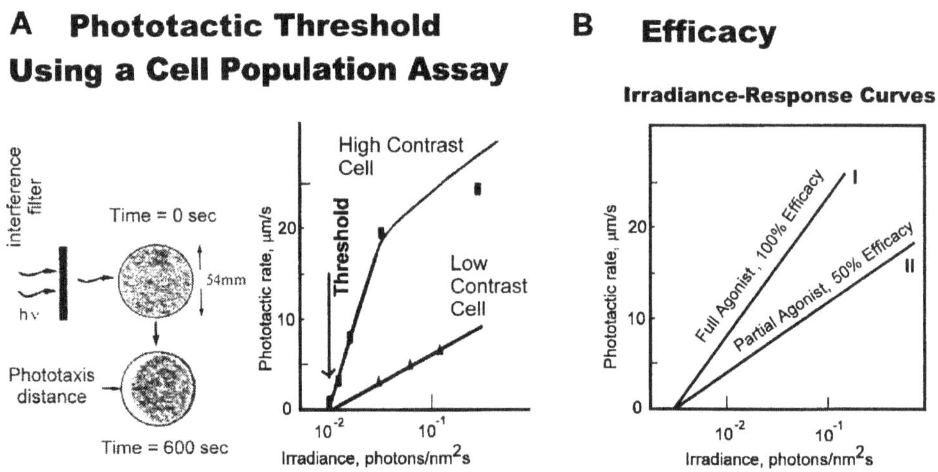

Figure 3. A. Method for measuring phototactic threshold using a population assay of free-swimming microorganisms. A different slope of the irradiance-response curves for high and low contrast cells are shown. The slope is proportional to the amplitude of the modulation.B. Efficacy for the same amplitude of light modulation, a cell with some lower-efficacy analogs (partial agonists) incorporated will show a lower slope compared to the native chromophore. Slope of II is one half of I.

free-swimming phototaxis referred to above, $T_e(\lambda)$ is almost 1.0 at all wavelengths and can be ignored. Response is then $\propto I_r\ A(\lambda)/I_{dark}$, which goes to zero as I_r becomes small.

The phototaxis population method taking advantage of Boltzmann transport. In this technique, phototactic cells are sufficiently concentrated; i.e. cells are colliding with each other that the negative phototaxis-swimming rate is about one-half that of non-colliding cells (Figure 3A). For *Chlamydomonas* a concentration of about 6×10^6 cells/ml is sufficient. Under this condition free swimming is limited by the frequent collisions of cells and a front forms as the cells remaining on the cell-diluted side towards the light bang into the cells in front. Since the solution next to the front is clear, this collision front is easy to see and its progress is easy to follow. The measurement involves a large number of cells, so this method has high signal to noise. This movement of a "shock" wave due to the freely swimming cells colliding with the more packed cells in front is described by Boltzmann in his consideration of particle transport. It leads to a linear rate correlation with the log of photon irradiance, which is convenient for doing threshold action spectra. The slope is dependent on the modulation contrast as the cell rotates and can be separately plotted as a measure of the intracellular screening [35]. Nultsch et al. [36] also found the rate of free-swimming phototaxis for *Chlamydomonas* was linear with log of irradiance.

The slope of the irradiance-response curve is also dependent on the size of the cell response for a given modulation. This may be less than normal when the native chromophore is substituted by an analog, which has lower efficacy for producing the phototaxis response. This is quite rare, but has been observed (see below). Efficacy, in this case, is defined as the ratio of the slopes of the phototactic rate per log photon irradiance of test analog compared to that of the native chromophore (Figure 3B). Lower efficacy or partial agonism can occur if the temporal response is altered.

Free-Swimming Tracking. This technique has the advantage that one knows what each individual cell does, such as positive or negative taxis or both, however, even with a video system it is quite difficult to count as many cells as can be conveniently accommodated by the simpler Boltzmann-transport population method. Consequently, the signal to noise is typically lower and the threshold determination less accurate. The main advantages are that it can be used when only a low concentration of cells is available, when cells are transparent, or when the cells have poor motility.

In the modulated situation of free-swimming rotating cells, if the cells have a uniform direction of phototaxis then the following procedure may be best (Figure 4). Subtract the number of cells going away from the light from the number towards the light and then divide by the total number of cells counted to give the modulation function. Typically, this parameter plots as a straight line as a function of log irradiance [9]. This example has an assumption that has to be verified, namely a particular distribution of swimming velocities as a function of intensity. If at low light intensities the velocities have random orientations, but show increasingly bimodal orthogonal alignment with the light beam at higher intensities then one could compare cells aligned within some angles to those not aligned [37]. This analysis can also be applied to crawling cells. For example, the crawling of *Amphistegena* [6] could be reanalyzed and plotted this way (Figure 5D).

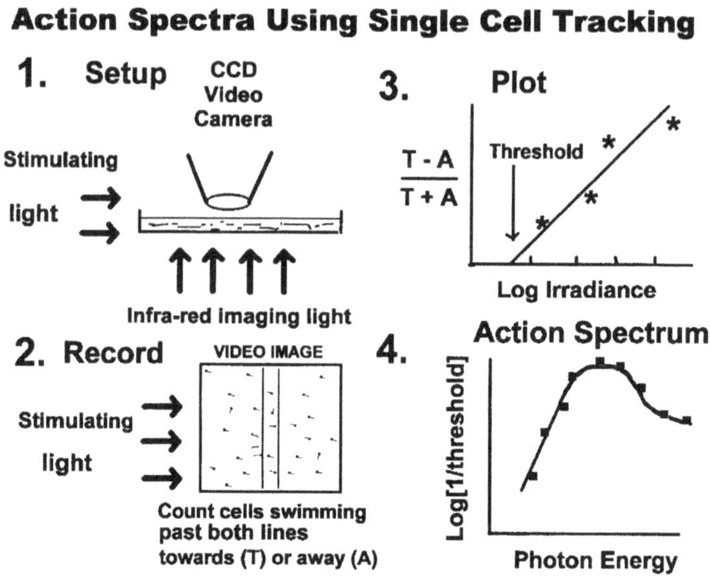

Figure 4. Method for measuring phototactic thresholds and action spectra using tracking of individual free-swimming organisms.

Application to phototropism. The same approach could be used for action spectroscopy of phototropism such as that of the sporangiophores of *Phycomyces*. Using the Fast Orthogonal Search program [38] or past experience one can choose the primary response frequency (1/600 s) or a ten-minute period in the case of the light-growth response which underlies the phototropism of *Phycomyces* sporangiophores. Then the cells can be stimulated at this period, for example, light on for 5 min and off for 5 min, and the amplitude of the peak-to-peak modulation of light-growth rate used as the response to plot as a function of log irradiance.

Other instances from the literature. In the stramenopiles: The aggregation rate for chloroplasts in *Vaucheria* [39] (Figure 6B) is nicely linear. Phototropic bending angle for *Vaucheria* is also linear and the reciprocal of threshold gives flavoprotein like action spectrum (Figure 6F) [40]. In the plasmodial slime molds: The initial rate of clearing within the plasmodium of *Physarum polycephalum*, called photoavoidance, increases linearly with log photon irradiance making it possible for Ueda et al. [41] to use the reciprocal of threshold as their measure of "action" (Figure 6E). Some multistate or multipigment complexity seems to exist at high light intensities as the amplitudes of the peak response varies with wavelength. Other examples of linearity with log photon irradiance are phototaxis of the aveolate *Stentor coeruleus* [42], rate of phototaxis of the red algae *Porphyridium cruentum* [43], the percentage of protonemata of the fern *Adiantum capillus-veneris* showing diastrophe [44], and phototropism of the rhizoid of the green algae, *Boergesenia forbesii* (Cladophorophyceae) [45]. Diastrophe is the state in which the chloroplasts gather above and below the central axis of the protonema and none remain along the sidewalls.

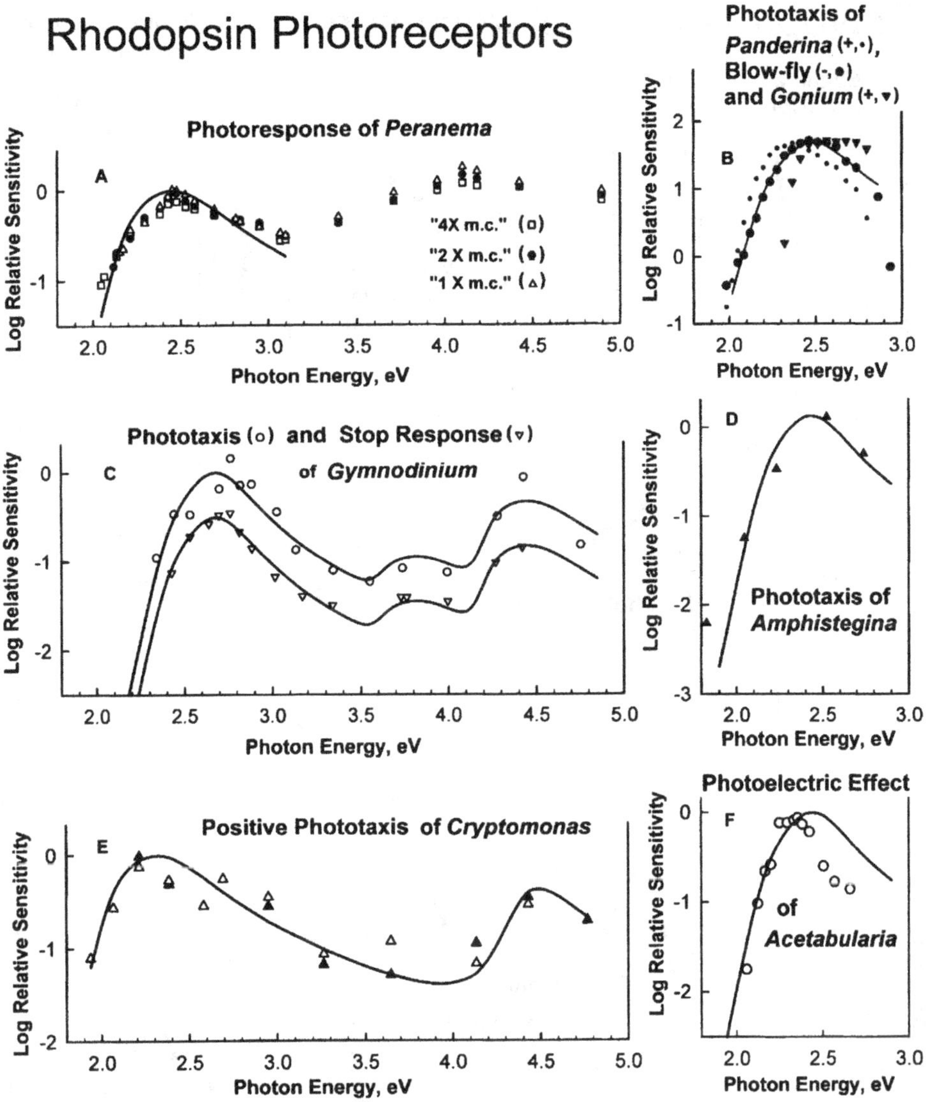

Figure 5. Rhodopsin-like action spectra replotted from published data using the methods described in the text.A. The reaction time of the ecclitic response of *Peranema trichophorum* [54,55] (peak at 2.45 ± 0.01 eV). The original plots were transformed using irradiance-response information in the paper. They show spectral broadening due to spectrally impure light from a prism-dispersed source.B. Positive phototaxis of the chlorophyte *Panderina morum* (2.42 ± 0.01 eV), negative phototaxis of the blow-fly larvae (2.485 ± 0.005 eV) and positive phototaxis of the chlorophyte *Gonium pictorale* (2.63 ± 0.01 eV), [18].C. Phototaxis and ecclitic or stop response of the dinoflagellate *Gymnodinium splendens* Lebour [61] (2.68 ± 0.01 eV).D. Crawling phototaxis of *Amphistegina radiata* [6] (2.43 ± 0.02 eV). The lowest energy point (678 nm) was eliminated from the curve fitting since colored (not blocked interference) filters give erroneous results two orders of magnitude below peak sensitivity.E. Positive phototaxis of *Cryptomonas rostratiformis* [79] (2.32 ± 0.02 eV).F. Photoelectric effect of *Acetabularia crenulata* [60] (2.40 ± 0.02 eV). The smooth curves are the rhodopsin standard curves shifted to their respective maxima with a half spectral width at half maximum (τ) of 0.21 eV.

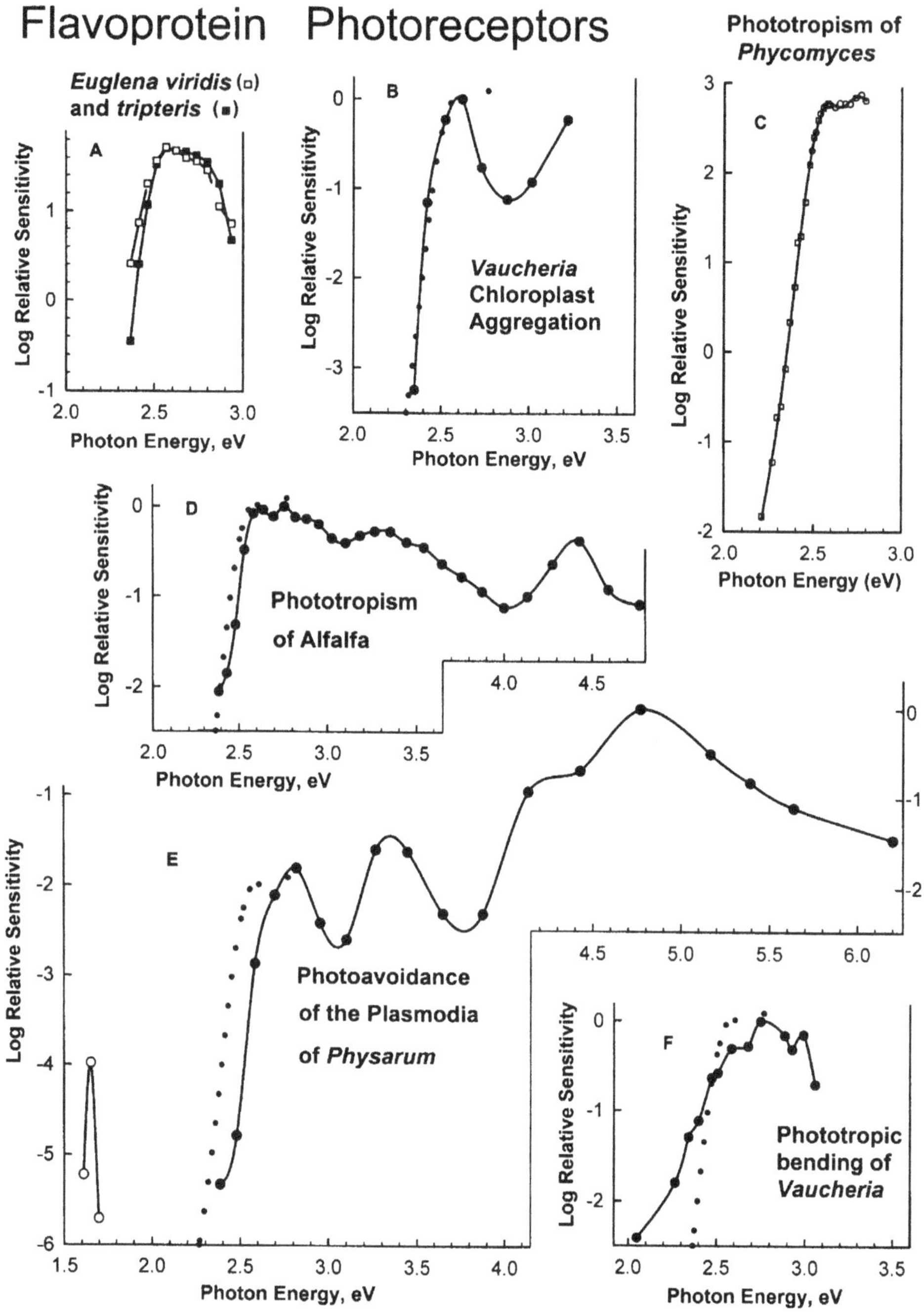

Figure 6. Flavoprotein/pterin-like action spectra replotted from published literature using methods described in the text.A. Phototaxis of *Euglena viridis* □ and *tripteris* ■ [18], B. Chloroplast aggregation in *Vaucheria* [39], C. Phototropism in the sporangiophores of *Phycomyces blakesleeanus* [78], D. Phototropism in alfalfa [48], E. Photoavoidance of the plasmodia of *Physarum polycephalum* [41], F. Phototropic bending of *Vaucheria geminata* [40]. To make comparisons easier, the curve in C has been dotted in B, D, E and F and all plots are on the same scale.

Examples of incorrect analysis in spite of the linearity between response and the log of irradiance. Non-threshold criterion responses were used in spite of the slopes of irradiance-response curves varying with wavelength. In Song et al. [42] action spectrum of *Stentor coeruleus*, they used 50% phototaxis (number "away"/total) as their criteria of response. Nultsch and Schuchart [43] plotted the half maximum of peak response of the red algae, *Porphyridium cruentum* (replotted in Figure 12 using thresholds). Yatsuhashi et al. [44] used 30% of the protonemata of the fern *Adiantum capillus-veneris* showing diastrophe response. Marangoni et al. [46] used unspecified criteria from plots of response versus irradiance for *Fabrea salina* phototaxis (replotted using thresholds in Figure 7C). Kadota et al. [47] used 10° as a response criterion for the phototropic response of the protonemata. In *Boergesenia forbesii* rhizoid phototropism Ishizawa and Wada [45] used 25°. Baskin and Iino (1987) treated part of their data using the threshold, but not all, making it difficult to compare the two parts of the spectrum (Figure 6D). Although it has long been a requirement for determination of action spectra that the irradiance-response curves be parallel in order to use finite-response criteria, the cause and the correction of this problem has often not been considered.

Application to crawling phototaxis on or near a surface. Potentially the same method could probably be applied to crawling cells such as the response of the pseudopods of amoebae. The variability of the action spectra reported for the movement responses of *Amoeba proteus* is quite astonishing ([49] and references therein, [50]).

A modulation procedure can be used in any situation in which the continuous temporal behavior of an organism can be monitored. However, when the integral or final response, e.g. Northern Blots, are used as the response for the photoinduction of a gene the methods are different. This situation is outside the scope of this review.

Advantages of this approach. Both of the above techniques yield a straight-line fit of response with the log of irradiance. For this reason, the threshold can be determined from measurement of responses at only three or four irradiances above threshold. A complete photon-irradiance-response curve going to high intensities is unnecessary [35]. Further, these straight-line fits need not be parallel as required in classical action spectroscopy [24].

This innovation enables an accurate action spectrum of cellular behavior over a wide spectral range to be finished in a few hours and in principle minutes with the proper equipment. For most microorganisms, this is important since the spectra sensitivity of their behavior is only stable for a day or hours. One may then average the complete spectra obtained from different days allowing for their different sensitivities. This improved speed of measurement has significantly extended application of this technique, as discussed below.

3.4.4 Assays that give linear fits with photon irradiance or hyperbolic tangent curves with log photon irradiance

Traditionally, these methods have been favored in spite of the difficulty of applying them to free-swimming phototaxis. Threshold is much more difficult to determine because the photon irradiance-response curve has to be fit by a hyperbolic tangent or related curve $[(S/(S+S_{1/2}) \equiv (1/(1+S_{1/2}/S)$, S being the irradiance and $S_{1/2}$ being the irradiance that gives 50% response] (see [26]). To get an "operational" threshold, one has to determine the tangent line at the inflection point or midpoint of the curve. To know the midpoint or inflection point requires, in general, many points of data apart from all the points to define the curve. In addition, to define this curve one has to make measurements at much higher intensities than near threshold with the consequence that

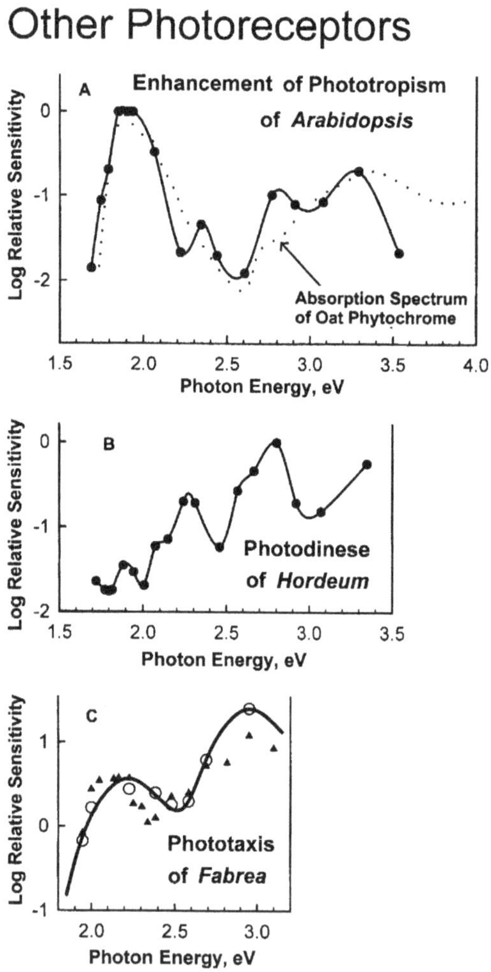

Figure 7. Action spectra due to other receptor pigments replotted from published literature using methods described in the text. A. *Arabidopsis thaliana* enhancement of phototropism [150], dotted curve is the absorption spectrum of oat phytochrome (Figure 13.33, [151]), B. *Hordeum vulgare* photodinese [152] (pigments unknown), C. *Fabrea salina* phototaxis [46], solid curve is sum of two rhodopsin curves, $\tau = 0.21$ eV, photon energy maxima at 2.22 ± 0.01 eV and 2.95 ± 0.02 eV fit to ○ from thresholds determined from response per log irradiance plot using data of Figure 5, ▲ from Figure 6.

frequently additional pigments may become involved and the situation is no longer simple.

Use of a criterion response, e.g. $S_{1/2}$, other than threshold, for the rotating cell, introduces an error due to the modulation of the screening. The value of using threshold for action spectra plots is clearly seen when data for criterion response were replotted as threshold response. For example, the action spectra for *Dunaliella* phototaxis [51] is replotted along with the original in the right most curve of Figure 8. *Fabrea salina* phototaxis [46] is replotted along with the original in Figure 7C. *Chlamydomonas* phototaxis [36] is replotted in Figure 17 of Foster and Smyth [22]. The replotted points fit parts of the rhodopsin standard curve.

Non-threshold action spectra, such as equal photon and equal response action spectra and their analysis are discussed at length in Foster and Smyth [22].

If there is no wavelength dependent modulation, more than one photochemical reaction, strong absorption of pigments or significant screening of receptor pigments then a particularly simple situation may prevail. Plots of $R[I(\lambda)]$ versus $\ln[I_r(\lambda)]$ at several important wavelengths will be needed to establish whether the above conditions hold. If these curves are superimposable except with respect to translation along the log irradiance axis then a simple method may be used. A further requirement is that it must be possible to measure the size of response with good precision in the linear range of the response function. If all requirements are met, only one parameter is needed

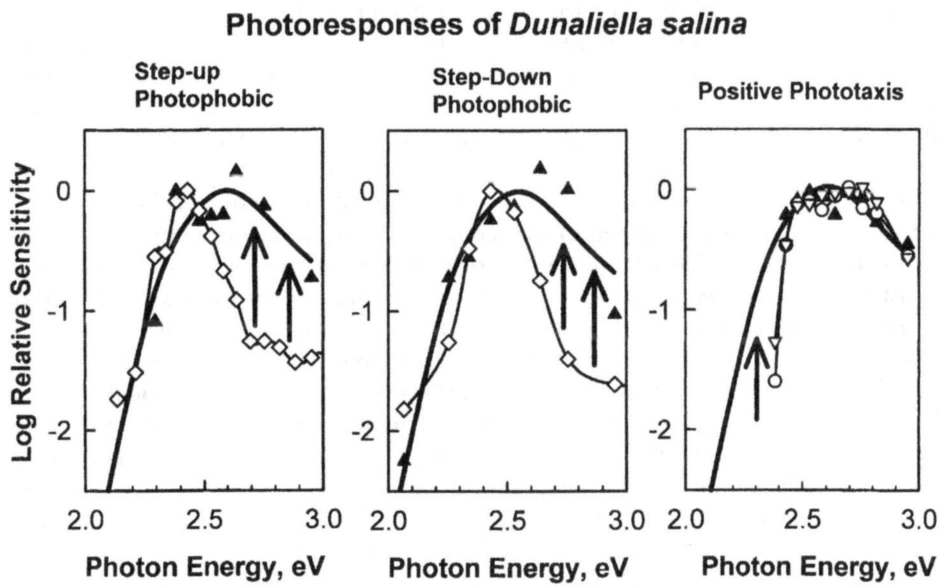

Figure 8. Threshold sensitivity versus finite criteria for response. Step-up (peak at 2.60 ± 0.04 eV), step down (2.55 ± 0.02 eV) and positive phototaxis (2.61 ± 0.02 eV) responses for *Dunaliella salina* show a rhodopsin-like pigment [51] using the methods of the text. The 'threshold' method values are in filled symbols, the original analysis is in open symbols traced by a thin line. The thick line shows a common rhodopsin-curve fit with the arrows indicating how the original spectrum was distorted by using an above threshold criterion for response.

from the analysis, as there is only one wavelength dependent parameter in the response equation, since $R[I(\lambda)] \propto n_r \cdot \varphi_r \cdot \sigma_r(\lambda) \cdot I_r(\lambda) \cdot \tau$. The ratio $R[I(\lambda)]/I_r(\lambda) \propto n_r \cdot \varphi_r \cdot \sigma_r(\lambda) \cdot \tau$ may be measured to determine relative values of $\varphi_r \cdot \sigma_r(\lambda)$. Although more measurements are desirable to improve the precision, in principle only one measurement of $R[I(\lambda)]$ at a suitable $I_r(\lambda)$ will be necessary to determine one point of the spectrum.

3.4.5 How to obtain an action spectrum from the irradiance-response curves

The x-intercept of the irradiance-response curve is the photon irradiance at the threshold for response. The reciprocal of this threshold irradiance is the sensitivity. An action spectrum is obtained by plotting the negative of the log of this photon irradiance or the log sensitivity as a function of the photon's energy (Figure 4). Particularly noteworthy in this plot is that the low photon-energy cutoff for the receptor is a straight line. It is always straight unless there is self-absorption or something else that is distorting it. Therefore, one may use its nonlinearity in combination with an absorption measurement to correct for self-absorption. The abscissa should always be linear with respect to energy. Preferentially the ordinate should be logarithmic for reasons already discussed. Examples of action spectra replotted in standard form are shown in Figures 5, 6 and 7.

Frequently curves are plotted which relate to photosensitivity as a function of wavelength, but the measured variable is not photon irradiance and hence the data values are not proportional to quantum efficiency. Therefore, these curves are not true action spectra and will not in general be comparable to absorption spectra. For example, the percentage of phototactic zoospores versus wavelength is plotted for the phototaxis of the zoospores of the brown alga *Pseudochorda gracilis* Kawai et al. [52]. The duration of photic suppression of protoplasmic movement is plotted for *Amoeba proteus* [53]. The reaction time to turning on the light is plotted for the rapid deflection of the anterior end (an ecclitic response) of *Peranema trichophorum* as a function of the wavelength [54,55]. The rate of change of cell concentration with wavelength was measured for *Cryptomonas* [56]. For *Paramecium bursaria,* Matsuoka and Nakaoka [57] plotted the number of cells accumulated in light as a function of wavelength. They appeared to use a constant energy stimulus of 0.05 W/cm^2 and not a constant photon stimulus. As we have discussed, this is not appropriate since pigments count photons. However, they used a photodiode to measure the light intensities. Since its response is approximately constant with respect to incident photons counts, if they did not correct for the difference in the photon energy for each wavelength their data may fortuitously be correct. Fabczak et al. [58] plotted the reciprocal of the latencies of ciliary reversal in response to light. Häder and Melkonian [59] plotted the fraction of *Euglena mutabilis* cells gliding toward the light source. While these plots indicate regions of spectral sensitivity they are **not** action spectra and hence are even more difficult to interpret. In each case a calibration of the response measure against a varied photon-irradiance stimuli must be made and that correction applied. Sometimes it is given in the paper, but not applied. For example, Schilde [60] plots the early receptor potential as a function of wavelength for a constant short stimulus and gives a separate plot of the early receptor potential as a function of the magnitude of the pulse stimulus. One can then be confident that the only error is that due to the wavelength dependence of the screening of the receptor. Forward [61] plotted the percentage of cells showing positive phototaxis of *Gymnodinium* in his Figure 2b and the calibration curve in Figure 3 enabling the calculation of an action spectrum (Figure 19, [22]) (Figure 5C). Gaertner [62] plotted the percentage of chloroplasts oriented as a function of wavelength and gave a calibration curve so that an action spectrum could be drawn.

3.4.6 The problem of multiple pigments

Stramenopiles and euglenoids typically have multiple pigments for free-swimming phototaxis, namely flavoproteins and pterins. The flavoproteins have been well described and the presence of pterins [63] suggested by the action spectra. After selective inhibition of one system, e.g. KI to inhibit the flavoprotein response, the threshold action spectrum may be done to reveal the second pigment. A priori using threshold responses one cannot tell how many pigments may be contributing to the response. In most cases, the absorption cross-sections will be different, which can be determined from the shape of the saturating irradiance-response curves. Study of the fluorescence of isolated cilia of *Euglena gracilis* at low temperature has found three groups of pigments; possibly pterins absorbing at 3.54, 3.18, and 2.99 eV (350, 390 and 415 nm); flavoproteins absorbing at 2.48–2.76 eV (450–500 nm), and unknown fluorescent pigments absorbing at 2.25–2.38 eV (520–550 nm) [64]. At physiological temperatures, the action spectrum is already four orders of magnitude down by 2.25 eV.

3.4.7 Adaptation leading to non-stationarity of response

If the response is not constant with time, i.e. is non-stationary, either because of cell development or because of light history, then it is more difficult to use that response. An example is the photophobic response of *Chlamydomonas.* One has to wait sufficiently long to return to the dark-adapted state before restimulating. It is best that all cells in each experiment have the same history prior to the stimulation.

3.4.8 Multi-state photoreversible pigments such as phytochrome or protostome rhodopsins

Near threshold, the responses are as already described, the concentration of pigments generated by the near threshold stimuli being too low to be significantly excited. What is quite different and important to observe if one suspects a photoreversible pigment is that, the irradiance-response curves extending far from threshold will not be similar as a function of photon energy. Typically, the irradiance-response curves will be bell shaped with different absolute peaks. If this occurs then multi-wavelength experiments need to be carried out, with a minimum of one wavelength per anticipated pigment or pigment form. Some recommendations for these experiments can be found in Hartmann [23] and Schäfer et al. [24].

3.4.9 Self-screening

Optical densities (OD) of two can be observed at selected wavelengths through a small cell, leading to problems. In *Euglena gracilis* the presumptive photoreceptor (the PAB) is strongly wavelength dependent as well as dichroic with an OD of about 1.4 to 1.5 (see section below) at 2.76 eV (450 nm). The simplest way to treat this problem is to do the threshold action spectrum as described, plot it and then correct with different assumptions. For example, one might assume the pigment is uniformly distributed in the structure. Knowing Beer's law and with a measured absorption spectrum for the structure one can calculate the correction. Beer's law is that absorption is directly proportional to the concentration, which implies that in an isotropic uniform absorber the light intensity will decay exponentially going through. In the case of the PAB the correction would be that self-screening at 2.75 eV would reduce the response by about 3.5 times.

3.4.10 The importance of actinic light modulation

One cannot overemphasize the importance of light modulation in experiments. While free-swimming cells self modulate, one must do it under other circumstances. To measure response to light under other circumstances, e.g. for enzyme assays, the light should always be modulated (but seldom is) to give the stimulus-dependent response independent of the steady-state dark activation that also occurs. This technical oversight in rhodopsin studies has limited the dynamic range for the assay from the steady-state dark activation of about 3–5% to 100% response [17].

3.4.11 Problems of dichroic receptors

Again, *Euglena gracilis* is an example of the problem. For initial experiments, this can be ignored if the stimulating light is unpolarized. Any polarizing elements in the optical path can be a problem, such as use of monochromators, mirrors, or Polaroid filters. Subsequently, the effects of polarization are important and different polarizations should be used with held cells for which one has control of cell orientation. *Euglena* uses dichroism to enhance its mechanism of contrast modulation. Using polarized light oriented in different directions relative to its photoreceptor will then give details of this dichroism. Potentially this data will give the relative dipole orientations for different spectral transitions. Complications can be turned into useful information.

3.4.12 Null and relative action spectra

Delbrück and Shropshire [65] advocated null action spectra with strong arguments from measurement theory. Suppose *Phycomyces* sporangiophores (see Figure 9) are illuminated on one side by a test stimulus of photon energy, ν_{test}, and on the other side by $\nu_{standard}$. With the condition that either one alone will give a phototropic response, when the irradiances of the stimuli is balanced to give no response then the system has its own internal control. For a cell, it can be assumed that there is an equal quantity of pigment for each photon energy (or wavelength). This cannot be assumed for different cells or in particular between different days of different cells. Neither the amount of pigment nor that of different components in the sensory-response pathway must be constant from one experiment to the next. Further, the degree of signal amplification is not a factor. The curve of the ratio of these two photon energies versus the phototropic bending angle is unfortunately rather shallow so the method by itself is relatively

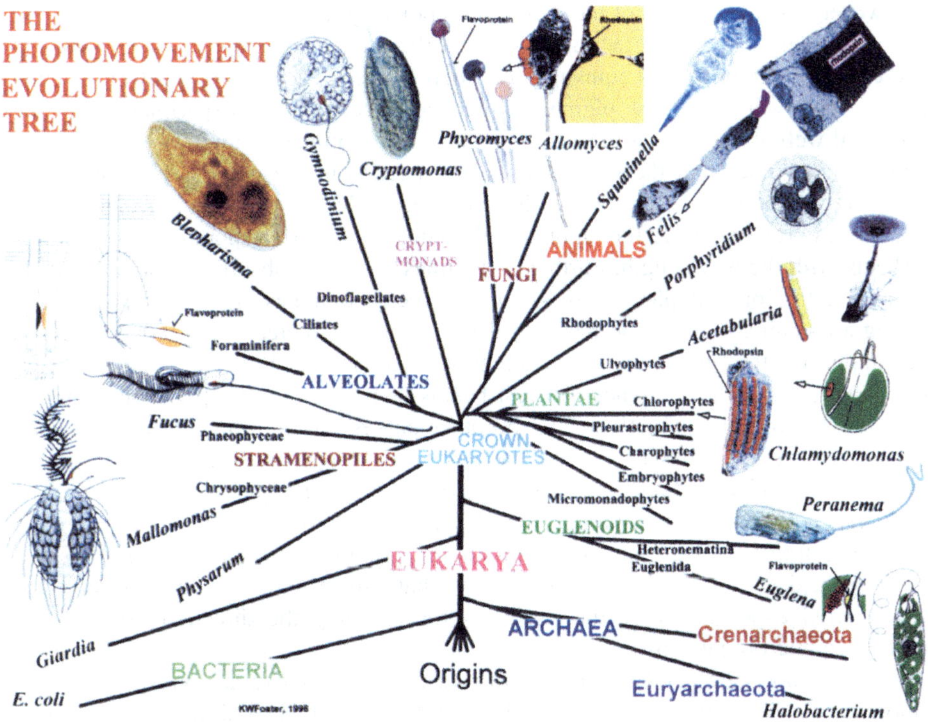

Figure 9. Evolutionary tree of some of the photomovement organisms (see text). Pictures of organisms modified from the following sources: the dinoflagellate *Gymnodinium* (Figure 137), the cryptomonad *Cryptomonas* (Figure 128), the chrysophyte (or synurophyte) *Mallomonas* body (Figure 127), the rotifer *Squatinella* (Figure 15) and the ciliate *Blepharisma* (Figure 325) from Patterson [153]; phaeophyte *Fucus* and cilium of *Mallomonas* from [154]; the stramenopile 'eyes' from Andersen [155], the euglenoid heteronematina *Peranema* from the web site //life-science.rutgers.edu/~triemer/peranema.htm (1998); chlorophyte *Chlamydomonas*, the chytridiomycete *Allomyces*, and mammal *Felis* rod cell with the eye detail from Figure 1 of Saranak and Foster [9]; the *Phycomyces* sporangiophores from Foster photograph; the euglenoid *Euglena gracilis* (Figure 11) from Leedale [156], the rhodophyte *Porphyridium* from Figure 1a of Gabrielson et al. [157] and the ulvophyte *Acetabularia* from Bonotto [158]. The branch lengths are drawn to be proportional to time since divergence (within the limitations of current knowledge). As for all constructed trees, the connections are necessarily speculative and the names are circa 1998.

insensitive. This can be corrected by modulating the ratio of the photon energies over a wide enough range that the null can be determined accurately. A similar method with *Phycomyces* light-growth responses have been carried out. Two wavelengths were alternated every 5 min (half the period that gives the maximal response to a modulated intensity), varying their relative irradiance as a function of time [66]. As discussed, for free swimming organisms Mast [18] used a reference beam perpendicular to the stimulus beam. For similar cells, Halldal [67] oriented the stimulating light and the reference beam in opposite directions. Because the response curve is shallow near null, maximum light intensity was used to obtain the least diffuse boundaries of response.

However, this method is not the method of choice for action spectra. The result will be inaccurate if there are interactions between different wavelengths, i.e. if responses at different wavelengths are not equivalent or independent of each other. This would be the case if there were multiple pigments involved. Because this experiment must be performed well above threshold, there is a danger of stimulating multiple pigments. In addition, it assumes that the geometry with respect to the receptor is equivalent and hence the distribution of light within the cell is equivalent at different wavelengths. In *Phycomyces*, for example, the light distribution in the aerial sporangiophore varies markedly with the wavelength. Further, it assumes screening absorption is independent of wavelength. For example, the free-swimming cell modulates the light by its rotation making a dynamic comparison between the oppositely directed light sources. Greatly oversimplifying the situation will give a hint of what happens. Suppose a cell is moving orthogonal to both the reference and test beams as per Halldal [67]. On one side the beam hits the receptor directly while the other passes through the cell attenuated by $A(\lambda)$. It will choose to swim toward the test beam if $I_{(test)} > k \cdot I_{(ref)} \cdot A(\lambda)_{(ref)}$ and the reference beam if $k \cdot I_{(ref)} > I_{(test)} \cdot A(\lambda)_{(test)}$. Taking the average yields an $I_{(test)} \equiv k \cdot I_{(ref)}/2 \cdot (A(\lambda)_{(ref)} + 1/A(\lambda)_{(test)})$, not the $I_{(test)} \equiv k \cdot I_{(ref)}$ needed for an accurate action spectrum.

Under appropriate conditions in a well-studied system [66], the relative-null method may give the desired more precise results. Note that null-response experiments are often the preferred experimental method such as for mapping the directivity of receptor antennas in microorganisms.

One may note the modest success of Mast's [18] relative action spectra (with orthogonal beams) replotted in Figures 5B and 6A, although Halldal [67] (with opposed beams) was less successful. Out of the range of maximum response, no intensity for some wavelengths could balance the constant reference light resulting in false zero responses. Consequently, with the additional absorption problem he obtained just peaks of response and distorted action spectra. Later Halldal [68] wisely switched to a threshold method to measure the action spectrum of *Platymonas* (replotted in Figure 18, [22]).

3.4.13 Photophobic responses and responses to pulse or step stimuli

These responses in eukaryotes are typically threshold level-crossing responses (like triggering an action potential) due to a large stimulus (a quick increase or decrease in light). Consequently, they have a distinctive character and uniformity (all-or-none). Burr [1] suggests that they deserve a distinctive name, the "ecclitic" response, derived from the Greek word εκκλισις, meaning a turning out of one's course. Ecclitic responses have a different use from phototaxis assays.

When to avoid an ecclitic response. Suppose one wanted to know whether the activity of *cis*-retinal was due to a small contamination of *trans*-retinal. Using the ecclitic response, it takes very little *trans*-retinal to saturate the response [69]. Hence, it might be tempting to suggest that a *trans*-retinal could be responsible for the activity of 11-*cis* retinal. However, with the threshold-phototaxis measurement one could observe that over many orders of magnitude of added pigment the response rises linearly with concentration of exogenous added chromophore. This is because the threshold is proportional to receptor concentration. Now to ask whether the 11-*cis* response is due to a *trans* contaminant, one only needs to measure the threshold of both at the same added concentration. The observation of only a slight difference in comparative sensitivity of the two would suggest that *trans* contamination could not be responsible for the 11-*cis* response and in fact, they have similar sensitivity.[1] As a rule, when the concentration of reconstituted pigment in the cell needs to be compared measure the threshold for phototaxis.

Advantages and disadvantages of pulse stimuli. The simple aspect is their typical dependence of response, $1 - \exp(-S)$, on the size of the stimulus, S, the photon irradiance (I_r) at the receptor times the activation cross section of the chromophore ($\varphi_r \sigma_r$). This expression holds reasonably for short light exposures that have fewer than two-photon hits on the same receptor molecule during the light exposure. Analysis becomes quite problematic at high photon irradiances, because of significant concentrations of multiple pigments formed having their own absorption spectra. The problem with action spectra based on pulse stimuli comes in that the intensity incident is not the intensity at the receptor. It may be filtered by screening pigments reducing the incident intensity by $T_e(\lambda)$, i.e. $I_r = I_i \cdot T_e(\lambda)$. Hence, response is proportional to $1 - \exp[-I_i \cdot T_e(\lambda) \cdot \varphi_r \sigma_r(\lambda)]$ and since φ_r is typically fairly constant one can get a reliable spectrum if one can contrive by orientation of stimulus to make $T_e(\lambda)$ constant and not variable with experimental trial. A virtue is the ability to calculate an absolute value for $\varphi_r \sigma_r(\lambda)$.

What if the beginning of the up or down step is not synchronized with the cell's orientation for either stimulus as happens in rotating cells? The consequence is that the size of the stimulus varies according to the distribution of possible cell attenuations depending on the cell orientation. The range of these attenuations is strongly wavelength dependent. For example, if the cell is comparatively transparent at a particular wavelength then the slope of the response to log photon irradiance curve will be highest. On the other hand, if there is a wide range of possible attenuations, from the light shining directly on the sensitive area to when it is attenuated by the chloroplast (typically reducing stimulus to 1%) the slope will be dramatically decreased. The fraction of situations when the light is direct will give response nearest threshold, so that extrapolating to threshold will again be the best analysis alternative.

---

[1] Note that the published [8] relative sensitivity of 11-*cis* to all-*trans* was slightly in error due to the fact that the all-*trans* extinction coefficient was used for both.

Step down stimuli. Barghigiani et al. [70] did an action spectrum with irradiance response curves for responses to a step-down stimulus. For the photocatalytic model light converts pigment from an activatable state to an inactivation state with interconversion in which the forward state is light dependent and the reverse is thermal [23]. This leads to the equation: $dp/dt = -k_1 I_r p + k_2(1-p)$, where p is the proportion of the photoreceptor pigment in the active form, I_r is the incident light intensity on the receptor, k_1 is the rate constant for chemical inactivation, and k_2 is the rate constant for dark regeneration. Setting $dp/dt = 0$ since before the step down one has presumably reached equilibrium, then $p = k_2/(k_2 + k_1 I_r) = 1/(1 + k_1 I_r/k_2)$. For a step down one might expect the response to be proportional to change in amount of pigment being activated, namely $k_1 I_r \cdot p$ or $k_2/(1 + k_2/k_1 I_r)$ which is of the classic form $(1/(1 + S_{1/2}/S)$ as discussed earlier. Since k_2 and φ_r are hopefully wavelength independent, $I_r = I_i \cdot T_e(\lambda)$ and $k_1 = \varphi_r \sigma_r$, then the response curves should be proportional to $1/[1 + 1/C \cdot \sigma_r I_i \cdot T_e(\lambda)]$ where $C = k_2/\varphi_r$. The action spectrum is a product of the pigment and the transmittance of its external screen. Barghigiani et al. fitted the curves instead to $1 - \exp(-k_I)$, while not recommended it is a reasonable approximation since the suggested equation has the same limit at low intensity and will give a reasonable action spectrum, particularly if $T_e(\lambda)$ is fairly constant. A step-up stimulus may be similarly analyzed.

Pulsed action spectra. They are done only when the action spectrum of this response or the average effect of screening over all orientations of the cell is desired. The ecclictic response is the only known cellular photoresponse for some cells. For rotating cells this is normally used only when the relative location of the receptor pigments and the screen are unknown. One can then measure the absorption spectrum of the cell and calculate the effect of stimulating over all orientations of the screen. This sounds simpler than it is and I have never seen anyone do it successfully. In defense of the method, when I considered working on *Chlamydomonas*, my decision hinged on whether it had rhodopsin or not. One day of doing the ecclitic action spectrum convinced me that a rhodopsin was the correct interpretation of Nultsch et al. [36] paper. For stationary cells, such as in phototropism or with normally swimming cells held on a micropipette most of the problems can be avoided and the advantages can be realized.

3.5 Presentation of action spectra and irradiance response curves

To be widely understood, vocabulary in photobiology should be as common with that outside the field as possible. Subscribing to this philosophy I recommend the terms suggested by Bell and Rose [71].

3.5.1 Units

Units of light intensities or photon irradiances have always been non-uniform. The standard SI unit is photons per meter^2, which leads to Ephotons/m^2 ($E = 10^{18}$) or equivalently 1.661 μmoles/m^2 for a typical intensity. Neither of these units is easy to visualize. If one rigorously knew the absorption capture cross-section of a pigment, then irradiance of a pigment could be expressed as fraction of its capture cross-section; either of absorption or activation, i.e. quantum efficiency for activation times the absorption cross section. The organisms actually respond in this unit, since this determines whether a stimulus is perceived as being large or small. However, at present the capture cross-sections for the pigments of phototactic algae are not exactly known, particularly as a function of wavelength. The activation cross-section may not be constant under different conditions. For this review I am making the following unconventional compromise, i.e. all units have been converted to photons/nm^2 for exposure and photons/nm^2 s for photon irradiance. These are equivalent to 10^{18} photons/m^2 and 10^{18} photons/m^2 s respectively. These values are easily comparable to the measured activation cross-sections, which we will express in nm^2. For reference, the activation cross-section of *Bos* rhodopsin with 11-*cis*-N-retinylidene as chromophore is about 0.01 nm^2, i.e. 100 photons/nm^2 would be enough photons to give each rhodopsin a single photon if equally distributed [72]. The common unit of μmole/m^2 is equivalent to 0.6022 photons/nm^2.

There would be less confusion for future generations if fewer energy units were adopted in biophysics and bioenergetics. The most appropriate candidate is the electron volt (eV). If there is an electric potential of 100 mV across a membrane, a unit charged species on the high-energy side has a potential energy of 100 meV. Temperature can be considered in terms of thermal energy (0 K ≡ –273.15°C, 8.6174×10^{-5} eV/K) giving 25.26 meV for 20°C and 26.73 meV for 37°C. Typically in physics the energy of photons are given in eV (1.23985/wavelength in μm). 500 nm equals 2.48 eV. Similarly thermal activation energies of receptors could be given in electron volts (1 eV = 23.06 kcal/mole = 96.48 kJ/mole). This would make it much easier to consider the different energies involved in a cell. Therefore, as an experiment in pedagogy, we use this single unit.

3.5.2 Why plot action spectra as log sensitivity versus photon energy

Photon energy is the natural variable to express the abscissa in electronic spectra. Pigment bands have a relatively simple shape as a function of photon energy. Further, the shape of the spectrum does not change to a first approximation when the maximum is shifted due to a changed local field. Use of a log ordinate insures that no matter what the sensitivity, the shape of the spectrum does not change. Further, the low energy cutoff, which invariably falls off exponentially with energy, is a straight line in this plot. The log ordinate also insures that the error of each measured point is similar no matter the position of the point in the spectrum. Use of an absolute value rather than a relative one keeps more of the information of the experiment (for example [8]). Later, this may be important for interpretation by readers, for example, was the experiment done on the low or high intensity receptor system. An absolute unit would be the negative of the logarithm of the threshold light irradiance.

"There has been considerable reluctance to plot spectra in the way just described merely because plots of linear wavelength and linear absorption are more familiar. But nostalgia is a poor reason for following a bad convention. There is no question that the plot described above is the correct one. George Wald in his Nobel lecture [73] even cited as one of the significant milestones in rhodopsin research the demonstration of Dartnall

[74] that this plot gives absorption spectra that are similar for nearly all rhodopsins. If action spectra were always plotted in this way much confusion in their interpretation would disappear from the literature" [35].

3.5.3 Why standardize the proportion of scales

The easier it is to compare action spectra and absorption spectra the easier it is to interpret results. Standardization of curves is a great help. The alternative of normalizing to a peak can be avoided by making each curve, no matter the scale, have the same standard shapes. If log sensitivity is plotted the curve may be simply slid up or down and the shape does not change. This is the reason that compilations of absorption spectra are normally plotted in this way [75]. Another virtue of this approach is that the absolute sensitivities can be presented in the same graph. Finally, to present many spectra in the same graph or to make comparisons it is helpful to plot the spectra with energy going one way or the other and with the same proportion of scale of log sensitivity to energy difference. For this review I have chosen to make 1 eV in photon energy on the abscissa have the same length as 2 $\log_{10}$ units of sensitivity on the ordinate.

3.5.4 Curves to fit rhodopsins

Saranak and Foster [15] have used the following convenient approximation to fit rhodopsin spectra. These parameters must be calculated for every action spectral peak. The shape consists of a Lorenzian multiplied by a Boltzmann equation. The Lorenzian is expected to be a precise fit on the high-energy side of the peak. The exponential low-energy tail provided by the Boltzmann equation is also anticipated.

$f(\omega) = \delta^2/[(\omega_{peak} - \omega)^2 + \delta^2] \times \{1 + \exp[\alpha(\omega_{peak} - \omega_{cutoff} - \omega)]\}^{-1}$ where ω, ω_{peak}, and ω_{cutoff} are the photon energies, δ is the half-width at half-maximum of the absorption curve and is normally considered to be related to the life time of the excited state, α is the slope of the low-frequency (energy) cutoff. α may be considered to be 1/kT where k = Boltzmann constant (0.08620 meV K^{-1}) and T, "effective temperature". For all rhodopsins, T is taken as 639 K and ω_{cutoff} as 0.286 eV as calculated for the standard rhodopsin curve (*Bos*) of Knowles and Dartnall [76]. δ varies between the 0.166 eV of *Halobacterium* rhodopsin [77] and the 0.31 eV of *Bos* [15]. The spectral peak, ω_{peak}, may vary over a wide range. Most microorganism rhodopsins (see Figure 5) have a δ of 0.21 eV.

3.5.5 Curve to fit flavoproteins

The same equations above with different parameters can be used to fit the action spectra for flavoproteins (see Figure 6C, data from [78]). This curve shows the sum of two slopes, due to the main peak at 2.75 eV (451 nm) and the narrower shoulder peak at 2.58 eV (481 nm) on the low energy slope.

3.5.6 Presentation of irradiance-response curves

A semilogarithmic plot of response as a function of log photon irradiance is the conventional presentation. Preferably, standard errors of the mean of each point should be shown.

3.5.7 Presentation of errors for action spectra

Most authors have failed to give error bars. When the averages are not provided all the data points should be shown. Forward [61] made the effort to give the standard deviations, but did not give the number of trials at each wavelength to calculate the standard error of the mean which is really needed. The papers with the most consistent use of error bars are those from Lipson's group. Lipson [26] and references therein to his group's papers give a discussion for the criterion case. His remarks can be easily adapted to the threshold determination and to the actual action spectra. If this is too much trouble one can get by with doing multiple trials of action spectra and then averaging these with analysis of the error as will be done by commercial software. Because of the inconstancy of cell sensitivity, this is probably a better approach then determining each irradiance-response curve individually and then hoping the sensitivity does not change from day to day.

3.6 Action spectral identification of receptor pigments of small animals and microorganisms

Many lessons may be learnt from the literature on the identification of receptor pigments.

3.6.1 The pigment responsible for light induction of carotenoid synthesis

This work is a classic example of using threshold action spectra for the identification of a pigment responsible for a particular function, such as the light control of carotenoid synthesis [16]. The assay for the synthesis was restoration of sensitivity for phototaxis using the native pigment. This study took advantage of a mutant that fails to synthesize carotene and retinal, but contains the apoprotein opsin. When retinal synthesis is induced by light, the retinal combines with opsin to form rhodopsin and the cells swim away from a source of light. The amount of light required to trigger a phototactic response is inversely proportional to the concentration of rhodopsin. Therefore, the decrease in amount of light necessary to generate that response can serve as a measure of the amount of retinal synthesized in cells after induction. The four steps of the procedure were analog incorporation, exposure to inducing light, dark incubation and phototaxis measurement. All-*trans*-7,8-dihydroretinal forms a rhodopsin in *Chlamydomonas* with peak absorption at 2.86 eV (434 nm). To minimize errors due to receptor screening, an optically thin layer of cells was exposed from above. The intensities and wavelengths of the inducing light were varied to obtain the intensity-response and action spectral curves. After at least 30 minutes of dark incubation the phototactic threshold was assayed at 2.27 eV (546 nm), which is well absorbed by the native light-induced pigment (2.45 eV, 505 nm peak), but very poor at exciting the blue-shifted analog (2.86 eV, 434 nm). The result was that the action spectrum of the light induced increase in sensitivity at 2.27 eV corresponded to the blue-shifted analog that was incorporated into the rhodopsin. This spectrum was shifted 0.41 eV (or –71 nm) from that of native rhodopsin. The action spectrum of the induced pigment was that of native rhodopsin. Light induction depended linearly on light exposure and rhodopsin concentration before the exposure. The conclusion from this work was that the activation of rhodopsin autoregulates carotene synthesis.

3.6.2 Test of the phycoerythrin hypothesis as phototaxis pigment for Cryptomonads

Erata et al. [79] confirmed that phycoerythrin is present in *Cryptomonas rostratiformis* Skuja and the action spectrum of Watanabe and Furuya [56] on which the phycoerythrin hypothesis was based. Then they tested this hypothesis by doing action spectra on different cryptomonads that do not have phycoerythrin. They found that *Chroomonas nordstedtii* Hansging species had the same action spectrum, but no phycoerythrin. In *Chroomonas coerulea* (Geotler) Skuja, a species with an eyespot, they found a different action spectrum. On the basis of the earlier work, Foster and Smyth [22] had suggested that phycoerythrin might lie in the intrathylakoid spaces of electron-dense material positioned internal to the eyespot as an example of a dielectric slab waveguide. Since cryptomonads are now shown to not use phycoerythrin and see toward their ventral side [80], this suggestion was wrong. Rather the eyespot blocks for the photoreceptor membranes on the opposite or ventral side ([79]. *Chroomonas mesostigmatica* (R. Andersen, personal communication) and *Rhodomonas stigmatica* [81] have their eyespots located about half way down the side of the cell, on the ventral side beneath the gullet. In each case the eyespot is associated with a flattened vesicle toward the ventral side. Given the direction of reception it seems likely that the darkened membrane about one-quarter wavelength ventral from the eyespot (in electron micrograph Figures 6–9 [79]) would be the most likely location of the photoreceptor responsible for phototaxis. How this receptor would communicate to control the cilia is not clear. Taking the advice of Foster and Smyth [22], Erata et al. [79] plotted the threshold action spectrum for the eyespot-less strains (replotted in Figure 5E). This action spectrum raises the interesting possibility that the pigment might be a rhodopsin.

3.6.3 The pigments responsible for the direction of phototaxis

In most instances, the pigments responsible for light tracking in phototaxis are different from the pigments determining the direction of taxis. In *Chlamydomonas,* it has been reported that photosynthesis is involved in determining the sign of taxis [82], but only a few wavelengths were used in their analysis. Photosynthesis does not seem to be the only factor. In a nonphotosynthetic organism, *Allomyces,* it is not known what determines the direction. In a few organisms, the direction does not change. The difficulty is that one is stimulating more than one receptor and the direction change does not occur at threshold but at some much higher intensity. In the seminal paper by Nultsch et al. [36], apparently there was large drop in phototaxis at the low energy end of the spectrum (in fact he did not include these irradiance-response curves in the paper) because, in hindsight, the cells reversed direction. The direction is probably controlled directly by membrane potential and relative ion concentrations, which are under the indirect control of many pigments, including chlorophylls. Since there is probably some advantage to the cell to optimize its photosynthesis or protect its photosynthesis apparatus from photooxidative damage, it would make teleological sense to use the photosynthetic system to regulate direction. On the other hand, it uses rhodopsin to track because this receptor is spectrally tuned to see through the green window left open by the transmission through chlorophylls.

3.6.4 Pigments in Fabrea salina *causing phototaxis*

Interestingly, if the data of Marangoni et al. [46] is recalculated using thresholds it is fit well by the sum of two rhodopsins (2.22 and 2.95 eV peaks with $\gamma = 0.21$ eV) as shown in Figure 7C. This shape of rhodopsin has been found for most eukaryotic rhodopsins that are not fungi or animals, which are broader. As discussed in detail below other evidence is required to demonstrate that they are rhodopsins since use of two such pigments controlling one response in a microorganism would be novel.

3.6.5 Pigment for light response of Blepharisma

Matsuoka et al. [83] used step-up stimuli to see if blepharismin, the main absorbing pigment in the cell, is also the pigment responsible for the ciliary reversal response of *Blepharisma japonicum.* They did action spectra for cells with the reduced and wavelength-shifted oxidized forms. However, the log irradiance-response (I-R) slopes vary strongly with wavelength. The action spectrum therefore depends strongly on the chosen response criteria. It is not clear why 90% saturation was used. Whether self absorption or another pigment is responsible for the I-R slope dependence on wavelength should be determined. Scevoli et al. [84] did a similar experiment measuring both latency and fraction of cells responding in the presence of the reduced form of blepharismin. Again, it is not clear whether the bulk absorption, the receptor pigment, or a combination is being measured. Both groups concluded that blepharismin was the responsible pigment. Only if the wavelength dependence of I-R slopes were due to self-absorption is it possible for blepharismin to be the photopigment. If another pigment is involved then it is likely that the bulk pigment is responsible for the dependence of the I-R slopes on wavelength. Hence, when the bulk pigment is suspected of being the receptor more work has to be done to demonstrate it. This is the same predicament as labeling gels with a "specific" antibody but the most abundant protein is labeled. Their results are in sharp contrast to that of Kraml and Marwan [85] who found a broad peak at about 3 eV (400 nm) having plotted the reciprocal of the time lag for the step-up photophobic response and extrapolated to threshold. Since the measurement started from darkness, it could have been measuring photoreactivation. Unfortunately, they did not plot log sensitivity or give enough irradiance-response curves to be clear what the full spectrum was. These papers do not resolve the nature of the photopigment.

3.6.6 Photobehavior pigments in Dunaliella salina

Wayne et al. [51] measured the step-up, step-down and phototactic responses of *Dunaliella salina*. One problem was that their step-up and step-down stimuli were not synchronized with the cell's orientation. The wider the range in intensities hitting the receptor (associated with local high attenuation) the shallower the slope of the response curves. They did not present action spectra obtained from the "threshold" method of analyzing fluence rate-response curves [22] because they could not locate the linear portion of the fluence rate-response curves. Instead, they chose the half-maximal response and got distorted action spectra. They then compared these action spectra to one obtained for phototaxis. In phototaxis action spectra, attenuation has the opposite net effect. The more contrast across the cell the greater is the modulation and hence the higher the slope of the irradiance-response curve. Again, threshold would have been the best to plot, rather than the half-maximal response, which is a product of the absorbing screen, mostly β-carotene next to the receptor. The authors concluded "the same photoreceptor pigment may be responsible for both photophobic responses whereas a second photoreceptor pigment is responsible for the phototactic response". Although their phototactic action spectrum was very similar to the flavoprotein spectrum (e.g. Figure 5A), they concluded this was not the pigment because it was not sensitive to KI and did not fluoresce. In spite of their spectral widths (γ of about 0.09 eV) being too narrow to be rhodopsins or carotenoproteins, they suggested they might be responsible for both pigments. The replotted and original action spectra are shown in Figure 8. The replotted spectra use the correct "threshold" method on the irradiance-response curves provided in their paper. The step-up (peak at 2.60 ± 0.04 eV), step-down (2.55 ± 0.02 eV) and phototaxis (2.61 ± 0.02 eV) action spectra are not significantly different with respect to shape (all have a γ of about 0.21 eV) and position (Figure 8). Therefore, they could have concluded the same rhodopsin was responsible for all three responses in this green alga as in *Chlamydomonas*. Because of the large variance at low stimulus levels sometimes it is best to extrapolate from the more precise measurements above threshold than to fall back on the incorrect traditional method.

3.6.7 Some action spectra do not suggest known pigments

An interesting example is the action spectrum of photodinesis of barley root hairs (Figure 7B). Photodinesis is the light-induced movement of the cytoplasm and intracellular organelles perhaps by plasma streaming or other means. The spectrum for gliding phototactic orientation of *Euglena mutabilis* may be similar [59]. A single or a combination of pigments could be responsible in each case.

3.7 Criteria for identification of rhodopsins

3.7.1 Consistency with a rhodopsin action spectrum

The spectrum should show appropriate peaks, one from about 2.2 eV (570 nm) to 3.6 eV (340 nm) and a second at 4.44 eV (279 nm). It should show the characteristic low-energy rhodopsin slope of about 11.8 orders/eV and a reasonable γ from the 0.166 eV of bacteriorhodopsin to the 0.31 eV of *Bos*. In many species, there are fluorescent pigments like flavins, which could act as antenna pigments to a rhodopsin. Hence, the slope of the low-energy cutoff of the action spectrum should be measured. Since the low-energy slope of rhodopsin is significantly less than for flavins (11.8 compared to 16.5 orders/eV, [35]), at low enough photon energies the characteristic slope of a rhodopsin should become apparent. The ability to measure sensitivity to 10^{-7} of the peak should make it possible to identify a rhodopsin in the presence of a flavoprotein.

3.7.2 Reversible blockage of response by inhibiting chromophore synthesis

Retinal needs to be present and a different retinal isomer may accumulate after light exposure. Finding inhibition while suggestive does not prove the case for the presence of rhodopsin. Finding lack of inhibition, however, strongly suggests some other pigment is responsible.

For example, inhibition of PAB formation by nicotine or hydroxylamine with light [86] results in loss of photoaccumulation in *Euglena*. However, this does not mean it is using rhodopsin as a chromophore for phototaxis. We demonstrated early in 1983 (unpublished) that inhibition of carotenoid synthesis by chemicals such as nicotine and norflurazon blocked phototaxis in *Chlamydomonas*. While this suggested to us that we might have a rhodopsin in *Chlamydomonas*, we did not consider this proof that rhodopsin was the photoreceptor. This was because there are many ways that a nonspecific inhibitor can affect cells. For example, in the case of *Euglena* induction of PAB formation could be rhodopsin dependent.

Allomyces zoospores released from mycelia grown in the presence of norflurazon 200 μM for three generations showed lower phototaxis sensitivity than the control. The loss of sensitivity was overcome by incorporation with retinal [9] suggesting a rhodopsin receptor.

3.7.3 Irreversible bleaching by light in the presence of hydroxylamine

This is not absolutely selective since other chromophores formed from aldehydes might have the same effect. Until proven otherwise the possibility that they actually belong to the same family of receptors should be considered. For example, peridinin (an aldehyde), or a retinal could be used in dinoflagellates as a chromophore. Recovery after reconstitution with retinal or its analog and comparison of action spectra would provide a stronger suggestion for that opsin.

3.7.4 Replacement by photon energy shifting analog

Recovery of activity with retinal after removal by chemical inhibition or bleaching in the presence of hydroxylamine is a preliminary indication for a retinal dependent receptor. This point may be proven by analog replacement. In this experiment, incorporation of the analog is confirmed by its red spectral shift in the site as compared to its absorption in solution. This is because it reacts in the binding site with the lysine *N*. Unincorporated exogenous analog cannot contribute photodynamically or thermally since they do not absorb at the test wavelengths of the analog rhodopsin. In addition, since the measurements are near threshold, the experiments are done at very low irradiance. Further, the native chromophore cannot be responsible for the response since the analog spectral response is shifted with few exceptions with respect to native chromophore. I first heard of this approach from Jose Luis Reissig[2] in 1967. This suggestion was the inspiration for the work of Otto et al. [87] testing flavoproteins as photoreceptors in *Phycomyces* and then Foster et al. [8] for *Chlamydomonas,* Saranak and Foster [9] for *Allomyces* and Häder and Lebert [88] for *Euglena*. Russo et al. [89] later advocated this approach as probably the easiest way to prove the nature of a pigment. The phototaxis recovery by the replacement with retinal analogs is probably the strongest indicator of a rhodopsin pigment and may be considered the primary evidence. Action spectroscopy was employed for this task.

This approach has been applied to *Chlamydomonas* phototaxis using negative-phototaxis threshold action spectra. The Boltzmann transport method [8,16] was used to involve large quantities of cells without error from multicellular absorption. Retinal analogs were incorporated in place of the normal chromophore in a mutant unable to make retinal. A consistent spectral shift corresponding to the electronic properties of the incorporated analogs was obtained showing the physiological role of the analogs in response [8]. A similar approach was used with the positively phototactic zoospores of *Allomyces* using an individual cell assay [9]. Retinal analogs β-apo-12′-carotenal (red shifting; [15]), octadienal, hexenal (blue shifting; [13]) were also incorporated into *Allomyces*. Either, irreversible bleaching with light in the presence of hydroxylamine was used to remove the native chromophore, or a carotenogenesis inhibitor was used to

[2] Jose Luis Reissig was a Research Professor at C.W. Post College, Long Island University, when, in 1975, he took a research sabbatical with Max Delbrück in Pasadena. His project was to test analogs of riboflavin that would be able to shift the action spectrum of *Phycomyces*. Max obtained many such analogs from a friend in Konstanz. Reissig was unlucky with the analogs he tried during his stay in Pasadena. He was born 1926 in Buenos Aires, Argentina and was trained at the Universidad de Buenos Aires, University of Michigan (B.Sc.) and Caltech (Ph.D. in 1951 with Sterling Emerson). He worked in Buenos Aires with Luis F. Leloir (Nobel Prize Laureate), at Cornell with A. Srb, in Edinburgh with C. Auerbach, in Copenhagen with M. Westergaard and in Paris with E. Wollman. He was Profesor Titular de Genetica (Full Professor of Genetics) in 1961 at the Universidad de Buenos Aires and enjoyed a comfortable and respected status in Argentina until he left the country in 1967, following the arrival of the dictatorship of General Ongania. He accepted what he quickly found to support himself and his family. In 1987 he took an early retirement from Post College. In our last exchange of letters (1989–1990) he was in Britain where he was a follower of a Buddhist monk (Enrique Cerdá-Olmedo personal communication).

reduce endogenous retinal [9]. Red-shifting analogs may give different sensitivity at wavelengths the accessory pigments do not absorb.

3.7.5 Measurement of activation cross section

By measuring the pulse response to high enough intensities it is possible to obtain the activation cross section, this should be about 0.01 nm^2 for a rhodopsin. Sineshchekov [90] measured the activation cross-section of rhodopsin in *Chlamydomonas* to be about 0.8$Å^2$ or 0.008 nm^2. This is consistent with a rhodopsin.

3.7.6 Other types of evidence

Depolarizations and hyperpolarizations have always been associated with rhodopsins, however, there is no reason to assume that they could not be associated with other pigments. Nevertheless, the photoelectric responses seen in *Paramecium bursaria* [91] and *Acetabularia* [60] may be indicative of a potential rhodopsin photoreceptor system. Probably the finding of retinal or antibody cross reactivity to rhodopsins is not helpful since with the divergence observed and the ubiquity of other functions for these and related compounds there is no assurance they relate to the response being studied.

3.8 The evolution of rhodopsin photoreceptors

3.8.1 An application to determining the evolutionary relationships of photoreceptors

Following in the footsteps of Richard Eakin [92] the evolution of rhodopsin-based vision may begin with distributed photoreceptors on the cell surface of unicellular creatures. Later these became plasma membrane patches as in green algae like *Chlamydomonas reinhardtii* [8,22] and in phototactic zoospores of chytridiomycetes like *Allomyces reticulatus* [9]. In *C. reinhardtii* the patch is positioned on the cell surface by the ciliary roots. These patches of photoreceptor control the motion of one or more cilia. The patch signals the cilium electrically with the same time course as in animal vision. Later these patches became physically associated with the cilium as in jellyfish and vertebrate eyes. Possibly, this occurred independently in diverse groups, since the cilia were responsible for locomotion in all the ancestral groups. The evolutionary relationships of discussed organisms are shown in Figure 9.

Although only bilateral animals have had their rhodopsin genes sequenced, rhodopsin is believed to be the universal visual pigment of the Animal Kingdom. In 1980 [22], Foster and Smyth suggested that the photoreceptor for phototaxis in green algae might be a rhodopsin situated in the plasma membrane overlying the eyespot. There was a possibility that dinoflagellates might use the carotenoid peridinin [93] or a rhodopsin as photoreceptor for phototaxis. Eakin [92] suggested that a rhodopsin in the plasma membrane might overlay the paraxonemal body of *Euglena*. In 1984 [8], Foster et al.

showed that a rhodopsin-like photoreceptor was indeed responsible for phototaxis in *C. reinhardtii*. In 1988 [94], Klein et al. obtained the amino-acid sequence of a G-protein-catalyzing receptor (GPCR) for cAMP in *Dictyostelium* that showed some recognizable similarity to a GPCR including rhodopsin. Unfortunately, the evolutionary position of dictyostelids remains a matter of controversy.

3.8.2 The current view

There is increasing acceptance of the view that archaea and eukarya rhodopsins are homologous [95], the result of a common ancestor formed by gene duplication [96]. The simplest hypothesis would be a continuous history of rhodopsin from the earliest eukaryotes. Initially, like in *Halobacterium*, the rhodopsins in eukaryotes most likely were unlocalized in the plasma membrane. Like in *Halobacterium* light stimulation resulted in an ecclitic response (a reversal of flagellar motors in response to light change) rather than phototaxis (sensing light direction). The rhodopsin-like action spectrum for the reaction time of the step-up ecclitic response of *Peranema trichoforum* (Figure 5A, calculated from the data of [54,55]) suggests this. *Peranema,* in the Heteronematales branch of the Euglenoid kingdom, does not have a PAB or eyespot. Note that the action spectrum is broader than expected for rhodopsins, presumably due to the spectral impurity of the stimulating light. During this period free-swimming positive and negative phototaxis pigments were not rhodopsins, but rather, in general, combinations of flavins and pterins as still seen today in euglenoids and stramenopiles (Foster and Saranak, in preparation). Later this situation changed and rhodopsin became involved in real phototaxis. An intriguing possibility, although not proven in any instance, is that rhodopsins became ubiquitous among the alveolates (see Figure 9). For example, among the foraminifera is the crawling phototaxis of *Amphistegena radiata* (Figure 5D, derived from [6]). In addition, some ciliates (Figure 7C) may be using rhodopsins to control their ciliary behavior. In dinoflagellates, specific eye structures could couple light efficiently to membrane receptors like rhodopsins (Figure 5C). Cryptomonads (Figure 5E) may also have acquired rhodopsins.

The above suggestion of rhodopsins in the alveolates may still be considered speculative, but some tolerance must be given until the work is repeated and extended. The different spectral characteristics observed may be due to different chromophores in the putative rhodopsins or experimental problems. In the known eukaryotic rhodopsins a variety of related chromophores are used, namely, 4-hydroxy-N-retinylidine, A1- and A2-*cis*-N-retinylidine, or *trans*-N-retinylidine. After all, the spectrum of *C. reinhardtii* phototaxis of Nultsch et al. [36] did not suggest rhodopsin to the original authors because of a dip at 460 nm and too sharp a cut-off at 540 nm. Reanalysis as a threshold action spectrum and review of the *Volvox* action spectrum [97] was the basis for the rhodopsin hypothesis [22] in green algae.

3.8.3 The light tracking pigments of free swimming phototaxis

Following the alveolates branching off the evolutionary tree (Figure 9) rhodopsin dominates for free-swimming phototaxis. There has been a clear switch from ciliary control by flavoprotein/pterins to rhodopsins between the stramenopiles and alveolates and the change is complete after that. Why this switch should have come about is not clear. The flavoprotein system is even more sensitive, just as fast and adapts more quickly than does the rhodopsin. It only misses high sensitivity in the "green window" left by chlorophyll absorption. Was it just an accidental loss of the particular flavin system with the niche later filled by rhodopsin? Alternatively, did the rhodopsin have a competitive advantage in an environment dominated by green bacteria?

In the plantae, the Chlorophytes, *Chlamydomonas* and *Volvox aureus* [97] (replotted for the stop response and threshold positive phototaxis response in [8]) show a clear rhodopsin action spectrum. The Ulvophyte, *Acetabularia,* produces transient depolarizing potentials in response to intense light flashes. Given the similarity of these early receptor currents and the evolutionary proximity to chlorophytes, it is likely that they use rhodopsin photopigments in spite of their action spectrum (Figure 5F) being only moderately similar to rhodopsins. The maximum sensitivity is near 530 nm similar to a green cone [60]. However, if a rhodopsin shape is assumed the real peak would shift to about 2.44 eV or 508 nm. The Pleurastrophyte, *Platymonas* (≡ *Tetraselmis*) *subcordiformis* [68], has a rhodopsin-like action spectrum for threshold positive and negative phototaxis (replotted in [8]). The Micromonadophyte (formerly Prasinophyceae), *Pyramimonas,* has a unique quarter-wave stack eyespot structure [22] probably implying a membrane receptor. Given their evolutionary position one may speculate that it also uses rhodopsin as its phototactic tracking receptor.

In the fungi, Robertson [98] measured the degree of phototaxis in 5 min of the zoospores of *Allomyces reticulatus* (a Blastocladiale Chytridiomycete) in response to light produced by a monochromator. He found that maximum response occurred between 470 and 525 nm with some response to 610 nm. Since this is consistent with a rhodopsin, Saranak and Foster [9] carried out the procedure for proof of identity of a rhodopsin as outlined above. Perhaps it is interesting that this fungal behavior has the same kinetic requirements as that of *Chlamydomonas* and animal vision. On the other hand, phototropism in the zygomycete, *Phycomyces,* is certainly not due to rhodopsin [7] and has a 2000 times slower response.

In the animals, Menzel and Roth [99] measured the action spectra of three positively phototactic rotifers, *Asplanchna priodonta, Polyarthra remata,* and *Filinia longiseta.* The relative sensitivities were measured using 9 wavelengths and 8 intensities. Both the *A. priodonta* and *F. longiseta* may have parabolic quarter-wave stack reflectors so that the light coupling may strongly depend on wavelength and have several pronounced maxima and minima. The action spectrum is necessarily a product of the efficiency of light coupling due to the antenna and location of the receptors and the absorption of the photoreceptor due to its orientation and its chromophore's extinction coefficient. Based on the spectra and the evolutionary position of the organisms, rhodopsins would be the most likely hypotheses. The predicted normal reflective maxima of the light gathering and beam shaping mirrors for *A. priodonta* and *F. longiseta* correspond well to the observed action spectra peak.

Outside of the animal/fungal/plantae branch, there are possible rhodopsins. In the Alveolate Kingdom Dinoflagellates such as *Gymnodinium splendens* (Figure 5C), *Gyrodinium dorsum, Peridinium balticum,* and *Gonyaulax,* ciliates like *Fabrea salina* (Figure 7C) and *Paramecium bursaria* and foraminiferins such as *Amphistegina radiata*

(Figure 5D) may have rhodopsins. Some Cryptomonads like *Cryptomonas rostratiformis* (Figure 5E) and *Chroomonas nordstedtii* may have rhodopsins. Finally, *Peranema trichophorum* (Figure 5A) of the Euglenoid Kingdom may (but not necessarily) have rhodopsin (also see Figure 9). Given the criterion of consistency with a rhodopsin spectrum has been met, further experiments may be suggested.

3.8.4 Amino-acid sequences of the rhodopsins

Green algal rhodopsins. Unfortunately, the amino acid sequence of any green algal rhodopsin is still unknown. A protein claimed to be "chlamyopsin" [100] belongs instead to those proteins, which are glycoproteins associated with photoreceptors and are generally characterized as filaments with extremely high lysine content (about one-sixth of amino acids) and leucine repeats. The best characterized is peripherin (about one-sixteenth of amino acids are lysine) in vertebrate rods. It is associated with the disc rim presumably with other proteins to accommodate the high curvature [101], but its mutation is a major cause of retinitis pigmentosa or various types of macular degeneration [102]. The mistaken identity was probably due to the assumption that only the photoreceptor protein would have this location specificity and ability to bind retinal. This assumption is not true for vertebrates, *Chlamydomonas* or *Volvox.* A premature conclusion was reached without testing the expressed products. The low conservation of rhodopsin sequence and the phylogenetic distance between green algae and animals makes finding the pigment by PCR extremely difficult.

Small animal rhodopsins. Unfortunately, the amino acid sequences have been on large organisms. It would be interesting to compare the rhodopsins of sponge larvae, rotifers, and jellyfish larvae (e.g. *Polyorchis*) with the vertebrate, mollusk and insect.

3.9 Criteria for identification of flavoproteins

The principles are the same as for identifying rhodopsins.

Consistent action spectrum. The relatively steep low-energy slope of 16.5 orders/eV [35] is characteristic for flavoproteins. They have a specific maximum at about 450 and 480 nm. The action spectra for *Phycomyces* [35,78] (Figure 6C) and alfalfa (Figure 6D, [48]) are particularly flavin like.

Blocked by inhibiting the chromophore's synthesis. In the case of *Phycomyces* double carotene synthesis, mutants have not changed the phototropic sensitivity implying the lack of involvement of carotene or retinoid receptor proteins.

Response inhibition. KI, a triplet-state quencher, and phenylacetate at mM concentrations should inhibit response. There is the caveat that chemical inhibitors are not necessarily specific. However, this inhibition does seem to be correlated with flavin-like spectra and not inhibit rhodopsin systems or the response due to pterins at as low concentrations.

The spectral shift of an analog. For flavins, the wavelength-shifting analog of choice seems to be roseoflavin, which causes a red shift. Roseoflavin was first applied by Otto et al. [87] to *Phycomyces* phototropism, and more recently, Häder and Lebert [88] applied it to *Euglena gracilis.*

Other types of evidence. Other types of evidence would include an appropriate activation cross-section for a flavin and appropriate transition dipole moments. In the case of oriented photoreceptors, it would be possible to determine the efficacy of different linear polarizations of light as a function of wavelength. This method may be used to test consistency with a flavoprotein, or, if different from expectation, exclude the possibility of a flavoprotein. Jesaitis [103] attempted this in *Phycomyces*. He found the orientations of the dipole moments were consistent with a flavoprotein hypothesis.

3.10 The evolution of flavoprotein photoreceptors

In spite of the zeal expressed for rhodopsin being a universal pigment controlling photobehavior [104–108] other pigments also have wide distribution. Probably the presence of retinal in an organism has been overemphasized as an indicator of rhodopsin having a role in phototaxis. Retinal plays an important role in development of many organisms. For example, retinal induces carotenogenesis in *Phycomyces* and is not involved in its phototropism [7,35].

Possibly, the most ancient photoreceptors associated with cilia and free-swimming phototaxis (Figure 9) are the pterin/flavoproteins of some euglenoids (*Euglena gracilis*) and stramenopiles (*Mallomonas*, *Fucus*) (Foster and Saranak, in preparation). These photoreceptors mediate behavior with the same temporal time course as the rhodopsins. Other free-swimming stramenopile action spectra like those of the Phaeophyceae, *Pseudochorda gracilis* [52] and *Ectocarpus siliculosus* [109] are also likely due to flavoproteins although their action spectra seems to be distinctly different from *Euglena* and cryptochrome spectra. The constant-photon action spectrum (at 10 photons/nm^2) of *Pseudochorda gracilis* has peaks at 420 and 460 nm without a peak at 380 nm. Unfortunately, since constant-stimuli action spectra are products of the screen and the receptor (see [22]) and the measure of response is probably very non-linear, it is quite difficult to estimate the absorption of the pigment other than to identify peaks of the active or screening pigment. Probably, the autofluorescence in the cilium reported by Kawai [110] in euglenoids, Chrysophyceae, Phaeophyceae, Synurophycease, Xanthophyceae and Prymnesiophyceae is an excellent indicator for flavin use as a chromophore in photobehavior. Equally, its absence in dinoflagellates, Cryptomonads, Micromonadophytes, and Chlorophyceae would imply that it is not used in these organisms. Since autofluorescence is not seen in the stramenopile groups Raphidophyceae, Eustigmatophyceae and Bacillariophyceae this could mean they are not phototactic or have a different receptor.

Flavoproteins are also responsible for phototropism in the zygomycete *Phycomyces blakesleeanus* (Figure 6C) (Foster, 1972), in the tips of the plant alfalfa (Figure 6D) [48], and the stramenopile *Vaucheria geminata* (Figure 6F) [40]. They dominate the photobehavior of fungi. Some could also be interacting with G-proteins for signal transduction. Flavoproteins also have many other duties such as being found as receptors to control plant guard cells, mictic female production in the rotifer *Notommata copeus* [111], *Physarum polycephalus* photoavoidance (Figure 6E) [41], and chloroplast movements in *Vaucheria* (Figure 6B) [39]. Taken together this suggests the ubiquitous use of flavoproteins (Figure 9) as receptors throughout the eukaryotes. They may have important applications in animals too.

3.11 A suggestion for the branching pattern of the evolutionary tree based on the distribution of photoreceptors and eye structures

In Figure 9, "The Photomovement Evolutionary Tree", the branch lengths are drawn proportional to time. The branch points have remained speculative due to the inability of rRNA sequence methods to resolve the separation of the crown eukaryotes. Accepting provisionally the behavioral photopigment information discussed above and related information suggests a possible branching. Beginning from the root of the tree and following the trunk leading to animals one first has the gene duplication creating rhodopsins between the Bacteria and Archaea branches. Next there is the acquisition of mitochondria between the *Giardia* and Euglenoid branches. The use of cilia associated flavoproteins (with pterins) dominates phototaxis in the Euglenoid, Haptophyte, and throughout the Stramenopile Kingdoms. The haptophytes (not shown) could be branching from the stramenopiles or from the trunk. Rhodopsin is probably diffusely present in some plasma membranes such as in *Peranema*. After the stramenopiles have branched off, flavoproteins lose control of the rapid photoresponses like that of free swimming phototaxis. Rhodopsin and the quarter-wave stack antennas [22] which optimize coupling light to membrane receptors begin their dominance in these responses in the Alveolate, which is continued throughout the Cryptomonads, Plantae, Fungi and Animal Kingdoms. Then, there is the change from tubular to flat cristae in the mitochrondria after the alveolates have branched off. The dominance of flavoproteins remains for slower photoresponses like phototropism in *Phycomyces* and Embryophytes. Then further along the trunk leading to animals, the acquisition of chloroplasts creates the Plantae branch including red and green algae and Embryophytes. The second cilium is abandoned in favor of single-cilium rear propulsion before the fungi/animal divergence. Finally, color vision arises in some animal branches.

3.12 Criteria for identification of other pigments

Pterins. Biochemical identification and consistency of action spectra seem the best avenues for detection.

Phytochromes, red and far-red. Characteristic of these pigments is the interconvertable forms. This leads to biphasic irradiance-response curves with the amplitudes of the maximum responses being wavelength dependent even after correction is made for differences in modulating amplitude. This is unlike the case for catalytic photoreceptors that would show the same maximum response irrespective of wavelength. Evidence of this interconvertabililty plus consistency with the absorption of these pigments is presently the primary criteria (Figure 7A).

Photosynthesis. If photosynthesis is involved in behavior, the action spectrum should be similar to that for oxygen evolution. Further, the threshold should be at the compensation point for photosynthesis. Generally, mutants can be found which interfere with both, and there are many chemical inhibitors as well.

3.13 Use of action spectra to characterize rhodopsin

3.13.1 Method of incorporation of analogs of retinal into rhodopsin

It is common knowledge in pharmacology that ligands with an extraordinary range of affinities (K_D from 10^{-11} to 10^{-2}), bind to G-protein catalyzing receptors (GPCR) which are related to rhodopsin. Not at all surprisingly, a wide range of retinal analogs (more than one hundred) activates rhodopsins. These have been primarily tested by recovery of phototaxis in *Chlamydomonas reinhardtii* and *Allomyces reticulatus.* These systems are particularly favorable because of the ease of removing their endogenous chromophores. In *Chlamydomonas,* the native chromophore may be removed by use of a mutant (FN68) that synthesizes only a small amount of retinal and in *Allomyces,* chromophores may be removed by simultaneous application of hydroxylamine and bleaching light.

Action spectroscopy of the threshold phototaxis response permits demonstration of the spectral shift of pigments in the receptor site as well as knowledge of the absolute sensitivity at the peak wavelength. The spectral shifts imply the incorporation of the analogs into the opsin-binding pocket to form an imine bond. An exception is the amide bond formed between the lysine *N* and the acid fluoride analog of retinal.

Particularly important in observing analog activity is the noise or background level. A statement that an analog does not show activity when the signal-to-noise ratio is very low leads to false negative conclusions. One has to say something like, “The response with the analog incorporated was not seen above the background or control level which was x% of the response of the native chromophore”. Further, it is necessary when there is no detected response to show the presence of the analog in the site. It is meaningless to conclude that an analog shows no response when there is no analog in the site.

There are many experimental details that influence the results: The purity, solubility, stability of the analogs, their distribution to and concentration at the binding site, the temperature, condition of the cells, the presence of anti-oxidant. In the initial experiments of Foster’s group the concentration was purposely held as high as possible to avoid false negative observations. They found that adding vitamin E as antioxidant increased the lifetime of potential response by a factor of about four. For all-*trans* retinal without vitamin E the response is gone 45 min after incorporation (Saranak, unpublished). They also found that a temperature of 25°C reduced the swimming of some strains of *Chlamydomonas*.

3.13.2 The Schiff-base counter-ion and receptor site of rhodopsin

Hundreds of retinal analogues have been incorporated into the retinal binding site of *Chlamydomonas* rhodopsin. Many were useful to probe the nature of the binding site. The data can be looked at from a different prospective. This is really an opportunity to study what a structured environment, such as can be engineered in a protein, does to the spectrum and other properties of a retinal-like molecule.

One can readily localize charge within the chromophore-binding site. Figure 10 is a plot of the spectral peak of photon energy, determined by action spectroscopy, versus the reciprocal of the length (number of carbon atoms) of the π-bonded conjugate chain of

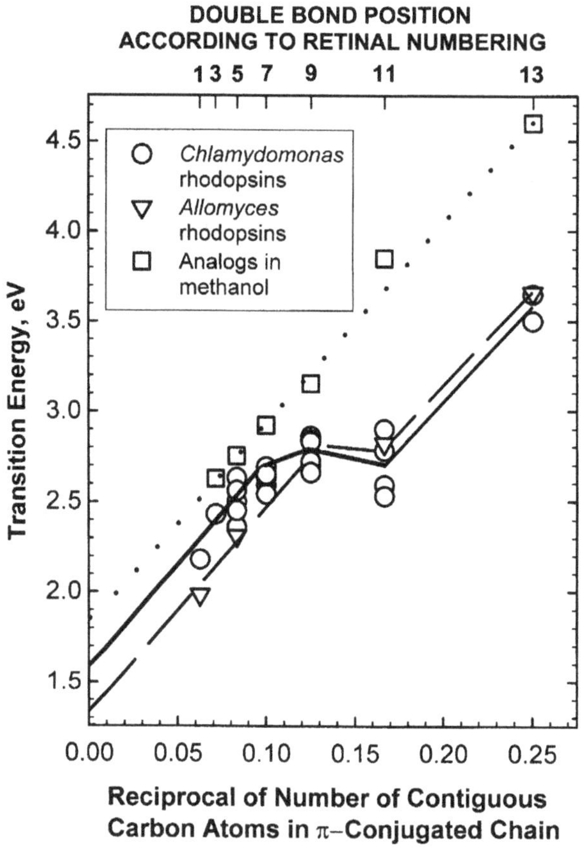

Figure 10. The effect of length of π-conjugated bonds of retinal analogs on action spectral peak. This particular example shows the effect of the counter ion near the 11–12 bond in *Allomyces reticulatus* [9] and *Chlamydomonas reinhardtii* [13] as well as the additional red-shift in *A. reticulatus* compared to in methanol solution. The double bond position 11 (the upper scale) corresponds to having the 11–12 double bond as according to standard chromophore numbering with the lysine *N* being at 16. The chromophore with the 1–2 double bond has four additional carbons in the conjugated chain. See text for more details.

the analog. The plot of the analogs in methanol solution is approximately linear as expected for such π-conjugated chains. However, bound in the site one can see distinctly different shifts as a function of the reciprocal length relative to the in-solution curve. Each chromophore is connected to a lysine *N* at a fixed position and to a first approximation is laid out identically along the pocket. Since the greatest spectral shift is seen for a charge group located at the end of the chain, it appears from the plot that a charged group in *Chlamydomonas* rhodopsin lies near the 10 or 11 carbon of *N*-retinylidene (using retinal numbering) [13]. This was the first experimental determination of the location of the counter ion in a rhodopsin. At the time, it was thought that it was next to the protonated Schiff base. Subsequently, using NMR spectroscopy and computer modeling, Han et al. [112] found that for *Bos* rhodopsin, the counter ion is slightly closer to the *N*-end. It was also possible to verify that, with respect to the native chromophore, there is a position which Foster's group referred to as the neutral zone, in which charged groups have no effect on shifting the spectral maximum. An electron donating group near the *N*-end red shifts (to lower photon energy), but near the ring end blue shifts (to higher photon energy) (see original reference for details, [13]). It has been noted that a number of amino acids contribute to the net resultant spectral shifts [113]. However, in Figure 10 only two shifts (implying at most three sites) are observed. One at 10–11 is due to the counter ion. A second is at the *N* imine bond. The third is at the end of the ring which shifts between the green sensitive *Allomyces* and the blue-green sensitive *Chlamydomonas* rhodopsin. Apparently, most of the amino acid residues contribute globally or indirectly to the effective charge of the three effective sites. Houjou et al. [114] gives a detailed discussion of the factors involved.

These results make it easy to understand how rhodopsins can have spectral maxima over a wide range of wavelengths, making color vision possible; either the two color system of many monkeys, the three color system of apes or the superior four color systems of chickens and bees. The different spectral shifts are made by changes in the local electron donor and acceptor properties along the length in the chromophore. The shifts are due to particular polar or charged groups with local effective (or partial) charges along the site. However, some authors think of the reaction as a dipole (chromophore) interacting with the parallel dipole of the protein or as the chromophore lying in a local effective electric field. It is now clear that by molecular engineering the rhodopsin one can study the effects of local field or environment on chromophores incorporated into the site.

Another aspect of the site is the penetrance to the site from the external medium. This can be studied with ions and zwitterions in solution, both of which have the potential to shift the spectrum. If a charge enters the site then one can measure the effect of charge on shifting the spectrum, on average 117 ± 7 meV in *Chlamydomonas* rhodopsin, (Saranak and Foster, unpublished). This compares to an estimated carboxylate shift of 220 meV.

Another aspect is the shape of the site, which can be measured by inserting analogs of different shape and then using action spectroscopy to determine if a pigment is formed and how readily the analog was incorporated from the peak sensitivity. Because the analogs are not in equilibrium with the binding site due to formation of the imine bond, one cannot get a true affinity. However, by measuring the lowest concentration that will give a measurable response one can get an idea of how easily a particular molecule can find the binding site. As in pharmacology, the native chromophores or ligands are not necessarily the best for distribution to the site, the highest in affinity or efficacy. How much gets in depends on the available concentration times its incorporation rate. In the case of *Chlamydomonas*, the binding site for retinal is about 4 carbons bonds longer than the native retinal's chain length. Further it has a restriction in the region of the 8–9 carbons (according to standard retinal numbering), known as the 9-methyl group steric constraint, as compounds with a β-ionone ring in this region have incorporated poorly yielding no detectable effect [15].

3.13.3 Initial mechanism of rhodopsin activation

Although the whole photocycle is important, a main interest has been in the activation process. Because, it was not clear from other assays (such as bleaching) what was important in vivo, an in vivo assay was developed by Foster's group to obtain results under physiological conditions. The in vivo assay has sufficient sensitivity that a single photon activating rhodopsin would be easily observed even if that activation resulted in the permanent inhibition of the receptor from further activation.

Foster's model for activation was derived from testing many series of retinal analogs. A model explaining the activity of all analogs tested provides a stronger basis for the understanding of receptor activation than any models purported to explain the activation of the native pigment.

Testing the isomerization hypothesis. This is at the heart of determining the appropriate paradigm for the mechanism of visual activation. In the first step of vision, a photon is absorbed by *N*-retinylidene, which lies in the regulatory site of rhodopsin. Upon photoexcitation, charge is redistributed as the electron density shifts toward the imine (C=N) end of the chromophore. The charge redistribution triggers double-bond isomerization of *N*-retinylidene, changes in the apparent pK_a of the imine nitrogen, proton motion in opsin, potentially the direct excitation of rhodopsin as well as bleaching (release of retinal from its binding site). Those phenomena, which are known to occur, lead by post hoc reasoning to the hypotheses that one or another of these events activates rhodopsin to initiate the visual cascade. Geometrical *cis-trans* isomerization, discovered by Hubbard and Wald [115], was put forward by Kropf and Hubbard [116] as the mechanism for activation of rhodopsin. To test these alternatives in *Chlamydomonas*, it seemed reasonable to incorporate isomerization-locked analogs of retinal in place of the native all-*trans* chromophore. Some analogs would of course be too bulky to fit right or would not orient correctly or in some other way be inefficient. We did not draw any conclusions from such analogs. As it turned out for each bond that may be individually locked and might otherwise undergo *cis-trans* isomerization, we found some locked analogs with normal activation using the action-spectral method [12]. We also found that response was due to their incorporation, because the spectra were always shifted from the native spectrum. Only the syn-anti bond between the 15th carbon of the *N*-retinylidene chromophore and *N* was not locked in any of the locked analogs.

From these experiments, it was concluded that blocking individual double-bond isomerizations does not prevent the activity of rhodopsin. Therefore, if isomerization is required it does not matter at what position along the chromophore that it occurs. Secondly, it was concluded that changes in molecular geometry of the chromophore might be unnecessary for protein activation. Of course, if as in *Bos* rhodopsin, 11-cis-N-retinylidene or at least the β-ionone portion, is acting like an antagonist then some latter step of activation might be blocked if it cannot change shape by isomerization. However, if a short chromophore without the β-ionone portion is used then activation will go to completion. In other words, the triggering part comes not from removal of the antagonism, but there is a distinct initiating process. We confirmed that incorporation of 11-*cis*-retinal inhibited the spontaneous activity of free opsin [117].

The 11–12 bond. Different analogs locking the 11–12 bond were incorporated into *Chlamydomonas,* four isomers of 7-member ringed analogs, two 6-member ringed analogs (13-*cis* and *trans* 9–12-phenyl), and two isomers of 5-member ringed analogs. The phototaxis sensitivity (reciprocal of threshold and proportional to the number of molecules incorporated) varied from normal to 50 times lower than all-*trans* retinal. This means, not surprisingly, that some relatively bulky analogs do not incorporate as well as the native. The 6-member ringed analogs showed somewhat lower than normal efficacy (87%). Even the 11-*cis* and 11,13-*dicis* 5-member rings that severely inhibit twisting, let alone isomerization, worked well. Analogs without an 11,12-π-conjugated bond also formed effective chromophores, namely, 11,12-dihydroretinal and n-hexenal (with action-spectral maxima about 3.65 eV or 340 nm). Several analogs were also tested for their ability to cause light-induced activation of retinal synthesis: *trans*-9–12-phenyl, n-hexenal and n-hexanal [118] and 11,12-dihydroretinal [16] do so. The five locked analogs tested by Takahashi et al. [119,120], 9,11-*dicis*(7)-, 11-*trans*(5), and 11-*cis* ret-7, 9,11-*dicis*-ret-7 and 9,12-phenyl-ret, were confirmed to be active.

The 13–14 bond. The above argument applies to the analogs that block the 13–14-bond isomerization. The normal isomerization is thought to be all-*trans* to 13-*cis* on the basis that all-*trans* is detected in *Chlamydomonas* kept in darkness [121] and 13-*cis* accumulates after light exposure. Foster's group found that five 13–14-bond locked analogs [four with 5-member rings, all-*trans*, 9-*cis*, 13-*cis* [12], and 11-*cis*], and one with the naphthalene group (naphthaldehyde) (Saranak and Foster, in preparation) would activate rhodopsin in vivo. The 9–12-phenyl locked analog inhibits isomerization about the 13–14 bond as well yet was seen to be active both by Foster et al. [12] and Takahashi et al. [119]. Their action spectral peaks were consistent with their electronic structure. Three activated, with normal sensitivities (measured as thresholds) and greater than 90% efficacy compared to the native analog. The others showed low phototaxis sensitivity and about 40% efficacy for all-*trans* and about 80% efficacy for the 13-*cis* analog. Not too surprisingly, with the very short 17–20 second stimulus used by other groups [119,120,122], they did not see the activity of this all-*trans* locked analog. With the same short stimuli, these authors also failed to see the activity of the 13-*cis*-locked analog. Unfortunately, they did not test the fully active analogs. In bacteriorhodopsin, use of the same 13–14-locked analogs does not results in proton pumping, but a large conformational change is seen by atomic force microscopy [123]. In *Chlamydomonas*, pumping protons is not required and only the conformational change of the protein is needed.

Since incorporation of an analog which locks the 11–12 bond blocked bathorhodopsin formation and subsequent bleaching in *Bos* rhodopsin [124], it would seem that the few picosecond period before the chromophore returns to its original ground state is all that is necessary to trigger the active state. It should be noted that charge separation is complete in less than 5 ps in bacteriorhodopsin [125]. The suggested nanosecond or longer period to trigger the protein is evidently not required.

Inactive chromophores. A variety of analog chromophores will incorporate into the site, compete with retinal incorporation and absorb light, but not activate rhodopsin. Just dumping energy (via absorbed photons) into the chromophore pocket is insufficient to form active rhodopsins. Both ketones and retinonitriles failed to activate *Chlamydomonas* rhodopsin to give a phototaxis response [14]. Analogs that bind to a *N* in the binding site but form an amine bond with the *N* instead of an imine also failed to activate *Bos* rhodopsin [126]. A similar analog in bacteriorhodopsin failed to result in a conformation change [123]. Since an amine bond is not a conjugated double bond, it effectively decouples the *N* from the chromophore. What is common in all these cases is the loss of the directional change in dipole moment that presumably triggers rhodopsin. Rousso et al. [123] restored a dipole moment change to the amine bound chromophore by adding an electron donor at the ring end of the chromophore and got a conformation change in bacteriorhodopsin. They concluded that an asymmetric charge distribution must be generated in the excited state to get a conformation change.

Minimum requirement for active analogs. In the alga *Chlamydomonas* the minimum acyclic chain that would form an active chromophore was n-hexenal with a spectral peak,[3] measured by action spectroscopy, of 3.65 eV (340 nm) [13]. This was repeated in the chytridiomycete *Allomyces* [9] as shown in Figure 11. Hence, members of both the plantae and fungal kingdoms have the same activation requirements. It would be interesting to know whether the animal kingdom is an exception.

With such a small chromophore, it is hard to imagine that it would be as effective in sterically triggering rhodopsin if a steric trigger were required with the native chromophore. The charge motion of the proton on the lysine *N* is likely to be quite small following activation since with the native chromophore the C=N bond is *anti* before and after isomerization at the 11-ene [127]. We also added a series of naphthalene analogs including naphthalene aldehyde (Saranak and Foster, in preparation). Again this chromophore was normally effective and the only bond that could "isomerize" is the *syn-anti* bond.

Several short analogs (*n*-hexenal, *n*-hexanal) were tested by Hegemann et al. [128] with *Chlamydomonas* strain CC2359 (not FN68) using potentially toxic concentrations (25 μM) and incorporated for 3 hours without vitamin E. This strain is a phototactic

[3] When the spectrum resulting from *n*-hexanal incorporation was published [13], it was thought that this was the spectrum of *n*-hexanal proper. However, our subsequent model predicts the theoretical location of its maximum well below 4.4 eV (280 nm), the limit of our measurements. What was observed was undoubtedly a result of polymerization of this chromophore. If sufficient time (days) is allowed for polymerization in addition to the ene peak at 3.7 eV (340 nm) a diene peak is seen at about 2.85 eV (440 nm).

mutant with an unusually narrow intensity range exhibiting phototaxis. They conclude that "hexenal or hexanal, did neither enter the rhodopsin binding site, nor influence its behavior in any way". More correctly, these analogs were not in the rhodopsin-binding site at the time of test. Perhaps the analogs had evaporated or oxidized by the time of the test. In the experience of Foster's group without vitamin E, the activity of all-*trans*-retinal is gone in about one hour. Further, these analogs are such small aldehyde molecules that they distribute everywhere in a short time. They were fully active in *Chlamydomonas* after 30 minutes of incorporation. From their work Hegemann et al. concluded that a chromophore with four double bonds is required at a minimum since the shortest they found active was dimethyl-octatrienal. Sineshchekov et al. [129] confirmed the activity of dimethyl-octatrienal, however, they saw no response to the shorter analog, citral, using 500 nm without vitamin E. Since citral incorporated into rhodopsin absorbs at about 340 nm, this negative result was not too surprising. Consequently, the analogs shorter than the trienal have only been properly tested by Foster's group.

Conclusion. In our view [13], the results are not consistent with the traditional formulation of the *cis-trans* isomerization hypothesis for the activation of rhodopsin.

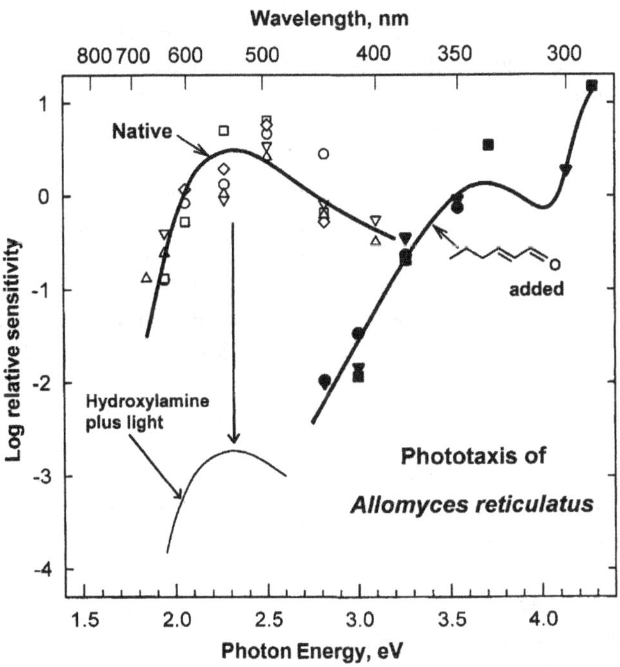

Figure 11. Recovery of phototaxis sensitivity by analog substitution in *Allomyces reticulatus.* A combination of hydroxylamine and light lowers its sensitivity by more than three orders of magnitude. With this low background analogs may be incorporated and tested for incorporation and shifting of the phototaxis action spectrum. The *n*-hexenal action spectrum is shown as an example.

According to the *cis-trans* hypothesis [116], light, by isomerizing the chromophore, destroys the fit of the chromophore with respect to its binding site in opsin and this results in visual excitation. This idea is similar to the discredited idea that ligands distort the receptors they enter causing activation, rather than the idea that their electron donation and polar properties are responsible for triggering or kicking conformation and electron/proton redistribution within the receptor. Their shape may of course help to stabilize the desired protein conformation. As discussed above, the controlling factor in spectral sensitivity is the protein surrounding the chromophore. It controls the absorption by influencing the distribution of electron density in the ground and excited states of the chromophore. Furthermore, when we restrict the geometrical changes, but keep the electronic aspects of the chromophore intact, we observe no loss of activity [12]. Therefore, it seems reasonable that the large change in the distribution of electron density on excitation of the chromophore [130] alters the protein's conformation and charge distribution [131]. For example, the highly polarizable amino acids thought to line the retinal binding pocket must also have their charge redistributed. These changes will propagate throughout the protein, altering the hydrogen-bonding network. Thus the electron redistribution, which is similar in all analogs at the C = N location, could directly excite opsin. We postulate that the electronic change switches or triggers the protein to the enzymatically active conformation. The implication is that any asymmetric π-electron system properly oriented producing a substantial dipole moment change with light absorption would activate rhodopsin. The rhodopsin case would appear to share some fundamental similarity to other studied photoreceptors, as shown in Table 1. A further implication is that the response is triggered extremely rapidly, probably in less than a few picoseconds.

Discrepancies in results. In spite of the confirmation of the activity of most analogs, criticism of Foster's model has rested on tests in *Chlamydomonas* of two compounds, already mentioned, that lock isomerization of the 13–14 bond with 5-member rings that hold the bond in the *cis* or *trans* position. Since Foster found the *trans* compound to have sufficiently low efficacy that others probably could not detect its activity in a short assay, the real discrepancy is the activity of the 13-*cis*-13–14-locked analog. Foster's group found this analog had 80% efficacy (see Figure 3B) and normal sensitivity compared to the native chromophore and therefore should have been detected by others.

As a general practice in research, discrepancies in results call for carefully examination of the technical details that might contribute. Some of these are shown in Table 2. The conditions used by Foster's group were derived by trial and error until reproducible results were obtained. These parameters are important and should be investigated before concluding that a particular compound fails to work.

A few comments are in order. The photophobic response of *Chlamydomonas* strain CC2359 was measured by Lawson et al. [69]. This strain is a phototactic mutant with an unusually narrow intensity range under which it exhibits phototaxis. Since the photophobic response requires the rapid development of a large signal, those with less than 100% efficacy or low affinity or distribution to the site were found inactive. Three other groups tested the 13-*cis*-13–14-locked analog. Zacks et al. [122], however, grew FN68 at 25°C (7°C higher than Foster), a temperature that makes FN68 inactive.

Further, their 10–16-day old cells were frequently too old. They measured phototaxis only within a 20-second period following a large change in light level [122]. Sineshchekov et al. [129] grew CC2359 at 28°C, even hotter. Sineshchekov et al. [129] reported that the analog failed to work at 22–24°C when added at 0.015 μM, probably because the threshold for response at that low concentration was below their level of detection. Takahashi et al. [119,120] also reported that the analog (at an estimated concentration of 0.125 μM) failed to show response in the 17-second period following a large change in light level. Since they found that all the 11–12-locked analogs showed response, the failure with the 13–14-locked 13-*cis* analog is surprising. Since all aldehyde chromophores are toxic to various degrees and subject to their own photoconversion, my group always added the antioxidant vitamin E (0.025%); none of these groups added an antioxidant. The other three 13–14-locked analogs that Foster's group found fully active were not tried.

For any analog substitution experiment, only Foster's population assay could be done sufficiently near threshold that response could be measured from rhodopsin molecules that can only respond once. This makes the important distinction between activation, an

Table 1. Comparison of photoreceptors

Chromophore	Flavin/pterin	p-courmaryl	*N*-retinylidene
Pigment	Flavoprotein + pterin in crystal or amorphous	Photoactive Yellow Protein	Rhodopsin
Required nature of chromophore	π-conjugated chain with three planar rings, polar, asymmetric and polarizable	π-conjugated chain, polar, asymmetric and polarizable	π-conjugated chain, polar, asymmetric and polarizable
Charged before light exposure	?	Negatively	Protonated N +, but will work if not protonated, provided chromophore remains asymmetric
Charged after light exposure	?	Neutral	Deprotonated, neutral
Isomerized on activation	Not possible	Yes, but not essential for activity	Yes, but not essential for activity
Post activation events	Energy migration to lowest energy chromophore; Electron transfer, Oxidized to reduced state	Chromophore is core of H-bonding network which is altered	Disulfide bond reduction, alteration of H-bonding network, Entropic irreversibility

Table 2. Parameters affecting measurement of 13-cis-13-14-locked 5-member ring retinal

Factors that influence the measurement of analog activity	Foster et al. (1989)	Lawson et al. (1991)	Takahashi et al. (1991)	Takahashi et al. (1992)	Zacks et al. (1993)	Sineshchekov et al. (1994)
Phototaxis assay	Population assay	Video Cell tracking	Video cell tracking	Video cell tracking		
Light intensity	Threshold at 480 nm (peak) of 0.0023 photons/nm^2 s		< 100 photons/nm^2 s	< 100 photons/nm^2 s	0.076–25.2 photons/nm^2 s	
Duration of stimulation	10 min of continuous light		17 s immediately following a step-up stimulus	10 s following a step up	20 s following a step up	
Pulse Response assay		Video cell tracking	Video cell tracking	Video cell tracking	Video cell tracking	Photoelectric response
Light exposure		0.9 or 3.6 photons/nm^2	< 1700 photons/nm^2	?	0.05–10 photons/nm^2	4.3 photons/nm^2
Wavelengths tested	Complete action spectrum	500 ± 20 nm	400 to 530 nm	423, 430, 455, 489 nm	500 ± 20 nm	500 nm broadband
Sensitivity of method	Saw activity down to 0.1% normal sensitivity and down to 40% efficacy	Failed to see activity of 9,12-phenyl	Saw activity of 9,12-phenyl and 11–12-locked analogs	Saw activity of trans–11–12-locked-5 member ring and 11-cis-7-member ring analogs	Failed to see activity of 6-cis-locked, 9,12-phenyl-locked	
Strain	FN68	CC2359	CC2359	FN68	FN68	CC2359
Age	7–10 days	10-14 days	?	?	10-16 days	?
Growth Temperature	18°C	?	?	?	25°C	28°C
Testing Temperature	20°C	?	21–22°C	23°C	?	22–24°C
Concentration of analog, μM	25 μM	Up to 1 μM	Up to 0.125 μM	?	0.002, 0.02, 0.08, 0.4 μM	0.015 μM
Added antioxidant	0.025% vitamin E	None	None	None	None	None
Incubation time	10 min to overnight	?	Maybe 180 min	Maybe 5–8 h	?	40 min

? = not mentioned in the publication
blank = not applicable

initiation event, and driving a full photocycle that is permitting both activation and regeneration of the system.

3.13.4 Requirement of detachment of the chromophore or protonation of the C = N bond

Foster et al. [12] showed that incorporation of the acid fluoride of retinal had an action spectral peak at 3.5 eV (354 nm) implying that there was no red-shifting protonation of the nitrogen. Of course, no protonation was expected since an amide bond is formed. This was the first demonstration that protonation was not required for rhodopsin activation. This lack of requirement was confirmed in *Bos* rhodopsin by Fahmy and Sakmar [132] using an ultraviolet absorbing mutant. Further, this chromophore does not detach from the opsin backbone. Therefore, hydrolysis of the chromophore during the photocycle is not necessary for activation or for completing a photocycle. This analog also showed normal efficacy and phototactic threshold, so that it had as good coupling to activation of the receptor as any other analog. Others have shown that retinal does not need to be attached either to *Bos* [133] or archaea rhodopsins [134,135], when the lysine side residue is replaced by glycine and the full *N*-retinylidene chromophore is incorporated.

3.13.5 The activation site of rhodopsin

Foster et al. [13] found that the shortest acyclic analog (*n*-hexenal) and naphthalene aldehyde incorporated into *Chlamydomonas* gave full response. *n*-hexenal was also fully active in *Allomyces* [9]. This shows that the protein only needs changes around the *N*-end of the chromophore. These changes could be the electron density shift along the chromophore and/or changes in the pK_a of the *N* or adjacent protein residues.

3.13.6 Determination of β-ionone ring conformation in an active site

The *N*-retinylidene chromophore in *Chlamydomonas* rhodopsin could be in the 6-s-*trans* or 6-s-*cis* conformation. Analogs were available in both conformations and could be incorporated in the chromophore-less mutant. One could measure the relative rate the two analogs enter the site. However, since rings held the conformations neither analog had the exact shape, flexibility, or other properties as the native. Hence, there is no reason to believe a choice made in this way would be correct. There is no perfect solution, but the action spectral approach may be the best. Since the analogs get in easily, the sites are not particularly restrictive, then the environments that determine the spectrum of each are probably very similar. Hence, one can compare the closeness of the spectra to that of the native chromophore. Since 8,16-methanoretinal (6-s-*trans*) shows a 0.04 eV shift from the native and 8,18-methanoretinal (6-s-*cis*) shows a 0.14 eV shift, probably the approximate isomeric shape is 6-s-*trans* [136]. This result has recently been confirmed in a more detailed study [137].

3.13.7 Summary of implications for visual activation

Following absorption of a photon, in vivo studies using the action spectra of microorganisms has shown that isomerization, detachment of the chromophore from the protein, protonation of the chromophore *N*, and attachment of the chromophore to the protein are not required to activate vision. On the other hand, there must be a dipole moment change with an asymmetric charge distribution on excitation, i.e. a π-conjugated chain made asymmetric by a heteroatom or strong electron donating group at one end. A π-conjugated chain connected to a nitrile, ketone or an amine bond to an *N* does not produce an activating chromophore. Indiscriminate dumping of photons into the regulatory site of rhodopsin does not work. However, a π-conjugated chain connected to an imine or amide bond can light-activate rhodopsin. The chromophore can be very short such as *N*-hexenylidene or *N*-naphthalylidene suggesting that the electronic change in the vicinity of the *N* is all that is absolutely necessary. These results imply that while normally the visual chromophore isomerizes, its accompanying shape change is not the property that is essential for triggering vision. Rather it is the charge separation or redistribution produced in the protein (not just the charge separation in the chromophore) that drives rhodopsin activation. This occurs in a time of a few picoseconds.

3.14 Application to determining the mechanism of retinal synthesis

Threshold-phototaxis action spectral peaks have been used to identify the in vivo products of the enzyme β-carotene dioxygenase of *Chlamydomonas*. With different substrates for the enzyme, it has been possible to clarify the in vivo cleavage of β-carotene into retinal, the first step in biosynthesis of retinoids [15]. The structure-activity relationships of the substrates and the enzymatic active site were also revealed. The activity of this enzyme was studied by using a mutation that lacks β-carotene synthesis, thus permitting the incorporation of synthetic carotenoids of various structures. Since retinal and related analogs form chromophores with opsin in *Chlamydomonas*, the action-spectral peaks of the phototaxis restored by carotenoid incorporation could be used to suggest the products formed by this enzyme that cleaves carotenoids. The data from a study of 12 different carotenoid analogs suggested that the physiologically relevant cleavage of β-carotene into retinal is central rather than excentric. This question arose because of the controversy created between the Goodman group [138,139] and the Krinsky group [140,141] on this question. In addition, these experiments provide new insight into the binding site of the enzyme. When apocarotenoids were substrates, the enzyme targeted the double bond located a constant distance away from the carbonyl group on the acyclic end and consequently, retinal was not produced. In hindsight, this was reasonable as it means that the binding site of the enzyme prefers the more polar group. β-Carotene (with rings at both ends) does not allow that preference. The distance away from the binding site (normally to split β-carotene in two) was preserved so that one can envisage an enzyme with a microguillotine (involving molecular O_2) at a fixed distance from a polar binding site.

Products could be inferred because of the unique spectral positions that different products have in *Chlamydomonas* rhodopsin. Both the products and the substrates that can directly incorporate into the opsin-binding site were detected by recovery of phototaxis. The technique had several important advantages. Foremost was the sensitivity to measure the spectra of a substrate or product that was only one thousandth the sensitivity of the native chromophore in *Chlamydomonas*, corresponding to perhaps 30 molecules incorporated per cell. Second, unlike chromatographic techniques, which necessarily involve assumptions about the nature of the products to be detected, this study did not have to make any such assumptions. The unanticipated finding of the preference for the polar end groups by the site located at a fixed distance from the cleavage site in the dioxygenase enzyme was hence possible.

3.15 Practical advise to maximize results with minimum cost and time

The number of data points needed for an action spectrum. This is analogous to the Nyquist criteria for observation of a particular frequency of response. To be believed it is necessary to have at least two points within any spectral feature such as a peak. If the spectrum is broad, fewer points are needed than if there are multiple peaks to describe. One must, however, keep in mind the distinguishing features of different pigments. For example, if a pigment could be a carotenoid or flavin these may be difficult to distinguish because their peak sensitivities may overlap. However, their low energy tails are easy to distinguish, the flavin being much sharper than the carotene. Hence, the effort to measure the cutoff to as much as three-orders of magnitude down from the peak will be rewarded with a definitive answer. Further, worth noting is that the precision required to distinguish similar peaks is much higher than to determine the linear slope (threshold plotted versus photon energy) of the low photon energy cutoff. Quinones et al. [142] were trying to distinguish between a carotene (zeaxanthin) and a flavoprotein receptor pigment for blue-light-induced enhancement of the "red-light stimulated chlorophyll-*a* fluorescence quenching" in cotton adaxial guard cells and light-grown corn coleoptiles. They compared these responses respectively to blue-light-stimulated stomatal opening and the phototropic curvature of oat coleoptiles. Had they extended their spectra a little further to lower energy they could have answered their question quickly. There must be at least twenty action spectra on *Euglena gracilis* phototaxis for which a similar extension could have excluded carotenes or retinoids as being responsible for either the positive or negative phototaxis.

Photon irradiance-response curves. Their shape has distinguished four classes. Those of the form $(S/(S + S_{1/2})$ which plotted against log irradiance is a hyperbolic tangent and related curves. S is the stimulus as discussed earlier. These are the most common ones seen in action spectra. They are analogous to the dose-response curves of ligands with their receptors. Examples are the integrated area and the peak-to-peak response to a pulse stimulus [143]. In this case, sometimes authors plot response versus irradiance (rather than log irradiance) and take the slope as a measure of spectral sensitivity.

There are those of the form $1 - \exp(-S)$. In this case, S is the photon irradiance times the activation cross section of the chromophore. This expression holds reasonably for short light exposures that have fewer than two photon hits on the same receptor molecule during the light exposure. Analysis becomes quite problematic at high photon exposures, because of significant concentrations of multiple pigments formed having their own absorption spectra.

Finally, there are those that are linear with log irradiance. These are much quicker to carry out, which is particularly important for action spectroscopy, because only a few points near threshold are needed. Several examples are worth repeating here. One, is the free-swimming phototactic rate of rotating (self-modulating) cells [8]. A second is the use of the directional modulation function [(toward − away)/(toward + away)] of free-swimming (self-modulating) cells [9]. Similarly, the peak-to-peak amplitude of the response to a stimulus modulated near the principal response frequency of the organism could be used. For example, the light-growth response of *Phycomyces* should be modulated at about 1.7×10^{-3} Hz.

Checking for accuracy of results. Plotting as I have suggested helps in detecting the errors in either labeling or recording of data. For example, Nultsch and Schuchart [43] would have probably checked out the labeled wavelengths in the irradiance-response curves for the red alga *Porphyridium* (Figure 12) before publishing it if they had used my suggested plots. My review and replot of their published data suggest that the labeled 534 nm might have been 504 nm which would be more reasonable.

3.16 The relative roles of absorption and action spectroscopy

3.16.1 A good application of absorption spectroscopy

Implications of measured optical density (OD) of isolated PAB. One of the most exciting developments with respect to *Euglena* phototaxis has been the isolation of the paraxenemal body (PAB) [144]. This has enabled its more accurate characterization. Foster and Smyth [22] estimated its OD to be about 0.20 from Wolken's [145] measurement of 7.1% (OD = 0.032) for the absorption through the PAB. This was corrected for scattering by assuming that the absorption at 410 nm is 0.58 times the absorption at 450 nm, as in flavoproteins and the relatively small size of the PAB (about 0.5 μm^2) relative to Wolken's measurement aperture (2-μm diameter). A measurement of 1.41 OD has been made by Gualtieri et al. [104] with an isolated PAB using a better instrument with a 0.5-μm diameter aperture. James et al. [105] confirmed this high value. They give the absorption for four different apertures, which extrapolated to the approximately $0.7 - \mu m^2$ cross-section of the PAB as determined by electron microscopy, of about 1.5 OD (the precise value is uncertain due to the partial presence of eyespot in the same field).

If a flavoprotein extinction coefficient (ε) of 10,000 M/liter/cm and range of thickness from 0.4 μm [146] to 0.7 μm [106] is assumed, the molecular weights would range from 0.4 to 0.7 kDa. Assuming a rhodopsin ε = 40–50,000 M/liter/cm (11-*cis* or all-*trans* retinylidene chromophore), the range of possible molecular weights is from 1.5 kDa to 3.5 kDa. [Molecular weight associated with each chromophore = density of protein × extinction coefficient of chromophore × pathlength/O.D. in appropriate units]

Since a rhodopsin would be at least 26 kDa, a rhodopsin protein crystal is an implausible suggestion. The action spectral evidence, dichroic nature of the crystal, and the availability of feasible energy transfer mechanisms [64] strongly suggest a flavin crystal. Such a quasicrystal may also have pterins and even some protein (riboflavin = 376 Da). The number of chromophores in the volume of the crystal (assuming 0.3 μm^3, 0.7 kDa per chromophore) would be about 4×10^8.

3.16.2 Behavioral response to determine how many receptor molecules used by the cell

Estimate of threshold. Foster and Smyth [22] estimated the minimum number of photoreceptor molecules was about 10^6 assuming no self-screening of receptor

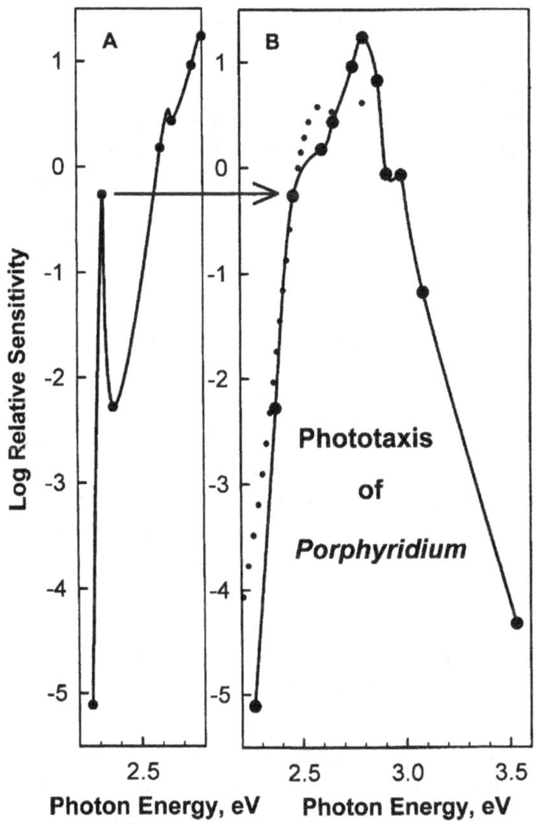

Figure 12. Importance of appropriate plotting as a check on accuracy of data. Phototaxis of the red alga *Porphyridium cruentum* (Ag.) Naegeli [43]. Apparently, as plotted in A there was an irradiance-response curve incorrectly labeled as '534 nm'. The figure is plotted from determining the thresholds of the irradiance-response curves in their Figures 8A and B. We will probably never know whether changing it to 504 nm as plotted in B is correct. The spectrum cuts off like a flavoprotein (the dotted curve is from Figure 6C), but does not have its typical absorption toward higher photon energy (see Figure 6D).

molecules (based on the information then available). This number is small relative to the number of chromophores estimated (maximum about 4×10^8) in the crystal. The high OD of the PAB means that self-screening must be included. The number of absorbed photons to get a response is not known. An upper limit can be calculated by the number of photons that could be absorbed during the integration time of the detector. Assuming 0.1 s for the integration time, 0.45 μm^2 cross-section in Creutz and Diehn [147] experiment, 2 mW/m^2 of 472 nm at threshold [(2 mJ/s)/(energy in joules of a 472 nm photon) = 0.002/(hc/λ, where h is Planck's constant, c is the speed of light and λ is the wavelength) = $0.002 J/s/(6.626 \times 10^{-34}\ J \cdot s \times 2.998 \times 10^8\ m/s/472 \times 10^{-9}\ m) = 4.75 \times 10^{15}$ photons/s], 96% absorbed assuming 1.455 OD for the PAB absorption. Multiplying together, this yields about 205 photons potentially absorbed at normal incidence. However, given the typical 15% efficiency of a crystal detector this reduces to 30 photons detected perpendicular to the longitudinal cell axis. The amount of pigment on the surface is about 5×10^4, much smaller than implied by the threshold. Therefore, it is likely that the PAB is acting as an antenna to another receptor. Sineshchokov et al. [64] report that the pterins and flavoproteins can transfer energy to the lowest energy species which does the final transduction to control behavior. It should be noted, however, that the pterins in the absence of the flavins might still be able to produce a response. While it has been speculated that a rhodopsin might be this final absorbing species, the action spectrum of the low energy tail does not pick up a trace of the shallower-sloped rhodopsin.

3.16.3 Action spectra to suggest the nature of the photopigment

Action spectra are complex in the case of *Euglena*. The effect of the dichroic crystal depends on the orientation of the cell that determines the observed behavioral response and therefore difficult to estimate. Since the absorption of the dichroic crystal is so great, there is additionally a correction for self-absorption. Nevertheless, the identification as a flavin/pterin system is clear and direct and has not changed since Mast's 1917 paper [18].

3.16.4 The problem of using an absorption spectrum in preference to an action spectrum for suggesting a receptor pigment

The absorption spectra of small objects are always difficult. Light scatter and the inability to dilute the pigment are prime problems. Both problems lead to filling in of any fine structure to the point that all one has left is a broad peak [148]. The problems are exemplified by the experience of Crescitelli et al. [149] who measured the transmission spectrum of the eyespot of *Chlamydomonas* in the mutant BF4/M18 which they kindly let us examine. In this mutant cell, one has a complete eyespot predominately of carotenes and a reduced amount of chlorophylls concentrated into small regions. The absorption through a single cell of 0.04 OD is about 1000 times that possible for the receptor itself for that probe size and is appropriate for the carotenes, which form the visible part of the eye. Probably this is β-carotene although in some strains it is replaced with a mixture of polar carotenes (Foster, unpublished). The spectrum measured, however, is smooth and similar to rhodopsin although rhodopsin is not what is being measured. The total absorption is as expected for the carotenes of eyespot. The absorption shows dichroism because the eyespot is birefringent. Using the same spectrometer as Crescitelli et al. [149], James et al. [105] also shows a broad peak for *Euglena*, again with the carotenoids and flavins overlapping and again the total absorption is reasonable for a rhodopsin. However, rhodopsin, even if it were present in small concentration, is **not** what is being measured. Absorption spectroscopy, unlike action spectroscopy, is not selective, it sees all the absorbers and scatterers. Gualtieri et al. [104] had similar problems. Each of the workers recorded something different in the red part of the spectrum [104,105,145]. Interestingly the absorption was unbleachable and not sensitive to hydroxylamine [105], as has always been found to the best of my knowledge for rhodopsin receptors. Barsanti et al. [86] found an effect on formation of the PAB if hydroxylamine was present. Perhaps a rhodopsin is involved in the light induction of this structure just as it is involved in the light induction of retinal synthesis in *Chlamydomonas* [16]. It is also not surprising that the bulk PAB photoreceptor for phototaxis is not rhodopsin.

3.17 Conclusion

Action spectra are not just to suggest the nature of a pigment in a cell. Action spectroscopy has entered the arsenal of useful and powerful spectroscopies, in particular, when in vivo conditions or high selectivity or sensitivity are desirable. Use of this technique argues for a new paradigm for visual excitation and has mapped out the receptor sites of rhodopsin and β-carotene dioxygenase. It has clarified the evolutionary distribution of rhodopsin and flavoprotein photoreceptors and suggested the evolutionary branching of the crown eukaryotes. Finally, action spectroscopy has helped to determine the pigments responsible for light regulation of behavior and gene expression. As biologists move deeper into studies of cellular regulation, and in particular the ubiquitous light regulation, I anticipate that action spectroscopy will continue to play an important role into the future.

Acknowledgements

The author would like to thank Drs. Jureepan Saranak and Jack Sullivan for making valuable suggestions and comments on the manuscript.

References

1. A.H. Burr (1984). Photomovement behavior in simple invertebrates. In: M.A. Ali (Ed.), *Photoreception and Vision in Invertebrates* (pp. 179–215). Plenum Press, New York.
2. S. Petridou, K. Foster, K. Kindle (1997). Light induces accumulation of isocitrate lyase mRNA in a carotenoid-deficient mutant of *Chlamydomonas reinhardtii*. *Plant Molecular Biology*, **33**, 381–392.

3. M. Watanabe (1995). Action spectroscopy: Photomovement and photomorphogenesis spectra. In: W.M. Horspool, P.-S. Song (Eds), *CRC Handbook of Organic Photochemistry and Photobiology* (pp. 1276–1288). Boca Raton, CRC Press.
4. F.T. Haxo, L.R. Blinks (1950). Photosynthetic action spectra of marine algae. *J. Gen. Physiol.*, **33**, 389–422.
5. C.W. Hawryshyn, F.I. Harosi (1991). Ultraviolet photoreception in carp: microspectrophotometry and behaviorally determined action spectra. *Vision Research*, **31**, 567–576.
6. A. Zmiri, D. Kahan, S. Hochstein, Z. Reiss (1974). Phototaxis and thermotaxis in some species of *Amphistegina* (Foraminifera). *J. Protozool.*, **21**, 133–138.
7. K.W. Foster (1977). Phototropism of coprophilous Zygomycetes. *Ann. Rev. Biophys. Bioeng.*, **6**, 419–443.
8. K.W. Foster, J. Saranak, N. Patel, G. Zarrilli, M. Okabe, T. Kline, K. Nakanishi (1984). A rhodopsin is the functioning photoreceptor for phototaxis unicellular eukaryote *Chlamydomonas*. *Nature*, **311**, 756–759.
9. J. Saranak, K.W. Foster (1997). Rhodopsin guides fungal phototaxis. *Nature*, **387**, 465–466.
10. C. Weber (1982). Electrical activities of a type of electroretinogram recorded from the ocellus of a jellyfish *Polyorchis penicillatus* (Hydromedusae). *J. Exp. Zool.*, **223**, 231–243.
11. K.W. Foster, J. Saranak, F. Derguini, V. Jayathirtha Rao, G.R. Zarrilli, M. Okabe, J.-M. Fang, N. Shimizu, K. Nakanishi (1988). Rhodopsin activation: a novel view suggested by in vivo *Chlamydomonas* experiments. *J. Amer. Chem. Soc.*, **110**, 6588–6589.
12. K.W. Foster, J. Saranak, F. Derguini, G.R. Zarrilli, R. Johnson, M. Okabe, K. Nakanishi (1989). Activation of *Chlamydomonas* rhodopsin in vivo does not require isomerization of retinal. *Biochemistry*, **28**, 819–824.
13. K.W. Foster, J. Saranak, P.A. Dowben (1991). Spectral sensitivity, structure, and activation of eukaryotic rhodopsins: Activation spectroscopy of rhodopsin analogs in *Chlamydomonas*. *J. Photochem. Photobiol. B: Biol.*, **8**, 385–408.
14. K. Nakanishi, F. Derguini, V. Jayathirtha Rao, G. Zarrilli, M. Okabe, T. Lien, R. Johnson, K.W. Foster, J. Saranak (1989). Theory of rhodopsin activation: Probable charge redistribution of excited state chromophore. *Pure Appl. Chem.*, **61**, 361–364.
15. J. Saranak, K.W. Foster (1994). The in vivo cleavage of carotenoids into retinoids in *Chlamydomonas reinhardtii*. *J. Exp. Bot.*, **45**, 505–511.
16. K.W. Foster, J. Saranak, G.R. Zarrilli (1988). Autoregulation of rhodopsin synthesis in *Chlamydomonas reinhardtii*. *Proc. Natl. Acad. Sci. USA*, **85**, 6379–6383.
17. Longstaff, R.D. Calhoon, R.R. Rando (1986). Deprotonation of the Schiff base of rhodopsin is obligate in the activation of the G-protein. *Proc. Natl. Acad. Sci. USA*, **83**, 4209–4213.
18. S.O. Mast (1917). The relation between spectral color and stimulation in the lower organisms. *J. Exp. Biol.*, **22**, 471–528.
19. L.N.M. Duysens (1970). Photobiological principles and methods. In: P. Halldal (Ed.), Photobiology of Microorganisms (pp. 1–16). Wiley-Interscience, New York.
20. W. Shropshire Jr. (1972). Action spectroscopy. In: K. Mitrakos, W. Shropshire Jr. (Eds), *Phytochrome* (pp. 161–181). Academic Press, New York.
21. L.O. Björn (1979). Photoreversibly photochromic pigments in organisms: properties and role in biological light perception. *Quart. Rev. Biophys.*, **12**, 1–23.
22. K.W. Foster, R.D. Smyth (1980). Light antennas in phototactic algae. *Microbiol. Rev.*, **44**, 572–630.

23. K.M. Hartmann (1983). Action spectroscopy. In: W. Hoppe, W. Lohmann, H. Markl, H. Ziegler (Eds), *Biophysics* (pp. 115–144). Springer-Verlag, New York.
24. Schäfer, L. Fukshansky, W. Shropshire Jr. (1983). Action spectroscopy of photoreversible pigment systems. In: W. Shropshire Jr., H. Mohr (Eds), *Photomorphogenesis* (pp. 39–67). Springer-Verlag, Berlin.
25. P. Galland (1987). Action spectroscopy. In: H. Senger (Ed.), *Blue Light Responses; Phenomena and Occurrence in Plants and Microorganisms* (pp. 37–52). CRC Press, Boca Raton, Florida.
26. E.D. Lipson (1991). Action spectroscopy. In: F. Lenci, F. Ghetti, G. Colombetti, D.-P. Häder, P.-S. Song (Eds), *Biophysics of Photoreceptors and Photomovements in Microorganisms* (pp. 293–309). Plenum, New York.
27. E.D. Lipson (1995). Action spectroscopy: Methodology. In: W.M. Horspool, P.-S. Song (Eds), *CRC Handbook of Organic Photochemistry and Photobiology* (pp. 1257–1266). CRC Press, Boca Raton.
28. T.P. Coohill (1995). Action spectroscopy: Ultraviolet radiation. In: W.M. Horspool, P.-S. Song (Eds), *CRC Handbook of Organic Photochemistry and Photobiology* (pp. 1267–1275). CRC Press, Boca Raton.
29. D-P. Häder (1987). Photosensory behavior in procaryotes. *Microbiol. Rev.*, **51**, 1–21.
30. I. Manton (1965). Some phyletic implications of flagellar structure in plants. *Adv. Bot. Res.*, **2**, 1–34.
31. R.A. Andersen, D.J.S. Barr, D.H. Lynn, M. Melkonian, O. Moestrup, M.A. Sleigh (1991). Terminology and nomenclature of the cytoskeletal elements associated with the flagellar/ciliary apparatus in protists. *Protoplasma*, **164**, 1–8.
32. R.A. Millikan (1916). A direct photoelectric determination of Planck's "h". *Phys. Rev.*, **7**, 355–388.
33. A. Einstein (1912). Thermodynamische Begründung des photochemischen Äquivalentgesetzes. *Ann. D. Phys.*, **37**, 832–838.
34. G.T. Fechner (1966). *Elemente der Psychophysik, 1860 (Available in English translation as: Elements of psychophysics*. Holt, Renehart and Winstin.
35. R.D. Smyth, J. Saranak, K.W. Foster (1988). Algal visual systems and their photoreceptor pigments. *Prog. Phycol. Res.*, **6**, 255–286.
36. W. Nultsch, G. Throm, I. von Rimscha (1971). Phototaktische Untersuchungen an *Chlamydomonas reinhardtii* Dangeard in homokontinuierlicher Kultur. *Arch. Mikrobiol.*, **80**, 351–369.
37. D-P. Häder, M. Lebert (1997). Behavioral mutants for *Euglena gracilis*: Functional and spectroscopic characterization. *J. Plant Physiol.*, **151**, 188–195.
38. K.H. Chon, M.J. Korenberg, N.H. Hostein-Rathlou (1997). Application of fast orthogonal search to linear and nonlinear stochastic systems. *Annals Biomed. Engin.*, **25**, 793–801.
39. M.R. Blatt (1983). The action spectra for chloroplast movements and evidence for blue light photo receptor cycling in the alga *Vaucheria*. *Planta*, **159**, 267–276.
40. H. Kataoka (1975). Phototropism in *Vaucheria geminata*. I. The action spectrum. *Plant Cell Physiol.*, **16**, 427–437.
41. T. Ueda, Y. Mori, T. Nakagaki, Y. Kobatake (1988). Action spectra for superoxide generation and UV and visible light photoavoidance in plasmodia of *Physarum polycephalum*. *Photochem. Photobiol.*, **48**, 705–709.
42. P.-S. Song, D.-P. Häder, K.L. Poff (1980). Phototactic orientation by the ciliate *Stentor coeruleus*. *Photochem. Photobiol.*, *32*, 781–786.
43. W. Nultsch, H. Schuchart (1980). Photomovement of the red alga *Porphyridium cruentum* (Ag.) Naegeli. II. Phototaxis. *Arch. Microbiol.*, **125**, 181–188.

44. H. Yatsuhashi, A. Kadota, M. Wada (1985). Blue- and red-light action in photoorientation of chloroplasts in *Adiantum* protonemata. *Planta*, **165**, 43–50.
45. K. Ishizawa, S. Wada (1979). Action spectrum of negative photo tropism in *Boergesenia forbesii*. *Plant Cell Physiol.*, **20**, 983–987.
46. R. Marangoni, S. Puntoni, L. Favati, G. Colombetti (1994). Phototaxis in *Fabrea salina* I. Action spectrum determination. *J. Photochem. Photobiol. B: Biol.*, **23**, 149–154.
47. Kadota, M. Koyama, M. Wada, M. Furuya (1984). Action spectra for polarotropism and phototropsim in protonemata of the fern *Adiantum capillus-veneris*. *Physiol. Plant.*, **61**, 327–330.
48. T.I. Baskin, M. Iino (1987). An action spectrum in the blue and UV for phototropism in alfalfa. *Photochem. Photobiol.*, **46**, 127–136.
49. T.W. James (1987). Photomechanical transduction in *Amoeba proteus*: An action spectrum. *J. Photochem. Photobiol. B: Biol.*, **1**, 203–211.
50. K. Lazowski, L. Kuznicki (1991). Influence of light of different colors on motile behavior and cytoplasmic streaming in *Amoeba proteus*. *Acta Protozool.*, **30**, 73–77.
51. R. Wayne, A. Kadota, M. Watanabe, M. Furuya (1991). Photomovement in *Dunaliella salina*: Fluence rate-response curves and action spectra. *Planta*, **184**, 515–524.
52. H. Kawai, M. Kubota, T. Kondo, M. Watanabe (1991). Action spectra for phototaxis in zoospores of the brown alga *Pseudochorda gracilis.*, *Protoplasma*, **161**, 17–22.
53. L. Hitchcock Jr. (1961). Color sensitivity of the *Amoeba* revisited. *J. Protozool.*, **8**, 322–324.
54. L.B. Shettles (1937). Response to light in *Peranema trichophorum* with special reference to dark-adaptation and light-adaptation. *J. Exp. Zool.*, **77**, 215–249.
55. L.B. Shettles (1938). Effect of ultraviolet light and X-rays on *Peranema trichophorum*. *J. Cell. Comp. Physiol.*, **12**, 263–272.
56. M. Watanabe, M. Furuya (1974). Action spectrum of phototaxis in a cryptomonad alga *Cryptomonas* sp. *Plant Cell Physiol.*, **15**, 413–420.
57. K. Matsuoka, Y. Nakaoka (1988). Photoreceptor potential causing phototaxis of *Paramecium bursaria*. *J. Exp. Biol.*, **137**, 477–485.
58. S. Fabczak, H. Fabczak, N. Tao, P.-S. Song (1993). Photosensory transduction in ciliates.I. An analysis of light-induced electrical and motile responses in *Stentor coeruleus*. *Photochem. Photobiol.*, **57**, 696–701.
59. D-P. Häder, M. Melkonian (1983). Phototaxis in the gliding flagellate, *Euglena mutabilis*. *Arch Microbiol.*, **135**, 25–29.
60. C. Schilde (1968). Schnelle photoelektrische Effekte der Alge *Acetabularia*. *Z. Naturforschg.*, **23b**, 1369–1376.
61. R.B. Forward, Jr. (1974). Phototaxis by the dinoflagellate *Gymnodinium splendens* Lebour. *J. Protozool.*, **21**, 312–315.
62. R. Gaertner (1979). Die Bewegung des *Mesotaenium*-Chloroplasten im Starklichtbereich. II. Aktionsdichroismus und Wechselwirkungen des Photorezeptors mit Phytochrom. *Z. Pflanzenphysiol.*, **63**, 428–443.
63. P. Galland, P. Keiner, D. Dörnemann, H. Senger, B. Brodhun, D.-P. Häder (1990). Pterin- and flavin-like fluorescence associated with isolated flagella of *Euglena gracilis*. *Photochem. Photobiol.*, **51**, 675–680.
64. V.A. Sineshchekov, D. Geiß, O.A. Sineshchekov, P. Galland, H. Senger (1994). Fluorometric characterization of pigments associated with isolated flagella of *Euglena gracilis*: evidence for energy migration. *J. Photochem. Photobiol. B: Biol.*, **23**, 225–237.
65. M. Delbrück, W. Shropshire Jr. (1960). Action and transmission spectra of *Phycomyces*. *Plant Physiol.*, **35**, 194–204.

66. P.A. Ensminger, H.R. Schaefer, E.D. Lipson (1991). Action spectra of the light-growth response of *Phycomyces*. *Planta*, **184**, 498–505.
67. P. Halldal (1958). Action spectra of phototaxis and related problems in Volvocales, *Ulva*-gametes and Dinophyceae. *Physiol. Plant.*, **11**, 118–153.
68. P. Halldal (1961). Ultraviolet action spectra of positive and negative phototaxis in *Platymonas subcordiformis*. *Physiol. Plant.*, **14**, 133–139.
69. M.A. Lawson, D.N. Zacks, F. Derguini, K. Nakanishi, J.L. Spudich (1991). Retinal analog restoration of photophobic responses in a blind *Chlamydomonas reinhardtii* mutant. *Biophys. J.*, **60**, 1490–1498.
70. Barghigiani, G. Colombetti, B. Franchini, F. Lenci (1979). Photobehavior of *Euglena gracilis*: Action spectrum for the step-down photophobic response of individual cell. *Photochem. Photobiol.*, **29**, 1015–1019.
71. C.J. Bell, D.A. Rose (1981). Light measurement and the terminology of flow. *Plant Cell Environ.*, **4**, 89–96.
72. H.J.A. Dartnall (1972). Photosensitivity. In: H.J.A. Dartnall (Ed.), *Photochemistry of Vision. Handbook of Sensory Physiology* (Vol. VII/1, pp. 122–145). Springer-Verlag, New York.
73. G. Wald (1968). Molecular basis of visual excitation. *Science*, **162**, 230–239.
74. H. Dartnall (1953). The interpretation of spectral sensitivity curves. *Brit. Med. Bull.*, **9**, 24–30.
75. DMS (1966) *UV Atlas of Organic Compounds, Published in collaboration with the Photoelectric Spectrometry Group, London and the Institut für Spektrochemie und Angewandte Spektroskopie*. Dortmund, Plenum Press, New York.
76. Knowles, H.J.A. (1977). Dartnall, The photobiology of vision. In: H. Davson (Ed.), *The Eye* (p. 76). Academic Press, New York.
77. N.A. Dencher (1978). Light-induced behavioural reactions of *Halobacterium halobium*; evidence for two rhodopsins acting as photopigments. In: S.R. Caplan, M. Ginzburg (Eds), *Energetics and Structure of Halophilic Microorganisms* (pp. 67–85). Elsevier/North Holland Biomedical Press, Amsterdam.
78. K.W. Foster (1972). The photoresponses of *Phycomyces*: Analysis using manual techniques and an automatic machine which precisely tracks and measures growth during programmed stimuli. Ph.D. Thesis, California Institute of Technology, Pasadena, CA.
79. M. Erata, M. Kubota, T. Takahashi, I. Inouye, M. Watanabe (1995). Ultrastructure and phototactic action spectra of two genera of cryptophyte flagellate algae, *Cryptomonas* and *Chroomonas*. *Protoplasma*, **188**, 258–266.
80. H. Kaneda, M. Furuya (1987). Effects of timing of flashes of light during the course of cellular rotation in phototactic orientation of individual cells of *Cryptomonas*. *Plant Physiol.*, **84**, 178–181.
81. D.R.A. Hill (1991). A revised circumscription of *Cryptomonas* (Cryptophyceae) based on examination of Australian strains. *Phycologia*, **30**, 170–188.
82. T. Takahashi, M. Watanabe (1993). Photosynthesis modulates the sign of phototaxis of wild-type *Chlamydomonas reinhardtii*. Effects of red background illumination and 3-(3′,4′-dichlorophenyl)-1,1-dimethylurea. *FEBS Lett.*, **336**, 516–520.
83. T. Matsuoka, S. Matsuoka, Y. Yamaoka, T. Kuriu, Y. Watanabe, M. Takayanagi, Y. Kato, K. Taneda (1992). Action spectra for step-up photophobic response in *Blepharisma*. *J. Protozool.*, **39**, 498–502.
84. P. Scevoli, F. Bisi, G. Colombetti, F. Ghetti, F. Lenci, V. Passaraelli (1987). Photomotile responses of *Blepharisma japanicum*.I. Action spectra determination and time-resolved fluorescence of photoreceptor pigments. *J. Photochem. Photobiol. B: Biol.*, **1**, 75–84.

85. M. Kraml, W. Marwan (1983). Photomovement responses of the heterotrichous ciliate *Blepharisma japonicum*. *Photochem. Photobiol.*, **37**, 313–319.
86. L. Barsanti, V. Passarelli, P. Lenzi, P.L. Walne, J.R. Dunlap, P. Gualtieri (1993). Effects of hydroxylamine, digitonin and triton X–100 on photoreceptor (paraflagellar swelling) and photoreception of *Euglena gracilis*. *Vision Res.*, **33**, 2043–2050.
87. M.K. Otto, M. Jayaram, R.M. Hamilton, M. Delbrück (1981). Replacement of riboflavin by an analogue in the blue light photoreceptor of *Phycomyces*. *Proc. Natl. Acad. Sci. USA*, **78**, 266–269.
88. D.-P. Häder, M. Lebert (1998). The photoreceptor for phototaxis in the photosynthetic flagellate *Euglena gracilis*. *Photochem. Photobiol.*, **68**, 260–265.
89. V.E.A. Russo, J.A.A. Chambers, F. Degli-Innocenti, T. Sommer (1984). Photomorphogenesis in microorganisms. In: G. Colombetti, F. Lenci, P-S. Song (Eds), *Sensory Perception and Transduction in Aneural Organisms* (pp. 231–249). Plenum Press, New York.
90. O.A. Sineshchekov (1991). Photoreception in unicellular flagellates: bioelectric phenomena in phototaxis. In: H.R.H. Douglas (Ed.), *Light in Biology and Medicine. Vol. II. Proceedings of the III. Congress of the European Society for Photobiology* (pp. 191–202). Budapest, Plenum Press, New York.
91. Y. Nakaoka, R. Tokioka (1988). Photoreceptor potential causing phototaxis in *Paramecium bursaria*. *J. Exp. Biol.*, **137**, 477–485.
92. R.M. Eakin (1982). Continuity and diversity in photoreceptors. In: J.A. Westfall (Ed.), *Visual Cells in Evolution* (pp. 91–105).
93. R.B. Forward, Jr. (1973). Phototaxis in a dinoflagellate: action spectra as evidence for a two-pigment system. *Planta*, **111**, 167–178.
94. P.S. Klein, T.J. Sun, C.L. Saxe, A.R. Kimmel, R.L. Johnson, P.N. Devreotes (1988). A chemotactic receptor controls development in *Dictyostelium discoideum*. *Science*, **241**, 1467–1472.
95. K. Seidel, B. Scharf, M. Gautel, K. Kleine, D. Oesterhelt (1995). The primary structure of sensory rhodopsin II: a member of an additional retinal protein subgroup is coexpressed with its transducer, the halobacterial transducer of rhodopsin II. *Proc. Natl. Acad. Sci.*, **92**, 3036–3040.
96. E.W. Taylor, A. Agarwal (1993). Sequence homology between bacteriorhodopsin and G-protein coupled receptors: exon shuffling or evolution by duplication? *FEBS*, **325**, 161–166.
97. K. Schletz (1976). Phototaxis bei *Volvox*. Pigmentsysteme der Lichtrichtungsperzeption. *Z. Pflanzenphysiol.*, **77**, 189–211.
98. J.A. Robertson (1972). Phototaxis in a new *Allomyces*. *Arch. Mikrobiol.*, **85**, 259–266.
99. R. Menzel, F. Roth (1972). Spektrale Phototaxis von Planktonrotatorien. *Experientia*, **28**, 356–357.
100. W. Deininger, P. Kroger, U. Hegemann, F. Lottspeich, P. Hegemann (1995). Chlamyrhodopsin represents a new type of sensory photoreceptor. *EMBO J.*, **14**, 5849–5858.
101. R.S. Molday, D. Hicks, L. Molday (1987). Peripherin: A rim specific membrane protein of rod outer segment discs. *Invest. Ophthalmol. Vis. Sci.*, **28**, 50–61.
102. M. Nakazawa, Y. Wada, Y. Chida, M. Tamai (1997). A correlation between computer-predicted changes in secondary structure and the phenotype of retinal degeneration associated with mutations in perhipherin/RDS. *Current Eye Research*, **16**, 1134–1141.
103. A.J. Jesaitis (1974). Linear dichroism and orientation of the *Phycomyces* photopigment. *J. Gen. Physiol.*, **63**, 1–21.
104. P. Gualtieri, L. Barsanti, V. Pasarelli (1989). Absorption spectrum of a single isolated paraflagellar swelling of *Euglena gracilis*. *Biochim. Biophys. Acta*, **993**, 293–296.

105. T.W. James, F. Crescitelli, E.R. Loew, W.N. McFarland (1992). The eyespot of *Euglena gracilis*: a microspectrophotometric study. *Vision Res.*, **32**, 1583–1591.
106. P. Gualtieri (1993). A biological point of view on photoreception (no-imaging vision) in algae. *J. Photochem. Photobiol. B: Biol.*, **18**, 95–100.
107. P.L. Walne, V. Passarelli, P. Lenzi, L. Barsanti, P. Gualtieri (1995). Isolation of the flagellar swelling and identification of retinal in the phototactic flagellate, *Ochromonas danica* (Chrysophyceae). *J. Eukaryot. Microbiol.*, **42**, 7–11.
108. P.L. Walne, P. Gualtieri (1994). Algal visual proteins: an evolutionary point of view. *Crit. Rev. Plant Sci.*, **13**, 183–197.
109. D.G. Müller, I. Maier, H. Müller (1987). Flagellum autofluorescence and photoaccumulation in heterokont algae. *Photochem. Photobiol.*, **46**, 1003.
110. H. Kawai (1992). Green flagellar autofluorescence in brown algal swarmers and their phototactic responses. *Bot. Mag. Tokyo*, **105**, 171–184.
111. P. Clement, E. Wurdak (1984). Photoreceptors and photoreceptions in rotifers. In: M.A. Ali (Ed.), *Photoreception and Vision in Invertebrates* (NATO ASI series 74, pp. 241–288). Plenum Press, New York.
112. S.M. Han, B.S. DeDecker, S.O. Smith (1993). Localization of the retinal protonated Schiff base counter-ion in rhodopsin. *Biophys.J.*, **65**, 899–906.
113. S.W. Lin, G.G. Kochendoerfer, K.S. Carrol, D. Wang, R. Mathies, T.P. Sakmar (1998). Mechanisms of spectral tuning in blue cone visual pigments. Visible and Raman spectroscopy of blue-shifted rhodopsin mutants. *J. Biol. Chem.*, **273**, 24583–24591.
114. H. Houjou, Y. Inoue, M. Sakurai (1998), Physical origin of the opsin shift of bacteriorhodopsin. Comprehensive analysis based on medium effect theory of absorption spectra. *J. Amer. Chem. Soc.*, **120**, 4459–4470.
115. R. Hubbard, G. Wald (1952). *Cis-trans* isomers of vitamin A and retinene in vision. *Science*, **115**, 60–63.
116. A. Kropf, R. Hubbard (1958). The mechanism of bleaching rhodopsin. *Ann. NY Acad. Sci.*, **74**, 266–280.
117. A. Surya, K.W. Foster, B.E. Knox (1995). Transducin activation by the bovine opsin apoprotein. *J. Biol. Chem.*, **270**, 5024–5031.
118. K.W. Foster, J. Saranak (1992). A 9-phenyl retinal analog and hexenal appear to normally activate the rhodopsin of *Chlamydomonas*. *Biophys. J.*, **61**, 527a.
119. T. Takahashi, K. Yoshihara, M. Watanabe, M. Kubota, R. Johnson, F. Derguini, K. Nakanishi (1991). Photoisomerization of retinal at 13-ene is important for phototaxis of *Chlamydomonas reinhardtii*: Simultaneous measurements of phototactic and photophobic responses. *Biochem. Biophys. Res. Comm.*, **178**, 1273–1279.
120. T. Takahashi, M. Kubota, M. Watanabe, K. Yoshihara, F. Derguini, K. Nakanishi (1992). Diversion of the sign of phototaxis in a *Chlamydomonas reinhardtii* mutant incorporated with retinal and its analogs. *FEBS Lett.*, **314**, 275–279.
121. F. Derguini, P. Mazur, K. Nakanishi, D.M. Starace, J. Saranak, K.W. Foster (1991). All-*trans*-retinal is the chromophore bound to the photoreceptor of the alga *Chlamydomonas reinhardtii*. *Photochem. Photobiol.*, **54**, 1017–1021.
122. D.N. Zacks, F. Derguini, K. Nakanishi, J.L. Spudich (1993). Comparative study of phototactic and photophobic receptor chromophore properties in *Chlamydomonas reinhardtii*. *Biophys. J.*, **65**, 508–518.
123. Rousso, E. Khachatryan, Y. Gat, I. Brodsky, M. Ottolenghi, M. Sheves, A. Lewis (1997). Microsecond atomic force sensing of protein conformational dynamics: Implications for the primary light-induced events in bacteriorhodopsin. *Proc. Natl. Acad. Sci. USA*, **94**, 7937–7941.

124. J. Buchert, V. Stefancic, K. Doukas, R.R. Alfano, R.H. Callender, J. Pande, H. Akita, V. Balogh-Nair, K. Nakanishi (1983). Picosecond kinetic absorption and fluorescence studies of bovine rhodopsin with a fixed 11-ene. *Biophys. J.*, **43**, 279–283.
125. R. Simmeth, G.W. Rayfield (1990). Evidence that the photoelectric response of bacteriorhodopsin occurs in less than 5 picoseconds. *Biophys. J.*, **57**, 1099–1101.
126. C.P. Govardhan, D.D. Oprian (1994). Active site-directed inactivation of constitutively active mutants of rhodopsin. *J. Biol. Chem.*, **269**, 6524–6257.
127. L. Huang, H. Deng, G. Weng, Y. Koutalos, T. Ebrey, M. Groesbeek, J. Lugtenberg, M. Tsuda, R.H. Callender (1996). A resonance Raman study of the C=N configuration of *octopus* rhodopsin, bathorhodopsin, and isorhodopsin. *Biochemistry*, **35**, 8504–8510.
128. P. Hegemann, P. Gärtner, R. Uhl (1991). All-*trans* retinal constitutes the functional chromophore in *Chlamydomonas* rhodopsin. *Biophys. J.*, **60**, 1477–1489.
129. O.A. Sineshchekov, E.G. Govorunova, A. Der, L. Keszthelyi, W. Nultsch (1994). Photoinduced electrical currents in carotenoid-deficient *Chlamydomonas* mutants reconstituted with retinal and its analogs. *Biophys. J.*, **66**, 2073–2084.
130. R. Mathies, L. Stryer (1976). Retinal has a highly dipolar vertically excited singlet state: implications for vision. *Proc. Natl. Acad. Sci. USA*, **73**, 2169–2173.
131. A. Lewis (1978). The molecular mechanism of excitation in visual transduction and bacteriorhodopsin. *Proc. Natl. Acad. Sci. USA*, **75**, 549–553.
132. K. Fahmy, T.P. Sakmar (1993). Light-dependent transducing activation by an ultraviolet-absorbing rhodopsin mutant. *Biochem.*, **32**, 9165–9171.
133. E.A. Zhukovsky, P.R. Robinson, D.D. Oprian (1991). Transducin activation by rhodopsin without a covalent bond to the 11-cis-retinal chromophore. *Science*, **251**, 558–560.
134. N. Friedman, S. Druckmann, J. Lanyi, R. Needleman, A. Lewis, M. Ottolenghi, M. Sheves (1994). A covalent link between the chromophore and the protein backbone of bacteriorhodopsin is not required for forming a photochemically active pigment analogous to the wild type. *Biochemistry*, **33**, 1971–1976.
135. U. Schweiger, J. Tittor, D. Oesterhelt (1994). Bacteriorhodopsin can function without a covalent linkage between retinal and protein. *Biochemistry*, **33**, 535–541.
136. K.W. Foster, J. Saranak, R. van der Stehen, J. Lugtenburg (1987). Retinal in *Chlamydomonas* rhodopsin is in a planar 6-s-trans conformation as shown by in vivo incorporation of 6–7 locked retinal analogs. *Invest. Ophthal. Vis. Sci.*, **28** (Suppl.), 253.
137. M. Sakamoto, A. Wada, A. Akai, M. Ito, T. Goshima, T. Takahashi (1998). Evidence for the archaebacterial-type conformation about the bond between the β-ionone ring and the polyene chain of the chromophore retinal in chlamyrhodopsin. *FEBS Lett.*, **434**, 335–338.
138. D.S. Goodman, H.S. Huang, T. Shiratori (1966). Mechanism of the biosynthesis of vitamin A from β-carotene. *J. Biol. Chem.*, **241**, 1929–1932.
139. D.S. Goodman, H.S. Huang, M. Kanai, T. Shiratori (1967). The enzymatic conversion of all-trans β-carotene into retinal. *J. Biol. Chem.*, **242**, 3543–3554.
140. Tang, X.-D. Wang, R.M. Russell, N.I. Krinsky (1991). Characterization of β-apo-13-carotenone and β-apo-14′-carotenal as enzymatic products of the exocentric cleavage of β-carotene. *Biochemistry*, **30**, 3829–3834.
141. X.-D. Wang, G.-W. Tang, J.G. Fox, N.I. Krinsky, R.M. Russell (1991). Enzymatic conversion of β-carotene into β-apo-carotenals and retinoids by human, monkey, ferret, and rat tissues. *Arch. Biochem. Biophys.*, **285**, 8–16.
142. M.A. Quinones, Z.M. Lu, E. Zeiger (1996). Close correspondence between the action spectra for the blue-light response of the guard-cell and coleoptile chloroplasts, and the spectra for blue light-dependent stomatal openings and coleoptile phototropism. *Proc. Natl. Acad. Sci. USA*, **93**, 2224–2228.

143. K.W. Foster, E.D. Lipson (1973. The light-growth responses of *Phycomyces*. *J. Gen. Physiol.*, **62**, 590–617.
144. P. Gualtieri, L. Barsanti, G. Rosati (1986). Isolation of the photoreceptor (paraflagellar body) of the phototactic flagellate *Euglena gracilis*. *Arch. Microbiol.*, **145**, 303–305.
145. J.J. Wolken (1977). *Euglena*: The photoreceptor system of phototaxis. *J. Protozool.*, **24**, 518–522.
146. P.A. Kivic, M. Vesk (1972). Structure and function in the Eugenoid eyespot apparatus: the fine structure, and response to environmental changes. *Planta*, **105**, 1–14.
147. C. Creutz, B. Diehn (1976). Motor responses to polarized light and gravity sensing in *Euglena gracillis*. *J. Protozool.*, **23**, 552–556.
148. L.N.M. Duysens (1956). The flattening of the absorption spectrum of suspensions as compared to that of solutions. *Biochim. Biophys. Acta*, **19**, 1–12.
149. F. Crescitelli, W. James, J.M. Erickson, E.R. Loew, N. McFarland (1992). The eyespot of *Chlamydomonas reinhardtii*: a comparative microspectrophotometric study. *Vision Res.*, **32**, 1591–1600.
150. A.-K. Janoudi, K.L. Poff (1976). Action spectrum for enhancement of phototropism by *Arabidopsis thaliana* seedlings. *Photochem. Photobiol.*, **56**, 655–659.
151. K.M. Hartmann, W. Haupt (1983). Photomorphogenesis. In: W. Hoppe, W. Lohmann, H. Markl, H. Ziegler (Eds), *Biophysics* (pp. 542–559). Springer-Verlag, New York.
152. M. Keul (1976). Action spectrum of photodinesis with barley root hairs *Hordeum vulgare*. *Z. Pflanzenphysiol.*, **79**, 40–52.
153. D.J. Patterson (1996). *Free-living freshwater protozoa: a colour guide* (p. 223). John Wiley, NY.
154. van den Hoek, D.G. Mann, H.M. Jahns (1996). *Algae: An Introduction to Phycology*. Cambridge Univ. Press.
155. R.A. Andersen (1989). Taxonomy, structure and ultrastructure: the Synurophyceae and their relationship to other golden algae. *Beiheft zur Nova Hedwigia*, **95**, 1–26.
156. G.F. Leedale (1967). *Euglenoid Flagellates* (p. 242). Prentice-Hall, Englewood Cliffs, NJ.
157. P.W. Gabrielson, D.J. Garbary, M.R. Sommerfeld, R.A. Townsend, P.L. Tyler (1990). Phylum Rhodophyta. In: L. Margulis, J.O. Corliss, M. Melkonian, D.J. Chapman (Eds), *Handbook of Protista, Jones and Bartlett Pub.* (pp. 102–118). Boston.
158. S. Bonotto (1988). Recent progress in research in *Acetabularia* and related Dasycladiales. *Prog. Phycol. Res.*, **6**, 59–235.

Photomovement
D.-P. Häder and M. Lebert, editors.

Chapter 4

Light responses in purple photosynthetic bacteria

Judith P. Armitage

Table of contents

Abstract

The vast majority of facultative photosynthetic bacteria are motile, most using flagella to swim about their environment. All species examined to date respond to changes in the intensity and wavelength of light and to changes in the chemical composition of their environment. It seems probable that bacteria are able to balance a wide range of sensory signals to move to and maintain themselves in optimum environments. While free swimming bacteria are almost certainly unable to respond to the direction of light, there is now evidence that colonies of bacteria, and possible bacteria in dense mats are able to respond to the direction as well as large changes in intensity. This may be responsible, balanced with other chemosensory signals, for maintaining the structures of some microbial mats. Recent studies of the mechanisms of signaling at a molecular level have shown that the sensory pathway from the receptor, which involved sensing changes in the rate of electron transfer, to the flagella motor are shared with the chemosensory pathway, thus allowing balancing of the different environmental signals. It has also become clear that, unlike many non-photosynthetic species, there are multiple sensory pathways in photosynthetic species, induced under different growth condition. This results in major differences in the range of stimuli sensed under different growth conditions and may help enhance the flexibility of metabolism seen in many purple bacteria.

4.1 Introduction

Motility, active movement around their environment, caused Leeuwenhoek to realize that the minute particles he could see down his microscope were in fact living organisms (see [1,2] for reviews). The first identification of bacteria, therefore, relied on their ability to swim. We now know that it takes about 2–5% of an average bacterial genome to code for flagella and their control. The metabolic cost of maintaining and expressing these genes means that if a bacterium is motile, it must provide it with some advantage for growth and survival [3]. All motile bacteria examined show changes in their swimming behavior in response to environmental change, but the stimuli to which a bacterium responds depends on the species. Thus *Escherichia coli* is repelled by sodium acetate while *Rhodobacter sphaeroides* is attracted; *Pseudomonas putida* is attracted to aromatic hydrocarbons invisible to most other species. A species is attracted to an optimum environment for growth, which will be different for an anaerobic fermenter and an obligate aerobe.

Bacteria can sense and respond to a wide range of stimuli including metabolites, terminal electron acceptors such as oxygen or nitrate, pH, temperature and, of course, light. A motile photosynthetic bacterium, particularly one which can grow with as much metabolic flexibility as the purple bacteria, can sense and respond to a very wide range of stimuli, and the nature of the response will change depending on the current growth conditions (see [4–8] for reviews). This discussion will deal with what we know about sensing in purple bacteria, in particular photosensing, but that cannot be separated from other sensory pathways and the consequences for these species in their natural environment.

I will, in general, describe the phenomena of bacteria changing swimming behavior in response to light as "photoresponses", both positive and negative. There is a great deal of confusion over the nomenclature pertinent to bacterial gradient responses. In the strict sense of the definition, bacteria do not show tactic responses as they are too small to directly sense gradients and orient themselves along them (see below). The words "chemotaxis" and "phototaxis" should really be saved for the oriented behavior of larger organisms [9]. Older literature tends to refer to the light-dependent responses of bacteria as "positive" or "negative photophobic" responses depending on whether the bacteria reverse out of light or dark respectively, and gradient dependent responses as klinokinesis with adaptation. However, the research community studying bacterial responses to chemical gradients (a large and powerful community with several thousand publications over the past few years) has abandoned the strict description of behavior and simply use "bacterial chemotaxis". It may not be accurate, but everyone knows what they mean! It has recently been suggested that "scotophobia" (fear of the dark) is a better word to describe the reversal response of bacteria when faced with a light/dark boundary [10]. In most cases this is correct, but every now and again a photosynthetic bacterium does not behave as expected (see later). Adding to the confusion, there is now some evidence that, unlike responses to chemical gradient, some species of photosynthetic bacteria, when growing as colonies on agar surfaces, actually show true phototaxis, the whole colony moving toward the light. Throughout this review, I will play safe (hopefully) and just use "photoresponses" to describe the response of bacteria to changes in light.

4.2 Historical perspective

In the middle of the nineteenth century microbiology was starting to develop a real scientific footing, particularly in Germany. Early in the century the extensive traveler and man of great curiosity, Gottfried Ehrenberg, had used an early microscope to see what were almost certainly bundles of flagella on a species of *Chromatium* [11]. He wrote a series of monographs during his life, culminating in one in 1883 describing "wave-shaped" flagella which he thought were required for swimming. The man who edited this review was the man who first really observed and described photoresponses in bacteria, Thomas W. Engelmann [12]. Engelmann isolated a bacterium from the Rhine close to his laboratory, which he named "*Bacterium photometricum*". He had started his observations on bacterial behavior on "putrefactive" bacteria and found that when the oxygen concentration on his slide started to fall, the cells swam to air bubbles, or the edge of the cover slip. The addition of carbonic acid, however, slowed them down and "caused great uneasiness" amongst them, probably the first description of tumbling behavior. He also observed that they accumulated around certain nutrients, suggesting a hunger response. He suggested that, taken with the responses to oxygen and carbonic acid, "breathing sensitivity" as he described it, bacteria must be animals with the same urges and needs as humans, i.e. the behavior confirmed the "unity of organic nature".

With "*Bacterium photometricum*" he took his behavioral experiments even further (we now assume from the descriptions that he was looking at a *Chromatium* species). One of the reasons we can be fairly sure he was working on *Chromatium* is his description of the pattern of swimming of the cells he was watching. He described them as swimming smoothly at 20 to 40 μm s^{-1}, rotating at 3–6 revolutions per second around their long axis with occasional brief reversals. This, together with the absorption spectrum and reversals when moving over a light/dark boundary, perfectly describes *Chromatium*. To look at the behavior of his bacteria, he placed an opaque disc with a circular hole between a laboratory burner and his microscope and focussed a light spot onto an otherwise dark and sealed microscope slide. Within 10 min the bacteria had filled the light spot. His next experiment was to use the gas light to shine a spectrum onto a layer of cells and see where they accumulated. He then compared this to the absorption spectrum of the living cells. Bacteria showed weak accumulation in the wavelengths between 810 and 570 nm and between 510 and 550 nm but the strongest accumulation was in the far-red, beyond the visible region at about 850 nm. The blue region and the visible red/orange region became relatively empty of cells. Observing the behavior of the swimming bacteria Engelmann saw that they appeared "frightened" when they went over a light/dark or a wavelength boundary such as yellow/red or far red/red and reversed back into the light or the specific wavelength. When moving in the opposite direction, i.e. dark/light or red/yellow, there was no change in behavior. Together these responses resulted in the accumulation in particular wavelengths of light.

Engelmann could not technically measure the absorption of his bacteria in the infra-red, but the observation that accumulation and absorption coincided in the visible regions of the light spectrum led him to suggest that his bacterial species also absorbed in the infra-red. A study into bacterial photoresponses therefore led to the first description of the absorption spectrum of bacteriochlorophyll! Engelmann speculated on the cause of the behavioral responses. He decided the responses were not caused by oxygen (which was still thought to be a by-product of bacterial photosynthesis) but by a chemical produced by light-dependent metabolism, perhaps starch. He concluded that it was produced in the light and decayed slowly in the dark, therefore bacteria incubated for a prolonged period in the dark swam quickly after the light was switched on, but if the light was then switched off would continue swimming for several minutes.

Engelmann did a wonderful job of describing the light dependent behavior of *Chromatium*, even the change in responses if oxygen was present, but little more was done on photosynthetic bacterial behavior, or indeed bacterial behavior in general, until almost 70 years later.

4.3 Bacterial motility

4.3.1 Motility and patterns of behavior

Flagella structure and operation. Flagellate bacteria swim by rotating semi-rigid helical flagella. Unlike eukaryotic flagella, they are usually made up of a polymer of a single protein, flagellin. Flagella are passive structures, rotated at their base by a series of rings in or close to the cytoplasmic membrane. Several motors from Gram negative species have now been examined genetically and structurally, and they all seem to have the same basic structure (Figure 1a) [3,13–15]. A pair of rings in the outer membrane, the L and

P rings, allow the flagellar rod to pass through to the inner membrane. The rod is fused to the MS ring which is in the inner membrane. Around the MS-ring is arranged a ring of eight or so membrane spanning protein complexes, the Mot complexes [16–19]. The

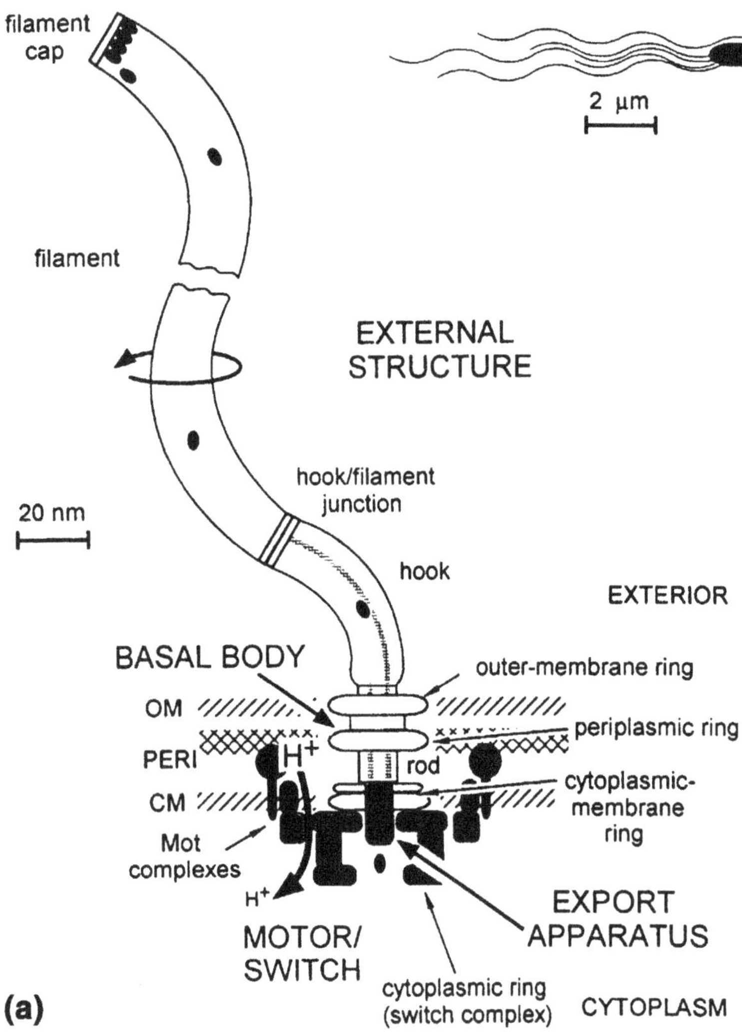

Figure 1. (a) A diagrammatic representation of a bacterial flagellum. OM outer membrane, PERI periplasmic space, CM cytoplasmic membrane. The flagellum is rotated at its base by the movement of protons between the Mot complex and the Motor/Switch. The insert shows a peritrichous bacterium swimming with a bundle of rotating filaments (from [3] with permission). (b) shows the pattern of swimming of representative phototrophic bacteria swimming over a light/dark boundary. A; a bacterium, such as *Chromatium* reverses when it swims over a boundary by reversing the rotational direction of the polar flagella. B; *Rhodospirillum rubrum* reverses the direction of both polar bundles while C; *Rhodobacter sphaeroides* stops when moving over the boundary, a response that can result in trapping in the dark rather than the light.

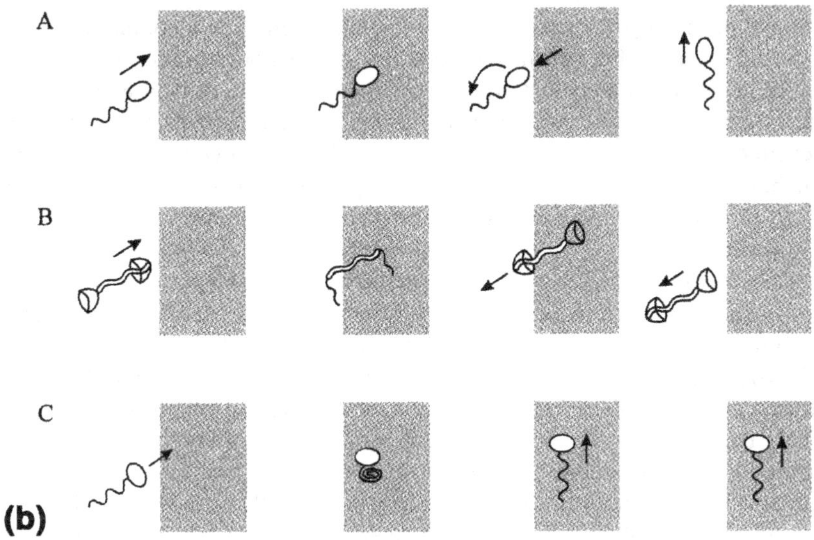

Figure 1. Continued.

Mot complexes are thought to be the site of proton translocation through the motor complex [20]. Protons, or in marine or alkalophilic species sodium ions, are transported, through the Mot protein complexes, down the ion gradient across the cytoplasmic membrane. One of the Mot proteins, MotB appears to have a peptidoglycan binding site and therefore provides the anchor, or stator, of the motor. The Mot complexes probably interact with a ring of proteins on the cytoplasmic face of the MS-ring, the C-ring, which is the rotor component of the motor. As the protons flow from the Mot complex to the C-ring and into the cytoplasm, the electrochemical energy is transformed into mechanical rotation [21]. The proton driven motor can rotate at about 300 Hz, using about 1000 protons per revolution, but the sodium driven motor can rotate at well over 1000 Hz [22,23], quite a feat for a 45 nm protein complex in a fluid membrane. This can drive swimming at speeds of around 25 $\mu m\ s^{-1}$, the average speed of most species studied in the lab, to over 100 $\mu m\ s^{-1}$ for many marine species. This is many body lengths per second through an environment which, for organisms of this size, is all viscosity. Although this may appear a large energy expenditure for a cell, the fact that there are only usually between one and eight flagella per cell means that the expenditure is usually no more that 1% of the cellular energy, and may be as little as 0.1% for cells in rich medium. This might change, however, for species growing under chemolithotrophic and oligotrophic conditions or for cells that under some growth conditions become multi-flagellate, and flagella synthesis may use a significant percentage of the cells' metabolic energy. The observation that most bacterial species living under energy limited conditions still swim, supports the view that swimming must provide a survival advantage.

Patterns of swimming. Flagella may be arranged as single filaments, as in *R. sphaeroides*; peritrichously, as in *E. coli* or as polar tufts, as in *Rhodospirillum rubrum.*

In fact single polar flagella or polar tufts of flagella seem to be a common arrangement on the purple bacteria. Few purple bacteria are peritrichous, which may indicate that stopping or reversals are the most efficient swimming pattern in their environment.

What is really remarkable about the rotating filaments is that, although the electrochemical ion gradient is always in one direction, the direction of filament rotation can switch, or the motor can stop, although the gradient remains constant. This means that e.g. *R. rubrum* or *C. vinosum* swim smoothly in one direction for a while and then suddenly reverse direction. *R. sphaeroides* on the other hand swims smoothly for a while and then stops. When it stops the flagellar filament changes conformation into a short-wavelength, large-amplitude form which rotates slowly, reorienting the cell for its next period of smooth swimming, when the functional filament reforms [24]. Peritrichously flagellate species tend to rotate their flagella in the same direction most of the time, causing the flagella to come together as a bundle and push the cell forward. Periodicaly the direction of rotation of some of the motors switch, causing the bundle to fly apart and the cell to "tumble" on the spot. When the flagella return to rotating in the same direction and reform a functional bundle the cell is usually pointing in a new direction. This pattern of smooth swimming interspersed with direction changing results in bacteria moving around their environment in a random three-dimensional pattern [25,26].

Some species of bacteria change their swimming patterns under different growth conditions, and this may be important for some photosynthetic species in sediments. *R. centenum* and several marine species such as *Vibrio parahaemalyticus* or *V. alginolyticus* swim in liquid using single polar flagella, but when placed on solid media they induce large numbers of lateral flagella that allow them to move over the agar surface [10]. In the case of the *Vibrio* species this also means a move from proton to sodium driven motors [27,28].

4.3.2 Methods for studying bacterial responses and their limitations

Studies on bacterial behavior are limited by the size of bacteria. Flagella are too small to see by normal microscopy and it is very difficult to study behavior in the natural environment. Most studies have been carried out on laboratory cultures of single species presented with a step-up or step-down in stimulus strength. This may bear little resemblance to the environmental gradients encountered in the natural environment. Very little work has been carried out on behavior in gradients, on the behavior of cells confronted with more than one gradient or on mixed population. Extrapolation from laboratory results to the field should therefore be made with caution [29].

Because of the problems involved in following single cells, Nossal adapted a method used to look at spermatazoa to follow swimming populations of bacteria [30]. This involves using the scattering of laser light to follow the formation and movement of bands of bacteria in liquid nutrient medium. Alternatively, if bacteria are placed in semi-solid medium containing a nutrient, as they metabolize the nutrient, this again creates a gradient which they follow. This results in a travelling band of bacteria and, if the medium is complex, it can result in several rings of travelling bacteria, each ring of bacteria responding to a particular nutrient gradient [31,32]. This occurs more slowly than in liquid medium and the medium is more manageable. Swarming in soft agar plates provides a qualitative measure of bacterial behavior, while measurement of light scattering is semi-quantitative. A number of equations have been developed to predict population behavior using this system. Mixtures of chemicals can result in the formation of complex, changing patterns of bacteria as they spread through the medium, the formation and changing of which may provide insights into the uses of chemical signals not only in bacterial movement but the developmental processes in higher organisms [33]. Another method for looking at population behavior was developed 100 years ago by Pfeffer. A capillary containing a chemoeffector is placed into a culture of bacteria. After a time the capillary is removed, the outside washed and the number of bacteria which have swum up the gradient into the capillary counted [34–36]. This again provides a semi-quantitative measure of the responses of bacteria to chemical gradient. In all these cases it is possible to measure the population movement, but difficult to quantify the gradient.

Few methods have been tried for looking at population behavior in natural environments. In general the behavior of bacteria isolated from the environment has been examined in the laboratory and has been extrapolated to the behavior of bacteria in the natural environment. One recent exception has been the measurement of lamina formation within 1 cm microcores made in natural mats filled with fine glass beads. The behavior of bacteria was examined in a section of the core directly on the microscope. Using the natural fluorescence of bacteriochlorophylls, Bachofen has been able to measure the movement *in situ* of purple non-sulfur bacteria throughout the day, as light intensity and other possible stimuli, such as sulfide or oxygen, change [37]. Following the movement of these phototrophs suggests that they may move substantial distances and the method provides an approach for looking at movement of bacteria under completely natural conditions. Some studies have also been made of the behavior of "clouds" of highly motile marine bacteria found close to surfaces [38,39] and the possible role of motility in mat formation [40,41], but these studies are in their infancy.

To analyze the effect of stimulus gradients on bacteria ideally you would like to follow the behavior of free-swimming single cells in real time. Unfortunately the small size of bacteria and their relatively fast swimming speed coupled with 3-dimensional swimming patterns makes this quite difficult. There are several computerized motion analysis systems which can identify a bacterium by its size and record its movement as an array of x,y coordinates with time, using video frame rates. Some new systems can follow up to 100 cells at any time while they are in the plane of focus of the microscope and average the behavior to produce mean swimming speeds and direction changing frequency (see [29] for detail). The population of cells can then be subjected to the release of a potential chemoeffector from a caged molecule by photoexcitation, or a flash of photosynthetic light of different intensities [41,42]. The population response and the time taken for adaptation can then be measured. This response to either a step-up or step-down of the stimulus and the resulting behavior then needs to be extrapolated to suggest a gradient response. This is at best an approximation, the steps may not be very physiological and the cells are tracked in 2 rather than 3 dimensions. A 3-dimensional tracker has been built, but is only able to track one cell at a time [25,43]. However, using this tracker it has been possible to identify and model different patterns of swimming in bacteria with single flagella such as *Pseudomonas* and *R. sphaeroides* and compare that to the swim/tumble behavior of *E. coli* [43,44]. Using this it has been shown that the transient reversals shown by *Pseudomonas* make it more suited to moving through media with obstacles, such as mud, compared to the swim/tumble pattern of *E. coli* which seems to be more efficient in liquid medium. This may have some relevance to the swimming pattern of purple bacteria, which tend to have polar tuft of flagella and also change direction by reversing.

More detailed analysis of behavior can come from examining the response of the flagellar motor itself by either following the wobble of the cell body which counter-rotates during swimming or using DIC microscopy [45,46]. Unfortunately, cells must be slowed by the addition of Ficoll because of the high rate of filament rotation, which is beyond the resolution of most data capture systems. Optical tweezers and electrorotation have been used to analyze the behavior of the motor itself, but the complex and very expensive equipment required means that it has not yet been extended to examine population behavior [47–49].

A classical method is to tether cells by their flagella or hooks using anti-flagella antibody and follow the behavior of the rotating cell body [50,51]. This can be done in a flow chamber and the environment controlled and changed. Computerized motion analysis then follows the rotational behavior of the cell and this is extrapolated to free-swimming responses. The obvious problems are that the cell body is subject to much greater viscous drag and subtle changes in flagellar behavior may be missed. In all these cases there is also the problem of measuring enough cells for the results to be statistically significant. If all cells in a population respond, then things are not too difficult, but with bacteria often only a percentage may respond, and these responses may vary. A large number of cells must be measured and this may be very time consuming. In the end it may only give a rough idea about what a bacterium might do in the natural environment when faced with a single change. Very few experiments have looked at multiple responses [51a]. The simultaneous photorelease of "caged" attractant and repellent indicated that when faced with a simultaneous step-up in a repellent and an attractant, the repellent signal may be faster, but the system is additive [42].

4.3.3 Bacterial photoresponses

Bacteria can apparently sense and respond to gradients of oxygen and chemicals but can bacteria sense a light gradient? This has been a major question since it became understood that they can sense gradients of chemicals and oxygen. Single celled bacteria have a major problem when it comes to sensing changes in their environment. It is worth remembering that the environment around a bacterium is very different from that experienced by larger organisms. A bacterium is so small that it experiences no inertia, but only viscosity, to the extent that when a bacterium stops, even though it may have been swimming at 50 $\mu m\ s^{-1}$, it will stop dead; within the diameter of a proton. This also has implications for gradient sensing, as there is no displacement of medium around a bacterium. Being only a few μm in length they cannot sense a spatial gradient i.e. they have no nose [52]. They are therefore unable to show responses in the same way as singled-celled eukaryotes; they cannot orient in the direction of a stimulus and move up or down the gradient. Instead they sense the environment "now" and compare the signal strength with that a few seconds before. If their world is improving they (usually) swim for longer in the positive direction, biasing a usual random pattern of swimming in a positive direction. In addition to the problem of gradient sensing is the problem of medium viscosity. The highly viscous medium for such small organisms with rotating filament means that in general they do not (cannot) swim in straight lines, but in gentle curves. This, combined with the constant buffeting, limits the overall distance a bacterium can move. Therefore, can a bacterium sense the direction of light or does it simply respond to the step-down or up in light intensity? Even swimming at 100 $\mu m\ s^{-1}$ a cell will need to swim several tens of millimeters to perceive a change of even 1% in light intensity, unless perhaps it is in a dense mat.

Photoresponses in free-swimming cells. It now seems unlikely that free-swimming bacteria in relatively clear medium can sense the direction of a light source. Many early experiments elegantly described bacteria as simply swimming back over a light/dark boundary by reversal. Swimming over a dark/light boundary had no effect; the cells just kept swimming and changing direction as though nothing had changed [53–57] (Figure 1b). These data all suggested that photosynthetic bacteria showed a simple reversal response when either swimming over a boundary or subjected to a step-down in light intensity, with no sense of direction, but this was tested more precisely recently.

In an experiment adapted from the earlier experiments on motile eukaryotic photosynthetic microbes, free-swimming *Rhodospirillum centenum* and *R. sphaeroides* were incubated anaerobically on a microscope slide [58]. A single fiber-optic fiber was inserted into the culture and a light beam shone through the culture. It presented the cells with a sharp dark/light gradient around the edge of the beam and a shallow gradient in intensity towards the source. There was no illumination other than this single light beam. A low-light sensitive camera was focussed onto the microscope stage and the pattern of accumulation measured by a computer program which quantified the change in light scattering within the light beam. Cells from a culture of the phototactic eukaryotic alga *Chlamydomonas* swam into the beam and then swam directly towards the source of the light. *R. centenum* on the other hand increased the light scattering within the beam over a period of a few minutes, but the increase in scattering was uniform along the beam with no increase towards the light source, consistent with the cells swimming over the dark/light boundary and then becoming trapped in the light. *R. centenum,* therefore, appeared to sense a large change in intensity at the edge of the light beam, but not the shallow change experienced along the light beam. Unexpectedly, *R. sphaeroides* culture did not accumulate within the light beam, but just outside. This is consistent with the cells swimming over the light/dark boundary and stopping (the direction changing response of *R. sphaeroides*). Presumably once stopped the next random change in direction is more likely to lead them out of the light than into the beam, clearing the light path and resulting in accumulation outside. It is this response which suggests that "scotophobia" may be unsuitable for light responses shown by some bacterial species.

Colonies of *R. centenum* grown on agar plates do respond to the direction of a light gradient (see below) [59]. To investigate whether single cells taken from a colony actively showing phototaxis could move towards the source of the light, swarming colonies were resuspended in water and subjected to the experiment described above. The individual cells from the swarming, phototactic colony showed no response other than to the light/dark boundary.

This suggests that individual free-swimming cells in a liquid medium do not sense the direction of light but sense a step-change in intensity or in useful wavelength.

Colony behavior of R. centenum. Like a number of other bacterial species (*Proteus, Vibrio alginolyticus, Azospirillum* and some *Salmonella* species) when *R. centenum* is grown on the correct concentration of agar, the cells develop a large number of lateral flagella, while still retaining their usual polar filament. Using these lateral filaments, cells are able to move over surfaces, but rather than the colony spreading across the surface as occurs in other species, the colony as a whole physically moves [10,59]. The colonies move towards infra-red light but are repelled by white light (Figure 2). Experiments where colonies were faced with angled light beams such that the intensity and direction were not necessarily coincident showed that the colonies did indeed move towards the light source [59]. If light is shone from two sides the colony will move at an angle between the light beams. Action spectra provided confusing data, as the wavelengths used for attraction and for repulsion are both absorbed by bacteriochlorophyll, suggesting that photosynthetic activity can result in two different signals.

If the individual cells within a colony are observed during active swarming, they are not oriented and moving towards the light, but swimming in large circles, swirling away from the edge of the colony. Measurement of oxygen within the colony showed that there is an oxygen gradient within the colony in addition to a gradient of light [60]. This leaves open the possibility that the colony response is a combination of the aerotactic and photoresponses of individual cells, each cell following a light gradient and then reversing in response to the increased oxygen levels close to the colony surface and this moves the colony forward. Alternatively the cells at the edge, in high light, oxygen and nutrient may excrete a metabolite which acts as an attractant for the cells in the inside of the colony, which will be in very low light, anaerobic, nutrient depleted conditions.

What is the role of this surface swarming behavior? It is possible that this type of behavior may help some bacterial species to layer within microbial mats. The light intensities within mats can be expected to be very low and under these circumstances there may be photoresponses, as the intensity may move below a photosynthetic threshold rapidly enough to allow photoresponses and this, combined with other electron transport sensing systems such as aerotaxis, may serve to maintain some species in their optimum position in mats (see later). As will be discussed later, the sensory pathway from both the photosensory and the chemosensory receptors is the same, which would allow the balancing of responses. It is intriguing that the attractant and repellent wavelengths identified in *R. centenum* are both absorbed by bacteriochlorophyll and the differential mechanisms involved in sensing remain to be elucidated.

Figure 2. Movement of a colony of *Rhodospirillum centenum* across an agar plate response to IR light shone from the right side of the plate. The colony was photographed at 0′, 60′, and 90′ by which time it had moved about 4 cm. Photographs kindly provided by H. Gest and C. Bauer.

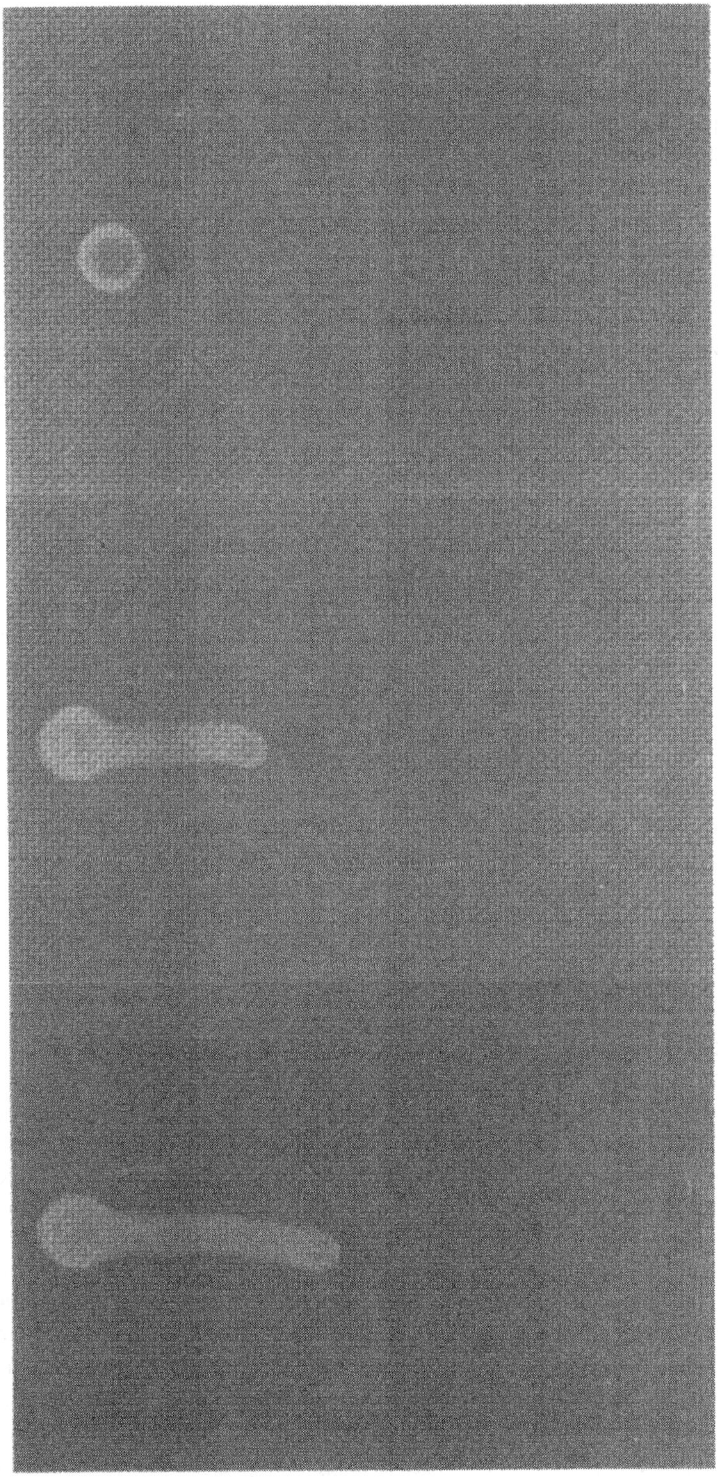

Light responses in non-photosynthetic species. It was shown many years ago that if free-swimming *E. coli* were subjected to a flash of intense blue light they tumbled [61–63]. If the flash was short enough they recovered but if a longer flash was given they continued tumbling for a long period and often finally died. It was initially thought that this "blue-light" response by a non-photosynthetic species was the result of oxidation of a flavin compound in the electron transport chain resulting in the inhibition of respiration and thus a repellent response. Recent work, however, has identified the repellent produced by the flash of blue light as being the product of photoxidation of porphyrins [64]. Whether responses to blue light of this intensity has any physiological role is uncertain, but a large number of bacterial species, photosynthetic and non-photosynthetic are repelled by blue light, probably the most damaging wavelength from the point of view of free radical dependant cell damage, and moving rapidly from these wavelengths may be more important that moving away from UV light [65]. Many photosynthetic bacteria are also repelled by blue light. The pathway for this response does not appear to involve the photosynthetic system, rather a specialized photosensory pigment, the photoactive yellow protein. This is discussed elsewhere (see chapter by Hellingwerf, this volume).

Recently several consortia of bacteria have been isolated from natural environments where a motile but non-photosynthetic species is surrounded by a cluster of non-motile photosynthetic bacteria, usually green sulfur bacteria. If this consortium is subjected to a step-down in light intensity, the consortium reverses, the non-photosynthetic bacterium reversing in response to light [66] (Figure 3). It is assumed that the signal must come from the cluster of photosynthetic bacteria surrounding the motile cell, but the nature of a signal which can pass between the species in less than 1 s is unknown. It is known, however, that *E. coli* can respond to changes in pH using the chemosensory pathway (see later) [67,68]. A change in light intensity would lead to a transient change in extracellular pH as photosynthetic activity changes and it is tempting to speculate that this change in pH could signal to the non-photosynthetic cell and cause reversals.

4.5 The role of photosynthesis in responses

There is a fascinating, if rather improbable, theory about the possible origins of photosynthesis from gradient-dependent phototaxis. It has been suggested that photosynthesis itself might be an evolutionary development from bacterial phototaxis [69]. The "black smokers" or hydothermal vents at the bottom of the oceans produce far-red radiation and recently there have been several reports that bacteria containing bacteriochlorophyll *a* and *b* have been found living around these vents, in the supposed pitch-black abysses of the deep oceans. The radiation spectrum is a balance between the radiance of the black-smoker and absorbance by the water, and shows two peaks; one at 800–950 and the other at 1000–1150 nm, close to the maxima of bacteriochlorophyll *a* and *b* absorbance, respectively. The vents provide the nutrients necessary for life, but all organisms living near must balance the steep temperature gradient and the concentration of potentially toxic compounds also present, to produce a signal that maintains them in their optimum environment. It is speculated that earliest micro-organisms (perhaps 3.8 billion years ago) used phototaxis to the intense radiation of the

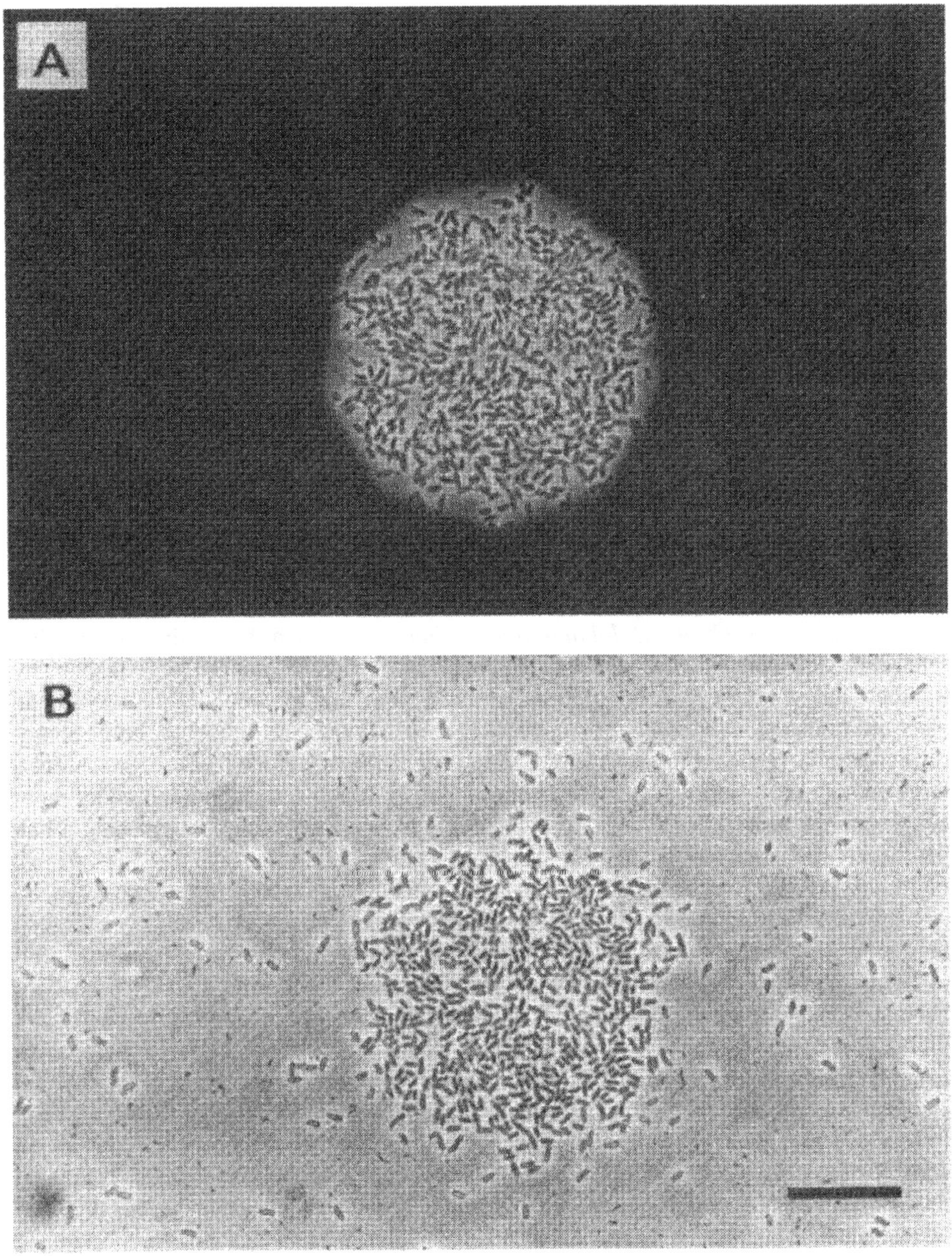

Figure 3. Response of a consortium of non-motile photosynthetic bacteria and motile non-photosynthetic species. The consortium accumulates in light of about 740 nm (see text for details). Very similar patterns of accumulation are seen with motile photosynthetic species. (Picture kindly provided by J. Overmann).

hydothermal vents to maintain themselves in their optimum position for growth and the pigments used eventually evolved into bacteriochlorophyll. Photosynthesis would, therefore, be a spin off from thermal detection by a chemotrophic early bacterium. Although the above speculation is entertaining, it should be treated with great caution. It seems more probable that the phototrophic organisms found around the thermal vents have found their way there from the surface layers of the ocean in much more recent times, rather than evolved there. Much more research needs to be carried out on the general bacterial population in deep oceans and comparisons carried out with populations in surface waters before conclusions about the origins of the populations can be made.

There has been evidence for many years that photosynthesis is essential for photoresponses in most bacterial photosynthetic species. The spectrum to which the bacteria respond is fairly coincident with the photosynthetic action spectrum [12,54] (Figure 4). In fact the lack of photoresponses has been used to isolate mutants in photosynthesis. Mutants in the reaction center proteins, but with the complete complement of light-harvesting bacteriochlorophyll and carotenoid pigments, were found to not only be unable to grow photosynthetically, but also had lost photoresponses [70]. This suggests that the pigments themselves cannot act as receptors and there are no photosynthesis-independent receptors. This is very different to the position found in archaeal species such as *Halobacterium salinarium,* which has retinal containing proteins, SR1 and SRII, dedicated to generating photoresponses (see chapter by Spudich, this volume). This is supported by the finding that inhibitors of photosynthetic electron transport, antimycin A, stigmatellin or myxathizol inhibited both photoresponses and photosynthetic electron transport, under conditions where there was a large enough Δp to allow continued swimming [71]. It was also found that under these conditions the bacteria showed normal chemotaxis, but not aerotaxis, suggesting a link between stimuli altering the electron transport activity [72].

Photosynthetic bacteria adapt to the light intensity in which they are grown by altering the relative concentration of light harvesting complexes and reaction centers, so that under low light the cells have a cytoplasm packed with invaginated membrane full of light harvesting pigments to catch all the photons that fall onto the cell, with light harvesting rather than electron transport being the limiting step in photosynthesis. Under high light, however, the cells adapt to the high photon radiance by having far fewer invaginations and a much smaller complement of light harvesting pigments and a fast electron transfer rate. As there are large numbers of photons falling on the cell, electron transport is rate limiting rather than light harvesting. This results in the photosynthetic activity of low light cells being saturated at a wide range of light intensities, while that of high light cells is only saturated at high light. When cells grown under the different sets of conditions were incubated under different intensities of light and then subjected to a step-down of 97% of that starting intensity, the high-light grown cells responded whatever the starting intensity, showing that the step-down must take them below a critical level of photosynthetic activity. On the other hand low-light grown cells showed no responses until the starting, and thus the final intensity, was extremely low. As the photosynthetic activity was saturated at even really low light intensities this again supports a direct link between the rate of photosynthetic activity and the photoresponse [71] and that there are no photoreceptors responding to an absolute change in light intensity. If photosynthetic electron transport remains saturated after a step-down there is no response, but the cells respond if the rate falls.

Recently, it has been shown that free-swimming *R. sphaeroides* incubated in low light show an increase in swimming speed in response to an increase in light intensity, even though the electrochemical proton gradient is saturated for flagellar motor activity [73].

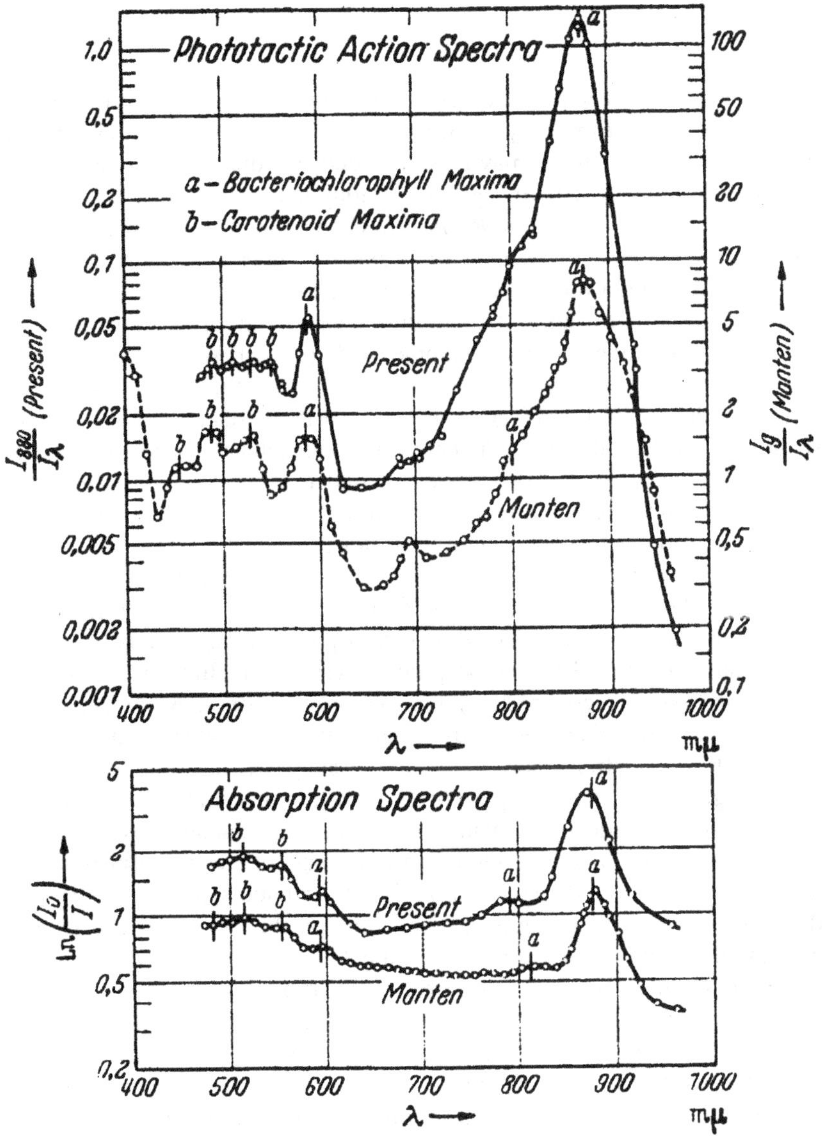

Figure 4. Action spectrum of photoresponses and photosynthetic absorbance of *Rhodospirillum rubrum*. The data compares that of Manten with Clayton. Taken from [57] with permission.

The increase in speed was not seen in cells grown in very low light, only in cells grown in "normal" or high light intensities. However all cells responded when the light intensity was returned to the prestimulus level. This again indicates that photosynthetic electron transport is involved in signaling photoresponses [73a].

4.5.1 Interaction between photoresponses and other electron transport dependent behavior

The purple non-sulfur bacteria can grown not only using photosynthetic electron transfer, but also using respiratory electron transfer and anaerobic respiration. There is evidence that some components may be shared between the different pathways and the pathways may compete. The pathways are certainly not mutually exclusive; respiratory electron transfer stops if pigmented *R. sphaeroides* is illuminated. If there is interaction between the different electron transfer pathways, how does this effect the responses of cells to light or oxygen or other terminal electron acceptors?

R. sphaeroides can grow anaerobically in the dark using terminal electron acceptors such as dimethylsulfoxide or trimethylamine oxide [74,75]. When growing on DMSO, the cells are also attracted by a gradient of DMSO and the cells show a step-down response to a reduction in DMSO concentration. DMSO reductase minus mutants show no response to DMSO, indicating that there is not a receptor independent of the terminal enzyme. If cells growing on DMSO are exposed to oxygen or light, however, both electron transport to the terminal reductase and responses to gradients or step-downs in DMSO concentration are lost [72]. This indicates that when electron are diverted from the DMSO to the terminal oxidase or to cyclic electron transfer there is no longer a response to DMSO, as there is no electron flow to the reductase. The response depends on active electron transport, and the electron transport pathways compete.

Similar results are found for the step-down response to light in the presence and absence of oxygen. Oxygen reduces the size of the step-down response to light considerably, suggesting that oxygen also reduces the rate of electron transport through the photosynthetic electron transfer chain, possibly by diverting electrons to the alternative, high-affinity oxidase [72,76]. This response could serve a physiological purpose as photosynthetically growing cells would be damaged by swimming into oxygenated environments and the reduction in the rate of electron transport caused by electrons being diverted to the terminal oxidase from photosynthetic electron transport would lead to reversals or to stopping (depending on the species) and bias the pattern of swimming back into an anaerobic environment.

The hypothesis that photoresponses are controlled by the rate of electron transfer is supported by recent data on *R. centenum*. This species swarms as a colony towards red light, even in the presence of oxygen. A mutant was isolated which swarmed much faster than the wild type and the mutation turned out to be in a gene responsible for transporting the oxidase Cu-cofactor, resulting in reduced respiration. This again would support the idea that an increased rate of electron transport causes an increased response (B. Rushing and C. Bauer, personal communication). Purple bacteria are therefore responding not to specific stimuli, but to changes in the rate of electron transfer.

4.5.2 The primary signal

The data outlined above strongly suggests that a change in the rate of electron transport causes the primary signal in the positive photoresponses of purple bacteria to changes in light intensity, rather than there being a specific photoreceptor. However, a change in the rate of electron transport can have two effects, it can alter the size of the electrochemical proton gradient (Δp) or it can alter the redox state of one of the electron transport components, either of which might possibly be the sensory signal. It is extremely difficult to separate the two events as any natural change in one automatically results in a change in the other parameter. There have been several suggestions in the past that bacteria may have a "protometer" sensing changes in the Δp, or even that the flagellar motor may itself act as a "protometer", as well as a proton driven rotor, sensing its own driving force [77,78]. Until recently, no mutants had been identified that might indicate the nature of the "protometer" or redox sensor, although redox sensor appear to be involved in other systems such as transcriptional regulation of nitrogen fixation and control of expression of aerobic/anaerobic pathways [79–83]. The genome sequence of *E. coli,* however, identified a gene coding for a protein with homology to both the chemosensory receptor proteins (see later) and the redox sensing NifL family of proteins. A mutation in this gene, named *aer,* resulted in a loss in accumulation around an air bubble while overexpression resulted in hypersensitivity to oxygen and the cells turning yellow as a result of increased intracellular FAD [80,84]. Aer, therefore, appears to be a redox sensing protein, sensing the rate of respiratory electron transport and signaling through a chemotaxis receptor-like domain to the chemosensory pathway. Given the normal mid-point potential of FAD, it has been suggested that it may sense a change in electron flow between the quinone and the dehydrogenases. Intriguingly, however, mutants in *aer* still showed responses to oxygen, but to lower concentrations, banding away from the minuscus rather than at the miniscus as is seen in wild type cells. Individual cells still showed a response to a step-up or step-down in oxygen, but much weaker. This indicated a second sensory receptor. Double mutants in one of the chemosensory receptors, Tsr, as well as Aer show no oxygen responses, indicating that this receptor, previously identified as being responsible for responses to serine, pH and temperature also senses oxygen. No redox prosthetic group has been identified as being coupled to this transmembrane protein and it has been suggested that the protein might itself sense the change in $\Delta\psi$, making it a protometer, but as yet there is no firm evidence to confirm this suggestion.

The photoresponses and oxygen responses of *R. sphaeroides* were used to try and identify whether this organism was likely to be sensing a change in Δp or a change in redox potential of an electron transport component. *R. sphaeroides* is the ideal organism for this measurement as it only responds (when grown anaerobically in low light) to a step-down in a stimulus [85], and it is possible to non-invasively measure the size of the Δp by using the absorption spectrum of the membrane bound carotenoids [86]. The addition of a small amount of the proton ionophore FCCP causes a reduction in the Δp without loss of swimming. It also effectively "clamps" the Δp by rapidly equalizing the proton gradient. FCCP not only reduces the Δp, it also causes an increase in the rate of electron transfer as the "back-pressure" of protons which usually controls the rate of electron transport is relieved by the FCCP. The addition of FCCP, therefore, caused the

Δp to fall and the electron transfer rate to increase. If *R. sphaeroides* senses Δp, then it should respond to the addition of FCCP, however there was no response to its addition, but a step-down response on its removal; conditions under which the Δp increased again, but the rate of electron transport decreased [71]. This strongly suggests a redox sensor rather than a "protometer" is involved in the primary photoresponse signal.

This was confirmed by looking at the response of a population of *R. sphaeroides* when the Δp was reduced either by FCCP or by a reduction in light. The experiments were carried out such that the rate of reduction was the same in both cases, but in the case of the reduction in light the electron transport rate would also decrease, whereas it would not when FCCP was added. The only population of cells to show a step-down response were those subjected to the reduction in light, the addition of FCCP had no effect on behavior although the Δp had fallen at the same rate as the darkened cells. The same result has recently been seen with free swimming cells, with low levels of electron transport inhibitors reducing the speed changes seen in response to a pulse of light, but FCCP having no effect on the response even though it reduced the Δp (Figure 5). The initial signal is therefore probably a redox change within the electron transfer chain, sensed by a sensory transducer.

The sensor has not been identified in *R. sphaeroides* but a non-phototaxis mutant has been isolated in *R. centenum* with a mutation in a gene which codes for a protein with homology to *E. coli* chemoreceptors. This mutant has lost all photoresponses, positive and negative, but remains normal for chemotaxis. (C. Bauer, H. Gest personal communication). Recent data show that the Reg(Prr) system know to control photosynthetic gene expression also controls expression of the chemosensory genes required for photoresponses and aerotactic responses in *R. sphaeroides*. This system senses respiratory electron flow (S. Romagnoli and J. P. Armitage unpublished).

4.6 Interaction with the chemosensory pathway

4.6.1 The chemosensory pathway

The chemosensory pathway in the enteric bacterium *E. coli* is perhaps the best understood sensory system in biology. Four sets of homologous membrane spanning sensory receptors sense the change in concentration of a limited number of chemoeffectors; serine, aspartate, maltose, galactose, and signal the change in periplasmic occupancy across the membrane to the conserved signaling domain on the cytoplasmic side of the membrane (for reviews see [87–89]) (Figure 6). A linker protein, CheW, transmits this change to a histidine protein kinase, CheA. CheA is a member of the histidine protein kinase (HPK) superfamily of proteins which are generally involved in responding to environmental changes, often by controlling transcription of specific operons [90–92]. In response to a reduction in receptor binding, the conformation of the cytoplasmic domain changes causing CheA to autophosphorylate at a conserved histidine using ATP and then transfers the phosphate to one of two response regulators. CheY is a small 14 kDa protein with the classical asparate pocket of a phospho-relay response regulator, but lacking the usual DNA binding domain. When phosphorylated, CheY-P can bind to the "switch" protein (FliM) on the cytoplasmic face of the flagellar motor and cause the normally CCW rotating flagellum to switch to CW, thus causing the cells to tumble. CheA-P can also transfer its phosphate to a second response regulator,

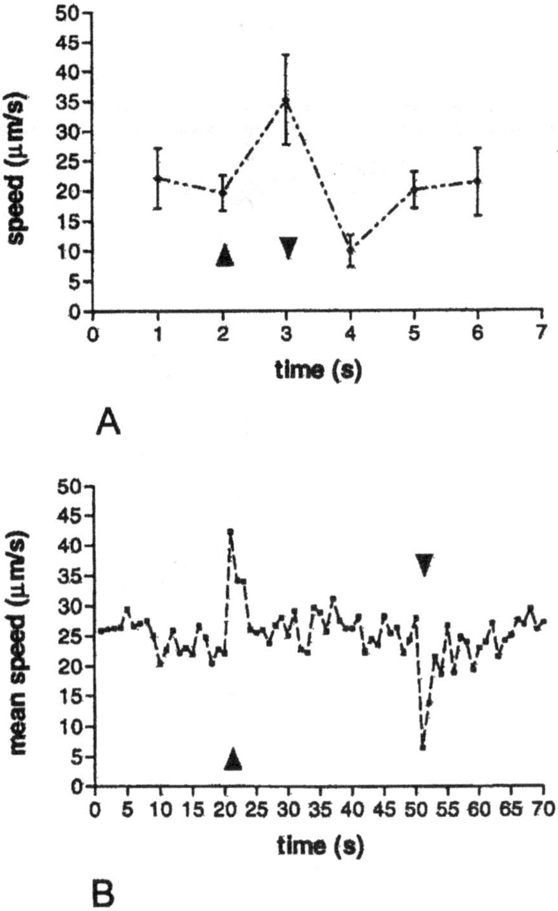

Figure 5. Response of free swimming *Rhodobacter sphaeroides* to a (A) one second flash of white light and (B) a 30 s flash of light. The swimming speed increases and rapidly adapts at the beginning the increase in intensity and the cells stop and adapt when the light is reduced again.

a methyl esterase enzyme, CheB. This enzyme works in conjunction with a methyl transferase, CheR, and together their activity is responsible for resetting the receptor into a non-signaling state, once the chemoeffector concentration stabilizes, by adding or removing methyl groups to glutamate residues on the cytoplasmic domain of the MCPs. CheR has constitutive activity while the activity of CheB is altered by its phosphorylation. The receptor methylation is therefore linked directly to the activity of CheA. CheB is a slow enzyme and the memory within the chemosensory signal probably come from the difference in timing of signal transmission, about 100 – 200 ms and receptor adaptation which takes many tens of seconds, dependent on the signal size. Signal termination is the result of increased dephosphorylation of CheY-P by an activator, CheZ which competes with the motor switch for CheY-P, but not CheY [93]. CheY-P has an inherent rate of dephosphorylation, but this is increased by binding to CheZ.

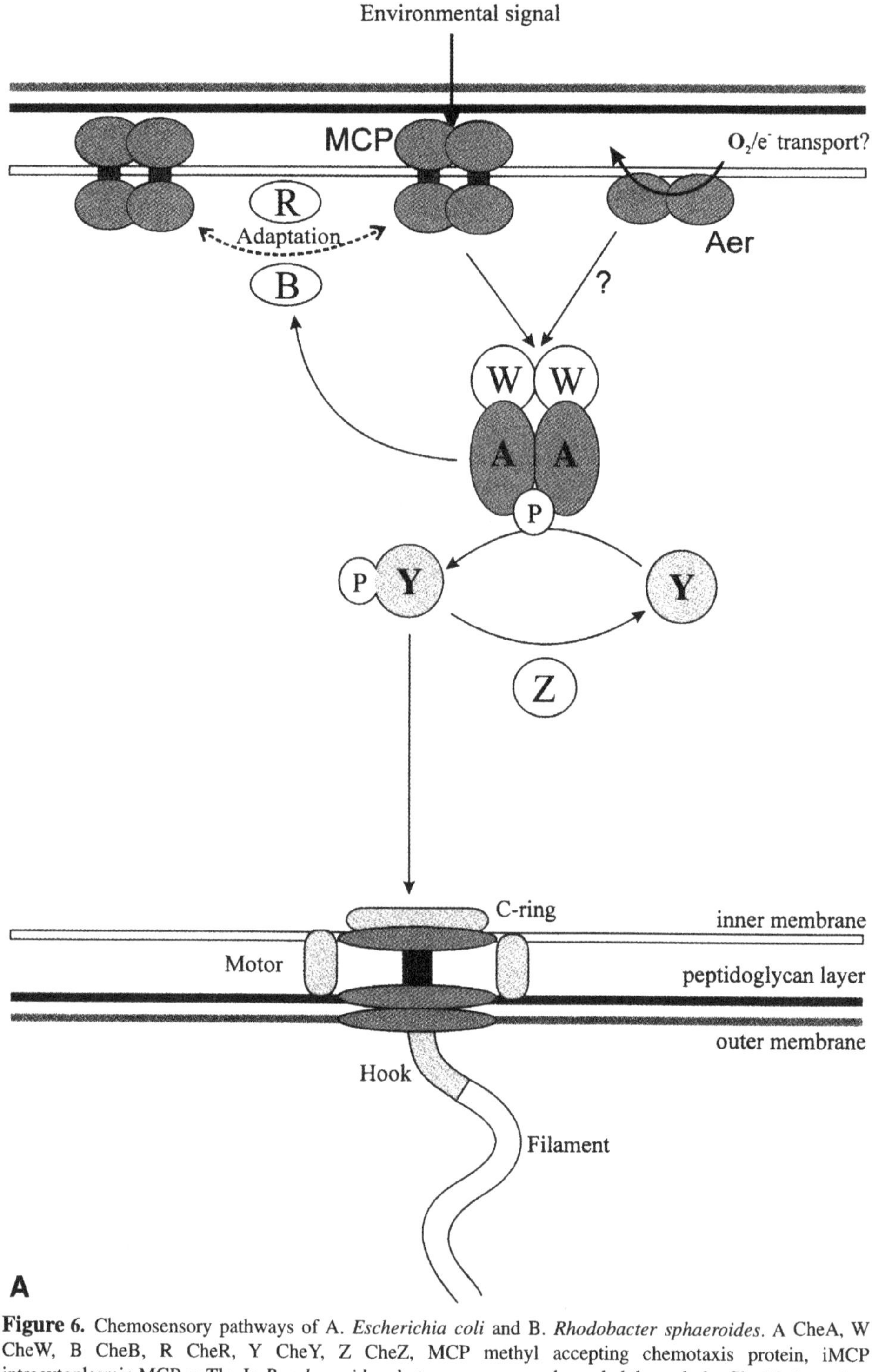

Figure 6. Chemosensory pathways of A. *Escherichia coli* and B. *Rhodobacter sphaeroides*. A CheA, W CheW, B CheB, R CheR, Y CheY, Z CheZ, MCP methyl accepting chemotaxis protein, iMCP intracytoplasmic MCP or Tlp. In *R. sphaeroides* photoresponses are channeled through the CheA2 dependent pathway, CheY4 and CheY5 are motor binding proteins, the other CheYs probably act as phosphate sinks.

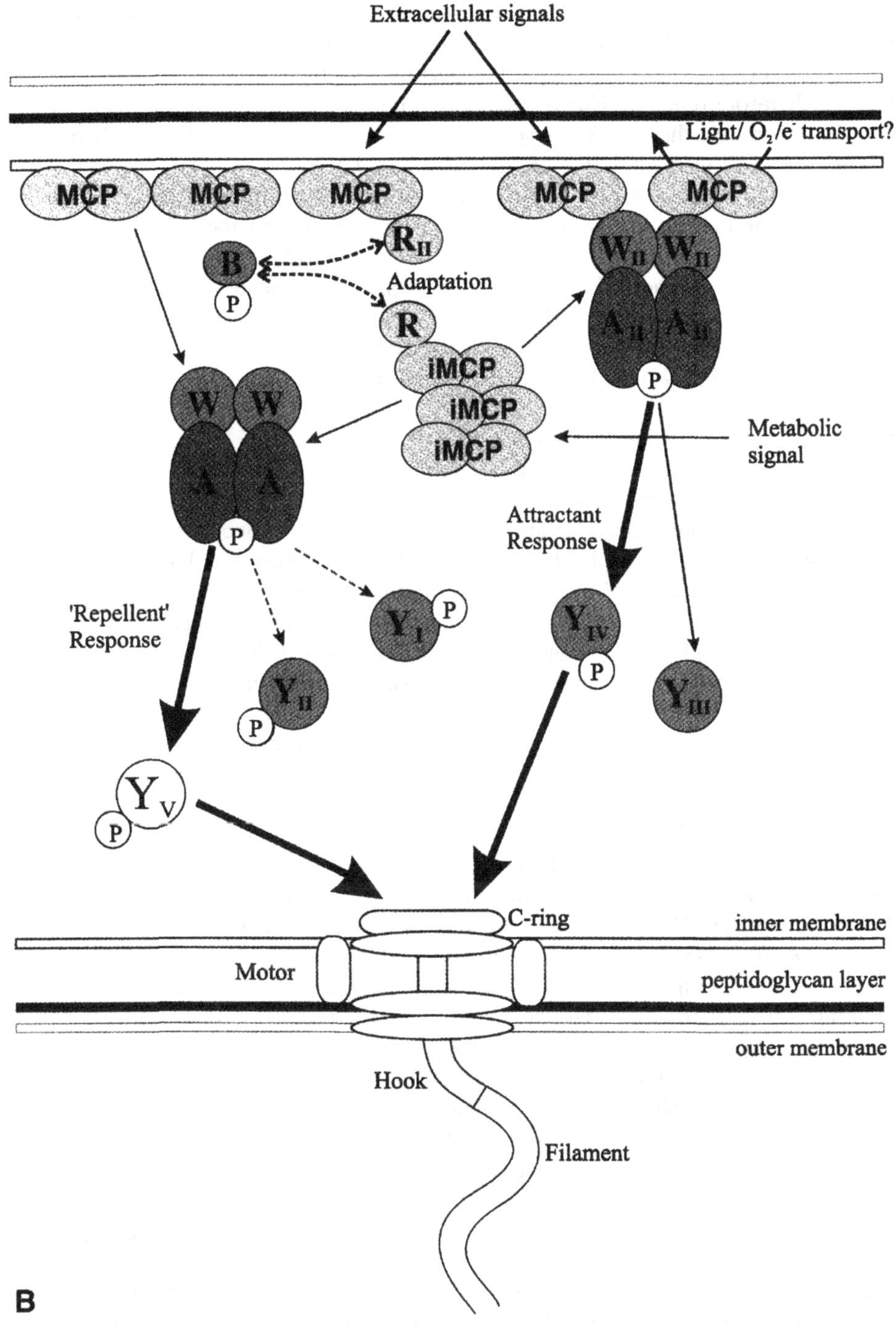

Figure 6. Continued.

Signals from the phosphoenol pyruvate (PEP)-dependent phosphotransferase sugar transport system, which phosphorylates sugars as they are transported into the cell, also interacts with CheA. Increased sugar transport reduces the level of CheA phosphorylation and therefore reducing CheY-P and thus tumbling [94]. Aer, the aerotaxis sensor also feeds through the CheA phospho-relay system [84]. Thus all identified tactic signals in *E. coli* feed through a single phospho-relay pathway. One exception is the metabolite, fumarate. There is evidence that fumarate can interact directly with the flagellar motor to induce switching [95,96]. The reason and the cause are unclear. Mutants in the fumarase gene show decreased tumbling rates, which can be overcome by the addition of extracellular fumarate (W. Marwan, personal communication). The site of interaction of fumarate with the motor is unknown, but it acts in the absence of all of the chemosensory genes. It has been suggested that this may represent a very early chemosensory pathway linked to the metabolic state of the cell.

4.6.1.1 R. sphaeroides and R. centenum

Does the same chemosensory pathway occur in the two photosynthetic members of the α-subgroup studied in any detail and how does it relate to chemosensing? Results suggest that the photoresponses (and other electron- transfer-dependent responses) do feed through a very similar phospho-relay pathways to control the flagellar motor, but the phospho-relay systems are more complex than that identified in enteric species.

Figure 7 shows the genes identified in *R. sphaeroides* involved in controlling flagellar behavior. Interestingly the operon organisation seen in *R. sphaeroides* is very closely related to that found in other α-subgroup species studied in detail, *Sinorhizobium meliloti, Agrobacterium tumefaciens* and *Caulobacter crescentus*, but is different from the organisation identified in *R. centenum* [4,97]. However, despite variations in the gene order, no member of the α-subgroup has yet been found to have a copy of *cheZ*, but all have at least two copies of *cheY*. In *R. centenum* one CheY is free while the other is fused to CheA [98]. The significance of this is currently unknown, but it may be that competition between the different CheYs for phosphate from CheA-P results in signal termination. This would depend on the different CheY-Ps having different affinities for the flagellar switch. This has not as yet been tested, but there is some evidence from the related *S. meliloti* that this may be the case [99]. Alternatively, it has been suggested that the lack of CheZ in the α-subgroup reflects the steepness of gradients found in natural environments in contrast to gradients found in the normal enteric environment. A more shallow gradient may require, it has been suggested, a more rapid signal termination and thus CheZ aids the natural dephosphorylation rate, which is fast enough for the bacteria in natural gradients. More research on binding and dephosphorylation kinetics, not to mention the steepness of natural gradients, is required before the reason for this difference is identified. It seems likely to be significant that while lacking CheZ, all the α-subgroup species have at least 2 copies of CheY.

Deletion or interruption of any of the genes of the *R. centenum* chemotaxis pathway results in the loss of colony phototaxis and a loss of free swimming chemotaxis, suggesting that both signals go through a single common pathway [97,98,100]. As with enteric species, disruption of *cheW*, *cheR* or *cheAY* resulted in smooth swimming while a *cheB* disruption caused tumbling. Interestingly, however, disruption of the *cheY* gene coding for a free CheY resulted in the loss of colony photoresponses, but only a minor

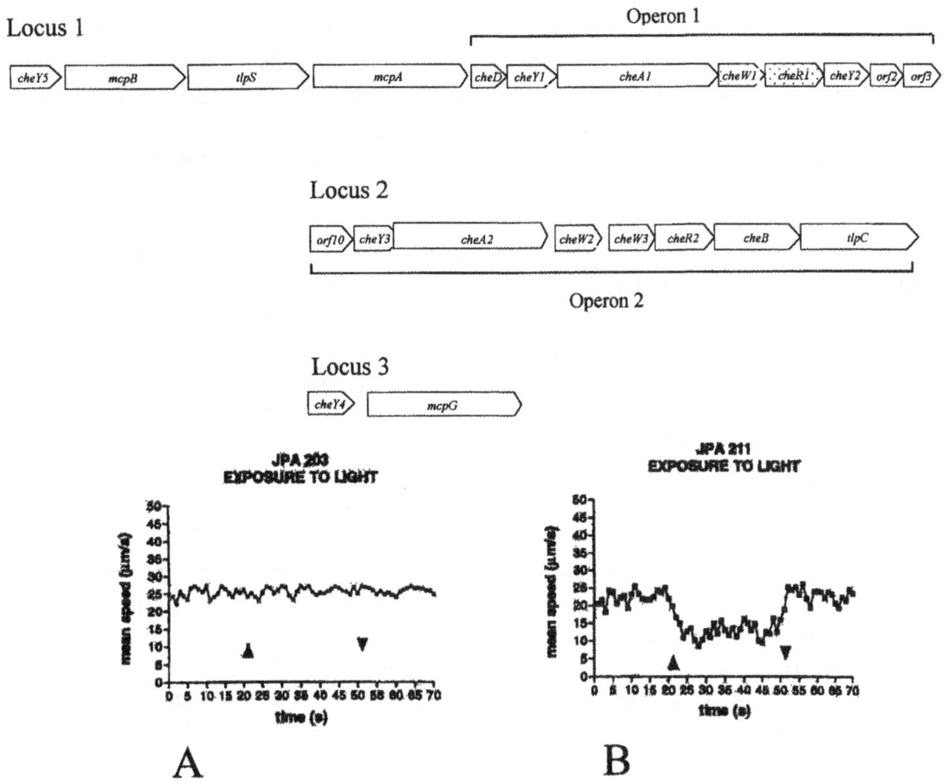

Figure 7. A. Organization of the genes coding for chemosensory genes in *R. sphaeroides*. Tlp-transducer-like protein (lacks the membrane-spanning domains, but retain the signaling domain of MCPs). B. Photoresponses of two mutants of *R. sphaeroides*. (A) A transposon insertion into *cheA2*, (B) an in-frame deletion of *cheA2*. The transposon causes the loss of down-stream genes while the in-frame deletion results only in the loss of CheA2. The loss of response suggests CheW2 is involved in signaling, while the inverted response suggests an additional signaling pathway.

change in tumbling frequency in free-swimming cells. This could suggest that there are different switch domains on the lateral motors used to move over agar surfaces, compared to the polar flagellum of the free-swimming cells. CheY might interact with the lateral flagella while CheAY might interact with the polar flagellum during smooth swimming. Both of the photosensory responses are mediated through the same phospho-relay system as chemotaxis. Whether the responses require methylation-dependent adaptation also remains to be identified, although Aer, responsible for aerotactic behavior in *E. coli* has no obvious methylation sites on the cytoplasmic domain, suggesting that perhaps adaptation of the receptors is not required for electron-transport dependent signaling. The mutational studies on *R. centenum* chemosensory genes also showed that the *che* genes are not required for swarmer differentiation, unlike many other swarmer species such as *Proteus mirabilis* or *Serratia marcescens* which lose the ability to differentiate if the Che pathway is disrupted.

The situation in *R. sphaeroides* has turned out to be far more complex than even that of *R. centenum*. Chemotaxis to many attractants (no repellents have been found) depends on transport and at least partial metabolism [101–103]. It was shown that glutamate transport mutants also lost chemotaxis to glutamate, showing that there were no independent MCP-like receptors for glutamate. Expression of a completely unrelated glutamate transporter from *E. coli* restored both transport and chemotaxis, suggesting that the process of transport was not linked to sensory signaling, but rather an intracellular receptor [104]. Similarly, mutants in the glucose-6-phosphate dehydrogenase gene lose taxis to sugars metabolized through the Entner-Doudoroff pathway, but not to sugars metabolized through the pathways Embden-Meyerhof, while transport of all sugars remains active [102]. The sugar responses are also independent of the PTS system shown to be involved in sugar taxis in *E. coli* (see earlier). The only sugar transported through a PEP-dependent PTS system in *R. sphaeroides* is fructose, the only sugar with a modified PTS system and unable to alter the activity of CheA.

Until recently, no MCP-like receptors had been identified. An operon coding for a set of genes with a high level of similarity to the *che* genes in enterics was identified. The operon was found to contain two copies of the *cheY* gene but no *cheB* or *cheZ*. Two genes were found with homology to MCP genes, but without the transmembrane domains, indicating cytoplasmic sensing. When one of these was mutated the responses lost were not to a single chemoeffector, as would be the case with *E. coli*, but to all chemoeffectors under aerobic but not anaerobic conditions [105]. Mutants were made in the *che* genes of this first operon, however, unlike *E. coli* or *R. centenum*, the mutants all showed almost normal patterns of swimming and normal photosensory and chemosensory responses. The only change in behavior was to some sugars, where responses were unexpectedly inverted. Taking mutants deleted for the complete operon, a second round of mutagenesis was carried out and the cells subjected to a phototaxis screen. Transposon mutants were placed in the bottom of a darkened test tube with a band of illumination around the center. Motile cells swam up the tube, but cells with normal photoresponses were delayed in the light region. Cells from the top of the tube were reinoculated at the bottom of another tube and the screen repeated 8 times. The mutants that still swam to the top of the tube were then plated onto swarm plates and mutants with altered photo- and chemoresponses identified. This screen identified a complete second *che* operon in *R. sphaeroides* [106] (Figure 7). This operon included two more copies of *cheW*, another *cheA*, *cheY*, *cheR* and the first copy of *cheB* and another apparent cytoplasmic sensory transducer. Mutations in these genes resulted in the loss of swarming on agar plates. When the $cheA_2$ gene was deleted in frame the mutants were found to still respond, but the responses were inverted i.e. when presented with a step-down in a chemoattractant the cells swam normally but when it was added they all stopped. Similarily, the cells stopped when given an increase in light and increased swimming speed when the light was reduced, confirming that signals from the photoreceptors do not go directly to the motor, but use the same sensory pathway as some chemosensory signals (Figure 7B). The fact that cells with both copies of CheA still show a response, albeit an inverted response, suggests there is yet another pathway. The inversion also indicated the involvement of an MCP-like transducer in both the photoresponse and the chemotactic response as inverted responses in *E. coli* have been produced when the methylation level of the MCPs is altered; CheB must be phosphorylated by CheA to demethylate the MCPs, if it is not active the receptors are overmethylated and send an aberrant signal to CheA. Interestingly the mutants cells still swam normally, stopping and starting when incubated in non-gradient condition. This is again completely different from the situation with enteric species where a mutation in any *che* gene results in either a smooth swimming or tumbly phenotype. This led to the identification of a 4th copy of *cheY*. What is clear from these mutant studies, however, is that the photoresponses integrate with the chemosensory pathway in all species to control the flagellar motor.

Further mutant studies have produced mutants which completely lack the photoresponse. If $cheA_2$ is deleted in frame, allowing expression of the $cheW_2$ gene in the same operon there is an inverted response, but if it is interrupted by a transposon, which also causes the loss of $cheW_2$, there is no response. CheW proteins link transducers to the CheA of the sensory pathway and this suggests there must be a transducer responsible for transmitting the signal from photosynthetic electron transport to the flagellar motor.

Why has no photosensory transducer been found in the mutational screens? *R. sphaeroides* gave one more unexpected result. Antibody produced to the highly conserved domain of one of the *E. coli* MCPs (Tsr, the serine receptor) was used to look for MCPs in *R. sphaeroides*. In cells grown anaerobically in high light, a weak signal was found to a protein at about 65 kDa using Western blots, and immunogold electron microscopy revealed a few gold particles in the membrane but a cluster inside the cell. When cells were grown aerobically, however, the hybridisation signal on Western blots increased almost 20 times, and immunoelectron microscopy revealed not only the small cytoplasmic cluster, but large numbers of gold particles at the cell poles [106a] and confirmed by GFP fusions [106b] (Figure 8). MCPs are found at the poles in *E. coli* and *C. crescentus*, probably allow the formation of the quaternary signaling complexes with CheW and CheA [107–109]. Mutants lacking *cheA* or *cheW* do not localize their MCPs at the poles. The differential expression of *mcp* genes under aerobic and anaerobic conditions is the first account of environmental regulation of the expression of chemosensory receptors and suggests that the behavioral responses of *R. sphaeroides* might be expected to be very different under aerobic and anaerobic conditions. Recent hybridisation studies suggest there may be as many as 12 *mcp*-like genes in *R. sphaeroides*, and the phenotypic studies suggest that they may be responsible for sensing different metabolic states under different growth condition. The expression levels of the two operons has shown that they are expressed at different levels under aerobic, anaerobic dark and photosynthetic conditions, implying that different pathways may be linked to different MCPs, expressed under different growth conditions. It has recently been found that MCP clustering in *E. coli* is essential for signaling, bringing together a critical number of CheA molecules for optimum signaling. One set of sensory genes may, therefore, code for proteins which are attached to the membrane bound aerobic MCPs and the other to the cytoplasmic MCPs. It remains to be discovered which of the MCP homologues is (are) involved in redox sensing.

Many of the purple non-sulfur bacteria may turn out to be similar, with a chemosensory pathway similar to that of enteric when growing heterotrophically, but expressing a different sensory system geared to the different metabolic requirements of the cell, and perhaps the change in membrane structure, when growing photo-

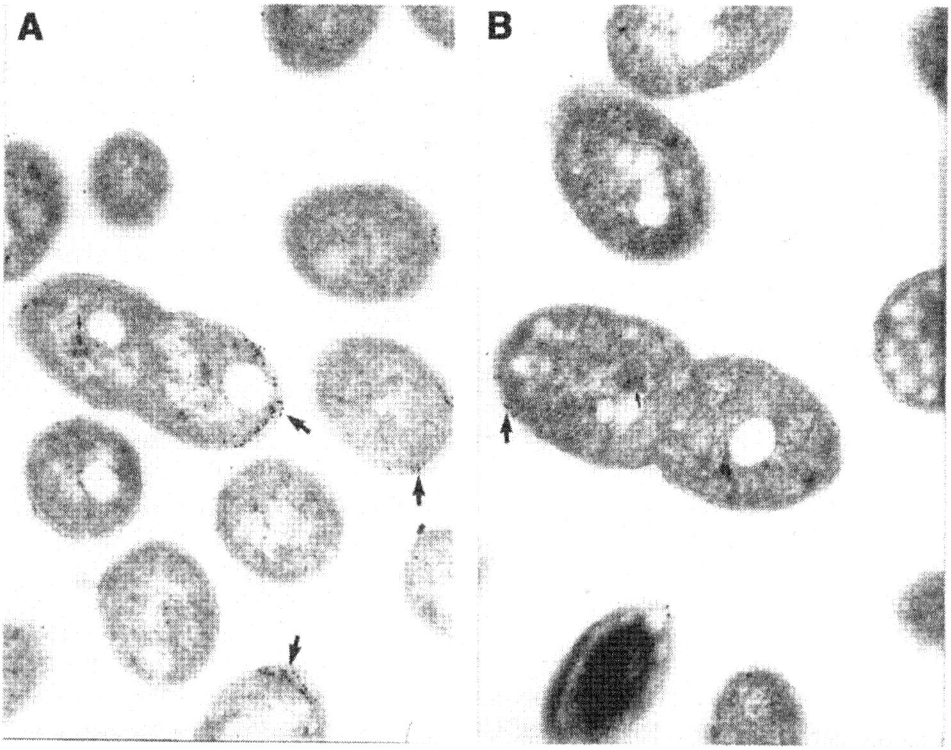

Figure 8. Electron micrographs showing immunogold labeling of MCP-like proteins in *R. sphaeroides* under different growth conditions.A. Aerobically grown *R. sphaeroides* with large cluster at the poles of the cell, but also cytoplasmic clusters.B. Anaerobically grown with fewer polar clusters but retaining cytoplasmic clusters.

heterotrophically, responding to changes in the metabolic state and in the redox state of electron transport chain components.

Summary

It is now clear that the photoresponses shown by purple bacteria are sensed as a result of changes in the rate of electron transport, and it is possible that, as respiratory and photosynthetic electron transport may share components that it is primarily part of a general response system to decreases in the rate of electron transport. The signal, when generated, probably feeds via an MCP-like transducer to a cytoplasmic sensory pathway shared with the chemosensory pathway. In this way the different environmental signals encountered by bacterial species in the "wild" can be balanced to produce a response that moves a bacterium to its optimum environment for growth.

Acknowledgements

I would like to thank the BBSRC, Wellcome Trust and NERC for funding research in the behavior of *R. sphaeroides* and C. Bauer, H. Gest and J. Overmann for unpublished data and provision of pictures.

Note added in proof. The genomes of *Rhodobacter sphaeroides* and *Rhodopseudomonas palustris* have been sequenced and suggest behavior is even more complex than detailed here, with another chemosensory gene locus and up to 29 receptors in *Rps. palustris*.

References

1. B.J. Ford (Ed.) (1991). *The Leeuwenhoek Legacy.* Biopress and Farrand Press, Bristol and London.
2. J.P. Armitage (1997). Three hundred years of bacterial motility. In: M.G. Ord, L.A. Stocken (Eds.), *Further Milestones in Biochemistry* (pp. 107–172). JAI Press Inc, Greenwich, Conn.
3. R.M. Macnab (1996). Flagella and motility. In: F.C. Neidhardt, R.I. Curtiss, J.L. Ingraham, E.C.C. Lin, G. Lowe, B. Magasanik, W.S. Reznikoff, M. Riley, M. Schaechter, H.E. Umbarger (Eds.), *Escherichia coli* and *Salmonella*: Cellular and Molecular Biology (pp. 123–145). ASM, Washington.
4. J.P. Armitage, R. Schmitt (1997). Bacterial chemotaxis: *Rhodobacter sphaeroides* and *Sinorhizobium meliloti* -variations on a theme? *Microbiology,* **143**, 3671–3682.
5. J.B. Stock, M.G. Surette (1996). Chemotaxis. In: F.C. Neidhardt, R.I. Curtiss, J.L. Ingraham, E.C.C. Lin, K.B. Low, B. Magasanik, W.S. Reznikoff, M. Riley, M. Schaechter, H.E. Umbarger (Eds.), *Escherichia coli* and *Salmonella: Cellular and Molecular Biology* (pp. 1103–1129). ASM, Washington.
6. J.P. Armitage (1997). Behavioural responses of bacteria to light and oxygen. *Arch. Microbiol.*, **168**, 249–261.
7. J.P. Armitage (1998). Motility responses towards light shown by phototrophic bacteria. In: M.X. Caddick, S. Baumberg, D.A. Hodgson, M.K. Phillips-Jones (Eds.), *Microbial Responses to Light and Time* (pp. 33–56). CUP, Cambridge, U.K.
8. H. Gest (1995). Phototaxis and other sensory phenomena in purple photosynthetic bacteria. *FEMS Microbiol. Rev.*, **16**, 287–294.
9. D.-P. Häder (1987). Photosensory behavior in prokaryotes. *Microbiol.Rev.*, **51**, 1–21.
10. L. Ragatz, Z.-Y. Jiang, C.E. Bauer, H. Gest (1994). Phototactic purple bacteria. *Nature*, **370**, 104–104.
11. G.S. Ehrenberg (1883). Die Infusionstrierchen als vollkommene Organismen. In: T.W. Engelmann (Ed.). Leipzig (p. 15).
12. T.W. Engelmann (1883). *Bakterium photometricum.* Ein Beitrag zur vergleichenden Physiologie des Licht-und Farbensinnes. *Pfuegers Arch. gesamte Physiol. Menschen Tiere*, **42**, 183–186.
13. C.J. Jones, S.-I. Aizawa (1991). The bacterial flagellum and flagellar motor: Structure, assembly and function. *Adv. Microb. Physiol.*, **32**, 109–172.
14. L. Shapiro (1995). The bacterial flagellum: From genetic network to complex architecture. *Cell*, **80**, 525–527.
15. S.C. Schuster, S. Khan (1994). The bacterial flagellar motor. *Annu. Rev. Biophys. Biomol. Struct.*, **23**, 509–539.

16. S.M. Block, H.C. Berg (1984). Successive incorporation of force generating units in the bacterial rotary motor. *Nature*, **309**, 470–472.
17. D.F. Blair, H.C. Berg (1988). Restoration of torque in defective flagellar motors. *Science*, **242**, 1678–1681.
18. D.F. Blair (1990). The bacterial flagellar motor. *Sem. in Cell Biol.*, **1**, 75–85.
19. S. Khan (1993). Gene to ultrastructure: The case of the flagellar basal body. *J. Bacteriol.*, **175**, 2169–2174.
20. D.F. Blair, H.C. Berg (1990). The MotA protein of *E. coli* is a proton-conducting component of the flagellar motor. *Cell*, **60**, 439–449.
21. N.R. Francis, G.E. Sosinsky, D. Thomas, D.J. DeRosier (1994). Isolation, characterization and structure of bacterial flagellar motors containing the switch complex. *J. Mol. Biol.*, **235**, 1261–1270.
22. M., Meister, G. Lowe, H.C. Berg (1987). The proton flux through the bacterial flagellar motor. *Cell*, **49**, 643–650.
23. Y. Magariyama, S. Sugiyama, K. Muramoto, Y. Maekawa, I. Kawagishi, Y. Imae, S. Kudo (1994). Very fast flagellar rotation. *Nature*, **371**, 752–752.
24. J.P. Armitage, R.M. Macnab (1987). Unidirectional intermittent rotation of the flagellum of *Rhodobacter sphaeroides*. *J. Bacteriol.*, **169**, 514–518.
25. H.C. Berg, D.A. Brown (1972). Chemotaxis in *Escherichia coli* analysed by three-dimensional tracking. *Nature*, **239**, 500–504.
26. H.C. Berg (1983). *Random Walks in Biology*. Princeton, New Jersey.
27. N. Hirota, Y. Imae (1983). Na^{+}-driven flagellar motors of an alkalophilic *Bacillus* strain YN-1. *J. Biol. Chem.*, **258**, 10577–10581.
28. T. Atsumi, L. McCarter, Y. Imae (1992). Polar and lateral flagellar motors of marine *Vibrio* are driven by different ion-motive forces. *Nature*, **355**, 182–184.
29. J.P. Armitage, H.L. Packer (1997). Bacterial motility and chemotaxis. In: D.R. Soll (Ed.), *Motion Analysis of Living Cells* (pp. 1–24). Wiley-Liss Inc, New York.
30. R. Nossal, S.H. Chen, C.C. Lai (1977). Use of laser scattering for quantitative determinations of bacterial motility. *Optical Comm.*, **4**, 35.
31. J.B. Armstrong, J. Adler, M.M. Dahl (1967). Non-chemotactic mutants of *Escherichia coli*. *J. Bacteriol.*, **93**, 390–398.
32. J. Adler (1969). Chemoreceptors in bacteria. *Science*, **166**, 1588–1597.
33. E.O. Budrene, H.C. Berg (1995). Dynamics of formation of symmetrical patterns by chemotactic bacteria. *Nature*, **376**, 49–53.
34. W. Pfeffer (1883). Locomotorische Richtungsbewegungen durch chemische Reize. *Berichte der Deutschen Botanischen Gesellschaft*, **1**, 524–533.
35. W. Pfeffer (1884). Locomotorische Richtungbewegungen durch chemische Reize. In: W. Pfeffer (Ed.), *Untersuchungen aus dem Botanische Institut Tubingen, Verlag von Wilhelm Engelmann* (pp.363–482). Leipzig.
36. J. Adler (1976). The sensing of chemicals by bacteria. *Scientific American*, **234**, 40–47.
37. A. Joss, K. Mez, B. Kanel, K.W. Hanselmann, R. Bachofen (1994). Measurement of fluorescence kinetics of phototrophic bacteria in their natural environment. *J. Plant. Physiol.*, **144**, 333–338.
38. J.G. Mitchell, L. Pearson, A. Bonazinga, S. Dillon, H. Khouri, R. Paxinos (1995). Long lag times and high velocities in the motility of natural assemblages of marine bacteria. *Appl. Environ. Microbiol.*, **61**, 877–882.
39. G.M. Barbara, J.G. Mitchell (1996). Formation of 30- to 40-micrometer-thick laminations by high-speed marine bacteria in microbial mats. *Appl. Environ. Microbiol.*, **62**, 3985–3990.

40. M. Huettel, S. Forster, S. Kloeser, H. Fossing (1996). Vertical migration in the sediment dwelling sulfur bacteria *Thioploca* spp. in overcoming diffusion limitations. *Appl. Environ. Microbiol.*, **62**, 1863–1872.
41. J.W. Barton, R.M. Ford (1997). Mathematical model for characterization of bacterial migration through sand cores. *Biotechnol. Bioeng.*, **53**, 487–496.
42. S. Khan, J.L. Spudich, J.A. McCray, D.R. Trentham (1995). Chemotactic signal integration in bacteria. *Proc. Natl. Acad. Sci. USA*, **92**, 9757–9761.
43. P.D., Frymier, R.M., Ford, H.C. Berg, P.T. Cummings (1995). Three-dimensional tracking of motile bacteria near a solid planar surface. *Proc. Natl. Acad. Sci. USA*, **92**, 6195–6199.
44. K.J. Duffy, R.M. Ford (1997). Turn angle and run time distribution characterize swimming behavior for *Pseudomonas putida*. *J. Bacteriol.*, **179**, 1428–1430.
45. G. Lowe, M. Meister, H.C. Berg (1987). Rapid rotation of flagellar bundles in swimming bacteria. *Nature*, **325**, 637–640.
46. S.M. Block, K.A. Fahrner, H.C. Berg (1991). Visualization of bacterial flagella by video-enhanced light microscopy. *J. Bacteriol.*, **173**, 933–936.
47. D.C. Fung, H.C. Berg (1995). Powering the flagellar motor of *Escherichia coli* with an external voltage source. *Nature*, **375**, 809–812.
48. S.M. Block, D.F. Blair, H.C. Berg (1989). Compliance of bacterial flagella measured with optical tweezers. *Nature*, **338**, 514–518.
49. R.M. Berry, L. Turner, H.C. Berg (1995). Mechanical limits of bacterial flagellar motors probed by electrorotation. *Biophys. J.*, **69**, 280–286.
50. H.C. Berg, S.M. Block (1984). A minature flow cell designed for rapid exchange of media under high-powered objectives. *J. Gen. Microbiol.*, **130**, 2915–2920.
51. M. Silverman, M. Simon (1974). Flagellar rotation and the mechanism of bacterial motility. *Nature*, **249**, 73–74.
51a. H.L. Packer and J.P. Armitrage (2000). The behavioural responses of *Rhodobacter sphaeroides* to linear gradients of the nutrients succinate and acetate. *Appl. Env. Microbiol.*, **66**, 5186–5191.
52. H.C. Berg, L. Turner (1995). Cells of *Escherichia coli* swim either end forward. *Proc. Natl. Acad. Sci. USA*, **92**, 477–479.
53. N. Pfennig (1968). *Chromatium akenii* (Thiorhodaceae). In: G. Wolf (Ed.), *Encyclopedia cinematographica, Institut für den Wissenschaftlichen Film* (pp. 3–9). Göttingen.
54. A. Manten (1948). Phototaxis in the purple bacterium *Rhodospirillum rubrum* and the relationship between phototaxis and photosynthesis. *Ant. van Leeuwen.*, **14**, 65–86.
55. R.K. Clayton (1958). On the interplay of environmental factors affecting taxis and mobility in *Rhodospirillum rubrum*. *Arch. Microbiol.*, **29**, 189–212.
56. R.K. Clayton (1957). Patterns of accumulation resulting from taxes and changes in motility of microorganisms. *Arch. Microbiol.*, **27**, 311–319.
57. R.K. Clayton (1953). Studies in the phototaxis of *Rhodospirillum rubrum*.I. Action spectrum, growth in green light and Weber-law adherence. *Arch. Microbiol.*, **19**, 107–124.
58. M.J. Sackett, J.P. Armitage, E.E. Sherwood, T.P. Pitta (1997). Photoresponses of the purple nonsulfur bacteria *Rhodospirillum centenum* and *Rhodobacter sphaeroides*. *J. Bacteriol.*, **179**, 6764–6768.
59. L. Ragatz, Z.-Y. Jiang, C.E. Bauer, H. Gest (1995). Macroscopic phototactic behavior of the purple photosynthetic bacterium *Rhodospirillum centenum*. *Arch. Microbiol.*, **163**, 1–6.
60. S. Romagnoli, A. Hochkoeppler, L. Damgaard, D. Zannoni (1997). The effect of respiration on the phototactic behavior of the purple nonsulfur bacterium *Rhodospirillum centenum*. *Arch. Microbiol.*, **167**, 99–105.

61. B.L. Taylor, D.E. Koshland Jr. (1975). Intrinsic and extrinsic light responses of *Salmonella typhimurium* and *Escherichia coli*. *J. Bacteriol.*, **123**, 557–569.
62. R.M. Macnab, D.E. Jr. Koshland (1974). Bacterial motility and chemotaxis: light-induced tumbling response and visualization of individual flagella. *J. Mol. Biol.*, **84**, 399–406.
63. B.L. Taylor, J.B. Miller, H.M. Warrick, D.E. Koshland Jr. (1979). Electron acceptor taxis and blue light effect on bacterial chemotaxis. *J. Bacteriol.*, **140**, 567–573.
64. X.-H. Yang, A. Sasarman, H. Inokuchi, J. Adler (1996). Non-iron porphyrins cause tumbling to blue light by an *Escherichia coli* mutant defective in *hemG*. *Proc. Natl. Acad. Sci. USA*, **93**, 2459–2463.
65. D.A. Hodgson, A.E. Berry (1998). Light regulation of carotenoid biosynthesis in *Myxococcus xanthus*. In: M.X. Caddick, S. Baumberg, D.A. Hodgson, M.K. Phillips-Jones (Eds), *Microbial responses to light and time* (pp. 185–212). CUP, Cambridge.
66. J.M. Frostl, J. Overmann (1998). Physiology and tactic response of the phototrophic consortium "*Chlorochromatium aggregatum*". *Arch. Microbiol.*, **169**, 129–135.
67. M. Kihara, R.M. Macnab (1981). Cytoplasmic pH mediates pH taxis weak-acid repellent taxis of bacteria. *J. Bacteriol.*, **145**, 1209–1221.
68. D.R. Repaske, J. Adler (1981). Changes in intracellular pH of *Escherichia coli* mediates the chemotactic response to certain attractants and repellents. *J. Bacteriol.*, **145**, 1196–1208.
69. E.G. Nisbet, J.R. Cann, C.L. Van Dover (1995). Origins of photosynthesis. *Nature*, **373**, 479–480.
70. J.P. Armitage, M.C.W. Evans (1981). The reaction centre in the phototactic and chemotactic responses of *Rhodopseudomonas sphaeroides*. *FEMS Microbiol. Letts.*, **11**, 89–92.
71. R.N. Grishanin, D.E. Gauden, J.P. Armitage (1997). Photoresponses in *Rhodobacter sphaeroides*: role of photosynthetic electron transport. *J. Bacteriol.*, **179**, 24–30.
72. D.E.Gauden, J.P. Armitage (1995). Electron transport-dependent taxis in *Rhodobacter sphaeroides*. *J. Bacteriol.*, **177**, 5853–5859.
73. D.M. Harrison, H.L. Packer, J.P. Armitage (1994). Swimming speed and chemokinetic response of *Rhodobacter sphaeroides* investigated by natural manipulation of the membrane potential. *FEBS Lett.*, **348**, 37–40.
73a. S. Romagnoli, J.P. Armitage (1999). Roles of chemosensory pathways in transient changes of swimming speed of *Rhodobacter sphaeroides* induced by changes in photosynthetic electron transport. *J. Bacteriol.*, **181**, 4825–4833.
74. T.C. Bonnett, P. Cobine, R.E. Sockett, A.G. McEwan (1995). Phenotypic characterisation and genetic complementation of dimethylsulfoxide respiratory mutants of *Rhodobacter sphaeroides* and *Rhadobacter capsulatus*. *FEMS Microbiol. Lett.*, **133**, 163–168.
75. S.J. Ferguson, J.B. Jackson, A.G. McEwan (1987). Anaerobic respiration in the Rhodospirillaceae: characterisation of pathways and evaluation of roles in redox balancing during photosynthesis. *FEMS Microbiol. Rev.*, **46**, 117–143.
76. J.P. Armitage, C. Ingham, M.C.W. Evans (1985). Role of the proton motive force in phototactic and aerotactic responses of *Rhodopseudomonas sphaeroides*. *J. Bacteriol.*, **163**, 967–972.
77. B.L. Taylor (1983). Role of proton motive force in sensory transduction in bacteria. *Ann. Rev. Microbiol.*, **37**, 551–573.
78. A.N. Glagolev (1984). Bacterial H^+-sensing. *Trends Biochem. Sci.*, **9**, 397–400.
79. J.H. Zeilstra-Ryalls, S. Kaplan (1995). Aerobic and anaerobic regulation in *Rhodobacter sphaeroides* 2.4.1: The role of the *fnr*L gene. *J. Bacteriol.*, **177**, 6422–6431.
80. A. Rebbapragada, M.S. Johnson, G.P. Harding, A.J. Zuccarelli, H.M. Fletcher, I.B. Zhulin, B.L. Taylor (1997). The Aer protein and the serine chemoreceptor Tsr independently sense

intracellular energy levels and transduce oxygen, redox, and energy signals for *Escherichia coli* behavior. *Proc. Natl. Acad. Sci. USA*, **94**, 10541–10546.
81. I.B. Zhulin, M.S. Johnson, B.L. Taylor (1997). How do bacteria avoid high oxygen concentrations? *Biosci. Rep.*, **17**, 335–342.
82. I.B. Zhulin, E.H. Rowsell, M.S. Johnson, B.L. Taylor (1997). Glycerol elicits energy taxis of *Escherichia coli* and *Salmonella typhimurium*. *J. Bacteriol.*, **179**, 3196–3201.
83. S. Hill, S. Austin, T. Eydmann, T. Jones, R. Dixon (1996). *Azotobacter vinelandii* NifL is a flavoprotein that modulates transcriptional activation of nitrogen fixation genes via a redox-sensitive switch. *Proc. Natl. Acad. Sci. USA*, **93**, 2143–2148.
84. S.I. Bibikov, R. Biran, K.E. Rudd, J.S. Parkinson (1997). A signal transducer for aerotaxis in *Escherichia coli*. *J. Bacteriol.*, **179**, 4075–4079.
85. H.L. Packer, D.E. Gauden, J.P. Armitage (1996). The behavioral response of anaerobic *Rhodobacter sphaeroides* to temporal stimuli. *Microbiology*, **142**, 593–599.
86. A.J. Clark, J.B. Jackson (1981). The measurement of membrane potential during photosynthesis and during respiration in intact cells of *Rhodopseudomonas capsulata* by both electrochromism and by permeant ion distribution. *Biochem. J.*, **200**, 389–397.
87. T.W. Grebe, J. Stock (1998). Bacterial chemotaxis: the five sensors of a bacterium. *Curr. Biol.*, **8**, R154-R157.
88. R.M. Macnab (1995). Flagellar switch. In: J.A. Hoch, T.J. Silhavy (Eds), *Two-component signal transduction* (pp. 181–199). ASM, Washington.
89. J.B. Stock, M.G. Surette, M. Levit, P. Park (1995). Two-component signal transduction systems: Structure-function relationships and mechanisms of catalysis. In: J.A. Hoch, T.J. Silhavy (Eds), *Two-component Signal Transduction* (pp. 25–51). ASM, Washington.
90. R.B. Bourret, K.A. Borkovich, M.I. Simon (1991). Signal transduction pathways involving protein phosphorylation in prokaryotes. *Annu. Rev. Biochem.*, **60**, 401–441.
91. J.S. Parkinson (1993). Signal transduction schemes of bacteria. *Cell*, **73**, 857–871.
92. L.A. Alex, M.I. Simon (1994). Protein histidine kinases and signal transduction in prokaryotes and eukaryotes. *Trends Genet.*, **10**, 133–138.
93. M. Eisenbach (1996). Control of bacterial chemotaxis. *Mol. Microbiol.*, **20**, 903–910.
94. J.W. Lengeler, K. Jahreis (1996). Phosphotransferase systems or PTSs as carbohydrate transport and as signal tranduction systems. In: W.N. Konings, H.R. Kaback, J.S. Lolkema (Eds), *Handbook of Biological Physics* (pp. 573–598). Elsevier Science BV, Amsterdam.
95. R. Barak, M. Eisenbach (1992). Fumarate or a fumarate metabolite restores switching ability to rotating flagella of bacterial envelopes. *J. Bacteriol.*, **174**, 643–645.
96. R. Barak, I. Giebel, M. Eisenbach (1996). The specificity of fumarate as a switching factor of the bacterial flagellar motor. *Mol. Microbiol.*, **19**, 139–144.
97. Z.Y. Jiang, H. Gest, C.E. Bauer (1997). Chemosensory and photosensory perception in purple photosynthetic bacteria utilize common signal transduction components. *J. Bacteriol.*, **179**, 5720–5727.
98. Z.Y. Jiang, C.E. Bauer (1997). Analysis of a chemotaxis operon from *Rhodospirillum centenum*. *J. Bacteriol.*, **179**, 5712–5719.
99. V. Sourjik, R. Schmitt (1998). Phosphotransfer between CheA, CheY1, and CheY2 in the chemotaxis signal transduction chain of *Rhizobium meliloti*. *Biochemistry*, **37**, 2327–2335.
100. Z.Y. Jiang, B.G. Rushing, Y. Bai, H. Gest, C.E. Bauer (1998). Isolation of *Rhodospirillum centenum* mutants defective in phototactic colony motility by transposon mutagenesis. *J. Bacteriol.*, **180**, 1248–1255.
101. C.J. Ingham, J.P. Armitage (1987). Involvement of transport in *Rhodobacter sphaeroides*. *J. Bacteriol.*, **169**, 5801–5807.

102. S.Y. Jeziore, P.A. Hamblin, W.C. Bootle, P.S. Poole, J.P. Armitage (1998). Metabolism is required for chemotaxis to sugars in *Rhodobacter sphaeroides*. *Microbiology*, **144**, 229–239.
103. P.S. Poole, J.P. Armitage (1989). Role of metabolism in the chemotactic response of *Rhodobacter sphaeroides* to ammonia. *J. Bacteriol.*, **171**, 2900–2902.
104. M.H.J. Jacobs, A.J.M. Driessen, W.N. Konings (1995). Characterization of a binding protein-dependent glutamate transport system of *Rhodobacter sphaeroides*. *J. Bacteriol.*, **177**, 1812–1816.
105. M.J. Ward, D.M. Harrison, M.J. Ebner, J.P. Armitage (1995). Identification of a methyl-accepting chemotaxis protein in *Rhodobacter sphaeroides*. *Mol. Microbiol.*, **18**, 115–121.
106. P.A. Hamblin, B.A. Maguire, R.N. Grishanin, J.P. Armitage (1997). Evidence for two chemosensory pathways in *Rhodobacter sphaeroides*. *Mol. Microbiol.*, **26**, 1083–1096.
106a. D.H. Harrison, J. Skidmore, J.P. Armitage, J.R. Maddock (1999). Localisation and environmental regulation of methyl-accepting chemotaxis proteins in *Rhodobacter sphaeroides*. *Mol. Microbiol.*, **31**, 885–892.
106b. G.H. Wadhams, A.C. Martin, J.P. Armitage (2000). Identification and localisation of a methyl-accepting chemotaxis protein in *Rhodobacter sphaeroides*. *Mol. Microbiol.*, **36**, 1222–1233.
107. M.R.K. Alley, J.R. Maddock, L. Shapiro (1992). Polar localization of a bacterial chemoreceptor. *Genes Dev.*, **6**, 825–836.
108. J.R. Maddock, M.R.K. Alley, L. Shapiro (1993). Polarized cells, polar actions. *J. Bacteriol.*, **175**, 7125–7129.
109. J.R. Maddock, L. Shapiro (1993b). Polar location of the chemoreceptor complex in the *Escherichia coli* cell. *Science*, **259**, 1717–1723.

Photomovement
D.-P. Häder and M. Lebert, editors.

Chapter 5

Color-sensitive vision by haloarchaea

John L. Spudich

Table of contents

Abstract

Halobacterium salinarum, a member of the Archaea domain, exhibits attractant or repellent motility responses to light gradients dependent on the color of the stimulating light. Two photoreceptors called sensory rhodopsins I and II (SRI and SRII) control a signal transduction pathway that mediates these responses. The sensory rhodopsins are seven-helix membrane proteins that are structurally and functionally similar to animal visual pigments. They couple retinal photoisomerization to receptor activation and are complexed with membrane-embedded transducer proteins (HtrI and HtrII) that modulate a cytoplasmic phosphorylation cascade controlling the cells' flagellar motors. The Htr proteins resemble the chemotaxis transducers from Eubacteria, such as *Escherichia coli*. The SR-Htr signaling complexes have facilitated studies of the biophysical chemistry of signal generation and relay, from the photobiophysics of initial excitation of the receptors to the final output at the level of the flagellar motor switch, shedding light on the fundamental principles of sensory transduction and more broadly the nature of dynamic interactions between membrane proteins. This review provides a brief historical perspective of the study of photosensory behavior of halobacteria and includes recent advances that have led to new insights into molecular mechanisms of signaling by membrane complexes.

5.1 The primitive visual system of halobacteria: behavioral physiology

Halobacteria, salt-loving archaeal prokaryotes, exhibit photoresponses to changes in light intensity and color by altering their swimming behavior. Motility responses of halobacteria to light were first described in the modern literature by Hildebrand and Dencher [1], who noted the opposite effects of orange (attractant) and near-UV/blue (repellent) light on swimming behavior. Individual cell tracking analysis clarified the behavioral mechanism [2], which is based on temporal light gradient-modulation of swimming reversal frequencies.

The behavioral physiology of phototaxis is understood from early visual cell tracking techniques and more recent computerized infrared video motion analysis [3]. Also a rapid population method for quantitating phototaxis accumulation and dispersion has been developed and applied to halobacteria ([4]; see also [5,6]). Several mathematical models have been presented that account for various aspects of the response kinetics (reviewed by [7]).

In the absence of stimulating gradients the cells exhibit smooth swimming runs of from ~2 to ~30 seconds duration interrupted by subsecond "swimming reversals". The reversals result from a switch in the leading end of the rod shaped cells and are characterized by a roughly 180° change in swimming direction. The alternation of runs and reversals produces a zigzag motility pattern and a random walk in three-dimensional space. An increase in the intensity of orange light or a decrease in that of blue light transiently inhibits swimming reversals. These responses are observed for changes in light intensity occurring in the sub-second to seconds range ("temporal gradients"). Conversely a temporal decrease in orange or increase in blue light enhances reversals.

Temporal light gradients are experienced by the swimming cell as time-dependent changes in light intensity generated by its translational motion in spatial light gradients. Therefore when the cells are swimming in spatial gradients their random walk is biased so that they swim longer distances when they are traveling up an orange light gradient or down a blue light gradient compared to the opposite directions. This bias accumulates the cells in regions of relatively intense orange light and disperses them from regions containing high intensities of near-UV or blue light.

The motility behavior described above is most strictly classified as positive and negative photoklinokinesis [8]. However most investigators of halobacterial photo-behavior refer to it as attractant and repellent phototaxis, by analogy with the term "chemotaxis", which is commonly applied to the similar biased random walk behavior of bacteria in gradients of attractant and repellent substances [9]. This review uses the latter terminology, but the readers should note that some authors restrict the term "phototaxis" to a behavior in which a cell responds to the direction of a light beam (such as is exhibited by eukaryotic algae), rather than to an intensity gradient.

5.2 Archaeal rhodopsins and the natural habitat of halobacteria

Halobacteria live in the Dead Sea, solar evaporation ponds, and other regions of near to fully saturated brine where solar radiation is intense. *Halobacterium salinarum*, the most studied species, takes advantage of the two principal roles played by light in the biosphere: as energy provider and as information carrier. *H. salinarum* membranes contain a family of four archaeal rhodopsins (Figure 1), photoactive proteins that are similar to our visual pigments in their structure and photochemistry: bacteriorhodopsin (BR; [10]) and halorhodopsin (HR; [11,12]) harvest solar energy by light-driven electrogenic transport of protons and chloride, respectively, across the cytoplasmic membrane. SRI [13,14] and SRII [15,16] are phototaxis receptors that use light energy to send signals to the flagellar motor via the transducer proteins HtrI [17] and HtrII [18,19], respectively ("*h*alobacterial *t*ransducers for sensory *r*hodopsin *I* and *II*").

The four archaeal rhodopsins and the three functions, proton transport, chloride transport, and phototaxis signaling, appear to account for retinal pigmentation and retinal-dependent functions in *H. salinarum*. Over 30 archaeal rhodopsins have been described and they all correspond in absorption spectrum and function to BR, HR, SRI or SRII [20–22]. Motility studies using retinal analogs to reconstitute phototaxis behaviour in *Chlamydomonas* algae suggest a protein similar to archaeal rhodopsins in this organism [23,24] Recently a photoreactive retinylidene protein, NOP-1, homologous to the archaeal rhodopsins, has been demonstrated in the filamentous fungus *Neurospora crassa*, definitively extending this family to the eukaryotic domain [25,26]. The long-lived photochemical reaction intermediates of heterologously expressed NOP-1 suggest a sensory rather than an ion pumping function.

Several reviews are available on BR [27–32] and HR [33,34]. Comprehensive reviews on the archaeal sensory rhodopsins covering all published work prior to 1988 [5] and covering all published work between 1988 and 1997 [35] are available, and a review covering both prokaryotic and eukaryotic microbial sensory rhodopsins has appeared [24]. The functions of electrogenic ion transport and sensory signaling are distinctly different; nevertheless recent work reveals that they both result from modifications of the same phototransduction mechanism. Three minireviews focus on comparison of the transport and signaling mechanisms by this family [36–38].

Detailed analysis of the cells' swimming behavior in their natural habitat are not available, but from their physiology in the laboratory a plausible scenario can be

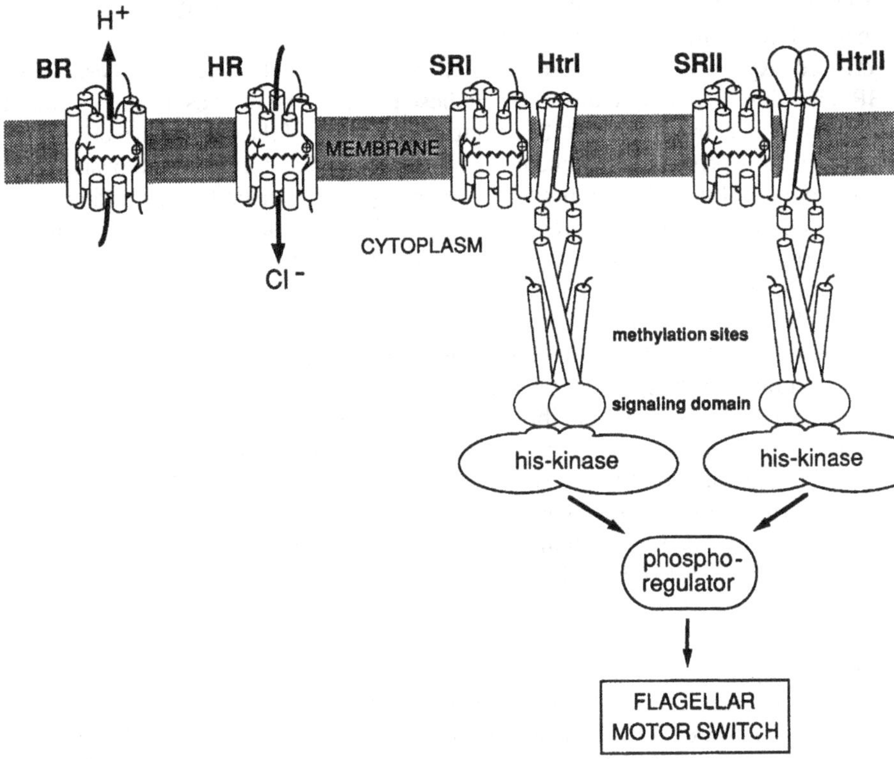

Figure 1. The four archaeal rhodopsins in *H. salinarum*. The transport rhodopsins BR (a proton pump) and HR (a chloride pump) are shown in addition to the sensory rhodopsins SRI and SRII with components in their signal transduction chains. Each rhodopsin consists of seven transmembrane α-helices enclosing a retinal chromophore linked through a protonated Schiff base to a lysine residue in helix G. The sensory rhodopsins are complexed to their corresponding transducer proteins HtrI and HtrII, which have conserved methylation and histidine kinase-binding domains that modulate kinase activity which in turn controls flagellar motor switching through a cytoplasmic phosphoregulator. The structures drawn for the Htr transducers are only approximate, since crystal structures are not available. The transducers are represented as dimers based on the dimeric structure of the homologous *E.coli* aspartate chemotaxis receptor Tar and on quantitative disulfide crosslinking into dimers observed following oxidation of HtrI containing engineered cysteine residues (104). The oligomeric state of the SRs, assumed to be monomeric in the drawing, is not known. The four-helix bundle structure of the transmembrane and methylation domains is based on the structures of the corresponding domains for Tar, for which extensive evidence exists. The relative positions of Htr and SR helices are arbitrary and chosen for illustration only. The depiction of SRs and Htrs as coupled physically within the membrane is based on recent transducer chimera analysis [123].

envisioned. *H. salinarum* grows at its maximum rate chemoheterotrophically in aerobic conditions. When oxygen and respiratory substrates are plentiful, *H. salinarum* cells would be expected to avoid sunlight and potential photooxidative damage. Accordingly, they synthesize the repellent receptor SRII (also known as phoborhodopsin) as their only rhodopsin. SRII absorbs blue-green light near the energy peak of the solar spectrum at the Earth's surface. Hence, its wavelength sensitivity is tuned to be maximally effective for seeking the dark.

Decreased oxygen tension suppresses SRII production and induces synthesis of BR and HR, enabling orange light absorbed by these pumps to be used as an energy source. Like respiratory electron transport, BR pumps protons out of the cell, contributing to the inwardly-directed proton electrochemical potential needed for ATP synthesis, active transport, and motility. HR is an inwardly-directed pump, transporting chloride into the cell. Like cation ejection, anion uptake hyperpolarizes the membrane positive-outside. Therefore, the electrogenic inward transport of chloride contributes to the membrane potential component of proton motive force without loss of cytoplasmic protons. HR therefore helps maintain pH homeostasis by avoiding cytoplasmic alkalization.

Production of SRI is induced along with BR and HR. SRI mediates attractant responses to orange light, guiding the cells into illuminated regions where the ion pumps will be maximally active. SRI exhibits a second signaling activity to ensure it will not perilously guide the cells into higher energy light. A long-lived photointermediate from orange light-activation of SRI, a species called S_{373}, absorbs near-UV photons and mediates a strong repellent response. The color-sensitive signals from SRI, therefore, attract the cells into a region containing orange light only if this region is relatively free of near-UV photons. When back in a rich aerobic environment, the *H. salinarum* cells turn off synthesis of BR, HR, and SRI and turn on SRII production.

Although the sensory rhodopsins are dedicated phototaxis receptors and are responsible for phototaxis under most conditions, some earlier work, especially action spectroscopy, suggested that BR mediated attractant responses. Studies of SR-deficient mutants have confirmed attractant responses to orange light due to light-driven proton pumping by BR [39,40]. The BR-mediated responses occur at high light intensities and are most evident in partially de-energized cells. Proton motive force ($\Delta\mu_{H^+}$) or membrane potential ($\Delta\Psi$) changes [41] have been suggested to be involved in aerotaxis, which occurs in *H. salinarum* [4] An aerotaxis sensor has also been identified recently which may respond to these changes [42]), and hence BR may contribute via this sensor. Alternatively, a hypothetical cellular device measuring proton motive force, called a "protometer", has been proposed as the sensor [40]. It is also possible that the BR-mediated responses may result from secondary consequences of electrogenic proton pumping (e.g. $\Delta\Psi$ changes) on metabolic or signal transduction pathways [43]. The difference between these possibilities may be only semantic, if one accepts as a "protometer" a component(s) with a different primary function(s) in the cell.

5.3 From physiology to molecular components

5.3.1 Discovery of the receptors

The behavioral studies in the 1970s demonstrated that the responses depend on retinal, and the action spectrum [1] for the attractant response matched closely the absorption spectrum of the light-driven proton pump bacteriorhodopsin (BR), at that time the only known retinal-containing pigment in *H. salinarum*. The existence of a second light-driven ion pump, the chloride transporter halorhodopsin (HR), was subsequently demonstrated [11,12]. In 1982, mutant strains (ion flux mutants, e.g. Flx3, Flx15) were isolated that lack both of the transport rhodopsins, yet retained both positive and negative phototactic responses [44]. The phenotype of Flx mutants predicted that yet undiscovered retinal-containing proteins that did not generate light-driven ion fluxes must exist and function as sensory rhodopsins mediating phototaxis. The search for the predicted sensory rhodopsins used Flx mutants because they lacked spectroscopic and functional interference by the transport rhodopsins.

The first sensory rhodopsin (now called SRI) was soon found by laser flash kinetic spectroscopy of Flx mutant membranes [13]. Further spectroscopic and behavioral studies established that SRI was unusual among photosensory receptors in its ability to discriminate color: the same SRI molecule is capable of producing attractant responses to orange light and repellent responses to new-UV light [14]. Photochromic reactions of the receptor protein's orange light-absorbing dark state and its near-UV light-absorbing long-lived photointermediate confer a simple color-discrimination capability allowing the cell to migrate into a spectral region optimal for energy capture by the transport rhodopsins while avoiding UV-blue light photooxidative damage.

Combined flash spectroscopy and behavioral analysis proved again fruitful and led to the detection of sensory rhodopsin II (SRII), also referred to as phoborhodopsin [15,16]. The existence of SRII was rapidly confirmed in a number of laboratories and shown to be a distinct protein from SRI and to mediate repellent phototaxis [45–47].

5.3.2 Cloning of receptor genes and identification of signal transduction components

The spectroscopic characterization of SRI and the availability of Flx mutants lacking the much more abundant BR and HR proteins enabled purification of SRI [48] and the use of protein sequence information to clone its gene, *sopI* (sensory opsin I, where opsin designates the apoprotein of a rhodopsin pigment) [49]. Biochemical analysis of phototaxis mutant membranes identified the second component in the SRI signaling pathway, a methyl-accepting transducer protein now known as HtrI [50]. HtrI was purified and its sequence used to clone its gene *htrI*, which was found to be immediately upstream of *sopI* and co-transcribed with the receptor apoprotein [17]. HtrI was found to modulate the SRI photocycle by altering its rate and pH dependence indicating physical association of the two proteins [51], and a number of biochemical and biophysical studies have further confirmed a molecular complexation between the receptor and its transducer in the light [52–54] and in the dark [55,56].

The gene encoding an SRII apoprotein was cloned from *Natronobacterium pharaonis* with a similar strategy and found to be paired with its transducer gene, as in the *htrI-sopI* operon [18]. The gene encoding *H. salinarum* SRII was obtained recently in a comprehensive cloning effort of the transducer gene family of this organism, because of its position adjacent to the *htrII* transducer gene [19,57].

The first of the transducers HtrI had been identified because of its carboxylmethylation, a reversible covalent modification that is characteristic of eubacterial transducers

[58]. The homology to the large eubacterial chemotaxis transducer family became definitive when the gene was cloned because of the conservation of the histidine kinase (CheA)-binding site ("signaling domain") in the cytoplasmic domain of the molecule [17]. In eubacterial chemotaxis, CheA phosphorylates a soluble protein CheY that binds to a flagellar motor switch complex controlling reorientation behavior. CheA and CheY comprise a two component regulatory system homologous to a large number of such systems controlling motility and transcriptional activity in Eubacteria, but previously not observed in the Archaea. Using oligonucleotide probes to the conserved regions of *cheA* genes, Rudolph and Oesterhelt [59] cloned an *H. salinarum* operon encoding homologs of CheA and CheY, as well as two other chemotaxis genes CheB and CheJ. This group also showed autophosphorylation of CheA and CheA dephosphorylation by CheY, as occurs in the *E. coli* counterparts [60].

5.4 Structure and function of SR-Htr molecular complexes

5.4.1 Structure of sensory rhodopsins

The gene-predicted sequences of the sensory rhodopsins indicate hydrophobic proteins with seven transmembrane segments forming a retinal binding pocket highly conserved with that of the transport rhodopsins (Figure 2). Crystal structures of SR proteins have not yet been accomplished, but an atomic resolution structure of BR (at 2.3 Å) is available from cryoelectron microscopy of two-dimensional crystals and x-ray crystallography of three-dimensional crystals [61–64]. The BR structure provides a good first approximation to the structures of SRI and SRII, because the transmembrane helices can be aligned without gaps while preserving the positions of the 22 residues in the retinal binding cavity [65].

The structure of the chromophores in sensory rhodopsins has been examined by retinal extraction, reconstitution with retinal isomers, and resonance Raman spectroscopy (summarized in detail in [35]). As in archaeal transport rhodopsins and in visual pigments, the sensory rhodopsin chromophores all contain a protonated Schiff base linkage at the attachment site of the retinal to a lysine residue in helix G.

The functional photoreactions of the archaeal rhodopsins each entails photoisomerization of the retinal from all-*trans* to 13-*cis*. However, a difference in isomer **exclusivity** of the unphotolyzed pigments has become evident through binding studies using retinal isomers and retinal analogues. The BR apoprotein, Bop, forms pigments with retinal, added as either the all-*trans* or 13-*cis* isomer, and in the dark the apoprotein catalyzes the isomerization in its chromophore into a mixture of all-*trans* and 13-*cis* isomers, the latter accompanied by isomerization about the C=N bond as well [29]. In contrast, SRI apoprotein (SopI) does not form a retinylidene pigment with 13-*cis* retinal [66] or with 13-*cis*-locked retinal (a rigid ring includes the 13–14 double bond; [67]). Both the Bop and SopI apoproteins form pigments with all-*trans* 13-desmethyl retinal and in BR the chromophore thermal equilibrium largely favors the 13-*cis* form with this analogue [68]. Even in this case SopI does not thermally isomerize the chromophore to 13-*cis*. The SRII from *Natronobacterium pharaonis* (pSRII) has also been shown to exhibit exclusive binding to the all-*trans* isomer [69]. Therefore, archaeal sensory rhodopsins, like animal visual pigments [70], and as opposed to archaeal transport rhodopsins, exhibit isomer exclusivity. The structural basis of the isomer-binding selectivity is likely to result from the greater steric restriction in the sensory rhodopsin binding pockets compared to those of BR and HR [43,66,71,72]. This property may be physiologically relevant in avoiding thermal noise and for this reason be a property of the sensory as opposed to the transport pigments.

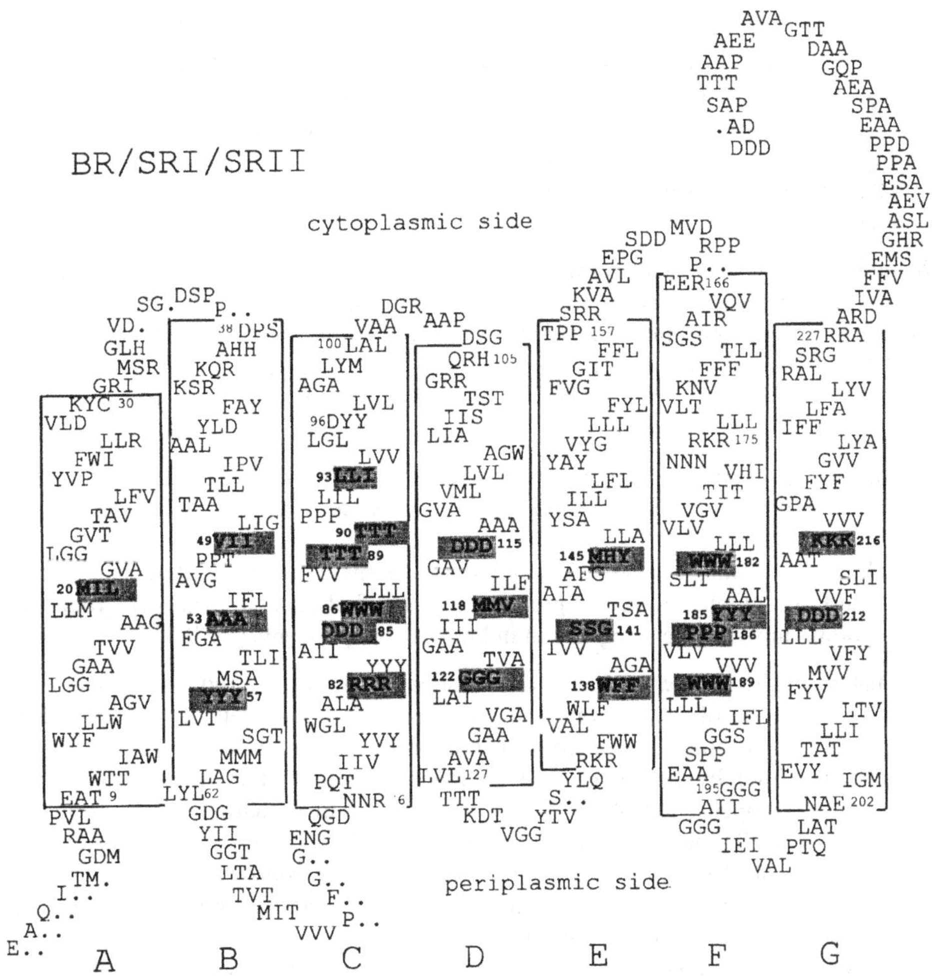

Figure 2. Alignment of the BR, SRI, and SRII primary sequences in a two-dimensional folding topology. Hydropathy analysis of the primary sequences in each case indicates the presence of seven transmembrane α-helices (designated A through G). The helix boundaries have been drawn based on those of BR. Position numbers refer to the residues in BR. Residues corresponding to the retinal-binding pocket conserved among the archaeal transport and sensory rhodopsins are shaded.

5.4.2 Color regulation

Interactions between the retinal and its protein environment cause the absorption maxima of the rhodopsins to strongly deviate from the absorption maximum of a protonated Schiff base model compound in methanol in the presence of Cl^-. The shift to longer wavelength caused by protein-retinal interactions and expressed as a wavenumber difference has been designated the "opsin shift" [73]. The opsin shift has been investigated in detail for BR [27,74–76] and it is likely that the similar absorption spectra and opsin shifts of HR and SRI and likely to result from the same color regulation mechanism as in BR. Three contributing factors have been identified:

1. the positive charge on the protonated Schiff base nitrogen is only **weakly** stabilized by a complex counterion provided by the protein environment,
2. the protein forces the C_5—C_6 single bond in the retinal to be 6*s-trans* allowing the ring and chain to adopt a co-planar conformation, and
3. a third factor is evident from a chromophore analogue in which the other two factors are eliminated [76].

While it has been suggested that the weak counterion accounts for two-thirds of the BR opsin shift and the co-planarization for most of the rest, a recent study indicates that the contribution of the third factor is at least 40% [76]. The physical basis of this important third factor is not yet clear. Two possibilities involving protein dipoles have been suggested based on the observation that photoexcitation results in polarization of the retinal [77]:

1. polar groups around the polyene chain of retinal may reduce the excited state-ground state energy difference by stabilizing the excited state or destabilizing the ground state, and
2. **polarizable** protein side chains may reduce the energy of the excited state. Further support for dipole effects is provided by the finding that hydroxyl groups are important in wavelength regulation in human cone pigments [78,79].

The nature of the counterion has been investigated in archaeal rhodopsins and visual pigments because of its relevance to the opsin shift as well as to Schiff base deprotonation, an important event for proton translocation by BR and signaling (discussed below) by archaeal sensory rhodopsins and human rod rhodopsin. The counterionic compensation of the Schiff base unit positive charge in BR is primarily from ionized Asp85, but involves also Arg82 and Asp212 [80–82]. These 3 residues are conserved in all SRI and SRII sequences. Asp85 is substituted by Thr in the corresponding position in HR, and a functionally important Cl^- apparently substitutes for the missing carboxylate counterion [83].

The mechanism of color regulation in SRII, whose short wavelength absorption maximum differs from that of the other three archaeal rhodopsins, has been examined using a wide range of chromophore analogues [84]. Ring/chain co-planarization in the 6-*s-trans* conformation is sufficient to explain nearly all of the opsin shift of 2200 cm^{-1} in SRII. The relatively small opsin shift of SRII indicates its lack of a significant contribution from factor (3) discussed above. The lesser influence of this factor, suggested to be retinal interactions with protein dipoles, may lead to a reduction in the inhomogeneous broadening of the absorbance bands, revealing the underlying vibrational fine-structure in SRII that is diffused in the other pigments. In this regard, it is notable that a serine residue in helix E conserved among the more red shifted pigments BR (Ser141), HR, and SRI, is substituted by glycine in each of the known SRII sequences (Gly128 in *H. salinarum* SRII). This substitution, which is within the retinal-binding pocket, may be largely responsible for the more blue shifted absorption of SRII if the hydroxyl group contributes to the opsin shift.

5.4.3 Photochemical reaction cycles

Photoisomerization of the retinal from all-*trans* to 13-*cis* initiates the functional thermal protein transitions of each of the archaeal rhodopsins. The photoexcitation of the dark state of SRI, SR_{587}, initiates a cyclic series of transitions (a photochemical reaction cycle, or more commonly "photocycle") containing three resolved intermediate states, $SR_{587} \rightarrow S_{610} \rightarrow S_{560} \rightarrow S_{373} \rightarrow SR_{587}$ [5]. The only long-lived intermediate detected in the SRI photocycle is S_{373} (800 ms in isolated membranes and 1.2 seconds in energized cells at 23°C). The rate of formation of S_{373} is 3000 times higher than its rate of decay (Figure 3). Therefore S_{373} the accumulates in physiologically active concentrations after

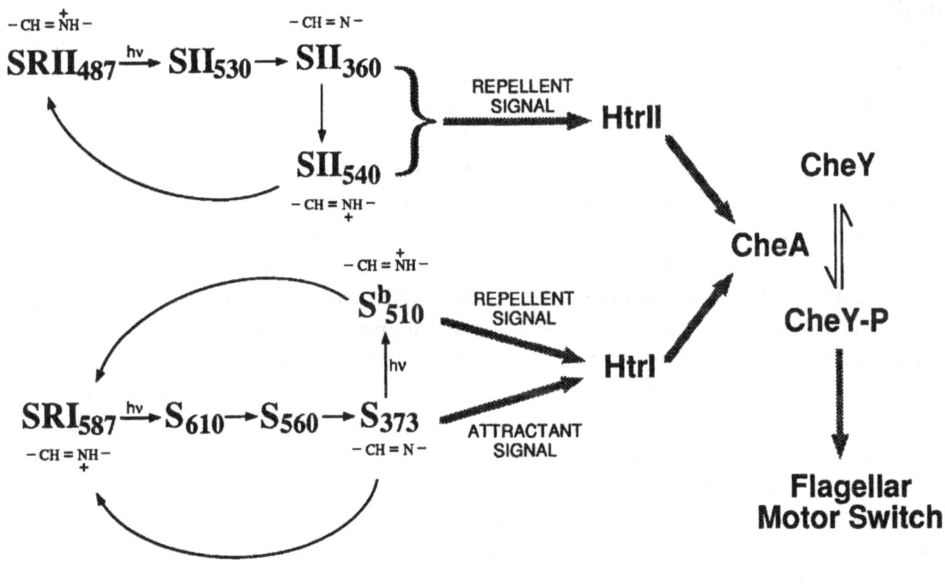

Figure 3. Photochemical reaction cycles of *H. salinarum* sensory rhodopsins I and II and their coupling to the flagellar motor. Arrows with hν indicate light reactions. Subscripts are the wavelength maxima observed for the pigments or calculated for their photointermediates from flash photolysis data. The state of protonation of the Schiff base is indicated. Approximate first order half-lives at room temperature for SRI intermediates S_{610}, S_{560}, S_{373}, and Sb_{510} are 90 μs, 270 μs, 800 ms, and 80 ms, respectively, and for SII_{530}, SII_{360}, and SII_{540}, 160 μs, 120 ms, and 330 ms, respectively. The indicated signaling states (see text) in each photocycle transmit signals through HtrI and HtrII to CheA. CheA controls the extent of phosphorylation of CheY, and phosphorylated CheY induces swimming reversals. Redrawn from [35].

a flash of light or in the photosteady state under continuous orange illumination. Near-UV light photoreconverts S_{373} ten times more rapidly to SR_{587} via the intermediate S^{b}_{510}. In the cell, the orange and near-UV light photoreactions of SR_{587} and S_{373} generate attractant and repellent responses, respectively. The activation of as few as 1–2 S_{373} molecules is sufficient to elicit a repellent response [14]. Hence, the photointerconversion of the two receptor forms by light (photochromicity) and the opposite cellular signals produced by these photoreactions provide the cell with a simple color-discriminating capability with exquisite sensitivity.

The photocycles of SRII (Figure 3) and pSRII have been studied in intact membranes and in detergent at physiological and at cryogenic temperatures. Using the nomenclature from the analogous species in BR, relatively long-lived M and O-like intermediates have been observed for both pigments at room temperature (summarized in [35]).

5.4.4 Receptor signaling states

The lifetimes of photocycle intermediates have been measured by laser flash kinetic absorption spectroscopy and modified by replacement of retinal with chromophore analogues in vivo using retinal-deficient strains of *H. salinarum*. The motility responses mediated by the modified SRs has been measured by computerized cell tracking and motion analysis [3]. In an early study of SRI, an acyclic retinal analogue decreased the photocycling rate and increased the phototaxis sensitivity of the cells, indicating that signaling is governed by the lifetime of a photocycle intermediate(s) rather than by the frequency of photocycling [85]. This result provided a quantitative method for the determination of spectral states producing phototaxis signals in vivo. The photocycle is measured for each modified SR and concentrations of photocycle intermediates integrated over time are calculated. These values are compared with the sensitivity of the cells containing the modified SR derived from fluence-response curves obtained by cell tracking and motion analysis. In the case of SRI the attractant response to orange light is proportional to the concentration of the unprotonated Schiff base species in the photocycle, S_{373}, indicating that this intermediate is the attractant signaling state [86].

As discussed above, photoexcitation of S_{373} results in a repellent response by the cells. This repellent response to S_{373} stimulation is stronger than the attractant response to SR_{587} and is proportional to the concentration of S_{373} in the photosteady state generated by orange light [14]. Therefore, S_{373} plays a dual role as orange light attractant signaling state and near-UV repellent receptor. Since either a decrease in orange light or an increase in near-UV light in an orange background results in a reduction of S_{373} levels, the strong near-UV response has been suggested to be caused by the rapid disappearance of the attractant signaling state [87]. However, simultaneous stimulation with orange and near-UV light produces a strong repellent response, even though it results in a net increase in S_{373} concentration. Therefore, a distinct signaling state with a repellent effect has been concluded to be produced by S_{373} photoexcitation. This conclusion has been confirmed genetically by isolation of an SRI mutant (D201N) that does not produce attractant signals to orange light, but still mediates wild-type near-UV repellent responses to S_{373} excitation [88]. These findings necessitate a model in which two distinct signaling states are formed by photoexcitation of SR_{587} and S_{373}. In a published model [35] it is tentatively assumed that S^b_{510} is the repellent signaling state, since it is the only intermediate observed and its lifetime is compatible with a signaling function.

Although three distinct states (the unstimulated state and the attractant and repellent signaling states) are necessary to explain the photoresponses, shuttling between only two structurally distinct conformations of SRI may produce these three states. Based on mutant and suppressor analysis, a mechanism requiring only two SRI conformations has been proposed in which SR_{587} is an equilibrium mixture of the two conformations, and this equilibrium is shifted in opposite directions in the S_{373} and S^b_{510} states [37], and this proposal has received strong support from suppressor analysis [89].

Signaling state analysis using retinal analogues has also been applied to SRII [90]. As was found for SRI, signaling efficiency correlated with photocycle duration. The results indicated that the signaling conformation of the protein is formed in the M intermediate and persists through the lifetime of O.

5.4.5 Relationship of sensory rhodopsin and transport rhodopsin mechanisms

The transport and sensory rhodopsins established that energy and sensory transducing proteins can evolve from a common progenitor and share much of their detailed mechanism despite their different functions [36]. A key shared feature between BR and SRII appears to be an interhelical salt-bridge (between Asp85 in BR (Asp73 in SRII) on helix C and the retinylidene Schiff base on helix G) that is released by photoisomerization of retinal [38,91]. In BR disruption of the salt-bridge by proton transfer is known to contribute to a conformational change resulting in a movement of helix F towards the periphery of the protein on the cytoplasmic side ([92]; reviewed in [37]). This opening of the structure likely facilitates uptake of the transported proton through the cytoplasmic channel in the BR pumping cycle. Disruption of the homologous salt-bridge in SRII by the mutation D73N constitutively activates the receptor, and a residual light-induced phototaxis response by cells carrying D73N shows that a second consequence of photoisomerization that does not require deprotonation of the Schiff base, also contributes to the signaling state. Transducer-free SRI uses the same salt-bridge mechanism to carry out light-driven proton transport, but interaction with its transducer disrupts the salt bridge in the dark by raising the apparent pK of Asp76 on helix C from 7.2 to 8.5 [36]. The purpose of this effect may be to poise the receptor in a partially activated state, which is then able to produce opposite signals depending on the color of the stimulus light, an idea supported by genetic evidence [89]. In a current model for signaling [38] the salt-bridge-controlled conformational change and a second consequence of photoisomerization are used to modulate interaction with the transmembrane regions of the Htr proteins.

The activating role of Asp73 is analogous to that of Glu113 in human rod rhodopsin [93]. Glu113, on helix C, forms a salt-bridge with the Schiff base on helix G, and serves as the proton acceptor during photoconversion to the G protein-activating state metarhodopsin-II_{380}. Disruption of the Glu113-protonated Schiff base salt bridge by mutagenic replacement of Glu113 with Gln constitutively activates the rhodopsin apoprotein. This observation supports the notion that in rhodopsin the counterion-Schiff base salt-bridge constrains the protein in an inactive conformation and the constraint is released by its light-induced disruption. The similar observation in SRII argues for a remarkable generality of this mechanism in retinylidene receptors from archaea to man.

Consistent with dark-activation by the substitution of Asn for Asp73 in SRII, cells carrying this mutation exhibit a strongly reduced, but still detectable, taxis response. Therefore salt-bridge disruption appears to be sufficient for shifting the conformational equilibrium toward the signaling conformation, but another consequence of photo-isomerization of retinal must also contribute to this shift. This result strengthens the analogy to visual pigments, in which proton transfer from the Schiff base to Glu113 is an important factor in stabilizing the G protein-activating state [93] and to which other determinants also contribute significantly [94,95].

5.4.6 Structure of the transducers

The transducer HtrI, the second protein in the SRI signaling pathway, was first identified by mutant analysis [50,96]. Based on its genetic association with SRI and its reversible carboxylmethylation, which is characteristic of eubacterial chemotaxis transducers [58], the protein was proposed to function as a transducer for SRI, relaying signals from the receptor to cytoplasmic components controlling the flagellar motor. HtrI, was isolated from SDS-PAGE gels, and partial sequence information used to identify and clone its gene, *htrI* [17]. Like the eubacterial transducers (methyl-accepting chemotaxis proteins or MCPs), the HtrI protein contains two transmembrane helices and a strongly conserved cytoplasmic region involved in binding of a histidine kinase (see Figure 1), and flanking regions containing carboxylmethylation sites.

The proposed role of HtrI as an SRI transducer has been confirmed in a number of investigations. The *htrI* gene and the *sopI* gene encoding the SRI apoprotein are part of an operon under control of a single promoter and expression of the *htrI-sopI* pair was shown to restore phototaxis in a mutant containing a deletion in the *htrI-sopI* region [17,53]. The most definitive genetic evidence is that deletion of the region encoding the methylation and signaling domain of HtrI, although not affecting the proper folding and membrane association of the shortened protein, prevents restoration of SRI phototaxis [97]. Furthermore, biochemical and spectroscopic evidence shows that SRI and HtrI are physically associated in the membrane (see below). It has been useful in mutagenesis experiments that cotranscription of *htrI* and *sopI* is not required for their functional association, because *htrI* chromosomally expressed from its native promoter and *sopI* expressed from a plasmid or from the *bop* locus on the chromosome produce an active complex [98,99].

After *htrI* was cloned, related genes have been identified in halobacteria. The *htrI* gene from *H. vallismortis* was cloned and the predicted protein sequence found to be 57% identical to *H. salinarum* HtrI [100]. The first *htrII* genes were identified in *H. vallismortis* and *N. pharaonis* [18], and *htrII* from *H. salinarum* was cloned recently [57], defining a second class of phototaxis transducers. As is *H. salinarium htrI*, these four *htr* genes are positioned immediately upstream of *sop* genes. By use of probes to the signaling domain-encoding region of *htrI,* a family of 13 halobacterial taxis transducers have been cloned [57]. One integral membrane transducer, HtrXI, has been demonstrated to function in aerotaxis [42], and soluble transducers, HtrVIII [101] and Car [102], in chemotaxis.

Alignment of the phototaxis Htr sequences with eubacterial chemotaxis transducers reveals a number of conserved features. They each have two transmembrane helices (TM1 and TM2). In the case of Tar [103] the two helices have been demonstrated to dimerize into a four-helix bundle, and a similar dimerization has been shown to occur in HtrI by crosslinking of engineered sulfhydryls [104]. There are several regions of conservation: the most strongly conserved is the signaling domain of approximately 60 residues, implicated in binding the histidine kinase, CheA. Recognizably conserved methylation sites flank the signaling domain. A region of weak but detectable homology contained within approximately 40 residues at the cytoplasmic end of TM2, called the linker region in eubacterial MCPs, has been found to be important in relaying SRI signals in HtrI mutagenesis studies [89,105].

The first transmembrane helix TM1 is located close to the N-terminal end in the archaeal phototaxis transducers and in the eubacterial chemotaxis transducers. In the *E. coli* chemotaxis transducer TM2 is found approximately 150 residues beyond TM1, defining a periplasmic ligand-binding domain. This domain of the aspartate receptor has been crystallized and its aspartate-binding site studied [103]. In the HtrI protein the TM2 sequence is adjacent to that of TM1 and there is little or no periplasmic domain. Similarly, the *N. pharaonis* HtrII contains a periplasmic domain of less than 20 residues. This difference in structure presumably reflects the lack of a periplasmic effector-recognition site in these phototaxis transducers, which detect instead conformational changes of the SR receptors via interactions within the membrane (see below). The primary sequence of HtrII from *H. salinarum* however does predict a large periplasmic domain of ~250 residues, consistent with it functioning as a chemotaxis receptor as well as a phototaxis receptor [106].

In all of the transducers the highly conserved signaling domain of about 60 residues is flanked by methylation helices, involved in chemotaxis adaptation in *E. coli*. The level of sequence conservation of these regions is considerably lower than that of the signaling domain, however sequence alignment has revealed candidate methylation sites in HtrI and the other transducers. In vivo radiolabeling with tritiated methionine, followed by SDS-PAGE and autofluorography has revealed carboxylmethylation on the halobacterial transducers (summarized in [35]). Stimulus-induced demethylation of HtrII occurs, whereas photostimulation causes no detectable changes in methylation extent in HtrI [107]. Mutagenesis studies of the *H. salinarum* phototaxis transducers indicates that a single pair of Glu residues at positions 265 and 266 is responsible for all of the HtrI methylation observed, and similarly a single Glu pair at positions 513 and 514 is responsible for nearly all of the methyl-labeling in HtrII [107]. Cells containing the unmethylatable transducers are still able to perform phototaxis and adapt to light stimuli, and methanol release assays show that methyl group turnover is still induced in response to SRI and SRII photostimulation. Furthermore pulse-chase experiments with *in vivo* ^{3}[H]-methyl labeled cells demonstrate that repetitive stimulation of the SRI-HtrI (unmethylatable mutant form) complex induces methyl turnover on the other transducers visible by SDS-PAGE. These results suggest that changes in the level of methylation of other Htr proteins (there are at least 13 known in *H. salinarum*) play a role in the adaptation to phototaxis stimuli. Since elimination of the primary methylation sites in HtrI or HtrII by mutagenesis does not greatly perturb phototaxis responses mediated through SRI or SRII, methylation changes may play a fine-tuning rather than essential role in phototaxis.

5.4.7 SR-Htr interaction and the signaling process

The SRI protein forms a molecular complex with HtrI both in the dark and in the light. Therefore, stimulus relay from the signaling states of SRI to HtrI does not involve protein association/dissociation, but rather structural changes within the complex, unlike signaling from visual rhodopsin to the G-protein transducin. The complexation has been directly demonstrated by the co-purification of HtrI with SRI in his-tagged SRI affinity chromatography (E.N. Spudich and J.L. Spudich, unpublished results). As a result of its interaction with the receptor, HtrI influences various properties of SRI both in its SR_{587} and S_{373} states (summarized in [35]). The presence of HtrI partially shields the chromophore in the unphotolyzed state SR_{587} from attack by hydroxylamine. In addition, the apparent pK_a of the Schiff base is above 12 in the presence of HtrI and 9.5 in its absence. Also the pK_a of Asp76, the protonation state of which can be monitored by the blue to purple transition, is shifted (from 7.2 to 8.5) by the presence of HtrI.

The above effects are those evident in the dark. The effects of HtrI on events during the photocycle were first detected by S_{373} decay measurements. The formation and decay of S_{373} involve deprotonation and reprotonation of the Schiff base in SRI, respectively. In the absence of HtrI these proton transfers result in light-driven electrogenic pumping of protons across the membrane [108] when Asp 76 is ionized. In contrast, in SRI complexed with HtrI these proton transfer reactions occur within the complex, because no changes in proton concentration are detected in the medium and the reactions are independent of external pH [109]. Removal of HtrI by mutation causes reprotonation of the Schiff base (i.e. S_{373} decay) to become highly pH-dependent ([51], and transient stoichiometric proton release is detected during the photocycle [52]. The protonation kinetics are first order and the rate constant is proportional to external proton concentration. The slope of this pH dependence is significantly less than one (0.36), suggesting a complex coupling of proton transfer events in SRI to the bulk pH [110]. When HtrI is present in sub-stoichiometric amounts, both pH-independent (HtrI-complexed) and pH-dependent (HtrI-free) photocycling SRI species are observed [52,54] as would be expected from a stable complexation of a fraction of the SRI molecules with the available HtrI. The binding of HtrI to SRI also alters the temperature dependence of S_{373} decay [111].

The rate of flash-induced deprotonation is also greatly affected by HtrI binding [56] and the yield of S_{373} is larger in the complex because of the suppression of thermal branching reactions from the S_{610} and S_{560} states to SR_{587} [24]. Also in bovine rod rhodopsin an increase in flash-yield of the deprotonated Schiff base species (Meta-II_{380}) is observed upon binding of its transducer (the G-protein transducin) [112]. In free SRI in membranes as well as in purified detergent-solubilized SRI [113] the formation of S_{373} occurs in μs times in the purple form ($t_{1/2} = 10$ μs at 18°C, [56]; $t_{1/2} = 3$ to 5 μs at 23°C, I. Szundi and R.A. Bogomolni, unpublished results), in which Asp76 is the proton acceptor [114]. In the blue form of free SRI Asp76 is not ionized and the rate of S_{373} formation is reduced 1000-fold ($t_{1/2}$ > 10 ms). In the SRI-HtrI complex (Asp76 protonated, blue form, measured at pH values 5 to 8 in membranes and in purified complex) neither the fast (10 μs) nor the slow (> 10 ms) rate is detected; rather a first order rate of 300 μs is observed. Note that in the complex (blue form) Asp76 is not available as a proton acceptor, and therefore HtrI interaction facilitates deprotonation of the Schiff base in this state of the protein, since without HtrI the blue form exhibits the > 10 ms rate. HtrII also modulates the photochemical reaction cycle of SRII in halobacterial membranes [115,116].

The earliest demonstration of single photon-driven proton pumping by HtrI-free SRI used pH and TTP^+ electrodes and membrane envelope vesicles [108]. Measurements at high light intensities with pH electrodes have been applied to whole cells and revealed a second mode of proton translocation interpreted as a two-photon cycling between the intermediates S^b_{510} and S_{373} [117]. The presence of both one- and two-photon pumping modes in SRI was confirmed by measurements with *H. salinarum* membranes attached to black lipid membranes (BLM; [110]). In this study proton pumping was reported for membranes containing wild-type SRI-HtrI complex, although in previous work, HtrI was reported to block proton release [52] and pumping [108]. The blockage of SRI-catalyzed proton translocation by HtrI was recently confirmed by pH-electrode measurements using membrane vesicles [116]. Since the BLM method is more sensitive, although not quantifiable in terms of protein-specific activity, it is possible that HtrI reduces pumping by SRI to a level below the detection limit of the other methods (~5%). Alternatively, a small fraction of HtrI-free SRI may be present in the membranes, and responsible for the BLM signals. Supporting this latter possibility are the flash photolysis data [110] of complex-containing membranes used for the BLM measurements, that show a slow (> 10 ms) phase in S_{373} formation with an amplitude of 2–5% which vanishes above pH 7 (i.e. at a pH where Asp76 becomes ionized in free SRI), as is characteristic of HtrI-free but not for HtrI-complexed SRI.

Analysis of these effects of HtrI on SRI in the context of current understanding of the BR pumping mechanism has led to the following interpretation: Like BR, SRI contains both a cytoplasmic and extracellular channel capable of proton conduction. Alternate access of these channels to the Schiff base during the photocycle permits proton release and uptake on opposite sides of the membrane, hence producing vectorial proton translocation. HtrI increases the pK_a of the gatekeeper for the extracellular channel, Asp76, thereby preventing it from accepting the Schiff base proton. The cytoplasmic channel is also blocked by the interaction with HtrI, although the mechanism is less clear. Either the proton movement or the structural changes during the switch in accessibility of the Schiff base (or both) may generate the receptor phototaxis signals.

The proton acceptor in the purple form of HtrI-free SRI has been identified as Asp76 by FTIR difference absorption spectroscopy on wild type and D76N mutant membranes [114]. However, the fate of the proton after its release from the Schiff base during S_{373} formation in the complex is unclear. The proton acceptor is not Asp76, since it is already protonated in SR_{587}. The FTIR light-dark difference spectra display signals in the carboxylate region that indicate perturbation, but not protonation, of a carboxylate group [114,118,119], which suggests the involvement of some other group. Site-specific mutagenesis identified another residue in SRI, His166, that appears to be critical to this process as well as to reprotonation of the Schiff base during the recovery phase of the SRI photocycle [120]. His166 was found to be important in phototaxis signaling and to play a role either as a structurally important residue or possibly as a participant in Schiff base proton transfers.

Polarization anisotropy in the SRI photocycle kinetics demonstrated a large effect of HtrI on the rotational diffusion rate of SRI and the angle of the chromophore with respect to the membrane plane (Bogomolni, unpublished results). A rotational diffusion time of SRI in native membranes of about 200 μs was measured, whereas HtrI-free SRI exhibited a much shorter rotation time (< 10 μs). The 200 μs time is significantly slower than that expected from a 25 kDa membrane protein in a lipid bilayer environment but is comparable to the rotational times observed for small aggregates of HR or BR. The transport rhodopsins occur in aggregated or oligomeric states in the membrane and strong evidence for this aggregation are their visible circular dichroism (CD) spectra which shows a negative-positive band expected from exciton interaction between proximal retinal chromophores. In contrast, SRI in the native state yields a CD spectrum devoid of exciton coupling [121]. The 200 μs correlation time would be consistent with an SRI molecule in a more massive complex in the membrane. In addition, according to the residual of the anisotropy HtrI induces an approximately 5° tilt of the SRI chromophore.

5.4.8 Mapping the region of receptor-transducer interaction

The portion of HtrI necessary for conferring pH-independence to the S_{373} decay process was localized by deletion analysis to the *N*-terminal 147 residues containing the two transmembrane helices and the linker region [122]. Within this fragment, substitution with neutral amino acids either accelerates (Glu56, Asp86, Glu87, or Glu108) or retards (Arg70, Arg84, or Arg99) S_{373} decay [105]. Opposite effects on the rate cancel in double mutants containing one replaced acidic and one replaced basic residue. The effect of substitution of Glu56 depends on the electronegativity of the residue introduced. These results indicate that electrostatic interactions of these residues with SRI or with other HtrI residues are involved in the coupling of HtrI to the SRI photoactive site.

Further work with chimeras indicates SRI-HtrI interaction occurs within the transmembrane domains of the two proteins [123]. Viewed from the outside of the cell, four distinct domains of the Htr proteins are evident:

1. a periplasmic region which is small (< 5 residues) in HtrI and large (~250 residues) in HtrII,
2. a membrane domain formed by two transmembrane hydrophobic helices TM1 and TM2),
3. a hydrophilic "linker" region of ~200 residues extending from the membrane surface to
4. the methylation and signaling domain homologous to the domains of eubacterial chemotaxis receptor/transducers that control the kinase activity.

The methylation and signaling domain of HtrI were found to be dispensable for the control of SRI photoreaction kinetics in a truncated transducer [97] and more extensive deletion analysis [122] established, as noted above, that the *N*-terminal 147 residues of HtrI, which contain the two transmembrane helices and ~90 residues of the cytoplasmic linker, are sufficient for interaction with SRI. Deletion of the entire linker region in that study and in an independent investigation [124] resulted in loss of the spectroscopically detectable interaction with SRI. This negative result does not distinguish whether the cytoplasmic portion is required for receptor interaction or alternatively for proper folding or stability of the partial transducer proteins.

Chimeras between HtrI and HtrII were constructed to overcome this limitation, since they are full-length transducer molecules more likely to fold properly [123]. Phototaxis responses of cells expressing the chimeras together with SRI or SRII were analyzed by motion analysis, and membranes isolated from these cells were studied by flash photolysis. The results demonstrated that the presence of the two transmembrane helices of HtrI in a chimera is necessary and sufficient for functional transducer complexation with SRI; i.e. for wild-type SRI photoreactions and attractant and 2-photon repellent phototaxis responses. Additionally, as previously demonstrated [122] chaperone-like facilitation of SRI folding or stability by HtrI was shown to also depend only on the two transmembrane helices of HtrI in chimeric transducers. Similarly, the two transmembrane helices of HtrII specify interaction with the repellent receptor SRII according to motility analysis and laser flash spectroscopy. The results support that the membrane domains of the receptor/transducer complexes, consisting of the 7 helices of the receptor interacting with the 4-helix bundle of the transducer dimer, produce SRI- and SRII-specific signals to the flagellar motor via interchangeable cytoplasmic domains.

5.4.9 From signaling complex to the flagellar motor

Early cell tracking studies and mutant analysis established that an integrated signal from phototaxis and chemotaxis receptors modulates the flagellar motor switch. A cluster of genes designated *cheY*, *cheB*, *cheA* and *cheJ* has been cloned from *H. salinarum* [59,60,125]. The first three genes are homologous to their counterparts in the *E. coli* chemotaxis system and it was shown that CheA has autophosphorylation activity and CheY stimulates its dephosphorylation, as expected from phosphotransfer to CheY. Deletion of either *cheA* or *cheY* results in a smooth swimming phenotype, as in *E. coli*. This observation fits the expectation from the *E. coli* paradigm that phospho-CheY causes swimming reversals upon binding to the flagellar motor switch in *H. salinarum*. The *cheJ* gene is not homologous to any of the known *E. coli* chemotaxis genes, but it is homologous to *B. subtilus cheC* [126]. A homolog of the *E. coli cheW* gene is found upstream and a homolog of *cheR* downstream of the *cheYBAJ* operon (K. Jung and J.L. Spudich, unpublished), whereas homologs of *cheR* and *cheZ* have not been identified. The interaction of CheY-P with the flagellar switch complex differs in *E. coli* and *H. salinarum*: It biases the *E. coli* flagellar motor to rotate clockwise (causing tumbles), whereas it induces a change in the direction of rotation (causing swimming reversals) regardless of the initial direction in *H. salinarum* [60].

An additional component has been proposed as part of the phototaxis signal transduction chain: fumarate ("switch factor") binding protein (FBP, [127]). The existence of FBP has been deduced from biochemical experiments showing that

(a) fumarate is released to the cytoplasm when reversal-inducing stimuli are delivered through SRI or SRII [128,129], and
(b) fumarate restores stimulus-induced reversals in a non-reversing mutant at the level of one or a few molecules per cell [127].

Fumarate is required for switching the direction of flagellar rotation in cytoplasm-free envelopes of *E. coli* [130]. It may act by lowering the activation energy for switching and may connect the bacterial metabolic state to tactic behavior [131].

5.5 Perspectives

The SR-Htr signaling complexes have taken their place as an opportune system to study the chemistry of signal transduction, and, more broadly, as a model system for understanding the nature of dynamic interactions between membrane proteins. Structural features and their dynamics are beginning to be revealed by molecular spectroscopy, and information with near-atomic resolution is within reach by crystallographic techniques. Study of the SR-Htr molecular complexes benefits from the close similarities between SRs and BR, one of the few membrane proteins undergoing extensive structure/function analysis at the atomic level, and between Htrs and the *E. coli* Tar for which partial crystallographic and extensive genetic and biochemical information is available. At the functional level, the processes of receptor activation and signal relay in visual pigments and archaeal sensory rhodopsins have been mutually informative. Future are likely to involve the use of overexpressed components for in vitro structure/function studies and for crystallography on two-dimensional or three-dimensional lattices of the complex, and exploitation of the molecular genetic and molecular biophysical tools that are now in place.

Acknowledgments

Work referred to performed in the author's laboratory has been supported primarily by NIH, NSF, and the University of Texas at Houston.

References

1. D. Hildebrand, N. Dencher (1975). Two photosystems controlling behavioural responses of *Halobacterium halobium*. *Nature*, **257**, 46–48.
2. J.L. Spudich, W. Stoeckenius (1979). Photosensory and chemosensory behavior of *Halobacterium halobium*. *J. Photobiochem. Photobiophys.*, **1**, 43–53.
3. T. Takahashi (1992). Automated measurement of movement responses in Halobacteria. In: D.-P. Häder (Ed.), *Image analysis in biology* (pp. 315–328). CRC Press, Boca Raton.
4. W. Stoeckenius, E.K. Wolff, B. Hess (1988). A rapid population method for action spectra applied to *Halobacterium halobium*. *J. Bacteriol.*, **170**, 2790–2795.
5. J.L. Spudich, R.A. Bogomolni (1988). Sensory rhodopsins of halobacteria. *Annu. Rev. Biophys. Biophys. Chem.*, **17**, 193–215.
6. D. Oesterhelt, W. Marwan (1993). Signal transduction in halobacteria. In: M. Kates (Ed.), *The Biochemistry of Archaea (Archaebacteria), New Comprehensive Biochemistry Vol. 26* (pp. 173–187). Elsevier Science Publishers B.V., Amsterdam.

7. D. Petracchi, S. Lucia, G. Cercignani (1994). Photobehaviour of *Halobacterium halobium*: proposed models for signal transduction and motor switching. *J. Photochem. Photobiol. B: Biol.*, **24**, 75–99.
8. D.-P. Häder (1987). Photosensory behavior in procaryotes. *Microbiol Rev.*, **51**, 1–21.
9. H.C. Berg (1983). *Random walks in biology* (pp. 172). Princeton University Press, Princeton, NJ.
10. D. Oesterhelt, W. Stoeckenius (1973). Functions of a new photoreceptor membrane. *Proc. Natl. Acad. Sci. USA*, **70**, 2853–2857.
11. A. Matsuno-Yagi, Y. Mukohata (1977). Two possible roles of bacteriorhodopsin; a comparative study of strains of *Halobacterium halobium* differing in pigmentation. *Biochem. Biophys. Res. Comm.*, **78**, 237–243.
12. B. Schobert, J.K. Lanyi (1982). Halorhodopsin is a light-driven chloride pump. *J. Biol. Chem.*, **257**, 10306–10313.
13. R.A. Bogomolni, J.L. Spudich (1982). Identification of a third rhodopsin-like pigment in phototactic *Halobacterium halobium*. *Proc. Natl. Acad. Sci. USA*, **79**, 6250–6254.
14. J.L. Spudich, R.A. Bogomolni (1984). The mechanism of colour discrimination by a bacterial sensory rhodopsin. *Nature*, **312**, 509–513.
15. T. Takahashi, H .Tomioka, N. Kamo, Y. Kobatake (1985). A photosystem other than PS370 also mediates the negative phototaxis of *Halobacterium halobium*. *FEMS Microbiol. Lett.*, **28**, 161–164.
16. H. Tomioka, T. Takahashi, N. Kamo, Y. Kobatake (1986). Flash spectrophotometric identification of a fourth rhodopsin-like pigment in *Halobacterium halobium*. *Biochem. Biophys. Res. Comm.*, **139**, 389–395.
17. V.J. Yao, J.L. Spudich (1992). Primary structure of an archaebacterial transducer, a methyl-accepting protein associated with sensory rhodopsin I. *Proc. Natl. Acad. Sci. USA*, **89**, 11915–11919.
18. R. Seidel, B. Scharf, M. Gautel, K. Kleine, D. Oesterhelt, M. Engelhard (1995). The primary structure of sensory rhodopsin II: a member of an additional retinal protein subgroup is coexpressed with its transducer, the halobacterial transducer of rhodopsin II. *Proc. Natl. Acad. Sci USA*, **92**, 3036–3040.
19. W. Zhang, A. Brooun, M.M. Müller, M. Alam (1996). The primary structures of the Archaeon *Halobacterium salinarium* blue light receptor sensory rhodopsin II and its transducer, a methyl-accepting protein. *Proc. Natl. Acad. Sci. USA*, **93**, 8230–8235.
20. J. Otomo, H. Tomioka, Y. Urabe, H. Sasabe (1992). Properties and the primary structures of new bacterial rhodopsins. In: J.L. Rigaud (Ed.), *Structures and functions of retinal proteins, Colloque INSERM* (Vol. 221, pp. 105–108). John Libbey Eurotext, London.
21. J. Soppa, J. Duschl, D. Oesterhelt (1993). Bacterioopsin, haloopsin, and sensory opsin I of the halobacterial isolate *Halobacterium* sp. strain SG1: three new members of a growing family. *J. Bacteriol.*, **175**, 2720–2726.
22. Y. Mukohata (1994). Comparative studies on ion pumps of the bacterial rhodopsin family. *Biophys. Chem.*, **50**, 191–201.
23. K.W. Foster, J. Saranak, N. Patel, G. Zarilli, M. Okabe, T. Kline, K. Nakanishi (1984). A rhodopsin is the functional photoreceptor for phototaxis in the unicellular eukaryote *Chlamydomonas*. *Nature*, **311**, 756–759.
24. J.L. Spudich, D.N. Zacks, R.A. Bogomolni (1995). Microbial sensory rhodopsins: photochemistry and function. *Israel J. Chem.*, **35**, 495–513.
25. J.A. Bieszke, E.L. Braun, L.E. Bean, S. Kang, D.O. Natvig, K.A. Borkovich (1999). The nop–1 gene of *Neurospora crassa* encodes a seven transmembrane helix retinal-binding protein homologous to archaeal rhodopsins. *Proc. Natl. Acad. Sci. USA*, **90**, 8034–8039.

26. J.A. Bieszke, E.N. Spudich, K.L. Scott, K.A. Borkovich, J.L. Spudich (1999). A eukaryotic protein, NOP–1, binds retinal to form an archaeal rhodopsin-like photochemically reactive pigment. *Biochemistry*, **38**, 14138–14145.
27. R.A. Mathies, S.W. Lin, J.B. Ames, W.T. Pollard (1991). From femtoseconds to biology: mechanism of bacteriorhodopsin's light-driven proton pump. *Ann. Rev. Biophys. Biophys. Chem.*, **20**, 491–518.
28. K.J. Rothschild (1992). FTIR difference spectroscopy of bacteriorhodopsin: toward a molecular model. *J. Bioenerg. Biomembr.*, **24**, 147–167.
29. T.G. Ebrey (1993). Light energy transduction in bacteriorhodopsin. In: M. Jackson (Ed.), *Thermodynamics of membranes, receptors and channels* (pp. 353–387). CRC Press, New York.
30. M.P. Krebs, H.G. Khorana (1993). Mechanism of light-dependent proton translocation by bacteriorhodopsin. *J. Bacteriol.*, **175**, 1555–1560.
31. J.K. Lanyi (1998). Understanding structure and function in the light-driven proton pump bacteriorhodopsin. *J. Struct. Biol.*, **124**, 164–178.
32. U, Haupts, J, Tittor, D. Oesterhelt (1999). Closing in on bacteriorhodopsin: progress in understanding the molecule. *Annu Rev Biophys Biomol Struct.*, **28**, 367–399.
33. J.K. Lanyi (1990). Halorhodopsin: a light-driven electrogenic chloride transport system. *Physiol. Rev.*, **70**, 319–330.
34. D. Oesterhelt (1995). Structure and function of halorhodopsin. *Israel J. Chem.*, **35**, 475–494.
35. W.D. Hoff, K.-H. Jung, J.L. Spudich (1997). Molecular Mechanism of Photosignaling by Archaeal Sensory Rhodopsins. *Annu. Rev. Biophys. Biomolec. Struct.*, **26**, 223–258.
36. J.L. Spudich (1994). Protein-protein interaction converts a proton pump into a sensory receptor. *Cell*, **79**, 747–750.
37. J.L. Spudich, J.K. Lanyi (1996). Shuttling between protein conformations: The common mechanism for sensory transduction and ion transport. *Curr. Opin. Cell Biol.*, **8**, 452–457.
38. J.L. Spudich (1998). MicroReview: Variations on a molecular switch: transport and sensory signaling by archaeal rhodopsins. *Molecular Microbiology*, **28**, 1051–1058.
39. B. Yan, S.W. Cline, F. Doolittle, J.L. Spudich (1992). Transformation of a Bop$^-$Hop$^-$Sop-I$^-$Sop-II$^-$ *Halobacterium halobium* mutant to Bop$^+$: effects of bacteriorhodopsin photoactivation on cellular proton fluxes and swimming behavior. *Photochem. Photobiol.*, **56**, 553–561.
40. S.I. Bibikov, R.N. Grishanin, A.D. Kaulen, W. Marwan, D. Oesterhelt, V.P. Skulachev (1993). Bacteriorhodopsin is involved in halobacterial photoreception. *Proc. Natl. Acad. Sci. USA*, **90**, 9446–9450.
41. S.I. Bibikov, V.P. Skulachev (1989). Mechanisms of phototaxis and aerotaxis in *Halobacterium halobium. FEBS Lett.*, **243**, 303–306.
42. A. Brooun, J. Bell, T. Freitas, R.W. Larsen, M. Alam (1998). An archaeal aerotaxis transducer combines subunit I core structures of eukaryotic cytochrome c oxidase and eubacterial methyl-acceptingchemotaxis proteins. *J. Bacteriol.*, **180**, 1642–1646.
43. B. Yan, A. Xie, G.U. Nieuhaus, Y. Katsuta, J.L. Spudich (1993). Steric constraints in the retinal binding pocket of sensory rhodopsin I. *Biochemistry*, **32**, 10224–10232.
44. E.N. Spudich, J.L. Spudich (1982). Control of transmembrane ion fluxes to select halorhodopsin-deficient and other energy transduction mutants of *Halobacterium halobium. Proc. Natl. Acad. USA*, **79**, 4308–4312.
45. E.N. Spudich, S.A. Sundberg, D. Manor, J.L. Spudich (1986). Properties of a second sensory receptor protein in *Halobacterium halobium. Proteins*, **1**, 239–246.
46. E.K. Wolff, R.A. Bogomolni, P. Scherrer, B. Hess, W. Stoeckenius (1986). Color discrimination in halobacteria: spectroscopic characterization of a second sensory receptor

covering the blue-green region of the spectrum. *Proc. Natl. Acad. Sci. USA*, **83**, 7272–7276.

47. W. Marwan, D. Oesterhelt (1987). Signal formation in the halobacterial photophobic response mediated by a fourth retinal protein (P480). *J. Mol. Biol.*, **195**, 333–342.
48. E.S. Schegk, D. Oesterhelt (1988). Isolation of a prokaryotic photoreceptor: sensory rhodopsin from halobacteria. *EMBO J.*, **7**, 2925–2933.
49. A. Blanck, D. Oesterhelt, E. Ferrando, E.S. Schegk, F. Lottspeich (1989). Primary structure of sensory rhodopsin I, a prokaryotic photoreceptor. *EMBO J.*, **8**, 3963–3971.
50. E.N. Spudich, C.A. Hasselbacher, J.L. Spudich (1988). A methyl-accepting protein associated with bacterial sensory rhodopsin I. *J. Bacteriol.*, **170**, 4280–4285.
51. E.N. Spudich, J.L. Spudich (1993). The photochemical reactions of sensory rhodopsin I are altered by its transducer. *J. Biol. Chem.*, **268**, 16095–16097.
52. K.D. Olson, J.L. Spudich (1993). Removal of the transducer protein from sensory rhodopsin I exposes sites of proton release and uptake during the receptor photocycle. *Biophys. J.*, **65**, 2578–2585.
53. E. Ferrando-May, M. Krah, W. Marwan, D. Oesterhelt (1993). The methyl-accepting transducer protein HtrI is functionally associated with the photoreceptor sensory rhodopsin I in the archaeon *Halobacterium salinarium*. *EMBO J.*, **12**, 2999–3005.
54. M. Krah, W. Marwan, A. Vermeglio, D. Oesterhelt (1994). Phototaxis of *Halobacterium salinarium* requires a signalling complex of sensory rhodopsin I and its methyl-accepting transducer HtrI. *EMBO J.*, **13**, 2150–2155.
55. B. Yan, E.N. Spudich, M. Sheves, G. Steinberg, J.L. Spudich (1997). Complexation of the signal transducing protein HtrI to sensory rhodopsin I and its effect on thermodynamics of signaling state deactivation. *J. Phys. Chem.*, **101**, 109–113.
56. K.-H. Jung, E.N. Spudich, P. Dag, J.L. Spudich (1999). Transducer Binding and Transducer Mutations Modulate Photoactive Site Deprotonation in Sensory Rhodopsin I. *Biochemistry*, **38**, 13270–13274.
57. W. Zhang, A. Brooun, J. McCandless, P. Banda, M. Alam (1996). Signal transduction in the Archaeon *Halobacterium salinarium* is processed through three subfamilies of 13 soluble and membrane-bound transducer proteins. *Proc. Natl. Acad. Sci. USA*, **93** 4649–4654.
58. J.B. Stock, M.G. Surette (1996). Chemotaxis. In: F.C. Neidhardt (Ed.), *Escherichia coli* and *Salmonella typhimurium* (pp. 1103–1129). ASM Press, Washington D.C.
59. J. Rudolph, D. Oesterhelt (1995). Chemotaxis and phototaxis require a CheA histidine kinase in the archaeon *Halobacterium salinarium*. *EMBO J.*, **14**, 667–673.
60. J. Rudolph, N. Tolliday, C. Schmitt, S.C. Schuster, D. Oesterhelt (1995). Phosphorylation in halobacterial signal transduction. *EMBO J.*, **14**, 4249–4257.
61. N. Grigorieff, T.A. Ceska, K.H. Downing, J.M. Baldwin, R. Henderson (1996). Electron-crystallographic refinement of the structure of bacteriorhodopsin. *J. Mol. Biol.*, **259**, 393–421.
62. E. Pebay-Peyroula, G. Rummel, J.P. Rosenbusch, E.M. Landau (1997). X-ray structure of bacteriorhodopsin at 2.5 Angstroms from microcrystals grown in lipidic cubic phases. *Science*, **277**, 1676–1681.
63. Y. Kimura, D.G. Vasylyev, A. Miyazawa, A. Kidera, M. Matsushima, K. Mitsuoka, K. Murata, T. Hirai, Y. Fujiyoshi (1997). Surface of bacteriorhodopsin revealed by high-resolution cryoelectronic crystallography. *Nature*, **389**, 206–211.
64. H. Luecke, H.-T. Richter, J.K. Lanyi (1998). Proton transfer pathways in bacteriorhodopsin at 2.3 Å resolution. *Science*, **280**, 1934–1937.
65. R. Henderson, J.M. Baldwin, T.A. Ceska, F. Zemlin, E. Beckmann, K.H. Downing (1990). Model for the structure of bacteriorhodopsin based on high-resolution electron cryo-microscopy. *J. Mol. Biol.*, **213**, 899–929.

66. B. Yan, K. Nakanishi, K., Spudich, J.L. (1991). Mechanism of activation of sensory rhodopsin I: Evidence for a steric trigger. *Proc. Natl. Acad. Sci. USA*, **88**, 9412–9416.
67. B. Yan, T. Takahashi, R. Johnson, F. Derguini, K. Nakanishi, J.L. Spudich (1990). All-*trans*/13-*cis* isomerization of retinal is required for phototaxis signaling by sensory rhodopsins in *Halobacterium halobium*. *Biophysical J.*, **57**, 807–814.
68. W. Gärtner, P. Towner, H. Hopf, D. Oesterhelt (1983). Removal of methyl groups controls the activity of bacteriorhodopsin. *Biochemistry*, **22**, 2637–2644.
69. J. Hirayma, N. Kamo, Y. Imamoto, Y. Shichida, T. Yoshizawa (1995). Reason for the lack of light-dark adaptation in *pharaonis* phoborhodopsin: reconstitution with 13-*cis*-retinal. *FEBS Lett.*, **364**, 168–170.
70. R.R. Birge (1990). Nature of the primary photochemical events in rhodopsin and bacteriorhodopsin. *Biochim. Biophys. Acta*, **1016**, 293–327.
71. M. Ariki, Y. Shichida, T. Yoshizawa (1987). Low temperature spectrophotometry on the photoreaction cycle of sensory rhodopsin. *FEBS Lett.*, **225**, 255–258.
72. J. Hirayama, Y. Imamoto, Y. Shichida, T. Yoshizawa, A.E. Asato, R.S. Liu, N. Kamo (1994). Shape of the chromophore binding site in pharaonis phoborhodopsin from a study using retinal analogs. *Photochem. Photobiol.*, **60**, 388–393.
73. K. Nakanishi, R. Crouch (1995). Application of artificial pigments to structure determination and study of photoinduced transformations of retinal proteins. *Israel J. Chem.*, **35**, 253–272.
74. H.J. De Groot, S.O. Smith, J. Courtin, E. Van den Berg, C. Winkel, J. Lugtenburg, R.G. Griffin, J. Herzfeld (1990). Solid-state ^{13}C and ^{15}N NMR study of the low pH forms of bacteriorhodopsin. *Biochemistry*, **29**, 6873–6883.
75. A. Albeck, N. Livnah, H. Gottlieb, M. Sheves (1992). ^{13}C NMR studies of model compounds for bacteriorhodopsin: factors affecting the retinal chromophore chemical shifts and absorption maximum. *J. Am. Chem. Soc.*, **114**, 2400–2411.
76. B. Yan, J.L. Spudich, P. Mazur, S. Vunnam, F. Derguini, K. Nakanishi (1995). Spectral tuning in bacteriorhodopsin in the absence of counterion and coplanarization effects. *J. Biol. Chem.*, **270**, 29668–29670.
77. R.A. Mathies, L. Stryer (1976). Retinal has a highly dipolar vertically excited state: implications for vision. *Proc. Natl. Acad. Sci. USA*, **73**, 2169–2173.
78. S.L. Merbs, J. Nathans (1993). Role of hydroxyl-bearing amino acids in differentially tuning the absorption spectra of the human red and green cone pigments. *Photochem. Photobiol.*, **58**, 706–710.
79. A.B. Asenjo, J. Rim, D.D. Oprian (1994). Molecular determinants of human red/green color discrimination. *Neuron*, **12**, 1131–1138.
80. A. Der, S. Szara, R. Toth-Boconadi, Z. Tokaji, L. Keszthelyi, W. Stoeckenius (1991). Alternative translocation of protons and halide ions by bacteriorhodopsin. *Proc. Natl. Acad. Sci. USA*, **88**, 4751–4755.
81. T. Marti, S.J. Rosselet, H. Otto, M.P. Heyn, H.G. Khorana (1991). The retinylidene Schiff base counterion in bacteriorhodopsin. *J. Biol. Chem.*, **266**, 18674–18683.
82. T. Marti, H. Otto, S.J. Rosselet, M.P. Heyn, H.G. Khorana (1992). Anion binding to the Schiff base of the bacteriorhodopsin mutants Asp-85->Asn/Asp-212->Asn and Arg-82->Gln/Asp-85->Asn/Asp-212->Asn. *J. Biol. Chem.*, **267**, 16922–16927.
83. J. Sasaki, L.S. Brown, Y.S. Chon, H. Kandori, A. Maeda, R. Needleman, J.K. Lanyi (1995). Conversion of bacteriorhodopsin into a chloride ion pump. *Science*, **269**, 73–75.
84. T. Takahashi, B. Yan, P. Mazur, F. Derguini, K. Nakanishi, J.L. Spudich (1990). Color regulation in the archaebacterial phototaxis receptor phoborhodopsin (sensory rhodopsin II). *Biochemistry*, **29**, 8467–8474.

85. D.A. McCain, L.A. Amici, J.L. Spudich (1987). Kinetically resolved states of the *Halobacterium halobium* flagellar motor switch and modulation of the switch by sensory rhodopsin I. *J. Bacteriol.*, **169**, 4750–4758.
86. B. Yan, J.L. Spudich (1991). Evidence that the repellent receptor form of sensory rhodopsin I is an attractant signaling state. *Photochem. Photobiol.*, **54**, 1023–1026.
87. W. Marwan, S.I. Bibikov, M. Montrone, D. Oesterhelt (1995). Mechanism of photosensory adaptation in *Halobacterium salinarium. J. Mol. Biol.*, **246**, 493–499.
88. K.D. Olson, X.-N. Zhang, J.L. Spudich (1995). Residue replacements of buried aspartyl and related residues in sensory rhodopsin I: D201N produces inverted phototaxis signals. *Proc. Natl. Acad. Sci. USA*, **92**, 3185–3189.
89. K.-H. Jung, J.L. Spudich (1998). Suppressor mutation analysis of the sensory rhodopsin I/ transducer complex: Insights into the color-sensing mechanism. *J. Bacteriology*, **180**, 2033–2042.
90. B. Yan, T. Takahashi, R. Johnson, J.L. Spudich (1991). Identification of signaling states of a sensory receptor by modulation of lifetimes of stimulus-induced conformations: The case of sensory rhodopsin II. *Biochemistry*, **30**, 10686–10692.
91. E.N. Spudich, W. Zhang, M. Alam, J.L. Spudich (1997). Constitutive signaling of the phototaxis receptor sensory rhodopsin II from disruption of its protonated Schiff base-Asp73 salt bridge. *Proc. Natl. Acad. Sci. USA*, **94**, 4960–4965.
92. S. Subramaniam, M. Gerstein, D. Oesterhelt, R. Henderson (1993). Electron diffraction analysis of structural changes in the photocycle of bacteriorhodopsin. *EMBO J.*, **12**, 1–8.
93. R. Rao, D.D. Oprian (1996). Activating mutations of rhodopsin and other G protein-coupled receptors. *Annu. Rev. Biophys. Biomol. Struct.*, **25**, 287–314.
94. K.P. Hofmann, S. Jäger, O.P. Ernst (1995). Structure and function of activated rhodopsin. *Isr.J. Chem.*, **35**, 339–355.
95. K. Fahmy, F. Siebert, T.P. Sakmar (1995). Photoactivated state of rhodopsin and how it can form. *Biophys. Chem.*, **56**, 171–181.
96. E.N. Spudich, T. Takahashi, J.L. Spudich (1989). Sensory rhodopsins I and II modulate a methylation/demethylation system in *Halobacterium halobium* phototaxis. *Proc. Natl. Acad. Sci. USA*, **20**, 7746–7750.
97. V.J. Yao, E.N. Spudich, J.L. Spudich (1994). Identification of distinct domains for signaling and receptor interaction of the sensory rhodopsin I transducer, HtrI. *J. Bacteriol.*, **176**, 6931–6935.
98. E. Ferrando-May, B. Brustmann, D. Oesterhelt (1993). A C-terminal truncation results in high-level expression of the functional photoreceptor sensory rhodopsin I in the archaeon *Halobacterium salinarium. Mol. Microbiol.*, **9**, 943–953.
99. M.P. Krebs, E.N. Spudich, H.G. Khorana, J.L. Spudich (1993). Synthesis of a gene for sensory rhodopsin I and its functional expression in *Halobacterium halobium. Proc. Natl. Acad. Sci. USA*, **90**, 3486–3490.
100. T. Kitajima, U. Mukohata (1995). Cloning and sequencing of the HtrI gene from *Haloarcula vallismortis*: the HtrI is more similar to HtrI from *Halobacterium salinarium* than HtrII of *Haloarcula vallismortis*. In: *Bacterial Rhodopsins Structure, Function and Evolution* (pp. 153–157). Nagoya University International Symposium, Nagoya.
101. A. Brooun, W. Zhang, M. Alam (1997). Primary structure and functional analysis of the soluble transducer protein HtrXI in the archaeon Halobacterium salinarium. *J. Bacteriol.*, **179**, 2963–2968.
102. K.F. Storch, J. Rudolph, D. Oesterhelt (1999). Car: a cytoplasmic sensor responsible for arginine chemotaxis in the archaeon *Halobacterium salinarum. EMBO J.*, **18**, 1146–1158.
103. M.V. Milburn, G.G. Prive, D.L. Milligan, W.G. Scott, J. Yeh, J. Jancarik, D.E. Koshland Jr., S.H. Kim (1991). Three-dimensional structures of the ligand-binding domain of the

bacterial aspartate receptor with and without a ligand. *Science*, **254**, 1342–1347.
104. X.-N. Zhang, J.L. Spudich (1998). HtrI is a dimer whose interface is sensitive to receptor photoactivation and His-166 replacements in sensory rhodopsin I. *J. Biol. Chem.*, **273** 19722–19728.
105. K.-H. Jung, J.L. Spudich (1996). Protonatable residues at the cytoplasmic end of transmembrane helix-2 in the signal transducer HtrI control photochemistry and function of sensory rhodopsin I. *Proc. Natl. Acad. Sci. USA*, **93**, 6557–6561.
106. S. Hou, A. Brooun, H.S. Yu, T. Freitas, M. Alam (1998). Sensory rhodopsin II transducer HtrII is also responsible for serine chemotaxis in the archaeon *Halobacterium salinarum. J. Bacteriol.*, **180**, 1600–1602.
107. B. Perazzona, J.L. Spudich (1999). Identification of methylation sites and effects of phototaxis stimuli on transducer methylation in *Halobacterium salinarum. J. Bacteriol.*, **181**, 5676–5683.
108. R.A. Bogomolni, W. Stoeckenius, I. Szundi, E. Perozo, K.D. Olson, J.L. Spudich (1994). Removal of transducer HtrI allows electrogenic proton translocation by sensory rhodopsin I. *Proc. Natl. Acad. Sci. USA*, **91**, 10188–10192.
109. K.D. Olson, P. Deval, J.L. Spudich (1992). Absorption and photochemistry of sensory rhodopsin-I: pH effects. *Photochem. Photobiol.*, **56**, 1181–1187.
110. U. Haupts, E. Bamberg, D. Oesterhelt (1996). Different modes of proton translocation by sensory rhodopsin I. *EMBO J.*, **15**, 1834–1841.
111. B. Yan, E.N. Spudich, M. Sheves, G. Steinberg, J.L. Spudich (1996). Complexation of the signal transducing protein HtrI to unactivated sensory rhodopsin I and its effect on thermodynamics of deactivation. *J. Phys. Chem.*, **101**, 109–113.
112. A. Pulvermuller, K. Palczewski, K.P. Hofmann (1993). Interaction between photoactivated rhodopsin and its kinase: stability and kinetics of complex formation. *Biochemistry*, **32**, 14082–14088.
113. M.P. Krebs, E.N. Spudich, J.L. Spudich (1995). Rapid high-yield purification and liposome reconstitution of polyhistidine-tagged sensory rhodopsin I. *Protein Expression and Purification*, **6**, 780–788.
114. P. Rath, E.N. Spudich, D.D. Neal, J.L. Spudich, K.J. Rothschild (1996). Asp76 is the Schiff base counterion and proton acceptor in the proton translocating form of sensory rhodopsin I. *Biochemistry*, **35**, 6690–6696.
115. J. Sasaki, J.L. Spudich (1998). The transducer protein HtrII modulates the lifetimes of sensory rhodopsin II photointermediates. *Biophys. J.*, **75**, 2435–2440.
116. J. Sasaki, J.L. Spudich (1999). Proton circulation during the photocycle of sensory rhodopsin II. *Biophys. J.*, **77**, 2145–2152.
117. U. Haupts, C. Haupts, D. Oesterhelt (1995). The photoreceptor sensory rhodopsin I as a two-photon-driven proton pump. *Proc. Natl. Acad. Sci. USA*, **92**, 3834–3838.
118. O.Bousché, E.N. Spudich, J.L. Spudich, K.J. Rothschild (1991). Conformational changes in sensory rhodopsin I: Similarities and differences with bacteriorhodopsin, halorhodopsin, and rhodopsin. *Biochemistry*, **30**, 5395–5400.
119. P. Rath, K.D. Olson, J.L. Spudich, K.J. Rothschild (1994). The Schiff base counterion of bacteriorhodopsin is protonated in sensory rhodopsin I: Spectroscopic and functional characterization of the mutated proteins D76N and D76A. *Biochemistry*, **33**, 5600–5606.
120. X.-N. Zhang, J.L. Spudich (1997). His–166 is critical for active site proton transfer and phototaxis signaling by sensory rhodopsin I. *Biophysical J.*, **73**, 1516–1523.
121. C.A. Hasselbacher, J.L. Spudich, T.G. Dewey (1988). Circular dichroism of halorhodopsin: Comparison with bacteriorhodopsin and sensory rhodopsin I. *Biochemistry*, **27**, 2540–2546.
122. B. Perazzona, E.N. Spudich, J.L. Spudich (1996). Deletion mapping of the sites on the HtrI

transducer for sensory rhodopsin I interaction. *J. Bacteriol.*, **178**, 6475–6478
123. X.-N. Zhang, J. Zhu, J.L. Spudich (1999). The specificity of interaction of archaeal sensory rhodopsins with their cognate transducers is determined by the transmembrane helices. *Proc. Natl. Acad. Sci. USA*, **273**, 19722–19728.
124. M. Krah, W. Marwan, D. Oesterhelt (1994). A cytoplasmic domain is required for the functional interaction of SRI and HtrI in archaeal signal transduction. *FEBS Lett.*, **353**, 301–304.
125. J. Rudolph, D. Oesterhelt (1996). Deletion analysis of the *che* operon in the Archaeon *Halobacterium salinarium. J. Mol. Biol.*, **258**, 548–554.
126. J. Kirby, M.M. Saulmon, C.J. Kristich, G.W. Ordal (1999). CheY-dependent methylation of the asparagine receptor, McpB, during chemotaxis in *Bacillus subtilus. J. Biol. Chem.*, **274**, 11092–11100.
127. W. Marwan, W. Schaefer, D. Oesterhelt (1990). Signal transduction in *Halobacterium* depends on fumarate. *EMBO J.*, **9**, 355–362.
128. W. Marwan, D. Oesterhelt (1991). Light-induced release of the switch factor during photophobic responses of *Halobacterium salinarium. Naturwissenschaften*, **78**, 127–129.
129. M. Montrone, W. Marwan, H. Grunberg, S. Musseleck, C. Starostzik, D. Oesterhelt (1993). Sensory rhodopsin-controlled release of the switch factor fumarate in *Halobacterium salinarium. Mol. Microbiol.*, **10**, 1077–1085.
130. R. Barak, I. Giebel, M. Eisenbach (1995). The specificity of fumarate as a switch factor of the bacterial flagellar motor. *Mol. Microbiol.*, **19**, 139–144.
131. M. Eisenbach (1996). Control of bacterial chemotaxis. *Mol. Microbiol.*, **20**, 903–910.

Notes added in proof

1. Archaeal rhodopsin homologs have now been demonstrated in eubacteria as well as in eukaryotic microbes (O. Beja, L. Aravind, E.V. Koonin, M.T. Suzuki, A. Hadd, L.P. Nguyen, S.B. Jovanovich, C.M. Gates, R.A. Feldman, J.L. Spudich, E.N. Spudich, E.F. DeLong, (2000) Bacterial rhodopsin: evidence for a new type of phototrophy in the sea. *Science*, **289**, 1902-1906), and a comprehensive comparative analysis of microbial and higher animal retinylidene proteins has appeared (J.L. Spudich, C.-H. Yang, K.-H. Jung, E.N. Spudich, (2000) Retinylidene Proteins: Structures and Functions from Archaea to Humans. *Annual Reviews Cell & Dev. Biol.*, **16**, 365-392).
2. The complete sequence of the *Halobacterium* genome has been obtained and reveals 16 Htr transducers in addition to HtrI and HtrII (W.V. Ng, S.P. Kennedy, G.G. Mahairas, B. Berquist, M. Pan et al. (2000) Genome sequence of *Halobacterium* species NRC-1. *Proc. Natl. Acad. Sci. USA*, **97**, 12176-12181).
3. An intermediate-resolution crystallographic projection structure of SRII has been produced from photoactive 2-D crystals of the protein (E.R.S. Kunji, E.N. Spudich, R. Grisshammer, R. Henderson, J.L. Spudich (2000) Electron Crystallographic Analysis of Two-dimensional Crystals of Archaeal Sensory Rhodopsin II: A 6.9-Å Projection Structure. *J. Mol. Biol.,* in press). The structure shows that the helix positions match the 7-helix arrangement of the archaeal transport rhodopsins rather than that of the eukaryotic visual pigments. The structural similarity of SRII to the transport rhodopsins further supports models in which the transport and signalling mechanisms derive from the same retinal-driven changes in protein conformation.

Further support for this hypothesis has also been obtained from electron paramagnetic resonance spectroscopy of SRII (A.A. Wegener, I. Chizhov, M. Engelhard M, H.J. Steinhoff (2000) Time-resolved detection of transient movement of helix F in spin-labelled pharaonis sensory rhodopsin II. *J. Mol. Biol.* 2000, **301**, 881-891).

4. Laser-flash spectroscopic analysis of the SRI back-photoreaction revealed previously undetected species with maximal absorption near 410 nm and 550 nm (T.E. Swartz, I. Szundi, J.L. Spudich, R.A. Bogomolni (2000) New photointermediates in the two photon signaling pathway of sensory rhodopsin I. *Biochemistry*, **39**, 15101-15109). Therefore these species as well as S^{b}_{510} are candidates for two-photon-induced repellent signalling states of SRI.

Photomovement
D.-P. Häder and M. Lebert, editors.

Chapter 6

Photoactive yellow protein, a photoreceptor from purple bacteria

Wim Crielaard, Remco Kort and Klaas J. Hellingwerf

Table of contents

Abstract

In members of the Archaea positive and negative phototactic responses are mediated via retinal-containing sensory rhodopsin photoreceptors, according to a mechanism that is similar to the mechanism of enterobacterial chemotaxis. In Bacteria the situation is less well resolved, even though the accumulation of photosynthetic bacteria in a light spot is one of the most extensively studied tactic responses of prokaryotes. Only recently, however, has it been reported that in this family of organisms (i.e. in the purple- or proteobacteria) another type of phototactic response occurs: blue light, of physiological intensities, evokes a repellent response. The photoreceptor that presumably mediates this response is the Photoactive Yellow Protein (PYP), a member of the xanthopsins. This family of photoreceptors consists of 4-hydroxy-cinnamate containing proteins, for which rich detail concerning structure and function is available.

In this contribution we will review the structure and function of PYP, and the initial molecular genetic studies aimed to further characterize the signal transduction chain responsible for the photoresponses mediated through PYP.

6.1 Introduction

Besides twitching motility [1], with use of fimbriae, gliding motility (e.g. [2]), and floatation regulated by the buoyancy of the cell [3], flagella-based swimming is one of the ways that prokaryotes (i.e. members of the domains of the Archaea and the Bacteria; [4]) have developed to move towards a more optimal environment. The archaetype of this response is the chemotactic response in *Escherichia coli* (for reviews see [5–7]. Net migration of organisms through this type of response is caused by a random walk of short "runs", spaced by "tumbles" in which the flagella either rotate in opposite direction or pause. In this pattern the length of the runs is biased by the chemical and physical stimuli from the environment of the cell, through specific receptors, called methyl-accepting chemotaxis proteins. Tactic migration in which light signals are processed, in a mechanism that is very similar to the mechanism of chemotaxis in *E. coli*, occurs in representatives of the halophilic branch of the Archaea (for a recent review see [8]). Despite earlier discussions about terminology (see below), this process is now generally referred to as phototaxis; in many of these organisms both phototactic attractant and repellent responses can be discriminated. Both responses compete and are integrated at the level of the intracellular signal transduction machinery.

Light can be detected by its intensity, color and direction. Particularly the latter aspect, in the past, has led to the introduction of rather complicated terminology regarding the characterisation of tactic responses of micro-organisms. It is generally agreed upon that purple bacteria are only capable of sensing light-intensity and color. Recently, however, it was reported [9,10] that colonies of *Rhodospirillum centenum* are capable of sensing the **direction** of light, which therefore was claimed to be the first report of true phototaxis in prokaryotes [11]. To further substantiate this possibility, motion analysis of single cells is required to exclude that colony migration towards the light is a result of sensing differences in light intensity caused by shading within a colony, rather than sensing of the direction of illumination by individual cells. Recent studies with single cells exposed to a light beam emitted from an optical fiber, however, have indicated that *Rsp. centenum* accumulated uniformely in all parts of the beam, whereas the eukaryote *Chlamydomonas reinhardtii* was capable of swimming towards the light source, showing true phototaxis [12]. In this review, we will use the term "phototaxis" quite loosely, i.e. to refer to processes in which (individual) bacteria show a net migration, in response to changes in their ambient light climate. Both a positive (attractant) and a negative (repellent) phototactic response could be observed in *Rsp. centenum*, depending on the light intensity used [10]. At low light intensity a positive phototactic response was recorded, with a wavelength-dependence that suggested that this response is mediated through the photosynthesis pigments. The tactic response of *Rsp. centenum*, recorded at high light intensities, which caused the cells to migrate away from the light source, was elicited mainly by light in the wavelength region between 550 and 600 nm. Also in *Rsp. centenum* a gene cluster has been identified that encodes Che signal transduction component, that mediate phototaxis as well as chemotaxis [13].

In 1993 we reported a new type of light-induced repellent response in the halophilic purple-sulphur bacterium *Ectothiorhodospira halophila* [14]. The initial observation that led to these studies was that whereas in a light spot of red- or infrared light (i.e. light that can be absorbed by the photosynthesis machinery) cells of this species accumulate, a different response is observed with blue light. With light of the latter color cells accumulate rather at the edge of the spot, indicating that besides an attractant response (selectively elicited by red light), these cells additionally display a repellent response towards blue light. Subsequent motion analyses of *E. halophila* cells showed a relative increase in the number of reversals of swimming direction, upon a step-up in the intensity of incoming blue light, in the physiological range of light intensities. This response to blue light showed adaptation, with kinetics that are similar to the kinetics of adaptation in chemotaxis of enterobacteria. Because it was known at that time that a low-abundance, highly absorbing photoactive protein was present in *E. halophila* (i.e. photoactive yellow protein (PYP); see further below), the wavelength dependence of this blue-light response was subsequently investigated. These experiments revealed that light with a wavelength longer than 500 nm did not elicit a measurable increase in the probability of directional switching of the cells, whereas a maximal effect was observed with light of 440 nm (see Figure 1). So, in contrast to the attractant response in this [15] and other purple bacteria, this new repellent response is not dependent on the photosynthetic machinery as the primary photoreceptor. On the contrary, its wavelength dependence matches the absorption spectrum of PYP, which makes this latter protein the designated candidate for the photoreceptor of this new repellent response.

6.2 Structure and function of the xanthopsins

The first member of the xanthopsins (a family of photoreceptors containing 4-hydroxy-cinnamate) the photoactive yellow protein from *E. halophila* was discovered in 1985 by T.E. Meyer [16]. Subsequently, photoactive yellow proteins were isolated from *Rhodospirillum salexigens* and *Chromatium salexigens*, which are also halophilic phototrophic purple bacteria [17,18]. PYP from *E. halophila* is an 125 amino acids (14 kDa) water-soluble protein and displays a main absorption band at 446 nm

(ε_{max} = 45.5 mM^{-1} cm^{-1} [16]), due to its thiol-ester linked 4-hydroxy-cinnamate chromophore [19–21]. The three-dimensional structure of the ground-state of PYP is now available (in the databases) at 1.4 Å resolution [22] and the structure of the protein in solution has been obtained at high resolution with ^{1}H-NMR [23].

Although the chromophore in PYP has a completely different chemical structure, its photocycle strongly resembles that of the archaeal sensory rhodopsins (e.g. [24]). After absorption of a blue photon, PYP enters a cyclic chain of reactions (the photocycle, see Figure 2). In this photocycle the ground state (λ_{max} 446 nm, pG) is converted (within nanoseconds) into a red-shifted intermediate (λ_{max} 465 nm, pR), followed by the formation of a relatively long-lived intermediate (λ_{max} 355 nm, pB), and recovery of the ground state [25,26]. In the pB state the chromophore is protonated and isomerized to the *cis* configuration [27]. Very recently, using picosecond transient absorption spectroscopy, two new photocycle intermediates were identified between pG and pR [28], with 3 and 220 ps lifetimes and a λ_{max} of 510 nm. In pG the anionic chromophore is in hydrogen-bonding contact with the buried and protonated E46 and with Y42 [22]. Interestingly, the pKa of both the chromophore and of E46 are strongly shifted by the protein environment, to a lower and higher value, respectively. FTIR analyses [29,30] revealed that the hydrogen bond between the chromophore and E46 remains intact in pR at low temperature. Therefore, photo-isomerisation of the double bond of the chromophore most likely takes place by rotation across both the C7-C8-double bond and the C-S single bond that links the chromophore to the apoprotein (i.e, a two-bond

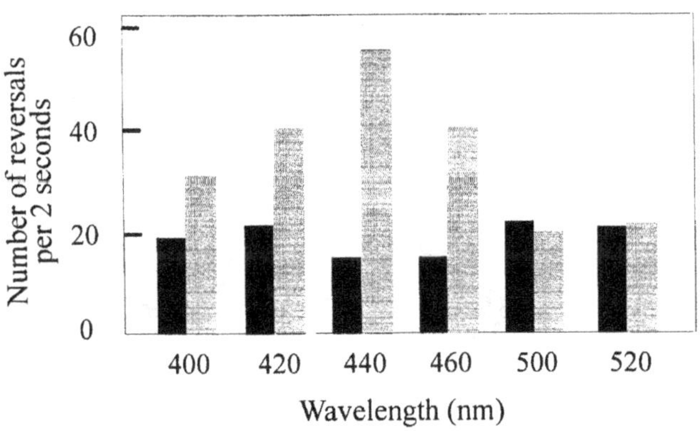

Figure 1. Wavelength dependence of the repellent response towards blue light in *Ectothiorhodospira halophila*. A cell suspension, containing a high proportion of highly motile bacteria, was observed in an anaerobic capillary with a microscope equipped with a video camera. Twenty four frames were recorded per second on a VHS videotape. Cells were observed in green light (>540 nm), saturating for photosynthesis. For the determination of the wavelength dependence of the (step-up) repellent response, cells were incubated for at least 30 min (anaerobically) in yellow-green light (negligible irradiation below 500 nm) in order for the cells to become adapted to this light regime. Side illumination (photostimuli) on the capillary was provided via an optical fiber. Narrow bandwidth (9 nm) interference filters (400 420, 440, 460, 500 and 520 nm) were used to select different wavelengths. The experiments were carried out on a single cell suspension within a period of 10 min. The total number of reversals during 2 s before (black bars) and after (gray bars) the step-up is plotted for each wavelength. Adapted from [14].

isomerization process, see also [31]). Subsequently, the long-living intermediate pB is formed. To achieve this, the chromophore must take up a proton, while E46 most likely deprotonates. In the pB state the protein (presumably) must significantly change its conformation, to generate a signal for the initiation of a phototactic response. Different techniques have yielded different estimates of the extent of this conformational transition. Time-resolved X-ray diffraction experiments with PYP crystals [32] have led to the conclusion that this conformational transition is mainly confined to the chromophore binding pocket. Spectroscopic and thermodynamic analyses of the photocycle transitions of PYP in solution [33], on the other hand, have led to a model that describes formation of the pB state as a partial unfolding of the protein, in which a much larger part of the protein is involved. Recently, this model has gained more evidence by an NMR study of the long-lived pB intermediate, which reveals a large degree of disorder in this intermediate [34].

Finally, the ground state of PYP (i.e. pG) is recovered after re-isomerisation and deprotonation of the chromophore and protein refolding. Thus, photo-isomerisation and proton transfer are essential also in the PYP photocycle, like in the photocycle of the sensory rhodopsins, which explains at least part of the strong similarity of their photochemical properties. It remains to be established whether the partial reactions in the recovery process of the pG state of PYP proceed independently or in one concerted reaction. The detailed characterization of the spatial structure of the signaling state of

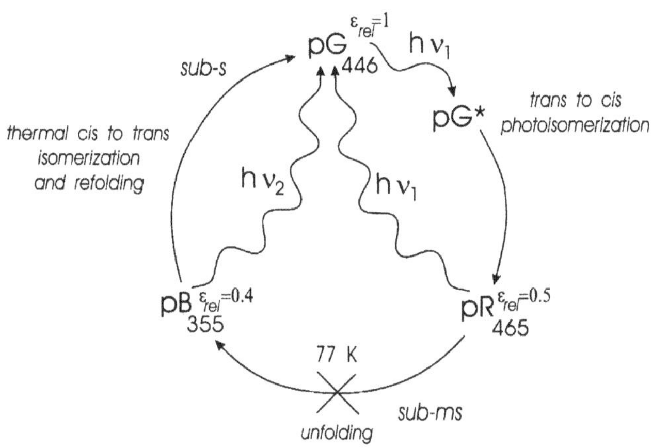

Figure 2. Schematic representation of the photocycle of the *Ectothiorhodospira halophila* photoactive yellow protein. pG, pR and pB are the ground state, the red-shifted and the blue-shifted intermediate of PYP, respectively. For each, the wavelength of maximal absorbance is indicated by a subscript; the relative extinction coefficient by a superscript. pG* is the excited state of PYP, formed after absorption of a blue photon. These intermediates interconvert through thermal (straight lines) or light-induced (wavy lines) reactions. The approximate time scale of the thermal reactions is indicated. The subscripts 1 and 2 for $h\nu$ refer to a blue and a UV photon, respectively. The photocycle of PYP is blocked at a temperature of 77 K or below, after formation of pR. The formation of pB is paralleled by a partial unfolding of the protein. Note that recently new photocycle intermediates have been identified between pG* and pR [28] and in the light-induced reaction between pB and pG (Hendriks and Hellingwerf, unpublished results).

PYP forms a major challenge for future research. The recently initiated time-resolved X-ray experiments, with extended time resolution [31], as well as the already mentioned ^{1}H-NMR experiments [34] are promising approaches towards this goal.

This biophysical characterization of PYP is greatly facilitated by the possibility to heterologously express (see below) and reconstitute PYP [35]. An additional advantage is that the reconstitution can also be carried out with a large range of chromophore analogues [36]. Using a so called "triple bond" (4-hydroxyphenylpropiolic acid) analogue and a "locked" chromophore (through reconstitution with 7-hydroxy-coumarin-3-carboxylic acid, in which a covalent bridge is present across the C7-C8-vinyl bond ([37], see the legend of Figure 3 for a detailed explanation), the latter possibility has provided further evidence that rotation of the carbonyl group of the thioester-linked chromophore of PYP (see above), is of critical importance for photoactivation of PYP. Other reconstitution experiments revealed that by varying the chemical structure of the chromophore of PYP a considerable (more than 40 nm) redshift of its absorbance maximum can be obtained [36,38]. However, such shifts are not sufficient to match the putative photoreceptor that mediates the negative phototactic response in *Rsp. centenum*. Consequently, it is not likely that a PYP homologue plays a role in this latter process.

6.3 Molecular genetics of PYP mediated responses in photosynthetic bacteria

Studies of the genes involved in negative phototaxis, mediated by a photoactive yellow protein, were initiated through the cloning of PYP from *E. halophila* [20,39]. Subsequently, cloning and sequencing of the gene encoding *Rsp. salexigens* PYP was also described by Kort et al. [39]. The latter PYP also contains, like the *E. halophila* PYP [39], the chromophore *trans* 4-hydroxy cinnamic acid, as could be demonstrated with high performance capillary zone electrophoresis. Additionally, evidence was presented [39] for the presence of a gene encoding a PYP homologue in *Rhodobacter sphaeroides*. Recently the genetic region encoding this gene has been cloned and sequenced ([40], see Figure 4). Sequence analyses showed that this *pyp* gene encodes a 124 amino acid protein with 48% identity to the other three known xanthopsins (see also Figure 5). Downstream from this *pyp* gene, a number of adjacent open reading frames were identified, including a gene encoding a CoA-ligase homologue (pCL). The basic structure of the chromophore binding pocket in *Rb. sphaeroides* PYP has been conserved, as was shown by analyzing its 3D structure, constructed by homology-based molecular modeling. In agreement with the presence of PYP in *Rb. sphaeroides*, we were able to detect the chromophoric group of PYP, 4-hydroxy cinnamic acid, in intact cells of *Rb. sphaeroides*. This chromophore could be isolated from phototrophically grown cells only [39]. These latter findings for *Rb. sphaeroides* are strongly reminiscent of the characteristics of the purple sulfur bacterium *Rhodospirillum salexigens*, in which protein-attached chromophore, as well as the PYP homologue, could only be identified in cells grown anaerobically in the light [39].

Previously, analysis of the flanking regions of the *E. halophila pyp* gene also showed an ORF (in this organism directly downstream of the *pyp* gene), encoding a CoA-ligase

A

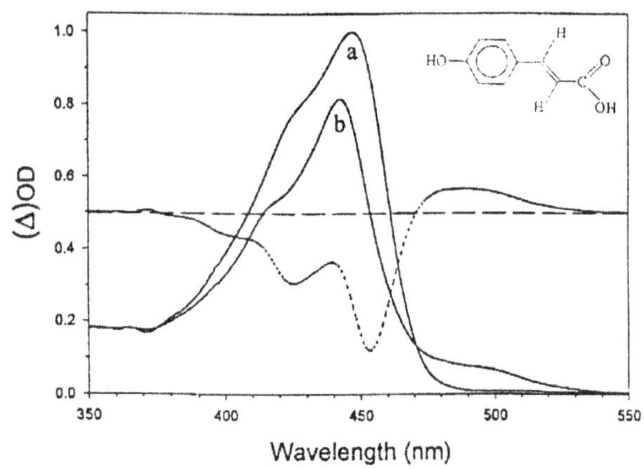
a
b
HO
H
H
C
O
OH
(Δ)OD
Wavelength (nm)

B

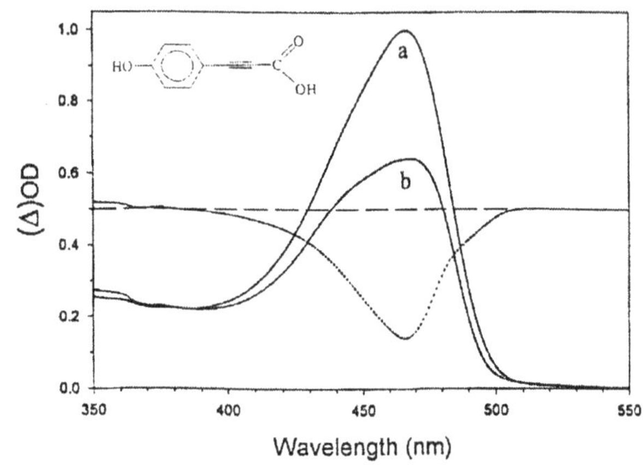
a
b
HO
C
O
OH
(Δ)OD
Wavelength (nm)

C

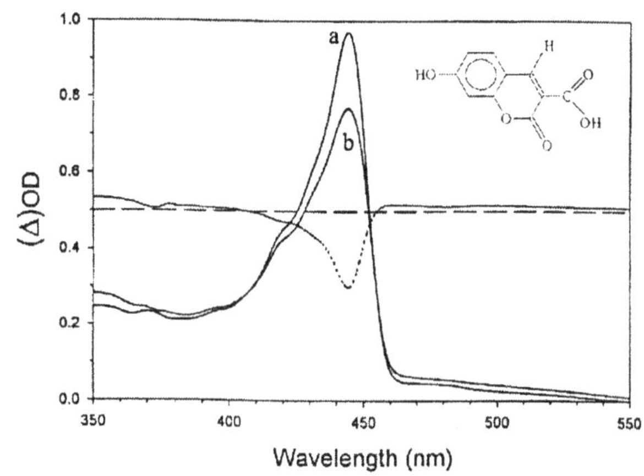
a
b
HO
H
C
O
OH
(Δ)OD
Wavelength (nm)

Figure 3. Light-induced formation of pR and pR-like intermediates in PYP reconstituted with the indicated chromophores (A: 4-hydroxy-cinnamic acid; the native chromophore; B: 4-hydroxyphenylpropiolic acid; the "triple bond" chromophore; C: 7-hydroxycoumarin-3-carboxylic acid the "locked" chromophore). Whereas PYP absorbs maximally at 446 nm the hybrids absorb maximally at 464 and 443 nm, respectively. Samples containing 20 μM PYP (hybrid) were frozen in 1 cm acrylic cuvettes in the dark. After recording the dark spectra (a traces) on a diode array spectrophotometer, each sample was illuminated in the cryostat for 20 min, to induce pR formation (b traces). The absorbance in both spectra was set to zero at 600 nm for background subtraction. The dotted line in each panel represents the difference spectrum between traces a (pG) and b (pR). Because the triple bond (panel B) cannot undergo *cis/trans* photoisomerization and because of the presence of the lock across the C7-C8-vinyl double bond in the other hybrid (panel C) it was expected that these two hybrids would not be able to photocycle. Clearly like the native protein (panel A), both are able. Both hybrids, upon photoexcitation, display authentic photocycle signals in terms of the presence of ns to μs intermediates; the "triple bond" hybrid, in addition, goes through a blue-shifted-like intermediate state (not shown), with very slow kinetics (adapted from [37]).

homologue, suggesting a plant-like conversion of 4-hydroxy cinnamic acid to its CoA derivative before linkage to apoPYP. The pathway of biosynthesis of 4-hydroxy cinnamic acid has been extensively studied in higher plants [41], but no information is available on the conservation of this pathway in *Rb. sphaeroides*, *E. halophila* or any other member of the Bacteria. In higher plants, the two enzymes of central importance in the metabolic conversions relevant for 4-hydroxy cinnamic acid are: phenylalanine ammonia lyase (PAL), which catalyses the reaction from either phenylalanine or tyrosine to 4-hydroxy cinnamic acid, and *p*-coumaryl:CoA ligase (pCL, or 4-hydroxy cinnamyl:CoA ligase), which activates 4-hydroxy cinnamic acid through a covalent coupling to CoA, via a thiol ester bond [41].

The *pyp* gene from *E. halophila* was used for heterologous overexpression in both *E. coli* and *Rb. sphaeroides*, aimed at the development of a holoPYP overexpression system. In both organisms the protein could be immunologically detected, but its yellow color was not observed [36,38]. Subsequent molecular genetic construction of a histidine-tagged version of PYP led to its 2500-fold overproduction in *E. coli* and allowed the rapid purification of the heterologously produced apoprotein. As already mentioned, holoPYP can be reconstituted by the addition of 4-hydroxy cinnamic acid anhydride to apoPYP as was first shown by Imamoto and co-workers [35].

From the preliminary analysis of the *pyp* gene cluster of *Rb. sphaeroides* (Figure 4), evidence for the presence of a *pyp* gene plus the pCL-based activation of the chromophore is now also available for a member of the α-group of the proteobacteria.

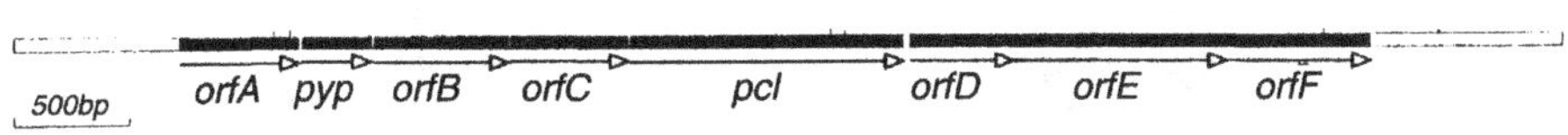

Figure 4. The *Rhodobacter sphaeroides* "*pyp* gene cluster". Schematic representation of the 6.5-kb DNA fragment from *Rb. sphaeroides*, containing Orfs A-F, *pyp* and *pcl* genes (see text for details, modified from [40]).

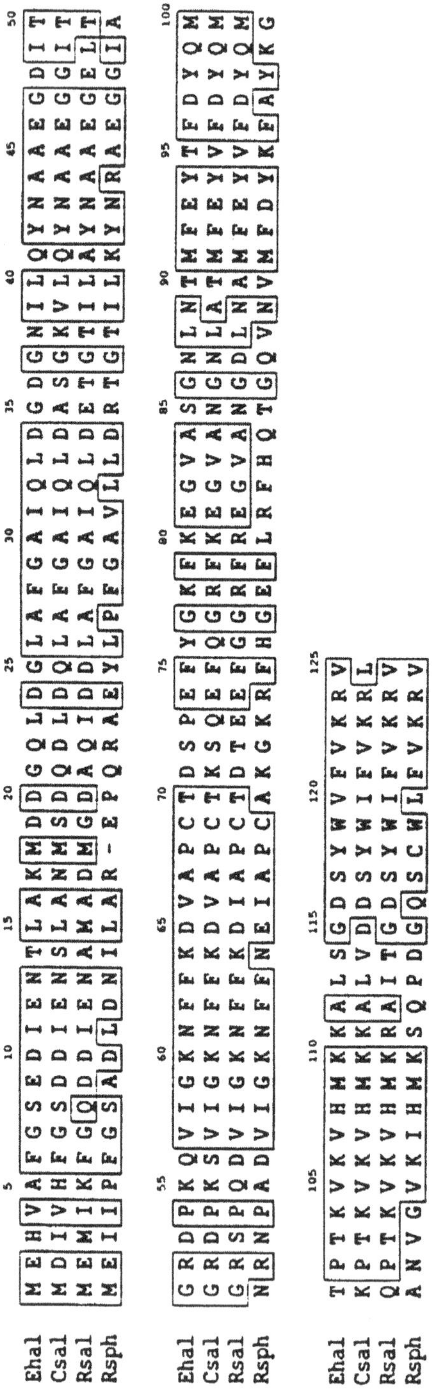

Figure 5. Multiple sequence alignment of the 4 known members of the xanthopsins. Abbreviations: Ehal = *Ectothiorhodospira halophila*, Csal = *Chromatium salexigens*, Rsal = *Rhodospirillum salexigens*, Rsph = *Rhodobacter sphaeroides*.

Genetic characterization of the PYP-initiated signal transduction pathway has been initiated and the identification of a "*pyp* gene cluster" in the genetically well characterized *Rb. sphaeroides*, will significantly facilitate this process.

6.4 Signal transduction

The signal transduction pathway from PYP to the flagella, largely remains to be uncovered. Most likely the blue-shifted intermediate pB is the signaling state of PYP, since it is by far the most stable transient intermediate, and it has a characteristic conformation change, allowing interactions with a hypothetical transducer (see [24] for a review). The amino acid residues important for signaling may be localized in a region of PYP, homologous to the PAS-core domain [42]. This latter domain, identified in PYP by Lagarias and co-workers [43] is found in many proteins throughout all three kingdoms, were it functions in signaling and signal transduction. PYP can be seen as the structural prototype for the three-dimensional fold of the PAS domain superfamily [42]. The homologous PAS-core region ranges from residues 29 to 69 and contains a number of conserved residues including three glycines (G37, G51 and G59) which may act as hinges in the dynamics of the protein in the processes of formation of the signaling state [44]. When it has been demonstrated conclusively that PYP is a photoreceptor involved in negative phototaxis, the assumption that the PYP transducer is a homologue of the methyl-accepting chemotaxis proteins, like the transducers of SRI and SRII (i.e. HtrI and HtrII) gains strength, since this would provide a straight forward mechanism for adaptation, as observed for free swimming *E. halophila* cells in their blue light response [14]. In that case, also further downstream, a two-component regulatory system may be involved, consisting of homologues of the kinase CheA and the response regulators CheY and CheB, just like in chemotaxis in *E. coli.*

6.5 Outlook

In the field of microbiology, besides the photoactive yellow protein, several new types of photoreceptor have been characterized during the past few years, in several classes of organisms. Examples are a rhodopsin [45] and a phytochrome [46] in cyanobacteria and the rhodopsins and flavin-containing photoreceptors in unicellular eukaryotes [47,48]. There will be a lot of excitation in the detailed characterization of the structure of these photoreceptors and in finding out their role in cellular physiology. It will be even more exciting to characterize those photoreceptors, for instance for tactic migration in gliding motility and for circadian synchronization, of which we so far only know that they exist.

References

1. J. Henrichsen (1983). Twitching motility. *Annu. Rev. Microbiol.*, **37**, 81–93.
2. D.-P. Häder (1987). Photosensory behaviour in prokaryotes. *Microbiol. Rev.*, **51**, 1–21.

3. J.F. Imhoff (1992). Taxonomy, phylogeny and general ecology of anoxygenic phototrphic bacteria. In: N.H. Mann, N.G. Carr (Eds), *Photosynthetic Prokaryotes* (pp. 53–92). Plenum Press, New York.
4. C.R. Woese (1987). Bacterial evolution. *Microbiol. Rev.*, **51**, 221–271.
5. J.P. Armitage (1992). Behavioural responses in bacteria. *Ann. Rev. Physiol.*, **54**, 683–714.
6. J.S. Parkinson (1993). Signal transduction schemes of bacteria. *Cell*, **73**, 857–871.
7. A.M. Stock, S.L. Mowbray (1995). Bacterial chemotaxis: a field in motion. *Curr. Opinion Struct. Biol.*, **5**, 744–751.
8. W.D. Hoff, K.-H. Jung, J.L. Spudich (1997). Molecular mechanism of photosignaling by archaeal sensory rhodopsins. *Ann. Rev. Biophys. Biomol. Struct.*, **26**, 221–256.
9. L. Ragatz, Z.-Y. Jiang, C.E. Bauer, H. Gest (1994). Phototactic purple bacteria. *Nature*, **370**, 104.
10. L. Ragatz, Z.-Y. Jiang, C.E. Bauer, H. Gest (1995). Macroscopic phototactic behaviour of the purple photosynthetic bacterium *Rhodospirillum centenum*. *Arch. Microbiol.*, **163**, 1–6.
11. H. Guest (1995). Phototaxis and other sensory phenomena in purple photosynthetic bacteria. *FEMS Microbiol. Rev.*, **16**, 287–294.
12. M.J. Sackett, J.P. Armitage, E.E. Sherwood, T.P. Pita (1997). Photoresponses of the purple non-sulfur bacteria Rhodospirillum centenum and Rhodobacter sphaeroides. *J. Bacteriol.*, **171**, 6271–6278.
13. Z.Y. Jiang, H. Gest, C.E. Bauer (1997). Chemosensory and photosensory perception in purple photosynthetic bacteria utilize common signal transduction components. *J. Bacteriol.*, **179**, 5720–5727.
14. W.W. Sprenger, W.D. Hoff, J.P. Armitage, K.J. Hellingwerf (1993). The eubacterium *Ectothiorhodospira halophila* is negatively phototactic, with a wavelength dependence that fits the absorption spectrum of the photoactive yellow protein. *J. Bacteriol.*, **175** 3096–3104.
15. E. Hustede, M. Liebergesell, H.G. Schlegel (1989). The photophobic response of various sulphur and nonsulphur purple bacteria. *Photochem. Photobiol.*, **50**, 809–815.
16. T.E. Meyer (1985). Isolation and characterization of soluble cytochromes, ferredoxins and other chromophoric proteins from the halophilic phototrophic bacterium *Ectothiorhodospira halophila*. *Biochim. Biophys. Acta*, **806**, 175–183.
17. T.E. Meyer, J.C. Fitch, R.G. Bartsch, G. Tollin, M.A. Cusanovich (1990). Soluble cytochromes and a photoactive yellow protein from the moderately halophilic purple phototrophic bacterium *Rhodospirillum salexigens*. *Biochim. Biophys. Acta*, **1016**, 364–370.
18. M. Koh, G. Van Driessche, B. Samyn, W.D. Hoff, T.E. Meyer, M.A. Cusanovich, J.J Van Beeumen (1996). Sequence evidence for strong conservation of the photoactive yellow proteins from the halophilic phototrophic bacteria *Chromatium salexigens* and *Rhodospirillum salexigens*. *Biochemistry*, **35**, 2526–2534.
19. W.D. Hoff, P. Düx, K. Hård, B. Devreese, I.M. Nugteren-Roodzant, W. Crielaard, R. Boelens, R. Kaptein, J. Van Beeumen, K.J. Hellingwerf (1994). Thiol ester-linked *p*-coumaric acid as a new photoactive prosthetic group in a protein with rhodopsin-like photochemistry. *Biochemistry*, **33**, 13959–13962.
20. M. Baca, G.E.O. Borgstahl, M. Boissinot, P.M. Burke, D.R. Williams, K.A. Slater, E.D. Getzoff (1994). Complete chemical structure of photoactive yellow protein: Novel thioester-linked 4-hydroxycinnamyl chromophore and photocycle chemistry, *Biochemistry*, **33**, 14369–14377.
21. W.D. Hoff, B. Devreese, R. Fokkens, I.M. Nugteren-Roodzant, J. Van Beeumen, N. Nibbering, K.J. Hellingwerf (1996). Chemical reactivity and spectroscopy of the thiol ester-

linked *p*-coumaric acid chromophore in the photoactive yellow protein from *Ectothiorhodospira halophila*. *Biochemistry*, **35**, 1274–1281.

22. G.E.O. Borgstahl, D.R. Williams, E.D Getzoff (1995). 1.4 Å Structure of photoactive yellow protein, a cytosolic photorecptor: unusual fold, active site and chromophore. *Biochemistry*, **34**, 6278–6287.
23. P. Düx, G. Rubinstenn, G.W. Vuister, R. Boelens, F.A.A. Mulder, K. Hård, W.D. Hoff, A. Kroon, W. Crielaard, K.J. Hellingwerf, R. Kaptein (1998). Solution structure and backbone dynamics of the photoactive yellow protein. *Biochemistry*, **37**, 12689–12699.
24. K.J. Hellingwerf, W.D. Hoff, W. Crielaard (1996). Photobiology of microorganisms: how photosensors catch a photon and use it to initialize signalling. *Mol. Microbiol.*, **21**, 683–693.
25. T.E. Meyer, E. Yakali, M.A. Cusanovich, G. Tollin (1987). Properties of a water soluble, yellow protein isolated from a halophilic phototrophic bacterium that has photochemical activity analogous to sensory rhodopsin. *Biochemistry*, **26**, 418–423.
26. W.D. Hoff, I.H.M. Van Stokkum, H.J. Van Ramesdonk, M.E. Van Brederode, A.M. Brouwer, J.C. Fitch, T.E. Meyer, R. Van Grondelle, K.J. Hellingwerf (1994). Measurement and global analysis of the absorbance changes in the photocycle of the photoactive yellow protein from *Ectothiorhodospira halophila*. *Biophys. J.*, **67**, 1691–1705.
27. R. Kort, H. Vonk, X. Xu, W.D. Hoff, W. Crielaard, K.J. Hellingwerf (1996). Evidence for *trans-cis* isomerization of the *p*-coumaric acid chromophore as the photochemical basis of the photocycle of photoactive yellow protein. *FEBS Lett.*, **382**, 73–78.
28. L. Ujj, S. Devanathan, T.E. Meyer, M.A. Cusanovich, G. Tollin, G.H. Atkinson (1998). New photocycle intermediates in the photoactive yellow protein from *Ectothiorhodospira halophila*: picosecond transient absorption spectroscopy. *Biophys. J.*, **75**, 406–412.
29. A. Xie, W.D. Hoff, A.R. Kroon, K.J. Hellingwerf (1996). Glu46 donates a proton to the 4-hydroxycinnamate anion chromophore during the photocycle of photoactive yellow protein. *Biochemistry*, **35**, 14671–14678.
30. Y. Imamoto, K. Mihara, O. Hisatomi, M. Kataoka, F. Tokunaga, N. Bojkova, K. Yoshihara (1997). Evidence for proton transfer from Glu-46 to the chromophore during the photocycle of photoactive yellow protein. *J. Biol. Chem.*, **272**, 12905–12908.
31. A. Perman V. Srajer, Z. Ren, T. Teng, C. Pradervand, T. Ursby, D. Bourgeois, F. Schotte, M. Wulff, R. Kort, K.J. Hellingwerf, K Moffat (1998). Energy transduction on the nanosecond time scale: early structural events in a xanthopsin photocycle. *Science*, **279**, 1946–1950.
32. U.K. Genick, G.E.O. Borgstahl, K. Ng, Z. Ren, C. Pradervand, P.M. Burke, V. Srajer, T.-Y. Teng, W. Schildkamp, D.E. McRee, K. Moffat, E.D. Getzoff (1997). Structure of a protein photocycle intermediate by millisecond time-resolved crystallography. *Science*, **275**, 1471–1475.
33. M.E. Van Brederode, W.D. Hoff, I.H.M, Van Stokkum, M.-L.Groot, K.J. Hellingwerf (1996). Protein folding thermodynamics applied to the photocycle of the photoactive yellow protein. *Biophys. J.*, **71**, 365–380.
34. G. Rubinstenn, G.W. Vuister, F.A Mulder, P.E. Düx, R. Boelens, K.J. Hellingwerf, R. Kaptein (1998). Structural and dynamic changes of photoactive yellow protein during its photocycle in solution. *Nature Struct. Biol.*, **5**, 568–570.
35. Y. Imamoto, T. Ito, M. Kataoka, F. Tokunaga (1995). Reconstitution photoactive yellow protein from apoprotein and *p*-coumaric acid derivates. *FEBS Lett.*, **374**, 157–160.
36. A. Kroon, W.D. Hoff, H. Fennema, J. Gijzen, G.-J. Koomen, J.W. Verhoeven, W. Crielaard, K.J. Hellingwerf (1996). Spectral tuning, fluorescence and photoactivity in hybrids of photoactive yellow protein, reconstituted with native and modified chromophores. *J. Biol. Chem.*, **271**, 31949–31956.

37. R. Cordfunke, R. Kort, A. Pierik, B. Gobets, G.J. Koomen, J.W. Verhoeven, K.J. Hellingwerf (1998). Trans/cis (Z/E) photoisomerization of the chromophore of photoactive yellow protein is not a prerequisite for the initiation of the photocycle of this photoreceptor protein. *Proc. Natl. Acad. Sci. USA*, **95**, 7396–7401.
38. S. Devanathan, U.K. Genick, E.D. Getzoff, T.E. Meyer, M.A. Cusanovich, G. Tollin (1997). Preparation and properties of a 3,4-dihydroxycinnamic acid chromophore variant of the photoactive yellow protein. *Arch. Biochem. Biophys.*, **340**, 83–89.
39. R. Kort, W.D. Hoff, M. van West, A.R. Kroon, S.M. Hoffer, K.H. Vlieg, W. Crielaard, J.J. van Beeumen, K.J. Hellingwerf (1996). The Xanthopsins: a new family of eubacterial blue-light photoreceptors. *EMBO J.*, **15**, 3209–3218.
40. R. Kort, M.K. Phillips-Jones, D.M. van Aalten, A. Haker, S.M. Hoffer, K.J. Hellingwerf, W. Crielaard (1998). Sequence, chromophore extraction and 3-D model of the photoactive yellow protein from *Rhodobacter sphaeroides*. *Biochim. Biophys. Acta*, **11**, 1–6.
41. K. Hahlbrock, D. Scheel (1989). Physiology and molecular biology of phenylpropanoid metabolism. *Annu. Rev. Plant Physiol. Plant Mol. Biol.*, **40**, 347–369.
42. J.-L. Pellequer, K.A. Wager-Smith, S.A. Kay, E.D. Getzoff (1998). Photoactive yellow protein: A structural prototype for the three-dimensional fold of the PAS domain super family. *Proc. Natl. Acad. Sci. USA*, **95**, 5884–5890.
43. D.M. Lagarias, S.-H. Wu, J.C. Lagarias (1995). Atypical phytochrome gene structure in the green alga *Mesotaenium caldariorum*. *Plant Mol. Biol.*, **29**, 1127–1142.
44. D.M.F. Van Aalten, W.D. Hoff, J.B.C. Findlay, W. Crielaard, K.J. Hellingwerf (1998). Concerted motions in the photoactive yellow protein. *Protein Engineering*, **11**, 873–879.
45. J.H. Geerdink, A. Haker, H.C.P. Matthijs, W.D. Hoff, K.J. Hellingwerf, L.R. Mur, A rhodopsin as photoreceptor in chromatic adaptation of the cyanobacterium *Calothrix* sp. In: P. Mathies et al. (Eds), *Photosynthesis: from Light to Biosphere* (Vol. I, pp. 303–306). Kluwer Academic Publishers, Dordrecht, The Netherlands.
46. J. Hughes, T. Lamparter, F. Mittmann, E. Hartmann, W. Gärtner, A. Wilde, T. Börner (1997). A prokaryotic phytochrome. *Nature*, **386**, 663.
47. V. Campuzano, P. Galland, M.I. Alvarez, A.P. Eslava (1996). Blue-light receptor requirement for gravitropism, autochemotropism and ethylene response in *Phycomyces*. *Photochem. Photobiol.*, **63**, 686–694.
48. J. Saranak, K.W. Foster (1997). Rhodopsin guides fungal phototaxis. *Nature*, **387**, 465–466.

Photomovement
D.-P. Häder and M. Lebert, editors.

Chapter 7

Light perception and signal modulation during photoorientation of flagellate green algae

Georg Kreimer

Table of contents

Abstract

Many flagellate green algae possess a single specialized optical device, the eyespot apparatus, for detecting light direction and intensity. Despite the great variety in movement patterns and cell shapes observed in green algae, the design principles of their eyespot apparatuses are similar and produce a highly directional photoreceptor. In conjunction with helical motion the eyespot functions as a combined absorbance screen/ interference reflector producing a modulated signal at the location of the retinal-based photoreceptor. It carries information about the orientation of the cell relative to the stimulus direction and the light intensity. Molecular dissection of the different components of this system has recently started. However, compared to other visual systems, little is yet known about structural components and the signaling cascade(s) initiated upon photoreceptor excitation. This review covers general aspects (location, ultrastructure, interaction with microtubular flagellar roots, phylogeny) and discusses functions of the eyespot apparatus in signal perception and modulation. Additionally, recent progress in identification of putative signaling and structural elements will be summarized.

7.1 Introduction

Since the first microscopic studies of algae by Leuwenhoek and Hooke, the peculiar photobehavior of many flagellate algae has attracted the interest of scientists from different fields. Yet, little is known about the mechanisms facilitating and regulating these fine-tuned and highly adaptable basic algal movement responses. In principle, the accumulation or dispersal of flagellate algae from certain environmental areas can result from three types of photobehavior [1]:

1. Photokinesis, in which the steady-state swimming velocity or the frequency of directional changes is affected by the light intensity.
2. Photophobic or photoshock responses to sudden changes in light intensity. These are transient, stereotyped alterations in motion in response to light stimuli exceeding certain threshold values. Often they involve a short reversal of the direction of movement due to a change in the flagellar beat type, and rapid adaptational processes can occur. Like kinesis this response is independent of the direction of light.
3. Phototactic responses, which depend on both intensity and direction of the light stimulus. Orientation both towards (positive phototaxis) and away from the light source (negative phototaxis) are exhibited by the cells. Phototaxis often results from short, subtle differences in the beat of the flagella without a general change of the flagellar beating mode.

The basic concepts of algal photoorientation, postulating a photoreceptive organelle and signal transducing and processing chains, which couple the flagellar responses to the event of photon absorption, have already been put forward early (for review see Haupt, this volume). The evolving concepts were closely coupled to advances in the understanding of the motion patterns of the cells and the interpretation of the function of an conspicuous "organelle", the eyespot or stigma. Already Ehrenberg [2] noted that most flagellate algae exhibiting photoorientation possess a red to orange pigmented spot, which he termed eyespot. He suggested that the eyespot is the photosensitive organelle involved in the different photoresponses. Later in the 19th and up to the middle of the 20th century this function of the eyespot was controversially discussed by e.g. Strasburger, Engelmann, Mast, Buder and Halldal. Now at the end of the 20th century, evidence that the eyespot, as it is seen in the light microscope, is not the location of the photoreceptor(s) in the different algal groups is overwhelming. Electron microscopy has revealed that the most conspicuous components of the eyespot sensu stricto are carotenoid-rich lipid globules. These are the major light-modifying structures in complex multicomponent "organelles", termed eyespot apparatuses, which enable the cells to determine the light direction and are therefore often compared to directional antennas. Their function is closely adapted to the rotational movement observed in many flagellates [for review see 3–9].

Conspicuous eyespot apparatuses are present in all major algal phylogenetic lineages, reflecting the importance of a precise photobehavior for algae. However, the origin of eyespots is, in contrast to that of chloroplasts, generally regarded as polyphyletic. Consequently, they differ in several features; e.g. ultrastructure and spatial relationship to the chloroplasts and flagellar apparatus has led early to their grouping in several classes [3]. Also the photoreceptor localizations apparently differ. For example, in green algae the photoreceptor of phototactic and photophobic responses is localized most likely in the membranes overlying the eyespot, whereas in *Euglena* the paraxonemal body is the most probable localization (see chapters by Deininger and Hegemann, Gualtieri, and Lebert, this volume). However, algal eyespot apparatuses share also many features, which probably reflect minimal functional demands of singular structures used for detecting direction and intensity of light. In general, they must operate over a wide range of light intensities and incidences, and they must discriminate sufficiently between the wavelengths used by the photoreceptor(s) and a fluctuating, diffuse background illumination in the aqueous environment. In addition, the whole light detecting and signaling system must be adapted to the general locomotion pattern of the cells in order to obtain signals containing directional information. This has apparently favored the evolution of light modifying structures, which enhances the contrast at the photoreceptor location in different ways [4,7]. Their optimization was probably also strongly influenced by the changed adaptive requirements after acquisition of chloroplasts in the different algal groups. Molecular phylogenetic evidence supports the view that these endosymbiotic events have happened several times in the algal lineages [10]. The observation that algae often retained their photobehavior even after loss of photosynthesis underlines the importance of gaining the ability to detect the direction and quality of light. In some dinophytes the eyespot is even the only remnant of a first endosymbiont after acquisition of a new photosynthetic endosymbiont. Except for the Dinophyta, where different types of eyespots are found, only one characteristic type of eyespot apparatus is observed in each group of the major algal lineages [3,9–11]. Functional comparisons should therefore be mainly restricted to phylogenetically related phyla and care should be taken when generalizing signaling mechanisms.

The currently best studied group regarding both eyespot and flagella function during photoresponses are the green algae. This review is restricted to the eyespot of unicellular green algae and their photoresponses. For recent comprehensive reviews on motility and photoorientation in multicellular Volvocales see Hoops [12] and Kirk [13].

7.2 Location and structure of the eyespot apparatus

The functional green algal eyespot is ultrastructurally complex and involves local specializations from different compartments (see below). The term eyespot apparatus has therefore been introduced to differentiate between the whole light receiving and modulating "organelle", the eyespot sensu stricto (i.e. the plate(s) of carotenoid-rich lipid globules, which modulates the light signal), and the eyespot membranes overlying the lipid globules [5,14]. The needs of a light detecting "organelle" in a unicellular organism for functioning as a directional antenna in conjunction with cell rotation during locomotion are perfectly reflected by both, ultrastructure and position of the eyespot apparatus. Although the basic structure of green algal eyespot apparatuses is very similar, different specializations with respect to ultrastructure, localization and association with microtubular flagellar roots are well documented. Since also different swimming behaviors are known from green algae [15,16], any basic model of phototaxis and signal modulation by their eyespot apparatus should at least not be in contradiction with the different specializations of this "organelle" described below and the general moving patterns.

7.2.1 Location and general features

The eyespot sensu stricto is located within a chloroplast and exhibits no close association with a flagellum (type A eyespots according to Dodge [3]). In most species it is easily seen in the light microscope due to the high carotenoid content of the lipid globules (Figures 1, 2). These are specifically enriched in β,β-carotene, β,ψ-carotene and lycopene [17,18]. Absorption spectra in aqueous media of fractions enriched in intact eyespot apparatuses or eyespot globules from *Spermatozopsis similis* exhibit a strong absorption up to 550 nm ([19], Renninger, Backendorf, Kreimer and Planta, in press, Figure 7). As will be outlined below (see Section 7.5) this specialized pigmentation is important for the light-modifying properties of the eyespot. Usually only one eyespot is present in each cell. When multiple eyespots occur, they are in close proximity forming apparently a functional unit (e.g. in *Pyramimonas* species, [5,20]), or they are oriented along the axis of the flagellar beat (e.g. zoospores of *Microthamnion kuetzingianum*, [21]). Two eyespots can, however, be observed in cells prior/during cytokinesis. In green algae the eyespot can be distributed either by division, in a semiconservative fashion or de novo synthesis. Complex eyespot/cytoskeletal interactions prior to/during basal body segregation and cell division are often intimately involved in this process [5,7,22,23]. The eyespot covers an area between 0.3 to 10 μm^2. Eyespot to cell body surface relations strongly depends on the total size of the alga and its eyespot. In general, a considerable surface area appears to be used by flagellate green algae for detecting the direction and intensity of light. In e.g. *Chlamydomonas reinhardtii* (strain g1, [24]) the average eyespot area is 1.3 μm^2. This represents about

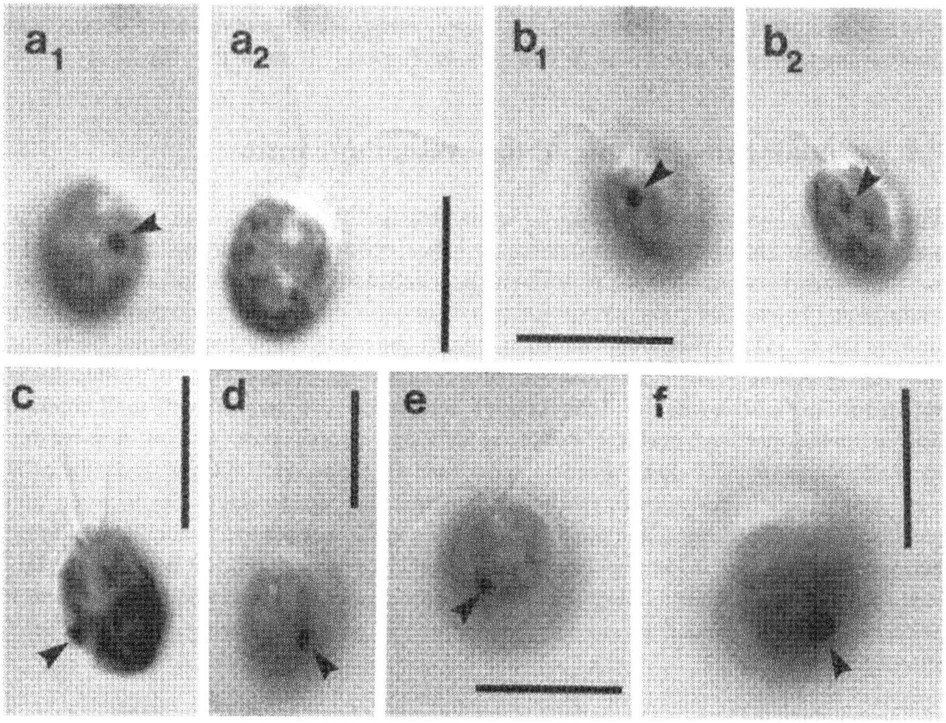

Figure 1. Differential interference contrast images of different strains of *Chlamydomonas reinhardtii* demonstrating the variability of eyespot apparatus positions. All strains exhibit strong negative phototactic orientation. (a–c) Strain 806; 1 and 2 denote different optical sections from the same cell. (d–e) Gametes of strain 595 and (f) vegetative cell of strain 593. Living cells were immobilized and imaged as described in [26]. Arrow head, eyespot apparatus. Scale bars = 10 μm.

1% of the cell body surface. In gametes of *Chlamydomonas* the eyespot is often even slightly larger than in the vegetative cells. Some species posses very large eyespots, e.g. the prasinophyte *Tertaselmis desikacharyi* [25]. Here the eyespot area can reach 7 μm^2. Based on a simplified ellipsoidal cell shape thus about 2% of the cell surface are covered by the light detecting "organelle" in this alga. This relatively large surface area used for

Figure 2. Differential contrast images of *Haematococcus pluvialis* (a–d) and *Spermatozopsis similis* (e–g). Note the different shapes and positions of the eyespot apparatuses (arrow heads) within the chloroplast and with respect to the flagellar apparatus. Note also protrusion of the eyespot apparatus of *H. pluvialis* (c) beyond the cell surface. Large arrows in (a, b) point to the flagella. Small arrows in (c, d) point to the flagellar canals in the thick wall of *H. pluvialis*. *H. pluvialis* cells were imaged in a living, immobilized state, whereas *S. similis* was fixed and imaged without immobilization. In ($a_{1,2}$) and ($g_{1,2}$) different focus planes of the same cells are shown. Scale bars = 10 μm.

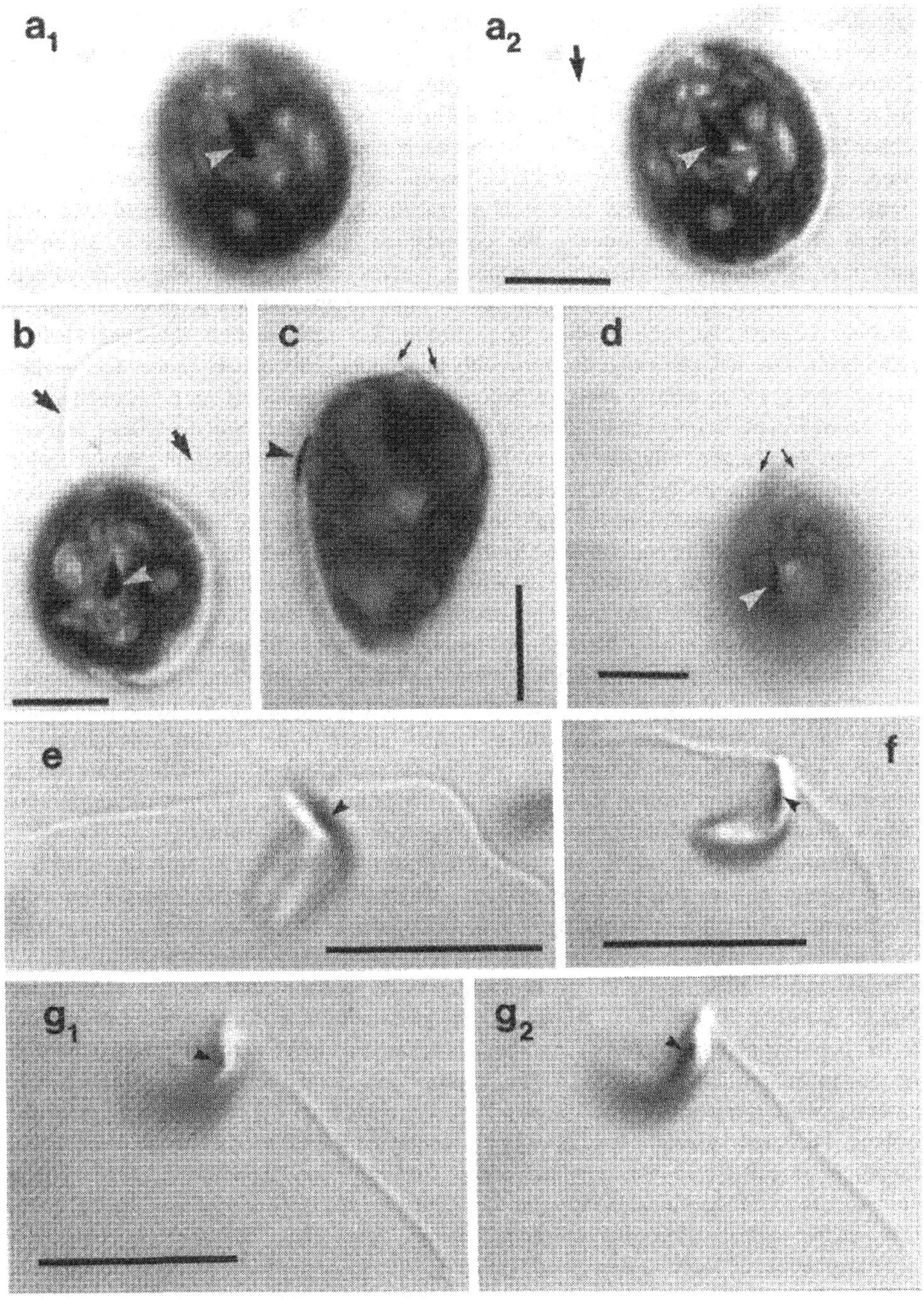
a1
a2
b
c
d
e
f
g1
g2

light detection might well explain the extremely high light sensitivity of some green algae, which is additionally increased by the presence of a "reflector" behind the photoreceptor [4,7,26]. Besides the area of the eyespot apparatus, its overall shape, surface geometry and location within the cell affects its properties as a directional light sensor (see Section 7.5). Its shape differs between species and ranges from ovoid (most often, e.g. *C. reinhardtii*, Figure 1) to comma-shaped (e.g. *Haematococcus spp.* Figure 2a–c). The longitudinal axis is often parallel to the cells longitudinal axis. The eyespot is always located towards the cell surface, most often roughly in a central portion of the chloroplast. However, in some species (e.g. *S. similis*, Figure 2e–g) it is located in a lobe of the chloroplast close to the flagellar apparatus. Eyespots can slightly protrude beyond the cell surface (e.g. Figure 2c) and exhibit different surface geometries. The surfaces range from roughly straight/slightly convex (many *Chlamydomonas* spp.) to complex surfaces with concave depressions (e.g. *Hafniomonas reticulata*, Figure 9a). Specializations in form and surface structure are often observed in species with larger cells. The functional significance of these different geometries for photobehavior has not yet been studied in detail. However, its impact on the efficiency of detecting and modulating the light signal may be significant, especially when the tracking direction of the cell deviates from the direction of light incidence (see Section 7.5). For instance, the probability of absorption of oblique light rays by dichroic-oriented receptors will be enhanced in curved, elongated protruding eyespot apparatuses like that of *Haematococcus*. This should theoretically favor the acquisition of the light source from any initial orientation and should allow a very efficient tracking. Also the overall position of the eyespot apparatus in the cell is important for these properties. Within the cell the eyespot apparatus is located either in the anterior, median (most often) or posterior region. Although its position is characteristic for each species, some variability is observed within a population; e.g. in cultures of *C. reinhardtii* (strain 806) cells with eyespots in all three positions can be observed (Figure 1a–c).

As summarized in Figure 3, eyespot position also differs in relation to the insertion of the flagella and, as will be described later, consequently also in their association with flagellar roots. In a few prasinophytes (e.g. *Mesostigma viride*) and *Pedinomonas* spp. the eyespot apparatus is localized opposite of the flagellar insertion (Figure 3a,b), whereas in most green algae the eyespot apparatus is localized more or less lateral to the flagella insertion (Figure 3a,c). However, irrespective of their localization with respect to the flagellar insertion eyespots are always located roughly perpendicular to the axis of cell movement (Figure 3, [5]). This is important for its function as a directional antenna, which performs spatial scans of the environment [4]. In addition, when the eyespot is located laterally it is always closer to one flagellum than to the other. This flagellum is called *cis*, whereas the other is called *trans* flagellum [27,28]. Where studied, also a defined location in relation to the plane of flagellar beat was resolved for the eyespot apparatus. Most commonly, when seen from the front, a clockwise displacement by 20–45° is described [5,22,29].

Many algae exhibit helical swimming and, where analyzed, rotation of the cell body is reported irrespective whether the cells are uni-, bi- or quadri-flagellate (for reviews see [15,16,30]). *C. reinhardtii,* for instance, rotates counterclockwise around its longitudinal axis when seen from behind and one helical turn corresponds roughly to one cell rotation [29,31]. Thus, in principal, the eyespot apparatus can either face towards the inside or the outside of the helix. Both situations have been reported for *C. reinhardtii* [28,31]. How can this discrepancy be explained? When analyzing a moving cell different velocities must be considered; e.g. *Chlamydomonas* as it moves on a helix has both a translational (linear) and a rotational (angular) velocity. A change in the orientation of a cell surface point, e.g. the eyespot apparatus, of a helically moving cell with respect to the axis of the helix can result from a change in the direction of either of these velocities with respect to the cell body [32]. For *Chlamydomonas* the direction of the translational velocity with respect to the cell body is roughly constant when swimming in the breaststroke mode. Crenshaw, based on the experimental data from different studies [28,29,31,33], suggested that *C. reinhardtii* might change the rotational

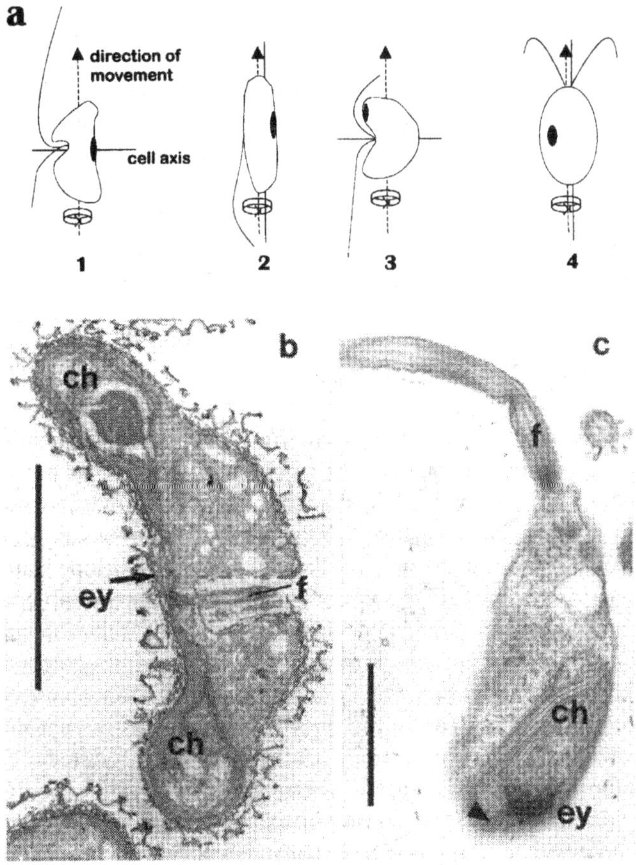

Figure 3. (a) Schematic drawing of the basic different positions of green algal eyespot apparatuses with respect to the direction of movement and the cell axis. 1 *Mesostigma viride*, 2 *Pediomonas* spec., 3 *Nephroselmis olviaceae*, 4 *Chlamydomonas*-type cell. (b, c) Electron micrographs of *M. viride* (b) and *Spermatozopsis similis* (c). Note the close hexagonal packing of the eyespot lipid globules. ey eyespot apparatus, ch chloroplast. Arrow head in (c) microtubular flagellar root. Scale bars: b 5 μm, c 1 μm. Figure (a) is redrawn with modifications after Melkonian and Robenek [5].

velocity with respect to its body. He suggested a model in which unicellular organisms moving on helical paths modulate components of their rotational velocity as functions of the stimulus intensity and thereby orient themselves in the stimulus field [34]. In permeabilized cell models of *C. reinhardtii* eyespot position on the helical path changes presumably as a consequence of the Ca^{2+}-dependent change in flagellar dominance. At 10^{-9} M Ca^{2+} it is located at the outer edge and at 10^{-7} M Ca^{2+} at the inner edge of the helix [28]. Assuming that these and the changes in flagellar beat of the *cis*- and *trans*-flagellum observed by Rüffer and Nultsch [33] upon photostimulation indeed alter the direction of the cell's rotational velocity, Crenshawś model might well explain the differences observed in the eyespot orientations relative to the axis of the helix. They might reflect different internal free Ca^{2+}-concentrations induced by distinct light intensities used during the recordings. Rüffer and Nultsch [31] point to the often observed disturbance of a smooth swimming path by the strong light necessary for high speed filming in their analysis. Recently, also Schaller et al. [35] suggested that eyespot position might change due to altered flagellar dominance. They additionally suggested that the eyespot is either raked outwards/backward (positive phototaxis) or inward/forward (negative phototaxis) relative to the swimming direction and can thus gain an additional directivity/shielding component (see Section 7.5). However, although these hypotheses are quite intriguing, in general too little is yet known about the directivity of the eyespot apparatus with this respect. Here closer analysis of more species during positive and negative phototactic orientation are clearly needed.

7.2.2 Ultrastructure of the eyespot apparatus

The functional green algal eyespot apparatus involves local specializations from two different subcellular compartments [5]. The most conspicuous part are one to several layers of carotenoid-rich lipid globules within the chloroplast, which are often associated with a thylakoid (the eyespot sensu stricto, Figure 4a,d,e, Figure 9a). They are closely linked to specialized areas of the chloroplast envelope and the adjacent plasma membrane, which are called eyespot membranes. Eyespot apparatuses can be isolated as structurally intact units [19] pointing to a rather stable linkage between the different components. None of the structural proteins of the green algal eyespot apparatus has been identified yet. It is also not known how association and the conspicuous constant spacing of the different constituents of the eyespot apparatus is facilitated. Although complex interactions with cytoskeletal elements do occur (see below), protein/protein and protein/lipid interactions of eyespot specific components will most likely dominate. Recently, the identification of some of these components has started. Both, biochemical approaches (isolation and characterization of eyespot globules from *S. similis,* Renninger, Backendorf, Kreimer and Planta, in press) and molecular biological approaches (e.g. analysis of the different *ey* mutant strains of *C. reinhardtii*; Roberts and Dieckmann, personal communication) will hopefully lead to the identification of some of these proteins.

Although closely associated, the eyespot membranes are separated by a constant space of 10 to 40 nm and are often attached to each other by a fuzzy fibrillar/granular material. In *M. viride* this space appears to be larger (50–53 nm) and contains pronounced granular material, which appears to be ordered in two rows (Figure 4a, [5]). In addition often microtubular roots/bands pass through this space, as shown in the

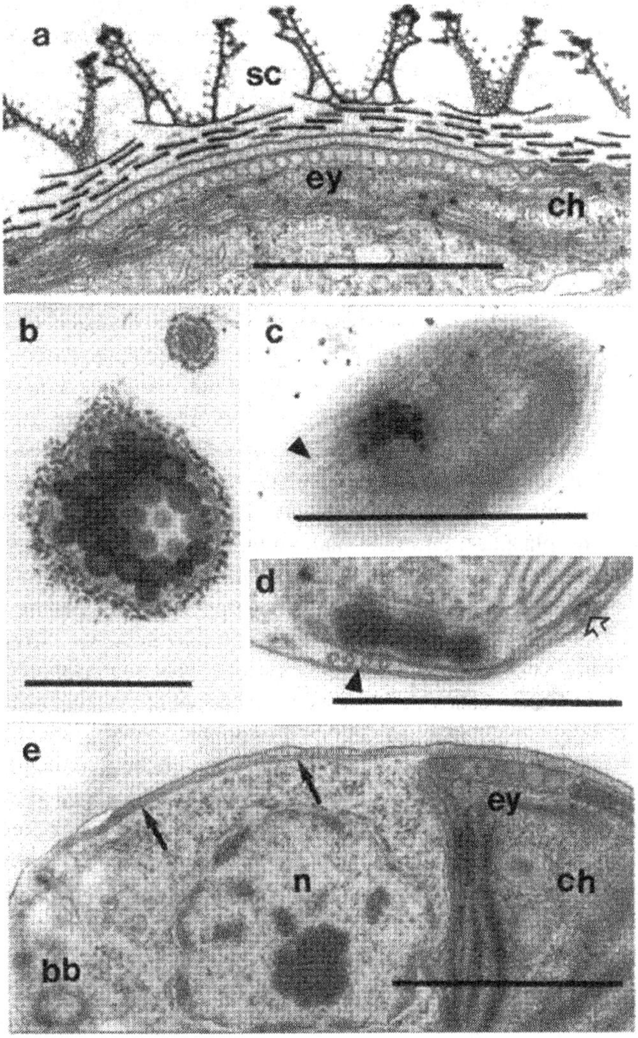

Figure 4. Thin sections through eyespot apparatuses of different Chlorophyceae. (a) Cross section of the eyespot apparatus of *Mesostigma viride*. Note the enlarged space between the chloroplast envelope and the plasma membrane, in which two rows of regularly spaced granules are visible. (b) Tangential section through the eyespot apparatus of *Pseudonephroselmis* spec. Note the hexagonal globule shape and the constant spacing between the globules. (c–e) Different thin sections through *Spermatozopsis similis* showing associations of the microtubular flagellar roots with the eyespot apparatus: oblique (c), cross (d) and longitudinal (e) sections. Arrow heads, 2s microtubular flagellar root; open arrow, 2d microtubular flagellar root; arrows, microtubular flagellar root; bb basal body, ch chloroplast, ey eyespot apparatus, sc scales. Scale bars: a–c, e 1 μm; d 0.5 μm.

tangential and cross sections (Figures 3c and 4c–e). Occasionally parts of these microtubular bands are co-isolated with the eyespot apparatuses (Kreimer, unpublished observation). A thickened appearance of the plasma membrane in the eyespot area in some thin sections pointed already early to specializations of the eyespot membranes [36,37]. Later freeze fracture analysis of different eyespot apparatuses demonstrated intramembrane particle specializations in two of the eyespot membranes, i.e. the plasma membrane and the outer chloroplast envelope [5,38–40]. Whereas large particles (16 nm to 20 nm) typically for the plasma membrane and chloroplast envelope were missing, smaller particles were particularly enriched in the area of the eyespot. The size distribution of the particles in the plasma membrane patch center around 8 nm to 12 nm and in the outer chloroplast envelope around 2 nm to 6 nm. Whereas the particle density in all investigated species was high in the plasma membrane (7,200 to 10,000 particles/μm^2), larger variations were observed in the outer chloroplast envelope (900 to 7,100 particles/μm^2). It was suggested that these particles may represent signaling components involved in photoreception of flagellate green algae or even the photoreceptor itself [5,38,39,41]. This assumption is sustained by both, the close match of the particle density and the minimal calculated number of photoreceptors per eyespot based on retinal extraction (~30,000 [42]) and recent localization of chlamyopsin, the putative photoreceptor, by immunofluorescence [43]. However, it has not yet been clearly shown in which of the eyespot membranes the receptor is localized. The observations that in *Mesostigma* only the plasma membrane patch exhibited a high intramembrane particle density [5] and that this alga shows positive phototaxis under low light conditions (M. Marin, personal communication), lends indirect support to the plasma membrane as major location of signaling elements involved in photoresponses. Irrespective of its precise localization, placing of the photoreceptor in front of highly pigmented and reflective lipid globules increases the directivity of the whole system (see Section 7.5).

Pigmented granules are often conspicuous components of light detecting organelles in different algae and ciliated protozoa [3,7,44]. Also in green algae they form the most prominent portion of the eyespot apparatus. Within the chloroplast they are usually highly ordered and form distinct layers. Analysis of more than 70 green algal taxa revealed a remarkable constant globule size, ranging from 80 to 130 nm for most species [5]. In some species the sizes are more variable (e.g. *Nephroselmis spp.*, *Pseudonephroselimis*, *Pyramimonas virginica*, Figure 4b, [5,20,45]), and diameters up to 200 nm are observed. However, one should always consider that globule sizes measured from thin sections can be affected by the plane of the sections and might also depend on the fixation method employed. The globule number per eyespot is more variable. The total number ranges from 30 in the smallest eyespots to about 2000 in eyespots of vegetative cells of *Volvox* (summarized in [5]). On average about 130 globules per μm^2 are observed. In cross/longitudinal sections the globules appear slightly compressed (Figure 4a,d,e). Tangential sections usually reveal their close hexagonal packing (Figures 3c, 4b,c). Only seldom pentagonal or heptagonal packing is observed. However, in some prasinophytes the globule size and also the packing density appears more variable (for review see [46]). In *S. similis* the close association of the globules is even maintained to a high degree in isolated eyespot apparatuses [19]. Even purified isolated eyespot globules retain, at least partially, their characteristic shape indicating the presence of material which stabilizes the globules (Renninger, Backendorf, Kreimer and Planta, in press). The eyespot globules are separated by a constant space of about 7 to 8 nm [5,36]. Depending on the fixation procedure employed, it appears either electron dense or translucent. This extractability by the conventional fixation procedures points to the presence of proteins/lipids in this space. This assumption is supported by freeze fracturing of eyespot apparatuses of vegetative cells and zoospores. In some specimens (e.g. Figure 4 in [38], Figure 2 in [39]) the fracture plane passes through the plane of the lipid globules. In such cases intramembrane particles can be identified which exactly outline the shape of the globules. The size of these particles appears somewhat variable, but fits to the dimensions of the space reported from thin sectioning. SDS-PAGE analysis revealed specific enrichment of several proteins in a fraction enriched in eyespot lipid globules from *S. similis* (Renninger, Backendorf, Kreimer and Planta, in press). At least some of these proteins might be involved in maintaining the highly ordered eyespot plate in this green alga.

According to the number of globule layers present single-, double- and multi-layered eyespots are distinguished. Each individual globule layer is usually subtended by a thylakoid. Based on globule-layer arrangement and association with thylakoids three basic subtypes have been introduced [4,5]:

1. The globule layer is subtended by a single thylakoid (Figure 9a). In eyespots with more than one layer this thylakoid is associated only with the upper globule layer. An additional constant space between the thylakoid and the following layer is present. This arrangement is typically found in the *Chlamydomonadales* and *Volvocales*.
2. Two globule layers are associated with one thylakoid. This type occurs in many double-layered eyespots.
3. No thylakoid is present between the layers and they are continuous with each other. However, again exceptions are found in some prasinophytes (see [46] for review).

In most species the globule layers are regularly spaced. Analyses of the spacing of the layers in the different subgroups have revealed extremely constant values. The combined thickness of one globule layer and the space between the next layer are 160 to 180 nm (first subtype), 150 to 220 nm (second subtype), and 180 to 220 nm (third subtype). In single-layered eyespots the distance from the inner surface of the globules to the plasma membrane is 120 to 200 nm. This distance is slightly reduced in multi-layered eyespots (120 to 140 nm [4,5]). Again these dimensions might be slightly different in vivo, as fixation can lead to either swelling or shrinkage of structures. However, as shown in Figure 9a, an extremely periodic spacing of the different components is one of the most striking features of multi-layered green algal eyespot apparatuses. As will be outlined later (see Section 7.5) it is one basis for contrast enhancement at the presumptive location of the photoreceptor in the plasma membrane patch overlying the globule layers.

7.3 Interactions with basal bodies and microtubular flagellar roots

Conspicuous interactions between flagellar microtubular roots (usually the 2s root) and the eyespot apparatus occur in many, but not all, green algae. In those species where these interactions occur the microtubules are attached to both, the plasma membrane and the outer chloroplast envelope in the eyespot region (Figure 4d). These interactions are apparently not stable because they are only very seldom observed in isolated eyespot apparatuses [19]. As the signal generated in the region of the eyespot apparatus finally affects the flagella and neither the flagella nor the basal bodies are identical in green algal cells, these interactions will also briefly be considered here.

Basal bodies in green algae, as well as in other algae, undergo a developmental cycle and need more than one cell cycle for maturation, i.e. they differ in age and developmental status within one cell ([47,22,48], reviewed in [49]). A biflagellate cell thus contains a younger basal body (no. 2), which was formed during the preceding cell cycle and an older basal body (no. 1) formed at least one cell cycle earlier than basal body no. 2. Hence during the next division the younger basal body becomes the older basal body of the following cell cycle, because it is distributed to the progeny cell with a newly formed basal body, which is the new basal body No. 2. As a consequence also the flagella undergo transformation, which is most obvious in species with flagella of different size or beat pattern. Thus, depending on their developmental status, the flagella probably also differ in specific components. This e.g. explains well their different Ca^{2+} sensitivity [28] and the different beat patterns observed for the two flagella of *Nephroselmis olivacea*, which change during maturation [47,49]. Due to this flagellar/basal body developmental cycle and the subtle differences of flagellar behavior in response to changes in the concentration of free Ca^{2+} positional stability and constant interaction with the developmentally identical flagellum during interphase is essential for a fixed reaction pattern of the cell to light stimuli.

In green algae it is established that the eyespot apparatus is associated in many cases with the immature basal body No. 2 [5,22,23,50,51]. It is not yet known how absolute orientation and positioning, irrespective of the mode of eyespot apparatus distribution during cytokinesis, is sensed. Cytoskeletal elements are obvious candidates [22,23,51]. For example, prior/during cytokinesis of *S. similis* the interaction between the eyespot apparatus and the 2s flagellar root is lost and complex, not stable interactions with different roots are observed for the newly formed eyespot plate. Interactions with the 1s root occur first at the location where the new eyespot plate is assembled prior to chloroplast division. After translocation of the new eyespot towards the anterior part of the cell its interphase association with the 2s root is established [23]. It is, however, not yet known whether the 1s root induces assembly of the eyespot plate or if the interaction occurs after its assembly. Clearly, the inherent asymmetry of the basal bodies and interactions of the different roots emerging from the mature/immature basal body might be one key to sense and mediate the absolute orientation within asymmetric cells [22,23].

During interphase pronounced associations between the eyespot apparatus and microtubules or microtubular roots/bands are well documented for many green algae (for review see [5]). The nomenclature used here follows the new terminology for cytoskeletal elements associated with the flagellar apparatus of protists suggested by Andersen et al. [52]. Figure 5 summarizes the different known associations for green algae, which range from no interactions to complex intercalations of the eyespot with the microtubular roots. Associations of the microtubular roots (2s, 2d) with the eyespot apparatus are observed in the majority of green algae. In those cases where no associations are reported two different groups of eyespot locations with respect to the insertion of the flagella were differentiated [5]. In the first group the eyespot is located opposite of the flagellar insertion. This is found in the Mamiellales (*Mantoniella*, *Mamiella*) and the prasinophyte *Mesostigma* (Figure 5a). In the second group with missing root associations the eyespot is located lateral to the flagellar insertion (Figure 5d). This is found in members of the Chlorodendrales (*Tetraselmis spp.*, *Scherffelia dubia*) and in Pyramimonadales (e.g. *Pyramimonas parkae*, *P. olivacea*). However, in some Chlorodendrales a rhizoplast (Ca^{2+}-modulated, contractile fibrous root emerging from the basal bodies) attaches to the plasma membrane close to the eyespot (e.g. [53]) or directly connects to the eyespot as in *Halosphera minor* [54]. A conspicuous variability with respect to the eyespot/flagellar apparatus localization occurs in the genus *Pyramimonas*. Here four different orientations are observed (summarized in [20]). Whereas in most green algae two flagellar roots (s, d) emerge from each basal body, *Nephroselmis spp.* possess only three roots [45,50]. From basal body No. 2 only a broad 2d root emerges (Figure 5b), which runs over the eyespot in a curved path, and the 2s root is missing. However, in most green algae the eyespot is

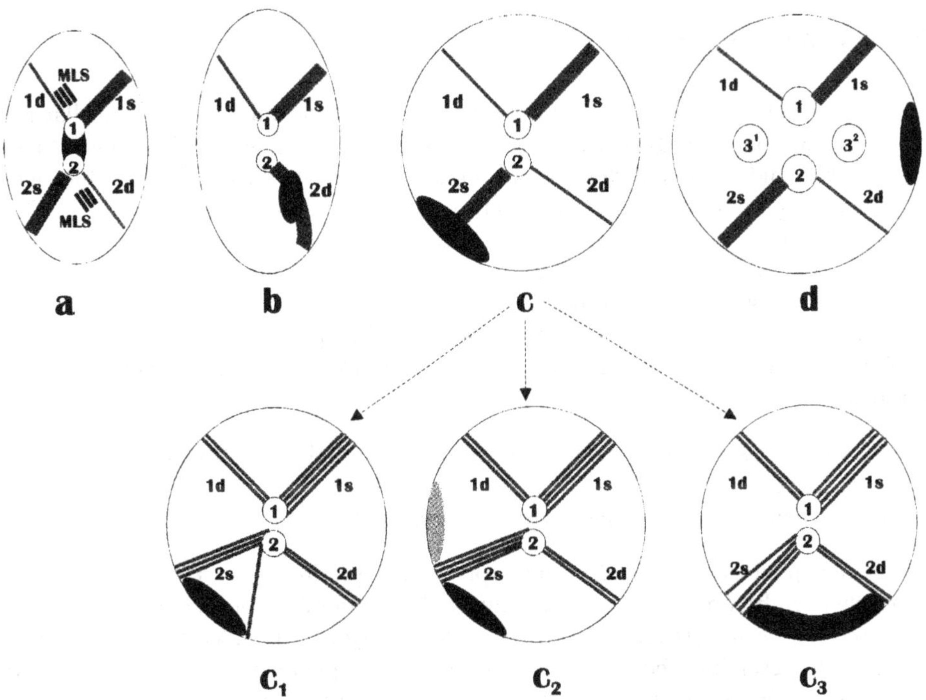

Figure 5. Schematic representation of different eyespot apparatus associations with respect to microtubular flagellar roots (1d, 1s, 2s, 2d) and the basal bodies during interphase. (a) *Mesostigma viride*, (b) Pseudoscourfieldiales (*Nephroselmis olivacea*), (c) advanced Chlorophyceae: C_1 most Chlorophyceae, C_2 most Ulovophyceae and Trebouxiophyceae, C_3 some female gametes of siphonalean green algae, (d) some Chlorodendrales and Pyramimonadales. Basal body numbering: 1 developmentally old and 2 developmentally young basal body. Further explanations are given in the text.

associated with the 2s flagellar root (Figures 3c,d, 5c). As depicted in Figure $5c_{1-3}$ different specifications of this type are known. In the most often observed type, e.g. in many *Chlamydomonas*-type cells, the eyespot is intercalated between the 2s root (Figure $5c_1$). In reproductive cells of Ulvophyceae and Trebouxiophyceae the eyespot is associated with one edge of the 2s root (Figure $5c_2$), whereas in gametes of siphonalean green algae the lipid globules extend from the edge of the 2s to the edge of the 2d root (Figure $5c_3$). Also in uniflagellate *Pedinomonas* species, where the eyespot apparatus is localized opposite to the insertion of the flagellum (Figure $3a_2$), the eyespot is intercalated between two micotubular flagellar roots [55]. It is widely accepted that the eyespot apparatus/cytoskeletal interactions are important for proper positioning of the light sensing organelle in relation to the flagellar apparatus. Occasionally, it was suggested that microtubular flagellar roots might be involved in signal transduction towards the flagella. However, as several phototactic green alga clearly lack these interactions (see above), roots cannot be involved in the general phototactic signaling process.

7.4 Eyespot phylogeny within the green algae

Among other features, the different eyespot positions and microtubular flagellar root systems have been used to deduce green algal phylogeny and their relation to the charophytes and embryophytes as well as to identify the ancestral green flagellate. Especially regarding the latter two controversial suggestions have been made. With respect to the eyespot apparatus one scenario assumes a posterior located eyespot, positioned on the cleavage plane with a single-layer of globules and not associated with microtubular flagellar roots for the ancestral green flagellate (reviewed in [46,56]). Until recently it was not possible to use molecular phylogenetic analysis to tackle the evolution of the different specializations within the green algal lineage for two reasons:

1. no eyespot possessing flagellate member of the charophyte/embryophyte lineage was known, and
2. too few genera of the prasinophytes had been analyzed.

Molecular phylogenetic studies now place the prasinophyte *Mesostigma* to the charophyte/embryophyte lineage and almost all genera of the prasinophytes have been analyzed. These analyses revealed that the prasinophytes form at least four independent lineages representing the early divergence in the phylogeny of the Chlorophyta [45,56,57]. This allowed for the first time the comparison of the evolution of the eyespot position, ultrastructure and its different flagellar root interactions within the green algae with a phylogenetic tree based on complete nuclear-encoded small subunit rRNA coding regions (Figure 6 [45]). This analysis revealed that the assumptions about primitive eyespot features (posterior location, single-layered, not associated with microtubular roots) coincide well with the molecular data. However, no clear evolutionary trend with respect to the other characteristics is evident. Each monophyletic genus of the prasinophytes has apparently evolved separately different characteristics of the eyespot

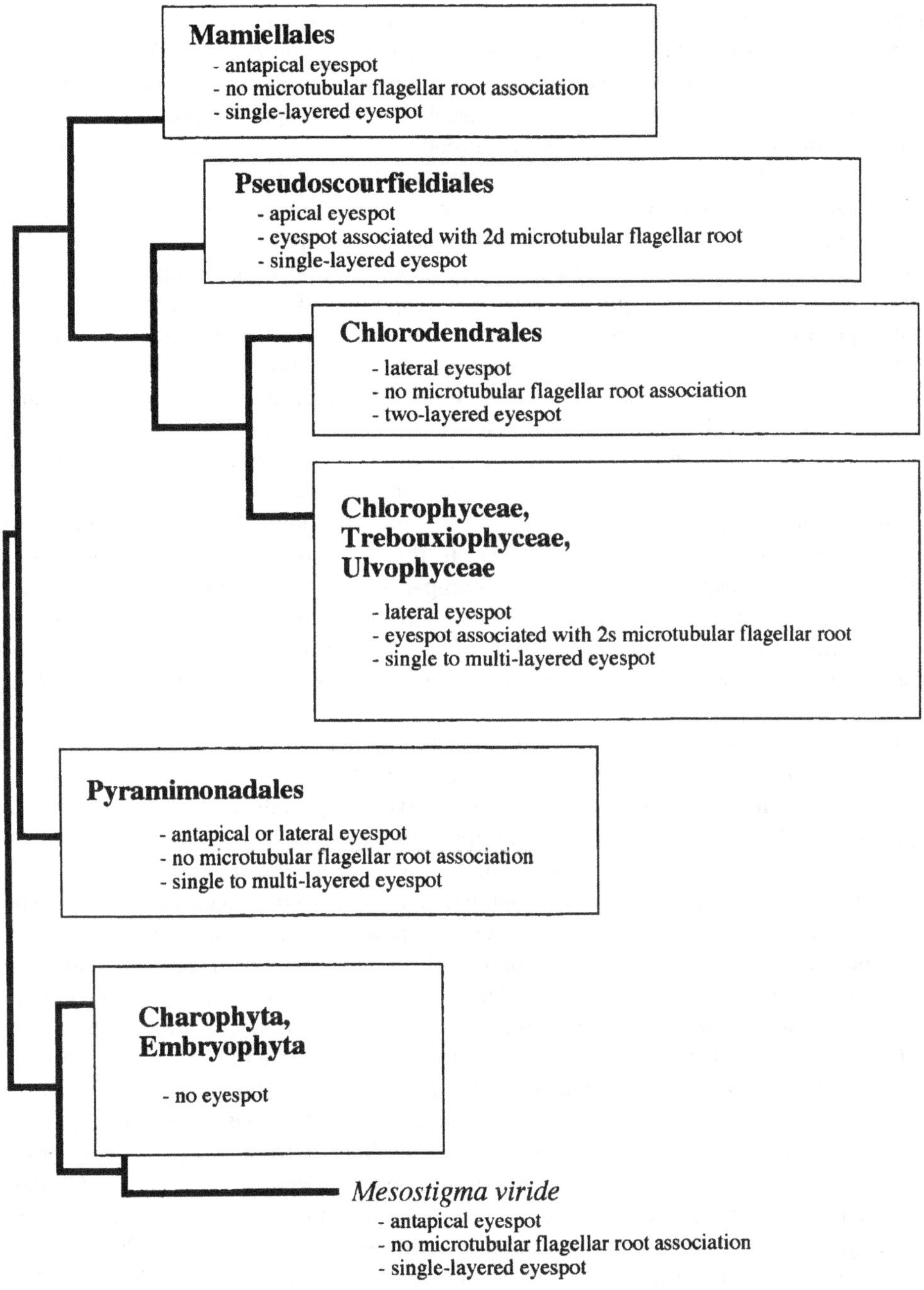

Figure 6. Simplified phylogenetic tree of the green algae (based on nuclear-encoded SSU rRNA sequence comparisons) combined with different structural/positional eyespot characteristics. The tree is rooted with *M. viride* and some charophytes and Embryophyta. Please note that not all members of the different shown green algal lineages possess eyespots. For detailed trees including the analyzed genera see [45,56,57].

apparatus. E.g. in the Pyramimonadales antapical eyespot position with a direct connection of a rhizoplast to the eyespot occurs (*Halosphaera minor* [54]) as well as variable lateral positions with no connections are observed (*Pyramimonas* spp. [20]). In addition, in the genus *Pyramimonas* miscellaneous eyespot designs and multiple eyespots are found, pointing to a rather broad evolutionary tendency with respect to positioning of the eyespot apparatus in this group. This might be interpreted as a result of the lack of flagellar root interactions. However, in the more advanced Chlorodendrales, which also usually lack a direct flagellar root-eyespot interaction in interphase, such a variability is not observed and the lateral eyespot position is rather stable. Still, during cytokinesis there might be transient eyespot/cytoskeletal interactions ensuring stable positioning in both groups. Melkonian and Robenek [5] suggested that lateral eyespot position in conjunction with the 2s root association has evolved late. This situation occurs in the advanced lineages; i.e. the Ulvophyceae, Trebouxiophyceae and Chlorophyceae, which represent monophyletic lineages [58]. The conclusion of Melkonian and Robenek is confirmed by this analysis. However, it should be noted that interactions with albeit different (2d) roots have evolved separately also in the *Nephroselmis* lineage. As supported by this tree, also multi-layered eyespots have evolved independently at least twice. Due to the interesting evolutionary position of green algae with respect to higher plants, comparison with their signaling pathways for light detection promises interesting insights into the evolution of light detecting systems in the future. However, currently still too little is known at the molecular level, e.g. of the putative receptor only two sequences are known (*C. reinhardtii* and *Volvox* [43,59]).

It is obvious that eyespot position and swimming behavior must be closely coupled in order to allow determination of the direction of light in such an effective way as is displayed by many green algae. Like the various specializations of the eyespot apparatuses, different swimming patterns apparently evolved independently in the green algal lineages [15]. Because the general motion pattern of *Mesostigma* is similar to that of *Nephroselmis* (Marin, personal communication), current molecular phylogenetic analysis supports the view that the ancient swimming behavior of green algae might have been a sidewards motion. Previously backward swimming was suggested as most primitive [15,60,61]. These authors also proposed that forward swimming is more advanced than backward/sideward motion, which is in accordance with the phylogeny [56,57]. Currently it is believed that the symmetrical (flagellar) beat is more primitive than the asymmetric (ciliary) beat [15]. However, some prasinophytes, e.g. *Nephroselmis olivacea* [15,47], exhibit a heterodynamic flagellar beat, i.e. the two flagella beat with different waveforms. A similar behavior is known from the glaucocystophyte *Cyanophora paradoxa* [62]. Because the molecular phylogenetic data support the idea that green algae and glaucocystophytes share a common origin [10], close analysis of the flagellar beat of *Mesostigma* could shed light on the question which flagellar beat form is ancient. A heterodynamic beat as the more primitive form might better explain the parallel evolution of different diverse beat forms observed in the green algal lineage and the necessary regulatory elements than a flagellar beat form. In summary, apparently a close co-evolution of the eyespot apparatus, the flagellar apparatus and the swimming patterns necessary for the fine tuned and highly adaptable phototactic responses has occurred in the different lineages of the green algae.

7.5 Signal generation and modulation

As will be discussed in the following paragraphs a signal encoding information about the cells orientation relative to the direction of the light stimulus is generated by the eyespot apparatus. Perception and modulation of the signal as well as the actual intensity encountered by the photoreceptor is affected by several parameters. These include the general motion pattern of the cell with its rolling around the axis of motion, the ultrastructure and location of the eyespot apparatus, and also the spectral properties of the whole "optical" system [4,5,7,8,32,34]. The transient photophobic response is less affected since no spatial information is required. Here only the light intensity must exceed a certain threshold within a given time window to trigger the stereotyped response (for review see Govorunova and Sineshchekov, this volume). However, to obtain directional and intensity information needed for phototaxis, high constraints must be fulfilled by the optical system and the signaling cascade(s) involved. For instance, correct phasing of the phototactic response of the cell is only possible when signal generation, processing, response and return to a new excitable state are adapted to the cell's rotational speed; e.g. for *C. reinhardtii* these values vary between 1 and 2.5 Hz [63]. Further complexity arises from multiple effects of different external and internal factors on photoorientation (e.g. [64–67]) and the presence of rapid adaptational processes. Desensitization towards photophobic stimuli occurs within seconds and does not affect the cells ability for precise phototaxis [68–72]. This is the basis for the occurrence of precise phototactic orientation over several orders of light intensities (~ 1 to 10^6 erg cm^{-2} s^{-1} [73]). Desensitization mechanisms are not yet known, but probably involve modifications of the receptor and/or ion channels involved in signal transmission [7]. Elements under discussion include reduced membrane excitability rather than photoreceptor bleaching [72], Ca^{2+}-modulated protein kinases/phosphatases and light-modulated heterotrimeric G-proteins [74–76].

In addition to the directivity already gained by its ultrastructure, the spectral properties of the whole optical system further increases the overall directivity and sensitivity. Here absorption, reflection and interference of phototactic active light are important parameters [4,7,26]. Both, sensitivity of the behavioral responses and the corresponding primary electrical events originating from the region of the eyespot apparatus peak at about 500 nm (reviewed by [59], see chapter by Foster, this volume). Absorption by the carotenoid-rich eyespot globules occurs in the same, but somewhat broader, spectral region as revealed by in vivo microspectrophotometry (e.g. [77,78]) and absorption spectra of isolated eyespot apparatuses and globules in aqueous solutions (Figure 7, [19], Renninger, Backendorf, Kreimer and Planta, in press). This close spectral match in conjunction with cell rotation makes the green algal eyespot apparatus highly directional. Efficient photoreceptor excitation is only possible when light falls from outside of the cell on the eyespot membranes (Figure 8a). Light passing through the cell will be absorbed and scattered back by the cell constituents and the eyespot carotenoids/globules, thus reducing excitation of the receptor. This concept of periodic shading/illumination of the photoreceptor has been proposed early by several authors (for review see [4,5]). Flash experiments proved that a response can indeed only be evoked with high probability when the eyespot is oriented towards the light source (e.g. [79]). Directivity of the eyespot is further increased by the dichroic orientation of the retinal chromophore within the membrane. In *H. pluvialis* and *C. reinhardtii* the chromophore is oriented almost parallel to the plane of the membrane [80,81].

Early experiments with multiple light sources by Buder [82] point towards a relatively large half-beam width of the eyespot apparatuses. Foster and Smyth [4] gave values of about 60°, which allows the tracking of a diffuse light source and prevents its rapid loss. Green algae align their swimming path and not the antenna direction, i.e. the direction of maximal sensitivity of the eyespot apparatus, with the direction of light during phototaxis (Figure 8b). Thus under continuous illumination cells will encounter a stimulus intensity, which is modulated in a regular fashion by their helical motion and relative orientation towards the light source [4]. The shape of the modulated signal and the duration of the light/dark period depends on several parameters, which will be considered later in more detail. In a simplified view the stimulus intensity will vary with the angular deviation of the swimming path from the light source in the following way: a maximal signal intensity and modulation is expected when the cells swim perpendicular to the light source and the signal decreases with an increased alignment of the swimming path either towards or away from the light source. As photoreceptor excitation in green algae leads to predominantly Ca^{2+}-carried inwardly directed currents in the eyespot region (for review see [59] and Govorunova and Sineshchekov, this volume), different local increases in the cytoplasmic free Ca^{2+} concentration are

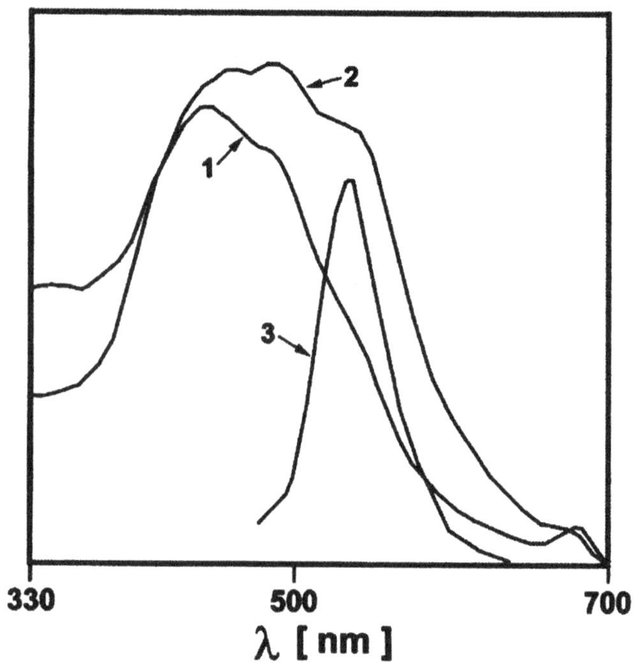

Figure 7. Absorption spectra of fractions enriched in (1) isolated eyespot apparatuses and (2) isolated eyespot globules from *Spermatozopsis similis* in aqueous solutions. For comparison the wavelength dependence of eyespot reflectivity of *Chlamydomonas reinhardtii* (3, redrawn from [78]) has been included. Note that the spectra are not normalized.

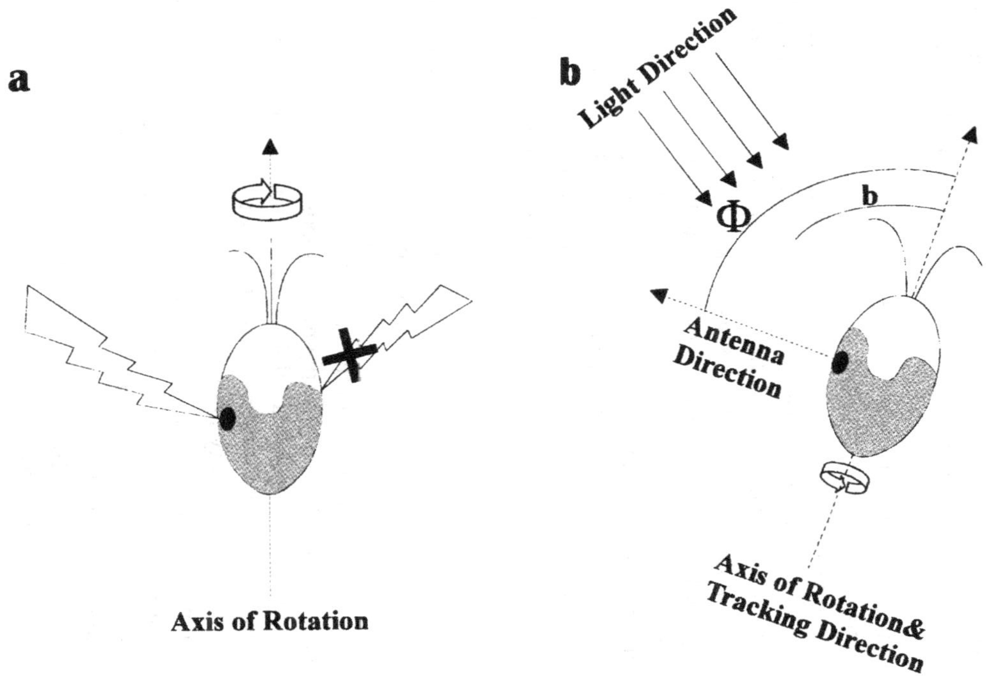

Figure 8. (a) Schematic drawing illustrating the concept of periodic illumination/shading of the photoreceptor. Only light falling from outside on the surface of the eyespot apparatus will lead to photoreceptor excitation in the eyespot membranes overlying the carotenoid-rich globules. Light falling through the cell will be absorbed/scattered by the cell constituents and the eyespot sensu stricto. In addition, back reflection will contribute to the generation of a periodically modulated signal in conjunction with cell rotation. (b) Scheme illustrating often used angles for defining the scanning parameters of the eyespot apparatus. The scan angle (ϕ) is defined as the angle between the tracking direction and the antenna direction. The tracking angle (b) is the angle between the tracking and light direction. For detailed definitions of additional antenna parameters not used in this review (see the reviews [4,8,32,34]).

probably triggered in response to the changed stimulus intensity. It is widely accepted that green algae thus start to orient themselves towards the light source in response to a Ca^{2+}-dependent, light evoked signal, which finally affects the flagellar beat. E.g. the *cis*- and *trans*-flagellum of *C. reinhardtii* exhibit complementary beat changes to step-up/step-down light stimuli ([33], for review see [9,83]). The basis for this behavior is the Ca^{2+}-mediated flagellar dominance. Mutants lacking this dominance are not phototactic [28,84,85]. During tracking of the light source, however, the cells probably tend to minimize the signal sensed by the receptor. In other words, once aligned with the light direction deviation from it can presumably be sensed by the appearance or increase of the same signal.

The eyespot enhances the precision of the orientation by increasing the front-to-back contrast [86,87]. Total contrast values of up to eight have been measured by means for rhodopsin-triggered photocurrents [88,89]. Especially in those green algae where the eyespot is exposed in thin lobes of the chloroplast (e.g. *S. similis*, Figure 2e–g) or only small parts of the cell body/chloroplast shield the photoreceptor (e.g. *M. viride,* Figure 3b), the importance of the eyespot globules as a shielding device becomes evident. In some species placement of the pyrenoid body or starch behind the eyespot apparatus additionally increases photoreceptor shielding during cell rotation. In addition, as will be outlined below, the reflective properties of the eyespot sensu stricto also contributes greatly to an optimal front to back contrast. Recently, the major role of the eyespot globules in light attenuation by absorption and scattering has been questioned by Schaller and Uhl [78] for *C. reinhardtii*. However, the strong attenuation of argon-ion laser light (principal lines at 488 nm and 514 nm) by the eyespot of different *Chlamydomonas* strains (Figure 1a–c, [87]) as well as in other green algae (Figure 2, [7,26]) does not support these results. Even in the closest vicinity of the eyespot the attenuation of the argon-ion laser by the chloroplast does not approach that observed in the eyespot area. In addition, other in vivo microspectrophotometry analyses [77] do support the major contribution of the eyespot carotenoids to the overall shielding properties for blue green light. Also the pigment composition and absorption data of isolated eyespot apparatuses/purified globules underline these shielding properties and clearly show considerable absorption up to 540 to 550 nm ([17–19], Figure 7). Changes in the spectral sensitivity in different eyespot mutants additionally stress the importance of an intact eyespot for a perfectly adapted screen for the retinal-based receptor [87,90]. However, the cell body and photosynthetic pigments clearly contribute to these general screen properties and generate already sufficient contrast modulation to allow photoorientation. This has early been nicely demonstrated in different eyespot mutants (e.g. [86]). To unravel the impact of these different components in more detail further analyses of colorless algae and eyespot mutants are desirable.

Both, directivity and sensitivity of the eyespot apparatus is further enhanced by reflection of phototactic active light (Figure 9b). This phenomenon has been long known and occurs in all green algal eyespots so far analyzed irrespective of the number of globule layers [4,7,26,78,82,91]. Signal strength increases with the integrity of the globule layers [87]. Maximal reflection occurs when the eyespot surface is roughly perpendicular to the light source and decreases at oblique angles of light incidence. No or only weak reflection signals are detectable when the cells are swimming directly towards the light source, i.e. at extremely acute angles between eyespot surface and light incidence. Focusing of the reflected light has been observed in eyespots with convex surfaces [26,91]. In addition already minor changes in the angle of incidence lead to color modulation of the reflected light. An impressive video demonstrating these properties can be found on the optical disk produced by Pickett-Heaps and Pickett-Heaps [92].

In their thorough analysis of the optical principles of algal light antennas Foster and Smyth [4] were the first to postulate that green algal eyespots might take advantage of reflection to increase both light intensity and illumination time at the photoreceptor location. They compared the exactly spaced alternating globule layers with high refractive index and spaces with low refractive indices observed in multi-layered eyespots (Figure 9a) with a combined interference reflector and wave guide, which is screened at one end. Calculated patterns of light intensity distribution for monochromatic light (480 nm) within a multi-layered eyespot coincides with that predicted for an quarter-wave interference reflector. The position of the first absolute maximum was predicted to be close to the plasma membrane (Figure 9c). Analysis of eyespot reflection by confocal microscopy provided the first experimental support for the hypothesis of quarter-wave interference reflection in green algal eyespots. The measured intensity distribution patterns in vertical optical sections matched closely the theoretical intensity distribution. In addition, for a given light intensity, the intensity of reflected light at the location of the plasma membrane in multi-layered eyespots was found to be about twice that of single/double layered eyespots [26]. The placement of a reflector with similarities to a slab wave guide behind the photoreceptor thus contributes in several ways to an increased directivity and sensitivity. First, for light falling through the cell, back reflection will additionally to absorption suppress excitation of the photoreceptor. The curved arrangement of the globule layers also helps to trap light at acute angles of incidence, which will be absorbed by the high amounts of carotenoids in the globules. Secondly, for light falling from the outside on the eyespot apparatus, interference reflection will lead to a specific intensity increase within the absorption range of the retinal-based receptor and an increased exposure time of the receptor. Thus the probability of photon absorption, which is proportional to the effective photoreceptor concentration multiplied by the exposure time and the intensity, should be considerably be increased. Reflection thus extends the sensitivity to lower fluence rates, where it probably has the most pronounced effects on the efficiency of photon absorption. A more detailed treatment of the physical principles of eyespot apparatuses is given in the review of Foster and Smyth [4]. However, the eyespot sensu stricto should and cannot be regarded as an ideal reflector.

Mast [91] noted that blue and green light is specifically reflected by green algal eyespots. This fits well with the calculations of Foster and Smyth [4] and the absorption spectrum of the retinal-based receptor [42]. Recently, the spectral dependence of eyespot reflection of *C. reinhardtii* cells captured on micropipettes has been determined. In these measurements reflectivity was found to peak at 530 to 550 nm [78]. A maximal reflection index for anomalous dispersions is predicted for wavelengths where the absorbance changes with wavelength are greatest [93]. The eyespot reflectivity peak coincides exactly with a shoulder observed in the absorption spectrum of isolated green algal eyespot apparatuses (Figure 7). This shoulder can be attributed to eyespot carotenoids as its relative abundance increases significantly in purified globules. However, reflection signals are also observed in the blue/green region (see e.g. Figure 9b, [7,26,87]). In non-immobilized cells of *Eudorina* Pickett-Heaps and Pickett-Heaps [92] neatly demonstrated that already small positional changes of the colony lead

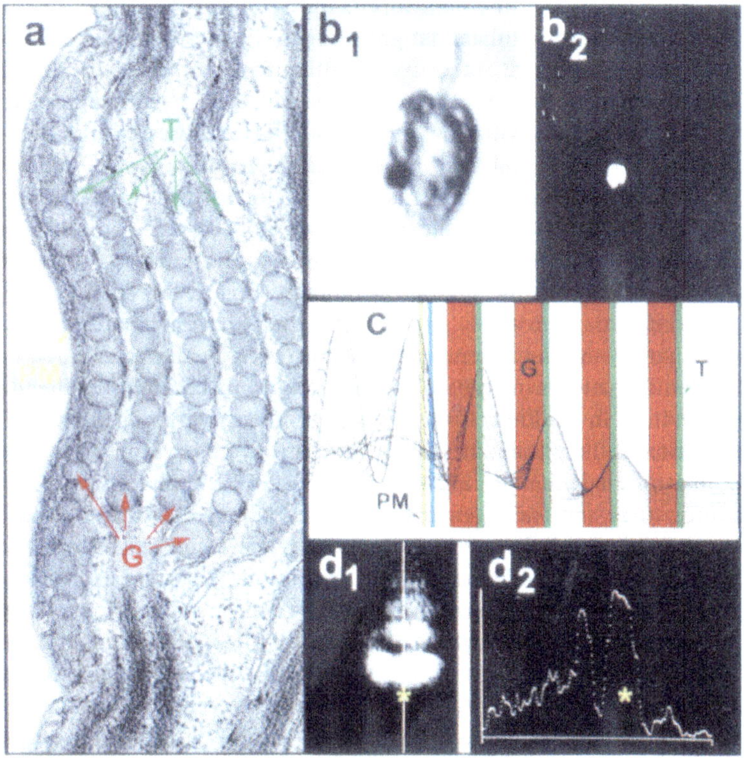

Figure 9. Eyespot reflectivity and the concept of quarter-wave interference reflection. (a) Cross section through the eyespot apparatus of *Hafniomonas reticulata* demonstrating the regularly spacing of the carotenoid-rich globules (G) with an high refractive index and the low refractive spaces between them including the underlying thylakoids (T) in multi-layered eyespot apparatuses of green algae. The plasma membrane (PM) patch overlying the eyespot senso stricto is most likely the location of the photoreceptor. Note the partially concave surface of the eyespot. (b) Optical section through a living, immobilized cell of *Tetraselmis chuii*, (b_1) differential interference contrast image and (b_2) simultaneously taken reflection image from the same focus plane, (c) calculated pattern of light intensity within and near a multi-layered eyespot apparatus for 480 nm light with different angles of incident (0° to 70°). The light falls from outside (left side). Further details for the calculations are given in the legend of Fig. 8 in [4]. Abbreviations and color code as in (a). (d) Vertical optical section (d_1 reflection mode of the confocal microscope) through the central portion of the eyespot apparatus of a living cell of *H. reticulata*. The intensity profile (d_2) has been collected along the indicated line. Yellow asterisks denote the region where the plasma membrane is positioned. Figure (c) has been redrawn and modified from [4], whereas Figures (a, b, d) have been modified from [26].

to a modulation of the color of the reflected/diffracted light, creating a rainbow effect ranging from blue to yellow/red. Changes in the pattern of reflectance for three wavelengths (400 nm, 500 nm, 600 nm) has recently been calculated for different scanning and tracking angles (for definition of these angles see Figure 8b) and revealed interesting and important properties of the eyespot [8]. Briefly, reflectance for perpendicular light incidence was predicted to be optimal at 500 nm. At acute angles the eyespot reflectance is predicted to exhibit a blue shift, whereas light of increasing wavelengths should be less reflected. Surprisingly, Schaller and Uhl [78] observed only a small blue shift in the reflectivity spectrum upon increasing angles of light incidence. The reasons for this discrepancy are not yet known, but might lie in the size of the eyespot apparatus and an increasing role of diffraction at oblique angles of analysis.

The above described angular dependence of the preferentially reflected light color might well explain differences observed for action spectra obtained at intermediate/high and low light intensities in *C. reinhardtii* [8]. The former typically exhibit a prominent blue-side peak and a steep drop in the red region (e.g. [94]). In contrast those obtained at low light levels and a recently published action spectrum solely obtained for dim flash induced orientational changes (i.e. the onset of phototactic orientation) agree well with the Dartnall nomogram of rhodopsin (e.g. [35,95]). Suppression in the long wavelength portion of the action spectra at high intensities might be ascribed to the predicted changes in reflection/absorption properties of the eyespot at acute angles of light incidences. At low intensities the swimming path of cells is usually not as straight as under high light conditions. In addition, the eyespot position in cells with low intracellular Ca^{2+} is towards the outer edge of the helix [28], which might lead to a slightly prolonged illumination time. Thus the probability of photon capture by cells with a roughly normal orientation of the eyespot apparatus with respect to the light source can be expected to be higher than under increased light intensities. These spectra will thus mainly reflect the primary orientational response. In other words they will be dominated by the wavelength dependence of the receptor and that of the eyespot sensu stricto at normal light incidence. Because light of 530 nm to 550 nm is preferentially reflected in these orientations (see above) only a slight distortion of the rhodopsin spectrum should occur. Photon absorption probability should even be increased by extending the intensity peak up to 550 nm where the receptor can still absorb. In this context it is important that already short flashes or even a single photon is sufficient to evoke an orientational response [35,79,96]. Thus under conditions where absorption of any photon in the relevant range is important for an orientational response, the overall optical properties are apparently fine tuned towards this demand. At higher intensities the swimming path will be increasingly aligned and consequently more oblique rays will hit the eyespot surface. Under these conditions light trapping by the globule layers and its absorption by the highly concentrated carotenoids will increase. The globules strongly absorb up to 550 nm (Figure 7). In parallel the residual reflected light will encounter a blue shift [8]. At acute angles the predicted changes in the reflection/absorbance properties of the eyespot should therefore change the spectral composition and intensity encountered by the receptor in such a way that a relative blue dominance and a suppression towards the long wavelength tail occurs. Despite the minimized exposure of the eyespot surface in well aligned algae, a shift to the receptor peak absorbance also here might lead to an increased photoreceptor excitation probability upon small deviations from the tracking direction. The eyespot properties under oblique illumination might also help to suppress receptor excitation by diffuse light, further optimizing tracking. Therefore action spectra obtained at higher light intensities might reflect mainly the wavelength dependence of optimal tracking. As discussed by Kreimer [7] color modulation may, however, also be important for a photocycle of the receptor. Experimental data supporting a photocyle also in green algae are not yet published, although retinal analogues which affect the lifetime of the active signaling state in bacterial rhodopsin also affect the sign of phototaxis [97].

The shape of the modulated signal, modulation contrast and the duration of light/dark periods will also be affected in a complex manner by parameters as eyespot position and surface geometry or the pitch angle of the helical swimming path. In some algae the latter can change as a function of light intensity, thus additionally complicating model calculations and predictions. Pure positional effects are easier accessible to model calculations. At identical tracking angles a sinusoidal signal modulation is predicted for both, cells with median or anteriorly located eyespot apparatuses. Dark and light periods, however, differ [8]. With an exactly median positioned eyespot apparatus both periods are equal. Already upon a slight decrease in the scan angle (90° to 70°) the phases become unequal. For cells exhibiting positive taxis, illumination time of the receptor is likely to increase, whereas during negative taxis the light period is probably reduced. For eyespot positions close to the flagellar base, as e.g. observed in *S. similis* (Figure 2e–g), even a modulated permanent illumination of the receptor for a positive phototactic cell can be envisaged when tracking and scan angle are similar. An inward or outward raking of the eyespot, as suggested by Schaller et al. [35], might furthermore modulate the effective illumination/shading times during cell rotation. The above described differences in illumination period for positive and negative phototactic cells might encode information of the cells actual swimming direction. Due to complex effects of different parameters on the sign of phototaxis (e.g. ionic composition of the medium, age of the culture, photosynthetic activity and the stage of the cell cycle [64,67,73,98]), experimental analysis of this hypothesis will be complex. Available experimental evidence emphasizes the substantial role of additional factors in determination of the sign of phototactic behavior. For example it is well documented that preillumination affects the sign of photoresponse. Preillumination as well as photosynthetic electron transport can affect the resting membrane potential and might thereby also control the general excitability and the sign of phototactic behavior. Irrespective of the involved mechanism(s), the concept of periodic shading/illumination and reduced photoreceptor excitation upon alignment of swimming and light direction can explain how unicellular green algae might sense whether they swim towards or away from the light source, pending the eyespot is not located perfectly on the cells equator. Long lasting, Ca^{2+}- and light intensity-dependent currents saturating at already low light intensities reported for *Haematococcus* [80] and recently also for *Volvox* [99] might well encode this information.

7.6 Signaling

Excitation of the photoreceptor in green algae activates depolarizing currents in the eyespot region, which are mainly carried by Ca^{2+} (see Govorunowa and Sineschekov and Deininger and Hegemann, this volume). In this respect the green algal visual signaling system is similar to invertebrate photoreception, where most systems depolarize [100]. The primary Ca^{2+} signal is processed rapidly (< 140 ms) and, depending on the intensity, lead to two basic differential responses of the flagella (for review see [83,101]). Phototaxis is based on subtle differences in the normal asymmetric, breaststroke type beat pattern of the flagella and the stop or shock response is characterized by a transient switch to a symmetrical, undulating waveform that causes the cell to move backwards. The shock response lasts for about 500 to 1000 ms. In contrast changes in flagellar beat amplitude and frequency in response to dim flashes last only up to about 80 ms [102]. The shock response is only initiated by a massive increase in intraflagellar free Ca^{2+} upon activation of voltage-gated Ca^{2+} channels in the flagellar membrane. When the depolarizing stimulus exceeds a certain threshold, these channels are activated in an all-or-none manner [89,103]. Flagellar currents have mainly been characterized by capacitive measurements on whole cells (summarized by Govorunowa and Sineshchekov, this volume). Only recently the first green algal flagellar ion channels have been characterized by the patch clamp technique at the single channel level [104]. In flagellar membrane preparations of *S. similis* they described a small conductance Ca^{2+}-permeable cation channel and two large conductance channels, one permeable for Cl^- and the other for potassium. Mutant analysis strongly support the idea that different ion channels are activated in response to low and high light stimuli. Mutants of *C. reinhardtii* (*ppr1* to *4*) were found to be defective in channel(s) involved in the generation of flagellar currents observed during photoshock. No effects on currents originating from the eyespot region or in the phototactic response are exhibited by these mutants [105]. Recordings of flagellar currents under low light conditions are unfortunately not yet published. Apparently only very few flagellar channels are activated under these conditions. Thus, as no evidence for two different receptors involved in the photoresponses exists, the signaling cascade initiated upon rhodopsin excitation is believed to be branched [7,76,80,88,89,106]. Signal transduction during photoshock presumably involves no biochemical amplification steps and can be explained solely by electrical signal spread (reviewed by [59]). The situation, however, appears to be different for phototaxis.

The series of electrical events clearly differ with light intensity. The first detectable inward current is graded with light intensity and is mainly carried by Ca^{2+}. High light intensities, which evoke photoshock responses, trigger this current on time scales between 30 to 50 μs [69,99,107–110]. Based on these fast activation involvement of biochemical amplification steps in photoresponses of green algae has been questioned and has lead to the suggestion that the rhodopsin itself forms a light-gated ion channel complex [43,89]. However, as already pointed out earlier [7,80], low light responses point to the presence of signal amplification mechanisms and do not support this proposal. Detailed kinetic analysis of photocurrents in *C. reinhardtii* suggests that bleaching of a single rhodopsin may activate more than one Ca^{2+} channel in the eyespot region ([109], Uhl, personal communication). Already early population measurements on *C. reinhardtii* pointed to rising times of the photoreceptor current in the ms range upon low intensity flashes [111]. Also for *H. pluvialis* low light electrical responses, which peak in the ms range and saturate at low intensities, have been reported [80,88]. Optimized single cell measurements on *C. reinhardtii* and *Volvox carteri* demonstrate that the delay time and the time needed for reaching the maximum of the photoreceptor currents shift to several milliseconds below 10^{18} photons m^{-2}. Furthermore the intensity dependence of the rising phase of the photoreceptor currents is clearly biphasic in both species [99,112]. Green algal electrical responses to rhodopsin excitation under low light conditions thus clearly fall in the range reported for fast invertebrate second messenger visual systems (e.g. [100,113]). The time scale of green light-induced electrical responses (about 50 μs to several ms) in these algae is remarkable and can best be explained by close coupling of rhodopsin with the channels [69]. In conjunction with the limited space at which the primary signal is perceived and processed, i.e. the eyespot membranes, this would allow the observed graded electrical responses at varying light intensities. Coupling might either occur by direct rhodopsin-channel interactions or additional signaling elements.

In analogy to known visual systems, heterotrimeric G-proteins are good candidates for a coupling between the rhodopsin and the ion channels. Different lines of evidence point to the presence of putative heterotrimeric G-proteins in preparations of eyespot apparatuses of *C. reinhardtii* and *S. similis* (for review see [7]). Low intensities of green light and Ca^{2+} modulate the activity of at least one of the putative G_α-subunits in *S. similis* [76]. However, currently the messenger(s) mediating activation of the ion channels in the plasma membrane region of the eyespot is (are) still elusive. Because rapid green light-induced changes in the IP_3 level occur in both *C. reinhardtii* and *S. similis* (Brünjes and Kreimer, unpublished results), it is tempting to propose similarities to the invertebrate visual system. IP_3 is assumed to play a major role in many invertebrate photoreceptors (reviewed in [100]). Interestingly one of the putative G_α-proteins in eyespot preparations exhibits cross-reactivity towards antibodies directed against the G_q-subtype [76]. In invertebrate photoreceptors this G_α-subtype couples phospholipase c to rhodopsin-activation [100,114]. Alternatively, the G-protein(s) may directly activate the ion channels or may be involved in light adaptation by controlling the opening probabilities of ion channels. Such mechanisms are well documented in vertebrates/invertebrates and higher plants. Involvement in other rhodopsin-controlled processes must also be considered [76]. The previously postulated biochemical amplification cascade under low light conditions, although now gathering increasing experimental support, thus still awaits further conclusive experimental evidence.

Unfortunately the green algal opsin sequence data allow no clear conclusion with respect to their putative reaction partners. They exhibit principal differences to all other known opsins (for review see Deininger and Hegemann, this volume). No homology to bacteriorhodopsin or bacterial sensory rhodopsins exists and the overall homologies to the known opsins are surprisingly low. Nevertheless, in some regions homologies to motifs of invertebrate opsins are evident. All so far known eukaryotic opsins belong to the group of G-protein coupled receptors, which possess a DRY/ERY consensus motif involved in G-protein binding. This motif is missing in green algae. However, only the arginine is absolutely conserved in this receptor subfamily [115]. Also in the first identified putative plant G-protein coupled receptor only the arginine is conserved in this position [116], whereas in the green algal opsin sequences an exchange to lysine has occurred. In other subfamilies of G-protein coupled receptors this motif is even completely absent [115]. Thus lack of this motif per se does not exclude interactions with G-proteins.

Several lines of evidence now point to complex cellular control mechanisms in green algal phototransduction. Identification of the intracellular messenger system(s) involved will be a future challenge, which requires the combined application of molecular genetic, biochemical and electrophysiological techniques. The presence of rapid adaptational processes [69–72] will add additional complexity to these analyses. In comparison to visual signaling cascades of invertebrates, these studies will contribute to our understanding how similar biological problems have been solved by unicellular plants.

Acknowledgements

Continuing support by the DFG is acknowledged. Special thank is also given to M. Melkonian for stimulating discussions, comments on the manuscript and for introducing me to this fascinating field of algal cell biology. I also want to express my gratitude to B. Marin for late night discussions and for providing some of his EM negatives and help with the phylogenetic tree. Gift of *C. reinhardtii* strains 595 and 593 by O.A. Sineshchekov and 806 by P. Hegemann is acknowledged. Thanks is also given K.H. Linne-von Berg for critical reading of the manuscript and to all who provided pre-prints and reprints of their recent work. Special thanks is also given to the different members of my laboratory for stimulating discussions, help and fun during the last years.

References

1. B. Diehn, M. Feinleib, W. Haupt, E. Hildebrand, F. Lenci, W. Nultsch (1977). Terminology of behavioral responses of motile microorganisms. *Photochem. Photobiol.*, **26**, 559–560.
2. C.G. Ehrenberg (1838). *Die Infusionsthierchen als vollkommene Organismen*. Leipzig.
3. J.D. Dodge (1973). *The Fine Structure of Algal Cells*. Academic Press, London.
4. K.W. Foster, R.D. Smyth (1980). Light antennas in phototactic alga. *Microbiol. Rev.*, **44**, 572–630.
5. M. Melkonian, H. Robenek (1984). The eyespot apparatus of flagellated green algae: a critical review. In: F.E. Round, D.J. Chapman (Eds), *Progress in Phycological Research* (Vol. 3, pp. 193–268). Biopress, Bristol.
6. W. Nultsch, D.-P. Häder (1988). Photomovement in motile microorganisms. II. *Photochem. Photobiol.*, **47**, 837–869.
7. G. Kreimer (1994). Cell biology of phototaxis in flagellate algae. *Int. Rev. Cytol.*, **148**, 229–310.
8. P. Hegemann, H. Harz (1998). How microalgae see the light. In: M.X. Caddick, S. Baumberg, D.A. Hodgson, M.K. Phillip-Jones (Eds), *Microbial Responses to Light and Time. Society for General Microbiology Symposium 56* (pp. 95–105). Cambridge University Press.
9. H. Kawai, G. Kreimer (2000). Sensory mechanisms: Phototaxes and light perception in algae. In: B.S.C. Leadbeater, J.C. Green (Eds), *The Flagellates: Unity, Diversity and Evolution* (pp. 124–146). Taylor and Francis.
10. M. Melkonian (1996). Phylogeny of photosynthetic protists and their plastids. *Ver. Dtsch. Zool. Ges.*, **89**, 71–96.
11. J.D. Dodge (1984). The functional and phylogenetic significance of dinoflagellate

eyespots. *BioSystems*, **16**, 259–267.
12. H.J. Hoops (1997). Motility in the colonial and multicellular Volvocales: structure, function, and evolution. *Protoplasma*, **199**, 99–112.
13. D.L. Kirk (1998). *Volvox. Molecular-genetic Origins of Multicellularity and Cellular Differentiation*. Cambridge University Press, Cambridge.
14. M. Melkonian, H. Robenek (1981). The eyespot apparatus of green algae: structure, development and function. *Phycologia*, **20**, 110.
15. I. Inouye, T. Hori (1991). High-speed video analysis of the flagellar beat and swimming patterns of algae: possible evolutionary trends in green algae. *Protoplasma*, **164**, 54–69.
16. S.F. Goldstein (1992). Flagellar beat patterns in algae. In: M. Melkonian (Ed.), *Algal Cell Motility* (pp. 99–153). Chapman and Hall, New York, London.
17. I. Ohad, I. Goldberg, R. Broza, S. Schuldiner, E. Gan-Zvi (1969). Changes in lipid and pigment composition and photosynthetic activity during formation of chloroplast lamellae in a mutant of *Chlamydomonas reinhardtii* y-1. In: H. Metzner (Ed.), *Progress in Photosynthesis Research* (Vol. 1, pp. 284–295). International Union of Biological Sciences, Tübingen.
18. M. Grung, G. Kreimer, M. Calenberg, M. Melkonian, S. Liaaen-Jensen (1994). Carotenoids in the eyespot apparatus of the flagellate green alga *Spermatozopsis similis*: Adaptation to the retinal-based photoreceptor. *Planta*, **193**, 38–43.
19. G. Kreimer, U. Brohsonn, M. Melkonian (1991). Isolation and partial characterization of the photoreceptive organelle for phototaxis of a flagellate green alga. *Eur. J. Cell Biology*, **55**, 318–327.
20. T. Hori, Ø. Moestrup, L.R. Hoffman (1995). Fine structural studies on an ultraplanctonic species of *Pyramimonas*, *P. virginica* (Prasinophyceae), with a discussion of subgenera within the genus *Pyramimonas*. *Eur. J. Phycol.*, **30**, 219–234.
21. M.W. Watson (1975). Flagellar apparatus, eyespot and behavior of *Microthamnion kuetzingianum* (Chlorophyceae) zoospores. *J. Phycol.*, **11**, 439–448.
22. J.A. Holmes, S.K. Dutcher (1989). Cellular asymmetry in *Chlamydomonas reinhardtii*. *J. Cell Sci.*, **94**, 273–285.
23. K.-F. Lechtreck, I.B. Reize, M. Melkonian (1997). The cytoskeleton of the naked green flagellate *Spermatozopsis similis* (Chlorophyta): flagellar and basal body developmental cycle. *J. Phycol.*, **33**, 254–265.
24. G.J. Pazour, O.A. Sineshchekov, G.B. Witman (1995). Mutational analysis of the phototransduction pathway of *Chlamydomonas reinhardtii*. *J. Cell Biol.*, **131**, 427–440.
25. B. Marin, K. Hoef-Emden, M. Melkonian (1996). Light and electron microscope observations on *Tetraselmis desikacharyi* sp. nov. (Chlorodendrales, Chlorophyta). *Nova Hedwigia*, **112**, 461–475.
26. G. Kreimer, M. Melkonian (1990). Reflection confocal laser scanning microscopy of eyespots in flagellated green algae. *Eur. J. Cell Biology*, **53**, 101–111.
27. B. Huang, Z. Ramanis, S.K. Dutcher, D.J.L. Luck (1982). Uniflagellar mutants of *Chlamydomonas*: Evidence for the role of basal bodies in transmission of positional information. *Cell*, **29**, 745–753.
28. R. Kamiya, G.B. Witman (1984). Submicromolar levels of calcium control the balance of beating between the two flagella in demembranated models of *Chlamydomonas*. *J. Cell Biol.*, **98**, 97–107.
29. U. Rüffer, W. Nultsch (1985). High-speed cinematographic analysis of the movement of *Chlamydomonas*. *Cell Motil. Cytoskeleton*, **5**, 251–263.
30. M.A. Sleigh (1991). Mechanisms of flagellar propulsion. A biologistś view of the relation between structure, motion and fluid mechanics. *Protoplasma*, **164**, 45–53.
31. U. Rüffer, W. Nultsch (1987). Comparison of the beating of cis- and trans-flagella of

Chlamydomonas cells held on micropipettes. *Cell Motil. Cytoskeleton*, **7**, 87–93.
32. H.C. Crenshaw (1993). Orientation by helical motion – I. Kinematics of the helical motion of organisms with up to six degrees of freedom. *Bull. Math. Biol.*, **55**, 197–212.
33. U. Rüffer, W. Nultsch (1991). Flagellar photoresponses of *Chlamydomonas* cells held on micropipettes: II. Change in flagellar beat pattern. *Cell Motil. Cytoskeleton*, **18**, 269–278.
34. H.C. Crenshaw (1993). Orientation by helical motion – III. Microorganisms can orient to stimuli by changing the direction of their rotational velocity. *Bull. Math. Biol.*, **55**, 231–255.
35. K. Schaller, R. David, R. Uhl (1997). How *Chlamydomonas* keeps track of the light once it has reached the right phototactic orientation. *Biophys. J.*, **73**, 1562–1572.
36. H.J. Arnott, R.M. Brown Jr. (1967). Ultrastructure of the eyespot and its possible significance in phototaxis of *Tetracystis excentrica*. *J. Protozool.*, **14**, 529–539.
37. P.L. Walne, H.J. Arnott (1967). The comparative ultrastructure and possible function of eyespots: *Euglena granulata* and *Chlamydomonas eugametos*. *Planta*, **77**, 325–353.
38. M. Melkonian, H. Robenek (1980). Eyespot membranes of *Chlamydomonas reinhardtii*: a freeze-fracture study. *J. Ultrastruc. Res.*, **72**, 90–102.
39. M. Melkonian, H. Robenek (1980). Eyespot membranes in newly released zoospores of the green alga *Chlorosarcinopsis gelatinosa* (Chlorosarcinales) and their fate during zoospore settlement. *Protoplasma*, **104**, 129–140.
40. H. Robenek, M. Melkonian (1981). Comparative ultrastructure of eyespot membranes in gametes and zoospores of the green alga *Ulva lactuca* (Ulvales). *J. Cell Sci.*, **50**, 149–164.
41. W. Nultsch (1983). The photocontrol of movement of *Chlamydomonas*. In: D.J. Cosens, D. Vince-Pruce, (Eds), *The Biology of Photoreception, Society for Experimental Biology* (pp. 521–539). Great Britain.
42. M. Beckmann, P. Hegemann (1991). In vitro identification of rhodopsin in the green alga *Chlamydomonas*. *Biochemistry*, **30**, 3692–3697.
43. W. Deininger, P. Kröger, U. Hegemann, F. Lottspeich, P. Hegemann (1995). Chlamyrhodopsin represents a new type of sensory photoreceptor. *EMBO J.*, **14**, 5849–5858.
44. H.-W. Kuhlmann (1998). Photomovements in ciliated protozoa. *Naturwissenschaften*, **85**, 143–154.
45. B. Marin (1996). *Die frühe Evolution der Grünalgen: Vergleichende Untersuchungen ultrastruktureller Merkmale, sowie phylogenetische Sequenzanalysen von kern- und plastidenkodierten ribosomalen RNA-Sequenzen*. Ph.D. Dissertation, Universität zu Köln, Köln, Germany.
46. S.D. Sym, R.N. Pienaar (1993). The class Prasinophyceae. In: F.E. Round, D.J. Chapman (Eds), *Progress in Phycological Research* (Vol. 9, pp. 281–376). Biopress, Bristol.
47. M. Melkonian, I.B. Reize, H.R. Preisig (1987). Maturation of a flagellum/basal body requires more than one cell cycle in algal flagellates: studies on *Nephroselmis olivacea* (Prasinophyceae). In: W. Wiessner, D.G. Robinson, R.C. Starr (Eds), *Algal Development, Molecular and Cellular Aspects* (pp. 102–113). Springer, Berlin, Heidelberg, New York, Tokyo.
48. Ø. Moestrup, T. Hori (1989). Ultrastructure of the flagellar apparatus in *Pyramimonas octopus* (Prasinophyceae) II. Flagellar roots, connecting fibers, and numbering of individual flagella in green alga. *Protoplasma*, **148**, 41–56.
49. P.L. Beech, K. Heimann, M. Melkonian (1991). Development of the flagellar apparatus during the cell cycle in unicellular algae. *Protoplasma*, **164**, 23–37.
50. Ø. Moestrup, H. Ettl (1979). A light and electron microscopical study of *Nephroselmis olivacea* (Prasinophyceae). *Opera Bot.*, **49**, 1–39.
51. K.P. Gaffal, S. el-Gammal, G.J. Friedrichs (1993). Computer-aided 3D-reconstruction of

the eyespot-flagellar/basal apparatus-contractile vacuoles-nucleus-associations during mitosis of *Chlamydomonas reinhardtii*. *Endocyt. Cell Res.*, **9**, 177–208.
52. R.A. Andersen, D.J.S. Barr, D.H. Lynn, M. Melkonian, Ø. Moestrup, M.A. Sleigh (1991). Terminology and nomenclature of the cytoskeletal elements associated with the flagellar/ciliary apparatus in protists. *Protoplasma*, **164**, 1–8.
53. H. Robenek, M. Melkonian (1979). Rhizoplast-membrane associations in the flagellate *Tetraselmis cordiformis* Stein (Chlorophyceae) revealed by freeze-etching and thin sections. *Arch. Protistenkd.*, **122**, 340–351.
54. T. Hori, I. Inouye, T. Horiguchi, G.T. Boalch (1985). Observations on the motile stage of *Halosphera minor* Ostenfeld (Prasinophyceae) with special reference to the cell structure. *Bot. Mar.*, **28**, 529–537.
55. H. Ettl, I. Manton (1964). Die feinere Struktur von *Pedinomonas minor* Korschikoff. *Nova Hedwigia*, **8**, 421–451.
56. M. Melkonian, B. Marin, B. Surek (1995). Phylogeny and evolution of the algae. In: R. Arai, M. Kato, Y. Doi (Eds), *Biodiversity and Evolution* (pp. 153–176). The National Science Museum Foundation, Tokyo.
57. T. Nakayama, B. Marin, H.D. Kranz, B. Surek, V.A.R. Huss, I. Inouye, M. Melkonian (1998). The basal position of scaly green flagellates among the green algae (Chlorophyta) is revealed by analyses of nuclear-encoded SSU rRNA sequences. *Protist*, **149**, 367–380.
58. T. Friedl (1997). The evolution of the green algae. In: D. Bhattacharya (Ed.), *The Origin of Algae and Plastids* (pp. 87–101). Springer, Wien.
59. P. Hegemann (1997). Vision in microalgae. *Planta*, **203**, 265–274.
60. M.A. Sleigh (1981). Flagellar beat patterns and their possible evolution. *Biosystems*, **14**, 423–431.
61. M. Melkonian (1983). Functional and phylogenetic aspects of the basal apparatus in algal cells. *J. Submicroscop. Cytol.*, **15**, 121–125.
62. K. Heimann, I.B. Reize, M. Melkonian (1989). The flagellar developmental cycle in algae: flagellar transformation in *Cyanophora paradoxa* (Glaucocystophyceae). *Protoplasma*, **148**, 106–110.
63. W. Nultsch (1988). Untersuchungen zum Bewegungs- und Reaktionsverhalten des Flagellaten *Chlamydomonas reinhardtii*, Sitzungsberichte der Wissenschaftlichen Gesellschaft an der Johann-Wolfgang-Goethe-Universität Frankfurt am Main Bd. 24, Franz Steiner Verlag, Wiesbaden, Stuttgart.
64. R.L. Stavis, R. Hirschberg (1973). Phototaxis in *Chlamydomonas reinhardtii*. *J. Cell Biol.*, **59**, 367–377.
65. W. Nultsch (1979). Effect of external factors on phototaxis of *Chlamydomonas reinhardtii*. III. Cations. *Arch. Microbiol.*, **123**, 93–99.
66. N. Morel-Laurens (1987). Calcium control of phototactic orientation in *Chlamydomonas reinhardtii*: Sign and strength of response. *Photochem. Photobiol.*, **45**, 119–128.
67. T. Takahashi, M. Watanabe (1993). Photosynthesis modulates the sign of phototaxis of wild-type *Chlamydomonas reinhardtii*. *FEBS Lett.*, **336**, 516–520.
68. P. Hegemann, B. Bruck (1989). Light-induced stop responses in *Chlamydomonas reinhardtii*: Occurrence and adaptation phenomena. *Cell Motil. Cytoskeleton*, **14**, 501–515.
69. O.A. Sineshchekov, F.F. Litvin, L. Keszthelyi (1990). Two components of photoreceptor potential in phototaxis of the flagellated green alga *Haematococcus pluvialis*. *Biophys. J.*, **57**, 33–39.
70. D.N. Zacks, F. Derguini, K. Nakanishi, J.L. Spudich (1993). Comparative study of phototactic and photophobic receptor chromophore properties in *Chlamydomonas reinhardtii*. *Biophys. J.*, **65**, 508–518.

71. D.N. Zacks, J.L. Spudich (1994). Gain setting in *Chlamydomonas reinhardtii*: Mechanism of phototaxis and the role of the photophobic response. *Cell Motil. Cytoskeleton*, **29**, 225–230.
72. E.G. Govorunova, O.A. Sineshchekov, P. Hegemann (1997). Desensitization and dark recovery of the photoreceptor current in *Chlamydomonas reinhardtii*. *Plant Physiol.*, **115**, 633–642.
73. M.E. Feinleib, G.M. Curry (1971). The relationship between stimulus intensity and oriented phototactic response (topotaxis) in *Chlamydomonas*. *Physiol. Plant.*, **25**, 346–352.
74. L. Linden, G. Kreimer (1995). Calcium modulates rapid protein phosphorylation/dephosphorylation in isolated eyespot apparatuses of the green alga *Spermatozopsis similis*. *Planta*, **197**, 343–351.
75. U. Schlicher, L. Linden, M. Calenberg, G. Kreimer (1995). G proteins and Ca^{2+}-modulated protein kinases of a plasma membrane-enriched fraction and isolated eyespot apparatuses of *Spermatozopsis similis* (Chlorophyceae). *Eur. J. Phycol.*, **30**, 319–330.
76. M. Calenberg, U. Brohsonn, M. Zedlacher, G. Kreimer (1998). Light- and Ca^{2+}-modulated heterotrimeric GTPases in the eyespot apparatus of a flagellate green alga. *Plant Cell*, **10**, 91–103.
77. F. Crescitelli, T.W. James, J.M. Erickson, E.R. Loew, W.N. McFarland (1992). The eyespot of *Chlamydomonas reinhardtii*: a comparative microspectrophotometric study. *Vision Res.*, **32**, 1593–1600.
78. K. Schaller, R. Uhl (1997). A microspectrophotometric study of the shielding properties of eyespot and cell body in *Chlamydomonas*. *Biophys. J.*, **73**, 1573–1578.
79. J.S. Boscov, M.E. Feinleib (1979). Phototactic response of *Chlamydomonas* to flashes of light. II. Response of individual cells. *Photochem. Photobiol.*, **30**, 499–505.
80. O.A. Sineshchekov (1991). Electrophysiology of photomovements in flagellated algae. In: F. Lenci, F. Ghetti, G. Colombetti, D.-P. Häder, P.S. Song (Eds), *Biophysics of Photoreceptors and Photomovements in Microorganisms* (pp. 191–202). Plenum Press, New York.
81. K. Yoshimura (1994). Chromophore orientation in the photoreceptor of *Chlamydomonas* as probed by stimulation with polarised light. *Photochem. Photobiol.*, **60**, 594–597.
82. J. Buder (1917). Zur Kenntnis der phototaktischen Richtungsbewegungen. *Jahrb. Wiss. Bot.*, **58**, 105–220.
83. G.B. Witman (1993). *Chlamydomonas* phototaxis. *Trends Cell Biol.*, **3**, 403–408.
84. C.J. Horst, G.B. Witman (1993). *ptx1*, a nonphototactic mutant of *Chlamydomonas*, lacks control of flagellar dominance. *J. Cell Biol.*, **120**, 733–741.
85. U. Rüffer, W. Nultsch (1997). Flagellar photoresponses of ptx1, a nonphototactic mutant of *Chlamydomonas*. *Cell Motil. Cytoskeleton*, **37**, 111–119.
86. N. Morel-Laurens, M.E. Feinleib (1983). Photomovement in an "eyeless" mutant of *Chlamydomonas*. *Photochem. Photobiol.*, **37**, 189–194.
87. G. Kreimer, C. Overländer, O.A. Sineshchekov, H. Stolzis, W. Nultsch, M. Melkonian (1992). Functional analysis of the eyespot in *Chlamydomonas reinhardtii* mutant *ey* 627, *mt-*. *Planta*, **188**, 513–521.
88. O.A. Sineshchekov (1991). Photoreception in unicellular flagellates: bioelectric phenomena in phototaxis. In: R.H. Douglas (Ed.), *Light in Biology and Medicine* (Vol. 2, pp. 523–532). Plenum Press, New York.
89. H. Harz, C. Nonnengäßer, P. Hegemann (1992). The photoreceptor current of the green alga *Chlamydomonas*. *Philos. Trans.R. Soc. London Ser.*, **B 338**, 39–52.
90. O.A. Sineshchekov, E.G. Govorunova, F.F. Litvin (1989). Role of photosynthetic apparatus and stigma in the formation of spectral sensitivity of phototaxis in flagellated green algae.

Biofizika, **34**, 255–258.

91. S.O. Mast (1927). Structure and function of the eyespot in unicellular and colonial organisms. *Arch. Protistenkd.*, **60**, 197–220.
92. J. Pickett-Heaps, J. Pickett-Heaps (1995). *Cellular evolution in the green algae. Volvocales, Tetrasporales and Chlorococcales*. Cytographics Production, Cytographics Pty. Ltd., Ascot Vale.
93. E. Hecht (1974). *Optics*. Addison-Wesley Publishing Company.
94. W. Nultsch, T. Throm, I. von Rimscha (1971). Phototaktische Untersuchungen an *Chlamydomonas reinhardii* Dangeard in homokontinuierlicher Kultur. *Arch. Mikrobiol.*, **80**, 351–69.
95. K.W. Foster, J. Saranak, N. Patel, G. Zarilli, M.Okabe, T. Kline, K. Nakanishi (1984). A rhodopsin is the functional photoreceptor for phototaxis in the unicellular eukaryote *Chlamydomonas*. *Nature*, **311**, 756–759.
96. P. Hegemann, W. Marwan (1988). Single photons are sufficient to trigger movement responses in *Chlamydomonas reinhardtii*. *Photochem. Photobiol.*, **48**, 99–106.
97. T. Takahashi, M. Kubota, M. Watanabe, K. Yoshihara, F. Derguini, K. Nakanishi (1992). Diversion of the sign of phototaxis in a *Chlamydomonas reinhardtii* mutant incorporated with retinal and its analogs. *FEBS Lett.*, **314**, 275–279.
98. R. Dolle, J. Pfau, W. Nultsch (1987). Role of calcium ions in motility and phototaxis of *Chlamydomonas reinhardtii*. *J. Plant Physiol.*, **126**, 467–473.
99. F.-J. Braun (1997). Untersuchungen von Photorezeptorströmen und intrazellulären Ionenkonzentrationen in *Chlamydomonas reinhardtii* und *Volvox carteri*, Ph.D. Dissertation, Universität Regensburg, Regensburg, Germany.
100. R. Ranganathan, D.M. Malicki, C.S. Zuker (1995). Signal transduction in *Drosophila* photoreceptors. *Annu. Rev. Neurosci.*, **18**, 283–317.
101. G. Kreimer (1995). Flagellar shock responses in green algae. *Bot. Acta*, **108**, 169–171.
102. E.M. Holland, H. Harz, R. Uhl, P. Hegemann (1997). Control of phobic behavioral responses by rhodopsin-induced photocurrents in *Chlamydomonas*. *Biophys. J.*, **73**, 1395–1401.
103. K. Yoshimura, C. Shingyoji, K. Takahashi (1997). Conversion of beating mode in *Chlamydomonas* flagella induced by electric stimulation. *Cell Motil. Cytoskeleton*, **36**, 236–245.
104. K. Hill, R. Hemmler, P. Kovermann, M. Calenberg, G. Kreimer, R. Wagner (2000). A Ca^{2+}-and voltage-modulated flagellar ion channel is a component of the mechanoshock response in the unicellular green alga *Spermatozopsis*, *Biochim. Biophys. Acta*, **1466**, 187–204.
105. A. Matsuda, K. Yoshimura, O.A. Sineshchekov, M. Hirono, R. Kamiya (1999). Isolation and characterization of novel *Chlamydomonas* mutants that display phototaxis but not photophobic response. *Cell Motil. Cytoskeleton*, *41*, 353–362.
106. P. Kröger, P. Hegemann (1994). Photophobic responses and phototaxis in *Chlamydomonas* are triggered by a single rhodopsin photoreceptor. *FEBS Lett.*, **341**, 5–9.
107. F.F. Litvin, O.A. Sineschekov, V.A. Sineschekov (1978). Photoreceptor electric potential in the phototaxis of the alga *Haematococcus pluvialis*. *Nature*, **271**, 476–478.
108. H. Harz, P. Hegemann (1991). Rhodopsin-regulated calcium currents in *Chlamydomonas*. *Nature*, **351**, 489–491.
109. C. Beck (1996). Lokalisation und Eigenschaften lichtinduzierter Ionenströme in *Chlamydomonas reinhardtii*, Ph.D. Dissertation, Universität München, München, Germany.
110. E.-M. Holland, F.-J. Braun, C. Nonnengässer, H. Harz, P. Hegemann (1996). The nature of rhodopsin-triggered photocurrents in *Chlamydomonas*. I. Kinetics and influence of divalent cations. *Biophys. J.*, **70**, 924–931.
111. O.A. Sineshchekov, E.G. Govorunova, A. Der, L. Keszthelyi, W. Nultsch (1992).

Photoelectric responses in phototactic flagellated algae measured in cell suspension. *J. Photochem. Photobiol. B: Biol.*, **13**, 119–134.

112. F.-J. Braun, P. Hegemann (1998). Two light-activated conductances in the eye of the green alga *Volvox carteri*. *Biophys. J.*, **76**, 1668–1678.
113. S.B. Laughlin, M. Weckström (1993). Fast and slow photoreceptors – a comparative study of the functional diversity of coding and conductances in the Diptera. *J. Comp. Physiol. A*, **172**, 593–609.
114. T. Suzuki, K. Narita, K. Yoshihara, Nagai K., Y. Kito (1995). Phosphatidyl inositol-phospholipase C in squid photoreceptor membrane is activated by stable metarhodopsin via GTP-binding protein, Gq. *Vision Res.*, **35**, 1011–1017.
115. L. Birnbaumer, M. Birnbaumer (1995). G proteins in signal transduction. In: M. Shinitzky (Ed.), *Biomembranes: Signal Transduction across Membranes* (Vol. 3, pp. 153–252). VHC-Verlagsgesellschaft, Weinheim.
116. S. Plakidou-Dymock, D. Dymock, R. Hooley (1998). A higher plant seven-transmembrane receptor that influences sensitivity to cytokinins. *Curr. Biol.*, **8**, 315–324.

Photomovement
D.-P. Häder and M. Lebert, editors.

Chapter 8

Algal eyes and their rhodopsin photoreceptors

Peter Hegemann and Werner Deininger

Table of contents

Abstract

The eyes of green flagellate algae as *Chlamydomonas reinhardtii* and *Volvox carteri* constitute the simplest and most common visual system found in nature. The eyes contain optics, photoreceptors and elementary components of a signal transduction chain. Rhodopsin serves as the photoreceptor, as it similarly does in animal vision. Upon light stimulation, its all-*trans*-retinal chromophore isomerizes into 13-*cis*.

At low light the rhodopsin activates via an amplification system non selective cation conductance within the eyespot area. In addition, the rhodopsin itself is thought to constitute a Ca^{2+}/H^{+} conductance, which dominates at high light levels. The Ca^{2+} and H^{+} influx into the eyespot region triggers flagellar currents which control flagellar beating and thereby orientation of the algae in light. The identification of proteins contributing to this signaling system just begun with the isolation and cloning of opsins from *Chlamydomonas* and *Volvox*. These plant opsins are highly charged and not typical 7-helix receptors. But still, algal opsins and their genes show striking homologies to the animal counterparts. This relation is discussed here in some detail.

8.1 Algal eyes

Green algae such as *Chlamydomonas, Haematococcus, Dunaliella, Volvox* and several other members of the *Chlorophycean* class are among the smallest organisms which possess eyes [1,2]. These eyes have diameters in the range of only 1 μm. Most single-celled species are biflagellate and swim with their flagella in forward direction. The cells rotate counter-clockwise around their swimming axis, and during rotation the eye scans the environment for light intensity and color. The photoreceptor receives a modulated light signal and the flagellar beat frequency, the beating space, and the beating mode are modulated accordingly. This signaling system comprises a set of rhodopsin-, transmitter-, and voltage-regulated conductances, which are located in the eyespot membrane and all the way along the flagellar membrane (reviewed in [3], see chapter by Sineshchekov and Govorunova, this volume). The swimming direction is changed until the received signal exhibits a minimal degree of intensity- and color modulation [4,5], see chapter by Kreimer, this volume).

The location of the eye is of fundamental importance for the reception of an interpretable light signal. In most *Chlorophyceaen* genera the orange eye is located between flagella and the cell equator. In *Chlamydomonas* the eye is positioned almost precisely equatorial. The location slightly varies during the cell cycle. During helical swimming the eye advances the flagellar beating plane by 20 to 40°. The angle corresponds well to the time needed for signal transduction from the eye to the flagella. The eyespot is attached to the microtubular rootlet (MTR) that emanates from the *cis* basal body and extends towards the distal end underneath the plasmalemma [6]. The *cis*-basal body is the one that developed during the last cell division. It becomes the *trans*-basal body in all subsequent generations [7], which implies that the eye must be disconnected from the MTR during division.

In the colonial alga *Volvox carteri* the 2000 to 4000 somatic cells are very similar to *Chlamydomonas* gametes. Somatic cells are "working cells" and imbedded in the cell matrix at the surface of the colony, whereas the 16 large reproductive cells (gonidia) have no direct contact with the extracellular medium in the adult colony. The individual cells are fixed within the spheroid, whereas the whole spheroid rotates around the axis of propagation. The flagella of the individuals beat synchronized and in an almost precisely parallel fashion. The flagella of different cells beat at different angles but all beat towards the colonial posterior pushing the colony to the anterior side along the axis of rotation [8]. The cells in the front part of the colony possess larger and more sensitive eyes than those at the rear end [9,8].

Modulation of light incidence does not automatically imply that the received light signal is proportionally modulated. Signal modulation would be minimal in fully transparent cells with randomly distributed photoreceptors. Hence, light direction could not be detected in such a cell. However, the algal photoreceptors are localized in the eye and an optical system provides the directivity. The optical system operates on the basis of light reflection and constructive interference [4,10]. It consists of layered carotenoid-rich lipid globules which are closed, hexagonally packed. The layers reflect the light, and the reflected waves interfere if the difference of their light paths is a multiple of $\lambda/2$, thus producing maxima and minima of interference waves [1]. At perpendicular light incidence the major maximum is $\lambda/2$ outside the outermost layer. Therefore, the eyespot overlaying part of the plasmalemma has been favored as the ideal location of the photoreceptor [4]. Reflection and interference increases with the number of layers. Although interference reflection dominates the optical system of most algal eyes, other factors as absorption of the chloroplast or the orientation of the receptor chromophore parallel to the plane of the eyespot membranes [11] also contribute to the directivity of the eye.

It should be kept in mind, that the absolute light sensitivity of a visual system is primarily determined by the number of photoreceptor molecules, whereas the spatial resolution is determined by the optics or, in other words, the optics is of great importance for path-finding, but only of low importance for non-directional phobic responses.

Since reflection and especially interference depend on the color of the light, green algal eyes are well-suited only for a certain color range, which is smaller than the range of a lens system. Interference occurs at a tilted angle at wavelengths shorter than the optimum, whereas it is simply reduced for longer wavelengths [5,12]. Optimal light incidence is rare, just as it is in any other visual system. Thus, the light signal is color modulated if cells do not swim parallel to the light.

8.2 *In vivo* identification and characterization of algal rhodopsins

Because phototaxis includes directivity, the optical properties of the eye and other shading pigments contribute to phototaxis action spectra to a greater extent than they contribute to spectra for phobic responses. Moreover, phototaxis is measured in continuous light where adaptation phenomena distort the action spectra. Therefore, only phototaxis action spectra constructed from recordings at low irradiance reveal the nature of the behavioral photoreceptor in green algae (see chapter by Foster, this volume).

Threshold phototaxis action spectra for *Chlorophycaeen* algae are rhodopsin-shaped with maxima between 460 and 560 nm, whereas high intensity spectra are more complex. In contrast, action spectra for flash-induced responses are all rhodopsin shaped and peak between 490 and 520 nm [13,14]. They are to a much lower degree distorted by eye pigments and other light absorbing cell material.

The existence of a rhodopsin-type photoreceptor in *Chlorophyceae* was experimentally verified by reconstituting blind retinal-deficient *Chlamydomonas* cells. Phototaxis and photophobic responses were restored after addition of all-*trans* retinal. Other mono-*cis* isomers were much less effective and their ability to restore phototaxis was most likely caused by all-*trans* retinal impurities. Several groups contributed to a detailed *in vivo* characterization of the chromophore in that they took these blind cells and added a variety of retinal analogs that were either unable to isomerize special C–C bonds or were devoid of methyl groups, or lacked the entire ring. The behavior was monitored by using several independent test systems (reviewed by [15,16]). The main conclusions from these reconstitution experiments are briefly summarized as follows:

1. Chlamyrhodopsin contains an all-*trans*, 6-S-*trans* retinal chromophore very much like bacterial rhodopsins, whereas in animal rhodopsins, 11-*cis* retinal or the derivatives 11-*cis*-3-hydroxy, 11-*cis*-4-hydroxy or 11-*cis* 3,4-dehydroretinal form the functional chromophore.
2. The algal chromophore undergoes a 13-*trans* to *cis* isomerization during illumination. The isomerization is converted into a conformational change of the protein via the 13-methyl group.
3. The retinal is in a planar conformation across the C6–C7 single bond, which links the *trans* conformation of the polyene chain to the ionone ring.
4. The ring is not a functionally essential component. It accelerates reconstitution, but a chromophore with only three conjugated double bonds plus methyl groups also restores photosensitivity.

Unlike most animal rhodopsins, the chlamyrhodopsin chromophore is easily accessible by hydroxylamine in light. Hydroxylamine bleaches the chromophore by cleavaging the retinylidene linkage and forming retinaloxime. After washing out the hydroxylamine, retinal restores the chromophore and the cell's ability to carry out phototaxis and phobic responses [17].

The presence of all-*trans* retinal in *Chlamydomonas* cells was verified by extracting retinoids from the cells and by analyzing them using HPLC. Due to the large amounts of carotenoids in the eyes and photosynthetic units of wild type cells, the early experiments were carried out again with the white strain FN68, which synthesizes only small amounts of retinal but the full amount of opsin protein. Upon illumination with green light, the few functional rhodopsin molecules induce retinal synthesis and the complete reconstitution of all rhodopsin within several minutes [18]. From these resensitized cells all-*trans* retinal was first extracted and spectroscopically identified [19]. Later, also wild type cells were extracted. In light-grown cells, besides all-*trans*-retinal, small amounts of the 13-*cis* and 11-*cis* isomers were found [20,21]. Kreimer et al. [22] identified all-*trans* and 11-*cis* retinal in the chlorophycean alga *Spermatozopsis similis* thus confirming the previous findings on *Chlamydomonas*.

8.3 Chlamyopsin purification and localisation

Chlamyopsin was, as the first member of the algal opsin family, biochemically identified by labeling the protein in a *Chlamydomonas* retinal-deficient mutant. ^{3}H-retinal was added to the cells at concentrations that were just sufficient to fully reconstitute the phototactic sensitivity. A 30 kDa protein appeared as the only labeled retinal protein of the total membrane fraction [19]. Unfortunately, no procedure could be developed to enrich rhodopsin-containing membranes from these white cells. Therefore, eyespot membranes were purified from wild type cells. The opsin absorption was identified in these fractions using differential microspectrophotometry [19]. However, the excess of carotenoids was so high that the chromophore could not be characterized in any further detail until today. After treating the membranes with detergent, the retinal of the chromophore could be exchanged for ^{3}H-retinal and the enriched opsin visualized by fluorography. The opsin was finally purified to homogeneity from this source [23].

Polyclonal antibodies against the purified chlamyopsin were used for the localization of the opsin in fixed and permeabilized cells [23]. The opsin appeared as a sharply localized spot of 1 μm in diameter near the equatorial position where in living cells the eyespot is seen. The opsin localization was fully conserved in white retinal-deficient cells which are unable to develop a visible eyespot pigmentation. Although the eyespot pigmentation disappears early in the division cycle [24,25], the opsin spot of the mother cell persists until the end of the first division cycle and is still visible at the time when the two new opsin spots are beginning to be formed in the daughter cells. The opsin as the eyespot structure [26] is seen very close to the cleavage furrow. The daughter opsin spots initially also lie near the cleavage furrow before they migrate to the distal sides of the cells, where later the pigmented eyespots are seen. Shortly after cleavage, in the daughter cell containing the maternal opsin the opsin concentration is larger, whereas the opsin spot of the sister reaches its full size only later on [23]. This observation lead to the suggestion, that in each division cycle the opsin molecules of the mother cell are at least partially conserved and reused in the functional eye that develops in one of the daughter cells, whereas opsin is *de novo* synthesized in the other daughter cell (suggested by [27]). Anti-chlamyopsin antibodies also identified the opsin in eyespot

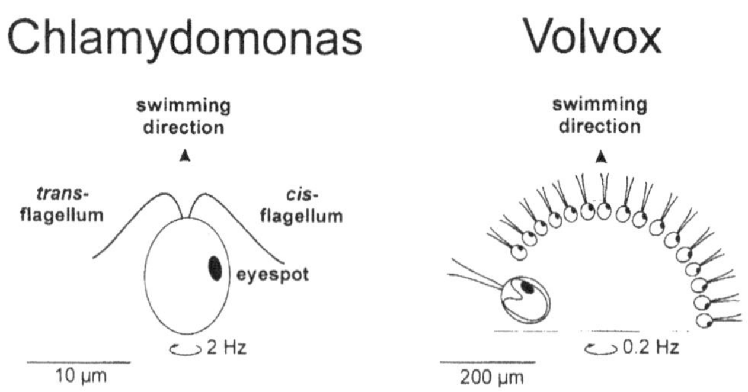

Figure 1. Schematic representations of the unicellular chlorophyceae *Chlamydomonas reinhardtii* and the multicellular species *Volvox carteri* (drawing by F.-J. Braun, modified).

preparations of the chlorophycaean alga *S. similis* [28]. The molecular weight of 32 kDa is slightly higher than that of chlamyopsin. The similisopsin is of special interest because in eyespot preparations of this alga a GTPase activity was discovered, which is light-regulated. Light regulation is suppressed by addition of the anti-chlamyopsin antibodies. This is the only biochemical evidence so far, that algal rhodopsins control a G-protein activity ([28,29], see also chapter by Kreimer, this volume).

8.4 Algal opsin genes

For the identification of the chlamyopsin gene RNA was isolated from non-synchronized cells, that were grown at low light levels. These cells showed the highest light sensitivity, contained the largest number of rhodopsin molecules and were likely to accumulate the most chlamyopsin RNA. Two overlapping opsin cDNA fragments were amplified by RACE-PCR [23]. Later, the complete chlamyopsin gene (*cop*) was sequenced from an EMBL3 clone [30,31]. Meanwhile, using the chlamyopsin cDNA as a probe, the opsin gene and cDNA from the colonial alga *Volvox carteri* (*vop*) were received from a cDNA and a genomic library [32]. The gene structure of both opsins is very similar (Figure 2). The coding regions are interrupted by 7 introns with 6 introns at the same position. Only intron 7 is located in *Volvox* 31 bp further downstream. Within the chlamyopsin gene introns are especially variable in size ranging between 63 and 955 bp. In the volvoxopsin gene the largest intron comprises only 371 base pairs. Compared to the opsin genes of higher eukaryotes [33,34], the algal genes contain three additional introns (Figure 3). However, it is remarkable that the algal introns I2 and I4 are positionally conserved in opsin genes from vertebrates and arthropodes (Figure 3). At the position of the algal I7, an insertion is found in all animal opsin genes. It follows that the algal I2, I4 and I7 predate the plant animal divergence [35]. The animal C-terminal sequences, extending the algal opsins, are separated by an additional intron and are especially divergent in composition and length, suggesting that they constitute relatively late evolutionary achievements.

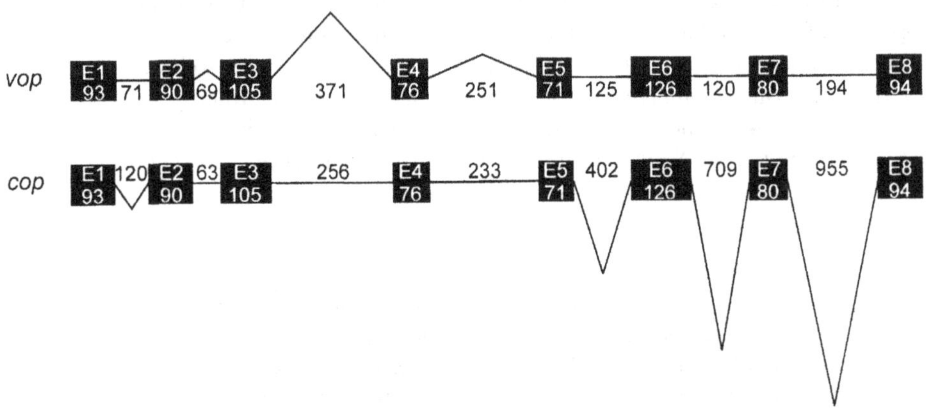

Figure 2. Exon-intron distribution within the two algalopsin genes *cop* and *vop*. The numbers indicate exon (E) and intron sizes.

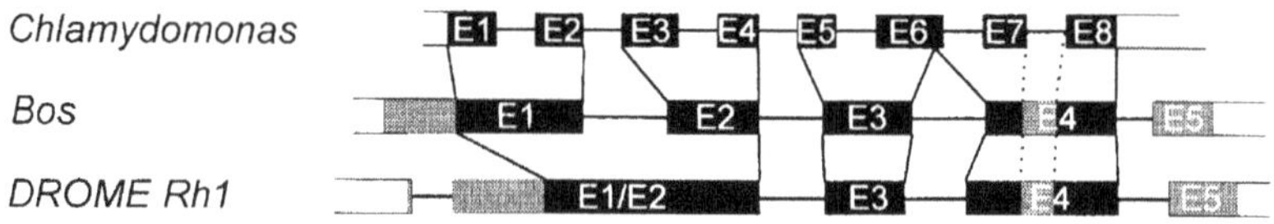

Figure 3. Comparison of the intron-exon arrangement within the *chlamyopsin, bovine opsin (bos)* and *drosophila rh1* (*Drome rh1*) genes.

The coding regions of the two algal opsins show strong codon biases, with preferentially G or C at the third position. This bias is partially responsible for the efficiency of gene expression [31,36]. Due to this criterium, *cop* and *vop* belong to the group of moderately expressed genes [37].

8.5 The opsin proteins

The algal opsin genes code for proteins with molecular weights of around 26 kDa. Algal opsins do not possess a hydrophilic C-terminal extension. But, already Weiss et al. [38] have shown that a rhodopsin without such a tail is functional.

Volvoxopsin and chlamyopsin are 65% identical and certainly belong to the same protein family. The algal opsins are unique in many respects. First of all, the proteins are highly charged and transmembrane segments can hardly be identified. K/R-K/R could define the ends of transmembrane segments because positive charges are qualified to interact with the lipid headgroups, but, hydrophobicity and hydropathy plots only identify 2 to 4 segments that are long enough to define transmembrane helices. Thus, the topography of the algal rhodopsins is totally unclear.

The algal sequences were compared to 25 archaean opsin sequences including bacteriorhodopsin, halorhodopsin and the sensory rhodopsins I and II. No significant homology was found. Besides the retinal binding lysine, no amino acid of known functional importance is conserved in the algal sequences, although the chromophoric properties of algal and archaean rhodopsins are similar. However, algal opsins do show sequence homology to animal opsins, with a slightly higher relation to invertebrate than to vertebrate opsins. Compared to animal opsins, algal opsins are truncated at both ends. But, from the 74 amino acids which are identical in the 7 invertebrate sequences presented in Figure 4, 18 are still present in both algal sequences. Another 18 positions

Figure 4. Comparison of the amino acid sequences of chlamyopsin (Chlam), volvoxopsin (Volvo) and various opsins from invertebrates: *Calliphora vicina* (Calli), *Drosophila melanogaster* Rh6 (Drom6), *Sphrodomantis sp.* (Sphrod), *Procambarus clarkii* (Proc), *Limulus polyphemus* (Limul). *Paroctopus* (Octop), Todarodes (Todar). Consesus sequences are labeled in red, amino acids which are identical in one alga and one or more invertebrates are marked in blue. Amino acids identical only in the two algae are underlayed in green, positions with homologous amino acids in all sequences are indicated by arrows. Assumed transmembrane segments in vertebrate receptors are shaded. Homologous amino acids are defined as explained by Tang et al. [43].

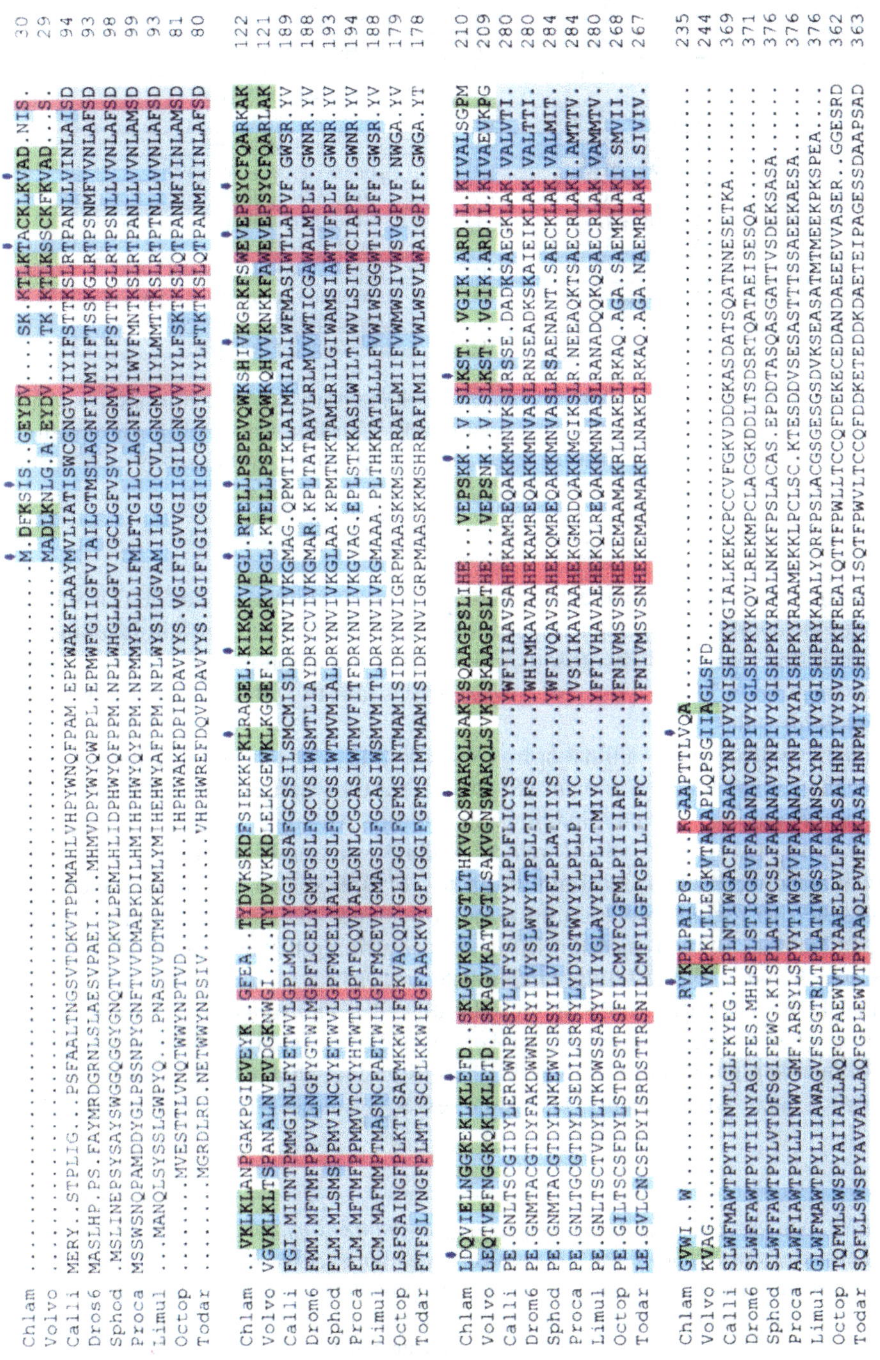

	T. pacificus	algae	location	function
1	L_{65}	Cop L_{17}/Vop L_{18}	H1/CL1	G-Protein interaction
2	I_{73}	Cop V_{25}/Vop V_{26}	H2	structural/G-Protein interaction
3	G_{104}	Cop G_{50}/Vop G_{53}	EL1	structural
4	Y_{111}	Cop Y_{55}/Vop Y_{56}	H3	proposed counter-ion of the protonated schiff-base
5	Y_{220}	Cop Y_{169}/Vop Y_{168}	H5	structural
6	L_{259}	Cop L_{200}/Vop L_{199}	CL3/H6	G-Protein interaction
7	F_{270}/W_{274}	Cop W_{213}/W_{215}	H6	conserved structural motif
8	K_{305}	Cop K_{225}/Vop K_{226}	H7	retinal binding site
9	$HEK_{230-232}$	Cop $HE_{179-180}$ Vop $HE_{180-181}$	CL3	G-Protein binding (conserved in all invertebrate VP)
10	K_{261}	Cop K_{202}/Vop K_{201}	H6	lysine at the beginning of TM6
11	W_{166}	Cop W_{108}/Vop W_{107}	H4	proposed interaction with the chromophore in invertebrates
12	A_{308}	Cop A_{229}	H7	chromophore-binding

Figure 5. List of amino acids, which are of functional importance in animal rhodopsins and are conserved in the algal opsin sequences. Abbreviations: H1-H7 helix1–helix7, CL 1–3 cytoplasmic loops 1–3 EL 1 and 2 extracellular loops 1 and 2. Cop Chlamyopsin, Vop Volvox opsin. After [44,45].

show functionally related amino acids in all 9 sequences (Figure 4 and 5). This means that half of the amino acids, which are identical in the invertebrates, are still functionally conserved in the algae. Chlamyopsin is most related to calliphora-opsin, including 54 identical positions (23%), whereas volvoxopsin is mostly related to DROME rh6, covering 49 identities (21%).

Within the quite divergent animal opsin sequences the cytoplasmic H1–2 loop (Figure 6A) and the retinal binding domain (Figure 6B) are of particularly high similarity. Although the putative retinal binding sites of the two algal opsins are relatively dissimilar, the H1–2 loop and the retinal binding site show the highest degree of conservation between algal and animal opsins (Figures 4 and 6). Besides the gene structure, the two regions provide the strongest evidence that algal and animal opsins developed from a common ancestor. But still, it is striking that within the algal group exon 8, which codes for the retinal binding site, underwent a much faster change than the exons 1–7. One reason might be that the unicell *Chlamydomonas* is a fast signal transducer whereas the spheroidal alga *Volvox* needs a signaling system with slow transduction and higher amplification [39]. This difference should reveal changes in the immediate retinal environment. The algal exon 8 allows the construction of a

Figure 6. A Alignment of the H1–2 loop regions from *Drosophila* opsin Rh1 (Drome 1), *Octopus* opsin (octopus), *bovine* (Bos) in comparison with the related region of the algal opsins Cop und Vop. B Amino acid alignment of the retinal-binding sites from 20 different retinal proteins. C Comparison of the pore loop-like region with pore loops of several K^+ channels. Completely conserved amino acids are underlayed with red boxes, amino acid identities and conservative exchanges between algal opsins and some other opsins are blue shaded. In C amino acid identities in Cop and Vop are underlayed with green boxes.

(a)

Volvox	TK.KTLKS.SC
Chlamy	SK.KTLKT.AC
DROME1	T.TKSLRTPAN
Octopu	SKTKSLQTPAN
bovine	HK.K.LRTPLN

(b)

1. Sensory Rhodopsin II	MYTIVVVYLDFISKVAFVAFAVLGADA	Archaea
2. Sensory Rhodopsin I	GAALTYAFLDVLAKVPYVYFFYARRQA	Archaea
3. Halorhodopsin	ITSWAYSGMDIVAKYLFAFLLLRWVVN	Archaea
4. Bacteriorhodopsin	IETLLFMVLDVSAKVGFGLILLRSRAI	Archaea
5. Chlamyopsin	VKP LPAIPGKGAAPTTLVQA	Algae
6. Volvoxopsin	VKPKLTLEGKVTAKAPLQPSGIIAGLS	Algae
7. Octopus dofleini	VTPYAAELPVLFAKASAIHNPIVYSVS	Invertebrata
8. Todarodes pacificus	VTPYAAQLPVMFAKASAIHNPMIYSVS	Invertebrata
9. Limulus polyphemus	LTPLATIWGSVFAKANSCYNPIVYGIS	Invertebrata
10. Procambarus clarkii	LSPVYTIWGYVFAKANAVYNPIVYAIS	Invertebrata
11. Sphodromantis sp.	ISPLATIWCSLFAKANAVYNPIVYGIS	Invertebrata
12. Drosophila melanogaster rh6	LSPLSTICGSVFAKANAVCNPIVYGLS	Invertebrata
13. Melanopsin Xenopus laevis	LTPYSKTVPAVIAKASAIYNPIIYGII	Vertebrata
14. Calliphora vicina rh1	LTPLNTIWGACFAKSAACYNPIVYGIS	Invertebrata
15. Drosophila melanogaster rh1	LTPLNTIWGACFAKSAACYNPIVYGIS	Invertebrata
16. Ancient Opsin Salmo salar	LDPRLAAAPAFFSKTAAVYNPVIYVFM	Vertebrata
17. Procottus jettelesi	FGPIFMTLPAFFAKTSAIYNPLIYICM	Vertebrata
18. Grenn-sensitive Opsin Gallus gallus	FTATLMAVPAFFSKSSSLYNPIIYVLM	Vertebrata
19. Pinealopsin Petromyzon marinus	IQPVLASLPSYFSKTATVYNPIIYVFM	Vertebrata
20. Peropsin Homo sapiens	IPPPMAIIAPLFAKSSTFYNPCIYVVA	Vertebrata

(c)

Volvox	VEPSNKVSLKSTVGIKARDLKIVAEV
Chlamy	VEPSKKVSLKSTVGIKARDLKIVALS
TOK1	ALYFCTVSL.LTVGLGDILPKSVGAK
dSlo	CVYFLIVTM.STVGYGDVYCETVLGR
mSlo	CVYLLMVTM.STVGYGDVYAKTTLGR
Kv1.1	AFWWAVVSM.TTVGYGDMYPVTIGGK

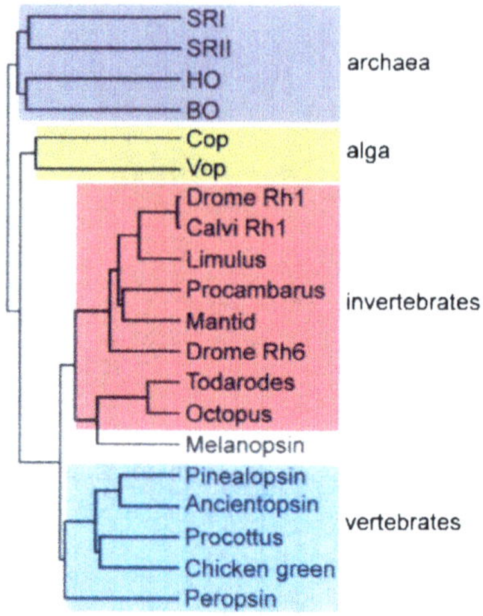

Figure 7. A dendrogram derived from the alignment of the retinal binding sites, corresponding to exon 8 of the algal genes.

dendrogram. Therein algal opsins and animal opsins form a superfamily, from which archaean opsins are separated (Figure 7).

The algal sequences do not give us any clear information about the interaction of the rhodopsins with G-proteins. The tripeptide DRY or ERY at the cytoplasmic border of the third TM segment in G-protein coupled receptors is a characteristic feature for rhodopsin-G-protein interaction [40]. This triplet is missing in the two algal receptors. On the other hand, the sequence comprising the third cytoplasmic loop CL3 between the transmembrane segments 5 and 6 in invertebrate opsins also plays a dominant role for G-protein recognition. CL3 differs in length but is similar in invertebrate and algal opsins (Figures 4 and 5). The algal sequences contain the motive S/TKK-S/TKS/T twice, beginning at position Cop_{14} and Cop_{184}. These segments are part of the cytoplasmic loops CL1 and CL3 in several invertebrates and some vertebrates. The amino acids all interact with cytoplasmic proteins as G-proteins, kinases and arrestins [41].

Algal opsins exhibit characteristics that suggest their direct participation in a receptor-ion channel complex. The lysine-rich sequences interspaced by hydrophobic amino acids (Cop_{58-80}) are reminiscent of S4 stretches of voltage-gated or cGMP-gated channels. In addition, the sequence VSLKSTVGI ($Cop_{187-195}$), which is identical in both algal opsins reminds us to the pore loop region of voltage-gated potassium channels (Figure 6C). Since flash-induced photoreceptor currents begin at high flash energies with a delay below 50 μs (see chapter by Govorunova, this volume), we had concluded that the rhodopsin is either directly linked to the ion channel or by itself forms a light-regulated channel complex [42]. Such a complex is likely to be composed of 4 to 5 subunits. In intact cells, the large intramembrane particles within the plasmalemma overlying the eyespot have been interpreted as such multimeric units [1]. If the rhodopsins are coupled to G-proteins and, in addition, form by themselves ion channel complexes, this would elegantly explain the activation of two different conductances [39].

The motif EVEPSKKV is part of the microtubule associated protein–2 from human. The identical 8-amino acid sequence is in chlamyopsin part of the CL3 like domain and adjacent to the putative pore loop region. It might link the chlamyopsin to a microtubule associated protein that attaches it to the microtubule rootlet and keeps the eyespot at its specific position.

References

1. M. Melkonian, H. Robenek (1984). The eyespot apparatus of flagellated algae: A critical review. *Progr. Phycol. Res.*, **3**, 193–268.
2. G. Kreimer (1994). Cell biology of phototaxis in flagellated algae. *Int. Rev. Cytol.*, **148**, 229–310.
3. O. Sineshchekov, E. Govorunova (1999). Rhodopsin-mediated phototsensing in green flagellate algae. *Trends Plant Sci.*, **4**, 58–63.
4. K.W. Foster, R.D. Smyth (1980). Light antennas in phototactic algae. *Microbiol. Rev.*, **44**, 572–630.
5. P. Hegemann, H. Harz (1998). How microalgae see the light. In: M.X. Caddick, S. Baumberg, D.A. Hodgson, M.K. Phillip-Jones (Eds), *Microbial Responses to Light and Time (Soc. Gen. Microbiol. Symp.*, pp. 95–105). Cambridge University Press.
6. M. Melkonian (1984). Flagellar apparatus ultrastructure in relation to green algal classification. In: D.E.G. Irvine, D.M. John. (Eds), *Systematics of Green Algae* (pp. 73–120). Academic Press.
7. K.P. Gaffal (1988). The basal body-root comples of *Chlamydomonas reinhardtii* during mitosis. *Protoplasma*, **143**, 118–129.
8. H.J. Hoops (1997). Motility in the colonial and multicellular Volvocales: structure, function and evolution. *Protoplasma*, **199**, 99–112.
9. H. Sakaguchi, K. Iwasa (1979). Two photophobic responses in *Volvox carteri*. *Plant Cell Physiol.*, **20**, 909–916.
10. G. Kreimer, M. Melkonian (1990). Reflection confocal laser microscopy of eyespots in flagellated green algae. *Eur. J. Cell Biol.*, **53**, 101–111.
11. K. Yoshimura (1994). Chromophore orientation in the photoreceptor of *Chlamydomonas* as probed by stimulation with polarized light. *Photochem. Photobiol.*, **60**, 594–597.
12. M. Land (1972). The physics and biology of animal reflectors. *Progr. Biophys. Mol. Biol.*, **24**, 75–106.
13. K. Schletz (1976). Phototaxis bei *Volvox*. Pigmentsysteme der Lichtrichtungsperzeption. *Z. Physiol.*, **77**, 189–211.
14. R. Uhl, P. Hegemann (1990). Probing visual transduction in a plant cell. Optical recording of rhodopsin-induced structrual changes from *Chlamydomonas reinhardtii*. *Biophys. J.*, **58**, 1295–1302.
15. K. Nakanishi, R. Crouch (1995). Application of artificial pigments to structure determination and study of photoinduced transformants of retinal proteins. *Israel J. Chem.*, **35**, 253–272.
16. J.L. Spudich, D.N. Zacks, R.A. Bogomolni (1995). Microbial sensory rhodopsins: Photochemistry and function. *Israel J. Chem.*, **35**, 495–513.
17. P, Hegemann, U. Hegemann, K.W. Foster (1988). Reversible bleaching of *Chlamydomonas reinhardtii* rhodopsin *in vivo*. *Photochem. Photobiol.*, **48**, 123–128.
18. K.W. Foster, J. Saranak, G. Zarrilli (1988). Autoregulation of rhodopsin synthesis in *Chlamydomonas reinhardtii*. *Proc. Natl. Acad. Sci. USA*, **85**, 6379–6383.

19. M. Beckmann, P. Hegemann (1991). In *vitro* identification of rhodopsin in the green alga *Chlamydomonas*. *Biochemistry*, **30**, 3692–3697.
20. P. Hegemann, W. Gärtner, R. Uhl (1991). All-trans retinal constitutes the functional chromophore in *Chlamydomonas* rhodopsin. *Biophys. J.*, **60**, 1477–1489.
21. F. Derguini, P. Mazur, K. Nakanishi, D.M. Starace, J. Saranak, K.W. Foster (1991). All-trans-retinal is the chromophore bound to the photoreceptor of the green alga *Chlamydomonas reinhardtii*. *Photochem. Photobiol.*, **54**, 1017–1021.
22. G. Kreimer, F-J. Marner, U. Brohsonn, M. Melkonian (1991). Identification of *11-cis* and *all-trans* retinal in the photoreceptive organelle of a flagellate green alga. *FEBS Lett.*, **293**, 49–52.
23. W. Deininger, P. Kröger, U. Hegemann, F. Lottspeich, P. Hegemann (1995). Chlamyrhodopsin represents a new type of sensory photoreceptor. *EMBO J.*, **14**, 5849–5858.
24. H. Ettl (1976). Über den Teilungsverlauf des Chlaoroplasten bei *Chlamydomonas*. *Protoplasma*, **88**, 75–84.
25. J.A. Holmes, S.K. Dutcher (1989). Cellular asymmetry in *Chlamydomonas reinhardtii*. *J. Cell Sci.*, **94**, 273–286.
26. K.P. el Gaffal, S. Gammal, G.J. Friedrichs (1993). Computer- aided 3D-reconstitution of the eyespot-flagellar/basal apparatus-contractile vacuoles-nucleus-associations during mitosis of *Chlamydomonas reinhardtii*. *Endocytol. Cell Res.*, **9**, 177–208.
27. D.L. Kirk (Ed.) (1998). *Volvox, Developmental and cell biology series*. Cambridge University Press.
28. M. Calenberg, U. Brohsonn, M. Zedlacher, G. Kreimer (1998). Light and Ca^{2+}-modulated heterotrimeric GTPases in the eyespot apparatusses of the flagellate green algae, the photoperceptive organelles for phototaxis and photoshock. *Plant Cell*, **10**, 91–103.
29. U. Schlicher, L. Linden, M. Calenberg, G. Kreimer (1995). G-Proteins and Ca^{2+}-modulated protein kinases of a plasma membrane enriched fraction and isolated eyespot apparatusses of *Spermatozopsis similis* (Chlorophyceae). *Eur. J. Phycol.*, **30**, 319–330.
30. M. Fuhrmann (1996). *Aufbau und Sequenz des Chlamyopsingens. Diplomthesis*. University Regensburg.
31. M. Fuhrmann, W. Oertel, P. Hegemann (1999). A synthetic gene coding for the green fluorescent protein (GFP) is a versatile reporter in *Chlamydomonas reinhardtii*. *Plant Journal*, **19**, 353–361.
32. E. Ebnet, M. Fischer, W. Deininger, P. Hegemann (1999). Volvoxrhodopsin, a light regulated sensory photoreceptor of the colonial alga *Volvox carteri*. *Plant Cell*, **11**, 1–12.
33. J. Nathans, D.S. Hogness (1983). Isolation, sequence analysis and intron-exon arrangement of the gene encoding bovine rhodopsin. *Cell*, **34**, 807–814.
34. C.S. Zucker, A.F. Cowman, G.M. Rubin (1985). Isolation and structure of a rhodopsin gene from *D. melanogaster*. *Cell*, **40**, 851–858.
35. M. Marchioni, W. Gilbert (1986). The triose phosphate isomerase gene from maize: Introns antedate the plant-animal divergence. *Cell*, **46**, 133–141.
36. T. Ikemura (1985). Codon usage and t-RNA content in unicellular and multicellular organisms. *Mol. Biol. Evol.*, **2**, 13–34.
37. R. Schmitt, S. Fabry, D.L. Kirk (1992). In search of the molecular origins of cellular differentiation in *Volvox* and its relatives. *Int. Rev. Cytol.*, **139**, 189–265.
38. R. Weiss, S. Osawa, W. Shi, C.D. Dickerson (1994). Effect of carboxy terminal truncation on the stability and G-coupling activity of bovine rhodopsins. *Biochemistry*, **33**, 7587–7593.
39. F.-J. Braun, P. Hegemann (1999). Two light activated conductances in the eye of the green alga *Volvox carteri*. *Biophys. J.*, **76**, 1668–1678.

40. K. Fahmy, T. Sakmar (1995). Regulation of the rhodopsin-transducin interaction by a highly conserved carboxylic acid. *Biochemistry*, **32**, 7229–7236.
41. T. Sakmar, K. Fahmy (1995). Properties and photoactivity of rhodopsin mutants. *Israel. J. Chem.*, **35**, 325–337.
42. H. Harz, C. Nonnengäßer, P. Hegemann (1992). The photoreceptor current of the green alga *Chlamydomonas. Phil. Trans. R. Soc. Lond. B.*, **338**, 39–52.
43. L. Tang, T.G. Ebrey, S. Subramanian (1995). Sequences and structure of retinal proteins. *Israel J. Chem.*, **35**, 193–209.
44. M.D. Hall, M.A. Hoon, N, Ryba, P. Ju, J.D.D. Pottinger, J.N. Keen, H.R. Saibil, B.C. Findlay (1991). Molecular cloning and primary structure of squid (*Loligo phorbesi*) rhodopsin, a phospholipase C directed G-protein linked receptor. *Biochem. J.*, **274**, 35–40.
45. W. Gärtner, W. Towner (1995). Yearly review: Invertebrate pigments and chromophore protein interactions. *Photochem. Photobiol.*, **62**, 1–16.

Photomovement
D.-P. Häder and M. Lebert, editors.

Chapter 9

Electrical events in photomovement of green flagellated algae

Oleg A. Sineshchekov and Elena G. Govorunova

Table of contents

Abstract

Phototaxis and photophobic responses in green flagellated algae are mediated by a rhodopsin-type photoreceptor. Its photoexcitation triggers a rapid cascade of electrical phenomena in the cell membrane. The photoinduced electrical responses in green flagellated algae can be recorded extracellularly from an individual cell by a suction pipette technique, or from a cell suspension. Photoexcitation leads to the onset of a photoreceptor current (PC) across the patch of the cell membrane overlaying the eyespot. The PC consists of at least two components activated by different mechanisms. The first mechanism most likely involves translocation of ions across the membrane either by the rhodopsin itself, or through an ion channel directly coupled to it. The second mechanism of PC generation appears to operate via a cascade of biochemical amplification. Membrane depolarization induced by PC leads to the unbalanced motor response of the flagella, which is the basis for phototaxis. If depolarization exceeds a critical level, a voltage-gated flagellar current (FC) is triggered across the flagellar membrane, which gives rise to the photophobic response of the cell. FC, and, upon prolonged light stimulation, PC are associated with depolarization-activated K^+ currents across the cell membrane. PC generation is the earliest so far detectable event in the signal transduction pathway. Therefore, investigation of PC allows in vivo probing the photoreceptor function and the role of the phototaxis directional antenna. The processes of energy metabolism provide the negative feedback loop for the light control of behavior in green flagellated algae by regulation of the phototaxis sign.

9.1 Introduction

Motile microorganisms tend to concentrate in the area of optimal illumination. This is particularly important for phototrophic species, most of which display very prominent photobehavior. Three types of photomovement could be phenomenologically distinguished in microorganisms: photokinesis, photophobic response, and phototaxis [1]. Light-induced changes in the linear velocity or the frequency of directional change is termed photokinesis. A transient change in linear velocity (normally a stop response) followed by a change of direction upon an abrupt change in the light intensity is called photophobic, or photoshock response. The ability of a cell to actively adjust the swimming path with respect to the light incidence is defined as phototaxis. This classification, however, does not reflect the nature of sensory transduction mechanisms. Independently on their phenomenological type, the photomotile responses could be based on photodynamic action of light, light energy accumulation, or specialized photoreception [2,3].

Photokinesis was reported in several species of green flagellates including *Chlamydomonas* [4], but has not been yet investigated in detail. In photosynthetic eukaryotes it is thought to be regulated by the processes of photosynthetic energy conversion (for review see [5]). On the contrary, phototaxis and photophobic response in green flagellates are mediated by a specific photoreceptor system, separate from the photosynthetic apparatus. Analysis of action spectra for phototaxis and photophobic response in *Platymonas* [6,7], *Chlamydomonas* [8] and *Volvox* [9] pointed to the involvement of rhodopsin-type photoreceptors [10]. Reconstitution of phototaxis in "blind" *Chlamydomonas* mutant by incorporation of exogenous retinoids further corroborated this hypothesis [11]. Recently, photoreceptor rhodopsins were identified in vitro in *Chlamydomonas* and *Volvox* (for review see chapter by Hegemann and Deininger, this volume). A single rhodopsin species was detected in *Chlamydomonas* retinal-deficient cells upon reconstitution with ^{3}H-retinal, indicating that phototaxis and photophobic response ought to be mediated by the same photoreceptor [12]. However, physiological and biochemical data in favor of separate photoreceptors for phototaxis and photophobic response were also presented [13]. Green flagellated algae are the only presently known photosynthetic eukaryotes to possess rhodopsin-type proteins similar to those involved in animal vision.

Phototaxis is the most complex behavioral response in green flagellates. They usually carry two or four flagella. Beating of the flagella in a so-called "breast-stroke" (ciliary) style propels the cell in the direction of its flagella-bearing end [14]. Many green flagellates follow helical swimming path that results from combination of rolling along the longitudinal axis of the cell and rotation in the plane of flagellar beating [15]. In *Chlamydomonas*, a lateral component of the 3-dimentional flagellar beating accounts for the former [16,17], whereas differences in the flagellar waveforms and/or beat frequencies might be the reason for the latter [17–19]. A distinct, highly specialized photoreceptor apparatus is employed for tracking the direction of the light. The eyespot, or stigma, was the first part of this apparatus recognized by light microscopy [20,21]. In *Chlorophyceae*, the eyespot is a part of the chloroplast and consists of one to several layers of carotenoid granules usually sandwiched between thylakoid membranes (for review see [10,22,23]). It has a lateral position with respect to the flagellar apparatus and is often shifted out of the plane of flagella beating [17]. The structural association between the eyespot and flagellar roots was found in several species of green algae (for review see [22]). The photoreceptor is confined to the membranes overlaying the eyespot. Most probably rhodopsin molecules are imbedded in the plasma membrane [24], although their localization to the outer chloroplast envelope was also suggested [25].

The hypothesis of periodic shading/illumination of the photoreceptor was already suggested in early studies on phototaxis [20,21]. The eyespot plays a major role in this process, although chloroplast absorption also contributes to it [9,26–28]. According to a recently developed view, the shading function of the eyespot is based not only on its absorption, but also on the interference of the light reflected by the eyespot layers with different refractive indices [10]. Observation of reflection patterns produced by the eyespots of various structures by confocal laser scanning microscopy supported the above hypothesis [29]. Illumination of the photoreceptor is maximum when the light is normal to the outer surface of the eyespot, and minimum when the light strikes from the back side. The eyespot and associated structures form a directional antenna which scans the environment during the helical swimming path. The signal received by the antenna is nearly constant when the axis of the helix is parallel to the light direction, but becomes periodic when the cell deviates from it. The periodic signal is processed to make a corrective motor response to re-align the swimming path with the light direction. Positive phototaxis is swimming towards the light source, negative phototaxis is swimming away from it. Processes of energy metabolism control the sign of phototaxis, although photo-orientation itself is not directly linked to them.

A photophobic response is induced by a change in stimulus intensity regardless of its direction. It is called the step-up or step-down photophobic response, depending on whether it is elicited by the increase or decrease in the light intensity. The photophobic response may consist of a brief cessation of flagella beating, as in *Volvox* [30], or the stop can be accompanied by a temporary transition from "breast-stroke" flagellar beating to flagellar undulation resulting in backward swimming of the cell, as in *Chlamydomonas* [31,32]. The photophobic response does not require the presence of the eyespot, since it is not affected in an eyespot-deficient *Chlamydomonas* mutant [26]. Phototaxis can be observed independently from the photophobic response at least in unicellular green flagellates [33,34]. Moreover, positive phototaxis is saturated at stimulus intensities lower than those necessary to saturate the stop-up photophobic response, as it has been shown in *Chlamydomonas* [35] and *Dunaliella* [36]. On the other hand, the dependence of phototaxis on desensitization of the photophobic response was found [37].

The light stimulus perceived by the photoreceptor molecules in the eyespot region of the cell needs to be transduced to an intracellular signal and transmitted to the motor apparatus. The possible involvement of electric processes in photosensory transduction in flagellated algae was initially proposed on the basis of indirect evidences, such as the dependence of photobehavior on the ionic composition of external medium [31,38–40], the influence of external electric fields on phototaxis [41], and control of flagella motion by electric current injection [42]. However, directly measuring the photoinduced electrical responses in green flagellates did not succeed until suitable experimental techniques have been developed. Using these techniques, it has been found that photoexcitation triggers a cascade of rapid electrical phenomena in the cell membrane [43–45]. The electrical responses are the earliest so far detectable events in signal transduction chains for photomovement in green flagellates.

So far, photosensory transduction has been investigated by electrophysiological methods only in a limited number of green flagellates, namely, *Haematococcus, Chlamydomonas, Polytomella, Spermatozopsis, Hafniomonas* and *Volvox*. Nevertheless, the common features of the photoelectric cascades found in these microorganisms suggest that the same basic scheme holds for the whole group of chlorophycean flagellates, although only partial comparative analysis of different species is possible at present. This review comprises the description of the cascade of the photoinduced electrical events, its role in regulation of photobehavior, and application of the electrophysiological approach to investigation of photosensory transduction in green flagellates.

9.2 Methods

9.2.1 Intracellular recording

Unicellular green flagellated algae usually do not exceed 10–20 μm in diameter, hence intracellular recording from them is rather problematic. Nevertheless, spontaneous and photoinduced changes of the membrane potential could be measured by this technique in *Haematococcus* [43]. Spontaneous changes appeared as spikes of complex kinetics, under certain conditions correlating to flagellar re-orientation. The onset of light caused slow hyperpolarization of the plasma membrane and fast high-amplitude electrical signal on the chloroplast membranes. The action spectrum of these responses coincided with that of photosynthesis. Similar photosynthetically driven changes of the plasma membrane potential was later measured in *Dunaliella* [46]. Specific effects of the blue-green light responsible for phototaxis could not be detected after the cell was impaled, but appeared as rapid potential changes when the recording microelectrode was pressed against the cell surface [43]. These observations led to the conclusion that phototaxis in green flagellates involves the generation of local photoinduced currents across the cell membrane which are highly sensitive to the cell damage. Therefore, methods for extracellular measurements of these currents have been developed.

9.2.2 Extracellular recording

Let us assume that photoexcitation leads to the onset of a local electrical current across a small patch of the cell membrane, as shown in Figure 1. The electrical circuit is closed via the rest of the cell membrane and the saline external medium. The part of the photoinduced current that flows through the external resistance R_{ext} produces a potential difference ΔV, which is picked up by the electrodes and can be measured by a voltage amplifier. Alternatively, the signal can be measured by a current to voltage converter, the output voltage of which is proportional to the input current. Using this type of instrument has two advantages:

1. the increase in R_{ext} does not change the signal amplitude, but leads to a decrease of the noise,
2. the influence of the external capacitance C_{ext} on the kinetics of the output signal is minimized, since the input voltage is kept at the constant level.

However, ΔV produced by the cells is small, which means that the influence of C_{ext} on the kinetics of the signal recorded by a voltage amplifier is negligible. Consequently, kinetics of rhodopsin-mediated responses recorded extracellularly by a voltage amplifier and by a current to voltage converter is practically the same, although the signals were termed “potentials” or “currents” according to the type of instrument used in a particular study. In this review, we will always refer to the extracellular responses as “currents” for the sake of clarity and in order to emphasize their origin from transmembrane currents.

Two methods for extracellular recording were developed. The response of an individual cell can be detected if R_{ext} is increased by sucking an individual cell into a tip of a glass micropipette, and the electrical signal between the inside and outside the pipette is measured. Alternatively, the currents from many synchronously excited cells can be picked up by the electrodes properly immersed in suspension of freely swimming cells.

Single-cell recording (suction pipette technique). The development of the suction pipette technique significantly facilitated investigation of photoelectric responses involved in phototaxis in green flagellates [44]. It has been observed that one component of the electrical signal changes its sign from positive to negative upon drawing the eyespot into the pipette (Figure 2). (Note that here the current flowing into the pipette is termed positive, which differs from the generally used sign convention). Therefore, it could be concluded that photoexcitation induces an inward current associated with the

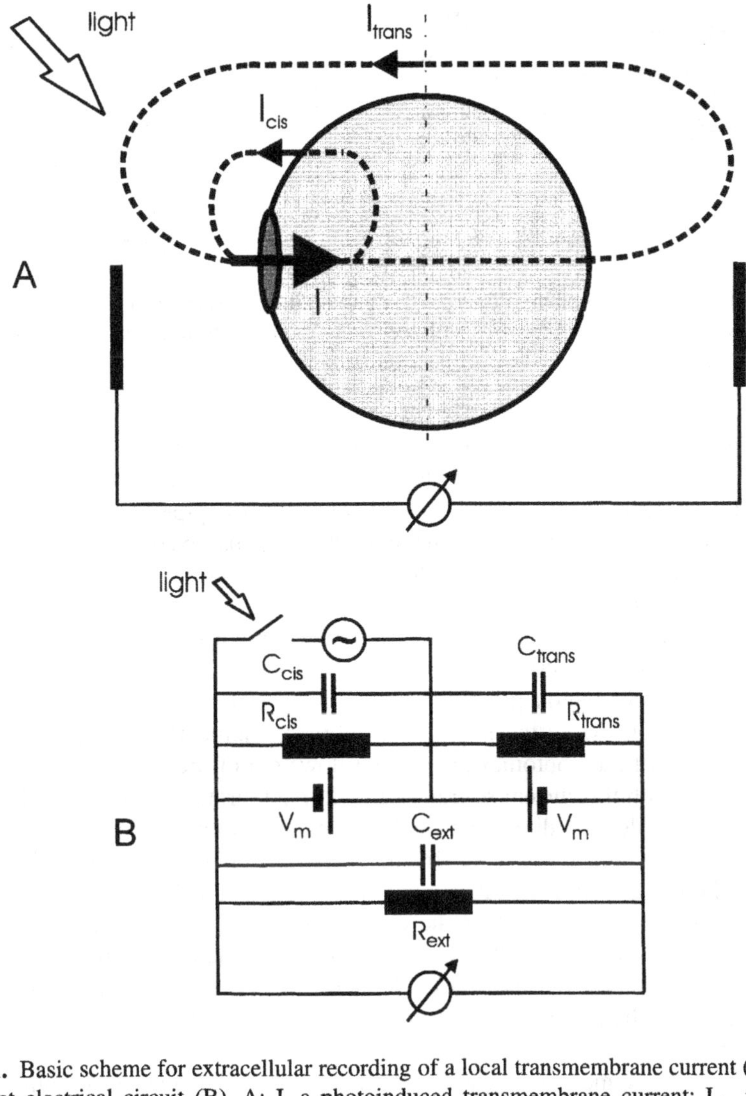

Figure 1. Basic scheme for extracellular recording of a local transmembrane current (A), and its equivalent electrical circuit (B). A: I, a photoinduced transmembrane current; I_{cis}, part of the current closing the circuit via the half of the plasma membrane that contains the current source shown as a small ellipse; I_{trans}, part of the current closing the circuit via the other half of the plasma membrane. B: C_{cis} and R_{cis}, the capacitance and the resistance, respectively, of the half of the plasma membrane containing the current source; C_{trans} and R_{trans}, the capacitance and the resistance, respectively, of the other half of the plasma membrane; C_{ext} and R_{ext}, the capacitance and the resistance, respectively, of the extracellular medium; V_m, the resting membrane potential. (Modified from [45]).

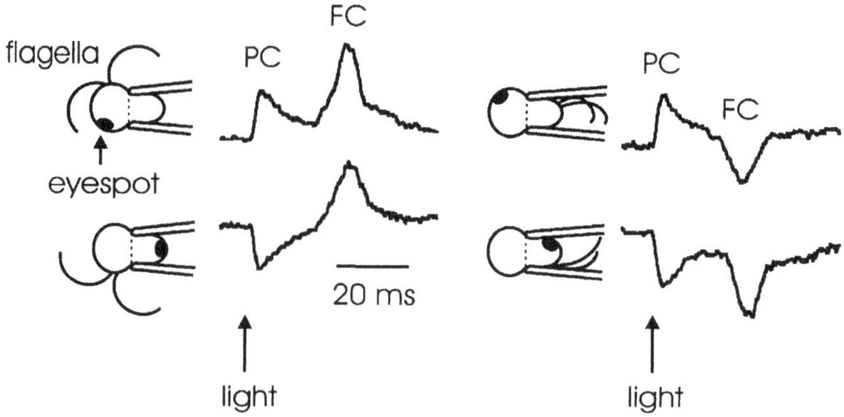

Figure 2. The dependence of the PC and FC signs on the position of the eyespot and flagella inside or outside the suction pipette. The traces were recorded from *Haematococcus pluvialis*; arrow indicates the time of the excitation flash. (Modified from [68]).

eyespot region of the cell. The origin of another component of the signal could be interpreted as an inward current in flagellar region of the cell, since its sign was determined by position of the flagella outside or inside the pipette in the same manner (Figure 2). Initially, these two components of the response were termed the "primary potential difference", or PPD, and the "regenerative response", or RR [45]. According to the localization of the current sources, these terms were later substituted by the "photoreceptor potential" [47], or the "photoreceptor current", or PC [48], and the "flagellar current", or FC [48], respectively.

When a cell is sucked into a pipette, the amplitude of the recorded signal is influenced by the ratio between the surfaces of the membrane fragments inside and outside the pipette, since part of the photoinduced current flows out of the cell on the same side of the glass barrier with the current source and is shunted through the external fluid (I_{cis} in Figure 1). Most stable recording can usually be achieved when approximately 1/3 of the protoplast is drawn into the pipette. In this case, 2/3 of the photoinduced current is shunted, if the current source is in a larger portion of the cell membrane outside the pipette, and 1/3 of it is shunted, if the current source is in a smaller portion of the membrane inside. Accordingly, the shunt current is minimum when the current source is completely separated from the rest of the cell membrane. This situation is achieved for PC recording when only the eyespot was sucked into the pipette of a smaller tip diameter [49].

The surfaces of the membrane fragments inside and outside the suction pipette are of the same order of magnitude. Therefore, clamping the potential of the membrane fragment inside the pipette to a constant level cannot be achieved. The resistance of the seal between the cell surface and the glass of the pipette is usually in the order of 100 MOhm, which is too low for the currents from single channels to be resolved from the background noise. Multilayered cell walls of a complex glycoprotein composition found in green flagellates preclude formation of gigaseals necessary for the single-channel recording. Digestion of the cell wall by autolysin [50] or other lytic enzymes, its perforation by microsurgery, or using cell wall-deficient mutants did not so far lead to gigaseal formation. However, a light-dependent single channel current was briefly reported in *Chlamydomonas* [51].

A similar suction pipette technique has been established independently for recording light-induced responses in photoreceptor cells of vertebrate retina [52]. Vertebrate photoreceptors maintain in darkness a steady ionic current which flows into the outer segment of the cell and out of its inner segment (for review see e.g. [53]). When the outer or the inner segment of the cell is sucked into the pipette, light-induced decrease in the circulating current due to the decrease in membrane conductance in the outer segment is observed. In contrast, photoresponses in flagellates are thought to result from the increase rather than decrease in the membrane conductance, similar to what was found in most invertebrate photoreceptors (for review see [54,55]).

Application of the suction pipette technique is limited to relatively large flagellates with elastic cell walls, such as *Haematococcus* [45], or cell wall-deficient mutants of *Chlamydomonas reinhardtii* [48]. Strains of *Chlamydomonas* with the wild-type cell walls are suitable for this technique only after treatment with autolysin [56]. Recently, the suction pipette technique has been successfully applied to somatic cells of a "dissolver" mutant of *Volvox carteri* that lacks extracellular matrix [57]. Another disadvantage of the suction pipette technique is deformation of the protoplast by sucking, which may affect the results of recording due to, for instance, stimulation of mechanoreceptors of the cell [58,59]. To overcome these problems, a method for photoelectric measurements in suspension of freely swimming microorganisms has been introduced [60,61].

Recording from cell suspensions. Delivery of a short excitation flash leads to generation of the photoinduced currents by individual cells in suspension, as it is shown in Figure 1. To take advantage of simultaneous recording from many cells, the currents from the individual cells should not compensate each other. This can be achieved by two modifications of the method.

In the unilateral mode of measurements, the excitation light is delivered along the line connecting the electrodes immersed in suspension of non-oriented cells (Figure 3a, left). The number of quanta captured by the photoreceptor of an individual cell is determined by the angle of the light incidence on the eyespot surface. Upon photoexcitation, the cells oriented with their eyespots towards the light source generate maximum PC, whereas the cells in the opposite orientation generate minimum current. Therefore, the electrodes pick up the difference signal, the sign of which is determined by the direction of the currents in the cells oriented with their eyespots towards the light source. To maintain the sign convention accepted for the suction pipette studies, this sign is considered positive. FC is an all-or-nothing response that appears when the membrane depolarization exceeds a critical level (see below). Therefore, the probability of FC generation is higher in the cells oriented with their eyespots towards the excitation flash, which gives rise to a non-compensated signal. The sign of FC measured in the unilateral mode depends on the angular distance between the eyespot and the flagella. For instance, this distance is slightly more than 90° in most strains of *Chlamydomonas reinhardtii,* which results in a small negative FC recorded in the unilateral mode (Figure 3a, right).

In the pre-oriented mode of measurements, the currents from individual cells are aligned due to pre-orientation of cells by a directional stimulus, such as light, gravitation, etc. The electrodes are placed one after another along the direction of the orienting stimulus, whereas the excitation flash is delivered at 90° to this direction (Figure 3b, left). Under these conditions, only currents from oriented cells contribute to the recorded signal, whereas the difference signal from non-oriented cells is not

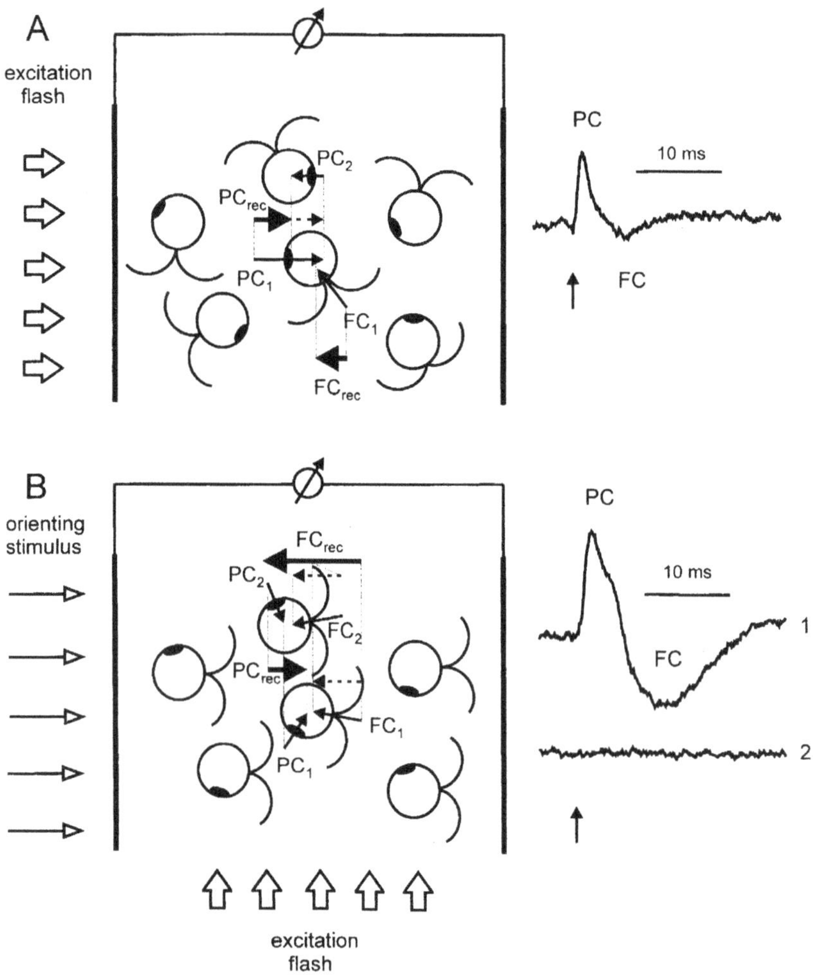

Figure 3. Two modifications of the suspension method for photoelectric measurements: unilateral mode (A) and pre-oriented mode (B). PC_1, PC_2, FC_1 and FC_2, photocurrents generated by individual cells in suspension. PC_{rec} and FC_{rec}, resultant current picked up by the electrodes. The traces at right were recorded from *Chlamydomonas reinhardtii* strain 495 (+); arrow indicates the time of the excitation flash. 1, the signal measured in pre-oriented mode in the presence of the orienting stimulus, 2, the signal recorded by the same set-up in the absence of the orienting stimulus. (Modified from [64]).

detected. The amplitude of the signal picked up by the electrodes is proportional to the cosine of the angle between the direction of the current and the line connecting the electrodes. Consequently, the amplitude of the signal recorded in the pre-oriented mode depends on a degree of orientation of the cells in suspension and can be used for an instant estimation of it. Most green flagellates swim with their flagella forward in helices of quite narrow cone angles, which means that the direction of FC in oriented cells is almost parallel to the line connecting the electrodes, whereas the direction of PC is close to perpendicular to this line. This accounts for the increased contribution of FC to the signal kinetics observed in the pre-oriented mode as compared to the unilateral mode (Figure 3b, right).

The suspension method for photoelectric measurements is not limited by the cell size or structure, and therefore can be applied to a broader range of microorganisms than the suction pipette technique [23,62]. Particularly important is that it is applicable to a large number of *Chlamydomonas* mutants [63–65]. A high signal-to-noise ratio and the possibility of recording the signals under fully physiological conditions constitute further advantages of this method. However, the complex origin of the signals collected from many millions of cells requires careful quantitative analysis of the results for their correct interpretation. Another problem of this method is that recording slow components of the photoelectric cascade is disturbed by the motion of cells in suspension.

9.3 The cascade of rhodopsin-mediated photoelectric responses and their role in regulation of photobehavior

The earliest step in the rhodopsin-mediated signal transduction chain in green flagellates is the PC generation. Localization of the PC to the eyespot region of the cell was directly shown by recording the signal upon illumination of different parts of the cell with a microbeam [66,67]. It was further confirmed by recording the PC from "excised eyes" – eyespot-containing vesicles detached from the cells [49,57]. Furthermore, the PC in *Volvox* could only be recorded from vegetative cells, and totally disappeared during their conversion into gonidia in parallel with the disappearance of the visible eyespots [57]. The action spectrum for PC generation clearly coincided with the action spectra for photoaccumulation, phototaxis and photophobic response of the cells [45,48, 57]. Both PC and phototaxis can be suppressed by the same range of chemical agents [45,48]. On the other hand, the PC is insensitive to DCMU, an inhibitor of photosynthesis [45].

The PC recorded from a cell sucked into a pipette upon excitation with a pulse of continuous light consists of a transient peak which decays to a lower stationary level and dissipates with a time constant of several tens of milliseconds after switching off the light (Figure 4). These two phases were originally defined as the "primary potential difference", or PPD, and the "late potential difference", or LPD [60,68]. The term "stationary photoreceptor current", or P_{St}-current for the stationary phase of the extracellularly recorded signal was suggested later [57] and will be used in this review. The P_{St}-current has a major physiological importance, since microorganisms deal with gradual rather than pulsed light stimuli in their natural habitat. However, investigation of the signal kinetics requires using short flashes to avoid the influence of light

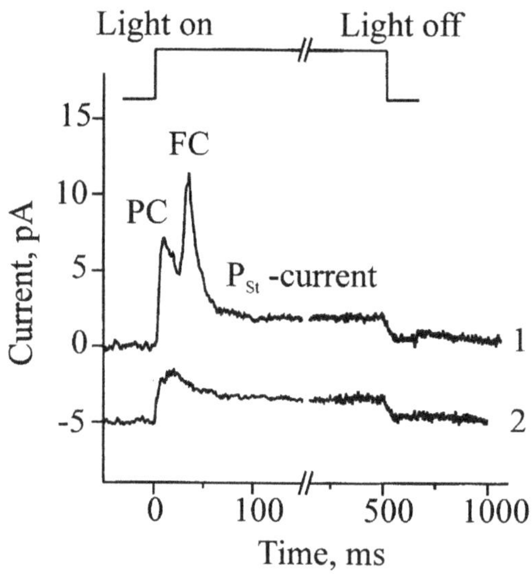

Figure 4. Photoelectric responses elicited by a pulse of continuous light at two different fluence rates. *Haematococcus pluvialis*; suction pipette technique; the duration of the light pulse is shown above. 1 the trace recorded at stimulus intensity above the threshold for FC triggering, 2 the trace recorded from the same cell at under-threshold stimulus intensity.

adaptation. The flash-induced PC also appears to be a multicomponential process, as it will be discussed in a separate section below.

The role of PC in phototaxis was examined by optical monitoring of flagellar beating in a cell fixed on a micropipette, which was undertaken in parallel to recording electrical responses. To simulate periodic illumination of the photoreceptor in a freely swimming microorganism, a *Haematococcus* cell was exposed to a modulated light stimulus of low intensity [60,68–71]. Periodic changes in P_{St}-current amplitude were the only photoelectric responses observed under these conditions. Step-up stimulus induced the increase in the beat frequency of the *cis*-flagellum (the one closest to the eyespot), and the decrease in the beat frequency of the *trans*-flagellum, whereas a step-off stimulus caused the opposite responses. The observed beat frequency changes were accompanied by slight changes in beating curvature, also opposite in the two flagella of the cell. Such unbalanced motor responses of the two flagella would lead to the correction of the swimming path in a freely swimming cell, i.e. to phototaxis. Therefore, it could be concluded that gradual changes in the amplitude of the P_{St}-current constitute the initial step in the signal transduction chain for phototaxis. Complex photoinduced changes in flagellar beat pattern and frequency were investigated by high-speed microcinematography in *Chlamydomonas* cells held on micropipettes [72,73]. Low intensity flashes only resulting in generation of a transient PC caused brief transient changes in the pattern of flagella beating [19]. These changes apparently correspond to the flash-induced changes in the direction of freely swimming cells monitored by videomicroscopy and motion analysis, which are regarded as the elementary motile responses for phototaxis [34,74].

A spike-like FC superimposes the PC if the intensity and/or duration of the light stimulus exceeds a certain threshold [45]. The delay time of the FC decreases upon increase of stimulus intensity from 100 to 5 milliseconds, but its peak amplitude does not change significantly. Such "all-or-none" appearance of the FC resembles action potentials found in other excitable membranes. A close time correlation between FC and a switch from "breast-stroke" style of flagella beating to undulation was noticed in a cell fixed on a micropipette [45]. In a freely swimming cell, this switch is characteristic for the photophobic response [31]. Therefore it could be concluded that the FC is the driving force for the photophobic response of the cell [48,60,68]. Recently this notion has been directly proven by parallel recording photoelectric currents and flagella beating from the same cell [19]. Furthermore, no FC could be detected in *Chlamydomonas* mutants *ptx2* and *ptx8* [65] and *ppr1-ppr4* [75] lacking the photophobic response. Interestingly, no FC, at least at room temperature, could be found in a "dissolver" mutant of *Volvox* [57]. The photophobic response in the multicellular *Volvox* involves only a stop of flagellar beating without a switch to the undulation beating mode [30], which can apparently explain the lack of FC in this organism.

Spontaneous spikes of similar kinetics can sometimes be observed in the dark or under continuous illumination [45]. Spontaneous spikes recorded by the suction pipette technique could be correlated to those measured by the intracellular microelectrodes [43] and to spontaneous photophobic responses in freely swimming cells. It has been proposed that control of the cell movement by the processes of energy metabolism involves regulation of the frequency of spontaneous photophobic responses [71,76]. Simultaneous recording of electrical responses and flagellar beating directly proved that the spontaneous FC leads to a switch from "breast-stroke" beating style to flagellar undulation, as does the photoinduced FC [19]. The highest probability of the spontaneous FC was observed just after the end of the flash-induced photophobic response, or at 10–20 mM external K^+, i.e. in a presumably depolarized cell [19].

The integral comprised by the PC before the beginning of the FC is nearly constant over a wide range of stimulus intensities and is insensitive to the substitution of Ca^{2+} by Ba^{2+} [60,68,77]. This indicates that a certain amount of charges should enter the cell to initiate the FC. When the cell is hyperpolarized by the photosynthetically active red background illumination [43], this integral increases more than twice [60,68]. Therefore it could be concluded that the FC is activated by the PC-induced depolarization of the membrane to a certain level. Thus, electrophysiological studies have shown that phototaxis and photophobic response share not only a single photoreceptor species, but also involve common initial steps in the signal transduction chain identified as PC generation. In *Chlamydomonas* mutants *ptx2* and *ptx8* which lack photophobic response, phototaxis is also inhibited, although the PC is not affected [65]. This shows the existence of an element in the signal transduction chain downstream from the PC generation common for the phototaxis and the photophobic response pathways, which is absent in these mutants.

The spike-like FC is accompanied by a transient current of an approximately 20-fold smaller peak amplitude and a peak time of a few hundreds of milliseconds under physiological conditions [77]. This current is also localized to the flagellar membrane and is therefore defined as the "slow flagellar current" (F_s), to be distinguished from the spike-like "fast flagellar current" (F_f). The functional role of F_s is presumably related to the control of the duration of backward swimming during the photophobic response [49,78]. In this review, we will still refer to the spike-like flagellar current as FC, except when discussing the differences between F_f and F_s.

Both PC and FC are carried mostly by Ca^{2+} ions under physiological conditions. However, generation of F_s and P_{St}-current seems to be linked to K^+ efflux that counterbalances depolarization of the membrane caused by these long-lasting inward currents [57,78]. Furthermore, transient K^+ currents (KC) are triggered by the membrane depolarization induced by F_f [79]. Their direction can be inward or outward depending on the electrochemical driving force for K^+. The outward K^+ current observed at low external K^+ concentrations accelerates restoration of the resting membrane potential after depolarization of the cell by photoexcitation.

Three major steps could be identified so far in the cascade of photoelectric processes, each one including several individual components. The primary step comprises photoreceptor currents driven by rhodopsin photoexcitation (PC). They occur in the eyespot region of the cell and are involved in both phototaxis and photophobic response. The PC generation leads to membrane depolarization, which above a certain threshold triggers voltage-activated electrical currents across the flagellar membrane (FC). These currents give rise to the photophobic response of the cell and represent the second step of the cascade. The membrane depolarization caused by either PC (under continuous light) or FC initiates voltage-activated transmembrane K^+ currents, which are not associated with specific motor responses, but rather play a role in photosensory adaptation. These currents constitute the third step of the electrical cascade. The role of the electric processes in regulation of photobehavior in green flagellated algae is shown in Figure 5. It has to be taken into account that the suggested scheme is certainly oversimplified, since the detailed investigation of individual components of the electrical cascade and their cause-effect relationships is still in progress.

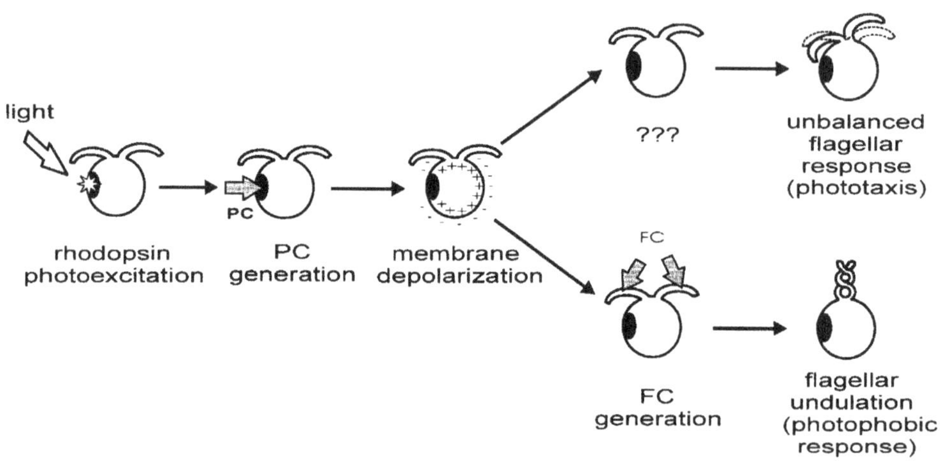

Figure 5. An overview of the rhodopsin-mediated signal transduction chains for phototaxis and photophobic response in green flagellated algae.

9.4 Photoreceptor currents

9.4.1 Response components

The PC can be induced by a flash over a range of intensities that covers at least five orders of magnitude. The delay time between the excitation flash and the onset of PC considerably varies within this range. The delay time of PC evoked by a high-energy flash is only limited by the time resolution of the measuring equipment (Figure 6, inset; [47,49]). On the other hand, the onset of PC generated in response to a low-intensity flash can be delayed up to a few milliseconds (Figure 6, curve 1). Particularly long delay times of about 10 ms were observed in *Polytomella* (unpublished observations) and *Volvox* [57]. The observed 1,000-fold variation of the PC delay times implies that different mechanisms for PC generation might operate at high and low flash intensities.

Upon a decrease in flash intensity, the PC becomes slower in both rise and decay, its peak time increases from several hundreds of microseconds to a few milliseconds (Figure 6). Two components in both rise and decay of a laser flash-induced PC recorded from *Haematococcus* by the suction pipette technique could be distinguished [47]. Only the first component of the rise had virtually no delay even at low temperature, whereas the delay time of the second component of the rise increased from 120 to 400 microseconds with the decrease in stimulus intensity. Red background illumination did

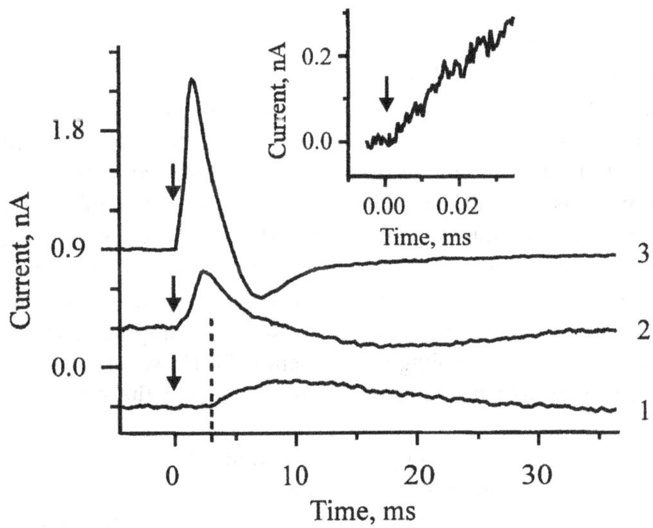

Figure 6. Photoelectric responses evoked in cell suspension at different flash energies. Arrows indicate the time of the flash. Main figure: *Chlamydomonas reinhardtii* strain 495 (+). Photoflash excitation, relative stimulus intensity: 1 0.01%, 2 2.5%, 3 100%. Dashed line shows the response delay in trace 1. Inset: The onset of PC recorded with an improved time resolution in *Chlamydomonas reinhardtii* strain 516/white–3 supplemented with 10^{-8} M all-*trans* retinal. Laser flash excitation, 500 nm, 2 $J \times m^{-2}$. The laser artifact is digitally subtracted.

not influence the first component of the PC rise, but increased the amplitude of the second component. The two kinetic components found in the PC rise indicate that two different processes contribute to PC generation, which could be defined as "early PC" and "late PC" (or, alternatively, "fast PC" and "slow PC").

The extremely short delay time and stability of the first response component indicated that its generation precedes biochemical steps of the signal transduction chain. In animal photoreceptors, the light-induced conformational changes of the rhodopsin molecule are accompanied by an intramolecular charge redistribution. The rhodopsin molecules are highly ordered, and therefore all the photoinduced movements of charges within individual rhodopsin molecules occur in parallel and cause a displacement current across the cell membrane. Membrane potential changes induced by this current were shown to follow the kinetics of the rhodopsin photochemical conversion and were referred to as "early receptor potential" (ERP), to be distinguished from the delayed "late receptor potential" (LRP) caused by light-induced ion movement across the photoreceptor membrane (for review see [80]). It has been proposed that the fast PC component found in *Haematococcus* might also reflect light-induced charge movements within photoreceptor molecules, although other possible explanations for the biphasic kinetics of the signal rise could not be ruled out [47]. It was calculated for animal photoreceptors, that a large number of rhodopsin molecules (> 1,000,000) should be photoconverted within the integration time of the cell membrane to make the ERP resolved from the background noise [81]. However, flagellated algae contain only about 3,000 to 400,000 photoreceptor molecules per cell [10] which disfavors the suggested above correlation of the fast PC component in *Haematococcus* with ERP in animal visual transduction.

The decay of PC in *Haematococcus* could be fitted by two exponentials with time constants of 2.5–6 and 14–32 milliseconds [47]. The relative amplitudes of the decay components depended on experimental conditions and could change during the course of experiment in a particular cell, similar to that of the components of the signal rise. However, establishing the possible correlation between the components of the rise and decay of the signal was difficult, since the capacity of the cell membrane and the pipette should be taken into account for quantitative deconvolution of the signal kinetics. Recently, complex decay kinetics of the flash-induced PC has been reported in *Volvox* [57]. Acceleration of the decay kinetics of the flash-induced PC with the increase in photon exposure correlated to the increase in the peak amplitude [49], which might indicate that PC inactivation is voltage-dependent [77]. However, it cannot be excluded that the PC decay can also be determined by decreasing the driving force for Ca^{2+} [49,77].

The existence of two response components could also be concluded from the analysis of the dependence of the peak amplitude of the flash-induced PC on stimulus intensity [68]. The stimulus-response dependence measured over a full range of stimulus intensities is clearly biphasic [61]. The saturation intensity of the first phase can be up to three orders of magnitude lower than that of the second phase (Figure 7).

The extremely high saturation level of the second phase implies that it is only limited by photoconversion of the photoreceptor pigment upon absorption of quanta. For photoprocesses of this type, the response amplitude should exponentially depend on the stimulus intensity [82]. Taking into account the existence of a low-saturating phase, the

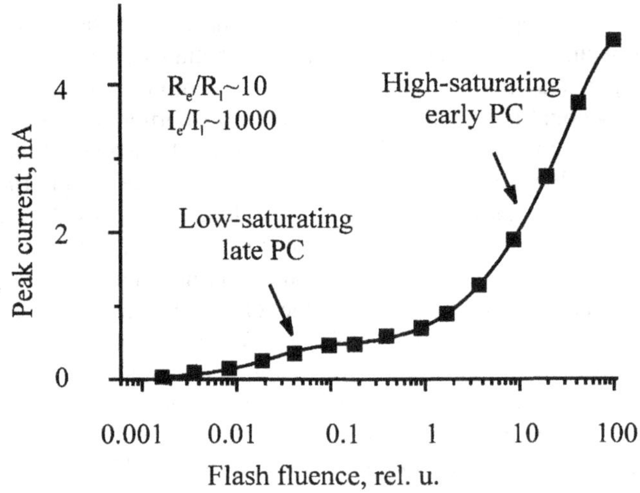

Figure 7. Dependence of the PC peak amplitude on the fluence of the excitation flash. *Chlamydomonas reinhardtii* strain 495 (+); photoelectric measurements in cell suspension. Solid line shows a computer fit to the Michaelis function for low-saturating PC and to exponential function for high-saturating PC.

high-saturating phase of the stimulus-response curve for the PC peak amplitude could be simulated by the exponential function $R = R_{sat1} + R_{sat2} \times (1 - e^{-\phi\sigma I})$, where R is the response amplitude, R_{sat1} is maximum amplitude of the low-saturating phase, and R_{sat2} is maximum amplitude of the high-saturating phase, ϕ is the quantum efficiency of response generation, σ is the optical cross section of a photoreceptor molecule, and I is photon exposure [68]. This fitting gives the value for R_{sat1} about 10 to 20% of R_{sat2}, and the value about 8×10^{-21} m^2 for the product of the quantum efficiency and the optical cross section, which is reasonable for a single absorbing molecule such as a retinal-containing protein.

If the photoresponse involves enzymatic or energy-transferring state, i.e. the active state of the photoreceptor pigment regulates the rate of a light-independently inactivated metabolic process, the dependence of the response amplitude on stimulus intensity is hyperbolic [82]. The peak amplitude of the photoelectric response from animal photoreceptors mediated by the enzymatic cascade of amplification could be simulated by a rectangular hyperbola, given by Michaelis equation $R/R_{sat} = I/(I + I_{50\%})$, where R is the response amplitude, R_{sat} is maximum response at saturation, I is photon exposure, and $I_{50\%}$ is the photon exposure yielding the half maximal response amplitude [83]. The fluence dependence of PC peak amplitude at lower intensities closely fits this equation (Figure 7).

Attempts were made to simulate the stimulus-response curve of the PC as a single-component process [57,77]. However, experimental data did not fit either the Michaelis function, or a more general equation including cooperative effects $R/R_{sat} = I^n/(I^n + I^n_{50\%})$, unless an inactivation mechanism reducing the response amplitude at saturating light intensities was proposed [77].

Calculation of the integral comprised by the PC in the high-saturating phase of the stimulus response curve shows that the number of elementary charges transported across the membrane upon absorption of one photon by the photoreceptor molecule is close to one. Taking into account a virtually instant onset of the current and its being limited by the photoconversion of the photoreceptor pigment, it can be suggested that the early PC originates from translocation of the ions across the membrane by the rhodopsin itself. Investigation of the ionic dependence of PC shows that it is mostly driven by the influx of Ca^{2+} ions (see below). The electrochemical driving force for Ca^{2+} in flagellates under physiological conditions is directed inward, and it is not known if the photoreceptor rhodopsin can transport the ions against the electrochemical gradient. Nucleotide sequences of *Chlamydomonas* opsin cDNA and *Volvox* opsin DNA indicate that the respective proteins contain many polar and charged amino acid residues which might form pores in the membrane [84,85]. On the other hand, intramembrane particles which might represent multimeric protein complexes, but are too small for individual rhodopsin molecules, were found in the eyespot region of *Chlamydomonas* by freeze-fracture electron microscopy [24]. Therefore, one of the possibilities is that the photoreceptor rhodopsin is closely associated with a low-conductance ion channel in 1:1 stoichiometry [77].

The delay time of the late PC indicate that it is likely limited by a turn-over time of a biochemical second messenger. Delay times ranging from 5 to 200 ms are typical for the onset of transmembrane ionic currents in animal visual systems mediated by an enzymatic cascade [54,80], and therefore might indicate the involvement of similar biochemical mechanisms for the signal amplification in flagellates. Under continuous illumination of 0.1 W m^{-2}, the number of quanta absorbed by the photoreceptor is around 10^3 per second, whereas the amplitude of P_{St} recorded under these conditions corresponds to about 10^7 elementary charges per second being transported across the membrane [60,68]. This means that generation of the late PC involves about 4 orders of signal amplification.

The mechanism for this amplification is not yet known, but recent results of biochemical and immunological studies on isolated eyespot apparatuses of *Chlamydomonas reinhardtii* [86,87] and *Spermatozopsis similis* [88] point to the presence of various molecular elements also found in enzymatic cascades of animal photosensory transduction (for review see [89,90]). Proteins with characteristics of subunits of animal heteromeric G proteins specifically associated with eyespot membranes were detected in such preparations [91–93]. Light-dependent cGMP hydrolysis was reported in a reconstituted system with bovine transducin and phosphodiesterase [86]. Furthermore, light-dependent GTPase activity with the action spectrum similar to that of rhodopsin absorption was found, which could be inhibited by antibodies raised against *Chlamydomonas* rhodopsin [94]. In addition, Ca^{2+}-dependent protein kinase and phosphatase activities were characterized in the same in vitro preparation [93,95].

Activation of the biochemical amplification cascade leading to late PC generation can be linked to early PC. For instance, it may result from the increase in intracellular Ca^{2+} concentration due to early PC generation (Figure 8a). Alternatively, rhodopsin photoexcitation may initiate two independent signal transduction mechanisms resulting in early PC and late PC generation (Figure 8b). This would mean that the photoreceptor rhodopsin from green flagellated algae simultaneously functions as the molecular device for translocation of the ions across the membrane, and as a catalyst of an enzymatic cascade, which so far has not been found in any other retinal-containing proteins involved in sensory transduction or energy transfer.

9.4.2 Ion selectivity of PC and molecular mechanisms for phototaxis

The PC can be inhibited by the removal of Ca^{2+} ions from the external medium or by the addition of La^{3+}, Cd^{2+}, ruthenium red and a number of organic blockers of Ca^{2+} channels (l-*cis*-diltiazem, verapamil, pimozide) [45,48,57,60]. Therefore, the PC is at least partially driven by an influx of Ca^{2+} ions across the photoreceptor membrane. Light-induced $^{45}Ca^{2+}$ uptake could be detected in *Chlamydomonas* cell wall-deficient mutant, although sensitivity and time resolution of such measurements were not high enough to directly prove the above hypothesis [96].

The dependence of the flash-induced PC on external Ca^{2+} concentration was studied by the suction pipette technique. Maximum peak amplitude was found below 10^{-6} M Ca^{2+}, which was explained by extremely tight binding of Ca^{2+} to the putative channel

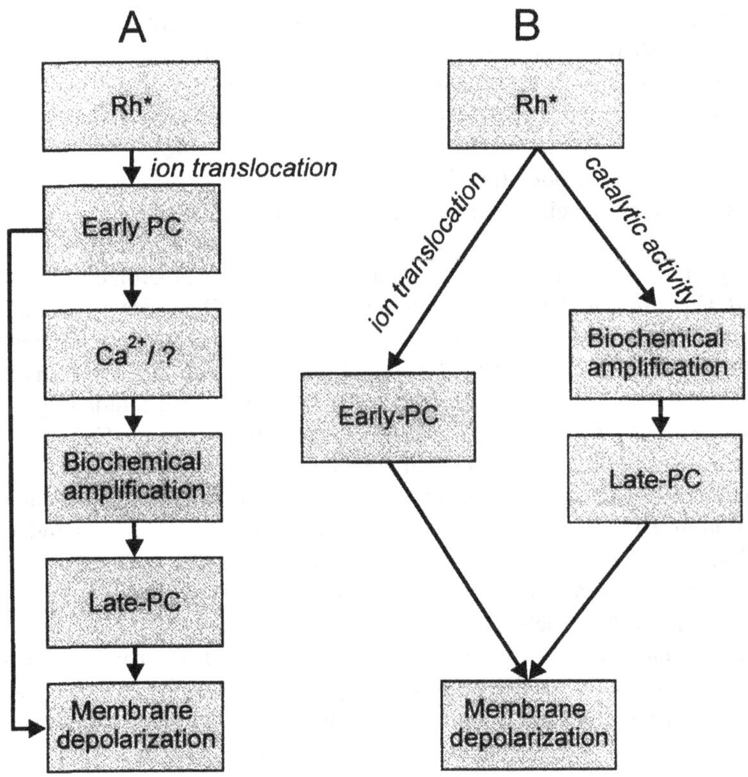

Figure 8. Schematic presentation of the two possible sequences of the primary events in the rhodopsin-mediated sensory transduction in green flagellated algae.

[49]. Only small changes in PC were observed after the replacement of external Ca^{2+} by Sr^{2+} or Ba^{2+}, indicating that the kinetics of PC decay is not limited by inactivation of the channel by cytosolic Ca^{2+} [49,77]. Small PC could also be measured after substitution of Ca^{2+} by Mg^{2+}, although Mg^{2+} at millimolar concentrations had an inhibitory effect [49].

Small flash-induced PC could be recorded in a Ca^{2+}-free buffer in the presence of EGTA [60] or BAPTA [49]. This means that PC also involves a Ca^{2+}-independent component. The influence of monovalent cations on the flash-induced PC was investigated by the suction pipette technique [78]. The PC was enhanced with a relative preference for $K^+ > NH_4^+ > Na^+$, when monovalent cations were added at concentrations of 20 or 40 mM to the bath solution containing 0.1 mM Ca^{2+}. However, removal of K^+ from the bath solution or its replacement by other monovalent cations had no effect. Therefore, it could be concluded that K^+ influx does not contribute to the flash-induced PC under physiological conditions. A small residual current was observed when NMG^+ was the only cation present in the external medium, which could only be explained by a H^+ influx or an anion efflux. The decrease in external pH significantly increases the PC recorded in *Volvox* [57].

The ionic selectivity of P_{St} is not yet completely examined. P_{St} is less sensitive to ruthenium red than the transient PC peak observed upon the onset of continuous illumination, although verapamil and l-*cis*-diltiazem equally inhibit both components of the current (unpublished observations). Removal of Ca^{2+} from the external medium led to a smaller suppression of P_{St} as compared to that of the transient PC peak [57]. P_{St} completely disappeared at 10 mM external K^+, which gave rise to the notion that at lower K^+ concentrations a non-localized K^+ efflux accompanies P_{St} and promotes it by stabilizing the membrane potential [57].

Phototaxis in green flagellated algae requires the presence of Ca^{2+} ions in the external medium, which cannot be fully substituted by Mg^{2+}, Sr^{2+} or Ba^{2+} [39,97,98]. Within a certain range, a decrease in the stimulus intensity can be compensated by the increase in external Ca^{2+} to yield the same phototactic rate [99]. Various calcium channel blockers inhibit phototaxis [96,100]. Experiments in reactivated, demembranated *Chlamydomonas* cell models revealed different sensitivities of *cis*- and *trans*-axonemes to Ca^{2+} concentration which might account for the unbalanced motor response of the two flagella necessary for phototaxis [15]. Prolonged incubation at a Ca^{2+} concentration below 10^{-8} M led to selective and reversible inactivation of the *trans*-axoneme, whereas the *cis*-axoneme was inactivated at 10^{-7} to 10^{-6} M Ca^{2+}. At an intermediate concentration of 10^{-8} M both axonemes remained active. This Ca^{2+}-dependent shift in flagellar dominance was not found in the cell models of a non-phototactic *ptx1* mutant deficient in two 75-kDa axonemal proteins [101]. *Ptx5, ptx6* and *ptx7* mutants also lack this shift and likely have defects in genes encoding axonemal proteins [65]. In *ptx1* mutant, photoinduced changes in beat amplitude and period occur in both flagella as in the *trans*-flagellum in the wild type [102]. Asymmetric phosphorylation of a 138-kDa protein of an inner dynein arm complex causes different sensitivity of the two axonemes to Ca^{2+} as recently shown by analysis of phototactic mutant strains [103].

The above data give rise to the notion that phototaxis involves alteration of intraflagellar Ca^{2+} concentration, although not so dramatic as necessary for photophobic responses. The PC is mostly carried by Ca^{2+} ions under physiological conditions.

However, diffusion of Ca^{2+} ions from the cell lumen to a narrow intraflagellar space seems to be unlikely [104], especially taking into account that transient changes in flagella beating could be detected in less than 50 ms after the onset of the PC [19]. On the other hand, it was suggested that flagellar root tubules linking the photoreceptor site to flagella might participate in transmission of the signal [105]. Nevertheless, the most likely hypothesis is that changes in intraflagellar Ca^{2+} concentration necessary for phototaxis are mediated by Ca^{2+} fluxes across the flagellar membrane. Possible mechanism for their onset is activation of specific Ca^{2+} channels or transporters in the flagellar membrane at membrane potentials more negative than those needed for triggering FC. These Ca^{2+} fluxes can be too small or non-electrogenic, which explains why they were not yet identified electrophysiologically in wild type cells. The *ptx3 Chlamydomonas* mutant may have a deficiency in the mechanism for regulation of intraflagellar Ca^{2+} concentration involved in phototaxis [65]. Suppression of phototaxis in this mutant can be explained neither by inhibition of the PC, which is only two-fold decreased as compared to wild type, nor by the deficiency in the axonemal sensitivity to Ca^{2+}. A photophobic response and a FC found in this mutant indicated that the voltage-dependent Ca^{2+} channels responsible for them were not affected.

9.5 Voltage-gated currents

9.5.1 Flagellar currents and molecular mechanisms for the photophobic response

The correlation found between FC generation and a switch from "breast-stroke" flagellar beating to flagella undulation unequivocally proved that FC is the trigger for the photophobic response of the cell [19,48,60,68]. In a narrow range of stimulus intensities, a switch to undulation could be observed in only the *cis*-flagellum [60,68,69]. Therefore, it could be concluded that each flagellum behaves as an individual excitable organelle and that FC generation is likely triggered by activation of the ion channels located to the membrane covering each flagellum. No FC could be recorded immediately after complete mechanical amputation of the flagella, and the time course of the signal recovery was found to correlate with that of flagellar regrowth [56]. If amputation was partial, the amplitude of the FC was proportional to the total length of the two flagella. In "bald" mutants of *Chlamydomonas reinhardtii* which only possess short flagella stubs, substantially decreased FC was recorded. When one flagellum was inside the pipette and the other one remained outside, two distinct FC peaks of opposite signs were detected. These data led to the conclusion that the ion channels involved in FC generation are evenly distribution along the whole length of flagella [56].

FC can be abolished by the removal of Ca^{2+} ions from the external medium [45,48]. It also disappears after the addition of ions of heavy metals [45,49], or a wide range of Ca^{2+} transport inhibitors such as ruthenium red, verapamil and pimozide [48,68]. These findings indicate that FC generation is a strictly Ca^{2+}-dependent process and is most probably driven by activation of voltage-gated Ca^{2+}-channels in the flagellar membrane. *Chlamydomonas* mutants *ppr1-ppr4* lack both FC and photophobic response, whereas phototaxis and Ca^{2+} dependence of axoneme beating are not affected [75]. This shows that *ppr* mutants are likely defective in the flagellar Ca^{2+} channels involved in the photophobic response.

Ion selectivity of FC has been studied by the suction pipette technique [49,78]. FC measured after substitution of Ca^{2+} by Sr^{2+} or Ba^{2+} in the external medium revealed permeability of the flagellar channels for these divalent cations. The kinetics of FC decay did not change in Ba^{2+}, indicating that FC inactivation is not caused by the increase in Ca^{2+} concentration in the intraflagellar space. A small conductance of the channels for Mg^{2+} was also found in the absence of Ca^{2+}, but no FC could be measured after the addition of millimolar concentrations of Mg^{2+} on top of 0.1 mM Ca^{2+}. Small FC could also be observed in the absence of Ca^{2+} upon the decrease in the bath pH from 6.8 to 5.8, which could be explained by partial permeability of the flagellar channels for H^+.

F_f and F_s could be separated not only by their kinetics, but also by their different ion specificity [49,77,78]. Substitution of Ca^{2+} by Ba^{2+} only slightly increased the peak amplitude of F_f, whereas that of F_s became several times larger. However, adding Ba^{2+} on top of Ca^{2+} did not enhance F_s. Saturation of F_s amplitude was already achieved at 10^{-6} M Ca^{2+}, whereas a much higher concentration of more than 1 mM was required to saturate the F_s amplitude in Ba^{2+}. These findings led to the conclusion that a low F_s amplitude in Ca^{2+} is caused by a Ca^{2+}-induced down-regulation of the channels responsible for F_s generation. F_s observed in Ba^{2+} was much more sensitive to inhibition by Cd^{2+} than F_f. Both F_s amplitude and the duration of the recovery of the average swimming speed during the photophobic response measured by a population assay increased from $Ba^{2+} > Sr^{2+} > Ca^{2+}$. Video recording of individual cells revealed extended spiraling observed for 2 to 8 s after a flash in Ba^{2+} that could only be observed upon stimulation with long light pulses or step-up stimuli in Ca^{2+}. It has been proposed that the time of spiraling corresponds to the rate of extrusion or sequestration of the divalent cations.

Ca^{2+} entry during the action potential in Characeae is accompanied by Cl^- efflux which enhances depolarization of the membrane [106]. Participation of Cl^- fluxes in the flagellar response has not been found yet.

Studies on isolated, reactivated flagellar apparatuses and axonemes have established that Ca^{2+} acts directly on the axoneme and at around 10^{-4} M induces a switch from the ciliary type stroke to undulation [16,107]. Several *mbo Chlamydomonas* mutants have been isolated which display neither a photophobic response nor a Ca^{2+}-induced change in the flagellar beating type [108]. Two axonemal proteins were identified by alteration of their phosphorylation at $> 10^{-6}$ M Ca^{2+}, one of which was deficient in the *mbo* mutants [109]. Therefore, this Ca^{2+}-dependent phosphorylation was suggested to be involved in initiating the photophobic response.

9.5.2 K^+ currents

Neither F_s nor P_{St}-current could be observed at high external K^+ [57,78]. It has been proposed that these long-lasting depolarizing currents are only possible if counterbalanced by a K^+ efflux down the gradient of electrochemical potential. The K^+ efflux repolarizes the cell and leads to increasing the intracellular concentration of the divalent cations necessary to alter flagella beating without a significant drop in the negative membrane potential. F_f triggered voltage-gated transient K^+ currents that were outward below 0.6 mM of external K^+, and inward above it [79]. The presumable K^+ efflux accompanying F_s was not sensitive to Cs^+ [78], whereas transient K^+ currents triggered by F_f could be suppressed by several mM TEA^+ or Ca^{2+} [79]. It is not clear at present, if various K^+ currents observed in *Chlamydomonas* are mediated by the same or different species of K^+ channels.

Investigation of the photoresponses measured when the pipette and the bath contained K^+ at different concentrations gave rise to the conclusion that the K^+ efflux accompanying F_s, as well as the transient K^+ currents triggered by F_f, are non-localized, i.e. evenly distributed over the cell membrane [78,79]. However, localization of the K^+ channels to flagellar membrane cannot be excluded, since correct interpretation of the results obtained by the suction pipette technique under asymmetric ionic conditions is not easy. For instance, voltage-gated K^+ channels responsible for the decay of the regenerative response in the ciliate *Paramecium* were found in ciliary membranes [110].

9.6 Application of the electrophysiological approach to the investigation of the photosensory transduction in green flagellates

9.6.1 The nature of the photoreceptor protein

Extremely low concentration of the photoreceptor protein in the presence of high amounts of photosynthetic pigments complicates the preparative isolation of rhodopsin and spectroscopic studies in green flagellated algae. Consequently, investigations of rhodopsin-mediated photobehavior have been undertaken for in vivo testing of the rhodopsin structure and function. However, recording rhodopsin-mediated electrical responses provides better time resolution than behavioral assays and is therefore free from possible contribution of down-stream elements of the signal transduction chain to the result of measurements. Action spectra for PC recorded by the suction pipette technique are least influenced by spectral characteristics of the eyespot and the chloroplast of the cell, as compared to action spectra of phototaxis. Therefore, PC action spectrum represents the closest match for the rhodopsin absorption spectrum. The PC action spectra in *Haematococcus, Chlamydomonas* and *Volvox* all have their maxima around 500 nm indicating similar rhodopsin photoreceptors in these species [45,48,57]. The overall shape of the PC spectrum could be fitted by Dartnall's standard curve for rhodopsin absorption [48].

The PC action spectrum in *Haematococcus* reveals a complex fine structure [60,68]. A similar but less pronounced fine structure was found in the absorption spectrum of the eyespot membranes isolated from *Chlamydomonas* cells [87], although it could originate from the absorption of the eyespot carotenoids. An action spectrum of photobehavior and an absorption spectrum of sensory rhodopsin II (SRII) in *Halobacterium* have similar maxima (487 nm) and structured band shapes [111], whereas absorption spectra of other bacterial and animal rhodopsins have simple bell shapes at room temperature. The fine structure of PSII absorption spectrum was explained by vibronic transitions in a single rhodopsin species which pointed to the high rigidity of the chromophore in the protein moiety [112]. Similar explanation could also be suggested for the PC action spectrum in *Haematococcus*, although contribution of more than one pigment species could not be entirely ruled out [3,60,68,70,71]. Only one retinal-binding protein was so far identified in *Chlamydomonas* by biochemical studies [12,87]. However, it was found that restoration of photophobic response in a carotenoid-deficient *Chlamydomonas* induced by the addition of all-*trans* retinal was inhibited by 13-*trans*-locked retinal, whereas phototaxis was unaffected [13]. This result might indicate the involvement of two separate photoreceptor species for photophobic response and phototaxis in flagellates. It is well documented that photobehavior in archaebacteria is mediated by two different sensory rhodopsins present in the same cell [113].

The product of optical cross section and quantum efficiency calculated from fitting the high saturation phase of the stimulus-response curve for PC peak amplitude by an exponential function was approximately 8×10^{-21} m^{-2} [60]. This value corresponds to a single rhodopsin molecule rather than to a pigment complex as in photosynthetic pigment units. Hydroxylamine, the agent known to induce light-dependent cleavage of the chromophore in retinal-containing proteins, caused a specific inhibition of PC [64,114]. No PC could be detected in "blind" carotenoid-deficient mutants of *Chlamydomonas* unless the cells were reconstituted by the addition of exogenous retinal or its analogs [64].

Measuring PC elicited by the polarized light stimulus in a cell sucked into a pipette enabled probing the chromophore orientation. Maximum response amplitude was found when the plane of polarization was parallel to the eyespot surface [60,70,115]. This result shows that transition dipole moments of chromophore molecules lie in the plane of the cell membrane, which is also typical for animal rhodopsins [116] and bacteriorhodopsin [117]. Contribution of the eyespot interference reflection to the observed effects of polarized light is unlikely, since it is minimal for the light beam parallel to the eyespot [10,29]. When the e-vector of the stimulus was perpendicular to the eyespot surface, its rotation around the axis of the incidence did not influence the PC amplitude [115]. This observation indicated that the orientation of the retinal polyene chains is not ordered within the plane of the membrane.

Restoration of phototaxis in "blind" carotenoid-deficient *Chlamydomonas* mutants by retinal and its analogs provided an in vivo evidence for a rhodopsin-type photoreceptor [11]. Phototaxis measured by a long-term population assay ("dish test") could be restored by the addition of a wide range of retinoid compounds including retinal analogs prevented from isomerization around $C_{13}=C_{14}$ double bond and short-chained aldehydes [118,119]. The results of these experiments gave rise to a new hypothesis for activation of eukaryotic rhodopsins not involving *cis/trans* double bond isomerization [120]. However, subsequent studies undertaken by video recording and motion analysis of individual cells [35,121] and by detecting flash-induced motile responses in cell suspensions by a light scattering assay [122] could not reproduce some of the results obtained by Foster's group. The details of the reconstitution experiments in *Chlamydomonas* were recently reviewed [123–125]. Taking into account possible reasons for discrepancies found between the results obtained by different techniques, it could be concluded that the natural chromophore for *Chlamydomonas* rhodopsin is all-

trans, 6-s-*trans* retinal that undergoes the all-*trans*/13-*cis* isomerization upon photoexcitation [35,121]. The functional chromophore requires the presence of at least three conjugated C = C double bonds in the polyene chain and a methyl group at C_{13} position [122]. However, investigation of chromophore requirements even by sophisticated behavioral assays could not rule out possible effects of the added compounds on downstream elements of the signal transduction chain and/or antenna function. Therefore, restoration of PC in suspensions of "blind" cells upon addition of exogenous chromophores was examined [64]. No PC could be measured after the addition of 13-*trans*-locked retinal, 13-*cis*-locked retinal, 13-demethyl-retinal and citral. On the other hand, the addition of all-*trans*-retinal, 9-demethyl-retinal and dimethyl-octatrienal resulted in the appearance of normal PC, which proved that restoration of phototaxis and photophobic response in "blind" mutants by these compounds was indeed due to reconstitution of the functional rhodopsin. However, full restoration of the ability of the "blind" cells to photo-orientation required longer time after the addition of retinal than restoration of PC measured in the same assay. This observation suggests that, besides reconstitution of the rhodopsin, additional mechanisms might contribute to restoration of phototaxis measured by the "dish test" on a time scale of minutes.

The recovery of PC after the saturation flash occurs within several hundred milliseconds thus indicating the upper limit for the turn-over time of the rhodopsin photocycle [60,68,70,126]. This value is of the same order as the turn-over time of the photocycles in bacterial sensory rhodopsins (for recent review see [124]). However, the processes of PC desensitization and dark recovery are clearly influenced by the membrane potential [79]. It was shown that the time course of PC desensitization correlated to the kinetics of the first-flash-induced signal. Generation of FC which is a voltage-activated process downstream from rhodopsin photoconversion nevertheless strongly influenced PC dark recovery. Furthermore, the recovery rate was clearly dependent on the external K^+ concentration. Therefore, the time course of PC recovery after the flash is likely determined by the rate of restoration of the resting membrane potential, rather than by rate of the rhodopsin photoconversion.

9.6.2 The phototaxis antenna function

Modulation of light incidence on the photoreceptor during helical swimming of the cell under lateral illumination was already suggested in early studies on phototaxis in flagellates [20,21]. The physiological significance of such modulation could be tested by measuring PC in a cell sucked into a pipette and illuminated at different angles. Maximum PC amplitude was observed when the eyespot was facing the light source, whereas the minimum one was found when the eyespot was turned away from it [48,70]. Spectral sensitivity of the difference between maximum and minimum PC amplitudes correlated to the absorption spectra of the eyespot and the chloroplast [70]. The estimated extinction coefficient of the cell is in the range of 0.3 at 500 nm, which can only account for the 2-fold decrease in the PC amplitude [77]. However, the difference between maximum and minimum PC amplitudes was 3-fold in *Haematococcus pluvialis* [60,68] which has only one layer of carotenoid granules in the eyespot [25], and 8-fold in *Chlamydomonas reinhardtii* [48] which has an eyespot consisting of 2–4 layers of carotenoid granules [127]. Studies by confocal laser scanning microscopy have shown that the intensity of the eyespot interference reflection correlates to the number of the layers of carotenoid granules in different species [29]. Electrophysiological data demonstrate that the eyespot interference reflection indeed plays a functional role in modulation of the physiological signal during helical swimming of the cell.

Spectral sensitivity of phototactic accumulation of the cells in *Chlamydomonas* mutants with defects in the eyespot structure is shifted to shorter wavelengths indicating that chloroplast absorption contributes to modulation of the photoreceptor illumination during the cell rotation [27]. PC from these mutants could be investigated by the suspension method. The PC amplitude measured in unilateral mode upon excitation with 500 nm light was much lower in the eyespot-deficient mutant than in wild type cells. However, only a small difference was observed at 440 nm light (within the absorption range of the Soret band of chlorophyll a). PC measured in the pre-oriented mode revealed no changes in spectral sensitivity, which indicates that the photoreceptor itself was not affected by this mutation [63].

Phototaxis in "blind" carotenoid-deficient mutants of *Chlamydomonas* can be restored by the addition of exogenous retinoid compounds [11]. Photoelectric measurements revealed that maximum PC in carotenoid-deficient mutants reconstituted with retinal is generated when the cell is turned away from the light source (Fig. 9;

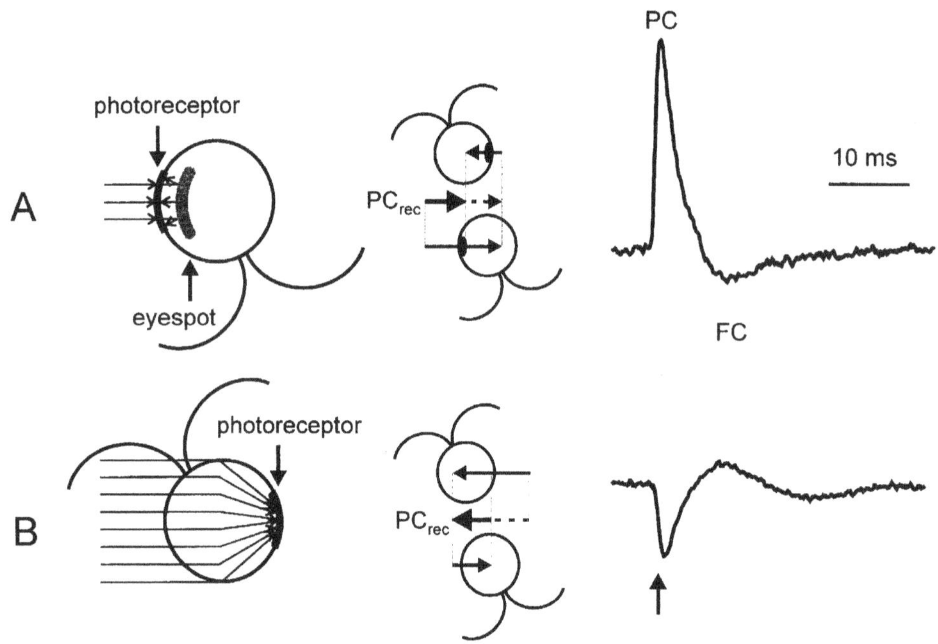

Figure 9. Phototaxis antenna function in green (A, strain 495(+)) and carotenoid-deficient (B, strain CC2359 with 10^{-8} M all-*trans* retinal) *Chlamydomonas reinhardtii.* Left schematic presentation of the mechanism for photoreceptor illumination, middle a scheme for photoelectric recording in unilateral mode of suspension measurements, right, photoelectric responses recorded from cell suspension, arrow shows the time of the excitation flash.

[64]). Cells of strain CC2359 used in electrophysiological studies lack eyespots as seen by both light and electron microscopy, even after their photobehavior has been restored by the addition of exogenous retinal [128]. The inverted dependence of the PC amplitude on the angle of light incidence shows that modulation of the photoreceptor illumination necessary for phototaxis likely occurs in this strain due to focusing of the light on the photoreceptor membrane by the transparent cell body ("lens effect"). Such a mechanism has been proposed earlier for phototropism in *Phycomyces* [129], but was not known in flagellates so far. Because of this mechanism of light focusing instead of the eyespot interference reflection, a periodic signal received by the photoreceptor during the rotation cycle in the carotenoid-deficient mutant would be phase-shifted by 180° as compared to the wild type. Consequently, the unbalanced motor response of flagella would also appear phase-shifted by the same angle, which would lead to turning of the cell in the opposite direction as compared to the wild type. Positive phototaxis found in retinal-reconstituted mutant cells under conditions when the wild type cells displayed negative phototaxis corroborated this hypothesis [64].

9.6.3 Regulation of phototaxis by the processes of energy metabolism

Phototaxis in green flagellated algae is mediated by a specific photoreceptor system separate from the photosynthetic apparatus. However, accumulation of the cells in the area of the optimal light conditions for phototrophic metabolism requires the involvement of a negative feedback loop in light control of photomovement. This loop is provided by regulation of phototaxis by photosynthesis and other processes of energy conversion. The energy-dependent mechanisms for regulation of photomovement are often found in prokaryotes and gliding unicellular eukaryotes (for review see [130–132]).

Most species of green flagellates are capable of both positive and negative phototaxis. Similar spectral sensitivity found for positive and negative phototaxis in the same species [6,8] indicated that a single receptor system is responsible for both processes. Therefore, the sign of phototaxis should be regulated at the level of the signal transduction chain, or at the effector level. The sign of phototaxis usually changes from positive to negative upon increase in the stimulus intensity [33] and depends on pre-irradiation [8,133–136], ionic composition of the medium [38,40,137], temperature [138], the nature of an exogenous chromophore [139], and a number of other factors. Different sensitivity of the two axonemes for Ca^{2+} [15] is supposed to play an important role in regulation of the phototaxis sign [140,141].

A fast switch from positive to negative phototaxis can be induced by the onset of the red background illumination that is not efficient for photo-orientation [27,142]. The spectral sensitivity of this phenomenon has a cut-off at 700 nm. It is saturated at light intensities known to saturate photosynthesis, and can be suppressed by DCMU. The sign of phototaxis was equally rapidly reversed after the red background illumination was simply turned off or switched to far-red light, which enabled to rule out a possible involvement of a phytochrome system.

It has been proposed that control of the phototaxis sign is linked to electrical processes in the cell membrane [135]. Measuring the resting membrane potential in

Haematococcus and *Dunaliella* by intracellular microelectrodes revealed that the onset of photosynthetically active red light induced gradual hyperpolarization of the plasmalemma that dissipated to the dark level after switching off the light [43,46]. These potential changes occurred on the same time scale as the changes in phototaxis sign observed upon switching on and off the red light [27]. Red background illumination increases the amplitude of PC [70], especially in the presence of KCN when photosynthesis is likely the main energy resource for keeping the resting membrane potential at a physiological level [47]. The increase in the electrochemical driving force for Ca^{2+} upon hyperpolarization of the cell membrane is likely the reason for the observed increase in PC amplitude, which apparently leads to the reversal of the phototaxis sign.

Upon the increase in stimulus intensity, the swimming path of individual *Chlamydomonas* cells becomes increasingly aligned with the stimulus direction throughout the whole range of intensities, whereas the net phototactic response of the population changes from positive to negative [33]. At intermediate stimulus intensities, individual *Haematococcus* and *Chlamydomonas* cells alternate between swimming towards and away from the light source, changing direction about every 25 s [143]. Strictly periodic electrical activity has been observed in the plasma membrane by an extracellular microelectrode deeply embedded into an invagination of the protoplast in *Haematococcus* [43,76]. The periodic electrical activity is regulated by the processes of photosynthesis and respiration [144–146]. It likely results from the activity of contractile vacuoles and can be optically monitored by recording the periodic local micromovements of the protoplast in a cell held on a micropipette [147]. It has been proposed that klinokinesis, i.e. spontaneous changes in the swimming direction is controlled by this activity [76]. The dependence of klinokinesis on red background illumination and its decomposition into two separate periodic processes corroborated this hypothesis [143]. Spontaneous spikes recorded by intracellular microelectrodes and suction pipette usually correlate to the impulses of periodic electrical activity and are probably the driving force for spontaneous changes in the swimming direction [76]. A phase shift between the two periodic electrical processes likely associated with the two flagella of the cell observed upon light stimulation [76] might control the switch from positive to negative phototaxis.

Figure 10 schematically shows the interaction of the two systems for light control of the behavior in green flagellated algae, one being responsible for alignment of the swimming path with the stimulus direction, and another regulating the sign of the response, i.e. swimming towards or away from the light source.

Acknowledgements

We thank F.-J. Braun and P. Hegemann for providing us with their manuscripts submitted for publication, and G. Kreimer, U. Rüffer and W. Nultsch for the reprints of their recent work. We are grateful to P. Hegemann for stimulating discussions that greatly contributed to this review. Critical reading of the manuscript by S.P. Balashov is highly acknowledged. We thank I.M. Altschuler for his help in preparation of the manuscript. This work was supported by INTAS-RFBR grant No. 95-1134 and RFBR grant No. 99-04-49015.

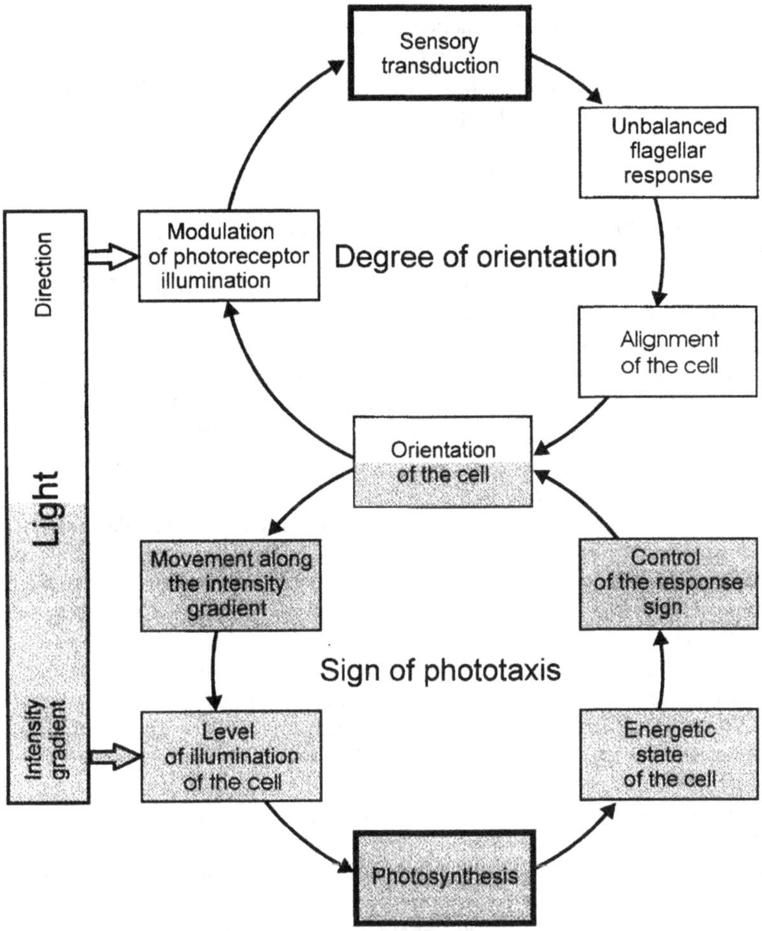

Figure 10. Two feedback loops in the light control of behavior in green flagellated algae.

References

1. B. Diehn, M. Feinleib, W. Haupt, E. Hildebrand, F. Lenci, W. Nultsch (1977). Terminology of behavioral responses of motile microorganisms. *Photochem. Photobiol.*, **26**, 559–560.
2. O.A.Sineshchekov, F.F. Litvin (1974). Phototaxis in microorganisms, its mechanism and relation to photosynthesis. *Usp. Sovr. Biol.*, **78**, 58–75 (In Russian).
3. O.A. Sineshchekov, F.F. Litvin (1982). Photoregulation of movement in microorganisms. *Usp. Mikrobiol.*, **17**, 62–87 (In Russian).
4. W. Nultsch, G. Throm (1975). Effect of external factors on phototaxis of *Chlamydomonas reinhardtii*. I. Light. *Arch. Mikrobiol.*, **103**, 175–179.
5. W. Nultsch (1975). Phototaxis and photokinesis. In: M.J. Carlyle (Ed.), *Primitive Sensory and Communication Systems. The Taxes and Tropisms of Microorganisms and Cells* (pp. 29–90). Academic Press, London, N.Y., San Francisco.
6. P. Halldal (1958). Action spectra for phototaxis and related problems in *Volvocales*, *Ulva*-gametes and *Dinophyceae*. *Physiol. Plant.*, **11**, 118–153.

7. P. Halldal (1961). Ultraviolet action spectra of positive and negative phototaxis in *Platymonas subcordiformis*. *Physiol. Plant.*, **14**, 133–139.
8. W. Nultsch, G. Throm, I. von Rimscha (1971). Phototaktische Untersuchungen an *Chlamydomonas reinhardtii* Dangeard in homokontinuerlicher Kultur. *Arch. Mikrobiol.*, **80**, 351–369.
9. K. Schletz (1976). Phototaxis bei *Volvox*. Pigmentsysteme der Lichtrichtungsperzeption. *Z. Pflanzenphysiol.*, **77**, 189–211.
10. K.-W. Foster, R.D. Smyth (1980). Light antennas in phototactic algae. *Microbiol. Rev.*, **44**, 572–630.
11. K.-W. Foster, J. Saranak, N. Patel, G. Zarilli, M. Okabe, T. Kline, K. Nakanishi (1984). A rhodopsin is the functional photoreceptor for phototaxis in the unicellular eukaryote *Chlamydomonas*. *Nature*, **311**, 756–759.
12. P. Kröger, P. Hegemann (1994). Photophobic responses and phototaxis in *Chlamydomonas* are triggered by a single rhodopsin photoreceptor. *FEBS Lett.*, **341**, 5–9.
13. D.N. Zacks, F. Derguini, K. Nakanishi, J.L. Spudich (1993). Comparative study of phototactic and photophobic receptor chromophore properties in *Chlamydomonas reinhardtii*. *Biophys. J.*, **65**, 508–518.
14. D.L. Ringo (1967). Flagellar motion and fine structure of the flagellar apparatus in *Chlamydomonas*. *J. Cell Biol.*, **33**, 543–571.
15. R. Kamiya, G.B. Witman (1984). Submicromolar levels of calcium control the balance of beating between the two flagella in demembranated models of *Chlamydomonas*. *J. Cell Biol.*, **98**, 97–107.
16. M. Bessen, R.B. Fay, G.B. Witman (1980). Calcium control of waveform in isolated flagellar axonemes of *Chlamydomonas*. *J. Cell Biol.*, **86**, 446–455.
17. U. Rüffer, W. Nultsch (1985). High-speed cinematographic analysis of the movement of *Chlamydomonas*. *Cell Motil. Cytoskeleton*, **5**, 251–263.
18. U. Rüffer, W. Nultsch (1987). Comparison of the beating of *cis*- and *trans*-flagella of *Chlamydomonas* cells held on micropipettes. *Cell Motil. Cytoskeleton*, **7**, 97–93.
19. E.M. Holland, H. Harz, R. Uhl, P. Hegemann (1997). Control of phobic behavioral responses by rhodopsin-induced photocurrents in *Chlamydomonas*. *Biophys. J.*, **73**, 1359–1401.
20. J. Buder (1917). Zur Kenntnis der phototaktischen Richtungsbewegungen. *Jahrb. Wiss. Bot.*, **58**, 105–220.
21. S.O. Mast (1927). The structure and function of the eye-spot in unicellular and colonial organisms. *Arch. Protiskenkd.*, **60**, 197–220.
22. M. Melkonian, H. Robenek (1984). The eyespot apparatus of flagellated green algae: a critical review. *Prog. Phycol. Res.*, **3**, 193–268.
23. G. Kreimer (1994). Cell biology of phototaxis in flagellated alga. *Int. Rev. Cytol.*, **148**, 229–310.
24. M.Melkonian, H. Robenek (1980). Eyespot membranes of *Chlamydomonas reinhardtii*: a freeze-fracture study. *J. Ultrastruct. Res.*, **72**, 90–102.
25. T. Ristori, G. Rosati (1983). The eyespot membranes of *Haematococcus pluvialis* Flotow *(Chlorophycea):* their ultrastructure and possible significance in phototaxis. *Monitore Zool. Ital.*, **17**, 401–408.
26. N.M.L. Morel-Laurens, M.E. Feinleib (1983). Photomovement in an "eyeless" mutant of *Chlamydomonas*. *Photochem. Photobiol.*, **37**, 189–194.
27. O.A. Sineshchekov, E.G. Govorunova, F.F. Litvin (1989). Role of photosynthetic apparatus and stigma in the formation of spectral sensitivity of phototaxis in flagellated green algae. *Biofizika*, **34**, 255–258 (In Russian).
28. K. Schaller, R. Uhl (1997). A microspectrophotometric study of the shielding properties of

eyespot and cell body in *Chlamydomonas*. *Biophys. J.*, **73**, 1573–1578.
29. G. Kreimer, M. Melkonian (1990). Reflection confocal laser scanning microscopy of eyespots in flagellated green algae. *Eur. J. Cell Biol.*, **53**, 101–111.
30. W.G. Hand, W. Haupt (1971). Flagellar activity of the colony members of *Volvox aureus* Ehrbg. during light stimulation. *J. Protozool.*, **18**, 361–364.
31. J.A. Schmidt, R. Eckert (1976). Calcium couples flagellar reversal to photostimulation in *Chlamydomonas reinhardtii*. *Nature*, **262**, 713–715.
32. U. Rüffer, W. Nultsch (1995). Flagellar photoresponses of *Chlamydomonas* cells held on micropipettes: III. Shock response. *Bot. Acta*, **108**, 255–265.
33. M.E.H. Feinleib, G.M. Curry (1971). The relationship between intensity and oriented phototactic response (topotaxis) in *Chlamydomonas*. *Physiol. Plantarum*, **25**, 346–352.
34. J.S. Boscov, M.E. Feinleib (1979). Phototactic response of *Chlamydomonas* to flashes of light. II. Response of individual cells. *Photochem. Photobiol.*, **30**, 499–505.
35. T. Takahashi, K. Yoshihara, M. Watanabe, M. Kubota, R. Johnson, F. Derguini, K. Nakanishi (1991). Photoisomerization of retinal at 13-ene is important for phototaxis of *Chlamydomonas reinhardtii*: simultaneous measurements of phototactic and photophobic responses. *Biochem. Biophys. Res. Commun.*, **178**, 1273–1279.
36. R. Wayne, A. Kadota, M. Watanabe, M. Furuya (1991). Photomovement in *Dunaliella salina*: fluence rate-response curves and action spectra. *Planta*, **184**, 515–524.
37. D.N. Zacks, J.L. Spudich (1994). Gain setting in *Chlamydomonas reinhardtii:* mechanism of phototaxis and the role of the photophobic response. *Cell Motil. Cytoskeleton*, **29**, 225–230.
38. P. Halldal (1957). Importance of calcium and magnesium ions in phototaxis of motile green algae. *Nature*, **179**, 215–216.
39. R.L. Stavis, R. Hirschberg (1973). Phototaxis in *Chlamydomonas reinhardtii*. *J. Cell Biol.*, **59**, 367–377.
40. H. Sakaguchi (1979). Effects of external ionic environment on phototaxis in *Volvox carteri*. *Plant Cell Physiol.*, **20**, 1643–1651.
41. I. Marbach, A.M. Mayer (1971). Effect of electric field on the phototactic response in *Chlamydomonas reinhardtii*. *Isr. J. Bot.*, **20**, 96–100.
42. K.M. Nichols, R. Rickmenspoel (1978). Control of flagella motion in *Chlamydomonas* and *Euglena* by mechanical microinjection of Mg^{2+} and Ca^{2+} and by electric current injection. *J. Cell Sci.*, **29**, 233–247.
43. O.A. Sineshchekov, V.K. Andrianov, G.A. Kurella, F.F. Litvin (1976). Bioelectric phenomena in unicellular flagellated alga and their relation to phototaxis and photosynthesis. *Fiziologiya rastenij*, **23**, 229–237 (In Russian).
44. O.A. Sineshchekov, V.A. Sineshchekov, F.F. Litvin (1978). Photo-induced bioelectric responses in phototaxis of a unicellular flagellated alga. *Dokl. Akad. Nauk S.S.S.R.*, **239**, 471–474 (In Russian).
45. F.F. Litvin, O.A. Sineshchekov, V.A. Sineshchekov (1978). Photoreceptor electric potential in the phototaxis of the alga *Haematococcus pluvialis*. *Nature*, **271**, 476–478.
46. D. Remis, W. Simonis, H. Gimmler (1992). Measurement of the transmembrane electrical potential of *Dunaliella acidophila* by microelectrodes. *Arch. Microbiol.*, **158**, 350–355.
47. O.A. Sineshchekov, F.F. Litvin, L. Keszthelyi (1990). Two components of photoreceptor potential in phototaxis of the flagellated green alga *Haematococcus pluvialis*. *Biophys. J.*, **57**, 33–39.
48. H. Harz, P. Hegemann (1991). Rhodopsin-regulated calcium currents in *Chlamydomonas*. *Nature*, **351**, 489–491.
49. E.M. Holland, F.-J. Braun, C. Nonnengässer, H. Harz, P. Hegemann (1996). The nature of rhodopsin-triggered photocurrents in *Chlamydomonas*. I. Kinetics and influence of divalent

cations. *Biophys. J.*, **70**, 924–931.

50. U.G. Schloesser, H. Sachs, D.R. Robinson (1976). Isolation of protoplasts by means of a "species-specific" autolysin in *Chlamydomonas*. *Protoplasma*, **88**, 51–64.
51. J.M Sullivan, K.-W. Foster (1987). Light-dependent ion channel activity in *Chlamydomonas*. *Biophys. J.*, **51**, 274a.
52. K.-W. Yau, T.D.Lamb, D. Baylor (1977). Light-induced fluctuations in membrane current of single toad rod outer segments. *Nature*, **269**, 78–80.
53. P.A. McNaughton (1990). Light response of vertebrate photoreceptors. *Physiol. Rev.*, **70**, 847–883.
54. K. Nagy (1991). Biophysical processes in invertebrate photoreceptors: recent progress and a critical overview based on *Limulus* photoreceptors. *Quart. Rev. Biophys.*, **24**, 165–226.
55. R. Ranganatan, D.M. Malicki, C.S. Zuker (1995). Signal transduction in *Drosophila* photoreceptors. *Annu Rev. Neurosci.*, **18**, 283–317.
56. C. Beck, R. Uhl (1994). On the localization of voltage-sensitive calcium channels in the flagella of *Chlamydomonas reinhardtii*. *Biophys. J.*, **125**, 1119–1125.
57. F.-J. Braun, P. Hegemann (1999). Two independent photoreceptor currents in the spheroid alga *Volvox carteri*. *Biophys. J.*, **76**, 1668–1678.
58. G. Kreimer, G.B. Witman (1994). Novel touch-induced, Ca^{2+}-dependent phobic response in a flagellate green alga. *Cell Motil. Cytoskeleton*, **29**, 97–109.
59. K. Yoshimura (1996). A novel type of mechanoreception by the flagella of *Chlamydomonas*. *J. Exp. Biol.*, **199**, 295–302.
60. O.A. Sineshchekov (1991). Electrophysiology of photomovements in flagellated algae. In: F. Lenci, F. Ghetti, G. Colombetti, D.-P. Häder, P.-S. Song (Eds), *Biophysics of Photoreceptors and Photomovements in Microorganisms (pp. 191–202). Plenum Press, New York.*
61. O.A. Sineshchekov, E.G. Govorunova, A. Der, L. Keszthelyi, W. Nultsch (1992). Photoelectric responses in phototactic flagellated algae measured in cell suspension. *J. Photochem. Photobiol. B: Biol.*, **13**, 119–134.
62. O.A. Sineshchekov, W. Nultsch (1992). *Signal transduction in rhodopsin-mediated photomovements in flagellates*. In: 5th Int. Conf. on Retinal Proteins, Dourdan.
63. G. Kreimer, C. Overländer, O.A. Sineshchekov, H. Stolzis, W. Nultsch, M. Melkonian (1992). Functional analysis of the eyespot in *Chlamydomonas reinhardtii* mutant *ey 627, mt⁻*. *Planta*, **188**, 513–521.
64. O.A. Sineshchekov, E.G. Govorunova, A. Der, L. Keszthelyi, W. Nultsch (1994). Photoinduced electric currents in carotenoid-deficient *Chlamydomonas* mutants reconstituted with retinal and its analogs. *Biophys. J.*, **66**, 2073–2084.
65. G.J. Pazour, O.A. Sineshchekov, G.B. Witman (1995). Mutational analysis of the phototransduction pathway of *Chlamydomonas reinhardtii*. *J. Cell Biol.*, **131**, 427–440.
66. O.A. Sineshchekov (1978). *The investigation of the photoelectric processes in phototaxis of a green alga*. Ph.D. Thesis, Moscow State University, Moscow (In Russian).
67. T. Ristori, C. Ascoli, R. Banchetti, P. Parrini, D. Petracci (1981). Localization of photoreceptor and active membrane in the green alga *Haematococcus pluvialis*. In: Proc. 6th Int. Cong. Protozool., Warsaw, p. 314.
68. O.A. Sineshchekov (1991). Photoreception in unicellular flagellates: bioelectric phenomena in phototaxis. In: R.H. Douglas (Ed.), *Light in Biology and Medicine* (Vol. 2, pp. 523–532). Plenum Press, New York.
69. O.A. Sineshchekov (1983). *Phototaxis in Microorganisms*. In: Proc. All-Union Conf. "Application of Lasers in Biology" (pp. 91–98). Moscow Univ. Press, Moscow. (In Russian).
70. O.A. Sineshchekov (1988). Phototaxis in microorganisms and its role in photosynthesis

regulation. In: I.N. Gogotov (Ed.), *Phototrophic Microorganisms* (pp. 11–18). Acad. Sci. U.S.S.R., Pushchino (In Russian).

71. O.A. Sineshchekov, F.F. Litvin (1988). The mechanisms of phototaxis in microorganisms. In: A.B. Rubin (Ed.), *Molecular Mechanisms of Biological Action of Optical Radiation* (pp. 412–427). Nauka, Moscow (In Russian).
72. U. Rüffer, W. Nultsch (1990). Flagellar photoresponses of *Chlamydomonas* cells held on micropipettes: I. Change in flagellar beat frequency. *Cell Motil. and Cytoskeleton*, **15**, 162–167.
73. U. Rüffer, W. Nultsch (1991). Flagellar photoresponses of *Chlamydomonas* cells held on micropipettes: II. Change in flagellar beat pattern. *Cell Motil. and Cytoskeleton*, **18**, 269–278.
74. P. Hegemann, W. Marwan (1988). Single photons are sufficient to trigger movement responses in *Chlamydomonas reinhardtii*. *Photochem. Photobiol.*, **48**, 99–106.
75. A. Matsuda, K. Yoshimura, O.A. Sineshchekov, M. Hirono, R. Kamiya (1998). Isolation and characterization of novel mutants of *Chlamydomonas* that display phototaxis but not photophobic response. *Cell Motil. Cytoskeleton*, **41**, 353–362.
76. O.A. Sineshchekov, V.V. Sudnitsin, F.F. Litvin (1984). Periodic electrical activity in a unicellular flagellated alga and its possible relation to klinophotokinesis. *Biofizika*, **29**, 643–648 (In Russian).
77. H. Harz, C. Nonnengässer, P. Hegemann (1992). The photoreceptor current in the green alga *Chlamydomonas*. *Phil. Trans. R. Soc. Lond. Ser. B*, **338**, 39–52.
78. A. Nonnengässer, E.-M. Holland, H. Harz, P. Hegemann (1996). The nature of rhodopsin-triggered photocurrents in *Chlamydomonas*. II. Influence of monovalent cations. *Biophys. J.*, **70**, 932–938.
79. E.G. Govorunova, O.A. Sineshchekov, P. Hegemann (1997). Desensitization and dark recovery of the photoreceptor current in *Chlamydomonas reinhardtii*. *Plant Physiol.*, **115**, 633–642.
80. R.A.Cone, W.L. Pak (1971). The early receptor potential. In: W.R. Loewenstein (Ed.), *Handbook of Sensory Physiology* (Vol. 1, pp. 345–365). Springer Verlag, Berlin.
81. D.G. Stavenga (1980). Short wavelength light in invertebrate visual sense cell: pigments, potentials and problems. In: H Senger (Ed.), *The Blue Light Syndrome* (pp. 5–24). Springer Verlag, Berlin.
82. K.M. Hartmann (1983). Action spectroscopy. In: W. Hoppe, W Lohmann, H. Markl, H. Ziegler (Eds), *Biophysics* (pp. 115–144). Springer Verlag, Berlin.
83. D.A. Baylor, T.D. Lamb, K.-W. Yau (1979). The membrane current of single rod outer segment. *J. Physiol.*, **288**, 589–611.
84. W. Deininger, P. Kroeger, U. Hegemann, P. Lottspeich, P. Hegemann (1995). Chlamyrhodopsin represents a new type of sensory photoreceptor. *EMBO J.*, **23**, 5849–5858.
85. W. Deininger, E. Ebnet, P. Hegemann (1996). Algal rhodopsins represent a new type of sensory photoreceptors. In: R. Schmitt, P. Hegemann, E. Harris (Eds), *Proc. 7th Int. Conf. Cell Mol. Biol. Chlamydomonas.* (p. S30). Regensburg Univ. Press, Regensburg.
86. I.L. Dumler, S.N. Korolkov, M.N. Garnovskaya, D.V. Parfenova, R.N. Etinghof (1989). The systems of photo- and pheromone signals transduction in unicellular eukaryotes. *J. Protein Chem.*, **8**, 387–389.
87. M. Beckmann, P. Hegemann (1991). In vitro identification of rhodopsin in the green alga *Chlamydomonas*. *Biochemistry*, **30**, 3692–3697.
88. G. Kreimer, U. Brohnsonn, M. Melkonian (1991). Isolation and partial characterization of the photoreceptive organelle for phototaxis of a flagellate green alga. *Eur. J. Cell Biol.*, **55**, 318–327.
89. L. Stryer (1986). Cyclic GMP cascade in vision. *Annu. Rev. Neurosci.*, **9**, 87–119.

90. U.B. Kaupp, K.-W. Koch (1992). Role of cGMP and Ca^{2+} in vertebrate photoreceptor excitation and adaptation. *Annu. Rev. Physiol.*, **54**, 153–175.
91. S.N. Korolkov, M.N. Garnovskaya, A.S. Basov, A.S. Chunaev, I.L. Dumler (1990). The detection and characterization of G-proteins in the eyespot of *Chlamydomonas reinhardtii*. *FEBS Lett.*, **270**, 132–134.
92. P.Hegemann, H. Harz, Photoreception in *Chlamydomonas*. In: J. Kurjan, B.J. Taylor (Eds), *Signal Transduction, Prokaryotic and Simple Eukaryotic Systems* (pp. 279–307). Academic Press, London.
93. U. Schlicher, L. Linden, M. Calenberg, G. Kreimer (1995). G proteins and Ca^{2+} -modulated protein kinases of a plasma membrane-enriched fraction and isolated eyespot apparatuses of *Spermatozopsis similis* (Chlorophyceae). *Eur. J. Phycol.*, **30**, 319–330.
94. M. Calenberg, U. Brohnsonn, M Zedlacher, G. Kreimer (1998). Light- and Ca^{2+}-modulated heteromeric GTPases in the eyespot apparatus of a flagellate green alga. *Plant Cell*, **10**, 91–103.
95. L. Linden, G. Kreimer (1995). Calcium modulates rapid protein phosphorylation/ dephosphorylation in isolated eyespot apparatuses of the green alga *Spermatozopsis similis*. *Planta*, **197**, 343–351.
96. P. Hegemann, K. Neumeier, U. Hegemann, E. Kuehnle (1990). The role of calcium in *Chlamydomonas* photomovement responses as analyzed by calcium channel inhibitors. *Photochem. Photobiol.*, **52**, 575–583.
97. R.L. Stavis (1974). The effect of azide on phototaxis in *Chlamydomonas reinhardtii*. *Proc. Natl. Acad. Sci. USA*, **71**, 1824–1827.
98. W. Nultsch (1979). Effect of external factors on phototaxis of *Chlamydomonas reinhardtii*. III. Cations. *Arch. Microbiol.*, **123**, 93–99.
99. R. Dolle, J. Pfau, W. Nultsch (1987). Role of calcium ions in motility and phototaxis of *Chlamydomonas reinhardtii*. *J. Plant Physiol.*, **126**, 467–473.
100. W. Nultsch, J. Pfau, R. Dolle (1986). Effects of calcium channel blockers on phototaxis and motility of *Chlamydomonas reinhardtii*. *Arch. Microbiol.*, **144**, 393–397.
101. C.J. Horst, G.B. Witman (1993). *Ptx1*, a nonphototactic mutant of *Chlamydomonas*, lacks control of flagellar dominance. *J. Cell Biol.*, **120**, 733–741.
102. U. Rüffer, W. Nultsch (1997). Flagellar photoresponses of *ptx1*, a nonphototactic mutant of *Chlamydomonas*. *Cell Motil. Cytoskeleton*, **37**, 111–119.
103. S.J. King, S.K. Dutcher (1997). Phosphoregulation of an inner dynein arm complex in *Chlamydomonas reinhardtii* is altered in phototactic mutant strains. *J. Cell Biol.*, **136**, 177–191.
104. W. Nultsch (1983). The photocontrol of movement of *Chlamydomonas*. In: D.J. Cosens, D. Vince-Price (Eds), *Soc. Exp. Biol. Symp. XXXVI* (pp. 521–539). Great Britain.
105. M Melkonian (1978). Structure and significance of cruciate flagellar root systems in green algae: comparative investigations in species of *Chlorosarcinopsis* (Chlorosarcinales). *Plant Syst. Evol.*, **130**, 265–292.
106. V.Z. Lunevsky, O.M. Zherelova, I.Y. Vostrikov, G.N. Berestovsky (1983). Excitation of *Characeae* cell membranes as a result of activation of calcium and chloride channels. *J. Membrane Biol.*, **72**, 43–58.
107. J.S. Hyams, G.C. Borisy (1978). Isolated flagellar apparatus of *Chlamydomonas*: characterization of forward swimming and alteration of waveform and reversal of motion by calcium ions in vitro. *J. Cell Sci.*, **33**, 235–253.
108. R.A. Segal, B. Huang, Z. Ramanis, D.J.L. Luck (1984). Mutant strains of *Chlamydomonas reinhardtii* that move backwards only. *J. Cell Biol.*, **98**, 2026–2034.
109. R.A. Segal, D.J.L. Luck (1985). Phosphorylation in isolated *Chlamydomonas* axonemes: a phosphoprotein may mediate the Ca^{2+}-dependent photophobic response. *J. Cell Biol.*, **101**,

1702–1712.

110. R. Eckert, P. Brehm (1979). Ionic mechanisms of excitation in *Paramecium*. *Ann. Rev. Biophys. Bioenerg.*, **8**, 353–383.
111. T. Takahashi, H. Tomioka, Y. Nakamori, N. Kamo, Y. Kobatake (1987). Phototaxis and the second sensory pigment in *Halobacterium halobium*. In: T. Kobayashi (Ed.), *Primary Processes in Photobiology* (pp. 101–108). Springer Verlag, Berlin, Heidelberg.
112. T. Takahashi, B. Yan, P. Mazur, F. Derguini, K. Nakanishi, J.L. Spudich (1990). Color regulation in the archaebacterial phototaxis receptor phoborhodopsin (sensory rhodopsin II). *Biochemistry*, **29**, 8467–8474.
113. R. Bogomolni, J.L. Spudich (1982). Identification of a third rhodopsin-like pigment in phototactic *Halobacterium halobium*. *Proc. Natl. Acad. Sci. USA*, **79**, 6250–6254.
114. O.A. Sineshchekov, E.G. Govorunova, F.F. Litvin (1991). The rhodopsin-like photoreceptor in *Haematococcus*: hydroxylamine effect on photoelectric responses and phototaxis of cells. *Sensornie Systemi*, **5**, 51–55 (In Russian).
115. K. Yoshimura (1994). Chromophore orientation in the photoreceptor of *Chlamydomonas* as probed by stimulation with polarized light. *Photochem. Photobiol.*, **60**, 594–597.
116. G. Wald, P.K. Braun, I.R. Gibbons (1963). The problem of visual excitation. *J. Opt. Soc. Am.*, **53**, 20–35.
117. M.P. Heyn, R.J. Cherry, U. Mueller (1977). Transient and linear dichroism studies on bacteriorhodopsin: determination of the orientation of the 568 nm all-*trans* retinal chromophore. *J. Mol. Biol.*, **117**, 607–620.
118. K.-W.Foster, J. Saranak (1988). Rhodopsin activation: a novel view suggested by in vivo *Chlamydomonas* experiments. *J. Am. Chem. Soc.*, **110**, 6589–6591.
119. K.-W. Foster, J. Saranak, F. Derguini, G.R. Zarrilli, R. Johnson, M. Okabe, K. Nakanishi (1989). Activation of Chlamydomonas rhodopsin in vivo does not require isomerization of retinal. *Biochemistry*, **28**, 819–824.
120. K.-W. Foster, J. Saranak, P.A. Dowben (1991). Spectral sensitivity, structure and activation of eukaryotic rhodopsins: activation spectroscopy of rhodopsin analogs in *Chlamydomonas*. *J. Photochem. Photobiol., B: Biol.*, **8**, 385–408.
121. M.A. Lawson, D.N. Zacks, F. Derguini, K. Nakanishi, D.L. Spudich (1991). Retinal analog restoration of photophobic responses in a blind *Chlamydomonas reinhardtii* mutant. *Biophys. J.*, **60**, 1490–1498.
122. P. Hegemann, W. Gärtner, R. Uhl (1991). All-*trans* retinal constitutes the functional chromophore in *Chlamydomonas* rhodospin. *Biophys. J.*, **60**, 1477–1489.
123. K. Nakanishi, R. Crouch (1995). Application of artificial pigments to structure determination and study of photoinduced transformations of retinal proteins. *Isr. J. Chem.*, **35**, 253–272.
124. J.L. Spudich, D. Zacks, R. Bogomolni (1995). Microbial sensory rhodopsins: photochemistry and function. *Isr. J. Chem.*, **35**, 495–513.
125. P. Hegemann (1997). Vision in microalgae. *Planta*, **203**, 265–274.
126. A. Der, R. Toth-Boconadi, L. Keszthelyi (1997). Electric currents evoked by double-flash excitation of the phototactic alga *Chlamydomonas reinhardtii*. *J. Photochem. Photobiol. B: Biol.*, **38**, 76–80.
127. H. Sager, G.E. Palade (1954). Chloroplast structure in green and yellow strains of *Chlamydomonas*. *Exp. Cell Res.*, **7**, 584–588.
128. M.A.Lawson, P. Satir (1994). Characterization of the eyespot regions of "blind" *Chlamydomonas* mutants after restoration of photophobic responses. *J. Euk. Microbiol.*, **41**, 593–601.
129. W. Shropshire Jr. (1962). The lens effect and phototropism of *Phycomyces*. *J. Gen. Physiol.*, **45**, 949–958.

130. W. Nultsch, D.-P. Häder (1979). Photomovement in motile microorganisms. *Photochem. Photobiol.*, **29**, 423–437.
131. W. Nultsch, D.-P. Häder (1988). Photomovement in motile microorganisms. II. *Photochem. Photobiol.*, **47**, 837–869.
132. D.-P. Häder (1986). Signal perception and amplification in photomovement of prokaryotes. *Biochim. Biophys. Acta*, **864**, 107–122.
133. P. Halldal (1960). Action spectra of induced phototactic response changes in *Platymonas*. *Physiol. Plant.*, **13**, 726–735.
134. A.M. Mayer (1968). *Chlamydomonas*: adaptation phenomena in phototaxis. *Nature*, **217**, 875–876.
135. I. Marbach, A.M. Mayer (1970). Direction of phototaxis in *Chlamydomonas reinhardtii* and its relation to cell metabolism. *Phycologia*, **9**, 255–260.
136. R. Uhl, P. Hegemann (1990). Adaptation in *Chlamydomonas* phototaxis.I. A light scattering apparatus for measuring the phototactic rate of microorganisms with high time resolution. *Cell Motil. Cytoskeleton.*, **15**, 230–244.
137. N. Morel-Laurens (1987). Calcium control of phototactic orientation in *Chlamydomonas reinhardtii*: sign and strength of response. *Photochem. Photobiol.*, **45**, 119–128.
138. H. Sakaguchi, K. Tawada (1977). Temperature effect on the photoaccumulation and photophobic response of *Volvox aureus*. *J. Protozool.*, **24**, 284–288.
139. T. Takahashi, M. Kubota, M. Watanabe, K. Yoshihara, F. Derguini, K. Nakanishi (1992). Diversion of the sign of phototaxis in a *Chlamydomonas reinhardtii* mutant incorporated with retinal and its analogs. *FEBS Lett.*, **314**, 275–279.
140. G.B. Witman (1993). *Chlamydomonas* phototaxis. *Trends Cell Biol.*, **3**, 403–408.
141. K. Schaller R., David, R. Uhl (1997). How *Chlamydomonas* keeps track of the light once it has reached the right phototactic orientation. *Biophys. J.*, **73**, 1562–1572.
142. T. Takahashi, M. Watanabe (1993). Photosynthesis modulates the sign of phototaxis of wild-type *Chlamydomonas reinhardtii*. *FEBS Lett.*, **336**, 516–520.
143. O.A. Sineshchekov, E.G. Govorunova (1991). Rhythmic motion activity of unicellular flagellated algae and its role in phototaxis. *Biofizika*, **36**, 603–608 (In Russian).
144. V.V. Sudnitsin, O.A. Sineshchekov, F.F. Litvin (1984). Influence of light and electrical current on the impulses of periodic activity in a unicellular flagellated alga. *Biofizika*, **29**, 842–844 (In Russian).
145. V.V. Sudnitsin, O.A. Sineshchekov, V.A. Boichenko, F.F. Litvin (1986a). Role of photosystem I of photosynthesis in regulation of the rhythm of periodic activity in a phytoflagellate. *Biofizika*, **31**, 430–433 (In Russian).
146. V.V. Sudnitsin, O.A. Sineshchekov, F.F. Litvin (1986). Connection of rhythmic activity in the flagellate *Haematococcus pluvialis* with respiration. *Biofizika*, **31**, 530–531. (In Russian).
147. O.A. Sineshchekov, V.V. Sudnitsin, F.F. Litvin (1988). Periodic micromovements in protoplast of unicellular flagellate *Haematococcus pluvialis*. *Biofizika*, **33**, 370–371 (In Russian).

Photomovement
D.-P. Häder and M. Lebert, editors.

Chapter 10

Rhodopsin-like-proteins: light detection pigments in *Leptolyngbya*, *Euglena*, *Ochromonas*, *Pelvetia*

Paolo Gualtieri

Table of contents

Abstract

This chapter deals with the photoreception strategy of microorganisms belonging to different phyla, such as *Leptolyngbya, Euglena, Ochromonas, Pelvetia*, which, according to the author's opinion use rhodopsin-like proteins. A brief introduction on the theoretical assumptions of the functioning of these proteins along with a survey of their distribution in nature precedes the description of the four cases. For each organism, the main characteristics of its photobehavior and the experiments which have strengthened the hypothesis of the presence of rhodopsin-like proteins are outlined.

10.1 Introduction

Proteins selected by evolution for photoreception possess stringent and specific functional characteristics, such as efficient photon-capturing ability, suitable spectral distribution, low thermal noise, and gain control [1]. A photoreceptive molecule must absorb a photon with high probability; hence it must have a high absorption coefficient. The capability of a visual system to detect a single photon is an adaptation for detecting light as low as possible. For a planktonic microorganism this represents an adaptive advantage: a photosynthetic organism in dim light can obtain more metabolic energy if it is able to detect low light intensities and move to more lighted, hence, more suitable areas for growth. Photon spectral distribution of the sun, i.e. the number of photons per frequency of spectral interval, has a maximum close to 1600 nm. Photoreceptive structures do not possess maximal photon-capturing power since they absorb mostly in the blue region. Since the more red-sensitive a photoreceptive pigment is, the higher is the rate of its thermal activation (dark noise), evolution takes into account the minimization of this drawback. Another requirement is that after absorbing a photon the molecule must pass the information on with high probability to the next stage in the detection chain; hence, it must have a high quantum efficiency. Although the photon has enough energy to be detected against the background of thermal activation, it does not have enough energy to transmit the signal reliably to the effector (e.g. flagellum) or to drive a response. One or more amplification stages are commonly required between the receptor and the effector. It is generally accepted that reception roles (chemoreception, mechanoreception, photoreception) are assigned to G-protein-linked receptors in eukaryotes [2]. Photoreceptors identified thus far are all characterized by seven membrane-spanning α-helices connected by extra- and intra-membranous loops [3], and interacting with G-proteins, which are therefore the first amplification factor. Exceptions to the presence of G-proteins, could be represented by the opsins of Archaebacteria [4] and *Chlamydomonas* [5], whose photoreceptive proteins are not even considered 7-transmembrane proteins.

Which are the proteins candidate for photoreception?

According to Mayr [6], living organisms are made up of macromolecules having extraordinary characteristics. Many of these macromolecules are so specific and unique for their ability to carry out a particular function, as rhodopsin-like proteins do in the photoreceptive process, to be present in animal and plant kingdoms every time this specific function is demanded. Notwithstanding this thought, rhodopsin(s) have been always considered a sole possession of animals. The recognition that rhodopsin is widely distributed outside the animal kingdom and also involved in non-visual purposes, such as melanosome migration and photic regulation of pupillary constriction [7] has been a surprising development of more recent years. In 1986 Appelbury and coworkers [8], on the basis of their observation that opsin-like genes can be identified in a wide variety of vertebrate, invertebrate and unicellular species, speculated that a primitive photopigment gene first evolved in unicellular organisms and was passed on, with some degree of sequence conservation, to a wide variety of present day species. Therefore, imaging eyes present in mollusks, arthropods and vertebrates, and no-imaging eyes found especially in lower species (microorganism photoreceptors), although anatomically very different, would employ similar visual transducers, namely rhodopsin-like proteins, which are 7 transmembrane α-helices receptors with retinal as chromophore. What is so special about this light absorbing group? First, the retinal-opsin complex has an intense absorption band whose maximum can be shifted into the visible region of the spectrum, over the entire range from 380 nm to 640 nm. Second, light isomerizes retinal very efficiently and rapidly. Moreover, its rate of isomerization in the dark is very low, about once in a thousand years. No other compound in nature comes close to matching the extremely high signal-to-noise ratio of retinal-opsin complex. Third, remarkable structural changes (movements of single α-helices) are produced by isomerization of retinal. Light is converted into atomic motion of sufficient magnitude to trigger a signal reliably and reproducibly.

Other candidate photoreceptive molecules, such as hypericin pigments and flavins (with pterin cofactor) are widely spread believed to act as photoreceptive proteins. In other chapters of this book the reader can find many examples and detailed explanations of their functions. Although the proposition that flavins could function as a near-UV/visible-light detector dates back to fifty years ago [9], at present a reliable biochemical identification of flavin photoreceptors lacks, and the identification of photoreceptor protein content is only based on action spectroscopy. However, Cashmore and his group, have recently reported mutant plants of *Arabidopsis* lacking CRY1 and CRY2 genes, which show no evidence of first-positive curvature, but still retain second-positive curvature [10,11]. In their opinion, the CRY1 and CRY2 genes might regulate the expression of a protein kinase with a putative redox-sensing domain.

This chapter deals with the photoreception strategy of microorganisms belonging to different phyla, and which, according to the author's opinion, use rhodopsin-like proteins.

10.2 *Leptolyngbya*

The genus *Leptolyngbya* sp. was created by Anagnostidis and Komarek [12] to accommodate a diverse set of very thin (< 3 μm) filamentous cyanobacteria previously included within the LPP-group B. This genus was accepted by Rippka and Herdman [13] who divided it into four clusters. These cyanobacteria show a very marked photobehavior.

Microscope observation reveals that the entire trichomes of this filamentous cyanobacterium move as a screw. If we allowed these cells to move in a semi-solid medium in a Petri dish with a source of light located in a precise spot, the trichomes grow from the mother colony fragments toward the light source. Microscope observation of the dish reveals that the trichomes move in parallel rows. By moving the dish to achieve a change in the light direction at a right angle to the original direction, a subsequent deviation of the growth direction is elicited, which reflects the light direction change. In Figure 1 we can see the 90° deviation in growth direction of the trichomes due to the corresponding dish movement (from position 1 to position 2). Therefore, these photosynthetic bacteria use light to grow toward optimal light intensities, with a sort of oriented movement with respect to the stimulus direction. Differently from the majority of motile cyanobacteria that respond to light, *Leptolyngbya* uses a peculiar steering mechanism for orientation towards light. Normally Oscillatoriacee are characterized by trichomes that are not able to change their direction and steer toward the light; however, some of them use a three steps forward-one step backward technique to gradually make their way toward the light [14].

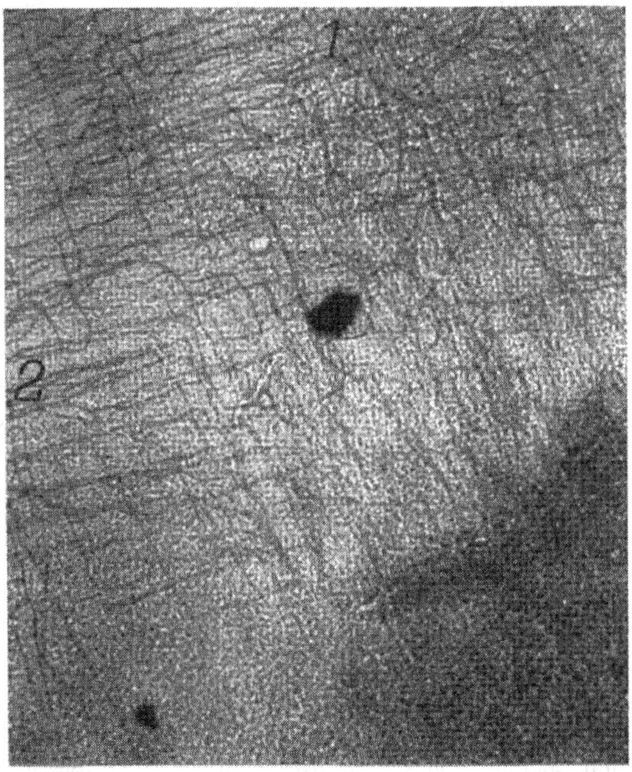

Figure 1. Photobehavior of *Leptolyngbya*: a Petri dish containing agarized growth medium was inoculated with the cyanobacterium, placed in a black box with a hole on the top, positioned over the center of the dish, and illuminated from above. By moving the dish to achieve a change in the light direction at a right angle to the original direction (from lamp 1 to lamp 2), *Leptolyngbya* orientated itself with respect to the new direction, i.e. the trichomes bent and grew at a right angle to the original direction. Scale bar 60 μm (courtesy Patrizia Albertano).

Nostocaceae, whose trichomes are made up of chains that are constricted where the cells join, are able to steer actively towards the light source. The trichomes may either turn end on to the light or glide as a U-shape with its center directed towards light [14]. *Leptolyngbya* (Oscillatoriacee) possesses more inflexible trichomes; nevertheless, it can change the direction of the trichome by bending it towards the light source at the level of the apical cell. This might suggest a visual system located in the apical cell, unlike other cyanobacteria where the whole trichome has been considered responsible for light detection [15]. These red *Leptolyngbya* possess an orange spot at the tip of the apical cell of every trichome, which resembles the eyespot or stigma of carotenoid -rich lipid globules almost invariably present in phototactic flagellate algae [16].

Electron microscopy revealed that this orange tip is characterized by osmiophilic globules of about 50 nm in diameter arranged in a peripheral cap extending 2–3 μm from the apex and with a possible layered pattern (Figures 2a and 2b). These globules are smaller than those present in eukaryotic algae, whose size is about 10 –30 nm [17].

Micro-spectrophotometric analysis of the tip of the apical cell of *Leptolyngbya* trichomes revealed a complex absorption spectrum. Two principal bands, centered at 456 nm and 504 nm respectively, were identified. Rhodopsin -like proteins (band at 504 nm) could be located inside the plasma membrane of the tip of the apical cell. On the basis of eukaryotic comparison, these proteins could be the photoreceptive pigment of this cyanobacterium, whereas carotenoids inside the globules (band at 456 nm) would have the screening role.

To verify the hypothesis of rhodopsin-like proteins, we treated *Leptolyngbya* with hydroxylamine (NH_2OH). This is a very commonly used reaction to identify retinal -based pigments [18], since NH_2OH selectively impairs rhodopsin based visual system. In the presence of NH_2OH, photo-orientation capability of cyanobacteria was completely impaired. The trichomes progressively lost their guidance mechanism and their growth pattern changed from parallel to more and more disordered. This photo-orientation impairment was reflected in the spectroscopic characteristics of the tip. In this case the spectrum did not show the band centered at 504 nm, whereas the band centered at 456 nm was present without significant difference with respect to the spectrum of untreated cells. This data is consistent with the hypothesis that the ability to sense light is via a rhodopsin-like protein located in the plasma membrane of the tip of the apical cell [16].

10.3 *Euglena*

Since the turn of the century the photosynthetic and photosensitive flagellate *Euglena gracilis* has provided an intriguing subject for photobiological studies. This flagellate dwells in natural shallow ponds, and uses sunlight as a source of energy and information. Its chloroplasts are the energy-supplying devices, whereas a simple but sophisticated system is used as light detector. This system consists of a locomotory flagellum, a stigma and a photoreceptor [19–21]. In general, as the cell rotates while swimming, the stigma comes between the light source and the paraflagellar swelling, thus modulating the light that reaches it, and regulating the steering of the locomotory flagellum. This configuration of stigma, swelling and flagellum, can be considered a

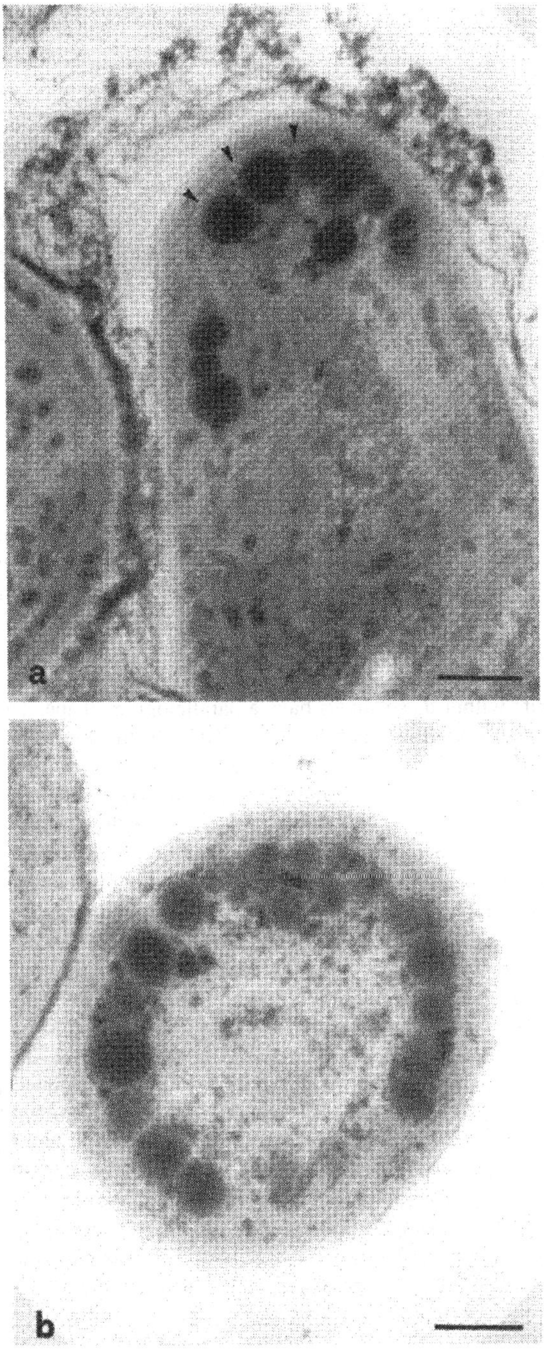

Figure 2. Transmission electron micrographs of a longitudinal (a) and a transverse (b) sections of the apical portion of a *Leptolyngbya* trichome. Both micrographs clearly show osmiophilic globules that resemble the stigma of eukaryotic algae. These globules lay in close contact with internal membranes (arrowheads). Scale bar 0.4 μm.

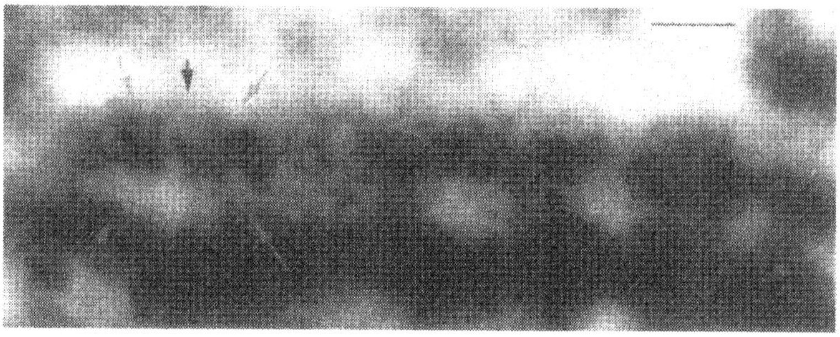

Figure 3. Freeze-fracture image of a cleavage plane perpendicular to a crystalline sheet of the photoreceptor of *Euglena*, showing integral proteins with their polar heads (black arrow) and the lipid bilayer (grey arrows). Scale bar 7 nm.

simple but complete visual system. However, the possible photoreceptive roles of the cytoplasmic stigma and the paraflagellar swelling of *E. gracilis* are still under debate, because of conflicting interpretations of the results obtained thus far by the different research groups working on this microorganism [22,23]. Nevertheless, our data provides evidence that the swelling is a rhodopsin-based photoreceptor.

The photoreceptor, a three-dimensional proteic crystal, is composed of a stack of crystalline sheets with a regular organization of component proteins. Fourier analysis suggests that the unit cells of proteic subunits within these sheets have a parallelogram shape. The dimensions of the monoclinic unit cell (or distorted hexagon, since $\beta = 107°$) of the sheet are about 50 Å × 40 Å. Figure 3 shows a cleavage plane perpendicular to one of the crystalline layers (for example, between the n-crystalline sheet and the $(n+1)$-crystalline sheet), although the cleft is not perfectly straight but somewhat curved. Several integral proteins are visible, showing their polar heads and the cylindrical hydrophobic cores. The layer seems to be constituted of the same protein, and the height of this integral membrane protein is about 70 Å, as a membrane should be. No other proteic structures or peripheral proteins can be seen between the layers. Lipids layers can be seen between proteins [24].

In 1989 Gualtieri and coworkers [25] and subsequently in 1992 Crescitelli and coworkers [26] measured the absorption spectrum of a single *Euglena* paraflagellar swelling. Due to the great similarity between these spectra and the absorption curve of rhodopsin α-band centered at 500 nm, both research groups suggested a pigment based on a rhodopsin-like protein. Successive experiments on inhibition of both the formation of the swelling and cellular photoorientation by means of hydroxylamine and nicotine showed that retinal is the chromophore present in the photoreceptor [27,28]. The extraction of retinal from intact cells [29] and from photoreceptors isolated from demembranated cells (unpublished results) has further strengthened the hypothesis of a rhodopsin-like protein. The estimated number of proteins, on the basis of the number of retinal molecules and of photoreceptor integrated optical density, is about 10^7 molecules. In 1997 Barsanti and co-workers reported the presence of a photochromic pigment in the paraflagellar swelling of *Euglena gracilis* which undergoes repeated and reversible fluorescence changes with a determinate kinetics [30]. According to these authors, the paraflagellar swelling possesses optical bistability, i.e. the non-fluorescent parent form (first conformer) of its molecules upon photo-excitation generates a fluorescent stable intermediate (second conformer) that can be photochemically driven back to the parent form. The quantum yields of the forward and reverse reactions are almost the same and close to unit [30]. These data have allowed us to identify the crystalline paraflagellar swelling as the true photoreceptor of *Euglena gracilis.*

How can *Euglena* detect light direction and successively steer? When light strikes the photoreceptor, a photoelectric signal is generated. Under the influence of this strong electric field, a displacement photocurrent could be propagated through paraxial rod filaments via charge transfer between rod proteins and, depending on the light intensity, through the cellular microtubular system and cytoskeleton since the function and structure of lattice-like components appear suited to propagate conformational waves [31]. The coherent excitation should have long-range effects, since the coiled paraxial rod filaments are hundreds of microns long. The component proteins of each of the seven paraxial rod filaments could, perhaps, be arranged around a hollow core, as in microtubules, each filament thus serving as a conduit for electron transport in transduction. We can suppose that the ATPase activity present in the flagellum [32] is located at the level of the goblet appendages of the paraxial rod. At these locations the mobile electrons should sink, thus activating ATPase proteins. The displacement of these goblet appendages along the axonemal microtubules should generate longitudinal waves of contraction along the paraxial rod. Its contractility can bend the flagellum sideways and/or forward during the well-known shock response [33,34]. In fact, the stiffening should swing the flagellum sideways, damping out some undulatory waves of the axoneme, whereas intermittent ATPase activity along the paraxial rod could augment other waves of the axoneme. This action turns the body away from or toward a source of light, depending on its fluence rate [24].

10.4 *Ochromonas*

In an effort to find a three-dimensional photoreceptive structure easier to isolate than the photoreceptor of *Euglena,* hence more promising for photoreceptive protein isolation and purification, we tested the freshwater, naked chrysophyte, *Ochromonas danica.* In this biflagellate, heterokont alga, the short flagellum bears a swelling in close juxtaposition to the chloroplastidic stigma. Although *Ochromonas* shows photo-behavior [35], there are no definitive data on the nature of its photoresponses or on the role or biochemistry of the flagellar swelling. We can hypothesize that *Ochromonas* uses a two-instant orientation mechanism as *Euglena* does. The long flagellum beats with uniplanar sine wave starting from the base, and its two rows of mastigonemes causes the pulling effect of its movement. The short flagellum directed laterally beats in helical waves. The cell thus regularly rotates along its major axis and scans the environment with its photoreceptor-stigma complex.

Ochromonas flagella lack the extensive paraxonemal material present in the *Euglena* flagellum; moreover, *Ochromonas* lacks either a cell wall or a pellicle, and this makes them fragile and more sensitive to the external environment than *Euglena* cells.

The *Ochromonas* photoreceptor is a three-dimensional proteic crystal with an organized structure of crystalline sheets similar to that of *Euglena*. A preliminary Fourier analysis of electron micrographs of photoreceptors suggests that the unit cells of proteic subunits within these sheets have a hexagonal shape. A hexagonal array of spots is clearly visible (Figure 4a,b). This cell could have a P^3 symmetry as reported by Henderson for bacterio-rhodopsin in *Halobacterium halobium* [36].

However, the extraction solution used for *Ochromonas* caused the cells, still motile, to round up; this shape appears to be highly labile when exposed to unfavorable conditions. The further addition of Ca^{2+} to the cells caused the acceleration of flagellar beating and their subsequent detachment at the level of insertion, i.e. the transition zone. The isolation yield of photoreceptors was quite high, but most cells burst and contamination by cellular debris was too high to allow so far an effective photoreceptive protein purification procedure.

We have identified all-trans retinal, corresponding roughly to 10^5 molecules/cell, in purified extracts of *Ochromonas* [35], and this result has been confirmed by another group (Hegemann, personal communication). On this basis, we have suggested that the photoreceptive system in *O. danica* use rhodopsin-like proteins, since retinal is the chromophore of rhodopsin-like proteins.

A green fluorescence emission, characteristic of blue-light absorbing pigments, is evident both in the short flagellum and in the swelling of *Ochromonas*. Such fluorescence has been observed also in other members of the Chrysophyceae, as well as in other algal groups [37,38]. While a green fluorescence has sometimes been suggested to be associated with flavins [39], a green emission has also been ascertained for rhodopsin photoproducts [40,41]. Therefore, the green fluorescence of the short flagellum and swelling of *Ochromonas* does not exclude the presence of a rhodopsin-like pigment.

Unlike *Euglena*, which also has a green fluorescent swelling associated with a flagellum, the situation in *Ochromonas* is complicated by the fact that both the flagellum and its swelling have a green emission, making the precise localization of the pigment more difficult. We have tried to demonstrate a possible photocycle of the O*chromonas* photoreceptor as for E*uglena*, but so far the attempts have been unsuccessfully.

10.5 *Pelvetia*

Fucoid algae, including species of *Pelvetia* and *Fucus* (Phaeophyceae), are a model system for studying the de novo formation and expression of cellular polarity. Newly fertilized *Pelvetia* zygotes (50–100 μm) have no morphological or biochemical polarity, except for the point of sperm entry, but 10 to 12 hours after fertilization they develop an axis that is expressed as localized growth or germination at one end. The subsequent first mitotic division at about 18 hours after fertilization is perpendicular to the developmental axis, partitioning the zygote into two progeny cells that develop very differently: a smaller rhizoid cell containing the growth site and a larger thallus cell. The rhizoid cell becomes the adult holdfast, while the thallus cell becomes the vegetative and reproductive part, the thallus. The site of germination can be determined by a number of external gradients (reviewed by [42]), including unilateral blue light to which the cells are exquisitely sensitive (Figure 5). The light need not to be present continuously. A 90-minute exposure to unilateral light several hours before germination is sufficient

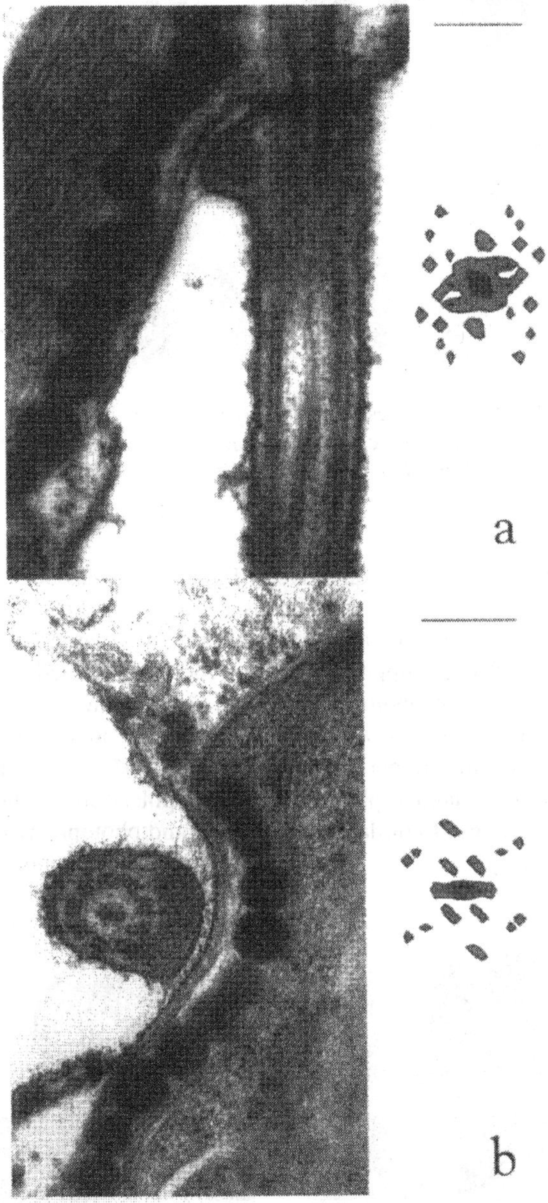

Figure 4. *Ochromonas* photoreceptive apparatus in both longitudinal (a) and transverse (b) sections. The axoneme of the longer flagellum, which bear the photoreceptor (arrow), is well visible facing the row of stigma globules. Next to each section, the Fast Fourier Transform of the photoreceptor is shown (courtesy Giovanna Rosati).

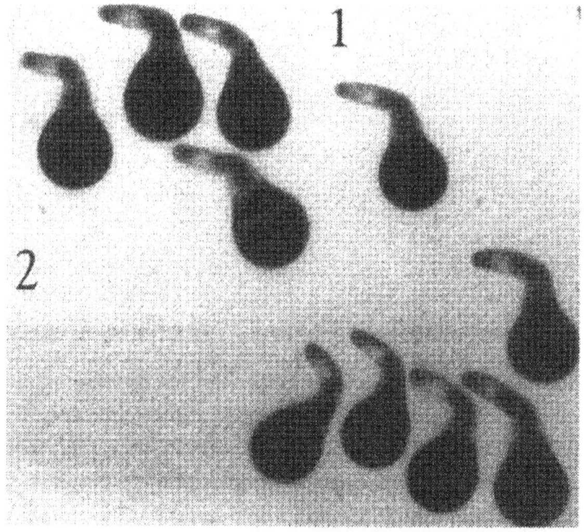

Figure 5. Germinated *Pelvetia* zygotes: a change by 90° (from 1 to 2) in the direction of light 24 h after germination led to the photoresponse of the rhizoids, i.e. they turned away from the light source (courtesy K.R. Robinson).

to polarize a population of zygotes effectively (Figure 6). It has been proposed that the formation of an intracellular Ca^{2+} gradient is essential to the early development of *Pelvetia* zygotes and embryos, including the processes of polarization, germination and rhizoidal growth. Evidence supporting this hypothesis includes measurement of differential Ca^{2+} flux during photopolarization [43], the measurement of an inward current at the nascent rhizoid [44] and the effects of Ca^{2+} ionophore gradients, which induced germination on the higher concentration side [45]. In addition, cyclic GMP (cGMP) has been demonstrated to play a role in photopolarization. The concentration of cGMP increased in response to unilateral blue light, and photopolarization was blocked by an inhibitor of guanylyl cyclase [46]. Recently, a high concentration of retinal has

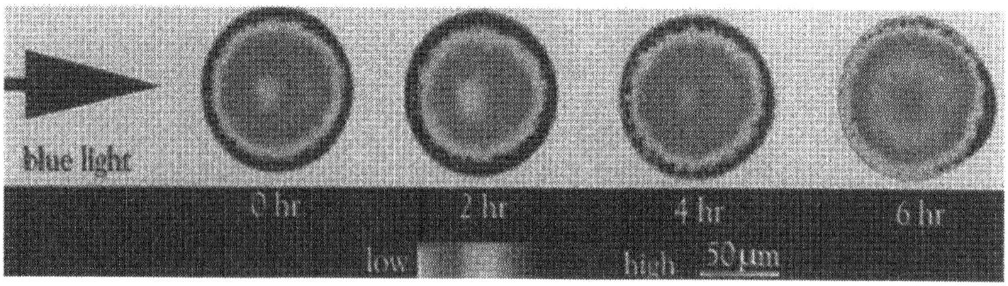

Figure 6. Images of a *Pelvetia* zygote injected with Rhodamine B dextran. Time 0 was defined as when the blue polarizing light was turned on, at 5.5 hours after fertilization. From left to right, the images were obtained at 0, 2, 4 and 6 h, at which time the zygote had germinated (courtesy K.R. Robinson).

been extracted and purified from *Pelvetia* zygotes, suggesting that rhodopsin-like molecules might be the photoreceptors also in this algal system [47]. The presence in *Pelvetia* of a rhodopsin-like protein acting as sensing protein, and the formation of an intracellular gradients of cGMP and Ca^{2+}, give to *Pelvetia* a biochemical pattern similar to that of higher Eukarya.

10.6 Conclusions

Life is characterized by the ability to reproduce, grow and develop, utilize energy, respond to environment, maintain homeostasis, and allow evolutionary adaptation. Therefore, it is not surprising that the basic proteins required to maintain these abilities, such as rhodopsin-like proteins, have been selected for the physical characteristics and conserved throughout the three domains of Bacteria, Archaea, Eukarya [48].

References

1. P.L. Walne, P. Gualtieri (1994). Algal visual proteins: An evolutionary point of view. *Crit. Rev. Plant Sci.*, **13**, 185–197.
2. L. Stryer (1988). *Biochemistry*. Freeman, New York.
3. M.J. Berridge (1993). Inositoltriphosphate and calcium signaling. *Nature*, **361**, 315–325.
4. W.D. Hoff, K.H. Jung J.L. Spudich (1997). Molecular mechanism of photosignalling by archaeal sensory rhodopsins. *Annu. Rev. Biophys. Biomol. Struct.*, **26**, 223–258.
5. W. Deininger, P. Kröger, U. Hegemann, F. Lottspeich, P. Hegemann (1995). Chlamyrhodopsin represents a new type of sensory photoreceptor. *EMBO J.*, **14**, 5849–5858.
6. E. Mayr (1982). *The growth of biological thought. Diversity, evolution and inheritance*. The Belknap Press of Harward University Press, Cambridge.
7. I. Provencio, G. Jiang, W.J. De Grip, W. Par Hayes, M.D. Rollag (1998). Melanopsin: An opsin in melanophores, brain, and eye. *PNAS (USA)*, **95**, 340–345.
8. R.L. Martin, C. Wood, W. Baehr, M. Applebury (1986). Visual pigment homologies revealed by DNA hybridization. *Science*, **232**, 1266–1269.
9. A.W. Galston (1950). Riboflavin, light and the growth of plants. *Science*, **111**, 619–624.
10. M. Ahmad, A.R. Cashmore (1993). *HY4* gene of *A. thaliana* encodes a protein with characteristics of a blue-light photoreceptor. *Nature*, **366**, 162–166.
11. M. Ahamad, J.A. Jarillo, O. Smirnova, A.R. Cashmore (1998). Cryptochrome blue-light photoreceptors of *Arabidopsis* implicated in phototropism. *Nature*, **392**, 720–723.
12. K. Anagnostidis, J. Komarek (1988). Modern approach to the classification system of cyanophytes. 3-Oscillatoriales. *Arch. Hydrobiol. Suppl.*, **80**(1–4), 327–472.
13. R. Rippka, M. Herdman (1992). *Pasteur Culture Collection of Cyanobacteria. Catalogue and Taxonomic Handbook* (Vol. I: Catalogue of Strains). Institut Pasteur, Paris.
14. R.W. Castenholz (1982). Motility and taxes. In: N.G. Carr, B.A. Whitton (Eds), *The Biology of Cyanobacteria, Botanical Monographs* (Vol. 19, pp. 413–439). Blackwell Scientific Publication, Oxford.
15. W. Nultsch, D.-P. Häder (1988). Photomovement in motile microorganisms – II. *Photochem. Photobiol.*, **47**, 837–869.
16. P. Albertano, L. Barsanti, V. Passarelli, P.A. Gualtieri (2000). Complex photoreceptive structure in the cyanobacterium *Leptolyngbya sp*, *Micron*, **1**, 27–34.
17. M. Melkonian, H. Robenek (1984). The eyespot apparatus of flagellate algae: a critical review. *Progr. Phycol. Res.*, **3**, 193–268.

18. A. Knowles, H.J.A. Dartnall (1977). Photochemistry of extracted visual pigments. In: H. Davson (Ed.), *The Eye* (pp. 289–345). Academic Press, New York.
19. P. Gualtieri, L. Barsanti, V. Passarelli, F. Verni, G.A. Rosati (1990). Look into the reservoir of *Euglena*: SEM investigation of the flagellar apparatus. *Micron Microscop. Acta*, **21**, 131–138.
20. G. Rosati, F. Verni, L. Barsanti, V. Passarelli, P. Gualtieri (1991). Ultrastructure of the apical zone of *Euglena gracilis*: photoreceptors and motor apparatus. *Electr. Microsc. Rev.*, **4**, 319–342.
21. F. Verni, G. Rosati, P. Lenzi, L. Barsanti, V. Passarelli, P. Gualtieri (1992). Morphological relationship between paraflagellar swelling and paraxial rod in *Euglena gracilis*. *Micron Microscop. Acta*, **23**, 37–44.
22. V.A. Sineshchekov, D. Geiss, O.A. Sineshchekov, P. Galland, H. Senger (1994). Fluorometric characterization of pigments associated with isolated flagella of *Euglena gracilis*: evidence or energy migration. *J. Photochem. Photobiol. B. Biol.*, **23**, 225–237.
23. D.-P. Häder, M. Lebert (1998). The Photoreceptor for the phototaxis in the photosynthetic flagellate *Euglena gracilis*. *Photochem. Photobiol.*, **63**, 260–265.
24. P.L. Walne, V. Passarelli, L. Barsanti, P. Gualtieri (1998). Rhodopsin: a photopigment for phototaxis in *Euglena gracilis*. *Crit. Rev. Plant Sci.*, **17**, 569–574.
25. P. Gualtieri, L. Barsanti, V. Passarelli(1989). Absorption spectrum of a single isolated paraflagellar swelling of *Euglena gracilis*. *Biochem. Biophys. Acta.*, **993**, 293–296.
26. T.W. James, F. Crescitelli, E.R. Loew, W.N. McFarland (1992). The eyespot of *Euglena gracilis*: a microspectrophotometric study. *Vision Research*, **32**, 1583–1591.
27. L. Barsanti, V. Passarelli, P. Lenzi, P. Gualtieri (1992). Elimination of photoreceptor (paraflagellar swelling) and the photoreception in *Euglena gracilis* by means of the carotenoid biosynthesis inhibitor nicotine. *J. Photochem. Photobiol.*, 135–144.
28. L. Barsanti, V. Passarelli, P. Lenzi, P.L. Walne, J.R. Dunlap, P. Gualtieri (1993). Effects of hydroxylamine, digitonin and Triton X-100 on photoreceptor (paraflagellar swelling) and photoreception of *Euglena gracilis*. *Vision Res.*, **33**, 2043–2050.
29. P. Gualtieri, P. Pelosi, V. Passarelli, L. Barsanti (1992). Identification of a rhodopsinic photoreceptor in *Euglena gracilis*. *Biochim. Biophys. Acta.*, **1117**, 55–59.
30. L. Barsanti, V. Passarelli, P.L. Walne, P. Gualtieri (1997). In vivo photocycle of the *Euglena gracilis* photoreceptor. *Biophys. J.*, **72**, 545–553.
31. J. Atema (1973). Microtubule theory of sensory transduction. *J. Theor. Biol.*, **38**, 181–190.
32. E. Piccinni, V. Albergoni, O. Coppellotti (1975). ATPase activity in flagella from *Euglena gracilis*. Localization of the enzyme and effect of detergents. *J. Protozool.*, **22**, 331–335.
33. H.S. Jennings (1906). *Behavior of the Lower Organisms*. Columbia University Press, New York.
34. S.O. Mast (1911). *Light and the Behaviour of Lower Organisms*. Wiley and Sons, New York.
35. P.L. Walne, V. Passarelli, P. Lenzi, L. Barsanti, P. Gualtieri (1995). Isolation of the flagellar swelling and identification of retinal in the phototactic flagellate, *Ochromonas danica* (Chrysophyceae). *J. Euk. Microbiol.*, **42**, 7–11.
36. R. Henderson (1977). The purple membrane from *Halobacterium halobium*,. *Annu. Rev. Biophys. Bioeng.*, **6**, 87–109.
37. D.G. Müller, I. Maier, H. Müller (1987). Flagellar autofluorescence and photoaccumulation in heterokont algae. *Photochem. Photobiol.*, **46**, 1003–1008.
38. A.W. Coleman (1988). The autofluorescent flagellum: A new phylogenetic enigma. *J. Phycol.*, **24**, 118–120.
39. P. Galland, H. Senger (1988). The role of flavin as photoreceptor. *J. Photochem. Photobiol. B*, **1**, 277–294.

40. A.V. Guzzo, G.L. Pool (1968). Visual pigment fluorescence. *Science*, **159**, 312–314.
41. A.V. Guzzo, G.L. Pool (1969). Fluorescence spectra of the intermediates of rhodopsins bleaching. *Photochem. Photobiol.*, **9**, 565–570.
42. L.F. Jaffe (1968). Localization in the developing *Fucus* egg and the general role of localizing currents. *Adv. Morphol.*, **7**, 295–328.
43. K.R. Robinson, L.F. Jaffe (1975). Polarizing fucoid eggs drive a calcium current through themselves. *Science*, **187**, 70–72.
44. R. Nuccitelli (1978). Ooplasmic segregation and secretion in the *Pelvetia* egg is accompanied by a membrane-generated electrical current. *Dev. Biol.*, **62**, 13–33.
45. K.R. Robinson, R. Cone (1980). Polarization of fucoid eggs by a calcium ionophore gradient. *Science*, **207**, 77–78.
46. K.R. Robinson, B.J. Miller (1997). The coupling of cyclic GMP and photopolarization of *Pelvetia* zygotes. *Dev. Biol.*, **187**, 125–130.
47. K.R. Robinson, R. Lorenzi, N. Ceccarelli, P. Gualtieri (1998). Retinal identification in *Pelvetia fastigiata*. *Biochem. Biophys. Res. Commun.*, **243**, 776–778.
48. W.E.G. Müller (1997). Evolution of Protozoa to Metazoa. *Theory Biosci.*, **116**, 145–168.

Photomovement
D.-P. Häder and M. Lebert, editors.

Chapter 11

Phototaxis of *Euglena gracilis* – flavins and pterins

Michael Lebert

Table of contents

Abstract

Photoorientation of unicellular organisms has been known for more than a century combining light perception, signal transduction and responses in a single cell. One of the model systems for the study of this phenomenon is *Euglena gracilis*, a unicellular, photosynthetic flagellate. *Euglena* uses different environmental stimuli to reach and stay in regions optimal for growth and reproduction. Among these stimuli light and gravity are the most important. Results of several decades clearly indicate that the photoreceptor molecules are located in a small organelle, the paraxonemal body (PAB). Detailed analysis revealed that the PAB shows a paracrystalline organization, i.e. molecules in this structure have a well defined orientation with respect to each other. This organization is the basis for detection of light direction, because the absorption probability depends on the orientation of the chromophores with respect to the light source. The role of the stigma, a red structure located close to the PAB is, according to the model, a screening device which absorbs light when the cell has a certain orientation to the light and by this means limits the light exposure of the PAB at the second predicted maximum of absorption probability. Overwhelming evidence indicates the presence of flavins and most likely pterins in the PAB acting as chromophoric groups of the photoreceptor molecules. Action spectra of phototaxis (i.e. the wavelength dependence of the sensitivity of the response) very closely resemble the absorption properties of flavins. Spectroscopic evidence shows the close relation between the absorption properties of the PAB and the known action spectra, thus reemphasizing the structure as the location of the photoreceptor molecules. Fluorimetric measurements indicate an energy transfer from pterins to flavins in the intact PABs. Several proteins containing pterins and flavins were identified in the structure. However, no final experimental evidence for the direct involvement of flavin and/or pterins was presented. Recently, a discussion was raised about the possibility of the involvement of a rhodopsin-like protein like in green algae. While the hypothesis can not be finally excluded, the critical interpretation of the presented evidences make the idea not very likely. In contrast to the knowledge about the chromophores not much is known about the elements of the signal transduction chain. Experimental evidence seems to exclude the direct involvement of the membrane potential as in the case of *Chlamydomonas* and other systems. Finally, it can be stated that while not very large, the field consists of a small number of very active groups and further progress is to be expected soon.

Abbreviations: PAB paraxonemal body, PFB paraflagellar body, PAR paraxonemal rod, FMN flavin mono nucleotide, FAD flavin adenine dinucleotide, LOV light oxygen voltage sensing domain.

11.1 Introduction

11.1.1 General

The following chapter discusses specific known or hypothesized characteristics of the phototaxis in *Euglena gracilis*, a unicellular, photosynthetic, fresh-water flagellate. Every cell is subject to the influence of many physical as well as chemical factors. The cell must react to these stimuli to survive. Cells in a tissue as well as single cells can and will experience sudden and at times dramatic changes in pH, temperature, light environment and the orientation in space, to name a few. Cells use specific or unspecific receptors, which interact with one or a few of these stimuli. As a result of this interaction the receptor changes its signaling state, possibly by changes of its conformation. This change in the signaling state triggers a signal transduction chain which in turn changes the activity of the effector, for example the beating pattern of a flagellum or the expression of a gene. In parallel to the above discussed excitation branch of a signal transduction, adaptation leads to a decrease or even an disappearance of the observable reaction [1]. Adaptation can be observed on all levels of the signal transduction (receptor, second messenger, effector; [2]). Frequently, transient reversible covalent modifications of the receptor proteins are involved [2–4]. In general, in most signal transduction chains reactions are observed during a change of the stimulus level rather than in response to the absolute stimulus level.

As a specialized case, protists, in contrast to multicellular organisms, combine signal perception, signal transduction and effector in a single cell [5]. This advantage was noted even by the earliest investigators [6–9] and consequently, some single cells (*Euglena gracilis*, *Chlamydomonas reinhardtii* and later the bacterium *Escherichia coli*) were established as model systems for signal transduction research.

11.1.2 The organism

Euglena gracilis is a unicellular, photosynthetic flagellate [10,11]. No sexual cycle is known. This observation and the finding that *Euglena* is octaploid [12] makes the induction of mutants difficult. The size and form are highly variable. The size ranges between 50 μm and 80 μm length and 8 μm to 12 μm width, depending on cell age and culture conditions [13]. The cell form varies between an almost perfect sphere to an extended cylinder [14–20]. The whole cell is covered with a slime sheath [21]. *Euglena* uses paramylon as storage substance (β-1,3-glucan; [22–25]). At the front end, outside the reservoir, but in close proximity to the paraxonemal body (see below) it posses a red structure, the stigma or eyespot [26]. The stigma is formed by globules (200–300 nm diameter) filled with carotenoids [27–35] and has no relationship in terms of origin or spatial orientation to chloroplasts [36,37]. Some evidence for the presence of flavins in the eyespot was presented by Sperling et al. [38]. In contrast to green algae (see related chapter by G. Kreimer, this volume), these globules are not organized in a rigid structure [9,36,39–44] and can not function as an interference reflector device as shown for stigmata in many green algae [43,44]. Most likely, the stigma functions as an screening device and is not directly involved in photoperception as proposed by earlier authors [45].

As a typical member of the order of Euglenophyceae, *Euglena* posses a small, bottle-like invagination at the anterior of the cell ("reservoir"). The 5 μm × 10 μm structure is connected to the medium only by a narrow channel ("pharynx"). The channel can be closed by a peripharyngial ring [46]. In contrast to other parts of the cell, in the reservoir the membrane is not associated with pellicle elements, but connected with a close system of longitudinally oriented microtubules. In the channel region also radial oriented microtubules were found [47,48]. The channel and reservoir membrane is connected by microfibrils to these microtubules. This functional unit might be related to mechanoreception in *Euglena* [49]. Associated with the reservoir is a system of contractile vacuoles (one main vacuole with several accessory vacuoles) for osmoregulation [48]. The content of the main vacuole is periodically released into the reservoir (every 20–30 s). At the bottom of the reservoir two flagella originate, of which only one leaves the reservoir [50]. Close to the contact point of the two flagella a small (1 μm × 0.7 μm × 0.7 μm; [37,41,52,51–54]) organelle can be found, the paraxonemal body (PAB; formerly called paraflagellar body (PFB); [55]). The paraxonemal membrane has a specialized structure [56]. The paraxonemal rod (PAR) connects the PAB to the cell matrix [57]. The PAR consists of several proteins [47,57,58]. The function of the structure is still unknown. The PAB has a paracrystalline organization [41,42]. Based on optical diffraction patterns Wolken [54] constructed a model of packed rods in a helical pattern of the PAB. Piccini and Mammi [59] revealed the presence of a paracrystalline structure of the PAB of monoclinic or slightly hexagonal cell units with the principal axes: $a = 8.9$ nm, $b = 7.7$ nm, $c = 8.3$ nm und $\beta = 110°$. According to Michel [60] this structure can be interpreted as a 3-D crystal of type I. Type I crystals are stacks of 2-D crystal arrays stabilized by interplane hydrophobic interactions and hydrophilic interactions with the surrounding aqueous environment. In the case of the PAB more than 100 layers of the 2-D crystals are thought to form the observed structure [61]. The 2-D type crystals can often be found in structures related to light energy conversion, for example the purple membrane in *Halobacterium salinarium*, with a very high concentration of bacterio-rhodopsin [62] or the photosynthetic membranes of higher plants [63] and purple bacteria [62]. In contrast to these specialized membrane structures, the PAB is the only example for a light detecting system.

The pellicle forms a layer below the cytoplasmic membrane and is composed of helical oriented parallel strips which can slide with respect to each other [64,65]. The sliding of strips explains the extreme variability of cell forms observed in *Euglena*, as well as the so called "euglenoid movement" [15,16,64,65]. Euglenoid movement can be observed under conditions where the cell has lost its trailing flagellum and moves over the surface by a sequence of body extensions and contractions. The pellicle is mainly formed by proteins (80%), fatty acids and carbohydrates [66].

The one emerging flagellum serves as a trailing flagellum. While variable, the velocity of the cells was measured as one to two cell length per second (50 μm s^{-1} to 100 μm s^{-1}; [67–73]) and depends on cell age and culture conditions.

During flagellar movement a helix-like structure of the flagellum can be observed. Waves are wandering from the base to the top of the flagellum with a frequency of approximately 20 Hz [74]. The corresponding force vectors result in a rotation of the cell body around its long axis with a frequency of 1–2 Hz [69]. The frequency, amplitude as well as the angle between cell body and flagellum can be experimentally influenced [75–77].

11.1.3 Ecology

Euglena as a freshwater aquatic organism belongs to the biggest ecosystem of the world, the aquatic ecosystem. Especially the marine ecosystems are an integral part of the global climate control. By the uptake of carbon dioxide and the production of oxygen, photosynthetic organisms are among the key players [78]. In addition, unicellular motile algae are primary producers in the food web. Both facts contribute to the growing concern regarding the potential hazardous effects of increased UV-B radiation due to ozone depletion. Related experiments clearly demonstrated, that the orientation reactions of motile organisms are primary targets for increased UV-B exposure [71,79–84].

The orientation reactions are used to reach and stay in horizons of the water column with light conditions which are optimal for growth and reproduction, but are not harmful by excessive solar irradiation for the sensitive photosynthetic apparatus [85]. Light and gravity are the two most important environmental factors used for orientation by the organisms and act synergistically or antagonistically on the net movement of the cells [85–89]. Any impairment of these reactions might result in reduced biomass production and as a consequence of the reduced photosynthetic activity will increase the atmospheric carbon dioxide concentration with an expected impact on global warming.

11.2 Photoresponses

Due to the evident relation between cell behavior and the stimulus, light responses attracted the attention of many researchers for more than 100 years. The first authors who realized the potential benefits of analyzing light responses of single cells were Cohn [90], Pfeffer [91], Jennings [92,93], Buder [8], Mast [94], Bolte [95] and Engelmann [96]. Since then *Euglena* has been a model system for related questions [97]. Following the definition given by Diehn et al. [98], three different responses can be distinguished: (1) Photokinesis, which describes the influence of the light intensity on the steady-state speed of an organism. A positive photokinesis is defined as an increase of the speed in the presence of light, while a negative photokinesis describes an decrease of the speed in the presence of light as compared to the dark control (see introductory chapter by Haupt, this volume). The response is independent of the light direction. (2) Photophobic response is defined as a transient response to a sudden change in light intensity (dI/dt). After the reaction the original behavior is resumed. As in photokinesis, the response is independent of the light direction. A great variation of reactions can be found, each characteristic for a species. These include delayed or sudden stops, back swimming, tumbling or change in direction of movement during forward locomotion. Instead of terms like "positive" or "negative" response the definition given by Haupt [99] will be used. A reaction to a sudden decrease or increase of light intensity are called step-down or step-up photophobic response. (3) Phototaxis is a orientation with respect to the light direction. An orientation toward a light source is called positive, away from the light source negative phototaxis; movement at an angle with respect to the light source is called dia-phototaxis. In contrast to the other responses it is a vectorial response. In the original definition by Diehn et al. [98] phototaxis also included an orientation based on trial and error ("biased random walk"; originally introduced for chemo-orientation in *Escherichia coli* [100,101]. In the following, the term phototaxis will only be used for active steering systems, following the limitation introduced by Nultsch and Häder [102]. In the older literature nowadays outdated terms are used for the different photoresponses. Positive and negative photophobotaxis are older terms for step-up and step-down photophobic responses, respectively. Positive and negative phototopotaxis are positive and negative phototaxis.

The main focus of this chapter will be on phototaxis of *Euglena gracilis*, and the other photoresponses will only be described shortly.

11.2.1 Methodology

In general, two methods can be distinguished: population methods and single cell measurements, both having advantages and disadvantages. Single cell measurements are based on direct, either microscopic or macroscopic observations of cells before, during and after stimulation. In the case of more than one reaction to the applied stimulus (i.e. phototaxis and photophobic responses), the observer can easily focus on the response of interest. Before the introduction of semi- or fully automatic computer-aided cell tracking systems measurements were often tedious and error prone and, due to limited observations statistical significance was hard to establish.

In contrast, statistical significance is easy to establish in population measurements. In these setups the reaction of a whole population with many individuals is measured. On the other hand, an intrinsic disadvantage of population methods is that the observed result is often the consequence of a mix of different reactions (photokinesis, photophobic responses and phototaxis; [103]). It might be difficult to distinguish between the contribution of each response to the measured parameter.

11.2.1.1 Population methods

Photokinesis. Depending on the reaction and the parameter measured different methods are applicable. Nultsch [104] used the spreading of cyanobacteria on semi-solid agar from a central inoculation with different irradiances after different times to estimate the photokinetic reaction value.

Photophobic responses. Light-trap systems are another example for a population method to measure photophobic responses. In these systems a continuous light field is projected in a liquid culture of homogenously distributed organisms. Depending on light quality and quantity cells entering the light field will either accumulate in the light-trap or leave the light trap. The general setup, originally invented by Engelmann [105], is still in use in many variations. The accumulation of cells can either be quantified after the completion of the reaction photographically or densitometrically [104,106]. Under some circumstances the continuous measurement of the accumulation is necessary. For this purpose single-beam [107–109] or double beam system are used. In double beam systems actinic light sources (for stimulation) and measuring beam (usually infrared light) are separated. One important example is the "phototaxigraph", invented by Lindes et al. [110] and improved by Diehn [111]. The latter system is an example for the limitation of the method and uncritical interpretation of data obtained (see below). A serious disadvantage of the light-trap method is a possible interference of the measurement with cells reacting phototactically to light scattered by cells already in the light field. While these limitations can be partially overcome [112,113] by sophisticated modifications, only a parallel microscopically examination of reacting cells can exclude a potential misinterpretation of data obtained with the light-trap method.

Phototaxis. Population methods measuring phototactic responses are all based on absorbance changes caused by the accumulation of cells. A simple system can be used to measure phototaxis in slow moving diatoms, cyanobacteria or slime molds. Organisms are inoculated in the middle of an agar plate and irradiated from one side [114,115]. A similar method can be applied for measuring phototaxis in single, motile cells. A homogeneous cell culture is placed in a Petri dish and illuminated from one side. The phototactic movement leads to a mass displacement which macroscopically clears one side of the Petri dish. The kinetics of the occurrence of this "clearing zone" can be taken as a measure for the phototactic activity. Again, as discussed above the observed reaction might or might not be solely caused by phototaxis and needs further validation.

The above mentioned "phototaxigraph" consists, like most of the systems of a cylindrical cuvette which is horizontally orientated. The cuvette is rotated around its long axis, to prevent an interference of the light reaction with gravitaxis (orientation toward or away from the center of gravity). Actinic light is impinging from above and below the system. The phototactic response is caused by light scattering from cells already in the actinic light field. The continuous comparison of the absorbance at >800 nm (phototactically inactive) inside and outside the actinic light zone is taken as a measure for the phototactic response. As discussed above, the measured response in this case is a mix of photokinetic, photophobic and phototactic responses [116,117].

In order to overcome these problems later versions uses actinic light entering the rotating cuvette parallel to its long axis [118]. The accumulation of cells is again measured by two measuring beams and comparing the absorption at > 620 nm. Later versions utilized actinic light impinging from 45° to reduce the light gradient caused by the absorption of the organisms [119,120].

11.2.1.2 Individual cell methods

Photokinesis. From the beginning of individual cell measurements researchers tried to apply semi-automatic or automatic methods. To access photokinesis the time needed for a given distance was measured directly under the microscope. The simplest enhancement included a video camera and a recorder. The reaction was first recorded and subsequently slowly replayed. Cell positions were marked on the screen on transparent foils which were analyzed later. This method was improved by Feinleib and Curry [118] who measured tracks taken from microphotographs with a constant exposure time. Similar methods were applied by various authors [121–124].

Photophobic responses. Phobic responses can easily be detected and quantified by microscopic observation [125–127]. Improved methods include the use of micro-cinematography [128], video recording [129] or the flying spot scanning apparatus [130,131]. With the advent of dedicated fast hardware efforts concentrated on the development of automatic analysis systems. In the first systems, video images were recorded and subsequently digitized. The positions of cells in the subsequent digitized images were determined. Using this spatial and temporal information, a path reconstruction can be obtained and sudden changes in the forward movement can be detected [132]. Due to faster hardware the same process can be done in real-time fully automatically. In the discussion of population methods potential problems in measurements and the subsequent analysis were pointed out. Automatic systems allow to measure the reaction of organisms in statistically significant numbers and by this means to combine the accuracy of single cell measurements with the statistical superiority of population methods. Nevertheless, the same restrictions as in population methods apply. Most modern systems allow, in parallel to the measurements the direct observation of the cells, which is a necessary control for the quality of the measurement. In addition, an analysis of the graphical representation of the raw data might be important to determine which of several potential responses was measured.

Phototaxis. As in measurements of photophobic responses researchers tried very early to automatically detect phototaxis. The same systems used for photophobic response analysis were also used for phototaxis research [129–131]. Systems using dedicated hardware for image processing were the next [133,134]. Later on, fast personal computers as well as flash analog-digital converters allowed to automate the analysis [135–143]. In the most advanced systems even three-dimensional analysis of the movement of organisms is possible [144,145]. In addition, on-line statistical analysis of the data is possible in parallel to the ongoing measurements [146]. However, the same restrictions as in all other cases discussed above apply: The direct observation of the cell behavior can never be substituted by pure numbers, which are prone to misinterpretations. Every conclusion must be paralleled by a careful analysis of the experimental procedure.

11.2.2 Photokinesis

Photokinesis was first described 120 years ago [147]. It is defined as the light-dependent change in velocity of movement. For *Euglena* reports indicate a weak positive photokinesis (increase of velocity in light; [148,149]) with a saturation at about 300 lx white light. Mast [150] reported a 10 to 15-min lag-period before the new steady state of velocity is reached, indicating an involvement of metabolic processes. Ascoli [70] found that the acceleration might be due to an increased beating frequency of the emerging flagellum. However, also the non photosynthetic relative *Astasia longa* shows photokinesis [151].

The wavelength dependence of this reaction is not yet clear. While some reports indicate the involvement of photosynthetic pigments (chl *b* and/or β-carotene; [70,116,148] others [152] reported a strong effect of blue light on photokinesis. Up to now, it remains unclear whether, like in prokaryotes (for summary see [153]), photokinesis is depending on photosynthesis or a specialized blue light photoreceptor is involved. Typical for photokinesis, no adaptation can be observed [148].

11.2.3 Photophobic responses

Photophobic responses ("Schreckbewegung") were first observed by Engelmann [96,105]. The older literature is summarized by Haupt [154], Feinleib and Curry [155], Diehn [156], Nultsch [116], Nultsch and Häder [157] and Nultsch and Häder [102].

The photophobic response is defined as a sudden behavioral response (stop or change in direction of movement) as a response to a rather steep increase ("step-up" response) or decrease ("step-down") in irradiance. In addition, a discrimination threshold must exist, which describes the minimal irradiance difference before and after the change in illumination [158]. Neither the steepness nor the duration of the change in irradiance nor the time after which a response is considered to be related to the irradiance change are well defined which makes comparison of results of different authors with different experimental protocols difficult.

In *Euglena*, like in many other motile microorganisms, step-down photophobic responses are observed at low irradiance levels and, separated by an "indifferent" irradiance range with no evident reactions, step-up photophobic responses at higher irradiance levels [111,159]. Doughty and Diehn developed a model for the step-down photophobic response, based on numerous inhibitor, ionophores, ion channel blockers, various pH and ion concentration studies performed by themselves and others [160–166]. The model is based on the assumption that upon irradiation the PAB transiently modulates the activity of a postulated flagellar Na^+/K^+ exchange pump. The resulting increase in the sodium concentration in the intraflagellar space will in turn open sodium-controlled calcium channels. The resulting increase in the intraflagellar calcium concentration is thought to eventually trigger a change in the beating pattern of the flagellum.

The action spectrum for the photophobic responses has typical characteristics of a flavin chromophore (see discussion below [167–169].

11.2.4 Phototaxis

11.2.4.1 General

In contrast to photokinesis and photophobic response, phototaxis depends on the direction of light. Phototaxis of unicellular flagellates has been known for more than hundred years [8,45,91,170,171]. In most cases, the spectral sensitivity is limited to the ultraviolet/blue-green spectral region (300 nm–550 nm). Much of the older literature is reviewed by Bowne and Bowne [172,173] and will only be mentioned shortly. *Euglena* shows a distinct positive phototaxis (movement toward a light source) at low irradiances (< 10 W m^{-2}) and a negative phototaxis at higher irradiances (movement away from a light source; Figure 1). Buder [8] devised a brilliant experiment to answer the question whether phototaxis is a result of a directional movement with respect to the light direction or is an orientation in the spatial gradient of light intensity. Light was supplied as a converging light beam by means of a biconvex lens. Phototactically responding cells were still moving in the direction of the light source even after they passed the focal point and encountered a corresponding decrease in the irradiance. Haupt [174] was the first to introduce the "one-" or "two-instant" mechanism concept which was later applied to other systems [129,175–178]. In the one-instant mechanism at least two photoreceptor molecules detect an intracellular light gradient by the comparison of the light absorption at two different locations at the same time. In contrast, the two-instant mechanism assumes a temporal comparison of the light absorption by one photoreceptor at different positions of the cell in respect to the light source. Besides a comparable fast movement, this mechanism requires a basic "memory" system. The latter mechanism seems to be implemented in fast, free moving single cells [150,156], while the former

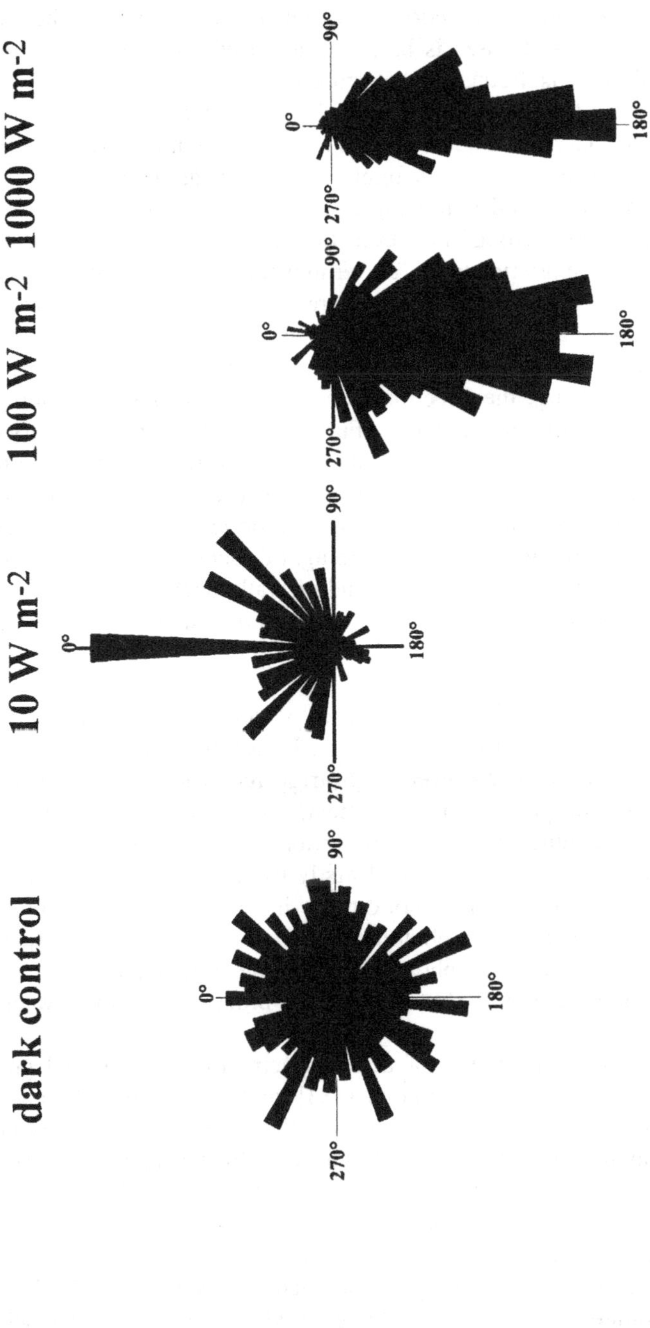

Figure 1. Circular histograms of the orientation of *Euglena gracilis* at different irradiances. Actinic light from 0°. Every histogram represents more than 1000 analyzed cell tracks (after [73]).

system might be used by relatively slow moving, gliding cells. A two-instant mechanism is also implemented in free moving prokaryotic cells which can, like eukaryotic cells, react to a great variety of environmental stimuli. In most cases misleadingly, these phenomena are called, depending on the stimuli photo-"taxis", chemo-"taxis" etc. In contrast, true taxis is based on the directional movement with respect to the stimuli source. Bacterial "taxis" in most cases is based on a biased random walk. The free movement of bacterial cells is frequently interrupted by a reversal of the forward movement (for example *Halobacterium salinarium*) or short tumbling periods (*Escherichia coli*). In most cases a frequent change in the direction of movement can be observed. In the case of a movement in a light or chemical gradient reorientational events are suppressed when the cell is moving in the "correct" direction or are more frequent when moving in the "wrong" direction. In general, moving in the "correct" direction is only favored by approximately 5% over movement in the "wrong" direction [100,101]. The molecular basis of this behavioral response is well characterized and established [179,180]. The observed population displacement with respect to the stimulus source is caused by a trial and error mechanism. In contrast, eukaryotic tactically oriented cells do change the direction of movement only in the case of passive disorientation with respect to the stimuli source. Further reasoning for a necessary clear distinction between the above discussed bacterial "taxis" and true taxis is given by an experiment of Häder et al. [181]. *Euglena* cell were placed on a rotating cuvette and light-depending orientation was analyzed by means of a motion analysis system [137]. Cells were capable of orienting with respect to the light direction for rotational speeds of up to 20° s^{-1}. If *Euglena* would use a biased random walk mechanism to move away or toward a light source the rotational displacement would not allow any orientation.

11.2.4.2 Photoreceptor

Action spectroscopy. Theoretical background and practical application of action spectroscopy is covered in depth in this volume (cf. Foster, this volume). In general, action spectroscopy uses the fact that every light-triggered reaction is based on the absorption of photons by the photoreceptor pigment. As a consequence, the spectral sensitivity of the reaction must reflect the absorption properties of the pigment(s) involved. In praxis, the analysis is complicated firstly by the possible involvement of shading pigments (i.e. pigments which are not directly involved in photoperception but absorb in the same spectral range as the photoreceptor pigment(s)) and secondly by the possible involvement of more than one photoreceptor pigment. Oltmann [182] was the first to measure an action spectrum of phototaxis with a broad maximum between 450 and 500 nm.

Figure 2 summarizes some examples of action spectra for negative and positive phototaxis. While Bünning and Schneiderhöhn [183] could not extend the analyzed wavelength region below 400 nm due to limitations of the light source used, both other examples show substantial sensitivity below 400 nm. All published action spectra agree in the finding that no reactions can be observed above 550 nm. Earlier reports of a sensitivity in the red region [148] might be due to blue-light-leaky filters used [167] or to the strong aerotaxis of *Euglena* [184]. Besides these limitations substantial differences can be observed between the published action spectra. Most likely, these differences can be accounted for by the methodology used. Bünning and Schneiderhöhn

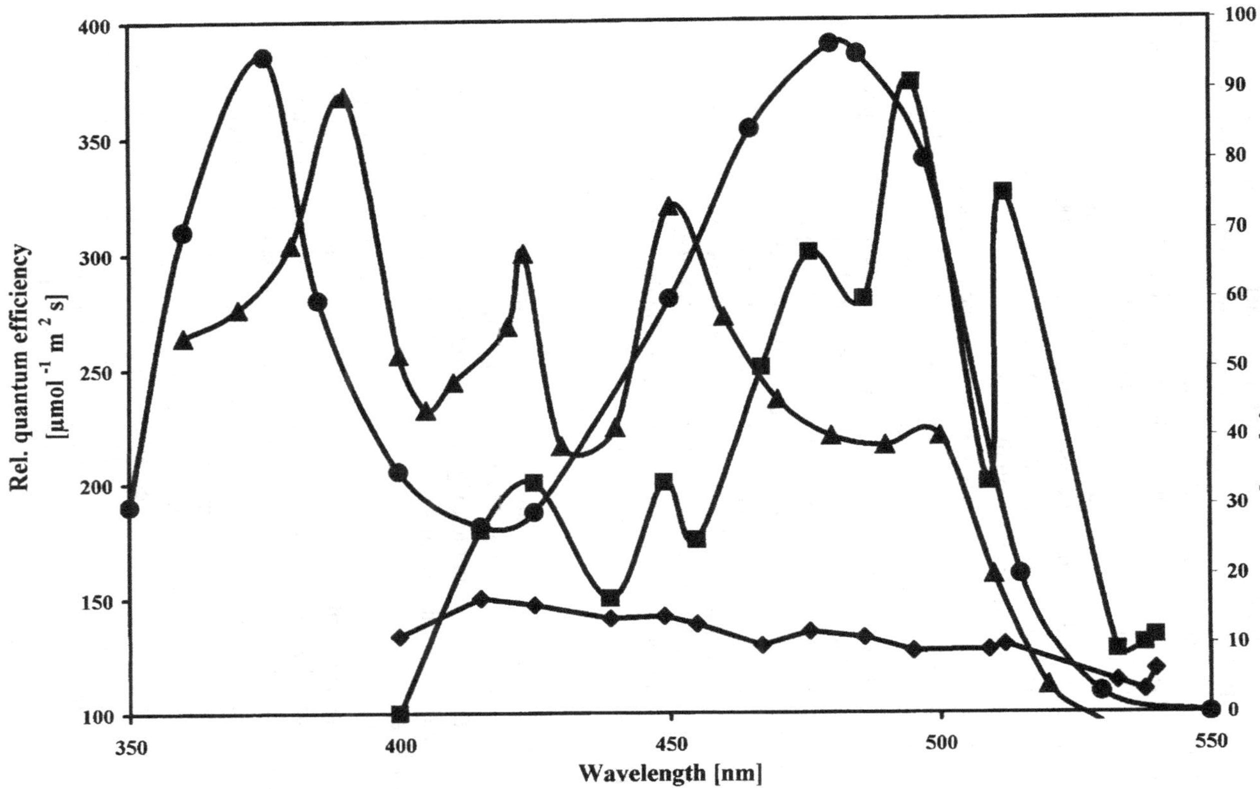

Figure 2. Action spectra of phototactic orientation. Negative (diamonds) and positive (squares) phototaxis after [183], negative phototaxis (triangles) after [190] and positive phototaxis (circles) after [167]. All figures redrawn from original data.

as well as Diehn or Checcucci et al. [185–187] used population methods which mainly quantified photoaccumulation (see also [188,189]). As discussed above photoaccumulations can be caused by a variety of factors including phototaxis, photophobic responses and photokinesis. In addition, it remains unclear whether the results were corrected for the quantum flux. If this is not the case, the results obtained reflect also the emission spectrum of the lamp used. Häder and Reinecke [190] used an automatic cell tracking system for the analysis which combines the high resolution of single cell measurements with the high number of cell tracks analyzed in population methods. With the above mentioned limitations in mind, all authors report a broad maximum between 440 and 500 nm. When measured in the wavelength region below 400 nm, a second major maximum is observed between 360 and 390 nm. The observed fine structures in the published action spectra differ substantially. Foster and Smyth [191] were the first to discuss the specific use of threshold action spectra for the identification of photoreceptor pigments. In the ideal case an action spectrum should be constructed from dose-effect curves. The linear part of the dose-effect curve must be extrapolated to the threshold value. Plotting the inverse of these values will yield the true absorption spectrum of the photoreceptor pigments involved, while the slope will give an indication of the screening pigments involved. Foster and Smyth [191] applied this approach on published data for phototactic orientation of *Chlamydomonas reinhardtii*. The obtained results were the first indication that a retinal protein [191] is involved in photoperception of this organism as later directly shown by experimental evidences [192–197].

Using this approach, data of Häder and Reinecke [190] were reanalyzed (Figures 3, 4). While the overall picture did not change substantially certain differences can be observed. Two major maxima can be detected (385 nm and 460 nm). Minor peaks can be seen at 410 nm and 490 nm. The maximum at 385 nm is substantially higher than the blue light maximum.

This action spectrum is very similar to the action spectrum of the step-down photophobic response published by Diehn [167]. He determined the action spectrum by combining results with the phototaxigraph combining data with longitudinally and transversely polarized light. For this reason the results are not directly comparable (see discussion above).

The absorption properties of the screening pigments (Figure 4) mainly reflect the absorption of the chloroplasts with chlorophyll and possibly carotenoids. Figure 5 shows a schematic drawing of cell orienting away or toward a light source. During negative phototaxis *Euglena* utilizes the rear end of the cell as additional screening device. The main absorbing organelles are the chloroplasts. The differences between the threshold action spectrum and the original spectrum are mainly due to the elimination of the distortion of the action spectrum by the screening pigments.

None of the action spectra published measured phototactic sensitivity below 300 nm while a sensitivity in this region might provide useful information [198]. All chromophores candidates discussed have a strong absorbance in the region below 300 nm. It would be interesting to know whether this absorbance could trigger phototaxis. Matsunaga et al. [199] measured an action spectrum for the step-up and step-down photophobic response which clearly in the UV-B and UV-C wavelength region (280–315 nm and below 280 nm, respectively) demonstrated a high sensitivity of the responses. The same needs to be shown for positive and negative phototaxis.

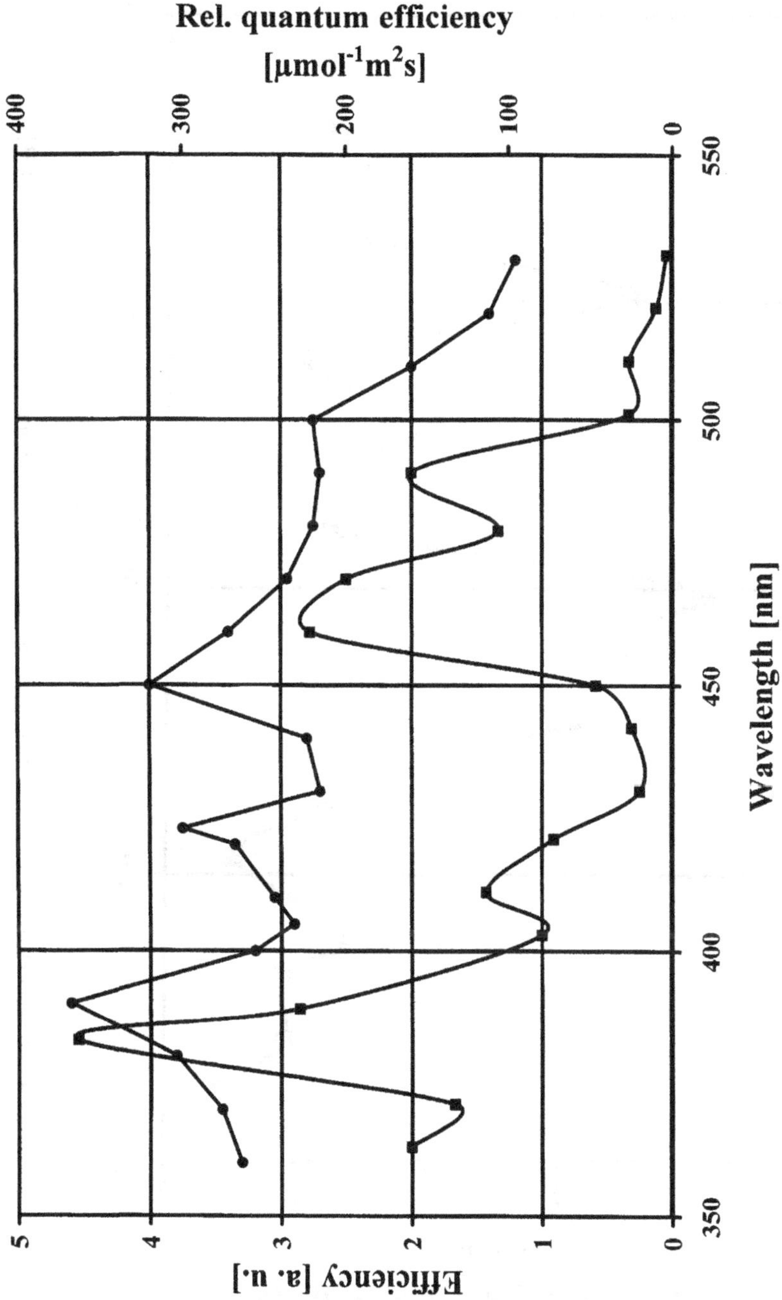

Figure 3. Threshold action spectrum of negative phototaxis of *Euglena gracilis* constructed according to Foster and Smyth [191] (squares) in comparison to the original spectrum published by Häder and Reinecke [190] (circles).

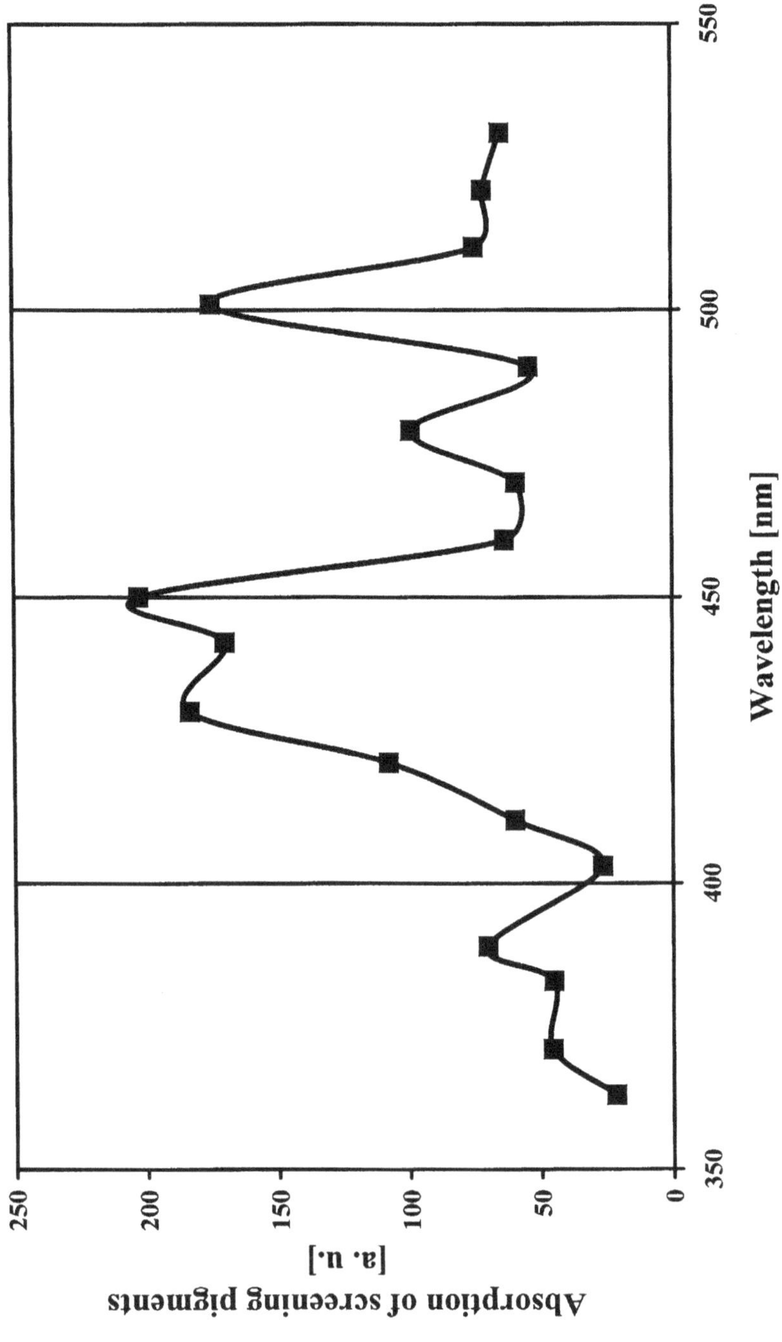

Figure 4. Absorption properties of screening pigments as analyzed according to Foster and Smyth [191].

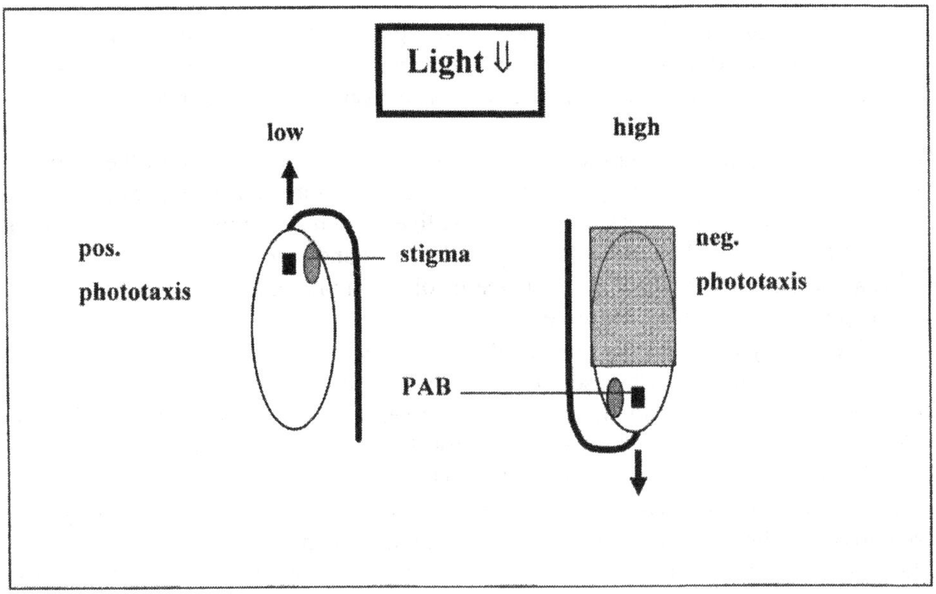

Figure 5. Schematic drawing of *Euglena gracilis* cells orienting toward or away from a light source.

Location of the photoreceptor. The potential role in light perception of the *P*ar*A*-xonemal *B*ody (PAB, formerly *P*ara*F*lagellar *B*ody, PFB [55]) instead of the stigma was proposed very early (see review by [155]). The electron dense material of the PAB is organized in a paracrystalline array [41,42,59]. The best evidence for the role of the PAB in phototactic orientation is that *Euglena* strains without PAB and stigma do not show negative or positive phototaxis [73,200,201]. Strains without stigma but with an intact PAB show normal negative phototaxis but no positive phototaxis [186,201–204]. Checcucci and coworkers [186] also determined the action spectrum for photo-accumulations in these stigma-less mutants which was identical to the wild type spectrum. Unfortunately, no mutants without PAB but with an intact stigma were isolated and characterized up to now. The close relative *Astasia longa,* which possesses a stigma but no PAB does not show any phototaxis [205–207]. In summary, it seems very likely that the PAB is the location of the photoreceptor molecules.

In contrast to the highly organized structures of the stigmata of green algae, the stigma of *Euglena gracilis* does not show a distinct structure. The globuli are not oriented with respect to each other, are surrounded by membranes and contain carotenoids [208]. In addition, the stigma is not part of the chloroplast. The positioning of the stigma, a critical factor for the correct function in *Chlamydomonas*, is variable with respect to the cytoplasmic membrane, but seems to be fixed with respect to the PAB and the reservoir membrane [44]. An interference reflection mechanism as in *Chlamydomonas* and other green algae seems not possible as judged by the available structural information. In the same line of evidence is the finding that the stigma of *Euglena* does not reflect light [9,44].

However, no direct evidences are available up to now which would support the hypothesis of the PAB as location of the photoreceptor molecules. An experiment where a flagellar response to a localized irradiation was measured was never published.

Photoreceptor pigment(s). Based on the above mentioned action spectra the involvement of one or more yellow pigments in photoperception of *Euglena gracilis* was suggested. Mainly flavins or carotenoids, most likely bound to proteins were discussed (54,61,99,102,154,157,174,191,209–226]. Later, Galland and coworkers introduced pterins as possible UV-absorbing substance involved in phototactic photoperception of *Euglena gracilis* ([227,228]; Figure 6).

Relatively early carotenoids were ruled out as possible candidates for photoperception. The main arguments are based on the photochemical properties of the molecules. The lifetime of the excited state of carotenoids is very short ($<10^{-13}$ s; [215,217,219,223] . The quantum efficiency for the singlet to triplet state (which has a longer lifetime; [229]) is extremely low ($\phi = 0.001$; [230]) and will not be sufficient to populate that state significantly [219,231]. For these reasons it is very unlikely that carotenoids are directly involved in phototactic photoperception.

Several authors reported the existence of flavins in the PAB based on microspectrofluorimetric measurements of single PABs [232–236]. The authors took the fluorescence emission at 520 nm when the organelle was excited between 400 and 500 nm as indication for the presence of flavins. Ghetti et al. [234] also measured fluorescence excitation spectra of PABs in vivo. The observations clearly supported the hypothesis of a flavin chromophore located in the PAB. Sineshchekov et al. [236] extended the approach and excited at different wavelengths. Their results support the idea that more than one chromophore (flavins and pterins) are located in the PAB.

Benedetti and Lenci [233] mentioned in their report a transient increase in fluorescence emission at 520 nm for several seconds followed by a gradual decrease of fluorescence emission intensity. The authors explain this observation by a possible interaction of flavin chromophores with the molecules of the PAB matrix (i.e. proteins, [224]) with an subsequent release of flavins from the structure due to a modification of the matrix. An alternative explanation is that the artificial high irradiation of the PAB leads to a saturation of the coupled signal transduction chain with a consequential increase of fluorescence emission of chromophores which could not channel their absorbed light energy to the transduction chain. Subsequently, photoreceptor molecules could be reversibly modified which causes the observed decrease of fluorescence emission.

Additional hints for the involvement of flavins in phototaxis came from quencher and inhibitor studies of excited flavins [237–241]. Unfortunately, the results presented by these authors are contradicting each other. While Mikolajczyk and Diehn observed no effect of potassium iodide, a potent quencher of excited flavins, on negative phototaxis but a strong effect on positive phototaxis, Lenci and coworkers reported a strong effect of the same substance on negative phototactic orientation. These differences in the observed behavior of *Euglena gracilis* might have several reasons: First of all, *Euglena* shows a strong circadian rhythm which not only impairs orientational responses [128,242–246] but several othr physiological reactions [13,107,148,185,188,201,202, 247–260]. In addition, culture age as well as culture conditions have an influence on phototactic orientation [261]. Even under standardized conditions, excluding any influence of circadian rhythms, phototactic efficiency as measured by the *r*-value ([262]; the *r*-value ranges between 0 and 1, zero indicating no orientation of a culture and one that all cells swim in the same direction) can vary by a factor of two from one culture to another (personal observation). Bünning and Schneiderhöhn [183] used the term

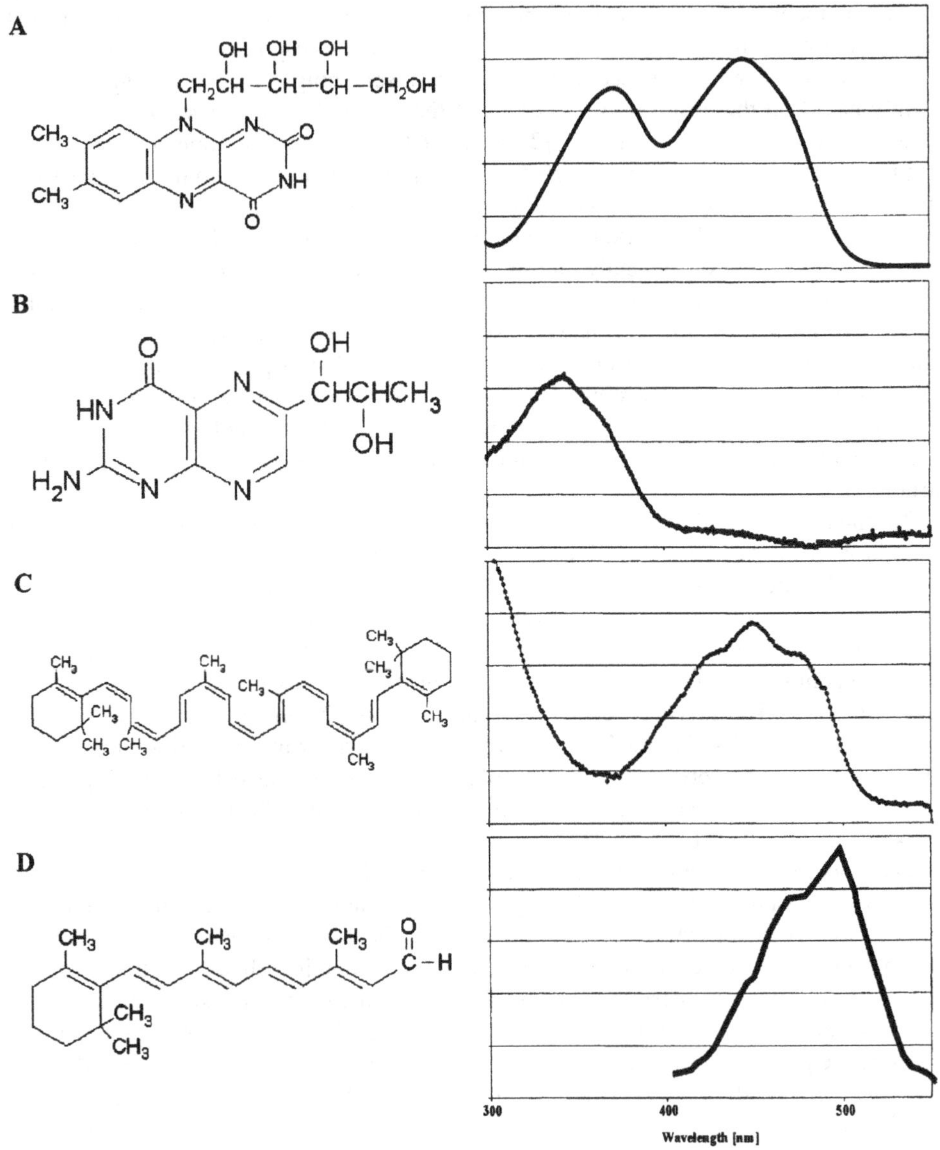

Figure 6. Relevant chromophores and the related absorption spectra. (A) riboflavin in water, (B) biopterin in water, (C) β-carotene in methanol and (D) absorption spectrum of sensory rhodopsin II of *Natronobacter sphaeroides* (after [346], modified).

"launisch" (unreliable) for the highly variable behavior of *Euglena*. As a consequence, the comparison of results of different authors requires a careful examination of the experimental conditions used.

Flavins were not only found in the PAB, but also in the stigma [29,38,263]. These findings were often discussed with respect to photoperception, but Heelis et al. [231] concluded from experiments with isolated stigmata that the ". . . photochemical activity of the pigments contained within the stigma of *Euglena* are too low to support suggestions of a direct role for this organelle in the photoresponses of this organism.". The main role of the stigma seems to be that of a screening device. This hypothesis is mainly based on the absorption properties of the organelle which absorbs in the same range as the phototactic sensitivity [27,33,35,264]. However, the transmittance of the stigma of 50% in green cells and 70% in dark-grown cells would only account for a small change in light intensity received by the proposed photoreceptor, the PAB [35]. In addition, Benedetti and coworkers did not observe a dependence of the absorption of the plane of polarization of the analyzing light. In contrast, the phototactic response shows a clear relation to the plane of polarizing light [111,265,266]. This, as well as the finding of a polarization of the fluorescence emission measured in intact PABs [185] effectively rules out the hypothesis of Bound and Tollin [265] of two perpendicularly oriented photoreceptor systems in the stigma. The chromophores involved in phototaxis must be fixed with respect to each other to explain the strong effect of polarization of actinic light, a requirement fulfilled by molecules in the PAB which show a quasi-crystalline structure.

Rosenbaum and Child [267] were the first to introduce a technique to isolate *Euglena* flagella with PABs still attached which makes the system accessible for biochemical analysis. The method was later enhanced by Gualtieri and co-workers [268]. Galland et al. [228] used this method to analyze the chromophore content of the PAB by fluorescence spectroscopy. They concluded from their results on the existence of flavins and pterins in the organelle. Comparing the fluorescence emission of preparations with and without PABs attached reveals that the bulk flavin and pterin fluorescence is associated with PABs (Figure 7) excluding the flagella as primary location of flavins in the preparations. Nevertheless, riboflavin-binding sites were also found in flagella of *Euglena* [269–271]. However, the bulk majority of the flavin-binding sites are associated with the PAB [271].

The maximum fluorescence emission of a given PAB preparation depends on the status of the sample. Solubilized samples show a largely enhanced fluorescence in comparison to intact PABs (Figure 8). This can be taken as an indication that the photoreceptor molecules are coupled to a signal transduction chain, and the absorbed light energy will be transferred to the chain. In the moment when the signal transduction chain is saturated (see [233] and discussion above) or the signal transduction chain is disrupted (e.g. by solubilization) the fluorescence emission will increase drastically. The low quantum yield of fluorescence in the PAB (0.005; [234]) as compared to free riboflavin (0.25) is in the same line of arguments. Lenci [272] discusses this finding as an indication of a de-excitation pathway other than radiative decay.

Sineshchekov et al. [236] showed by fluorescence techniques an energy transfer between pterins (UV-A absorbing pigment) and flavins. Besides fluorescence emission at 440 nm and 520 nm (depending on the excitation wavelength), the authors reported

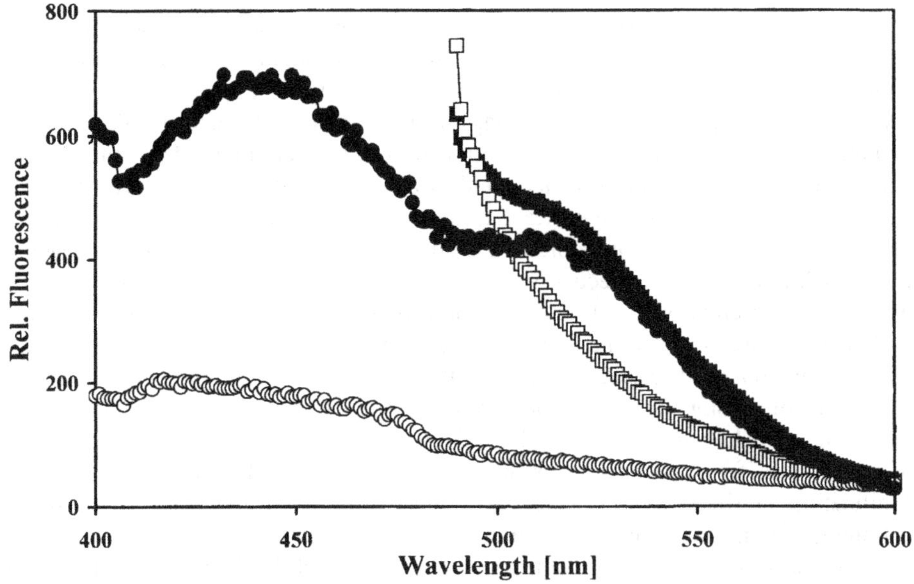

Figure 7. Comparison of fluorescence emission spectra of preparations with (filled symbols) and without (open symbols) PABs attached. Circles excitation at 360 nm, squares excitation 460 nm.

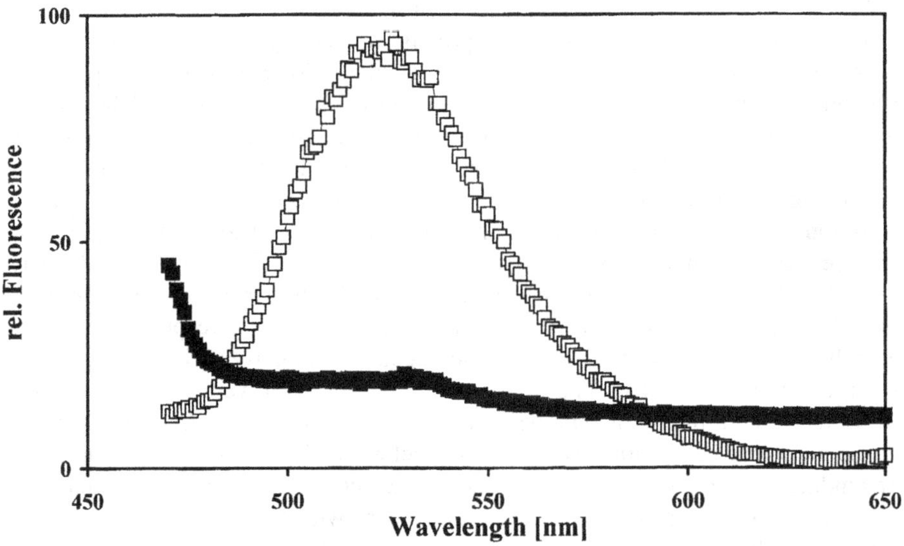

Figure 8. Comparison of fluorescence emission of intact (closed symbols) and solubilized (open symbols) PAB preparations. Excitation 450 nm.

a long wavelength fluorescence emission at 580 nm when the sample was irradiated with 520 nm light and an emission at 620 nm when irradiated at 550 nm. The short wavelength emission was attributed to pterins and the intermediate one to flavins. The nature of the long-wavelength emitting species remained unclear. The authors speculated about the existence of a yet unknown third pigment in the PAB but excluded rhodopsin due to the low fluorescence yield of almost all rhodopsins known [273]. An alternative hypothesis would be either that flavins acting as "core" species have special spectroscopic properties comparable to the special pair in the photosynthetic reaction center which have different spectroscopic properties than "normal" pigments due to the specific chemical environment in the reaction center protein complex. It might even be possible that the treatment of the flagella preparation in the experiments performed leads to the production of degraded proteins with changed spectroscopic properties. The involvement of a long-wavelength absorbing pigment seems not very likely, because none of the published action spectra showed any significant sensitivity beyond 550 nm. The energy transfer between pterins and flavins in intact PABs can be disrupted by solubilization of the organelle [236,274]. This indicates the need for close spatial positioning of both molecules for an effective energy transfer [275]. The latter point is further emphasized by the finding of a strong polarization of the fluorescence of the PAB as to be expected in the highly structured PAB [232,236]. Based on chemical experiments it could be shown that pterins are preferentially located at the outside of the PAB structure while flavins are inside the structure and are not accessible by reductants or oxidants [274]. The comparison of the action spectrum of negative phototaxis and the fluorescence excitation spectra of PABs revealed as close resemblance which emphasizes that the PAB is in fact the location of the photoreceptor molecules.

A more direct approach is the isolation and the biochemical characterization of the PAB. Unfortunately, the method yields only about 5% PABs of the possible total number [271]. In addition, the preparations are only stable for 24 h after which the fluorescence emission disappears and the proteins are rapidly degraded. However, it was possible to analyze preparations for protein content. Comparing flagella preparation with and without PABs still attached, four proteins could be identified. Three of them contained pterins (Mr 27, 27.5, 31.6 kDa) while one protein was associated with a flavin ((M_r 33.5 kDa) [276,277]). The same authors analyzed PAB proteins before and after exposure to UV radiation [278,279] which might be the consequence of the high absorption of flavins and pterins in the short ultraviolet wavelength region. After exposure cells did not show any phototactic orientation. The biochemical analysis revealed that proteins in the PAB were preferentially destroyed by UV. The authors concluded that the proteins in the PAB are important for signal perception in phototaxis. Recently the binding properties of flavins to flagella were analyzed ([28]; Dederichs and Hertel, unpublished results). The dissociation constant (K_d) were 6 nM for riboflavin and FMN (flavin mono nucleotide, see below), 30 nM for FAD (flavin adenine dinucleotide) and 12 nM for roseoflavin (see below) indicating a very specific binding. The binding-protein has a strong hydrophobic character. The apparent molecular weight of 32–35 kDa of this protein is in good agreement with the flavin-containing protein identified by Brodhun and Häder [276] to be part of the PAB photoreceptor complex.

A very successful applied strategy for analyzing signal transduction chains is mutant analysis [3,281]. While *Euglena* is a model system for environmental controlled signal transduction chains for more than hundred years and many mutants in several reactions are known (48,256,282–287) only a few mutants in phototaxis were isolated up to now. The induction of mutants in *Euglena* is difficult due to polyploidy [12]. As a result most mutations are non-chromosomal, i.e. affect mitochondrial or chloroplastic genomes. The mutants analyzed by Lebert and Häder [73] were not isolated in a dedicated approach but spontaneous mutants isolated by accident. Three out of four mutants were stable, while one mutant strain included some revertants [73]. All the mutants had not only lost the PAB (as judged by a fluorescence technique which takes the flavin fluorescence (520 nm) as an indication for the presence of an intact PAB) but also the stigma. None showed a normal negative or positive phototaxis. At very high irradiances the strains showed a weak diaphototaxis (orientation in an angle with respect to the light source). One of the mutant strains was analyzed in more depth. The biochemical analysis of this mutant revealed that no flavins were present in the PAB preparations while pterins were present in normal amounts [274]. This finding is a clear indication for the prominent role of flavins in photoperception in *Euglena gracilis*.

Another successfully applied technique for the analysis of a receptor system is chromophore substitution. This method was used for an in depth analysis of the chromophores involved in photoresponses of *Chlamydomonas reinhardtii* ([288–290]; cf. related chapters in this volume). A comparable approach was the use of roseoflavin to substitute the natural flavin in the PAB. Roseoflavin has a red-shifted absorption spectrum as compared to a typical flavin like riboflavin [291] with an maximum absorbance at 496 nm (in water) and tailing absorption up to 600 nm. Since the phototactic sensitivity ends between 520 nm and 550 nm (see section: action spectroscopy) roseoflavin is a useful molecule for a substitution experiment. In fact, roseoflavin-treated cells showed a shifted phototactic sensitivity closely related to the absorption spectrum of roseoflavin [274]. Interestingly, these cells did not show a clear negative or positive phototaxis but both orientational responses at the same time. The explanation of this behavior is based on the low absorbance of the chloroplasts in the rear end of the cells in the green spectral range. As a result, a low modulation of the light derived signal of the PAB can be achieved when moving away or toward the light source. Under normal conditions the reorientational movements of the cells are biased by the lower modulation of the light signal when the cell is moving away from the light source due to the high absorption of cellular content in the blue spectral region.

Summarizing it can be stated that clear evidences favor the PAB as location of the photoreceptor molecules. The primary role of the stigma is not in photoperception (for a more in depth discussion of the role of the stigma see below). Overwhelming evidence exists for the hypothesis that flavin is the primary chromophore involved in photoperception of phototaxis. Pterins very likely act as "light antenna" for the flavins. However, recently a discussion was raised about the possible role of rhodopsin in photoperception (see Gualtieri, this volume). The evidence will be discussed in the following.

The absorption spectrum of single PABs in vivo are quite different to the fluorescence excitation spectra measured [27,292] even when the excitation spectra are corrected for the instrument specific constants (i.e. emission spectrum of the lamp, transmission properties of excitation and emission monochromator etc.; [274,293]. The absorption spectrum shows a single peak centered around 500 nm. Unfortunately, none of the absorption spectra were measured below 400 nm which would be important for the interpretation of the data. However, the spectra were taken as evidence for the existence of a rhodopsin-like photoreceptor in *Euglena* which contains retinal as a chromophoric group. Alternatively, the spectrum was interpreted as a protein stabilized "red" flavin semiquinone [269]. This interpretation was rejected by Gualtieri [61] based on the circumstantial argumentation that under the experimental conditions used (pH < 5 and in air) the radical can not form. However, the microchemical environment of a chromophore in a most likely hydrophobic protein pocket can not easily be predicted and consequently, the interpretation of the absorption spectrum as an semiquinone can not be finally excluded.

The comparison of the action spectrum of negative phototaxis and the absorption properties of the PAB as published by Gualtieri et al. [292] shows only a very poor resemblance (Figure 9). In contrast, the action spectrum can easily be explained by the pterin/flavin hypothesis. The major peaks at 370 nm and 460 nm can be attributed to flavin. The high sensitivity in the UV region is due to the additional absorption of pterins which could channel the light energy to flavins via fluorescence energy coupling. The peak around 410 nm can be attributed to a semihydro flavin formed during the continuous illumination in the experimental procedure of measuring negative phototaxis. The fine structure in the blue region is typical for flavins in low temperature measurements where the motility of the molecules is more limited. A limited motion of flavins is to be expected in a paracrystalline structure like the PAB.

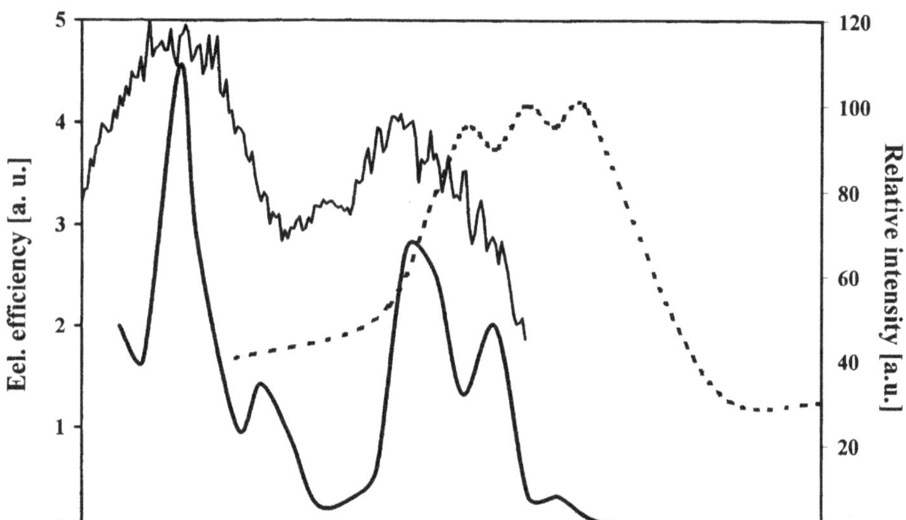

Figure 9. Comparison of the absorption properties of a single PAB (dashed line [292], modified), the action spectrum of negative phototaxis (heavy line, after [190], recalculated according to [191]) and the corrected fluorescence excitation spectrum of isolated PABs.

Alternatively it might be possible that the specific protein environment causes such distinct fine structure as in the case of phototropin (NPH1; [294]). Phototropin binds non-covalently flavin mono nucleotide (FMN). The chromophore binds to the LOV domain of the protein which controls the autophosphorylation of the molecule (see chapter by Iino, this volume for details).

When *Euglena* cells were treated with nicotine, an inhibitor of retinal biosynthesis [295], the photoreceptor organelle PAB disappeared and no photoaccumulation could be observed anymore [296]. When nicotine was removed from the medium cells showed photoaccumulation again after several generations. This recovery was paralleled by the new synthesis of PABs as shown by electron microscopy. As to be expected, photoaccumulation could not be restored by the simple addition of retinal to the medium because no PAB structure was present in the treated cells. In the discussion of population methods it was already mentioned that photoaccumulation can be the result of photokinesis, photophobic responses, phototaxis or even chemotaxis (i.e. aerotaxis) to oxygen produced by cells in the light field. In the published results no effort was made to measure one or several of the responses separately. This might be the reason that a careful analysis of the effect of nicotine on negative and positive phototaxis revealed that none of the responses were impaired [274] even after a prolonged treatment period of more than six month. Nevertheless, retinal could be identified in whole cell extracts by means of chemical methods (HPLC, mass and gas chromatography; [297]). The relevance of that finding is unclear due to the methodology used which did not exclude the accidental formation of retinal from β-carotene. The amount of retinal found would account for 2×10^7 rhodopsin molecules per cell, a number in good agreement with theoretical calculations by Foster and Smyth [191] based on fluence-response curves. It this number is correct and all of the rhodopsin is located in the PAB, no other photoreceptor molecule must be expected in the PAB.

Hydroxylamin is known to inhibit the formation of retinal photo-intermediates which appear after the absorption of a photon. In fact, the substance inhibits phototaxis in *Euglena* [298]. Finally, a photocycle of the hypothetical rhodopsin-like photoreceptor was published [299]. In the experiments cells were irradiated with UV (365 nm) or blue light (436 nm) and fluorescence emission was recorded above 397 nm or 470 nm, respectively. Under blue light no substantial emission could be recorded. When the cells were irradiated with UV light a green fluorescence emission was observed which gradually increased for up to 8 s and returned to its initial values after about 10 s when the cells were irradiated with blue light. This amounts to a total of 18 s for the completion of the photocycle. This would be the slowest photocycle recorded for a rhodopsin [300] with the exception of a bacterio-rhodopsin mutant [301]. The kinetic of the build-up of a stable intermediate (the signaling species?) must be comparable to the rotational speed of the cells (1–2 Hz, [69]). The results reported indicate a substantial increase of the fluorescent species in less than a second (20% to 60% of the maximum) which would well be in accordance with the rotational speed. An increase and gradually decrease of the fluorescence emission of the PAB was also reported by Benedetti and Lenci ([233]; see discussion above). However, the photocycle results are not compatible with the pterin/flavin hypothesis, because both chromophores normally do not produce stable intermediates while they are very often observed in rhodopsins especially in invertebrates [302,303]. Recently, a discussion was raised whether the LOV-domain in

NPH1 [294] could stabilize a UV absorbing photo intermediate of flavins but no clear experimental evidence was presented up to now (R. Hertel, personal communication). If the hypothesis of a stable intermediate could be substantiated it could explain the finding of Barsanti et al. [299] on the basis of a flavin/pterin system.

The obvious contradiction between the absorption properties of isolated PABs with a maximal absorption around 500 nm [292] and the absorption properties of the pigments involved in fluorescence emission with a maximal absorption around 360 nm is solved by the assumption of a secondary β-band of the resting state of the proposed rhodopsin-like protein centered in the UV-region [27,292,304]. Nevertheless, the existence of this β-band needs to be shown. In addition, it remains unclear why the excitation in the blue region alone did not yield any fluorescence while the absorption spectrum of single PABs shows substantial absorption in the wavelength range between 400 and 450 nm (between 40 and 50% of the maximum; Figure 9). In the context of the results discussed above it would be very helpful to firstly measure the absorption properties below 400 nm and secondly measure the absorption of single PABs with 360 nm background light. Under this conditions the absorption maximum around 500 nm should disappear and a new maximum centered around 440 nm should appear.

Theoretical considerations were also used for the substantiation of the rhodopsin hypothesis [61,305]. First of all, rhodopsin-like proteins can be found in a wide variety of organisms. Based on this observation Martin et al. [306] proposed the evolution of a primitive photopigment gene in a unicellular ancestor of the super-kingdoms [307] which was subsequentially passed to many recent species, not only acting as photoreceptors but also as chemo-receptors (i.e. β-adrenergic receptor; [308]). A common motif found in almost all rhodopsin species is the existence of 7 α helices ("7-helix-protein" superfamily). The finding that eukaryotic rhodopsin, in contrast to the prokaryotic ones, are in many known cases coupled to the signal transduction chains via G-proteins as well as a more complete sequence comparison between known 7-helix-proteins makes the hypothesis of a common ancestor not very likely but points more in the direction of a co-evolution [309,310]. As a consequence, an argumentation based on the fact that many organisms use rhodopsin as a photoreceptor and therefore all organisms use rhodopsin might be an unjustified generalization. Flavo-proteins were developed very early in evolution, too. They are present in all living systems with a wide variety of functions including metabolic pathways [308], DNA repair (photolyases; [311–314]); many of them carry a flavin/pterin chromophore system. Recently, it was found that a protein which inherits a photolyase motif is involved in phototropism [315] and most likely carries a flavin as chromophore [316]. An additional family of proteins was identified which controls blue light-dependent phototropism [294]. One of these proteins, NPH1, is mainly responsible for the blue light response and uses flavins as chromophoric groups. Flavin (as FMN) is bound to the LOV domain of the protein. The LOV domain is closely related to the PAS domain found in a number of proteins associated with light perception and signal transduction, including phytochromes [317], the photoactive yellow protein (PYP; [318]; see related chapters in this volume) and the oxygen sensor FixL [319]. Functions of the PAS/LOV domains include the mediation of protein-protein interactions and cofactor binding, functions essential for a coupling to signal transduction chains. It seems, that the PAS/LOV domain can be found in extremely diverse groups of organisms originated in all kingdoms. Thus, a flavin associated photoreceptor motif might be present in many organisms and an evolutionary case can be made for both proposed photoreceptor systems, rhodopsin and flavin/ pterin.

Another consideration is centered around the signal-to-noise ratios of rhodopsins versus flavins [61]. In fact, the thermal stability of rhodopsin, e.g. how often is a spontaneous isomerization of the retinal to be expected, is much higher than for a flavin. Gualtieri [61] estimated one event in one thousand years (rhodopsin) versus 10,000 events for flavins. However, this estimate neglects first of all that the signal-to-noise ratio of a signal transduction chain is not only determined by the thermal stability of the chromophores involved but by the thermal stability of the signal transduction chain as a whole. A simple experiment makes this problem obvious. When we encounter a prolonged stay in total darkness we suddenly start to "see" light flashes or "light noise". While it is still under discussion whether this observation is related to thermally driven isomerization of retinal in our visual system or to a synaptic "noise" phenomenon, it clearly indicates the necessity for a clear distinction between receptor noise and overall signal transduction noise. In addition, this effect can be interpreted as an extreme change in sensitivity of the signal transduction chain involved. This phenomenon is very common. In the hearing process of animals the threshold value was determined to be below or very close to thermal noise [320]. Another example is the tactic response of *Euglena gracilis* to gravity. In seems that the integration of the signal over a period of several rotations enables *Euglena* to detect the very small force induced by gravity [321]. More generally speaking, a vectorial stimulus such as light can be detected even in a very "noisy" environment by integration over some time. However, such an integration period needs to be shown in the case of positive and negative phototaxis of *Euglena gracilis.*

Summarizing, it can be stated that a century of research yielded a great amount of independent indications that flavins and very likely pterins are involved in phototactic light perception. The rhodopsin hypothesis, while not totally excluded seems not very likely in the light of the presented experimental findings.

Mechanism of phototactic orientation. The involvement of the stigma in the mechanism of phototactic orientation was proposed very early. The so-called shading hypothesis proposes a periodic shading of the PAB by the stigma during the rotation of cell when the organism moves at an angle with respect to the light source [8]. Each shading event of the PAB will lead to a reorientational movement of the cell. As a consequence the cell will move toward the light source at low irradiances (< 10 W m^{-2}; positive phototaxis) and due to a change in the reorientational movements of the flagellum away from the light source (negative phototaxis; [102,322]; Figure 1) at higher irradiances. If the hypothesis holds, reorientation would be based on repetitive photophobic responses. In both cases, for positive and negative phototaxis only a minimal modulation of light impinging on the PAB is to be expected. In the case of positive phototaxis cells are moving directly toward the light source, and the stigma will not shade the PAB during the rotation. In the case of negative phototaxis the PAB will be in addition shaded by the rear end of the cell which is filled with chloroplasts.

Several results could not be explained by the shading hypothesis. When a culture is irradiated with two perpendicular oriented light sources with a low irradiance level

($5\ W\ m^{-2}$) the population splits in two: half of the cell moving toward one light source the other toward the second light source. When the irradiance level of both light sources differs by more than 10% all cells move in the direction of the brighter light source [181]. At higher irradiance levels the cells are moving on the resultant of the vector addition of both irradiance levels. Threshold determinations for positive and negative phototaxis ($0.21\ W\ m^{-2}$ and $2.1\ W\ m^{-2}$, respectively; [322]) are not comparable to the much higher thresholds for step-up and step-down photophobic responses [167]. If phototaxis in *Euglena* is really based on repetitive photophobic responses then inhibitors of the photophobic responses should also impair phototaxis. None of the tested substances had any effect on phototaxis [261]. In addition, stigma-less mutants did not show positive phototaxis but oriented away from the light source (negative phototaxis; [186,201,202,204,323]. All discussed results made the validity of the original shading hypothesis very unlikely.

Bound and Tollin [265] and Creutz and Diehn [324] observed a distinct polarotaxis (orientation respect to polarized light) in *Euglena*. Diehn [111] concluded from comparable experiments the existence of two photoreceptor systems in *Euglena* (see discussion of chromophoric groups involved in photoperception). Further experiments resulted in a new model for the phototactic orientation [266]. Based on experiments with polarized light a dichroitic orientation of the photoreceptor molecules was shown. Such an dichroitic orientation was already predicted on the basis of the paracrystalline structure of the PAB which includes an rigid spatial arrangement of the chromophores with respect to each other (see above). The main absorption vector was determined to be turned approximately 50° with respect to the long axis of a cell. During the rotation of a cell the probability of absorbing a photon changes with the position of the chromophores with respect to the light source. During one rotation two positions with a maximal absorption probability are to be expected. The role of the stigma is most likely to suppress the second maximum of photon absorption. This interpretation would also include an explanation of the lack of positive phototaxis in stigma-less mutants. Positive phototaxis depends on the absorption of the stigma to suppress the second maximum of absorption by the PAB. While the stigma has a variable position in the cell it seems always to be in a strict position with respect to the PAB [44]. In the absence of the stigma this screening does not exist and, consequently, the cell will react two times during one rotation. These reaction will be at opposite positions thus resulting in no orientational movement of the cell at all. Negative phototaxis involves the screening of the PAB by the rear end of the cell filled with organelles. Consequently, a stigma is not an indispensable requisite for the negative tactic response.

Signal transduction. The knowledge about the signal transduction chain of phototaxis in Euglena is scarce. Flavins might be coupled via a cytochrome to the signal transduction chain, but no clear experimental evidence is available up to now [61,325,326]. As discussed above, all substances impairing the photophobic responses of Euglena did not have any effect on phototactic orientation [261] leaving the signal transduction as a block box. Some researchers proposed the involvement of the membrane potential in signal transduction [327,328] like in many other organisms [329,330]. While injection of negative current pulses as well as the manipulation of the ionic environment (Ca^{2+} and Mg^{2+}) did in fact change the flagellar beating pattern

[75–77,331,332] the application of external electrical fields did not have any effect on phototaxis in *Euglena* [261]. The galvanotaxis reported earlier is most likely due to chemotaxis or a passive alignment of the cells due to surface charges (existing or induced, [333,334]. Thus, the impact of changes in the membrane potential on phototaxis remains unclear. Recently, a report was published indicating the involvement of the membrane potential in the switch between positive and negative phototaxis [274]. $TPMP^+$, a lipophilic cation which reduces the membrane potential shifted the threshold irradiance between positive and negative phototaxis to lower irradiances levels. The influence of the ATP concentration seems to be limited to the direct impact on flagellar movement [335].

Older reports proposed a piezoelectric effect of a structure close to the PAB connected to the flagellum (the PAR?; [336]) triggered by the photoexcited receptor molecules. While the theory as well as the structure possibly involved was discussed controversially for quite some time it is still an open question [39,337–339].

Summarizing it can be stated that the elements involved in signal transduction are still unknown. The signal transduction chain is still a "black box".

11.3 Further remarks

The mystery of the phototactic orientation of *Euglena gracilis* is still not solved after more than hundred years of research. Neither the chromophoric group(s) involved in photoperception nor the elements of the signal transduction chain are identified. While the evidences for the involvement for flavins/pterins are overwhelming, a role of rhodopsins can not be finally excluded. In addition, the analysis of phototaxis is hampered by contradictory reports in the literature. Only recently a paradigm shift away from pure physiological, descriptive and often only qualitative studies to strictly quantitative ones as well as a biochemical approach can be observed. The whole field is in transition and starts following a strategy successfully applied in other systems (e.g. *Chlamydomonas*). This strategy includes a combined effort of physiologists, biochemists, biophysicists and molecular biologists. It will be the "Golden Triangle" i.e. structure, function and genetics that must be elucidated to solve the riddle. However, many issues must be addressed. These include a kinetic analysis of phototaxis, especially the primary events of the orientational response. Is there a lag-phase (integration period) like in gravitaxis of *Euglena* [321]? If yes, how long is it? What is the irradiance dependence? Some of these questions were already addressed but need a reevaluation [340–344]. Modern motion analysis systems are now capable of high resolution analysis in a quantitative way, thus all related questions can be answered fast. What is the exact relation between photophobic responses and phototaxis? Is there a single photoreceptor for all three responses (like in *Chlamydomonas* [345]) or independent ones? The biochemical approach will be more complicated. The isolation procedure of the PAB (the location of the photoreceptor as all researchers agree) must be optimized, to supply the amounts of protein needed for the necessary spectroscopic and protein chemistry experiments. These experiments include the raising of antibodies, sequencing (at least partially) to identify the genes involved. One way of optimizing the procedure might be the approach of Barsanti et al. [299] to use demembranized cell ghosts. The sequencing of proteins in the PAB will open a whole wealth of possibilities introduced by modern molecular biology, including knock-out mutants to verify the function of a given protein in the phototaxis signal transduction chain. It should not take another hundred years to solve the *Euglena*-enigma!

Acknowledgements

The author would like to thank D.-P. Häder, W. Marwan, P. Hegemann, J. Soppa for fruitful discussions and R. Hertel for supplying unpublished results.

References

1. P. Galland (1991). Photosensory adaptation in aneural organisms. *Photochem. Photobiol.*, **54**, 1119–1134.
2. G. Rispoli (1998). Calcium regulation of phototransduction in vertebrate outer rods. *J. Photochem. Photobiol. B: Biol.*, **44**, 1–20.
3. R.C. Stewart, F.W. Dahlquist (1987). Molecular components of bacterial chemotaxis. *Chem. Rev.*, **87**, 997–1025.
4. M. Alam, M. Lebert, D. Oesterhelt, G.L. Hazelbauer (1989). Methyl-accepting taxis proteins in *Halobacterium halobium. EMBO J.*, **8**, 631–639.
5. S. Aaronson (1963). Is there a protozoan analog of the metazoan nervous system? I. Acetylcholine and cholinesterase. *J. Protozool.*, **10** (Suppl.), 8.
6. W. Pfeffer (1886). Über chemotaktische Bewegungen von Bakterien, Flagellaten. *Volvocineen. Untersuchungen Bot. Inst. Tübingen*, **2**, 582–661.
7. R. Aderhold (1888). Beiträge zur Kenntnis richtender Kräfte bei der Bewegung niederer Organismen. *Jenaische Zeitschr. f. Med. u. Naturwisschft.*, **22**, 311–342.
8. J. Buder (1919). Zur Kenntnis der phototaktischen Richtungsbewegungen. *Jahrb. Wiss. Bot.*, **58**, 105–220.
9. S.O. Mast (1927). Structure and function of the eyespot in unicellular and colonial organisms. *Arch. Protistenk.*, **60**, 197–220.
10. D.E. Buetow (1968). Morphology and ultrastructure of *Euglena*. In: D.E. Buetow (Ed.), *The Biology of Euglena* (pp. 109–184). Academic Press, New York, London.
11. D.E. Buetow (1968). *The Biology of Euglena* (Vol. I: General biology and ultrastructure). Academic Press, New York, London.
12. H.Z. Hill, J.A. Schiff, H.T. Epstein (1966). Studies of chloroplast development in *Euglena*. XIV. Sequential interactions of ultraviolet light and photoreactivating light in green colony formation. *Biophys. J.*, **6**, 125–144.
13. M. Lebert, M. Porst, D.-P. Häder (1999). Circadian rhythm and gravitaxis in *Euglena gracilis. J. Plant Physiol.*, **155**, 344–349.
14. N. Kamiya (1939). Die Rhythmik des metabolischen Formwechsels der Euglenen. *Ber. Dtsche. Bot. Ges.*, **57**, 231–240.
15. E. Mikolajczyk (1972). Patterns of body movements of *Euglena gracilis. Acta Protozool.*, **11**, 317–331.
16. E. Mikolajczyk (1973). Effect of some chemical factors on the euglenoid movement in *Euglena gracilis. Acta Protozool.*, **12**, 133–143.
17. E. Mikolajczyk, B. Diehn (1976). Light-induced body movement of *Euglena gracilis* coupled to flagellar photophobic responses by mechanical stimulation. *J. Protozool.*, **23**, 144–147.

18. E. Mikolajczyk, B. Diehn (1978). Morphological alteration in *Euglena gracilis* induced by treatment with CTAB (Cetyltrimethylammonium bromide) and Triton X-100: correlations with effects on photophobic behavioral responses. *J. Protozool.*, **25**, 461–470.
19. E. Mikolajczyk, L. Kuznicki (1981). Body contraction and ultrastructure of *Euglena. Acta Protozool.*, **20**, 1–24.
20. J.M. Murray (1981). Control of cell shape by calcium in the euglenophyceae. *J. Cell Sci.*, **49**, 99–117.
21. A. Diskus (1955/56). Färbestudien an den Schleimkörperchen und Schleimausscheidungen einiger Euglenen. *Protopl.*, **45**, 460–477.
22. J. Schiller (1952). Über die Vermehrung des Paramylons und über Alterserscheinungen bei Eugleninen. *Österr. Bot. Zeitschr.*, **99**, 27 + 414–420.
23. E. Kamptner (1952). Eine polarisationsoptische Untersuchung an Paramylonkörnern von *Euglena* und *Phacus. Österr. Bot. Zeitschr.*, **99**, 556–588.
24. J.Z. Kiss, E.M. Roberts, R.M. Brown Jr., R.E. Triemer (1988). X-ray and dissolution studies of paramylon storage granules from *Euglena. Protopl.*, **146**, 150–156.
25. J.Z. Kiss, A.C. Vasconcelos, R.E. Triemer (1988). The intramembranous particle profile of the paramylon membrane during paramylon synthesis in *Euglena* (Euglenophyceae). *J. Phycol.*, **24**, 152–157.
26. E.G. Pringsheim (1937). Uber das Stigma bei farblosen Flagellaten. *Cytologia*, **1**, 234–255.
27. T.W. James, F. Crescitelli, E.R. Loew, W.N. McFarland (1992). The eyespot of *Euglena gracilis*: A microspectrophotometric study. *Vision Res.*, **32**, 1583–1591.
28. N. J. Krinsky, T.H. Goldsmith (1960). The carotenoids of the flagellated alga, *Euglena gracilis. Arch. Biochem. Biophys.*, **91**, 271–279.
29. P.G. Sperling, P.L. Walne, O. J. Schwarz, L.L. Triplett (1973). Studies on characterization of pigments from isolated eyespots of Euglenoid flagellates. *J. Phycol. Suppl.*, **9**, 20.
30. G.K. Strother, J. J. Wolken (1960). Microspectrophotometry of *Euglena*. Chloroplast and eyespot. *Nature*, **188**, 601–602.
31. G.K. Strother, J. J. Wolken (1961). In vivo absorption spectra of *Euglena*: chloroplast and eyespot. *J. Protozool.*, **8**, 261–265.
32. D.V. Heelis, W. Kernick, G.O. Philips, K. Davies (1979). Separation and identification of the carotenoid pigments of stigmata isolated from light-grown cells of *Euglena gracilis* strain. Z. *Arch. Microbiol.*, **121**, 207–211.
33. D.V. Heelis, P.F. Heelis, W.A. Kernick, G.O. Phillips (1980). The stigma of *Euglena gracilis* strain Z: an investigation into the possible occurrence of carotenoproteins and nucleic acids. *Cytobios*, **29**, 135–143.
34. T. Osafune, J.A. Schiff (1980). Stigma and flagellar swelling in relation to light and carotenoids in *Euglena gracilis* var. *bacillaris. J. Ultrstruct. Res.*, **73**, 336–349.
35. P.A. Benedetti, G. Bianchini, A. Checcucci, R. Ferrara, S. Grassi (1976). Spectroscopic properties and related functions of the stigma measured in living cells of *Euglena gracilis. Arch. Microbiol.*, **111**, 73–76.
36. P.A. Kivic, M. Vesk (1974). The structure of the eyespot apparatus in bleached strains of *Euglena gracilis. Cytobiol.*, **10**, 88–101.
37. E. Kronestedt, B. Walles (1975). On the presence of plastids and the eyespot apparatus in a porfiromycin-bleached strain of *Euglena gracilis. Protopl.*, **84**, 75–82.
38. P.G. Sperling, P.G. Pagni, P.L. Walne, E.L. Wehry (1976). Fluorometric evidence for flavins in isolated eyespots of *Euglena gracilis* var. *bacillaris. Photochem. Photobiol.*, **24**, 373–375.
39. P.L. Walne, H. J. Arnott (1967). The comparative ultrastructure and possible function of eyespots: *Euglena granulata* and *Chlamydomonas eugemetos. Planta*, **77**, 325–353.

40. J.D. Dodge (1969). A review of the fine structure of algal eyespots. *Brit. Phycol. J.*, **4**, 199–210.
41. P.A. Kivic, M. Vesk (1972). Structure and function in the euglenoid eyespot apparatus: the fine structure, and response to environmental changes. *Planta*, **105**, 1–14.
42. P.A. Kivic, M. Vesk (1972). Structure and function in the euglenoid eyespot. The probable location of the phototaxis photoreceptor. *J. Exp. Bot.*, **23**, 1070–1075.
43. G. Kreimer, M. Melkonian (1990). Reflection confocal laser scanning microscopy of eyespots in flagellated green algae. *Europ. J. Cell Biol.*, **53**, 101–111.
44. G. Kreimer (1994). Cell biology of phototaxis in flagellated algae. *Int. Rev. Cyt.*, **148**, 229–310.
45. R.H. France (1909). Untersuchungen über die Sinnesorganfunktion der Augenflecke bei Algen. *Arch. Hydrobiol.*, **4**, 37–48.
46. B. Schussnig (1960). *Handbuch der Protophytenkunde, Band II*. Gustav Fischer Verlag, Jena.
47. G.F. Leedale (1982). Ultrastructure. In: D.E. Buetow (Ed.), *The Biology of* Euglena, *Vol. III: Physiology* (pp. 1–27). Academic Press.
48. G. Rosati, L. Barsanti, V. Passarelli, A. Giambelluca, P. Gualtieri (1996). Ultrastructure of a novel non-photosynthetic *Euglena* mutant. *Micron*, **27**, 367–373.
49. E. Mikolajczyk, B. Diehn (1979). Mechanosensory response and mechanoreception in *Euglena gracilis*. *Acta Protozool.*, **18**, 591–602.
50. G.B. Bouck (1982). Flagella and the cell surface. In: D.E. Buetow (Ed.), *The Biology of* Euglena, *Vol. III: Physiology* (pp. 29–51). Academic Press, New York.
51. G.F. Rosati, L. Verni, V. Barsanti, V. Passarelli, P. Gualtieri (1991). Ultrastructure of the apical zone of *Euglena gracilis*: photoreceptors and motor apparatus. *Electron. Microsc. Rev.*, **4**, 319–342.
52. R. Ferrara, R. Banchetti (1976). Effect of streptomycin on the structure and function of the photoreceptor apparatus of *Euglena gracilis*. *J. Exp. Zool.*, **198**, 393–402.
53. P. Omodeo (1980). The photoreceptive apparatus of flagellated algal cells: Comparative morphology and some hypothesis on functioning. In: F. Lenci, G. Colombetti (Eds), *Photoreception and Sensory Transduction in Aneural Organisms* (pp. 127–154). Plenum Press, New York, London.
54. J.J. Wolken (1977). *Euglena*: The photoreceptor system for phototaxis. *J. Protozool.*, **24**, 518–522.
55. R.A. Andersen, D.J.S. Barr, D.H. Lynn, M. Melkonian, O. Moestrup, M.A. Sleigh (1991). Terminology and nomenclature of the cytoskeletal elements associated with the flagellar/ciliary apparatus in protists. *Protopl.*, **164**, 1–8.
56. H. Robenek, M. Melkonian (1983). Structural specialization of the paraflagellar membrane of *Euglena*. *Protopl.*, **117**, 154–157.
57. J.S. Hyams (1982). The *Euglena* paraflagellar rod: structure, relationship to other flagellar components and preliminary biochemical characterization. *J. Cell Sci.*, **55**, 199–210.
58. H.M. Ngô, G.B. Bouck (1998). Heterogeneity and a coiled prediction of trypanosomatid-like flagellar rod proteins in *Euglena*. *J. Eukaryot. Microbiol.*, **45**, 323–333.
59. E. Piccinni, M. Mammi (1978). Motor apparatus of *Euglena gracilis*: ultrastructure of the basal portion of the flagellum and the paraflagellar body. *Boll. Zool.*, **45**, 405–414.
60. H. Michel (1990). General and practical aspects of membrane protein crystallization. In: H. Michel (Ed.), *Crystallization of Membrane Proteins* (pp. 73–89). CRC Press, Boca Raton, FL.
61. P. Gualtieri (1993). *Euglena gracilis*: is the photoreception enigma solved? *J. Photochem. Photobiol. B: Biol.*, **19**, 3–14.
62. D. Oesterhelt (1989). Photosynthetic systems in prokaryotes: The retinal proteins of

halobacteria and the reaction centre of purple bacteria. *Biochem. Int.*, **18**, 673–694.
63. W. Kühlbrandt (1992). Two-dimensional crystallisation of membrane proteins. *Q. Rev. Biophys.*, **25**, 1–49.
64. T. Suzaki, R.E. Williamson (1986). Pellicular ultrastructure and euglenoid movement in *Euglena oxyuris* Schmarda. *J. Protozool.*, **33**, 165–171.
65. T. Suzaki, R.E. Williamson (1986). Cell surface displacement during euglenoid movement and its computer simulation. *Cell Motil. Cytoskel.*, **6**, 186–192.
66. G. Bricheux, G. Brugerolle (1987). The pellicular complex of Euglenoids. II. A biochemical and immunological comparative study of major epiplasmic proteins. *Protopl.*, **140**, 43–54.
67. T.L. Jahn, E.C. Bovee (1965). Movement and locomotion of microorganisms. *Ann. Rev. Microbiol.*, **19**, 21–58.
68. E.C. Bovee (1982). Movement and locomotion of *Euglena*. In: D.E. Buetow (Ed.), *The Biology of* Euglena, *Vol. III: Physiology* (pp. 143–168). Academic Press, New York.
69. C. Ascoli, M. Barbi, C. Frediani, A. Mure (1978). Measurements of *Euglena* motion parameters by laser light scattering. *Biophys. J.*, **24**, 585–599.
70. C. Ascoli (1975). New techniques in photomotion methodology. In: G. Colombetti (Ed.), *Biophysics of Photoreceptors and Photobehaviour of Microorganisms* (pp. 109–120). Lito Felici, Pisa.
71. D.-P. Häder, M. Häder, S.-M. Liu, W. Ullrich (1990). Effects of solar radiation on photoorientation, motility and pigmentation in a freshwater *Peridinium*. *BioSystems*, **23**, 335–343.
72. K. Vogel, R. Hemmersbach-Krause, C. Kühnel, D.-P. Häder (1993). Swimming behavior of the unicellular flagellate, *Euglena gracilis*, in simulated and real microgravity. *Micrograv. Sci. Technol.*, **5**, 232–237.
73. M. Lebert, D.-P. Häder (1997). Behavioral mutants of *Euglena gracilis*: Functional and spectroscopic characterization. *J. Plant Physiol.*, **151**, 188–195.
74. S.F. Goldstein (1992). Flagellar beat patterns in algae. In: M. Melkonian (Ed.), *Algal Cell Motility, Current Phycology* (Vol. 3, pp. 1–38). Chapman and Hall, NY.
75. K.M. Nichols, R. Rikmenspoel (1977). Mg^{2+}-dependent electrical control of flagellar activity in *Euglena*. *J. Cell Sci.*, **23**, 211–225.
76. K.M. Nichols, R. Rikmenspoel (1978). Control of flagellar motion in *Chlamydomonas* and *Euglena* by mechanical microinjection of Mg^{2+} and Ca^{2+} and by electric current injection. *J. Cell Sci.*, **29**, 233–247.
77. K.M. Nichols, R. Rikmenspoel (1980). Flagellar waveform reversal in *Euglena*. *Exp. Cell Res.*, **129**, 377–381.
78. D.-P. Häder (1997). Impact on marine ecosystems. In: C.S. Zerefos, A.F. Bais (Eds), *Solar Ultraviolet Radiation. Modelling, Measurements and Effects, NATO ASI Series* (Vol. I 52, pp. 209–231). Springer, Berlin, Heidelberg.
79. D.-P. Häder (1986). Effects of solar and artificial UV irradiation on motility and phototaxis in the flagellate, *Euglena gracilis*. *Photochem. Photobiol.*, **44**, 651–656.
80. D.-P. Häder (1986). The effect of enhanced solar UV-B radiation on motile microorganisms. In: R.C. Worrest, M.M. Caldwell (Eds), *Stratospheric Ozone Reduction, Solar Ultraviolet Radiation and Plant Life* (pp. 223–233). Springer, Berlin, Heidelberg, New York, Tokyo.
81. S. Gerber, D.-P. Häder (1993). Effects of solar irradiation on motility and pigmentation of three species of phytoplankton. *Env. Exp. Biol.*, **33**, 515–521.
82. D.-P. Häder, M. Häder (1988). Inhibition of motility and phototaxis in the green flagellate, *Euglena gracilis*, by UV-B radiation. *Arch. Microbiol.*, **150**, 20–25.
83. D.-P. Häder, M. Häder (1990). Effects of UV radiation on motility, photo-orientation and

pigmentation in a freshwater *Cryptomonas*. *J. Photochem. Photobiol. B: Biol.*, **5**, 105–114.
84. D.-P. Häder, S.-M. Liu (1990). Motility and gravitactic orientation of the flagellate, *Euglena gracilis*, impaired by artificial and solar UV-B radiation. *Curr. Microbiol.*, **21**, 161–168.
85. D.-P. Häder, K. Griebenow (1988). Orientation of the green flagellate, *Euglena gracilis*, in a vertical column of water. *FEMS Microbiol. Ecol.*, **53**, 159–167.
86. J.O. Kessler, N.A. Hill, D.-P. Häder (1992). Orientation of swimming flagellates by simultaneously acting external factors. *J. Phycol.*, **28**, 816–822.
87. B. Eggersdorfer, D.-P. Häder (1991). Phototaxis, gravitaxis and vertical migrations in the marine dinoflagellate, *Prorocentrum micans*. *FEMS Microbiol. Ecol.*, **85**, 319–326.
88. B. Eggersdorfer, D.-P. Häder (1991). Phototaxis, gravitaxis and vertical migrations in the marine dinoflagellates, *Peridinium faeroense* and *Amphidinium caterii*. *Acta Protozool.*, **30**, 63–71.
89. M. Lebert, D.-P. Häder (1996). How *Euglena* tells up from down. *Nature*, **379**, 590.
90. F. Cohn (1863). Verhalten der grünen mikroskopischen Pflanzen und Thiere zum Lichte. Jber. Schles. Gesellsch. *Vaterl. Cultur.*, **41**, 102–105.
91. W. Pfeffer (1881). *Pflanzenphysiologie*. Verlag Wilhelm Engelmann, Leipzig.
92. H.S. Jennings (1906). *Behavior of the lower organisms*. Indiana University Press, London, Bloomington.
93. H.S. Jennings (1910). *Das Verhalten der niederen Organismen*. Dtsch. von E. Mangold, Leipzig, Berlin.
94. S.O. Mast (1907). Light reactions in lower organisms. II *Volvox globator*. *J. Comp. Neur. Psychol.*, **17**, 99–180.
95. E. Bolte (1920). Über die Wirkung von Licht und Kohlensäure auf die Beweglichkeit grüner und farbloser Schwärmzellen. *Jahrb. Wiss. Bot.*, **59**, 287–324.
96. T.W. Engelmann (1882). Über Licht- und Farbenperception niedrigster Organismen. *Pfluegers Arch.*, **29**, 387–400.
97. J.J. Wolken (1961). Euglena – *An experimental organism for biochemical and biophysical studies*. Inst. of Microbiol., Rutgers, The State University.
98. B. Diehn, M. Feinleib, W. Haupt, E. Hildebrand, F. Lenci, W. Nultsch (1977). Terminology of behavioral responses of motile microorganisms. *Photochem. Photobiol.*, **26**, 559–560.
99. W. Haupt (1975). Locomotion. *Fortschr. Bot.*, **37**, 177–185.
100. H.C. Berg, D.A. Brown (1972). Chemotaxis in *Escherichia coli* analysed by threedimensional tracking. *Nature*, **239**, 77–79.
101. C. Berg (1975). Chemotaxis in bacteria. *Ann. Rev. Biophys. Bioeng.*, **4**, 119–136.
102. W. Nultsch, D.-P. Häder (1988). Photomovement in motile microorganisms II. *Photochem. Photobiol.*, **47**, 837–869.
103. R.K. Clayton (1964). Phototaxis in Microorganisms. In: A.C. Giese (Ed.), *Photophysiology* (pp. 51–77). Academic Press, New York, London.
104. W. Nultsch (1962). Der Einfluβ des Lichtes auf die Bewegung der Cyanophyceen. II. Photokinesis be *Phormidium autumnale*. *Planta*, **57**, 613–623.
105. T.W. Engelmann (1883). *Bakterium photometricum*. Ein Beitrag zur vergleichenden Physiologie des Licht- und Farbensinnes. *Pfluegers Arch.*, **30**, 95–124.
106. D.-P. Häder, W. Nultsch (1971). Untersuchungen zur Abgrenzung der Arten *Phormidium uncinatum* und *Phormidium autumnale*. *Schweiz. Ztschr. Hydrologie*, **33**, 567–577.
107. V.G. Bruce, C.S. Pittendrigh (1960). An effect of heavy water on the phase and period of the circadian rhythm in *Euglena*. *J. Cell. Comp. Physiol.*, **56**, 25–31.
108. K. Brinkmann (1966). Temperatureinflüsse auf die circadiane Rhythmik von *Euglena gracilis* bei Mixotrophie und Autotrophie. *Planta*, **70**, 344–389.

109. V.G. Bruce (1970). The biological clock in *Chlamydomonas reinhardi*. *J. Protozool.*, **17**, 328–334.
110. D.A. Lindes, B. Diehn, G. Tollin (1965). Phototaxigraph: recording instrument for determination of rate of response of phototactic microorganisms to light of controlled intensity and wavelength. *Rev. Sci. Inst.*, ***36***, 1721–1725.
111. B. Diehn (1969). Two perpendicularly oriented pigment systems involved in phototaxis of *Euglena*, *Nature*, **122**, 366–367.
112. D.-P. Häder, W. Nultsch (1973). Negative photo-phobotactic reactions in *Phormidium uncinatum* and *Phormidium autumnale*. Schweiz. *Z. Hydrol.*, **33**, 566–577.
113. D.-P. Häder (1974). Participation of two photosystems in the photo-phobotaxis of *Phormidium uncinatum*. *Arch. Microbiol.*, **96**, 255–266.
114. G. Drews (1959). Beiträge zur Kenntnis der phototaktischen Reaktionen der Cyanophceen. *Arch. Protistenk.*, **104**, 389–430.
115. W. Nultsch (1961). Der Einfluß des Lichtes auf die Bewegung der Cyanphyceen.I. Phototopotaxis von *Phormidium autumnale*. *Planta*, **56**, 632–647.
116. W. Nultsch (1975). Phototaxis and Photokinesis. In: M.J. Carlile (Ed.), *Primitive Sensory and Communication Systems* (pp. 29–90). Academic Press, London, New York, San Francisco.
117. H. Schuchart (1975). Phototaxis methodology. In: G. Colombetti (Ed.), *Biophysics of Photoreceptors and Photobehaviour of Microorganims* (pp. 85–108). Lito Felici, Pisa.
118. M.E. Feinleib, G.M. Curry (1967). Methods for measuring phototaxis of cell populations and individual cells. *Physiol. Plant.*, **20**, 1083–1095.
119. G. Throm (1968). Untersuchungen zum Reaktionsmechanismus von Phobotaxis and Kinesis an *Rhodospirillum rubrum*. *Arch. Protistenk.*, **110**, 313–371.
120. W. Nultsch, G. Throm (1975). Effect of external factors on phototaxis of *Chlamydomonas reinhardtii*. *Arch. Microbiol.*, **103**, 175–179.
121. D.M. Phillips (1972). Comparative analysis of mammalian sperm motility. *J. Cell Biol.*, **53**, 561–573.
122. K. Iwasa, A. Shimizu (1972). Motility of the diatom *Phaeodactylum tricornutum*. *Exp. Cell Res.*, **74**, 552–558.
123. G.K. Ojakian, D.F. Katz (1973). A simple technique for the measurement of swimming speed of *Chlamydomonas*. *Exp. Cell Res.*, **81**, 487–491.
124. W.R. Schneider, R.N. Doetsch (1974). Velocity measurements of motile bacteria by use of a video-tape recording technique. *Appl. Microbiol.*, **27**, 283–284.
125. E. Hildebrandt, N. Dencher (1974). Photophobische Reaktionen von *Halobacterium halobium* und ihre Beziehung zum Retinal-Protein-Komplex der Zellmembran. *Ber. Dtsch. Bot. Ges.*, **87**, 93–99.
126. E. Hildebrandt, N. Dencher (1975). Two photosystems controlling behavioural responses of *Halobacterium halobium*. *Nature (London)*, **257**, 46–48.
127. D.-P. Häder (1977). Influence of electrical fields on photophobic reactions in blue-green algae. *Arch. Microbiol.*, **114**, 83–86.
128. J.A. Schmidt, C. Creutz, B. Diehn (1974). Factors controlling the expression of phototaxis in *Euglena*. In: G. Schenck (Ed.), *Progress in Photobiology, Proc. of the VI. Intern. Congr. on Photobiol.* (p. 359). Frankfurt, Dtsch. Ges. f. Lichtforsch. eV.
129. M.E.H. Feinleib, G.M. Curry (1971). The relationship between stimulus intensity and oriented phototactic response (topotaxis) in *Chlamydomonas*. *Physiol. Plant.*, **25**, 346–352.
130. D. Davenport, C.A. Wright, D. Causley (1962). Technique for the study of the behavior of motile microorganisms. *Science*, **135**, 1059–1060.
131. W.G. Hand, R. Forward, D. Davenport (1967). Short-term photic regulation of a receptor

mechanism in a dinoflagellate. *Biol. Bull.*, **133**, 150–165.
132. W. Marwan, D. Oesterhelt (1990). Quantitation of photochromism of sensory rhodopsin I by computerized tracking of *Halobacterium halobium* cells. *J. Mol. Biol.*, **215**, 277–285.
133. D. Davenport, G.J. Culler, J.O.B. Greaves, Forward R.B., W.G. Hand (1970). The investigation of the behavior of microorganisms by computerized television. *IEEE Transaction Biomedical Engineering*, **17**, 230–237.
134. D. Davenport (1973). Studies in microorganismal behaviour by computerized television. In: A. Pérez-Miravete (Ed.), *Behaviour of Micro-organims* (pp. 106–116). Plenum Press, London, New York.
135. P. Gualtieri, G. Colombetti, F. Lenci (1985). Automatic analysis of the motion of microorganisms. *J. Microscopy*, **139**, 57–62.
136. T.D. Coates, J.T. Harman, W.A. McGuire (1985). A microcomputer-based program for video analysis of chemotaxis under agarose. *Computer Methods Programs Biomed.*, **21**, 195.
137. D.-P. Häder, M. Lebert (1985). Real time computer-controlled tracking of motile microorganisms. *Photochem. Photobiol.*, **42**, 509–514.
138. D.B. Dusenbery (1985). Using a microcomputer and videocamera to simultaneously track 25 animals. *Comput. Biol. Med.*, **15**, 169.
139. D.-P. Häder, K. Griebenow (1987). Versatile digital image analysis by microcomputer to count microorganisms. *EDV Med. Biol.*, **18**, 37.
140. P. Gualtieri, F. Ghetti, V. Passarelli, L. Barsanti (1988). Microorganism track reconstruction: an image processing approach. *Comput. Biol. Med.*, **18**, 57–63.
141. P. Gualtieri, P. Coltelli (1989). A digital microscope for real time detection of moving microorganisms. *Micron Microsc. Acta*, **20**, 99–105.
142. D.-P. Häder, K. Vogel (1991). Simultaneous tracking of flagellates in real time by image analysis. *J. Math. Biol.*, **30**, 63–72.
143. D.-P. Häder (1994). Real-time tracking of microorganisms. *Binary*, **6**, 81–86.
144. C. Kühnel-Kratz, D.-P. Häder (1993). Real time three-dimensional tracking of ciliates. *J. Photochem. Photobiol.*, **19**, 193–200.
145. C. Kühnel-Kratz, D.-P. Häder (1994). Light reactions of the ciliate *Stentor coeruleus* – a three-dimensional analysis. *Photochem. Photobiol.*, **59**, 257.
146. M. Lebert, D.-P. Häder (1999). Image analysis: A versatile tool for numerous applications. *G.I.T. Special edition: Imaging Microscopy*, **1**, 5–6.
147. E. Strasburger (1878). *Wirkung des Lichtes und der Wärme auf Schwärmsporen*. Jena.
148. J.J Wolken, E. Shin (1958). Photomotion in *Euglena gracilis*. I. Photokinesis. II. Phototaxis. *J. Protozool.*, **5**, 39–46.
149. M. Zhenan, R. Shouyu (1983). The effect of red light on photokinesis of *Euglena gracilis*. In: C.K. Tseng (Ed.), *Proceedings of the Joint China-U.S. Phycology Symposium* (pp. 311–321). Sci. Press Beijing, China.
150. S.O. Mast (1911). *Light and behavior of organisms* John Wiley & Sons, New York, Chapman & Hall Ltd., London.
151. D.-P. Häder, M.A. Häder (1989). Effects of solar UV-B irradiation on photomovement and motility in photosynthetic and colorless flagellates. *Environ. Exp. Bot.*, **29**, 273–282.
152. B. Diehn (1973). Phototaxis and sensory transduction in *Euglena*. *Science*, **181**, 1009–1015.
153. D.-P. Häder (1979). Photomovement. In: W. Haupt, M.E. Feinleib (Eds), *Encyclopedia of Plant Physiology, New Series, vol. 7 Physiology of Movements* (pp. 268–309). Springer, Berlin, Heidelberg.
154. W. Haupt (1959). Die Phototaxis der Algen. *Handbuch der Pflanzenphysiologie*, **17**(1), 318–370.

155. M.E. Feinleib, G.M. Curry (1971). The nature of the photoreceptor in phototaxis. In: *Handbook of Sensory Physiology, Vol. I.: Receptor Mechanisms* (pp. 366–395). Springer, Berlin, Heidelberg, New York.
156. B. Diehn (1973). Phototaxis in *Euglena*. 1. Physiological basis of photoreception and tactic orientation. In: A. Pérez-Miravete (Ed.), *Behaviour of Microorganisms* (pp. 83–90). Plenum Press, New York, London.
157. W. Nultsch, D.-P. Häder (1979). Photomovement of motile microorganisms. *Photochem. Photobiol.*, **29**, 423–437.
158. R. Clayton (1959). Phototaxis of purple bacteria. *Handbuch der Pflanzenphysiologie*, **17**, 371–387.
159. B. Diehn (1969). Phototactic response of *Euglena* to single and repetitive pulses of actinic light: orientation time and mechanism. *Exp. Cell Res.*, **56**, 375–381.
160. M.J. Doughty, B. Diehn (1979). Photosensory transduction in the flagellated alga, *Euglena gracilis*. I. Action of divalent cations, Ca^{2+} antagonists and Ca^{2+}-ionophores on motility and photobehavior. *Biochim. Biophys. Acta*, **588**, 148–168.
161. M.J. Doughty, B. Diehn (1982). Photosensory transduction in the flagellated alga, *Euglena gracilis*. III. Induction of Ca^{2+}-dependent responses by monovalent cation ionophores. *Biochim. Biophys. Acta*, **682**, 32–43.
162. M.J. Doughty, B. Diehn (1983). Photosensory transduction in the flagellated alga, *Euglena gracilis*. IV. Long term effects of ions and pH on the expression of step-down photobehaviour. *Arch. Microbiol.*, **134**, 204–207.
163. M.J. Doughty, B. Diehn (1984). Anion sensitivity of motility and step-down photophobic responses of *Euglena gracilis*. *Arch. Microbiol.*, **138**, 329–332.
164. M.J. Doughty, R. Grieser, B. Diehn (1980). Photosensory transduction in the flagellated alga *Euglena gracilis*. *Biochim. Biophys. Acta*, **602**, 10–23.
165. C. Barghigiani, G. Colombetti, F. Lenci, R. Banchetti, M.P. Bizzaro (1979). Photosensory transduction in *Euglena gracilis*: Effect of some metabolic drugs on the photophobic response. *Arch. Microbiol.*, **120**, 239–245.
166. P.L. Walne, F. Lenci, E. Mikolajczyk, G. Colombetti (1984). Effect of pronase treatment on step-down and step-up photophobic responses in *Euglena gracilis*. *Cell Biol. Intern. Rep.*, **8**, 1017–1027.
167. B. Diehn (1969). Action spectra of the phototactic responses in *Euglena*. *Biochim. Biophys. Acta*, **177**, 136–143.
168. C. Barghigiani, G. Colombetti, B. Tranchini, F. Lenci (1979). Photobehavior of *Euglena gracilis*: action spectrum for the stepdown photophobic response of individual cells. *Photochem. Photobiol.*, **29**, 1015–1019.
169. M.J. Doughty, B. Diehn (1985). Flavins as photoreceptor pigments for behavioral responses in motile microorganisms, especially in the flagellated alga, *Euglena* sp. *Structure and Bonding*, **41**, 45–70.
170. R.H. France (1908). Experimentelle Untersuchungen über Reizbewegungen und Lichtsinnesorgane der Algen. *Zt. Ausb. Entw. lehre*, 2, 29–43.
171. S.O. Mast (1914). Orientation in *Euglena* with some remarks on tropisms. *Biol. Zentralbl.*, **34**, 641–664.
172. S.W. Bowne Jr., G.D. Bowne (1967). Taxis in *Euglena*. *Exp. Cell Res.*, **47**, 545–553.
173. S.W. Bowne Jr., G.D. Bowne (1968). Phototaxis. *Naturwiss. Rdschau*, **21**, 394–395.
174. W. Haupt (1966). Phototaxis in plants. *Int. Rev. Cytol.*, **19**, 267–299.
175. J.S. Boscov (1974). *Responses of* Chlamydomonas *to single flashes of light*. Thesis, Tufts University.
176. M.E. Feinleib (1975). Behaviour models in flagellates. In: G. Colombetti (Ed.), *Biophysics of Photoreceptors and Photobehaviour of Microorganisms* (pp. 68–84). Lito Felici, Pisa.

177. M.E. Feinleib (1975). Phototactic responses of *Chlamydomonas* to flashes of light. I. Responses of cell populations. *Photochem. Photobiol.*, **21**, 351–354.
178. J.S. Boscov, M.E. Feinleib (1979). Phototactic response of *Chlamydomonas* to flashes of light – II. response of individual cells. *Photochem. Photobiol.*, **30**, 499–505.
179. S.L. Mowbray, M.O.J. Sandgren (1998). Chemotaxis receptors: a progress report on structure and function. *J. Struct. Biol.*, **124**, 257–275.
180. S.L. Mowbray (1999). Bacterial chemoreceptors: recent progress in structure and function. *Mol. Cells*, **9**, 115–118.
181. D.-P. Häder, M. Lebert, M.R. DiLena (1986). New evidence for the mechanism of phototactic orientation of *Euglena gracilis*. *Curr. Microbiol.*, **14**, 157–163.
182. F. Oltmanns (1917). Über Phototaxis. *Z. Bot.*, **9**, 257–338.
183. E. Bünning, G. Schneiderhöhn (1956). Über das Aktionsspektrum der phototaktischen Reaktionen von *Euglena*. *Arch. Microbiol.*, **24**, 80–90.
184. A. Checcucci, G. Colombetti, G. del Carratore, R. Ferrara, F. Lenci, (1974). Red light induced accumulation of *Euglena gracilis*. *Photochem. Photobiol.*, **19**, 223–226.
185. A. Checcucci, L. Favati, S. Grassi, T. Piaggesi (1975). The measurement of phototactic activitiy in *Euglena gracilis Klebs*. *Monitore Zool. Ital.*, **9**, 83–98.
186. A. Checcucci, G. Colombetti, R. Ferrara, F. Lenci (1976). Action spectra for photo-accumulation of green and colorless *Euglena*: evidence for identification of receptor pigments. *Photochem. Photobiol.*, **23**, 51–54.
187. A. Checcucci, G. Colombetti, R. Ferrara, F. Lenci (1976). Further analysis of the mass photoresponses of *Euglena gracilis* Klebs (Flagellata euglenoidina). *Monitore Zool. Ital. N.S.*, **10**, 271–277.
188. E. Bünning, J. Gössel (1959). Ergänzende Versuche über die phototaktischen Aktions-spektren von *Euglena*. *Arch. Microbiol.*, **32**, 319–321.
189. E. Bünning, M. Tazawa (1957). Über die negativ-phototaktische Reaktion von *Euglena gracilis*. *Arch. Microbiol.*, **27**, 306–310.
190. D.-P. Häder, E. Reinecke (1991). Phototactic and polarotactic responses of the photosynthetic flagellate, *Euglena gracilis*. *Acta Protozool.*, **30**, 13–18.
191. K.W. Foster, R.D. Smyth (1980). Light antennas in phototactic algae. *Microbiol. Rev.*, **44**, 572–630.
192. P. Hegemann, U. Hegemann, K.W. Foster (1988). Reversible bleaching of *Chlamydomonas reinhardtii* rhodopsin in vivo. *Photochem. Photobiol.*, **48**, 123–128.
193. K.W. Foster, J. Saranak, N. Patel, G. Zarilli, M. Okabe, T. Kline, K. Nakanishi (1984). A rhodopsin is the functional photoreceptor for phototaxis in the unicellular eukaryote *Chlamydomonas*. *Nature*, **311**, 756–759.
194. M. Beckmann, P. Hegemann (1991). *In vitro* identification of rhodopsin in the green alga *Chlamydomonas*. *Biochem.*, **30**, 3692–3697.
195. W. Deininger, P. Kröger, U. Hegemann, Lottspeich, F.P. Hegemann (1995). Chlamyrhodo-psin represents a new type of sensory photoreceptor. *EMBO J.*, 235849–5858.
196. F. Derguini, P. Mazur, K. Nakanishi, D.M. Starace, J. Saranak, K.W. Foster (1991). All-trans-retinal is the chromophore bound to the photoreceptor of the alga *Chlamydomonas reinhardtii*. *Photochem. Photobiol.*, **54**, 1017–1021.
197. P. Hegemann (1997). Vision in microalgae. *Planta*, **203**, 265–274.
198. P. Halldal (1967). Ultraviolet action spectra in algology. A review. *Photochem. Photobiol.*, **6**, 445–460.
199. S. Matsunaga, T. Hori, T. Takahashi, M. Kubota, M. Watanabe, K. Okamoto, K. Masuda, M. Sugai (1998). Discovery of signaling effect of UV-B/C light in the extended UV-A/ blue- type action spectra for step-down and step-up photophobic responses in the unicellular flagellate alga *Euglena gracilis*. *Protopl.*, **201**, 45–52.

200. E.G. Pringsheim (1948). The loss of chromatophores in *Euglena gracilis*. *New Phytol.*, **47**, 52–87.
201. J. Vavra (1962). Instability of the stigma in apochlorotic *Euglena gracilis* var. *bacillaris*. *J. Protozool. (Suppl.)*, **9**, 28–29.
202. I. Gössel (1957). Über das Aktionsspektrum der Phototaxis chlorophyllfreier Euglenen und über die Absorption des Augenflecks. *Arch. Microbiol.*, **27**, 288–305.
203. R. Ferrara (1975). General review on phototactic action spectra. In: G. Colombetti (Ed.), *Biophysics of Photoreceptors and Photobehaviour of Microorganisms* (pp. 121–145). Lito Felici, Pisa.
204. D.-P. Häder (1993). Simulation of phototaxis in the flagellate *Euglena gracilis*. *J. Biol. Phys.*, **19**, 95–108.
205. T. Suzaki, R.E. Williamson (1983). Photoresponse of a colorless euglenoid flagellate, *Astasia longa*. *Plant Sci. Lett.*, **32**, 101–107.
206. E. Mikolajczyk (1984). Photophobic responses in Euglenina. 1. Effects of excitation wavelength and external medium on the step-up response of light- and dark-grown *Euglena gracilis*. *Acta Protozool.*, **23**, 1–10.
207. E. Mikolajczyk (1984). Photophobic responses in Euglenina: 2. Sensitivity to light of the colorless flagellate *Astasia longa* in low and high viscosity medium. *Acta Protozool.*, **23**, 85–92.
208. P.P. Batra, G. Tollin (1964). Phototaxis in *Euglena*. I. Isolation of the eye-spot granules and identification of the eye-spot pigments. *Biochim. Biophys. Acta*, **79**, 371–378.
209. A.W. Galston (1950). Riboflavin, light and the growth of plants. *Science*, **111**, 619–624.
210. J.J. Wolken (1957). A comparative study of photoreceptors. *Transact. Acad. Sci., New York*, **19**, 315–327.
211. J. Wolken (1960). Photoreceptors: Comparative studies. In: M.B. Allen (Ed.), *Comparative Biochemistry of Photoreactive Systems* (pp. 145–167). Academic Press, New York, London.
212. W. Haupt (1980). Localization and orientation of photoreceptor pigments. In: F. Lenci, G. Colombetti (Eds), *Photoreception and Sensory Transduction in Aneural Organisms* (pp. 155–172). Plenum Press, New York, London.
213. S.W. Bendix (1960). Phototaxis. *Bot. Rev.*, **26**, 145–208.
214. S.W. Bendix (1960). Pigments in phototaxis. In: M.B. Allen (Ed.), *Comparative Biochemistry of Photoreactive Systems* (pp. 107–127). Academic Press, New York, London.
215. P.-S. Song, T.A. Moore, M. Sun, (1972). Excited states of some plant pigments. In: C.O. Chichester (Ed.), *The Chemistry of Plant Pigments* (pp. 33–74). New York.
216. A. Checcucci, R.A. Weale (Eds) (1973). *Primary Molecular Events in Photobiology*. Elsevier Sci. Publ., Amsterdam, London, New York.
217. P.-S. Song, T.A. Moore (1974). On the photoreceptor pigment for phototropism and phototaxis: is a carotenoid the most likely candidate? *Photochem. Photobiol.*, **19**, 435–441.
218. J.F. McKellar, G.O. Phillips, A. Checcucci, F. Lenci (1975). Excited states of flavins in photoreception processes. In: J.B. Birks (Ed.), *Excited States of Biological Molecules* Wiley, New York.
219. R.W. Bensasson (1975). Spectroscopic and biological properties of carotenoids. In: G. Colombetti (Ed.), *Biophysics of Photoreceptors and Photobehaviour of Microorganisms* (pp. 146–163). Lito Felici, Pisa.
220. F. Lenci (1975). Photochemistry of flavins. In: G. Colombetti (Ed.), *Biophysics of Photoreceptors and Photobehaviour of Microorganisms* (pp. 164–183). Lito Felici, Pisa.

221. A. Checcucci (1976). Molecular sensory physiology of *Euglena*. *Naturwiss.*, **63**, 412–417.
222. G. Colombetti, F. Lenci (1980). Identification and spectroscopic characterization of photoreceptor pigments. In: F. Lenci, G. Colombetti (Eds), *Photoreception and Sensory Transduction in Aneural Organisms* (pp. 173–188). Plenum Press, New York, London.
223. W. Shropshire Jr. (1980). Carotenoids as primary photoreceptors in blue-light responses. In: H. Senger (Ed.), *The Blue Light Syndrome* (pp. 172–186). Springer.
224. R. Banchetti, G. Rosati, F. Verni (1980). Cytochemical analysis of the photoreceptor in *Euglena gracilis* Klebs (Flagellata Euglenoidina). *Monitore Zool. Ital. (N.S.)*, **14**, 165–171.
225. P.-S. Song (1980). Primary photophysical and photochemical reactions: Theoretical background and general introduction. In: F. Lenci, G. Colombetti (Eds), *Photoreception and Sensory Transduction in Aneural Organisms* (pp. 189–210). Plenum Press, New York, London.
226. P. Galland, H. Senger (1988). The role of flavins as photoreceptors. *J. Photochem. Photobiol. B: Biol.*, **1**, 277–294.
227. P. Galland, H. Senger (1988). The role of pterins in the photoreception and metabolism of plants. *Photochem. Photobiol.*, **48**, 811–820.
228. P. Galland, P. Keiner, D. Dörnemann, H. Senger, B. Brodhun, D.-P. Häder (1990). Pterin- and flavin-like fluorescence associated with isolated flagella of *Euglena gracilis*. *Photochem. Photobiol.*, **51**, 675–680.
229. R. Bensasson, E.J. Land, B. Maudinas (1976). Triplet states of carotenoids from photosynthetic bacteria studied by nanosecond ultraviolet and electron pulse irradiation. *Photochem. Photobiol.*, **23**, 189–193.
230. E.J. Land (1975). Photochemistry of polyenes. *Photochem. Photobiol.*, **22**, 286–288.
231. D.V. Heelis, P.F., Heelis, F. Bradshaw, G.O. Phillips (1981). Does the stigma of *Euglena gracilis* play an active role in the photoreception processes of this organism? A photochemical investigation of isolated stigmata. *Photobiochem. Photobiophys.*, **3**, 77–81.
232. P.A. Benedetti, A. Checcucci (1975). Paraflagellar body (PFB) pigments studied by fluorescence microscopy in *Euglena gracilis*. *Plant Sci. Lett.*, **4**, 47–51.
233. P.A. Benedetti, F. Lenci (1977). In vivo microspectrofluorometry of photoreceptor pigments in *Euglena gracilis*. *Photochem. Photobiol.*, **26**, 315–318.
234. F. Ghetti, G. Colombetti, F. Lenci, E. Campani, E. Polacco, M. Quaglia (1985). Fluorescence of *Euglena gracilis* photoreceptor pigment: an in vitro microspectrofluorometric study. *Photochem. Photobiol.*, **42**, 29–33.
235. W. Schmidt, P. Galland, H. Senger, M. Furuya (1990). Microspectrophotometry of *Euglena gracilis*. *Planta*, **182**, 375–381.
236. V.A. Sineshchekov, D. Geiß, O.A. Sineshchekov, P. Galland, H. Senger (1994). Fluorometric characterization of pigments associated with isolated flagella of *Euglena gracilis*: evidence for energy migration. *J. Photochem. Photobiol.*, **23**, 225–237.
237. B. Diehn, B. Kint (1970). The flavin nature of the photoreceptor molecule for phototaxis in *Euglena*. *Physiol. Chem. Physics*, **2**, 483–488.
238. E. Mikolajczyk, B. Diehn (1975). The effect of potassium iodide on photophobic responses in *Euglena*: evidence for two photoreceptor pigments. *Photochem. Photobiol.*, **22**, 269–271.
239. G. Colombetti, D.-P. Häder, F. Lenci, M. Quaglia (1982). Phototaxis in *Euglena gracilis*: effect of sodium azide and triphenylmethyl phosphonium ion on the photosensory transduction chain. *Curr. Microbiol.*, **7**, 281–284.
240. F. Lenci, G. Colombetti, D.-P. Häder (1983). Role of flavin quenchers and inhibitors in the sensory transduction of the negative phototaxis in the flagellate, *Euglena gracilis*. *Current*

Microbiol., **9**, 285–290.
241. F. Lenci, D.-P. Häder, G. Colombetti (1984). Photosensory responses in freely motile microorganisms. In: G. Colombetti, F. Lenci (Eds), *Membrane and Sensory Transduction* (pp. 199–229). Plenum Press, New York.
242. M. Verworn (1889). *Psychophysiologischen Protistenstudien.* Fischer, Jena.
243. R. Pohl (1948). Tagesrhythmus im phototaktischen Verhalten in der *Euglena gracilis. Z. Naturf.*, **3b**, 367–374.
244. W.J. Kostir (1952). Factors which induce motility or passivity in *Euglena. Proc. Am. Soc. Protozool*, **3**, 5.
245. L.N. Edmunds Jr. (1984). Physiology of circadian rhythms in microorganisms. *Adv. Microb. Physiol.*, **25**, 61–148.
246. S.K. Petersen-Mahrt, N.G.A. Ekelund, S. Widell (1994). Influence of UV-B radiation and nitrogen starvation on daily rhythms in phototaxis and cell shape of *Euglena gracilis. Physiol. Plant.*, **92**, 501–505.
247. V.G. Bruce, C.S. Pittendrigh (1956). Temperature independence in a unicellular 'clock'. *Proc. Nat. Acad. Sci.*, **42**, 676–682.
248. V.G. Bruce, C.S. Pittendrigh (1958). Resetting the *Euglena* clock with a single light stimulus. *Am. Nat.*, **42**, 295–306.
249. Bruce, V.G. (1973). The role of the clock in controlling phototactic rhythms. In: A. Pérez-Miravete (Ed.), *Behaviour of Micro-organisms* (pp. 257–266). Plenum Press, London, New York.
250. E. Bünning (1973). *The Physiological Clock* (3rd ed.). English Univ. Press, London.
251. G. Tollin, M.J. Robinson (1969). Phototaxis in *Euglena.* V. Photosupression of phototactic activity by blue light. *Photochem. Photobiol.*, **9**, 411–416.
252. M.N. Verma (1967). Diurnal variation in a pond in Seoni. *India. Hydrobiol.*, **30**, 129–137.
253. J.R. Cook, P. Harris, D.S. Nachtwey (1974). Irreversible plastid loss in *Euglena gracilis* under physiological conditions. *Plant Physiol.*, **53**, 284–290.
254. W.G. Walther, L.N. Edmunds Jr. (1973). Studies on the control of the rhythm of photosynthetic capacity in synchronized cultures of *Euglena gracilis. Plant Physiol.*, **51**, 250–258.
255. D.L. Laval-Martin, D.J. Shuch, L.N. Edmunds Jr. (1979). Cell cycle-related and endogenously controlled circadian photosynthetic rhythms in *Euglena. Plant Physiol.*, **63**, 495–502.
256. I.A. Carrè, D.L. Laval-Martin, L.N. Edmunds Jr. (1989). Circadian changes in cyclic AMP levels in synchronously dividing and stationary-phase culture of the achlorophyllous ZC mutant of *Euglena gracilis. J. Cell. Sci.*, **94**, 267–272.
257. I.A. Carré, L.N. Edmunds Jr. (1992). cAMP dependent kinase in the algal flagellate *Euglena gracilis. J. Biol. Chem.*, **267**, 2135–2137.
258. K. Goto, D. Laval-Martin, L.N. Edmunds Jr. (1985). Biochemical modeling of an autonomously oscillatory circadian clock in *Euglena. Science*, **228**, 1284–1288.
259. I. Tong, I.A. Carré, L.N. Edmunds Jr. (1991). Circadian rhythmicity in the activities of adenylate cyclase and phosphodiesterase in synchronously dividing and stationary-phase culture of the achlorophyllous ZC mutant of *Euglena gracilis. J. Cell. Sci.*, **100**, 365–369.
260. I. Tong, L.N. Edmunds Jr. (1993). Role of cyclic GMP in the mediation of circadian rhythmicity of the adenylate cyclase – cyclic cAMP – phosphodiesterase system in *Euglena. Biochem. Pharm.*, **45**, 2087–2091.
261. D.-P. Häder, M. Lebert, M.R. DiLena (1987). Effects of culture age and drugs on phototaxis in the green flagellate, *Euglena gracilis. Plant Physiol. (Life Sci. Adv.)*, **6**, 169–174.
262. E. Batschelet (1981). Circular statistics in Biology. In: R. Sibson, J.E. Cohen (Eds),

Mathematics in Biology. Academic Press, London, New York.

263. C.J. Bartlett, P.L. Walne, O.J. Schwarz, O.H. Brown (1972). Large scale isolation and purification of eyespot granules from *Euglena gracilis* var. *bacillaris*. *Plant Physiol.*, **49**, 881–885.
264. L.G.M. Baas-Becking, P.A. Ross (1926). Notes on microspectra. I. The absorption spectrum of *Euglena*. *J. Gen. Physiol.*, **9**, 111–114.
265. K.E. Bound, G. Tollin (1967). Phototactic response of *Euglena gracilis* to polarized light. *Nature*, **216**, 1042–1044.
266. D.-P. Häder (1987). Polarotaxis, gravitaxis and vertical phototaxis in the green flagellate, *Euglena gracilis*. *Arch. Microbiol.*, **147**, 179–183.
267. J.L. Rosenbaum, F.M. Child (1967). Flagellar regeneration in protozoan flagellates. *J. Cell Biol.*, **34**, 345-364.
268. P. Gualtieri, L. Barsanti, G. Rosati (1986). Isolation of the photoreceptor (paraflagellar body) of the phototactic flagellate *Euglena gracilis*. *Arch. Microbiol.*, **145**, 303–305.
269. A. Nebenführ, A. Schäfer, P. Galland, H. Senger, R. Hertel (1991). Riboflavin-binding sites associated with flagella of *Euglena*: a candidate for blue-light photoreceptor? *Planta*, **185**, 65–71.
270. R. Neumann, R. Hertel (1994). Purification and characterization of a riboflavin-binding protein from flagella of *Euglena gracilis*. *Photochem. Photobiol.*, **60**, 76–83.
271. B. Brodhun, R. Neumann, R. Hertel, D.-P. Häder (1994). Riboflavin-binding sites in the flagella of *Euglena gracilis* and *Astasia longa*. *J. Photochem. Photobiol.*, **23**, 135–139.
272. F. Lenci (1994). Molecular basis of photoreception. In: R.C. Jennings, G. Zucchelli, F. Ghetti, G. Colombetti (Eds), *Light as an Energy Source and Information Carrier in Plant Physiology* (pp. 147–157). Plenum Press, New York, London.
273. V.A. Sineshchekov, F.F. Litvin (1987). Fluorescence of rhodopsins and its connection with the primary processes of light energy transformation. *Biofizika*, **32**, 540–555.
274. D.-P. Häder, M. Lebert (1998). The photoreceptor for phototaxis in the photosynthetic flagellate *Euglena gracilis*. *Photochem. Photobiol.*, **68**, 260–265.
275. T. Förster (1948). Intermolecular energy migration and fluorescence. *Ann. Phys.*, **2**, 55–75.
276. B. Brodhun, D.-P. Häder (1990). Photoreceptor proteins and pigments in the paraflagellar body of the flagellate *Euglena gracilis*. *Photochem. Photobiol.*, **52**, 865–871.
277. B. Brodhun, D.-P. Häder (1995). A novel procedure to isolate the chromoproteins in the paraflagellar body of the flagellate *Euglena gracilis*. *J. Photochem. Photobiol. B: Biol.*, **28**, 39–45.
278. D.-P. Häder, B. Brodhun (1991). Effects of ultraviolet radiation on the photoreceptor proteins and pigments in the paraflagellar body of the flagellate, *Euglena gracilis*. *J. Plant Phys.*, **137**, 641–646.
279. B. Brodhun, D.-P. Häder (1993). UV-induced damage of photoreceptor proteins in the paraflagellar body of *Euglena gracilis*. *Photochem. Photobiol.*, **58**, 270–274.
280. A. Dederichs (1998). *Flavin-bindende Proteine in* Phycomyces, Euglena *und* Chlamydomonas. Dissertation Uni. Freiburg i. Br.
281. J.J. Falke, R.B. Bass, S.L. Butler, S.A. Chervitz, M.A. Danielson (1997). The two-component signaling pathway of bacterial chemotaxis: A molecular view of signal transduction by receptors, kinases and adaptation enzymes. *Ann. Rev. Cell Dev. Biol.*, **13**, 457–512.
282. J.A. Schiff, H. Lyman, G.K. Russel (1971). Isolation of mutants from *Euglena gracilis*. *Meth. in Enzymol.*, **23**, 143–162.
283. J.A. Schiff, H. Lyman, G.K. Russel (1980). Isolation of mutants in *Euglena gracilis*: An addendum. *Meth. Enzymol.*, **69**, 23–29.

284. A. Shneyour, M. Avron (1975). Properties of photosynthetic mutants isolated from *Euglena gracilis*. *Plant Physiol.*, **55**, 137–141.
285. A. Shneyour, M. Avron (1975). A method for producing, selecting, and isolating photosynthetic mutants of *Euglena gracilis*. *Plant Physiol.*, **55**, 142–144.
286. R.E. Triemer (1986). Light and electron microscopic description of a colorless euglenoid, *Serpenomonas costata* n. g., n. sp. *J. Protozool.*, **33**, 412–415.
287. U. Brönstrup, W. Hachtel (1986). Protein synthesis in isolated mitochondria from a streptomycin-bleached mutant of *Euglena gracilis* Klebs. *J. Plant Physiol.*, **124**, 45–53.
288. K.W. Foster, J. Saranak, F. Derguini, G.R. Zarrilli, R. Johnson, M. Okabe, K. Nakanishi (1989). Activation of *Chlamydomonas* rhodopsin in vivo does not require isomerization of retinal. *Biochemistry*, **28**, 819–824.
289. K.W. Foster, J. Saranak, P.A. Dowben (1991). Spectral sensitivity, structure and activation eukaryotic rhodopsins: activation spectroscopy of rhodopsin analogs in *Chlamydomonas*. *Photochem. Photobiol.*, **8**, 385–408.
290. P. Hegemann, W. Gärtner, R. Uhl (1991). All-trans retinal constitutes the functional chromophore in *Chlamydomonas* rhodopsin. *Biophys. J.*, **60**, 1477–1489.
291. S. Kasai, R. Miura, K., Matsiu (1975). Chemical structure and some properties of roseoflavin. *Bull. Chem. Soc. Jpn.*, **48**, 2877–2880.
292. P. Gualtieri, L. Barsanti, V. Passarelli (1989). Absorption spectrum of a single isolated paraflagellar swelling of *Euglena gracilis*. *Biochim. Biophys. Acta*, **993**, 293–296.
293. U. Kopf, J. Heinze (1984). 2,7-Bis(diethylamino)phenazoxonium chloride as quantum counter for emission measurements between 240 and 700 nm. *Anal. Chem.*, **56**, 1931–1935.
294. Christie, M. Salomon, K. Nozue, M. Wada, W.R. Briggs (1999). LOV (light, oxygen, or voltage) domains of the blue-light photoreceptor phototropin (nph1): Binding sites for the chromophore flavin mononucleotide. *Proc. Natl. Acad. Sci. USA*, **96**, 8779–8783.
295. P.G. Sperling, A. Schimz (1980). Photosensory retinal pigments in *Halobacterium halobium*. *Biophys. Struct. Mech.*, **6**, 165–169.
296. L. Barsanti, V. Passarelli, P. Lenci, P. Gualtieri (1992). Elimination of photoreceptor (paraflagellar swelling) and photoreception in *Euglena gracilis* by means of the carotenoid biosynthesis inhibitor nicotine. *J. Photochem. Photobiol. B: Biol.*, **13**, 135–144.
297. P. Gualtieri, P. Pelosi, V. Passarell, L. Barsanti (1992). Identification of a rhodopsin photoreceptor in *Euglena gracilis*. *Biochim. Biophys. Acta*, **1117**, 55–59.
298. L. Barsanti, V. Passarelli, P. Lenci, P.L. Walne, J.R. Dunlap, P. Gualtieri (1993). Effects of hydroxylamine, digitonin and triton X-100 on photoreceptor (paraflagellar swelling) and photoreception of *Euglena gracilis*. *Vision Res.*, **33**, 2043–2050.
299. L. Barsanti, V. Passarelli, P.L. Walne, P. Gualtieri (1997). In vivo photocycle of the *Euglena gracilis* photoreceptor. *Biophys. J.*, **72**, 545–553.
300. R.A. Bogomolni, J.L. Spudich (1991). Archaebacterial rhodopsins: sensory and energy transducing membrane proteins. In: Sensory Receptors and Signal Transduction (pp. 233–255). Wiley-Liss.
301. J. Soppa, J. Otomo, J. Straub, J. Tittor, S. Meeßen, D. Oesterhelt (1989). Bacteriorhodopsin mutants of *Halobacterium* sp. GRB. II. Characterization of mutants. *J. Biol. Chem.*, **264**, 13049–13056.
302. K. Hamdorf, J. Schwemer (1975). Photoregeneration and the adaptation process in insect photoreceptors. In: A.W. Snyder, R. Menzel (Eds), *Photoreceptor Optics* (pp. 253–289). Springer, Berlin.
303. N.K. Franceschini (1983). In-vivo microspectrofluorimetry of visual pigments. In: D.J. Cosens, D. Vince-Price (Ed.), *The biology of photoperception. Society for Experimental Biology* (pp. 53–85). London.

304. D.G. Stavenga, R.P. Smits, B. Hoenders (1993). Simple exponential function describing the absorbance bands of visual pigment spectra. *Vision Res.*, **33**, 1011–1017.
305. P.L. Walne, V. Passarelli, L. Barsanti, P. Gualtieri (1998). Rhodopsin: A photopigment for phototaxis in *Euglena gracilis*. *Crit. Rev. Plant Sci.*, **17**, 559–574.
306. R.L. Martin, C. Wood, W. Baehr, M. Applebury (1986). Visual pigment homologies revealed by DNA hybridization. *Science*, **232**, 1266–1269.
307. L. Margulis (1993). *Symbiosis in Cell Evolution*. W.H. Freeman, New York.
308. L. Stryer (1988). *Biochemistry* (3rd ed,). W.H. Freeman and Company, New York.
309. J. Soppa (1994). Two hypotheses – one answer: Sequence comparison does not support an evolutionary link between halobacterial retinal proteins including bacterio-rhodopsin and eukaryotic G-protein-coupled receptors. *FEBS Letters*, **342**, 7–11.
310. P.A. Kivic, P.L. Walne (1983). Algal photosensory apparatus probably represent multiple parallel evolutions. *Biosystems*, **16**, 31–38.
311. M. Schuman-Jorns, B. Wang, S.P. Jordan, L.P. Chanderkar (1990). Chromophore function and interaction in *Escherichia coli* DNA photolyase: Reconstitution of the apoenzyme with pterin and/or flavin derivates. *Biochem.*, **29**, 552–561.
312. G. Payne, M. Wills, C. Walsh, A. Sancar (1990). Reconstitution of *Escherichia coli* photolyase with flavins and flavin analogues. *Biochem.*, **29**, 5706–5711.
313. L.P. Chanderkar, M.S. Jorns (1991). Effect of flavin structure and redox state on catalysis by and flavin-pterin energy transfer in *Escherichia coli* DNA photolyase. *Biochem.*, **30**, 745–754.
314. S. Kanai, R. Kikuno, H. Toh, H. Ryo, T. Todo (1997). Molecular evolution of the photolyase-blue-light photoreceptor family. *J. Mol. Evol.*, **45**, 535–548.
315. A.R. Cashmore (1998). The cryptochrome family of blue/UV-A photoreceptors. *J. Plant Res.*, **111**, 267–270.
316. C. Lin, D.E. Robertson, M. Ahmad, A.A. Raibekas, M.S. Jorns, P.L. Dutton, A.R. Cashmore (1995). Association of flavin adenine dinucleotide with the *Arabidopsis* blue light receptor CRY1. *Science*, **269**, 968–970.
317. I.B. Zhulin, B.L. Taylor (1997). PAS domain S-boxes in Archaea, Bacteria and sensors for oxygen and redox. *Trends Biochem. Sci.*, **22**, 331–333.
318. J.L. Pellequer, K.A. Wager-Smith, S.A. Kay, E.D. Getzoff (1998). Photoactive yellow protein: a structural prototype for the three-dimensional fold of the PAS domain superfamily. *Proc. Natl. Acad. Sci. USA*, **95**, 5884–5890.
319. M.A. Gilles-Gonzalez, G.S. Ditta, D.R. Helinski (1991). A haemoprotein with kinase activity encoded by the oxygen sensor of *Rhizobium meliloti*. *Nature*, **350**, 170–172.
320. T. Holton, A.J. Hudspeth (1986). The transduction channel of hair cells from the bull-frog characterized by noise analysis. *J. Physiol.*, **375**, 195–227.
321. M. Lebert, D.-P. Häder (1999). Negative gravitactic behavior of *Euglena gracilis* can not be described by the mechanism of buoyancy-oriented upward swimming. *Adv. Space Res.*, **24**, 851–860.
322. D.-P. Häder, G. Colombetti, F. Lenci, M. Quaglia (1981). Phototaxis in the flagellates, *Euglena gracilis* and *Ochromonas danica*. *Arch. Microbiol.*, **130**, 78–82.
323. R. Ferrara, S. Grassi, G. del Carratore (1975). An automatic homocontinuous culture apparatus. *Biotechnol. Bioengineer.*, **17**, 985–995.
324. C. Creutz, B. Diehn (1976). Motor responses to polarized light and gravity sensing in *Euglena gracilis*. *J. Protozool.*, **23**, 552–556.
325. F. Fong, J.A. Schiff (1978). Blue-light absorbance changes and phototaxis in *Euglena*. *Plant Physiol. (Suppl.)*, **61**, 74.
326. F. Fong, J.A. Schiff (1979). Blue-light-inducted absorbance changes associated with carotenoids in *Euglena*. *Planta*, **146**, 119–127.

327. O. Froehlich, B. Diehn (1974). Photoeffects in a flavin-containing lipid bilayer membrane and implications for algal phototaxis. *Nature*, **248**, 802–804.
328. E. Hildebrandt (1974). Effects of some inhibitors in *Euglena* photomotion. In: G. Schenk (Ed.), *Progress in Photobiology* (pp. 358). Proc. 6th Int. Congr. Photobiol, Dtsch. Ges. Lichtforsch. E.V., Frankfurt.
329. P.J. Simons (1981). The role of electricity in plant movements. *New Phytol.*, **87**, 11–37.
330. H. Harz, C. Nonnengässer, P. Hegemann (1992). The photoreceptor current of the green alga *Chlamydomonas*. *Phil. Trans.R. Soc. Lond. B*, **338**, 39–52.
331. K.M. Nichols, A. Jacklet, R. Rikmenspoel (1980). Effects of Mg^{2+} and Ca^{2+} on photoinduced *Euglena* flagellar responses. *J. Cell Biol.*, **84**, 355–363.
332. C. Tamponnet, J.-P. Rona, J.-N. Barbotin, R. Calvayrac (1988). Effects of high external calcium concentrations on etiolated *Euglena gracilis* Z cells and evidence of an internal membrane potential. *Biochim. Biophys. Acta*, **943**, 87–94.
333. F.W. Bancroft (1913). Heliotropism, differential sensibility, and galvanotropism in *Euglena*. *J. Exp. Zool.*, **15**, 383–428.
334. C. Ascoli, M. Barbi, C. Frediani, D. Petracchi (1978). Effects of electromagnetic fields on the motion of *Euglena gracilis*. *Biophys. J.*, **24**, 601–612.
335. B. Diehn, G. Tollin (1967). Phototaxis in *Euglena*. IV. Effect of inhibitors of oxidative and photophosphorylation on the rate of phototaxis. *Arch. Biochem. Biophys.*, **121**, 169–177.
336. E.C. Bovee, T.L. Jahn (1972). A theory of piezoelectric activity and ion movements in the relation of flagellar structures and their movements to the phototaxis of *Euglena*. *J. Theor. Biol.*, **35**, 259–276.
337. E. Piccini, V. Albergoni (1973). ATPase activity in isolated flagella from *Euglena gracilis*. *J. Protozool.*, **20**, 456–458.
338. E. Piccinni, V. Albergoni, O. Coppelotti (1975). ATPase activity in flagella from *Euglena gracilis*: Localization of the enzyme and effects of detergents. *J. Protozool.*, **22**, 331–335.
339. E. Piccinni, P. Omodeo (1975). Photoreceptors and phototactic programs in protista. *Boll. Zool.*, **42**, 57–79.
340. B. Diehn, G. Tollin (1966). Phototaxis in *Euglena*. III. Lag phenomena and the overall mechanism of the tactic response to light. *Photochem. Photobiol.*, **5**, 839–852.
341. T. Shimmen (1981). Quantitative studies on step-down photophobic response of *Euglena* in an individual cell. *Protopl.*, **106**, 37–48.
342. B. Diehn, J.R. Fonseca, T.R. Jahn (1975). High speed cinemicrography of the direct photophobic response of *Euglena* and the mechanism of negative phototaxis. *J. Protozool.*, **22**, 492–494.
343. M.J. Doughty (1991). A kinetic analysis of the step-up photophobic response of the flagellated alga *Euglena gracilis* in culture medium. *J. Photochem. Photobiol. B: Biol.*, **9**, 75–85.
344. M.J. Doughty (1993). Step-up photophobic response of *Euglena gracilis* at different irradiances. *Acta Protozool.*, **32**, 73–77.
345. P. Kröger, P. Hegemann (1994). Photophobic responses and phototaxis in *Chlamydomonas* are triggered by a single rhodopsin photoreceptor. *FEBS Lett.*, **341**, 5–9.

Photomovement
D.-P. Häder and M. Lebert, editors.

Chapter 12

Yellow-light sensing phototaxis in cryptomonad algae

Masakatsu Watanabe and Mayumi Erata

Table of contents

Abstract

Cryptomonad algae (cryptophyte algae), originated from a eukaryotic host cell and a eukaryotic endosymbiont by secondary endosymbiosis, are unicellular biflagellate, freshwater or marine, eukaryotic algae, unique among flagellate algae in having phycobiliproteins (Cr-phycoerythrin or Cr-phycocyanin) as photosynthetic accessory pigments. An eyespot is found only in some species (e.g. *Chroomonas coerulea*) although most of the cryptomonad species are phototactic. Cryptomonad phototaxis may be exclusively positive (*Cryptomonas* sp. CR-1, *Cryptomonas rostratiformis* and *Chroomonas nordstedtii*) or exclusively negative (*Chroomonas coerulea*) or positive at low fluence rates and negative at higher fluence rates (*Cryptomonas maculata*) or diaphototactic, i.e. orienting perpendicular to the direction of the light beam (*Cryptomonas* sp. S2). Action spectroscopy for positive phototaxis in *Cryptomonas* sp. CR-1 led to the discovery of its very unique yellow-light sensing with an action peak in the yellow light region at ca. 560 nm, which was later confirmed in the action spectrum for positive phototaxis in *Cryptomonas rostratiformis* and *Chroomonas nordstedtii*, together with a new action peak at 460 nm (the former showing also a UV peak at 280 nm), strongly suggesting that the photoreceptor mediating positive phototaxis is common among these species, and not phycobilins. Analyses using repeated pulses of stimulus light showed that light is sensed by the ventral side of *Cryptomonas* sp. CR-1 cells. The striking similarity of the action spectra for photoaccumulation and light-induced membrane depolarization in the ciliate *Paramecium* with that for the cryptomonad phototaxis seems to suggest considerable common nature of their unidentified photoreceptors. Analyses of phototactic responses of *Cryptomonas* sp. CR-1 cells to intermittent light stimuli with variable light durations and dark intervals revealed a striking dark interval dependence, suggesting also considerable similarity in the photoreceptor-signal transduction processes between cryptomonads and the green alga *Chlamydomonas*, in which retinal-binding protein(s) are supposed to work as the photoreceptor(s) for its green-light sensing phototaxis, suggesting that a dark interval dependence of phototaxis is theoretically expected on the basis of kinetic analyses of the photoreceptor current. The presence of Ca^{2+} is crucially needed for phototactic orientation and antagonized by K^+ ions, consistently with other phototactic organisms such as *Chlamydomonas, Euglena* and *Paramecium.* The ecological significance of cryptomonad phototaxis is demonstrated, in general accordance with other algae, as the means for the cells to locate themselves in photosynthetically advantageous light habitats and to avoid harmful strong light or UV radiation both horizontally and vertically. The molecular identification of the photoreceptor pigment(s) as well as the understanding of the ecological and biochemical meanings of the UV-sensing in *Cryptomonas* is a challenging subject of future studies.

12.1 Principal characteristics and phylogeny of cryptomonads

Cryptomonad algae (cryptophyte algae) are unicellular biflagellate, freshwater or marine, eukaryotic algae, unique among flagellate algae in having phycobiliproteins (Cr-phycoerythrin or Cr-phycocyanin) as photosynthetic accessory pigments [1,2]. Cell shapes (Figure 1, [3]) may be ovate, ellipsoidal or oblong, mostly dorsiventrally compressed; periplasts, found only in cryptomonads are the cell coverings typically consisting of multiple plates. A ventral groove, called furrow, mostly followed by a tubular invagination, called gullet, begins in the vestibular region of the cell and extends posteriorly. Large and small ejectosomes (trichocysts), extrusive organelles unique in cryptomonads, are found lining the furrow/gullet and in the cell surface region, respectively. Two flagella emerge from the opening of the gullet/furrow anteriorly. An eyespot is found only in a few species (e.g. *Chroomonas coerulea*) (Figure 2) [4] although most of the cryptomonad species are phototactic. Chloroplast colors are brownish, red, blue-green or olive green depending on the accessory pigments present. The phycobiliproteins are located in the intrathylakoidal lumen [5]. *Chilomonas* has a reduced chloroplast which lacks pigments. Chloroplasts are surrounded by a double

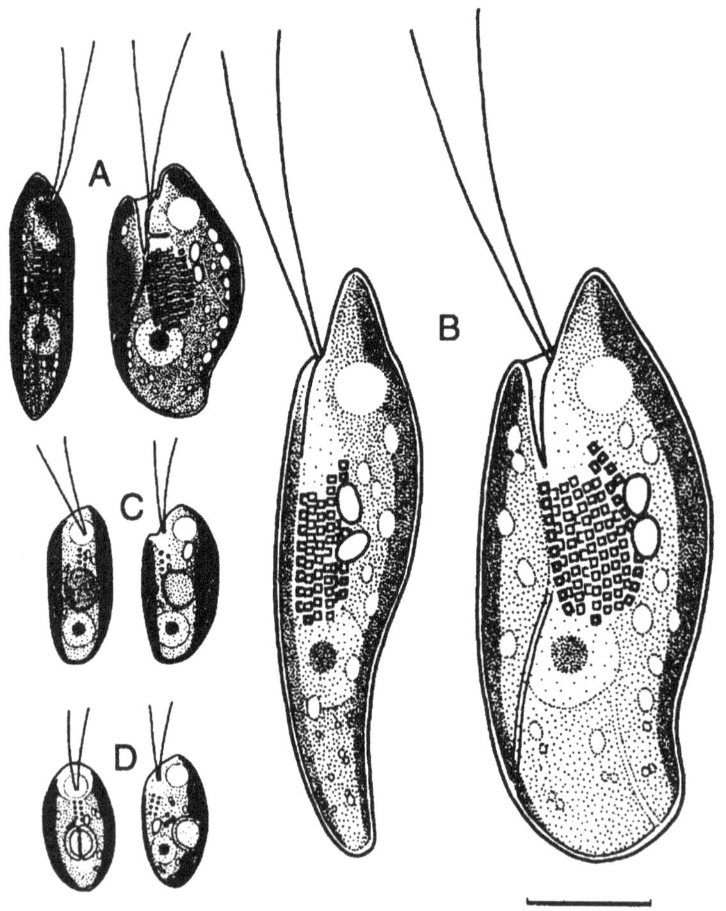

Figure 1. Cell shapes of some cryptomonad species. A *Cryptomonas* sp., B *Cryptomonas rostratiformis*, C *Chroomonas nordstedtii*, D *Chroomonas coerulea* (from [3]). Scale bar = 20 μm (A), 10 μm (B, C, D).

membrane called chloroplast ER, which surrounds the chloroplast, starch grains and a nucleomorph which is a vestigial nucleus remaining from a red algal endosymbiont [6]. *Goniomonas* lacks plastids and a nucleomorph [2,7].

Recent ultrastructural and molecular biological analyses have revealed that the cryptomonads have been originated from a eukaryotic host cell and a eukaryotic endosymbiont by secondary endosymbiosis [2,6–9]. The host nuclear encoded rRNA appears to be related to *Goniomonas*, whereas the nucleomorph encoded rRNA appears to be related to red algae [6,7,10].

12.2 Positive, negative and diaphototaxis

Häder et al. [11] reported that the marine flagellate *Cryptomonas maculata* (=*Rhodomonas maculata*) shows weak positive phototaxis at fluence rates below 15 W m^{-2} and a more pronounced negative phototaxis at fluence rates above 15 W m^{-2}, in contrast to the freshwater species *Cryptomonas* sp. CR-1, *Cryptomonas rostratiformis* and *Chroomonas nordstedtii*, for which only positive phototaxis is found [4,12]. *Chroomonas coerulea*, having an eyespot, is reported to show only negative phototaxis [4]. Häder et al. [13] next reported an interesting observation that in nitrogen deficient *Cryptomonas maculata* cells, which show a reduced pigment concentration with concomitant decrease in photosynthetic efficiency, negative phototaxis commences even at a fluence rate as low as 3.5 W m^{-2}, though being less prone to photobleaching.

Rhiel et al. [14] reported a very interesting discovery and characterization that a freshwater *Cryptomonas* sp. (strain S2), they had isolated from a pond near Marburg, shows an unusual phototactic response: at all effective fluence rates it orients perpendicular to the direction of the light beam (diaphototaxis). When grown under nitrogen and phosphorus deficiency, the degree of orientation increases though the absorption spectrum does not change considerably. Rhiel et al. [15] further showed that this diaphototaxis was considerably impaired when the rotation speed as well as the swimming speed was drastically decreased in the presence of high viscosity media (0.6% (w/v) methyl cellulose), indicating that the mechanism of light detection is dependent on a periodic shading or irradiation mechanism. They also demonstrated in this freshwater *Cryptomonas* sp. S2 a rather pronounced negative gravitaxis which is only partially modified by phototaxis.

12.3 Photoreceptor(s)

12.3.1 Action spectra

12.3.1.1 Cryptomonas *sp. CR-1 – The first study on cryptomonad phototaxis*

Even much before the idea of the secondary symbiotic origin of cryptomonads, their above-mentioned unique photosynthetic pigment composition among flagellate algae suggested the possibility of unique photoreceptor characteristics, leading Watanabe and Furuya [12] to determine the action spectrum for positive phototaxis in *Cryptomonas* sp. CR-1 and to the discovery of its very unique yellow-light sensing with an action peak at ca. 560 nm.

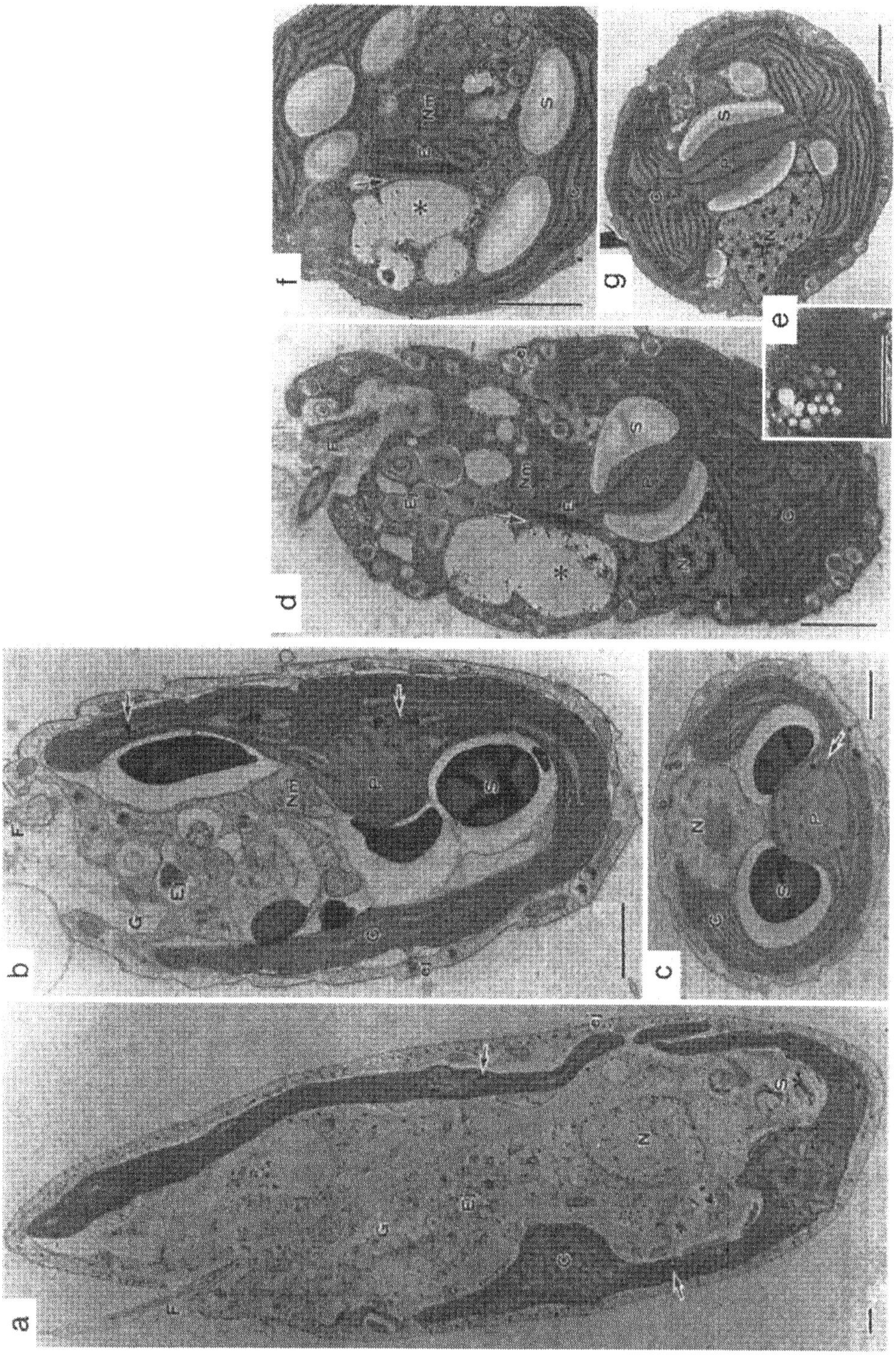

Figure 2. Transmission electron micrographs of *Cryptomonas rostratiformis*, *Chroomonas nordstedtii* and *Chroomonas coerulea*. Arrows indicate the osmiophilic globules. *C* Chloroplast, *E* eyespot, *Ej* large ejectosome (gullet-associated), *ej* small ejectosome (periplast-associated), *F* flagellum, *G* gullet, *N* nucleus, *Nm* nucleomorph, *P* pyrenoid, *S* starch grain. Bars: 1 μm. **a** *Cryptomonas rostratiformis*, longitudinal section. **b** *Chroomonas nordstedtii*, longitudinal section. **c** *Chroomonas nordstedtii*, transverse section. **d** *Chroomonas coerulea*, longitudinal section. Note two kinds of vacuoles adjacent to the eyespot disc (arrow and asterisk). **e** *Chroomonas coerulea*, grazing section through the eyespot disc. **f** *Chroomonas coerulea*, transverse section through the eyespot. **g** *Chroomonas coerulea*, transverse section through the pyrenoid, showing the H-shaped outline of the chloroplast-pyrenoid complex. (from [4]).

Action spectra have been determined for phototaxis in several flagellated algae of the Euglenophyta [16–18], Dinophyta [19–21], and Chlorophyta [19,22–24]. In all but two cases they have two features in common: a maximum sensitivity around 440–520 nm (blue-green), and very little, if any, sensitivity above 560 nm. The rare exceptions are *Prorocentrum micans* (Dinophyta) with a maximum sensitivity at about 570 nm (yellow) [19] and *Peridinium gatunense* with a maximum sensitivity at 640 nm (red) [25]. On the other hand, in filamentous cyanophytes, action spectra for phototaxis show two maximum peaks at 490 and 560 nm, the latter coinciding with the absorption maximum of phycoerythrin in these algae [26]. Because Cryptophyta is the only group, except Glaucophyta, of flagellated algae that has phycobiliproteins as photosynthetic accessory pigments, it seemed quite interesting to determine to which type the action spectrum for phototaxis of this group resembles. Attempts were therefore made to determine the action spectrum of the phototactic response in this alga.

An apparatus was designed, constructed and used to measure the phototactic response at the cell population level in this and several other following studies (Figure 3a). A phototactically inactive measuring beam of 750 nm was chosen to determine changes in cell concentration by a photomultiplier at the rear end of the transparent rectangular cuvette into which monochromatic stimulus light was projected horizontally. Some typical examples of recorded charts obtained with a cell suspension of *Cryptomonas* sp. CR-1 at 3.5×10^5 cells ml^{-1} are shown in Figure 3b. The slopes of these curves (rate of change in cell concentration in a 2-mm layer just inside of the rear side of the cuvette) were used as index of phototactic responses. The phototactic responses were generally linear with the logarithm of the stimulus fluence rate and never saturated in a fluence rate range below 1 Wm^{-2} at all tested wavelengths. Therefore the (equal quantum) action spectrum for phototaxis was determined at a fluence rate of 0.83 micromole m^{-2} s^{-1} e.g. 0.25 and 0.15 W m^{-2} at 400 and 680 nm, respectively).

The action spectrum obtained (Figure 4a) showed a single peak at 560 nm and a shoulder at 490 nm. Blue light was somewhat effective but less so than orange light. The longer the wavelength beyond 560 nm, the less the rate of phototactic response. Wavelengths longer than 640 nm had no effect.

It was shown that the action spectrum for phototaxis in *Cryptomonas* sp. CR-1 is very unique among flagellate algae and but similar in part to that for the filamentous cyanophyte *Phormidium uncinatum*, with maxima at 380–400, 480–500 and 550–560 nm. The fact that the action spectral peak wavelength and profile in the longer

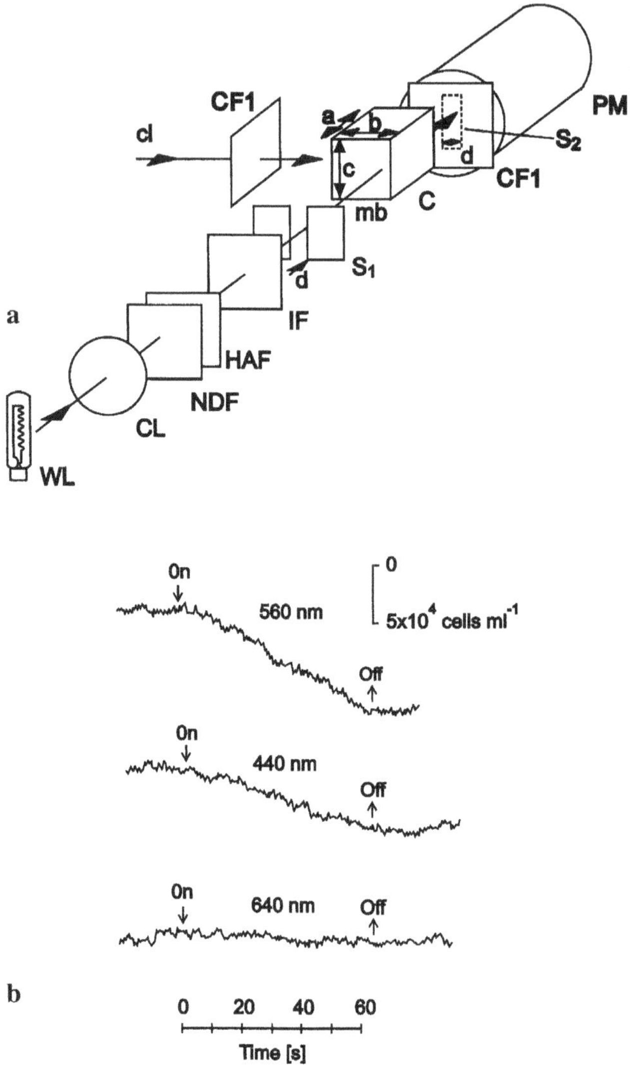

Figure 3. **a** Schematic representation of the apparatus for photometric measurement of population-level phototaxis. WL tungsten lamp. CL collimating lens. NDF neutral density filter. HAF heat absorbing filter. S1 knife-edge slit (d = 2 mm). mb measuring beam (2 mm wide, 15mm high). C transparent rectangular cuvette made of 1.5 mm thick polymethacrylate resin (a = 12 mm, b = 14 mm, c = 12 mm) with a 1.5 mm thick glass lid. CF1 sharp cut-off filter to keep light of wavelengths shorter than 690 nm out of the photomultiplier. S2 slit. CF2 appropriate sharp cut-off filter to remove stray light and the second order spectra from the stimulus light. sl monochromatic stimulus light from a spectro-irradiator. PM head-on type photomultiplier. **b** Time-course of stimulus light-induced cell concentration changes at the rear of the cell suspension relative to the stimulus light beam. Stimulus light fluence rate was 0.83 μmol m^{-2} s^{-1}. Initial cell concentration was 3.5×10^5 cells ml^{-1} (from [12]).

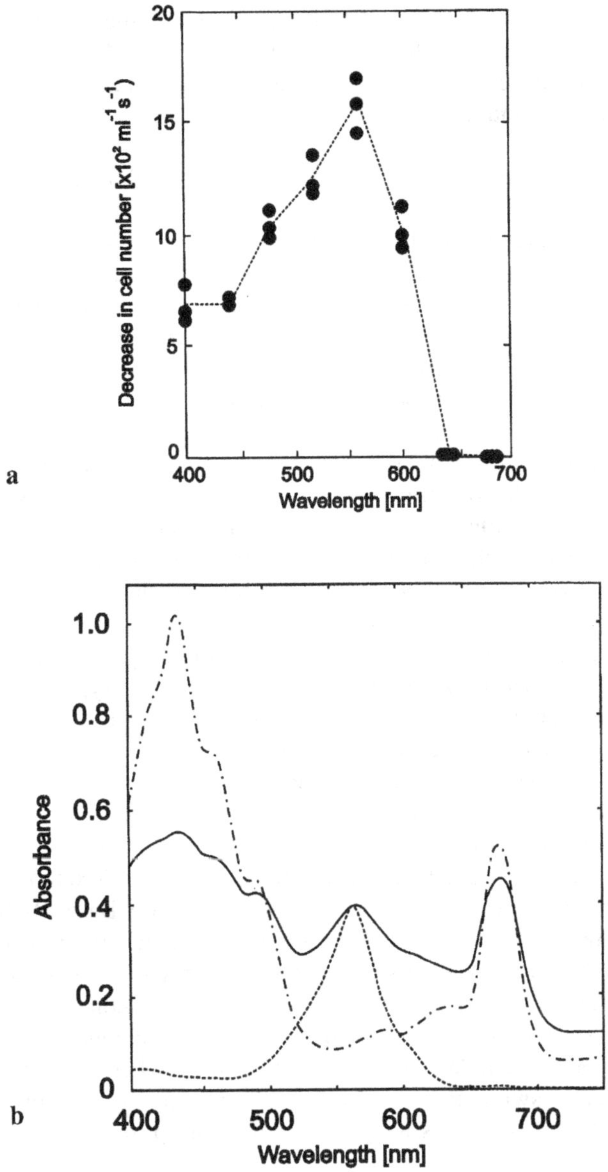

Figure 4. a Equal-quantum action spectrum for population-level positive phototaxis in *Cryptomonas* sp. CR-1 (at 0.83 μmol m^{-2} s^{-1}). Each point shows the value for a single measurement. **b** Absorption spectra of intact cells (solid line); of phycobiliproteins extracted with 0.1 M phosphate buffer, pH 6.5 (dotted line); and of the green pellet of disrupted cells after phosphate buffer extraction (dot-dashed curve). Curves are adjusted to show the absorbance of a 1.0 cm layer containing 1.8×10^6 cells ml^{-1} (from [12]).

wavelength region of *Cryptomonas* sp. CR-1 coincided with the absorption spectrum of the phycobiliproteins extracted with phosphate buffer (Figure 4b) urged further studies to determine whether or not the phycobiliproteins of this alga had a role as (one of) the photoreceptor(s) for phototaxis.

12.3.1.2 Cryptomonas rostratiformis, Chroomonas nordstedtii *and* Chroomonas coerulea

This peak position was confirmed by a recent, individual cell-level re-examination and extension of the action spectrum by Erata et al. [4] using the Okazaki Large Spectrograph [27,28] and computerized motion analysis systems with three species containing different phycobiliproteins (*Cryptomonas rostratiformis*, *Chroomonas nordstedtii* and *Chroomonas coerulea*) irrespective of whether the cellular phycobilins are phycoerythrin (absorption peak at 545–565 nm) (*Cryptomonas*) or phycocyanin (absorption peak at 645 nm) (*Chroomonas*).

Since cryptomonads of different genera contain different kinds of phycobilins such as Cr-phycocyanin and Cr-phycoerythrin, having different absorption peaks [29,30], action spectroscopy for the phototaxis of these different genera was expected to provide valuable information as to the photoreceptor identity. Two cryptomonad genera were chosen for this purpose, *Cryptomonas* and *Chroomonas*, the former having only Cr-phycoerythrin and the latter having only phycocyanin as their phycobilin pigment. The latter is the only genus which includes members with an eyespot, and therefore was expected to give information on the role of the eyespot in cryptomonad phototaxis, particularly its relation to the location of photoreceptor (cf. [31]), although most phototactic cryptomonad genera do not have an eyespot.

Phototaxis measurements were carried out using a computerized motion analysis system as described by Takahashi and Kobatake [32] and Takahashi et al. [33] (Figure 5). Three or five identical sets of the system were utilized simultaneously at different monochromatic wavelength positions in the range of 250 to 700 nm at the Okazaki Large Spectrograph (Figure 6) [27,28], National Institute for Basic Biology, Okazaki, Japan. For measurements, appropriate exposure times for the stimulus light and video frame-acquisition (Figure 5b) were selected for each species, following the results of preliminary experiments. Positive and negative phototactic indices used in this study were calculated as follows (cf. [34]): Positive phototactic index $= \{[X+] - 0.25\} \times (4/3)$, Negative phototactic index $= \{[X-] - 0.25\} \times (4/3)$, where X+ and X– represent the fractions of cells swimming toward and away from the stimulus light source, respectively. The factor (4/3) was adopted to adjust the positive value of the indices between 0 and 1.

The fluence-response curves for phototaxis in each of the three species are shown in Figure 7. *Cryptomonas rostratiformis* and *Chroomonas nordstedtii* showed positive phototaxis (Figure 7a–c). In both species, the index of positive phototaxis almost linearly correlates with the logarithm of the fluence rate of less than 5 μmol m^{-2} s^{-1}, while positive phototaxis was saturated or suppressed at each wavelength examined with more intense actinic light. For *Cryptomonas rostratiformis*, responses in the UV range (Figure 7b) were also measured; considerably high phototactic responses were observed at 260, 280 and 300 nm. In contrast *Chroomonas coerulea* showed remarkable negative phototaxis (swimming away from the light source) (Figure 7d,e). This response increased linearly with the logarithm of the fluence rate at fluence rates of more than 10 μmol m^{-2} s^{-1}. When the stimulus light was very weak (0.0002–1.0 μmol m^{-2} s^{-1}), the cells showed a weak response without exhibiting any positive phototaxis (Figure 7e).

Based on the results in Figure 7a-e, equal quantum action spectra for phototaxis were calculated in the three species at fluence rates where neither saturation nor suppression

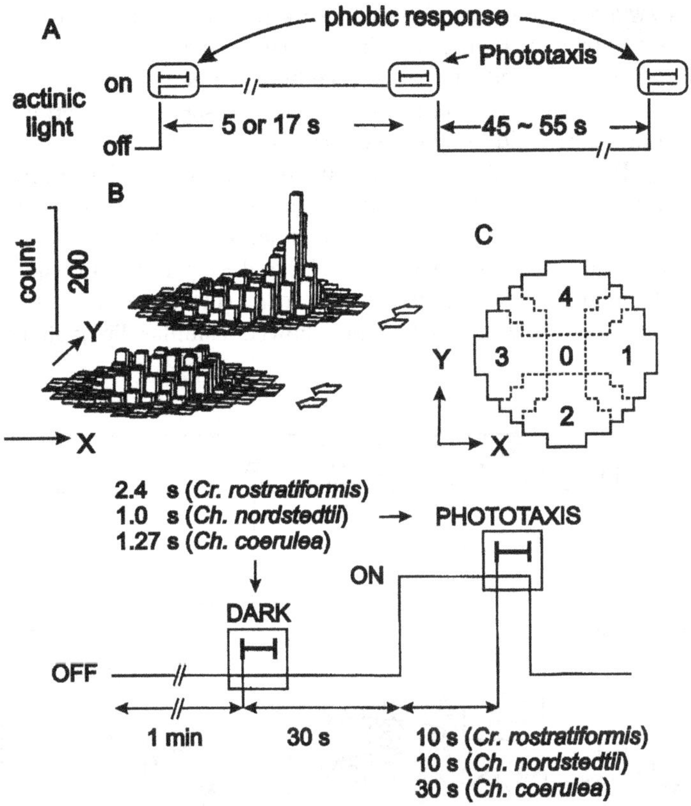

Figure 5. Measurement of individual cell-level photomovement responses by a computerized motion analysis system. **a** A An example of timing chart for stimulus irradiation and data acquisition. Swimming tracks of microorganisms were recorded in six consecutive digitized frames at a frame rate of 5 s^{-1} during the periods indicated by the horizontal bars. A computer software automatically detects the displacement of each individual cell track between two consecutive frames and the data were accumulated. B Histogram of the two-dimensional displacement per 200 ms of the cells responding positive-phototactically (upper) or without phototaxis (lower). Arrows indicate the direction of stimulus light. C The histogram was then divided into 5 subdomains. The phototaxis index was calculated as the integration of the counts over the domain 1 divided by that over domains 1, 2, 3 and 4. The magnitude of photophobic response was calculated from the number of cell tracks whose displacement falls in subdomain 0 after stimulus irradiation (from [33]) **b** Timing chart for stimulus irradiation and data acquisition optimized for each of the cryptomonad species examined. Horizontal bars indicate the time periods during which the digitized data of six consecutive frames were acquired. (from [4]).

is observed: in *Cryptomonas rostratiformis*, at 6.64 μmol m^{-2} s^{-1}, *Chroomonas nordstedtii* at 8.31 μmol m^{-2} s^{-1} and *Chroomonas coerulea* at μmol m^{-2} s^{-1}.

The action spectrum of *Cryptomonas rostratiformis* shows a peak at 560 nm (Figure 8a). Monochromatic light at wavelengths longer than 680 nm have almost no effect in eliciting a phototactic response. The spectrum shows a large drop at 460 nm which was not reported for *Cryptomonas* sp. CR-1. Except this point, the phototaxis action spectrum of *Cryptomonas rostratiformis* closely resembles that of *Cryptomonas* sp. CR-1 ([12]; also shown in Figure 8a). It also has a small peak in the UV range at 280 nm. It has been proposed that equal-quantum action spectra are generally more distorted due to screening pigment(s) than the action spectrum determined form the fluence rate at the threshold or the rate that gives the maximal response at each wavelength [35]. To examine this, reconstituting the threshold action spectrum was also tried for *Cryptomonas rostratiformis* (Figure 8b) using fluence rate-response curves shown in Figure 7a and b and those from another set of experiments (data not shown).

Basically, the reconstituted curves were similar to the equal-quantum action spectrum in Figure 8a. A small difference is that the former exhibited a dip in the phototactic sensitivity at about 480 nm whereas the latter showed a similar dip at about 460 nm.

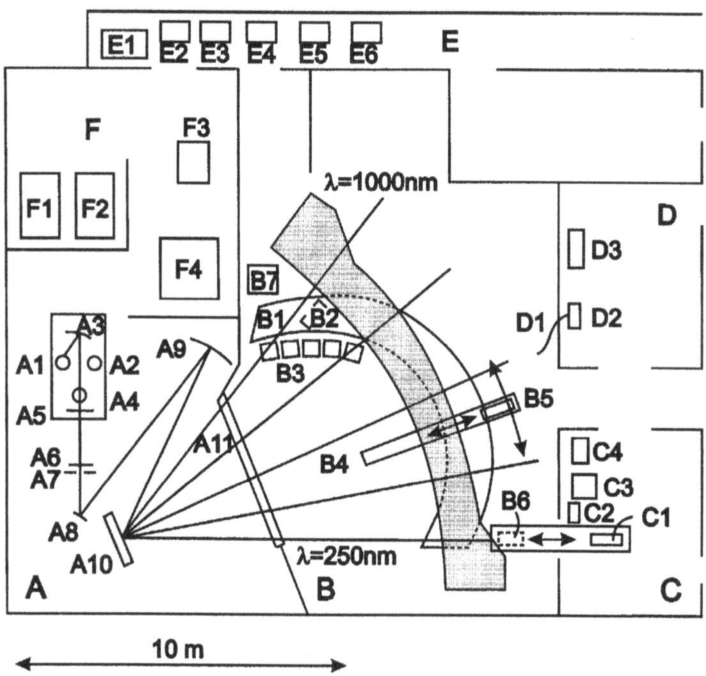

Figure 6. Spatial arrangement of the Okazaki Large Spectrograph (OLS) at the National Institute for Basic Biology (NIBB), Okazaki, Japan. A1 30 kW Xe lamp. A7 entrance slit. A9 condensing mirror. A10 plane grating (double-blazed at 250 and 500 nm). B1 irradiation stage. B2 sample box. B3-6 and C2 box transfer robot. C2 control panel. C3 CRT terminal. D1 optical fiber bundle. D2 its outlet. E1 host computer ([27]).

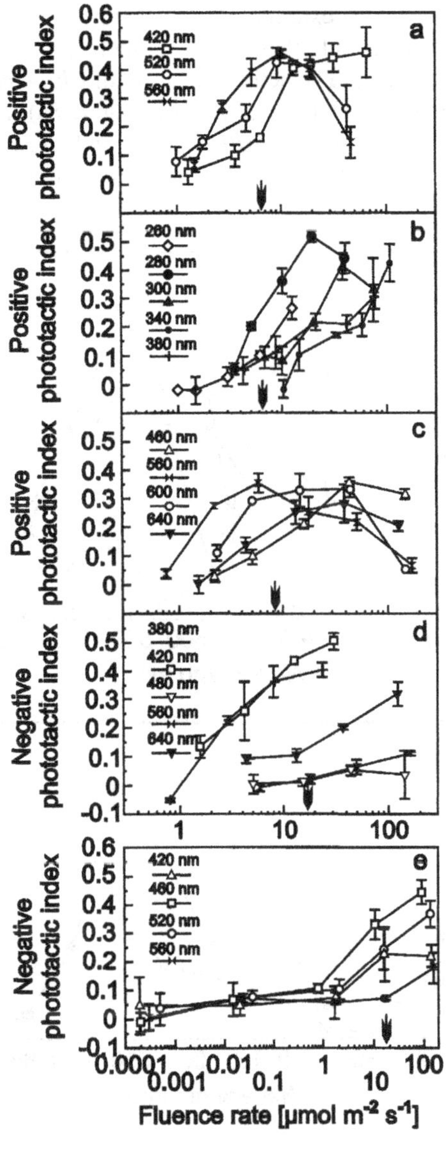

Figure 7. Fluence rate-response curves for phototaxis in three cryptomonad species. **a** Positive phototaxis of *Cryptomonas rostratiformis* in the visible region. **b** positive phototaxis of *Cryptomonas rostratiformis* in the UV region. **c** positive phototaxis of *Chroomonas nordstedtii* in the visible region. **d** negative phototaxis of *Chroomonas coerulea* in the visible region. **e** negative phototaxis of *Chroomonas coerulea* in the lower fluence rate range. Bars represent standard errors. The arrows on the horizontal axes indicate the fluence rate at which the equal-quantum action spectra in Figure 8 a were determined. (from [4]).

The two *Chroomonas* species, examined in this study, not only show opposite signs of their phototactic response but also differ in their action spectra in both peak wavelengths and spectral shapes. The action spectrum of *Chroomonas nordstedtii* shows a large peak at 560 nm and a smaller one at 460 nm (Figure 8a). These peak wavelengths agree with those of both *Cryptomonas rostratiformis* and CR-1, and the general shape of the spectrum resembles those of the two *Cryptomonas* strains. On the other hand, the

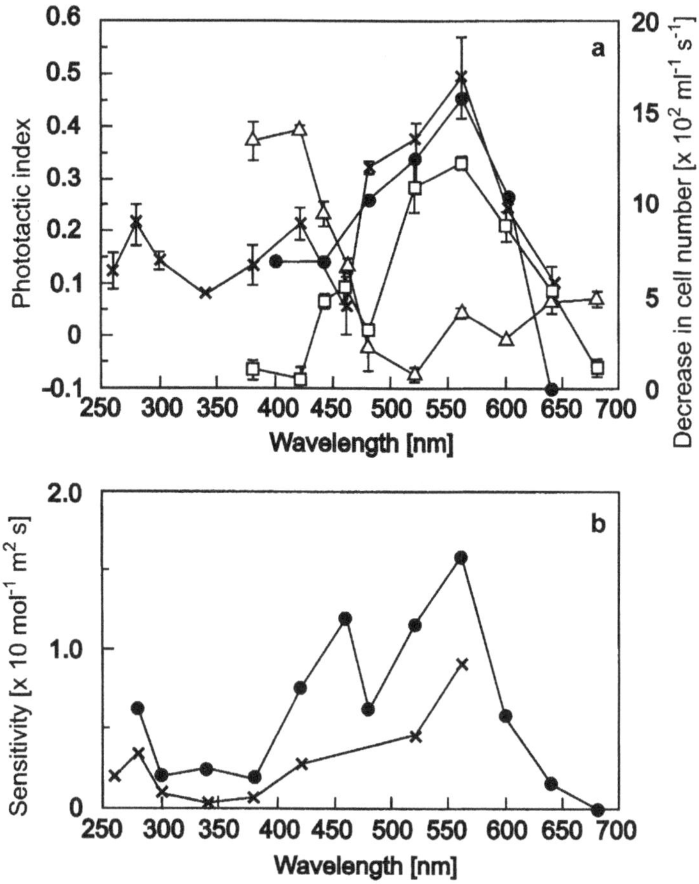

Figure 8. Action spectra for individual cell-level phototaxis in three cryptomonad species. **a** Equal-quantum action spectra: (crosses) positive phototaxis in *Cryptomonas rostratiformis* (at 6.64 μmol m^{-2} s^{-1}); (squares) positive phototaxis in *Chroomonas nordstedtii* (at 8.31 μmol m^{-2} s^{-1}); (triangles) negative phototaxis in *Chroomonas coerulea* (at 16.6 μmol m^{-2} s^{-1}). Bars indicate standard errors. (filled circles) Equal-quantum action spectrum for population-level positive phototaxis in *Cryptomonas* sp. CR-1, shown for comparison (from Figure 4a). **b** The threshold action spectrum reconstituted from two sets of fluence rate-response curves of positive phototaxis in *Cryptomonas rostratiformis*. The threshold fluence-rates were determined according to [35], pp. 621–622), from the fluence rate-response curves shown in Figure 7a and b (crosses), and from the curves obtained during another series of experiments (filled circles). (from [4]).

action spectrum for negative phototaxis of *Chroomonas coerulea* had a peak at 420 nm (Figure 8a). The results strongly suggest that the characteristics of the photoreceptor mediating positive phototaxis are common, at least among the two strains of *Cryptomonas* and *Chroomonas nordstedtii*.

It is well-known that cryptomonads contain phycobilin pigments, as do red algae and cyanobacteria, but there is an important difference between cryptomonads and the other two groups of organisms in their composition of phycobilins. Each organism of the cryptomonads contains only a single type of phycobilin, whereas each member of the latter two groups possesses several kinds (e.g. phycoerythrin, phycocyanin and allophycocyanin), comprising the pigment complexes on the thylakoid surface known as "phycobilisomes". Cryptomonads generally vary in plastid pigmentation from genus to genus. Most species of *Cryptomonas* show brownish pigmentation as a result of bearing phycoerythrin, whereas *Rhodomonas* has a bright red color, as it possesses another type of phycoerythrin [36]. In contrast, *Chroomonas* has a blue-green pigmentation, because it contains only phycocyanin.

Since the action spectroscopical study for phototaxis in *Cryptomonas* sp. CR-1 [12], the possibility of involvement of phycobilin in cryptomonad phototaxis has been postulated because there was a similarity between the absorption spectrum of the pigment and the phototactic action spectrum. If the phycobilins are universally related to the phototaxis of the cryptomonads, the similarity in spectral shape and peaks between phototactic action spectra and absorption spectra might be expected in all of the variously pigmented cryptomonads.

However, by comparison between the action spectra and the absorption spectra, shown in Figures 8 and 9, it is clear that the peaks in the absorption spectra do not coincide with those in the phototactic action spectra in *Cryptomonas* and *Chroomonas* species examined. This fact indicates that there is no obvious relationship between phototactic response and phycobilin pigmentation for at least these two *Chroomonas* species. The difference between the equal-quantum action spectrum and the threshold action spectrum may be ascribable to the screening effect of phycocyanin, which shows complex and relatively intense absorbance at 400–480 nm (Figure 9). Table 1 summarizes the above results together with those by Watanabe and Furuya [12] on *Cryptomonas* sp. CR-1. The presence or absence of an eyespot (cf. Figures 1 and 2) is possibly the only difference that may affect phototaxis.

In general, the eyespot is considered as an important organelle that confers the ability of many unicellular algae to sense precisely the direction of light [35], presumably reinforcing directivity of the light-sensing devices of cells by reflection [31,37–40] of incident light beams. Thus, it is possible that the negative phototaxis of *Chroomonas coerulea* results from intensified phototactic sensitivity due to the presence of an eyespot, since in the other two species a drop in the positive phototactic indices was observed at the highest fluence rates applied. However, *Chroomonas coerulea* did not show positive phototaxis even at very low stimulus fluence rates (Figure 7c), suggesting that the negative phototaxis of this species could not be accounted for by a simple increase in phototactic sensitivity.

Furthermore, the action spectral peak at around 420 nm differs from those of *Cryptomonas rostratiformis*, *Chroomonas nordstedtii*, described above, and *Cryptomonas* sp. CR-1 [12] seen at ca. 560 nm. Therefore the negative and positive phototaxis

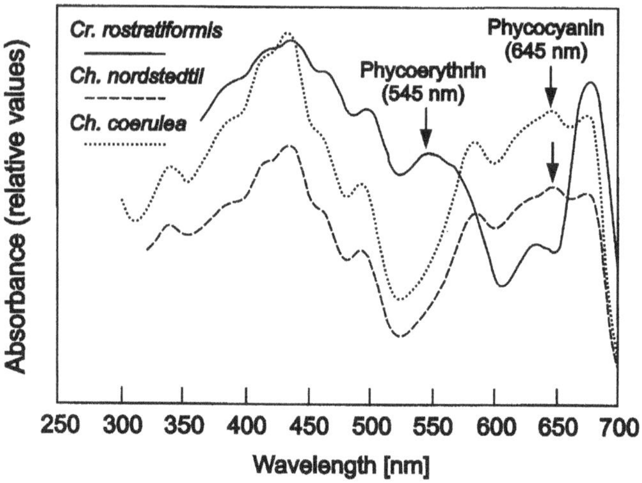

Figure 9. Absorption spectra of cell suspensions of *Cryptomonas rostratiformis*, *Chroomonas nordstedtii* and *Chroomonas coerulea* (form [4]).

observed above might be mediated by different photosensory systems.

The above results, taken together with the fact that DCMU, a potent inhibitor of photosynthetic electron flow from photosystem II (to which the phycobiliproteins supply photon energy as part of the antenna-pigment system) to photosystem I, did not block *Cryptomonas* phototaxis at 10–5 M whereas it completely blocked photosynthetic oxygen evolution [41], should reasonably rule out the possibility of involvement of phycobiliproteins as the photoreceptor pigment for cryptomonad phototaxis.

12.3.1.3 Comparison with Paramecium

The fact that the action spectra for photoaccumulation and light-induced membrane depolarization in *Paramecium*, a ciliate protozoon [42], appears strikingly similar to that for cryptomonad phototaxis [43] is of particular interest, considering the above-mentioned partially protozoic origin of cryptomonad: *Paramecium bursaria* shows photoaccumulation in a irradiated region. Intracellular recordings from both *Chlorella-*

Table 1. Phototactic response and other characteristics of cryptomonads (from [4])

	Cryptomonas rostratiformis	*Cryptomonas* CR-1[a]	*Chroomonas Nordstedtii*	*Chroomonas coerulea*
Phycobilin	phycoerythrin	phycoerythrin	phycocyanin	phycocyanin
Absorption max.	545 nm	565 nm	645 nm	645 nm
Eyespot	–	–	–	+
Phototaxis (sign)	positive	positive	positive	negative
Action max.	280, 560 nm	560 nm	560 nm	420 nm

[a] from [12]

containing and *Chlorella*-free cells showed that a step-increase in the light intensity induced a steady depolarization of membrane potential, and a step-down caused recovery to the original level. The action spectrum of the depolarization corresponded to that for photoaccumulation and showed two peaks at 420 and 560 nm (Figure 10a,b). The amplitude of the depolarization became larger as light intensity was increased.

The elucidation of the chemical nature of the hidden photoreceptors for phototaxis in both groups of organisms with a possible evolutionary relationship is an extremely interesting and challenging subject of study awaiting extensive future studies.

12.3.2 Photoreceptor localization

12.3.2.1 Species with and without eyespot

The photoreceptive site in *Chroomonas coerulea* is reasonably presumed to be somewhere near the eyespot. Structural and physiological comparisons between two species of *Chroomonas* with and without eyespot, i.e. *C. coerulea* and *C. nordstedtii*, would provide useful information on the photoreceptor localization in these two species of *Chroomonas* and also in other eyespot-less cryptomonad species such as those of *Cryptomonas*.

12.3.2.2 Cell rotation

A series of studies on phototaxis in *Cryptomonas* sp. CR-1 by Watanabe and Furuya [44,45], Uematsu-Kaneda and Furuya [46] and Kaneda and Furuya [47,48] have related the phase of cell rotation to the cell's sensitivity to the light stimulus, and these authors discussed the possible intracellular localization of the photoreceptive site.

Kaneda and Furuya [48] showed that light is sensed by the ventral side of *Cryptomonas* sp. CR-1 cells by infrared videomicrographical analysis of phototactic orientation of individual cells, rotating at a frequency of ca. 2 Hz, to repeated pulses of 20 μs duration at 500 ms intervals. Repeated flashes of light were provided with a combination of a strobe and an interference filter of ca. 570 nm transmission peak wavelength.

The light beam was perpendicular to the vertical side of the cuvette. The movement of individual cells in a flat, rectangular cuvette ($10 \times 10 \times 3$ mm) was recorded and measured by combination of infrared microscopic observation light, an infrared sensitive video camera mounted on a microscope, a time lapse video recorder, and a video monitor. *Cryptomonas* cells have an opening in a subapical position on the ventral side, from which the flagella emerge, so that the relationship between cell rotation and the helical path could be studied with relative ease (Figure 11).

Cryptomonas cells swim along a helix, their ventral side facing always toward the forward direction of the helix. The cell body always follows the flagella, so that it rotates once during each helical turn, as seen at high magnification Nomarski differential interference-contrast video-microscopy (Figure 11). The strobe was flashed at intervals of 500 ms since the percentage of cells rotating with a period of 475–525 ms was highest (ca. 40%). The position of each cell, at the time of the first flash, and the swimming track of the cell on the monitor TV screen, for a 3-s period from 1 s before the onset of the light stimuli until 2 s after their beginning, were photographed and analyzed.

The phase of a helical path at the time of the first flash was found to be crucial to the phototactic response of each cell. When only the cells that rotated with a period of 475 to 525 ms were selected and analysed, as many as 80% and 70% of the cells in phases

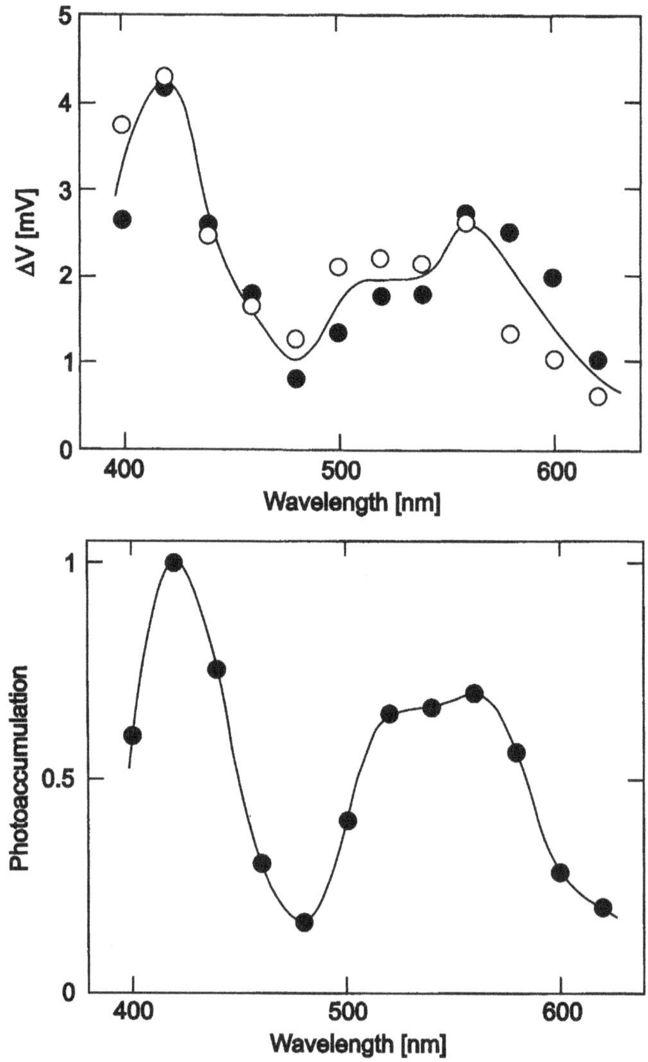

Figure 10. **a** Equal-fluence-rate action spectrum for light-induced depolarization of the cell membrane in the cilliate *Paramecium bursaria* (at 5 W m^{-2}). Eight light pulses of various wavelengths were applied and the amplitude of steady depolarization was measured. Resting potential was -27 ± 3 mV. Each point is the mean of three different specimens of *Chlorella*-containing (filled circles) and *Chlorella*-free (open circles). **b** Equal-fluence-rate action spectrum for photoaccumulation of the ciliate *Paramecium bursaria* (at 4 W m^{-2}). The number of accumulated cells was normalized with that at 420 nm. Accumulations are means of five measurements of Chlorella-containing cells. (from [42]).

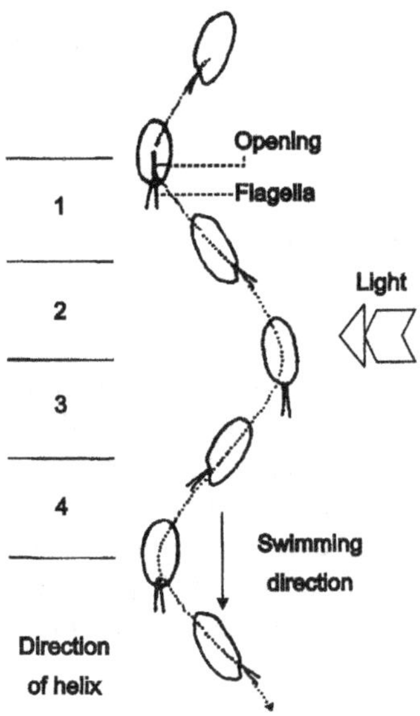

Figure 11. Schematic representation of the relationship between rotation and helical movement of a *Cryptomonas* sp. CR-1 cell. The cell has an opening in a subapical position on the ventral side, from which the flagella emerge. 1, 2, 3 and 4 (phases of rotation) represent quarters of a period (from [48]).

1 and 2 (Figure 11), respectively, showed a shift in the track toward the light source, whereas only 40% (about the same level of the dark control) of the cells either in phases 3 or 4 (Figure 11) showed such positive shift. This result clearly demonstrates that the cell must be irradiated during phases 1 or 2 of the rotation period for recognition of the direction of the stimulus light. As the light was incident on the ventral sides of the cells only during phases 1 and 2, the cells reoriented their courses toward the light source, predominantly, only when the ventral sides of the cells were exposed to the flashes of light. Taking this fact together with the observation by Kaneda and Furuya [47] that the anterior side of each cell must be illuminated for the recognition of the direction of stimulus light, it can be concluded that the cell can recognize the light direction only when the light is incident on the anterior ventral side of a cell.

The unidentified photoreceptor molecules may thus be located near the anterior ventral side of the cell (namely, near the opening), so it can absorb the light only when the light enters from the anterior ventral side. Each chloroplast which occupies a large volume of the *Cryptomonas* cell may absorb and attenuate light from directions other than the anterior ventral side of a cell.

12.4 Signal transduction

The signal transduction processes of phototaxis which lie between the light-requiring reaction of excitation of the photoreceptor molecules by the stimulus light and the final response of change in cell movement may be considered as dark reactions, possibly involving various second messengers such as calcium ions as well as specific regulatory proteins.

12.4.1 Temporal separation of light and dark reactions by intermittent light stimuli

Watanabe and Furuya [44] tried temporal separation of such light and dark reactions in the phototactic responses of *Cryptomonas* cells by using intermittent light stimuli with variable frequencies, light durations and dark intervals as well as fluence rates. When the dark interval was shorter than 60 ms, the so-called flicker fusion phenomenon was observed irrespective of the light duration: the phototactic responses of cell population were just linearly dependent on the logarithm of the total incident light fluence, with good reciprocity between duration and fluence rate. In contrast, when the dark interval exceeded 250 ms, the responses were remarkably reduced regardless of light duration and were not affected by increasing the fluence rate of the stimulus light pulses.

12.4.1.1 Frequency

Stimulus light of 0.5 W m^{-2} at 570 nm was chopped with a rotating sector at a frequency range between 0.125 and 32 Hz so that the duration of the light pulses was equal to that of the dark intervals. Responses of the cell population to the light pulses were compared with the response to continuous light of 0.25 W m^{-2}, the same total incident light fluence rate as that of the repeated stimuli. The data (Figure 12) clearly showed a frequency dependence of the phototactic response. The responses at 16 and 32 Hz were as large as those to the continuous stimulus, whereas those at low frequencies (2 and 1 Hz) were reduced to approximately 20% of that of the continuous stimulus. In the range between 0.5 and 0.125 Hz, the lower the frequency the larger were the responses. As the incident light fluence rate was equal in all cases, the diverse responses in Figure 12 are attributable to the different frequencies of the light-dark cycles.

12.4.1.2 Reciprocity

The effect of stimulus light dose on phototactic response was measured by varying the duration of light pulses (1, 2, 4 and 16 ms) in fixed cycles with a period of 32 ms. The data in Figure 13 show that the phototactic response is linearly dependent on the logarithm of the duration of the light pulses at this higher frequency.

The effects of varying the fluence rate and duration of each light pulse were then tested at different total energies and with various dark intervals, in fixed cycles with a period length of 32 ms. The phototactic responses correlated with the total doses given for a unit period of time, so that the Bunsen-Roscoe reciprocity law held.

12.4.1.3 Dark interval

Phototactic responses to repeated pulses of light with various durations (16 ms to 1 s) separated by dark intervals of different lengths (16 ms to 1 s) were measured to determine which of these two factor is essential in determining the level of phototactic responses (Figure 14).

All of the responses to repeated light pulses having 16-ms, 250-ms or 1-s duration were almost the same and were as large as that to the continuous light stimulus, provided that the dark interval was 250 ms or longer. It is evident that the length of the dark interval, rather than the duration of the light pulses, determines the phototactic response of cell populations of *Cryptomonas* sp. CR-1.

Next, it was examined whether there is any effect of fluence rate on the phototactic response to repeated light pulses of the lower frequency. Phototactic responses to light pulses of a constant frequency (about 1.85 Hz) (cycle length = 540 ms) with different dark intervals (0, 135, 270 or 405 ms) were measured using different fluence rates. Phototactic response to continuous light reached a plateau at the fluence rate of 0.17 W m^{-2} in this experiment. The reduction in phototactic responses with 270-ms and 405-ms dark intervals was not affected by increasing the fluence rate from 0.5 W m^{-2} to 2.0 W m^{-2}. Thus, the responses are essentially dependent on the length of dark interval, but not on the total incident light energy delivered per unit time at this lower frequency.

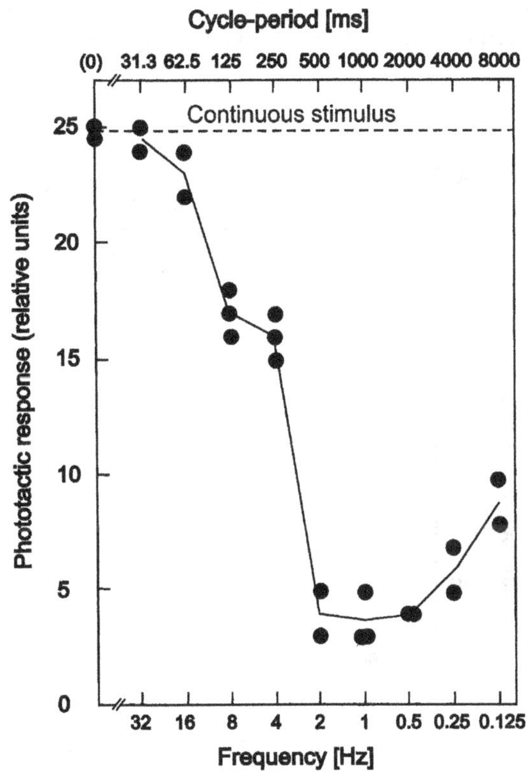

Figure 12. Effects of the frequency of repeated light pulses on the phototactic responses of cell populations of *Cryptomonas* sp. CR-1. Stimulus light of 0.5 W m^{-2} at 570 nm was chopped with a rotating sector so that the duration of the light pulses is equal to that of the dark intervals regardless of it frequency. Responses of the cell population to the light pulses were compared with the response to continuous light of 0.25 W m^{-2}, the total incident energy per unit time is the same as that of the repeated stimuli. (from [44]).

The phototactic response in *Cryptomonas* sp. CR-1 could thus be separated into two elementary processes, photoreception and the following dark reaction(s), using cyclic treatments consisting of light pulses and dark periods. In this respect, it resembles many other light-dependent processes, including photosynthesis [49], vision [50] and photomorphogenesis [51,52].

Although the presence of dark intervals of this critical length seemed to suggest some kind of decay of crucial biochemical or physico-chemical transient states in the photoreceptor-signal transduction processes, its identity has not adequately been addressed so far.

12.4.1.4 Dark interval in Chlamydomonas

In this connection, an extremely interesting and important photo-electrophysiological analysis was recently done by Yoshimura and Kamiya [53] on photoreceptor current (PRC) of the flagellate green alga *Chlamydomonas*, in which retinal-binding protein(s) are believed work as the photoreceptor for green-light sensitive phototaxis [33,54–57], in response to sinusoidal cyclic changes of the light fluence rate. The amplitude of PRC was highest at the frequencies of 1–5 Hz, overlapping the rotation frequency of the

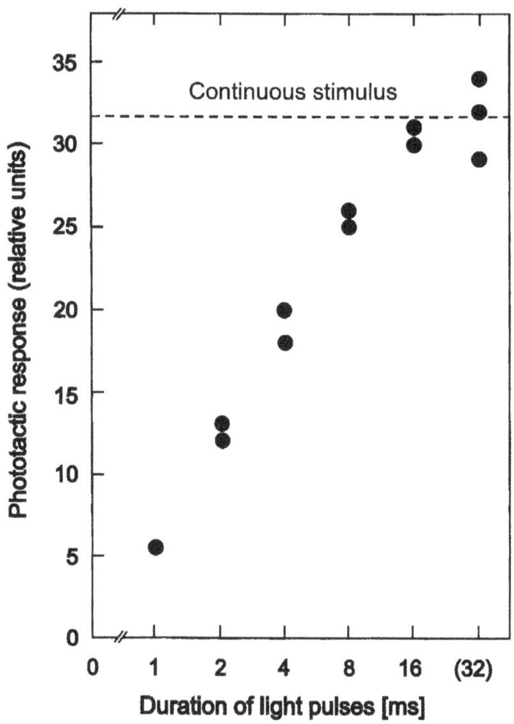

Figure 13. Effects of durations of light pulses at about 31 Hz (cycle length = 32 ms) on phototactic responses of *Cryptomonas* sp. CR-1. Stimulus light of 0.5 W m^{-2} at 570 nm was chopped with a rotating sector to give various combinations of durations of light pulses and dark interval. Each point shows the value of a single measurement (from [44]).

swimming cells. The PRC was generated only at the beginning of the light phase and soon decayed during the light phase. Further analysis indicated that desensitization, provably involving membrane depolarization [58], occurs within 10 ms after the onset of the light stimulus and has completely decayed ca. 1 s after the end of the light stimulus. The authors discuss that the inactivation and desensitization mechanism of PRC, providing the short PRC which is supposed to be important for the proper guidance of the swimming direction during phototaxis, optimizes the *Chlamydomonas* photoreceptor for the detection of a cyclic light change of 1–5 Hz caused by cell rotation.

Although the above analyses were done with an immotile mutant of *Chlamydomonas* with paralysed flagella for convenience of the electrophysiological measurements, the authors supposed that the degree of phototaxis should depend on the length of the dark period and not on that of the light period provided that the light period is long enough, provided their data on the PRC hold also in the behavioral response of swimming cells. Indeed, this is exactly what was observed by Watanabe and Furuya [44] for the phototaxis in *Cryptomonas* sp. CR-1! Unfortunately, the previous studies on the behavioral response of *Chlamydomonas* to repetitive flashes paid no attention to its dependency on the length of the light and dark period [59,60].

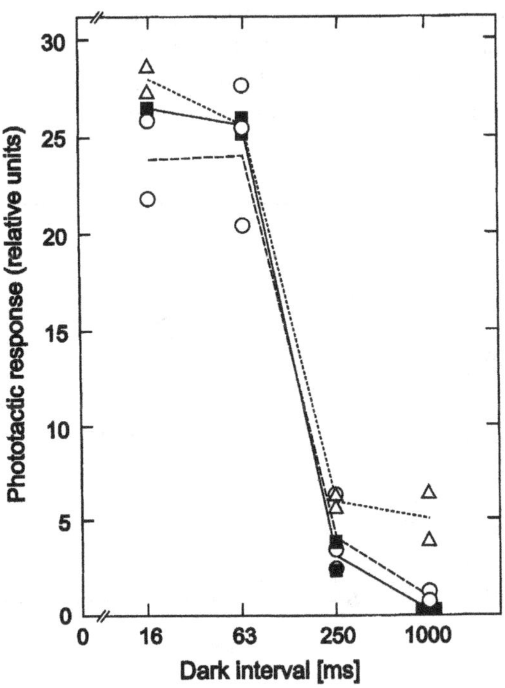

Figure 14. Effects of dark intervals on phototactic responses of cell populations in *Cryptomonas* sp. CR-1. Stimulus light of 0.5 W m^{-2} at 570 nm was chopped with a rotating sector to give various combinations of durations of light pulses and dark interval. Durations of the light pulses were 16 ms (circles), 250 ms (squares) or 1 s (triangles). Each point shows the value from a single measurement. (from [44]).

Such an obvious accordance between *Chlamydomonas* and *Cryptomonas* as to the dependence of phototaxis on the length of dark period and not on the length of the light period may indicate the existence of very similar photoreceptor(s) and signal transduction systems between these two phylogenetically and action spectroscopically very remote organisms.

Future molecular characterization of this inactivation process will provide invaluable insight into the mechanisms of phototactic light sensing not only in *Chlamydomonas* but also in other flagellate algae including cryptomonads.

12.4.2 Calcium ions

Uematsu-Kaneda and Furuya [75] and Kaneda and Furuya [76] examined effects of calcium and potassium ions on the phototaxis of *Cryptomonas* sp. CR-1 and obtained results consistent with those in other organisms such as *Chlamydomonas, Euglena* and *Paramecium*, the presence of Ca^{2+} being crucially needed for phototactic orientation and antagonized by K^+ ion.

Figure 15 shows the effects of the removal and addition of Ca^{2+} in the external medium on phototaxis in *Cryptomonas* CR-1 as measured at the population level using the photoelectrical measuring apparatus. Cells had been transferred from the growth medium to a basal medium that consisted of 5 mM MOPS (pH 7.0), 0.16 mM $MgCl_2$, 0.5 mM $CaCl_2$, 0.68 mM KCl and 0.16 mM NaCl, 4 h prior to the experiments. When calcium was removed from the medium by adding 1 mM EGTA, the phototactic

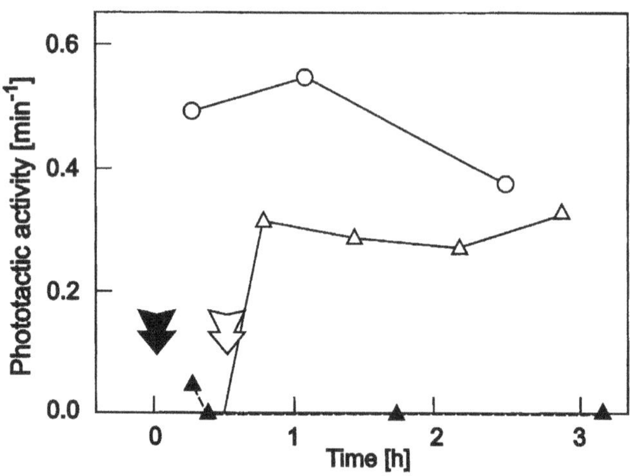

Figure 15. Effects of the removal and addition of Ca^{2+} on population-level phototaxis in *Cryptomonas* sp. CR-1. EGTA (in mM) was added to the basic medium at time 0 (black arrow), $CaCl_2$ (1 mM) was added 30 slater (white arrow), (open circles) the control in the basic medium. (closed triangles) time course in the presence of 1 mM EGTA. (open triangles) time course in the presence of both 1 mM EGTA and 1 mM $CaCl_2$. Each point is the value of a single measurement. (from [75]).

response was totally inhibited, but the swimming rate was not much affected. This effect of EGTA was partially reversed by the addition of 1 mM $CaCl_2$. When 15 mM KCl was added to the medium, phototaxis was greatly inhibited without significant influence on the swimming rate. The KCl-induced inhibition was partially removed by the addition of 15 mM $CaCl_2$ or $MgCl_2$.

The authors discuss the mechanism of this antagonistic effect of K^+ and Ca^{2+} on phototaxis in *Cryptomonas* and speculated that metal ions may alter the condition of the membrane as reported in squid giant axon [61] and in *Chara* cells [62]. In this connection, it is noteworthy that in the ciliate *Paramecium* increased KCl acts as a depolarizing stimulus that results in a transient increased conductance to Ca^{2+}, which produces an influx of Ca^{2+}, thus resulting in the increased concentration of Ca^{2+} in the cilia and cortex, activating the mechanism for ciliary reversal [63]. The same authors [76] later clearly demonstrated at the individual cell level that Ca^{2+} ions are essential for phototactic orientation (Figure 16a) and the change in swimming direction (Figure 16b) of the same organisms.

12.5 Ecological significance of cryptomonad phototaxis

12.5.1 Light habitats and UV avoidance

The ecological significance of algal phototaxis is generally considered as the means for the cells to locate themselves in photosynthetically advantageous light habitats and to avoid harmful strong light or UV light both horizontally and vertically [64–66].

A demonstrative field work in this connection was done by Smolander and Arvola [67] on seasonal variation in the diurnal vertical distribution of *Cryptomonas marssonii* in a small, highly humic lake in Finland during a summer season (between May and September) using a close-interval Blaker-type sampler. The results indicated that the cells were phototactic; they were typically concentrated at the surface or subsurface during daylight, whereas in darkness the highest cell concentrations were recorded in deeper water, usually near the upper limit of anoxia. During a dense cyanobacterial bloom in August the cells of *Cryptomonas marssonii* were also concentrated by day into the same water layer, where oxygen was depleted. However, the cells seemed to avoid totally anoxic water.

Damaging effects of solar and artificial UV light to the cells' motility, photo-orientation and pigmentation of *Cryptomonas* are well documented [68–73]. For example, in the freshwater *Cryptomonas* sp. S-2, the diaphototactic orientation is impaired within about 90 min of solar radiation. The percentage of motile cells and the average velocity of the swimming cells decreased within about the same exposure time. Since removing short wavelength UV light by means of an artificial ozone filter or UV cut-off filters increased the tolerated exposure time, the damaging effect seems to be caused by the solar UV-B component. Solar radiation also bleached the photosynthetic pigments of the cells as shown by absorption difference spectra [68]. The same authors later confirmed this conclusion by experiments with artificial UV-B radiation of 50 mW m^{-2} from a transilluminator [69].

Erata et al. [4] recognized a UV action peak (at 280 nm) in the action spectrum for positive phototaxis in *Cryptomonas rostratiformis* (Figure 6). Such intriguing apparent

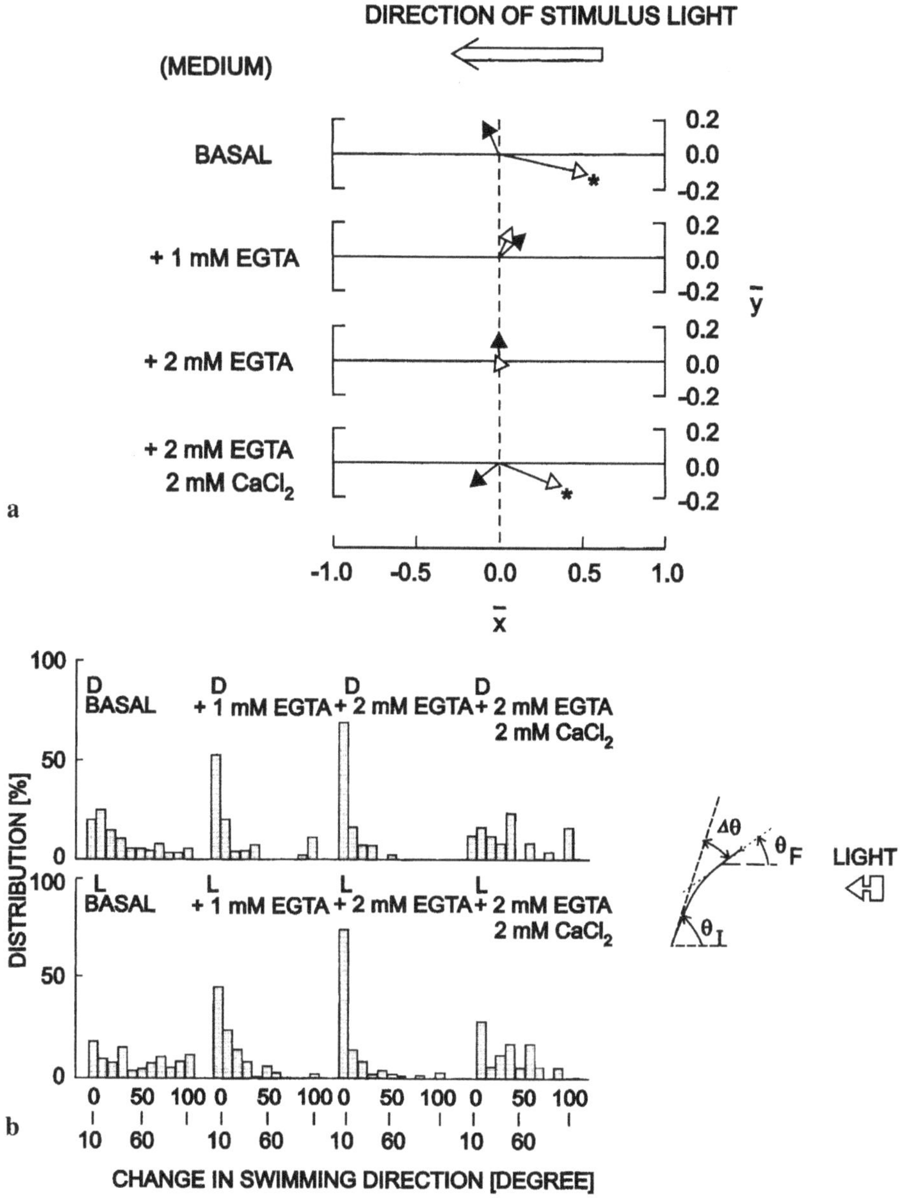

Figure 16. Effect of removal of Ca^{2+} by EGTA on the phototactic behavior of individual cells of *Cryptomonas* sp. CR-1. **a** Effect on the swimming direction of laterally illuminated cells. Each arrow represents the mean vector from 40 measurements. Black-headed arrow: samples not irradiated; white-headed arrow: those irradiated with the stimulus light; *: the result showing significant orientation toward the light source ($p < 0.05$). **b** Effect on the degree of change in swimming direction. Swimming tracks of cells for a 6-s period (schematically represented by the long arrow) were recorded. The absolute value of change in swimming direction ($\delta\phi$) of 50 cells was measured in each case. The distribution of changes is presented at intervals of 10°. D not irradiate. L irradiated with the stimulus light (from [76]).

suicide responses to UV radiation have been commonly observed in various flagellate algae such as the green alga *Platymonas* (= *Tetraselmis*) *subcordiformis* (at 270 nm) [22], the dinoflagellates *Gymnodinium splendens* (at 280 nm) [20], *Scrippsiella hexapraecingula, Peridinium foleaceum, Alexandrium hiranoi* and *Gymnodinium mikimotoi* (260–280 nm) [21], and the euglenoid *Euglena gracilis* (at 270 nm) [18], the ecological significance of these positive phototaxis toward UV light source remaining for future elucidation. In this connection it is noteworthy that a sensory perception of UV-B radiation by the ciliate *Blepharisma japonicum* is also reported [74].

12.6 Future prospects

As seen from the overview above, the most crucial, interesting and rewarding subjects of future studies are undoubtedly molecular identification, localization and signal transduction of the photoreceptor pigment(s). Extensive endeavors taking full advantage of the recent molecular biological and spectroscopic methodologies are awaited.

References

1. C. van den Hoek, D.G. Mann, H.M. Jahns (1995). *Algae: An Introduction to Phycology* (pp. 235–243). Cambridge University Press, Cambridge.
2. P. Kugrens (1999). Cryptomonad systematics – an algal enigma? In: J. Seckbach (Ed.), *Enigmatic Microorganisms and Life in Extreme Environments* (pp. 127–138). Kluwer Academic, Dordrecht.
3. M. Erata, M. Chihara (1987). Cryptomonads from the Sugadaira-moor, central Japan. *Bull. Sugadaira Mont. Res., Univ. Tsukuba*, **8**, 57–69.
4. M. Erata, M. Kubota, T. Takahashi, I. Inouye, M. Watanabe (1995). Utrastructure and phototactic action spectra of two genera of cryptophyte flagellate algae, *Cryptomonas* and *Chroomonas*. *Protoplasma*, **188**, 258–266.
5. E. Gantt (1979). Phycobiliproteins of Cryptophyceae. In: M. Levandowsky, S.H. Hunter (Eds), *Biochemistry and Physiology of Protozoa* (Vol. 1, pp. 121–137). Academic Press, New York.
6. S.E. Douglas, C.A. Murphy, D.F. Spencer, M.W. Gray (1991). Cryptomonad algae are evolutionary chimeras of two phylogenetically distinct unicellular eukaryotes. *Nature*, **350**, 148–151.
7. B. Marin, M. Klingberg, M. Melkonian (1998). Phylogenetic relationships among the Cryptophyta: analyses of nuclear-encoded SSU rRNA sequences support the monophyly of extant plastid-containing lineages. *Protist*, **149**, 265–276.
8. G.I. McFadden (1990). Evidence that cryptomonad chloroplasts evolved from photosynthetic eukaryotic endosymbionts. *J. Cell Sci.*, **95**, 303–308.
9. J.D. Watson, M. Gilman, J. Witkowsky, M. Zoller (1992). *Recombinant DNA* (pp. 447–448). Freeman, New York.
10. G.I. McFadden, P.R. Gilson, D.R.A. Hill (1994). *Goniomonas*: rRNA sequences indicate that this phagotrophic flagellate is a close relative of the host component of cryptomonads. *Eur. J. Phycol.*, **29**, 29–32.
11. D.-P. Häder, E. Rhiel, W. Wehrmeyer (1987). Phototaxis in the marine flagellate *Cryptomonas maculata*. *J. Photochem. Photobiol. B: Biol.*, **1**, 115–122.
12. M. Watanabe, M. Furuya (1974). Action spectrum of phototaxis in a cryptomonad alga, *Cryptomonas* sp. *Plant Cell Physiol.*, **15**, 413–420.

13. D.-P. Häder, E. Rhiel, W. Wehrmeyer (1988). Ecological consequences of photomovement and photobleaching in the marine flagellate *Cryptomonas maculata*. *FEMS Microbiol. Ecol.*, **53**, 9–18.
14. E. Rhiel, D.-P. Häder, W. Wehrmeyer (1988). Photoorientation in a freshwater *Cryptomonas*. species. *J. Photochem. Photobiol. B: Biol.*, **2**, 123–132.
15. E. Rhiel, D.-P. Häder, W. Wehrmeyer (1988). Diaphototaxis and gravitaxis in a freshwater *Cryptomonas*. *Plant Cell Physiol.*, **29**, 755–760.
16. B. Diehn (1969). Action spectra of the phototactic responses in *Euglena*. *Biochim. Biophys. Acta*, **177**, 136–143.
17. A. Checcucci, G. Colombetti, R. Ferrara, F. Lenci (1976). Action spectra for photoaccumulation of green and colorless *Euglena*: evidence for identification of receptor pigments. *Photochem. Photobiol.*, **23**, 51–54.
18. S. Matsunaga, T. Hori, T. Takahashi, M. Kubota, M. Watanabe, K. Okamoto, K. Masuda, M. Sugai (1998). Discovery of signaling effect of UV-B/C light in the extended UV-A/blue-type action spectra for step-down and step-up photophobic responses in the unicellular flagellate alga *Euglena gracilis*. *Protoplasma*, **201**, 45–52.
19. P. Halldal (1958). Action spectra of phototaxis and related problems in Volvocales, *Ulva*-gametes and Dinophyceae. *Physiol. Plant.*, **11**, 118–153.
20. R.B. Forward (1974). Phototaxis by the dinoflagellate *Gymnodinium splendens Lebour.J. Protozool.*, **21**, 312–315.
21. T. Horiguchi, H. Kawai, M. Kubota, T. Takahashi, M. Watanabe (1999). Phototactic responses of four marine dinoflagellates with different types of eyespot and chloroplast. *Phycol. Res.*, **47**, 101–107.
22. P. Halldal (1961). Ultraviolet action spectra of positive and negative phototaxis in *Platymonas subcordiformis*. *Physiol. Plant.*, **14**, 133–139.
23. W. Nultsch, G. Throm, I. Rimscha (1971). Phototaktische Untersuchungen an *Chlamydomonas reinhardii* Dangeard in homokontinuierlicher Kultur. *Arch. Mikrobiol.*, **80**, 351–369.
24. R. Wayne, A. Kadota, M. Watanabe, M. Furuya (1991). Photomovement in *Dunaliella salina*: fluence rate-response curves and action spectra. *Planta*, **184**, 515–524.
25. S.-M. Liu, D.-P. Häder, W. Ullrich (1990). Photoorientation in the freshwater dinoflagellate, *Peridinium gatunense* Nygaard. *FEMS Microbiol.*, **73**, 91–102.
26. W. Nultsch (1962). Phototaktische Aktionsspektren von Cyanophyceen. *Ber. Dt. Bot. Ges.*, **75**, 443–453.
27. M. Watanabe, M. Furuya, M. Miyoshi, Y. Inoue, I. Iwahashi, K. Matsumoto (1982). Design and performance of the Okazaki Large Spectrograph for photobiological. research. *Photochem. Photobiol.*, **36**, 491–498.
28. M. Watanabe (1991). High-fluence rate monochromatic light sources, computerized analysis of cell movements, and microbeam irradiation of a moving cell – current experimental methodology at the Okazaki Large Spectrograph. In: G. Colombetti, F. Lenci, D.-P. Häder, P.-S. Song (Eds), *Biophysics of Photoreceptors and Photomovements of Microorganisms* (pp. 327–337). Plenum, New York.
29. D.R.A. Hill, K.S. Rowan (1989). The biliproteins of the Cryptophyceae. *Phycologia*, **28**, 455–463.
30. M. Mimuro, N. Tamai, A. Murakami, M. Watanabe, M. Erata, M.M. Watanabe, M. Takutomi, I. Yamazaki (1998). Multiple pathways of excitation energy flow in the photosynthetic pigment system of a cryptophyte, *Cryptomonas* sp. (CR-1). *Phycol. Res.*, **46**, 155–164.
31. M. Melkonian, H. Robenek (1984). The eyespot apparatus of flagellated green algae: a critical review. In: F.E. Round, D.J. Chapman (Eds), *Progress in Phycological Research* (Vol. 3, pp. 193–268). Biopress, Bristol.

32. T. Takahashi, Y. Kobatake (1982). Computer-linked automated method for measurement of the reversal frequency in phototaxis of *Halobacterium halobium*. *Cell Struct. Funct.*, **7**, 183–192.
33. T. Takahashi, K. Yoshihara, M. Watanabe, M. Kubota, R. Johnson, F. Derguini, K. Nakanishi (1991). Photoisomerization of retinal at 13-ene is important for phototaxis of *Chlamydomonas reinhardtii*: simultaneous measurements of phototactic and photophobic responses. *Biochem. Biophys. Res. Commun.*, **178**, 1273–1279.
34. T. Takahashi, M. Kubota, M. Watanabe, K. Yoshihara, F. Derguini, K. Nakanishi (1992). Diversion of the sign phototaxis in a *Chlamydomonas reinhardtii* mutant incorporated with retinal and its analogs. *FEBS Lett.*, **314**, 275–279.
35. K.W. Foster, R.D. Smyth (1980). Light antennas in phototactic algae. *Microbiol. Rev.*, **44**, 572–630.
36. M. Erata, M. Chihara (1989). Re-examination of *Pyrenomonas* and *Rhodomonas* (Class Cryptophyceae) through ultrastructural survey of red pigmented cryptomonads. *Bot. Mag. Tokyo*, **102**, 429–443.
37. H. Kawai, D.G. Müller, E. Folster, D.-P. Häder (1990). Phototactic response in the gametes of the brown alga, *Ectocarpus siliculosus*. *Planta*, **182**, 292–297.
38. G. Kreimer, M. Melkonian (1990). Reflection confocal laser scanning microscopy of eyespots in flagellated green algae. *Eur. J. Cell Biol.*, **53**, 101–111.
39. G. Kreimer, U. Brohnsonn, M. Melkonian (1991). Isolation and partial characterization of the photoreceptive organelle for phototaxis of a flagellate green alga. *Eur. J. Cell Biol.*, **55**, 318–327.
40. G. Kreimer, H. Kawai, D.G. Müller, M. Melkonian (1991). Reflective properties of the stigma in male gametes of *Ectocarpus siliculosus* (Phaeophyceae) studied by confocal laser scanning microscopy. *J. Phycol.*, **27**, 268–276.
41. M. Watanabe, Y. Miyoshi, M. Furuya (1976). Phototaxis in *Cryptomonas* sp. under condition suppressing photosynthesis. *Plant Cell Physiol.*, **17**, 683–690.
42. K. Matsuoka, Y. Nakaoka (1988). Photoreceptor potential causing phototaxis of *Paramecium bursaria*. *J. Exp. Biol.*, **137**, 477–485.
43. M. Watanabe (1995). Action spectroscopy – photomovement and photomorphogenesis spectra. In: B. Horspool, P.-S. Song (Eds), *CRC Handbook of Organic Photochemistry and Photobiology* (pp. 1276–1288). CRC Press, Boca Raton.
44. M. Watanabe, M. Furuya (1978). Phototactic responses of cell population to repeated pulses of yellow light in a phytoflagellate *Cryptomonas* sp. *Plant Physiol.*, **61**, 816–818.
45. M. Watanabe, M. Furuya (1982). Phototactic behavior of individual cells of *Cryptomonas* sp. in response to continuous and intermittent light stimuli. *Photochem. Photobiol.*, **35**, 559–563.
46. H. Uematsu-Kaneda, M. Furuya (1982). Effects of viscosity on phototactic movement and period of cell rotation in *Cryptomonas* sp. *Physiol. Plant.*, **56**, 194–198.
47. H. Kaneda, M. Furuya (1986). Temporal changes in swimming direction during the phototactic orientation of individual cells in *Cryptomonas* sp. *Plant Cell Physiol.*, **27**, 265–271.
48. H. Kaneda, M. Furuya (1987). Effects of the timing of flashes of light during the course of cellular rotation on phototactic orientation of individual cells of *Cryptomonas*. *Plant Physiol.*, **84**, 178–181.
49. R. Emerson, W. Arnold (1932). Separation of the reactions in photosynthesis by means of intermittent light. *J. Gen. Physiol.*, **15**, 391–420.
50. W.A. van de Grind, O.-J. Grusser, H.-U. Lunkenheimer (1973). Temporal transfer properties of the afferent visual system. In: R. Jung (Ed.), *Handbook of Sensory Physiology* (Vol. VII/3, pp. 431–573). Springer, Berlin.

51. Y. Inoue, M. Furuya (1975). Perithecial formation in *Gelasinospora reticulispora*. IV. Action spectra for the photoinduction. *Plant Physiol.*, **55**, 1098–1101.
52. M. Wada, M. Furuya (1974). Action spectrum for the timing of photo-induced cell division in *Adiantum* gametophytes. *Physiol. Plant.*, **32**, 377–381.
53. K. Yoshimura, R. Kamiya (2001). The sensitivity of *Chlamydomonas* photoreceptor is optimized for the frequency of cell-body rotation. *Plant Cell Physiol.* (in press).
54. P. Hegemann, W. Gärtner, R. Uhl (1991). All-trans retinal constitutes the functional chromophore in *Chlamydomonas* rhodopsin. *Biophys. J.*, **60**, 1477–1489.
55. M.A. Lawson, D.N. Zacks, F. Derguini, K. Nakanishi, J.L. Spudich (1991). Retinal analog restoration of photophobic responses in a blind *Chlamydomonas reinhardtii* mutant. Evidence for an archaebacterial-like chromophore in a eukaryotic rhodopsin. *Biophys. J.*, **60**, 1490–1498.
56. G. Kreimer (1994). Cell biology of phototaxis in flagellated algae. *Int. Rev. Cytol.*, **148**, 229–310.
57. P. Hegemann (1997). Vision in microalgae. *Planta*, **203**, 265–274.
58. E.G. Guvorunva, O.A. Sineshchekov, P. Hegemann (1997). Desensitization and dark recovery of the photoreceptor current in *Chlamydomonas reinhardtii*. *Plant Physiol.*, **115**, 633–642.
59. M.E. Feinleib (1975). Phototactic response of *Chlamydomonas* to flashes of light.I. Response of cell populations. *Photochem. Photobiol.*, **21**, 351–354.
60. J.S. Boskov, M.E. Feinleib (1979). Phototactic response of *Chlamydomonas* to flashes of light – II. Response of individual cells. *Photochem. Photobiol.*, **30**, 499–505.
61. I. Tasaki, T. Takenaka, S. Yamagishi (1968). Abrupt depolarization and bi-ionic action potentials in internally perfused squid giant axons. *Amer. J. Physiol.*, **215**, 152–159.
62. T. Shimmen, M. Kikuyama, M. Tazawa (1976). Demonstration of two stable potential states of plasmalemma of *Chara* without tonoplast. *J. Membrane Biol.*, **30**, 249–270.
63. R. Eckert (1972). Bioelectric control of ciliary activity. *Science*, **176**, 473–481.
64. R.B. Forward (1976). Light and diurnal vertical migration: photobehavior and photophysiology of plankton. In: K.C. Smith (Ed.), *Photochemical and Photobiological Reviews* (Vol. 1, pp. 157–209). Plenum, New York.
65. D.-P. Häder (1979). Photomovement. In: W. Haupt, M.E. Feinleib (Eds), *Physiology of Movements. Encyclopedia of Plant Physiology, New Series* (Vol. 7, pp. 268–309). Springer, Berlin.
66. D.-P. Häder (1991). Effects of enhanced solar ultraviolet radiation on aquatic ecosystems. In: G. Colombetti, F. Lenci, D.-P. Häder, P.-S. Song (Eds), *Biophysics of Photoreceptors and Photomovements of Microorganisms* (pp. 157–172). Plenum, New York.
67. U. Smolander, L. Arvola (1988). Seasonal variation in the diel vertical distribution of the migratory alga *Cryptomonas marssonii* (Cryptophyceae) in a small, highly humic lake. *Hydrobiol.*, **161**, 89–98.
68. D.-P. Häder, M. Häder (1989). Effects of solar radiation on photoorientation, motility and pigmentation in a freshwater *Cryptomonas*. *Bot. Acta*, **102**, 236–240.
69. D.-P. Häder, M. Häder (1990). Effects of UV radiation on motility, photo-orientation and pigmentation in a freshwater *Cryptomonas*. *J. Photochem. Photobiol. B: Biol.*, **5**, 105–114.
70. D.-P. Häder, M. Häder (1991). Effects of solar and artificial U.V. radiation on motility and pigmentation in the marine *Cryptomonas maculata*. *Environ. Exp. Bot.*, **31**, 33–41.
71. I. Zündorf, D.-P. Häder (1991). Biochemical and spectroscopic analysis of UV effects in the marine flagellate *Cryptomonas maculata*. *Arch. Microbiol.*, **156**, 405–411.
72. S. Gerber, D.-P. Häder (1993). Effects of solar irradiation on motility and pigmentation of three species of phytoplankton. *Environ. Exp. Bot.*, **33**, 515–521.

73. S. Gerber, D.-P. Häder (1995). Effects of enhanced solar irradiation on chlorophyll fluorescence and photosynthetic oxygen production of five species of phytoplankton. *FEMS Microbiol. Ecol.*, **16**, 33–42.
74. F. Lenci, G. Checcucci, F. Ghetti, D. Gioffre, A. Sgarbossa (1997). Sensory perception and transduction of UV-B radiation by the ciliate *Blepharisma japonicum*. *Biochim. Biophys. Acta*, **1336**, 23–27.
75. H. Uematsu-Kaneda, M. Furya (1982). Effects of calcium and potassium ions on phototaxis in *Cryptomonas*. *Plant Cell Physiol.*, **23**, 1377.
76. H. Kaneda, M. Furya (1987). Effect of calcium ions on phototactic orientation of individual *Cryptomonas* cells. *Plant Sci.*, **48**, 31–35.

Photomovement
D.-P. Häder and M. Lebert, editors.

Chapter 13

Photo-stimulated effects on diatom motility

Stanley A. Cohn

Table of contents

Abstract

Like many unicellular algae, benthic diatoms have active motility to help regulate their access to necessary resources such as light. However, diatom cells are constrained within a hardened silicate-based cell wall, which restricts their movement to a characteristic gliding motion along the surface to which they are attached. The restriction on movement results in a motile system with severe spatial constraints in changing direction and orientation. This chapter outlines the historical work relating to the investigations on the generation and regulation of diatom movement, and in particular how this movement is affected by light (such as the photo-based response of generating direction changes at light/dark boundaries). Additionally, this article discusses recent work on the comparative differences between the motile and adhesive abilities of different diatom species, and how light and other environmental influences affect them. The environmental regulation of motility is crucial to the ecological success of diatoms as it influences both the cells' access to light as well as helps determine the segregation of diatom species into microniches within the algal and benthic communities. Diatom motility affects both the accessibility of cells to nutrients and light as well as the exposure of cells to loss due to predation or physical disturbance. Light-based motile responses of diatoms are thus likely to be fundamental components of the cells' strategies for long-term viability in aquatic ecosystems.

13.1 Introduction

Like any motile organism, diatoms must have the mechanisms to make their movements directed and productive, and use the movement to gain some ecological advantage. Since diatoms are photosynthetic, they must acquire light as their primary energy resource, so it is not surprising that they have developed light-based mechanisms to control their orientation and movement. However, unlike most motile protists, diatom cells are constrained within a relatively rigid silicate-based cell wall that prohibits generation of the cell membrane protrusions and extensions required for either ciliary or amoeboid motility. This means that they locally are restricted to a relatively two-dimensional plane (like amoebae), but that unlike amoebae they lack the ability for essentially omnidirectional reorientation within that plane. Instead they must rely upon a secondary system of cellular secretions for providing the connections between the cell and the underlying substratum. The location and movement of these secretions are restricted by the physical structure of the cell wall components and regulated by as yet unknown photodetectors in the cell.

13.2 Overview of diatom motility

Diatoms are a unique form of beautiful and intricately shaped algae, known for their characteristic golden pigmentation and hardened silica-based cell walls [1]. There are two major groups of diatoms: pennates, elongated forms that are usually sediment dwelling and often motile, and centrics, cells that are cylindrically symmetrical, often planktonic and generally non-motile. Significant bi-directional movements are generally restricted to those pennates in which a characteristic slit, known as the raphe, is present in the cell wall (Figure 1). The forms of pennate diatoms are plentiful, with hundreds of known genera described [2,3]. Each of these forms of pennate diatoms have distinctive shapes of the raphe fissure [4], and characteristic orientations and paths of cell movement [5,6]. Many such pennates have been observed to move relatively rapidly (ca. 1–20 μm/s), with movements that are responsive to environmental signals. While a few centric and araphid diatoms have been observed to have motility, the movement is slow and less responsive, and appears to be solely due to the force of mucilage secretion through pores [7,8].

The graceful and active gliding movements of raphid pennate diatoms have been observed and studied for over 100 years [9–13]. The early investigators were quite aware of the hardened cell wall surrounding the algae and knew that it must place some constraints on their movement. While the resolution of the early microscopes was excellent, most cytology was performed on fixed and stained specimens. The lack of modern optical methods such as phase or differential interference contrast meant that the ability to detect cellular structures was somewhat limited for live moving specimens, particularly relatively transparent cells such as diatoms. It is therefore not surprising that early hypotheses for the movement of diatoms included protoplasmic streaming, flagellar beating, propulsion of water, or even pressure due to evolution of oxygen gas. The modern development of high-resolution light and electron microscopy has rendered such models obsolete. Electron microscopy has confirmed that the entire cell protoplast is confined within the hardened cell wall; no direct membrane extensions or protrusions outside the cell wall (as required for either amoeboid or ciliary movement) have yet been observed [5,14].

Numerous early observations on diatoms have proved to be quite useful, and often involved detailed descriptions of attached ink, dye, or other particles to diatoms and determination of their positions during cell movement (see [5] for discussion). Many of these observations showed that the movements of these extracellular particles were localized in the area of the raphe, whose shape and location is characteristic for each species. However, the accompanying proposals for protoplasmic extensions through the raphe were purely circumstantial, and there is currently no evidence that mature raphes contain any cytoplasmic extensions to the exterior of the cell. Nonetheless, these early observations clearly showed the importance of the raphe in the ability of the diatom in making contacts with external surfaces, and in determining the general speeds and attributes of raphe-based movements. Current observations using video-microscopy of fluorescent beads attached to diatoms (e.g. [15]) validate the raphe as a primary site of cell/substratum attachment.

More recent microscopic observations show that the movement of diatoms is generally smooth and bidirectional along the apical axis of the cell [16], although the orientation of the path varies considerably between species [5,6,17] and may be related to the raphe shape. Upon closer observation, the relatively smooth gliding observed at lower magnifications is actually made up of a number of more saltatory movements (e.g. [18]), and the diatom is observed to frequently change direction along its path and reverse itself [5]. While the orientations of paths traveled generally have a consistent angle of curvature [6], cells can be observed to have relatively frequent reorientations

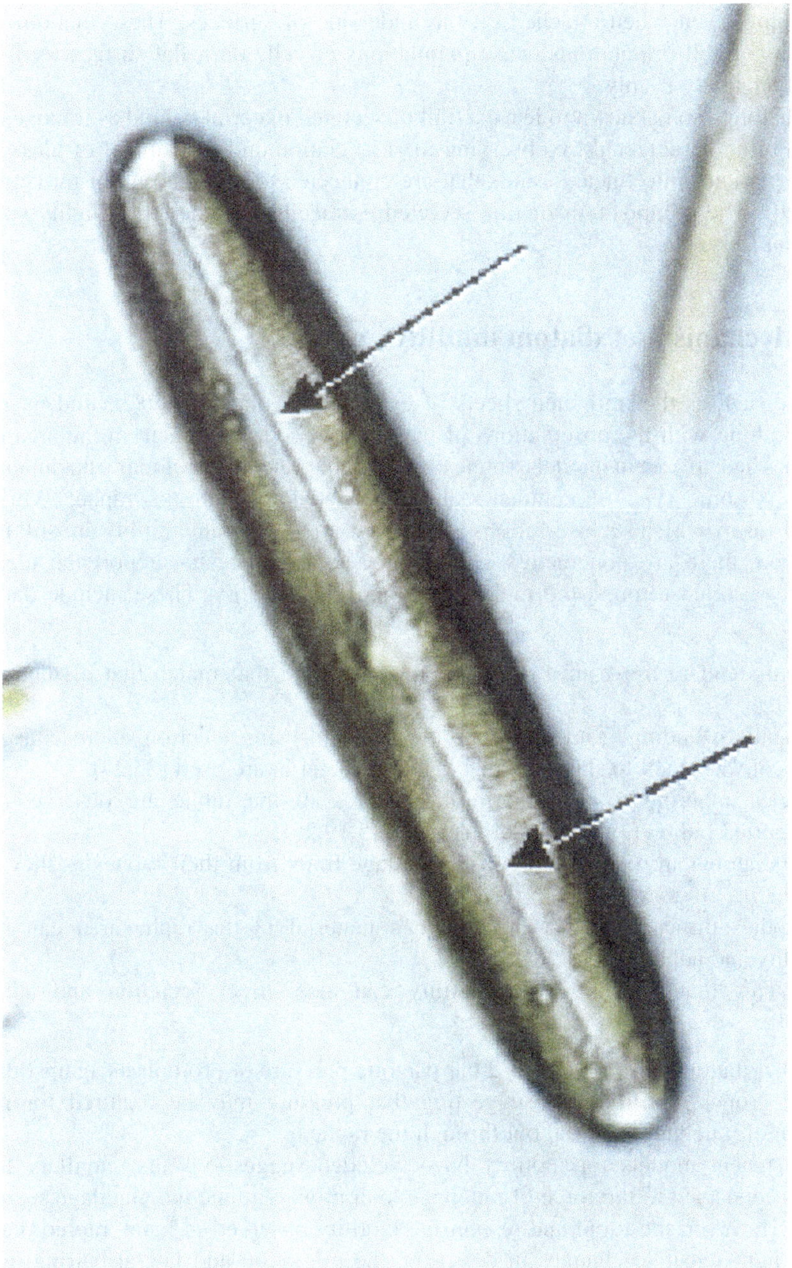

Figure 1. Light micrograph of *Pinnularia viridis* displaying the two slits in the cell wall (raphe branches, marked with arrows) through which mucilage material involved in generating cell movement is thought to be generated. Cell is approximately 150 μm in length.

and flipping, even when attached on the underside of surfaces. These random minor alterations of cell orientations cause populations of cells on a flat surface to disperse outward relatively evenly.

The diatom also is known to leave a trail of secreted material behind as it moves. This trail can often be detected by observing the association and attachment of glass beads or other particles into linear strands that are connected to the rear end of moving cells (e.g. [19]). The composition of this secreted extracellular material is highly species-dependent [20–22].

13.3 Mechanism of diatom motility

Evidence such as the trails and sheets of extracellular material left behind by motile diatoms, along with the observations of particle and cell attachment in the area of the raphe, has led to recent models suggesting that the agent of cellular attachment and motility is some type of cellular material secreted through the raphe. While the mechanisms by which such secretions would be used to generate motility are still poorly understood, there are nonetheless several lines of evidence to support the idea that mucilage secreted through the raphe is required for motility. These include data that indicate:

1. diatoms tend to move in a direction and curvature that match that of their raphe [5,17],
2. filaments extending from the raphe are detected using electron microscopy when protocols designed to stabilize mucilaginous material are used [23,24],
3. particles adhering to the diatom in the area of the raphe are observed to be transported bidirectionally along the raphe [5,19,25],
4. some diatoms are known to deposit mucilage trails from their raphes as they move [5,19,26],
5. antibodies directed against proteoglycan material in the raphe area can inhibit motility and adhesion [27,28], and
6. chemicals that interfere with motility can also affect secretion and adhesion [29,30].

Freshwater diatoms also require 1–2 bar osmotic pressure of protoplasts against the cell wall for proper motility [18] suggesting that pressure may be required for proper extrusion of mucilage material out through the raphe.

Some recent models for motility have included suggestions that capillary forces, possibly connected to the force of mucilage hydration, regulate the mucilage secretions [31,32]. However, the rapid and responsive motility observed in many raphid diatoms, such as light-sensitive changes in direction, the migration and lateral pairing of cells prior to sexual reproduction, and the ability of cells to pull themselves up from single points of contact [2,6] would seem to make such a mechanism unlikely.

One of the primary models developed over recent years centers on the possibility that the mucilage may be connected and actively transported along actin cables directly underneath the raphe. Under this model, the motile force could be produced by translocating membrane-bound attachment sites for the mucilage strands along the underlying actin filaments, effectively moving the mucilage strands down the length of the cell. If the mucilage strands were attached to the substratum, this would generate force between the substratum and the cell, thus pulling the cell along [24,33]. This movement would be similar to cytoplasmic streaming of membrane-bound vesicles or chloroplasts along actin cables in higher plant cells [34,35]. The actual motion would likely be generated by a force-producing "motor" protein, such as the myosins known to be present in many animal and algal cell types [36–38], some of which have known membrane-binding domains.

Bidirectional movement could be generated by the use of anti-parallel actin filaments, either by having actin filaments of each orientation contained within each actin cable, or by having each actin cable composed of uniformly oriented actin filaments with each cable being in an opposite orientation. Myosin has also been shown to be capable of directing movement along actin filaments in either direction [39,88], raising the possibility that any actin filament could be sufficient to support bidirectional movement. While actin-based motility for diatoms has not been unequivocally proven, latrunculin, a potent actin inhibitor from sea sponge, causes rapid and reversible inhibition of diatom motility [30,89], further supporting the primary importance of actin cables for motility. However, the direct connection of extracellular mucilage with the actin cables is not clear, as attached beads can move down the raphe and cross the central areas of cells where no raphe branch exists [15], and cells can retain substratum connections while flipping from one side of the cell to the other [6,40]. Rather than individual distinct actin-connected and membrane-associated mucilage strands generating the motility, the actin may instead be generating a flow of secreted raphe-based mucilage along the membrane. Thus, the mucilage would generally flow down the raphe canals, but could be pushed across the small areas of the cell wall (e.g. the central area and the tips of cells) where no raphe branch actually exists, by the force of the mucilage flow. In this way, cells may at times have a band of mucilage material that functionally surrounds the cell, providing a medium for cell movement.

13.4 Photo-regulated motility

In general, bacteria, algae and other motile protists are thought to generate chemotactic and phototactic movements by one of several general mechanisms [41,42] which include: regulation of the frequency of changes in cell direction (including changes at light/dark boundaries), changes in cell speed (photokinesis), or changes in cell orientation [43–45]. Functionally, these processes involve a diverse set of processes such as regulating the direction, orientation or speed of flagellar beating, control of cell secretions and attachments, modulating areas of cytoplasmic viscosity, or coordinating intracellular cytoskeletal connections with the membrane. However, because motile benthic diatoms require direct attachment with the substratum in order to move, they are restricted locally to movement in a two-dimensional plane. For structural reasons explained previously, diatom movement is predominantly constrained to forward and backward directions of movement, with the direction and orientation of movement related to the shape of their raphe [5,17].

Virtually all the current evidence suggests that motile pennate diatoms regulate their motility by detecting boundaries where changes in light intensity occur, and changing their direction of movement accordingly. Such responses are known as photophobic responses and can be stimulated by either increases (step-up) or decreases (step-down) in light intensity. Since the movement of diatoms is so closely connected to their substratum connections via mucilage, it is not yet clear whether the direction changes are generated by control over cell secretions (and their cellular attachments), rearrangements or reconnections to cytoplasmic cytoskeletal components, or both. Whatever the mechanism, diatoms show clear photophobic responses, shifting diatoms to move into areas of moderate light at light/dark boundaries, and shifting cells away from areas of very high intensity light [6,46]. In contrast, cells do not appear to actively alter their orientation or average speed with respect to light intensity or wavelength, and seem to rely primarily on regulating the frequency of direction change for generating their phototactic ability. As described below, the wavelengths and intensities responsible for generating these effects appear to be strongly species-specific.

Light regulation of some form is clearly evident in diatoms, and can be demonstrated by a number of observed photo-sensitive diatom motile behaviors. Diatoms are known to exhibit diurnal behaviors in which they surface in the sediment during early portions of the day and settle back down into the sediment at night (e.g. [47,48]). While this migratory behavior in the sediment is linked to diurnal light levels, the cells do appear to be able to retain diurnal migration patterns in the lab for several days, even under conditions of constant illumination. Diatoms also have been observed to accumulate in areas of moderate light and avoid areas of shade or high intensity light, which may be partially responsible for the movement of some diurnally migrating species back into the sediment hours before sunset.

Despite their ecological importance and widespread abundance, relatively little is known about the physiology behind diatoms' phototactic abilities. Groundbreaking studies in the understanding of the photo-based effects on diatom movements were carried out by Nultsch [49] who determined several important diatom behavioral responses. Using a modified microscope capable of exposing a sample of diatoms to a small area of light and independently measuring cell accumulation via spectrophotometric measurement of cell density, he measured a number of physical characteristics relating to light spot accumulation for the diatom *Nitzschia communis*. He determined an action spectrum for cell accumulation as a function of wavelength and light intensity, and showed how these behaviors can have characteristic peaks of activity. For example, the work indicated that the diatoms displayed a maximum sensitivity for light accumulation at light intensities in the range of 100–500 lx (ca. 2–10 μmol m^{-2} s^{-1}) and wavelengths around 370 nm. In addition, the data suggested that individual cells could be repelled by red light, a behavior that has been subsequently confirmed for other diatom species [6]. However, Nultsch's microscope set-up was designed to observe mass cell accumulation (density) in a defined area, and so was unable to analyze the behavior of individual cells.

Subsequent studies by Nultsch and Wenderoth [26,50–54] were designed to analyze the behavior of individual cells (*Navicula perigrina*) in more detail. Using a microscope in which a mask of defined area could be used to irradiate the cells, they were able to view the behavior of individual diatoms as they were exposed with light on different areas of the cell. They were able to show that irradiation at one end of a diatom with a moderate light level of 1000 lx (approximately 15–20 μmol m^{-2} s^{-1} for white light) will stimulate the cell to move in the direction of irradiation, and that diatoms can change direction based on the relative light exposure at the two ends. They found that moderate light irradiation of the trailing end of a cell caused a rapid change of direction, while irradiation at the leading end inhibited the cell from changing direction. Whole cell irradiations or irradiations of the center of the cell had little effect in stimulating directed cell movement. They speculated from these results that the cell moves in the direction of a detected light gradient. That is, they suggested that the cell determines which end is being exposed with more light, and generates movement in that direction.

13.5 Recent assays for photo-based movements in diatoms

Until recently, little had been done to build on Nultsch and Wenderoth's work. Our lab has recently undertaken a number of experiments to further understand the process of photo-stimulated movements in diatoms, the way in which these behaviors are species-specific, and how they contribute to the overall success of diatom communities. Our goal has been to understand several aspects of photo-based movements: the intensity and wavelength of light required to generate photo-responses; the differences in sensitivities between different species; and the effect of inter-species competition on the ability of cells to accumulate in light spots. In the process of undertaking these investigations, we have developed five types of light assays of diatom movement: the light boundary assay, the cell accumulation assay, the spot irradiation assay, the avoidance assay and the vertical migration assay. These assays, some of which were developed from previous microscopic motility assays [55–57] are all designed around the ability to microscopically observe individual cells, with each assay designed to help elucidate a different attribute of the motile process. In addition we have analyzed four different species of diatoms (*Craticula cuspidata*, *Nitzschia linearis*, *Stauroneis phoenicenteron*, and *Pinnularia viridis*) each with a distinctive set of motile characteristics. The information we have gained from each of these types of experiments is described below.

Light Boundary Assay. This assay is designed to help us determine the ability of cells to detect and respond directly to light/dark boundaries, i.e. the photophobic response. Generally we are assaying for the step-down photophobic response whereby cells change direction at a light/dark boundary and move back into the light. In the assay, we place individual cells into a circular field of light of defined size and intensity [6]. Typically we have a light spot with a diameter 3 to 5 times greater than the length of the cell we are working with (e.g. about a 350 μm diameter spot for *Craticula* cells which are about 100 μm long). Individual cells (usually dark adapted for 30 minutes prior to treatment) are placed into the center of the spot and allowed to move. Upon reaching the light/dark boundary, the cell is scored for its response as to whether or not it changes direction within one cell length of the boundary. A number of individual cells are measured (in our tests about 10–20 cells per data point), and the resulting percentage of cells responding is determined (Figure 2). Using a variety of filters we can alter the light intensity and wavelength, then compare the responses under different conditions and determine the attributes and thresholds of wavelength or light intensity that are required for the cell to respond to a boundary.

Because cells have a natural tendency to change direction occasionally, controls for this type of assay include some type of measurement of the basal rate of cell direction change for each wavelength/intensity condition being studied. In order to determine this basal rate of direction change for each of our treatment conditions, dark-adapted cells were placed in the center of an open field of light containing no light/dark boundary and allowed to start moving. The percentage of cells changing direction within once cell length of a "mock" boundary drawn only on the video monitor was then determined. The functional or net photo-based response for cells was then considered to be the frequency of cells reversing at the real light/dark boundary minus the frequency of cells reversing at the mock boundary.

What we discovered with this assay was not only a confirmation of the earlier work suggesting that a photophobic response appears to be the primary behavior responsible for generating diatom phototaxis, but that there were species specific responses [6] in light sensitivity. Three of the cell types (*Craticula cuspidata*, *Nitzschia linearis* and *Pinnularia viridis*) showed the same general trend, with a maximal step-down response

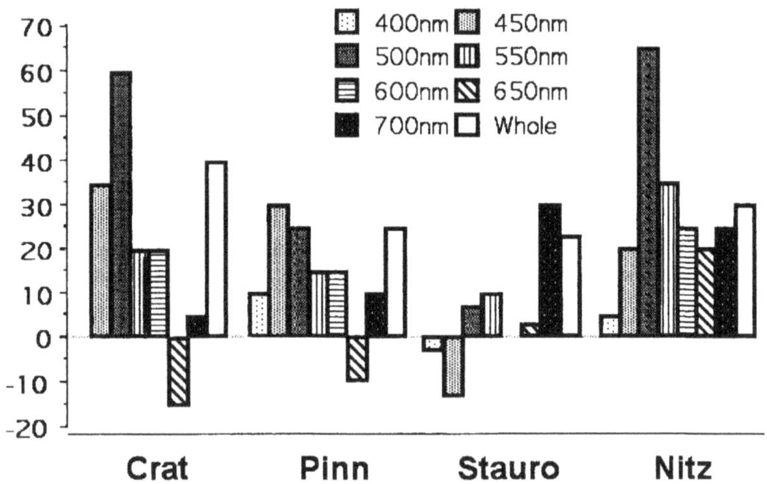

Figure 2. Wavelength-dependent sensitivity of four diatom cell types to direction reversal at light/dark boundaries. Cells were placed in the center of a 350–360 μm wide diameter spot of light at ca. 1 μmol m^{-2} s^{-1} irradiance. Cells were exposed with light wavelengths of 400 nm, 450 nm, 500 nm, 550 nm, 600 nm, 650 nm, or 700 nm generated by interference filters (shaded bars as per legend) or with full spectrum 100 W tungsten light (open bars). Moving cells were then scored to determine the percentage of cells which changed direction within one cell length of the light/dark boundary, moving back into the light. The control (no boundary) rate of direction reversal was determined by measuring the percentage reversing direction within one cell length of an artificial boundary drawn on an acetate film or the video screen representing the same spot size. This graph represents the net sensitivity of each of the cell types (Crat = *Craticula cuspidata*; Pinn = *Pinnularia viridis*; Stauro = *Stauroneis phoenicenteron*; Nitz = *Nitzschia linearis*) to the light/dark boundary test at each of the measured wavelengths (i.e. percent of cells changing direction at the actual boundary minus the percent of cells changing direction at the mock boundary). Note that the *S. phoenicenteron* cells have a distinctly different spectral response from that of the other cells (reproduced by permission of J. Phycol.).

at 450–500 nm blue-green light, while another species, *Stauroneis phoenicenteron*, showed a maximal response in the red range, at about 650 nm. In addition, the results suggested that *Craticula* and *Pinnularia* were somewhat repelled by red light, changing direction less frequently at the boundary than control cells (as suggested by Nultsch's earlier work) while *Stauroneis* was somewhat repelled by green light wavelengths around 400–450 nm.

By analyzing the actual light spot and the mock spot responses separately, we observed that the net response for cells was due to a combination of marked decreases in the basal rate of direction change at the responsive wavelengths, along with increased sensitivities to the light/dark boundary. That is, at the responsive wavelengths of light the basal underlying frequency of direction change is reduced, while the cells become more responsive to light/dark boundaries. The phototactic responses seem to be almost solely due to changes in the regulation of direction change, as no substantial wavelength-dependent changes in cell speed were observed.

We have also used a modification of the microscope assay to test for any irradiance-dependent photokinesis in the diatoms. In this test, the cells were placed into the center of a field of view with a wide open field diaphragm (i.e. no boundary) and measured the average speed of cell movement as a function of light irradiation. We detected no significant change in cell speed as a function of light irradiance within the range of 1 to 50 μmol s^{-1} m^{-2}, indicating that photokinesis is unlikely to be a strong modulator of diatom photo-accumulation.

Cell accumulation assay. This assay is designed to measure the functional ability for cell populations to accumulate in areas of increased light intensity. In contrast to the Nultsch studies, our assay measures the actual number of individual cells accumulating in a light area, and not the overall cell density via light absorption or scattering. Our assay consists of placing a culture of cells with a relatively uniform density on a small Petri plate. The plate is completely blackened to inhibit light exposure except for one small spot (usually 7 mm diameter) in the bottom of the Petri plate. A small spot of paper of the same size, which acts as a control area, is sealed onto the bottom of the Petri plate prior to the blackening of the outside of the plate. The Petri plate containing the cell culture is then illuminated from below with light of a known wavelength and intensity. The number of cells within the area of the light spot is counted (by transiently removing the top of the Petri plate) and compared to the number of cells in the unilluminated spot on the Petri plate (Figures 3, 4).

We have been able to use this assay to measure the rate at which different diatom species accumulate in the light spot under different conditions. We have found a number of interesting results of this work:

1. as with many of our other assays, the responses are species-specific, showing distinct activities for different species,
2. as suggested by the boundary assay work, light wavelength can directly cause the redistribution of diatom species, and
3. there may be competitive effects of species and density contribution in cell response.

The species-specific responses show up in investigations of both light intensity and of light wavelengths. In investigating four different species, we have found that each one has a characteristic response in terms of light sensitivity and wavelength (Figure 5). For example, we find in general that *Stauroneis phoenicenteron* cells accumulate in the light spots at a faster rate as light intensity increases. However, for two other species, *Nitzschia linearis* and *Craticula cuspidata*, the activity for the accumulation light response levels off in the range of 20–50 μmol m^{-2} s^{-1}. The fourth species, *Pinnularia viridis*, shows little ability to accumulate into light spots at any level of light intensity, a condition most likely due to its circular path of movement [6].

The four species also show distinct responses as to wavelength preferences. When we use filters to mimic a somewhat broader spectral range of light, more similar to what they might receive in a natural setting, we observe that at lower light levels (1–10 μmol m^{-2} s^{-1}) *Craticula* accumulates much more in green light than does *Stauroneis*, while the opposite is true in red light (although *Stauroneis* does accumulate somewhat in green light as the light intensity increases). *Nitzschia* and *Stauroneis* cells also appear to be quite sensitive to broad band blue light. While *Craticula* shows virtually no accumulation in red light at any intensity, *Nitzschia* accumulates well at

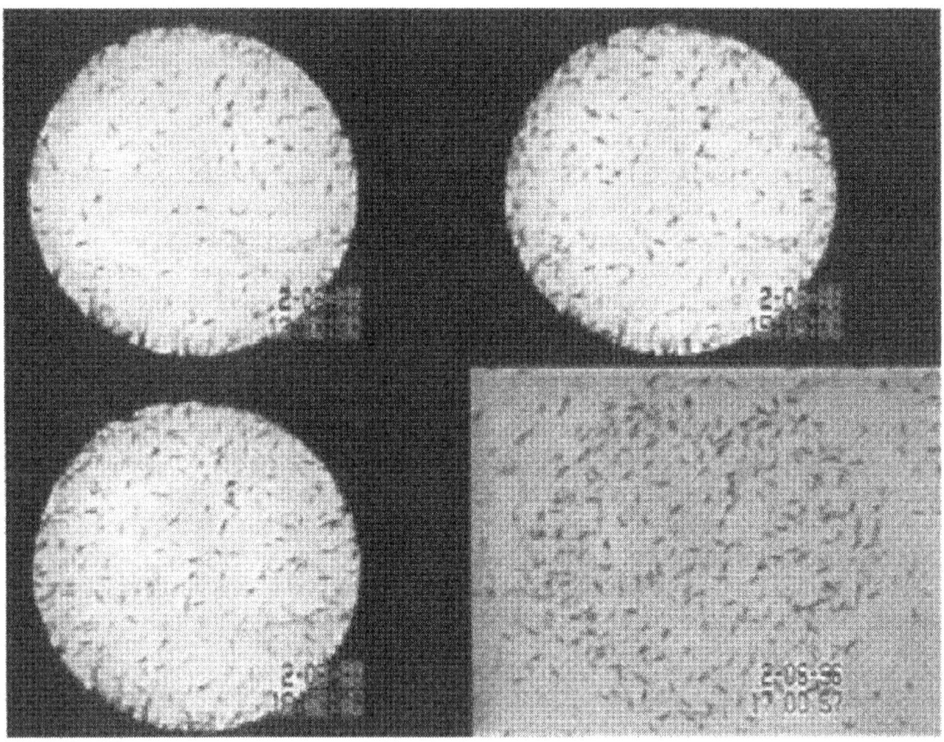

Figure 3. Time-lapse sequence of accumulation of *Craticula cuspidata* cells in a light spot. A culture of *C. cuspidata* cells was exposed with a 7 mm light spot, starting at the upper left frame. Cell densities observably increased after 2 hours (upper right panel) and 4 hours (lower left panel). Removal of the spot mask after 4 hours (lower right panel) reveals that cell density in the area of the former spot is greater than the surrounding area. Note in the last panel that cells are particularly aggregated near the former periphery of the light spot.

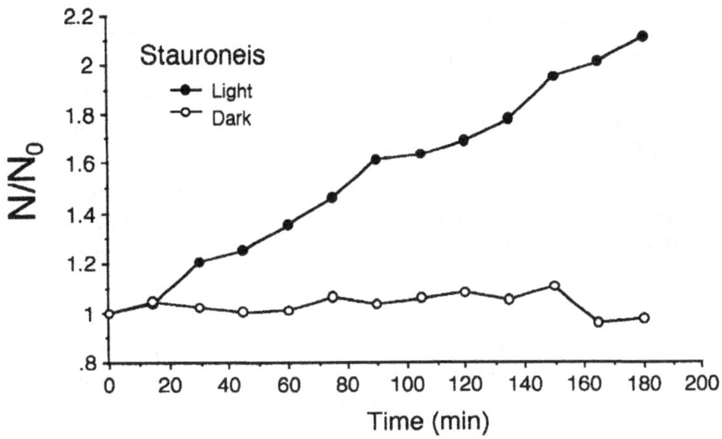

Figure 4. Cell accumulation in irradiated or unirradiated spots. Analysis of cell accumulation over time using assays as in Figure 3 can be used to determine the rates of cell accumulation. This graph represents the relative number of cells present in a 7 mm spot of full spectrum light of moderate irradiance (approx. 25 μmol m^{-2} s^{-1}; solid circles) or an unirradiated spot of the same size shielded from the light (open circles). Note that the slope of the accumulation under these conditions is essentially linear.

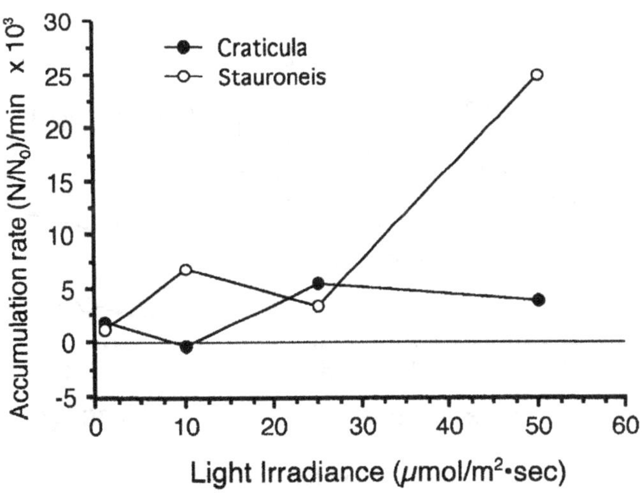

Figure 5. Rates of cell accumulation versus intensity for *Stauroneis phoenicenteron* and *Craticula cuspidata*. Analysis of the slopes from experiments as in Figure 4 were used to determine the rates of cell accumulation at different intensities of light for two species of diatoms. While *C. cuspidata* cells seem to plateau in sensitivity at about 20 μmol m^{-2} s^{-1}, *S. phoenicenteron* rates of accumulation continue to increase with light intensity.

both the red and blue ranges of light at higher light intensities. All of these results have correlated well with the results of our boundary assay studies.

Because of the species-specific differences in wavelength response, we have been able to use our accumulation assay to show that light irradiation can also be used to shift the distribution of species in an area (Figure 6). Using a modified version of our accumulation assay, we irradiated two identically sized spots on the same Petri plate with light of equal irradiance levels, but different wavelengths. Using a plate containing a mixed population of *Craticula cuspidata* and *Stauroneis phoenicenteron*, and irradiating the plate with one red spot and one green spot, we discovered that the relative abundance of *C. cuspidata* to *S. phoenicenteron* shifted dramatically in the two spots, with the *Craticula* accumulating in the green spot and *Stauroneis* accumulating in the red spot [58]. We are also in the initial stages of investigating the effect of cell-cell interactions on the rates of accumulation into the light spots by changing the density and species composition of the initial populations. Further work in this area will allow us to determine if the presence of certain cell species inhibits or enhances the ability of cells to accumulate into the light spots.

The cell accumulation assay has also allowed us to directly develop a mathematical model describing the rate of cell accumulation into light spots. By using three spots of different sizes on the same Petri plate, we have found that the rates of cell accumulation into the three spots differs significantly, with the rate of cell accumulation being inversely proportional to the radius of the spot (Figure 7). This result matches nicely with the idea that accumulation is due solely to the cells biasing their direction at the light/dark boundaries. Under such a model, the rate of cell increase into a spot (in cells/min) is proportional to the perimeter (and hence the radius) of the spot. However, because the area of the spot increases as the square of the radius, the larger the spot, the

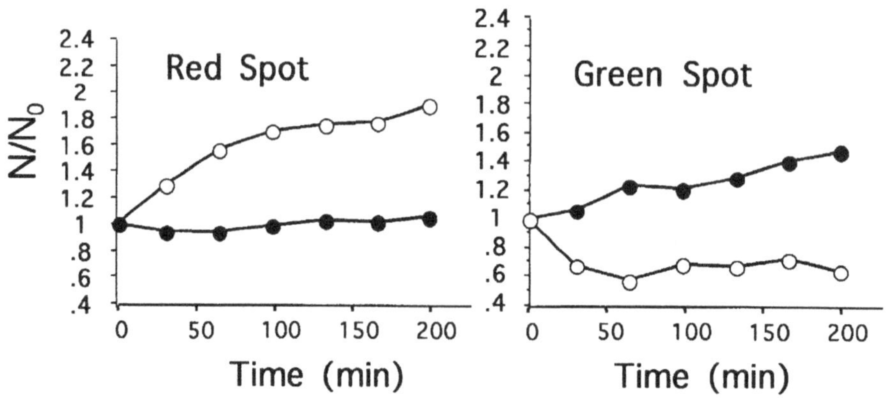

Figure 6. Comparison of cell accumulation for *Stauroneis phoenicenteron* and *Craticula cuspidata* in light spots at two different wavelengths. Areas of a mixed culture of *S. phoenicenteron* and *C. cuspidata* cells were exposed with two separate light spots, one with broad wavelength red light and the other with broad wavelength green light, both at irradiances of 5 μmol m^{-2} s^{-1}. *C. cuspidata* cells accumulated in the green spot but had almost no accumulation in the red spot. In contrast, *S. phoenicenteron* cells accumulated well in the red spot, but actually reduced in number within the green spot.

slower the increase in cell density (in cells/mm^2 per min). Under such a model, the rate of cell density increase within a spot can be given by the equation

$$C_{in} = \left(1 - \frac{k_1}{k_2}\right) C_{out} e^{-(2k_2/r)t} + \frac{k_1}{k_2} C_{out} \tag{13.1}$$

where C_{in} and C_{out} are the densities of cells inside and outside of the spot, k_1 is the rate constant for cells moving into a spot, k_2 is the rate constant for cells moving out of a spot, r is the radius of the spot, and t is time. Under this model, there are distinct parameters, the rate constants, that can be determined for each species under particular light conditions (unpublished, manuscript in preparation), using the fact that initial rates of accumulation are proportional to $k_1 - k_2$ and that cells reach a maximum density in the spot that is proportional to k_1/k_2. This model also predicts that the larger the spot, the longer it will take for the cells to accumulate into the spot to some maximal density, and that the time it takes to reach 1/2 the maximal density should be directly proportional to the radius of the spot.

Spot irradiation and avoidance assays. In the spot irradiation assay we used a technique to irradiate diatoms with high intensity light (ca. 500–2000 μmol m^{-2} s^{-1}), in

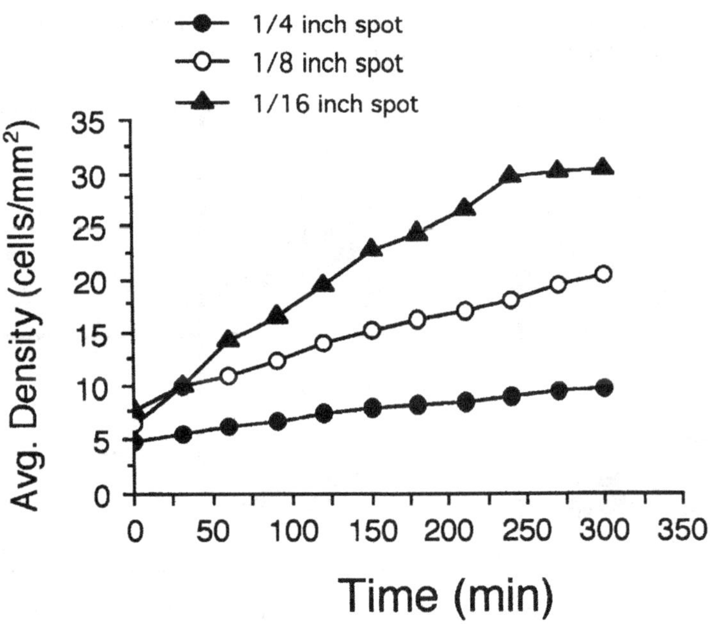

Figure 7. Comparison of cell accumulation for *Craticula cuspidata* cells in light spots at three different diameters. The same culture of *C. cuspidata* cells was exposed with full spectrum light at 30 μmol m^{-2} s^{-1} at three different locations, each with a different spot size (as noted in legend). Density of cells increased fastest in the spot with the smallest diameter. This correlates well with the mathematical model for cell accumulation, described in the text, where initial rates of density increase are inversely proportional to the spot diameter.

which we could precisely define the wavelength and spot size of the irradiation. This set-up, adapted from a previously designed epi-illumination irradiation set-up in the lab of Dr. J.D. Pickett-Heaps [59,60], allowed us to test the effect of irradiating specific locations on diatoms, as well as determining the relative effect of irradiations at different wavelengths. While similar to some types of experiments done previously [50,51,53], we were able to define the area of irradiation much more precisely. In addition, our work was with the high intensity photophobic response (i.e. step-up, out of the light) rather than the previously investigated step-down, into the light, response, allowing the two types of responses to be directly compared.

Our results showed a dramatic effect. With 100% reproducibility, *Craticula* cells irradiated on their leading end with high intensities of light (>500 μmol m^{-2} s^{-1} and exposures of ca. 1 s) quickly reversed direction (Figure 8). In contrast, cells exposed similarly at their trailing end never showed any change of direction, compared to 11% of control cells that changed direction. Analysis of the efficiency of each wavelength to stimulate direction change determined that the most efficient wavelength examined for this step-down response (i.e. 500 nm) is similar to the most responsive wavelength in our light/dark boundary assay. The similarity in the wavelength sensitivities suggests that the same light receptor or process may be involved in both the step-up and the step-down responses [46].

Multiple exposures on cells also showed that a first irradiation tended to inhibit the effect of a second irradiation. That is, an irradiation on the leading end of a cell caused a direction change, but appeared to partially inhibit the ability of a second irradiation at

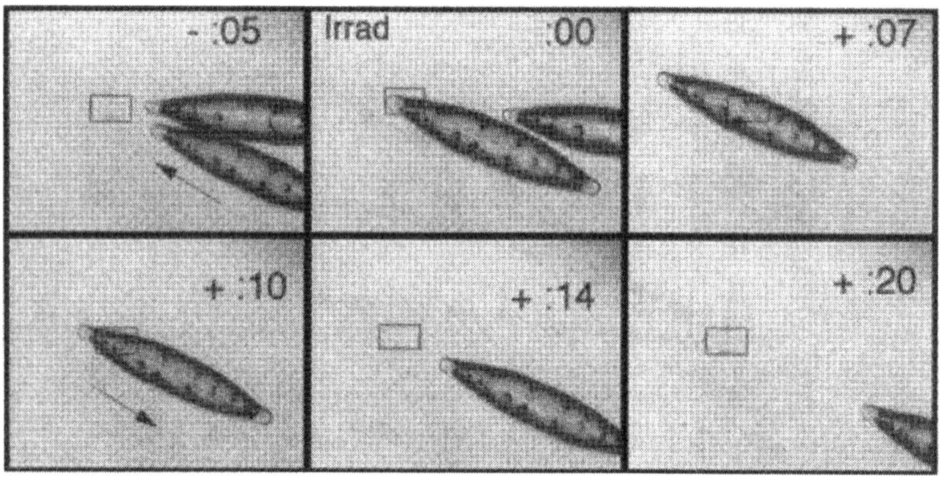

Figure 8. Time-lapse video sequence of high energy irradiation at the tips of *Craticula cuspidata* cells. The time relative to the irradiation (in seconds) is marked in the upper right of each panel; the site of the irradiation on the slide is marked with a box and placed in all the panels for relative position of cells. Five seconds prior to irradiation the cell is moving to the left toward the irradiation site. When the tip of the cell enters the irradiation site, it is exposed, using a timed shutter, with one second of very high intensity (ca. 1000 μmol m^{-2} s^{-1}) light. By seven seconds after irradiation the cell stops moving, and by ten seconds after irradiation the cell has changed direction. It continues moving in the opposite direction until it is out of the field of view.

the new leading end to stimulate another direction change. Similarly, an initial irradiation at the trailing end of a cell seemed to inhibit the stimulation of a direction change by a subsequent irradiation on the leading end. These effects appeared to be dependent on the relative durations of the first and second exposures, suggesting that light irradiation may be generating some type of secondary messenger molecule (e.g. [Ca^{2+}]) whose effects must be overcome by the second irradiation. Such regulation by second messengers is a common feature of many receptor-mediated motile processes [34,45,61] and there are indications that calcium may be involved in the diatom's ability to detect several types of environmental signals [90].

In order to better localize the area of the cell that is sensitive to the light exposure, we irradiated *Craticula cuspidata* cells with a small irradiation spot (ca. 17×20 μm) at several locations on the cell. Irradiation at the tips has 100% effectiveness in generating a direction change while irradiation at the edge (where large amounts of the chloroplasts are located) or the middle of the cell had little effect. These results suggest that the site of the photoreceptors is likely at the tips of these cells, near the ends of the raphe branches. Since the tips of these cells contain little or no chloroplasts, and because the most sensitive frequencies for the direction change are off of the maximal absorption peaks for the chlorophylls, we suspect that the receptors are not located in the chloroplasts themselves.

We have recently modified this assay to develop an avoidance response on a standard upright video-microscope, whereby a microscopic field is concurrently illuminated with low level light from below, and irradiated with a fixed spot of high irradiance from above using epi-illumination. The area of the epi-illumination spot is determined using a slide containing a film of fluorescent material or beads, and then marked directly over the video monitor, or superimposed on the video image using a computer. Individual cells can be placed in the field of view (outside of the high-energy irradiation spot) and allowed to move toward the high-energy spot, in much the same was as our boundary assay. We can then score the frequency of cells changing direction when encountering the high irradiance illumination, and compare it to cells passing through the same area when the epi-illumination is blocked. This set-up can generate higher irradiance than the controlled spot apparatus, so that by using neutral density filters and interference filters we can create epi-irradiation spots of 10 to 10^6 μmol m^{-2} s^{-1} at a variety of wavelengths.

By selectively altering the wavelength or intensity of the epi-illumination, we can determine the light characteristics required to stimulate an avoidance response, and directly compare the step-up response with the step-down response to better determine if they are generated by the same mechanisms in the cell. Initial tests on irradiance levels have indicated that a threshold light irradiance around 500–1000 μmol m^{-2} s^{-1} is required to stimulate the avoidance response for both blue (around 450 nm wavelength) and green (around 550 nm) irradiations.

Vertical migration studies. We have recently begun some studies assessing the ability of different types of cells to undergo light-stimulated vertical migrations. In these studies, cells are layered on the bottom of Petri dishes that have been blackened on the bottoms and sides to eliminate light exposure. The cells are then overlaid with two thin sheets of all cotton muslin, which act as a natural fibrous substrate over which the cells can migrate. The plates are then placed in a lighted incubator; since the bottoms and sides of the plates are black, the cells are effectively illuminated only from above. The number of cells present at each of the layers is then counted over time to determine the relative ability of the cells to migrate upward through the muslin material. This assay allows us to compare the abilities of different cells to move through a substrate and compare this to their photo-sensitivities as measured by the other three assays. In this way we can begin to assess the contribution of adhesion and cell shape to the ability of cells to undergo light-directed movement through a substrate.

While these studies are preliminary, it appears that the long and slender *Nitzschia* is the best at moving upward through the substrate, suggesting that its small cross-sectional area and long raphe length to cell volume ratio may aid in its ability to move through fibrous material. Such studies may provide an excellent model system, as many developing algal communities contain large populations of filamentous green and blue-green algae that act as substrates for cell attachment and movement. By controlling the wavelength and intensities of light in these assays, our future investigations may provide information on the process driving the spatial segregation of different diatom species that occurs during the development of some algal communities [62].

The results from previous work and the above assays all point to a process whereby accumulation of diatoms in the light is primarily due to an effect in which cells get caught in a light trap delineated almost solely by the relative intensities of light at the inside and outside boundaries of the spot. That is, there appears to be little ability for diatoms to be able to actively reorient themselves in particular directions. While a number of diatoms do in fact occasionally stop and swivel or pivot around and reattach to the substrate reoriented in a new direction, this appears to be a random event. For the most part, diatoms seem to wander about, changing direction and occasionally reorienting themselves, resulting in a relatively even dispersal of cells. However, once cells detect a significantly high light gradient (i.e. the light irradiance at one tip is significantly higher than the other tip) then the photophobic response is triggered. For moderate light levels this is a step-down response whereby the direction of cell movement is shifted toward the light, while at very high light intensities this is a step-up response in which the direction of movement is shifted away from the high-intensity light.

As such, the natural behavior of diatoms is to undergo a random dispersal of cells, in which cells occasionally come in contact with areas of greater light intensity. Once there, the cells become trapped (or more precisely biased in their direction of movement to stay) inside the area of light. This accounts for the relatively linear increase in cell density we find in the first few hours of our cell accumulation assay, as cells wander in at a relatively constant rate, and then have difficulty in leaving the spot. A similar linear rate of initial increase for cell density, after which a plateau of cell density was reached, was reported for the first few hours of cell accumulation in Nultsch's work [63], although the initial and actual cell densities were not reported, so the rate constants can not be determined from the data. The light trapping effect is supported by the video recordings of cell behavior during our light aggregation assay. These observations show that cells in the interior of the spot undergo relatively random alterations in cell direction and movements, while many cells situated at the boundary of the spot show the same behavior as cells in our boundary assay, i.e. they change direction at the light/dark boundary and move back into the light.

We believe this general behavior is also why *Pinnularia viridis* cells do not accumulate in light spots appreciably, even though they show a well-defined sensitivity to light when tested with our boundary assay. We suspect that the circular path curvature of *P. viridis* [5,6] causes cells to wander in a circle, and thus disperse over a surface at a much lower rate, with each cell effectively confined to a small area. This means that new *P. viridis* cells will likely wander into light spots much less frequently, and therefore have a much slower rate of accumulation. Those cells that make it in still get trapped, but a much lower percentage of the *P. viridis* cells outside the light spot will ever find their way in. These species-dependent differences in dispersal ability may have important implications in the light-dependent segregation of cells in algal communities.

It is important to note that almost all of the past and current studies on the light responses affecting diatom motility have dealt with the physiology of the movement and not the actual mechanism. While some advances have been made on understanding the diatom genes and proteins involved in cell adhesion [28], cell wall formation [64,65], silicon transport [66], and photosynthesis [67,68], almost no information is available on the protein or components responsible for light reception and transduction for the motility apparatus. The localization of the light sensitivity to the tips of the cell (often devoid of chloroplasts) and the shift in spectral sensitivity relative to the absorption or action spectra of diatom chloroplasts and pigments, suggest that the motility related photo-reception is independent from that used in photosynthesis. It suggests that there is a component or structure in the membrane or cytoplasm of diatom cells, near the tips of the underlying actin filaments and the ends of the raphe branches, where the actin-based motile generation can be modulated. While the light detection may involve any of the known types of plant photosensors [91], some of which may be involved in photo-sensitive gene expression [92], at this time the molecular basis for the reception and transduction system remains a mystery.

13.6 Ecology of photo-stimulated motility

Clearly, the photo-responsive behavior of diatoms must play a role in selective advantage for the cells if the response is to have ecological (and evolutionary) meaning. As important contributors to the overall primary production of many aquatic communities, it is clear that light-based responses would be among the most important for the cells [42,69]. There is already clear evidence that light intensity can affect diatom growth rates [70] and that light can affect the relative abundance and stratification of diatoms and other algae in various periphyton communities [71–74]. Conversely, community composition can affect the quality of the available light [75,76]. The ability to have some photo-regulation of cell movements is therefore likely to be advantageous to diatoms in terms of maximizing light for energy capture, regulating nutrient acquisition, and controlling their refuge from cell loss or predation.

For example, the observed abilities of diatoms to accumulate in areas of low to moderate light and avoid areas of extremely high light intensities, suggests that motility allows the cells to maximize light for photosynthesis, while avoiding high energy light which may damage the chloroplasts or bleach the photopigments. Just as importantly, perhaps, may be the ability for light to direct cells to areas most conducive to their long-term survival. The difference in spectral sensitivities of diatom cells, described earlier, allows for different species to migrate to different areas within algal mats. For example, *Craticula* cells, most sensitive to green, might be able to migrate into areas where there are abundant green algae and still use the available light wavelengths effectively. In contrast, the red-sensitive *Stauroneis* might migrate away from such areas, thus reducing the immediate pressure of interspecies competition of diatoms.

While the speed of the more rapid diatoms can often correspond to a movement of up to one cell length per second, in actual values (10–20 μm/s) this is slow compared to the movements of the protozoa, crustaceans, or mollusks which feed upon the diatoms. In terms of avoiding predation, it therefore seems much more likely that rather than a direct and immediate retreat from predators, the motility may provide a mechanism for helping the cells establish a long-term refuge strategy for avoiding them. The light-dependent vertical migration of diatoms may thereby serve several purposes. This may allow diatoms in sediment to actively move to the surface in the early daylight hours to accumulate more light, but resettle into the sediment at night where they may have access to greater local concentrations of settled organic nutrients. The motility of many diatoms appears to be affected by the chemical composition of the environment [77–79] so that the diurnal movements may have both chemotactic and phototactic components.

There is no doubt that the variety of environmental conditions within aquatic communities leads to a unique set of ecological trade-offs for the cells at each level in an algal community [42,80]. Light-regulated migrations to particular levels within the algal community (or sediment) may thus allow diatoms to balance the processes of maximizing their light accumulation while minimizing their time exposed to potential predators or loss from the surface due to water movements. Conversely, the protection provided by the overlying layers of other algae also produces shading and reduced light exposure and potential for energy production. The balance required between the needs for light exposure, protection from cell loss due to predation or water movement or disturbance, and nutrient acquisition both from the water column and from the sediment lead to particular pressures for cells at each stratum of a developing benthic community. Such localized conditions can easily lead to localized diatom populations that can best take advantage of these sets of conditions.

There is evidence that adhesion and motility of different diatom species may strongly influence the ability of cells to become localized to different areas in an algal community as it develops [62,81,82]. This stratification is likely due in large part to the different motile strategies that diatoms have within the community. Because this motility is related to the cells' ability to secrete and adhere to substrates, it seems likely that their stratification would be associated with both the adhesive properties of individual cell types with particular surfaces [83,84] as well as their motile responses, particularly to the different light regimes in the layers of an algal community [75,76,85,86]. Moreover, there is some evidence that the secreted material may itself change the cohesive properties of the sediment [87].

Investigations in our lab looking at the effects of temperature on motility and adhesion independently indicate that the two processes may not be coordinately regulated. In all four species tested, temperature causes a steady increase in motile speed until about 35–40°C, at which point the motility rapidly falls to zero. Adhesion, on the one hand, seems to have species-specific effects, where some species lose adhesion dramatically at higher temperatures, while others maintain a relatively constant net adhesion to the substratum. Such results suggest that any light-based responses may also be directed to only one of the two processes (e.g. the force generation mechanisms) and may not necessarily affect both motility and adhesion.

Other results from our lab [40,58] also suggest that there are indeed differential competitive abilities of diatoms. That is, the sensitivities of the different species to light intensity cause different species to have a motile advantage at different light levels. For example, in our tests using *C. cuspidata*, *P. viridis*, *N. linearis*, and *S. phoenicenteron* together, in which all four species are in a mixed population competing for the same light spot, we find that *N. linearis* clearly accumulates the fastest at low light levels, while *S. phoenicenteron* accumulates fastest at higher intensities (Figure 9).

The investigations in our lab described in the previous sections are therefore providing the initial information we will need for future comparison of lab and field studies to investigate a number of questions about the relationship between the distribution of diatoms within an algal community and their motile responses. These questions include:

- Are cells that have strong adhesion and resistance to water flow, but poor vertical migration likely to remain near the bottom of developing communities? Are such cells good at initial colonization? Are they most responsive to the full spectrum of light more likely seen for initial colonizers?
- Are cells which have moderate or poor adhesion ability likely to be found within deeper levels of an algal community or sediment, where there would be less cell loss

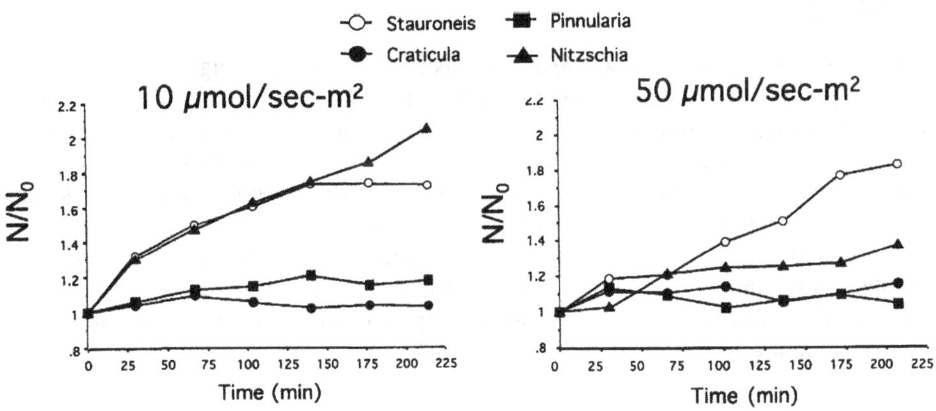

Figure 9. Comparison of cell accumulations for four diatom species placed in full-spectrum light spots. Four species of diatoms (*Craticula cuspidata*, *Pinnularia viridis*, *Stauroneis phoenicenteron* and *Nitzschia linearis*) were placed together in a mixed cell culture and allowed to accumulate in the same light spot as with the regular accumulation assays. At lower light levels (10 μmol m^{-2} s^{-1}; left panel) *Nitzschia* accumulates best over time, while *Stauroneis* accumulates fastest at higher light intensities (e.g. 50 μmol m^{-2} s^{-1}; right panel).

from the surface? Are such cells poor at initial colonization? Are they more sensitive to particular wavelengths of light as might be found within algal communities where overlying species absorb out particular wavelengths?

- Does the difference in motile sensitivities to light or wavelength correspond to the location or temporal appearance of diatoms within an algal community? For example, do the contrasting light responses of *Stauroneis* and *Craticula* cause these cells to exist within different areas of an algal community?
- Do cells with highest sensitivity to light intensity begin to migrate first (earliest in the day) followed by those that require higher light intensities for threshold of response? Do these sensitivities correlate with the amount of light found within various areas of an algal community (i.e. are the cells near the surface more or less sensitive to light compared to those near the bottom of the community where there may be less light intensity).
- To what degree are our measured characteristics of motility – speed, adhesion, sensitivity to light intensity and wavelength, path curvature – correlated with the observed behaviors of diatoms in natural communities?
- What are the relative contributions of light quality, substratum composition, nutrient and ionic composition of the water and species distribution within the community to ability of diatoms to regulate their motility and spatial orientation? To the diatom's ability to sense and respond to seasonal changes?

The relative scarcity of diatom research connecting lab data and field data on light responses suggests diatom ecology is likely to be a fruitful area of future research, particularly in the investigation of how light affects the ability of diatom to respond to other changes in its environment.

Summary

Motile pennate diatoms show clear and distinct reactions to light exposure, with evidence pointing to the generation of photophobic reactions, generated at the tips of the cells and biasing the direction changes of the cells, being the primary motile response. These photo-sensitive reactions show clear characteristic differences between diatom species with respect to both light intensity and light wavelength. While the molecular mechanisms driving these direction changes are still poorly understood, it is apparent that these light based reactions help allow the diatoms to differentiate themselves within the algal communities, and likely contributes to the relative success of diatoms by allowing different species to occupy different spatial and temporal domains within the algal communities.

Acknowledgments

The author would like to gratefully acknowledge the work of Samantha Dunbar, Roy Weitzell, and all the graduate and undergraduate students in my lab whose work contributed significantly to many of the experiments outlined in this article. Much of the work described was done with the help of NSF Grants IBN-9407279 and IBN-9982897.

This article was also supported by the DePaul University Research Council and The College of Liberal Arts & Sciences.

References

1. F.E. Round, R.M. Crawford (1990). The Bacillariophyta. In: L. Margulis, J. Corliss, M. Melkonian, D.J. Chapman (Eds), *Handbook of Protoctista* (pp. 574–599). Jones and Bartlett, Boston.
2. F.E. Round, R.M. Crawford, D.G. Mann (1990). *The Diatoms. Biology and Morphology of the Genera*. Cambridge University Press, Cambridge.
3. E. Fourtanier, J.P. Kociolek (1999). Catalogue of the diatom genera. *Diatom Research*, **14**, 1–190.
4. E.J. Cox (1977). Raphe structure in naviculoid diatoms as revealed by the scanning electron microscope. *Nova Hedwigia Beih.*, **54**, 261–274.
5. L.A. Edgar, J.D. Pickett-Heaps (1984), Diatom locomotion. In: F.E. Round, D.J. Chapman (Eds), *Progress in Phycological Research* (Vol. 3, pp. 47–88). Biopress Ltd., Bristol.
6. S.A. Cohn, R.E. Weitzell Jr. (1996). Ecological considerations of diatom cell motility: I. Characterization of motility and adhesion in four diatom species. *J. Phycology*, **32**, 928–939.
7. J.D. Pickett-Heaps, D.R.A. Hill, R. Wetherbee (1986). Cellular movement in the centric diatom *Odontella sinensis*. *J. Phycol.*, **22**, 334–339.
8. J.D. Pickett-Heaps, D.R.A. Hill, K.L. Blaze (1991). Active gliding motility in an araphid marine diatom *Ardissonea* (formerly *Synedra*) *crystallina*. *J. Phycol.*, **27**, 718–725.
9. T.W. Engelmann (1879). Ueber die Bewegungen der Oscillarien und Diatomeen. *Botanische Zeitung*, **37**, 49–56.
10. H.L. Smith (1888). A contribution to the life history of the Diatomaceae – part II. *Proceedings of the American Society of Microbiology*, **9**, 126–167.
11. O. Müller (1893). Die Ortsbewegung der Bacillariaceen betreffend. I. *Ber. Dt. Bot. Ges.*, **11**, 571–576.
12. R. Lauterborn (1896). *Untersuchungen über Bau, Kernteilung und Bewegung der Diatomeen*. W. Englemann Pub., Leipzig.
13. D.D. Jackson (1905). Movements of diatoms and other microscopic plants. *American Naturalist*, **461**, 287–291.
14. J.D. Pickett-Heaps, A.M. Schmid, L.A. Edgar (1990). The cell biology and phylogeny of diatom valve formation. In: F.E. Round, D.J. Chapman (Eds), *Progress in Phycological Research* (Vol. 7, pp. 1–168). Biopress Ltd, Bristol.
15. S.A. Cohn, D. Zelner, J. Crea, B. Wibisono, M. Silverman (1999). Analysis of diatom motility using light avoidance and fluorescent bead assays. *Mol. Biol. Cell*, **10(S)**, 264a.
16. M.A. Harper (1977). Chapter 8: Movements. In: D. Werner (Ed.), *The Biology of Diatoms* (pp. 224–249). University of California Press, Berkeley.
17. J. Bertrand (1992). Mouvements des diatomées. II – Synthèse des mouvements. *Cryptogamie Algol.*, **13**, 49–71.
18. S.A. Cohn, N.C. Disparti (1994). Environmental factors influencing diatom cell motility. *J. Phycol.*, **30**, 818–828.
19. R.W. Drum, J.T. Hopkins (1966). Diatom locomotion: an explanation. *Protoplasma*, **62**, 1–32.
20. K.D. Hoagland, J.R. Rosowski, M.R. Gretz, S.C. Roemer (1993). Diatom extracellular polymeric substances: Function, fine structure, chemistry, and physiology. *J. Phycol.*, **29**, 537–566.

21. Y. Wang, J. Lu, J-C. Mollet, M.R. Gretz, K.D. Hoagland (1997). Extracellular matrix assembly in diatoms (Bacillariophyceae). II. 2,6 Dichlorobenzonitrile inhibition of motility and stalk production in the marine diatom *Achnanthes longpipes*. *Plant Physiol.*, **113**, 1071–1080.
22. B.A. Wustman, M.R. Gretz, K.D. Hoagland (1997). Extracellular matrix assembly in diatoms (Bacillariophyceae). I. A model of adhesives based on chemical characterization and localization of polysaccharides from the marine diatom *Achnanthes longpipes* and other diatoms. *Plant Physiol.*, **113**, 1059–1069.
23. L.A. Edgar, J.D. Pickett-Heaps (1982). Ultrastructural localization of polysaccharides in the motile diatom *Navicula cuspidata*. *Protoplasma*, **113**, 10–22.
24. L.A. Edgar, J.D. Pickett-Heaps (1983). The mechanism of diatom locomotion.I. An ultrastructural study of the motility apparatus. *Proc. Roy. Soc. Lond. B.*, **218**, 331–343.
25. D. Werner (Ed.) (1977). *The Biology of Diatoms*. Univ. of Calif. Press, Berkeley.
26. K. Wenderoth (1983). *Phototaxis bei Desmidiaceen und Diatomeen, Biol. Film No. C1496* Institut für den Wissenschaftlichen Film, Göttingen, Germany.
27. J.L. Lind, C.J. van Vliet, A. Bacic, R. Wetherbee (1996). Characterization of molecules responsible for cell substrate adhesion in marine raphid diatoms. *Mol. Biol. Cell*, **7(S)**, 56a.
28. J.L. Lind, K. Heimann, E.A. Miller, C. van Vliet, N.J. Hoogenraad, R. Wetherbee (1997). Substratum adhesion and gliding in a diatom are mediated by extracellular proteoglycans. *Planta*, **203**, 213–221.
29. D.R. Webster, K.E. Cooksey, R.W. Rubin (1985). An investigation of the involvement of cytoskeletal structures and secretion in gliding motility of the marine diatom, *Amphora coffaeformis*. *Cell Motil.*, **5**, 103–122.
30. R. Wetherbee, J.L. Lind, N.C. Poulsen, T.P. Spurck (1996). Cell motility in marine raphid diatoms appears to be actin-based. *Mol. Biol. Cell*, **7(S)**, 560a.
31. R. Gordon, R.W. Drum (1970). A capillarity mechanism for diatom gliding locomotion. *Proc. Natl. Acad. Sci. (USA)*, **67**, 338–344.
32. R. Gordon (1987). A retaliatory role for algal projectiles, with implications for the mechanochemistry of diatom gliding motility. *J. Theor. Biol.*, **126**, 419–436.
33. L.A. Edgar, M. Zavortink (1983). The mechanism of diatom locomotion. II. Identification of actin. *Proc. Roy. Soc. Lond. B.*, **218**, 345–348.
34. D. Bray (1992). *Cell Movements*. Garland, New York.
35. R.E. Williamson (1992). Cytoplasmic streaming in characean algae: Mechanism, regulation by Ca^{2+}, and organization. In: M. Melkonian (Ed.), *Algal Cell Motility* (pp. 73–98). Chapman and Hall, New York.
36. T.M. Preston, C.A. King, J.S. Hyams (1990). *The Cytoskeleton and Cell Motility*. Blackie Publ., London.
37. J.W. LaClaire II, R. Chen, D.L. Herrin (1995). Identification of a myosin-like protein in *Chlamydomonas reinhardtii* (Chlorophyta). *J. Phycol.*, **31**, 302–306.
38. J.R. Sellers, H.V. Goodson, Motor proteins 2: myosin. In: P. Sheterline (Ed.), *Protein Profile* (Vol. 2, Issue 12, pp. 1323–1423). Academic Press, London.
39. Y.Y. Toyoshima (1991). Bidirectional movement of actin filaments along tracks of heavy meromyosin and native thick filaments. *J. Cell Sci. Suppl.*, **14**, 83–85.
40. S.A. Cohn, R.E. Weitzell Jr., A. Norris, M.J. Lazzarotto (1996). Comparative analysis of adhesion and photo-stimulated aggregation in pennate diatoms. *Mol. Biol. Cell*, **7(S)**, 232a.
41. D.-P. Häder, E. Hoiczyk (1992). Chapter 1. Gliding motility. In: M. Melkonian (Ed.), *Algal Cell Motility* (pp. 1–38). Chapman and Hall, New York.
42. R.J. Stevenson, M.L. Bothwell, R.L. Lowe (Eds) (1996). *Algal Ecology. Freshwater Benthic Ecosystems*. Academic Press, San Diego.

43. W. Nultsch, D.-P. Häder (1979). Photomovement of motile microorganisms. *Photochem. Photobiol.*, **29**, 423–437.
44. W. Nultsch, D.-P. Häder (1988). Photomovement in motile microorganisms – II. *Photochem. Photobiol.*, **47**, 837–869.
45. M. Melkonian (Ed.) (1992). *Algal Cell Motility.* Chapman and Hall, New York.
46. S.A. Cohn, T.P. Spurck, J.D. Pickett-Heaps (1999). High energy irradiation at the leading tip of moving diatoms causes a rapid change of cell direction. *Diatom Research*, **14**, 193–206.
47. F.E. Round, J.D. Palmer (1966). Persistent vertical migration rhythms in benthic microflora. II. Field and laboratory studies on diatoms from the bank of the river Avon. *J. Marine Biol. Assoc. UK.*, **46**, 191–224.
48. C.M. Happey-Wood, P. Jones (1988). Rhythms of vertical migration and motility in intertidal benthic diatoms with particular reference to *Pleurosigma angulatum*. *Diatom Research*, **3**, 83–93.
49. W. Nultsch (1971). Phototactic and photokinetic action spectra of the diatom *Nitzschia communis*. *Photochem. Photobiol.*, **14**, 705–712.
50. W. Nultsch, K. Wenderoth (1973). Phototaktische Untersuchungen an einzelnen Zellen von *Navicula peregrina* (Ehrenberg) Kützing. *Arch. Mikrobiol.*, **90**, 47–58.
51. K. Wenderoth (1979). Photophobische Reaktionen von Diatomeen im monochromatischen Licht. *Ber. Deutsch. Bot. Ges.*, **92**, 313–321.
52. K. Wenderoth (1982). *Photokinese und photophobische Reaktionen der Kieselalge* Navicula peregrina, *Biol. Film No. C1388.* Institut für den Wissenschaftlichen Film, Göttingen, Germany.
53. K. Wenderoth (1982). *Reaktion der Kieselalge* Navicula peregrina *auf Belichtung verschiedener Zellabschnitte, Biol. Film No. C1466.* Institut für den Wissenschaftlichen Film, Göttingen, Germany.
54. K. Wenderoth (1984). *Wirkungsspektrum der Step-down-Reaktion bei Diatomeen, Biol. Film No. C1520.* Institut für den Wissenschaftlichen Film, Göttingen, Germany.
55. S.A. Cohn, A.L. Ingold, J.M. Scholey (1987). Correlation between the ATPase and microtubule translocating activities of sea urchin egg kinesin. *Nature (Lond.)*, **328**, 160–163.
56. S.A. Cohn, A.L. Ingold, J.M. Scholey (1989). Quantitative analysis of sea urchin egg kinesin-driven microtubule motility. *J. Biol. Chem.*, **264**, 4290–4297.
57. S.A. Cohn, W.M. Saxton, R.J. Lye, J.M. Scholey (1993). Analyzing microtubule motors in real time. *Meth. Cell. Biol.*, **39**, 75–88.
58. S.A. Cohn, S.A. Dunbar, C. Skoczylas, J.A. Mucha (1997). Comparative analysis of diatom motility: phototactic behavior and sensitivity to ultraviolet radiation. *Mol. Biol. Cell*, **8(S)**, 386a.
59. O.G. Stonington, T.P. Spurck, J.A. Snyder, J.D. Pickett-Heaps (1989). UV-microbeam irradiations of the mitotic spindle.I. The UV microbeam apparatus. *Protoplasma*, **153**, 62–70.
60. T.P. Spurck, O.G. Stonington, J.A. Snyder, J.D. Pickett-Heaps, A. Bajer, J. Mole-Bajer (1990). UV-microbeam irradiations of the mitotic spindle. II. Spindle fiber dynamics and force production. *J. Cell Biol.*, **111**, 1505–1518.
61. W. Haupt (1982). Light mediated movement of chloroplasts. *Ann. Rev. Plant Phys.*, **33**, 205–233.
62. R.E. Johnson, N.C. Tuchman, C.G. Peterson (1996). Changes in the vertical micro-distribution of diatoms within a developing periphyton mat. *J. North Am. Benthol. Soc.*, **16**, 503–519.
63. W. Nultsch (1975). Phototaxis and photokinesis. In: M.J. Carlile (Ed.), *Primitive Sensory and Communication Systems* (pp. 29–90). Academic Press, New York.

64. N. Kröger, C. Bergsdorf, M. Sumper (1996). Frustulins: domain conservation in a protein family associated with diatom cell walls. *Eur. J. Biochem.*, **239**, 259–264.
65. N. Kröger, G. Lehmann, R. Rachel, M. Sumper (1997). Characterization of a 200-kDa diatom protein that is specifically associated with a silica-based substructure of the cell wall. *Eur. J. Biochem.*, **250**, 99–105.
66. M. Hildebrand, K. Dahlin, B.E. Volcani (1998). Characterization of a silicon transporter gene family in *Cylindrotheca fusiformis*: Sequences, expression analysis, and identification of homologs in other diatoms. *Mol. Gen. Genet.*, **260**, 480–486.
67. M. Eppard, E. Rhiel (1998). The genes encoding the light harvesting subunits of *Cyclotella cryptica* (Bacillariophyceae) constitute a complex and heterogeneous family. *Mol. Gen. Genet.*, **260**, 335–345.
68. P. Guenaeau, F. Morel, J. Laroche, D. Erdner (1998). The petF region of the chloroplast genome from the diatom *Thalassiosira weissflogii*: Sequence, organization and phylogeny. *Eur. J. Phycol.*, **33**, 203–211.
69. M.J. Caduto (1990). *Pond and Brook. A Guide to Nature in Freshwater Environments.* University Press of New England, Hanover.
70. A. Morel, L. Lazzara, J. Gostan (1987). Growth rate and quantum yield time response for a diatom to changing irradiences (energy and color). *Limnol. Ocenaogr.*, **72**, 1066–1084.
71. A.D. Steinman, C.D. McIntire (1986). Effects of current velocity and light energy on the structure of periphyton assemblages in laboratory streams. *J. Phycol.*, **22**, 352–361.
72. M.L. Bothwell, K.E. Suzuki, M.K. Bolin, F.J. Hardy (1989). Evidence of dark avoidance by phototrophic periphytic diatoms in lotic systems. *J. Phycol.*, **25**, 85–94.
73. R.J. Stevenson, C.G. Peterson (1989). Variation in benthic diatom (Bacillariophyceae) immigration with habitat characteristics and cell morphology. *J. Phycol.*, **25**, 120–129.
74. R.J. Stevenson, C.G. Peterson, D.B. Kirschtel, C.C. King, N.C. Tuchman (1991). Density-dependent growth, ecological strategies, and effects of nutrients and shading on benthic diatom succession in streams. *J. Phycol.*, **27**, 59–69.
75. B.B. Jørgensen, N.P. Revsbech, Y. Cohen (1983). Photosynthesis and structure of benthic microbial mats: Microelectrode and SEM studies of four cyanobacterial communities. *Limnol. Oceanogr.*, **28**, 1075–1093.
76. B.B. Jørgensen, D.J. DesMarais (1988). Optical properties of benthic photosynthetic communities: Fiber optic studies of cyanobacterial mats. *Limnol. Oceanogr.*, **33**, 99–113.
77. B. Cooksey, K.E. Cooksey (1980). Calcium is necessary for motility in the diatom *Amphora coffaeformis*. *Plant Physiol.*, **65**, 129–131.
78. B. Cooksey, K.E. Cooksey (1988). Chemical signal-response in diatoms of the genus *Amphora*. *J. Cell Sci.*, **91**, 523–529.
79. K.E. Cooksey (1981). Requirement for calcium in adhesion of a fouling diatom to glass. *Appl. Envir. Microbiol.*, **41**, 1378–1382.
80. W.M. Darley (1982). *Algal Biology: A Physiological Approach.* Blackwell Scientific Pub., London.
81. D.M. Patterson (1986). The migratory behavior of diatom assemblages in a laboratory tidal micro-ecosystem examined by low temperature scanning electron microscopy. *Diatom Res.*, **1**, 227–239.
82. P.V. McCormick, R.J. Stevenson (1991). Mechanisms of benthic algal succession in lotic environments. *Ecology*, **72**, 1835–1848.
83. N. Tanaka (1986). Adhesive strength of epiphytic diatoms on various seaweeds. *Bull. Jap. Soc. Sci. Fish.*, **52**, 817–821.
84. K. Becker (1996). Exopolysaccharide production and attachment strength of bacteria and diatoms on substrates with different surface tensions. *Microb. Ecol.*, **32**, 23–33.

85. C. Hudon, E. Bourget (1983). The effect of light on the vertical structure of epibenthic diatom communities. *Bot. Mar.*, **26**, 317–330.
86. L.-A. Hansson (1995). Diurnal recruitment patterns in algae: Effects of light cycles and stratified conditions. *J. Phycol.*, **31**, 540–546.
87. D.M. Patterson (1989). Short-term changes in the erodibility of intertidal cohesive sediments related to the migratory behavior of epipelic diatoms. *Limnol. Oceanogr.*, **34**, 223–234.
88. A.L. Wells, A.W. Lin, L.Q. Chen, D. Safer, S.M. Cain, T. Hasson, B.O. Carragher, R.A. Milligan, S.H. Lee (1999). Myosin VI is an actin-based motor that moves backwards. *Nature*, **401**, 505–508.
89. N.C. Poulsen, I. Spector, T.P. Spurck, T.F. Schultz, R. Wetherbee (1999). Diatom gliding is the result of an actin-myosin motility system. *Cell Motil. Cytoskel.*, **44**, 23–33.
90. A. Falciatore, M. Ribera d'Alcalà, P. Croot, C. Bowler (2000). Perception of environmental signals by a marine diatom. *Science*, **288**, 2363–2366.
91. K.J. Hellingwerf (2000). Key issues in the photochemistry and signalling-state formation of photosensor proteins. *J. Photochem. Photobiol. B-Biology*, **54**, 94–102.
92. C. Leblanc, A. Falciatore, M. Watanabe, C. Bowler (1999). Semi-quantitative RT-PCR analysis of photoregulated gene expression in marine diatoms. *Plant Mol. Biol.*, **40**, 1031–1044.

Photomovement
D.-P. Häder and M. Lebert, editors.

Chapter 14

Photomovement of microorganisms in benthic and soil microenvironments

Ferran Garcia-Pichel and Richard W. Castenholz

Table of contents

Abstract

Photomovement seems to play an important role in the ecology of natural populations of microorganims, particularly for phototrophs, living in sediments, microbial mats and soils. These (micro)habitats share optical and other physicochemical characteristics, very different from those of the plankton, which may allow photomovement to become of particular adaptive value. This is most obvious in the frequently encountered phenomenon of vertical migrations of benthic phototrophic microorganisms. These migrations involve the populational movement from or to the surface of the sediments (or mats) in response to, or in anticipation of, changes in environmental parameters. We provide a primer on the characteristics of sedimentary and soil light fields, review the information available on the phenomenon of vertical migration of microorganisms in these habitats and discuss the selective advantage provided to the organisms by light-responsive movements in nature.

14.1 Introduction

Massive daily movements of microalgae (cyanobacteria, diatoms) in shallow water and intertidal sediments, microbial mats and biofilms, resulting in conspicuous surface color changes are well known; solar irradiance seems to play a predominant role in the modulation of these short-distance vertical migrations. The daily migratory behavior of microorganisms in sedimentary environments seems indeed to be a widespread but poorly understood phenomenon. In order to fully understand these phenomena it would be desirable to integrate the observations on migrating behavior of microorganisms in nature with investigations of their photoresponses studied under controlled laboratory conditions. However, there has been only a few attempts to realize such goals. Some authors are of the opinion that the role of photoresponses in nature has yet to be determined [1]. This may be in fact a bit pessimistic, but certainly, when trying to ascertain the adaptive value of photophysiological mechanisms studied in the laboratory, it is common practice to invoke typical planktonic situations, and not those in sediments or soils. But the habitat offered to microoganisms in such microenvironments (topsoils, the sediments of marine and freshwater bodies of water) is under physical constraints very different from those acting in planktonic habitats. Mass transport in sediments, microbial mats and soils, for example, is largely governed by diffusion, as opposed to the turbulent mixing that predominates in the planktonic environment. In the light-exposed sediments of shallow waters and in soils, light attenuation is very strong and a function of both absorption and multiple scatter by both mineral particles and the biota. The resulting photic zones are extremely shallow, usually in the order of millimeters or less. In the water column of marine and freshwaters, the attenuation of incident light is comparatively weak and mostly due to absorption processes, scattering playing only a minor role; this results in photic zones typically in the range of 1–100 m. In view of the differences sketched above, it is perhaps not all too surprising that microorganisms dwelling in sedimentary and soil microenvironments should have developed a suite of specific adaptations, among them photomotility responses, geared differently than those providing adaptive value in planktonic systems. Thus, for example, it has been hypothesized that a motility mechanism based on runs and tumbles (such as that of *Escherichia coli*) is less efficient in porous media than one based on runs and random reversals -such as that of *Pseudomonas putida* [2]. The latter mechanism minimizes collisions with particles and should be of advantage in sediments and soils.

It is the purpose of this chapter to review the phenomenology of photomovements described for natural microenvironments in sediments, microbial mats and soils, and to try to unravel their ecophysiological significance based on the growing knowledge of microscale light fields and organismal photobiological responses.

14.2 Light fields: penetration, trapping, directionality and spectral composition

Visible light reaching benthic and soil environments is subject to intense attenuation due to the high density of mineral and biogenic particles. The use of fiber optic based microprobes [3–6] has enabled researchers in recent years to obtain small-scale measurements of the light fields and optical parameters within these environments [7–10]. In a survey of several such environments the measured photic depths for visible light (those depths where the scalar irradiance, E_0, or fluence rate, is attenuated to 1% of that incident) varied between 3.1 mm for quartz sand and 0.45 mm for silty muds [10]. In the ultraviolet (UV) at 310 nm, the corresponding depths were only 1.25 and 0.23 mm. The attenuation of E_0 is however not simply exponential, because multiple scattering plays a significant role in the attenuation process, along with absorption. In fact, as light penetrates into the sediment, soil or mat, and successive scatter events randomize the directionality of the photons, two non-intuitive, but important, phenomena take place: the light fields become increasingly diffuse and maxima in scalar irradiance (larger than that incident) build up close to the surface. Below this surficial zone where the E_0 maximum occurs, E_0 attenuates quasi exponentially. An abstracted representation of a typical E_0 vertical profile within a sedimentary environment is shown in Figure 1.

The spectral quality of E_0 is also strongly modified as it penetrates the sediment, mat or soil. A general pattern is that the shorter the wavelength the lesser the penetration, so that sunlight becomes redder with depth, and infrared spectral components penetrate the deepest. In the case of quartz sand, for example, the apparent attenuation coefficient, K_D, for scalar irradiance in the subsurface zone increases linearly with wavelength [10]. Specific characteristics of each microenvironment may somewhat alter this general pattern, so that the presence of iron minerals, for example, may contribute to an enhanced attenuation of UV and blue wavelengths [11] and the presence of populations of phototrophic microorganisms may impose strong absorption signatures due to their photosynthetic pigments on the overall spectrum of the scalar irradiance within the sediment [12,13].

The above is a succinct but unavoidably simplified account of the nature of light fields within benthic environments and soils, and the reader is urged to the literature cited above to obtain a more in-depth picture of the field. However, the features most relevant to the optical environment in which photomovements of microorganisms must operate have been mentioned. In short, we can think of two distinct optical zones, one close to

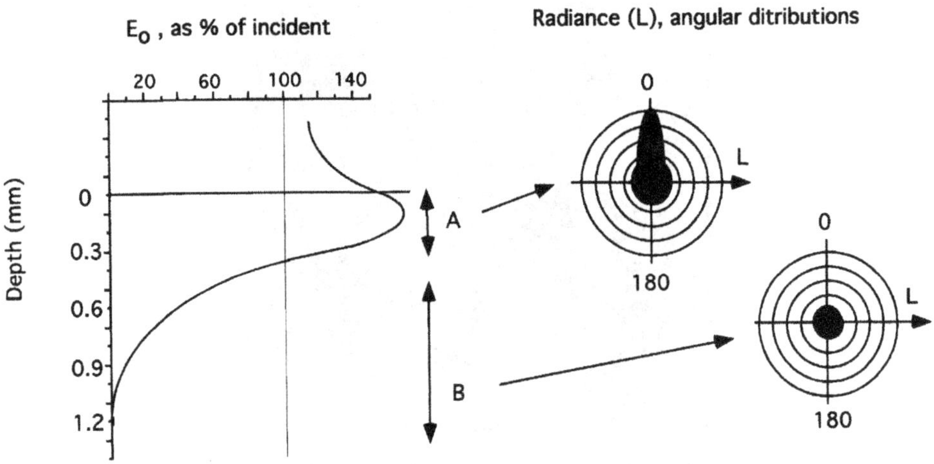

Figure 1. Generalized optical characteristics of the light field in sediments, mats and soils. Left: vertical profile of the scalar irradiance (E_o), expressed as percentage of incident (downwelling) irradiance. A maximum in E_o (a "light trap") occurs close to the surface (zone "A"). In the subsurface, Eo decays quasi exponentially with depth (zone "B"). Right: Angular distributions of radiance (L) in zones "A" and "B". In "A" the light field displays a strong vertical component, most photons travel downwards. In "B" the light field is close to diffuse.

the surface, where light trapping occurs, presenting maxima in scalar irradiance, high intensities and a downward directionality of the light, and a second subsurface zone where intensities decay exponentially and the light field is close to diffuse. When scattering is strong and absorption weak, the maxima due to light trapping and the depth of occurrence will be comparatively large, and the light will become diffuse close to the surface; this will be the case for visible and infrared wavelengths in sandy sediments without a well-developed phototrophic biota, or for red and infrared wavelengths in most sediment types and microbial mats. By contrast when absorption phenomena predominate, the light field will tend to maintain a strong downward component, maxima in scalar irradiance will be reduced in magnitude and will occur at the surface. This will be the case for UV and blue wavelengths in most cases, and for most wavelengths in the visible in those mats, sediments and soils with strong absorption (silts, feldspar sands, volcanic sands or any of these heavily colonized by phototrophic microbes).

14.3 Vertical migrations of microorganisms in sediments and soils

Migrations of microorganisms in sediments have been known for a long time, as in some cases they involve the appearance and disappearance of colored masses of organisms from the surface, bringing about conspicuous changes in the overall sediment appearance ([14–16] and see Figure 2). Similarly, the sudden appearance of cyanobacterial populations at the surface of wetted desert soils [11,17] has been

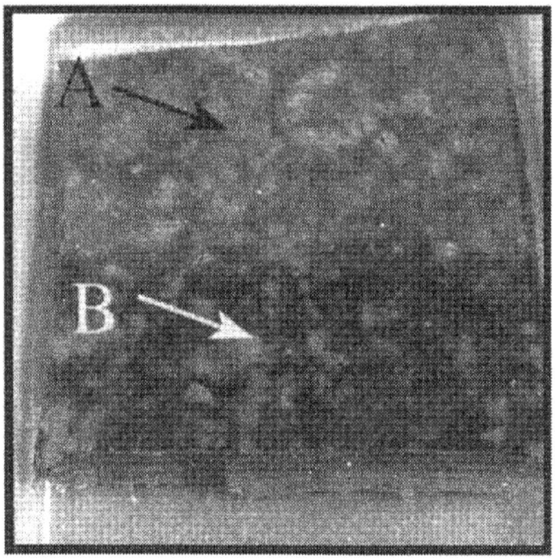

Figure 2. Excised, laminated microbial mat from a hypersaline pond near Guerrero Negro, Mexico. One half of the mat (upper area, A) was exposed to full solar irradiance throughout the morning, the lower half (B) was covered with a 10% transmittance screen for 1.5 h. The dark coloration in B is the result of oscillatorian cyanobacteria rapidly migrating to the surface in response to the change in light conditions. Reproduced from [65], with permission from the publisher.

observed. All of these migrations involve in most cases microalgae, most commonly diatoms or cyanobacteria, but also sometimes other algal types [18], and we will also focus most of our discussion on phototrophic microorganisms. However, photomovement is not relegated to these metabolic types, as other non-phototrophic bacteria, such as the sulfur bacterium *Beggiatoa* spp. and *Isosphaera pallida*, have been shown to display photomovement in laboratory experiments [19,20] and migrating behavior in nature [19,21]. It is a reasonable expectation that less obvious daily migrations occur in microbial types such as fermentative, sulfate-reducing and denitrifying bacteria as well, and perhaps that the application of novel techniques for the identification and quantification of microorganisms in the field will be able to unravel these in the near future. Recently, the vertical migration of anaerobic *Desulfovibrio* (sulfate-reducer) has been demonstrated as a probable negative aerotaxis [22].

Reports on migrating behavior of microorganisms, mostly qualitative, stem from a variety of environments including microbial mats in hot springs [16,23,24] and hypersaline waters [16,23,24], as well as in intertidal [25–27] and freshwater sediments [18,28] and in saline Antarctic ponds [29] indicating that the phenomenon is widespread. In soils, the presence of sustained migration has not yet been thoroughly described; but from anecdotal reports it is possible to infer that similar phenomena may occur in (wet) soils [11,17,30]. One should not forget, however, that in many other instances, virtually or totally non-motile organisms are able to build up functionally stable benthic populations, indicating that migrations are not a necessary adaptation of the microphytobenthos to the sedimentary habitat, and that other strategies of tolerating diurnal inhibitory conditions have evolved.

14.3.1 Cyanobacterial migrations

Benthic gliding filamentous cyanobacteria often display daily migrations. From the studies in which the extent of daily movements has been measured, the migrations of cyanobacteria involve a downward movement into the sediment as the sun rises in the sky clearing the sediment or mat to depths of around 0.5 mm below the surface [21,24,31]. This is reversed at dusk, when cyanobacterial populations return to the surface. Rhythmicity in the migration patterns of cyanobacteria in the absence of environmental cues has not been described. A quantitative description of cyanobacterial migration (*Spirulina* spp. and *Oscillatoria* spp.) in a microbial is reproduced in Figure 3. In one case, however, a migrating pattern more complicated than that depicted in Figure 3, involving a second night-time downward movement has been reported in a hot spring *Oscillatoria* [23]. It is thought that the nighttime behavior is a response to chemotactic rather than phototactic cues [32].

14.3.2 Anoxygenic photosynthetic bacteria

The swimming migration of benthic purple sulfur bacteria (*Chromatium* spp.) during daily cycles or after incubations under artificially imposed low light or darkness, has been described from marine sediments and hot spring microbial mats [33–35]. Although

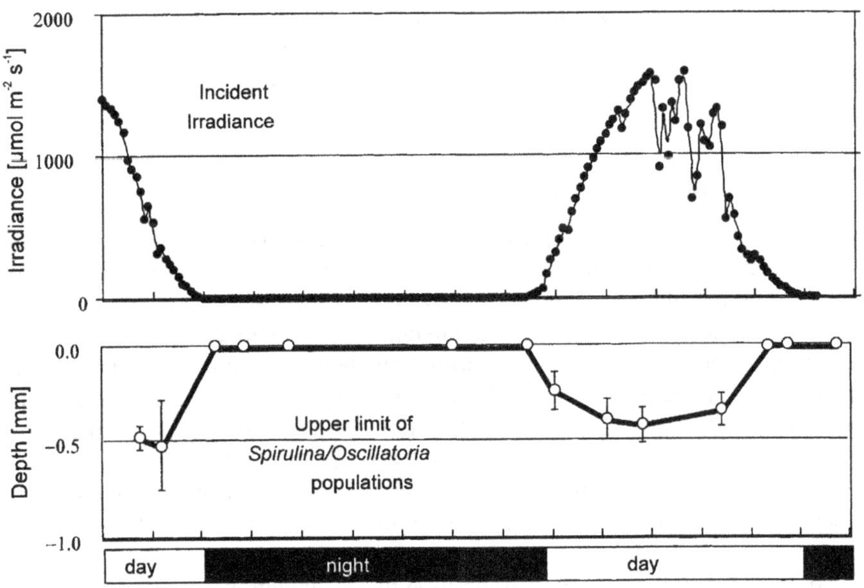

Figure 3. Daily migration of cyanobacteria in response to changes of light intensity (*Spirulina* cf. *subsalsa*, *Oscillatoria* sp.) in a hypersaline microbial mat from Baja California (Mexico). Upper graph depicts the incident solar radiation. Lower graph depicts the position of the upper edge of the cyanobacteria populations as determined from minicores. Bar at bottom indicates the periods of daytime and nighttime.

a full quantitative description has yet to be presented, these migrations involve the appearance of *Chromatium* populations at the surface of the mat at night and their return to the sediment interior at dawn.

14.3.3 Diatoms

Many pennate diatoms display active, relatively rapid gliding movements [36] and are usually much faster than those of cyanobacteria. The rhythmic movement of benthic diatoms in sediments has been well documented. It has been reported to involve upward migrations starting at dawn, continuing during daytime and downward migrations starting with waning light intensity before sunset [37–39], but this general pattern may be influenced by tidal rhythms in shallow marine sediments (Figure 4). Most of the studies on diatom migration are based on the lens paper technique [40], in which one counts the number of individuals trapped on a piece of tissue placed on the surface of the sediment. Alternatively, fluorescence based, non-invasive techniques have also been used recently [41] with similar results. With these techniques one obtains good quantitative data on the variability in biomass at the sediment surface, but there are no quantitative accounts on how deep the diatoms dive into the sediment, or on the depth-distribution of the biomass with time. Patterson [42] has demonstrated with low-temperature scanning electron microscopy that the bulk of diatom biomass within a tidal sediment was restricted to the upper 1.6 mm. The occasional presence of diatoms at greater depths has been attributed to physical turbulence and meiofaunal reworking of the sediments [39]. The migrations of diatoms may present a daily rhythmicity that is maintained for several cycles in the absence of environmental cues, both in tidally influenced sediments [41] and in tidally independent situations [43]. Thus, a circadian rhythm is a property of these eukaryotic microorganisms.

14.4 The role of photomovement in vertical migrations

Directly or indirectly, light cues play a significant role in driving vertical migrations of microorganisms in sediments and microbial mats. In populations of microbial phototrophs there is abundant evidence of the direct role of light responses. Commonly, for example, artificial manipulation of the magnitude of the incident radiation results in immediate population responses, and correlations between the depth of migration and the intensity of the incident radiation have been measured (see Figure 2). Additionally, whenever the phototrophic organisms involved in vertical migrations have been studied in the laboratory, they have been found to be light-responsive. Because changes in light intensity bring about virtually simultaneous variations in chemical parameters, it becomes almost impossible to disentangle the importance of each of the possible cues in the motility responses in nature. Oxygen partial pressures, for example, may vary upon a sudden darkening from supersaturation to anoxia within minutes [44]. Dramatic changes in the partial pressure of CO_2, and the concentration of sulfide at various depths occur as well. It is conceivable that the migrations are driven partly by chemotactic cues, a hypothesis that is not only supported by laboratory studies in non-phototrophic bacteria but also in cyanobacteria [32,45] and by the fact that the return of cyanobacteria to the surface of the sediments typically occurs also under complete darkness or at very

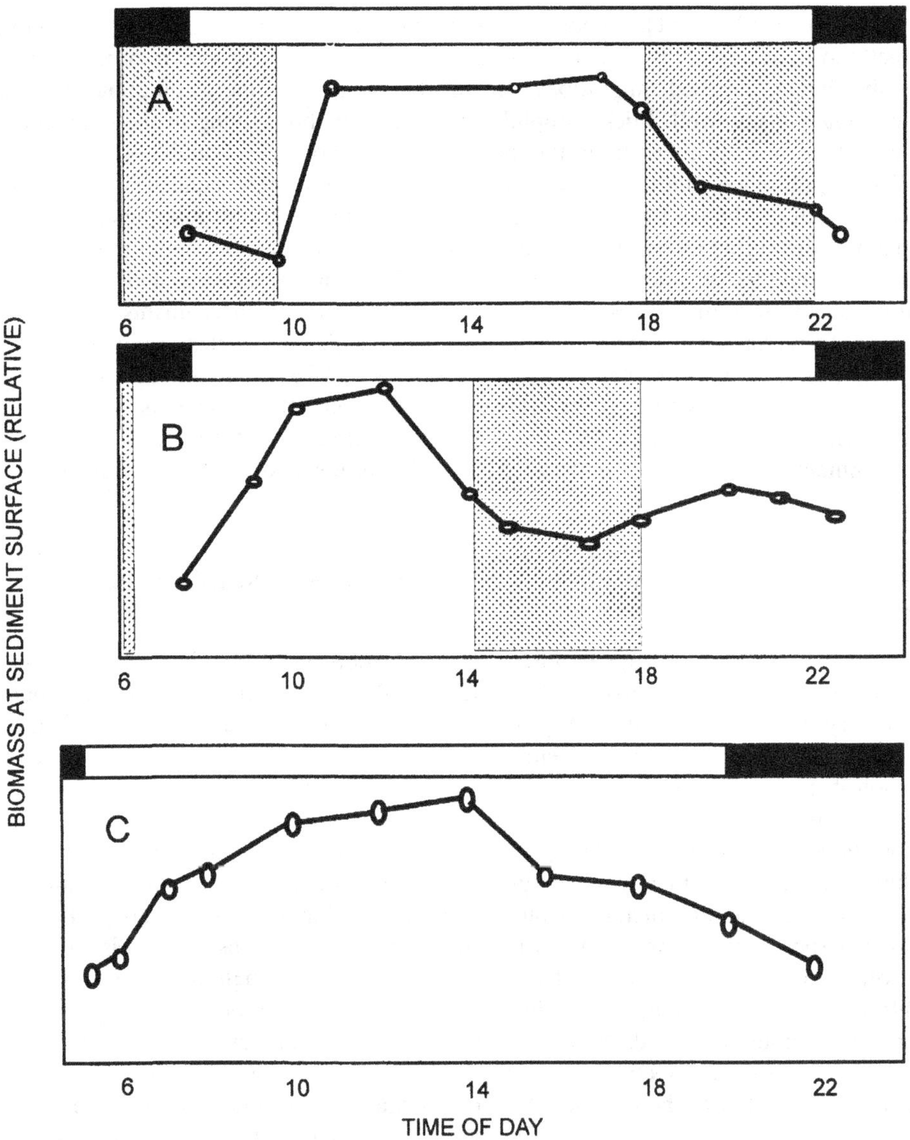

Figure 4. Daily and tidally influenced migration of sedimentary diatom communities. Data compiled and modified from [41] and [43]. Black bars on graphs indicate night periods, white bars daytime periods. The abscissa represents the variation of diatom biomass present at the surface of the sediment as inferred by Chl *a* fluorescence (A, B) or by diatoms trapped on lens paper (C). A, B: the same intertidal sediment studied on two days, differing in tidal period. Times of flooding are in gray background.C. migrations in a non-tidal pond. The diatom biomass at the surface mimics the curve of light intensity (not shown).

low irradiances [21,28,31]. Phototrophic microorganims are the most likely to respond directly to light cues, but the changes in the chemical microenvironment brought about by the light-controlled migrations are likely to affect other non-photosynthetic microorganisms as well. Microaerophilic organisms or those living in the oxic/anoxic interface, must react indirectly to the migrations of phototrophic populations.

The migrations of diatoms in sediments seem to be under the control of an internal clock, which can be set by both light and tidal cues, so that the populations are able to remain at the surface when light is available during low tides, but avoid the sediment surface at night and when high tides ebb or flow during daytime, and are able to anticipate the incoming tides (Figure 4). While the sensory mechanisms that allows diatoms to rephase and modulate their migrations with tidal cycles is unknown, it is certain that is a light-independent mechanism and probably coordinated with lunar cycles. Natural ecosystems are usually more complicated than a scientist would wish. However, the benthic flora may be in the position of integrating a variety of environmental signals to respond in a manner that brings about the highest increase in fitness.

14.5 Integrating knowledge of photoresponses and migratory behavior

Perhaps the simplest attempt to integrate information gained in the laboratory with observations in nature consists of elucidating the minimum speed of locomotion necessary to sustain migration. A good estimate of such minimum speed may help us ascertain, for example, which members in a complex community are likely to be responsible for the migrations, or alternatively, it may place a limit to the maximal distance that a given organism is able to migrate. Naturally migrating cyanobacteria show, under the microscope, gliding speeds above 0.5 $\mu m\ s^{-1}$. But even such apparently simple measurements may represent gross oversimplifications, as the physical nature of the sedimentary environment is not taken into account: diatoms of the genus *Gyrosigma*, showing speeds of 4.7 $\mu m\ s^{-1}$ when moving on an artificial substratum, display a net movement of only 0.17 $\mu m\ s^{-1}$, when moving not on, but through the same substratum [39], as movement through the sediment probably requires more tortuous paths than surface locomotion. In addition, most motile diatoms and cyanobacteria exhibit a periodic reversal of direction even in the absence of sensory cues, so that the net advance of a cell or trichome is less than the maximal rate of locomotion. Furthermore, positive and negative photokinesis (the dependence of locomotive speed on the light intensity reaching the cells) is known in many microorganisms, primarily diatoms and cyanobacteria, and shown to be driven by different wavelength bands in different microorganims [46,47].

Light properties that may be used by different sensing systems include intensity, polarization, color and directionality [48]. From the knowledge of sedimentary light fields and of physiological photoresponses, however, some generalized patterns may be expected or predicted. True phototaxis (the ability to move towards or away from the direction of light [47,49]), may not be of much help for microorganisms adapted to live at low light intensities in the subsurface of sediments, soils and mats, since the light fields may be close to diffuse deep below the surface. However, directed movements can still be of much use in microorganims dwelling at or close to the sediment surface, where the light fields contain a significant downward directionality. Surface dwelling cyanobacteria such as *Lyngbya* sp. from hot springs mats [50] and intertidal sediments [51] and the motile phases (hormogonia) of terrestrial *Nostoc* spp. from desert soils (Garcia-Pichel, unpublished) show this type of movement directed towards the light source. The bundle forming *Microcoleus chthonoplastes* is also able to display a "populational phototaxis", in that bundles of trichomes of this cyanobacterium are able to steer in the direction of the incoming light, whereas single trichomes are not apparently able to do so [52]. Typical upright growth is often displayed by surface dwelling *Scytonema* in terrestrial populations [53] and in intertidal carbonate stromatolites (L. Prufert-Bebout, pers. comm). However, photophobic responses in which the direction of movement is dependent on the detection of abrupt changes in light intensity [47,49] may often be used. This is most frequently realized by variation in the frequency of reversals in the direction of movement, so that, for example, an organism will reverse direction when sensing a sharp step-up in light intensity. This may involve only the first few leading cells in a motile filamentous cyanobacterium. The step-up photophobic response will result in a net accumulation of organisms in the regions of lower light intensity. In a step-down photophobic response, the organisms will tend to accumulate in the region of higher light intensity. Photophobic responses are the basis of photomovement in all flagellated bacteria [1], in most gliding cyanobacteria [49] and diatoms [54]. Step-up photophobic responses are probably very much involved in the migrations of gliding bacteria in natural microbial mats [19,28,31], although step-down photophobic responses have been most commonly studied in the laboratory. Photophobic responses are very well suited for determining directionality in the extremely steep gradients of either directed or diffuse light, and the sedimentary, soil and mat microenvironments may strongly select for this type of response. However, the use of photophobic responses within sedimentary light fields may not be problem-free as the changes in light intensity may be gradual and not always sudden [55]. If a population occupies the sediment surface during the night and a maximum in scalar irradiance occurs just below the surface when the sediment is illuminated (Figure 1), it is in principle possible that the population may become trapped at the surface during daytime, unable to move downward against a short, steep gradient of increasing light intensity, before the gradient starts to decrease with depth. If this theoretical consideration holds in nature, it would carry a selective advantage for benthic microorganims to respond preferentially to short-wavelength radiation, as the maxima in scalar irradiance are less pronounced and tend to occur closer to the surface. Indeed, most diatoms, but not all, show maximal responses to blue light [46,54]. Although pure-culture studies pointed out that the (step-down) photophobic response of cyanobacteria is driven preferentially by red wavelengths [47], the upward migrations of cyanobacteria in mats is preferentially prevented by short wavelengths, especially by UV radiation [29,31,56,57] and not by red (nor green) light.

14.6 Selective advantage of photomovements

Photomovements in phototrophic benthic microorganims may provide a means to avoid burial, to avoid photodamage and optimize photosynthetic performance or a means for microniche selection. These will be discussed separately in the following, but they should not be regarded as mutually exclusive functions.

14.6.1 Avoidance of burial

Most sediments will either slowly accrete by sedimentation of allochthonous matter, by mineral precipitation in situ, by growth of microorganisms, or by a mixture of all these processes. The importance of burial events in benthic communities becomes most evident in the formation of multilayered cyanobacterial mats. Each of the layers corresponds to what once were surface benthic communities, their buried remains lying, potentially active but dormant, well below the present photic zones in the mats [58]. But phototrophic organisms need to remain in the photic zones of the sediments, so that mechanisms for benthic microbial populations to avoid permanent burial and to recolonize new surfaces are of indisputable adaptive value. Even unicellular cyanobacteria such as the extremely halotolerant forms that build microbial mats in hypersaline waters or the extremely themophilic cyanobacteria of hot springs, display movements towards the light [59–61]. In some of these cyanobacteria, the gliding velocities (0.1–0.3 $\mu m\ s^{-1}$) measured [60] could possibly suffice for the populations to respond to fast changes in environmental conditions, but in many, the speeds are simply too slow to offer benefits for daily migrations; their importance can only be associated with long-term changes in light conditions such as those produced by slow burial or seasonal fluctuations in incident irradiance. The ability to develop specialized, light-responsive motile cells in otherwise nonmotile cyanobacteria, such as those grouped in the order Pleurocapsales or the formation of highly motile hormogonia in filamentous forms such as *Scytonema*, *Nostoc*, *Calothrix* and *Mastigocladus* [62], may also provide a means of founding new populations after catastrophic burial events. In a study of responses to burial with fast gliding cyanobacteria, Pentecost [28] measured the return of an oscillatorian cyanobacterium (gliding speed, ca. 1.2 $\mu m\ s^{-1}$) buried under 3 mm of silt. In the presence of optimal light conditions (ca. 4 $W\ m^{-2}$) it took 1.5 h for 50% of the filaments to return to the new sediment surface, and the return was almost complete within 3 h. In the dark, the cyanobacteria were also able to return to the surface but much more slowly and inefficiently: only ca 40% had returned to the surface within 3 h.

14.6.2 Optimization of photosynthetic performance

Phototrophic microorganims in nature are subjected to rapidly varying incident light intensities due to either the normal diel cycle or to cloud cover. Regulating their exposure by migrating through the sediment seems a plausible mechanism which will prevent them from having to break down and resynthesize potentially photosensitizing light-harvesting pigments in order to match incident radiation. Because the light intensity in sediments and microbial mats may vary by several orders of magnitude within short vertical distances, the possibility exists for motile benthic phototrophs to adapt into a "shade" type, with high contents of light harvesting pigments resulting in photosynthesis saturation at relatively low fluence rates, and to follow the corresponding saturating light intensities along the benthic light gradients. Indeed, this seems to be the strategy behind the fast migrations of the few benthic cyanobacteria studied so far. The photosynthetic performances of migrating field populations of cyanobacteria (*Microcoleus lyngbyaceus*, *Spirulina* cf. *subsalsa* and two *Oscillatoria sp.*) have been measured in several occasions [21,28,29,31]. Photosynthetic saturation occurred below or around 10% of incident radiation in all cases. In a single experiment in which the light fields, the photosynthetic performance, and the positioning of migrating populations of *Oscillatoria* and *Spirulina* spp. were determined concurrently, it was possible to calculate that the cyanobacteria operated around photosynthetic saturation, avoiding photoinhibitory intensities throughout the day [21].

14.6.3 Avoidance of photodamage

Exposure to excessive visible and/or ultraviolet may bring about considerable photodamage to microorganisms. Castenholz [16,63] considered the aggregating behavior of *Oscillatoria* cf. *terebrifromis* from hot spring mats (Figure 5) as a possible simple response to excessive radiation, whereby self-shading should minimize photodamage. After some initial experiments it had been hypothesized that UV radiation may play a significant regulatory role in the photomotility responses of benthic cyanobacteria [56]. In laboratory experiments with natural microbial mats from Solar Lake [57], evidence was obtained that UV-B radiation, at intensities typical for the natural setting was the most effective wavelength region promoting the downward retreat of populations of the cyanobacterium *Microcoleus chthonoplastes*. Later, in a comprehensive study of natural hypersaline benthic mats, Kruschel and Castenholz [31] could clearly demonstrate the importance of UV radiation in modulating the migratory behavior of both *Spirulina* cf. *subsalsa* and of *Oscillatoria* cf. *laetevirens*. Downward migration of *O.* cf. *laetevirens* was promoted by both visible light (intensities above 400 W m^{-2}) and by UV-A alone (intensities above 10 W m^{-2}). No retreats occurred with exposure to UV-B alone. Upward migration of either *S. subsalsa* or *O. laetevirens* occurred in the dark and under low visible intensities, (20 W m^{-2}) but it was prevented

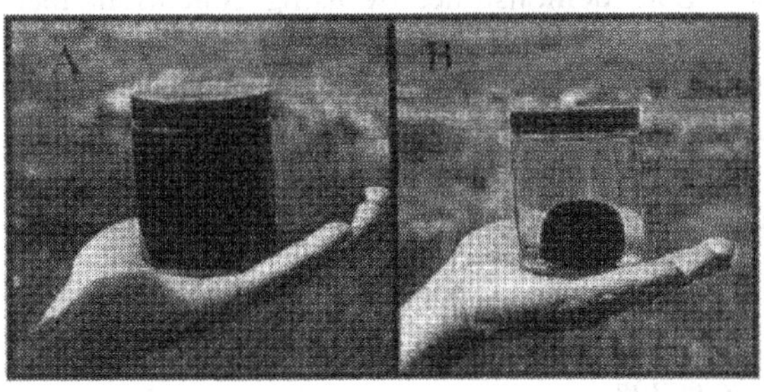

Figure 5. Aggregating (clumping) behavior of *Oscillatoria* cf. *terebriformis*. from Hunter's Hot Springs (Oregon, USA) A population of trichomes removed from the spring benthic mat and dispersed in spring water. B After 5 min in the light, trichomes have clumped into a ball, as a result of gliding motility. Temperature was approximately 40°C.

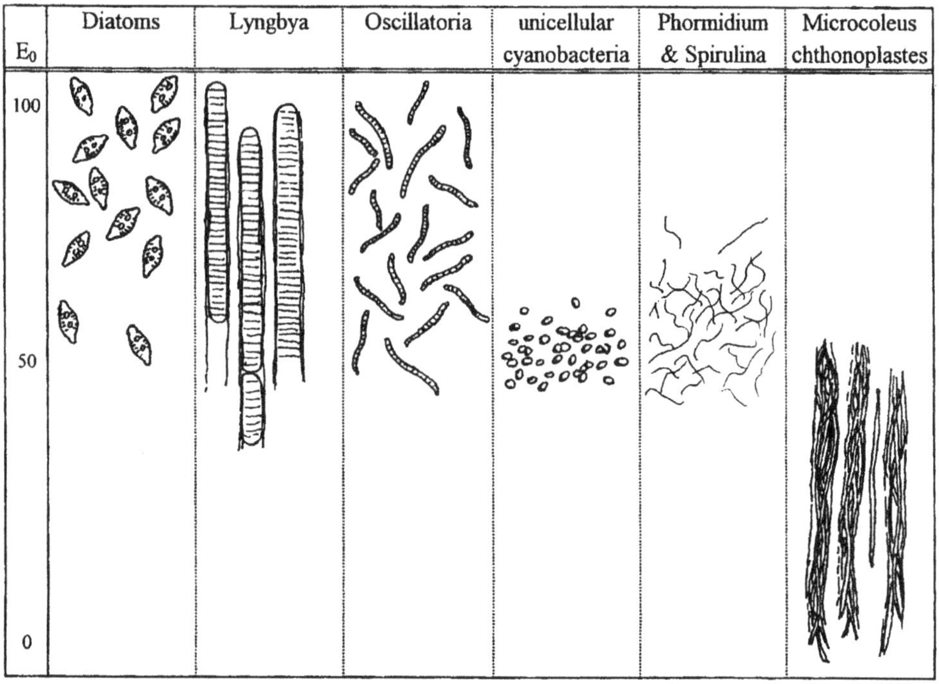

Figure 6. Spatial distribution reached through photomovements in several diatoms and cyanobacteria, typical inhabitants of benthic microbial mats, incubated within gradients of light intensity. Intensity (E0, left) varied between 100 μmol photons (PAR) m^{-2} s^{-1} and undetectable, the total distance covered by the gradient was ca. 3 cm. The distribution of the organisms was measured after 1 week of inoculation at the center of the gradient. The final distribution along the gradients reached, corresponds roughly to their natural distribution with depth in natural samples (from [51], with permission).

by very low intensities of UV-A (1.5 W m^{-2}) and perhaps also UV-B (0.1 W m^{-2}), whereas high intensities of visible light (100 W m^{-2}) were needed to attain similar results. The authors could demonstrate, by using concurrent measurements of UV-photodamage on the cyanobacteria and benthic light fields in the mats, that the migrating behavior represents a strategy which includes avoidance of damage by UV radiation and exposure to sufficient visible light for photosynthesis. Thus, this may be considered an optimization of photosynthetic performance. Recently, the possible influence of UV radiation on the photoresponses of marine benthic diatoms (*Gyrosigma balticum*) has also been recognized [64].

14.6.4 Microniche selection

In benthic microalgal communities one can often distinguish a layered distribution of specific organisms along the vertical gradients in light intensity and spectral distribution. Microbial mats and intertidal sandy sediments may exhibit, for example, a golden-brown diatom rich top zone, underlain by one to several distinct layers of cyanobacterial species. Using artificial microgradients of light intensity, Prufert-Bebout [51] has investigated the long-term photoresponses displayed by several diatoms and cyanobacteria isolated from intertidal mats and sediments. *Nitzchia scalpeliformis* and *Navicula* sp. (diatoms) and *Lyngbya aestuarii* (cyanobacterium), typical inhabitants of the top layers or "sun types", moved within the gradients towards the highest irradiance provided (100 μmol m^{-2} s^{-1}, PAR) while *Microcoleus chthonoplastes* moved within the gradient so that accumulation occurred at around 20 μmol m^{-2} s^{-1}. In general, the responses were such that they could explain the vertical distribution as found in natural settings, implying that photoresponses play a significant role in niche selection and the formation of multilayerd microalgal communities (Figure 6).

14.7 Conclusions and outlook

Up to date, our meager capacity to understand the photobiology of benthic migration in benthic systems is based on a few comprehensive studies and much indirect evidence. There is no question that a full understanding of the migrating behavior of microorganisms in terms of their photoresponses should involve concerted efforts leading to a sound description of the natural phenomenon, the optical and temporal circumstances in which it occurs, and the motility responses of the organisms involved, as studied under controlled conditions in the laboratory. Because microorganisms not only sense a range of environmental stimuli, but also are able to integrate the signals through common sensory pathways, the studies should not be relegated to photoresponses, but also to chemoresponses.

References

1. J.P. Armitage (1997). Behavioural responses of bacteria to light and oxygen. *Arch. Microbiol.*, **168**, 249–261.
2. K.J. Duffy, R.M.Ford (1997). Turn angle and run time characterize swimming behavior in *Pseudomonas putida*. *J. Bacteriol.*, **179**, 1428–1430.
3. B.B. Jørgensen, D.J. Des Marais (1986). A simple fiber-optic microprobe for high resolution light measurements. *Limnol. Oceanogr.*, **31**, 1376–1383.
4. B.K. Pierson, V.M. Sands, J.L. Frederick (1990). Spectral irradiance and distribution of pigments in a highly layered microbial mat. *Appl. Environm. Microbiol.*, **56**, 2327–2340.
5. C. Lassen, H. Ploug, B.B. Jørgensen (1992). A fiber-optic scalar irradiance microsensor: application for spectral light measurements in sediments. *FEMS Microbiol. Ecol.*, **86**, 247–254.
6. F. Garcia-Pichel (1995). A scalar irradiance fiber-optic microprobe for the measurement of ultraviolet radiation at high spatial resolution. *Photochem. Photobiol.*, **61**, 248–254.
7. B.B. Jørgensen, D.J. Des Marais (1988). Optical properties of benthic photosynthetic communities: fiber-optic studies of cyanobacterial mats. *Limnol. Oceanogr.*, **33**, 99–113.
8. M. Kühl, B.B. Jørgensen (1992). Spectral light measurements in microbenthic phototrophic communities with a fiber-optic microprobe coupled to a sensitive diode array detector. *Limnol. Oceanogr.*, **37**, 1813–1823.

9. M. Kühl, B.B. Jørgensen (1994). The light field of microbenthic communities: radiance distribution and microscale optics of sandy coastal sediments. *Limnol. Oceanogr.*, **39**, 1368–1398.
10. F. Garcia-Pichel, B.M. Bebout (1996). The penetration of ultraviolet radiation into shallow water sediments: high exposure for photosynthetic communities. *Mar. Ecol. Progress Ser.*, **131**, 257–262.
11. F. Garcia-Pichel, J. Belnap (1996). Microenvironments and microscale productivity of cyanobacterial desert crusts. *J. Phycol.*, **32**, 774–782.
12. B.B., Jørgensen, Y. Cohen, D.J.D. Marais (1987). Photosynthetic action spectra and adaptation to spectral light distribution in a benthic cyanobacterial mat. *Appl. Environm. Microbiol.*, **53**, 879–886.
13. B.K., Pierson, A. Oesterle, G.L. Murphy (1987). Pigments, light penetration and photosynthetic activity in the multi-layered microbial mats of Great Sippewissett Salt Marsh, Massachusetts. *FEMS Microbiol. Ecol.*, **45**, 365–376.
14. S. Winogradsky (1887). Über Schwefelbakterien. *Bot. Ztg.*, **45**, 489–610.
15. E. Fauré-Fremiet (1951). The tidal rhythm of the diatom *Hantzschia amphioxys*. *Biol. Bull.*, **100**, 17–33.
16. R.W. Castenholz (1968). The behavior of *Oscillatoria terebriformis* in hot springs. *J. Phycol.*, **4**, 132–139.
17. T.D. Brock (1975). Effect of water potential on a *Microcoleus* (Cyanophyceae) from a desert crust. *J. Phycol.*, **11**, 316–320.
18. C.M. Happey-Wood (1988). Vertical migration patterns of flagellates in a community of freshwater benthic algae. *Hydrobiologia*, **161**, 99–124.
19. D.C. Nelson, R.W. Castenholz (1982). Light responses in *Beggiatoa*. *Arch. Microbiol.*, **131**, 146–155.
20. S.J. Giovannoni, E. Schabtach, R.W. Castenholz (1987). *Isosphaera pallida*, gen. and comb. nov, a gliding, budding eubacterium from hot springs. *Arch. Microbiol.*, **147**, 276–284.
21. F. Garcia-Pichel, M. Mechling, R.W. Castenholz (1994). Diel migrations of microorganisms within a benthic, hypersaline mat community. *Appl. Environm. Microbiol.*, **60**, 1500–1511.
22. D., Krekeler, A. Teske, H. Cypionka (1998). Strategies of sulfate-reducing bacteria to escape oxygen stress in cyanobacterial mats. *FEMS Microbiol. Ecol.*, **25**, 89–96.
23. L.L. Richardson, R.W. Castenholz (1987). Diel vertical movements of the cyanobacterium *Oscillatoria terebriformis* in a sulfide-rich hot spring microbial mat. *Appl. Environm. Microbiol.*, **53**, 2142–2150.
24. R.W., Castenholz, B.B., Jørgensen, E. D'Amelio, J. Bauld (1991). Photosynthetic and behavioral versatility of the cyanobacterium *Oscillatoria boryana* in a sulfide-rich microbial mat. *FEMS Microbiol. Ecol.*, **86**, 45–58.
25. G.F. Whale, A.E. Walsby (1984). Motility of the cyanobacterium *Microcoleus chthonoplastes* in mud. *Br. Phycol. J.*, **19**, 117–123.
26. J. Pinckney, Y. Piceno, C.R. Lowell (1994). Short-term changes in the vertical distribution of benthic microalgal biomass in intertidal muddy sediments. *Diatom Res.*, **9**, 143–153.
27. A. Wulff, K. Sundbäck, C. Nilsson, L. Carlson, B. Jonsson (1997). Effect of sediment load on the microbenthic community of a shallow water sandy sediment. *Estuaries*, **20**, 547–558.
28. A. Pentecost (1984). Effects of sedimentation and light intensity on mat-forming Oscillatoriaceae with particular reference to *Microcoleus lyngyaceus* Gomont. *J. Gen. Microbiol.*, **130**, 983–990.
29. T.-L. Nadeau, C. Howard-Williams, R.W. Castenholz (1999). Effects of solar UV and visible irradiance on photosynthesis and vertical migration of *Oscillatoria* sp (Cyanobacteria) in an Antarctic microbial mat. *Aquat. Microb. Ecol.*, **20**, 213–243.

30. M.C. Davey, K.J. Clarke (1991). The spatial distribution of microalgae on Antarctic fellfield soils. *Antarctic Sci.*, **3**, 257–264.
31. C. Kruschel, R.W. Castenholz (1988). The effect of solar UV and visible irradiance on the vertical movements of cyanobacteria in microbial mats of hypersaline waters. *FEMS Microbiol. Ecol.*, **27**, 53–72.
32. L.L. Richardson, R.W. Castenholz (1989). Chemokinetic motility responses of the cyanobacterium *Oscillatoria terebriformis. Appl. Environm. Microbiol.*, **55**, 261–263.
33. B.B. Jørgensen (1982). Ecology of the bacteria of the sulphur cycle with special reference to anoxic-oxic interface environments. *Philos. Trans. R. Soc. London Ser. B*, **298**, 543–561.
34. C.-S. Jue (1990). *The effect of aerobic environments on* Chromatium *cf.* tepidum, *a thermophilic purple sulfur bacterium.* M.A. Thesis, University of Oregon Eugene, Oregon, USA.
35. M., Kühl, C. Lassen, B.B. Jørgensen (1994). Optical properties of microbial mats: Light measurements with fiber optic microprobes. In: L.J. Stal, P. Caumette (Ed.), *Microbial Mats. Structure, Development and Environmental Significance* (pp. 149–166). Springer-Verlag, Berlin.
36. M.A. Harper (1977). Movements. In: D. Werner, (Ed.), *The biology of diatoms* (pp. 224–249). University of California Press, Berkeley.
37. F.E. Round, J.D. Palmer (1966). Persistent vertical migration rhythms in benthic microflora. II. Field and laboratory studies on diatoms from the bank of the river Avon. *J. Mar. Biol. Assoc. UK*, **46**, 191–224.
38. C.M. Happey-Wood, P. Jones (1988). Rhythms of vertical migration and motility in intertidal benthic diatoms with particular reference to *Pleurosigma angulatum. Diatom Res.*, **3**, 89–93.
39. S.I. Hay, T.C. Maitland, D.M. Patterson (1993). The speed of diatom migration through natural and artificial substrata. *Diatom Res.*, **8**, 371–384.
40. J.W. Eaton, B. Moss (1966). The estimation of numbers and pigment content in epipelic algal populations. *Limnol. Ocean.*, **11**, 584–594.
41. J. Serôdio, J.M. d. Silva, F. Catarino (1997). Nondestructive tracing of migratory rhythms of intertidal benthic microalgae using *in vivo* chlorophyll *a* fluorescence. *J. Phycol.*, **33**, 542–553.
42. D.M. Patterson (1986). The migratory behaviour of diatom assemblages in a laboratory tidal micro-ecosystem studied by low-temperature scanning electron microscopy. *Diatom Res.*, **1**, 227–239.
43. F.E. Round, J.W. Eaton (1966). Persistent, vertical migration rhythms in benthic microflora. III the rhythm of epipelic algae in a freshwater pond. *J. Ecol.*, **55**, 609–615.
44. N.P. Revsbech, B.B. Jørgensen, T.H. Blackburn, Y. Cohen (1983). Microelectrode studies of the photosynthesis and O_2, H_2S and pH profiles of a microbial mat. *Limnol. Oceanogr.*, **28**, 1062–1074.
45. G. Malin, A.E. Walsby (1985). Chemotaxis of a cyanobacterium on concentration gradients of carbon dioxide, bicarbonate and oxygen. *J. Gen. Microbiol.*, **131**, 2643–2652.
46. W. Nultsch (1971). Phototactic and photokinetic action spectra of the diatom *Nitzschia communis. Photochem. Photobiol.*, **14**, 705–712.
47. D.-P. Häder (1987). Photosensory behavior in prokaryotes. *Microbiol. Rev.*, **51**, 1–21.
48. P. Hegemann, Vision in microalgae, Planta, 203 (1997) 265–274.
49. R.W. Castenholz (1982,). Motility and Taxes. In: N.G. Carr, B.A. Whitton (Eds), *The biology of cyanobacteria* (pp. 413–440). Blackwell Sci. Pub., Oxford, England.
50. C. Mrozek (1990). *Physical and chemical factors influencing the phototactic steering response in* Lyngbya *sp. (U-Sara.-L).* M.A. Thesis, University of Oregon Eugene, Oregon, USA.

51. L. Prufert-Bebout (1998). *Environmental controls on cyanobacterial distribution: the role of irradiance*. Ph.D. Thesis, Aarhus University, Aarhus, Denmark.
52. L. Prufert-Bebout, F. Garcia-Pichel (1994). Field and cultivated *Microcoleus chthonoplastes*: The search for clues to its prevalence in marine microbial mats. In: L.J. Stal, P. Caumette (Eds), *Microbial Mats. Structure, Development and Environmental Significance* (pp. 111–116). Spinger-Verlag, Heidelberg.
53. F.E. Fritsch (1922). The terrestrial algae. *J. Ecol.*, **10**, 220–236.
54. S.A. Cohn, R.E. Weitzell (1996). Ecological considerations of diatom cell motility.I. Characterization of motility and adhesion of four diatom species. *J. Phycol.*, **32**, 928–939.
55. N.B. Ramsing, L. Prufert-Bebout (1994). Motility of *Microcoleus chthonoplastes* subjected to different light intensities quantified by digital image analysis. In: L.J. Stal, P. Caumette (Eds), *Microbial Mats. Structure, Development and Environmental Significance, NATO ASI Series* (pp. 183–192). Springer Verlag, Heidelberg.
56. F. Garcia-Pichel, R.W. Castenholz (1994). On the significance of solar ultraviolet radiation for the ecology of microbial mats. In: L.J. Stal, P. Caumette (Eds), *Microbial Mats. Structure, Development and Environmental Significance* (pp. 77–84). Springer-Verlag, Heidelberg.
57. B.M. Bebout, F. Garcia-Pichel (1995). UVB-induced vertical migrations of cyanobacteria in a microbial mat. *Appl. Environm. Microbiol.*, **61**, 4215–4222.
58. B.B. Jørgensen, N.P. Revsbech, Y. Cohen (1983). Photosynthesis and structure of benthic microbial mats: microelectrode and SEM studies of four cyanobacterial communities. *Limnol. Ocean.*, **28**, 1075–1093.
59. R.D. Simon (1981). Gliding motility in *Aphanothece halophytica*: analysis of wall proteins in *mot* mutants. *J. Bacteriol.*, **148**, 315–321.
60. N.B. Ramsing, M.J. Ferris, D.M. Ward (1997). Light-induced motility of thermophilic *Synechococcus* isolates from Octopus Spring, Yellowstone National Park. *App. Environm. Microbiol.*, **63**, 2347–2354.
61. F. Garcia-Pichel, U. Nübel, G. Muyzer (1998). The phylogeny of unicellular, extremely halotolerant cyanobacteria. *Arch. Microbiol.*, **169**, 469–482.
62. W. Hernández-Muñiz and S.E. Stevens Jr. (1987). Characterization of the motile hormogonia of *Mastigocladus laminosus*. *J. Bacteriol.*, **169**, 218–223.
63. R.W. Castenholz (1967). Aggregation in a thermophilic *Oscillatoria*. *Nature*, **215**, 1285–1286.
64. G.J.C. Underwood, C. Nilsson, K. Sundbäck, A. Wulff (1999). Short-term effects of UVB radiation on chlorophyll fluorescence, biomass, pigments and carbohydrate fractions in a benthic diatom mat. *J. Phycol.*, **35**, 656–666.
65. R.W. Castenholz, F. Garcia-Pichel (1999). In: B.A. Whitton, M. Potts (Eds), *Ecology of Cyanobacteria: Their Diversity in Time and Space* (2000) (pp. 591–612). Kluwer Sci. Publ., Dordrecht, The Netherlands.

Photomovement
D.-P. Häder and M. Lebert, editors.

Chapter 15

Phytochrome as an algal photoreceptor

Gottfried Wagner

Table of contents

Dedicated to Professor Wolfgang Haupt on the occasion of his 80th birthday.

Abstract

Phytochrome in the cryptophytic kingdom of blue-green, red and green algae is described in this review, in terms of molecular properties and biological function. With the recent discovery of *Synechocystis* phytochrome, the concept of phytochrome being a light-regulated kinase was revitalized, even though phytochrome-dependent kinase activity in eukaryotic cells is still a matter of controversy. Possibly, the phytochrome signal transducing machinery has been adopted in plants through endocytobiosis of cyanobacteria and chloroplast development.

Phytochrome function in conjugating green algae, namely *Mougeotia* and *Mesotaenium,* has been studied in details. Unexpectedly, a domain at the C-terminal end of both pigments was discovered to be reminiscent of a microtubule-associated protein, not a trans-membrane protein. Also in *Mougeotia*, the cylindrical scaffold of microtubules was found to be light-regulated, mediated by calcium-calmodulin. Thus, the microtubular scaffold and associated proteins appear worthwhile to be considered as candidates to bridge the gap in the transduction chain between the formation of the "tetrapolar" phytochrome gradient in the *Zygnematales* and chloroplast orientation with respect to light. In a mono-molecular or multi-molecular "reaction unit", the blue-light photoreceptors in discussion need to be obeyed here as well, for competition to or co-action with phytochrome, both handling the chloroplast rotation through consistent control of the actin-myosin motor apparatus.

15.1 Introduction

In the course of plant evolution, phytochrome responses have been reported for blue-green, red and green algae ([1] for bibliography; [2]). In the blue-greens, i.e. cyanobacteria, the complete sequence of the *Synechocystis* chromosome has revealed a phytochrome-like sequence that yielded an authentic phytochrome when overexpressed in *Escherichia coli* [3]. Evidence for its physiological function is emerging: interruption or partial deletion of the PCC 6803 phytochrome gene yielded mutants unable to grow under blue light [4]. Also, the photoreceptor for positive phototaxis appears likely to be a phytochrome-like tetrapyrrole rather than chlorophyll *a* [5]. For red algae no genomic proof was performed as yet, but early evidence for the physiological function of phytochrome is given in *Porphyra tenera* [6,7].

15.2 Algal phytochrome physiology

The localization and dichroic orientation of the red/far red reversible photoreceptor and the factors which bring about the phytochrome responses (photoorientation) have been characterized in detail [8] in two species of conjugating green algae (*Zygnematales*, Figure 1). Also, phytochrome-mediated morphogenetic events were reported here [9–11], with recent inclusion of the close relatives *Spirogyra hyalina* and *Spirogyra crassa* [12,13]. The two species *Mesotaenium caldariorum* and *Mougeotia scalaris* (*Zygnematales*, Figure 2) have emerged as classical examples of photobiology in non-flagellates [8,14,15].

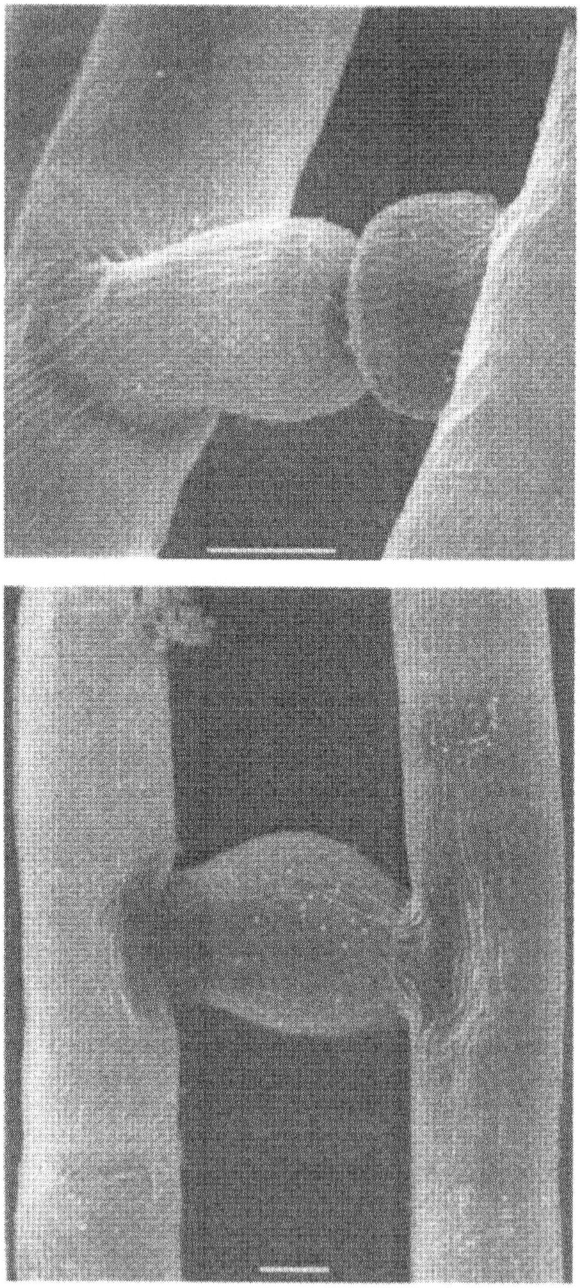

Figure 1. Scanning electron-microscopic view of two separate filaments of *Mougeotia scalaris* in vicinity to each other, with conjugation tube and zygospore formation proceeding half-way between two mating cells. Top: two protrusions grow out from two separate cells, to reach each other and to form a conjugation tube. Bar 25 mm. Bottom: A conjugation tube is formed and a zygospore inside has developed half way between the two now empty *Mougeotia* cell cylinders. Note loss of turgor here. Bar 20 μm (courtesy of U. Richter and A. Schmidt, unpublished).

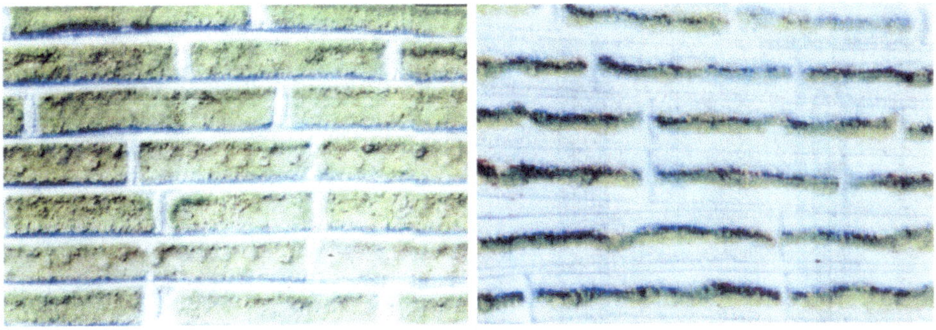

Figure 2. Micrograph of filaments of *Mougeotia scalaris*, oriented parallel to each other on a microscopic slide to form a closed layer. Left-hand side: The ribbon-shaped chloroplasts are in face position in order to fully collect low irradiance light, incident from above (= low irradiance response, syn. light-attractant response). Right-hand side: The ribbon-shaped chloroplasts are in profile position in order to avoid high irradiance light, incident from above (= high irradiance response, syn. light-avoidance response). (Billek and Wagner, unpublished).

Mougeotia is a filamentous green alga whose cells contain a single ribbon-like chloroplast nested between two large vacuoles and surrounded by a layer of cytoplasm [14]. The close relative, *Mesotaenium*, is a single-celled green alga but otherwise much alike *Mougeotia*. In both species, the chloroplast can rotate about its long axis and can respond to incident radiation by orienting perpendicular to the direction of light (Figures 2 and 3a, for further details see e.g. [16]). Phytochrome is implicated in the response because the systems are most sensitive to red light (see Figure 3b) and the red light effect can be reversed by far-red light. In addition to phytochrome, a blue light photoreceptor is at work in *Mougeotia* and *Mesotaenium*, leading to the same pattern of chloroplast orientation in low irradiance blue light ([17], face position, see Figure 2, left-hand side, and Figure 3a). High irradiance blue light leads to a light avoidance response with the chloroplast face parallel to the incident radiation (profile position, see Figure 2, right-hand side, and Figure 3a). In co-action, the blue light photoreceptor predominantly obeys light irradiance here, while phytochrome monitors light direction ([18] and references therein).

Using microbeams of red and far-red light, Wolfgang Haupt and his co-workers showed that phytochrome for chloroplast rotation is located near the periphery of the cytoplasm in the vicinity of the plasma membrane. When a microbeam of red light was directed at the cell surface under the microscope, the edge of the chloroplast adjacent to the microbeam rotated 90°, even though the chloroplast itself was not illuminated [19].

The plasma membrane localization of phytochrome specifically in *Mougeotia* was deduced from studies using microbeams of plane-polarized red and far-red light. Phytochrome is a dichroic pigment; that is, it has a preferred electrical vector orientation for the absorption of light. The highest absorption occurs when the electric vector is parallel to the absorbing vector of the pigment (see Figure 4 for demonstration of the phenomenon by means of rod-shaped crystals of dichroic bacteriorhodopsin). Haupt and his co-workers discovered that the ability of plane-polarized light to cause rotation depends on the plane of polarization relative to the long axis of the cell. This

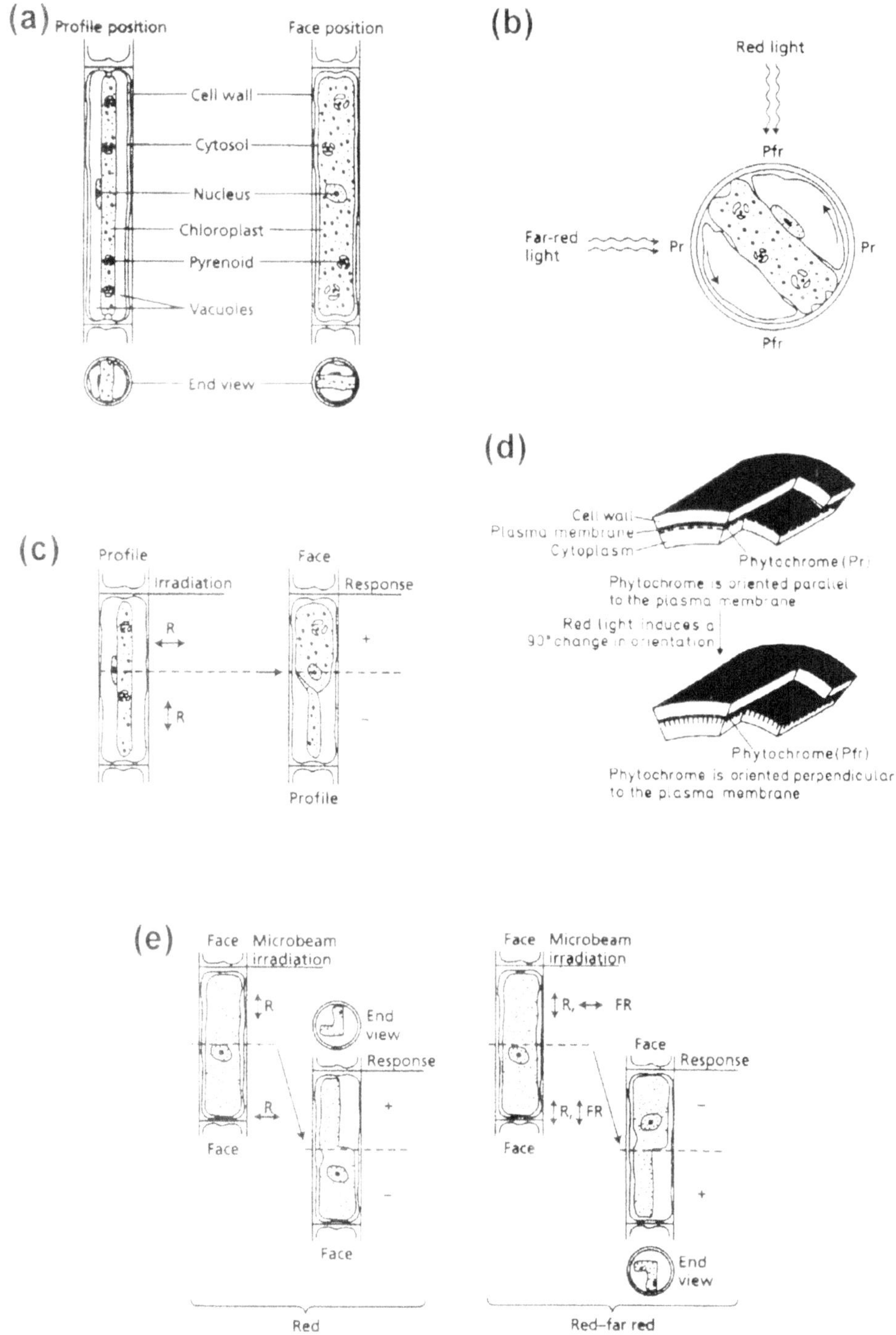
(a)
Profile position
Face position
Cell wall
Cytosol
Nucleus
Chloroplast
Pyrenoid
Vacuoles
End view
(b)
Red light
Pfr
Far-red light
Pr
Pr
Pfr
(c)
Profile
irradiation
Face
Response
R
R
+
-
Profile
(d)
Cell wall
Plasma membrane
Cytoplasm
Phytochrome(Pr)
Phytochrome is oriented parallel to the plasma membrane
Red light induces a 90° change in orientation
Phytochrome(Pfr)
Phytochrome is oriented perpendicular to the plasma membrane
(e)
Face
Microbeam irradiation
R
R
End view
Response
+
-
Face
Face
Red
Face
Microbeam irradiation
R, FR
R, FR
Face
Response
-
+
Face
End view
Red–far red

◄

Figure 3. Sketches of a *Mougeotia* cell as part of a multicellular filament; different chloroplast responses after different light treatments are shown as well. (a) A *Mougeotia* cell in top and end view, respectively. (b) The chloroplast rotation as the result of a tetrapolar phytochrome gradient. In cross orientation, microbeams of red and far-red light emphasize the Pfr/Pr-situation. (c) Partial irradiation of the *Mougeotia* cell cylinder by linearly polarized light (↔ orientation of the electrical vector E, see also Fig. 4) and partial response of the chloroplast (R red light). (d) Change in transition moment of phytochrome in the cortical layer of cytoplasm: Phytochrome (Pr) is putatively positioned surface-parallel versus phytochrome (Pfr) surface-perpendicular. (e) Experiment of photoreversibility of the chloroplast response as a result of change of orientation of the electrical vector (↔) and of wavelength (R red light, FR far red light), as indicated. The light treatments were given through microbeam irradiations at the right side cortical layer of cytoplasm, outside the chloroplast (modified after [8,16,20]).

phenomenon is depicted in the experiment shown in Figure 3c. Half of a cell was illuminated with plane-polarized red light vibrating transverse to the cell axis, while the other half was illuminated with light vibrating parallel to the cell axis. The chloroplast rotated from the profile to the face position only in response to red light vibrating in the transverse direction, producing a twist in the chloroplast. It was concluded that the active phytochrome has a defined orientation in the cell [8].

Later studies by the same laboratory imply that Pr molecules are arranged in left-handed spirals around the cell parallel to the cytoplasmic boundary while Pfr appears to be arranged perpendicularly to the cell wall.

What type of experiments have lead to the conclusion of specific phytochrome orientation in *Mougeotia*? The left and right hand sketches in Figure 3e show the result of two experiments through microbeam irradiations (black bars) at the right side cortical layer of cytoplasm, outside the chloroplast in two-dimensional projection of the cylindrical cell [20]. The two experiments start from the chloroplast oriented in face position. In the first experiment sketched at the left-hand side of Figure 3e, the cytoplasmic layer to the right is pulse-irradiated through slits of red light the plane-polarized radiation vibrating either parallel (upper light treatment) or perpendicular (lower light treatment) to the cell's long axis. In apparent contradiction to the results in Figure 3c, chloroplast reaction in this microbeam experiment is seen only when the red light was vibrating parallel to the cell's long axis (for explanation, see below). In the second experiment sketched at the right-hand side of Figure 3e, the cytoplasmic layer to the right is pulse-irradiated first through slits of red light the plane-polarized radiation vibrating parallel to the cell's long axis. Consecutively and without delay in this photoreversibility experiment, this first red light treatment is followed by a second pulse of light of different wavelength and plane of polarization. Here, the upper right side of the cytoplasmic layer is irradiated through a slit with far-red light with the plane-polarized radiation vibrating perpendicular to the cell's long axis, while the right side lower part of the cytoplasmic layer is irradiated through a slit with far-red light with the plane-polarized radiation vibrating parallel to the cell's long axis. Again, in apparent contradiction to the results in Figure 3c, the chloroplast reaction is seen here only when the plane-polarized radiation of both red and far-red light is vibrating parallel to the cell's long axis. The apparent contradiction is elegantly resolved by a model, understood

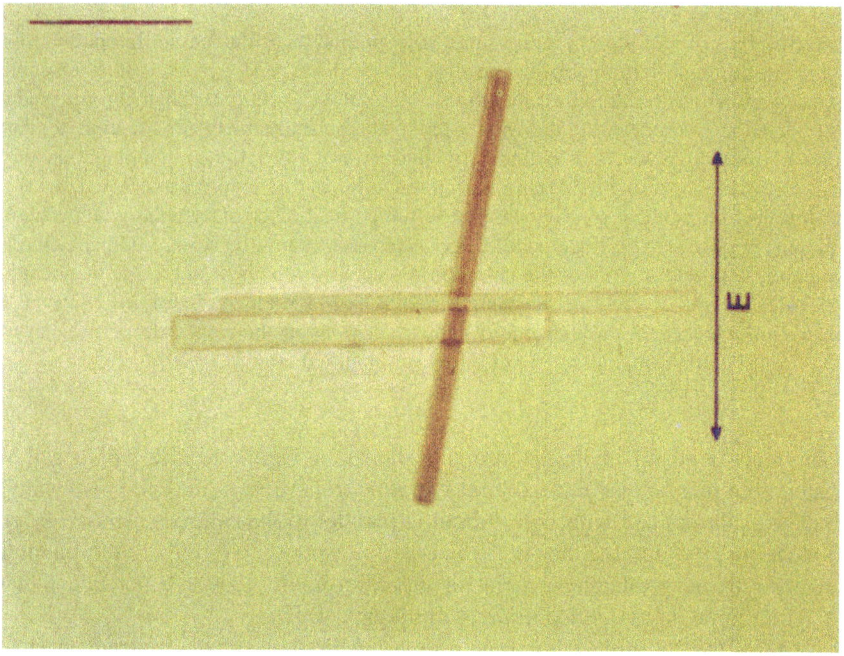
E

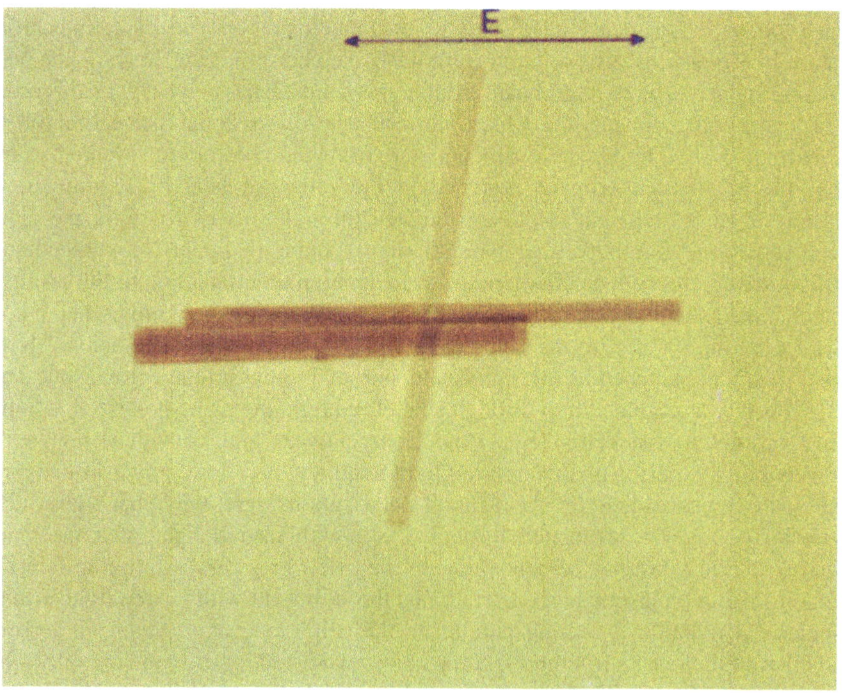
E

Figure 4. Demonstration of the phenomenon of absorption dichroism of linearly polarized light by a dichroic pigment. For the demonstration here, the retinal-protein bacteriorhodopsin has been used and grown in the shape of crystalline rods of up to 300 μm in length. Three of such rods were positioned here on a microscopic slide, two of them parallel to each other and one across. The microscope used was equipped with a linear polarizer of visible light. Bar 100 μm. Top: The plane of polarization was rotated almost parallel to the longitudinal axis of the single rod of crystalline bacteriorhodopsin (see E ↔), with full light absorption here but none by the two rods across. Bottom: The plane of polarization was rotated parallel to the longitudinal axis of the two rods of crystalline bacteriorhodopsin with full light absorption here, but none by the single rod positioned in perpendicular orientation (E electrical vector of the linearly polarized light used; ↔). In conclusion, the bacteriorhodopsin crystals show a preferred electrical vector orientation for the absorption of light parallel to the longitudinal crystal axis (courtesy of A. Schmidt, unpublished; see also [96]).

as the "Jaffe-Haupt model", of specific orientation of phytochrome in *Mougeotia* and possibly other systems [21], and change of the transition moment upon photoconversion of Pr to Pfr and vice versa (Figure 3d).

Finally, how does the chloroplast orientation in *Mougeotia* work? The effects described above are best accounted for by a scheme in which phytochrome is located on the plasma membrane of *Mougeotia* (Figure 3d) or on the microtubules forming a cylindrical scaffold underneath the membrane (Figure 5). It is difficult to imagine how

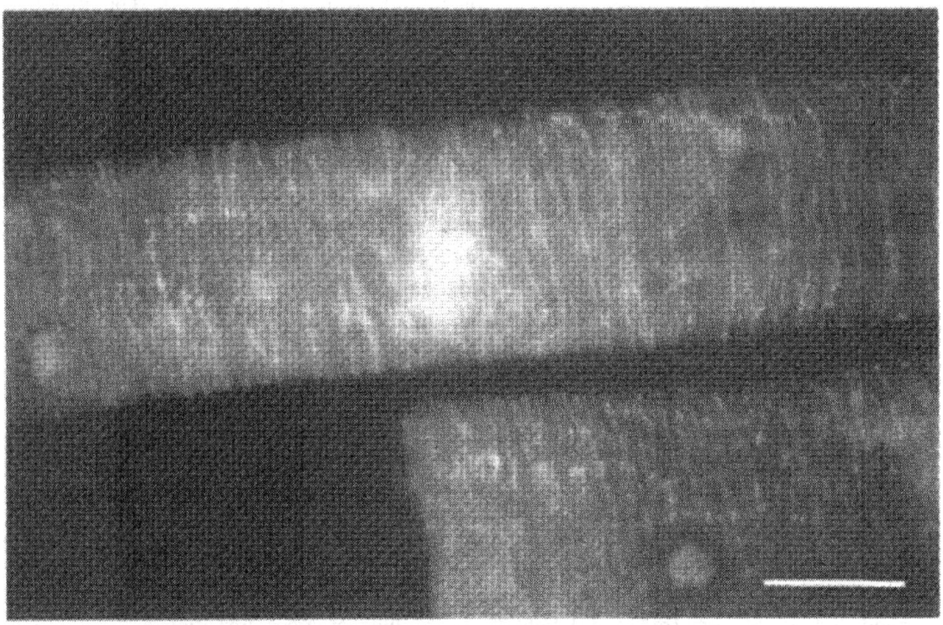

Figure 5. Immunofluorescence microscopy of cortical microtubules in *Mougeotia scalaris*. The microtubules run around the inner circumference of the cell membrane to form a cylinder such as the cell membrane. Bar 20 μm (from [59] with permission).

Pr and Pfr could maintain their respective orientation without being associated with some peripheral, relatively immobile structure such as the membrane or the cylinder of microtubules. Rotation occurs because the chloroplast moves away from a localized region of Pfr and toward a region of Pr. More specifically, the chloroplast movement occurs in response to a tetrapolar gradient of the active form of phytochrome, Pfr, forming a pigment pattern at the periphery of the cell cylinder (Figure 3b). The pigment gradient is translated into a gradient of actin-myosin interaction [22,23].

15.3 Molecular biology of prokaryotic phytochrome

Recent advances in molecular biology make the molecular analysis of algal phytochromes possible without protein purification, and the molecular species comprising these phytochrome gene families can be investigated.

Full length sequences of pro- and eukaryotic algal phytochromes have been reported for *Synechocystis* [3,24], *Mesotaenium* [25] and *Mougeotia* [26]. For *Synechocystis* phytochrome (Cph1) regions of strong similarity to plant phytochromes were found throughout the coding sequence whereas C-terminal homologies identify it as a likely sensory histidine kinase, a family to which plant phytochromes are related [24]. This, however, does not prove that the gene product is a genuine phytochrome. Phycocyanin levels prevent spectral photoreversibility measurements of phytochrome in cyanobacteria, so Hughes and coworkers [3] investigated the putative Cph1 gene product by expression-cloning in *E. coli* using the vector pQE12. The clone was engineered to express a C-terminal polyhistidine tag for nickel-affinity purification. The product, finally, was eluted as a homogenous apoprotein which could be concentrated to a 5–10 mg ml^{-1} solution.

Plant phytochrome apoproteins auto-catalytically attach linear tetrapyrrol chromophores such as phycocyanobilin (PCB) [27], abundant in the cytoplasm of cyanobacteria. Indeed, the *Synechocystis* PHY apoprotein attached purified PCB, producing a visibly photochromic holoprotein (Figure 6a). The reconstituted *Synechocystis* holo phytochrome was analyzed spectrophotometrically after exposure to saturating monochromatic 657 nm (red) and 731 nm (far-red) irradiation (Figure 6b). The spectra are reminiscent of plant phytochrome-PCB adducts with absorbance maxima at 658 and 702 nm after red and far-red irradiation, respectively, and an isosbestic point at 677 nm [28]. Hughes and coworkers [3] proudly state: ". . . there remains no barrier in principle to obtaining crystals for X-ray diffraction analysis of phytochrome molecular structure".

→

Figure 6. Photochromicity of expression-cloned and reconstituted *Synechocystis* phytochrome holoprotein. (a) Stoichiometric amounts of phycocyanobilin were added to PHY. After auto-assembly (20 min in darkness) the sample was divided and each portion irradiated with 731 nm (far-red, left) or 657 nm (red, right) light. Note the blue or green transition associated with phytochrome photoconversion. (b) Absorbance characteristics of the reconstituted phytochrome after irradiation with saturating red (R) or far-red (FR) light, and the calculated difference spectrum (from [3] with permission).

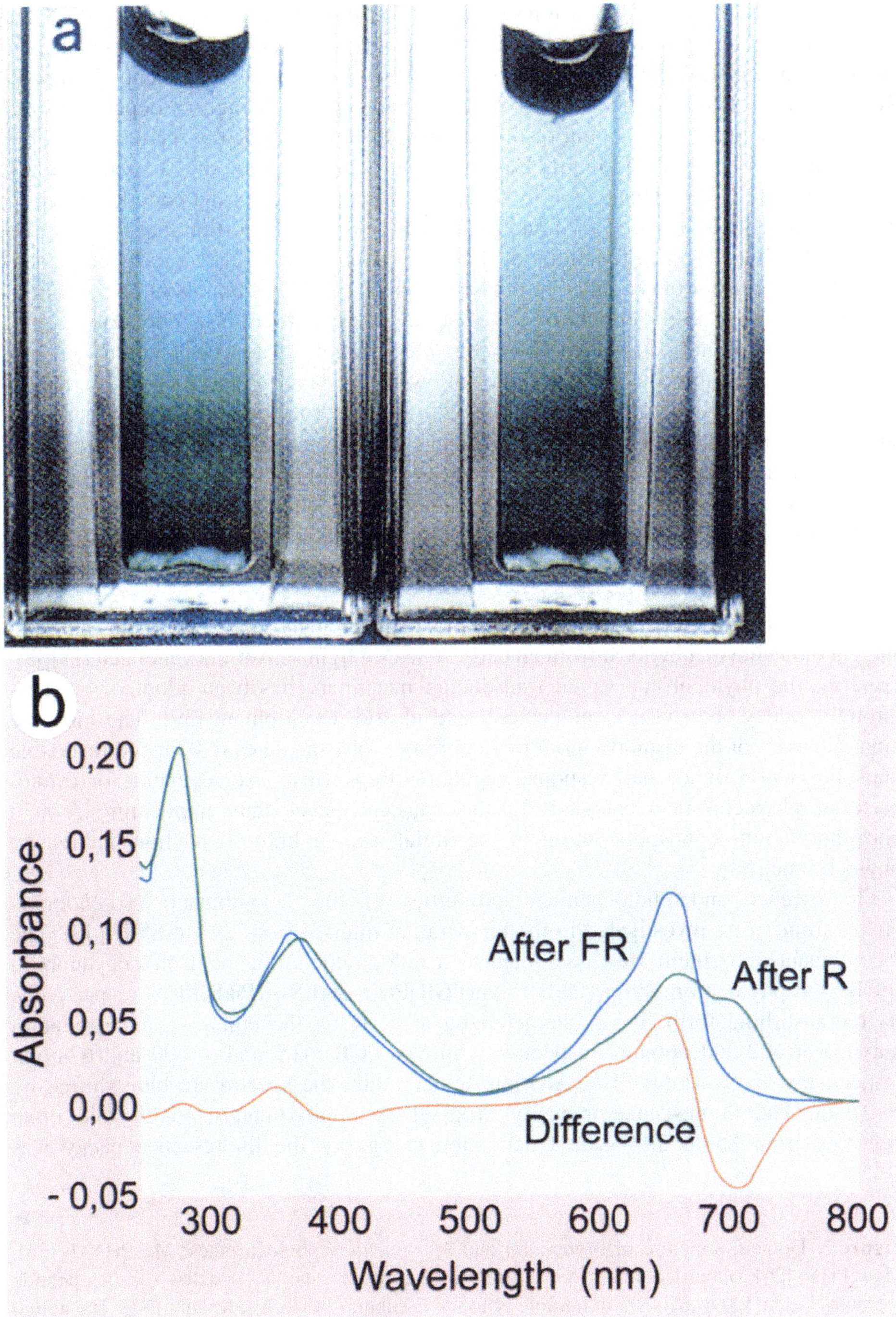
a
b
0,20
0,15
0,10
0,05
0,00
- 0,05
Absorbance
After FR
After R
Difference
300
400
500
600
700
800
Wavelength (nm)

Nevertheless, the three-dimensional structure of phytochrome from algae, mosses or higher plants remains obscure to date, even though successful experiments of phytochrome crystal growth were reported for the first time at the recent "European Symposium on Photomorphogenesis" [29,30]. For the time being, we depend on less powerful techniques of structural resolution such as low angle X-ray scattering [31,32] and homology modeling [33–36]. Based on the coordinates of the β-subunit of C-phycocyanin [37], e.g. *Mougeotia* phytochrome was modeled for the peptide stretch of Arg-320 to Ser-335 (Figure 7). An approximately 300 residue portion that is important for plant phytochrome function is missing from the *Synechocystis* sequence, immediately in front of the putative kinase region [24]. The recombinant apoprotein is soluble and can easily be purified to homogeneity by affinity chromatography. Phycocyanobilin and similar tetrapyrroles are covalently attached within seconds (Figure 6a), an autocatalytic process followed by slow conformational changes culminating in red-absorbing phytochrome formation. Spectral absorbance characteristics are remarkably similar to those of plant phytochromes (Figure 6b), although the conformation of the chromophore is likely to be more helical in the *Synechocystis* phytochrome. According to size-exclusion chromatography the native recombinant apoproteins and holoproteins elute predominantly as 115- and 170-kDa species, respectively. Both tend to form dimers *in vitro* and aggregate under low salt conditions.

Lagarias and co-workers conclude from biochemical analyses that phytochrome is an ancient molecule that evolved from an early state as a light sensor in cyanobacteria [38]. Possibly, the phytochrome signal transducing machinery has been adopted in plants through endocytobiosis of cyanobacteria and plastid development [39]. The histidine kinase activity of the cyanobacterial phytochrome Cph1 mediates red, far-red reversible phosphorylation of a small response regulator, Rcp1 (response regulator for cyanobacterial phytochrome), encoded by the adjacent gene, thus implicating protein phosphorylation-dephosphorylation in the initial step of light signal transduction by phytochrome [38].

Fluorescence and photochemical properties of the recombinant *Synechocystis* phytochrome were investigated in the temperature interval from 293 to 85 K [40]. The recombinant apoprotein was assembled to a holophytochrome with phycocyanobilin (PCB) and phytochromobilin (PϕB), Syn(PCB)PHY and Syn(PϕB)PHY, respectively. Its red-absorbing form, Pr, is characterized at 85 K by the emission and excitation maxima at 682 and 666 nm, respectively, in Syn(PCB)PHY and at 690 and 674 nm, respectively, in Syn(PϕB)PHY. At room temperature, the spectra are blue shifted by 5–10 nm. The fluorescence intensity dropped by approximately 15–20 fold upon warming from 85 to 293 K, and activation energy of the fluorescence decay was

Figure 7. Domain structure of *Mougeotia* and *Mesotaenium* phytochromes: MesPHY1b [25], MougPHY [26] (modified after [53]). *Mougeotia* phytochrome is modeled for the peptide segment of Arg-320 to Ser-335 in homology to the β-subunit of C-phycocyanin [37]. The amino acids Gly 321-Val 322 in the chromophore binding domain (-Arg-Gly-Val-His-Gly-Cys-) are characteristic of the zygnematophycean phytochromes known to date [26]. The chromophore is also displayed (courtesy of Ch. Betzel, unpublished).

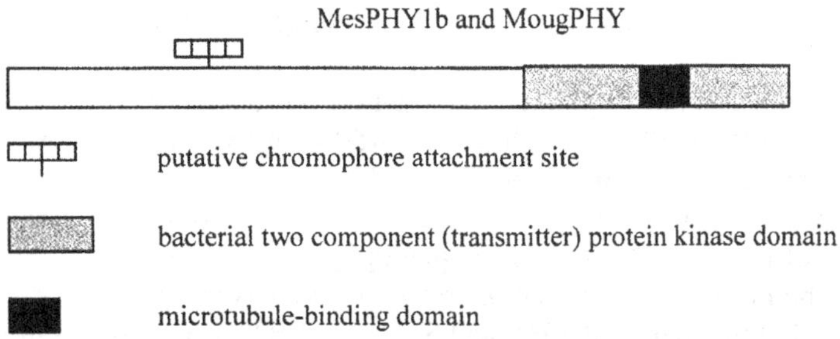
MesPHY1b and MougPHY
putative chromophore attachment site
bacterial two component (transmitter) protein kinase domain
microtubule-binding domain

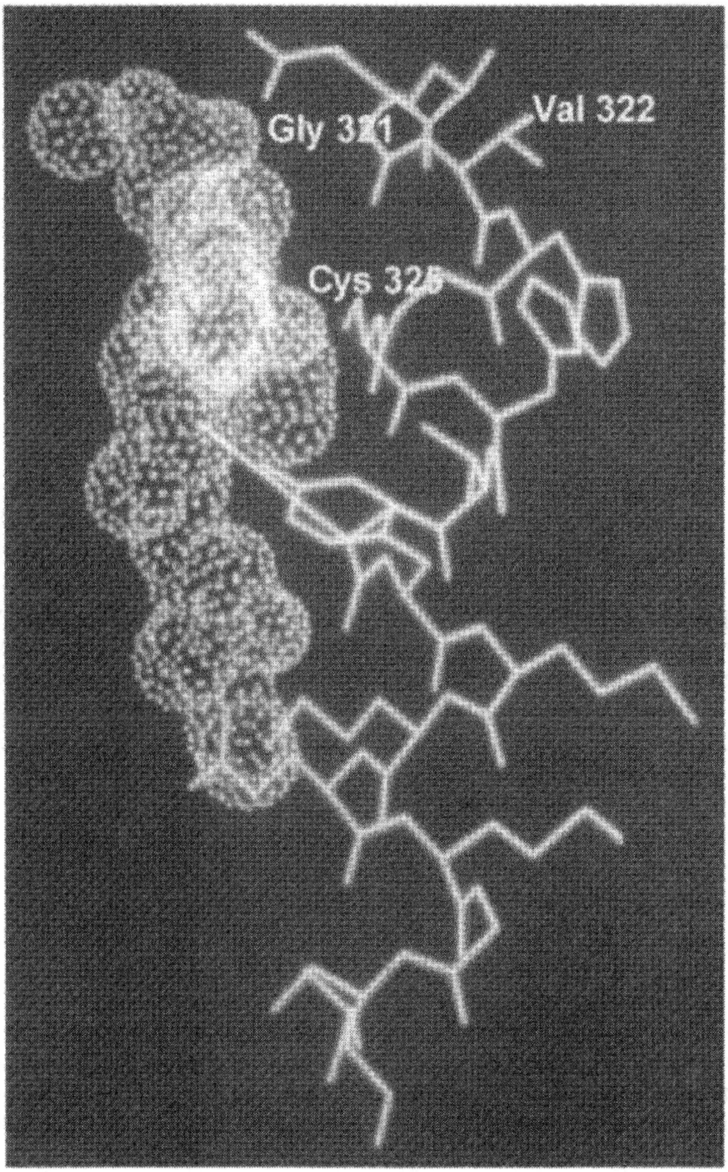
Gly 321
Val 322
Cys 325

estimated to be ca. 5.4 and 4.9 kJ mol^{-1} in Syn(PCB)PHY and Syn(PϕB)PHY, respectively.

Phototransformation of Pr upon red illumination was observed at temperatures above 160–170 K in Syn(PCB)PHY and above 140–150 K in Syn(PϕB)PHY with a 2–3 nm shift of the emission spectrum to the blue and increase of the intensity in its shorter wavelength region [40]. This was interpreted as a possible formation of the photoproduct of the meta-Ra type of the plant phytochrome. At ambient temperatures, the extent of the Pr phototransformation to the far-red-absorbing form, Pfr, was ca. 0.7–0.75 and 0.85–0.9 for Syn(PCB)PHY and Syn(PϕB)PHY, respectively. Fluorescence of Pfr and of the photoproduct similar to lumi-R was not observed. With respect to the photochemical parameters, Syn(PCB)PHY and Syn(PϕB)PHY are similar to each other and also to a small fraction of PHYA (PHYA″) and to PHYB. The latter were shown to have low photochemical activity at low temperatures in contrast to the major PHYA pool (PHYA′), which is distinguished by the high extent (ca. 50%) of Pr phototransformation at 85 K. These photochemical features are interpreted in terms of different activation barriers for the photoreaction in the Pr excited state.

The Raman spectra of the prokaryotic phytochrome suggest far-reaching similarities in chromophore configuration and conformation between the Pfr forms of *Synechocystis* phytochrome and the plant phytochromes (*e.g.* PHYA from oat), but also some differences, such as torsions around methine bridges and in hydrogen bonding interactions, in the Pr state [41]. *Synechocystis* phytochrome (PCB) undergoes a multistep photoconversion reminiscent of the PHYA Pr→Pfr transformation but with different kinetics. The first process resolved is the decay of an intermediate with red-shifted absorption (relative to the parent state) and a 25 μs lifetime. The next observable intermediate grows in with 300 (±25) μs and decays with 6–8 ms. The final state (Pfr) is formed biexponentially (450 ms, 1 s). When reconstituted with PϕB, the first decay of this *Synechocystis* phytochrome is biexponential (5 and 25 μs). The growth of the second intermediate is slower (750 μs) than that in the PCB adduct whereas the decays of both species are similar. The formation of the Pfr form required fitting with three components (350 ms, 2.5 s, and 11 s). H/D exchange in *Synechocystis* phytochrome (PCB) delays, by an isotope effect of 2.7, both growth (300 μs) and decay rates (6–8 ms) of the second intermediate. This effect is larger than values determined for PHYA (ca. 1.2) and is characteristic of a rate-limiting proton transfer. The formation of the Pc intermediate state of the PCB adduct of *Synechocystis* phytochrome shows a deuterium effect similar as PHYA (ca. 1.2). Activation energies of the second intermediate in the range 0–18°C are 44 (in H_2O/buffer) and 48 kJ mol^{-1} (D_2O), with essentially identical pre-exponential factors.

15.4 Molecular biology of eukaryotic algal phytochrome

Apart from difference spectra similar to those of higher plant phytochrome, [42–47], immunological approaches with more or less universal antibodies provided the first molecular indication that lower plants contained phytochromes with properties and characteristics similar to those of higher plants [48,49]. In addition to mosses and ferns, red (*Corallina, Gelidium, Porphyra)*, brown (*Cytoseira*) and green macroalgae (*Chara,*

Chaetomorpha, *Enteromorpha*, *Ulva*) were investigated [6,7,44,50]. Although immunoblots of red and brown algae showed stained bands in the expected M_r region, early efforts to find phytochrome DNA in red and brown algae were unsuccessful. Nevertheless, immunoblots are not necessarily irrelevant [2]. Currently, algae showing phragmoplast-formation such as *Mesotaenium*, *Mougeotia* and *Chara* [51] are the lowest eukaryotic plants known definitely to harbor phytochromes ([25,26,52]; see Figure 8).

Phytochrome gene expression was found to be light-regulated in *Mougeotia* and *Mesotaenium* (for review, see [53]). In *Mougeotia*, the level of phytochrome mRNA was

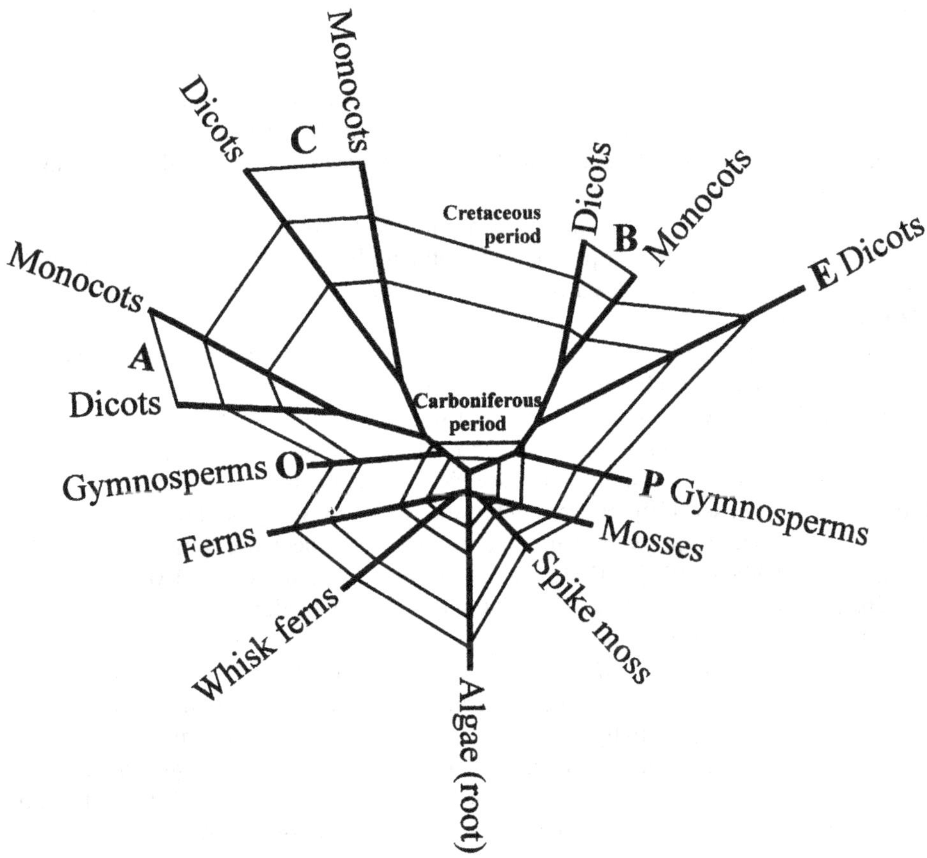

Figure 8. Phylogenetic tree of phytochrome with full length nucleotide sequences in schematic representation. Land plants deviated possibly 450 million years ago from charophycean green algae, defined as the root [97]. The root is surrounded by isobars when the carboniferous and cretaceous periods in organic evolution began and ended, respectively. During the course of plant speciation, events of duplication occurred of the single phytochrome gene in algae, resulting in the classes of O and P phytochrome genes in gymnosperms and A, B, C, E in angiosperms (from [98], with permission).

relatively high after 5 d of dark adaptation [54] as is the level of phytochrome protein monitored through immunolocalization of cytosolic phytochrome in *Mougeotia* [55]. The level of phytochrome-specific mRNA declined to less than 10% of the initial value after 5 min red-light irradiation, and the red-light effect could be reversed by a subsequent far-red light treatment [26]. These data, together with the presence of an evidently expressed single phytochrome gene in this alga indicate that phytochrome gene expression is autoregulated [26]. Phytochrome transcript and phytochrome protein levels in *Mesotaenium* are also light-regulated [56], although it is not known which gene is involved as all the genes are very similar in sequence, and RNA blot analysis cannot distinguish the expression pattern of each gene [25,45].

Functional domains in *Mesotaenium* MesPHY1b [25] and *Mougeotia* MougPHY [26] phytochromes have been predicted from similarities of the deduced amino acid sequences to known sequences of various functional proteins, including a C-terminal module homologous to bacterial transmitter modules (Figure 7). This finding appears consistent with the respective proposal by Schneider-Poetsch and his colleagues [57,58]. In *Mesotaenium* phytochrome, red/far-red reversible Ser/Thr protein kinase activity was shown which implicates PHY-mediated protein phosphorylation in the light signal transduction [99].

The sequence of the *Mougeotia* phytochrome indicated no hydrophobic transmembrane domains [26]. Instead, a possible microtubule binding domain was found in the C-terminal 16-mer 3-fold repeats in both *Mougeotia* and *Mesotaenium* phytochromes ([26]; see Figure 7).

With regard to the *Mougeotia* cortical microtubules, there is another interesting aspect to be reported: High irradiance blue light (HIBL), but not red light, were found to diminish indirect immunofluorescence of anti-tubulin bound to cortical microtubules in *Mougeotia*, while the overall microtubular pattern remained constant (Figure 5). The decrease in intensity of immunofluorescence is not the result of quenching of the fluorescence label used [59], but may inter alia reflect masking of anti-tubulin binding sites. If this conclusion were true, the fluorescence decrease might reflect MT-modification in *Mougeotia* in HIBL, e.g. by MT-associated proteins (MAP; [22,26]).

Preparative scale formation of algal protoplasts and controlled osmotic cell lysis have permitted separation of intact organelles from the phytochrome-enriched soluble protein fraction of *Mesotaenium caldariorum*. Kidd and Lagarias [45] have utilized the observation that red light-absorbing (Pr) and far-red light-absorbing (Pfr) forms of phytochrome are differentially retained on an anion exchange matrix to purify *Mesotaenium* phytochrome to apparent homogeneity. *Mesotaenium* phytochrome preparations with A650/A280 ratios greater than 0.78 exhibit a single 120-kDa band on silver-stained sodium dodecyl sulfate-polyacrylamide gels. Immunoblot analyses using a cross-reactive pea phytochrome monoclonal antibody reveal that, first, the 120-kDa band represents the full-length polypeptide, second, phytochrome is predominantly localized in the algal cytoplasm, and third, there are 150,000 to 250,000 phytochrome molecules per cell. Steric exclusion high pressure liquid chromatography analysis under non-denaturing conditions indicates that *Mesotaenium* phytochrome has an apparent molecular mass of 355 kDa. The absorption maxima for Pr and Pfr are 650 and 722 nm, respectively. Both are blue-shifted compared with those of phytochromes from dark-grown angiosperm tissue (see below). The molar absorption coefficient for

Mesotaenium Pr at 650 nm is 86,800 (±2800) liter mol^{-1} cm^{-1}, which is lower than that of higher plant phytochromes.

Detailed studies were undertaken also by the laboratory of Lagarias to determine whether this blue shift is due to a chromophore other than phytochromobilin or reflects a different protein environment for the phytochromobilin prosthetic group. Using reversed phase high performance liquid chromatography, Lagarias and co-workers [60] showed that soluble protein extracts prepared from algal chloroplasts contain the enzyme activities for ferredoxin-dependent conversions of biliverdin IXα to (3Z)-phytochromobilin and (3Z)-phytochromobilin to (3Z)-phycocyanobilin. In vitro assembly of recombinant algal apophytochrome was undertaken with (3E)-phytochromobilin and (3E)-phycocyanobilin. The difference spectrum of the (3E)-phycocyanobilin adduct was indistinguishable from that of phytochrome isolated from dark-adapted algal cells, while the (3E)-phytochromobilin adduct displayed red-shifted absorption maxima relative to purified algal phytochrome. These studies indicate that phycocyanobilin is the immediate precursor of the green algal phytochrome chromophore and that phytochromobilin is an intermediate in its biosynthesis in *Mesotaenium*.

15.5 Chloroplast orientation in *Mougeotia* and *Mesotaenium*

15.5.1 Mechanics of the movement

For spatial information to be adequately transformed into movement, driven by the actomyosin motor apparatus in *Mougeotia* and *Mesotaenium*, the primary signal stored in the photoreceptor memory of active phytochrome or the blue light pigment, must be transformed into the corresponding pattern of structurally fixed component of the motor apparatus (vectorial information). In principle, immobile structure at the cell periphery is given either by the plasmalemma or by the microtubules in close vicinity (see above).

Alignment of the actin filaments at sites, where chloroplast edge and cortical cytoplasm merge, shows that parts of the actomyosin-generated force is oriented in radial direction within the cylindrical cell. Chloroplast orientation, however, results from force in tangential direction. Hence, chloroplast movement requires successive lateral shift of the actin filaments which presumably proceeds in a statistical way, biased by sensor pigment-modulated plasmalemma (or cytoskeletal) anchorage sites to actin [22]. Consequently, the velocity of *Mougeotia* and *Mesotaenium* chloroplast rotational movement does not result merely from the velocity of actin-myosin interaction as is the case in translational movements, e.g. cytoplasmic streaming [22,61]. Rather, rotational movement may be dominated by biased, lateral movement of actin filaments along an increasing gradient of plasmalemma anchorage sites, even though the chloroplast in *Mougeotia* rotates in almost constant angular velocity (around 1 mrad s^{-1} [62]). Actually, the velocity of the movement of the *Mougeotia* chloroplast turns out to be two to three orders of magnitude slower than the translational actin-based intracellular movements (for review, see [63]). Furthermore, as shown by electron microscopy or after rhodamin phalloidin staining of formaldehyde-fixed cells, a special F-actin organization of seemingly single filaments is evident for walled *Mougeotia* in contrast to actin bundles in identically treated *Mougeotia* protoplasts [64]. It may be speculated that bundles of actin filaments would not allow enough freedom for the single actin filament to undergo the biased, lateral step-wise movement as implied in the model. Much alike in the high irradiance response, Mineyuki and coworkers [65] identified seemingly single cytoplasmic actin filaments in *Mougeotia*, using fluorescein-conjugated phalloidin, which emerge from the advancing front of the moving chloroplast. In cells fixed at 1–3 min after the onset of light induction, fluorescent filaments appeared at several sites along the leading edge of the moving chloroplast: these filaments were first short, then they grew longer and usually oriented themselves normal to the cell axis when the cells had been fixed at 5–10 min after the onset of irradiation. The filaments extended from the front and rear margins of the moving chloroplast, but vanished within 30 min after the chloroplast had reached its final position [65].

15.5.2 Calcium effects

Calcium appears to be involved in the chloroplast responses, even though specific sites of action are not identified as yet. Namely, the six paradigmas of L.F. Jaffe and R.E. Williamson [66] to identify calcium as second messenger in *Mougeotia* or *Mesotaenium* have not been met. Schönbohm and coworkers [67] did an extensive study of the effect of Ca^{2+} entry blockers on the light-induced chloroplast movements and on the chloroplast adhesion in *Mougeotia*. The organic inhibitors diltiazem and nifedipine, and the inorganic inhibitors Ruthenium Red, La^{3+} and Co^{2+} did not affect the low and high irradiance movements of the chloroplast (see also [68]). Only at toxic concentrations or after long-term incubations of 4 to 7 days, which resulted in unspecific side effects, the chloroplast movement was slightly inhibited [67]. This conclusion is supported by another evidence: Serlin and coworkers [69,70] have used the patch-clamp technique to determine whether or not phytochrome regulates ion channel activity in the plasmalemma of *Mougeotia* protoplasts. Consistent with the pharmacological results of Schönbohm et al. [67] and of Wagner and Grolig [68], the patch-clamp data of Serlin and coworkers [70] indicate that ". . . it is unlikely that channel activation is part of the mechanism leading to chloroplast rotation. . . in *Mougeotia*".

In extension of this detailed study, the same laboratory characterized the escape times (time required for loss of photoreversibility) to compare the transduction pathways involved in phytochrome-mediated chloroplast rotation and K^+ channel activation, respectively [71]. The escape time for chloroplast rotation was 2.5 min after red light irradiation (red and far-red light irradiations were 30 s). For channel activation, shorter red and far-red light irradiations (10 s) had to be used to obtain an escape time of 20 s. The difference in the escape times suggests that there is relatively rapid divergence in the transduction pathways leading from phytochrome activation to each of the two responses in the same cellular system, namely chloroplast rotation and channel activation. Because channel activation occurs 2 to 4 min after irradiation while the escape time is 20 s, it is unlikely that phytochrome acts directly on the channel.

Calcium-binding globules in *Mougeotia* [72] have been identified and analyzed by means of histochemistry using different techniques [73]. The globules form a member of the large family of physodes in lower plants [74]. The calcium-binding physodes are abundant at the chloroplast edges, where they accumulate even more during reorientational movement. A Ca^{2+} depletion upon fixation at different K^+/Na^+-ratios resulted in selective uptake of potassium, not sodium. Consistent with earlier findings [72], calcium-binding by the polyphenolic physode matrix does not depend merely on electric charge but also on the presence of protonated/deprotonated phenolic groups, together with ester-linked carbonyl oxygen, which seem to be good candidates for a co-ordinate type of calcium-binding.

Calmodulin (CaM) as the major calcium target is the product of a single gene in *Mougeotia* and reflects this alga being a calciophilic organism [75]. The calcium-binding affinity of *Mougeotia* calmodulin is diminished by three major amino acid differences compared e.g. with maize calmodulin, i.e. at position 26 within calcium-binding loop I and at positions 99 and 105 within and adjacent to calcium-binding loop III, respectively. Other amino acid replacements are seen as well. The indole ring of the rare amino acid 105-Trp in *Mougeotia* calmodulin, unique to date to all native calmodulins known at this position, is able to form H-bonds and often turns out to be part of a functional group. In *Mougeotia* calmodulin, as judged from the molecular model (Figure 9), this 105-Trp together with 141-Phe and 92-Phe possibly form a hydrophobic stack followed by conformational change of two Met residues in close neighborhood. As a result, Ca^{2+} affinity of *Mougeotia* calmodulin is diminished fivefold compared to maize calmodulin; furthermore, affinity of calcium-activated *Mougeotia* calmodulin in the cyclic 3′,5′nucleotide phosphodiesterase standard test is diminished 20 fold compared to maize calmodulin. Thus, in sum of the two changes in affinity, a possibly 100 fold difference in physiological function of *Mougeotia* calmodulin may result, compared to more classical calmodulins [75].

Based on the fluorescent calcium sensitive dye, indo-1, a fast (3–6 s) increase of cytoplasmic free calcium in *Mougeotia* has been detected upon irradiation by UV/blue light of 365 nm or 450 nm in a dose as used for induction of high irradiance chloroplast movement (HIBL; [76]). This cytoplasmic calcium increase turned out independent of the external calcium concentration (EGTA-buffered media of pCa 8 versus pCa 3) and probably reflects calcium released from internal stores such as the calcium-binding physodes. Either calcium by itself or calcium-activated calmodulin, isolated from *Mougeotia* and *Mesotaenium* [77,78], should be able to pass the presently undefined Ca^{2+} signal onto cytoskeletal regulatory proteins, e.g. by activation of protein kinases. In an in vitro assay, *Mougeotia* cytoplasmic extract phosphorylates rabbit myosin light chain [79]. Using the synthetic peptide KM-14 as a substrate, Roberts [80] has detected and partially purified a calcium-dependent protein kinase from *Mougeotia* and *Mesotaenium*, possibly a member of the large family of calcium-dependent protein kinases (CDPKs) [81,82]. The *Mougeotia* kinase was stimulated 40 fold by calcium with half-maximal stimulation occurring at 1.5 μM. The calmodulin-depleted enzyme was fully active and calcium-dependent, and was not stimulated further by exogenous calmodulin nor by the calcium effectors phosphatidylserine and diacylglycerol. Inhibition of the reorientational movement of the *Mougeotia* chloroplast by the antagonists trifluoperazin and the W-compounds [83] suggest involvement of calmodulin and/or CDPK, disregarding induction of the transductional chain by blue or red light [79].

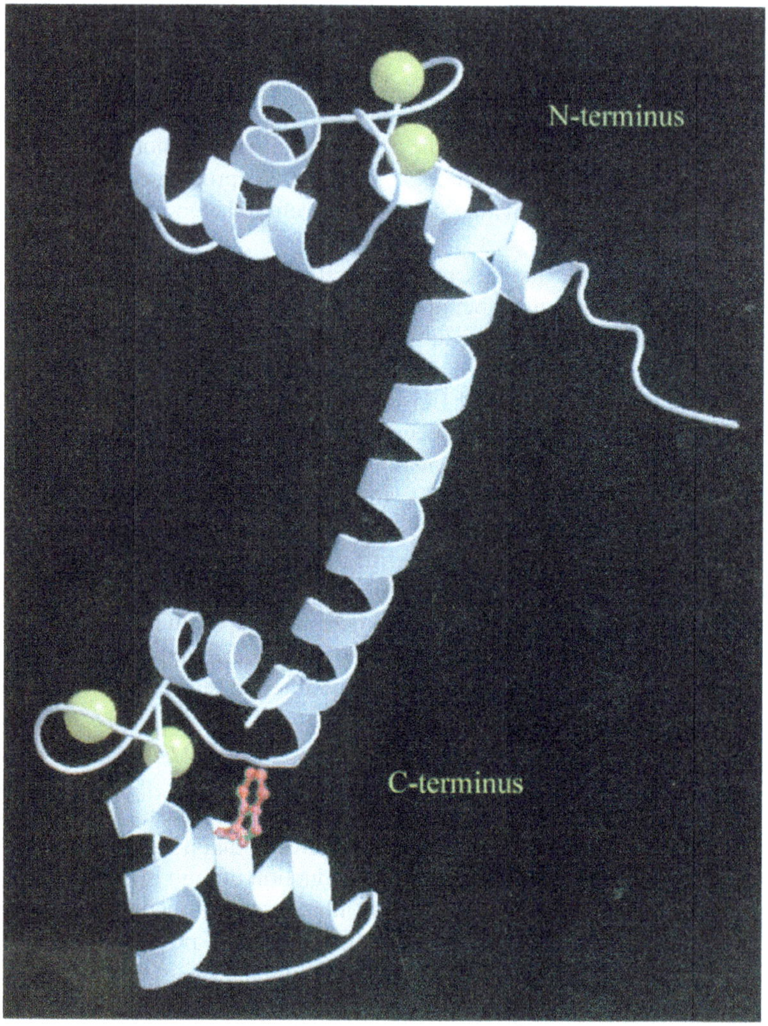
N-terminus
C-terminus

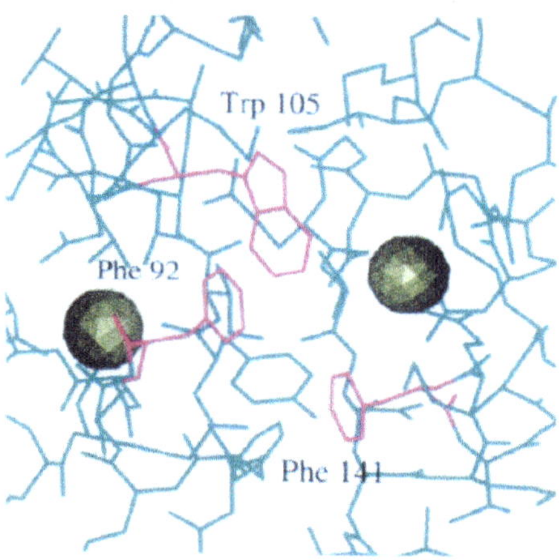
Trp 105
Phe 92
Phe 141

Figure 9. Molecular modeling of *Mougeotia* calmodulin in ribbon and string representation. At the top, the scheme is as follows: α-helical regions are gray ribbons, loops gray strings, the calcium atoms yellow spheres. The tryptophane residue 105 is indicated in red. (Courtesy of Ch. Betzel and M. Perbandt, unpublished). Bottom: Molecular modeling of the region around 105-tryptophane (Trp 105) in the deduced amino acid sequence of *Mougeotia* calmodulin. Neighboring side chains of Phe 92 and Phe 141 are candidates to form a hydrophobic ring interaction. The balls indicate the Ca^{2+} atoms (from [75], with permission).

15.5.3 Microtubules

Cortical microtubules in living plant cells are highly dynamic and turn over more rapidly than comparable structures in animal cells [84,85]. The MTs depolymerize after rise of cytoplasmic [Ca^{2+}], particularly in the presence of calmodulin [86,87]. In *Mougeotia* cells, the MTs in cortical array are located just beneath and in close contact to the plasmalemma and, as shown in *Mougeotia* protoplasts, are sensitive to MT-depolymerizing and MT-stabilizing agents such as colchicine/isopropyl-N-phenyl carbamate and taxol, respectively [88–90]. In the absence of calmodulin, the MTs adhering to plasma membrane ghosts of *Mougeotia* protoplasts are rather stable and depolymerize upon raising the [Ca^{2+}] only after pretreatment with Triton X-100 [91], while the presence of calmodulin possibly enhances the calcium sensitivity, as deduced from higher plant cells [92]. Most interestingly, in *Mougeotia* cells, MT-depolymerizing drugs speed up red-light-mediated chloroplast movement from a response time of more than 20 min to less than 10 min [93].

Compiling the reported effects of microtubuli-depolymerizing agents, CaM/CDPK-inhibitors and increased [Ca^{2+}] on the kinetics of *Mougeotia* chloroplast movement (Figure 10), we suggested the following hypothesis [59,76]: First, a physically or chemically induced rise of free cytoplasmic [Ca^{2+}] causes a possibly calmodulin-mediated microtubule/MAP modification and thus an ease in lateral shearing force to develop through actin/myosin interaction in *Mougeotia*. Second, as a consequence, the calcium increase in the cytosol, may it result from the external medium or from internal stores, will speed up the chloroplast rotation (Figure 10). In accordance, a calcium-elicited microtubule/MAP modification possibly leads also to the increased ease of chloroplast displacement in longitudinal direction in high-irradiance blue-light, as reported by Schönbohm [94]. Proof of the MT/MAP-function of calcium/calmodulin in *Mougeotia* as a scalar factor is in good progress [59,75], but calcium targets other than this may still await identification in this alga.

Epilogue

No information is available to date on the most challenging question of the vectorial signal of chloroplast orientation in *Mougeotia* or *Mesotaenium* (Jaffe-Haupt model). In face of the small cell dimensions of these organisms, this question is difficult to analyze; however, based on preliminary results, the vectorial signal here appears to be different from calcium.

Another question, difficult to answer at the present time and out of the scope of this review, are the molecular properties, location and function of the blue-light

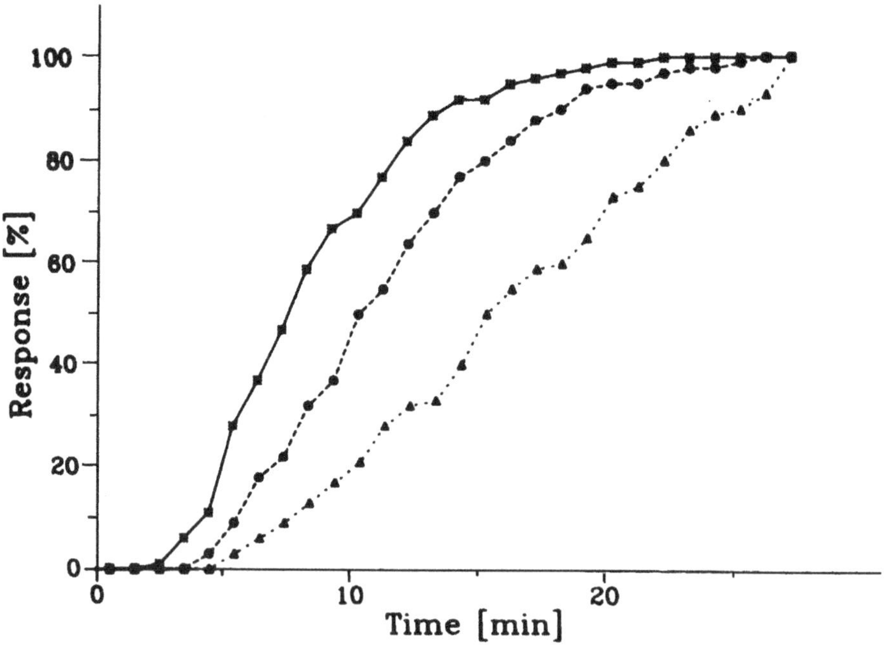

Figure 10. Velocity of chloroplast rotation in *Mougeotia scalaris*, mediated by different light regimes. Left-hand curve: UV-A-mediated chloroplast movement (time span to reach half maximal response ($R_{0.5}$) = 8.2 min, $n = 107$). Central curve: Blue-light-mediated chloroplast movement ($R_{0.5} = 11.0$ min, $n = 114$). Right-hand curve: Red-light-mediated chloroplast movement ($R_{0.5} = 15.4$ min, $n = 97$). The differences in $R_{0.5}$ are highly significant ($p < 0.0001$) and reflect three different levels of $[Ca^{2+}]_c$ (from [76], with permission).

photoreceptor(s) in the plant systems discussed in competition to or co-action with phytochrome (Herrmann and Kraml, 1997 and references therein [100]). A cryptochrome gene, photoregulated in its transcription, has been identified recently in *Mougeotia scalaris* [95].

Acknowledgements

Stimulating discussion with W. Haupt, M. Kraml, K.D. Brunner and C. Zörb during the course of writing this review is thankfully acknowledged. Work in the author's laboratory was financially supported by the Deutsche Agentur für Luft- und Raumfahrtangelegenheiten (FKZ 50 WB 9163) and the Deutsche Forschungsgemeinschaft (Wa 265/14–1).

References

1. D.J. Correl, J.L. Edwards, W. Shropshire (Eds) (1977). *Phytochrome: A Bibliography with Author, Biological Materials, Taxonomic, and Subject Indexes of Publications prior to 1975*. Smithonian Institution Press, Washington, D.C.

2. H.A.W. Schneider-Poetsch, Ü. Kolukisaoglu, D.H. Clapham, J. Hughes, T. Lamparter (1998). Non-angiosperm phytochromes and the evolution of vascular plants. *Physiologia Plantarum*, **102**, 612–622.
3. J. Hughes, T. Lamparter, F. Mittmann, E. Hartmann, W. Gärtner, A. Wilde, T. Börner (1997). A prokaryotic phytochrome. *Nature*, **386**, 663.
4. A. Wilde, Y. Churin, H. Schubert, T. Börner (1997). Disruption of a *Synechocystis* sp. PCC 6803 gene with partial similarity to phytochrome genes alters growth under changing light qualities. *FEBS Lett.*, **406**, 89–92.
5. J.-S. Choi, Y.-H. Chung, Y.-J. Moon, C. Kim, M. Watanabe, P.-S. Song, C.-O. Joe, L. Bogorad, Y.M. Park (1999). Photomovement of the gliding cyanobacterium *Synechocystis* sp. PCC 6803. *Photochem. Photobiol.*, **70**, 95–102.
6. M.J. Dring (1967). Phytochrome in red alga, *Porphyra tenera. Nature (London)*, **215**, 1411–1412.
7. H.G. Rentschler (1967). Photoperiodische Induktion der Monosporenbildung bei *Porphyra tenera* Kjellm. (*Rhodophyta-Bangiophyceae*). *Planta*, **76**, 65–74.
8. W. Haupt (1982). Light-mediated movement of chloroplasts. *Ann. Rev. Plant Physiol.*, **33**, 205–233.
9. H. Neuscheler-Wirth (1970). Photomorphogenese und Phototropismus bei *Mougeotia. Z. Pflanzenphysiol.*, **63**, 238–260.
10. M.J. Dring (1988). Photocontrol of development in algae. *Ann. Rev. Plant Physiol. Plant Mol. Biol.*, **39**, 157–174.
11. M. Wada, A. Kadota (1989). Photomorphogenesis in lower green plants. *Ann. Rev. Plant Physiol. Plant Mol. Biol.*, **40**, 169–191.
12. S.B. Agrawal, B.R. Chaudhary (1994). Effect of certain environmental factors on zygospore germination of *Spirogyra hyalina. Folia Microbiol.*, **39**, 291–295.
13. F. Grolig (1998). Nuclear centering in *Spirogyra*: force integration by filaments along microtubules. *Planta*, **204**, 54–63.
14. G. Senn (1908). *Die Gestalts- und Lageveränderung der Pflanzen-Chromatophoren* (pp. 1–397). Engelmann, Leipzig.
15. G. Mosebach (1958). Zur Phototaxis von *Mougeotia* (*Mesocarpus*). *Planta*, **52**, 3–46.
16. L. Taiz, E. Zeiger (1998). *Plant Physiology* (2nd ed.), Box 17.1 (after revision by G. Wagner) *Mougeotia: A chloroplast with a twist.* Sinauer Assoc., Inc., Publishers, Sunderland, Mass. (USA), pp. 493–494.
17. H. Gabrys, T. Walczak, W. Haupt (1984). Blue-light-induced chloroplast orientation in *Mougeotia*. Evidence for a separate sensor pigment besides phytochrome. *Planta*, **160**, 21–24.
18. H. Herrmann, M. Kraml (1997). Time-dependent formation of Pfr-mediated signals for the interaction with blue light in *Mesotaenium* chloroplast orientation. *J. Photochem. Photobiol.*, **37**, 60–65.
19. W. Haupt, R. Scheuerlein (1990). Chloroplast movement. *Plant Cell Environ.*, **13**, 595–614.
20. M. Kraml (1994). Light direction and polarization. In: R.E. Kendrick, G.H.M. Kronenberg (Eds), *Photomorphogenesis in Plants* (2nd ed., pp. 417–445). Kluwer Academic Publishers, Dordrecht, The Netherlands.
21. M. Wada, F. Grolig, W. Haupt (1993). Light-oriented chloroplast positioning. Contribution to progress in photobiology. *J. Photochem. Photobiol. B: Biol.*, **17**, 3–25.
22. F. Grolig, G. Wagner (1988). Light-dependent chloroplast orientation in *Mougeotia* and *Mesotaenium*: biased by pigment-regulated plasmalemma anchorage sites to actin filaments? *Botanica Acta*, **101**, 2–6.
23. G. Wagner (1996). Intracellular movement. *Progress in Botany*, **57**, 68–80.

24. T. Lamparter, F. Mittmann, W. Gärtner, T. Börner, E. Hartmann, J. Hughes (1997). Characterization of recombinant phytochrome from the cyanobacterium *Synechocystis*. *Proc. Nat. Acad. Sci. (USA)*, **94**, 11792–11797.
25. D.M. Lagarias, S.H. Wu, J.C. Lagarias (1995). Atypical phytochrome gene structure in the green alga *Mesotaenium caldariorum*. *Plant Mol. Biol.*, **29**, 1127–1142.
26. A. Winands, G. Wagner (1996). Phytochrome of the green alga *Mougeotia*: cDNA sequence, autoregulation and phylogenetic position. *Plant Mol. Biol.*, **32**, 589–597.
27. T.D. Elich, J.C. Lagarias (1989). Formation of a photoreversible phycocyanobilin-apophytochrome adduct *in vitro*. *J. Biol. Chem.*, **364**, 12902–12908.
28. W. Rüdiger (1987). Phytochrome: the chromophore and photoconversion. *Photobiochem. Photobiophys. Suppl.*, 217–227.
29. J. Hughes, E. Hartmann, T. Lamparter (1999). Cyanobacterial phytochrome as a model for phytochrome function. In: E. Hartmann, K.-J. Appenroth, T. Börner, H.-P. Haschke, J. Hughes, T. Lamparter, E. Schäfer, B. Weisshaar, D. Matzkuhn (Orgs), *European Symposium on Photomorphogenesis, Book of Abstracts* (p. 6.2). Berlin.
30. C.-M. Park, J.-I. Kim, J.Y. Shim, Z. Luka, P.-S. Song (1999). Trials and tribulation of phytochrome crystallization. In: E. Hartmann, K.-J. Appenroth, T. Börner, H.-P. Haschke, J. Hughes, T. Lamparter, E. Schäfer, B. Weisshaar, D. Matzkuhn (Orgs), *European Symposium on Photomorphogenesis, Book of Abstracts* (p. 12.2). Berlin.
31. S. Tokutomi, M. Nakasako, J. Sakai, M. Kataoka, K.T. Yamamoto, M. Wada, F. Tokunaga, M. Furuya (1989). A model for the dimeric molecular structure of phytochrome based on small-angle X-ray scattering. *FEBS Lett.*, **247**, 139–142.
32. M. Nakasako, M. Wada, S. Tokutomi, K.T. Yamamoto, J. Sakai, M. Kataoka, F. Tokunaga, M. Furuya (1990). Quaternary structure of pea phytochrome I dimer studied with small angle X-ray scattering and rotary-shadowing electron microscopy. *Photochem. Photobiol.*, **52**, 3–12.
33. J.L. Gabriel, J.K. Hoober (1991). Molecular modeling of phytochrome. *J. Theor. Biol.*, **151**, 541–556.
34. S. Tokutomi, T. Sugimoto, M. Mimuro (1992). A model for the molecular structure and orientation of the chromophore in a dimeric phytochrome molecule. *Photochem. Photobiol.*, **56**, 545–552.
35. Z.A. Luka, I.D. Volotovski (1993). Modelling of the structure of phytochrome. *Biophysics*, **38**, 1053–1058.
36. W. Parker, P. Goebel, C.R. Ross, P.-S. Song, J.J. Stezowski (1994). Molecular modeling of phytochrome using constitutive C-phycocyanin from *Fremyella diplosiphon* as a putative structural template. *Bioconjugate Chem.*, **5**, 21–30.
37. M. Duerring, G.B. Schmidt, R. Huber (1991). Isolation, crystallization, crystal structure analysis and refinement of constitutive C-phycocyanin from the chromatically adapting cyanobacterium *Fremyella diplosiphon* at 1.66 Å resolution. *J. Mol. Biol.*, **217**, 577–592.
38. K.C. Yeh, S.H. Wu, J.T. Murphy, J.C. Lagarias (1997). A cyanobacterial phytochrome two-component light sensory system. *Science*, **277**, 1505–1508.
39. S. Hanelt, B. Braun, S. Marx, H.W.W. Schneider-Poetsch (1992). Phytochrome evolution: A phylogenetic tree with the first complete sequence of phytochrome from a cryptogamic plant (*Selaginella martensii* Spring). *Photochem. Photobiol.*, **56**, 751–758.
40. V. Sineshchekov, J. Hughes, E. Hartmann, T. Lamparter (1998). Fluorescence and photochemistry of recombinant phytochrome from the cyanobacterium *Synechocystis*. *Photochem. Photobiol.*, **67**, 263–267.
41. A. Remberg, I. Lindner, T. Lamparter, J. Hughes, C. Kneip, P. Hildebrandt, S.E. Braslavsky, W. Gärtner, K. Schaffner (1997). Raman spectroscopic and light-induced

kinetic characterization of a recombinant phytochrome of the cyanobacterium *Synechocystis*. *Biochemistry*, **36**, 13389–13395.
42. R. Grill, H. Schraudolf (1981). *In vivo* phytochrome difference spectrum from dark grown gametophytes of *Anemia phyllidis* L. Sw. treated with norflurazon. *Plant Physiol.*, **68**, 1–4.
43. P. Lindemann, S.E. Braslavsky, E. Hartmann, K. Schaffner (1989). Partial purification and initial characterization of phytochrome from the moss *Atrichum undulatum* P. Beauv. grown in the light. *Planta*, **178**, 436–442.
44. F. López-Figueroa, P. Lindemann, S.E. Braslavsky, K. Schaffner, H.A.W. Schneider-Poetsch, W. Rüdiger (1989). Detection of a phytochrome-like protein in macroalgae. *Bot. Acta*, **102**, 178–180.
45. D.G. Kidd, J.C. Lagarias (1990). Phytochrome from the green alga *Mesotaenium caldariorum*. Purification and preliminary characterization. *J. Biol. Chem.*, **265**, 7029–7035.
46. H. Oyama, K.T. Yamamoto, M. Wada (1990). Phytochrome in the fern *Adiantum capillus-veneris* L.: Spectrophotometric detection in vivo and partial purification. *Plant Cell Physiol.*, **31**, 1229–1238.
47. T. Lamparter, S. Podlowski, F. Mittmann, H.A.W. Schneider-Poetsch, E. Hartmann, J. Hughes (1995). Phytochrome from protonemal tissue of the moss *Ceratodon*. *J. Plant Physiol.*, **147**, 426–434.
48. H.A.W. Schneider-Poetsch, H. Schwartz, R. Grimm, W. Rüdiger (1988). Cross-reactivity of monoclonal antibodies against phytochrome from *Zea* and *Avena*. *Planta*, **173**, 61–72.
49. H.A.W. Schneider-Poetsch, G. John, B. Braun (1990). The distribution of a phytochrome-like protein in the fern *Psilotum nudum*. An immunoblotting analysis of an early ancestor of all vascular plants. *Bot. Acta*, **103**, 235–239.
50. F. López-Figueroa, P. Lindemann, S.E. Braslavsky, K. Schaffner, H.A.W. Schneider-Poetsch, W. Rüdiger (1990). Detection of some conserved domains in phytochrome-like proteins from algae. *J. Plant Physiol.*, **126**, 484–487.
51. H. Sawitzky, F. Grolig (1995). Phragmoplast of the green alga *Spirogyra* is functionally distinct from the higher plant phragmoplast. *J. Cell Biol.*, **130**, 1359–1371.
52. H.Ü. Kolokisaoglu, S. Marx, C. Wiegmann, S. Hanelt, H.A.W. Schneider-Poetsch (1995). Divergence of the phytochrome gene family predates angiosperm evolution and suggests that *Selaginella* and *Equisetum* arose prior to *Psilatum*. *J. Mol. Evol.*, **41**, 329–337.
53. M. Wada, T. Kanegae, K. Nozue, S. Fukuda (1997). Cryptogam phytochromes. *Plant Cell Environ.*, **20**, 685–690.
54. A. Winands, G. Wagner, S. Marx, H.A.W. Schneider-Poetsch (1992). Partial nucleotide sequence of phytochrome from the zygnematophycean green alga *Mougeotia*. *Photochem. Photobiol.*, **56**, 765–770.
55. C. Hanstein, F. Grolig, G. Wagner (1992). Immunolocalization of cytosolic phytochrome in the green alga *Mougeotia*. *Bot. Acta*, **105**, 55–62.
56. L.Z. Morand, D.G. Kidd, J.C. Lagarias (1993). Phytochrome levels in the green alga *Mesotaenium caldariorum* are light-regulated. *Plant Physiol.*, **101**, 97–104.
57. H.A.W. Schneider-Poetsch, B. Braun (1991). Proposal on the nature of phytochrome action based on the C-terminal sequences of phytochrome. *J. Plant Physiol.*, **137**, 576–580.
58. H.A.W. Schneider-Poetsch (1992). Signal transduction by phytochrome: phytochromes have a module related to the transmitter modules of bacterial sensor proteins. *Photochem. Photobiol.*, **56**, 839–846.
59. B. Al-Rawass, F. Grolig, G. Wagner (1997). High irradiance blue light affects cortical microtubules in the green alga *Mougeotia scalaris*. *Plant Cell Physiol.*, **38**, 882–886.

60. S.H. Wu, M.T. McDowell, J.C. Lagarias (1997). Phycocyanobilin is the natural precursor of the phytochrome chromophore in the green alga *Mesotaenium caldariorum*. *J. Biol. Chem.*, **272**, 25700–25705.
61. F. Grolig (1990). Actin-based organelle movements in interphase *Spirogyra*. *Protoplasma*, **155**, 29–42.
62. N. Ytow, T. Yamada, S. Ishizaka (1992). Mechanics of the chloroplast rotation in *Mougeotia*: Measurement of angular velocity by laser diffractometry. *Cell Motil. Cytoskel.*, **23**, 102–110.
63. R.E. Williamson (1993). Organelle movements. *Ann. Rev. Plant Physiol. Plant Mol. Biol.*, **44**, 181–202.
64. F. Grolig, K. Weigang-Köhler, G. Wagner (1990). Different extent of F-actin bundling in walled cells and protoplasts of *Mougeotia scalaris*. *Protoplasma*, **157**, 225–230.
65. Y. Mineyuki, H. Kataoka, Y. Masuda, R. Nagai (1995). Dynamic changes in the actin cytoskeleton during the high-fluence rate response of the *Mougeotia* chloroplast. *Protoplasma*, **185**, 222–229.
66. R.E. Williamson (1992). Cytoplasmic streaming in characean algae: mechanism, regulation by Ca^{2+}, and organization. In: M. Melkonian (Ed.), *Algal Cell Motility* (pp. 39–72). Chapman and Hall, New York and London.
67. E. Schönbohm, J. Meyer-Wegener, E. Schönbohm (1990). No evidence for Ca^{2+} influx as an essential link in the signal transduction chains of either light-oriented chloroplast movements or Pfr-mediated chloroplast anchorage in *Mougeotia*. *J. Photochem. Photobiol.*, **5**, 331–341.
68. G. Wagner, F. Grolig (1985). Molecular mechanisms of photoinduced chloroplast movements. In: G., Colombetti, F. Lenci, and P.S. Song (Eds), *Sensory Perception and Transduction in Aneural Organisms* (pp. 281–298). Plenum Press, New York-London.
69. R.R. Lew, B.S. Serlin, C.L. Schauf, M.E. Stockton (1990). Red light regulates calcium-activated potassium channels in *Mougeotia* plasma membrane. *Plant Physiol.*, **92**, 822–830.
70. R.R. Lew, B.S. Serlin, C.L. Schauf, M.E. Stockton (1990). Calcium activation of *Mougeotia* potassium channels. *Plant Physiol.*, **92**, 831–836.
71. B.S. Serlin, R.R. Lew, F. Krasnoshtein, J. Krol, K.D. Sumida (1996). Phytochrome activation of K^+ channels and chloroplast rotation in *Mougeotia*: The escape times. *Plant Cell Physiol.*, **37**, 175–179.
72. F. Grolig, G. Wagner (1989). Characterization of the isolated calcium-binding vesicles from the green alga *Mougeotia scalaris*, and their relevance to chloroplast movement. *Planta*, **177**, 169–177.
73. A. Tretyn, F. Grolig G. Magdowski, G. Wagner (1996). Selective binding of Ca^{2+}, Zn^{2+}, Cu^{2+} and K^+ by the physodes of the green alga *Mougeotia scalaris*. *Folia Histochem. Cytobiol.*, **34**, 103–108.
74. A. Tretyn, F. Grolig, G. Magdowski, G. Wagner (1992). Electron microscopic characterization of calcium-binding physodes in the green alga *Mougeotia scalaris*. *Histochemistry*, **97**, 487–492.
75. C. Zörb, K.D. Brunner, M. Perbandt, C. Betzel, G. Wagner (1998). Cloning, recombinant expression and characterization of wild type-105-Trp-calmodulin of the green alga *Mougeotia scalaris*. *Bot. Acta*, **111**, 1–8.
76. U. Russ, F. Grolig, G. Wagner (1991). Changes of cytoplasmic free Ca^{2+} in the green alga *Mougeotia scalaris* as monitored by indo-1, and their effect on the velocity of chloroplast movements. *Planta*, **184**, 105–112.

77. G. Wagner P. Valentin, P. Dieter, D. Marmé (1984). Identification of calmodulin in the green alga *Mougeotia* and its possible function in chloroplast reorientational movement. *Planta*, **162**, 62–67.
78. S. Jacobshagen, D. Altmüller, F. Grolig, G. Wagner (1986). Quantification of *Mesotaenium* calmodulin by improved cyclic nucleotide phosphodiesterase test. In: A.J. Trewavas (Ed.), *Molecular and Cellular Aspects of Calcium in Plant Development* (pp. 201–217). Plenum Press, New York-London.
79. G. Wagner, F. Grolig, D. Altmüller (1987). Transduction chain of low irradiance response of chloroplast orientation in *Mougeotia* in blue or red light. *Photobiochem. Photobiophys., Suppl.*, 183–189.
80. D.M. Roberts (1989). Detection of a calcium-activated protein kinase in *Mougeotia* by using synthetic peptide substrates. *Plant Physiol.*, **91**, 1613–1619.
81. D.M. Roberts, A.C. Harmon (1992). Calcium-modulated proteins: targets of intracellular signals in higher plants. *Ann. Rev. Plant Physiol. Mol. Biol.*, **43**, 375–414.
82. D.M. Roberts (1993). Protein kinases with calmodulin-like domains: novel targets of calcium-signals in plants. *Curr. Opinion Cell Biol.*, **5**, 242–246.
83. G. Wagner, U. Russ, H. Quader (1992). Calcium, a regulator of cytokeletal activity and cellular competence. In: D. Menzel (Ed.), *The Cytoskeleton of the Algae* (pp. 411–424). CRC Press, Boca Raton-Ann Arbor-London-Tokyo.
84. G.O. Wasteneys, B.E.S. Gunning, P.K. Hepler (1993). Microinjection of fluorescent brain tubulin reveals dynamic properties of cortical microtubules in living plant cells. *Cell Motil. Cytoskeleton*, **24**, 205–213.
85. J.M. Hush, P. Wadsworth, D.A. Callaham, P.K. Hepler (1994). Quantification of microtubules dynamics in living plant cells using fluorescence redistribution after photobleaching. *J. Cell Sci.*, **107**, 775–784.
86. D.H. Zhang, P. Wardworth, P.K. Hepler (1990). Microtubule dynamics in living dividing plant cells: confocal imaging of microinjected fluorescent brain tubulin. *Proc. Natl. Acad. Sci. (USA)*, **87**, 8820–8824.
87. R.J. Cyr (1991). Calcium/calmodulin affects microtubule stability in lysed protoplasts. *J. Cell Sci.*, **100**, 311–317.
88. H.J. Marchant, E.R. Hines (1979). The role of microtubules and cell-wall deposition in elongation of regenerating protoplasts of *Mougeotia*. *Planta*, **146**, 41–48.
89. E. Nogales, S.G. Wolf, I.A. Khan, R.F. Ludueña, K.H. Downing (1995). Structure of tubulin at 6.5 Å and location of the taxol-binding site. *Nature*, **375**, 424–427.
90. M.A. Melan (1998). Use of fluorochrome-tagged taxol to produce fluorescent microtubules in solution. *Biotechniques*, **25**, 188–192.
91. T. Kakimoto, H. Shibaoka (1986). Calcium-sensitivity of cortical microtubules in the green alga *Mougeotia*. *Plant Cell Physiol.*, **27**, 91–101.
92. N.A. Durso, R.J. Cyr (1994). A calmodulin-sensitive interaction between microtubules and a higher plant homolog of elongation factor-1 alpha. *Plant Cell*, **6**, 893–905.
93. B.S. Serlin, S. Ferrell (1989). The involvement of microtubules in chloroplast rotation in the alga *Mougeotia*. *Plant Sci.*, **60**, 1–8.
94. E. Schönbohm (1987). Movement of *Mougeotia* chloroplasts under continuous weak and strong light. *Act. Physiol. Plant.*, **9**, 109–135.
95. K.D. Brunner, C. Zörb, Ü. Kolukisaoglu, G. Wagner (2000). Light-regulated transcription of a crytochrome gene in the green alga *Mougeotia scalaris*. *Protoplasma*, **214**, 194–198.
96. H. Michel, D. Oesterhelt (1980). Three-dimensional crystals of membrane proteins: Bacteriorhodopsin. *Proc. Natl. Acad. Sci. (USA)*, **77**, 1283–1285.
97. R. Moore, W.D. Clark, D.S. Vodopich (1998). *Botany* (2nd ed.). WCB/McGraw-Hill: McGraw-Hill Companies, pp. 1–919.

98. H.A.W. Schneider-Poetsch, M. Schmidt (1998). *Mit dem Gen eines pflanzlichen Sehpigments 450 Millionen Jahre in die Vergangenheit geschaut.* Forschung in Köln – Berichte aus der Universität, **2**, 34–39.
99. K.C. Yeh, J.C. Lagarias (1998). Eukaryotic phytochromes: Light-regulated serine/threonine protein kinases with histidine kinase ancestry. *Proc. Natl. Acad. Sci. U.S.A.*, **95**, 13976–13981.
100. H. Hermann, M. Kraml (1997). Time-dependent formation of Pfr-mediated signals for the interaction with blue light in *Mesotaenium* chloroplast orientation. *J. Photochem. Photobiol.*, **37**, 60–65.

Photomovement
D.-P. Häder and M. Lebert, editors.

Chapter 16

Keeping in tune with time: entrainment of circadian rhythms

J. Woodland Hastings

Table of contents

Abstract

The circadian biological clock is an endogenous cellular/biochemical regulatory mechanism that controls activity and many other processes in relation to the time of day, but its timing is usually not exact. This is corrected by a daily phase shift in a process called entrainment, mostly attributable to the light-dark cycle. The underlying biochemical mechanism of circadian rhythms can be explained in terms of a set of genes and their products whose activities oscillate due to autoregulatory negative feedback loops, with positive elements as well. While genes contributing to such mechanisms appear to be numerous and different, the mechanism itself as well as some elements of the genes appear to be conserved or to have evolved convergently. Many aspects of the effects of light on the circadian system and the way in which entrainment is achieved are well known [1–3]. These will be summarized in conjunction with an account of the current knowledge of the mechanisms of photoreception, along with evidence that cryptochrome may function as a circadian photoreceptor. In the two cases reported, light causes a phase shift in the molecular oscillation by induction of a clock gene in one, and in the other leads to the loss of proteins involved in the feedback loop. New molecular information is appearing rapidly, and some may well be outdated before this volume appears. As another reviewer has noted recently [4], summarizing knowledge in the circadian field at this time is like shooting a moving target. Stay tuned.

16.1 Circadian clocks and molecular mechanisms

The environmental day/night cycle provides an ecological niche related to time of day, to which we can attribute the origin and evolution of endogenous timing mechanisms [5,6]. For many organisms, activity varies rhythmically with time of day, the maximum occurring at a specific time of the cycle (acrophase). While the acrophase for a particular rhythm may differ in different species, the regulation of the timing is typically indirect in the sense that it involves an internal biological mechanism, rather than being a direct response to light. The mechanism responsible is referred to as the circadian clock.

It is not only activity that is regulated by this clock mechanism; many diverse physiological processes are also controlled, and different processes may themselves have different acrophases in a given organism. The phenomenon may be viewed as a daily differentiation, and indeed, as in development, the daily program and the overt rhythms involve the regulation of gene expression, either at the transcriptional or translational levels, or both [7].

Even though one must say that we still do not have a full understanding of the biochemical mechanisms responsible for the clock and the pathways leading to and from it, progress has been dazzling over the past decade (reviewed in [4,8–12]. Genes in which mutations lead to the loss of rhythmicity or to a change in the free running period (FRP) of a rhythm have been isolated and extensively characterized from several organisms, including *Drosophila* [13], *Neurospora* [7], mammals [14,15] and *Synechococcus* [16,17].

Such "clock" genes, while lacking extensive sequence similarities per se, are elements in functionally similar circadian mechanisms. Several code for protein products that are able to repress the synthesis of their respective messages, such that there is a daily circadian cycle of synthesis of clock-gene message and protein. The feedback loop includes positive elements as well, paired transcriptional activators, the action of which is blocked by the negative elements. Other "clock" genes block or alter circadian expression in other ways, and the existence of additional loops is emerging as studies continue, making a schematic representation terribly complicated. A simplification is attempted in Figure 1.

The existence of functionally similar genes and regulation has led to the conclusion that similar molecular strategies have evolved for the construction of biological clocks in different species, even if they have evolutionarily different origins [18]. In fact, certain clock genes appear to be conserved across species [19,20], which may account for the similarity in mechanisms.

16.2 The conventional clock paradigm is inadequate

The circadian mechanism has been modeled by distinguishing three major components: input, central oscillator and output, but there is little direct experimental evidence in support of this paradigm. Although the model is still often used in the interpretation of experiments, its value now seems limited, if not negative.

Molecular components of input and output pathways had generally been considered **not** to be clock components, in the expectation that their absence would not alter the

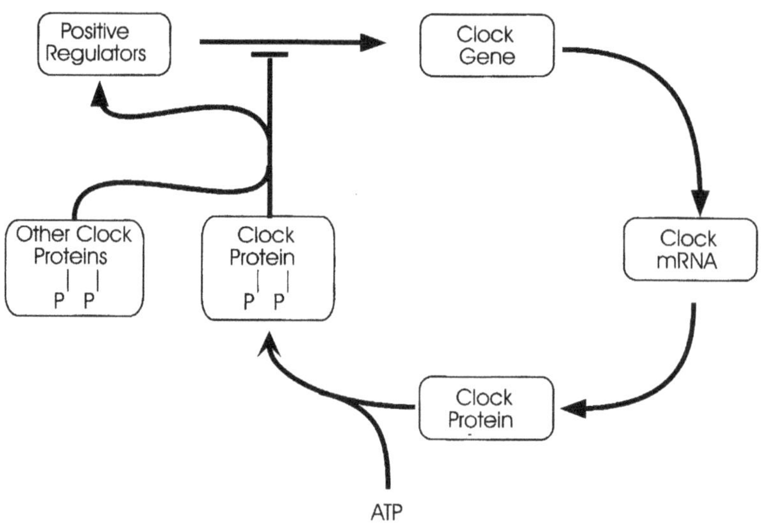

Figure 1. Schematic representation of components and pathways of the expression and regulation of a prototypical clock gene such as *per* and *tim*. Transcription of the clock gene is driven (designated by arrowhead) by positive regulators acting on the clock gene promoter, while the protein product of the gene enters the nucleus, and as a dimer with another clock protein, feeds back to represses its own transcription (designated by a T-head), and also up regulates the positive regulators. For detailed schemes of the individual systems, see Dunlap [4].

functioning of the clock. The analogy to a mechanical clock in which the removal of the hands or the reset knob has no effect on the timing of the clock itself has been persuasive for the view that the input and output components have no role in the timing mechanisms as such.

An explicit dissent from this view was put forward by Roenneberg and Merrow [21], based on results indicating that the clock genes *frq* (*Neurospora*) and *per* (*Drosophila*) are closely connected to their circadian entrainment pathways. They proposed a model with an autoregulatory gene and its products on an input pathway, feeding into a separate oscillator. In support of this, it was found that *frq9*, a null mutant of the *Neurospora* clock gene frequency, could be entrained by temperature but not by light cycles [22]. They interpreted this to mean that the *frq* gene product is involved in the light input pathway and that in *frq9* a circadian mechanism remains functional but not light entrainable. Several other recent findings support this kind of model, complicated though it be. For example, recent studies of two mouse cryptochrome genes, *mCry1* and *mCry2*, which are most probably involved in photoreception, indicate that they also act in the negative limb of the clock feedback loop [23].

The modeling of the clockworks and associated components is even more complex because of evidence from several systems that there may be more than one circadian oscillator, the hypothetical central component conceptualized in the clock paradigm. In humans maintained in isolation, the circadian rhythms of sleep/wake and body temperature were observed to desynchronize (spontaneously) and to then exhibit different FRPs, indicative of two oscillators [24,25]. In the alga *Gonyaulax polyedra* maintained in constant dim light (LL), different rhythms have been reported to dissociate and exhibit different FRPs [26–28]. In these cases transient phase coupling is suggestive of interaction between the two oscillators. There is also evidence for a circadian clock for photoperiodism distinct from the clock for circadian rhythmicity [29]. Thus, the question of how many circadian clocks there may be in a given organism, or how the clockworks are organized to give results described above, remains open.

16.3 Key clock properties

In organisms ranging from bacteria to mammals the existence of a circadian biological clock is inferred by an in vivo rhythm with a free running period (FRP; measured under constant laboratory conditions) of about (circa) one day (dian). A second essential feature is that the FRP of the rhythm is not greatly different at different but constant temperatures within a physiological range. This has been interpreted as a strong indication that the period, thus the accuracy of the timing, is of functional importance.

A third key property is entrainment, whereby the internal clock, as measured by a rhythm, which typically has a FRP slightly different from the 24 hour day, is synchronized to it, resulting in a stable phase relationship between the two. There are several environmental factors that change in a regular fashion over the daily cycle, and some of those (other than light), such as temperature, have been shown to act as zeitgebers (time givers) and to have the ability to **entrain**. But the most important and best studied entraining factor, which will be considered here, is light.

Entrainment to a 24 hour light/dark cycle followed by a free run in constant dim light is illustrated by the data of Figure 2, which are based on the rhythm of bioluminescence in the unicellular dinoflagellate *Gonyaulax*. Circadian rhythms may also be entrained by light dark cycles with cycle lengths different from 24 hours (e.g. 23 or 25 h, but there are limits, which will be discussed below.

It should also be noted that the term entrainment does not refer to "training" but to a train of events. In particular, a direct response to an environmental cycle does not constitute entrainment if it fails to alter the phase, as evaluated by a subsequent free run in constant conditions. Such responses, which have been incorrectly taken to be indicative of entrainment by some authors, are referred to as "masking effects" [6,30].

16.4 Actions of light on circadian systems: tonic and phasic effects

Two different kinds of light effects may be distinguished; this is done with organisms maintained under constant laboratory conditions (either constant light or darkness) so as to evaluate effects on the circadian mechanism.

If the light intensity is changed for a short time (minutes or a few hours), the phase of the circadian rhythm may be changed, either advanced or delayed, and by a time greater than the duration of the light pulse. Such an effect is considered to be discrete,

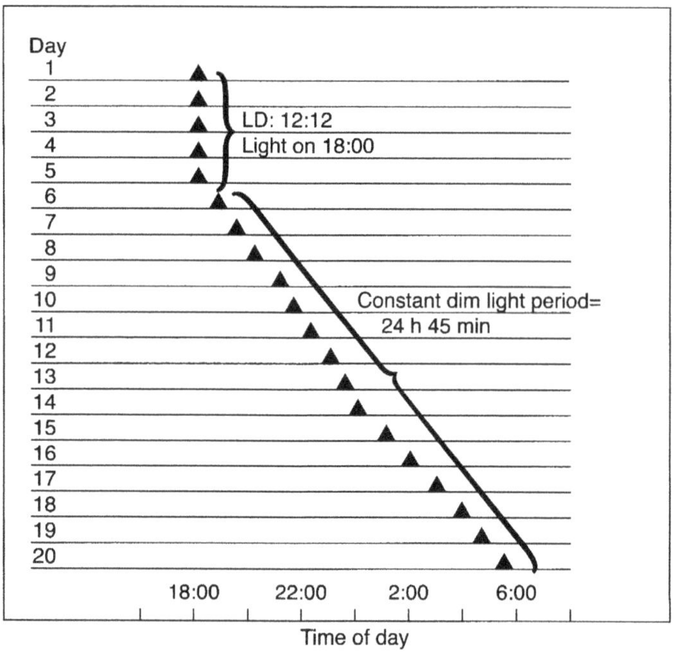

Figure 2. Plot showing the time of day at which maximum bioluminescence occurred in a culture of the dinoflagellate *Gonyaulax polyedra* subjected to a light-dark cycle for the first five days, during which time the clock is entrained, followed by being maintained in constant dim light, where the rhythm exhibits a free run with a period of 24.75 h. This uses the conventional way of plotting circadian data, placing sequential days beneath each other. From [106].

or *phasic* in nature, analogous to altering the position and thus the phase of a swinging pendulum by striking it once.

If the intensity (fluence rate) of the constant light under which the organisms are being maintained is different, the free running period (FRP) of the circadian rhythm may be different. This action may be viewed as occurring continuously, or *tonically*, like air resistance on a swinging pendulum. A different gas or a different partial pressure of a gas would provide a different resistance.

Aschoff [31] noted that in some (typically night active) species the period is longer at higher light intensities whereas in day active species it is shorter at higher intensities. In the unicellular dinoflagellate *Gonyaulax polyedra* (but so far in no other organism) it has been found that light of different spectral compositions can have qualitatively different effects on the period of the luminescence rhythm [32]. With an increasing intensity of blue light, the FRP becomes shorter, while with red light it becomes longer. This indicates that two different photoreceptors are involved in the light transduction to the *Gonyaulax* clock.

As elaborated below, entrainment of circadian rhythms can be well explained by phasic effects, but some role for tonic effects cannot be altogether excluded.

16.5 Daily phase shifts required for entrainment

Under full cycles in natural or laboratory (e.g. 12 h light-12 h dark; LD 12:12) conditions, the period of a circadian rhythm is exactly 24 h, even though the FRP may be greater or less than 24 h. This entrainment is analogous to resetting an inaccurate clock each day, but the convention for what constitutes an advance or a delay in a circadian clock differs from that used in mechanical clocks, often creating confusion. In circadian systems, the convention is that if the free running circadian period is longer than 24 h, the phase in the circadian clockworks must be **advanced** to an earlier time each cycle for entrainment, and conversely the phase has to be **delayed** if the FRP is shorter. Also by convention, advances in the circadian clockworks are positive, and delays negative. When entrained, the amount of phase shift each cycle is thus given by the equation

$$\text{FRP} - \text{T} = \text{phase shift},$$

where T is the period of the entraining cycle.

The physiological mechanism of entrainment has features that are similar in phylogenetically diverse species, both plant and animal, allowing us to generalize in describing the phenomenon. Thus, even if the different circadian systems originated independently, mechanistically similar systems have evolved.

16.6 Single exposures to light evoke phase shifts: the phase response curve (PRC)

An explanation for phase shifts needed for entrainment was provided by the discovery that single light pulses could result in phase shifts in the circadian oscillator, and that such shifts are different in both sign and magnitude when applied at different times in the circadian cycle [33,34]; reviewed in Johnson [35].

In the experimental protocol with the rhythm of bioluminescence in *Gonyaulax* many individual cell cultures are subjected to an entraining light-dark cycle and then transferred to and maintained in constant dim light or darkness. At two to four hour intervals thereafter some cultures are removed and exposed to light for one to three hours and then returned to the constant conditions. The bioluminescence rhythms of all cultures are subsequently measured and their phases determined [33,36].

The result, illustrated diagrammatically in Figure 3B, shows that light pulses given at a time corresponding to early night phase result in delays, increasing in magnitude to about 12 h around mid-night phase, while pulses given in late night phase result in advances, starting at about 12 h shifts and becoming less approaching day phase. Pulses given during the day phase are generally without effect or less effective; this is sometimes referred to as the "dead zone".

16.7 The PRC depends on intensity and duration of light exposure: type 0 and type 1 resetting

The phase shifts illustrated in Figure 3B depend upon the time in the cycle and are maximally 12 hour advances or delays. Potentially, then, with such a PRC the rhythm can be reset to any of all possible new phases. This is referred to as strong (type 0) resetting [35,37].

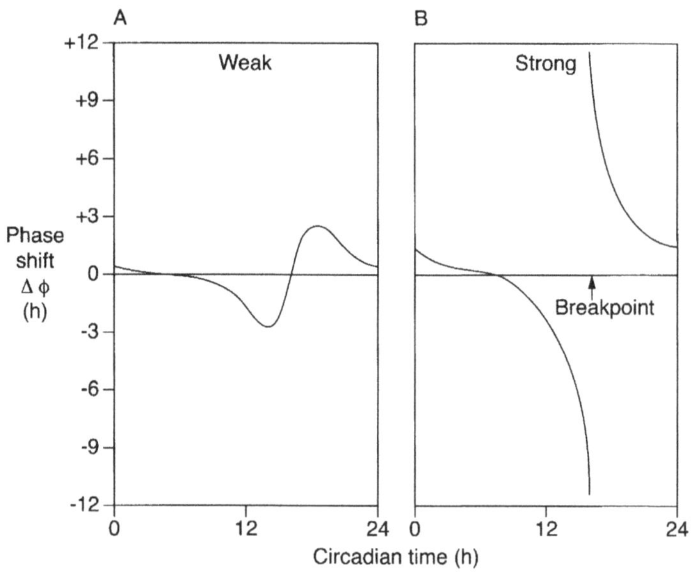

Figure 3. Schematic phase response curves for phase shifting by light of an organism maintained in constant conditions except for a brief exposure to light. Abscissa, circadian time of the exposure; ordinate, phase shift in hours. Panel A weak, type 1 shifts, and Panel B, strong, type 0 phase shifts. After [30]. See Johnson [35, 107].

The magnitude of phase shifts also depends upon the intensity (fluence rate) of the light and the duration of the exposure [33]. At lower fluence rates and shorter durations phase shifts are smaller, and the PRC has a quite different character, as shown in Figure 3, trace B. This is referred to as weak (type 1) resetting. The different responses can be modeled in a very satisfactory way by a limit cycle [38,39], the details of which are not discussed in this paper. For the present consideration, it should be noted that in both cases there are delays and advances at the onset and termination, respectively, of the night phase of the circadian clockworks.

16.8 The PRC can account for entrainment

If the FRP (e.g. 25 h) of a rhythm is **greater** than the entraining period (e.g. 24 h) of a light-dark cycle, the night (light sensitive) phase of the circadian clockworks (as determined by the PRC) will extend beyond the dark portion and receive an exposure to light at a time and for a duration such that a one hour phase advance will occur. If the FRP is even longer (e.g. 26 h), the exposure of the light sensitive phase will occur even earlier, at a point on the PRC where the phase shift is even greater, with the duration of the exposure also being greater. By the same reasoning, if the FRP is **shorter** than 24 h, the onset of night phase of the circadian clockworks will begin before the end of the light period and the light exposure will result in a delay in the phase of the clockworks, such that the equation above is satisfied.

Entrainment may thus be viewed as repetitive resetting by either advances or delays, each equal in magnitude to the difference between the entraining cycle length (T) and the FRP. Although this model invokes only a single phase shift per cycle, an entraining cycle with a short dark period (e.g. LD, 18:6) will involve both delays and advances, since the light-sensitive period of the circadian clockworks would be longer than the duration of darkness in the cycle. The equation would thus give a net phase shift, positive or negative, depending on the FRP, but would include both delays and advances.

In this model, entrainment by light is considered to be discrete, or phasic in nature, as discussed above. However, dusk and dawn are not sharp transitions, and it is clear that light may act continuously on the circadian system, as shown by the effect of intensity on the FRP, as illustrated in Figure 4 for the *Gonyaulax* rhythm of bioluminescence. The extent to which a continuous action of light may have a role, if any, in entrainment is not clear. But the success of the discrete model and the evidence below from experiments with skeleton photoperiods and single pulses per cycle suggests that entrainment is largely attributable to a discrete phase-shifting action of light.

16.9 Light resets clock genes: the PRC at the molecular level

The discovery and analysis of clock genes has provided the beginnings of our understanding of the molecular mechanism of circadian systems. If components thus far

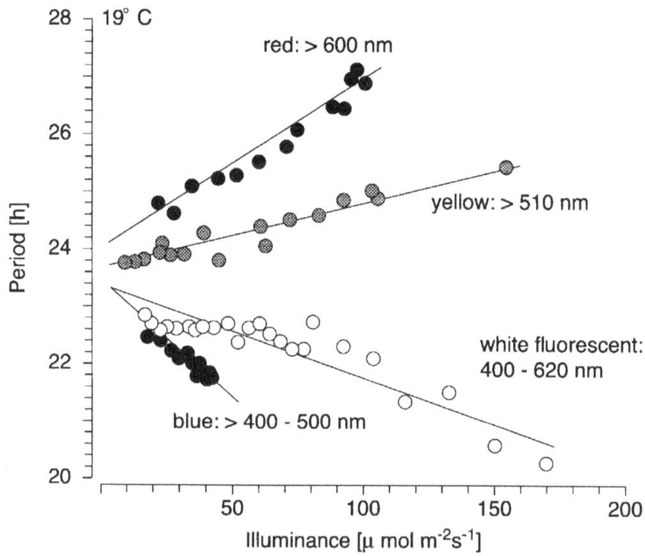

Figure 4. Effects of light intensity (abscissa) on the length of the free running period (FRP) for the circadian rhythm of bioluminescence in *Gonyaulax*, showing the different functions obtained for light of different spectral qualities. In blue light the FRP becomes shorter with increased intensity while in red light it becomes longer. From [32].

identified are elements of the circadian oscillator, then light induced phase shifting should act at that level.

Indeed, this has been shown in several systems, first in *Neurospora*, where brief (1 to 2 min) exposure to light induces the *frq* transcript [40], acting rapidly through the positive elements WC-1 and WC-2 [41]. Both *frq* mRNA and FRQ protein exhibit a circadian rhythm, the pattern of which is shown in Figure 5 along with the PRC for phase shifts by light. Induction of *frq* by light at a time prior to its normal peak would advance the phase, whereas induction after the usual peak time would set the phase back and thus delay the rhythm. This is consistent with the observed PRC for the intact organism. The finding provides a major advance in our knowledge of the mechanism of phase shifting and entrainment.

In *Drosophila* the mechanism involves the clock proteins TIM and PER, which in the normal cycle dimerize and function to repress their own transcription (Figure 1). Light causes a prompt decrease in the level of TIM protein, and thereby PER as well, since dimerization is required for stability [42–45]. These proteins are increasing in the late day and early evening so their inactivation brings their levels back to a low point, a delay. Conversely, light during the decreasing phase hastens the decline and advances the rhythm.

In mammals the light appears to induce clock genes *per1* and *per2*, and the phasing and direction of shift is similar to that in *Neurospora* [46,47]. In all three cases thus far examined light acts to increase or decrease the levels of negative elements, thereby rephasing the molecular cycle. How this is translated into physiological changes is still to be established.

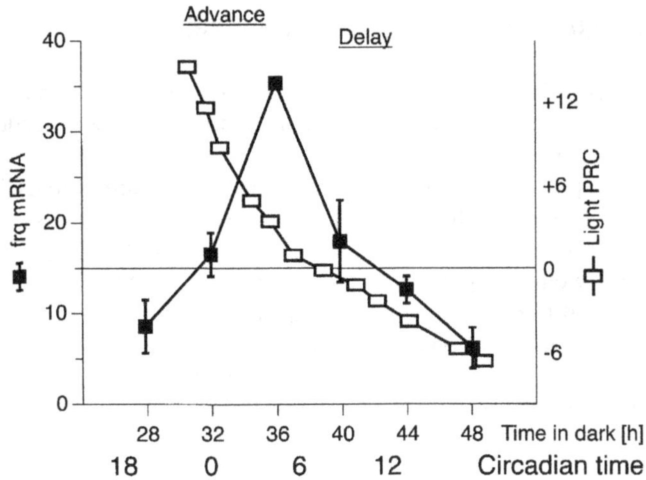

Figure 5. The phase of the circadian rhythm of *frq* mRNA relative to the PRC for resetting the conidiation rhythm in *Neurospora* by light. Exposure to light of a culture growing in DD causes a rapid increase in *frq* mRNA at all times in the cycle, such that the phase of the RNA rhythm is reset to the circadian time of its normal peak, thereby resetting the phase of the rhythm. PRC from [83]; RNA rhythm from [40].

16.10 Entrainment does not require light and dark cycles: skeleton photocycles

In nature entrainment involves cycles of alternating light and dark, and such cycles imposed under laboratory conditions are similarly effective. However, entrainment of the circadian clock does not require the full light dark cycle; it may be accomplished equally well by exposing the organism to a "skeleton photoperiod", or "skeleton photocycle" – two brief light exposures per day, effectively simulating dawn and dusk. Thus, for example, *Drosophila* maintained in constant dark (DD) except for the brief light exposures, can be entrained to 24 h by two 15 minute light pulses spaced every 12 h, or at alternating intervals of 10 h and 14 h, or 8 h and 16 h, etc. [1] (Figure 6).

As with complete photocycles, entrainment to other period lengths is also possible using two light pulses per cycle; for example a rhythm can be entrained to a 23 h skeleton cycle by light pulses spaced every 11.5 h. Skeleton cycles may also consist of a background of LL with two dark (or dim) intervals per 24 h, thus skeleton scotocycles.

If the two light pulses are spaced about 12 hours apart then the dark phase of the rhythm may fall in either of the dark intervals, depending on the phase of the rhythm at the time when the skeleton cycle was introduced. But if the skeleton timing is far removed from 12:12, thus 16:8, for example, the night phase of the rhythm will self position itself, always finally occurring in the longer of the two intervals. In *Drosophila*, the range of "bistability", where the night phase of the rhythm may occupy either interval, lies between 10.3 h and 13.7 h, outside of which it will always occupy the longer interval [1] (Figure 6).

16.11 Single light pulses per cycle can entrain

A skeleton photocycle does not need two exposures to light per cycle in order to entrain; a single will suffice, given that only a single phase shift per cycle is needed in order to satisfy the entrainment equation.

Such single light pulse experimental regimes have been referred to as T cycles [48], where T is the cycle length. However, full light-dark cycles have also been called T cycles, where T is the cycle length, so the use of the term "single pulse T cycle" or "skeleton T cycle" would be preferable. If T = 24 h and the FRP is greater than 24 h, then an advance equal to FRP − T h is required each cycle. In this case the rhythm will position itself so that its night phase immediately precedes the light pulse. When the FRP is less than 24 h delays are required, and the night phase of the rhythm will be located right after the time at which the light pulse occurs.

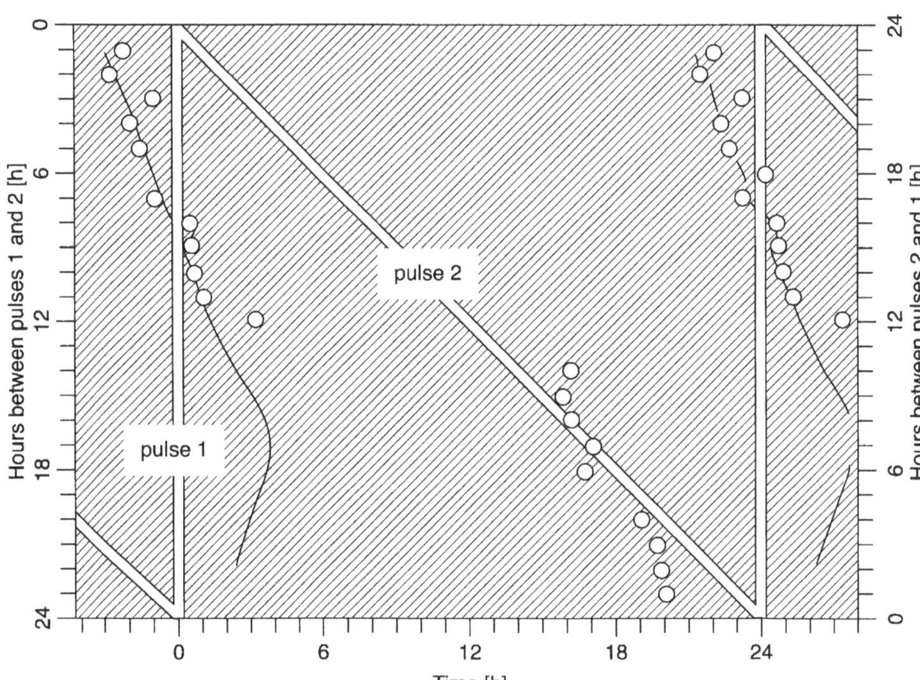

Figure 6. Phasing of the circadian rhythm of eclosion in *Drosophila* subjected to different skeleton photoperiods following prior entrainment in a light-dark cycle, with all cultures having the same initial phase, with the original day phase, as determined by the prior entrainment, starting at 0 hours on the abcissa. Each imaginary horizontal line represents a culture subjected to two 15 min light pulses per cycle, spaced as indicated by the vertical and slanting lines, and the self selected phase of the rhythm in each is indicated by a circle. In the cultures where the time between pulses 1 and 2 was 12 hours or less the initial phase was retained; with longer times between the pulses the night phase of the clock positioned itself so as to fall at the end of the longer interval. After [1].

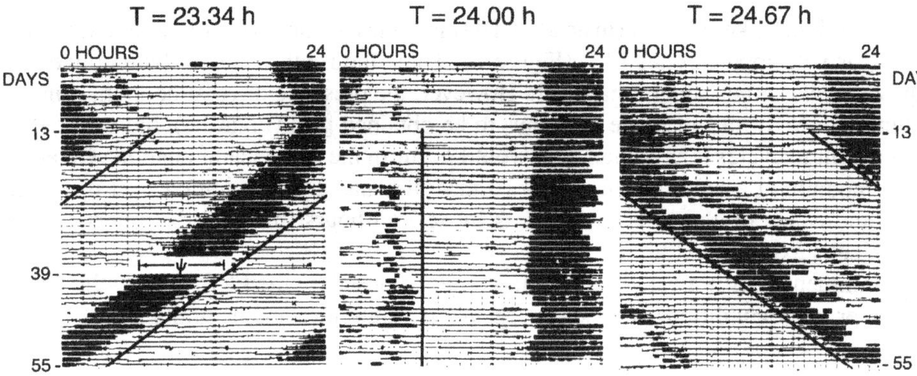

Figure 7. Activity rhythms of hamsters subjected to entrainment by single light pulses per cycle. The entraining cycles were shorter (23.33 h) and longer (24 h and 24.6 h) than the FRP, which was about 23.9 h. By self selection the animals position their activity rhythms so that they occur just prior to the light pulse for T = 23.33, providing an advance reset, and just after the light pulse in the other cases, giving delay resets. After [48].

As with full and skeleton cycles, single pulse T cycles with periods greater or less than 24 h also entrain. This is illustrated for the hamster in the experiments of Figure 7, where the T cycles in one experiment are shorter than the FRP (which in this case was about 23.9 h), and in the others longer, one by only 0.1 h. As can be seen, the night phases, characterized by running activity, are self positioned, respectively, immediately before and after the time at which the pulses occur.

As reported by many workers, circadian systems may be exquisitely sensitive to light. An impressive example is the squirrel monkey, which was shown to be entrained by a single one-second exposure to light per cycle [49].

16.12 Limits of entrainment

How different in length can an imposed light-dark (or skeleton) cycle be and still succeed in entraining a circadian rhythm? In principle, the limits of entrainment may be inferred from the PRC. In the extreme case, with a type 0 PRC with advance and delay phase shifts of up to 12 hours (Figure 3), entrainment could be accomplished with cycles ranging from 12 to 36 h; the former would be entrained by a 12 h advance each cycle and the latter by 12 h delays.

In *Gonyaulax*, where the PRC shows full 12 h phase advances following an appropriate pulse of bright light (Figure 8), entrainment to 12 hour (LD6:6) cycles has been demonstrated to occur with bright light but to break down in dim light cycles, showing the importance of light intensity on the magnitude of phase shifting [50].

In many organisms, however, the light intensity and other conditions may be such that the maximum phase shifts obtainable experimentally are much less than 12 hours, and in many systems only type 1 PRCs are obtained. Moreover, PRCs may have features idiosyncratic to the organism. For example, the PRC for the *Gonyaulax* luminescence rhythm with cells kept in LL (instead of DD) includes advances up to 12 hours, but no delays, or very modest ones [36] (Figure 8). Thus, the rhythm cannot be entrained to imposed cycles much longer than 24 h. However, the presence of creatine in the culture medium allows for a full range of delays to be obtained [51] (Figure 8), indicating that entrainment to cycles longer than 24 h would occur in the presence of creatine. Creatine in the medium at a concentration of ~ 10 mM also has a strong tonic effect upon the bioluminescence rhythm; the FRP is less by several hours [52], but this effect occurs only in cells exposed to blue light.

Entrainment does not occur beyond the limits imposed by the PRC, and rhythms free run under such conditions; for some organisms light-entrainment limits may be 24 h

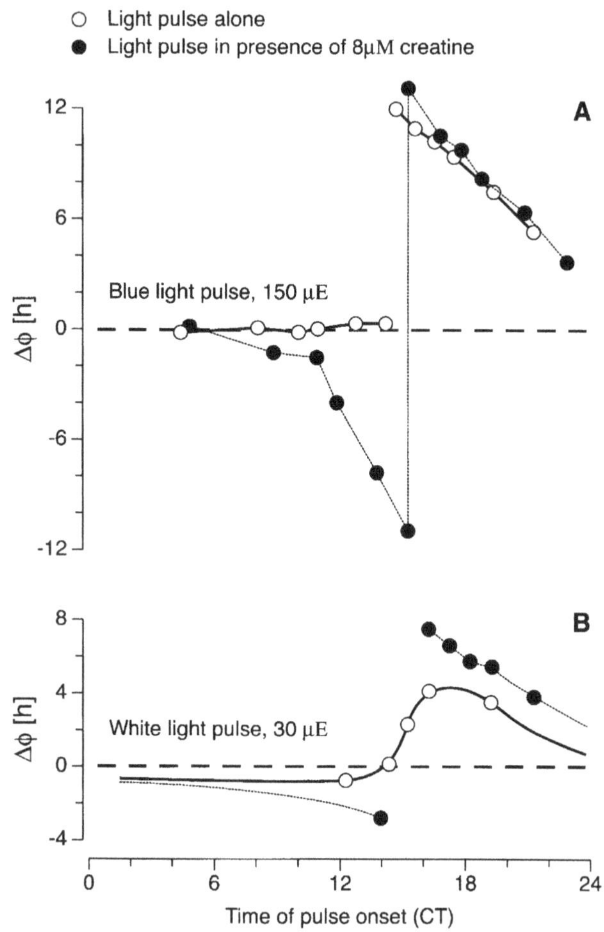

Figure 8. Creatine in the medium increases magnitude of light-induced phase shifts. PRCsare shown for (A) strong blue light pulses and (B) weak white light pulses with (filled circles) and without (open circles) 8 mM creatine. From [51]. Not shown are experiments documenting that light-induced phase shifts are completely blocked in the presence of the protein kinase inhibitor, 6-DMAP.

plus or minus only a few hours. With higher frequency light dark cycles that are well outside the range of entrainment (e.g. hour- or minute-long cycles), circadian clocks free run with an FRP near their FRP in constant light or constant dark [6]. However, the imposed high frequency cycle may have a demonstrable effect on the exact value of FRP that is not attributable simply to the tonic action of light on the FRP, but to the integral of the many, albeit very small, phase shifts that occur, some advances and some delays, as was shown in *Gonyaulax* [53].

16.13 Protein phosphorylation and the PRC

An explanation for the effect of creatine in speeding up the clock and evoking light-induced phase delays has not been obtained, but it may very well relate to an effect on protein phosphorylation. Creatine is involved as a shuttle molecule in phosphate transfer, and while its presence has not been demonstrated in *Gonyaulax*, it could have an action in this regard. This possibility is supported by the finding that an inhibitor of protein kinases, 6-dimethyl amino purine (6-DMAP; at mM concentrations), completely blocks phase shifts induced by light in both the presence and absence of creatine [54].

Staurosporine, another and more specific protein kinase inhibitor, increases the magnitude of light-induced phase shifting (at nM concentrations), both delays and advances, and in the absence of creatine [55] (Figure 9). It is postulated that the difference in the actions of 6-DMAP and staurosporine is because they act on different target kinases. Both inhibitors have a strong tonic effect on the FRP, increasing it from about 23 h to more than 30 h, but pulses of the inhibitors do not cause phase shifts. Many other protein kinase inhibitors are without effect on the *Gonyaulax* circadian system, so the targets may be inferred to be specific.

Inhibitors of protein serine/threonine phosphatases also have important effects on circadian rhythms, but they do not simply counter the effects of kinase inhibitors [56]. Okadaic acid administered chronically has a strong tonic effect on the FRP, increasing it to as much as 30 h (concentration dependent), while cantharadin results in a decrease of the FRP of about 2 h and calyculin is without effect. All three, however, cause a pronounced increase in light-induced phase shifts, also altering the time in the cycle when the maximum shift occurs. These several effects are postulated to involve steps in the circadian light transduction pathway; indeed, light transduction pathways are known to involve protein phosphorylation in other systems.

16.14 Circadian photoreception does not require vision

Light entrainment evidently involves photoreception and input pathways to the clock, but circadian photoreceptors and pathways are not the same as those used in vision, or indeed for sensory photoreception in general, in any of the several cases analyzed so far. This was first and very definitively shown by the pioneering work of Menaker [57] in sparrows; photoentrainment was not lost in birds blinded by removal of the eyes. By masking experiments it was shown that photoreception occurred in the brain [58], even in the absence of the pineal (an extra retinal photoreceptor organ in many species), but there does not appear to be a more specific localization, nor has the photoreceptor molecule been definitively identified.

It has also been shown in many other non mammalian vertebrates, as well as in invertebrates and unicellular organisms, that non visual or non sensory reception is primarily responsible for circadian entrainment [59,60]. In cultured chicken and lizard pineal glands, melatonin synthesis exhibits an entrainable circadian rhythm. In the silkmoth, entrainment occurs by the action of light directly on the brain [61], and the eye of the snail, *Bulla*, surgically trimmed to remove all retinal elements, retains a circadian rhythm and is entrained by light cycles [62]. *Drosophila* deprived of dietary carotenoids can have their visual sensitivity decreased by 10 orders of magnitude, with no decrement whatsoever in sensitivity to photic entrainment [63], suggesting that a rhodopsin is not involved in circadian photoresponses (however, see [64]). Also, genetic ablation of the eyes or mutations in visual transduction pathways of *Drosophila* do not block circadian entrainment [65]. In the alga *Chlamydomonas*, the photoreceptor pigment for phototaxis exhibits absorption peaking around 500 nm [66], but

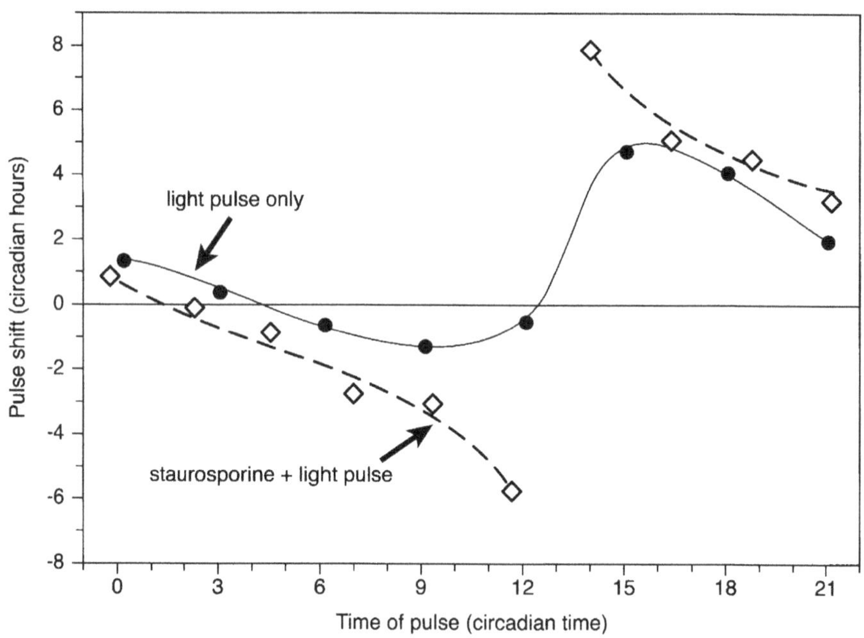

Figure 9. Staurosporine, a protein kinase inhibitor, enhances the responses to light pulses. The PRCs were obtained by subjecting cultures (maintained otherwise in dim white light) to three hour pulses of bright blue light (120 $\mu E\ m^{-2}\ s^{-1}$) at 3-h intervals, starting about 30 h after transfer from LD to LL. Solid line, light pulse only. Dashed line, 20 nM staurosporine added to cultures at time of transfer to LL. Phase shifts and times for light pulses are given in circadian hours, thus compensating for the period-lengthening effect caused by staurosporine. From [55].

light-induced phase shifting of the circadian system has a very different action spectrum, with peaks at longer wavelengths [67,68].

The photoperiodic response in insects, which is regulated by a circadian clock, can also occur in the absence of the functional organized photoreceptors, eliminated either by destroying or shielding the structures in question [69]; again, the brain is the site of photic input for the photoperiodic responses. Photoperiodic photoreception in the lizard is similarly mediated by extraretinal light receptors [70].

In mammals, by contrast, early attempts to demonstrate an extraretinal pathway for entrainment of circadian rhythms were unsuccessful [71], and have remained so, with the exception of a single recent report (next paragraph). In all experimental mammals tested eyes are required, but vision is not. In mice homozygous for retinal degeneration (*rd/rd*), all rods and most cones are lost by 150 days, along with all electrophysiological and behavioral responses to bright light [72], but such animals retain unabated circadian responses to light and entrainment, indistinguishable from wild type mice. The same is true for transgenic animals in which both rods and cones have been eliminated [73]. Still, the circadian light input must be ocular, because the removal of the eyes or severing of the optic nerve eliminates photoentrainment. These and other experiments indicate that photoreceptors other than rods and cones must be responsible for circadian light responses [74]. The cryptochromes may be such molecules (see below).

The exception was reported by Campbell and Murphy [75], who exposed humans to light in the popliteal region (behind the knee) at different times of day, in protocols which evoked type 1 delay and advance phase shifts of the circadian system. The finding remains to be confirmed, and if so to be explained in terms of the receptor and pathway involved. Similar exposures have been reported not to suppress melatonin during the night phase, diagnostic of the action of light on the circadian system [76,77].

Most blind humans who have no conscious perception of light are not entrained by light, such that their internal circadian clock "free runs" even when they maintain a 24 h activity-rest schedule [78–80]. Such persons often experience cyclic bouts of insomnia as their circadian pacemakers move out of and then back in phase with the 24-h day. Some totally blind individuals are actually entrained by light entering the eyes, as demonstrated by the suppression of nocturnal melatonin secretion by light, preventable by covering the eyes [79]. These individuals may be comparable to mice with degenerate retinas, with vision lost but ocular circadian photoreceptors intact. A few totally blind individuals have been identified as entrained, but not by light; in those cases no evidence for extraocular photoreception was obtained, and it was concluded that non photic stimuli are responsible [81].

16.15 Circadian photoreceptors are primarily blue-sensitive

Although the biochemical identity of a photoreceptor pigment for circadian responses is not yet established for certain in any individual species, action spectra and other considerations indicate that blue light is the most effective in most. In *Neurospora* the circadian system is sensitive to blue light but blind in the red [82], as shown also by the action spectrum for phase shifting [83]. Such action spectra for *Drosophila* [84] and hamsters [85] also show effectiveness only in the blue region.

However, in higher plants [86] and *Gonyaulax* the circadian system is sensitive to both blue and red, as shown in the action spectrum for phase shifting in *Gonyaulax* [87] (Figure 10). Attempts to identify phytochrome genes that might be responsible for the red responses in *Gonyaulax*, as in higher plants [88], have not been successful. Blue light is much more effective than red [89], and the different effects on the FRP of increasing intensities of blue and red light [32] (Figure 4) indicate that two photoreceptors and two input pathways are indeed involved in phase shifting, but their respective roles are not known.

The action of creatine, described in connection with its enhancement of phase shifting by light, also includes a tonic effect on the FRP, which is shortened in a concentration dependent fashion [52]. Its effect occurs only in blue light [51]. Phase response curves have been reported to differ for red and blue light [90], but the effect of creatine, whose action occurs in blue but not red light, was not determined. Allopurinol, known as an inhibitor of xanthine oxidase, lengthens the FRP of the bioluminescence rhythm, and inhibits phase advances during the second half of the subjective night, but acts on blue light effects only [91]. However, the effect of allopurinol in the presence of creatine, thus on the creatine-induced phase delays, was also not determined.

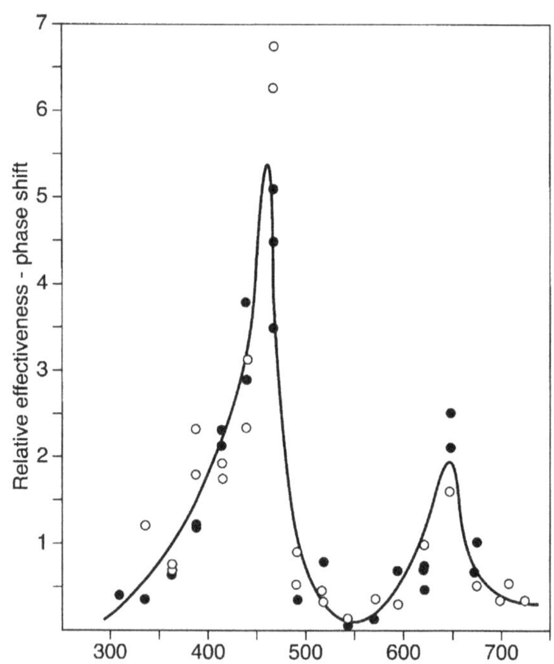

Figure 10. Action spectrum for shifting the phase of the bioluminescence rhythm in *Gonyaulax* by single 3-hour exposures to light of different wavelengths. From [87]. Different symbols give data from two independent experiments.

16.16 Cryptochrome: a newly identified photoreceptor is also a part of the molecular clock

Over the past 50 years a host of light effects, studied extensively in plants, have been found to be due to light in the blue region of the spectrum. Continued experimentation generated a protracted and relentless debate concerning the identity of the photoreceptor- whether a flavin, a carotenoid or some other molecule, with the name cryptochrome referring to blue light action on the cryptogams, but evidently appropriate also to its elusive nature. The gene coding for a molecule with appropriate properties was first isolated and characterized in *Arabidopsis* on the basis of the failure of a mutant impaired in *CryI* (also referred to as *HY4*) to respond to blue light suppression of hypocotyl elongation [92,93].

CRY1, the protein encoded by the gene, was found to be a 75 kDa flavoprotein, and shown to share sequence similarity with type I DNA photolyase, a blue light-dependent DNA repair flavoenzyme [94,95]. A second member of the *Arabidopsis* cryptochrome family, CRY2, also mediates blue light-induced shortening of the hypocotyl; overexpression of either gene results in hypersensitivity to blue light [96].

That cryptochromes function as photoreceptors for a number of different processes in plants is well established [93,97–99]. Their involvement in circadian control in plants has been described [88] but is less extensively investigated.

The possibility that such genes might occur in animals was proposed by Sancar and colleagues, who analyzed and isolated human genes having sequence similarities to photolyases but lacking DNA repair activity [100]. With knowledge that the *Arabidopsis* cry genes were also homologous to photolyase, they took the bold step of naming the human genes *cry1* and *cry2*, and proposing that the protein hCRY1 and hCRY2 function as blue light photoreceptors in humans, and that they function as the photoreceptors in the entrainment of circadian rhythms [101]. Mice lacking *cry2* have significantly longer FRPs and exhibit larger phase shifts to light pulses [102]. This last finding suggests that the protein has a function in the clock mechanism; if it served only as a photoreceptor it would be expected that the response to light should be reduced.

A more definitive demonstration that the proteins are actually involved in the clock mechanism came from the report that while the FRPs are altered in mice lacking either *cry1* or *cry2*, mutant mice lacking both proteins are altogether arhythmic in activity [103]. At the molecular level, Reppert and colleagues showed that the two genes affect molecular cycling, acting in the negative limb of the feedback loop of the clock [23]. Reporter gene assays showed that mCRY1 or mCRY2 alone abrogates CLOCK-:BMAL1-E box mediated transcription, the positive activator in the mammalian system (Figure 1). Thus, these proteins are definitely involved in the clock mechanism; the role of these proteins in circadian photoreception in mammals remains to be clarified.

Studies with *Drosophila* indicate that cryptochrome is present and definitely functions in both light reception and in molecular cycling of PER and TIM proteins [104, 105]. A rhythm mutant was identified as a null mutation in a *Drosophila*'s version of the blue light receptor cryptochrome. The mutation was mapped to a position within the ORF of a cryptochrome encoding gene, named cry^b. CRY was found to be a novel member of the cryptochrome/6–4 photolyase protein family, lacking photolyase activity but, as with the other cases analyzed, binding both flavin and pterin chromophores.

Mutant flies show relatively normal rhythmic behavior despite arhythmic expression of PER and TIM, but the mutant exhibits poor synchronization to light-dark cycles and shows no response to brief light pulses. Convincing evidence of a role in light reception was the finding that photosensitivity increased in a strain overexpressing CRY. Altogether, these results support the contention of Roenneberg and Merrow [21] that molecules involved in input pathways can be an integral part of the clock mechanism.

References

1. C.S. Pittendrigh (1965). On the mechanism of the entrainment of a circadian rhythm by light cycles. In: J. Aschoff (Ed.), *Circadian Clocks* (pp. 277–297). North Holland, Amsterdam.
2. C.S. Pittendrigh (1981). Circadian systems: entrainment. In: J. Aschoff (Ed.), *Handbook of Behavioural Neurobiology* (Vol. 4, pp. 95–124). Biological Rhythms, Plenum Press, NY.
3. M.W. Young (2000). The tick-tock of the biological clock. *Scientific American*, March, in press.
4. J.C. Dunlap (1999). Molecular bases for circadian clocks. *Cell*, **96**, 271–290.
5. T. Roenneberg, R.G. Foster (1997). Twilight times: Light and the circadian system. *Photochem. Photobiol.*, **66**, 549–561.
6. L.N. Edmunds (1988). *Cellular and molecular bases of biological clocks: Models and mechanisms of circadian time keeping*. Springer, NY.
7. J.C. Dunlap (1996). Genetic and molecular analysis of circadian rhythms. *Annu. Rev. Genetics*, **30**, 579–601.
8. P. Sassone-Corsi (1997). PERpetuating the PASt. *Nature*, **389**, 443–444.
9. A.J. Millar (1997). Circadian rhythms: PASing time. *Curr. Biol.*, **7**, R474–476.
10. J.C. Hall, M. Rosbash (1998). Mutations and molecules influencing biological rhythms. *Annu. Rev. Neurosci.*, **11**, 373–393.
11. S.M. Reppert (1998). A clockwork explosion. *Neuron*, **21**, 1–4.
12. M.W. Young (1998). The molecular control of circadian behavioral rhythms and their entrainment in *Drosophila*. *Annu. Rev. Biochem.*, **67**, 135–152.
13. J.C. Hall (1998). Genetics of biological rhythms in *Drosophila*. *Adv. Genetics*, **38**, 135–184.
14. M. Vitaterna, D. King, A.-M. Chang, J. Kornhauser, P. Lowrey, D. McDonald, W. Dove, L. Pinto, F. Turek, J.T. Takahashi (1994). Mutagenesis and mapping of a mouse gene, *clock*, essential for circadian behaviour. *Science*, **264**, 719–725.
15. J.S. Takahashi, L. Pinto, M. Vitaterna (1994). Forward and reverse genetic approaches to behavior in the mouse. *Science*, **264**, 1724–1733.
16. S.S. Golden, M. Ishiura, C.H. Johnson, T. Kondo (1997). Cyanobacterial circadian rhythms. *Annu. Rev. Plant Physiol. Plant Mol. Biol.*, **48**, 327–354.
17. M. Ishiura, S. Kutsuna, S. Aoki, H. Iwasaki, C. Andeersson, A. Tanabe, S. Golden, C.H. Johnson, and T. Kondo (1998). Expression of a gene cluster kaiABC as a circadian feedback process in cyanobacteria. *Science*, **281**, 1519–1523.
18. J.C. Dunlap (1998). Common threads in eukaryotic circadian systems. *Curr. Opin. Genet. Dev.*, **8**, 400–406.
19. S. Sun, U. Alsbrecht, O. Zhuchenko, J. Bailey, G. Eichele, C. Lee (1997). Rigui, a putative mammalian ortholog of the *Drosophila* period gene. *Cell*, **90**, 1003–1011.
20. H. Tei, H. Okamura, Y. Shigeyoshi, C. Fukuhara, R. Ozawa, M. Hirose, Y. Sakaki (1997). Circadian oscillation of a mammalian homologue of the *Drosophila* gene period. *Nature*, **389**, 512–516.

21. T. Roenneberg, M. Merrow (1998). Molecular circadian oscillators: an alternative hypothesis. *J. Biol. Rhythms*, **13**, 167–179.
22. M. Merrow, M. Brunner, T. Roenneberg (1999). Assignment of circadian function for the *Neurospora* clock gene *frequency*. *Nature*, **399**, 584–586.
23. K. Kume, M. Zylka, S. Sriram, L. Shearman, D. Weaver, X. Jin, E. Maywood, M. Hastings, S. Reppert (1999). mCRY1 and mCRY2 are essential components of the negative limb of the circadian clock feedback loop. *Cell*, **98**, 193–205.
24. J. Aschoff (1965). Circadian rhythms in man. *Science*, **148**, 1427–1432.
25. R. Wever (1979). *The circadian system of man*. Springer-Verlag, New York.
26. F. Von der Heyde, A. Wilkens, L. Rensing (1992). The effect of temperature on the circadian rhythms of flashing and glow in *Gonyaulax polyedra*: Are the two rhythms controlled by two oscillators? *J. Biol. Rhythms*, **7**, 115–123.
27. T. Roenneberg, D. Morse (1993). Two circadian oscillators in one cell. *Nature*, **362**, 362–364.
28. D.M. Morse, J.W. Hastings, T. Roenneberg (1994). Different phase responses of two circadian oscillators in *Gonyaulax*. *J. Biol. Rhythms*, **9**, 263–274.
29. D.S. Saunders (1990). The circadian basis of ovarian diapause regulation in *Drosophila melanogaster*: is the period gene causally involved in photoperiodic time measurement? *J. Biol. Rhythms*, **5**, 315–331.
30. M. Moore-Ede, F. Sulzman, C. Fuller (1982). *The clocks that time us*. Harvard University Press, Cambridge, MA.
31. J. Aschoff (1960). Exogenous and endogenous components in circadian rhythms. *Cold Spring Harbor Symposia Quant. Biol.*, **25**, 11–28.
32. T. Roenneberg, J.W. Hastings (1988). Two photoreceptors control the circadian clock of a unicellular alga. *Naturwissenschaften*, **75**, 206–207.
33. J.W. Hastings, B.M. Sweeney (1958). A persistent diurnal rhythm of luminescence in *Gonyaulax polyedra*. *Biol. Bull.*, **115**, 440–458.
34. P. DeCoursey (1960). Daily rhythm of light sensitivity in a rodent. *Science*, **131**, 33–35.
35. C.H. Johnson (1992). Phase response curves: what can they tell us about circadian clocks? In: T. Hiroshige, K. Honma (Eds), *Circadian Clocks from Cell to Human* (pp. 209–246). Hokkaido University Press, Sapporo, Japan.
36. C.H. Johnson, J.W. Hastings (1989). Circadian phototransduction: Phase resetting and frequency of the circadian clock of *Gonyaulax* cells in red light. *J. Biol. Rhythms*, **4**, 417–37.
37. A.T. Winfree (1980). *The geometry of biological time*. Springer-Verlag, New York.
38. A.T. Winfree (1987). When time breaks down, Princeton University Press, Princeton.
39. P. Lakin-Thomas (1995). A beginner's guide to limit cycles, their uses and abuses. *Biol. Rhythm Research*, **26**, 216–232.
40. S.K. Crosthwaite, J.J. Loros, J.C. Dunlap (1995). Light-inducing resetting of a circadian clock is mediated by a rapid increase in *frequency* transcript. *Cell*, **81**, 1003–1012.
41. S.K. Crosthwaite, J.J. Dunlap, J.J. Loros (1997). *Neurospora wc-1* and *wc-2*: transcription, photoresponses, and the origins of circadian rhythmicity. *Science*, **276**, 763–769.
42. M. Myers, K. Wager-Smith, A. Rothenfluh-Hilfiker, M. Young (1996). Light-induced degradation of TIMELESS and entrainment of the *Drosophila* circadian clock. *Science*, **271**, 1736–1740.
43. H. Zeng, Z. Qian, M. Myers, M. Rosbash (1996). A light-entrainment mechanism for the *Drosophila* period protein contributes to the circadian clock. *Neuron*, **380**, 129–135.
44. C. Lee, V. Parikh, T. Itsukaichi, K. Bae, I. Edery (1996). Resetting the *Drosophila* clock by photic regulation of PER and a PER-TIM complex. *Science*, **271**, 1740–1744.

45. M. Hunter-Ensor, A. Ousley, A. Sehgal (1996). Regulation of the *Drosophila* protein TIMELESS suggests a mechanism for resetting the circadian clock by light. *Cell*, **84**, 677–685.
46. Y. Shigeyoshi, K. Taguchi, S. Yamamoto, S. Takeida, L. Yan, H. Tei, S. Moriya, S. Shibata, J.J. Loros, J.C. Dunlap, H. Okamura (1997). Light-induced resetting of a mammalian circadian clock is associated with rapid induction of the *mPer1* transcript. *Cell*, **91**, 1043–1053.
47. L. Shearman, M. Zylka, D. Weaver, L. Kolakowski, S. Reppert (1997). Two *period* homologs: circadian expression and photic regulation in the suprachiasmatic nuclei. *Neuron*, **19**, 1261–1269.
48. J.A. Elliott (1976). Circadian rhythms and photoperiodic time measurement in mammals. *Fed. Proc.*, **35**, 2339–2346.
49. F.M. Sulzman, C.A. Fuller, M.C. Moore-Ede (1979). One second light pulses entrain circadian rhythms in a diurnal primate. *Physiologist*, **224**, 66.
50. J.W. Hastings, B.M. Sweeney (1959). The *Gonyaulax* clock. In: R.B. Withrow (Ed.), *Photoperiodism in Plants and Animals* (pp. 567–584). A.A.A.S. Press, Washington, D.C.
51. T. Roenneberg, W. Taylor (1994). Light-induced phase responses in *Gonyaulax* are drastically altered by creatine. *J. Biol. Rhythms*, **9**, 1–12.
52. T. Roenneberg, H. Nakamura, J.W. Hastings (1988). Creatine accelerates the circadian clock in the unicellular alga *Gonyaulax polyedra*. *Nature*, **334**, 432–434.
53. H. Broda, V.D. Gooch, W. Taylor, N. Aiuto, J.W. Hastings (1986). Acquisition of circadian bioluminescence data in *Gonyaulax* and an effect of the measurement procedure on the period of the rhythm. *J. Biol. Rhythms*, **1**, 251–263.
54. J. Comolli, W. Taylor, J.W. Hastings (1994). An inhibitor of protein phosphorylation stops the circadian oscillator and blocks light-induced phase shifting in *Gonyaulax polyedra*. *J. Biol. Rhythms*, **9**, 13–26.
55. J. Comolli, J.W. Hastings (1999). Novel effects on the *Gonyaulax* circadian system produced by the protein kinase inhibitor staurosporine. *J. Biol. Rhythms*, **14**, 10–18.
56. J. Comolli, W. Taylor, J. Rehman, J.W. Hastings (1996). Inhibitors of serine/threonine phosphoprotein phosphatases alter circadian properties in *Gonyaulax polyedra*. *Plant Physiol.*, **111**, 285–291
57. M. Menaker (1968). Extraretinal light perception in the sparrow, I: Entrainment of the biological clock. *Proc. Natl. Acad. Sci. USA*, **59**, 414–421.
58. J.P. McMillan, H. Keatts, M. Menaker (1975). On the role of eyes and the brain photoreceptors in the sparrow: entrainment to light cycles. *J. Comp. Physiol.*, **102**, 251–256.
59. M. Menaker (1989). Extraretinal photoreception. In: K.C. Smith (Ed.), *The Science of Photobiology* (2nd ed., pp. 215–230). Plenum Press, NY.
60. T. Yoshikawa, T. Oishi (1998). Extraretinal photoreceptions and circadian systems in nonmammalian vertebrates. *Comp. Biochem. Physiol. B Biochem. Mol. Biol.*, **119**, 65–72.
61. J.W. Truman, L.M. Riddiford (1998). Neuroendocrine control of ecdysis in silkmoths. *Science*, **67**, 1624–1626.
62. G.D. Block, D.G. McMahon (1984). Cellular analysis of the *Bulla* ocular circadian pacemaker system III. Localization of the circadian pacemaker. *J. Comp. Physiol.*, **155A**, 387–395.
63. W.F. Zimmerman, T.H. Goldsmith (1971). Photosensitivity of the circadian rhythm of visual receptors in carotenoid-depleted *Drosophila*. *Science*, **172**, 1167–1169.
64. C.S. Zucker (1996). The biology of vision in *Drosophila*. *Proc. Natl. Acad. Sci. USA*, **93**, 571–576.
65. M.S. Dushay, M. Rosbash, J.C. Hall (1989). The disconnected visual system mutations in

Drosophila melanogaster drastically disrupt circadian rhythms. *J. Biol. Rhythms*, **4**, 1–27.

66. K.W. Foster, J. Saaranak, N. Patel, G. Zarilli, M. Okabe, T. Kline, K. Nakanashi (1984). A rhodopsin is the functional photoreceptor for phototaxis in the unicellular eukaryote *Chlamydomonas*. *Nature*, **311**, 756–759.
67. C.H. Johnson, T. Kondo, J.W. and J.W. Hastings (1991). Action spectrum for resetting the circadian phototaxis rhythm in *Chlamydomonas*. II: Illuminated Cells. *Plant Physiology*, **97**, 1122–1129.
68. T. Kondo, C.H. Johnson, J.W. Hastings (1991). Action spectrum for resetting the circadian phototaxis rhythm in *Chlamydomonas*. I: Cells in darkness. *Plant Physiology*, **95**, 197–205.
69. D.S. Saunders (1981). Insect photoperiodism: the clock and the counter. *Physiol. Ent*, **6**, 99–116.
70. H. Underwood, B. Goldman (1987). Vertebrate circadian and photoperiodic systems; role of melatonin and the pineal gland. *J. Biol. Rhythms*, **2**, 279–315.
71. R.J. Nelson, T. Zucker (1981). Absence of extraocular photoreception in diurnal and nocturnal rodents exposed to direct sunlight. *Comp. Biochem. Physiol.*, **69A**, 145–148.
72. I. Provencio, S. Wong, A. Lederman, S.M. Argamaso, R.G. Foster (1994). Visual and circadian responses to light in aged retinally degenerate mice. *Vision Res.*, **34**, 1799–1806.
73. M. Freedman, R. Lucas, B. Soni, M. von Schantz, M. Munoz, Z. David-Gray, R. Foster (1999). Regulation of mammalian circadian behavior by non-rod, non-cone, ocular photoreceptors. *Science*, **284**, 502–504.
74. R.G. Foster (1998). Shedding light on the biological clock. *Neuron*, **20**, 829–832.
75. S.S. Campbell, P.J. Murphy (1998). Extraocular circadian phototransduction in humans. *Science*, **279**, 396–399.
76. S.W. Lockley, D.J. Skene, K. Thapan, J. English, D. Ribeiro, I. Haimov, S. Hampton, B. Middleton, M. von Schantz, J. Arendt (1998). Extraocular light exposure does not suppress plasma melatonin in humans. *J. Clin. Endocrin. Metab.*, **83**, 3369–3376.
77. M. Hébert, S. Martin, C.I. Eastman (1999). Nocturnal melatonin secretion is not suppressed by light exposure behind the knee in humans. *Neurosci. Letts.*, **274**, 127–130.
78. R.L. Sack, A.J. Lewy, M.L. Blood, L.D. Keith, H. Nakagawa (1992). Circadian rhythm abnormalities in totally blind people: incidence and clinical significance. *J. Clin. Endoc. Metab.*, **72**, 127–134.
79. C.A. Czeisler, T. Shanahan, E. Klerman, H. Martens, D. Brotman, J. Emens, T. Klein, J. Rizzo (1995). Suppression of melatonin secretion in some blind patients by exposure to bright light. *New England J. Med.*, **332**, 6–11.
80. S.W. Lockley, D.J. Skene, J. Arendt, H. Tabandeh, A.C. Bird, R. Defrance (1997). Relationship between melatonin rhythms and visual loss in the blind. *J. Clin. Endoc. Metab.*, **82**, 3763–3770.
81. E.B. Klerman, D.W. Rimmer, J.-J. Dijk, R.E. Kronauer, J.F. Rizzo, C.A. Czeisler (1998). Non photic entrainment of the human circadian pacemaker. *Am. J. Physiol.*, **274**, R991-R996.
82. M.L. Sargent, W.R. Briggs (1967). The effects of light on a circadian rhythm of conidiation in *Neurospora*. *Plant Physiol.*, **42**, 1504–1510.
83. S. Dharmananda (1980). *Studies of the circadian clock of* Neurospora crassa*: light-induced phase shifting*. Ph.D. thesis, Department of Biology, University of California, Santa Cruz.
84. D.D. Frank, W.F. Zimmerman (1969). Action spectra for phase shifts of a circadian rhythm in *Drosophila*. *Science*, **163**, 688–689.
85. J.S. Takahashi, P.J. DeCoursey, L. Bauman, M. Menaker (1984). Spectral sensitivity of a novel photoreceptive system mediating entrainment of mammalian circadian rhythms.

Nature, **308**, 186–188.
86. P.J. Lumsden (1991). Circadian rhythms and phytochrome. *Annu. Rev. Plant Physiol., Plant Mol. Biol.*, **42**, 351–371.
87. J.W. Hastings, B.M. Sweeney (1960). The action spectrum for shifting the phase of the rhythm of luminescence in *Gonyaulax polyedra. J. Gen Physiol.*, **43**, 697–706.
88. D.E. Somers, P.F. Devlin, S.A. Kay (1998). Phytochromes and cryptochromes in the entrainment of the *Arabidopsis* circadian clock. *Science*, **282**, 1488–1490.
89. T. Roenneberg, J.W. Hastings (1991). Are the effects of light on the phase and period of the *Gonyaulax* clock mediated by different pathways? *Photochem. Photobiol.*, **53**, 525–533.
90. T. Roenneberg, T.-S. Deng (1997). Photobiology of the *Gonyaulax* circadian system. I. Different phase response curves for red and blue light. *Planta*, **202**, 494–501.
91. T.-S. Deng, T. Roenneberg (1997). Photobiology of the *Gonyaulax* circadian system. II. Allopuriol inhibits blue-light effects. *Planta*, **202**, 502–509.
92. M. Ahmad, A.R. Cashmore (1993). *HY4* gene of *A. thaliana* encodes a protein with characteristics of a blue-light photoreceptor. *Nature*, **366**, 162–166.
93. A.R. Cashmore, J. Jarillo, Y.-J. Wu, D. Liu (1999). Cryptochromes: blue light receptors for plants and animals. *Science*, **284**, 760–765.
94. K. Malhotra, S. Kim, A. Batschauer, L. Dawut, A. Sancar (1995). Putative blue-light photoreceptors from *Arabidopsis thaliana* and *Sinapsis alba* with a high degree of sequence homology to DNA photolyase contain the two photolyase cofactors but lack DNA repair activity. *Biochemistry*, **34**, 6892–6899.
95. P.D. Hoffman, A. Batschauer, J.B. Hays (1996). *PHH1*, a novel gene from *Arabidopsis thaliana* that encodes a protein similar to plan blue-light photoreceptors and microbial photolyases. *Mol. Gen. Genet.*, **253**, 259–265.
96. C. Lin, H. Yang, H. Guo, T. Mockler, J. Chen, A.R. Cashmore (1998). Enhancement of blue-light sensitivity of *Arabidopsis* seedlings by a blue light receptor cryptochrome 2. *Proc. Natl. Acad. Sci. USA*, **95**, 2686–2690.
97. M. Ahmad, A.R. Cashmore (1996). Seeing blue: the discovery of cryptochrome. *Plant Mol. Biol.*, **30**, 851–861.
98. H. Guo, H. Yang, T. Mockler, C. Lin (1998). Regulation of flowering time by *Arabidopsis* photoreceptors. *Science*, **279**, 136–1363.
99. M. Ahmad, J. Jarillo, O. Smirnova, A.R. Cashmore (1998). Cryptochrome blue-light photoreceptors of *Arabidopsis* implicated in phototropism. *Nature*, **392**, 720–723.
100. D. Hsu, X. Zhao, S. Zhao, A. Kazantsev, R.-P. Wang, T. Todo, Y.-F. Wei, A. Sancar (1996). Putative human blue-light photoreceptors hCRY1 and hCRY2 are flavoproteins. *Biochemistry*, **35**, 13871–13877.
101. Y. Miyamoto, A. Sancar (1998). Vitamin B2-based blue-light photoreceptors in the retino hypothalamic tract as the photoactive pigments for setting the circadian clock in mammals. *Proc. Natl. Acad. Sci. USA*, **95**, 6097–6102.
102. R.J. Thresher, M. Vitatema, Y.K. Miyamoto, A. Kazantsev, D. Hsu, C. Petit, C. Selby, L. Dawut, O. Smithies, J.S. Takahashi, A. Sancar (1998). Role of mouse cryptochrome blue light photoreceptor in circadian photoresponses. *Science*, **282**, 1490–1494.
103. G.T.J. van der Horst, M. Muijtjens, K. Kobayashi, R. Takano, S.-I. Kanno, M. Takao, J. de Wit, A. Verkerk, A. Eker, D. van Leenen, R. Buijs, D. Bootsma, J. Hoeijmakers, A. Yasui (1999). Mammalian Cry1 and Cry2 are essential for maintenance of circadian rhythms. *Nature*, **398**, 627–630.
104. R. Stanewsky, M. Kaneko, P. Emery, B. Beretta, K. Wager-Smith, S.A. Kay, M. Rosbash, J.C. Hall (1998). The cryb mutation identifies cryptochrome as a circadian photoreceptor in *Drosophila. Cell*, **95**, 681–692.
105. P. Emery, W. So, M. Kaneko, J.C. Hall, M. Rosbash (1998). CRY, a *Drosophila* clock and

light-regulated cryptochrome, is a major contributor to circadian rhythm resetting and photosensitivity. *Cell*, **95**, 669–679.

106. J.W. Hastings (1964). The role of light in persistent daily rhythms. In: A.C. Giese (Ed.), *Photophysiology* (Vol. I, pp. 333–361). Academic Press, New York.
107. C.H. Johnson (1990). *PRC An atlas of Phase Response Curves for circadian and circadial rhythms*. Department of Biology, Vanderbilt University, Nashville, TN.

Photomovement
D.-P. Häder and M. Lebert, editors.

Chapter 17

Photomovement in ciliates

Francesco Lenci, Francesco Ghetti and Pill-Soon Song*

Table of contents

* This paper is dedicated to the memory of Prof. Renzo Nobili and Prof. Nicola Ricci who introduced F. Ghetti and F. Lenci to the study of ciliates and always were close to them with their scientific wisdom and precious friendship.

Abstract

The principal behavioral responses of ciliates to photic stimuli are briefly illustrated, with special attention to step-up photophobic reactions and phototaxis in *Blepharisma japonicum*, *Stentor coeruleus* and *Fabrea salina*. In *Fabrea* both the action spectrum for phototaxis and preliminary immunochemical and biochemical evidences suggest that a rhodopsin-type chromophore is the photoreceptor pigment. Action spectra for step-up photophobic reactions, phototaxis and membrane receptor potentials in *Blepharisma* and *Stentor* point to a hypericin-type photosensing chromophores: blepharismin and oxyblepharismin for red and blue *Blepharisma* cells, respectively, and stentorin for *Stentor*. These unique photoreceptor pigments are also phototoxic for the ciliates themselves and play a defensive role against predators. Concerning the molecular mechanisms at the basis of the light-initiated signaling process, several experimental lines of evidence indicated that photobehavioral responses in these ciliates are modulated by protonophores, ammonium chloride and extracellular pH. These findings brought to hypothesize that a proton translocation occurs in the photoreceptor apparatus. Recent spectroscopic results suggest that a light-induced electron transfer from the first singlet excited state of the pigment could be the very primary step of the photoreaction. In the physiological molecular environment, such an electron transfer could generate a transient intracellular pH gradient, which in turn could cause opening of Ca^{2+} channels. Ca^{2+} influx would trigger the stop and subsequent reversal of ciliary beating, i.e. the photomotile response. A working hypothesis including the photoactivation of a G-protein is finally presented.

17.1 Introduction

A wide variety of eukaryotic and prokaryotic microorganisms exploit light as an energy source to drive vital metabolic processes. Most of these microorganisms utilize light also as an information signal for controlling their motile behavior in order to get into environmental niches in which the illumination conditions are best for their growth.

Even though ciliated protozoa do not harvest and convert light energy directly for their metabolism, some of them perceive and react to photic stimuli. Light can in fact be an environmental cue for ciliates to gather into habitats which can be unfavorable for their predators and propitious for their preys and, in general, for food [1–4]. Also for ciliates, the main photobehavioral responses are photophobic reactions and phototaxis (for definitions see W. Haupt's Chapter in this volume). Photokinesis was suggested to occur only in a limited number of cases (see below). Step-up/step-down photophobic responses are elicited by a sudden increase/decrease in photon flux density and, usually, consist in a transient alteration in the activity of the motor apparatus. Phototaxis results from the detection of the direction of light propagation and is defined as positive or negative according to whether the oriented movement is toward or away from the source and as diaphotaxis, or transverse phototaxis, when the microorganism moves at an angle with respect to the light beam. Positive (negative) photokinesis consists of an increase (decrease) of the average cell speed caused by an increase in photon flux density [5].

In this chapter we will focus our attention mainly on those ciliates for which at least preliminary data are available on the molecular mechanisms for sensory perception and transduction of photic stimuli: the heterotrichous ciliates *Blepharisma japonicum*, *Fabrea salina* and *Stentor coeruleus*. Among them, of particular interest are *Blepharisma* and *Stentor*, in which the very same endogenous pigment is phototoxic for the ciliate itself [6,7], plays a defensive role against predators [8] and acts as the photosensing chromophore [9–16]. This almost unique behavior of an organism producing a pigment that photodamage not only its predators but the organism itself, is a puzzle for the protozoologists studying evolutionary processes.

The pigments of *Stentor* and *Blepharisma*, stentorin and blepharismins, respectively, have recently been found to have a molecular structure very similar to that of hypericin [17–20]. This class of hypericin-like photoreceptor molecules is unique among the photobiological light sensors, and investigations of their role in photosensory transduction should broaden our understanding of the basic nature of stimulus-response systems at the single cell level.

17.2 Photomotile responses

Stentor coeruleus. Already at the beginning of this century it was observed that *Stentor* (Figure 1) showed light avoiding reactions bringing the cells to gather in the shady or dark areas of culture vessels [21–23]. However, it was not until the late 1970s that this light-avoiding behavior was characterized as being due to a step-up photophobic reaction and a negative phototactic response [13,14].

A sudden increase in light intensity causes *Stentor* to stop swimming, reverse the direction of its ciliary beating and tumble until adaptation occurs and forward movement

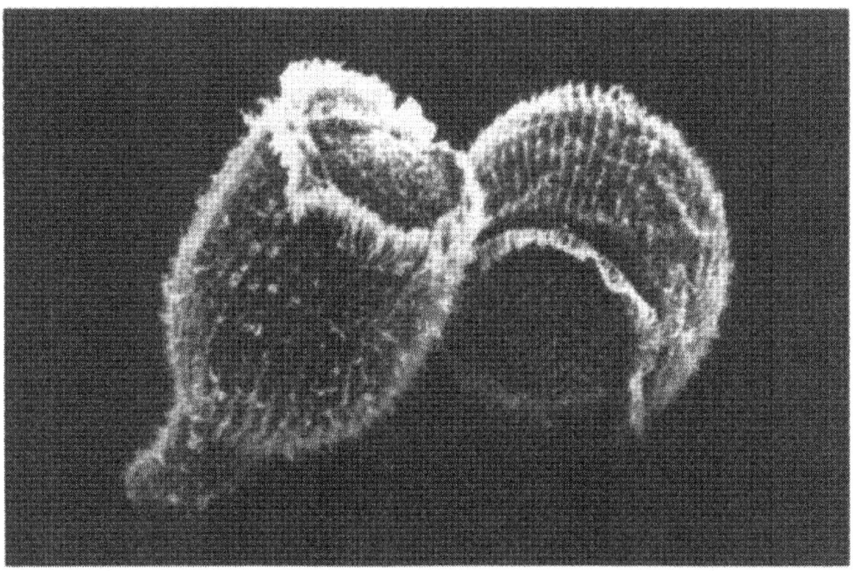

Figure 1. Scanning electron micrograph of two *Stentor coeruleus* cells.

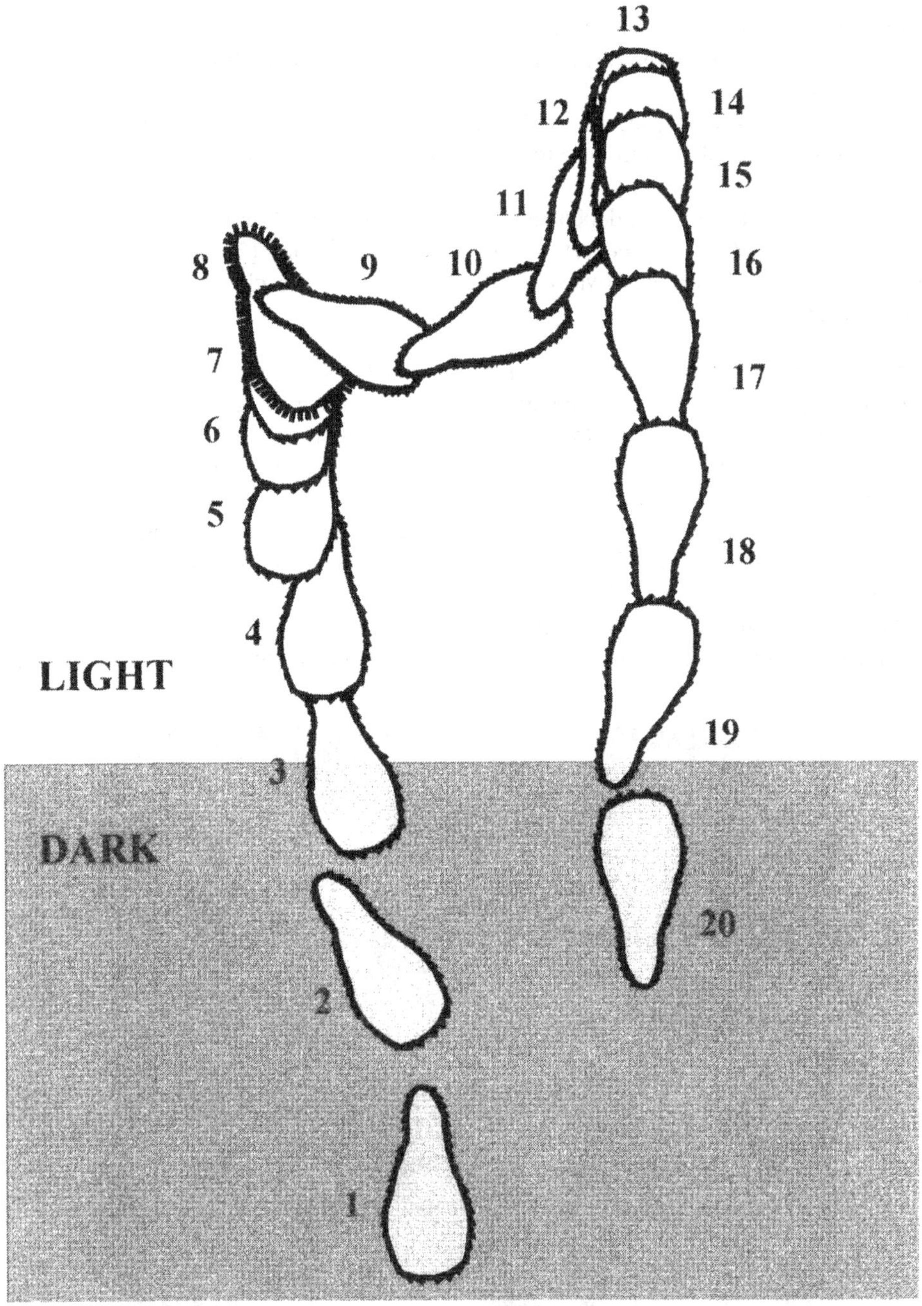

Figure 2. Schematic representation of step-up photophobic reaction at light-dark border for *Blepharisma japonicum*. A similar scheme holds in the case of *Stentor coeruleus*.

is resumed. When the cells cross a dark-light border (Figure 2), the final outcome of this maneuver is a change in the direction of swimming which can allow the cells to escape from the lighted region.

In addition to this step-up photophobic reaction, *Stentor* is also able to detect the direction of the incident light and to swim away from the light source along the light beam [14].

Whether or not there is a dependence of swimming speed on photon flux density in *Stentor* is not well established yet: Song et al. [14] and Song [24] did not observe any photokinetic effect, whereas Iwatsuki [25] reported that the ciliate exhibits a positive photokinesis.

Blepharisma japonicum. As mentioned previously, the red heterotrichous ciliate *Blepharisma* (Figure 3) has been an interesting subject of studies in the field of photobiology because of its endogenous pigment blepharismin. In fact, if *Blepharisma* is exposed to relatively strong light, in the presence of oxygen, blepharismin acts as a strong photosensitiser and readily causes its death [6,26].

Similarly to the case of *Stentor*, *Blepharisma* has been shown to accumulate in shaded regions [6,26] as a result of step-up photophobic responses, negative phototaxis and positive photokinesis [10,27,28].

The step-up photophobic reaction of *Blepharisma*, closely resembling the response exhibited by *Stentor*, consists of a stop of forward movement, followed by backward

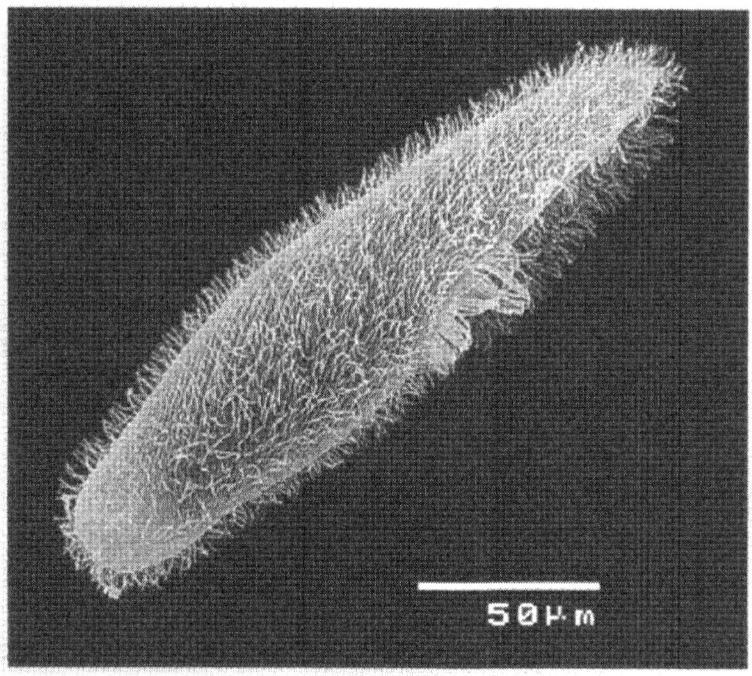

Figure 3. Scanning electron micrograph of *Blepharisma japonicum* (kindly provided by Prof. F. Verni, Ecology, Ethology and Evolution Dept., Pisa University).

swimming along a bent trajectory, so that the cell's longitudinal axis changes its direction (Figure 2). If the rise in light intensity results from entering into an illuminated area and the cell, moving backward, enters the dark area, it resumes forward swimming in a new direction [10]. Under continuous illumination the light avoiding maneuver is followed by a tumbling and it is repeated until the cell adapts to the new illumination conditions [27]. The step-up photophobic reaction of *Blepharisma* occurs with a time lag with respect to the stimulus application which decreases with increasing photon flux density and depends on the actinic wavelength [10,12,27]. Local stimulation of a *Blepharisma* cell showed that only the anterior part of the cell is photosensitive, even though the cell is homogeneously pigmented; therefore, it was suggested that the photoreceptor pigment mediating the photoresponse is localized in the anterior end [29].

Matsuoka and Taneda [30] reported that light-adapted cells of *Blepharisma* are capable of perceiving a reduction in photon flux density, to which they react with a transient suppression of ciliary reversal and an increase in swimming speed.

Fabrea salina. Whereas *Stentor* and *Blepharisma* react to photic stimuli to escape from the lighted region (step-up photophobic responses) or to swim away from the light source (negative phototaxis), *Fabrea* (Figure 4) tends to gather into illuminated spots. Under unilateral illumination *Fabrea* shows an evident positive phototaxis, swimming toward the light source in a manner approximately parallel to the light propagation direction [31–34]. It has also been reported that when phototactic light is suddenly switched off, cells exhibit a step-down photophobic reaction, consisting of a cessation of forward swimming followed by a short backward movement [35].

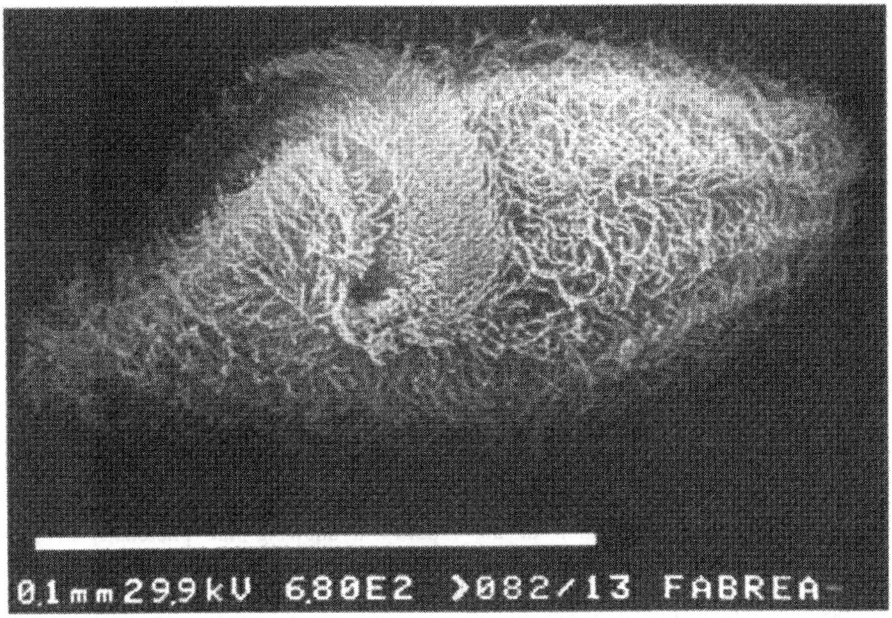

Figure 4. Scanning electron micrograph of *Fabrea salina*. (Kindly provided by Prof. F. Verni, Ecology, Ethology and Evolution Dept., Pisa University).

In these three ciliates, like in most other freely motile microorganisms, positive and negative phototaxis may actually result not from a truly oriented movement toward or away from the light source, but rather from a series of step-down or step-up photophobic reactions, giving rise to cell photoaccumulation or photodispersal, respectively (for a comprehensive discussion of this issue in the case of *Blepharisma* and *Stentor*, see D. Wood's chapter this volume).

In *Stentor* and *Blepharisma* the question whether they display a true negative phototactic response was answered by Song et al. [14] and Tsuda and Matsuoka [28] by means of specifically designed experiments. In both experimental set-ups, the cells were irradiated horizontally in unilateral direction using an actinic beam focused in the middle of the experimental chamber. In such a light beam, *Stentor* and *Blepharisma* moved away from the light source in the convergent portion of the beam between the light source and the image plane as well as in the divergent portion of the beam beyond the image plane, finally gathering far from the light source (Figure 5). A contribution of either step-up photophobic reaction or positive photokinesis should have caused accumulation in both the low photon density portions of the beam. In conclusion, these experiments show that the sensory signal for these ciliates is the light direction propagation and not the light gradient.

A comprehensive computer-assisted on-line image analysis of motion parameters of *Fabrea* under unilateral light stimulation and in the dark provided the convincing evidence that this ciliate swims toward the light source not because it is able to directly perceive the direction of light propagation, but rather because its mean free path is higher when it moves toward the light source (Colombetti, personal communication).

The contribution of photokinesis to photoaccumulation and/or photodispersal in ciliates still remains an open question, also because a misleading and confusing

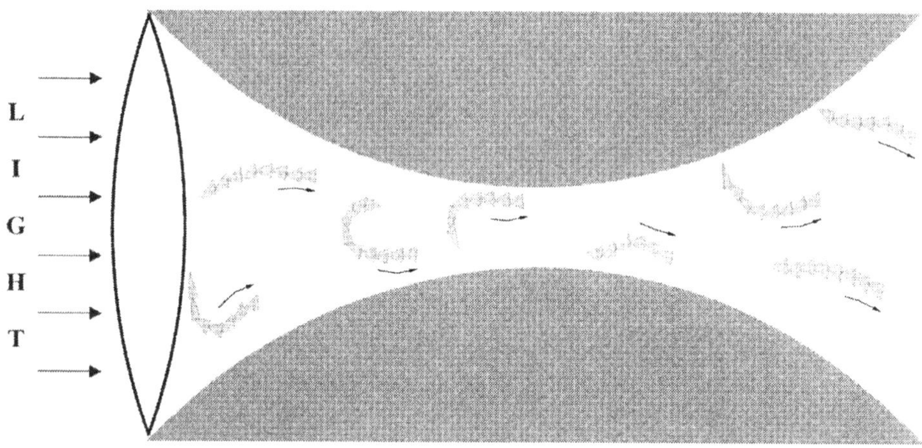

Figure 5. Schematic representation of the experiment showing that phototaxis in *Stentor coeruleus* results from perception of light propagation direction and not the light gradient (redrawn from [14]). The cells, irradiated horizontally in unilateral direction using an actinic beam focused in the middle of the experimental chamber, moved away from the light source both in the convergent and the divergent portion of the beam. A similar scheme holds in the case of *Blepharisma japonicum* [28].

terminology has often been used. For instance, in some cases photodispersal from non-directionally lighted regions has been plainly attributed either to pure photokinesis or negative phototaxis, without taking into account the occurrence of step-up photophobic responses.

In any case, if *Blepharisma* and *Stentor* do exhibit photokinesis, the crucial question to clarify concerns the driving force for positive photokinesis. Should photokinesis in these ciliates be linked to photophosphorylation, as they are non-photosynthetic, it would be interesting to explore the hypothesis that cellular free ATP level is modulated by light.

Loxodes striatus. The yellow ciliate *Loxodes* was shown to orient itself by means of a gravitactic reaction, going upward or downward depending on the illumination conditions and on the oxygen partial pressure in the medium [36–39].

From an ecological point of view, this capability of perceiving O_2 implies that for *Loxodes* both light and partial oxygen pressure are environmental cues which allow the cells to gather in the region where most of their predators, like crustacean plankton, usually stay far from [36,37,39].

Other ciliates. In a number of ciliates (e.g. *Chlamydodon mnemosyne*, various species of *Ophryoglena*, *Ichthyophtirius multifilius*, *Porpostoma notatum*) photobehavioral responses (mainly positive and negative phototaxis) markedly depend on the cell life-cycle. Most of these microorganisms also possess subcellular organelles which play a crucial role in allowing them to perceive light direction and orient their movement toward or away from the light source. The field of behavioral responses and morphological properties of the photoreceptor apparatus in the aforementioned single cell organisms has been thoroughly investigated and reviewed by Kuhlmann and co-authors [1,2,40–42].

17.3 Photosensory transduction

Stentorin and blepharismin are mainly localized in pigment granules ([43–45] and references therein). In both *Stentor* and *Blepharisma* the granules, distinct in nature from mitochondria [46,47], are about 500 nm in diameter, membrane-limited and arranged in strings parallel to the ciliary rows, spread all over the cellular body. In Figure 6 an electron micrograph of *Stentor* granules is shown. A honeycomb-like structure, made up of a folded membrane, has also been suggested to be contained in *Blepharisma* pigment granules [45].

17.3.1 Photosensors: pigment chromophores

Stentor coeruleus. The photoreceptive function of stentorin has been established on the basis of the resemblance of the action spectra for the photophobic and phototactic responses in *Stentor* to the absorption spectrum of the whole cells and of the extracted pigment (Figure 7) [13,14,48,49]. The action spectra for the membrane receptor potentials, for the light-induced receptor potential and ciliary stroke reversal are also consistent with the absorption spectra of the pigment stentorin ([16,50,51], see also D. Wood's chapter in this volume).

It was also found that a colorless mutant, as well as a caffeine-bleached *Stentor*, lost their photosensitivity, further supporting that stentorin is the photoreceptor [24].

Stentorin has a very similar absorption spectrum to that of hypericin, with a red shift of 7 nm for the longest wavelength absorption peak (from 588 nm for hypericin to 595 nm for free stentorin). In recent studies, the structure of stentorin has been identified as a derivative of hypericin (octahydroxyl-diisopropylnaphtholdianthrone, Figure 8, [17,52]).

Blepharisma japonicum. Under relatively low intensity irradiation (about 3.5 W/m^2), the red form of blepharismin progressively converts into a grey-blue form, oxyblepharismin, apparently neither toxic nor phototoxic for the cell [26,53]. A comparison of the action spectra for step-up photophobic responses of red and blue *Blepharisma* cells with blepharismin and oxyblepharismin (Figure 9) clearly indicates that these

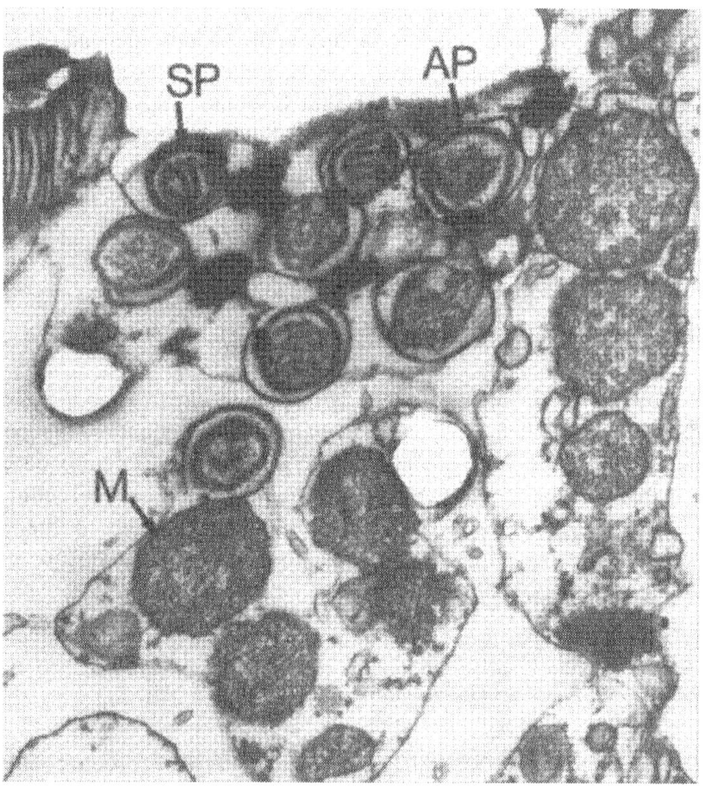

Figure 6. Transmission electron micrograph of *Stentor coeruleus* granules.

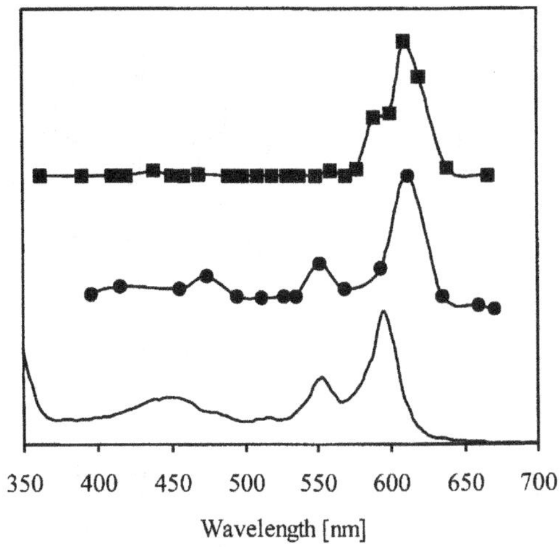

Figure 7. Action spectra of step-up photophobic response (circles: [94]) and phototaxis (squares: [14]) in *Stentor coeruleus* compared with the absorption spectrum of the photoreceptor chromophore stentorin (continuous line). It should be noted that the *in vivo* absorption spectrum of stentorin is red-shifted of about 10 nm.

pigments are the photoreceptor pigments for this motile reaction [9,11,12,54]. It remains to be seen whether or not the blepharismin phototransformation to oxyblepharismin is part of its photo-cycle or both species independently serve as photodetector in the photosensory transduction of *Blepharisma*.

The action spectrum for the membrane receptor potentials of red *Blepharisma* cells also correlated with the absorption spectra of blepharismin ([50,51], see D. Wood's chapter in this volume).

It was also observed that *Blepharisma* is capable of perceiving and transducing UV-B radiation (280–315 nm) as an environmental sensory stimulus and that blepharismin plays the role of sensing pigment for both visible and UV-B radiation [55].

The chemical structure of blepharismin has recently been deduced by mass spectrometry, NMR and FTIR spectroscopic methods. Data are consistent with the structure, 2,4,5,7,2′,4′,5′,7′-octahydroxy-6,6′-diisopropyl-1,1′-*p*-hydroxybenzylidene benzodianthrone [18], as shown Figure 8. This structure is in agreement with another structural assignment reported by Maeda et al. [19] who also elucidated the chemical structures of several blepharismin derivatives.

As shown in Figure 8, the bridged structure in blepharismin introduces a distortion of the upper and lower halves of the naphthoanthraquinone groups, thus causing a blue shift of the absorbance maximum as well as the broadening of the absorbance bands, relative to hypericin and stentorin. A semi-empirical calculation predicted that the bridged blepharismin ring system is 36 kcal/mol higher in energy than the hypericin ring system. This energy difference is consistent with what one would expect to lose from the loss of aromaticity of one aromatic ring. Also, the structure assigned accounts for the

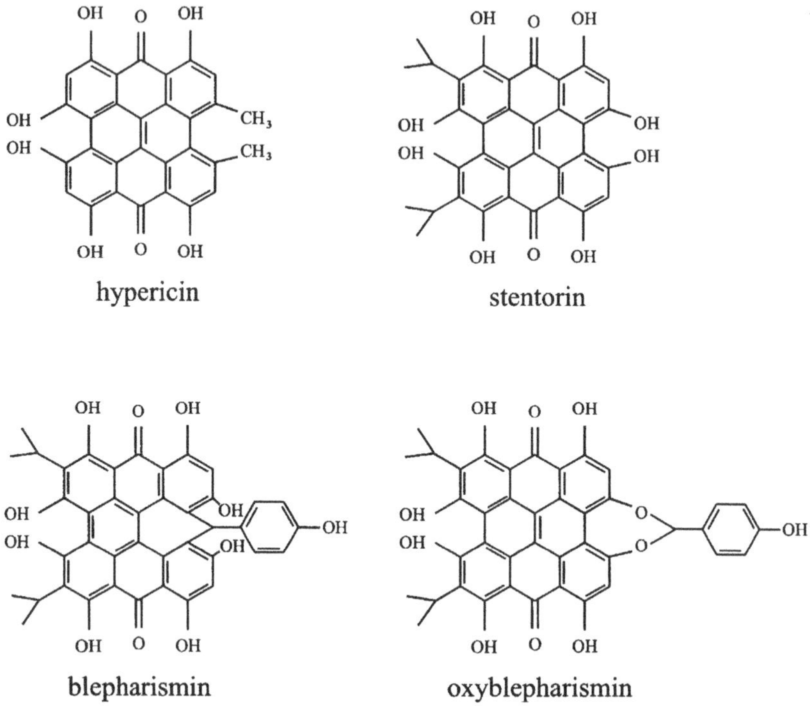

(A)

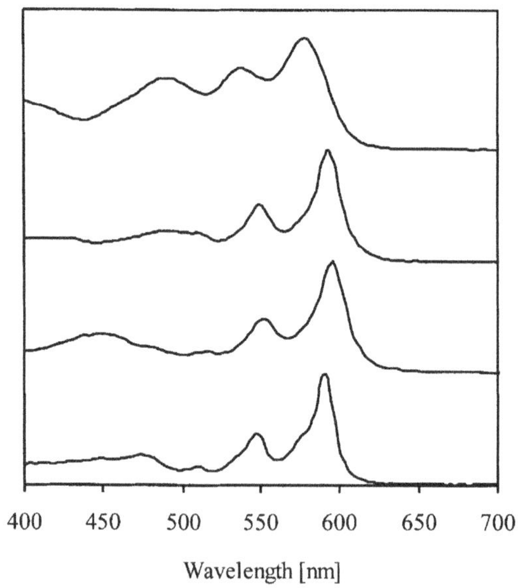

(B)

Figure 8. (A) Structural formulas of hypericin, stentorin [17], blepharismin [18,19] and oxyblepharismin [20]. (B) Optical absorption spectra of, from the bottom, hypericin, stentorin, oxyblepharismin and blepharismin.

facile photochemical transformation of blepharismin to oxyblepharismin, with loss of two hydrogen atoms.

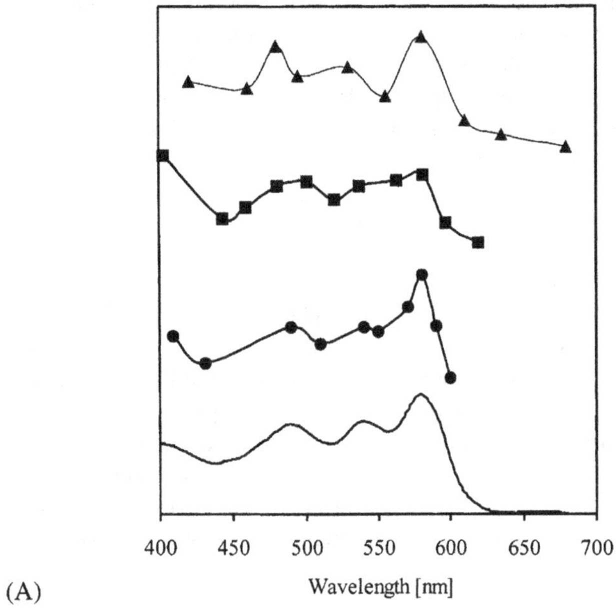

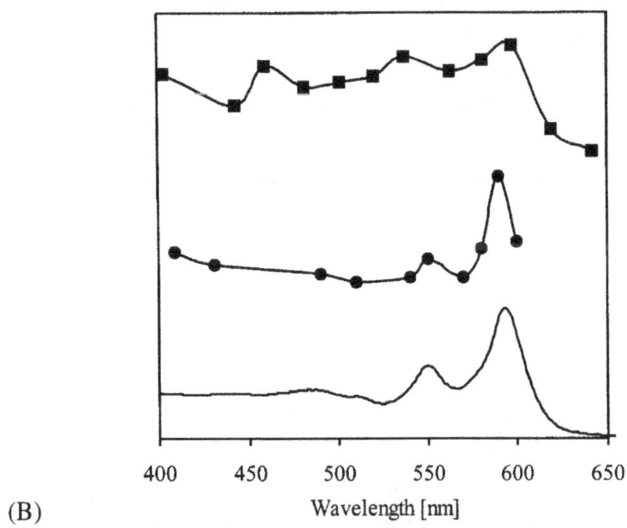

Figure 9. (A) Action spectra of step-up photophobic response (circles: [9], squares: [11]) and for the membrane receptor potentials (triangles: [99]) in *Blepharisma japonicum* red cells compared with the absorption spectrum of blepharismin (continuous line). (B) Action spectra of step-up photophobic responses (circles: [9], squares: [11]) in *Blepharisma japonicum* blue cells compared with the absorption spectrum of oxyblepharismin (continuous line).

Very recently, the structure of oxyblepharismin has also been determined by Spitzner et al. [20]. The proposed molecular structure, resulting from photoinduced rearrangement and irreversible dehydrogenation of blepharismin, accounts for the resemblance between its absorption spectrum and those of hypericin and stentorin (Figure 8).

In the landscape of photoreceptor pigments for photomotile responses of microorganisms (e.g. rhodopsins, flavins, pterins, *p*-coumaric acid, see, e.g. [56] and S. Braslawsky's chapter in this volume) the hypericin analogs stentorin, blepharismin and oxyblepharismin constitute a new unique class of photopigments (Figure 8). The blepharismin structure is particularly unique among the hypericin compounds because of its conjugation-disrupting bridge carbon, which accounts for its distinct absorption spectrum. It remains to be seen if ciliates other than *Stentor* and *Blepharisma* and even some non-ciliate cells make use of the hypericin-based pigment molecules for photoreception.

Fabrea salina. The action spectrum for positive phototaxis of *Fabrea* [32] (Figure 10) is very similar to that of *Paramecium bursaria*, in which experimental evidences suggest the presence of a rhodopsin-like pigment [57,58]. The inhibition of phototaxis by hydroxylamine (which is known to interfere with rhodopsin) and the detection, by means of immunochemical techniques, of a rhodopsin-like molecule on the plasma membrane [59] suggested that the phototactic response at 420 nm is mediated by a rhodopsin-like pigment [32]. Microspectrofluorometry and confocal microscopy data suggest, however, that also a hypericin-like pigment is present in *Fabrea* [60,61].

Loxodes striatus. The action spectrum of *Loxodes* photomotile reactions, determined using the rate of escape from oxic water to quantify photoresponsiveness, suggested a flavin as the photoreceptor responsible for the observed behavior. It was also shown that upon irradiation of a crude extract of the pigment, superoxide anion (O_2^-) was photodynamically produced, as reasonably expected by a flavin which is an efficient

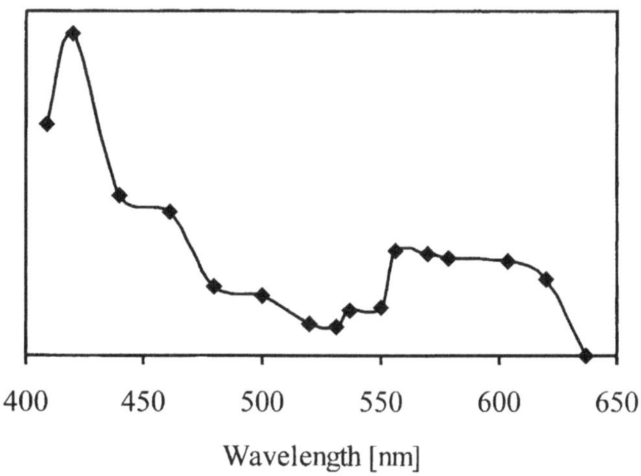

Figure 10. Action spectrum of positive phototaxis in *Fabrea salina* (redrawn from [32]).

photosensitiser. On the basis of these findings it has been proposed that *Loxodes* can perceive as a signal the superoxide anions intracellularly produced both biochemically in the dark and photodynamically under illumination [36–39].

17.3.2 Photoreceptor proteins

Stentor coeruleus. Two distinct forms of stentorin chromoproteins can be chromatographically isolated. These are stentorin I and II. Their absorption spectra are very similar with respect to peak shape. However, the absorption peaks of stentorin II in the visible region are red-shifted of about 10 nm. As the chromophores in stentorin I and II are similar in structure, the difference in peak position implies that the chromophore environment is responsible for the spectral shift. Stentorin I is strongly fluorescent while stentorin II is only weakly or non-fluorescent. Both stentorin I and II displayed anomalous behaviors on SDS-PAGE gels. For example, the apparent molecular mass of stentorin I was strongly dependent on the gel concentration. Stentorin II did not enter the gel. These anomalies suggested that stentorins might be glycosylated since the dependence of the apparent molecular weight on gel concentration is characteristic of glycoproteins [62]. However, neither stentorin I nor stentorin II were found to be glycosylated [63].

Stentorin I is not larger than cytochrome *c* (12.5 kDa). Detergent-solubilized stentorin I, previously thought to be a 100 kDa chromoprotein [64], is very likely a chromophore-detergent complex. Stentorin II differs from stentorin I in that it can be precipitated by TCA while stentorin I cannot. Detergent-solubilized stentorin II behaves like a large complex of molecular mass greater than 500 kDa. Stentorin II can be further resolved into stentorin II-A and stentorin II-B by hydrophobic interaction chromatography [65]. It contains the chromophore covalently bound to an approximately 50 kDa protein as determined by SDS-urea-PAGE. The amino acid composition of the approximately 50 kDa blue-green band from SDS-urea-PAGE of stentorin II-B is known [65]. Because the N-terminus is blocked, direct sequencing from the N-terminus is not feasible. However, two partial sequences, (1) NPFTAELVETA and (2) SILAADESTGTIG, were obtained for stentorin II-B tryptic fragments. Even though these sequences are too short to search for homologous proteins, several proteins exhibiting high homology to these sequences can be found. Two highly homologous gene products from photoreceptor cells were considered among those identified. The first is the gene product from *Drosophila ninaC*, which encodes a protein containing 1,501 amino acid residues with domains homologous to protein kinases and myosin [66]. The similarity of sequence 1 to –(279)HPFLTELIENE(289)– of *ninaC* is 73%, while its identity is 55%. The similarity of sequence 2 is 50% and the identity 42% to –(596)KVLAAILNIGNI(607)– of *ninaC*. The homologous fragment for sequence 1 is located in the kinase domain (17–282). There is a total of 10.2% of aspartate and glutamate residues in this kinase domain. The homologous fragment for sequence 2 is located in the myosin domain (329–1,053). This myosin domain contains 13% aspartate and glutamate residues.

Another protein homologous to the limited N-terminal sequences of stentorin II-B digest is a sodium-calcium exchanger from bovine rod photoreceptors [67]. Sequence 1 exhibits 60% similarity and 40% identity to –(995)QPLSLEWPET(1004)– of this protein, while sequence 2 displays 75% similarity and 42% identity to –(1056)VWWAHQVGETIG(1067)–. This 1,199 amino acid protein contains 17% aspartate and glutamate residues. Hydrophobicity analysis provides 12 hydrophobic segments in this protein so that 12 transmembrane segments in topology are defined. It also contains 6 N-linked glycosylation sites near the N-terminus. An unusual feature is the presence of 26 acidic residues which cluster in the segment –(967)EDEEEEDEEEE-DEEEEEEEEEEEEENE(994)–. The homologous segment (995–1004) for sequence 1 is located following this acidic segment. That of sequence 2 is located near the C-terminus.

Blepharisma japonicum. The blepharismin pigment was reported to be non-covalently bound to an apoprotein of molecular mass 38 kDa by Gioffré et al. [68] and Yamazaki et al. [69], whilst it was found to be covalently bound to a 200 kDa protein by Matsuoka et al. [70]. The large discrepancy between these two values of molecular weight has not been resolved. Since both these pigment protein complexes are spectroscopically indistinguishable, and both are consistent with the action spectra of *Blepharisma* photobehavior, then other properties of the pigment proteins must be examined in order to determine their functional role. The reported molecular mass value of 200 kDa for the blepharismin holoprotein seems anomalously large, in comparison to the 50 kDa stentorin IIb subunit. However, proteins with molecular masses comparable to the 200-kDa blepharismin holoprotein are not uncommon. Thus, it remains to be seen which of the two extremely different molecular mass proteins represent the functional blepharismin.

17.3.3 Signal trigger: primary photoprocesses

Photomotile responses of microorganisms such as ciliates involve a chain of chemical and/or physical events triggered by the photoexcited photoreceptor molecule. The nature and efficiency of the photomovement initiation trigger is linked to the primary photoprocess of the photoreceptor molecules. It was suggested that stentorin II, rather than stentorin I, is the photosensor because the former displays very low fluorescence quantum yield with ultrafast decay in picosecond time scale [71,72]. A picosecond pump-probe spectroscopy study indicated that a unique initial photoprocess occurred within 3 ps in excited stentorin II but not in the free chromophore species, hypericin or stentorin I [72]. Blepharismin in its protein bound form also exhibits a picosecond process [69].

Unlike rhodopsin, photoactive yellow protein (PYP) and phytochromes, both stentorin and blepharismin do not exhibit a photochemical transformation cycle that can be readily detected by spectrophotometry. The chemical parent compound hypericin for stentorin and blepharismin is an efficient photosensitiser, mediating various photodynamic actions in vivo and in vitro via singlet oxygen [7,73]. It has been suggested that singlet oxygen functions as a signal messenger for the photomovements in some microorganisms, such as *Anabaena variabilis, Physarum polycephalum* and *Loxodes* [74]. However, at least in *Stentor*, singlet oxygen does not seem to play a significant role in the photomovement transduction pathway, since stentorin II is a poor singlet oxygen generator, unlike its free chromophore and hypericin [75]. A singlet oxygen quencher did not specifically inhibit the photophobic response of *Blepharisma,* also suggesting that the active oxygen is not a photosensory transducer component for this organism [76]. What then is the primary photochemistry involved in triggering the photosensory transduction pathway in *Stentor* and *Blepharisma*? We can only speculate, since there are not enough data to answer this question.

Stentor coeruleus. Stentorin can mediate electron transfer processes in its excited state [77]. For example, the stentorin model chromophore, hypericin, can serve as an electron acceptor from an electron donor with sufficient reducing potential. Hypericin undergoes two reversible one-electron reductions at –0.87 V and –1.26 V vs. normal hydrogen electrode [78]. Additionally, an oxidation is observed at ca. +0.90 V in dimethylsulfoxide (DMSO). These formal potentials are consistent with the visible absorbance maximum of hypericin in DMSO at 599 nm (=2.1 eV). Given this information, we estimate that the excited state potentials for hypericin (Hyp) are approximately –0.90 V for the formation of hypericin cation radical Hyp^+ (due to uncertainty of these numbers, the oxidation potential of the excited state hypericin can be as high as +1.2V [69,77])

$$Hyp + h\nu \rightarrow Hyp^* \rightarrow Hyp^+ + e^-$$

and +0.90 V for the formation of hypericin anion radical Hyp^-,

$$Hyp + h\nu \rightarrow Hyp^* + e^- \rightarrow Hyp^-$$

where the electron e^- is supplied by an electron donor including the ground state Hyp itself.

Thus, in the excited state, hypericin can be a good electron donor as well as a good electron acceptor, depending on the redox potentials of the donor/acceptor pairs present in solution. Yamazaki et al. [79] demonstrated that *p*-benzoquinone, which functions as a classical electron acceptor, quenched the fluorescence of hypericin. The quenching is due to electron transfer from hypericin to *p*-benzoquinone at a diffusion-controlled rate (1.43×10^{10} M^{-1} s^{-1}). The electron donor property of hypericin in its excited state is consistent with the picosecond absorption difference measurements of hypericin showing that upon excitation in ~5 ps a new species having a transient absorption in the red/far-red region was formed. This species was not present in ground state absorption [80,81]. The long wavelength absorption, which may have resulted from a solvated electron, is faded by the electron scavenger solvent, acetone [80,81]. Weiner and Mazur [82] reported the formation of cation and anion radicals upon photolysis of hypericin.

Photolysis of stentorin II yields radical species, apparently producing stentorin cation radicals, in a manner similar to the scheme shown for the excited state hypericin [77]. Electron transfer probably occurs from the excited stentorin chromophore to a suitable acceptor/amino-acid residue. An efficient electron transfer process may account for the quenching of the fluorescence from the excited state stentorin. The rate for this process can be estimated from the short fluorescence decay lifetime. This value (1×10^{11} M^{-1} s^{-1}) is close to the diffusion limited rate constant for electron transfer from the excited state hypericin to *p*-benzoquinone. This reaction is likely a reversible process in native stentorin.

Hypericin can also be photooxidized in the presence of dithiodiethanol with a quantum yield of ca. 0.001. Photooxidation of hypericin with dithiodiethanol

($HOCH_2CH_2$–S–S–CH_2CH_2OH) generates mercaptoethanol (HS-CH_2CH_2OH) and oxy-hypericin [83]. Similarly, photooxidation of stentorin (HO-ST-OH) would likely produce a new quinone form of oxy-stentorin (O = ST = O) according to the following scheme:

$$\text{HO–ST–OH} + h\nu \rightarrow \text{HO–ST–OH}^*$$

$$\text{HO–ST–OH}^* + \text{R–S–S–R} \rightarrow \text{HO–ST–OH}^+ + [\text{R–S–S–R}]^- \rightarrow \text{O} = \text{ST} = \text{O} + 2\text{H}^+ + 2\text{R–S}^-$$

In native stentorin, two electrons and two protons resulting from the similar reactions reduce the suitable amino acid residue(s) such as cystine. The potential for the oxidation of mercaptoethanol is +0.02 V at pH 8 [84], while a potential of +0.6 V is necessary for the reduction of oxy-hypericin in DMSO [78]. Thus, a spontaneous reduction of oxy-hypericin in the ground state is feasible. Similarly, an oxy-stentorin can be reduced to stentorin to complete the photocycle, with the possible vectorial release of protons from the stentorin-bound pigment granule to the cytoplasm [85]. It remains to be seen if electron/proton transfer to and from the stentorin chromophore is accelerated by the apoprotein to initiate the functionally viable "photocycle" in *Stentor.* The redox potential of the cysteine-cystine system is estimated to be –0.21 V at pH 7.0 [86]. This redox potential would enhance the photooxidation of hypericin with respect to dithiodiethanol-mercaptoethanol system (–0.9 V at pH 10 for oxidation).

Blepharisma japonicum. Electron transfer to and from the excited state of blepharismin is likely to play a signal initiating role also in the photosensory transduction process of *Blepharisma*. To clarify this point, the fluorescence quenching of oxyblepharismin by electron acceptors with different reduction potentials has been studied in DMSO and in ethanol. It was shown that ground-state and excited state complexes (like fluorescent exciplexes) are not formed between the fluorophore and the quenchers. In DMSO, the bimolecular quenching rate constants, k_q, calculated on the basis of best fitting procedures, clearly show that the quenching efficiency decreases with the quencher negative reduction potential, E^0. The k_q (M^{-1} s^{-1}) and E^0 (V) values are, respectively, 7.8×10^9 and –0.134 for 1,4-benzoquinone, 8.9×10^9 and –0.309 for 1,4-naphthoquinone, 2.4×10^9 and –0.8 for nitrobenzene, 0.009×10^9 and –1.022 for azobenzene, 0 and –1.448 for benzophenone. These findings point to the fact that, upon formation of the encounter complex between oxyblepharismin and the quencher, an electron is released from excited oxyblepharismin to the quencher, similarly to what happens in stentorin. It was suggested that, in the pigment granules, such a light-induced electron transfer from excited blepharismins to a suitable electron acceptor, which can be coupled to a proton release into the cytoplasm [85], finally triggers the sensory transduction process in *Blepharisma* [87].

17.3.4 Signal generation and amplification

Motile behavior in *Stentor* and *Blepharisma* is controlled by ciliary strokes and their ultimate reaction to light is ciliary reversal. Therefore, it is important to look at what governs the ciliary movement. Membrane potentials control the ciliary movement. Evidence has been gathered indicating that ciliary activity is controlled primarily through electrical properties of the cell membrane [88,89]. The membrane is believed to control ciliary beatings by regulating the internal ionic environment, especially the intracellular calcium ions [90]. The motile mechanoresponse of *Stentor* is controlled by ion permeability of the cell membrane that determine resting, action, and mechanoreceptor potentials in the ciliate cell [16,91]. Similarly, the light responses in *Stentor* have been investigated using electrophysiological approaches. The photic receptor potential appeared after the onset of light stimulation of *Stentor* with a delay as long as 0.5 s at temperatures below 10°C. This contrasts markedly with the receptor potentials produced by mechanical and electrical stimulation [92]. For example, mechanical stimulus initiates a membrane depolarisation within 10 ms of the stimulus onset, which within another 20 ms triggers an action potential [91]. These differences in time course indicate that there are at least kinetic and possibly biochemical differences in the generation of the photic and mechanical receptor potentials.

The photomovement responses in *Stentor* are pH dependent, with lower responses at acidic pH and higher responses at neutral and basic pHs [93,94]. It was also demonstrated that heavy water (D_2O) enhanced the photophobic response and inhibited the phototactic response in *Stentor* [95].

Upon addition of protonophores, the photoresponses of *Stentor* decreased significantly. It is hypothesized that the protonophores would dissipate the pH gradient across the membrane, which could be generated by electron/proton transfer from the excited state stentorin [13,93]. As a result of the primary photoprocess of stentorin, a transient intracellular pH change may be coupled directly or indirectly to open the voltage-sensitive calcium channels. Indeed, in a model system consisting of hypericin embedded in phospholipid liposomes, a light-induced pH drop within the liposome has been reported [96]. An intracellular pH drop can also be triggered by the photoexcitation of hypericin or hypocrellin incorporated in 3T3 mouse fibroblast cells [97,98]. These model studies suggest that the primary photoprocess can lead to an intracellular pH change. Such a transient pH change could serve as the initial signal for the subsequent signal transduction pathway involved in the photomovement of the ciliate cells.

The membrane resting potential in ciliate cells is usually negative so that the cell interior is electrically negatively charged with respect to the extracellular compartment. *Stentor*, as most organisms, maintains a negative resting potential of about –50 to –55 mV across the cell membrane [51,91]. Upon excitation of the cell with a step-up light stimulus, a graded membrane receptor potential is generated, followed by an all-or-none action potential [16,99]. A pH-dependent action potential has also been observed by Kim [64]. In addition, Ca^{2+} ionophores such as calimycin, α-phosphatidic acid and Ca^{2+} blockers, including Ruthenium Red and methoxyverapamil, specifically inhibit both the step-up photophobic and negative phototactic responses in *Stentor* [100]. Although it is obvious in *Paramecium* as well as in other species, no membrane hyperpolarisation has been observed in *Stentor.* As in most ciliated protozoa, it seems that locomotion in *Stentor* is regulated by membrane-limited Ca^{2+} fluxes [101,102]. Also, as will be discussed later, actinic light signals are amplified by the influx of Ca^{2+} ions across the cell membrane via depolarisation by voltage-dependent calcium channels.

Perhaps the most convincing evidence for the generation of a transient pH change as an early intracellular transduction signal comes from the photoresponses of the ciliate

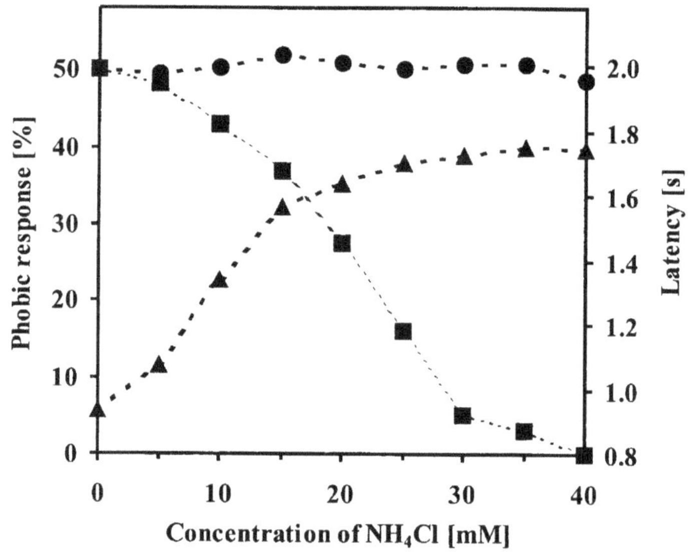

Figure 11. Effect NH_4Cl on the percentage (squares) and the latency (triangles) of the photophobic response in *Blepharisma japonicum*. No effect was noticed when NH_4Cl was replaced by NaCl (circles). (Redrawn from [94]).

cells to protonophores and exogenous ammonium chloride [94]. Ammonium chloride serves as a membrane permeable weak acid that lowers the intracellular pH. Figure 11 shows the effect of ammonium chloride on the photosensitivity and the photophobic response of *Blepharisma.* From these data, it was suggested that the artificially lowered intracellular pH in the presence of NH_4Cl makes inefficient any light-induced pH drop as the early signal for the subsequent transduction cascade. *Stentor* shows similar responses to ammonium chloride [94].

As described earlier, electrophysiological studies suggested a temporal correlation between light-induced membrane potential and photomovement responses, indicating that the membrane electrical events are involved in the photosensory transduction in *Stentor. Blepharisma* exhibits similar time courses correlating the membrane potential changes with the photophobic response. These results are not unexpected since modification of the motile responses by ionophores and ion channel blockers ($TPMP^+$, CCCP, FCCP, calimycin, and verapamil, etc.) as well as by extracellular ions (Ca^{2+}, H^+) have been observed before [103]. These agents are known to disrupt or alter cell membrane potential and are therefore expected to influence the photomotile responses in *Stentor.*

As already described, the final motile response in the photomovement of *Stentor* consists of at least the following steps: a stop reaction of its forward swimming, followed by brief backward swimming when light intensity is strong enough, and resumption of forward swimming. When higher light intensity than that required to elicit a receptor potential exceeding the threshold is applied, *Stentor* cells react with a receptor potential which rapidly triggers the action potential, and after the decay of the action potential, a plateau of depolarisation follows. With stimulation of light of similar intensity, *Stentor* reacted by altering its swimming direction with a stop reaction and a period of backward swimming. The latency (delay time between the initial onset of light stimulus and the appearance of action potential) of the stop reaction is shorter and the duration of backward swimming is longer with increasing light intensity. There is a temporal, and possibly functional connection between the light-induced membrane potential changes and the phobic response [50,51]. On the basis of electrophysiological measurements [99,104], a similar transduction model can be proposed for the photomovement response of *Blepharisma*.

Further investigation of the ionic basis underlying the light-induced membrane potential changes is warranted to shed light on our understanding of the photosensory transduction in *Stentor* and *Blepharisma*. One direction of future work could be the application of patch clamp techniques [105,106]. Patch clamp techniques [107] have allowed the measurement of the currents in single ionic channels. This technique has been successfully applied to single channel current recording in plasma membrane blisters of *Paramecium* [108]. Only recently, it has been applied to identify cGMP-dependent ion channels in *Stentor* cells [109]. It seems promising to apply it to investigate ion channel activities in *Stentor* and *Blepharisma* cells. Experiments on single channel activities may reveal the ion channels involved, their conductance and opening probability. Even more, if patch clamping of the intact cell is successful, it should be possible to observe the ion channel activity change with light stimulation in *Stentor* and *Blepharisma*.

17.4 Signal transduction models

In the previous section we discussed the possible role of an intracellular pH change (ΔpH) as an early transducing signal generated by the photoexcitation of photoreceptors stentorin and blepharismin in the pigment granules of *Stentor* and *Blepharisma*, respectively. A signaling role of the ΔpH in the photomovement of the ciliates warrants direct experimental confirmation. In the proposed transduction scheme, the light signal is amplified in terms of Ca^{2+} ion influx from the extracellular medium, which elicits structural changes in the ciliary contractile axonemes for the reversal of ciliary stroke. Figure 12 presents a schematic model for the photosensory transduction pathway for the photomovement responses such as the step-up photophobic response in *Stentor* and *Blepharisma* on the basis of photophysiological and electrophysiological results highlighted above [103,104,110].

The biochemical mechanisms underscoring each of the ionic events depicted in this scheme are virtually unknown. Here, we can only speculate about the possible biochemical mechanisms based on indirect physiological observations. Pertussis toxin blocks the inhibition of adenylate cyclase by catalysing the covalent modification of the inhibitory G protein [111]. The G-protein activator fluoroaluminate sensitized the ciliate cells, *Stentor* and *Blepharisma*, to the actinic light stimuli and enhanced their photophobic responses. Incubation of both ciliates in the presence of 8-bromo-guanosine 3′,5′-cyclic monophosphate (8-Br-GMP) resulted in specific inhibition of their photophobic responses, while lengthening the latency for the generation of action potentials, thus desensitizing the cells to the actinic light stimuli [99,112]. IBMX, an inhibitor of cGMP-dependent phosphodiesterase [113], and *l-cis*-diltiazem, a potent blocker of cGMP-dependent ion channels [114], also inhibited the photophobic response of *Stentor* and *Blepharisma* [99,110,112]. On the other hand, 6-anilino-5,8-quinolinedione (LY83583) significantly stimulated their photophobic responses, presumably due to the lowering of cellular cGMP. These results suggest that G-protein and cGMP/GMP play a signaling role in the photosensory transduction in *Stentor* and *Blepharisma*. (It is not clear how the "dark" activation of G-protein and the lowering of cellular cGMP by fluoroaluminate and LY83583, respectively, led to the enhanced photosensitivity of the ciliates. Since the cells treated with these drugs did not exhibit

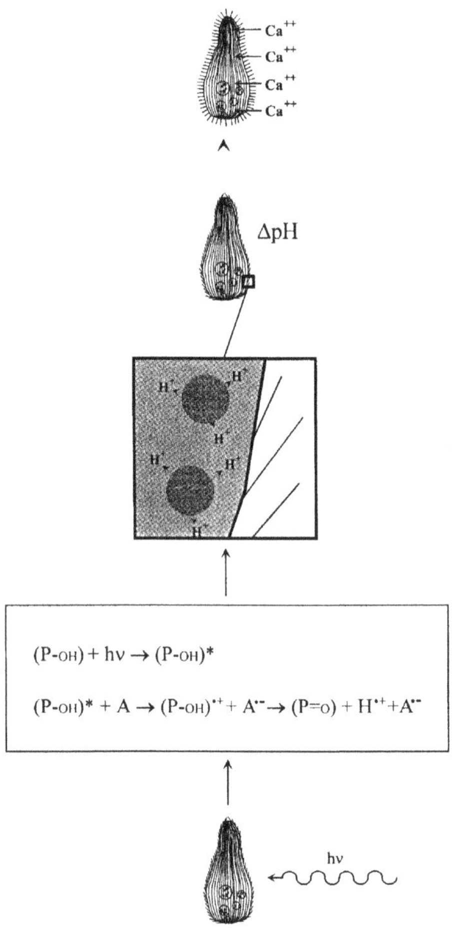

Figure 12. Ionic model for photomovement of ciliates. The granule-embedded chromophore after releasing an electron from its first excited singlet state to a suitable acceptor, can give raise to a proton translocation process into the cytoplasm, thus causing an intracellular pH decrease. The molecular processes linking this ΔpH with the opening of Ca^{2+} channels are depicted in Figure 13.

ciliary stroke reversals in the dark, these drugs cause a net increase in the capability of the ciliate cells to show photophobic responses).

In summary, the photoexcitation of stentorin in the pigment granule is proposed to generate a transient intracellular pH change, which can serve as a cellular signal associated with the depolarizing receptor potential. Subsequently, the initiating signal appears to be amplified by a sudden trigger of Ca^{2+} ion influx into the cell. The involvement of these photo-signal transduction events in the ciliate cells is consistent with the inhibitory effects of protonophores and Ca^{2+} channel blockers. It has also been suggested that a heterotrimeric G-protein plays an important role as a signal transducer in *Stentor* as well as *Blepharisma*. Thus, these cells provide an analogy to "visual" excitation system reminiscent of the rhodopsin-coupled transducin of higher animals. In both cases, G-protein activates a phosphodiesterase as an effector molecule, which leads to the lowering of cellular cGMP level. In the vertebrate visual system, the light-induced, rhodopsin-mediated lowering of intracellular cGMP level transiently closes cGMP-gated cation channels. However, in *Drosophila,* the light-induced, rhodopsin-mediated activation of the phospholipase C results in the opening of the cation channels, transient receptor potential in the plasma membrane, and influx of Ca^{2+} ions (for a review, see [115]). No experiments have been performed to identify a phospholipase C isoform, the activation of which may elicit influx of Ca^{2+} ions for the photomovement responses in the ciliate cells. The scheme proposed here combines vertebrate effector cGMP-dependent phosphodiesterase and invertebrate effector phospholipase C. This hypothetical model is also based on limited physiological evidence for the photophobic

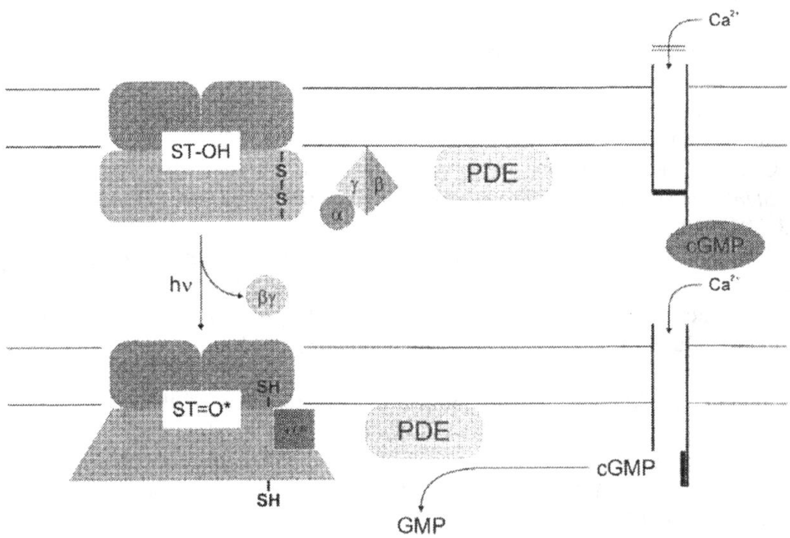

Figure 13. Hypothetical schematic model for photosensory transduction pathway in *Stentor coeruleus*. Cystine residues (-S-S-) of stentorin apoprotein can be the suitable acceptors of the electrons released from photoexcited stentorin (St-OH), thus allowing the release of protons into the cytoplasm; the subsequent cascade activation of G-protein (α-[β-γ]) and phosphodiesterase (PDE) causes the lowering of cGMP levels and the opening of Ca^{2+}-channels which eventually induces ciliary stroke reversal.

responses in *Stentor* and *Blepharisma,* but the fact that the light-induced action potentials in the ciliates are depolarizing, is consistent with the influx of Ca^{2+} ions.

Thus, the working hypothesis includes the photoactivation sequence stentorin → G-protein → phosphodiesterase → lowering of cGMP → opening of Ca^{2+}-channel → ciliary stroke reversal. This hypothesis is illustrated in a schematic model in Figure 13. A similar chain of events initiated by blepharismins is likely to be operative also in *Blepharisma*.

Acknowledgements

This work was supported by Kumho Petrochemical Co., LTD, and by USPHS-NIH (GM36956). This is Kumho Life & Environmental Science Laboratory Publication No. 14.

References

1. H.-W. Kuhlmann (1998). Photomovements in ciliated protozoa. *Naturwissenschaften*, **85**, 143–154.
2. H.-W. Kuhlmann (1998). Do phototactic ciliates make use of directional antennas to track the direction of light? *Europ. J. Protistol.*, **34**, 244–253.
3. R.W. Pierce, J.T. Turner (1992). Ecology of planktonic ciliates in marine food webs. *Reviews in Aquatic Sciences*, **6**, 139–181.
4. N. Ricci (1990). The behavior of ciliated protozoa. *Anim. Behav.*, **40**, 1048–1069.
5. B. Diehn, M.E. Feinleib, W. Haupt, E. Hildebrand, F. Lenci, W. Nultsch (1977). Terminology of behavioral responses of motile microorganisms. *Photochem. Photobiol.*, **26**, 559–560.
6. A.C. Giese (1973). *Blepharisma: The Biology of a Light-Sensitive Protozoan* (p. 366). Stanford University Press, Stanford, CA.
7. K.C. Yang, R.K. Prusti, P.-S. Song, M. Watanabe, M. Furuya (1986). Photodynamic action in *Stentor coeruleus* sensitized by endogenous pigment stentorin. *Photochem. Photobiol.*, **43**, 305–310.
8. A. Miyake, T. Harumoto, B. Salvi, V. Rivola (1990). Defensive function of pigment granules in *Blepharisma japonicum*. *Europ. J. Protistol.*, **25**, 310–315.
9. G. Checcucci, G. Damato, F. Ghetti, F. Lenci (1993). Action spectra of the photophobic response of the blue and red forms of *Blepharisma japonicum*. *Photochem. Photobiol.*, **57**, 686–689.
10. M. Kraml, W. Marwan (1983). Photomovement response of the heterotrichous ciliate *Blepharisma japonicum*. *Photochem. Photobiol.*, **37**, 313–319.
11. T. Matsuoka, S. Matsuoka, Y. Yamaoka, T. Kuriu, Y. Watanabe, M. Takayanagi, Y. Kato, K. Taneda (1992). Action spectra for step-up photophobic response in *Blepharisma*. *J. Protozool.*, **39**, 498–502.
12. P. Scevoli, F. Bisi, G. Colombetti, F. Ghetti, F. Lenci, V. Passarelli (1987). Photomotile responses of *Blepharisma japonicum* I: Action spectra determination and time-resolved fluorescence of photoreceptor pigments. *J. Photochem. Photobiol. B: Biol.*, **1**, 75–84.
13. P.-S. Song, D.-P. Häder, K.L. Poff (1980). Step-up photophobic response in the ciliate, *Stentor coeruleus*. *Arch. Microbiol.*, **126**, 181–186.
14. P.-S. Song, D.-P. Häder, K.L. Poff (1980). Phototactic orientation by the ciliate, *Stentor coeruleus*. *Photochem. Photobiol.*, **32**, 781–786.

15. N. Tao, L. Deforce, M. Romanowski, S. Meza-Keuthen, P.-S. Song, M. Furuya (1994). *Stentor* and *Blepharisma* photoreceptors: Structure and function. *Acta Protozool.*, **33**, 199–211.
16. D.C. Wood (1976). Action spectrum and electrophysiological responses correlated with the photophobic response of *Stentor coeruleus*. *Photochem. Photobiol.*, **24**, 261–266.
17. N. Tao, M. Orlando, J.-S. Hyon, M. Gross, P.-S. Song (1993). A new photoreceptor molecule from *Stentor coeruleus*. *J. Am. Chem. Soc.*, **115**, 2526–2528.
18. Checcucci, G., R.K. Shoemaker, E. Bini, R. Cerny, N. Tao, J.-S. Hyon, D. Gioffre, F. Ghetti, F. Lenci, P.-S. Song (1997). Chemical structure of blepharismin, the photosensor pigment for *Blepharisma japonicum*. *J. Am. Chem. Soc.*, **119**, 5762–5763.
19. M. Maeda, H. Naoki, T. Matsuoka, Y. Kato, H. Kotsuki, K. Utsumi, T. Tanaka (1997). Blepharismin 1–5, novel photoreceptor from the unicellular organism *Blepharisma japonicum*. *Tetrahedron Lett.*, **38**, 7411–7414.
20. D. Spitzner, G. Höfle, I. Klein, S. Pohlan, D. Ammermann, L. Jaenicke (1998). On the structure of oxyblepharismin and its formation from blepharismin. *Tetrahedron Letters*, **39**, 4003–4006.
21. E.B. Holt, F.S. Lee (1901). The theory of phototactic response. *Am. J. Physiol.*, **4**, 460–481.
22. H.S. Jennings (1904). *Contributions to the study of the behavior of lower organisms* (pp. 1–256). Publications of the Carnegie Institution of Washington No. 16.
23. S.O. Mast (1906). Light reactions in lower organisms. I. *Stentor coeruleus*. *J. Expl. Zool.*, **3**, 359–399.
24. P.-S. Song (1981). Photosensory transduction in *Stentor coeruleus* and related organisms. *Biochim. Biophys. Acta*, **639**, 1–29.
25. K. Iwatsuki (1992). *Stentor coeruleus* shows positive photokinesis. *Photochem. Photobiol.*, **55**, 469–471.
26. A.C. Giese (1981). The photobiology of *Blepharisma*. *Photochem. Photobiol. Rev.*, **6**, 139–180.
27. T. Matsuoka (1983). Negative phototaxis in *Blepharisma japonicum*. *J. Protozool.*, **30**, 409–414.
28. T. Tsuda, T. Matsuoka (1994). The cells of *Blepharisma* can detect light direction. *Microbios.*, **77**, 153–160.
29. T. Matsuoka (1983). Distribution of photoreceptors inducing ciliary reversal and swimming acceleration in *Blepharisma japonicum*. *J. Exp. Zool.*, **225**, 337–340.
30. T. Matsuoka, K. Taneda (1992). Step-up and step-down photoresponses in *Blepharisma*. *Zoological Science*, **9**, 529–532.
31. G. Colombetti, R. Bräucker, H. Machemer (1992). Photobehavior of *Fabrea salina*: responses to directional and diffused gradient-type illumination. *J. Photochem. Photobiol. B.: Biol.*, **15**, 253–257.
32. R. Marangoni, S. Puntoni, L. Favati.G. Colombetti (1994). Phototaxis in *Fabrea salina*. I. Action spectrum determination. *J. Photochem. Photobiol. B: Biol.*, **23**, 149–154.
33. R. Marangoni, A. Battistini, S. Puntoni, G. Colombetti (1995). Temperature effects on motion parameters and the phototactic reaction of the marine ciliate *Fabrea salina*. *J. Photochem. Photobiol. B: Biol.*, **30**, 123–127.
34. S. Puntoni, R. Marangoni, D. Gioffrè, G. Colombetti (1998). Effects of Ca^{2+} and K^+ on motility and photomotility of the marine ciliate *Fabrea salina*. *J. Photochem. Photobiol., B: Biol.*, **43**, 204–208.
35. G. Colombetti, R. Marangoni, H. Machemer (1992). Phototaxis in *Fabrea salina*. *Med. Biol. Env.*, **20**, 93–100.

36. T. Fenchel, B.J. Finlay (1984). Geotaxis in the ciliated protozoon *Loxodes*. *J. Exp. Biol.*, **110**, 17–33.
37. T. Fenchel, B.J. Finlay (1986). Photobehavior of the ciliated protozoon *Loxodes*: taxic, transient, and kinetic responses in the presence and absence of oxygen. *J. Protozool.*, **33**, 139–145.
38. B.J. Finlay, T. Fenchel (1986). Photosensitivity in the ciliated protozoon *Loxodes*: pigment granules, absorption and action spectra, blue light perception, and ecological significance. *J. Protozool.*, **33**, 534–542.
39. B.J. Finlay, T. Fenchel, S. Gardener (1986). Oxygen perception and O2 toxicity in the freshwater ciliated protozoon *Loxodes*. *J. Protozool.*, **33**, 157–165.
40. H.-W. Kuhlmann, R. Hemmersbach-Krause (1993). Phototaxis in the "stigma"-forming ciliate *Nassula citrea*. *J. Photochem. Photobiol. B: Biol.*, **21**, 191–195.
41. H.-W. Kuhlmann, R. Bräucker, A.G. Schepers (1997). Phototaxis in *Porpostoma notatum*, a marine scuticociliate with a composed crystalline organelle. *Europ.J. Protistol.*, **32**, 295–304.
42. M. Selbach, D.-P. Häder, H.-W. Kuhlmann (1999). Phototaxis in *Chlamydodon mnemosyne*: determination of the illuminance-response curve and the action spectrum. *J. Photochem. Photobiol. B: Biol.*, in press.
43. I.-H. Kim, J.S. Rhee, J.W. Huh, S. Florell, B. Faure, K.W. Lee, M. Kahsai, P.-S. Song, N. Tamai, T. Yamazaki, I. Yamazaki (1990). Structure and function of the photoreceptor stentorins in *Stentor coeruleus*.I. Partial characterization of the photoreceptor organelle and stentorins. *Biochim. Biophys. Acta*, **1040**, 43–57.
44. F. Ghetti (1991). Photoreception and photomovements in *Blepharisma japonicum*. In: F. Lenci et al. (Eds), *Biophysics of Photoreceptors and Photomovements in Microorganisms* (pp. 257–265). Plenum Press, New York.
45. T. Matsuoka, T. Tsude, M. Ishida, Y. Kato, M. Takayanagi, T. Fujino, S. Mizuta (1994). Presumed photoreceptor protein and ultrastructure of the photoreceptor organelle in the ciliated protozoa, *Blepharisma*. *Photochem. Photobiol.*, **60**, 598–604.
46. P.B. Weisz (1950). On the mitochondrial nature of the pigmented granules in *Stentor* and *Blepharisma*. *J. Morphol.*, **86**, 177–184.
47. F. Inaba, R. Nakamura, S. Yamaguchi (1958). An electron-microscopic study on the pigment granules of *Blepharisma*. *Cytologia (Tokyo)*, **23**, 72–79.
48. E.B. Walker, T.Y. Lee, P.S. Song (1979). Spectroscopic characterization of the *Stentor* photoreceptor. *Biochim. Biophys. Acta*, **587**, 129–144.
49. I.-H. Kim, R.K. Prusti, P.-S. Song, D.-P. Häder, M. Häder (1984). Phototaxis and photophobic responses in *Stentor coeruleus*. *Biochim. Biophys. Acta*, **798**, 298–304.
50. H. Fabczak, P.B. Park, S. Fabczak, P.-S. Song (1993). Photosensory transduction in ciliates. II. Possible role of G-protein and cGMP in *Stentor coeruleus*. *Photochem. Photobiol.*, **57**, 702–706.
51. S. Fabczak, H. Fabczak, N. Tao, P.-S. Song (1993). Photosensory transduction in ciliates.I. An analysis of light-induced electrical and motile responses in *Stentor coeruleus*. *Photochem. Photobiol.*, **57**, 696–701.
52. D.W. Cameron, A.G. Riches (1995). Synthesis of stentorin. *Tetrahedron Lett.*, **36**, 2331–2334.
53. F. Ghetti, G. Checcucci, F. Lenci, P.F. Heelis (1992). A laser flash photolysis study of the triplet states of the red and the blue forms of *Blepharisma japonicum* pigment. *J. Photochem. Photobiol. B: Biol.*, **13**, 315–321.
54. T. Matsuoka, Y. Watanabe, Y. Sagara, M. Takayanagi, Y. Kato (1995). Additional evidence for blepharismin photoreceptor pigment mediating step-up photophobic response of unicellular organism, *Blepharisma japonicum*. *Photochem. Photobiol.*, **62**, 190–193.

55. F. Lenci, G. Checcucci, F. Ghetti, D.Gioffrè, A. Sgarbossa (1997). Sensory perception and transduction of UV-B radiation by the ciliate *Blepharisma japonicum*. *Biochim. Biophys. Acta*, **1336**, 23–27.
56. K.J. Hellingwerf, W.D. Hoff, W. Crielaard (1996). Photobiology of microorganisms: How photosensors catch a photon to initialize signalling. *Mol. Microbiol.*, **21**, 683–693.
57. Y. Nakaoka, R. Tokioka (1988). Photoreceptor potential causing phototaxis in *Paramecium bursaria*. *J. Exp. Biol.*, **137**, 477–485.
58. Y. Nakaoka, R. Tokioka, T. Shinozawa, J. Fujita, J. Usukura (1991). Photoreception of *Paramecium* cilia: localization of photosensitivity and binding with anti-frog-rhodopsin IgG. *J. Cell. Sci.*, **99**, 67–72.
59. A. Podestà, R. Marangoni, C. Villani, G. Colombetti (1994). A rhodopsin-like molecule on the plasma membrane of *Fabrea salina.*, *J. Euk. Microbiol.*, **41**, 565–569.
60. R. Marangoni, R. Cubeddu, P. Taroni, G. Valentini, R. Sorbi, E. Lorenzini, G. Colombetti (1996). Microspectrofluorometry, fluorescence imaging and confocal microscopy of an endogenous pigment of the marine ciliate *Fabrea salina*. *J. Photochem. Photobiol. B: Biol.*, **34**, 183–189.
61. R. Marangoni, L. Gobbi, F. Verni, G. Albertini, G. Colombetti (1996). Pigment granules and hypericin-like fluorescence in the marine ciliate *Fabrea salina*. *Acta Protozoologica*, **35**, 177–182.
62. Y.P. See, G. Jackowski (1989). Estimating molecular weights of polypeptides by SDS gel electrophoresis. In: T.E. Creighton (Ed.), *Protein Structure. A Practical Approach* (pp. 1–21). IRL Press, Oxford.
63. N. Tao (1994). *Stentorin, the photoreceptor molecule and signal transduction in ciliate* Stentor coeruleus. Ph.D. Dissertation, University of Nebraska, Lincoln, NE.
64. I.-H. Kim (1988). *Characterization and primary processes of stentorin and its function in* Stentor coeruleus. Ph.D. Dissertation, University of Nebraska-Lincoln, Lincoln, NE.
65. R. Dai, T. Yamazaki, I. Yamazaki, P.-S. Song (1995). Initial spectroscopic characterization of the ciliate photoreceptor. *Biochim. Biophys. Acta*, **1231**, 58–68.
66. C. Montell, G.M. Rubin (1988). The Drosophila *ninaC* locus encodes two photoreceptor cell specific proteins with domains homologous to protein kinases and the myosin heavy chain head. *Cell*, **52**, 757–772.
67. R. Reilander, A. Achilles, U. Friedel, G. Maul, F. Lottspeich, N.J. Cook (1992). Primary structure and functional expression of the Na/Ca,K-exchanger from bovine rod photoreceptors. *EMBO J.*, **11**, 1689–1695.
68. D. Gioffré, F. Ghetti, F. Lenci, C. Paradiso, R. Dai, P.-S. Song (1993). Isolation and characterization of the presumed photoreceptor protein of *Blepharisma japonicum*. *Photochem. Photobiol.*, **58**, 275–279.
69. T. Yamazaki, I. Yamazaki, Y. Nishimura, R. Dai, P.-S. Song (1993). Time-resolved fluorescence spectroscopy and photolysis of the photoreceptor blepharismin. *Biochim. Biophys. Acta*, **1143**, 319–326.
70. T. Matsuoka, Y. Murakami, Y. Kato (1993). Isolation of blepharismin-binding 200 kDa protein responsible for behavior in *Blepharisma*. *Photochem. Photobiol.*, **57**, 1042–1047.
71. P.-S. Song, I.-H. Kim, S. Florell, N. Tamai, T. Yamazaki, I. Yamazaki (1990). Structure and function of the photoreceptor stentorins in *Stentor coeruleus*. II. Primary photoprocess and picosecond time-resolved fluorescence. *Biochim. Biophys. Acta.*, **1040**, 58–65.
72. S. Savikhin, N. Tao, P.-S. Song, W. Struve (1993). Ultrafast pump-probe spectroscopy of the photoreceptor stentorins from the ciliate *Stentor coeruleus*. *J. Phys. Chem.*, **97**, 12379–12386.
73. N. Duran, P.-S. Song (1985). Hypericin and its photodynamic action. *Photochem. Photobiol.*, **43**, 677–680.

74. F. Ghetti, G. Checcucci, F. Lenci (1992). Photosensitized reactions as primary molecular events in photomovements of microorganisms. *J. Photochem. Photobiol. B: Biol.*, **15**, 185–198.
75. R. Dai, P.-S. Song, J.L. Anderson, M. Selke, C.S. Foote (1992). *Hypericin and silkworm chlorophyll metabolites as anti-retroviral and anti-tumor photosensitizers.* IUBMB Conference on Biochemistry and Molecular Biology of Diseases Abstracts, Nagoya, Japan.
76. G. Checcucci, F. Lenci, F. Ghetti, P.-S. Song (1991). A videomicroscopic study of the effect of a singlet oxygen quencher on *Blepharisma japonicum* photobehavior. *J. Photochem. Photobiol. B: Biol.*, **11**, 49–55.
77. T.A. Wells, A. Losi, R. Dai, M. Anderson, J. Redepenning, P. Scott, S.-M. Park, J. Golbeck, P.-S. Song (1997). Electron transfer quenching and photoinduced EPR of hypericin and the ciliate photoreceptor stentorin. *J. Phys. Chem.*, **101**, 366–372.
78. J. Redepenning, N. Tao (1993). Measurement of formal potentials for hypericin in dimethylsulfoxide. *Photochem. Photobiol.*, **58**, 532–535.
79. T. Yamazaki, N. Ohta, I. Yamazaki, P.-S. Song (1993). Excited-state properties of hypericin: Electronic spectra and fluorescence decay kinetics. *J. Phys. Chem.*, **97**, 7870–7875.
80. F. Gai, M.J. Fehr, J.W. Petrich (1993). Ultrafast excitaed-state processes in the antiviral agent hypericin. *J. Am. Chem. Soc.*, **115**, 3384–3385.
81. F. Gai, M.J. Fehr, J.W. Petrich (1994). Role of solvent in excited-state proton transfer in hypericin. *J. Phys. Chem.*, **98**, 8352–8358.
82. L. Weiner, Y. Mazur (1992). EPR studies of hypericin. photogeneration of free radicals and superoxide. *J. Chem. Soc. Perkin Trans.*, **2**, 1439–1442.
83. R. Dai (1994). *Further characterization of the ciliate photoreceptor (stentorin) and the initial photoprocess.* Ph.D. Dissertation, University of Nebraska, Lincoln, NE.
84. J.H. Zagal, C. Paez (1989). Catalytic electrooxidation of 2-mercaptoethanol on a graphite electrode modified with metal-phthalocyanines. *Electrochim. Acta*, **34**, 243–247.
85. R.I. Cukier, D.G. Nocera (1998). Proton-coupled electron transfer. *Annu. Rev. Phys. Chem.*, **49**, 337–369.
86. W.W. Cleland (1964). Dithiothreitol, a new protective reagent for SH groups. *Biochemistry*, **3**, 480–482.
87. N. Angelini, A. Quaranta, G. Checcucci, P.-S. Song, F. Lenci (1998). Electron transfer fluorescence quenching of *Blepharisma japonicum* photoreceptor pigments. *Photochem. Photobiol.*, **68**, 864–868.
88. R. Eckert, Y. Naitoh (1972). Bioelectric control of locomotion in the ciliates. *J. Protozool.*, **19**, 237–243.
89. Y. Naitoh, R. Eckert (1974). The control of ciliary activity in protozoa. In: M.A. Sleigh (Ed.), *Cilia and Flagella* (pp. 305–352). Academic Press, London.
90. R. Eckert (1972). Bioelectric control of ciliary activity. *Science*, **176**, 473–481.
91. D.C. Wood (1982). Membrane permeabilities determining resting, action and mechanoreceptor potentials in *Stentor coeruleus*. *J. Comp. Physiol.*, **146**, 537–550.
92. D.C. Wood (1991). Electrophysiology and photomovement of *Stentor.* In: F. Lenci, F. Ghetti, G. Colombetti, D.-P. Häder, P.-S. Song (Eds), *Microorganisms Biophysics of Photoreceptors and Photomovements* (pp. 281–191). Plenum Press, New York.
93. E.B. Walker, M.J. Yoon, P.-S. Song (1981). The pH dependence of photosensory responses in *Stentor coeruleus* and model system. *Biochim. Biophys. Acta*, **6340**, 289–308.
94. H. Fabczak, S. Fabczak, P.-S. Song, G. Checcucci, F. Ghetti, F. Lenci (1993). Photosensory transduction in ciliates. Role of intracellular pH and comparison between *Stentor coeruleus* and *Blepharisma japonicum*. *J. Photochem. Photobiol. B: Biol.*, **21**, 47–52.

95. K. Iwatsuki, P.-S. Song (1985). Deuterium oxide (D_2O) enhances the photosensitivity of *Stentor coeruleus*. *Biophys., J.*, **48**, 1045–1048.
96. M.J. Fehr, M.A. McCloskey, J.W. Petrich (1995). Light-induced acidification by the antiviral agent hypericin. *J. Am. Chem. Soc.*, **117**, 1833–1836.
97. F. Sureau, P. Miskovsky, L. Chinsky, P.Y. Turpin (1996). Hypericin-induced cell pyhotosensitization involves an intracellular pH decrease. *J. Am. Chem. Soc.*, **118**, 9484–9487.
98. R. Chaloupka, F. Sureau, E. Kocisova, J.W. Petrich (1998). Hypocrellin A photosensitization involves an intracellular pH decrease in 3T3 cells. *Photochem. Photobiol.*, **68**, 44–50.
99. S. Fabczak, H. Fabczak, P.-S. Song (1993). Photosensory transduction in ciliates. III. The temporal relation between membrane potentials and photomotile response in *Blepharisma japonicum*. *Photochem. Photobiol.*, **57**, 872–876.
100. P.-S. Song (1983). Protozoan and related photoreceptors: Molecular aspects. *Annu. Rev. Biophys. Bioeng.*, **12**, 35–68.
101. R.K. Prusti, P.-S. Song, D.-P. Häder, M. Häder (1984). Caffeine-enhanced photomovement in the ciliate, *Stentor coeruleus*. *J. Photochem. Photobiol.*, **40**, 369–375.
102. K. Iwatsuki, P.-S. Song (1989). The ratio of extracellular Ca^+ to K^+ ions ratio affects the photoresponses in *Stentor coeruleus*. *Comp. Biochem. Physiol.*, **92A**, 101–106.
103. P.-S. Song, K.L. Poff (1989). Photomovement. In: K.C. Smith (Ed.), *The Science of photobiology* (pp. 305–346). Plenum Press, New York.
104. S. Fabczak, H. Fabczak, M. Walerczk, J. Sikora, B. Groszynska, P.-S. Song (1996). Ionic mechanisms controlling photophobic responses in the ciliate *Blepharisma japonicum*. *Acta Protozool.*, **35**, 245–249.
105. N. Neher, B. Sakmann (1992). The patch clamp technique. *Sci. Amer.*, **No. 3**, 44–51.
106. O.P. Hamill, A. Marty, E. Neher, B. Sakmann, F.J. Sigworth (1981). Improved patch-clamp techniques for high-resolution current recording from cells and cell-free membrane patches. *Pflügers Archiv: Eur. J. Physiol.*, **391**, 85–100.
107. E. Neher, B. Sakmann (1976). Single channel currents recorded from membrane of denervated frog muscle fibres. *Nature*, **260**, 799–802.
108. Y. Saimi, B. Martinac (1989). Calcium-dependent potassium channel in *Paramecium* studied under patch clamp. *J. Membrane Biol.*, **112**, 79–89.
109. P. Koprowski, M. Walerczyk, B. Groszynska, H. Fabczak, A. Kubalski (1997). Modified patch-clamp method for studying ion channels in *Stentor coeruleus*. *Acta Protozool.*, **36**, 121–124.
110. S. Fabczak, H. Fabczak, P.-S. Song (1994). Ca^{2+} ions mediate the photophobic response in *Blepharisma* and *Stentor*. *Acta Protozool.*, **33**, 93–100.
111. L. Stryer (1988). *Biochemistry* (p. 296). Freeman, W.H. and Company, New York.
112. H. Fabczak, N. Tao, S. Fabczak, P.-S. Song (1993). Photosensory transduction in ciliates. IV. Modulation of the photomovement response of *Blepharisma japonicum* by cGMP. *Photochem. Photobiol.*, **57**, 889–892.
113. N.D. Goldberg, M.K. Haddox (1977). Cyclic GMP metabolism and involvement in biological regulation. *Annu. Rev. Biochem.*, **46**, 823–896.
114. L.W. Haynes (1992). Block of the cyclic GMP-gated channel of vertebrate rod and cone photoreceptor by *l-cis*-diltiazem. *J. Gen. Physiol.*, **100**, 783–801.
115. K. Scott, C. Zuker (1997). Lights out: deactivation of the phototransduction cascade. *Trends Biochem. Sci.*, 350–354.

Photomovement
D.-P. Häder and M. Lebert, editors.

Chapter 18

Electrophysiology and light responses in *Stentor* and *Blepharisma*

David C. Wood

Table of contents

Abstract

The blue-green ciliate, *Stentor coeruleus*, swims away from light sources and collects in dimly lighted areas. This behavioral outcome is termed photodispersal. Photodispersal could be produced by one or more of the three light-sensitive behavioral responses these ciliates exhibit – photokinesis, photophobic responses and phototaxis. Of these photophobic responding appears to be the primary contributor to photodispersal, since: (1) photophobic responses, but not photokinesis, is observed at fluence rates just sufficient to produce photodispersal, (2) manipulations which alter or eliminate photophobic responses also eliminate photodispersal, and (3) photophobic responses has the necessary property of being sensitive to the direction of the incident light.

Photophobic responses begin with a reversal in the direction of ciliary beat thereby stopping the ciliate's forward swimming. Ciliary reversal, as viewed by videomicroscopy, can be induced by small membrane depolarizations (<10 mV) which do not produce action potentials. Even when action potentials are elicited from a cell, it reverses its cilia before the action potential is initiated. Thus, despite the all-or-none character of ciliary reversal it is not elicited by all-or-none action potentials as previously supposed.

Illumination of *Stentor* produces a graded membrane depolarization, the photic receptor potential, observed with amplitudes from 0 to 20 mV and long onset latencies from 100 to 200 ms. Current pulses applied during the course of the photic receptor potential indicate that the membrane resistance is not changed during these receptor potentials as is observed for receptor potentials generated by the opening or closing of ionic channels. This result is corroborated by the slow rate of receptor potential rise and the small changes in current observed when voltage-clamped cells are illuminated. The alternative mechanism for generating receptor potentials is by transmembrane electrogenic pumping.

18.1 Introduction

Many of the species in the ciliate genus *Stentor* have been given their species name because of their obvious coloration – *Stentor niger, Stentor amethystinus, Stentor coeruleus,* and *Stentor rubra* [1]. The closely related heterotrich *Blepharisma* is also colored and is even capable of changing its normal pink coloration to blue when grown under brightly illuminated and oxygenated conditions [2]. Ciliates of these genera are photosensitive, a characteristic not shared by the many genera of colorless ciliates. It is an obvious inference that the photosensitivity of these heterotrichs is attributable to their pigmentation and many lines of evidence support this conclusion. For instance, 10 action spectra for photosensitive behavioral responses produced by *Stentor* and *Blepharisma* have been constructed and all but one are well correlated with the appropriate pigment absorption spectrum [3–12]. Further, *Stentor niger* has an action spectrum that differs from that of *Stentor coeruleus* in the direction predicted by the difference in their absorption spectra [13]. Similarly, the action spectrum of pink *Blepharisma japonicum* is well matched by the absorption of its "pink" pigment, while blue *Blepharisma japonicum* have an action spectrum that is shifted to agree with the absorption of its "blue" pigment [9]. Additionally, *Stentor coeruleus* bleached of most of its pigment by caffeine treatment are less sensitive to light than are normally pigmented cells [3,14]. These many correlations leave little doubt that light absorption by the principal pigments in these organisms initiates their photosensitive behavioral responses.

Three photosensitive responses have been described for *Stentor* and *Blepharisma*: photokinesis, photophobic responses, and phototaxis. Photokinesis is exhibited by these species because they swim more rapidly when brightly illuminated than when dimly illuminated [15,16]. A *Stentor* or *Blepharisma* exhibits a photophobic response if, upon illumination, it undergoes the following stereotyped behavioral sequence: an abrupt cessation of forward swimming, backward swimming and then renewed forward swimming. Finally, these cells exhibit phototaxis when they gradually reorient the direction of their forward swimming so as to be moving progressively more and more away from the direction in which the actinic light is traveling [5]. In the final phase of this behavior they are swimming directly away from the light source. These three photosensitive responses, singly or in combination, result in *Stentor* and *Blepharisma* collecting in the darker areas of an unevenly illuminated vessel, a phenomenon termed photodispersal. However, the relative importance of these three behaviors in the production of photodispersal is not obvious and remains somewhat controversial.

It is the purpose of this review to summarize what is known about the cellular mechanisms by which *Stentor* and *Blepharisma* transduce light energy absorbed by their respective principal pigments into photophobic responses. The cellular mechanisms producing photokinesis and phototaxis are not considered here for several reasons, not the least of which is that very little is known about them. Secondly, photokinesis and phototactic behaviors are not closely linked in time to light onset, whereas an initial photophobic response generally occurs within 5 s of light onset. This short latency makes the correlation between light onset, the recorded transmembrane potentials and the resultant overt behaviors much easier to study. Thirdly, changes in transmembrane potential appear to be necessary for the production of photophobic responses and may be important in the production of photokinesis and phototaxis as well. Transmembrane potentials are best recorded with microelectrodes which also immobilize the cell being studied. However, a meaningful electrophysiological analysis can only occur if the behavioral responses and transmembrane potentials produced by an immobilized cell can be uniquely related to the behavioral responses of a swimming cell. Since both immobilized and swimming cells exhibit ciliary reversal and ciliary reversal initiates photophobic responses in swimming cells, the circumstances necessary for an electrophysiological analysis of photophobic responses are present. Conversely, the behavioral processes that produce photokinesis and phototaxis are presently unknown, so transmembrane potentials recorded from immobilized cells cannot be correlated with them. In sum then, this review will attempt to replace the sequence of arrows diagramed below with known cellular mechanisms.

$$\text{Light} \rightarrow \text{photopigment absorption} \rightarrow \text{photophobic response}$$

18.2 The cellular basis of photophobic responding

During normal forward swimming the somatic cilia of *Stentor* and *Blepharisma* produce their power stroke in a direction that is obliquely toward their posterior and this motion propels these organisms along a helical forward swimming path. A variety of stimuli – light, certain chemicals and electrical shocks – can elicit a sudden reorientation in the direction of the ciliary power stroke to an anterior pointing direction, a response termed ciliary reversal. This reorientation results in a sudden cessation in forward motion due to the high level of viscous drag acting on these small organisms [17]. The ciliary reversal may be brief, producing no more than a short hiatus in forward swimming, or the cilia may beat toward the anterior of the cell for a period of time during which the cell is propelled backward. In the later case, ciliary beating toward the posterior of the cell is eventually resumed and with it forward swimming, albeit in a direction different from the forward swimming direction exhibited prior to ciliary reversal. If light initiates this behavioral sequence, it is termed a photophobic response.

In *Stentor* and *Blepharisma* observation of these ciliary responses is facilitated by the fact that they possess two types of cilia as their order name "Heterotrichida" indicates. The somatic cilia of both of these organisms are aligned in long rows called kineties that run the length of the cell. Additionally they possess composite ciliary structures, called membranelles, each of which contains 3 adjacent rows of cilia with 20 to 25 cilia in each row [18]. These 60 to 75 cilia beat as a single unit. *Stentor* possesses a single row of about 250 membranelles that circumscribes the anterior surface (frontal field) of the cell. These membranelles cease their normal pattern of metachronous beating and synchronously assume a forward pointing direction at the onset of a period of ciliary reversal. This ciliary response is easily discerned in the appropriate lighting conditions.

The direction of the ciliary power stroke is determined by the concentration of Ca^{2+} in the intraciliary lumen. Using *Paramecium* models, i.e. *Paramecium* whose membranes had been permeabilized by treatment with weak detergents, Naitoh and Kaneko [19] and Nakaoka et al. [20], showed that the orientation of the ciliary power stroke shifted from backward to forward when the concentration of Ca^{2+} to which these cells were exposed was raised to above 1 μM. Similarly, Matsuoka et al. [21] demonstrated that a reversal in ciliary beat direction occurred in *Blepharisma* models when they were exposed to concentrations of Ca^{2+} above 1 μM. While the exact mechanism by which a rise in intraciliary Ca^{2+} alters the direction of the ciliary power stroke has not been completely elucidated, it appears to involve the binding of Ca^{2+} to calmodulin. The Ca^{2+}-calmodulin complex is then thought to bind to the axoneme or to influence axonemal phosphorylation and thereby redirect the orientation of ciliary beating.

$$\text{Light} \rightarrow \text{photopigment absorption} \rightarrow [Ca^{2+}]_i \rightarrow \text{ciliary reversal}$$
$$\rightarrow \text{photophobic response}$$

18.3 The ciliary action potential

The increase in intraciliary Ca^{2+} that produces ciliary reversal results from depolarization of the ciliary membrane. Using simultaneous intracellular recording and microphotography Eckert and Naitoh [22] and Eckert [23] demonstrated convincingly that reorientation of nonbeating cilia in *Paramecium* was correlated with a transient regenerative membrane depolarization that overshot 0 mV and had many of the properties of an action potential. Subsequently, Machemer and Eckert [24] and Machemer [25] used high speed cinematography coupled with intracellular recording to quantitatively correlate the direction and frequency of ciliary beating with changes in transmembrane potential. A similar, but qualitative, correlation between action potential production and ciliary reversal has been noted by those researchers who have recorded intracellularly from *Stentor* [3,4,10,27–28] and *Blepharisma* [11,29], but has not been studied using cinematography. In *Paramecium* these transient membrane depolarizations, while clearly regenerative, vary in amplitude with the strength of the eliciting stimulus. Similarly, when *Stentor* are in the pear-shaped swimming form, the recorded transient depolarizations vary in amplitude with stimulus intensity [3]. On the other hand, *Stentor* produce true all-or-none action potentials if the recordings are made when the ciliate is in the extended sessile form [30]. Deciliation of *Paramecium* results in cells that are no longer capable of producing these regenerative responses [31,32] and the amplitude of these responses increases as the cilia grow back to their normal size [33]. Therefore, the ion channels responsible for regenerative response production are located in the ciliary membrane. On the other hand, we have recently observed that deciliated *Stentor* are still capable of producing action potentials and therefore must have ion channels capable of action potential production located both in the ciliary membrane and in other areas of the plasma membrane (Wood, unpublished).

As in *Paramecium*, the peak of *Stentor*'s and *Blepharisma's* action potential becomes more positive as the extracellular concentration of Ca^{2+} increases indicating that an influx of Ca^{2+} is instrumental in action potential production [29,30]. The dependence of *Stentor's* action potential peak on extracellular Ca^{2+} concentration is well described by the prevailing biophysical model for the generation of steady state transmembrane potentials, the Goldman-Hodgkin-Katz equation, if it is assumed that the permeability of Ca^{2+} increases 116 fold above resting levels during action potential production. On the other hand changing the extracellular concentration of other potentially significant

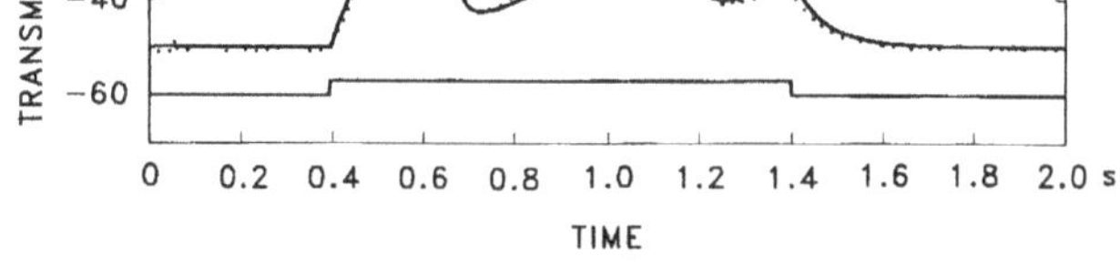

Figure 1. Action potentials elicited by a rectangular outward current pulse of 2 nA that was passed through one of two intracellular microelectrodes (bottom trace). The other microelectrode was used to record transmembrane potential (dotted line trace). Superimposed on this recorded trace is the output of the mathematical model developed to simulate *Stentor* action potentials using parameters in the equations that optimize the fit (solid line trace).

ions, K^+, Na^+, Mg^{2+} and Cl^-, does not significantly alter the peak amplitude of the action potential. The importance of Ca^{2+} influx in action potential generation is also supported by the fact that 50 μM La^{3+}, a Ca^{2+} channel blocker, blocks action potential production.

The membrane mechanisms responsible for action potential production in *Stentor* have been studied using two microelectrode voltage-clamp methodology [34]. A 30 to 60 mV step depolarization above a holding potential of –50 mV elicits an early inward current followed by a prolonged outward current from the cell being studied in the presence of normal culture medium. Introduction of 50 μM La^{3+} into the test chamber eliminates all the inward current normally elicited by these voltage steps. Thus, the early inward current is carried by Ca^{2+} and the channels through which this Ca^{2+} flows are opened by a modest degree of membrane depolarization. On the other hand, membrane potential does not appear to control the closing of these Ca^{2+} channels. By using a current microelectrode filled with 2 M CsCl outward current is blocked, indicating it is carried by K^+. This procedure isolates the inward Ca^{2+} current. The Ca^{2+} current can then be seen to inactivate during a maintained depolarization, but the rate of this inactivation does not increase monotonically as the amplitude of the depolarizing voltage step increases. Instead, the rate of the inactivation first increases and then decreases with larger and larger voltage steps, a pattern similar to that of the Ca^{2+} current itself. This correlation suggests that the Ca^{2+} current is inactivated by the increase in intracellular Ca^{2+} concentration that the Ca^{2+} current itself produces [35]. The conclusion that Ca^{2+} current inactivation is Ca^{2+}-dependent is reinforced by the observation that inward Ba^{2+} currents, generated when Ba^{2+} is substituted in the extracellular medium for Ca^{2+}, do not inactivate. Similarly, the outward K^+ current that repolarizes the cell membrane during the later stages of the action potential is also Ca^{2+}-dependent, since it is not elicited when extracellular La^{3+} is combined with intracellular BAPTA to maintain intracellular Ca^{2+} concentrations at a very low level. Likewise, when Ba^{2+} is substituted for extracellular Ca^{2+}, no outward currents are produced in the range of physiological transmembrane potentials. Thus, the processes that terminate the action potential – Ca^{2+} channel inactivation and K^+ channel activation – are both Ca^{2+}-dependent.

We have recently generated a mathematical model incorporating the channel characteristics outlined above to describe the ionic currents generated during the voltage clamp experiments. In this model it is assumed that the relevant ion channels are found in the ciliary membrane and that Ca^{2+} accumulates in the intraciliary lumen to produce Ca^{2+} channel inactivation and K^+ channel activation. This model adequately describes the form and amplitude of action potentials observed during normal current clamp recordings (Wood, unpublished). Three features of this model and the data on which it is based are relevant to the mechanisms involved in photophobic responses. Firstly, the amount of Ca^{2+} that enters a cilium during an action potential is sufficient to raise the intraciliary Ca^{2+} concentration to more than 50 μM, whether this estimate is based on calculations used in the mathematical model, determination of the amount of Ca^{2+} required to charge the ciliary membrane to a 65–75 mV action potential peak, or integration of the currents observed during voltage clamp recordings. Thus, the Ca^{2+} entering the cilium during an action potential is more than sufficient to produce ciliary reversal. Since action potentials in swimming *Stentor* vary in amplitude it is to be expected that the amount of Ca^{2+} entering the cilium also varies and hence the strength and duration of the ciliary reversal would be expected to vary, as has been observed [36]. Secondly, the Ca^{2+} channels that produce action potentials and hence ciliary reversal are voltage-dependent. Therefore, the cellular mechanism that transduces the absorption of light into action potentials and ciliary reversals must operate by producing membrane depolarization. Thirdly, both Ca^{2+} channel inactivation and K^+ channel activation are Ca^{2+}-dependent; therefore, whenever the Ca^{2+} concentration in the intraciliary lumen is high, as is the case during ciliary reversal, K^+ channels will be open and Ca^{2+} channels closed. With the ion channels in this state, action potentials and hence photophobic responses cannot be produced. In accord with this expectation, Colombetti et al. [37] observed that *Stentor* do not produce photophobic responses during periods of ciliary reversal.

$$\text{Light} \rightarrow \text{photopigment absorption} \rightarrow \text{action potential} \rightarrow [Ca^{2+}]_0 \rightarrow \text{ciliary reversal}$$
$$\rightarrow \text{photophobic response}$$

18.4 The photic receptor potential

As deduced above, light must act upon *Stentor* and *Blepharisma* to produce a membrane depolarization in order to trigger an action potential and ciliary reversal. Such a light-induced membrane depolarization, termed a photic receptor potential, has been recorded from both *Stentor* and *Blepharisma* [3,10,11,28,29]. This depolarization has several rather unusual characteristics. The most notable of these is the slowness of its onset – typically photic receptor potentials do not begin until 100 to 500 ms after light onset. This contrasts strongly with the brief 0.5 ms delay seen between mechanical stimulation and mechanoreceptor potential initiation as observed in *Stentor* [38]. Secondly, photic receptor potentials once initiated rise rather slowly to their peak with maximum rates of rise on the order of 20 mV/s; in contrast, mechanoreceptor potentials and action potentials rise at rates 40 times faster. Further, photic receptor potentials frequently follow very irregular time courses exhibiting small "bumps" and large waves during extended periods of illumination (Figure 2). Finally, the amplitude of photic receptor potentials exhibits considerable adaptation during extended periods of illumination and recovery of responsiveness occurs only very slowly (e.g. 5 min) after even short periods (e.g. 5 s) of illumination [10,11].

Despite the slowness and irregularity of these photic receptor potentials their properties strongly suggest that they are instrumental in the production of action potentials and photophobic responses. The most convincing evidence supporting this conclusion derives from the observation that action potentials are always initiated at the peak of the receptor potential depolarization as is evident in Figure 2 in which an action potential is produced near the peak of each of 3 waves of receptor potential depolarization. Additionally, the action spectra of the photic receptor potentials of both *Stentor* and *Blepharisma* have peaks at the same wavelengths as the peaks in their photophobic response action spectra [10,11]. Finally, while the initiation of the photic receptor potential occurs with a relatively long latency after light onset, photophobic responses occur with even longer latencies. These correlations suggest the photic

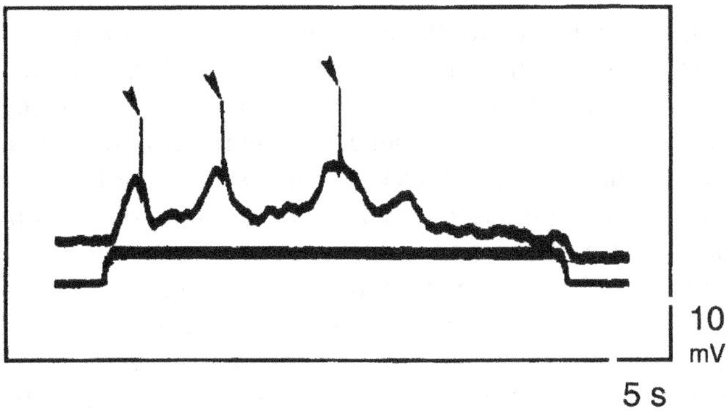

Figure 2. An example of a photic receptor potential recorded from an intracellular vacuole. The record shows waves of depolarization as are commonly seen during long periods of illumination. An action potential (arrowheads) was triggered by each of these waves. The bottom trace records the output of a photocell monitoring the stimulus light.

receptor potential is involved in the elicitation of action potentials and photophobic responses. Additional data suggest that the photic receptor potential is sufficient to elicit action potentials. For instance, photic receptor potentials that exceed 8 to 15 mV are sufficient to elicit action potentials [10], while depolarizations produced by outward current pulses have a similar threshold for eliciting action potentials (14.7 . . . 2.6 mV) [39]. Further, subthreshold photic receptor potentials can be summed with subthreshold mechanoreceptor potentials to elicit action potentials (Wood, unpublished results). However, mechanoreceptor potentials and photic receptor potentials are generated by entirely different cellular mechanisms and share only one common property – they both are depolarizing potentials. Therefore, the depolarization produced by the summed receptor potentials must be sufficient to elicit an action potential.

Light → photopigment absorption → receptor potential → action potential

→ $[Ca^{2+}]_i$ → ciliary reversal → photophobic response

18.5 The membrane mechanism producing photic receptor potentials

Only two basic mechanisms can be employed by cells to generate changes in transmembrane potential, such as receptor potentials and action potentials. The first and most common of these mechanisms requires a change in the membrane permeability of at least one ion species. This change alters the transmembrane flux of that ion producing a minuscule change in intracellular ionic concentration that, nevertheless, is sufficient to produce a substantial change in the transmembrane potential. Receptor potentials generated by this mechanism necessarily involve a change, generally an increase, in membrane conductance. The second mechanism capable of generating changes in transmembrane potential involves the pumping of specific ion species across the cell membrane, where the energy required to drive the relevant pump is derived from the metabolism of intracellular high energy compounds or from a transmembrane electrochemical gradient for some other ion or compound. Such a mechanism is termed an electrogenic pump and does not involve an appreciable change in membrane conductance. *Stentor* appears to employ an electrogenic pump for the generation of its photic receptor potential. The most direct evidence supporting this conclusion comes from comparison of the amplitude of the membrane hyperpolarizations produced by small inward current pulses before and during a photic receptor potential (Figure 3). The amplitude of the hyperpolarization produced divided by the current amplitude used in the test pulse gives the membrane resistance, the reciprocal of the membrane conductance. Five cells that produced receptor potentials exceeding 8 mV had the same membrane resistance before (15.6 ± 2.9 MΩ) and during (15.6 ± 2.8 MΩ) illumination (Wood, unpublished results). From a theoretical perspective (i.e. using either the Goldman-Hodgkin-Katz or parallel conductance equations), the membrane resistance during the period of illumination should have dropped to at least 0.95 of the resting membrane resistance to generate an 8 mV receptor potential if the receptor potential was generated by an increase the membrane permeability for Ca^{2+}. This calculated change in membrane resistance is significantly different from the observed change ($p < 0.05$; *t*-test) supporting the conclusion that the photic receptor potential is not produced by a change in membrane conductance. This conclusion is also suggested by the slow rate of rise of the photic receptor potential; the maximum observed rate of rise, 22 mV/s, could be produced across the average 8.8 nF membrane capacitance of *Stentor* by a sustained current of only 0.2 nA. While receptor currents of such small magnitude could be produced by either a conductance change mechanism or an electrogenic pump, they are uncharacteristic of receptor currents produced in large ciliates by stimuli activating a

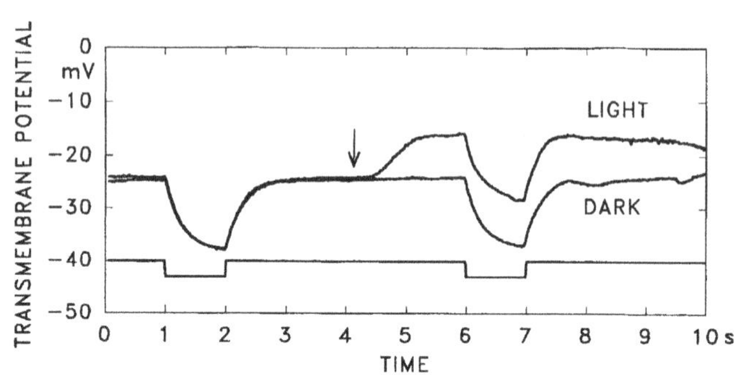

Figure 3. An example of the recordings used to assess whether the membrane resistance (conductance) changes during the production of a photic receptor potential. During each trace two 1 nA inward current pulses lasting 1 s were passed through one of two intracellular microelectrodes (bottom trace) recorded by the second microelectrode (top 2 records). The voltage deflection produced by these current pulses was used to determine the membrane resistance. During one of the voltage traces (marked DARK) the cell remained in the dark throughout the recording period. During a subsequent recording period (marked LIGHT) the cell was illuminated (at the arrow) with 14 W/m^2 of white light. The voltage deflection and hence the membrane resistance was as large during the receptor potential elicited by the light as before the cell was illuminated.

conductance change mechanism. For instance, mechanical stimuli elicit inward receptor currents in *Stentor* and *Paramecium* that exceed 10 nA. Further, photic receptor potential amplitudes are not influenced by changes in the steady state potential at which they are elicited. On the other hand, transient potentials generated by changes in membrane conductance in general and mechanoreceptor potentials in *Stentor* in particular are influenced by the transmembrane potential at which they are generated [28,30]. These lines of evidence indicate that photic receptor potentials are not generated by a conductance change mechanism but rather by an electrogenic pumping mechanism.

Unfortunately, there has not been a biochemical characterization of this electrogenic pumping mechanism. However, Fabczak et al. [29] have observed that the amplitude of the photic receptor potential increases as the extracellular concentration of Ca^{2+} increases suggesting that Ca^{2+} ions are being electrogenically pumped into the cell. It is also possible to speculate about the pump's cellular location. *Stentor* is about twice as sensitive to illumination of its anterior surface as to illumination of its posterior surface and in *Blepharisma* the difference is even more pronounced [6,15,40,41]. However, this difference in surface sensitivity cannot be due to localization of the receptor potential since any changes in transmembrane potential generated at one locus within the cell are also present at all other loci due to the long length constant of these cells. The distribution of pigment granules across the cell surface also appears to be fairly uniform. However, Matsuoka et al. [42] have observed that there are different forms of the photopigment in *Blepharisma,* some of which are unevenly distributed across the cell surface. But, this uneven distribution of photopigment forms does not provide a simple explanation for the differential surface photosensitivity, since the photosensitive anterior of *Blephasrisma* contains but one form of the photopigment and this form is also present in the insensitive posterior of the cell. Therefore, the authors are forced to hypothesize that the other photopigment forms actively inhibit the signal generated by the functionally active photopigment and thereby produce the observed differential photosensitivity. Another conceptually simpler explanation for the difference in surface photosensitivity is that the electrogenic pumping mechanism is localized only in the anterior portions of these cells.

$$\text{Light} \rightarrow \text{photopigment absorption} \rightarrow \text{electrogenic pump} \rightarrow \text{receptor potential}$$
$$\rightarrow \text{action potential} \rightarrow [Ca^{2+}]_i \rightarrow \text{ciliary reversal} \rightarrow \text{photophobic response}$$

18.6 The functional characteristics and significance of the photoreceptor system of *Stentor* and *Blepharisma*

Transmembrane potentials that arise with long latencies and have slow rise times are generally produced by cellular mechanisms that involve a sequence of chemical intermediates, as is the case for visual stimulus transduction in the vertebrate retina, for example. Generally, the sequence of chemical intermediates also produces a large amplification of the input signal. For example, a single photon of light absorbed by a rod in the retina produces a change in the transmembrane flux of millions of Na^+ ions. The long latencies and slow rise times which characterize *Stentor* and *Blepharisma* photic receptor potentials suggest they are also generated by a sequence of chemical intermediates. However, the available data suggest that these chemical intermediates need not amplify the signal. The fluence rate of 610 nm light required to elicit a photophobic response in 50% of an illuminated population of cells is reported to be 0.1 W/m^2 [5]. If this light impinges on only the frontal field of a 250 μm *Stentor* for 1 s to produce the photophobic response, then 3.2×10^{-9} joules of light energy are absorbed by the cell's pigment granules prior to the photophobic response assuming their absorbance is 0.458 as reported by Fabczak et al. [10]. However, a cell producing a 15 mV suprathreshold receptor potential and having a 17 MΩ membrane resistance need use only 13.2×10^{-12} joules to generate this potential. Thus, the light energy available to the cell is in excess of 200 times that needed to generate the photic receptor potential.

The relative insensitivity of *Stentor* and *Blepharisma* to light probably reflects the function that these photoresponses serve in their behavioral repertoire. Most photosensory systems both in invertebrates and vertebrates appear to have evolved for the purpose of detection – detection of potential prey, of predators, of possible mates, etc. To perform these functions effectively these photosensory systems need to be very sensitive. In contrast, the photoreceptive systems of *Stentor* and *Blepharisma* appear to have evolved to serve a protective function, since the feeding and reproductive behaviors of these organisms are not dependent on pigmentation. Indeed, one very successful member of the genus, *Stentor polymorphus*, is not pigmented. In the case of *Blepharisma* it has been proposed that the pigment acts as a screen to reduce the injurious effects of far UV radiation [2]. It is also possible that the pigment conveys a selective advantage to these ciliates because it becomes toxic to other organisms that have ingested it upon their exposure to light. Therefore, the pigment may serve as a deterrent to predation. Whatever the selective advantage conveyed by the pigment it comes with a price. Both *Stentor* and *Blepharisma* can be killed by the photodynamic action of their own pigmentation if they are continuously exposed to bright light for from minutes to hours [2,43]. Consequently, the photodispersal behavior exhibited by these organisms is adaptive because it removes them from brightly illuminated areas and hence from injury induced by their own pigment. Given that the function of the photoreceptor systems in *Stentor* and *Blepharisma* are primarily for protection and not for detection, it is evident that these systems need sense and react to light intensities that are only somewhat less intense than those that are injurious. Therefore, the photosensory mechanisms these organisms have evolved to serve this somewhat unique protective function may also be somewhat unique.

References

1. V. Tartar (1961). *The Biology of* Stentor. Pergamon, New York.
2. A.C. Giese (1973). Blepharisma, *The biology of a light-sensitive protozoan.* Stanford University Press, Stanford, CA, USA.
3. D.C. Wood (1976). Action spectrum and electrophysiological responses correlated with the photophobic response of *Stentor coeruleus. Photochem. Photobiol.*, **24**, 261–266.
4. P.-S. Song, D.-P. Häder, K.L. Poff (1980). Step-up photophobic response in the ciliate, *Stentor coeruleus. Arch. Microbiol.*, **126**, 181–186.

5. P.-S. Song, D.-P. Häder, K.L. Poff (1980). Phototactic orientation by the ciliate, *Stentor coeruleus*. *Photochem. Photobiol.*, **32**, 781–786.
6. M. Kraml, W. Marwan (1983). Photomovement responses of the heterotrichous ciliate *Blepharisma japonicum*. *Photochem. Photobiol.*, **37**, 313–319.
7. I.-K. Kim, R.K. Prusti, P.-S. Song, D.-P. Häder, M. Häder (1984). Phototaxis and photophobic responses in *Stentor coeruleus*. Action spectrum and role of Ca^{2+} fluxes. *Biochim. Biophys. Acta*, **799**, 298–304.
8. P. Scevoli, F. Bisis, G. Colombetti, F. Lenci, V. Passarelli (1987). Photomobile responses of *Blepharisma japonicum* I. Action spectrum determination and time-resolved fluorescence of photoreceptor pigments. *J. Photochem. Photobiol. B Biol.*, **1**, 75–84.
9. G. Checcucci, G. Damato, F. Ghetti, F. Lenci (1993). Action spectra of the photophobic response of blue and red forms of *Blepharisma japonicum*. *Photochem. Photobiol.*, **57**, 686–689.
10. S. Fabczak, H. Fabczak, N. Tao, P.-S. Song (1993). Photosensory transduction in ciliates.I. An analysis of light-induced electrical and motile responses in *Stentor coeruleus*. *Photochem. Photobiol.*, **57**, 696–701.
11. S. Fabczak, H. Fabczak, P.-S. Song (1993). Photosensory transduction in ciliates. III. The temporal relation between membrane potentials and photomobile responses in *Blepharisma japonicum*. *Photochem. Photobiol.*, **57**, 872–876.
12. T. Matsuoka, Y. Murakami, Y. Kato (1993). Isolation of Blepharismin-binding 200 kDa protein responsible for behavior in Blepharisma. *Photochem. Photobiol.*, **57**, 1042–1047.
13. M. Taffrau (1957). Les facteurs essentiales du phototropisme chez le Cilie heterotriche *Stentor niger*. *Bull. Soc. Zool. France*, **82**, 354–356.
14. R.K. Prusti, P.-S. Song, D.-P. Häder, M. Häder (1984). Caffeine-enhanced photomovement in the ciliate, *Stentor coeruleus*. *Photochem. Photobiol.*, **40**, 369–375.
15. T. Matsuoka (1983). Distribution of photoreceptors inducing ciliary reversal and swimming acceleration in *Blepharisma japonicum*. *J. Exp. Zool.*, **225**, 337–340.
16. K. Iwatsuki (1992). *Stentor coeruleus* shows positive photokinesis. *Photochem. Photobiol.*, **55**, 469–471.
17. E.M. Purcell (1977). Life at low Reynolds number. *Amer. J. Physics*, **45**, 3–11.
18. J.T. Randall, S.F. Jackson (1958). Fine structure and function in *Stentor polymorphus*. *J. Cell Biol.*, **4**, 807–830.
19. Y. Naitoh, H. Kaneko (1972). Reactivated triton-extracted models of *Paramecium*: Modification of ciliary movement by calcium ions. *Science*, **176**, 523–524.
20. Y. Nakaoka, H. Tanaka, F. Oosawa (1984). Ca^{2+}-dependent regulation of beat frequency of cilia in *Paramecium*. *J. Cell Sci.*, **65**, 223–231.
21. T. Matsuoka, Y. Watanabe, T. Kuriu, T. Arita, K. Taneda, M. Ishida, T. Suzuki, Y. Shigenaka (1991). Cell models of *Blepharisma*: Ca^{2+}-dependent modification of ciliary movement and cell elongation. *Europ. J. Protistol.*, **27**, 371–374.
22. R. Eckert, Y. Naitoh (1970). Passive electrical properties of *Paramecium* and problems of ciliary coordination. *J. Gen. Physiol.*, **55**, 467–483.
23. R. Eckert (1972). Bioelectric control of ciliary activity. *Science*, **176**, 473–481.
24. H. Machemer, R. Eckert (1973). Electrophysiological control of reversed ciliary beating in *Paramecium*. *J. Gen. Physiol.*, **61**, 293–316.
25. H. Machemer (1974). Frequency and directional responses of cilia to membrane potential changes in *Paramecium*. *J. Comp. Physiol.*, **92**, 293–316.
26. V. Chen (1972). *The electrophysiology of* Stentor polymorphus*: An approach to the study of behavior.* Ph.D. thesis, SUNY at Buffalo.
27. D. Mergenheimer (1971). Membrane potentials in *Stentor coeruleus*. *Protoplasma*, **72**, 359–365.

28. D.C. Wood (1973). Stimulus specific habituation in a protozoan. *Physiol. Behav.*, **11**, 349–354.
29. S. Fabczak, H. Fabczak, M. Walerczyk, J. Sikora, B. Groszynska, P.-S. Song (1996). Ionic mechanisms controlling photophobic responses in the ciliate *Blepharisma japonicum*. *Acta Protozool.*, **35**, 245–249.
30. D.C. Wood (1982). Membrane permeabilities determining resting, action and mechanoreceptor potentials in *Stentor coeruleus*. *J. Comp. Physiol.*, **146**, 537–550.
31. A. Ogura, K. Takahashi (1976). Artificial deciliation causes loss of calcium-dependent responses in *Paramecium*. *Nature*, **264**, 170–172.
32. K. Dunlap (1977). Localization of calcium channels in *Paramecium caudatum*. *J. Physiol. (Lond.)*, **271**, 119–133.
33. H. Machemer, A. Ogura (1979). Ionic conductances of membranes in ciliated and deciliated *Paramecium*. *J. Physiol. (Lond.)*, **296**, 49–60.
34. D.C. Wood (1995). Action potentials generated by I_{Ca} and $I_{K(Ca)}$ in a primitive ciliate. *Abstr. Soc. Neurosci.*, **21**, 65.
35. P. Brehm, R. Eckert (1978). Calcium entry leads to inactivation of calcium channel in *Paramecium*. *Science*, **202**, 1203–1205.
36. M. Doughty (1990). A kinetic analysis of a step-up photosensory response of the ciliate *Stentor coeruleus*. *Can. J. Microbiol.*, **36**, 414–418.
37. G. Colombetti, F. Lenci, P.-S. Song (1982). Effects of K^+ and Ca^{2+} ions on motility and photosensory responses of *Stentor coeruleus*. *Photochem. Photobiol.*, **36**, 609–611.
38. D.C. Wood, L. Kilderry (1997). The actin cytoskeleton is involved in mechanical stimulus transduction in the ciliate, *Stentor*. *Abstr. Soc. Neurosci.*, **23**, 480.
39. D.C. Wood (1991). Electrphysiology and photomovement of *Stentor*. In: F. Lenci, G. Colombetti, D.-P. Häder, P.-S. Song (Eds), *Biophysics of Photoreceptors and Photomovements in Microorganisms* (pp. 281–291). Plenum, London.
40. D.C. Wood, R. Marinelli (1991). Doublet *Stentor* do not display photodispersal. *Photochem. Photobiol.*, **53**, 353–357.
41. C. Kühnel-Kratz, D.-P. Häder (1993). Real time three-dimensional tracking of ciliates. *J. Photochem. Photobiol. B: Biol.*, **19**, 193–200.
42. T. Matsuoka, M. Sato, M. Maeda, H. Naoki, T. Tanaka, H. Kotsuki (1997). Localization of blepharismin photosensors and identification of a photoreceptor complex mediating the step-up photophobic response of the unicellular organism, *Blepharisma*. *Photochem. Photobiol.*, **65**, 915–921.
43. K.-C. Yang, R.K. Prusti, E.B. Walker, P.-S. Song, M. Watanabe, M. Furuya (1986). Photodynamic action in *Stentor coeruleus* sensitized by endogenous pigment stentorin. *Photochem. Photobiol.*, **43**, 305–310.

Photomovement
D.-P. Häder and M. Lebert, editors.

Chapter 19

Genetic analysis of phototaxis in *Dictyostelium*

Paul R. Fisher

Table of contents

Abstract

Dictyostelium discoideum slugs are doubtless the simplest multicellular eukaryotes in which classical and molecular genetic approaches to the study of the signal transduction pathways controlling photomovement are possible. The slugs are formed by chemotactic aggregation of starving amoebae, so that it is a simple matter to generate, from clonally-derived mutant cell lines, hundreds of genetically identical multicellular individuals for study. Adding to the already extensive knowledge of the unicellular stages of the *D. discoideum* life cycle, classical and molecular genetics have begun to unravel transduction of signals controlling phototaxis (and thermotaxis) in the slugs. Distributed over all seven genetic linkage groups are at least 20, but possibly as many as 55 genes of importance for slug behavior. The encoded proteins appear from pharmacological studies and mutant phenotypes to govern transduction pathways involving extracellular cAMP signals and corresponding receptors, heterotrimeric and small GTP-binding proteins, the intracellular second messengers cyclic AMP, cyclic GMP, Ca^{2+} and IP_3 as well as cytoskeletal proteins such as Actin Binding Protein-120 and myosin II. Pathways from the photo- and thermoreceptors converge first with each other and thence with those from extracellular tip activation (cyclic AMP) and inhibition (Slug Turning Factor and/or ammonia and/or adenosine) signals that control slug movement and morphogenesis.

19.1 Introduction

The first cellular slime mould, *Dictyostelium mucoroides*, was discovered by Brefeld [1], but was initially misclassified with the acellular slime moulds in the belief that the multicellular stage was a plasmodium – a unicellular, multinucleate syncytium arising from cell fusion during the aggregation of starving amoebae. Only in 1880 was it realized that these organisms exhibit an asexual life cycle that uniquely involves the formation by aggregation of a truly multicellular organism (Figure 1), renamed a pseudoplasmodium as a result of its originally mistaken identity [2]. The species *Dictyostelium discoideum* was first reported in 1935 by Raper [3] and has since become the best known of the cellular slime moulds because of its popularity as a model organism for studies of eukaryotic cell and molecular biology.

As described by Raper [4], the *D. discoideum* pseudoplasmodium is formed by a combination of starvation-induced cell differentiation and complicated morphogenetic movements, beginning with chemotactic[1] aggregation to form a mound of between 10^4 and 10^5 cells (Figure 1). This is followed by tip formation and upwards extension of the

[1] That aggregation occurs by chemotaxis was demonstrated by beautiful experiments in which aggregation centers on one side of a dialysis membrane were shown to attract aggregation competent amoebae migrating on the opposite side of the membrane [11] and cells downstream of an aggregation center in a slow current migrated towards the center while upstream cells did not [12]. The attractant was identified as cAMP twenty years later [13,14].

aggregate to form a standing "finger" that falls over and crawls away – the pseudoplasmodium. It migrates at average speeds from 0.2 to 2.0 mm/h through a slime sheath composed of cellulose and protein which it synthesizes, secretes and leaves behind in collapsed form as a trail. This behavior in combination with its elongated shape, lends the organism a superficial resemblance to a garden slug (Figure 2) and has resulted in the modern preference for the term "slug" when referring to it. The slug stage in the life cycle, interposed as it is between aggregation and culmination, may last from less than an hour to nearly two weeks depending on the strain and on conditions such

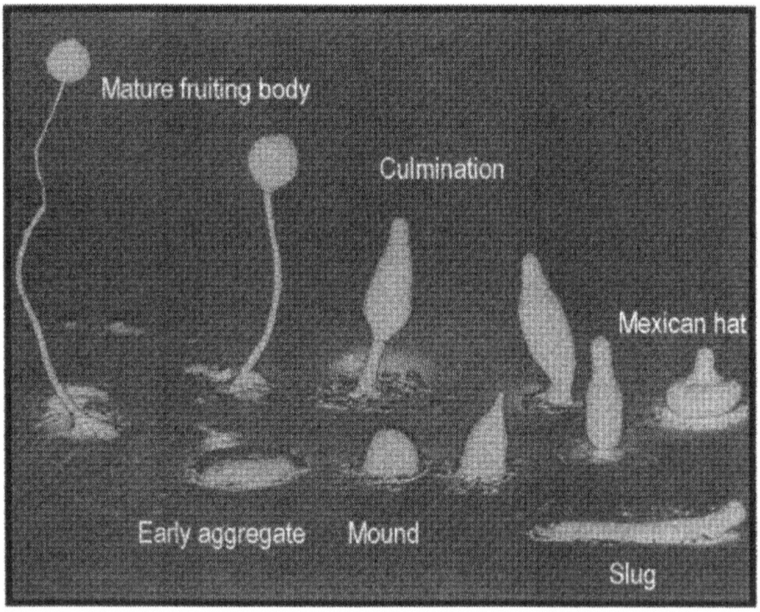

Figure 1. Stages in the life cycle of *Dictyostelium discoideum*. A scanning electron micrograph montage of various stages in the life cycle is shown. The original grey scale image was kindly made available by M.J. Grimson and R.L. Blanton, Dept. Biological Sciences, Texas Tech University. Amoebae multiply by growth on a bacterial food source (ingested by phagocytosis) accompanied by mitotic division. When the food source is exhausted, starvation induces differentiation to an aggregation competent form – amoebae that synthesize, secrete and are attracted by cAMP (other slime mould species use different attractants). After chemotactic aggregation during which amoebae are recruited into the aggregate in long streams of cells, a tip forms on the aggregation mound and is extended upwards to form a finger-like structure that falls onto its side to form the migrating slug. After a variable migration period the tip stops and the remainder of the slug is drawn up beneath it to begin culmination. Anterior cells funnel downwards towards the substratum and differentiate to form the stalk which, as it forms, is climbed by the other cells to finally form a droplet of spores atop the stalk. Each spore is derived from a single amoeba by differentiation and, under suitable conditions such as the presence of a bacterial food supply, will germinate to yield a single vegetative amoeba. Excluding the period of migration the whole cycle takes about 24 hours – roughly 8 hours to aggregation, 16 hours to slug formation, 24 hours to completion of culmination.

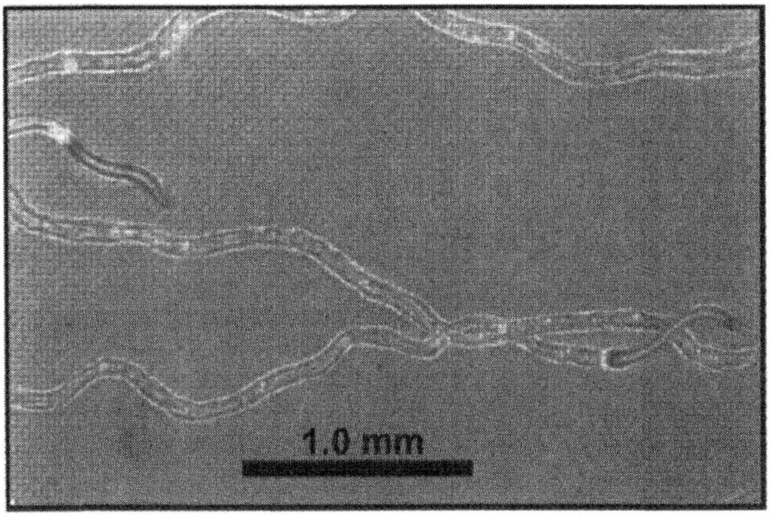

Figure 2. *Dictyostelium discoideum* slugs migrating towards a light source. The figure (from Figure 2 of [24]) shows trails of four slugs migrating on water agar towards a light source that was to the right of the image. Two slugs have already left the field of view, one has just entered on the left and one is about to leave on the right. Each slug is about 1 mm long and migrates at about 1 mm/h through an extracellular matrix of cellulose and protein which forms a so-called slime sheath and which collapses behind it to leave a trail. The image was taken by low magnification phase contrast microscopy.

as temperature, humidity, pH and ionic strength. At some variable point in time, apparently under the control of the slug tip, migration ceases and a series of complex morphogenetic movements leads to fruiting body formation.

To a first approximation the fruiting body consists of a droplet of spores atop a conical stalk mounted on a basal disc. The stalk cells are derived almost entirely from the anterior 20% or so of the slug amoebae, the spores from the bulk of the remainder and the basal disc from cells (1–2%) collected at the rear of the slug. The slug thus contains three major cell types (the rearguard cells, the prespore cells and the anterior, prestalk cells) that are spatially separated in the slug and predestined to form the three major morphological features of the fruiting body. This much has been known since the elegant experiments of Raper and Bonner who first demonstrated it using tissue transplants on slugs made from amoebae that had been color-tagged with vital stains [5,6] or by feeding them differently pigmented bacteria [4]. However, more recent analysis of the patterns of gene expression and histological staining in slugs and culminants has revealed greater underlying complexity – several subclasses of prestalk cells, each with its own distinct spatial distribution and morphogenetic fate map, some with representatives scattered through the prespore region – the so-called anterior-like cells (for a recent review, see [7]).

The study of photosensory (and thermosensory) responses in *Dictyostelium discoideum* began with the reports by Raper [4] and Bonner [8] that the slugs migrate with great precision both towards a lateral light source (Figure 2) (white light intensities as low as 1 $\mu W/m^2$ [9]) and towards the warmth in astonishingly shallow temperature gradients (as little as 0.04°C/cm [10]). In the soil environment these behaviors would take the slug to the surface where the drop in humidity and the overhead light would induce culmination, i.e. fruiting body formation [4,15,16]. Many years were to elapse before Häder and colleagues found that the individual postvegetative amoebae also are phototactic [17,18] and thermotactic [19]. Even now these responses remain little studied, despite recent isolation of the putative amoebal photoreceptor [20–22].

Because slugs are visible to the naked eye and leave slime trails, their phototactic (and thermotactic) behavior (recently reviewed by [23] and [24]) is much more experimentally tractable than that of the individual amoebae and accordingly has been the subject of more intensive research using a combination of behavioural, pharmacological and genetic approaches. In the remainder of this article I describe first the key features of slug turning behavior that have been discovered so far and then focus specifically on the pharmacological and genetic analysis of phototaxis.

19.2 Key features of slug turning behavior

Most of the key features of signal transduction controlling slug phototaxis and thermotaxis, as illustrated in Figure 3, were proposed more than a decade ago [25] and may be summarized as follows:

- *Dictyostelium* slugs sense light and temperature gradients separately.
- Photosensory and thermosensory transduction pathways converge early, upstream of most signal transduction components which accordingly function to regulate both phototaxis and thermotaxis.
- Sign reversals in slug turning responses occur in both phototaxis and thermotaxis.
- Slug behavior is controlled by the slug tip.
- Slug turning is mediated by transient, lateral shifts in slug tip position as determined by a temporary, local imbalance between the coupled processes of tip autoactivation and autoinhibition. Light intensity and temperature gradients across the slug tip cause turning responses by altering the balance between tip activation and inhibition, for example by stimulating differential production of tip inhibitor. This modulation of the tip activation/inhibition system occurs because, after converging, signals from the photoreceptor and thermoreceptor interact with the same central components as the tip activation and inhibition signals. Sign reversals can be economically explained because the turning behavior of the slug will depend upon whether tip activation or inhibition dominates the response to light or temperature gradients.
- Tip activation signals are carried by cAMP waves, now known to be scroll-shaped, while tip inhibition signals may be borne by one or more of the molecules NH_3, adenosine and the originally proposed Slug Turning Factor (STF). The inhibition signals may act primarily by influencing the pacemaker frequency associated with the cAMP waves.
- The central signal transduction components include second messengers (cAMP, inositol polyphosphates and the originally proposed cGMP and Ca^{2+}) along with

associated proteins, as well as signalling proteins (including heterotrimeric and small GTP-binding proteins) and cytoskeletal proteins (including ABP120). Recent work has shown that the mitochondria may also be involved. The grouping of these various components into a single central box in Figure 3 reflects the fact that it has not yet been determined exactly how they interact with one another. For some of these the evidence that they play a role is pharmacological (inositol polyphosphates and Ca^{2+}), but for most there is genetic evidence (i.e. based on mutant phenotypes) that they are essential for normal slug behavior. For this reason the number and identities of the

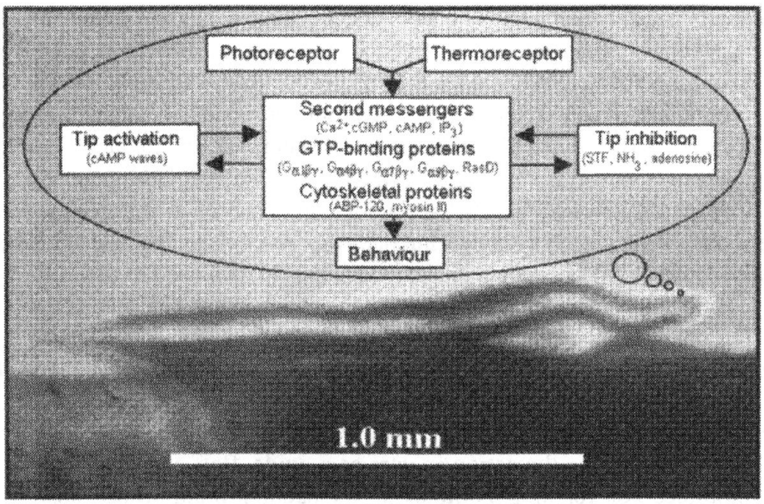

Figure 3. Simple model for *Dictyostelium* slug phototransduction and thermotransduction pathways. A lateral view of a slug migrating on a water agar surface is shown. The "thought bubbles" emanating from the tip indicate that the tip controls slug behavior via the indicated pathways. Signals from photoreceptor and thermoreceptor converge early and thence control the concentrations of the intracellular second messengers cAMP, cGMP, Ca^{2+} and possibly IP_3. The evidence for inositol polyphosphate (IP3) involvement is the pharmacological effect of Li^+ whose target could also (or instead) be glycogen synthase kinase 3 (GSK3). Heterotrimeric G proteins and the small GTP-binding protein RasD are involved in transducing the signals. These in turn modulate the tip activation and inhibition signals that determine the position of the slug tip. Transient lateral imbalances between tip activation and inhibition induced by this means by light and temperature gradients cause temporary lateral shifts in tip position and thence slug turning because the slug "follows its nose". Depending upon whether tip activation or inhibition dominates the response the slug turns either towards or away from the light source, or up or down the temperature gradient. Sign reversals in slug turning responses result from switches in the balance between control by tip activation and inhibition. This explains direction-dependent sign reversals in phototaxis that cause bidirectional phototaxis (see text) and temperature-dependent sign reversals in thermotaxis that cause movement towards the warmth or the cold depending on the temperature. Tip activation signals are believed to be carried by 3-dimensional spiral scroll waves of extracellular cAMP analogous to the 2-dimensional cAMP waves that mediate aggregation (see Figure 5). Candidate tip inhibitors are STF (Slug Turning Factor), ammonia and adenosine.

central signal transduction components are discussed separately in the section on genetic analysis.

I now examine each of these key features in more detail.

19.2.1 Dictyostelium *slugs sense light and temperature gradients separately*

In view of the extreme sensitivity of slugs to thermal gradients, one possible explanation of slug phototaxis was that it resulted from thermotactic responses to local warming. Slug phototaxis should therefore have been most sensitive to far red and infrared light where local warming would be most effective. However, in 1964 Francis [26] reported an action spectrum for slug phototaxis that disproved this – slug phototaxis was most sensitive to blue light (around 430 nm) and exhibited a second broad peak of sensitivity to longer wavelengths (around 550–600 nm). The spectrum was elegantly confirmed by Poff et al. [27] by measuring the angle of deviation of the migrating slug paths from a line bisecting the directions towards two lateral light sources of different wavelengths. Since then, photobiologically rigorous fluence-response curves for slug phototaxis have been obtained with light of many different wavelengths and used to determine the slug phototaxis action spectrum more accurately [9]. As a result the blue light peak was resolved into two peaks at 420 nm and 440 nm, while the broad peak at longer wavelengths was resolved into peaks at 560 nm and 610 nm. The fluence response curves also confirmed that phototaxis occurred at light intensities so low that local warming would be insufficient to create thermal gradients detectable even by *Dictyostelium* slugs. Using the action spectra as a guide, Poff and colleagues isolated a putative photoreceptor for slug phototaxis – a haem protein whose absorption and photooxidation action spectra were similar to the slug phototaxis action spectrum [27–29].

Further evidence that phototaxis does not result from thermotactic turns towards the slug's warmer side came from demonstration that phototaxis relies on a lens effect – light refracted at the roughly cylindrical surface of the slug is focused onto its distal side, so that when a slug turns towards a lateral light source, it is turning away from its most intensely illuminated side! The evidence that this is so came from three kinds of experiments, again beginning with the pioneering work of Francis [26]:

1. When vertical beams of light were used to illuminate one side of the slug tip, the slug turned away from the illuminated side [26,30,31].
2. Because of the high refractive index of mineral oil, light rays refracted at the surface of a slug immersed in it would diverge rather converge. Slugs treated in this manner exhibited negative phototaxis, turning away from instead of towards the light source [32].
3. When slugs were stained heavily with neutral red, so that the light was absorbed on their proximal sides rather than focused onto their distal sides, they migrated away from the light source rather than towards it [31,33]. In the UV range, slugs are negatively phototactic because strong absorption of the light defeats the lens effect [34].

Separate photosensing and thermosensing by *Dictyostelium* slugs was verified genetically by the isolation of mutants defective in phototaxis but unaffected in positive thermotaxis [10] and vice versa [35].

19.2.2 Photosensory and thermosensory transduction pathways converge early

Although light and warmth are sensed separately, the signal transduction pathways converge early. This conclusion was originally based on the fact that metabolites (Slug Turning Factor, STF) secreted by the amoebae at high cell density [36], several pharmacological agents [37] and independent mutations [38] that impair phototaxis also perturb positive thermotaxis and vice versa [39]. It has since been verified manifold by the effects of additional pharmacological agents on both phototaxis and thermotaxis [40,41], by detailed genetic and phenotypic analysis of a large collection of mutants [35,42] and by the discovery of cGMP responses to both light and heat stimuli that are altered in many mutants [43]. Classical genetics has identified on the seven *Dictyostelium* genetic linkage groups about a dozen genes (from an estimated total of around 20) that, when mutant, cause deficient slug phototaxis [33,35,42,44]. All but one [10] of the reported phototaxis mutations affect thermotaxis as well and the genetic analysis confirmed that both phenotypes almost always map to the same linkage group. In only a single mutant (from more than 20 that have been genetically analysed) is there evidence that the two phenotypes may have arisen from two independent mutations in different genetic loci. In three cases (*phoE* on linkage group III,[2] *phoJ* on linkage group IV†, *phoK* on linkage group V†) it has been verified by complementation tests involving independent mutant alleles of the same locus that the thermotaxis and phototaxis phenotypes are caused by the same mutation [35,42].

19.2.3 Sign reversals in slug turning responses occur in both phototaxis and thermotaxis

Until the early 1980s orientation by slugs either in response to a lateral light source or to temperature gradients was believed simply to consist of positive taxis. Then in 1980 Whitaker and Poff [45] reported the unexpected finding that at temperatures several degrees below the growth temperature slugs migrate towards the cold (negative thermotaxis). The temperature at which the transition from positive to negative thermotaxis occurred was dependent on the temperature to which the amoebae had been exposed during growth. Fisher and Williams [38] confirmed this behavior and found as well that at temperatures above the growth temperature the accuracy of thermotaxis similarly declined until, in mutants and also in wildtype slugs under certain conditions [37], a transition to negative thermotaxis occurred.

---

[2] The present physical map of the *D. discoideum* genome includes 6 chromosomes. Except for linkage group VII (chromosome 5) and linkage group V (not assigned to a physical chromosome) there is a one-to-one correspondence between the numbering of the genetic linkage groups (Roman numerals) and the physical chromosomes (Arabic numerals) [47,48]).

Although slugs migrate away from ultraviolet light [34], negative phototaxis by slugs in response to light in the visible range has not been reported except in the special case where the slugs have been heavily stained (see earlier). However, Fisher and Williams [46] found that slug phototaxis is actually bidirectional in that slugs migrate not directly towards the light, but at an angle ($\pm\alpha$) either side of the direction towards the light source (Figure 4). This happens because slugs whose current direction of travel is at an angle less than $\pm\alpha$ subsequently turn away from the light, while those whose current direction of travel is at an angle greater than $\pm\alpha$ turn towards the light. The angle at which the transition occurs becomes the preferred direction of travel ($\pm\alpha$). Thus slug turning behavior in both thermotaxis and phototaxis is characterized by sign reversals that depend on temperature in thermotaxis, and on the direction of migration in phototaxis [46].

In wildtype slugs under most conditions α is sufficiently small that bidirectional phototaxis is indistinguishable from unidirectional phototaxis towards the light. However, as illustrated in Figure 4, it is more evident (because α is greater) in mutants [35,43,46], in the presence of fluoride ions [37], EGTA (a Ca^{2+} chelator) [37] or metabolites (Slug Turning Factor – see below) excreted during slug formation at high cell densities [35,43,46]. In general these same factors caused shifts towards the growth temperature of the transition temperatures at which sign reversals occur in thermotaxis, the exception being fluoride ions which shifted transition temperatures away from the

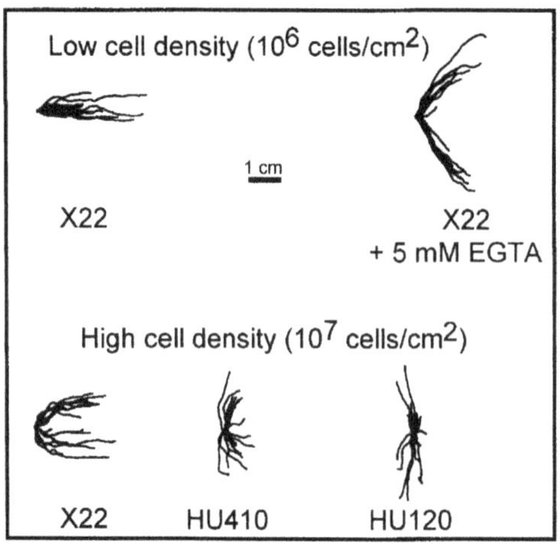

Figure 4. Bidirectional phototaxis by *Dictyostelium discoideum* slugs. Modified from Figure 2 of [25]. The effects on bidirectional phototaxis on water agar of mutations (wildtype X22 *versus* mutants HU410 and HU120) and environmental conditions (X22 at low density *versus* high density or with EGTA). Digitized trails were plotted so that the light source was to the right of the figure. In each case slug trails are plotted from a common origin although in reality slugs were formed and migrated from centres scattered over the 1 cm^2 area onto which starving cells were spread initially at the indicated cell densities. In wildtype X22 slugs at low density without EGTA bidirectional phototaxis is not observable because the two preferred directions ($\pm\alpha$) are too close together. Bidirectional phototaxis is evident in the mutants, in the presence of EGTA and at high cell densities (close to the origin) because α is greater under these conditions.

growth temperature. Any model of photosensory and thermosensory signal transduction in slugs must include an explanation of these sign reversals.

19.2.4 Slug behavior is controlled by the slug tip

That the slug tip controls slug behavior and polarity was first shown by Raper nearly 60 years ago [4]. Thus removal of the tip or of small amounts of tissue from it caused migration to cease until an intact tip was reestablished, whereas an excised tip was able to continue migration on its own in an unimpeded fashion. Control by the slug tip was confirmed specifically for phototaxis by demonstration that slugs respond to vertical microbeams of light only if they are shone on the tip [30,31]. Tip transplants between slugs of a wildtype strain and a phototaxis mutant showed that phototaxis was wildtype if the tip was wildtype and vice versa [25].

Since almost all reported mutations causing impaired phototaxis also result in thermotaxis deficiencies [25,35,38,42], it had always been assumed that thermotaxis, like phototaxis and morphogenesis in general, would also be tip-controlled. However Hashimoto and Matsui [49] reported a curious result from tip transplants between slugs formed from amoebae grown at 19°C and 25°C. At 20°C the former slugs ("+" slugs) were positively thermotactic, while the latter (" – " slugs) were negatively thermotactic as expected from the original report of negative thermotaxis and thermal adaptation by Whitaker and Poff [45]. Contrary to expectation, in heterologous transplants the presence of either a tip or a tail from a " – " slug resulted in negative thermotaxis. One possible explanation is that in the heterologous transplants, cells from the tails of " – " slugs take over the tip and thereby control the behavior. Another possibility is that the tip obtains its information about the growth temperature from a global signal emanating from the " – " cells irrespective of their location in the slug. The results of Hashimoto and Matsui [49], while unexpected, are thus not incompatible with tip control of thermotactic behavior.

19.2.5 Slug turning is mediated by modulation of the coupled processes of tip autoactivation and autoinhibition

The idea that slug turning might be mediated by coupled tip autoactivation and autoinhibition was first proposed by Hans Meinhardt [50] who inferred it from the spontaneous formation of tips in *Dictyostelium* aggregates and the subsequent generation of slugs with clear tip/body polarity [4][3] Matching the observed morphological polarity of slugs in the form of the tip/body pattern is a very pronounced behavioral polarity – the slug follows where the tip leads and in the absence of external directional stimuli tends to persist migrating in the original, randomly chosen direction [51]. Linked autocatalytic and inhibitory pathways are a general and necessary feature

[3] It should be noted that the polarity spoken of here is influenced by but is not equivalent to the spatial patterns of cell types controlled by cell cycle position at the onset of starvation, cell sorting and morphogens such as DIF (reviewed by [56]).

of physical, chemical and biological systems that spontaneously form such patterns from macroscopically homogeneous starting conditions. In *Dictyostelium* slugs, direct experimental evidence for tip to tail gradients of both tip inhibition and activation was obtained from experiments in which secondary tip formation was scored after transplantation of tissue to and from different points along the slug length [52–54]. Kopachik [55] verified the secretion by aggregates of a tip inhibitor and showed that it is a small molecule able to diffuse through a dialysis membrane. While tip inhibition signals may be carried by global concentration gradients of tip inhibitor, the tip activation signal is almost certainly encoded in spatiotemporal cAMP waves (see below).

Slug turning behavior during phototaxis and thermotaxis can be explained if light or temperature gradients induce lateral differences in tip activation or inhibition signals. According to this view, slug turning is a consequence of small lateral shifts in tip position that are induced during photosensory or thermosensory transduction by modulation of the tip activation/inhibition system. As explained in detail by Fisher et al. [25], one prediction of this activation/inhibition model for slug polarity and behavior is that slugs would turn towards or away from light or heat, depending on whether tip activation or inhibition dominated the response. In thermotaxis this would depend upon the temperature so that close to the growth temperature tip activation on the warm side of the slug would cause positive thermotaxis. At temperatures where negative thermotaxis occurs, tip inhibition on the warm side would result in slugs turning towards the cold. In phototaxis, the lens effect focuses light onto the distal side of the slug tip and the focusing advantage increases with the angle of deviation from the direction towards the light. At large angles, tip inhibition on the distal side would result in turns towards the light, but at low angles with small focusing advantage tip activation on the distal side would cause slugs to turn away from the light. The activation/inhibition model thus provides an economical explanation for the observed sign reversals in slug turning behavior during phototaxis and thermotaxis.

An alternative model for slug turning based on differential speeds has also been proposed. The idea is that turning results from a tractor mechanism in which cells on one side of the slug move forward faster than cells on the other side. Provided that the cells retain their relative positions this would result in a turn. The original evidence for this mechanism came from experiments suggesting, contrary to earlier work [4,26,32], that the average speed of migrating slugs was greater in the light than in the dark [30]. However, more detailed quantitative studies showed that average slug speed is in fact unaffected by light, by temperature gradients or by STF (Slug Turning Factor), a slug repellent and candidate tip inhibitor [36,57]. More recently it has been reported that slugs (and aggregating cells) migrate faster in the presence of optimal concentrations of NH_3, also a slug repellent [58–60] and candidate tip inhibitor controlling slug orientation [61,62]. Miura and Siegert (unpublished data) recently revisited the question. In some very elegant time-lapse filming experiments they found (unexpectedly) that slugs slow down in the presence of ammonia. Be that as it may, the effects of ammonia on slug speed can provide only indirect evidence for a differential speeds mechanism for slug turning in phototaxis. Miura and Siegert therefore filmed slugs during phototaxis. They found that slugs turned towards the light within a few minutes and later, after a delay of at least 20 minutes, accelerated (first the tip, later the rear) and elongated.[4] These results refute a differential speeds hypothesis for turning, since the acceleration occurred only after the turn.

Additional difficulties for the tractor mechanism of slug turning are posed by the fact that in the slug tip where turns are initiated, the cells rotate around the slug axis at right angles to the direction of slug movement (Figure 5) and they can and do change their relative positions [63]. Takahashi et al. [64] reported time lapse video evidence that during slug turns the tip cells change their relative positions in a manner consistent with the tip-shifting and not the differential speeds hypothesis. Miura and Siegert (unpublished data) found changes in the motion and relative positions of tip cells during

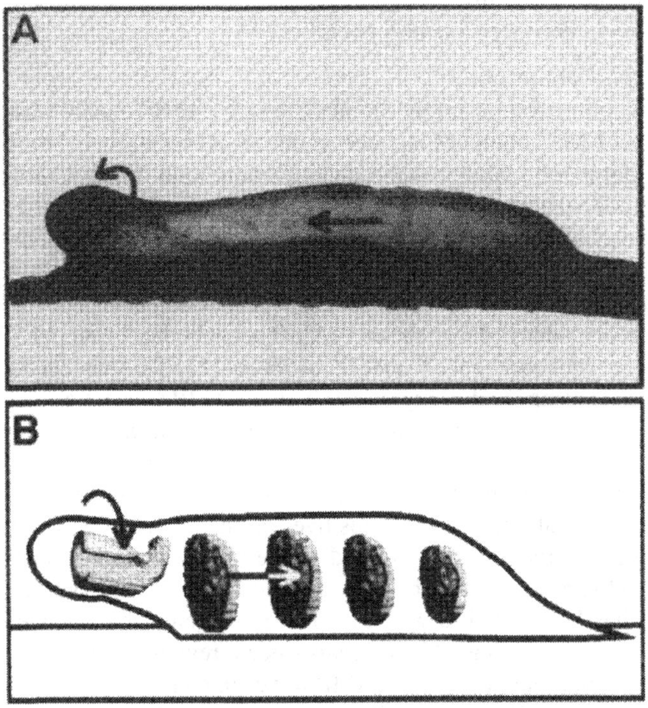

Figure 5. Cell movement and signal propagation in *Dictyostelium discoideum slugs.* Modified from panels A and B of Plate 8 from [93]. Original color image kindly provided by F. Siegert, Ludwig Maximilians University, Munich. **A.** Side view of a neutral red stained slug with arrows indicating the overall direction of cell movement as observed by tracking neutral red stained vesicles in the cells. **B.** A model showing the mode of signal propagation that explains the observed pattern of cell movement. Shown are the scroll wave in the tip rotating counter to the direction of movement of the cells, the twisting of the scroll wave at the rear of the tip region and its conversion into a train of planar waves propagating rearwards through the rest of the slug.

[4] Because it was accompanied by an increase in slug length, this acceleration during phototaxis may have been obscured in earlier studies by the corrections that were made for slug size. These corrections were necessary because larger, longer slugs migrate faster.

phototactic turns that were suggestive of changes in the geometry of the tip activation signal (twisting of the scroll wave) and consistent with the tip-shifting hypothesis. Recently Bonner [65] found a way to make flat (one cell thick) slugs at a glass-mineral oil interface thus allowing him to observe the movement of the individual cells in the slug. Consistent with the tip-shifting hypothesis, he found that turns are not associated with speed differences between amoebae on the two sides of the slug, but with lateral shifting of the "high point" around and towards which the active motility by the anterior cells is directed. The "high point" was recognizable as the region of highest cell motility in these two-dimensional slugs.

19.2.6 Tip activation signals are carried by cAMP waves, while the tip inhibitor may be NH_3, adenosine and/or Slug Turning Factor (STF)

19.2.6.1 The tip activation signal

The idea that tip autoactivation might be mediated by cAMP signals was always very attractive because cAMP stimulates its own synthesis and secretion in aggregating cells. By 1983 when Meinhardt elaborated the tip activation/inhibition model for *Dictyostelium* morphogenesis, there was already a great deal of evidence that cAMP waves analogous to those seen during aggregation might carry the tip activation signal controlling slug polarity and behavior [50]. Thus it was known that slug tips secrete cAMP [66,67] and by that means can attract aggregation competent cells [68], that slugs have cAMP receptors, adenylyl cyclase and cAMP phosphodiesterase [69,70], that there is an overall tip to tail gradient of cAMP in slugs [71,72], that anterior cells sort chemotactically to slug tips and to regions of higher cAMP concentration [73–76], that files of slug cells move in a pulsatile fashion in three dimensional spirals analogous to the two dimensional spiral waves seen in aggregation [77–79], that high cAMP concentrations cause slugs to lose polarity and disintegrate into smaller aggregates [80,81] and that a mutant, temperature sensitive for chemotactic aggregation, was also impaired in slug tip formation when shifted to the restrictive temperature [82]. Since then these early observations have been confirmed and extended, while much of the molecular detail has been filled in with the cloning of genes that encode many of the proteins involved in transducing cAMP signals (recent reviewers include [83–89]).

Recent studies of the role of putative cAMP waves in controlling slug morphogenesis were those of Siegert and Weijer and colleagues who showed that rotating waves of cell movement occur both in mounds during tip formation and subsequently in the tips of migrating slugs [63,90]. As illustrated in Figure 5, cells in the slug anterior rotated around the tip axis as if responding to a three-dimensional scroll wave, while cells in the rear (prespore) region moved forward in the direction of slug migration in a periodic fashion as if responding to planar, rearward propagating waves [63]. Slug models based on the properties of the cAMP signal relay system of aggregating cells were shown to be able to generate the observed behavior if there was a tip to tail step gradient in the excitability of the cells at the prestalk/prespore boundary [91,92]. Siegert and Weijer with their colleagues recently reported that microinjection of cAMP pulses into tips could generate optical density waves that interact with the natural waves annihilating them [93]. This is the first direct in vivo evidence that the waves are in fact cAMP mediated. Several known inhibitors of cAMP signal relay in aggregating cells were shown by Darcy and Fisher [41] to impair slug phototaxis and thermotaxis, providing in vivo evidence that these behaviors are indeed controlled by cAMP signals analogous to those that drive aggregation. Genetic confirmation of the role of extracellular cAMP signals was provided by demonstration of the importance of the CAR3 and CAR4 cAMP receptors (see later).

19.2.6.2 The tip inhibitors

Slug turning in phototaxis or thermotaxis can be explained by differential tip activation and/or inhibitor secretion across the slug tip in response to light or temperature gradients. Fisher et al. [44] reported the isolation in crude extracts of a small ($< 500\ M_w$) molecule (Slug Turning Factor, STF) with the properties expected of the slug tip inhibitor in phototaxis – a slug repellent whose secretion is stimulated by light and which impairs phototaxis (and thermotaxis) at high, uniform concentrations. Thus STF is a candidate tip inhibitor. Similar reasoning and analogous results led Bonner and coworkers to conclude that ammonia played an equivalent role to that proposed for STF in slug phototaxis [61] and thermotaxis [62]. It had been known for many years that culminating fruiting bodies use gradients of a gaseous repellent to orient away from each other and the substratum, and towards a mound of activated charcoal [94]. The identity of this gas as NH_3 was indicated by the demonstration that ammonia is a repellent of slugs and culminating fruiting bodies [58–60,95]. Because other small amines also act as slug repellents, it is possible that the action of NH_3 is mediated by alkalinization of an intracellular compartment [60]. The fact that slugs will orient towards the acid side of very steep pH gradients may be consistent with this [16]. However Schaap et al. [96] observed transient inhibition by ammonia and long term inhibition by weak acids of adenylyl cyclase activity, results that are not consistent in a simple way with an alkalinization mechanism for some of the actions of ammonia. As well as perturbing signal relay and morphogenesis [97,98], ammonia antagonizes cAMP binding to its receptor [99] so that its action might be more directly on the cAMP waves carrying the tip activation signal.

Although there is agreement that ammonia functions as a slug tip inhibitor controlling barrier avoidance by slugs and culminating fruiting bodies, a specific role for it in phototaxis and thermotaxis has been questioned [100]. To account for slug phototaxis and thermotaxis, light and temperature gradients must induce across the slug tip sustained gradients of tip inhibitor that are sufficient to cause turning. The secreted inhibitor must thereby form biologically significant concentration gradients in (STF) or around (NH_3) the slug tip. Slugs formed at high cell densities should accordingly be disoriented during phototaxis by the presence in the surrounding agar (STF) or air (NH_3) of high concentrations of these tip inhibitors. It has been shown that disorientation of slugs during phototaxis at high cell densities does occur but is entirely due to STF, with no contribution from NH_3 [100].

Because of the very rapid diffusion of gaseous NH_3 (15 ms to diffuse an average distance equal to the width of a slug), only steady state gradients maintained by sustained differences in secretion could work. An average slug secretes a total of 2–20 fmol of NH_3 per second [61], sufficient only to maintain a steady state gradient in air which is 10 to 100 fold shallower [100] than the minimum that slugs can detect in chemotaxis [60]. Achievable steady state gradients would actually be even shallower, because the differences in secretion on opposite sides of the slug tip can only be a fraction of the total. It seems that lateral ammonia gradients leading to slug turns may be of significance only when free diffusion of the gas is restricted by close proximity of a barrier.

Haser and Häder [101] reported that individual slug paths are straighter during phototaxis and during migration in darkness if activated charcoal is present in the agar or in the lid of the Petri dish. However, the global pattern of slug migration was little affected by the charcoal, apart from a relatively small difference attributable to the different optical properties of charcoal and water agar [102]. A major influence on the small scale straightness of slug trails is the time frequency and magnitude of spontaneous turning events combined with the speed of migration [51]. If slug migration is slower, trails appear less straight because spontaneous turns are compressed into shorter distances. The straighter trails on charcoal agar could be due partly to the fact that slugs migrate 30–60% faster when activated charcoal is present in the agar [57]. It is unknown if activated charcoal in the lid has the same effect on speed as when it is in the agar. In any case the results of Haser and Häder suggest charcoal adsorption of a volatile metabolite that either slows slug migration or increases the spontaneous turning rate. In both cases, it should be noted, the effect is the opposite of what might have been expected for adsorption of NH_3 in its role as tip inhibitor. Similarly surprising is the fact that if slugs are placed in a gentle air stream, they migrate downwind [101] not upwind as might have been expected to occur because of ammonia accumulation on the downwind side. The directional migration of slugs under the influence of gases clearly exhibits some as yet poorly understood features.

The third molecule that has been considered as a possible tip inhibitor controlling phototaxis is adenosine [103,41]. It is an antagonist of cAMP signaling whose synthesis is coupled to cAMP by virtue of it being a product of cAMP hydrolysis by phosphodiesterase and 5′-nucleotidase [104–106]. Adenosine impairs slug phototaxis and thermotaxis at concentrations consistent with the K_d (~ 350 μM) of the abundant low affinity adenosine binding sites in *Dictyostelium* membranes [107,41]. This supports the suggestion that adenosine has a morphogenetic role in slugs opposing that of cAMP [103]. However the concentrations at which it is active are significantly higher than those at which STF is active, so that it may not be the authentic tip inhibitor controlling phototaxis [41].

Whichever of the candidate tip inhibitors (one or more) proves to play the definitive role in slug phototaxis and thermotaxis, it must do so by modulating the twisted scroll waves of cAMP in the tip. One possible mechanism would be that the inhibitor causes a lateral gradient in excitability resulting in an altered morphology for the wave with lateral twisting or bending of the wave filament. Recent observations by Miura and Siegert (unpublished data) on cell movements in the tip during slug turning in phototaxis are consistent with altered signal geometry.

19.2.7 The central signal transduction components in phototaxis and thermotaxis are shared with other signalling pathways controlling morphogenesis

Except for STF, which has been little studied, a great deal of evidence has accumulated that the candidate tip activator (extracellular cAMP) and inhibitor (ammonia and adenosine) molecules controlling phototaxis and thermotaxis are important morphogens regulating the behavior and differentiation of individual cells in the multicellular stages of the *Dictyostelium* life cycle (for a recent review, see [83]). If slug turning behavior is indeed mediated by modulation of the tip autoactivation/inhibition system and lateral shifts in tip position, then the central signalling molecules in phototaxis should be common to behavior and morphogenesis. Direct genetic evidence for this was obtained by Darcy et al. [35,42,43] who found that many phototaxis-deficient mutants exhibited one of two major abnormalities in fruiting body formation – either the formation of stumpy fruiting bodies with little or no stalk, or of multiply tipped culminants with several stalks but few if any spores. Since the prestalk cells reside in the tip, these phenotypes are suggestive respectively of impaired tip activation and inhibition pathways.

Identification of the central signalling molecules involved in these pathways requires *in vivo* perturbation of the activities or concentrations of the molecules concerned. This can be done either pharmacologically by observing the effects of (hopefully) specific inhibitors, or genetically by observing the effects of specific mutations. Both approaches have been used in the study of *Dictyostelium* slug phototaxis and the results confirm that many of the central signalling molecules involved in phototaxis and thermotaxis are also important for normal multicellular morphogenesis. In the following sections I summarize pharmacological and genetic evidence for the identities of central signalling molecules in slug behavior that so far indicates roles for the cAMP receptors CAR3 and CAR4, the heterotrimeric G proteins $G_{\alpha1\beta\gamma}$, $G_{\alpha4\beta\gamma}$, $G_{\alpha7\beta\gamma}$ and $G_{\alpha8\beta\gamma}$, the small GTP binding protein RasD, the signalling protein glycogen synthase kinase 3 (GSK3), the second messengers Ca^{2+}, cAMP, cGMP and inositol polyphosphates along with associated proteins, the cytoskeletal protein ABP-120 and the mitochondria.

19.3 Pharmacological analysis

Pharmacological data has been used to support postulated roles in phototaxis for extracellular cAMP signals, for heterotrimeric G proteins and for the intracellular second messengers Ca^{2+} and inositol polyphosphates. Thus, Darcy and Fisher [41] reported that phototaxis (and thermotaxis) are impaired by agents that perturb extracellular cAMP signalling, namely ammonium salts, caffeine, adenosine and the slowly hydrolysable cAMP analogue, cAMPS. The conclusion that extracellular cAMP signals participate in the control of slug phototaxis has since been further supported by the phenotypes of mutants lacking the CAR3 or CAR4 cAMP receptors (see below).

In support of a role for heterotrimeric G proteins, Darcy and Fisher [40] found that slug phototaxis is impaired by pertussis toxin which specifically ADP-ribosylates and inactivates some classes of G-protein α subunit. Dohrmann et al. [37] also reported that fluoride ions have dramatic effects on slug behavior and, by analogy with other organisms, suggested adenylyl cyclase as the possible target. In fact, with hindsight it is clear that the cyclase in other organisms is not the direct target, but the G-protein α subunit which activates it. Presumably, one or more G-proteins are the targets for fluoride ions in phototaxis. Deranged phototaxis in specific *Dictyostelium* G-protein mutants has now confirmed that these proteins play essential roles in slug behavior (see later).

As evidence for a role for Ca^{2+} signalling, Dohrmann et al. [37] reported the detrimental effects of EGTA (Ca^{2+} chelator) and addition of extra Ca^{2+} on slug phototaxis and thermotaxis. This conclusion remains reasonable, especially since the slug tips do exhibit high levels of intracellular membrane-associated and free Ca^{2+} as well as unusual rotating Ca^{2+} waves [108–110].

Darcy and Fisher [40] reported pharmacological evidence from the inhibitory effects of lithium ions that inositol polyphosphate signalling may play a central role in phototaxis and thermotaxis. At the time it was widely believed that several enzymes involved in inositol polyphosphate metabolism were the relevant physiological targets for Li^+ in both mammalian cells and in *Dictyostelium*. Indeed the half-maximal effect of Li^+ on slug behavior occurred at concentrations as low as 0.25 mM, which compared well with the observed half-maximal inhibition of several *Dictyostelium* inositol polyphosphate phosphatases at Li^+ concentrations from 0.24 mM (Ins(1,4,5)P_3 phosphatase) to 2.5 mM [112]. When the gene encoding phospholipase C (whose activity liberates Ins(1,4,5)P_3) was disrupted [113], phototaxis and thermotaxis were unaffected, suggesting that the Li^+ effects may not have resulted from perturbed inositol polyphosphate metabolism after all (Fraser and Fisher, unpublished). However there are alternative phospholipase C-independent routes for Ins(1,4,5)P_3 production and other inositol polyphosphates also are known to play signalling roles [114]. Nonetheless, recent work in *Dictyostelium* and in other organisms suggests that glycogen synthase kinase 3 (GSK3) may in fact be the relevant biological target for Li^+, not inositol polyphosphatases as originally thought [115]. This means that until appropriate mutants have been tested, both the inositol polyphosphatases and GSK3 must be considered as potentially relevant targets for Li^+ in phototaxis.

19.4 Genetic analysis

Although pharmacological analysis has proved useful as a means of in vivo perturbation of the activities of potentially relevant signalling molecules in phototaxis, it has suffered from unavoidable uncertainties regarding drug access and specificity. This problem is less acute in genetic analysis based on the study of mutant phenotypes. As described below, both classical and molecular genetic techniques have been brought to bear on the functional dissection[5] of the signalling pathways involved in *Dictyostelium* slug behavior.

19.4.1 Classical genetics

From the early 70s through to the late 80s a range of classical genetic techniques and markers were developed for *D. discoideum* based on the parasexual genetic cycle in

[5] For the purposes of genetic analysis, a phototaxis gene is operationally defined as one that encodes a product whose activity at wildtype levels is necessary for normal phototaxis. This necessarily pragmatic definition clearly includes genes whose importance for phototaxis may be only indirect. Such pleiotropy is nonetheless informative, as illustrated by the effects of disruption of the mitochondrial large ribosomal RNA gene (see later).

which diploids (14 genetic linkage groups) resulting from rare fusion of haploid cells are selected and subsequently haploidized to yield progeny bearing new assortments of parental linkage groups [116,117]. This allows straightforward assignment of mutant genes to linkage groups because markers on different linkage groups reassort (segregate) into haploid progeny independently of each other. The diploids are sufficiently stable that their phenotypes can be studied in complementation tests, allowing assignment of mutant alleles to loci. Ordering of markers on chromosomes is possible (but tedious) based on the low frequency of mitotic crossing over in the diploid state. Large scale chromosome rearrangements also occur and can be detected genetically [118–121]. Diploids homozygous at all loci and isogenic to corresponding haploids can be isolated and studied to determine the effect of the diploid state *per se* on particular phenotypes [122].

19.4.1.1 Isolation and phenotypic characterization of phototaxis mutants

Although the first phototaxis deficient (Pho$^-$) mutant was reported in 1970 [123] and others were isolated and studied phenotypically in the early 80s [38,39,46], the first systematic genetic study of phototaxis was reported only recently [35,42,43]. Over 20 mutants were characterized phenotypically in relation to phototactic and thermotactic behavior, morphological defects and the newly discovered cGMP responses to light and warmth (Table 1) [43]. All but one of the phototaxis mutants reported to date have been impaired in positive thermotaxis as well, demonstrating early convergence of the signal transduction pathways from the photoreceptor and thermoreceptor (Figure 6). Two thermotaxis mutants isolated after chemical mutagenesis have been reported to be unaltered in phototaxis. One fails to exhibit negative thermotaxis at low temperatures [39], while the other shows the more commonly observed defect in positive thermotaxis at temperatures around the growth temperature [35]. The latter mutant, HPF228, and a *carC* disruptant (see later) are the only known strains with defective positive thermotaxis but normal phototaxis and they reveal the existence of gene products involved in thermotransduction prior to convergence of the thermosensory and photosensory pathways. Taken together the data demonstrate that, although light and warmth are sensed separately, turns towards them are under the control of the same signal transduction pathway.

The mutant phenotypes have provided genetic evidence that cGMP is an important intracellular second messenger in signal transduction during phototaxis and thermotaxis. Recent work revealed a relatively slow (peak after about 1 minute, compared to 10 seconds for cAMP stimuli) cGMP response to light and to warmth (Figure 7) [43]. The response was either positive (a 2-fold increase) or negative (a 2-fold decrease) depending on the strength of the stimulus (light intensity or warming rate) so that cGMP responses to these stimuli exhibit sign reversals analogous to the behavioral sign reversals. In the parental strain X22, the cGMP responses to light and warmth remained unaltered during aggregation and in the slug stage. Because of the technical difficulty of perfoming cGMP assays at the slug stage, the mutants were therefore characterized with respect to cGMP responses in postvegetative amoebae. Altered responses in the mutants may therefore better reflect signal transduction pathways at this stage of development than in slugs. Nonetheless, most phototaxis mutants showed impairment of the positive response and in some this was accompanied by impairment of the negative response,

Table 1. Genes important for *Dictyostelium* phototaxis and thermotaxis. Gene (locus) and mutant allele designations follow the standard style for this organism – an italicized, lower case, three letter code indicating the primary phenotype or protein associated with mutant alleles of that gene, followed by an italicized, upper case letter indicating the specific gene (locus). The corresponding phenotype is indicated by the corresponding non-italicized code, with the first letter capitalized and a superscript to indicate mutant ($^{-}$) phenotype. An italicized arabic number following the locus designation indicates the particular, usually mutant, allele and is normally simply a unique number assigned at the time of mutant isolation. A hyphen in place of an upper case letter indicates that the mutant allele has not been assigned to a specific locus e.g. by complementation tests. Mutant alleles were mapped to linkage groups by the parasexual genetic cycle. Primary phenotypes or proteins corresponding to genotype designations are *pho* – impaired phototaxis; *the* - impaired thermotaxis; *pka* – protein kinase A (regulatory subunit *pkaR,* catalytic subunit *pkaC*); *gelA* – gelation factor 120 (ABP-120, 120 kDa actin binding protein); *ecm* – extracellular matrix protein (*ecmA*, extracellular matrix protein A); *stp* – stumpy fruiting body morphology (little or no stalk); *mtp* – multiple tips on aggregates/fruiting structures; *mig* – short migration path in phototaxis/thermotaxis experiments; *mhcA* – myosin II heavy chain; *carC,D* – 7-transmembrane domain cAMP receptors CAR3,4; *gpbA* – G protein β subunit; *gpaA,G,H* – G protein α subunits 1,7,8; *rasD* – small GTP-binding protein of the Ras subfamily; *rnlA* – mitochondrial large subunit rRNA; *hspA* – chaperonin 60. Genes in the upper portion of the table were identified by classical genetics, those in the middle and lower portions by molecular genetic methods. Genes in the lower portion encode essential mitochondrial proteins.

Gene	Linkage group	Mutant allele	Mutant	Parental strain	Mutant phenotypes			
					Thermo-taxis	cGMP responses	Morphology & migration	Comment
phoA	IV	*phoA2315*	HU407	DU584	–	**III**	+	Requires *phoB2316*
phoB	VII	*phoB2316*	HU407	DU584	–	**III**	+	Requires *phoA2315*
phoC	VI	*phoC2307*	HPF9	X22	–	**III**	Mtp^{-}	*mtp-2302* on III *mtp-2303* on VI
phoD	IV	*phoD2317*	HPF3	HU1628	–	**III**	+	
phoE	III	*phoE2305*	HPF7	X22	–	**I**	Stp^{-}	*stpA2300* on III
		phoE2308	HPF11	X22	–	**I**	Stp^{-}	*stpA2301* on III
phoF	VI	*phoF2306*	HPF8	X22	–	**III**	Mtp^{-}	*mtpA2300* on III *mtpB2301* on VI

Table 1. Continued.

phoG	V	*phoG2309*	HPF12	X22	–	**III**	Stp⁻	*stpB2302* on V
phoH	III	*phoH355*	HU409	X22	–	**IV**	+	Small slugs
phoI	II	*phoI2301*	HPF2	X22	–	**IV**	+	
phoJ	IV	*phoJ2304*	HPF6	X22	–	**I**	Mtp⁻,Mig⁻	*mtpD2304* + *migA2300* on IV
		phoJ2313	HPF16	X22	–	**III**	Mtp⁻,Mig⁻	*mtpD2305* + *migA2301* on IV
phoK	V	*phoK2300*	HPF1	X22	–	**III**	+	
		phoK2311	HPF14	X22	–	**III**	+	
theA	II	*theA2300*	HPF13	X22	–	**III**$_W$,**IV**$_L$	Mig⁻	*migB2302* on II, probable unmapped *pho* mutation
theB	VII	*theB2301*	HPF228	X22	–	**IV**$_W$,**I**$_L$	Mig⁻	*migC2303* on VII
pho-	II	*pho-2319*	HPF229	X22	–	**I**	+	Complementation tests for linkage group II loci is difficult in X22 background.
	II	*pho-2303*	HPF5	X22	–	**II**	+	
	II	*pho-2318*	HPF230	X22	–	**II**	+	
	II	*pho-2314*	HPF17	X22	–	**IV**	+	
pkaR	III	*ecmA/pkaR* fusion	EcmA:Rm	AX2	?	?	Mig⁻ (small slugs + slow migration), culmination arrested.	
gelA	I	*gelA1000*	HG1264	AX2–214	–	?	Mig⁻ (delayed development, premature culmination), amoeboid motility slightly deranged.	
mhcA	IV	Triple Ala	HG1555	AX2	+/–	?	Ala replace the 3 phosphorylatable Thr in the tail domain. Null mutants arrest at tip formation.	
carC	III	disruption	*carC*⁻	AX3	–	?	Phototaxis slightly or not affected.	
carD	III	disruption	*carD*⁻	AX3	–	?	Reduced prestalk (tip) region with mislocalized prespore cells.	

Table 1. Continued.

Gene	Linkage group	Mutant allele	Mutant	Parental strain	Mutant phenotypes		
					Thermo-taxis	cGMP responses	Comment
gpbA	II	*gpbA*$^{++}$	HPF354	LW6	–	**?**	Overexpresser of wildtype $G_{\dagger}$ in $G_{\dagger}$-null background.
		gpbA1	HPF355		**?**	**?**	Temperature-sensitive
		gpbA2	HPF356		**?**	**?**	mutant alleles in a $G_{\dagger}$-null background.
gpaA	IV	disruption	gα1 null	AX3	+/–	**?**	Overexpression of continuously active form inhibits tip formation.
gpaD	IV	disruption	gα4 null	AX3	–	**?**	Transduces signals from folate and pterin receptors.
gpaG	III	disruption	LW1	JH10	–	**?**	
gpaH	IV	disruption	LW3	JH10	+	**?**	
rasD	VII	disruption	*rasD*$^{-}$	AX3	–	**?**	Overexpression of continuously active form yields Mtp^{-} phenotype.
rnlA	mtDNA	disruption	HPF231	AX2	–	**I**	Slightly slower growth.
hspA	VI	antisense inhibition	HPF334	AX2	–	**?**	Slow growth.

In the table the phenotypes of the mutants are indicated as follows. Thermotaxis: +, wildtype; –, impaired; +/–, slightly or not impaired; **?**, unknown. cGMP responses in postvegetative amoebae: **I**, wildtype; **II,** cGMP responses to light and warmth abolished; **III**, cGMP increase after strong stimuli abolished; **IV,** cGMP responses to strong and weak stimuli inverted (see Figure 7). Subscripts L and W indicate phenotypes specifically relating to the light or warmth stimulus. Absence of a subscript indicates responses to both stimuli combined. Morphology and migration: +, wildtype; –, mutant for the indicated phenotype.

while a few showed wild-type cGMP responses (Figure 7, Table 1) [43]. One mutant, defective in thermotaxis only (HPF228, *theB2301*), showed altered cGMP responses to warmth but wild-type responses to light. The results indicate that signals from the photoreceptor and thermoreceptor converge before regulation of the cGMP responses and orientation behavior in both phototaxis and thermotaxis. This is the first of the second messengers to have been systematically studied in this way in a collection of

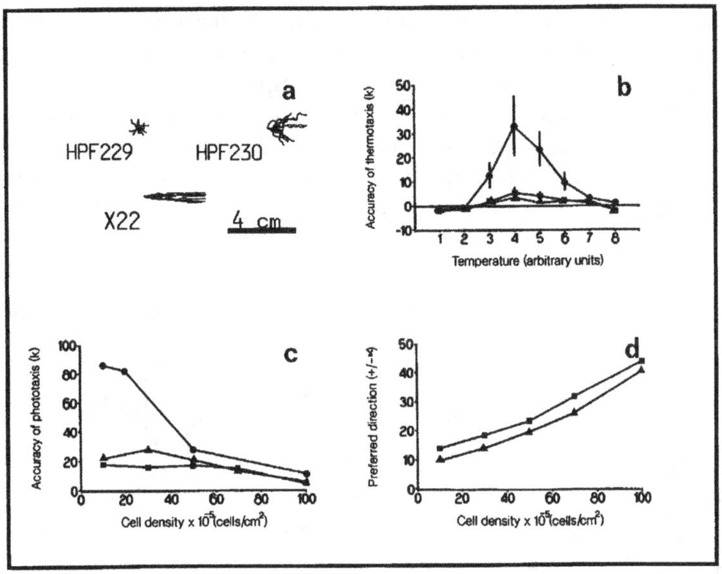

Figure 6. Phenotypes of *Dictyostelium discoideum* slug phototaxis mutants. Reproduced from Figure 1 of [35]. The phenotypes of wildtype strain X22 (●) and two phototaxis mutants HPF229 (■) and HPF230 (▲) are shown as an example. (a) Digitized trails in phototaxis on charcoal agar (water agar supplemented with 0.5% (w/v) activated charcoal to absorb stray light) are shown after plotting from a common origin with the direction towards the light source to the right of the figure. (b) Thermotaxis by slugs migrating in an 0.2°C/cm gradient at temperatures ranging from about 14°C (T = 1) to about 28°C (T = 8). Growth temperature was 21°C. The accuracy of thermotaxis (κ) is a statistical measure of how concentrated individual directions are around the mean direction (towards the warmth or towards the cold). The convention is followed that positive values reflect movement towards the warmth and negative values movement towards the cold. Vertical bars represent 90% confidence limits. The scale of the Y-axis required to accommodate the very accurate thermotaxis at T = 4 by the wildtype slugs obscures the fact that at T = 2, both mutants had already switched to significant negative thermotaxis (towards the cold) while the wildtype slugs showed no significant orientation at this temperature. (c) Accuracy of phototaxis on charcoal agar *versus* cell density at which the slugs were formed. Phototaxis becomes less accurate at high cell density because of the excretion into the medium of high levels of Slug Turning Factor (STF) which interferes with phototaxis. (d) Preferred directions of migration (+ α) either side of the light source in phototaxis by the two phototaxis mutants *versus* cell density. The wildtype X22 is not shown as α was 0° at all cell densities. At high cell densities phototaxis by the X22 slugs was bimodal close to the inoculation site where STF levels are highest, but this was obscured by readjustment of the preferred directions as slugs migrated away from this region.

phototaxis mutants. A similar study on chemotaxis mutants confirmed the already considerable evidence for a role for cGMP in signal transduction during chemotactic aggregation [124].

Many of the mutant phototaxis (*pho*) loci are associated with morphological defects in the fruiting bodies and/or with short migration (*mig*) paths of slugs [35,42] (Table 1). The short migration paths may be due to delayed development, slower movement and/or earlier "decisions" by the slugs to cease migration and culminate to form a fruiting body. The two major types of morphological abnormality observed in association with mutant *pho* loci are the formation of short, stumpy fruiting bodies containing spores but little or no stalk (stumpy, *stp*) and the formation of fruiting bodies with abnormal, multiple stalks (derived during culmination from multiple tips) and few or no spores (multiple tips, *mtp*). In the great majority of cases the morphological and slug migration defects mapped to the same linkage group as the mutant *pho* alleles and thus were probably caused by pleiotropic effects of these alleles [35,42]. In the case of *phoE/stpA* on chromosome III and *phoJ/mtpD* on linkage group IV this was almost certainly so, as

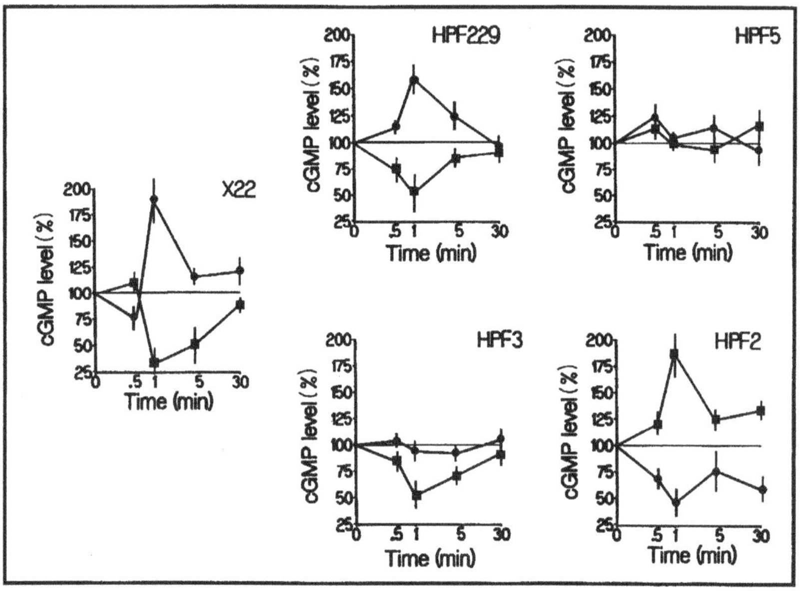

Figure 7. cGMP response phenotypes of phototaxis and thermotaxis mutants. Reproduced from Figure 5 of [43]. Cyclic GMP responses to a weak (■) or strong (●) combined stimulus of light/heat in amoebae of the parental strain X22 or a phototaxis mutant from each of the four observed phenotypic classes. HPF229 – Class **I**, wildtype cGMP response; HPF5 – Class **II**, cGMP responses to both strong and weak stimuli abolished; HPF3 – Class **III**, cGMP increase after a strong stimulus impaired, but decrease after a weak stimulus normal; HPF2 – Class **IV**, cGMP responses inverted i.e. cGMP decreases after a strong stimulus and increases after a weak stimulus (the inverse of the wildtype responses). Strong stimuli were empirically defined with respect to light as 65 lux "white" light from a tungsten bulb and with respect to warming as a warming rate of 0.033°C/min. Weak stimuli were empirically defined as 33 lux light and 0.017°C/min warming rate.

independently isolated mutants in these loci showed the same combination of behavioral and morphological phenotypes mapping to the same complementation group on the same chromosome [35]. The conclusion to be drawn from this genetic analysis is that the proteins involved in signal transduction during slug phototaxis and thermotaxis also play critical roles in normal morphogenesis and differentiation into the final fruiting body [35] as predicted from the tip activation/inhibition model for slug turning behavior [25].

19.4.1.2 Genetic analysis of mutant pho loci

The genes responsible for the behavioral and morphological defects of phototaxis mutants have been mapped and assigned to about a dozen loci distributed over all seven genetic linkage groups (Table 1), including the elusive linkage group V which had previously evaded genetic detection, despite clear cytogenetic evidence of its existence [35,42,118–122,125,126]. Note, however, that recent physical mapping of the genome failed to detect a chromosome corresponding to linkage group V [47,48]. Since the cytogenetics and genetic experiments were performed in different genetic backgrounds from the physical mapping work, the discrepancies may be explained by chromosome fusion/splitting events in the different lineages.

It is clear from the genetic analysis to date that the number of genes that can be mutated to yield viable cells able to aggregate and form slugs with defective phototaxis is most probably around 20, but could be as high as 55 [35]. This conclusion is based on three lines of evidence:

1. The frequency with which Pho^- mutants are isolated (1/400) under standard conditions of chemical mutagenesis is about 25 fold higher than expected for a single gene [35,42].
2. The frequency with which Pho^- mutants are isolated (1/600) by non-targeted disruption with integrating plasmid DNA is about 20 fold higher than expected for a single gene [127].
3. The frequency of genes represented by single mutants only (6/12) in a collection of assigned loci, yielded a maximum likelihood estimate of 17 genes in the pool of genetically detectable *pho* loci. Three loci (one of them on linkage group V) were found to be each represented by two independent mutants. This frequency is also consistent with perhaps 20 *pho* loci that can be detected by mutant isolation and genetic analysis [35,42].

Twenty phototaxis genes would be an underestimate if some chromosomal *pho* loci are also necessary for aggregation (e.g. *gpbA*, see below) or cell growth and division. Such loci would not be detectable by standard methods for phototaxis mutant isolation. At least twenty *pho* loci might be expected based on the complexity of known signal transduction pathways. There is pharmacological and/or genetic evidence that cAMP [41,128], cGMP [43], IP_3 [40] and Ca^{2+} [37], G-protein ([40]; Fraser and Fisher, unpublished) and Ras (Fraser and Fisher, unpublished) signalling pathways are involved in slug phototaxis. This being so, one rapidly arrives at quite large estimates of the number of genes that might be necessary for normal phototransduction and behavior.

The genetic analysis of phototaxis revealed a Pho^- phenotype in the multiply marked "tester" strain HU407 used in crosses with the phototaxis mutants [35]. This phenotype was due to mutant alleles of two loci, *phoA* and *phoB*, on linkage groups IV and VII respectively, both of which were required for expression of the Pho^- phenotype. The simplest explanation is that the products of these two loci may be able to substitute for each other functionally. This could occur, for example, if the two genes were homologs encoding similar proteins. Such multigene families are common in *Dictyostelium* as in other eukaryotes, with the cAMP receptor genes *(carA-D)* [129], adenylyl cyclase genes *(acaA, G)* [130], G protein α subunit genes (*gpaA-H*) [131] and the large actin gene family [132] being notable examples that could include genes necessary for phototaxis (see below).

19.4.2 Molecular genetics

19.4.2.1 Phototaxis genes identified by targeted mutagenesis

Because genes encoding so many signal transduction and cytoskeletal proteins in *Dictyostelium* have been cloned and corresponding mutants isolated, it is possible to determine if they play essential roles in slug behavior by examining the phototaxis phenotypes of the mutants. Using this approach, the first cloned genes known to be important for slug phototaxis have recently been identified.

cAMP receptor genes. The evidence that extracellular cAMP waves carry the tip activation signal was described in a previous section. Four cAMP receptor genes (carA-D encoding the 7-transmembrane domain receptors CAR1–4 respectively) have been cloned, sequenced and disrupted [129,133–136]. The CAR1 receptor mediates chemotactic aggregation and disruption of carA therefore prevents multicellular development. The carB gene seems likely to be important for slug behavior since it is expressed in tip cells and its disruption arrests development at the point of tip formation. Unfortunately this means that slugs are not formed so that these mutants cannot be tested for aberrant slug behavior. However carC and carD disruptants will form slugs and they are impaired in slug orientation behavior (see Table 1), confirming the central role played by extracellular cAMP signals in slug phototaxis and thermotaxis (Fraser and Fisher, unpublished). CAR4 is expressed preferentially in tip cells and null mutants produce slugs that have reduced prestalk (tip) and enhanced prespore regions, with some of the prespore cells mislocalizing into the stalk at culmination [136]. CAR3 is expressed in all cells in the slug but preferentially in prespore/spore cells at culmination [137].

pkaR *encoding the regulatory subunit of cAMP-dependent protein kinase.* Another gene shown to be important for phototaxis is pkaR which encodes the regulatory subunit of protein kinase A (PKA). Harwood and Williams and coworkers showed that overexpression of a cAMP-insensitive mutant form (Rm) of the inhibitory subunit of protein kinase A (pkaR) results in cells lacking active protein kinase A, because cAMP fails to relieve the heterodimeric holoenzyme from inhibition [138,139]. Constitutive expression leads to defects in cAMP signal relay (cAMP synthesis and secretion in response to cAMP) but not chemotaxis, so that the cells are unable to aggregate or form slugs. Expression of the mutant pkaR under the control of the ecmA promoter restricts its synthesis to anterior (PstA) cells and allows aggregation and slug formation. However the slugs that form exhibit impaired phototaxis as well as deranged multicellular development [128]. Since cAMP normally fulfils its second messenger role by regulating protein kinase A, the phototaxis deficient phenotype of these slugs indicates a second messenger role for cAMP in phototaxis. It is also possible that the effects of phenotypic protein kinase A deficiency are indirect, resulting from slug tip abnormalities due to effects on cell differentiation and morphogenetic movements within the slug [128].

Genes encoding cytoskeletal proteins. Increases in cytosolic Ca^{2+} can occur by influx from the extracellular medium or by release from intracellular stores. In other eukaryotes the second messenger inositol 1,4,5-triphosphate ($InsP_3$) triggers Ca^{2+} release from the endoplasmic reticulum and this is so for *Dictyostelium* as well [140]. At present the evidence for Ca^{2+} and inositol polyphosphate signalling in slug behavior is pharmacological – EGTA [37] (Ca^{2+} chelator) and Li^+ ions [40] (inhibitor of several key enzymes in inositol polyphosphate cycling) impair phototaxis and thermotaxis when present in the extracellular medium during slug formation and migration. Of the known Ca^{2+} binding proteins that might be relevant, several have been examined genetically for roles in phototaxis. They are the Ca^{2+}-regulated actin binding proteins (for recent review see [141]) – the capping/severing protein severin and the actin crosslinkers α-actinin, fimbrin (or plastin) and the 34 kDa actin-bundling protein. Mutants lacking these proteins are unaffected in development and behavior at the unicellular and multicellular stages, including slug phototaxis and thermotaxis [33,44]. Thus the role of Ca^{2+} as a second messenger in slug behavior may not be primarily via regulation of these particular actin-binding proteins. Other cytoskeletal proteins that have been shown not to be required for normal phototaxis are hisactophilin and protovillin (Schleicher and Fisher, unpublished).

One cloned gene encoding a cytoskeletal protein that did prove to be important for phototaxis is *gelA* (or *abpC* [48]) encoding the 120 kDa F-actin cross-linking gelation factor (or ABP-120) [142]. In the course of classical genetic studies of chemically-induced mutants affected in the gene encoding this protein, Wallraff and Wallraff [33] discovered that the mutants exhibit a phototaxis deficient phenotype and that the slugs migrate shorter distances than wild-type slugs, at least partly because aggregation and slug formation is slower while cessation of migration and fruiting body formation occurs sooner. The possibility that the slugs also migrate more slowly has not been tested. Fisher et al. [44] found similar defects in mutants in which the ABP120 gene had been disrupted. Both the chemically-induced mutants and the disruptants are also deficient in thermotaxis [44]. In cAMP-stimulated aggregation competent amoebae, incorporation of actin into the triton-resistant cytoskeleton is defective [142–144]. Mutant amoebae are reported to exhibit basically normal chemotaxis and motility [143] although they move slightly more slowly and exhibit subtle abnormalities in cell shape, in extension of pseudopodia [144] and in the periodicity of cAMP waves during early development [145].

The 120 kDa protein is not one of the Ca^{2+} regulated actin binding proteins, so that the means by which signals from receptors might interact with this cytoskeletal protein remain unknown. However, the related human protein ABP-280 is apparently regulated by phosphorylation [146,147] and is itself necessary for SAPK (stress-activated protein kinase) activity in protein phosphorylation cascades that are initiated by the extracellular signals lysophosphatidic acid and tumor necrosis factor α [148]. Both proteins contain an N-terminal actin-binding domain of the α-actinin/spectrin family followed by a rod domain consisting of multiple repeated segments that have been shown (in ABP-120) to exhibit immunoglobulin-like folding [149] and (in ABP-280) to interact with other proteins such as SEK-1 (which phosphorylates SAPK) [148,150,151]. It was therefore suggested that each repeated immunoglobulin-like fold of ABP-120 (and ABP-280) might mediate specific binding interactions with particular proteins [44]. Thus *Dictyostelium* ABP-120 could interact specifically with other proteins in the photosensory and thermosensory transduction pathways including the receptors, could be necessary for specific protein phosphorylation cascades in transduction of signals from these receptors to the cytoskeleton, and could itself be regulated by phosphorylation.

The *mhcA* gene on chromosome 4 encodes the myosin II heavy chain, another cytoskeletal protein which proved to play a role in slug phototaxis. Disruption or antisense inhibition of *mhcA* impaired [152] but did not abolish motility and chemotaxis by individual *Dictyostelium* amoebae [153,154]. However, myosin II is required for slug formation [155] so that slug behavior cannot be tested in myosin II-deficient strains (unlike ABP-120 which is essential for multicellular morphogenesis only in the simultaneous absence of α-actinin). Nonetheless multicellular development does proceed in a myosin II null mutant that has been rescued by expression of a mutant myosin in which alanines have replaced the three phosphorylatable threonines in the C-terminal tail region of the protein [156]. Phosphorylation of the myosin at these sites in the wildtype protein inhibits myosin thick filament assembly and association with the triton-insoluble cytoskeleton [156,157]. Although slugs are formed by a mutant strain which has the three threonine to alanine substitutions (HG1555, kindly provided by G. Gerisch), their phototaxis is impaired suggesting that myosin phosphorylation and disassembly from the cytoskeleton play important roles in phototaxis (Fraser and Fisher, unpublished).

Genes encoding heterotrimeric and small GTP-binding proteins. Darcy and Fisher [40] reported that phototaxis is impaired by pertussis toxin, which specifically ADP-ribosylates the α subunits of some classes of heterotrimeric GTP-binding proteins (G proteins). Together with the effects of fluoride ions on phototaxis and thermotaxis [37], these results provided pharmacological evidence that one or more G proteins play important signalling roles in slug behavior. This conclusion has now been confirmed genetically. In *Dictyostelium* 8 different G protein α subunit genes (gpaA-H) [131,158–162] and a single β subunit gene (gpbA on chromosome 2) [163] have been reported. Disruption mutants have been isolated for all except $G_{\alpha 6}$ and of these, the strains deficient in $G_{\alpha 1,4,5,7 \text{ or } 8}$ are able to form slugs. Testing these mutants (kindly provided by P. Devreotes and R. Firtel) for phototaxis and thermotaxis showed that $G_{\alpha 5}$ is not required for either of these behaviors, while each of $G_{\alpha 1,4,7,8}$ are needed, to different degrees, for one or both of them (Fraser and Fisher, unpublished; Table 1). $G_{\alpha 1}$ and $G_{\alpha 4}$ appear to play major roles in phototaxis since their absence had the greatest impact on this behavior.

Like many other gene products that are essential for normal slug phototaxis, some of the G protein α subunits are also important for multicellular morphogenesis. Thus Richard Firtel and coworkers reported that overexpression of $G_{\alpha 1}$(G45V) a continuously active mutant $\alpha 1$ subunit prevented prestalk cells from forming a tip [164,165]. In chimaeric slugs formed from mixtures of wildtype and mutant cells, the cells expressing the mutant α subunit were excluded from the tip. These results are those that might be expected if $G_{\alpha 1}$ transduces tip inhibition signals. Overexpression of the wildtype $\alpha 1$ subunit has more subtle effects on expression of genes in the tip region that are also consistent with this hypothesis [164]. $G_{\alpha 4}$ is known to transduce signals from folate-specific receptors for chemotaxis in vegetative amoebae and from pterin receptors later in development [166], raising the possibility that *Dictyostelium* slugs may use pterins as extracellular signals in the multicellular stages. *Dictyostelium discoideum* produces dictyopterin, a natural isomer of L-biopterin, throughout development but maximally in growth phase and postvegetative amoebae [167,168]. The possibility that dictyopterin plays a signalling role in slug behavior has not been tested and it is possible that the $\alpha 4$ subunit transduces signals from other, as yet unknown receptors. The $G_{\alpha 4}$ null mutant exhibits deranged multicellular morphogenesis beyond the slug stage [166]. No overt phenotypes have been reported to result from disruption of the gene encoding either the $\alpha 7$ or the $\alpha 8$ subunit, although overexpression of $G_{\alpha 7}$ causes abnormal morphogenesis beginning at the slug stage [169].

Verification of the essential role of heterotrimeric G proteins was also obtained using expression vectors encoding either wildtype or one of two mutant temperature-sensitive G_β subunits kindly provided by T. Jin and P. Devreotes [170]. Overexpression of the wildtype β subunit in a G_β-null strain rescued development [170], but the slugs exhibited slightly or moderately deranged phototaxis and thermotaxis (Fraser and Fisher, unpublished). Similar phenotypes resulted from overexpression at the permissive temperature (21°C) of either of the temperature-sensitive β subunits (encoded by the mutant alleles *gbpA1* and *gbpA2*), but development was not rescued at the restrictive temperature (27°C). If these strains were allowed to develop at the permissive temperature and then shifted to 27°C, phototaxis became severely disoriented (Fraser and Fisher, unpublished). These results confirm genetically that heterotrimeric G proteins are essential for normal slug behavior.

The *ras* genes encode a family of small GTP-binding proteins that play essential signalling roles in metazoan growth and development. Continuously active mutant Ras proteins have been implicated in deregulation of growth in malignant transformation [171]. *Dictyostelium discoideum* has 6 different *ras* genes [172–176]. The first of these to be discovered, *rasD* (originally called *Ddras*) on chromosome 6, is expressed in tip cells and when a continuously active mutant form of it (RasD[G12T]) is overexpressed, multitipped aggregates form that contain mostly prestalk cells and fail to proceed to culmination [177,178]. RasD is involved in transduction of cAMP signals that activate ERK2 (Extracellular signal Regulated Kinase 2) in aggregation competent cells [179,180]. The phenotypes associated with RasD[G12T] expression suggest that the wildtype RasD protein may function in transduction of tip activation signals. That it plays an important role in slug phototaxis and thermotaxis was shown by the fact that these responses are impaired in a *rasD* null strain, kindly provided by G. Weeks (Fraser and Fisher, unpublished).

19.4.2.2 A role for mitochondria revealed by nontargeted gene disruption

Recent developments in the molecular genetics of *Dictyostelium discoideum* make an alternative route to identification of genes important for phototaxis possible. Instead of proceeding from a known protein to the gene and thence via disruptants to the mutant phenotype, nontargeted disruption of genes important for phototaxis allows research to begin with a mutant phenotype, using the sequences inserted into the gene as a molecular tag to facilitate its isolation.

Two methods for nontargeted gene disruption have been described – REMI (Restriction Enzyme Mediated Insertion) [181] and nontargeted, recombinational integration of plasmid (RIP) [127]. In both methods a plasmid vector is used that can be inherited stably in *Dictyostelium* only if integrated into the genome. REMI has not yet been used to isolate phototaxis deficient mutants, but RIP disruption of *pho* genes was reported recently [127]. The disrupted gene was recovered from one of the mutants [182] and subsequently shown to be *rnlA*, the mitochondrial large subunit rRNA gene [183]. That disruption of this gene was responsible for the phototaxis deficient phenotype was demonstrated by experiments in which portions of the gene were used to target plasmid insertion into the same locus by homologous recombination. Southern blotting confirmed that plasmid DNA was inserted into a minority of the mitochondrial genomes in cells of both the original mutant and the targeted disruptants.

Because of the essential role of the large subunit rRNA in protein synthesis [184], it is expected that disruption of this gene by plasmid insertion would result in defective mitochondria lacking the mitochondrially encoded proteins essential for normal function. Thus these mutants may have revealed a role for the mitochondria in signal transduction. This is consistent with recent reports of mitochondrial Ca^{2+} transients that arise in concert with cytosolic Ca^{2+} responses to hormonal stimuli in mammalian cells [185], and with the observation that impairing mitochondrial electron transport functions modulates cytosolic Ca^{2+} waves [186]. Of the various second messengers that seem to play a role in slug phototaxis, only Ca^{2+} is known to mediate communication between the mitochondria and the cytosol in this way.

While all of the mitochondrial mutants exhibited severe defects in both phototaxis and thermotaxis, only some were impaired in growth. This suggests that signal transduction is more sensitive than other cellular activities to the presence of defective mitochondria and may help explain the pathology of human mitochondrial diseases [187]. Most of these disorders are characterized by the presence in cells of a subpopulation of defective, mutant mitochondria. The most commonly affected tissues are heart, muscle and central nervous system, which share not only high energy demands but also heavy dependence on signal transduction to mediate their normal activities.

Most mitochondrial proteins are encoded by nuclear genes, disruption of which would be expected to be lethal since all the mitochondria are supplied by the products of these genes and would all be affected equally. However antisense inhibition of expression of such genes need not be complete, with the level of inhibition depending on the amount of the antisense RNA made and thus on the number of copies of the antisense RNA expressing plasmid. In accordance with this it was found recently that although targeted disruptants of the nuclear gene for the essential mitochondrial protein chaperonin 60 could not be isolated, antisense-inhibited transformants could (Kotsifas, de Lozanne and Fisher, unpublished). Chaperonin 60 is the product of the *hspA* gene on chromosome 5 (genetic linkage group VI) and plays an essential role in correctly folding proteins newly imported into the mitochondrial matrix [188]. As was the case with the mitochondrial *rnlA* disruptants, all antisense-inhibited transformants were deficient in phototaxis and thermotaxis, but only a subset (those with the highest level of antisense inhibition) were also deficient in growth. These results confirmed that signal transduction for slug phototaxis and thermotaxis is more sensitive than cell growth and division to mitochondrial insufficiency.

The discovery of mitochondrial mutants deficient in phototaxis illustrates the ability of nontargeted gene disruption techniques to reveal the unexpected. Proceeding from mutant phenotype to gene to protein, rather than in the opposite direction, does not require a priori selection of a gene of interest and ensures from the outset that the gene under study really is necessary for the wild-type phenotype. The route from selected protein to mutant can harbour some surprises at the end, when disruptants fail to show the expected phenotype. A case in point is the phospholipase C gene, the disruption of which did not result in any of the expected mutant phenotypes but instead led to inability of spores to abort germination in unfavorable conditions ([113]; van Haastert, pers. comm.).

19.5 Conclusion

Genetic analysis is uniquely powerful in functional dissection of complex biological processes. Its application to phototaxis (and thermotaxis) by *Dictyostelium discoideum* slugs in combination with behavioral, physiological and pharmacological studies led to elucidation of the key features illustrated in Figure 2. The main contribution so far of the classical and molecular genetic analysis has been to show that a minimum of about 20 genes is needed for wildtype slug behavior, that many genes are important both for orientation behavior and for multicellular morphogenesis as expected from the tip activation/inhibition model for slug turning, that the photosensory and thermosensory transduction pathways converge early so that most genes are required for both, and that the central signalling components include the second messengers cAMP and cGMP, the cAMP receptors CAR3 and CAR4, the GTP-binding proteins $G_{\alpha1\beta\gamma}$, $G_{\alpha4\beta\gamma}$, $G_{\alpha7\beta\gamma}$, $G_{\alpha8\beta\gamma}$ and RasD, the cytoskeletal proteins ABP120 and myosin II and perhaps the mitochondria. Perhaps surprisingly, gene disruptions revealed that normal phototaxis and thermotaxis are highly sensitive to the presence of a subpopulation of defective mitochondria, but insensitive to the absence of the two major Ca^{2+}-regulated actin binding proteins. The power of the genetic approach is this ability to reveal the unexpected. It will therefore be interesting to see what surprises await us from future genetic dissection of *Dictyostelium* slug behavior.

Acknowledgements

I am grateful to colleagues who kindly consented to my citation of their unpublished results. This work was supported by grants from the Australian Research Council. My thanks go to Kota Miura, Florian Siegert and John Bonner for helpful email discussions on the mechanism of slug turning.

References

1. O. Brefeld (1869). *Dictyostelium mucoroides.* Ein neuer Organismus aus der Verwandschaft der Myxomyceten. *Abhandlungen der Senckenbergischen Naturforschenden Gesellschaft Frankfurt*, **7**, 85–107.
2. M.P. van Tieghem (1880). Sur quelques Myxomycetes a plasmode agrege. *Bull. Soc. Bot. Fr.*, **27**, 317–322.
3. K.B. Raper (1935). *Dictyostelium discoideum*, a new species of slime mold from decaying forest leaves. *J. Agric. Res.*, **50**, 135–147.
4. K.B. Raper (1940). Pseudoplasmodium formation and organization in *Dictyostelium discoideum. J. Elisha Mitchell Sci. Soc.*, **56**, 241–282.
5. J.T. Bonner (1944). A descriptive study of the development of the slime mold *Dictyostelium discoideum. Am. J. Bot.*, **31**, 175–182.
6. J.T. Bonner, M.K. Slifkin (1949). A study of the control of differentiation: the proportions of stalk and spore cells in the slime mold *Dictyostelium discoideum. Am. J. Bot.*, **36**, 727–734.
7. J. Williams (1997). Prestalk and stalk cell heterogeneity in *Dictyostelium.* In: Y. Maeda, K. Inouye, I. Takeuchi (Eds), Dictyostelium *A Model System for Cell and Developmental Biology* (pp. 293–304). Universal Academy Press Inc., Tokyo.
8. J.T. Bonner, W.W. Clarke, C.L. Neely, M.K. Slifkin (1950). The orientation to light and the extremely sensitive orientation to temperature gradients in the slime mold *Dictyostelium discoideum. J. Cell. Compar. Physiol.*, **36**, 149–158.
9. K.L. Poff, D-P. Häder (1984). An action spectrum for phototaxis by pseudoplasmodia of *Dictyostelium discoideum. Photochem. Photobiol.*, **39**, 433–436.
10. K.L. Poff, M. Skokut (1977). Thermotaxis by pseudoplasmodia of *Dictyostelium discoideum. Proc. Natl. Acad. Sci. USA*, **74**, 2007–2010.
11. E.H. Runyon (1942). Aggregation of separate cells of *Dictyostelium* to form a multicellular body. *The Collecting Net*, **17**, 88–88.
12. J.T. Bonner (1947). Evidence for the formation of cell aggregates by chemotaxis in the development of the slime mold *Dictyostelium discoideum. J. Exp. Zool.*, **106**, 1–26.
13. T.M. Konijn, J.G.C. van de Meene, J.T. Bonner, D.S. Barkley (1967). The acrasin activity of adenosine-3′,5′-cyclic phosphate. *Proc. Natl. Acad. Sci. USA*, **58**, 1152–1154.
14. T.M. Konijn, D.S. Barkley, Y.Y. Chang, J.T. Bonner (1968). Cyclic AMP: a naturally occurring acrasin in the cellular slime molds. *Am. Naturalist*, **102**, 225–233.
15. P.C. Newell, A. Telser, M. Sussman (1969). Alternative developmental pathways determined by environmental conditions in the cellular slime mold *Dictyostelium discoideum. J. Bacteriol.*, **100**, 763–768.
16. J.T. Bonner, A. Hay, D.G. John, H.B. Suthers (1985). pH affects fruiting and slug orientation in *Dictyostelium discoideum. J. Embryol. Exp. Morph.*, **87**, 207–213.
17. D-P. Häder, K.L. Poff, Light-induced accumulations of *Dictyostelium discoideum* amoebae (1979). *Photochem. Photobiol.*, **29**, 1157–1162.
18. D-P. Häder, K.L. Poff (1979). Photodispersal from light traps by amoebas of *Dictyostelium discoideum. Exp. Mycol.*, **3**, 121–131.
19. C.B. Hong, D.R. Fontana, K.L.Poff (1983). Thermotaxis in *Dictyostelium discoideum* amoebae and its possible role in pseudoplasmodium thermotaxis. *Proc. Natl. Acad. Sci. USA*, **80**, 5646–5649.

20. H.P. Vornlocher, D.P. Häder (1992). Biochemical and spectroscopic investigations of the presumed photoreceptor for phototaxis in amoebae of the cellular slime mold, *Dictyostelium discoideum*. *Acta Protozoologica*, **31**, 159–168.
21. T. Schlenkrich, M. Porst, D-P. Häder (1995). A rapid simple method for the isolation and characterization of the photoreceptor of *Dictyostelium discoideum* amoebae. *FEBS Lett.*, **364**, 276–278.
22. T. Schlenkrich, P. Fleischmann, D-P. Häder (1995). Biochemical and spectroscopic characterization of the putative photoreceptor for phototaxis in amoebae of the cellular slime mould *Dictyostelium discoideum*. *J. Photochem. Photobiol. B: Biol.*, **30**, 139–143.
23. J.T. Bonner (1994). The migration stage of *Dictyostelium*: behaviour without muscles or nerves. *FEMS Microbiol. Lett.*, **120**, 1–8.
24. P.R. Fisher (1997). Genetics of phototaxis in a model eukaryote, *Dictyostelium discoideum*. *Bioessays*, **19**, 397–407.
25. P.R. Fisher, U. Dohrmann, K.L. Williams (1984). Signal processing in *Dictyostelium discoideum* slugs. *Modern Cell Biol.*, **3**, 197–248.
26. D.W. Francis (1964). Some studies on phototaxis of *Dictyostelium*. *J. Cell. Compar. Physiol.*, **64**, 131–138.
27. K.L. Poff, W.L. Butler, W.F. Loomis (1973). Light-induced absorbance changes associated with phototaxis in *Dictyostelium*. *Proc. Natl. Acad. Sci. USA*, **70**, 813–816.
28. K.L. Poff, W.F. Loomis, W.L. Butler (1974). Isolation and purification of the photoreceptor pigment associated with phototaxis in *Dictyostelium discoideum*. *J. Biol. Chem.*, **249**, 2164–2168.
29. K.L. Poff, W.L. Butler (1974). Spectral characteristics of the photoreceptor pigment of phototaxis in *Dictyostelium discoideum*. *Photochem. Photobiol.*, **20**, 241–244.
30. K.L. Poff, W.F. Loomis (1973). Control of phototactic migration in *Dictyostelium discoideum*. *Exp. Cell Res.*, **82**, 236–240.
31. D-P.Häder, U. Burkart (1983). Optical properties of *Dictyostelium discoideum* pseudoplasmodia responsible for phototactic orientation. *Exp. Mycol.*, **7**, 1–8.
32. J.T. Bonner, F.E. Whitfield (1965). The relation of sorocarp size to phototaxis in the cellular slime mold *Dictyostelium purpureum*. *Biol. Bull.*, **128**, 51–57.
33. E. Wallraff, H.G. Wallraff (1997). Migration and bidirectional phototaxis in *Dictyostelium discoideum* slugs lacking the actin cross-linking 120 kDa gelation factor. *J. Exp. Biol.*, **200**, 3213–3220.
34. D-P. Häder (1985). Negative phototaxis of *Dictyostelium discoideum* pseudoplasmodia in UV radiation. *Photochem. Photobiol.*, **41**, 225–228.
35. P.K. Darcy, Z. Wilczynska, P.R. Fisher (1994). Genetic analysis of *Dictyostelium* slug phototaxis mutants. *Genetics*, **137**, 977–985.
36. P.R. Fisher, E. Smith, K.L. Williams (1981). An extracellular chemical signal controlling phototactic behaviour by *D. discoideum* slugs. *Cell*, **23**, 799–807.
37. U. Dohrmann, P.R. Fisher, M. Brüderlein, K.L. Williams (1984). Transitions in *Dictyostelium discoideum* behaviour: influence of calcium and fluoride on slug phototaxis and thermotaxis. *J. Cell Sci.*, **65**, 111–121.
38. P.R. Fisher, K.L. Williams (1982). Thermotactic behaviour of *Dictyostelium discoideum* slug phototaxis mutants. *J. Gen. Microbiol.*, **128**, 965–971.
39. M.J. Schneider, D.R. Fontana, K.L. Poff (1982). Mutants of thermotaxis in *Dictyostelium discoideum*. *Exp. Cell Res.*, **140**, 411–416.
40. P.K. Darcy, P.R. Fisher (1989). A role for G-proteins and inositol phosphate signalling in *Dictyostelium discoideum* slug behaviour. *J. Gen. Microbiol.*, **135**, 1909–1915.
41. P.K. Darcy, P.R. Fisher (1990). Pharmacological evidence for a role for cyclic AMP signalling in *Dictyostelium discoideum* slug behaviour. *J. Cell Sci.*, **96**, 661–667.

42. P.K. Darcy, Z. Wilczynska, P.R. Fisher (1993). Phototaxis genes on linkage group V in *Dictyostelium discoideum*. *FEMS Microbiol. Lett.*, **12**, 87–89.
43. P.K. Darcy, Z. Wilczynska, P.R. Fisher (1994). The role of cGMP in photosensory and thermosensory transduction in *Dictyostelium discoideum*. *Microbiol.*, 140 1619–1632.
44. P.R. Fisher, A.A. Noegel, M. Fechheimer, F. Rivero, J. Prassler, G. Gerisch (1997). Photosensory and thermosensory responses in *Dictyostelium* slugs are specifically impaired by absence of the F-actin cross-linking gelation factor (ABP-120). *Curr. Biol.*, **7**, 889–892.
45. B.D. Whitaker, K.L. Poff (1980). Thermal adaptation of thermosensing and negative thermotaxis in *Dictyostelium*. *Exp. Cell Res.*, **128**, 87–93.
46. P.R. Fisher, K.L. Williams (1981). Bidirectional phototaxis by *Dictyostelium discoideum* slugs. *FEMS Microbiol. Lett.*, **12**, 87–89.
47. W.F. Loomis, D.L. Welker, J. Hughes, D. Maghakian, A. Kuspa (1995). Integrated maps of the chromosomes in *Dictyostelium discoideum*. *Genetics*, **141**, 147–157.
48. A. Kuspa, W.F. Loomis (1996). Ordered yeast artificial chromosome clones representing the *Dictyostelium discoideum* genome. *Proc. Natl. Acad. Sci. USA*, **93**, 5562–5566.
49. Y. Hashimoto, K. Matsui (1988). Thermotaxis and the role of the *Dictyostelium discoideum* slug tip. *Cell Struct. Funct.*, **13**, 189–191.
50. H. Meinhardt (1983). A model for the prestalk/prespore patterning in the slug of the slime mold *Dictyostelium discoideum*. *Differentiation*, **24**, 191–202.
51. P.R. Fisher, W.N. Grant, U. Dohrmann, K.L. Williams (1983). Spontaneous turning behaviour by *Dictyostelium discoideum* slugs. *J. Cell Sci.*, **62**, 161–170.
52. A.J. Durston (1976). Tip formation is regulated by an inhibitory gradient in the *Dictyostelium discoideum* slug. *Nature*, **263**, 126–129.
53. A.J. Durston, F. Vork (1977). The control of morphogenesis and pattern in the *Dictyostelium discoideum* slug. In: P. Cappucinelli, J. Ashworth (Eds), *Development and Differentiation in the Cellular Slime Moulds* (pp. 17–26). Elsevier/North Holland, Amsterdam.
54. H.K. MacWilliams (1982). Transplantation experiments and pattern mutants in cellular slime mold slugs. In: S. Subtelny, P.B. Green (Eds), *Developmental Order: Its Origin and Regulation* (pp. 463–483). Alan R.Liss Inc., New York.
55. W.J. Kopachik (1982). Size regulation in *Dictyostelium*. *J. Embryol. Exp. Morphol.*, **68**, 23–35.
56. W.F. Loomis (1993). Lateral inhibition and pattern formation in *Dictyostelium*. *Curr. Topics Dev. Biol.*, **28**, 1–46.
57. E. Smith, P.R. Fisher, W.N. Grant, K.L. Williams (1982). Sensory behaviour in *Dictyostelium discoideum* slugs: phototaxis and thermotaxis are not mediated by a change in slug speed. *J. Cell Sci.*, **54**, 329–339.
58. J.T. Bonner, H.B. Suthers, G.M. Odell (1986). Ammonia orients cell masses and speeds up aggregating cells of slime molds. *Nature*, **323**, 630–632.
59. I.N. Feit, R.B. Sollitto (1987). Ammonia is the gas used for the spacing of fruiting bodies in the cellular slime mould *Dictyostelium discoideum*. *Differentiation*, **33**, 193–196.
60. T. Kosugi, K. Inouye (1989). Negative chemotaxis to ammonia and other weak bases by migrating slugs of the cellular slime moulds. *J. Gen. Microbiol.*, **135**, 1589–1598.
61. J.T. Bonner, A. Chiang, J. Lee, H.B. Suthers (1988). The possible role of ammonia in phototaxis of migrating slugs of *Dictyostelium discoideum*. *Proc. Natl. Acad. Sci. USA*, **85**, 3885–3887.
62. J.T. Bonner, D. Har, H.B. Suthers (1989). Ammonia and thermotaxis: further evidence for a central role of ammonia in the directed cell mass movements of *Dictyostelium discoideum*. *Proc. Natl. Acad. Sci., USA*, **86**, 2733–2736.

63. F. Siegert, C.J. Weijer (1992). Three-dimensional scroll waves organize *Dictyostelium* slugs. *Proc. Natl. Acad. Sci. USA*, **89**, 6433–6437.
64. H. Takahashi, T. Umeda, K. Inouye (1997). How does a slug turn?. In: The 1997 International *Dictyostelium* Conference, pp. 176–176.
65. J.T. Bonner (1998). A way of following individual cells in the migrating slugs of *Dictyostelium discoideum. Proc. Natl. Acad. Sci. USA*, **95**, 9355–9359.
66. J. Rubin, A. Roberston (1975). The tip of the *Dictyostelium discoideum* pseudoplasmodium as an organizer. *J. Embryol. Exp. Morph.*, **33**, 227–241.
67. J. Rubin (1976). The signal from fruting body and conus tips of *Dictyostelium discoideum. J. Embryol. Exp. Morphol.*, **36**, 261–271.
68. J.T. Bonner (1949). The demonstration of acrasin in the later stages of the development of the slime mold *Dictyostelium discoideum. J. Exp. Zool.*, **110**, 259–271.
69. E.J. Henderson (1975). The cyclic adenosine 3′,5′-monophosphate receptor of *Dictyostelium discoideum.* Binding characteristics and variation of binding levels during the life cycle. *J. Biol. Chem.*, **250**, 4730–4736.
70. C.L. Rutherford, R.D. Taylor, R.K. Merkle, L.T. Frame (1982). Cellular pattern formation: *Dictyostelium discoideum* as a system for a biochemical approach. *Trends Biochem. Sci.*, **7**, 108–111.
71. P. Pan, J.T. Bonner, H.J. Wedner, C.W. Parker (1974). Immunofluorescence evidence for the distribution of cyclic AMP in cells and cell masses of the cellular slime molds. *Proc. Natl. Acad. Sci. USA*, **71**, 1623–1625.
72. M. Brenner (1977). Cyclic AMP gradient in migrating pseudoplasmodia of the cellular slime mold *Dictyostelium discoideum. J. Biol. Chem.*, **252**, 4073–4077.
73. Y. Maeda (1977). Role of cyclic AMP in the polarized movement of the migrating pseudoplasmodium of *Dictyostelium discoideum. Dev. Growth Differ.*, **19**, 201–205.
74. J. Sternfeld, C.N. David (1981). Cell sorting during pattern formation in *Dictyostelium. Differentiation*, **20**, 10–21.
75. W.J. Kopachik (1982). Orientation of cells during slug formation in *Dictyostelium. Wilh. Roux's Arch. Dev. Biol.*, **191**, 348–354.
76. S. Matsukama, A.K. Durston (1979). Chemotactic cell sorting in *Dictyostelium discoideum. J. Embryol. Exp. Morphol.*, **50**, 243–251.
77. A.J. Durston, F. Vork (1979). A cinematographical study of the development of vitally stained *Dictyostelium discoideum. J. Cell Sci.*, **36**, 261–279.
78. R.I. Clark, T.L. Steck (1979). Morphogenesis in *Dictyostelium*: an orbital hypothesis. *Science*, **204**, 1163–1168.
79. R.L. Clark, G.S. Retzinger, T.L. Steck (1980). Novel morphogenesis in Ax-3, a mutant strain of the cellular slime mould *Dictyostelium discoideum. J. Gen. Microbiol.*, 319–331.
80. M. Nestle, M. Sussman (1972). The effect of cyclic AMP on morphogenesis and enzyme accumulation in *Dictyostelium discoideum. Dev. Biol.*, **28**, 545–554.
81. R.P. George (1977). Disruption of multicellular organization in the cellular slime molds by cyclic AMP. *Cell Diff.*, **5**, 293–300.
82. S.L. Barclay, E.J. Henderson (1982). Thermosensitive development and tip regulation in a mutant of *Dictyostelium discoideum. Proc. Natl. Acad. Sci. USA*, **79**, 505–509.
83. J.D. Gross (1994). Developmental decisions in *Dictyostelium discoideum. Microbiol. Rev.*, **58**, 330–351.
84. R.A. Firtel (1995). Integration of signalling information in controlling cell-fate decisions in *Dictyostelium. Genes. Dev.*, **9**, 1427–1444.
85. J. Williams (1995). Morphogenesis in *Dictyostelium*: new twists to a not-so-old tale. *Curr. Opin. Genet. Devel.*, **5**, 426–431.
86. C.D. Reymond, P. Schaap, M. Veron, J.G. Williams (1995). Dual role of cAMP during

Dictyostelium development. *Experientia*, **51**, 1166–1174.

87. P.J.M. van Haastert (1995). Transduction of the chemotactic cAMP signal across the plasma membrane of *Dictyostelium* cells. *Experientia*, **51**, 1144–1154.
88. W.F. Loomis (1996). Genetic networks that regulate development in *Dictyostelium* cells. *Microbiol. Rev.*, **60**, 135–150.
89. J.Y. Kim, P. van Haastert, P.N. Devreotes (1996). Social senses: G-protein-coupled receptor signalling pathways in *Dictyostelium discoideum. Chem. Biol.*, **3**, 239–243.
90. J. Rietdorf, F. Siegert, C.J. Weijer (1996). Analysis of optical density wave propagation and cell movement during mound formation in *Dictyostelium discoideum. Dev. Biol.*, **177**, 427–438.
91. O. Steinbock, F. Siegert, S.C. Müller, C.J. Weijer (1993). Three-dimensional waves of excitation during *Dictyostelium* morphogenesis. *Proc. Natl. Acad. Sci. USA*, **90**, 7332–7335.
92. T. Bretschneider, F. Siegert, C.J. Weijer (1995). Three-dimensional scroll waves of cAMP could direct cell movement and gene expression in *Dictyostelium* slugs. *Proc. Natl. Acad. Sci. USA*, **92**, 4387–4391.
93. F. Siegert, C.J. Weijer (1997). Control of cell movement during multicellular morphogenesis. In: Y. Maeda, K. Inouye, I. Takeuchi (Eds), *Dictyostelium* A Model System for Cell and Developmental Biology (pp. 425–436). Universal Academy Press Inc. Tokyo.
94. J.T. Bonner, M.R. Dodd (1962). Evidence for gas-induced orientation in the cellular slime molds. *Dev. Biol.*, **5**, 344–361.
95. J.T. Bonner (1993). Proteolysis and orientation in *Dictyostelium* slugs. *J. Gen. Microbiol.*, **139**, 2319–2322.
96. P.Schaap, R. Brandt, S. van Es (1995). Regulation of *Dictyostelium* adenylyl cyclases by morphogen-induced modulation of cytosolic pH or Ca^{2+} levels. *Dev. Biol.*, **167**, 179–188.
97. G.B. Williams, E.M. Elders, M. Sussman (1984). Modulation of the cAMP relay in *Dictyostelium discoideum* by ammonia and other metabolites: possible morphogenetic consequences. *Dev. Biol.*, **105**, 377–388.
98. F. Siegert, C.J. Weijer (1989). Digital image processing of optical density wave propagation in *Dictyostelium discoideum* and analysis of the effects of caffeine and ammonia. *J. Cell Sci.*, **93**, 325–335.
99. P.J.M. van Haastert (1985). The modulation of cell surface receptors from *Dictyostelium discoideum* by ammonium sulphate. *Biochim. Biophys. Acta*, **845**, 254–260.
100. P.R. Fisher (1991). The role of gaseous metabolites in phototaxis by *Dictyostelium discoideum* slugs,. *FEMS Microbiol. Lett.*, **77**, 117–120.
101. H. Haser, D-P. Häder (1992). Orientation and phototaxis in pseudoplasmodia of an axenic strain of the cellular slime mould, *Dictyostelium discoideum. Exp. Mycol.*, **16**, 119–131.
102. P.R. Fisher, K.L. Williams (1981). Activated charcoal and orientation behaviour by *Dictyostelium discoideum* slugs. *J. Gen. Microbiol.*, **126**, 519–523.
103. P. Schaap, M. Wang (1986). Interactions between adenosine and oscillatory cAMP signalling regulate size and pattern in *Dictyostelium. Cell*, **45**, 137–144.
104. P.C. Newell (1982). Cell surface binding of adenosine to *Dictyostelium* and inhibition of pulsatile signalling. *FEMS Microbiol. Lett.*, **13**, 417–421.
105. P.C. Newell, F.M. Ross (1982). Inhibition by adenosine of aggregation centre initiation and cyclic AMP binding in *Dictyostelium. J. Cell Sci.*, **89**, 123–127.
106. A. Theibert, P.N. Devreotes (1984). Adenosine and its derivatives inhibit the cAMP signalling response in *Dictyostelium discoideum. Dev. Biol.*, **106**, 166–173.
107. P.J.M. van Haastert (1983). Binding of cAMP and adenosine derivatives to *Dictyostelium discoideum* cells. Relationships of binding, chemotactic and antagonistic activities. *J. Biol. Chem.*, **258**, 9643–9648.

108. Y. Maeda, M. Maeda (1973). The calcium content of the cellular slime mold, *Dictyostelium discoideum*, during development and differentiation. *Exp. Cell Res.*, **82**, 125–130.
109. U.K. Tirlapur, J. Gross, V. Nanjundiah (1991). Spatial variation of sequestered calcium in the multicellular stage of *Dictyostelium discoideum* as assayed by chlortetracycline fluorescence. *Differentiation*, **48**, 137–146.
110. S. Saran, H. Nakao, M. Tasaka, H. Iida, F.I. Tsuji, V. Nanjundiah, I. Takeuchi (1994). Intracellular free calcium level and its response to cAMP stimulation in developing *Dictyostelium* cells transformed with jellyfish apoaequorin cDNA. *FEBS Lett.*, **337**, 43–47.
111. A.B. Cubitt, R.A. Firtel, G. Fischer, L.F. Jaffe, A.L. Miller (1995). Patterns of free calcium in multicellular stages of *Dictyostelium* expressing jellyfish aequorin. *Development*, **121**, 2291–2301.
112. M.M. van Lookeren Campagne, C. Erneux, R. van Eijk, P.J.M. van Haastert (1988). Two dephosphorylation pathways of inositol 1,4,5-triphosphate in homogenates of the cellular slime mould *Dictyostelium discoideum. Biochem, J.*, **254**, 343–350.
113. A.L. Drayer, J. van der Kaay, G.W. Mayr, P.J.M. van Haastert (1994). Role of phospholipase C in *Dictyostelium* - formation of inositol 1,4,5-triphosphate and normal development in cells lacking phospholipase C activity. *EMBO J.*, **13**, 1601–1609.
114. P. van Dijken, J.C.T. Bergsma, H.S. Hiemstra, B. Devries, J. van der Kaay, P.J.M. van Haastert (1996). *Dictyostelium discoideum* contains three inositol monophosphatase activities with different substrate specificities and sensitivities to lithium. *Biochem. J.*, **314**, 491–495.
115. C.M. Hedgepeth, L.J. Conrad, J. Zhang, H.C. Huang, V.M.Y. Lee, P.S. Klein (1997). Activation of the wnt signalling pathway – a molecular mechanism for the lithium action. *Dev. Biol.*, **185**, 82–91.
116. D.L. Welker, K.L. Williams (1982). A genetic map of *Dictyostelium discoideum* based on mitotic recombination. *Genetics*, **102**, 691–710.
117. W.F. Loomis (1987). Genetic tools for *Dictyostelium discoideum. Methods Cell Biol.*, **28**, 31–45.
118. D.L. Welker, K.L. Williams (1981). Genetic and cytological characterization of fusion chromosomes of *Dictyostelium discoideum. Chromosoma*, **82**, 321–332.
119. D.L. Welker, K.L. Williams (1985). Translocations in *Dictyostelium discoideum. Genetics*, **109**, 341–364.
120. K.L. Williams, G.E. Robson, D.L. Welker (1980). Chromosome fragments in *Dictyostelium discoideum* obtained from parasexual crosses between strains of different genetic background. *Genetics*, **95**, 289–304.
121. D.L. Welker, B.A. Metz, K.L. Williams (1982). Chromosome rearrangements in *Dictyostelium discoideum. Genetics*, **102**, 711–723.
122. D.L. Welker, K.L. Williams (1980). Mitotic arrest and chromosome doubling using thiabendazole, cambendazole, nocadazole, and benlate in the slime mould *Dictyostelium discoideum. J. Gen. Microbiol.*, **116**, 397–407.
123. W.F. Loomis (1970). Mutants in phototaxis of *Dictyostelium discoideum. Nature*, **227**, 745–746.
124. H. Kuwayama, S. Ishida, P.J.M. van Haastert (1993). Non-chemotactic *Dictyostelium discoideum* mutants with altered cGMP signal transduction. *J. Cell Biol.*, **123**, 1453–1462.
125. I.M. Zada-Hames (1977). Analysis of karyotype and ploidy of *Dictyostelium discoideum* using colchicine induced metaphase arrest. *J. Gen. Microbiol.*, **99**, 201–208.
126. G.E. Robson, K.L. Williams (1977). The mitotic chromosomes of the cellular slime mould *Dictyostelium discoideum*: a karyotype based on Giemsa banding. *J. Gen. Microbiol.*, **99**,

191–200.

127. Z. Wilczynska, P.R. Fisher (1994). Analysis of a complex plasmid insertion in a phototaxis-deficient transformant of a *Dictyostelium discoideum* selected on a *Micrococcus luteus* lawn. *Plasmid*, **32**, 182–194.
128. J.T. Bonner, J.G. Williams (1994). Inhibition of cAMP-dependent protein kinase in *Dictyostelium* prestalk cells impairs slug migration and phototaxis. *Dev. Biol.*, **164**, 325–327.
129. C.L. Saxe III, R.L. Johnson, P.N. Devreotes, A.R. Kimmel (1991). Expression of a cAMP receptor gene of *Dictyostelium* and evidence for a multigene family. *Genes Dev.*, **5**, 1–8.
130. G.S. Pitt, N. Milona, S. Borleis, K.C. Lin, R.R. Reed, P.N. Devreotes (1992). Structurally distinct and stage-specific adenylyl cyclase genes play different roles in *Dictyostelium* development. *Cell*, **69**, 305–315.
131. L. Wu, P.N. Devreotes (1991). *Dictyostelium* transiently express eight distinct G-protein α-subunits during its developmental program. *Biochem. Biophys. Res. Comm.*, **179**, 1141–1147.
132. P. Romans, R.A. Firtel (1985). Organization of the actin multi-gene family of *Dictyostelium discoideum* and analysis of variability in the protein coding regions. *J. Mol. Biol.*, **186**, 321–335.
133. P.S. Klein, T.J. Sun, C.L. Saxe III, A.R. Kimmel, R.L. Johnson, P.N. Devreotes (1988). A chemoattractant receptor controls development in *Dictyostelium discoideum*. *Science*, **241**, 1467–1472.
134. C.L. Saxe III, G.T. Ginsburg, J.M. Louis, R. Johnson, P.N. Devreotes, A.R. Kimmel (1993). CAR2, a prestalk cAMP receptor required for normal tip formation and late development of *Dictyostelium discoideum*. *Genes Dev.*, **7**, 262–272.
135. R.L. Johnson, C.L. Saxe III, R. Gollop, A.R. Kimmel, P.N. Devreotes (1993). Identification and targeted gene disruption of CAR3, a cAMP receptor subtype expressed during multicellular stages of *Dictyostelium* development. *Genes Dev.*, **7**, 273–282.
136. J.M. Louis, G.T. Ginsburg, A.R. Kimmel (1994). The cAMP receptor CAR4 regulates axial patterning during late development of *Dictyostelium*. *Genes Dev.*, **8**, 2086–2096.
137. Y.M. Yu, C.L. Saxe (1996). Differential distribution of cAMP receptors cAR2 and cAR3 during *Dictyostelium* development. *Dev. Biol.*, **173**, 353–356.
138. A.J. Harwood, N.A. Hopper, M.N. Simon, S. Bouzid, M. Veron, J.G. Williams (1992). Multiple roles for cAMP-dependent protein kinase during *Dictyostelium* development. *Dev. Biol.*, **149**, 90–99.
139. A.J. Harwood, N.A. Hopper, M.N. Simon, D.M. Driscoll, M.Veron, J.G.Williams (1992). Culmination in *Dictyostelium* is regulated by the cAMP-dependent protein kinase. *Cell*, **69**, 615–624.
140. G.N. Europe-Finner, P.C. Newell (1986). Inositol 1,4,5-triphosphate induces calcium release from a non-mitochondrial pool in amoebae of *Dictyostelium*. *Biochim. Biophys. Acta*, **887**, 335–340.
141. M. Schleicher, B. Andre, C. Andreoli, L. Eichinger, M. Haugwitz, A. Hofmann, J. Karakesiglou, M. Stockelhuber, A.A. Noegel (1995). Structure/function studies on cytoskeletal proteins in *Dictyostelium* amoebae as a paradigm. *FEBS Lett.*, **369**, 38–42.
142. M. Brink, Gerisch, G. Isenberg, G. Noegel, A.A. Segall, J.E. Wallraff, E.M. Schleicher (1990). A *Dictyostelium* mutant lacking an F-actin cross-linking protein, the 120-kD gelation factor. *J. Cell Biol.*, **111**, 1477–1489.
143. A.A. Noegel, B. Leiting, W. Witke, C. Gurniak, C. Harloff, H. Hartmann, E. Weissmüller, M. Schleicher (1989). Biological roles of actin-binding proteins in *Dictyostelium discoideum* examined using genetic techniques. *Cell Motil. Cytoskel.*, **14**, 69–74.
144. D. Cox, J. Condeelis, D. Wessels, D. Soll, H. Kern, D.A. Knecht (1992). Targeted

disruption of the ABP-120 gene leads to cells with altered motility. *J. Cell Biol.*, **116**, 943–955.

145. F. Rivero, B. Köppel, B. Peracino, S. Bozzaro, F. Siegert, C.J. Weijer, M. Schleicher, R. Albrecht, A.A. Noegel (1996). The role of the cortical actin cytoskeleton – F-actin crosslinking proteins protect against osmotic stress, ensure cell size, cell shape and motility, and contribute to phagocytosis and development. *J. Cell Sci.*, **109**, 2679–2691.
146. Y. Ohta, J.H. Hartwig (1995). Actin filament cross-linking by chicken gizzard filamin is regulated by phosphorylation *in vitro*. *Biochemistry*, **34**, 6745–6754.
147. Y. Ohta, J.H. Hartwig (1996). Phosphorylation of actin-binding protein 280 by growth factors is mediated by p90 ribosomal protein S6 kinase. *J. Biol. Chem.*, **271**, 11858–11864.
148. A. Marti, Z. Luo, C. Cunningham, Y. Ohta, J. Hartwig, T.P. Stossel, J.M. Kyriakis, J. Avruch (1997). Actin-binding protein-280 binds the stress-activated protein kinase (SAPK) activator SEK-1 and is required for tumor necrosis factor-α activation of SAPK in melanoma cells. *J. Biol. Chem.*, **272**, 2620–2628.
149. P. Fucini, C. Renner, C. Herberhold, A.A. Noegel, T.A. Holak (1997). The repeating segments of the F-actin cross-linking gelation factor (ABP-120) have an immunoglobulin-like fold. *Nature Struct. Biol.*, **4**, 223–230.
150. Y. Ohta, T.P. Stossel, J.H. Hartwig (1991). Ligand sensitive binding of actin-binding protein to immunoglobulin G Fc receptor (FcγIR). *Cell*, **67**, 275–282.
151. R.K. Andrews, J.E.B. Fox (1991). Interaction of purified actin-binding protein with the platelet membrane glycoprotein Ib-IX complex. *J. Biol. Chem.*, **266**, 7144–7147.
152. D. Wessels, D.R. Soll, D. Knecht, W.F. Loomis, A. de Lozanne, J. Spudich (1988). Cell motility and chemotaxis in *Dictyostelium* amoebae lacking myosin heavy chain. *Dev. Biol.*, **128**, 164–177.
153. A. de Lozanne, J.A. Spudich (1987). Disruption of the *Dictyostelium* myosin heavy chain gene by homologous recombination. *Science*, **236**, 1086–1091.
154. D.A. Knecht, W.F. Loomis (1987). Antisense RNA inactivation of myosin heavy chain gene expression in *Dictyostelium discoideum*. *Science*, **236**, 1081–1085.
155. E. Shelden, D.A. Knecht (1995). Mutants lacking myosin II cannot resist forces generated during multicellular morphogenesis. *J. Cell Sci.*, **108**, 1105–1115.
156. T.T. Egelhoff, R.J. Lee, J.A. Spudich (1993). *Dictyostelium* myosin heavy chain phosphorylation sites regulate myosin filament assembly and localization *in vivo*. *Cell*, **75**, 363–371.
157. E.R. Kuczmarski, S.R. Tafuri, L.M. Parysek (1987). Effect of heavy chain phosphorylation on the polymerization and structure of *Dictyostelium* myosin filaments. *J. Cell Biol.*, **105**, 2989–2997.
158. M. Pupillo, A. Kumagai, G.S. Pitt, R.A. Firtel, P.N. Devreotes (1989). Multiple alpha subunits of guanine nucleotide-binding proteins in *Dictyostelium*. *Proc. Natl. Acad. Sci. USA*, **86**, 4892–4896.
159. A. Kumagai, M. Pupillo, R. Gunderson, R. Miake-Lye, P.N. Devreotes, R.A. Firtel (1989). Regulation and function of Galpha protein subunits in *Dictyostelium*. *Cell*, **57**, 265–275.
160. A. Kumagai, J.A. Hadwiger, M. Pupillo, R.A. Firtel (1991). Molecular genetic analysis of two Galpha protein subunits in *Dictyostelium*. *J. Biol. Chem.*, **266**, 1220–1228.
161. J.A. Hadwiger, T.M. Wilkie, M. Strathmann, R.A. Firtel (1991). Identification of *Dictyostelium* Galpha genes expressed during multicellular development. *Proc. Natl. Acad. Sci. USA*, **88**, 8213–8217.
162. A.B. Cubitt, F. Carrel, S. Dharmawardhane, C. Gaskins, J. Hadwiger, P. Howard, S.K.O. Mann, K. Okaichi, K. Zhou, R.A. Firtel (1992). Molecular genetic analysis of signal transduction pathways controlling multicellular development in *Dictyostelium*. *CSH Symp.*

Quant. Biol., **57**, 177–192.
163. P. Lilly, L.J. Wu, D.L. Welker, P.N. Devreotes (1993). A G-protein beta-subunit is essential for *Dictyostelium* development. *Genes Dev.*, **7**, 986–995.
164. S. Dharmawardhane, A.B. Cubitt, A.M. Clark, R.A. Firtel (1994). Regulatory role of the G-alpha-1 subunit in controlling cellular morphogenesis in *Dictyostelium. Development*, **120**, 3549–3561.
165. J. Rietdorf, F. Siegert, S. Dharmawardhane, R.A. Firtel, C.J. Weijer (1997). Analysis of cell movement and signalling during ring formation in an activated Galpha1 mutant of *Dictyostelium discoideum* that is defective in prestalk zone formation. *Dev. Biol.*, **181**, 79–90.
166. J.A. Hadwiger, S. Lee, R.A. Firtel (1994). The G♀ subunit G♀4 couples to pterin receptors and identifies a signaling pathway that is essential for multicellular development in *Dictyostelium. Proc. Natl. Acad. Sci. USA*, **91**, 10566–10570.
167. R. Klein, R. Thiery, I. Tatischeff (1990). Dictyopterin, 6-(D-*threo*–1′,2′-dihydroxypropyl)-pterin, a new natural isomer of L-biopterin. Isolation from vegetative cells of *Dictyostelium discoideum* and identification. *Eur. J. Biochem.*, **187**, 665–669.
168. M. Gütlich, K. Witter, J. Bourdais, M. Veron, W. Rödl, I. Ziegler (1996). Control of 6-(D-*threo*–1′,2′-dihydroxypropyl) pterin (dictyopterin) synthesis during aggregation of *Dictyostelium discoideum. Biochem. J.*, **314**, 95–101.
169. L.J. Wu, C. Gaskins, K.M. Zhou, R.A. Firtel, P.N. Devreotes (1994). Cloning and targeted mutations of G♀7 and G♀8, two developmentally regulated G protein ♀-subunit genes in *Dictyostelium. Mol. Biol. Cell*, **5**, 691–702.
170. T. Jin, R.D.M. Soeda, J. Liu, A.R. Kimmel, P.N. Devreotes, P. Schaap (1998). Temperature-sensitive Gβ mutants discriminate between G protein-dependent and -independent signaling mediated by serpentine receptors. *EMBO J.*, **17**, 5076–5084.
171. F. McCormick (1994). Activators and effectors of *ras* p21 proteins. *Curr. Opin. Gen. Dev.*, **4**, 71–76.
172. C.D. Reymond, R.H. Gomer, M.C. Mehdy, R.A. Firtel (1984). Developmental regulation of a *Dictyostelium* gene encoding a protein homologous to mammalian Ras protein. *Cell*, **39**, 141–148.
173. S.M. Robbins, J.G. Williams, K.A. Jermyn, G.B. Spiegelman, G. Weeks (1989). Growing and developing *Dictyostelium* cells express different *ras* genes. *Proc. Natl. Acad. Sci. USA*, **86**, 938–942.
174. S.M. Robbins, V.V. Suttorp, G. Weeks, G.B. Spiegelman (1990). A *ras*-related gene from the lower eukaryote *Dictyostelium* that is highly conserved relative to the human *rap* genes. *Nucl. Acids Res.*, **18**, 5265–5269.
175. J. Daniel, G.B. Spiegelman, G. Weeks (1993). Characterization of a third *ras* gene, *rasB*, that is expressed throughout the growth and development of *Dictyostelium discoideum. Oncogene*, **8**, 1041–1047.
176. J. Daniel, J. Bush, J. Cardelli, G.B. Spiegelman, G. Weeks (1994). Isolation of two novel *ras* genes in *Dictyostelium discoideum* – evidence for a complex, developmentally regulated *ras* gene subfamily. *Oncogene*, **9**, 501–508.
177. C.D. Reymond, R.H. Gomer, W. Nellen, A. Theibert, P.N. Devreotes, R.A. Firtel (1986). Phenotypic changes induced by a mutated *ras* gene during the development of *Dictyostelium* transformants. *Nature*, **323**, 340–343.
178. S.A. Louis, G.B. Spiegelman, G. Weeks (1997). Expression of an activated *rasD* gene changes cell fate decisions during *Dictyostelium* development. *Mol. Biol. Cell*, **8**, 303–312.
179. M.L.W. Knetsch, S.J.P. Epskamp, P.W. Schenk, Y.W. Wang, J.E. Segall, B.E. Snaar-Jagalska (1996). Dual role of cAMP and involvement of both G-proteins and ras in

regulation of ERK2 in *Dictyostelium discoideum*. *EMBO J.*, **15**, 3361–3368.

180. L. Aubry, M. Maeda, R. Insall, P.N. Devreotes, R.A. Firtel (1997). The *Dictyostelium* mitogen-activated protein kinase 2 ERK2 is regulated by ras and cAMP-dependent protein kinase (PKA) and mediates PKA function. *J. Biol. Chem.*, **272**, 3883–3886.
181. A. Kuspa, W.F. Loomis (1992). Tagging developmental genes in *Dictyostelium* by restriction enzyme-mediated integration of plasmid DNA. *Proc. Natl. Acad. Sci. USA*, **89**, 8803–8807.
182. T.L. McMahon, Z. Wilczynska, C. Barth, D.J. Fraser, L. Pontes, P.R. Fisher (1996). Replicon rescue: a novel strategy to clone the genomic DNA flanking insertions of integrating shuttle vector DNA. *Nucl. Acids Res.*, **24**, 4096–4097.
183. Z. Wilczynska, C. Barth, P.R. Fisher (1997). Mitochondrial mutations impair signal transduction in *Dictyostelium discoideum* slugs. *Biochem. Biophys. Res. Comm.*, **234**, 39–43.
184. H.F. Noller (1993). Peptidyl transferase: protein, ribonucleoprotein, or RNA? *J. Bacteriol.*, **175**, 5297–5300.
185. R. Rizzutto, C. Bastianutto, M. Brini, M. Murgia, T. Pozzan (1994). Mitochondrial Ca^{2+} homeostasis in intact cells. *J. Cell Biol.*, **126**, 1183–1194.
186. S. Jouaville, F. Ichas, E.K. Holmuhamedov, P. Camacho, J.D. Lechleiter (1995). Synchronization of calcium waves by mitochondrial substrates in *Xenopus laevus* oocytes. *Nature*, **377**, 438–441.
187. L.I. Grossman, E.A. Shoubridge (1996). Mitochondrial genetics and human disease. *Bioessays*, **18**, 983–991.
188. J.P. Hendrick. F.U. Hartl (1995). The role of molecular chaperones in protein folding. *FASEB J.*, **9**, 1559–1569.

Photomovement
D.-P. Häder and M. Lebert, editors.

Chapter 20

Photomovement and photomorphogenesis in *Physarum polycephalum*: targeting of cytoskeleton and gene expression by light

Wolfgang Marwan

Table of contents

Abstract

Plasmodia of *Physarum polycephalum* are giant single cells that respond to UV and visible light by photomovement and photomorphogenesis. Light sensing is mediated by the *Physarum* phytochrome and at least three additional blue and UV light photoreceptors that converge into signal transduction pathways to control basic cellular functions. One pathway targets the cytoskeleton resulting in changes of the cell shape, the periodicity of protoplasmic streaming and the direction of cell migration during photomovement. Other pathways target gene expression and mediate photomorphogenetic responses like sporulation of starving plasmodia. The developmental program of sporulation occurs upon irreversible commitment and includes the activation of a gene expression cascade, differential regulation of the cell cycle and changes in cell motility including its regulation by external stimuli. Factors found to be involved in signal transduction are Ca^{2+}, cyclic nucleotides and metabolic intermediates.

Phylogentically, *Physarum* is more close to the protozoa than to any other phylum and its sensory physiology displays motifs found in both, plant and animal cells: the synergism of phytochrome and blue light photoreceptors in *Physarum* is a recurring theme in higher plant development and the sensory control of differentiation, cell cycle and cell motility is fundamental in developing animal cells as well.

Over the years, a considerable body of mainly physiological data on photosensing in *Physarum* plasmodia, namely photoreceptors involved, signal transduction mechanisms and downstream cellular targets has accumulated. This review puts these data together to draw a clearcut picture of the *Physarum* photobiology. It is meant to provide an up-to-date basis for new start-ups employing the genetic tools nowadays available for *Physarum polycephalum*, gene replacement and time-resolved somatic complementaion analysis. These tools together with the extreme experimental manipulatability of plasmodial cells make *Physarum* a unique cellular model which promises not only answers to old and new questions in photobiology and but also new general insight into central pathways of cell control.

20.1 Introduction

Physarum polycephalum plasmodia are giant multinuclear single cell organisms that respond to light by phototaxis and photomorphogenesis. *Physarum* plasmodia have attracted much attention by cell biologists. The perfect synchrony of all nuclei with respect to cell cycle and differentiation made *Physarum* a standard research object in the early days of cell cycle research [1]. Plasmodia in addition provide unusual possibilities for experimental manipulation because of their macroscopic size (a single plasmodium can cover surfaces ranging from parts of a millimeter to several square meters. K.-E. Wohlfarth-Bottermann and R. Stiemerling grew a 5.54 m^2 plasmodium (still a single cell) and entered the Guinness Book of Records in 1989 with the largest protozoon in the world [2]. For laboratory investigations, plasmodia can be either examined as a whole, or pieces of any size can be excised and analyzed separately since each of them stays alive and still behaves like a, yet smaller, single cell. In fact, this also works the other way round: plasmodia or pieces excised from them spontaneously fuse as they get into contact with each other, mix by vigorous protoplasmic streaming and thereby form a homogeneous single celled plasmodium with a new identity.

Due to the pioneering work of a small group of researchers, *Physarum* is nowadays a genetic organism (for reviews see [3–7]) that can be grown axenically [8,9] and is susceptible to targeted gene replacement [5,10].

Plasmodia develop from a diploid zygote formed by mating of two haploid amoebal cells. Upon continuous growth, the zygote undergoes multiple nuclear divisions finally resulting in a huge protoplasmic mass that contains millions of nuclei. Starving plasmodia differentiate into a highly motile network of strands. Driven by vigorous and a highly rhythmic protoplasmic streaming, they can migrate on solid surfaces over distances of many centimeters per day. A plasmodium is not an amorphous protoplasmic mass. It exhibits perfect polarity as migrating animal cells do. Polarity is morphological and functional. A migrating plasmodium is differentiated into a protruding front and a veined network that contains a considerable part of the protoplasm that follows the front (Figure 1).

Plasmodia are highly responsive to environmental stimuli. They are able to sense light, chemicals, temperature gradients, humidity and may be other stimuli. Visible light and UV control the motility behavior (phototaxis and photoavoidance response), cell differentiation (sporulation and spherulation) and metabolic processes (for detailed reviews see [11,12] and references therein).

This review tries to give a comprehensive overview over the photosensory physiology of *Physarum polycephalum* with special emphasis on photomovement and sporulation. These two photoresponses may be controlled by the same photoreceptors and both responses involve the cytoskeleton as molecular effector system in various ways.

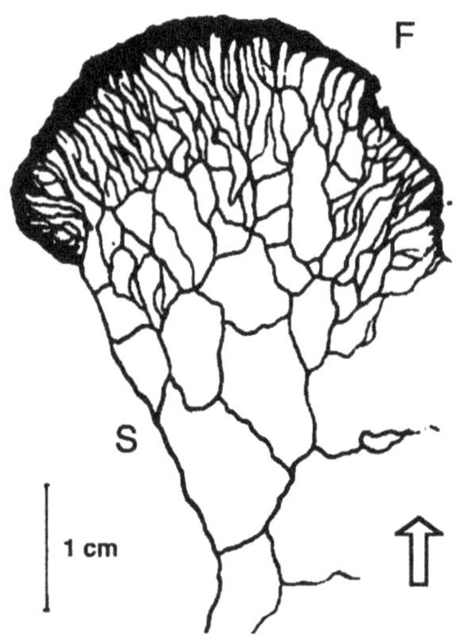

Figure 1. Migrating plasmodium of *Physarum confertum*. The plasmodium is polarized into a migrating front (F) and a network of plasmodial strands (S). The strands and the front form a protoplasmic continuum. The arrow indicates the direction of migration (from Wohlfarth-Bottermann & Stockem: [61]).

20.2 Life cycle of *Physarum polycephalum* and its use for genetics

During its life cycle, *Physarum polycephalum* proceeds through several well-defined states of cellular differentiation ([5], and references therein; Figure 2). The life cycle may start when haploid amoebae hatch from mature spores. The amoebae look like soil amoebae, feed on bacteria and multiply by mitosis as the "true" soil amoebae do. When maintained as single clones in the laboratory they can be grown for indefinite time. Cells of two amoebal clones of different mating type can fuse to give a diploid zygote. Upon

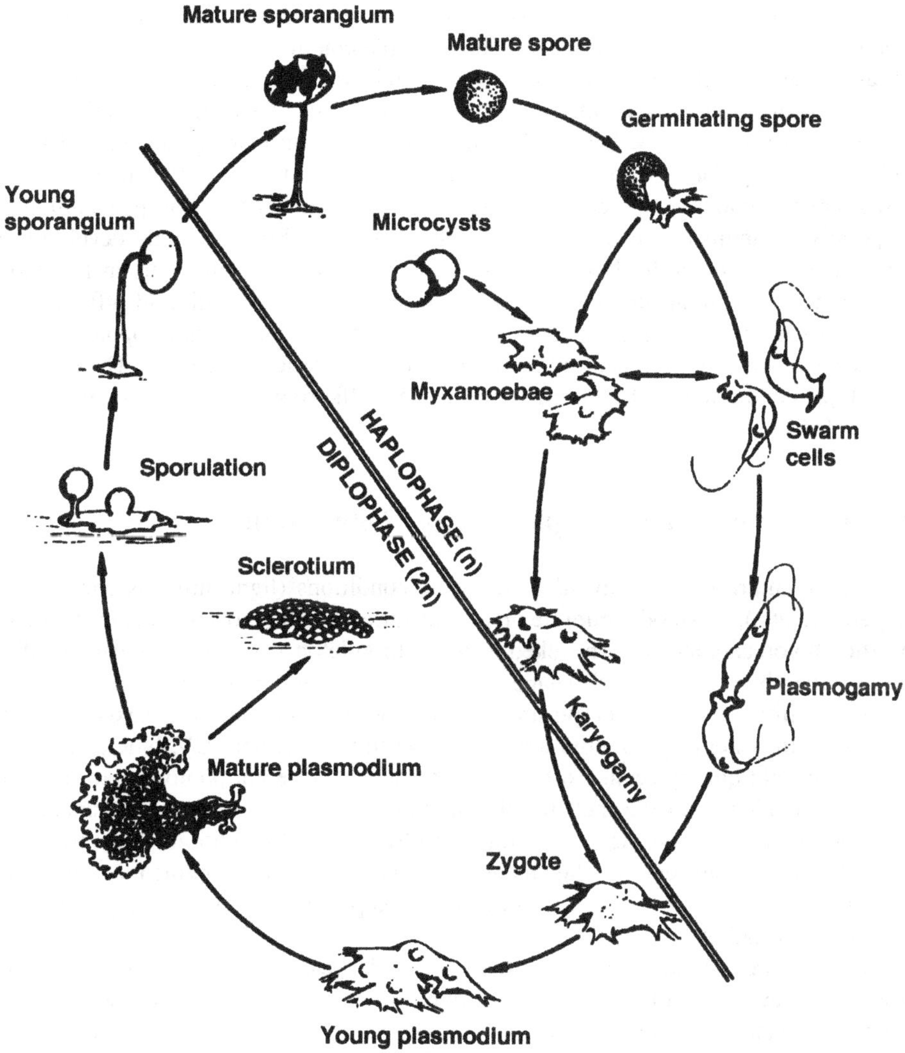

Figure 2. Life cycle of a typical myxomycete like *Physarum polycephalum*. Reprinted from [62].

continuous growth and multiple nuclear divisions, the zygote develops into a macroscopic, diploid and multinuclear plasmodium. The plasmodium continues growth as long as nutrition is available. Mature, i.e. starving plasmodia can encapsulate by forming macrocysts, so-called sclerotia or in an alternative developmental pathway they can sporulate. Sclerotization especially occurs upon desiccation. Sporulation of starving plasmodia is triggered by UV or visible light ([11,12] and references therein) or by heat shock [13]. During sporulation the entire plasmodial mass cleaves and differentiates into fruiting bodies each of which finally contains hundreds of haploid spores. Although the plasmodium is the most obvious (and an obligatory) stage, under natural conditions it may represent only a short episode in the life cycle of an acellular slime mould. Upon a closer look, many amoebae found in the soil might turn out to be slime mold amoebae, capable for plasmodium development under suitable conditions.

Plasmodial mutants of *Physarum* can be obtained by mutagenesis of uninuclear, haploid amoebae which are subsequently regenerated to plasmodia by mating or by apogamic development (see below). Plasmodia then contain many genetically identical nuclei. Screening for mutants is facilitated by a temperature-sensitive mutation that allows haploid amoebae to directly develop into a multinuclear haploid plasmodium at the permissive temperature (apogamic development [14,15]). Apogamic development allows to screen for both, dominant and recessive non-lethal mutations in a haploid background. i.e. without the need for screening the homozygous diploid offspring. At the non-permissive temperature, amoebae capable for apogamic development nevertheless can be crossed with haploid amoebae of other genetic background. The resulting diploid plasmodium can be sporulated and the offspring subjected to segregation analysis [15].

20.3 Photomovement and phototaxis of plasmodia

In search of nutrition or to avoid unfavorable conditions (light, noxious compounds, temperature etc.), plasmodia migrate on a solid surface. Photomovement is mediated by specific photoreceptors that are linked to the cytoskeleton via a biochemical signaling pathway thereby regulating cell shape and cytoplasmic streaming. Two effects of light on the motility of *Physarum* plasmodia have been described: photoavoidance and phototaxis. In general, plasmodia are responsive to far-UV, near-UV, blue and far-red light. Each light quality seems to be sensed through a different photoreceptor protein. This is concluded from the fact that the individual light responses can be uncoupled from each other by choosing the appropriate physiological condition. In contrast to a wide-spread misinterpretation, the characteristic intensive yellow pigment of *Physarum polycephalum* wild-type isolates is not a photoreceptor neither triggering photomovement nor sporulation.

Plasmodia are able to discriminate light and dark areas and to move to a dark place leaving not even a part of the plasmodium in the light. This response is called "photoavoidance". Plasmodia are also able to sense the direction of a light source and can directly migrate toward or away from it. This response is called "phototaxis". It is not clear whether photoavoidance and phototaxis use the same photoreceptors and signal transduction mechanisms although the action spectra look similar. Signal transduction in phototaxis and/or photoavoidance seems to involve Ca^{2+} released from intracellular stores into the cytoplasm, cyclic nucleotides and the mitochondrial citric acid cycle.

20.3.1 The mechanism of plasmodial motility and how it is measured

Plasmodial motility and cytoplasmic streaming. Plasmodia can migrate on a solid surface. As in animal cells, migration is based on the asymmetric activity of the cytoskeleton, however the mechanism appears somewhat different. The plasmodial protoplasm moves by continuous and highly rhythmic streaming through the protoplasmic veins. These veins are differentiated into an ectoplasmic tube and an endoplasmic core which represents the fluid protoplasm. The ectoplasmic tube consists of sheets and fibrils of actomyosin embedded in longitudinal and radial orientation into the ectoplasmic gel (Figure 3). The actomyosin structures are connected to a complex system of plasmalemma invaginations. The ectoplasmic tube is attached to the substrate via pseudopods ([16] and references therein). By reversible contraction of the longitudinal and the radial actomyosin fibrils the endoplasm is pumped back and forth

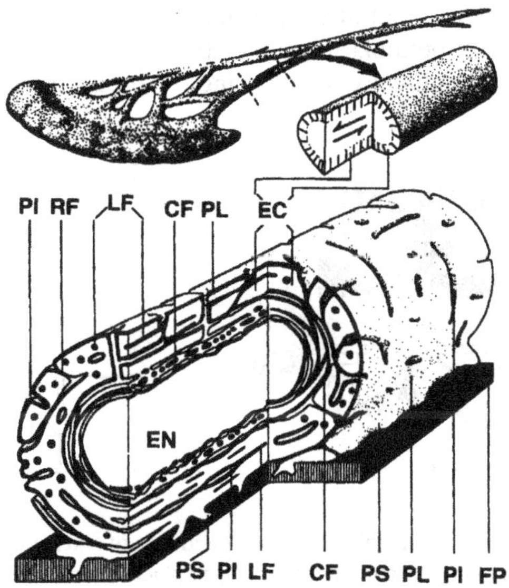

Figure 3. Subcellular structure of a plasmodial vein of *Physarum polycephalum*. The three-dimensional structure (bottom) of a protoplasmic vein as part of a migrating plasmodium (top) is shown. The vein is differentiated into an ectoplasmic gel and an endoplasmic sol. The sol (fluid protoplasm) is pumped through the lumen rhythmically forth and back (as indicated by the arrows). Actomyosin fibrils oriented in longitudinal, radial and circular direction are restricted to the ectoplasmic region. Depending on the size of the plasmodium, a vein can vary from about 0.1 to 1 mm in diameter. CF circular fibrils, LF longitudinal fibrils, RF radial fibrils, EC ectoplasm, EN endoplasm, PI plasmalemma invagination, PL plasmalemma, PS pseudopodium, FP Filter paper (reprinted from Fleischer & Wohlfarth-Bottermann [63]).

[16–18]. Since these contractions occur in an oscillatory manner, the phenomenon is called shuttle streaming. If there is a net protrusion of the cytoplasm into the front part, the plasmodium moves in forward direction.

Four different parameters of plasmodial motility have been analyzed:

1. migration of the entire plasmodium by observing traces left on the agar surface of the substratum,
2. measuring the pressure difference generated between the two parts of an isolated plasmodial strand,
3. measuring the tensional force generated by an isolated plasmodial strand, and
4. measuring the change in local thickness of a plasmodium as a function of time by light scattering.

Measuring protoplasmic pressure by the Kamiya double chamber method. That protoplasmic streaming indeed is mediated by contraction of the actomyosin fibrils was demonstrated by the classic experiments of Kamiya [19]. A protoplasmic strand is excised form a plasmodium and placed into the narrow gap connecting two chambers leaving no air in the gap (Figure 4a, upper part). The actomyosin fibrils reorganize in the plasmodium and the cytoplasm flows between the two parts rhythmically forth and back. One of the chambers is connected to a regulatable barometer. When the protoplasm streams from one part of the plasmodium into the other, the pressure in the chamber changes as the protoplasmic volume does. The pressure exerted by the plasmodium is measured indirectly by adjusting the pressure within the chamber to a

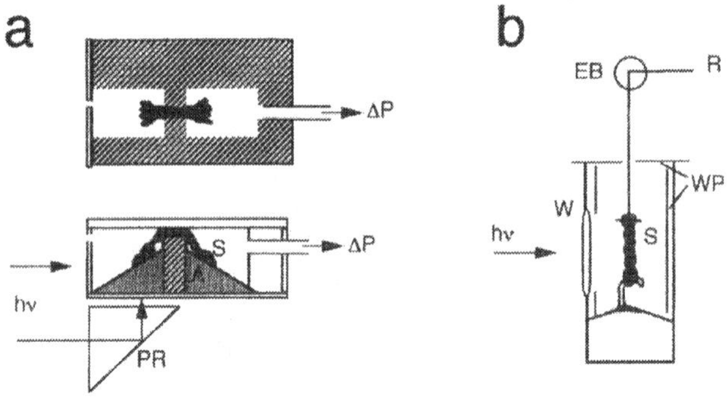

Figure 4. Devices to measure the biomechanical activity of excised plasmodial strands. (a) Kamiya double chamber. The plasmodial strand (S) is placed into two chambers partially filled with agar gel (A). The strand completely fills the narrow gap between the two chambers to make an air-tight seal. The pressure of the right chamber is adjusted by the operator to stop the protoplasmic flow through the gap between the two parts of the plasmodium. These changes in pressure (ΔP) are recorded by a barometer as a function of time. To measure light responses of a plasmodial strand, actinic light (hν) is projected into the left chamber from the side and from below through an optical prism (PR). (b) Moist chamber for measuring the isometric tension. A plasmodial strand (S) is hung vertically into a chamber kept moist by wet filter paper (WP). The isometric tension exerted by the plasmodial strand is measured by an electrobalance (EB) connected to a strip chart recorder (R). Actinic light (hν) is applied through a small window (W) (redrawn from Hato et al. [21]).

level just sufficient to stop the cytoplasmic streaming through the gap. The pressure necessary to suppress streaming, changes with time in an oscillatory manner and reflects the pressure exerted onto the endoplasm by contraction of the ectoplasmic fibrils. These contractions mediate the shuttle streaming activity of the protoplasm in the unperturbed cell. A plot of the pressure difference between the two chambers against the time is called "dynamoplasmogram", and a typical example is shown in Figure 7. The plasmodial response to light can be measured just by projecting a light spot into one of the two chambers (see below).

Measuring the isometric tension exerted by a plasmodial strand. The isometric tension exerted by a plasmodial strand can be measured by a method similar to that described by Kamiya et al. [20]. The strand portion of the slime mold is hung vertically into a moist chamber as shown in Figure 4b. The change in isometric tension is detected by an electro-balance and displayed on a strip chart recorder. The isometric tension changes in an oscillatory manner with a period of 1.5 to 3 min and an amplitude of 2 to 5 mg [21]. The plasmodial response to light again is measured by projecting actinic light into the chamber.

Measuring changes in the local plasmodial thickness by computerized video analysis. If the spatio-temporal pattern of the protoplasmic streaming in an intact plasmodium is of interest, the thickness of the plasmodium at defined points can be measured. In practice this is done by recording the stray light resulting from an infrared observation light source by a CCD camera coupled to a computer via a video frame grabber [22].

20.3.2 Phototaxis

How plasmodia do phototaxis. Phototaxis in general is defined as a movement response relative to the direction of the incident light [23]. The organism hence responds to light as a vectorial stimulus.

When plasmodia of *Physarum polycephalum* are migrating over a solid agar surface that is exposed to lateral light, the cells exhibit phototaxis [24]. Positive phototaxis occurs in response to low, negative phototaxis in response to high light intensity. The response pattern of an individual plasmodium during negative phototaxis depends on the direction of the stimulus light relative to the direction in which the plasmodium is migrating at the moment when it is exposed to the stimulus [24]. If the moving front of the plasmodium is hit at an angle of $>>0$ and $<90°$ within the plane of the substrate, the plasmodium performs a characteristic U-turn. The moving front gradually changes direction to move away from the light source until the light stimulus hits the plasmodium at an angle of about 120° (Figure 5a). The cellular response differs, if the light stimulus hits the plasmodium face on. Then the cell is repolarized and the ratio between forward and backward cytoplasmic streaming is reversed. More cytoplasm is transported to the back of the plasmodium where a new moving front is created (Figure 5b). The occurrence of these two photoresponses is not only interesting with respect to their underlying mechanisms for light detection and cell polarity reorganization. It is also a nice and clear example for true phototaxis in the sense of its definition by Diehn

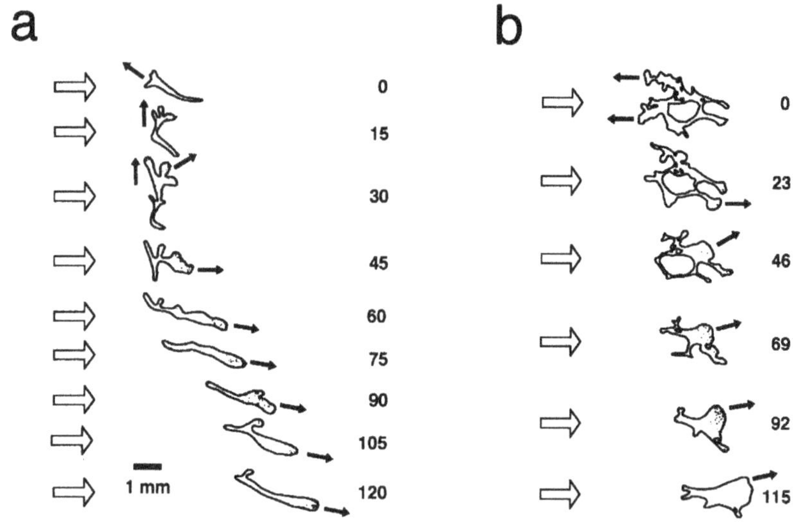

Figure 5. Phototactic orientation of small migrating plasmodia to lateral light. (a) U-turn induced by light hitting the moving front of the plasmodium at a certain angle with respect to the initial movement direction. (b) Repolarization of the movement direction by light hitting the plasmodium face on. Thick arrows indicate the direction of the stimulus light, thin arrows the direction of plasmodial movement. Numbers indicate the time in minutes after onset of the stimulus light (redrawn from Häder & Schreckenbach [24]).

et al. [23]. Dependence of the response on the angle of the light direction rather than on its respective intensity demonstrates that the response in controlled by the light direction rather than by small differences in light intensity in a shallow light intensity gradient (which often can not be completely excluded in such types of experiments).

Action spectra define the photoreceptors of phototaxis. In order to characterize the photoreceptor(s) involved in plasmodial phototaxis, Häder and Schreckenbach [24] measured stimulus response curves at wavelengths ranging from 300 to 700 nm. The fluence rate-response curves follow the same pattern at all wavelengths: low fluence rates cause positive and high fluence rates negative phototaxis [24]. From the fluence rate-response curves it seems that positive and negative phototaxis are mediated by different photoreceptor systems as indicated by a different wavelength-dependence of the relative photon effectiveness for each of the two responses.

An action spectrum for the negative phototaxis was obtained by measuring the phototactic orientation as a function of wavelength at constant energy fluence rate. The action spectrum indicates a pronounced effectiveness of near-UV and blue light but low or insignificant response in the region between 500 and 700 nm [24]. An important finding of the authors is that the action spectrum obtained from the deeply yellow colored wild type strain did not differ significantly form that obtained from an albino mutant strain. The albino strain was even five times more light sensitive as compared to the yellow wild type. This excluded the abundant yellow pigment as active photoreceptor species. Otherwise the efficiency of the signal transduction pathway would have to be at least 500 fold higher in the albino strain as compared to the yellow wild type and this seems extremely unlikely.

Signal transduction in phototaxis involves Ca^{2+} fluxes. The signal transduction mechanisms involved in phototaxis were analyzed by employing pharmacological evidence. Calcium transport blockers (ruthenium red or lanthanum ions) inhibit negative phototaxis. The same effect occurs in response to caffeine or phosphatidic acid that are known to enhance calcium fluxes [25]. Stimulation with light causes drastic changes of the cytoplasmic Ca^{2+} concentration [26] and a transient hyperpolarization of the plasma membrane [25]. The light-induced Ca^{2+} release seems to occur from intracellular stores (e.g. mitochondria of ER) since rhythmical Ca^{2+} fluxes across the membrane have been excluded [25,26]. Ca^{2+} release from the mitochondria is an attractive hypothesis since the turnover rate of metabolic intermediates in these organelles seems to be crucial for the photomovement responses observed in excised plasmodial strands ([27] and see below).

20.3.3 Photoavoidance response

The photoavoidance response and how it is quantified. Plasmodia are able to escape light even if it falls perpendicular onto the surface they are migrating on. In contrast to phototaxis, this response is called photoavoidance because it occurs independently from the direction of the incident light [28]. Bialczyk [28] measured an exact action spectrum for the photoavoidance response of young plasmodia of *Physarum nudum*. Round plasmodia (about 1 cm in diameter) without an oriented frontal part, placed on an agar surface, were partially irradiated with a light spot from above. Dependent on the light intensity, the plasmodia eventually avoided the light area and migrated into the dark. The probability that an individual plasmodium responded by avoidance was proportional to the logarithm of the fluence rate of the monochromatic stimulus light. Perfectly parallel fluence rate-response curves were obtained for different wavelengths that were used to calculate the wavelength-dependence of the relative photon effectiveness. An action spectrum with distinct peaks in the near-UV (375 nm) and the blue (452 nm) was obtained. The cells were not responsive to wavelengths above 500 nm. At 452 nm, a fluence rate of 0.2 W/m^2 caused half maximal response.

Whether plasmodia avoid or seek light depends on their physiological state. Whether a photomovement response of *Physarum nudum* plasmodia is positive or negative depends on both, the light intensity and the physiological state of the plasmodium. Young plasmodia avoid light. In contrast, mature plasmodia in which vegetative growth has been terminated, migrate towards light of low intensity but avoid light of high intensity [28–30]. Note that the light avoidance response can be abolished if plasmodia competent for spherulation are exposed for 24 h to blue light of intermediate intensity [31].

Four photoreceptors are distinguished by action spectrometry. An exact action spectrum for the photoavoidance response of *Physarum polycephalum* plasmodia was worked out by Ueda et al. [22]. The authors irradiated about a quarter of a 5 cm plasmodium with a monochromatic light beam from above and recorded its response with a high sensitivity video camera under infared (maximum at 920 nm) safe light. The rate of the avoidance was determined from the initial linear increase of the brightness level after irradiation caused by translocation of protoplasm away from the irradiated area. A plot of the rate of photoavoidance against the logarithm of the fluence rate yields perfectly linear curves at all wavelengths measured. The fluence rate-response curves were used to construct an action spectrum by plotting 1/threshold fluence rate against the wavelength. A threshold spectrum is expected to be congruent to the absorption spectrum (wavelength-dependence of the absorption cross section) of the active photoreceptor species, even if the photoreceptor kinetics are not known [32]. The threshold action spectrum exhibits distinct peaks at 260, 370, 460 and 750 nm and a shoulder at 290 nm (Figure 6). The nonuniform slope of the fluence rate-response curves suggests that more than one photoreceptor species or even antagonistically active photochemical intermediates may be involved. This is supported by the finding that the response to far or near UV and to blue light were persistent whereas the response to far-red light was transient lasting only about 5 minutes [22]. That different photoreceptors for blue and far-red light are involved in photomovement responses is also suggested by intersecting fluence rate-response curves of different slope that result when the isometric tension of excised plasmodial strands was measured in response to monochromatic light at different wavelengths [21]. The authors conclude that phytochrome or a photoreceptor similar to phytochome may be involved in the photomovement response [21]. Further evidence for different photosystems for UV and blue light is that upon starvation the plasmodial response to near UV decreases 15 fold while the response to the other wavelengths of maximal effectiveness remains constant [33]. Shifting the temperature of nourished plasmodia from 25°C to 31°C selectively reduces the sensitivity to blue light 12 fold while the sensitivities to UV-A and UV-C remain constant [33]. The action spectrum for the photoavoidance response shown in Figure 6 was measured using an

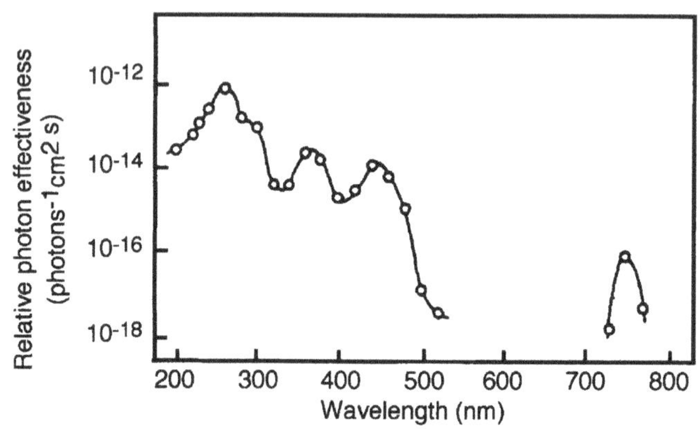

Figure 6. Action spectrum for the photoavoidance response of *P. polycephalum*. The relative photon effectiveness at each wavelength was obtained from fluence rate response curves and defined as 1/threshold fluence rate required to evoke a minimal response (replotted from Ueda et al. [22]).

albino strain of *Physarum polycephalum*, which again excludes the bulk yellow pigment of wild-type plasmodia from being the active photoreceptor.

Superoxide is generated by actinic blue light. Interestingly, light effective in causing the photoavoidance response also generates superoxide [22]. The action spectrum for superoxide generation in the blue and UV region is very similar to that for the photoavoidance response. From this similarity, the authors assume that superoxide generation occurs by the photochemical reactivity of the photoreceptors, but leave the question open whether or not superoxide is in the main pathway of signal transduction. Photobiologists know that similar or even identical action spectra never prove that two effects are mediated by one and the same photoreceptor.

Why is the response to far-red light not reported throughout the literature? While all authors found pronounced photomovement responses to UV and blue light [21,22,24,27,28,34,35], only some described responses to far-red light (21,22,33) while others found that there is no response to far-red light ([24,28] (for *Physarum nudum*)). Although these findings seem clearly contradictory at the first glance, they could be easily explained by different physiological conditions of the plasmodia under investigation. Plasmodia used for photomovement experiments are usually transferred from an ill-defined growth substrate (oat flakes) onto a non-nutrient agar gel on which they migrate and continually starve. Hence, experiments are carried out with plasmodia that might be in different states of starvation from study to study. The blue light photoreceptor that triggers sporulation is constitutively present and can even be activated in non-starved plasmodia. In contrast only starved plasmodia are susceptible to far-red light induction of sporulation [13]. Therefore it might well be that the responsiveness to far-red light depends on the degree of starvation in the case of photomovement as well. Note in this context that only mature plasmodia of *P. nudum* migrate towards dim light while young plasmodia do not [29,30]. Growth stage-dependence of photoreceptor equipment may be a general phenomenon. The photorecptors sensory rhodopsin-I and sensory rhodopsin-II mediate the photophobic response of *Halobacterium salinarum* ([35] and see the chapter by Spudich in this volume). The blue light photoreceptor sensory rhodopsin-II is constitutively present while sensory rhodopsin-I, the photoreceptor for orange light and near UV is strongly induced during growth of the culture. The efficiency of the signal transduction pathway that connects sensory rhodopsin-I to the flagellar motor in parallel changes in a complex manner [36]. This is the reason why action spectra recorded with cells from cultures at different stages of growth look completely different, a fact which produced considerable confusion in the literature for some years.

Signal transduction by cyclic nucleotides. Ueda et al. [37] found that cAMP and cGMP levels increased upon illumination with UV or blue light. They postulate that an increased concentration of cyclic nucleotides leads to an enhanced local development of contractile fibrils which squeeze the protoplasmic sol from the area resulting in photoavoidance [37]. Together with the finding that Ca^{2+} is involved in phototaxis [25] this nicely fits into the general schemes of mammalian signal transduction pathways. Nevertheless, finding the small molecules is only the first step and it would be interesting to see the proteins that are involved in the photocontrol of plasmodial motility.

20.3.4 Photomovement responses of excised plasmodial strands

Phototaxis, photoavoidance and photomovement responses of excised plasmodial strands may emerge from different mechanisms. The response of isolated strands or the local response of the protoplasm of an intact plasmodium frequently are called "phototactic" in the literature. Although it seems plausible that photoavoidance response and phototaxis are based on differential regulation of the shuttle streaming activity, no direct evidence has been presented.

Measuring the shuttle streaming activity of intact or excised plasmodial strands provides the advantage of a higher resolution in time than quantifying the migration of a plasmodium either with respect to the direction (phototaxis) or the intensity (photoavoidance) of the incident light. High resolution in time also allows detection of short and transient responses to light which might not be detectable if the path of an intact plasmodium is followed. However, one should keep in mind that phototaxis and photoavoidance of intact plasmodia and photomovement responses of (excised) plasmodial strands may emerge from different biochemical and biomechanical mechanisms. Therefore the findings on these three types of response should not be mixed up.

Actinic light biasses the protoplasmic shuttle streaming to one direction and modulates the underlying biochemical oscillation. Photomovement responses are also evident at the level of the excised plasmodial strand by recording either a dynamoplasmogram or the isometric tension. For recording a dynamoplasmogram, a plasmodial strand is put into a modified Kamiya double chamber (Figure 4a and see above). Stimulus light is applied to one chamber while the other chamber is kept dark. In the absence of any light stimulus there is an oscillatory change in pressure difference produced by the protoplasmic shuttle streaming. The pressure oscillates symmetrically around a baseline. A zero baseline indicates that the plasmodium does not migrate from one chamber into the other. When the stimulus light is set on, oscillation continues however with a shift in phase. In addition, the pressure now oscillates around a baseline of negative pressure indicating a net movement of the protoplasm from one chamber into the other. When the stimulus light is switched off, the oscillation and its baseline returns to the prestimulus level (Figure 7; [21]). This was an important experiment immediately suggesting a mechanism for the photoavoidance response: Translocation of the plasmodial protoplasmic mass is achieved by biassing the net direction of the oscillatory shuttle streaming away from the light stimulus. The experiment also shows, that irradiation can shift the phase of the oscillation. This immediately suggests a direct interference of the photoreceptors and the signaling pathway with the components mediating the biochemical oscillation.

Antagonistic red and far-red light effectiveness suggests that a phytochrome controls the shuttle streaming activity. By measuring dynamoplasmograms of isolated plasmodial

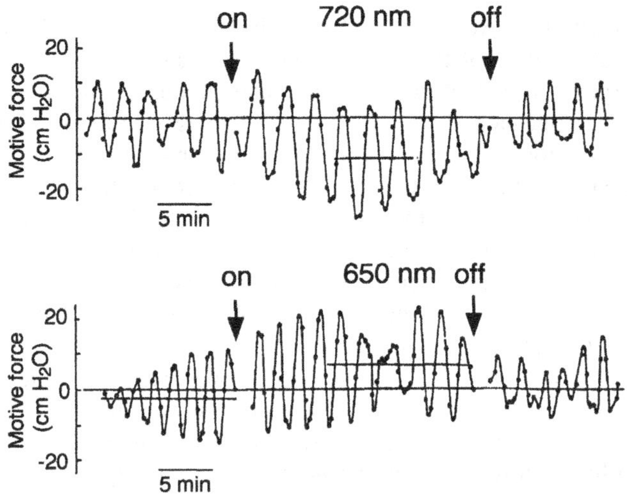

Figure 7. Dynamoplasmograms indicating the response of an excised plasmodial strand to far-red and red light, as recorded with the Kamiya double chamber. Horizontal lines indicate the maximal baseline offset of the force oscillation. The response is negative to far-red (720 nm) and blue-green (500 nm; not shown) and positive to red (650 nm) light (redrawn from Hato et al. [21]).

strands, Hato et al. [21] found that in addition to the effectiveness of blue-green (500 nm) light there is a pronounced response to red and far-red light. Far-red light (720 nm) caused a negative response as blue did, but red light (650 nm) provoked a positive response of comparable strength (Figure 7). All responses are completely reversible when the light is switched off again. From the antagonistic effectiveness of red and far-red light, the authors concluded that a phytochrome may trigger the photoavoidance response [21]. An important criterion that may discriminate a phytochrome from other red light photoreceptors (e.g. rhodopsins) is its ability for reversible photoconversion. This has not been shown for the photoavoidance response. It has however been shown for the induction of sporulation which can be triggered by far-red light and depends on a phytochrome as the avoidance response seems to do ([38] and see below). Genetic analysis of the *Physarum* photosensory system is just at the beginning. Looking at blind mutants should soon reveal whether photomovement and photomorphogenesis are mediated by the same photoreceptors.

It is surprising that the red/far-red antagonism is not observed by measuring the isometric tension. Blue-green (510 nm) light increases the isometric tension as far-red (720 nm) does. In contrast, red light causes only a very small, if any, decrease in isometric tension [21]. This again demonstrates that it is important to keep in mind which physical parameter is meant when talking about plasmodial "phototaxis".

Is signal transduction mediated by mitochondrial citric acid cycle intermediates? That the blue-light mediated photomovement response in plasmodia is coupled to respiration [27] sounds trivial since respiration is expected to be the main energy source in migrating plasmodia. However, it is interesting that the effect depends on the chemical nature and the composition of the supplemented substrate. This suggests that it might not be the energetic aspect of respiration which modulates the plasmodial photoresponse. In excised plasmodial strands that are submerged in a salt solution (note that strands are not submerged in the tensiometric measurements of Hato et al. [21]) blue light causes a prolongation of the period of the contraction-relaxation cycle as determined tensiometrically [27]. Out of different substrates for respiration added to the bath solution, only a combination of α-ketoglutarate and pyruvate (used in a molar ratio of 10 : 1 and applied in the presence of inhibitors of glycolysis) caused a prolongation of the period and an increase of the amplitude of the force oscillations and thus an enhancement of the photoresponse. Even more interesting, application of either α-ketoglutarate or pyruvate alone in most cases caused irregular oscillations which did not permit to determine any period. The authors conclude that pyruvate- and α -ketoglutarate-dehydrogenase complexes functioning in mitochondrial respiration are involved in the blue-light response of plasmodia [27]. Certainly, one can think of many other possible explanations for the reported phenomena that are equally likely. Because of his own work the author of this chapter recalled that changes in the cytoplasmic level of fumarate are involved in photo- and chemotaxis of *Halobacterium salinarum* and *Escherichia coli* [39–43]. Fumarate was shown to directly interact with the switch complex of the bacterial flagellar motor [44] and can even replace CheY in intact cells at low temperatures by lowering the free energy difference between the clockwise and counterclockwise states of the motor [45]. Metabolically induced changes in the cytoplasmic level of fumarate can even cause chemotaxis in the absence of the two-component phosphorylation cascade [42]. However to prove or disprove a specific effect of fumarate as signaling molecule in *Physarum* is not easy given the complexity of the metabolic network and because of the lack of a defined molecular target to look at.

20.4 Photomorphogenesis: photoinduction of sporulation

Starving plasmodia of *P. polycephalum* sporulate when they are exposed to light. This photomorphogenesis is mediated by a branched signal transduction pathway which integrates far-red and blue light, heat shock and the nutritional status of the cell [13]. Far-red light is sensed by the *Physarum* phytochrome (see below), blue light by a separate not yet identified blue light photoreceptor. Integration of far-red and blue light, both sensed by specific photoreceptors is responsible for many photomorphogenetic effects in higher plants as well [46]. In plants the pathways connecting phytochrome and cryptochrome to downstream events like changes in the transcriptional activity are under intensive investigation (for recent reviews see [47,48]). At present it is not clear whether or not the same photoreceptors that mediate photomovement do also trigger sporulation. The action spectra measured for the two types of photoresponse do not argue against this possibility. However, blind or color blind mutants should help to answer this question.

Sporulation of *Physarum* plasmodia is an all-or-none-response. The plasmodial protoplasm is completely converted into fruiting bodies if the induction was of sufficient strength. Alternatively, not even a single fruiting body is formed. With respect to the developmental decision the plasmodium behaves as a single cell in displaying perfect synchrony like it does in cell cycle control. Morphogenesis is driven by a cascade of differentially expressed genes and involves the mitotic and meiotic maturation of haploid spores.

In this section, recent results on the photoreceptors and on the physiological events involved in sporulation are described. It also reviews most recent results on mutants with defects in the signaling pathway and on their functional characterization by time-resolved somatic complementation analysis as a new experimental approach to investigate complex pathways.

20.4.1 The Physarum phytochrome system

Starved plasmodia can be induced by a far-red light pulse of about 1 mmol/m^2 to sporulation. The action spectrum for the induction of sporulation was calculated from the slopes of fluence response curves. It has a small maximum in the blue at about 464 nm and a pronounced maximum in the far-red region at about 738 nm (Figure 8; [38]). Far-red light induction is completely reversible by red light provided that the two light pulses follow each other within a short period of time (Figure 9 and see below). The action spectrum for the photoreversion was determined by preirradiating plasmodia with a short far-red light pulse and subsequently exposing them to a second pulse of monochromatic light, the wavelength of which was variable. The action spectrum of the photoreversion of the induction by far-red light was also calculated from the slope of fluence-response-curves [38]. The action spectra of both, the photoinduction and the

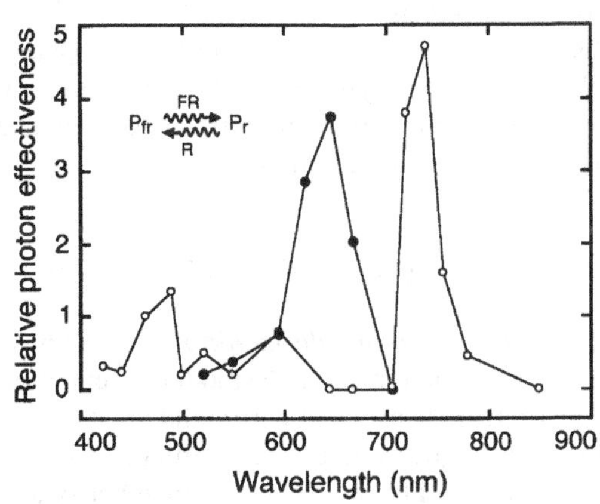

Figure 8. Action spectrum for the induction (open circles) and inhibition of photoinduced sporulation (closed circles). Sporulation was induced by exposing plasmodia grown and starved for six days in complete darkness to a one hour pulse of monochromatic light. The action spectrum for the inhibition of sporulation was measured by exposing plasmodia, that were preirradiated with a far-red light pulse of a fluence just sufficient to cause 80 to 90% sporulation, to a second pulse of different wavelength. Both spectra were obtained from fluence response curves (from Starostzik & Marwan [38]).

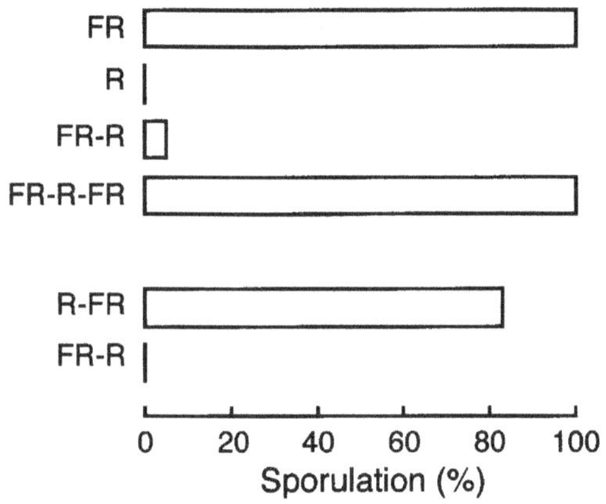

Figure 9. Induction of sporulation in starving plasmodia depends on the sequence of irradiation with far-red (FR) and red (R) light. FR and R were applied as 10 and 20 min pulses respectively (from Starostzik and Marwan [38]).

photoreversion of sporulation are similar to photoconversion spectra of P_{fr} and P_r respectively. With respect to the required photon fluence of 1 mmol/m^2, the FR-R reversibility, the validity of the reciprocity law and the time-dependent loss of reversibility (see below), photoinduction of sporulation is similar to a low fluence response in plants [49]. However, the activity of the *Physarum* phytochrome seems to be reversed as compared to the plant phytochromes. In plants, phytochrome is synthesized in the dark in its red-absorbing form P_r which is regarded to be physiologically inactive (for a recent review see [50]). By red light, P_r is converted to P_{fr} which mediates the photomorphogenetic responses of the plant. In contrast, it seems that the *Physarum* phytochorme is synthesized in the dark in its P_{fr} form and the photoconversion to P_r causes sporulation. In *Physarum* it remains to be established whether P_r induces or P_{fr} represses sporulation.

20.4.2 Sensory integration of multiple signals by a branched signaling pathway

Evidence for a blue light photoreceptor: plasmodia remember blue light exposure for a couple of days. When growing, non-starved plasmodia are exposed to a blue light pulse, sporulation does not occur. If light-exposed plasmodia are starved without further light induction, they remember the light stimulus and sporulate as soon as they have adopted a sufficient level of competence. Plasmodia only remember blue, not far-red light and the photoinduction can not be reverted by a subsequent red light pulse. This demonstrates that in addition to phytochrome, at least one independent blue light photoreceptor (cryptochrome) must be present. It cannot be excluded that the phytochrome is synthesized during the starvation period. The memory effect for blue light demonstrates that starvation is not necessary to induce cryptochrome biosynthesis. Hence, a photoreceptor signal and a starvation signal have to be present simultaneously and are processed by sensory integration to cause cellular commitment. If either of the two signals is quenched, sporulation is prevented ([13,51] and see below). This fact is symbolized by introducing a logic AND gate into the minimal model for the sensory control of sporulation (Figure 10).

Two points of no return indicate irreversible steps in the signaling pathway to cellular commitment. The induction of starved plasmodia to sporulation can be reverted in two different ways:

1. by a red light pulse that reverts P_r to P_{fr}, thereby abolishing the photoreceptor signal, and
2. by feeding glucose to the plasmodium which quenches the starvation signal.

In each case, the reversibility is gradually lost with increasing time span following the photoinduction. The photoreversibility of the phytochrome photoactivation is completely lost at about 2 h after the far-red pulse [13] visualized as time-dependent formation of the X-signal (Figures 10, 11). The loss of photoreversibility of the phytochrome induction is observed in plants as well [52].

The probability that refeeding of an induced plasmodium will prevent sporulation is time-dependent as well and becomes close to zero at five hours after an inductive light pulse [13,51,53]. The time course of the formation of the morphogenetic signal defined in this way is shifted by 2 to 3 h as compared to the time course for X-signal formation ([13]; Figure 11). The morphogenetic signal commits the cell to sporulation. Committed cells will sporulate even if they are refed.

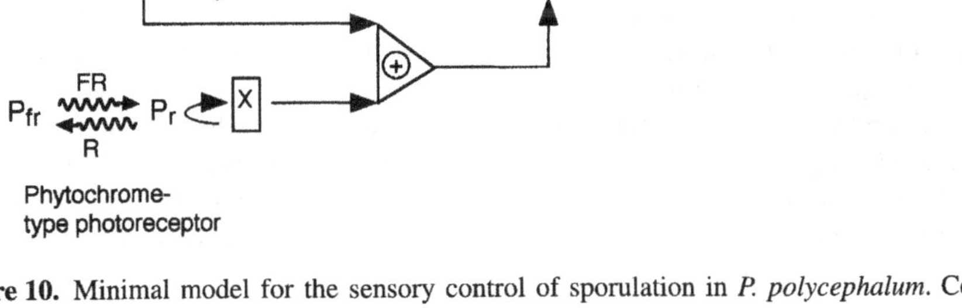

Figure 10. Minimal model for the sensory control of sporulation in *P. polycephalum*. Cellular commitment to sporulation depends on the formation of a sufficient amount of morphogenetic signal. It is formed, if the starvation signal and a second (photoreceptor-) signal are present at the same instant of time. The AND gate type sensory integration of these two stimuli is symbolized by a triangle. P_{fr}, P_r *Physarum* phytochrome in its far-red (FR) or red light (R) absorbing state, respectively, X X-signal conferring time-dependent loss of reversibility of the phytochrome induction, Y intermediate formed upon blue light induction and mediating the memory of starving plasmodia for blue light.

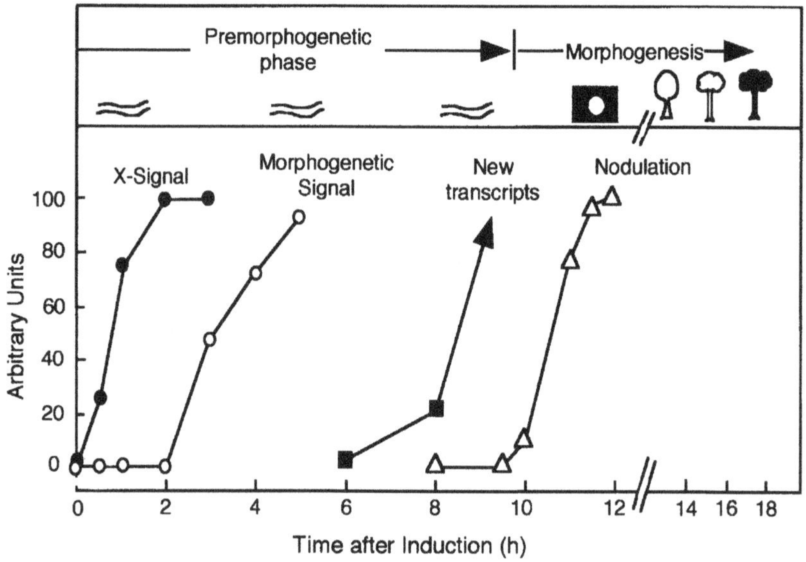

Figure 11. Time-dependent events following induction of sporulation. During the premorphogenetic phase, formation of the X-signal and the morphogenetic signal mark irreversible steps of the signaling pathway. Early differentially regulated transcripts can be detected by differential display RT-PCR well before the cascade-like expression of intermediate and high abundant at the beginning of the nodulation stage (cf. Fig. 12) starts.

Formation of the morphogenetic signal can also be detected by a different experimental approach [54]. As mentioned above, non-starved and therefore uncompetent plasmodia do not sporulate when exposed to blue light. When a competent plasmodium is exposed to blue or far-red light and immediately after the inductive light pulse fused to an uncompetent plasmodium, the resulting plasmodium does not sporulate. With increasing delay time elapsed between photoinduction and fusion, the probability for sporulation increases until it finally reaches 100% [13,51]. The kinetics for the formation of the morphogenetic signal in both types of experiment are identical within the experimental error.

The commitment point is glucose-sensitive independent of whether the plasmodium was induced via phytochrome, blue light or by heat shock. Hence, the irreversible step which becomes obvious as "point of no return" is located downstream of the point of integration of photoreceptor-, heat shock and starvation signals (Figure 10).

20.4.3 The gene expression cascade downstream of the developmental switch

Following the inductive light pulse, there is a premorphogenetic phase of about 10 h. Visible morphogenesis then starts by cleavage of the plasmodial strands into nodular structures that culminate and finally form the fruiting bodies (Figure 12). The entire protoplasmic mass is completely converted into fruiting bodies. Dependent on the size of the plasmodium, tens to hundreds of individual fruiting bodies are formed. The length of the premorphogenetic phase is constant irrespective whether the photon exposure delivered by the inductive pulse is near threshold or highly saturating.

Distinct morphogenetic intermediate stages are accompanied with a cascade-like differential expression of translatable mRNAs [55–57]. These mRNAs were translated in vitro in the presence of radioactive methionine and the translation products separated on a two dimensional polyacrylamide gel. Differentially occurring spots were identified and their presence evaluated over the entire period of morphogenesis (Figure 13). Out

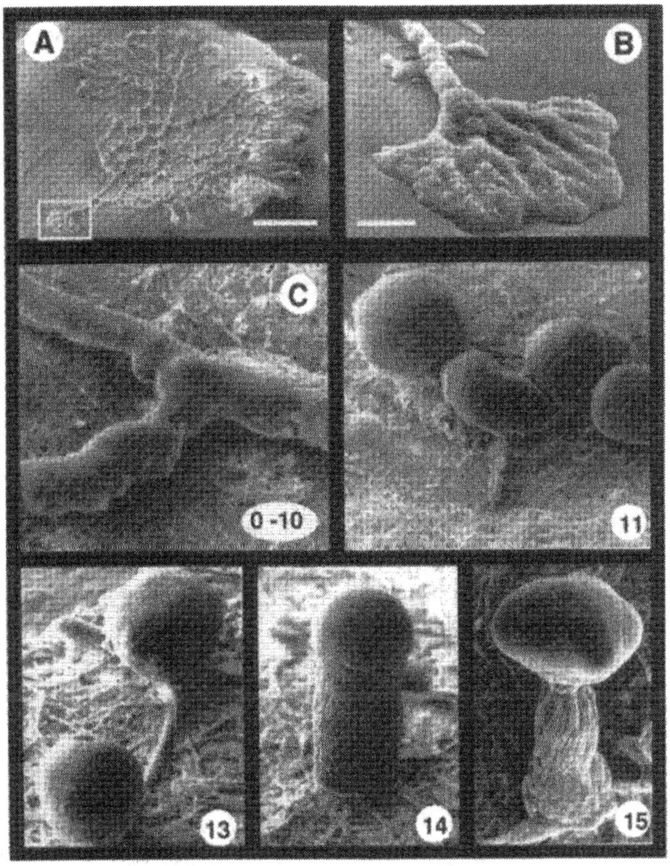

Figure 12. Scanning electron micrographs of a *P. polycephalum* plasmodium before and during sporulation. (A) veined network of a starving, migrating plasmodium. (B) Plasmodial pseudopodium as marked by the rectangle in (A). (C) Plasmodium before and at different stages of morphogenesis. After light induction there is a premorphogenetic phase of about ten hours (at 0–10 h) without changes in the plasmodial morphology. Morphogenesis starts when the plasmodial strands cleave into nodular structures (at 11, 13 h) that culminate (at 14 h) and reshape to form the mature fruiting bodies (at 15 h). The specimen was supported by a piece of filter paper. Bars correspond to 1 mm in (A) and to 100 μm in (B) and (C). Electron micrographs were taken by Th. Schreckenbach and M. Claviez (unpublished results).

of these differentially expressed mRNAs, the highly abundant messages of tubulins were identified [55,56]. These cytoskeletal proteins are presumably involved in culmination. During the premorphogenetic phase, early transcripts can be detected by a more sensitive method, Differential Display RT-PCR. The early transcripts are associated with the point of no return that is passed at about six hours after induction (Marwan and Cashmore, unpublished results).

20.4.4 Genetic dissection of the signaling pathway by somatic complementation analysis of photomorphogenetic mutants

Classes of mutants. The branched structure of the signal transduction pathway controlling sporulation greatly facilitates the definition of early signaling mutants. Each color blind mutant (which is sensitive to e. g. far-red light but not to blue or vice versa) must be defective in an early step. In a pilot screen for sporulation mutants, haploid amoebae capable for apogamic development were subjected to UV mutagenesis.

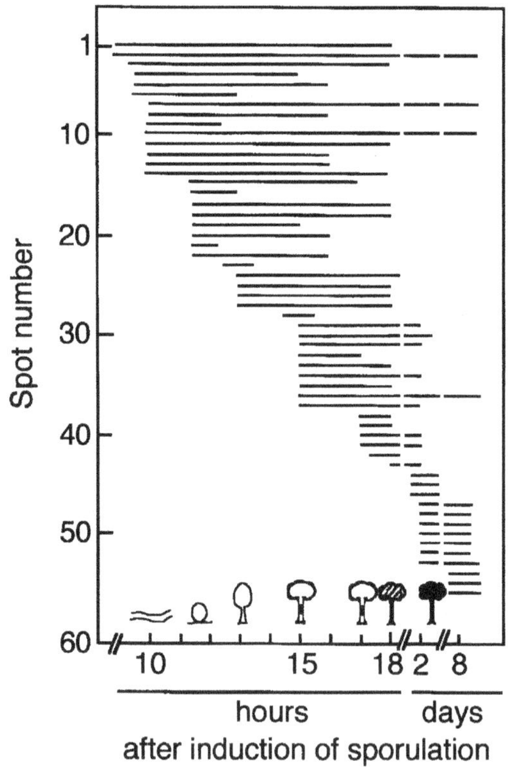

Figure 13. Sporulation in *P. polycephalum* is driven by a cascade of differentially regulated genes. RNA was isolated from different morphogenetic intermediates (bottom) and translated in vitro. Translation products were separated on a two-dimensional gel and spots not appearing in the dark control were identified. The horizontal bars indicate the time interval in which each species is expressed (redrawn from Putzer et al. [55]).

Individual clones were regenerated to haploid multinuclear plasmodia. The plasmodia were grown up, starved and screened for defects in the photosensory control of sporulation. Four phenotypes were obtained in the pilot screen: Blue blind, general blind, light-independent sporulating and a nodulation stage-arrest mutant [58].

Light-independent mutants sporulate as soon as they get competent by starvation. The probability for sporulation to occur grows with increasing time of starvation. The mutants behave similar to wild-type plasmodia that have been exposed to blue light, returned to complete darkness and then starved for several days. Although the mutants behave like having a constitutively activated photoreceptor, the mutation could be at any step upstream the point of integration of photoreceptor and starvation signal.

Time-resolved somatic complementation analysis. Analysis of genetic interactions between genes involved in signal transduction pathways usually generate static models, in most cases leaving the assignment of gene products to a well-defined regulatory function at the system level a difficult task to be solved. The fact that plasmodia spontaneously fuse as they get into contact provides an easy possibility to explore the potential of time-resolved somatic complementation analysis to study the architecture and function of signal processing regulatory networks in vivo. Experimentally, the approach is based on the fusion of cells carrying mutations in a signaling pathway to wild-type cells or to other mutants. The conceptual difference from genetic complementation analysis is that the signaling pathway is triggered in one of the complementing partners at a defined time before the cells are fused. From the concentration- and time-dependence of the complementation efficiency observed in the resulting heterokaryons, mutations could be functional characterized in detail even before the isolation of the gene is envisaged [59]. This in addition provides the possibility to identify those mutants out of a big collection that display a well-defined regulatory phenotype. This new concept should allow to analyze the structure and dynamics of a complete signaling pathway. The usefulness of the approach will be explained using two examples.

Time-resolved somatic complementation allows to discriminate mutations in the signaling pathway from photoreceptor or chromophore biosynthesis mutants. One example may explain the concept. If a color blind mutant (which is unable to sporulate by induction with light of a certain wavelength) is irradiated with light, no sporulation occurs. However, if the mutant is irradiated and after light induction fused to a non-induced plasmodium in the dark, sporulation may occur. This demonstrates that the mutant is equipped with a functional photoreceptor and tells that the phenotype must be due to a mutation in the signaling pathway [58].

Nos1 is a mutant which is general blind. If the mutant is irradiated with either far-red or blue light it does not sporulate. When a wild-type plasmodium is irradiated with light and immediately after irradiation the Nos1 plasmodium is fused to the wild-type, sporulation of the heterokaryon does not occur. This means that the mutation is dominant negative. However the dominant repression of sporulation is overcome with increasing time elapsed after light induction until fusion. This may be interpreted as time-dependent by-passing of the defective signaling intermediate in the heterokaryon. The kinetics of loss of dominance in turn reflect the formation kinetics of the intermediate that is downstream of the one which generates the dominant block [58].

Time-resolved somatic complementation analysis should provide enough functional insight into the pathway to order the signaling intermediates (defined by mutations) according to their hierarchy in controlling cell differentiation.

20.5 Spherulation of microplasmodia: a skotomorphogenesis that is inhibited by blue light

Spherulation of starving small submersed plasmodia (microplasmodia) is a default developmental pathway that is inhibited by light. As in sporulation, photoinhibition of spherulation also involves a change in the motility behavior of the plasmodia. In this respect light influences plasmodial motility in a duplicate way. Directly through photomovement responses and indirectly through processes associated with the photomorphogenetic responses, sporulation and inhibition of spherulation.

When small plasmodia grown submersed in liquid shaken culture are transferred to starvation medium and shaking is continued, they differentiate into resting structures (microsclerotia or spherules). Early events in this differentiation pathway include decreases in both, plasmodial motility and protein synthesis. Synthesis of plasmodial proteins is reduced and spherulation-specific proteins are expressed instead [31]. Spherule formation is inhibited by blue light with maximal effectiveness at 450 nm (16 W/m^2). Blue light also inhibits the synthesis of starvation-induced proteins at the mRNA level [60]. It also inhibits the loss of plasmodial motility [31]. In addition, the light avoidance response is lost. Hence, blue light causes the inhibition of a morphogenetic default pathway. Note that the same principle seems to be true for photomorphogenesis in *Arabidopsis*, where light inhibits the default pathway of de-etiolement.

20.6 Conclusions and outlook

Physarum is a lower eukaryote with an interesting photobiology. Light controls cell movement and morphogenesis, phytochrome and blue light photoreceptors converge into pathways that diverge again to control cytoskeleton, ion channels, metabolic reactions and gene expression. Mutants can be isolated, genes replaced and physiological and biochemical experiments performed at the single cell level. Time resolved somatic complementation analysis works by fusing plasmodial mutants thereby combining signaling pathways at different states of activation. This introduces the dimension of time into the analysis of genetic networks and expands the experimental possibilities provided by conditional mutagenesis in exploring central pathways of cell control.

Acknowledgements

The author gratefully acknowledges financial support from the Deutsche Forschungsgemeinschaft and from the Fonds der Chemischen Industrie.

References

1. H.P. Rusch, W. Sachsenmaier, K. Behrens, V. Gruter (1966). Synchronization of mitosis by the fusion of the plasmodia of *Physarum polycephalum*. *J. Cell Biol.*, **31**, 204–209.

2. K. Hausmann (1998). Nachruf auf Karl-Ernst Wohlfarth-Bottermann (1923–1997). *Biologie in unserer Zeit*, **28**, 60–61.
3. J. Dee (1982). Genetics of *Physarum polycephalum*. In: H.C. Aldrich, J.W. Daniel (Eds), *Cell Biology of Physarum* and *Didymium* (Vol. 1, pp. 211–251). Academic Press.
4. J. Dee (1987). Genes and development in *Physarum*. *Trends Genet.*, **3**, 208–213.
5. T.G. Burland, L. Solnicá-Krezel, J. Bailey, D.B. Cunningham, W.F. Dove (1993). Patterns of inheritance, development and the mitotic cycle in the protist *Physarum polycephalum*. *Adv. Microb. Physiol.*, **35**, 1–69.
6. J. Bailey (1995). Plasmodium development in the myxomycete *Physarum polycephalum*: genetic control and cellular events. *Microbiol.*, **141**, 2355–2365.
7. J. Bailey (1997). Building a plasmodium: development in the acellular slime mould *Physarum polycephalum*. *BioEssays*, **19**, 985–992.
8. J.W. Daniel, H.H. Baldwin (1964). Methods for culture of plasmodial myxomycetes. *Methods Cell Physiol.*, **1**, 9–41.
9. J. Dee, J.L. Foxon, R.W. Anderson (1989). Growth, development and genetic characteristics of *Physarum polycephalum* amoebae able to grow in liquid axenic medium. *J. Gen. Microbiol.*, **135**, 1567–1588.
10. T.G. Burland, D. Pallotta (1995). Homologous gene replacement in *Physarum*. *Genetics*, **139**, 147–158.
11. L. Rakoczy (1980). Effect of blue light on metabolic processes, development and movement in true slime molds. In: H. Senger, (Ed.) *The Blue Light Syndrome* (pp. 570–583). Springer Verlag, Heidelberg.
12. T. Schreckenbach (1984). Phototaxis and photomorphogenesis in *Physarum polycephalum* plasmodia. In: H. Senger (Ed.), *Blue Light Effects in Biological Systems* (pp. 463–475). Springer-Verlag, Berlin.
13. C. Starostzik, W. Marwan (1995). Functional mapping of the branched signal transduction pathway that controls sporulation in *Physarum polycephalum*. *Photochem. Photobiol.*, **62**, 930–933.
14. A.G. Wheals (1970). A homothallic strain of the myxomycete *Physarum polycephalum*. *Genetics*, **66**, 623–633.
15. R.W. Anderson (1979). Complementation of amoebal-plasmodial transition mutants in *Physarum polycephalum*. *Genetics*, **91**, 409–419.
16. K.E. Wohlfarth-Bottermann (1979). Oscillatory contraction activity in *Physarum*. *J. Exp. Biol.*, **81**, 15–32.
17. K.E. Wohlfarth-Bottermann (1962). Weitreichende fibrilläre Protoplasmadifferenzierungen und ihre Bedeutung für die Protoplasmaströmung.I. Elektronenmikroskopischer Nachweis und Feinstruktur. *Protoplasma*, **54**, 514–539.
18. S. Hatano, F. Oosawa (1966). Isolation and characterization of plasmodium actin. *Biochim. Biophys. Acta*, **127**, 488–498.
19. N. Kamiya (1942). Physical aspects of protoplasmic streaming. In: W. Seifriz (Ed.), *The structure of protoplasm* (pp. 199–244). Iowa State University Press, Ames, Iowa.
20. N. Kamiya, R.D. Allen, R. Zeh (1972). Contractile properties of the slime mold strand. *Acta Protozool.*, **11**, 113–123.
21. M. Hato, T. Ueda, K. Kurihara,, Y. Kobatake (1976). Phototaxis in true slime mold *Physarum polycephalum*. *Cell Struct. Funct.*, **1**, 269–278.
22. T. Ueda, Y. Mori, T. Nakagaki, Y. Kobatake (1988). Action spectra for superoxide generation and UV and visible light photoavoidance in plasmodia of *Physarum polycephalum*. *Photochem. Photobiol.*, **48**, 705–709.
23. B. Diehn, M. Feinleib, W. Haupt, E. Hildebrand, F. Lenci, W. Nultsch (1977). Terminology of behavioral responses of motile microorganisms. *Photochem. Photobiol.*, **26**, 559–560.

24. D.-P. Häder, T. Schreckenbach (1984). Phototactic orientation in plasmodia of the acellular slime mold *Physarum polycephalum*. *Plant Cell Physiol.*, **25**, 55–61.
25. D.-P. Häder (1985). Role of calcium in phototaxis of *Physarum polycephalum*. *Plant Cell Physiol.*, **26**, 1411–1417.
26. F. Achenbach, M.H. Weisenseel (1981). Ionic currents transverse the slime mold *Physarum*. *Cell Biol. Int. Rep.*, **5**, 375–379.
27. W. Korohoda, Z. Shraideh, Z. Baranowsky, K.E. Wohlfarth-Bottermann (1983). The blue-light reaction in plasmodia of *Physarum polycephalum* is coupled to respiration. *Planta*, **158**, 54–62.
28. J. Bialczyk (1979). An action spectrum for light avoidance by *Physarum nudum* plasmodia. *Photochem. Photobiol.*, **30**, 301–303.
29. L. Rakoczy (1963). Application of crossed light and humidity gradients for the investigation of slime molds. *Acta Soc. Bot. Polon.*, **32**, 393–403.
30. J. Bialczyk, L. Rakoczy (1975). Phototaxis of the *Myxomycetes*. Behaviour of old plasmodia in white light. *Bull. Acad. Polon. Sci.*, **23**, 571–575.
31. T. Schreckenbach, C. Verfuerth (1982). Blue light influences gene expression and motility in starving microplasmodia of *Physarum polycephalum*. *Eur. J. Cell Biol.*, **28**, 12–19.
32. K.W. Foster, R.D. Smyth (1980). Light antennas in phototactic algae. *Microbiol. Rev.*, **44**, 572–636.
33. T. Nakagaki, S. Umemura, Y. Kakiuchi, T. Ueda (1996). Action spectrum for sporulation and photoavoidance in the plasmodium of *Physarum polycephalum*, as modified differentially by temperature and starvation. *Photochem. Photobiol.*, **64**, 859–862.
34. Y. Mori, K. Matsumoto, T. Ueda, Y. Kobatake (1986). Spatio-temporal organization of intracellular ATP content and oscillation patterns in response to blue light by *Physarum polycephalum*. *Protoplasma*, **135**, 31–37.
35. W.D. Hoff, K.-H. Jung, J.L. Spudich (1997). Molecular mechanism of photosignaling by archaeal sensory rhodopsins. *Annu. Rev. Biophys. Biomol. Struct.*, **26**, 223–258.
36. J. Otomo, W. Marwan, D. Oesterhelt, H. Desel, R. Uhl (1989). Biosynthesis of the two halobacterial light sensors P480 and SR and variation in gain of their signal transduction chains. *J. Bacteriol.*, **171**, 2155–2159.
37. T. Ueda, Y. Mori, T. Nakagaki, Y. Kobatake (1988). Changes in cAMP and cGMP concentration, birefringent fibrils and contractile activity accompanying UV and blue light photoavoidance in plasmodia of an albino strain of *Physarum polycephalum*. *Photochem. Photobiol.*, **47**, 271–275.
38. C. Starostzik, W. Marwan (1995). A photoreceptor with characteristics of phytochrome triggers sporulation in the true slime mould *Physarum polycephalum*. *FEBS Lett.*, **370**, 146–148.
39. W. Marwan, W. Schäfer, D. Oesterhelt (1990). Signal transduction in *Halobacterium* depends on fumarate. *EMBO J.*, **9**, 355–362.
40. W. Marwan, D. Oesterhelt (1991). Light-induced release of the switch factor during photophobic responses of *Halobacterium halobium*. *Naturwiss.*, **78**, 127–129.
41. M. Montrone, W. Marwan, H. Grünberg, S. Mußeleck, C. Starostzik, D. Oesterhelt (1993). Sensory rhodopsin-controlled release of the switch factor fumarate in *Halobacterium salinarium*. *Mol. Microbiol.*, **10**, 1077–1085.
42. M. Montrone, D. Oesterhelt, W. Marwan (1996). Phosphorylation-independent bacterial chemoresponses correlate with changes in the cytoplasmic level of fumarate. *J. Bacteriol.*, **178**, 6882–6887.
43. M. Montrone, M. Eisenbach, D. Oesterhelt, W. Marwan (1998). Regulation of switching frequency and bias of the bacterial flagellar motor by CheY and fumarate. *J. Bacteriol.*, **180**, 3375–3380.

44. R. Barak, M. Eisenbach (1992). Fumarate or a fumarate metabolite restores switching ability to rotating flagella of bacterial envelopes. *J. Bacteriol.*, **174**, 643–645.
45. K. Prasad, S.R. Caplan, M. Eisenbach (1998). Fumarate modulates bacterial flagellar rotation by lowering the free energy difference between the clockwise and counterclockwise states of the motor. *J. Mol. Biol.*, **280**, 821–828.
46. H. Mohr (1994). Coaction between pigment systems. In: R.E. Kendrick, G.H.M. Kronenberg (Eds), *Photomorphogenesis in plants* (2nd ed., pp. 353–372). Kluwer Academic Publishers, Dordrecht.
47. A. Mustilli, C. Bowler (1997). Tuning in to the signals controlling photoregulated gene expression in plants. *EMBO J.*, **16**, 5801–5806.
48. C.R. Andersson, S.A. Kay (1998). COP1 and HY5 interact to mediate light-induced gene expression. *BioEssays*, **20**, 445–448.
49. A.L. Mancinelli (1994). The physiology of phytochrome action. In: R.E. Kendrick, G.H.M. Kronenberg (Eds), *Photomorphogenesis in plants* (2nd ed., pp. 211–269). Kluwer Academic Publishers, Dordrecht.
50. P.H. Quail (1997). The phytochromes: a biochemical mechanism for signaling in sight? *BioEssays*, **19**, 571–579.
51. B. Starostzik, W. Marwan (1994). Time-resolved detection of three intracellular signals controlling photomorphogenesis in *Physarum polycephalum. J. Bacteriol.*, **176**, 5541–5543.
52. A. Batschauer, P.M. Gilmartin, F. Nagy, E. Schäfer (1994). The molecular biology of photoregulated genes. In: R.E. Kendrick, G.H.M. Kronenberg (Eds), *Photomorphogenesis in plants* (2nd ed., pp. 559–593). Kluwer Academic Publishers, Dordrecht.
53. A. Chapman, J.G. Coote (1982). Sporulation competence in *Physarum polycephalum* CL and the requirement for DNA replication and mitosis. *J. Gen. Microbiol.*, **128**, 1489–1501.
54. A. Hildebrandt (1986). A morphogen for the sporulation of *Physarum polycephalum* detected by cell fusion experiments. *Exp. Cell Res.*, **167**, 453–457.
55. H. Putzer, C. Verfuerth, M. Claviez, T. Schreckenbach (1984). Photomorphogenesis in *Physarum*: Induction of tubulins and sporulation-specific proteins and of their mRNAs. *Proc. Natl. Acad. Sci. USA*, **81**, 7117–7121.
56. A.K. Werenskiold, B. Poetsch, F. Haugli (1988). Cloning and expression of a β tubulin gene of *Physarum polycephalum. Eur. J. Biochem.*, **174**, 491–495.
57. R. Martel, A. Tessier, D. Pallotta, G. Lemieux (1988). Selective gene expression during sporulation of *Physarum polycephalum. J. Bacteriol.*, **170**, 4784–4790.
58. A. Starostzik, W. Marwan (1998). Kinetic analysis of a signal transduction pathway by time-resolved somatic complementation of mutants. *J. Exp. Biol.*, **201**, 1991–1999.
59. W. Marwan, C. Starostzik (1997). Somatische Komplementationsanalyse: ein neuer Schlüssel zum Verständnis zellulärer Regulationsprozesse? *Biospektrum*, **3**, 25–27.
60. H. Putzer, K. Werenskiold, C. Verfuerth, T. Schreckenbach (1983). Blue light inhibits slime mold differentiation at the mRNA level. *EMBO J.*, **2**, 261–267.
61. K.E. Wohlfarht-Bottermann, W. Stockem (1972). Comparative studies on actomyosin-thread models of muscles and of myxomycete plasmodia. Their significance in the contractile mechanism of primitive motile systems. *Acta Protozool.*, **11**, 39–64.
62. C.J. Alexopoulos, C.W. Mims (1979). *Introductory Mycology* (3rd ed., p. 69). John Wiley and Sons.
63. M. Fleischer, K.E. Wohlfarth-Bottermann (1975). Correlation between tension force generation, fibrillogenesis and ultrastructure of cytoplasmic actomyosin during isometric and isotonic contractions of protoplasmic strands. *Cytobiol.*, **10**, 339–365.

Photomovement
D.-P. Häder and M. Lebert, editors.

Chapter 21

Genetics of *Phycomyces* and its responses to light

Enrique Cerdá-Olmedo and Luis M. Corrochano

Table of contents

Abstract

Phycomyces blakesleeanus is best known for its sensitive and precise responses to many stimuli, including light. The genetics of this fungus was developed as a research tool, but its many singularities render it separately attractive. Recessive mutants are isolated from multinucleate cells and mutations assigned to genes by complementation and recombination. *Phycomyces* heterokaryons offer unique experimental possibilities, for instance in the study of gene function in vivo and the causes of cell death. Behavioral mutations modify the effects of light on growth (phototropism), development (photophorogenesis), and metabolism (photocarotenogenesis). Many behavioral mutants suffer metabolic alterations, such as defects in the biosynthesis of carotenes, flavins and pterins. The study of mutant phenotypes provides abundant information on the nature of the photoreceptors and other signal transducers and on their organization in sensory pathways. An overall impression of parsimony arises from the application of the same genes for different purposes; this combinatorial gene usage seems to be the groundwork of the elaborate behavior of the fungus.

21.1 Introduction

The Zygomycete fungus *Phycomyces blakesleeanus* is a prototype of the use of light as a source of information by organisms that do not use it as a significant source of energy. *Phycomyces*, like many organisms from microbes to animals [1], relies chiefly on blue light for this purpose. The effects of light and other stimuli on *Phycomyces* have been investigated since the early days of fungal physiology [2–4]. The first century of research has been summarized and examined critically in a book [5] which is also the recommended source for methods and details. There are specialized reviews on the major effects of light: phototropism ([6], Galland, this volume), photomorphogenesis [7] and photocarotenogenesis [8]. The genetics of *Phycomyces* were tackled first by Hans Burgeff [9], but did not become a tool in photobiology until 1967, when the first behavioral mutants were isolated in the laboratory of Max Delbrück at the California Institute of Technology.

21.2 Morphology and development

The life cycles of *Phycomyces* (Figures 1 and 2) consist in successions of stages different in morphology, biochemistry, physiology and behavior. The mycelium grows and expands indefinitely as long as the food supply and other external factors permit; acid media (optimum pH 3.3) and certain detergents limit growth and result in the formation of distinct colonies. *Phycomyces* is strictly a filamentous fungus, and no conditions are known that will allow it to grow as a yeast; other Mucorales, including several *Mucor* species, and many other fungi of different groups are dimorphic: environmental cues lead them to shift between alternative states, as filamentous fungi and yeasts [10,11]. The sexual cycle of *Phycomyces* requires the confluent growth of mycelia of the two mating types, often called sexes. It leads to the formation of resting structures, the zygospores, which, after a dormancy of several months, require only water to grow and develop.

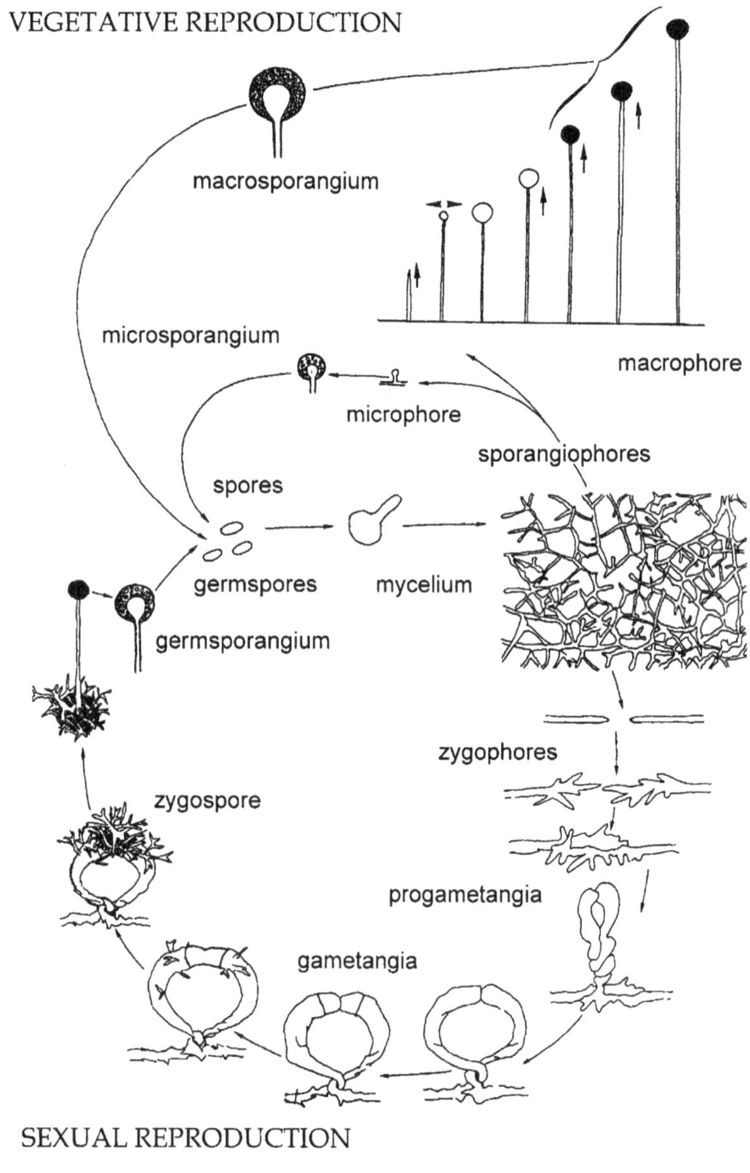

Figure 1. The life cycles of *Phycomyces*. Dormant spores, activated and incubated under suitable conditions, germinate to produce a mycelium, which is a branched cell of indefinite extension. Out of the mycelium grow two kinds of sporangiophores: macrophores and microphores. The stages of macrophore development are drawn and their growth is indicated by arrows. The sporangia contain spores, which complete the asexual cycle. Sexual reproduction starts when mycelia of opposite mating type grow near each other; both develop thicker hyphae, called zygophores, in response to chemical signals from the other. Progametangia are the zygophores of opposite mating type that come in contact, twist around each other, and finally adopt the shape of tongs. Gametangia are the two cells, separated at the ends of the progametangia, that fuse to form the zygospore. The zygospore is decorated with appendages, blackens, thickens, and rests for months. Germination of the zygospore gives rise to a germsporangiophore, a germsporangium, and germspores, all very similar, in terms of morphology and physiology, to their vegetative counterparts.

Three phases in the life of *Phycomyces* have received considerable attention to determine their dependence on external signals and on the internal conditions of the cells. Signals for spore germination augur suitable growth conditions and assure the ecological succession of saprophytes. Two kinds of sporangiophores are formed depending on the chances for spore dispersal. Sexual reproduction is preceded by an exchange of chemical signals between the mycelia of opposite sex before they make contact. Light plays a major role in sporangiophore formation, but not in the other two major transitions.

21.3 Genetics

21.3.1 Karyology of the life cycles

Phycomyces is a coenocytic organism (Figure 3). Mycelia, sporangiophores and young sporangia form a continuum inhabited by many nuclei, traversed by strong cytoplasmic

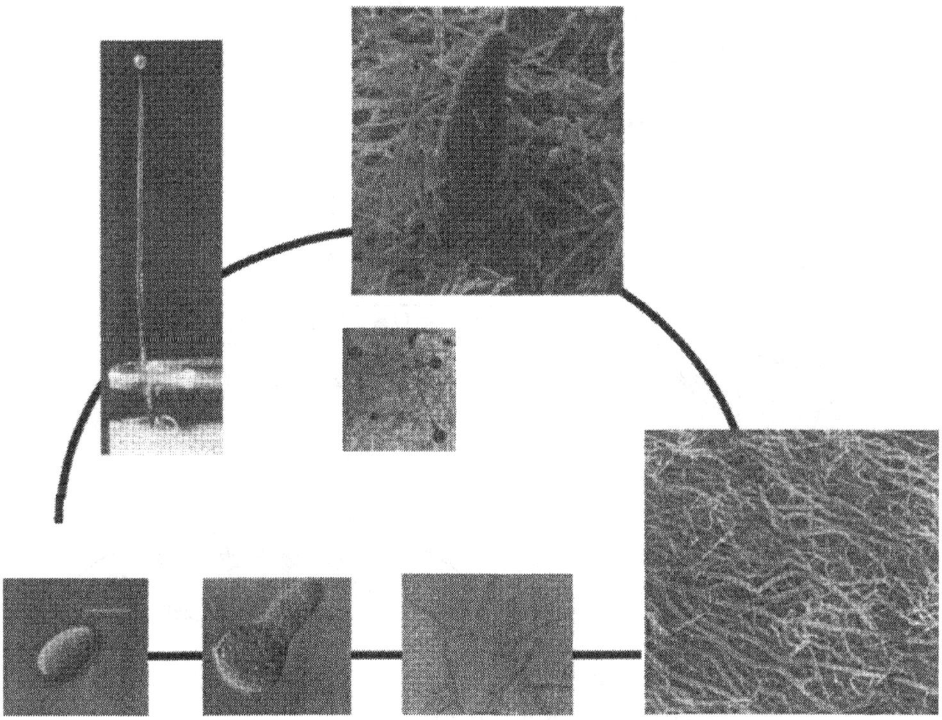

Figure 2. Structures of the vegetative cycle. The spores are about 9 μm long. Their germination produces one to three germinating tubes that grow into a branched mycelium; most mycelial hyphae are about 10 to 20 μm in diameter. Macrophores and microphores grow out of the mycelium into the air. Macrophores become several cm long; their sporangia, some 0.5 mm across, contain about 10^5 spores. Microphores are about 1–3 mm long; their sporangia, some 0.1 mm across, contain about 10^3 spores (photographs courtesy H.H. Heunert, W. Schröder and F. Gutiérrez-Corona).

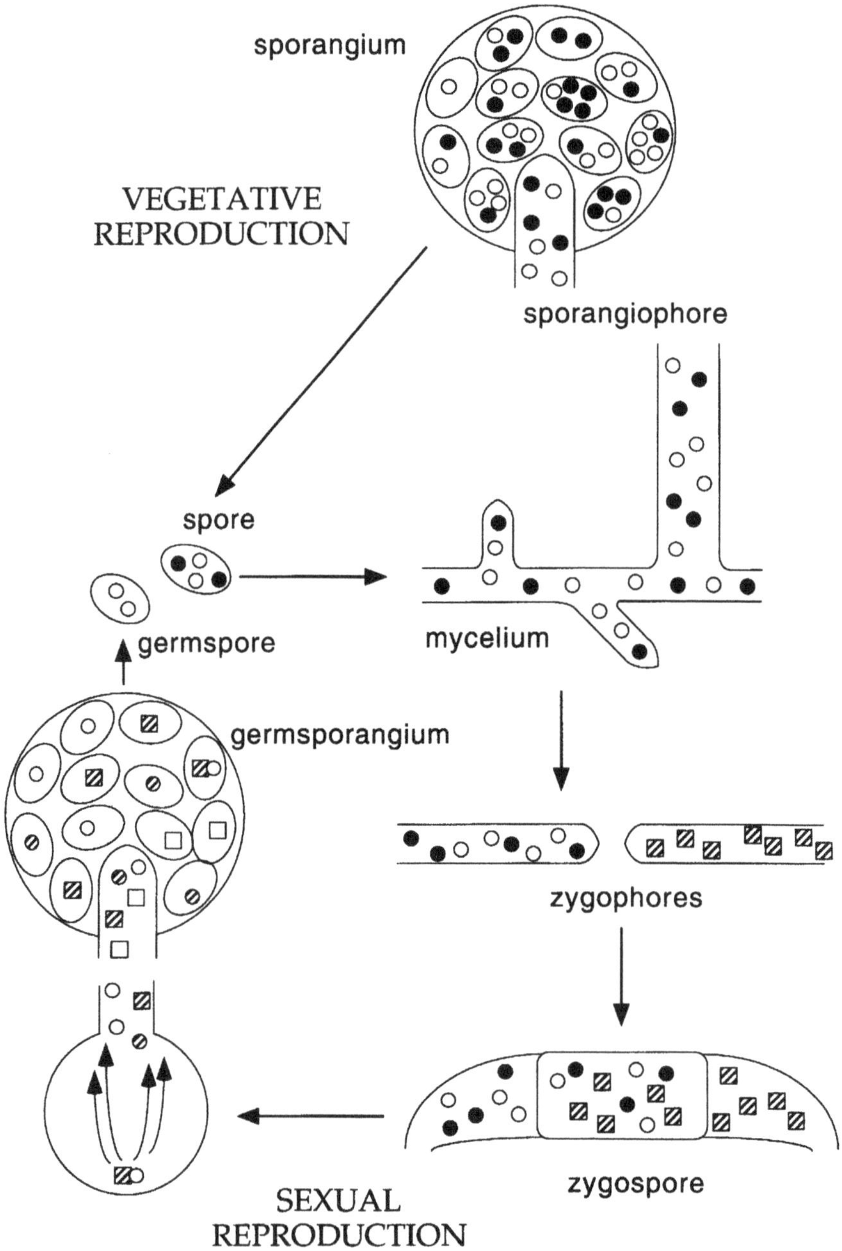

Figure 3. Karyology of the life cycles. *Phycomyces* is nearly always a coenocytic organism. Vegetative spores encase random samples of nuclei from the young sporangium; only a tiny proportion of the spores are uninucleate (0.3% in the standard wild type). The nuclei need not be genetically identical; a heterokaryon with two kinds of nuclei (open and closed circles) will form heterokaryotic spores with both kinds of nuclei and homokaryotic spores with either kind. The zygospore contains thousands of nuclei of both mating types (here represented by circles and squares), but in most zygospores a single diploid gives rise to all the progeny. The germspores are multinucleate, but derive from uninucleate primordia.

currents and not interrupted by septa or cell walls. Nearly all the spores are multinucleate; most of the spores of the standard laboratory strain, NRRL1555, contain three or four nuclei and only 0.3% are uninucleate [12]. The distribution of nuclei per spore is practically constant, as it is affected only slightly by external conditions. Uninucleate spores tend to be smaller and rounder than multinucleate spores and can be purified partially by sedimentation [13], but the method is cumbersome and has not entered common use. About 70% of the spores are uninucleate in a mutant [14], but most of the nuclei in this mutant are abnormal, probably polyploid.

After sporulation, mycelia are not wholly devoid of life, but contain propagula, small masses of cytoplasm which can resume growth if conditions allow. The number of nuclei in the propagula is highly variable, but some have only a few nuclei.

The sexual cycle entails the fusion of multinucleate gametangia so that many nuclei of both sexes come together in the zygospore. The nuclei seem to disappear without hints of karyogamy or meiosis and are seen again when the zygospore germinates and forms a multinucleate sporangiophore and a multinucleate sporangium. These structures of the sexual cycle are similar to their vegetative counterparts and their sexual origin is expressed by the prefix germ-. The germspores are multinucleate, but average fewer nuclei than vegetative spores.

21.3.2 Heterokaryons

Because *Phycomyces* cells contain many nuclei, they are homokaryons or heterokaryons, depending on whether the nuclei are genetically identical or not [15]. Heterokaryons can be obtained from homokaryons by several methods, two of which are commonly used. The simplest one [16] consists in subculturing mycelial pieces taken from the confluence of mycelia of opposite sex. Natural heterokaryons, formed early in the sexual cycle, grow out as vegetative mycelia. The presence of nuclei of opposite sex gives these *intersexual heterokaryons* a peculiar look: they are more colorful (β-carotene) and have fewer sporangiophores than normal mycelia and they produce special twisted structures (*pseudophores*). The other method [17] is more general, because it can be used for strains of the same sex, but requires some manual ability: two macrophores from different strains are cut, joined by their wounds, and allowed to regenerate together and sporulate.

The proportions of different kinds of nuclei in heterokaryons remain constant through mycelial expansion, sporangiophore development, and subculture of mycelial fragments. Random samples of nuclei are walled up into spore primordia and the mature spores are formed without further nuclear division. Heterokaryons form homokaryotic spores of both possible kinds and heterokaryons with various nuclear mixtures. Different kinds of spores formed by a heterokaryon with two kinds of nuclei, A and B, can be predicted to occur with the following frequencies:

$$f(\text{homokaryons A}) = \Sigma\, p^n f(n);$$

$$f(\text{homokaryons B}) = \Sigma\, (1-p)^n f(n);$$

$$f(\text{heterokaryons}) = 1 - A - B,$$

where p is the proportion of nuclei of type A in the original heterokaryon and $f(n)$ is the frequency of spores having n nuclei. In the standard wild type $f(1)=0.003$, $f(2)=0.090$, $f(3)=0.420$, $f(4)=0.410$, $f(5)=0.074$, and $f(6)=0.003$ [12].

The nuclear proportion p in the original heterokaryon can be calculated from the observed frequencies of different kinds of spores by solving the equations above, which can be done by a simple graphical method.

Homokaryons can be obtained, not only from spores of heterokaryons, but from mycelial pieces with few nuclei, so that there is a good chance that they are all identical. Suspensions of heterokaryotic mycelia broken in a blender or shaken in a stirrer grow into colonies of different nuclear composition, including homokaryons [137]. Pieces of old heterokaryotic mycelia subcultured on fresh agar produce patches with different nuclear proportions, reflecting statistical variations in the nuclear composition of the propagula. A pure culture from a single individual can be easily obtained from mixed cultures of homokaryons and heterokaryons by picking up a macrophore and laying it on fresh agar. A macrophore derives from a single spore or mycelial piece because *Phycomyces* mycelia do not fuse (*anastomose*) spontaneously. Mycelial pieces do not always provide pure cultures because they may consist of intertwined hyphae of different individuals.

Cultures from single nuclei can be obtained directly by selecting for mutants resistant to 5-carbon-5-deazariboflavin (1 mg/L). These mutants occur spontaneously at the unexpected frequency of about one in a million spores. The mutation is extremely recessive because it involves the loss of riboflavin permease [18,137].

21.3.3 *Isolation of mutants*

Phycomyces was one of the first organisms in which mutants were reported, both spontaneous [9] and radiation-induced [19], but the extreme scarcity of uninucleate spores is a serious hindrance in the search for recessive mutants. Spontaneous recessive mutations can be isolated only if they are frequent and have a selectable phenotype. A simple solution to this problem is to increase both the frequency of the mutations and the frequency of spores with a single viable nucleus by exposing the spores to a lethal mutagen. About 35% of the survivors are functionally uninucleate and express recessive mutations following exposure of spores to *N*-methyl-*N*′-nitro-*N*-nitrosoguanidine to a survival level of 1 to 5% [20].

A further improvement is to apply the mutagen to spores of a heterokaryon that contains a minority of nuclei resistant to deazariboflavin. A large majority of the resistant survivors originate from functionally uninucleate spores and can express new recessive mutations. It is recommended to inoculate plates with about fifty thousand survivors each, allow them to grow, harvest separately the spores from each plate, and lyophilize them. The resulting spore suspensions can be used to look for any kinds of mutants in the indefinite future and have the advantage of having selected out many undesirable traits, such as slow growth or failure to sporulate [20].

The genetic nomenclature of *Phycomyces* follows the recommendations of Demerec et al. [21]. Strains, whether wild or obtained in the laboratory, are named with one or more uppercase letters followed without interruption by a number. Genes are named with three lowercase letters that refer to their general function followed without interruption by an uppercase letter to specify the gene. Mutant alleles are designated with the name of the gene followed without interruption by a number; when the specific gene is unknown, the uppercase letter is replaced by a hyphen. The genotype of a strain is a list of the alleles that differ from those of the wild type and the symbols (+) or (−) for the mating type. Heterokaryosis is indicated by using the names or the genotypes of their components separated by an asterisk.

For example, the three letter code *mad* is used for genes and mutations related to phototropism; *car* for biosynthesis of carotene, *dar* for resistance to deazariboflavin, *rib* for biosynthesis of riboflavin. The gene for phytoene dehydrogenase is called *carB*, and one of the mutant alleles of this gene is *carB10*. This allele is found in strain C5, the original mutant obtained directly from the wild type and in additional strains derived from C5 by recombination or new mutation. The name of an allele is sometimes abbreviated to the name of the gene: for example, a *carB* strain is a strain that carries a mutant allele of gene *carB*.

The isolation of mutants from heterokaryons offers information on their dominance or recessivity. Mutants in genes *carR* and *carB* are red and white, respectively, owing to the loss of different enzymes of the carotene pathway, but their heterokaryons contain both enzymes and are yellow. Recessive mutants isolated from the spores of a *carB* * *carR* heterokaryon must be red or white; yellow mutants isolated in this way contain dominant mutations. The proportion of dominant and recessive mutations can be estimated from the color distribution and the nuclear proportion in the original heterokaryon.

21.3.4 Complementation

Mutations can be assigned to genes by the complementation test in *Phycomyces* heterokaryons, much as it is done in other organisms. Two mutations are said to complement when the heterokaryon of two mutant strains, each carrying one of the mutations, exhibits an approximately wild phenotype. Tentatively, mutations that fail to complement are assigned to the same gene and non-complementing mutations, to different genes. An essential control is needed: the heterokaryon of the wild type and a strain carrying both mutations must be similar to the wild type. The double mutant that is needed for this "cis" control may be difficult to obtain and this control is replaced usually by two separate tests for recessivity: the heterokaryons of the wild type and each of the two mutants must be similar to the wild type. The cis control is more stringent than the separate tests, but the results are expected to differ only in the case of some unlikely interactions between the mutant gene products.

Complementation between recessive mutations of the same gene seems to be rare in *Phycomyces,* with no well documented case, although it is reportedly frequent in other fungi. When this *intragenic complementation* occurs, a gene is defined by a set of mutations that do not complement any mutations of the gene, some of which may complement each other.

A *madJ* mutation [22] is the example in *Phycomyces* of an extreme rarity: a recessive mutation that does not complement recessive mutations of other genes. A possible explanation is given below (p. 611).

Dominance and complementation are not qualitative, but quantitative concepts. The heterokaryons of the wild type and a mutant allele are rarely identical to the wild type. Complementing heterokaryons are expected to resemble the wild type more than the mutants, not to be identical to it. *Phycomyces* offers unique experimental possibilities to obtain heterokaryons with practically any values of the nuclear proportion, to grow them to any desired size or stage, and to determine quantitatively their phenotype and their nuclear composition. Phenotypic measurements of the heterokaryons can be expressed as a function of the proportion of nuclei that carry a certain allele. Such functions may give considerable information on the action of the corresponding gene products. For example, the structure and operation of the carotenogenic biosynthetic pathway was established from carotene analyses of heterokaryons of *car* mutants [23–25]. The interactions between gene products responsible for phototropism were deduced from threshold measurements in heterokaryons [26].

Another unique application of *Phycomyces* heterokaryons is to elucidate the mechanism of action of lethal agents. The frequency of heterokaryosis among the survivors following the exposure of a heterokaryon to a lethal agent depends on the extent and kind of the damage inflicted to the nuclei. Quantitative estimates may be obtained for the contribution of four major classes of lethal events to the overall damage [27]; the results with various agents [27–29] provide hard evidence to avoid mistaking the morphological and metabolic consequences of death for its causes.

Recessive lethal alleles can be maintained indefinitely in heterokaryosis. Quantitative estimates of recessivity and complementation may give hints about gene function. If the lethal effect is limited to some growth stage, homokaryons for the lethal allele may be grown and analyzed directly. For example, if the allele blocks spore germination, homokaryons can be obtained by mechanical disruption of a heterokaryon; if the allele blocks sporulation, they can be obtained from spores of a heterokaryon.

21.3.5 Recombination

New combinations of the alleles of two strains are formed during their sexual reproduction. Zygospores remain dormant for at least two months and the duration of dormancy depends on the strains [30,31]. Zygospore germination is relatively synchronous: practically all the zygospores germinate within a month after the first [32]. The number of viable germspores per germsporangium varies from none to tens of thousands, depending on the cross, and is particularly high in crosses between isogenic strains with a common genetic background [33]. The colonies grown from germspores are usually homokaryotic [32,34]. Germspores are multinucleate, but, unlike vegetative spores, are formed from uninucleate primordia.

Genotypes are distributed very irregularly among the germspores in a germsporangium. When the germspores from many germsporangia are pooled and analyzed for the segregation at two genes, the most common result is that the two parental and the two recombinant combinations are about equally frequent, each group making up about one fourth of the progeny. Germspores do not contain nuclei that have avoided the recombination process (apomictic nuclei); they would exhibit parental allele combinations, and these are not over represented in pools of germsporangia and are completely absent from many individual germsporangia.

Only about 3% of the gene pairs are linked (when the parents differ in alleles of those genes, the frequency of recombinants is significantly lower than 50%). Recombination is so active that two mutations in the same gene recombine with detectable frequency (in the order of 1%) [34]. Mutations that recombine more frequently are in different genes and this is confirmed by complementation. Linkage allows the formation of a genetic map with 11 linkage groups, which may correspond to as many different chromosomes [35,36]. Many pairs of genes in the same chromosome segregate independently, as expected if multiple recombination events occur between them.

In most zygospores, a single pair of nuclei of different sex engender the nuclei in the germspores of the germsporangium. This is proven by crossing a heterokaryon with a homokaryon: most germsporangia contain markers from the nucleus in the homokaryon and only one of the nuclei in the heterokaryon [32]. Only one diploid nucleus, rarely more, divides to produce the progeny. The thousands of nuclei in the young zygospore presumably provide materials for the multiplication of the chosen few and their descendants in the germsporangiophore and the germspores.

When the parental strains differ at many genes, the germspores in a germsporangium represent many genetic combinations, always more than the four that would be produced by meiosis in a single diploid [137]. This rebuts the simple hypothesis of a single meiosis of a single diploid followed by mitotic multiplication of the meiotic products. The diploid nucleus must multiply before meiosis. The descendants might undergo independent meioses, but no hints of meiosis have been found by genetic analyses and searches for synaptonemal complexes.

A small fraction of the germspores, some 2% on the average, produce intersexual heterokaryotic colonies that are diploid for the sex marker and usually for other markers as well. Upon subsequent growth, these colonies develop into heterokaryons that contain complex mixtures of haploid and partially diploid nuclei [137]. These and other results support an alternative to meiosis that involves mitotic divisions only, but accompanied by frequent recombination between homologous chromosomes and random chromosome losses.

21.4 Photoresponses

All growing stages of *Phycomyces* respond to external stimuli, but not to the same stimuli or at the same time. Particularly conspicuous are the effects of blue light on the growth of the macrophores, the development of new sporangiophores and the synthesis of the yellow pigment, β-carotene. These responses can be measured quantitatively and extend over enormous fluence ranges, the upper thresholds being 10^7 to 10^{10} times larger than the lower thresholds.

21.4.1 Growth responses of the macrophores

Macrophores grow in stages I and IV, that is, before the appearance of sporangia and after the formation of spores. Most experiments deal with older macrophores that grow at about 1 μm/s for many hours [37] by extension of their "growing zone". This is roughly a transparent cylinder about 0.1 mm in diameter and about 3 mm long, located below the sporangium. Growth lacks any defined orientation in the absence of gravity and other stimuli applied to the growing zone; asymmetrical stimuli modify the direction of growth, either towards the source of stimulus (positive tropism) or away from it (negative tropism); stimuli applied symmetrically around the macrophore axis cause a transient change in growth velocity (mecism, sometimes called "growth response" in a restricted sense). Many stimuli cause growth responses: visible, ultraviolet and ionizing radiations, wind, gravity, pressure, various chemicals and the presence of obstacles near the growing zone of the macrophores. Research has concentrated on phototropism, gravitropism and avoidance of obstacles (probably an "autochemotropism", a tropism mediated by a chemical released from the macrophore).

The action spectra for the growth responses of *Phycomyces* depend very much on the actual experimental conditions, but phototropism is caused by wavelengths up to a little over 500 nm. The most effective wavelengths for positive phototropism are about 450 nm, with thresholds about 1 nW/m^2 for constant illuminations and about 3 $\mu J/m^2$ for pulses. Macrophores grow away from ultraviolet sources, with a major peak at 280 nm and a threshold of about 0.2 nW/m^2. The neutral wavelength that marks the limit between positive and negative tropism is about 310 nm [38].

The growth responses of *Phycomyces* extend up to 10 W/m^2; to respond to such a wide range, *Phycomyces* has two photosystems, optimized for different intervals of fluence rates. The photosystem for dim light, working for example at 10 nW/m^2, elicits two weak responses: a quick response, completed within half an hour after the onset of illumination, with bending rates 1 to 3°/min, and a delayed response, which starts one hour after the onset of illumination, with bending rates about 1°/min. The photosystem for bright light, working for example at 1 mW/m^2, elicits the same quick response and a strong delayed response, starting 40 min after the onset of illumination and bending at 3 to 4°/min [39].

Macrophores respond to fluence changes and illumination asymmetries; under constant, symmetrical illumination, they assume their standard cruising speed, which is independent of the absolute fluence. Once adapted to a bright illumination, they take a long time before they can react to a much weaker one; it is said that they have to adapt to the new conditions. One way to follow the kinetics of adaptation is to shift macrophores from a bright symmetrical illumination to darkness for a certain interval of time and then to a weak unilateral illumination; phototropism will be observed only if the interval was long enough to allow the cells to adapt. The adaptation is faster if darkness is replaced by a very dim illumination, called subliminal because it does not produce tropism by itself [40]. The most effective wavelength for the dim illumination is 485 nm, with a threshold of about 3 pW/m^2, but wavelengths up to 680 nm are quite effective [41]. This response is noteworthy for its extremely low threshold and for the effectiveness of the green to red spectral range. Another, weak, effect of red light is the inhibition of phototropism to a prior blue-light stimulus [42].

21.4.2 Photomorphogenesis

The existence of two kinds of sporangiophores, macrophores and microphores [43], is explained by their role in spore dispersion. The size of the macrophores and their sensitivity to various stimuli allow the spores to come out of the decaying organic matter into open air, where they can stick to or be eaten by animals. Microphores may be seen as economy devices made under severe competition or poor chances for long-range dispersal. Both kinds of sporangiophores are formed in the dark under standard laboratory conditions, but their presence is modified by many environmental circumstances, such as illumination, temperature, and the availability of asparagine, zinc, oxygen, retinol, and other chemicals [44,45].

Blue light stimulates the production of macrophores and inhibits that of microphores (Figure 4). Many cases of photomorphogenesis are known in other fungi [46,47] and, of

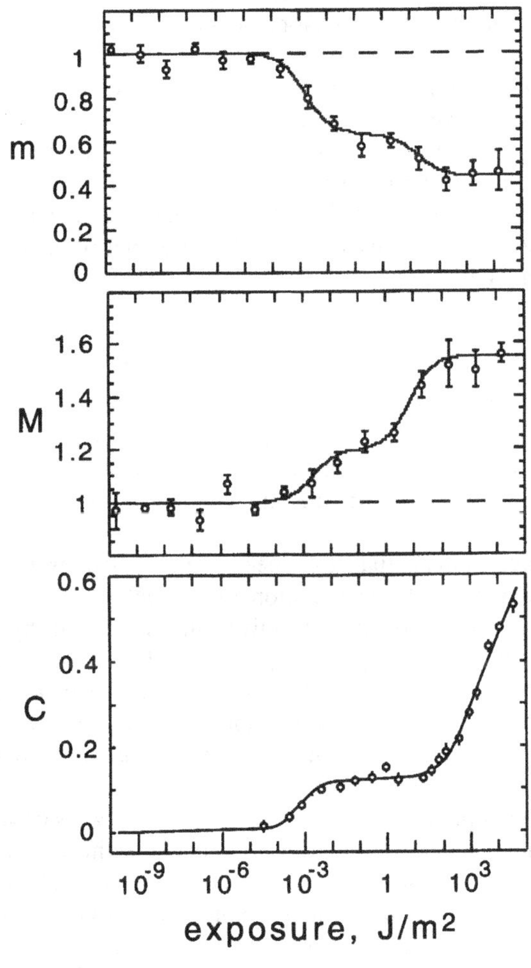

Figure 4. Inhibition of microphorogenesis, stimulation of macrophorogenesis and stimulation of β-carotene accumulation by blue-light pulses. Variables m and M are the number of microphores and the dry mass of macrophores, respectively, in illuminated cultures relative to those found in the dark. Variable C is the β-carotene absorbance in illuminated cultures minus that in dark cultures. The continuous lines represent computer-fitted algebraic expressions [49,57].

course, in plants and algae, but dim light usually has no effect. Thus, the photomorphogenetic responses of *Neurospora* and *Trichoderma* have thresholds of about 4 J/m^2. The striking trait of *Phycomyces* is its sensitivity, with thresholds under 100 $\mu J/m^2$. The effect of light depends on the product of the exposure time (between 12 s and 3 h) and the fluence rate. *Phycomyces* can be said to count and remember the number of photons received over a long time. The threshold corresponds to the arrival of one photon per μm^2 every 20 min.

The competence to detect light is not a constitutive property of the mycelium, but a developmental trait [48]. Four different kinds of detectors are used, because each of the two responses depends on two photosystems geared to different fluence ranges. The four responses can be distinguished by their thresholds and action spectra [49]. Because the competence periods of different detectors do not necessarily coincide in time, the relationships between responses and pulse fluences are two-step curves (Figure 4); microphores disappear completely under continuous illumination following a single-step stimulus-response curve with a threshold of about 10 nW/m^2 [48].

Light influences two other developmental transitions. The formation of the sporangium on the young sporangiophore is accelerated by blue light with a threshold of about 1 mW/m^2 [44]. The completion of the sexual cycle is inhibited by very bright light [50]. The response is peculiar because of its optimal wavelength around 360 nm and the threshold about 1 W/m^2.

21.4.3 Photocarotenogenesis and other responses to light

Illumination changes the chemical composition of *Phycomyces*, as shown by analyses under various conditions [51]. Bright light produces small changes in some metabolite concentrations and enzyme activities of the citric-acid cycle and related pathways [52–55]; the largest effect was on the glyceraldehyde 3-phosphate dehydrogenase activity [56].

The only chemical effect that has been investigated from the viewpoint of photobiology is the increased accumulation of the yellow pigment β-carotene. The biosynthesis of carotenoids is almost universally regulated by light [8]. Photocarotenogenesis resembles other photoresponses of *Phycomyces* in its low absolute threshold (about 10 $\mu J/m^2$) and the two-step fluence response curve (Figure 4). Photocarotenogenesis does not occur throughout mycelial growth, but is restricted to a competence period that coincides roughly with that for photophorogenesis when tested under the same culture conditions [57].

The biological role of photocarotenogenesis in *Phycomyces* is obscure. The protective effect against oxidative stress and radiations, particularly those absorbed by carotenes, is not conciliated easily with the very low threshold and the brief competence period of the *Phycomyces* response or with the photoinhibition of carotene production in a related fungus, *Blakeslea trispora* [58]. The ability to make carotenes does not modify the fitness of *Phycomyces* when exposed to various radiations [59,60].

Phycomyces has a strong capacity to repair ultraviolet-induced damage by photoreactivation with longer wavelengths [27]. Spores formed in the light photoreactivate much better than those formed in the dark [61], probably because the production of DNA photolyase is induced by light.

Bright light (about 1 W/m^2) inhibits mycelial growth [62], particularly in the presence of photosensitizers [63]. This is more likely to be a consequence of photodamage than a sensory response.

21.5 Behavioral mutants

Many wild strains of *Phycomyces blakesleeanus* and the related species *P. nitens* have been isolated throughout the world. One of them, NRRL1555, is the standard strain for physiology and genetics. Many mutants and recombinants have been obtained and are kept in the laboratories that work with this organism. An official strain collection was set up at the Institute of Genetic Ecology, Tohoku University, Sendai, Japan, under the care of Prof. T. Ootaki.

In *Phycomyces*, as in all organisms, the information acquired by sensors must be conveyed to effectors by specific sensory pathways. These are formally similar to biochemical pathways, but represent flows of information, not metabolites. The signal transducers that compose the pathways (sensors, effectors, and intermediate transducers) may be affected by mutations in the genes that govern them. Genetic analysis can identify the elements of sensory pathways and their relationships and produce a sensory pathway chart [64].

21.5.1 The piloboloides mutants

Behavioral mutants have been isolated by many different criteria. The earliest ones are the "piloboloides" (*pil*) mutants [9,65], characterized by the bulging growing zone of their macrophores, which reminds of the sporangiophores of a related fungus, *Pilobolus*. The altered morphology affects the focusing of light in the growing zone. Blue light causes a positive phototropism when the diameter of the growing zone does not exceed 0.2 mm, and a negative one when it becomes larger, up to about 0.5 mm [66,67]. The *pil* mutants, most of which belong to a single gene, *pilA* [68], confirm that *Phycomyces* does not respond to the direction of light, but to its distribution in the growing zone, a result that was reached from experiments with the wild type [69,70].

21.5.2 Pigment mutants

Mutants lacking β-carotene [71,138] are easy to find because their mycelia are not yellow, but red or white. Mutants in gene *carB* over accumulate phytoene. Mutants in the bifunctional gene *carRA* have various phenotypes: the *carR* mutants over accumulate lycopene; the *carA* mutants, deficient in phytoene synthase have traces of β-carotene; and the *carRA* mutants have traces of lycopene. The most complete block of the pathway is found in double mutants *carB carRA*, which are defective for the last seven reactions in the pathway and have no detectable carotenes. These mutants exhibit normal phototropism [64,72], thus negating the possibility that β-carotene is the photoreceptor for this response, an old conjecture based on rough similarities between the respective action and absorption spectra.

β-Carotene or a derived compound has an essential role in photophorogenesis [73]; the production of macrophores and microphores in *carB* and *carRA*, mutants is not modified by blue light pulses up to at 10 kJ/m^2 [74]. The role of β-carotene cannot be fulfilled by phytoene (abundant in the *carB* mutants), lycopene (abundant in the *carR* mutants) or retinol (added to the medium). The mutants exhibit reduced photomorphogenetic responses under bright continuous light.

A requirement of β-carotene for photocarotenogenesis is suggested by the observation that *carB carR* double mutants, which are defective for the last six reactions in the pathway, do not increase their phytoene content in response even to very bright light [57]. The same is true for the lycopene content of the *carR* mutants, but not for the phytoene content of *carB* mutants. This is a puzzling result that deserves reexamination.

Carotene overproducing mutations occur in genes *carS* [75], *carF* [76], and *carD* [77]. The *carS* and *carF* mutants are deficient in photocarotenogenesis [57,76,78]. The *carS* gene product plays a central role in the regulation of carotene biosynthesis as a mediator of end-product inhibition and other regulations of the pathway [75,79].

The *carA* mutants, which have much less β-carotene than the wild type, show diminished photophorogenesis and photocarotenogenesis. These defects are not due to the carotene shortage, because the same defects were found in *carA carS* recombinants which have more carotene than the wild type [57,74,80]. The *carA* mutants have defects both in the biosynthesis of carotene and its regulation [25,138].

The *carC* mutants, which were isolated because they have less β-carotene than the wild type, are partially defective in photocarotenogenesis. The analysis of double mutants indicates that *carC* acts before *carS* in the sensory pathway [81].

Pigments present in the growing zone affect phototropism in an indirect way because they modify the distribution of light [82]. A carotene overproducer shows negative tropism to blue light [83]. The role of β-carotene extends beyond its involvement in the responses to light. Many *car* mutants differ from the wild type in the numbers of macrophores and microphores produced in the dark [74,84]. This suggests that phorogenesis is influenced by the amount of β-carotene present in the cell. β-Carotene has an essential role in the sexual cycle as precursor of the sexual hormones; the *car* mutants lacking β-carotene fail to stimulate their partners and do not enter the sexual cycle [85]. The formation of zygospores is blocked completely or partially in *car* mutants with abnormal β-carotene concentrations. The causes are unclear: the *carS* mutants are sexually incapable while a *carF* mutant of the same carotene content produces some zygospores [76].

The riboflavin auxotrophic mutants *ribA, ribB,* and *ribD* are essentially normal for phototropism [86,87]; this does not eliminate riboflavin and its derivatives as candidate photoreceptors, if the concentrations required for normal growth are higher than those required for normal vision. The candidacy was very strongly supported by the observation that a *ribB* mutant fed riboflavin and its analogue roseoflavin responds to light of wavelength 529 nm, characteristic of roseoflavin absorption [86]. Riboflavin may not be the only chromophore for phototropism in *Phycomyces.* A *ribC* mutant has a reduced positive phototropism to wavelengths around 390 nm; its chemical analysis suggests that 6,7-dimethyl-8-ribityllumazine or an analogue, possibly a pterin, is required for phototropism [87].

21.5.3 Phototropism mutants

The *mad* mutants were isolated for defective phototropism of their macrophores. Most of them failed to turn downward or sideward towards a light source [64,88,89]. Hypertropic mutants built wavy macrophores as these grew between a pair of sources that were lit alternatively under conditions that did not allow the wild type to respond [90].

The *mad* mutants define ten unlinked genes, named in order of discovery from *madA* to *madJ* [22,91–95]. There are two major classes of *mad* mutants. Mutants in genes *madA*, *madB*, *madC*, and *madI* are class-1 mutants, defective in phototropism, but not in other tropisms. Informally they are said to be "night blind", because they do not respond to dim light. They respond to a certain illumination as the wild type would respond to a weaker one: mutant and wild-type stimulus-response curves look similar, with those of the mutants shifted to higher fluence rates.

Mutants in genes *madD, madE, madF, madG,* and *madH* are class-2 mutants, defective in gravitropism and avoidance of obstacles, as well as in phototropism. The recessive hypertropic mutants fall in gene *madH* [96]. Mutants in genes *madD, madE, madF* and *madG* are called "stiff" because they respond more slowly than the wild type, sometimes so slowly that they appear blind to all fluence rates.

21.5.4 Negative tropisms to ultraviolet C radiation

Tropisms towards blue sources and away from ultraviolet ones have been attributed to the same photoreceptors and effectors. Ultraviolet radiation is expected to be absorbed by chromophores that absorb blue light and by the apoproteins bound to them. The opposite directions of blue and ultraviolet tropisms were attributed to the strong ultraviolet absorbency of gallic acid in the macrophores [97]. Mutants affected in known *mad* genes were isolated for being unable to turn away from an ultraviolet C source [98]. The ultraviolet responses of the *mad* mutants, except *madG*, are defective, but not as much as their responses to blue light [99]. Unexpectedly, two kinds of mutants, *uvi* [100] and *hba* [101] have established the existence of a specific ultraviolet tropism. The former were isolated because they exhibit normal positive tropism towards blue sources, but defective negative tropism away from ultraviolet sources; they are believed to mark gene products specifically used for the ultraviolet response, not necessarily a wholly separate pathway. The *hba* mutants failed a colorimetric test for gallic acid; although their macrophores were almost free of this phenol, they turned away from ultraviolet sources as usual, and thereby they contradicted the absorbance hypothesis.

21.5.5 Mutants for other responses

The *pim* mutants were isolated because they produce microphores under constant illumination that completely prevents their appearance in the wild type [102]. These mutants are normal for phototropism and photomacrophorogenesis, but at least one of them seems to be defective in photocarotenogenesis. The *pim* mutants represent a clear genetic separation of two sensory pathways for the regulation of sporangiophore development.

The *pic* mutants defective in photocarotenogenesis were isolated because of their pallid color when grown in the light and their normal color when grown in the dark. The two alleles that were studied belong to different genes, *picA* and *picB* [80].

Gravitropism is a very slow response (turning velocity 0.1 to 0.3°/min), in comparison with phototropism (up to 5°/min, [39]) and the avoidance of obstacles (up to 8°/min, [103]). Mutants that are even slower are easy to obtain. Some of them lack [104] certain protein crystals present in normal macrophores [105], which are likely to act as statoliths. Other mutants with slow gravitropism contain the crystals [139]. The contrary phenotype occurs in a *carB* mutant that was isolated because of its white color, but exhibits a fast gravitropism [37,90,106], presumably because of an additional *geo* mutation. This mutant shows that slow gravitropism is an evolutionary preference: the different velocities presumably allow the cell to set priorities in the case of contradictory information from different sensors. The bending rates of other tropisms have been set by evolution as well, as shown by the isolation of the *madH* mutants, whose phototropism and avoidance are about twice as fast as those of the wild type [90].

21.6 A chart of the sensory pathways

The genetic analysis of the *Phycomyces* sensory pathways can be summarized in a chart (Figure 5) that establishes the signal pathways from receptors to effectors and indicates known mutational blocks. The chart is based on the phenotypes of the behavioral

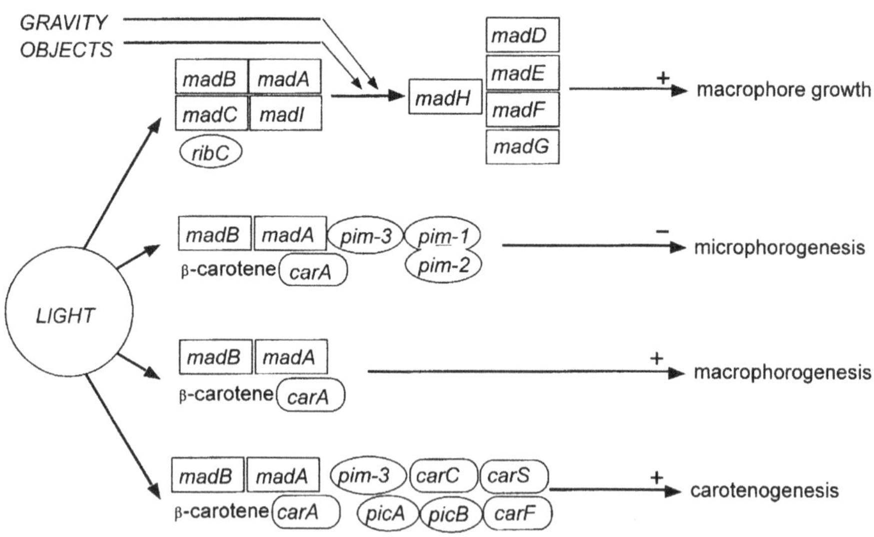

Figure 5. The sensory pathways of *Phycomyces* link the stimuli (left) and the responses (right). The flow of information is hindered by mutations at the genes given in italics or by lack of β-carotene.

mutants, and particularly on their pleiotropism (Table 1). The chart is necessarily partial because the search for some behavioral mutants is incomplete and the search for others has not even begun.

Sensory responses require that the energy absorbed by the sensor be less than the energetic cost of the response; this definition excludes from sensory physiology such responses as the accumulation of glucose in illuminated plants and photoreactivation after exposure to ultraviolet C radiation. Another exclusion should be simple photodamage; for example, an enzyme that contains riboflavin as a cofactor may be inactivated by blue light and this may lead to a modification of the chemical composition of the cell. Such a phenomenon would occur equally well in the behavioral mutants and in the wild type and probably would require strong illumination. An example is the inhibition of mycelial growth by light in the presence of quinacrine [107]. Photodamage and sensory responses are not mutually exclusive, because a product of the damage could be used by the cell as a signal in a sensory process.

21.6.1 Photoreceptors and other early transducers

All tropisms may be expected to share the genes required to regulate growth. Genes *madD, madE, madF, madG* and *madH*, defined by class-2 mutants, belong to this group because their mutants affect the tropisms of the macrophores to various stimuli. By contrast, genes *madA, madB, madC, madI,* and *ribC*, defined by class-1 mutants, are presumed to govern photoreception and other early steps of phototropism. Biophysical and biochemical analyses point to an involvement of these genes in photoreception.

The action spectra of class-1 mutants differ from that of the wild type in the overall effectiveness; those of *madB* and *madC* [108,109], and *ribC* [87] mutants seem to differ also in the shape (relative wavelength dependence). Because these mutants are relatively frequent, each is likely to suffer the loss of activity of a gene product, rather than specific modifications of its activity; the changes in the shape of the action spectra probably represent the loss of elements in a complex photoreceptor that contains several chromophores, including riboflavin, as seen above. The *madI* mutants have a consistent loss of an unidentified pterin [110], which could thus be one of the chromophores in the photoreceptor. The *madA* mutants have less flavin than the wild type, but flavin supplementation does not improve their defective phototropism [111]. The *madA* gene product has been proposed to interact physically with the *madB* and *madC* gene products. Nearly all the *mad* gene products seem to interact physically, as if they formed part of an aggregate [26,109,112,113], let us say, a "sensosome".

The night-blind mutants are defective in both photosystems used by the wild type to respond to different fluence rates. The photosystem for bright light is characterized by a strong delayed response which was not found in *madA*, *madB,* and *madC* mutants exposed up to 1 W/m^2; two weak responses, similar to those expected from the dim-light photosystem, were found in the mutants, but only when the illumination was much brighter than required in the wild type [114].

The *mad* mutants adapt to darkness. Mutants in gene *madA* behave exactly as the wild type when light fluences are measured not in absolute units, but relative to the respective thresholds; *madB* and *madH* mutants adapt more slowly and more rapidly than the wild

Table 1. Genetic analyses of the sensory responses of *Phycomyces*. The cells contain the references for research on the role of the gene on the left in the response above. Boxed cells indicate evidence of involvement, i.e. mutants in the gene are defective in the response. Non-boxed cells indicate that the responses are apparently normal in the mutants.

	Phototropism	Photomecism	Photomacro-phorogenesis	Photomicro-phorogenesis	Photocaroteno-genesis	Geotropism	Avoidance
madA	[64] [92] [26] [125] [108] [126] [115] [109] [114] [111] [98]	[127] [115] [128] [129] [112] [130]	[64] [73] [107] [84] [74]	[107] [84] [74]	[131] [80] [132]	[98] [99]	[64] [98] [99]
madB	[64] [92] [26] [125] [108] [126] [115] [109] [114] [98]	[128] [115] [113] [129] [112] [130]	[133] [64] [73] [107] [84] [74]	[107] [84] [74]	[64] [131] [80] [132]	[64] [98] [99]	[64] [98] [99]

Table 1. Continued.

	Phototropism	Photomecism	Photomacro-phorogenesis	Photomicro-phorogenesis	Photocaroteno-genesis	Geotropism	Avoidance
madC	[64] [92] [26] [93] [125] [88] [108] [126] [109] [114] [98] [134]	[128] [115] [113] [129] [112] [130] [135] [134]	[64] [93] [73] [107] [74]	[107] [74]	[64] [131]	[64] [88] [98] [99]	[64] [93] [88] [99]
madD	[64] [92] [26] [93] [125] [88] [99]	[127] [128] [129] [112]	[64] [93] [73] [84] [74]	[107] [84] [74]	[64] [131]	[64] [88] [99]	[64] [93] [88] [99]
madE	[64] [92] [26] [125] [89] [99]	[127] [128] [129] [112] [124]	[64] [73] [74]	[74]	[64]	[64] [89] [99]	[64] [89] [99]

Table 1. Continued.

	Phototropism	Photomecism	Photomacro-phorogenesis	Photomicro-phorogenesis	Photocaroteno-genesis	Geotropism	Avoidance
madF	[64] [125] [89] [98] [99]	[128] [129] [112]	[73] [74]	[74]	[98]	[89] [98] [99]	[64] [89] [98] [99]
madG	[125] [98] [99]	[129] [112] [135]	[73] [74]	[74]	[98]	[98] [99]	[98] [99]
madH	[90] [126] [115] [109] [136]	[136] [113] [135] [124]	[90] [74]	[74]		[90]	[90]
madI	[89] [111] [98]					[89] [99]	[89] [99]
madJ	[98] [99]		[102]	[102]		[98] [99]	[99]
No β-carotene	[64] [72]		[73] [84] [74]	[84] [74]	[57]		
carA	[64] [126]	[128]	[73] [84] [74]	[84] [74]	[80] [78] [57]		

Table 1. Continued.

	Phototropism	Photomecism	Photomacro-phorogenesis	Photomicro-phorogenesis	Photocaroteno-genesis	Geotropism	Avoidance
carA carS			[74]	[74]	[80] [57]		
carS			[84] [74]	[84] [74]	[80] [78] [57]		
carD			[74]	[74]	[57]		
carC			[74]	[74]	[80] [78] [57]		
picA	[80]		[74]	[74]	[80]		
picB	[80]		[74]	[74]	[80]		
pim-1	[102]		[102]	[102]	[102]		
pim-2	[102]		[102]	[102]	[102]		
pim-3	[102]		[102]	[102]	[102]		

type, respectively [115]. Since *madB* is considered a photoreceptor gene, at least part of the adaptation must occur early in the signal pathway.

Very little is known about the genetics of a related photoresponse: the effect of subliminal light on adaptation; the effectiveness of long wavelengths imply special photoreceptors, but the requirement of gene *madB* [116] brings it to the common fold of *Phycomyces* sensory photoresponses.

Genes *madA*, *madB* and *carA* are needed for the mycelial photoresponses, probably for early functions, as judged from the "night-blind" stimulus-response relationships in mutants of these genes. The phenotypes of the double mutants [74] suggest that the *madA* and *madB* gene products work in succession, not in parallel, and that the *carA* gene product interacts with them.

The requirement for *madA* and *madB* in all the photoresponses of *Phycomyces* suggests that all photoreceptors share some elements, and the specific requirements for *madC*, *madI*, *carA*, and β-carotene point to structural differences. Two *pim* mutants represent at least one function involved in photomicrophorogenesis only. A function shared by this response and photocarotenogenesis is defective in another *pim* mutant.

The same picture of different photoreceptors with common and specific elements is suggested by action spectroscopy. The three mycelial photoresponses of *Phycomyces* that have been studied in some detail (photomicrophorogenesis, photomacrophorogenesis and photocarotenogenesis), just as the growth responses, depend on two photosystems each, which exhibit fluence-dependent responses in different fluence ranges. The six action spectra for the mycelial photoresponses [49,57] are similar, but not identical, suggesting minor differences in the corresponding photoreceptors and their environment. The available mutants are probably deficient in the two photosystems, as in the case of phototropism, but this is not proven for most of them.

The available information is compatible with a simple proposal: carotene is synthesized in lipid and protein globules [117,118] that contain β-carotene and the products of the *car* genes. At least some of these globules are converted to photoreceptors and signal transducers by the incorporation, during a certain period of development, of the products of *madA*, *madB*, and other genes.

Blue light causes acidification of the mycelial cytoplasm (by 0.3 pH units) and hyperpolarization across the membrane, probably caused by the activation of a proton ATPase. The hyperpolarization does not occur in *madA*, *madB*, and *madC* mutants and is thus an early consequence of the activity of the photoreceptor [119].

The three mycelial photoresponses described above depend on genes *madA* and *madB*, but not on other *mad* genes. The same gene dependence is found in responses that occur in the macrophores. One example is the hastening of the development of sporangia on macrophores exposed to very bright blue light [44]. Another is the improved photoreactivation of ultraviolet damage in spores that have been produced in illuminated cultures, presumably due to the photoinduction of DNA photolyase [61].

21.6.2 Regulation of macrophore growth

Mutants in genes *madD, madE, madF, madG* and *madH* affect simultaneously all the tropisms that have been tested: phototropism, gravitropism and avoidance. The growth-rate fluctuations seen in unstimulated macrophores of *madD, madE, madF* and *madG* mutants [120] indicate that the corresponding genes are involved in the regulation of growth. The abnormal growth fluctuations are missing in *madH* mutants and in mutants of early genes; the *madH* mutants are hypertropic for all the stimuli.

Growth involves the weakening of the cell wall of the growing zone, which causes an immediate longitudinal extension, and the reinforcement of the cell wall with new materials [121,122]. Because the strength of the wall is due essentially to its chitin, the growth responses are expected to cut chitin fibers by the action of chitinase and then to reinforce the wall with new chitin synthesis. In keeping with this expectation, the number of free chitin-chain ends in the wild type increases shortly after illumination and chitin synthetase activity increases a little later; both are late effects of light, missing in *madB* and *madE* mutants [123].

Most *mad* mutants are defective both in phototropism and photomecism, but not necessarily to the same extent. This result concords with the view that phototropism results from local variations in growth velocity and additional factors needed for integrated detection and response.

Gene *madJ* is identified by a single mutant almost totally blind for blue light. It has a very complex phenotype: defective gravitropism and avoidance, slightly defective ultraviolet C tropism, normal photocarotenogenesis [98,99], normal photomacrophorogenesis and defective photomicrophorogenesis [102]. Before trying to place *madJ* in the sensory pathway chart one should consider the puzzling genetic results: in heterokaryosis the *madJ* mutant genetically complements various *mad* mutants, but not *madD*, *madE*, and *madF* mutants; in crosses, the *madJ* mutation recombines freely with *madD*, *madE*, and *madF* [22,98]. The *madJ* gene product may be a regulatory RNA confined to its native nucleus and needed for the expression of genes *madD*, *madE*, and *madF*.

The action spectra of several class-2 mutants are reported to differ slightly in shape from that of the wild type: phototropism in *madI* [110], *madF* and *madJ* [99] and photomecism in *madE* [124]. It is possible that some genes affect both sensors and effectors, but such a conclusion is premature in view of the slight differences, which may not be significant, and the inherent difficulties of action spectroscopy in this case, where the results depend markedly on the experimental conditions and geometrical considerations [70].

21.6.3 Combinatorial genetics of behavior

The use of a gene product for different purposes gives an impression of parsimony: some elements are common to several pathways. Blue-light receptors that require the *madA* and *madB* gene functions combine with different effectors to carry out various photoresponses, such as tropism, carotenogenesis, and morphogenesis. Different stimuli, such as blue light, gravity and obstacles, share the effectors that regulate macrophore growth. The product of gene *carA* is required for all the mycelial photoresponses. Several genes seem to be specific for a single response.

The photoresponses of *Phycomyces* may have evolved from an early photosystem that included the products of genes *madA* and *madB*. The improvement and the diversification of the responses to light required the recruitement of other existing gene products for new functions. It is tempting to assume that the growth photoresponses acquired their extraordinary sensitivity through the introduction of the *madC* gene product and the mycelial photoresponses did the same with the *carA* gene product and β-carotene. Genes with multiple functions are subject to contradictory evolutionary pressures. The usual solution of this problem is gene duplication followed by separate specialization. To estimate the extent of this phenomenon in *Phycomyces* behavior we must wait for the gene sequences to be known. An alternative evolutionary possibility is to improve the capacity of a gene product to interact with various partners for different tasks. This way seems to have been exploited by *Phycomyces* to create its complex set of behaviors.

References

1. H. Senger (Ed.) (1987). *Blue light responses: phenomena and occurrence in plants and microorganisms*. CRC Press, Boca Raton FL, USA.
2. J.B. Carnoy (1870). Recherches anatomiques et physiologiques sur les champignons. *Bull. Soc. R. Bot. Belg.*, **9**, 157–321.
3. J. Sachs (1879). Über Auschliessung der geotropischen und heliotropischen Krümmungen während des Wachsens. *Arb. Botan. Inst. Würzburg*, **2**, 209–225.
4. J. Massart (1888). Recherches sur les organismes inférieurs. 1. La loi de Weber vérifiée pour l'héliotropisme du champignon. *Bull. Acad. R. Sci. Belg.* (Ser. 3), **16**, 590–601.
5. E. Cerdá-Olmedo, E.D. Lipson (1987). *Phycomyces*. Cold Spring Harbor Laboratory Press, New York.
6. E. Cerdá-Olmedo, V. Martín-Rojas (1996). Phototropism in *Phycomyces*. In: C. Jennings, G. Zucchelli, F. Ghetti, G. Colombetti (Eds), *Light as an energy source and information carrier in plant physiology* (pp. 293–299). Plenum Press, New York.
7. L.M. Corrochano, E. Cerdá-Olmedo (1992). Sex, light, and carotenes: the development of *Phycomyces*. *Trends Genet.*, **8**, 268–274.
8. J. Avalos, E.R. Bejarano, E. Cerdá-Olmedo (1993). Photoinduction of carotenoid biosynthesis. *Meth. Enzymol.*, **214**, 283–294.
9. H. Burgeff (1912). Über Sexualität, Variabilität und Vererbung bei *Phycomyces nitens*. *Ber. Dtsch. Bot. Ges.*, **30**, 679–685.
10. M. Orlowski (1991). *Mucor* dimorphism. *Microbiol. Rev.*, **55**, 234–258.
11. H.D. Madhani, G.R. Fink (1998). The control of filamentous differentiation and virulence in fungi. *Trends Cell Biol.*, **8**, 348–353.
12. M. Heisenberg, E. Cerdá-Olmedo (1968). Segregation of heterokaryons in the asexual cycle of *Phycomyces*. *Mol. Gen. Genet.*, **102**, 187–195.
13. P. Reau (1972). Uninucleate spores of *Phycomyces*. *Planta*, **108**, 153–160.
14. T. Ootaki, N. Orihara, S. Umemura, S. Yabe, A. Miyazaki (1991). A mutant of *Phycomyces blakesleeanus* with a high proportion of uninucleate spores. *Trans. Mycol. Soc. Japan*, **32**, 87–102.
15. H. Burgeff (1914). Untersuchungen über Variabilität, Sexualität und Erblichkeit bei *Phycomyces nitens* Kuntze. *Flora*, **107**, 259–316.
16. W. Gauger, M.I. Peláez, M.I. Alvarez, A.P. Eslava (1980). Mating type heterokaryons in *Phycomyces blakesleeanus*. *Exp. Mycol.*, **4**, 56–64.
17. T. Ootaki (1973). A new method for heterokaryon formation in *Phycomyces*. *Molec. Gen. Genet.*, **121**, 49–56.

18. M. Delbrück, T. Ootaki (1979). An unstable nuclear gene in *Phycomyces*. *Genetics*, **92**, 27–48.
19. H. Dickson (1932). The effects of x–rays, ultraviolet light, and heat in producing saltants in *Chaetomium cochliodes* and other fungi. *Ann. Botany*, **46**, 389–405.
20. M.I.G. Roncero, C. Zabala, E. Cerdá-Olmedo (1984). Mutagenesis in multinucleate cells: the effects of N-methyl–N′-nitro-N-nitrosoguanidine on *Phycomyces* spores. *Mutat. Res.*, **125**, 195–204.
21. M. Demerec, E.A. Adelberg, A.J. Clark, P.E. Hartman (1966). A proposal for a uniform nomenclature in bacterial genetics. *Genetics*, **54**, 61–76.
22. V. Campuzano, P. Galland, A.P. Eslava, M.I. Alvarez (1995). Genetic characterization of two phototropism mutants of *Phycomyces* with defects in the genes *mad I* and *mad J*. *Curr. Genet.*, **27**, 524–527.
23. M.D. De la Guardia, C.M.G. Aragón, F.J. Murillo, E Cerdá-Olmedo (1971). A carotenogenic enzyme aggregate in *Phycomyces*: evidence from quantitative complementation. *Proc. Nat. Acad. Sci. USA*, **68**, 2012–2015.
24. C.M.G. Aragón, F.J. Murillo, M.D. De la Guardia, E. Cerdá-Olmedo (1976). An enzyme complex for the dehydrogenation of phytoene in *Phycomyces*. *Eur. J. Biochem.*, **63**, 71–75.
25. F.J. Murillo, S. Torres-Martínez, C.M.G. Aragón, E. Cerdá-Olmedo (1981). Substrate transfer in carotene biosynthesis in *Phycomyces*. *Eur. J. Biochem.*, **119**, 511–516.
26. J.R. Medina, E. Cerdá-Olmedo (1977). Allelic interactions in the photogeotropism of *Phycomyces*. *Exp. Mycol.*, **1**, 286–292.
27. E. Cerdá-Olmedo, P. Reau (1970). Genetic classification of the lethal effects of various agents on heterokaryotic spores of *Phycomyces*. *Mutat. Res.*, **9**, 369–384.
28. C. Pueyo, E. Cerdá-Olmedo (1981). Causes of cell death following exposure to different fungicides and heat. *Exp. Mycol.*, **5**, 112–119.
29. E. Cerdá-Olmedo, V. Martín-Rojas, B. Cubero (1996). Causes of cell death following ultraviolet B and C exposures and the role of carotenes. *Photochem. Photobiol.*, **64**, 547–551.
30. M. Orejas, T. Suárez, A.P. Eslava (1985). Helper strains for shortening the dormancy in *Phycomyces blakesleeanus*. *Curr. Genet.*, **9**, 369–372.
31. M.J.F. Sarabia, A.P. Eslava, M.I. Alvarez (1988). Influence of parental strains on the germination of *Phycomyces blakesleeanus* zygospores. *Genet. Res.*, **52**, 91–95.
32. E. Cerdá-Olmedo (1975). The genetics of *Phycomyces blakesleeanus*. *Genet. Res.*, **25**, 285–296.
33. M.I. Alvarez, A.P. Eslava (1983). Isogenic strains of *Phycomyces blakesleeanus* suitable for genetic analysis. *Genetics*, **105**, 873–879.
34. A.P. Eslava, M.I. Alvarez, P.V. Burke, M. Delbrück (1975). Genetic recombination in sexual crosses of *Phycomyces*. *Genetics*, **80**, 445–462.
35. M. Orejas, M.I. Peláez, M.I. Alvarez, A.P. Eslava (1987). A genetic map of *Phycomyces blakesleeanus*. *Mol. Gen. Genet.*, **210**, 69–76.
36. M.I. Alvarez, E.P. Benito, V. Campuzano, A.P. Eslava (1993). Genetic loci of *Phycomyces blakesleeanus*. In: S.J. O'Brien (Ed.), *Genetic maps* (6th ed., pp. 3.120–3.126). Cold Spring Harbor Laboratory Press, Cold Spring Harbor, NY.
37. K. Bergman, P.V. Burke, E. Cerdá-Olmedo, C.N. David, M. Delbrück, K.W. Foster, E.W. Goodell, M. Heisenberg, G. Meissner, M. Zalokar, D.S. Dennison, W. Shropshire (1969). *Phycomyces*. *Bact. Rev.*, **33**, 99–157.
38. G.M. Curry, H.E. Gruen (1959). Action spectra for the positive and negative phototropism of *Phycomyces* sporangiophores. *Proc. Nat. Acad. Sci. USA*, **45**, 797–804.
39. P. Galland, E.D. Lipson (1987). Blue–light reception in *Phycomyces* phototropism:

evidence for two photosystems operating in low– and high–intensity ranges. *Proc. Nat. Acad. Sci. USA*, **84**, 104–108.
40. P. Galland, L.M. Corrochano, E.D. Lipson (1989). Subliminal light control of dark adaptation kinetics in *Phycomyces* phototropism. *Photochem. Photobiol.*, **49**, 485–491.
41. X.-Y. Chen, Y.-Q. Xiong, E.D. Lipson (1993). Action spectrum for subliminal light control of adaptation in *Phycomyces* phototropism. *Photochem. Photobiol.*, **58**, 425–431.
42. G. Löser, E. Schäfer (1986). Are there several photoreceptors involved in phototropism of *Phycomyces blakesleeanus*? Kinetic studies of dichromatic irradiation. *Photochem. Photobiol.*, **43**, 195–204.
43. R.M. Thornton (1972). Alternative fruiting pathways in *Phycomyces*. *Plant Physiol.*, **49**, 194–197.
44. V.E.A. Russo, P. Galland, M. Toselli, L. Volpi (1980). Blue light induced differentiation in *Phycomyces blakesleeanus*. In: H. Senger (Ed.), *The Blue Light Syndrome* (pp. 563–569). Springer, Berlin.
45. L.M. Corrochano, E. Cerdá-Olmedo (1988). Photomorphogenesis in *Phycomyces*: dependence on environmental conditions. *Planta*, **174**, 309–314.
46. T. Kumagai (1988). Photocontrol of fungal development. *Photochem. Photobiol.*, **47**, 889–896.
47. L.M. Corrochano, E. Cerdá-Olmedo (1991). Photomorphogenesis in *Phycomyces* and in other fungi. *Photochem. Photobiol.*, **54**, 319–327.
48. L.M. Corrochano, E. Cerdá-Olmedo (1990). Photomorphogenesis in *Phycomyces*: competence period and stimulus-response relationships. *J. Photochem. Photobiol.*, **B5**, 255–266.
49. L.M. Corrochano, P. Galland, E.D. Lipson, E. Cerdá-Olmedo (1988). Photomorphogenesis in *Phycomyces* fluence–response curves and action spectra. *Planta*, **174**, 315–320.
50. Y. Yamazaki, H. Kataoka, A. Miyazaki, M. Watanabe, T. Ootaki (1996). Action spectra for photoinitiation of sexual development in *Phycomyces blakesleeanus*. *Photochem. Photobiol.*, **64**, 387–392.
51. A.F. Barrero, J.E. Oltra, J.A. Poyatos (1996). Acidic metabolites from *Phycomyces blakesleeanus*. *Phytochemistry*, **42**, 1427–1433.
52. G. Sandmann, W. Hilgenberg (1980). The light–dependent intermediary metabolism of *Phycomyces blakesleeanus* Bgff. II. Formation and degradation of oxalacetic acid. *Z. Pflanzenphysiol.*, **96**, 203–210.
53. R. Garcés, J.R. Medina (1985). Light–dependent decrease in alcohol dehydrogenase activity of *Phycomyces*. *Exp. Mycol.*, **9**, 94–98.
54. J. Rua, L.B. Rodríguez-Aparicio, F. Busto, J. Soler (1987). Effect of light on several metabolites of carbohydrate metabolism in *Phycomyces blakesleeanus*. *J. Bacteriol.*, **169**, 904–907.
55. M. Díez, J. Rua, F. Busto, J. Soler (1988). Analysis of the effect of light on carbohydrate metabolism in *Phycomyces sporangiophores* and mycelia. *Int. J. Biochem.*, **20**, 801–806.
56. L.B. Rodríguez-Aparicio, J. Rua, D. De Arriaga, J. Soler (1987). Light–induced effects of several enzymes of carbohydrate metabolism in *Phycomyces blakesleeanus*. *Int. J. Biochem.*, **19**, 1211–1215.
57. E.R. Bejarano, J. Avalos, E.D. Lipson, E. Cerdá-Olmedo (1991). Photoinduced accumulation of carotene in *Phycomyces*. *Planta*, **183**, 1–9.
58. R.P. Sutter (1970). Effect of light on β-carotene accumulation in *Blakeslea trispora*. *J. Gen. Microbiol.*, **64**, 215–221.
59. E. Cerdá-Olmedo, V. Martín-Rojas, B. Cubero (1996). Causes of cell death following ultraviolet B and C exposures and the role of carotenes. *Photochem. Photobiol.*, **64**, 547–551.

60. V. Martín-Rojas, A. Gómez-Puerto, E. Cerdá-Olmedo (1996). Lack of protection by carotenes against gamma–radiation damage in *Phycomyces*. *Radiat. Environ. Biophys.*, **35**, 193–197.
61. P. Galland (1996). Ultraviolet killing and photoreactivation of *Phycomyces* spores. *Microbiol. Res.*, **151**, 9–17.
62. R.M. Thornton (1973). New photoresponses of *Phycomyces*. *Plant Physiol.*, **51**, 570–576.
63. M.J. Carlile (1962). Evidence for a flavoprotein photoreceptor in *Phycomyces*. *J. Gen. Microbiol.*, **28**, 161–167.
64. K. Bergman, A.P. Eslava, E. Cerdá-Olmedo (1973). Mutants of *Phycomyces* with abnormal phototropism. *Mol. Gen. Genet.*, **123**, 1–16.
65. K. Koga, T. Ootaki (1983). Growth of the morphological, piloboloid mutants of *Phycomyces blakesleeanus*. *Exp. Mycol.*, **7**, 148–160.
66. K. Koga, T. Sato, T. Ootaki (1984). Negative phototropism in the piloboloid mutants of *Phycomyces blakesleeanus* Bgff. *Planta*, **162**, 97–103.
67. T. Ootaki, T. Tsuru (1993). Diphasic negative phototropism of sporangiophores of the piloboloid mutant of *Phycomyces blakesleeanus*. *Exp. Mycol.*, **17**, 103–108.
68. K. Koga, T. Ootaki (1983). Complementation analysis among piloboloid mutants of *Phycomyces blakesleeanus*. *Exp. Mycol.*, **7**, 161–169.
69. W. Shropshire (1962). The lens effect and phototropism of *Phycomyces*. *J. Gen. Physiol.*, **45**, 949–958.
70. L. Fukshansky (1993). Intracellular processing of a spatially non–uniform stimulus: case study of phototropism in *Phycomyces*. *J. Photochem. Photobiol. B: Biol.*, **19**, 161–186.
71. E. Cerdá-Olmedo (1985). Carotene mutants of *Phycomyces*. *Meth. Enzymol.*, **110**, 220–243.
72. D. Presti, W.-J. Hsu, M. Delbrück (1977). Phototropism in *Phycomyces* mutants lacking β-carotene. *Photochem. Photobiol.*, **26**, 403–405.
73. P. Galland, V.E.A. Russo (1979). Photoinitiation of sporangiophores in *Phycomyces* mutants deficient in phototropism and in mutants lacking β-carotene. *Photochem. Photobiol.*, **29**, 1009–1014.
74. L.M. Corrochano, E. Cerdá-Olmedo (1990). Photomorphogenesis in behavioral and colour mutants of *Phycomyces*. *J. Photochem. Photobiol. B: Biol.*, **6**, 325–335.
75. F.J. Murillo, E. Cerdá-Olmedo (1976). Regulation of carotene synthesis in *Phycomyces*. *Mol. Gen. Genet.*, **148**, 19–24.
76. B.J. Mehta, L.M. Salgado, E.R. Bejarano, E. Cerdá-Olmedo (1997). New mutants of *Phycomyces blakesleeanus* for β-carotene production. *Appl. Environ. Microbiol.*, **63**, 3657–3661.
77. L.M. Salgado, E.R. Bejarano, E. Cerdá-Olmedo (1989). Carotene–superproducing mutants of *Phycomyces blakesleeanus*. *Exp. Mycol.*, **13**, 332–336.
78. J.L. Revuelta, A.P. Eslava (1984). Photoregulation of carotenogenesis in *Phycomyces*. *Curr. Genet.*, **8**, 261–264.
79. E.R. Bejarano, F. Parra, F.J. Murillo, E. Cerdá-Olmedo (1988). End-product regulation of carotenogenesis in *Phycomyces*. *Arch. Microbiol.*, **150**, 209–214.
80. I. López-Díaz, E. Cerdá-Olmedo (1980). Relationship of photocarotenogenesis to other behavioral and regulatory responses in *Phycomyces*. *Planta*, **150**, 134–139.
81. J.L. Revuelta, A.P. Eslava (1983). A new gene (*carC*) involved in the regulation of carotenogenesis in *Phycomyces*. *Mol. Gen. Genet.*, **192**, 225–229.
82. T. Ootaki, K. Koga, H. Ito, T. Tsuru (1991). Diameter and amount of β-carotene determine the maximal bending angle of *Phycomyces* sporangiophores. *Bot. Mag. (Tokyo)*, **104**, 323–340.
83. T. Ootaki, K. Koga, S. Oosawa, R. Okazaki, T. Tsuru (1988). Optical properties of

piloboloid and β-carotene–overproducing mutants of *Phycomyces blakesleeanus* responsible for their negative phototropism. *Exp. Mycol.*, **12**, 313–324.
84. F. Gutiérrez–Corona, E. Cerdá-Olmedo (1988). Genetic determination of sporangiophore development in *Phycomyces. Dev. Genet.*, **9**, 733–741.
85. R.P. Sutter (1975). Mutations affecting sexual development in *Phycomyces blakesleeanus. Proc. Nat. Acad. Sci. USA*, **72**, 127–130.
86. M.K. Otto, M. Jayaram, R.M. Hamilton, M. Delbrück (1981). Replacement of riboflavin by an analogue in the blue–light photoreceptor of *Phycomyces. Proc. Nat. Acad. Sci. USA*, **78**, 266–269.
87. S. Tillmanns, H. Senger, P. Galland (1995). Reduced near–UV sensitivity in *Phycomyces* mutants affected in the biosynthesis of 6,7-dimethyl-8-ribityllumazine. *Photochem. Photobiol.*, **62**, 588–595.
88. M.I. Alvarez, T. Ootaki, A.P. Eslava (1983). Mutants of *Phycomyces* with abnormal phototropism induced by ICR–170. *Mol. Gen. Genet.*, **191**, 507–511.
89. M.I. Alvarez, A.P. Eslava, E.D. Lipson (1989). Phototropism mutants of *Phycomyces blakesleeanus* isolated at low light intensity. *Exp. Mycol.*, **13**, 38–48.
90. E.D. Lipson, I. López-Díaz, J.A. Pollock (1983). Mutants of *Phycomyces* with enhanced tropisms. *Exp. Mycol.*, **7**, 241–252.
91. A.P. Eslava, M.I. Alvarez, E.D. Lipson, D. Presti, K. Kong (1976). Recombination between mutants of *Phycomyces* with abnormal phototropism. *Mol. Gen. Genet.*, **147**, 235–241.
92. T. Ootaki, E.P. Fischer, P. Lockhart (1974). Complementation between mutants of *Phycomyces* with abnormal phototropism. *Mol. Gen. Genet.*, **131**, 233–246.
93. T. Ootaki, T. Kinno, K. Yoshida (1977). Complementation between *Phycomyces* mutants of mating type (+) with abnormal phototropism. *Mol. Gen. Genet.*, **152**, 245–251.
94. E.D. Lipson, D.T. Terasaka, P.S. Silverstein (1980). Double mutants of *Phycomyces* with abnormal phototropism. *Mol. Gen. Genet.*, **179**, 155–162.
95. V. Campuzano, J.M. Díaz-Mínguez, A.P. Eslava, M.I. Alvarez (1990). A new gene (*madI*) involved in the phototropic response of *Phycomyces. Mol. Gen. Genet.*, **223**, 148–151.
96. I. López-Díaz, E.D. Lipson (1983). Genetic analysis of hypertropic mutants of *Phycomyces. Mol. Gen. Genet.*, **190**, 318–325.
97. M. Delbrück, W. Shropshire (1960). Action and transmission spectra of *Phycomyces. Plant Physiol.*, **35**, 194–204.
98. V. Campuzano, P. Galland, H. Senger, M.I. Alvarez, A.P. Eslava (1994). Isolation and characterization of phototropism mutants of *Phycomyces* insensitive to ultraviolet light. *Curr. Genet.*, **26**, 49–53.
99. V. Campuzano, P. Galland, M.I. Alvarez, A.P. Eslava (1996). Blue-light receptor requirement for gravitropism, autochemotropism and ethylene response in *Phycomyces. Photochem. Photobiol.*, **63**, 686–694.
100. V. Martín-Rojas, H. Greiner, T. Wagner, L. Fukshansky, E. Cerdá-Olmedo (1995). Specific tropism caused by ultraviolet C radiation in *Phycomyces. Planta*, **197**, 63–68.
101. D. Weinkove, J.A. Poyatos, H. Greiner, E. Oltra, J. Avalos, L. Fukshansky, A.F. Barrero, E. Cerdá-Olmedo (1998). Mutants of *Phycomyces* with decreased gallic acid content. *Fungal Genet. Biol.*, **25**, 196–203.
102. R. Flores, E. Cerdá-Olmedo, L.M. Corrochano (1998). Separate sensory pathways for photomorphogenesis in *Phycomyces. Photochem. Photobiol.*, **67**, 467–472.
103. J.F. Lafay, J. Matricon, C. Bodéré (1975). The avoidance response of *Phycomyces*: distance dependence of the response. *Physiol. Vég.*, **13**, 259–263.
104. C. Schimek, P. Eibel, F. Grolig, T. Horie, T. Ootaki, P. Galland (1999). Gravitropism in *Phycomyces*: a role for sedimenting protein crystals and floating lipid globules. *Planta*, **210**, 132–142.

105. J.J. Wolken (1969). Microspectrophotometry and the photoreceptor of *Phycomyces*. *J. Cell. Biol.*, **43**, 354–360.
106. T. Ootaki, K. Ito, M. Abe, G. Lazarova, A. Miyazaki, T. Tsuru (1995). Parameters governing gravitropic response of sporangiophores in *Phycomyces blakesleeanus*. *Mycoscience*, **36**, 263–270.
107. I. López-Díaz, E. Cerdá-Olmedo (1981). Light–controlled phorogenesis and mycelial growth in *Phycomyces* mutants. *Curr. Genet.*, **3**, 23–26.
108. P. Galland (1983). Action spectra of photogeotropic equilibrium in *Phycomyces* wild type and three behavioral mutants. *Photochem. Photobiol.*, **37**, 221–228.
109. P. Galland, E.D. Lipson (1985). Modified action spectra of photogeotropic equilibrium in *Phycomyces blakesleeanus* mutants with defects in genes *madA*, *madB*, *madC*, and *madH*. *Photochem. Photobiol.*, **41**, 331–335.
110. N. Hohl, P. Galland, H. Senger, A.P. Eslava (1992). Altered pterin patterns in photoreceptor mutants of *Phycomyces blakesleeanus* with defective *madI* gene. *Bot. Acta*, **105**, 441–448.
111. N. Hohl, P. Galland, H. Senger (1992). Altered flavin patterns in photobehavioral mutants of *Phycomyces blakesleeanus*. *Photochem. Photobiol.*, **55**, 247–255.
112. R.C. Poe, P. Pratap, E.D. Lipson (1986). System analysis of *Phycomyces* light–growth response: double mutants. *Biol. Cybern.*, **55**, 105–113.
113. A. Palit, P. Pratap, E.D. Lipson (1986). System analysis of *Phycomyces* light–growth response. Photoreceptor and hypertropic mutants. *Biophys. J.*, **50**, 661–668.
114. A. Palit, P. Galland, E.D. Lipson (1989). High- and low-intensity photosystems in *Phycomyces* phototropism: effects of mutations in genes *madA*, *madB*, and *madC*. *Planta*, **177**, 547–553.
115. P. Galland, V.E.A. Russo (1984). Threshold and adaptation in *Phycomyces*. Their interrelation and regulation by light. *J. Gen. Physiol.*, **84**, 119–132.
116. P. Galland, M. Orejas, E.D. Lipson (1989). Light–controlled adaptation kinetics in *Phycomyces*: evidence for a novel yellow–light absorbing pigment. *Photochem. Photobiol.*, **49**, 493–499.
117. G.J.P. Riley, P.M. Bramley (1976). The subcellular distribution of carotenoids in *Phycomyces blakesleeanus* C115 *car-42 mad-107*(-). *Biochim. Biophys. Acta*, **450**, 429–440.
118. G.J.P. Riley, P.M. Bramley, (1982). The biosynthesis of carotenes in cell organelles of *Phycomyces blakesleeanus* C115 *carS42 mad-107*(-). *Cytobios*, **34**, 97–104.
119. J. Weiss, M.H. Weisenseel (1990). Blue light–induced changes in membrane potential and intracellular pH of *Phycomyces* hyphae. *J. Plant Physiol.*, **136**, 78–85.
120. P.A. Ensminger, E.D. Lipson (1992). Growth rate fluctuations in *Phycomyces* sporangiophores. *Plant Physiol.*, **99**, 1376–1380.
121. J.K.E. Ortega, R.I. Gamow (1976). An increase in mechanical extensibility during the period of light–stimulated growth. *Plant Physiol.*, **57**, 456–457.
122. J.K.E. Ortega, K.J. Manica, R.G. Keanini (1988). *Phycomyces*: turgor pressure behavior during the light and avoidance growth responses. *Photochem. Photobiol.*, **48**, 697–703.
123. L. Herrera-Estrella, J. Ruiz-Herrera (1983). Light response in *Phycomyces blakesleeanus*: evidence for roles of chitin biosynthesis and breakdown. *Exp. Mycol.*, **7**, 362–369.
124. P.A. Ensminger, E.D. Lipson (1991). Action spectra of the light–growth response in three behavioral mutants of *Phycomyces*. *Planta*, **184**, 506–509.
125. E.D. Lipson, D.T. Terasaka (1981). Photogeotropism in *Phycomyces* double mutants. *Exp. Mycol.*, **5**, 101–111.
126. P. Galland, V.E.A. Russo(1984). Light and dark adaptation in *Phycomyces* phototropism. *J. Gen. Physiol.*, **84**, 101–118.

127. K.W. Foster, E.D. Lipson (1973). The light growth response of *Phycomyces*. *J. Gen. Physiol.*, **62**, 590–617.
128. E.D. Lipson (1975). White noise analysis of *Phycomyces* light growth response system. III. Photomutants. *Biophys. J.*, **15**, 1033–1045.
129. R.C. Poe, P. Pratap, E.D. Lipson (1986). System analysis of *Phycomyces* light–growth response: single mutants. *Biol. Cybern.*, **55**, 99–104.
130. A. Palit, E.D. Lipson (1989). System analysis of *Phycomyces* light–growth response in single and double night–blind mutants. *Biol. Cybern.*, **60**, 385–394.
131. M. Jayaram, L. Leutwiler, M. Delbrück (1980). Light–induced carotene synthesis in mutants of *Phycomyces* with abnormal phototropism. *Photochem. Photobiol.*, **32**, 241–245.
132. G. Raugei, U. Dohrmann, U. Pohl, V.E.A. Russo (1982). Kinetics of photoaccumulation of β-carotene in *Phycomyces blakesleeanus*. *Planta*, **155**, 296–300.
133. K. Bergman (1972). Blue–light control of sporangiophore initiation in *Phycomyces*. *Planta*, **107**, 53–67.
134. P. Galland, S. Amon, H. Senger, V.E.A. Russo (1995). Blue–light reception in *Phycomyces*: red–light sensitization in *madC* mutants. *Bot. Acta*, **108**, 344–350.
135. A. Palit, P. Pratap, E.D. Lipson (1989). System analysis of *Phycomyces* light–growth response: *madC*, *madG*, and *madH* mutants. *Biophys. J.*, **55**, 519–526.
136. P. Galland, A. Palit, E.D. Lipson (1985). *Phycomyces*: phototropism and light-growth response to pulse stimuli. *Planta*, **165**, 538–547.
137. B. Mehta, E. Cerdá-Olmedo (2001). Intersexual partial diploids of *Phycomyces*. *Genetics*, (in press).
138. N. Arrach, E. Cerdá-Olmedo, J. Avatos (2001). A single gene for lycopene cyclase, phytoene synthase and regulation of carotene biosynthesis in *Phycomyces*. *Proc. Nat. Acad. Sci. USA*, **98**, 1687–1692.
139. P. Eibel, C. Schimek, V. Fries, F. Grolig, T. Schapat, W. Schmidt, H. Schneckenburger, T. Ootaki, P. Galland (2000). Statoliths in *Phycomyces*. Characterization of octahedral protein crystals. *Fungal Genet. Biol.*, **29**, 211–220.

Photomovement
D.-P. Häder and M. Lebert, editors.

Chapter 22

Phototropism in *Phycomyces*

Paul Galland

Table of contents

Abstract

The single-celled giant sporangiophore of *Phycomyces* reacts phototropically to far-UV, near-UV and blue light. Prerequisite for phototropism is the lens-like property of the growing zone by which unilateral light is focussed at the distal side of the sporangiophore. Physiological experiments and mathematical models of intensity and excitation profiles across the growing zone suggest dichroic photoreceptors in the vicinity of the cell wall, probably the plasmalemma. Action spectroscopy and the photophysiology of behavioral mutants indicate that a cryptochrome (flavin-like blue-light photoreceptor) mediates phototropism and other light responses of this fungus. Mutant analysis furthermore shows that the far-UV reception is mediated by a separate photoreceptor which interacts antagonistically with the cryptochrome. Analyses of sensory adaptation (range adjustment) and of far-UV reception provide evidence that the cryptochrome generates upon excitation a red light-absorbing intermediate – probably a flavosemiquinone – which can be detected spectroscopically in the growing zone, i.e. the light sensitive zone of the sporangiophore. Even though the red light-absorbing photoreceptor intermediate is unable to elicit directly phototropism it can modulate sensory adaptation to blue light and influence the interaction between the far-UV receptor and the cryptochrome. The analysis of the photoreception of *Phycomyces* suggests that one needs to reckon with the possibility that cryptochromes in general possess the ability to generate red-light absorbing intermediates and thus the potential to mediate red-light responses.

22.1 Introduction

In comparison to the classical plant objects of phototropism such as the grass coleoptiles and seedlings, the *Phycomyces* sporangiophore is in several ways unique. The sporangiophore is a single coenocytic cell, which, for simplicity, can be regarded as a thin, water-filled tube elongating under a turgor pressure of about 3 bar at an astonishing rate of 2–3 mm h^{-1}. Light and gravity perception, growth modulation and phototropism all occur in the small growing zone below the sporangium. The many complexities of signal transduction are thus restricted to this fragile transparent cylinder of 2 mm length and 100 μm diameter, whose growth and twist is modulated by unilateral light and gravity in a way that manifests as tropic bending.

During the early decades of this centuries it was not uncommon to investigate *Phycomyces* phototropism side by side with that of higher plants. From these and later investigations it became apparent that the physiological principles underlying the blue-light perception of *Phycomyces* and higher plants share a similar logic structure (comparative review [1]). The many similarities which *Phycomyces* phototropism shares with that of plants justifies the use of this organism as a model organism.

Despite of these similarities there evolved over the years a certain dichotomy in the type of scientific approaches and in the emphasis which was placed on certain scientific problems. Much if not most of these differences was the legacy of Max Delbrück who endeavored in the mid-fifties to launch *Phycomyces* as the "phage" of sensory physiology. The mutant approach has been introduced during the late sixties in the laboratory of Max Delbrück a time long before the advent of molecular biology which made the generation of mutants and their analysis mandatory. The Delbrück laboratory has also emphasized more than any other plant physiological laboratory the topic of adaptation and fostered this field through concepts which are of general applicability [2]. In the history of general physiology the work by Delbrück and Reichardt may be regarded as a singular occasion during which a serious attempt was made to bridge the gap between the photophysiology of animal and plant research.

Later work, which was initiated either in the Delbrück group or in its vicinity, pioneered such methods as system analysis, which was extensively applied by Lipson and associates [3], and the application of pressure probes and "governing equations" [4,5]. During the last decade a great deal of efforts went into the mathematical treatment of optical problems related to *Phycomyces* phototropism. These studies by Fukshansky and coworkers have provided the mathematical tools to predict the light distribution inside plant organs or the *Phycomyces* sporangiophore and thus to predict the excitation profiles that represent the input signals for the transduction chain.

The *Phycomyces* literature has been reviewed in regular intervals during the past decades. Older literature is critically discussed in previous review articles [6,7]. For physical principles regarding optics, photoreceptor dichroism and excitation profiles as well as for a critical discussion of problems pertaining to action spectroscopy the reader is referred to the reviews by Fukshansky [8,9].

22.2 The sporangiophore and its growing zone

The *Phycomyces* sporangiophore represents a giant aerial hypha which emerges from the mycelium approximately two days after spore germination. During the following 2 days it grows rapidly at an average rate of about 2–3 mm h^{-1}. During this time, approximately at the third day, it develops at its apex a spherical case, the sporangium, which contains about 10^5 spores (Figure 1). Growth, twist and bending occur in the 2–3 mm long growing zone. In the immature stage (stage 1 without sporangium) the growing zone extends up to the tip; in the mature stage (stage 4 with sporangium) the growing zone extends from 0.1 mm below the sporangium to about 2 to 3 mm below it. The vacuole of the sporangiophore extends into the growing zone. In young stage-4 sporangiophores it ends in the lower portion of the growing zone, while it extends through the entire growing zone in older sporangiophores. Most of the photophysiology has been done with stage-4 sporangiophores (Figure 1). If not otherwise stated, the results described in this chapter refer to this material.

The growing zone represents the sensitive and also the reactive zone of the sporangiophore. At its upper part it is continually formed anew and at the bottom part it is converted into non-reactive material. Growth is caused by the turgor of the sporangiophore (about 3 bar) and concomitant loosening of cell wall and new synthesis of chitosan and chitin, the latter one being synthesized in chitosomes [10]. In stage 4, not however in stage 1, growth is accompanied by twist. Stage-4 sporangiophores rotate at a rate of about 2–5° min^{-1} (Figure 2). The direction of rotation is clockwise when viewed from above [11,12].

22.3 Growth, light and dark reactions of the sporangiophore

22.3.1 Sporangiophore growth

The steady-state growth rate of the sporangiophore is about 20–40 $\mu m\ min^{-1}$ (Figure 2A). The growth rate is not completely steady but rather fluctuates giving rise

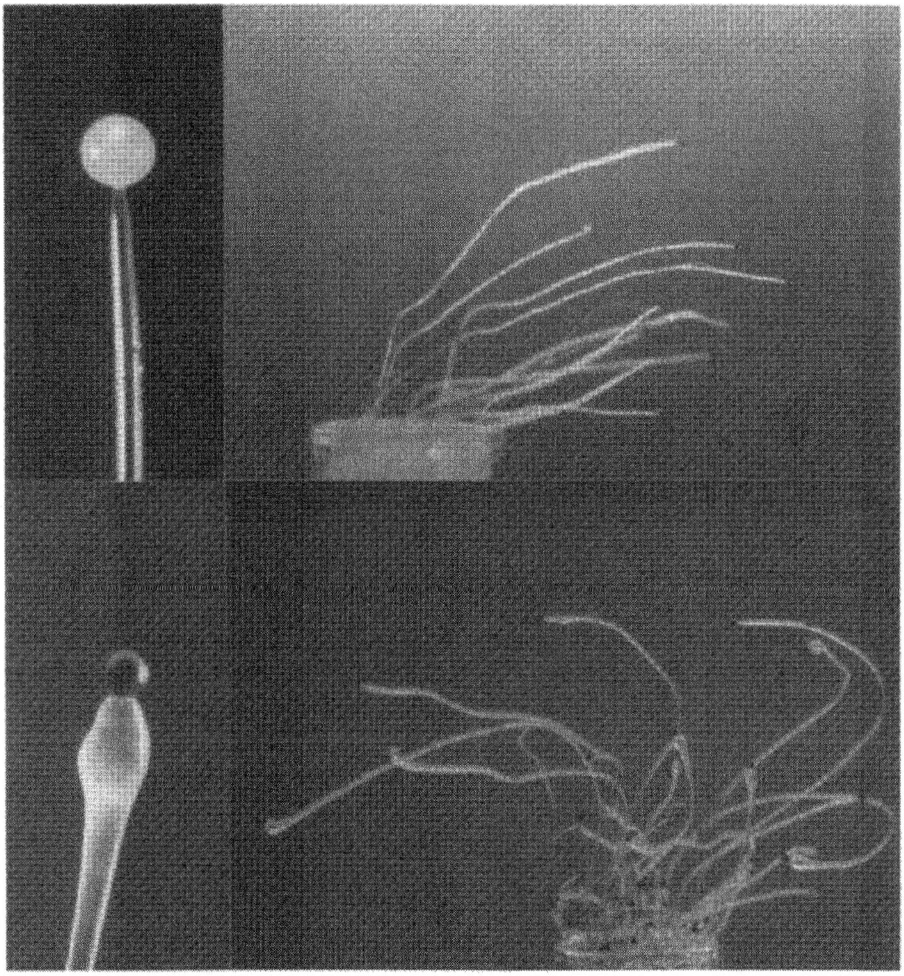

Figure 1. Sporangiophores of *Phycomyces blakesleeanus*. Top left: stage-2 sporangiophore of the wild type. The growth in this stage is temporarily arrested; the brilliant yellow color of the sporangium is caused by β-carotene. Top right: stage-4 sporangiophores of the wild type bending toward unilateral light from the right side. Bottom left: stage-4 sporangiophore of a piloboloid mutant. The growing zone is spherically enlarged. Bottom right: stage-4 sporangiophores of a piloboloid mutant bending away from unilateral light from the right side. The diameter of the sporangia is about 0.5 mm. Photographs by Tamotsu Ootaki.

to sudden bursts [13–15]. These fluctuations are at the basis of the sensory responses described below. A frequency spectrum of the fluctuations shows several small peaks between 0.3 and 10 mHz and a maximum at 10 mHz. Some of the phototropism mutants have an altered frequency spectrum [13].

The longitudinal stretch of the growing zone is accompanied by a twist (rotation) which is clockwise when viewed from above. The steady-state twist rate is in the order of 2–5° min^{-1} (Figure 2B). Like the growth rate also the twist rate fluctuates but, surprisingly, no correlation appears to exist between stretch and twist fluctuations [14]. The absence of correlation may be due to the fact that twist and stretch are unequally distributed in the growing zone [16].

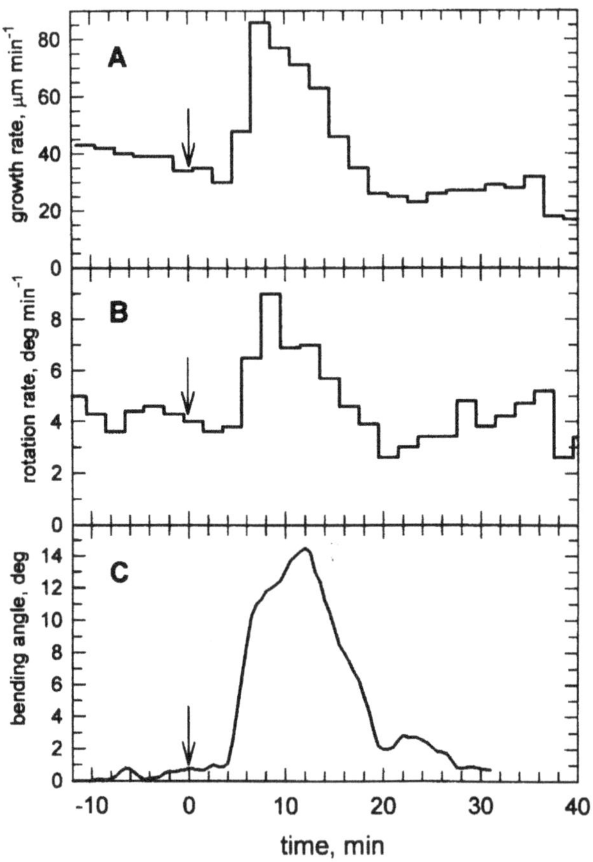

Figure 2. Light-growth response (A), light-twist response (B) and phototropism (C) of sporangiophores in response to a pulse of light at time 0. The data for A and B are from the same sporangiophore which was dark adapted and then exposed from 4 sides to a short pulse of white light (duration not specified; 160 mcs equivalent to about 6.4×10^{-3} Jm^{-2} of blue light) (modified after Oort, 1931). The phototropic bending of a dark-adapted sporangiophore (C) was elicited with a pulse of unilateral blue light (30 s, 450 nm, 2.4×10^{-3} J m^{-2}) (after [21]).

As in other plant and fungal cells the growth of the sporangiophore depends on the net rate of water uptake and the rate of cell wall extension as well as on the interrelated transpiration. By means of a pressure probe the relation between these as well as other parameters have been measured [17–19]. To do this one requires beside special measuring devices so-called "governing equations", which describe the processes which control cell enlargement, the net relative water uptake and the relative rate of extension of the cell wall chamber. The general purpose of governing equations is to determine the behavior of parameters under study when at the same time other parameters change simultaneously. In the field of physics the Maxwell equations in electricity and magnetism are an example for governing equations [4,5].

For stage-4 sporangiophores the volume of a 2-mm long growing zone amounts to about 16 nL, the volumetric transpiration rate is about 1.4 nL min^{-1}, the turgor pressure to 0.31 MPa ($\approx$ 3 bar) and the growth rate is about 33 μm min^{-1} [18]. The latter value corresponds to a volumetric growth rate of 0.528 nL min^{-1}. The rate for water uptake for young stage-4 sporangiophores is about 1.2 nL min^{-1} or more (Foster in: [6]). The water uptake of a sporangiophore ranges from about 1 to 7 times of its volumetric growth rate. Apparently, much of the water which is taken up, is lost again by transpiration and is not used exclusively for the enlargement of the growing zone. One should, however, keep in mind that transpiration is not restricted to the growing zone [18].

22.3.2 Light-growth response and light-twist response

The light-growth response consists of a transient acceleration of the growth rate in response to a pulse or a step-up stimulus [20,21]. Typically, the latency is about 3–4 min, then the growth rate increases and may display a transient deceleration before the normal growth rate is resumed (Figure 2A). Even after saturating stimuli the response lasts not much longer than 40–50 min. The dynamic range of the response equals that of phototropism, i.e. between 10^{-9} and 10 W m^{-2} for near-UV and blue light [20]. A step-down or pulse-down of the fluence rate elicits a dark-growth response, i.e. a transient decrease of the growth rate. The duration depends on the exact prestimulus and stimulus conditions. As a rule the response ceases after about 30–40 min [20,22]. A light-growth response is not accompanied by alterations of the turgor and must, therefore, be explained in terms of altered cell-wall mechanical properties [23] and light-induced chitin synthase activity [24].

The light-growth response has been extensively analyzed in the context of system identification and system analysis techniques. Instead of stimulating the sporangiophores with well-defined step- or pulse stimuli it is treated with so-called white noise, i.e. a continuous randomized variation of light intensity. The growth rate is monitored on an automated tracking machine and the growth rate is cross-correlated to the stimulus. The general input-output relation is obtained with a mathematical formalism (Wiener kernels) which allows the logic representation of the "black box" [3,25]. More modern system identification techniques employ still other stimuli, so-called sum-of-sinusoids [26]. Beside characterizing the wild type, the system analysis methods was also very helpful for the description of behavioral mutants. One significant outcome of these mutant studies was that the single gene products do not act sequentially along a linear transduction chain but rather interact with each other in a molecular complex [27].

The light-growth response is accompanied by a concomitant light-twist response meaning that the rotation rate is transiently increased after a light stimulus (Figure 2B) [11,12]. Elongation and twist appear to be somewhat independent from each other, because the increase of twist rate begins 2–3 min after the increase of the elongation [12].

When sporangiophores are stimulated with a pulse of or by continuous unilateral light the phototropic response is accompanied by transient light-growth and light-twist responses even though they are practically never monitored in phototropism experiments. Since phototropism, light-growth response and light-twist response are transient after a pulse of light (Figure 2), one might conclude that the transitoriness of pulse-induced phototropism is caused by the transient growth and twist responses. This is, however, not the case. With continuous unilateral light phototropism persists indefinitely (e.g. Figure 4) while the ensuing light-growth and light-twist responses are nevertheless transient (see below).

22.3.3 Gas responses

Sporangiophores, not however the mycelium, produce ethylene. An increase of exogenous ethylene elicits a transient increase of the growth rate which resembles that caused by light [28].

Sporangiophores can sense the presence of nearby obstacles without touching them (avoidance response). They bend away from obstacles which are as close as 0.1 to 4 mm provided that the air is humid [29]. No such humidity dependence is found for phototropism. The molecular mechanism of the avoidance response is not understood but it appears clear that a kind of "gas radar" is operating. Ethylene had been proposed as a contender for the gas X, because it can abolish the avoidance response when applied exogenously [28]. The ethylene hypothesis had subsequently been challenged on the grounds that the ethylene concentration, which is required to inhibit the avoidance response, exceeds by several orders of magnitude the amounts given off by the sporangiophore [30].

22.4 Gravitropism

The ability to respond to gravity is essential for the straight growth of *Phycomyces*. This can be best seen under microgravity. In a satellite sporangiophores grow and bend in a completely random fashion [31]. Horizontally placed sporangiophores begin to bend upward after about 30 min and they continue to grow slowly upward until a vertical position is reached after 10–12 hours [32]. Though gravitropism occurs in darkness as well as in light, there is a complex interaction between gravitropism and phototropism which is little understood (see below). As with phototropism [33], also for gravitropism the bending rate depends on the diameter of the sporangiophore. A smaller diameter correlates with increased bending rates [34].

Statoliths which are ubiquitous in plants have been identified only very recently [35,36]. These are paracrystalline protein bodies, so-called octahedral crystals, which had been described earlier without recognizing their statolith function [37,38]. Mutants which lack the octahedral crystals are partially defective in negative gravitropism; while the wild type bends completely upward the crystal-lacking mutants reach only a 45° position [35,36]. The residual gravitropism of the crystal-lacking mutants may be caused by flexure. It can, however, be excluded that flexure, which horizontal sporangiophores experience under their own weight, represents the exclusive cause of negative gravitropism. Since sporangiophores which are placed in water do bend upwards it appears likely that an internal sedimenting particle elicits the response [39,40].

22.5 Phototropism

22.5.1 Lens effect

A prerequisite for positive phototropism is an unequal light distribution in the sporangiophore. This is achieved through the so-called lens effect of the growing zone. The optics of the growing zone resembles that of a transparent cylinder. Because of the lens properties of the growing zone unilateral light is focussed into a narrow band on the distal side of the sporangiophore (Figure 3) whose granular fine structure becomes apparent on microphotographs [41]. That the unequal intensity distribution across the sporangiophore is causing phototropism has been verified in 1918 by the classical immersion experiment of Buder. When the sporangiophore is submerged in a medium with higher refractive index than that of the growing zone the converging lens changes into a diverging one and as a result, negative phototropism occurs [42,43]. The inversion of phototropism occurs only with visible light. If the same immersion experiment is done with far-UV light, which elicits negative phototropism, then no inversion of the bending direction is achieved [44].

The intensity pattern across the growing zone has been calculated on the basis of the refractive indices of cell wall, cytoplasm and vacuole [45,46] taking into account multiple internal reflections and concomitant interference patterns (Figure 3) ([47]; reviews: [8,9]). The calculated intensity pattern was experimentally confirmed by direct measurements with fiber optics [41]. Apparent discrepancies between theory and measurements could be resolved by calculating the artifactual contribution of the acceptance angle of the fiber optics [9].

It should be emphasized that the intensity pattern in the sporangiophore is not identical with the actual (photoreceptor) excitation profile. Though the actual excitation depends on the intensity distribution it depends in addition also on other important parameters such as adaptation, location of the photoreceptor and photoreceptor dichroism as well as wavelength. Fukshansky and coworkers have treated this problem in depth and have generated numerous excitation profiles for the mentioned parameters [8,9]. The knowledge of intensity and excitation profiles are powerful and necessary tools for the localization of the photoreceptor and for the interpretation of action spectra.

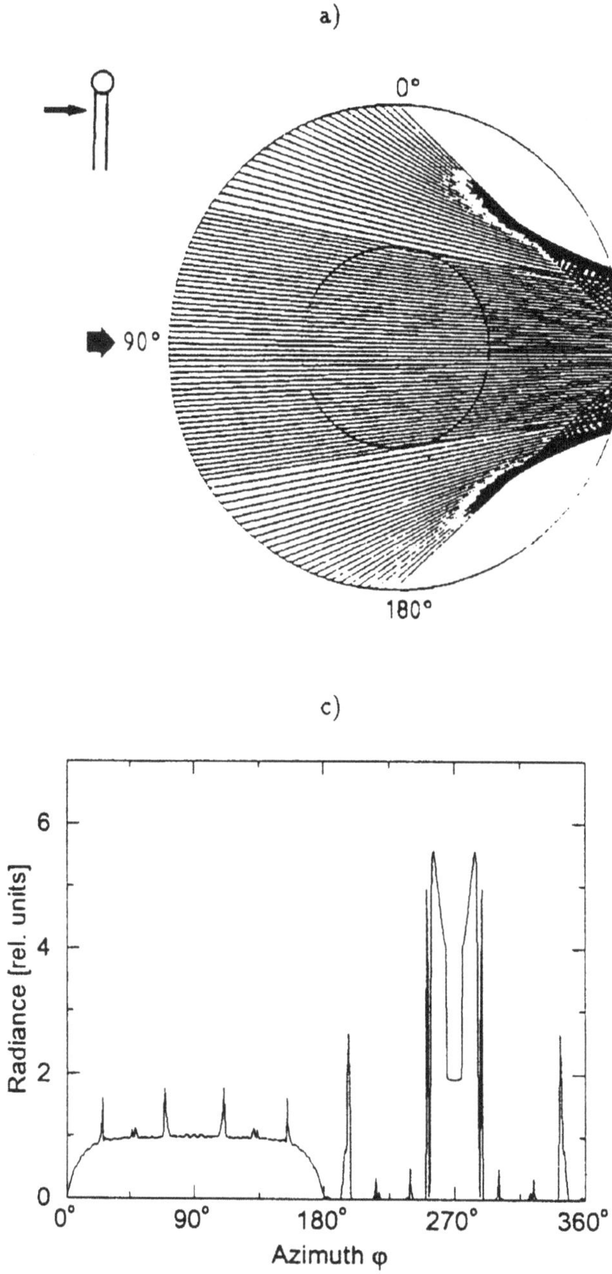

Figure 3. Top: Light propagation across a vertical growing zone under unilateral irradiation. The internal circle indicates the tonoplast. Bottom: The corresponding intensity profiles at the site of the plasma membrane. The light propagation and the intensity profiles were calculated for non-polarized light and a sporangiophore in air (courtesy [66]).

22.5.2 General properties of Phycomyces *phototropism*

For wavelengths above 300 nm phototropism is positive, i.e. the sporangiophore bends toward the light source (Figures 2C, 5). As explained above, the lens effect of the growing zone, which causes the focal band and thus the internal intensity difference between proximal and distal sides, is responsible for this behavior. The minimal intensity difference between proximal and distal sides which can still be perceived is in the order of 8% (discrimination threshold). This can be conveniently shown by placing the sporangiophore between two light sources of slightly different intensities [48,49].

Below 300 nm phototropism is, however, negative: the sporangiophore bends away from the unilateral light (Figure 5). The reasons for the negative bending are not fully understood. The proposal that gallic acid in the vacuole of the growing zone attenuates far-UV light and counteracts thus the lens effect [6,50], appears no longer valid. Calculations show that gallic acid could not completely abolish focussing of far-UV light [8,44] so that one should actually expect positive phototropism. Very special irradiation regimes, which include steps-down from adapting white light to unilateral far-UV light, elicit indeed positive phototropism in response to far-UV light [44]. Usually, however, i.e. under steady-state irradiation, only negative phototropism is observed.

22.5.3 Phototropism paradox

Phototropic bending is generated by differential growth rates at the proximal and distal sides. During steady state bending the growth rate at the distal side exceeds the average growth rate by up to 6–16% while that of the proximal side is by the same amount below average [51,52]. The fundamental paradox of phototropism consists in the fact that this growth-rate differential can be maintained permanently even though the related light-growth responses cease after about 40 min. Another way to formulate this paradox is that phototropism does not display adaptation while the related light-growth response does have adaptation. One can indeed maintain phototropism “indefinitely” by placing the sporangiophore on a “tropostat”, a simple device with which the geometry of the unilateral light and the growing zone can be kept constant [53]. Another way to manifest permanent phototropism is achieved by placing an obliquely irradiated sporangiophore on a slowly rotating turntable which compensates for the tropic response (Figure 4). The phototropism paradox is not unique to *Phycomyces* but is found even in grass coleoptiles [1].

It had been attempted to explain the phototropism paradox with the so-called “carrousel model” [54]. According to this model, permanent, i.e. non-adapting phototropism, is explained by the rotation of the sporangiophore. It is assumed that photoreceptors move along with the rotating cell wall from the flanks into the bright focus band where they experience a step-up causing a local light-growth response. As new photoreceptors continually move into the focal band there would occur a continual sequence of local light-growth responses so that phototropism would never adapt. Several observations contradict this elegant hypothesis: stage-1 sporangiophores do not rotate and yet they display phototropism. Even among stage-4 sporangiophores one can obtain single specimens which do not rotate but which nevertheless have normal phototropism (Galland and Lipson, unpublished).

To cope with the phototropism paradox it was postulated that "adaptation" is averaged across the sporangiophore [55]. Also more recent mathematical treatments of the problem rely on non-local signal processes [8].

22.5.4 Dynamic range

The *Phycomyces* sporangiophore has a huge dynamic range akin to that of the human eye. It reacts to blue light between 10^{-9} and 10^{2} W m^{-2} (Figure 5) [6,7,56]. This holds

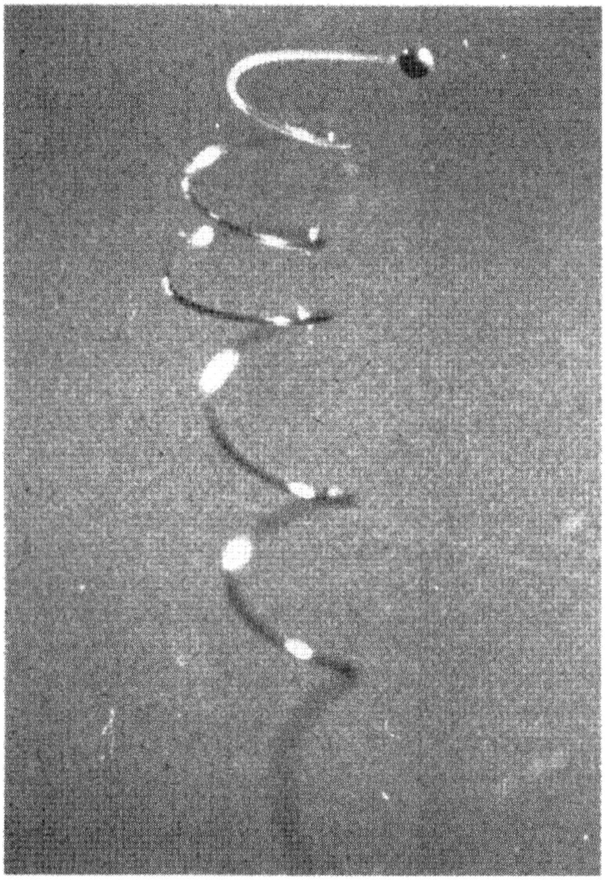

Figure 4. Phototropism of a sporangiophore mounted on a turntable which rotates at 3° min^{-1} counterclockwise when viewed from above. To generate the helix one needs to begin the experiment with a bent sporangiophore which is irradiated unilaterally with light originating from the direction of the observer. This way the counterrotation of the turntable compensates the phototropism in the direction of the light. The experiment demonstrates that phototropism does not adapt and can be maintained "indefinitely", in this experiment for 22 hours. Photograph by D.S. Dennison (from [6]).

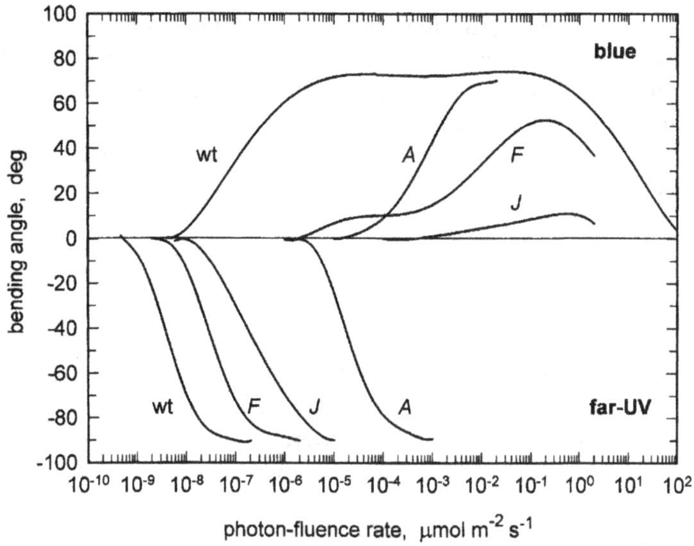

Figure 5. Photon fluence-rate response curve for photogravitropic equilibrium of *Phycomyces* wild type (wt) and various behavioral mutants (genotype *mad*). The capital letters indicate the corresponding *mad* genes. Far-UV = 279 nm. Blue light = 461 nm for low and middle fluence-rate ranges and broad-band blue-light for the high fluence-rate range (modified after [7,78]).

for so-called photogravitropic equilibrium experiments in which sporangiophores are exposed for 6–8 hours to unilateral light so that near threshold the gravitropic and the phototropic stimuli nearly balance each other. The bending rate of fully adapted sporangiophores critically depends on the intensity range. In the low- and middle-intensity range (10^{-7}–10^{-3} W m^{-2} of blue light) the bending rate is between 0.1 and 1.5° min^{-1}. In the high-intensity range (10^{-3}–1 W m^{-2}) the bending rate is between 1.5–3° min^{-1}. Above 1 W m^{-2} the bending rate decreases again and approaches zero near 10^2 W m^{-2} (Figure 5) [57]. Near threshold the number of excited photoreceptors is limiting; in the high-intensity region the photoreceptor populations at the proximal and distal sides are all in the excited states so that no differential growth response and thus no bending can occur.

22.5.5 Pulse-induced phototropism

In contrast to the work done with plants rather few data on pulse-induced phototropism are available for *Phycomyces*. The bending in response to a unilateral pulse of light is transient (Figure 2C). It commences about 4 min after the stimulus, a maximum bending angle of about 10–20° is reached after 10–20 min and then the sporangiophores returns to the original position. The response may be terminated after about 30 min. Fluence-response curves for pulse-induced phototropism were determined for blue light for the wild type and hypertropic mutants [21]. The threshold for the response is in the wild type about 10^{-6} J m^{-2} for stage 4 and 10^{-7} J m^{-2} for stage 1 [58]. The fluence-response curve of the hypertropic mutant displays three prominent peaks which are reminding of the first, second and third positive bending responses of grass coleoptiles [21].

For pulse-induced phototropism and also for the light-growth response reciprocity is obeyed for a limited range of stimulus durations. At 3×10^{-5} J m^{-2} (450 nm) reciprocity fails for pulses shorter than 0.06 seconds and an optimum response is obtained for a pulse duration of 2 ms [21].

22.5.6 *Photogravitropic equilibrium*

When sporangiophores are exposed for 6–8 hours to unilateral light they assume a photogravitropic equilibrium angle (Figure 5). In the vicinity of the absolute threshold the unilateral light stimulus and the gravitropic stimulus balance each other so that the resulting bending angle represents, as it were, a compromise between the two conflicting stimuli. The situation is, as will be shown below, however, more complex.

Polar angle. The polar angle is the bending angle in the plane of bending independently whether or not the sporangiophore bends in the plane of the incident light. Practically all fluence rate-response curves published display this angle.

For near-UV or blue light the absolute thresholds for photogravitropic equlibrium are near 10^{-9} W m^{-2} [59]. Above 10^{-7} W m^{-2} the corresponding fluence rate-response curves have reached a plateau value which is typically near 70° [59,60]. Above 10 W m^{-2} the bending angle decreases again (Figure 5). This behavior is explained by the assumption that the photoreceptors at the proximal and at the distal sides of the sporangiophore are all in the excited state so that no differential signal is perceived.

Beside the gravitropic contribution, the photogravitropic equilibrium angle is also determined by the optical path length. Because the optical path length increases in a bending sporangiophore the light attenuation increases concomitantly. That this factor must not be neglected has been shown recently by experiments in which the phototropic equilibrium angle was determined on a clinostat which eliminates the contribution of gravitropism [46,60,61]. If the optical path length surpasses a critical value then phototropism becomes negative. This occurs in the piloboloid mutants (see below), whose growing zones are spherically enlarged [62,63].

Aiming-error angle. For understanding the fluence rate-response curves of photogravitropic equilibrium it is important to distinguish between the polar angle and the so-called aiming error angle. The plane of bending is not necessarily identical with the plane of the incident light. Near threshold, i.e. 10^{-9} W m^{-2}, the plane of bending deviates about 100° clockwise from the plane of incident light when viewed from above. This clockwise deviation is called the aiming error. For elevated fluence rates causing a maximal polar angle, i.e. some 80°, the aiming error angle approaches zero [64]. To manifest an aiming error sporangiophores need to be adapted for a long time, usually longer than 6 hours, to low fluence rates. If fluence rate-response curves for photogravitropic equilibrium are obtained for insufficiently adapted sporangiophores (less than 6 hours exposure to unilateral light) then no aiming error is apparent even near threshold (Galland, unpublished).

The reason for the aiming error remains obscure. It must be considered as an intrinsic property of tropism as such as it is found also for the avoidance response [65] and gravitropism (Galland, unpublished). The aiming error has been attributed to the helical growth of the sporangiophore, i.e. its continual twist in a clockwise direction. The direction of rotation coincides indeed with the direction of the aiming error. Fukshansky and coworkers [66] have proposed a very satisfying explanation which rests on the light excitation pattern inside the growing zone. Because of photoreceptor dichroism (radial orientation of the receptors) the excitation pattern in the focal band is asymmetric. This asymmetry causes a deviation which is opposite to the direction of the helical growth of the cell wall. Near threshold the twist of the cell wall dominates so that the aiming error is clockwise. Above 10^{-7} W m^{-2} the contribution of the asymmetric excitation profile is strong enough to counteract the twist of the cell wall such that both forces compensate each other. This way the aiming error becomes zero.

22.5.7 Phototropic reversal

During steady-state bending of the sporangiophore the bending direction is reversed if the fluence rate of the actinic light is suddenly changed [67,68]. The reversal occurs always in the plane of bending irrespective from which side the step-up of light was given. Even a decrease of the light intensity [67] or sudden darkness [69] causes a phototropic reversal. Tropic reversal is not unique to phototropism; as a matter of fact, it occurs even for the avoidance response [69]. Light can also induce a tropic reversal during an avoidance response [69,70] and gravitropism [6]. Because all tropic responses have the phenomenon of reversal in common, it had been associated with the output of the signal chain [69].

22.5.8 Interaction between phototropism and gravitropism

Light and gravity interact in a complex manner which is little understood. In centrifugation experiments with vertically placed sporangiophores light above 10^{-2} W m^{-2} inhibited gravitropism while there was a slight light-induced enhancement in the middle intensity range [71]. In contrast to this observation is the finding that high-intensity light of 40 W m^{-2} apparently does not inhibit negative gravitropism of horizontally placed sporangiophores [72].

To understand the interaction between the two stimuli is essential for the interpretation of fluence rate-response curves for photogravitropic equilibrium. It remains still unclear if and to what extent the threshold is determined by the interaction of these stimuli. An increase of the threshold for photogravitropic equilibrium occurs when the gravi-stimulus is increased (Dennison in: [6]). A decrease of the gravi-stimulus by way of clinostating does not decrease the threshold but increases the slope of the fluence rate-response curve when the experiment is done with actinic light at 454 nm (Galland unpublished).

22.5.9 Cytological and biochemical studies on phototropism

The biochemical investigations on growth and phototropism are fragmentary. It is known that irradiation causes a transient increase of chitin synthase activity [24]. In stage 1 a redistribution of small apical vesicles, which probably consist of chitosomes and glycoproteins, occurs after unilateral irradiation [73]. Interestingly, the number of vesicles was elevated at the proximal to the light rather than at the distal side. This may appear puzzling in view of the fact that omnilateral irradiation causes a transient increase of chitin synthase activity [24]. It appears likely that the vesicle material is incorporated faster into the distal cell wall so that the steady-state vesicle number is smaller at that side.

22.6 Phototropism mutants

22.6.1 Mad mutants

The first mutants with reduced phototropic sensitivity were isolated by coworkers of Max Delbrück (Figure 5) ([74]; The abbreviation *mad* stands for Max Delbrück and not, as often assumed, for their abnormal behavior). Presently there exist ten different types of *mad* mutants, *madA* – *madJ*. Morphologically and even with respect to growth rate they are indistinguishable from the wild type. All of them have been characterized genetically by complementation analyses and for most of them the location of the corresponding genes on the linkage map is known [75,76]. So far none of these genes have been isolated and the function of the corresponding proteins remain unknown.

To bring order into the complex phenotypes of the *mad* mutants they can be most conveniently classified into two major groups according to their capacity or inability to display normal gravitropism. Class-1 mutants have elevated photogravitropic thresholds and normal gravitropism. Class-2 mutants have elevated photogravitropic thresholds and also greatly reduced gravitropism. A more detailed classification is feasible and needs to take into account even photodifferentiation.

Class-1 mutants: madA. The photogravitropic threshold of *madA* mutants is raised approximately 10^4-fold in comparison to that of the wild type (Figure 5) [64,74]. The threshold for the light-growth response is also elevated [20] while the avoidance response, negative gravitropism [74] and ethylene response [77] are normal.

The action spectrum for photogravitropic equilibrium is similar in shape to that of the wild type [64] so that it remains unclear whether or not the photoreceptor itself or else a step "behind" the receptor is affected. It should be stressed that the far-UV sensitivity is affected to about the same extent as the sensitivities to wavelengths above 300 nm (Figure 5) [78]. The phototropic dark-adaptation kinetics of these mutants are slower than those of the wild type and also the light-adaptation kinetics are altered [57].

The mutants are defective for photodifferentiation as the thresholds for light-induction of macrophores [79,80], for light-suppression of microphores [80], for photosporangiogenesis [81,82] and photoaccumulation of β-carotene [83] are elevated. The respective threshold shifts are, however, different from the one for photogravitropic threshold.

Sporangiophores of *madA* mutants contain only 10% of riboflavin of the wild type [84]. It appears unlikely that this defect is responsible for the loss of light sensitivity as an extra supplement of exogenous riboflavin does not raise the threshold [84].

Class-1 mutants: madB. The photogravitropic threshold of these mutants is raised approximately 10^6-fold [59,74]. The light-growth response is affected by about the same factor [73] while the avoidance response, negative gravitropism [74] and ethylene response [77] are normal.

The action spectrum for photogravitropic equilibrium is altered in that the prominent near-UV peak present in the wild type is missing [59]. The far-UV sensitivity is reduced to about the same extent as the blue-light sensitivity [78]. The phototropic dark- and light adaptation kinetics are slower than those of the wild type [57].

The defects in photodifferentiation are qualitatively similar though greater than those of the *madA* mutants [79–83]. The sporangiophores of the *madB* mutants have altered pterin patterns [84]. It remains unclear, however, whether or not the abnormal patterns are related to the altered behavior.

Class-1 mutants: madC. In *madC* mutants the threshold for photogravitropic equilibrium is raised about 10^6-fold, while gravitropism, avoidance response [74] and ethylene-growth response [77] are unaffected.

The photogravitropic action spectrum for this mutant is abnormal in that the near-UV peak is suppressed [59]. The sensitivity to far-UV light is reduced to about the same extent as that to blue light [78]. The phototropic light adaptation kinetics of *madC* mutants are greatly disturbed [57]. Taken together, the observations indicate that the photoreceptor system itself is defective.

All responses of photodifferentiation remain unaffected in *madC* mutants [79–83].

A puzzling feature, which is unique for *madC* mutants, is the fact that red light above 600 nm, which is itself phototropically ineffective, causes a partial suppression of the phototropic deficiency, i.e. red light lowers the phototropic threshold for blue light about 100-fold [85]. Other *mad* mutants or the wild type do not display this behavior. From this observation it was concluded that the blue-light receptor of *Phycomyces* contains a red-light absorbing intermediate and that *madC* mutants contain an elevated level of this receptor intermediate [85].

Double mutants in the combination *madAmadB* or *madBmadC* are completely blind [86]. Surprisingly, a double mutant *madAmadC* is no more affected than a single *madC* mutant [86]. In fact, even the entire action spectrum for photogravitropic equilibrium of the *madAmadC* double mutant resembles closely that of the *madC* single mutant [59]. The *madC* mutation is thus epistatic over the *madA* mutation. From this observation it was concluded that the photoreceptor system of *Phycomyces* must possess a branched input, i.e. two receptors, one of which would contain the *madA* and the *madC* gene products [59].

The flavin pattern of *madC* sporangiophores is more or less normal while the pattern of pterins displays substantial alterations [87]. In the *madAmadC* double mutant the flavin pattern is like that of the single *madC* mutant. In the *madA* single mutant the flavins were reduced (see above). Thus even on the biochemical level the *madC* mutation is epistatic over the *madA* mutation [84].

Class-1 mutants: madI. These mutants are only mildly affected in phototropism as the threshold for photogravitropic equilibrium is raised only 10 to 100-fold depending on the particular allele [88,89]. The mutants have a normal avoidance response and normal gravitropism [78,88]. They have not yet been analyzed with respect to photo-differentiation.

The action spectrum for photogravitropic equilibrium is specifically distorted in the near-UV indicating thus a defect of the photoreceptor system. This defect corresponds with an abnormal pattern of pterins in the sporangiophores [90]. The sensitivity to far-UV light is reduced to about the same extent as to blue light [78].

The strains L150, L152 and L154 have not yet been fully characterized genetically [88]. They have generally a similar phenotype as the *madI* mutants and display mildly altered photogravitropic action spectra [91]. The action spectrum for the light-growth response of strain L150 is similar to that of the wild type [92].

Class-2 mutants: mad D, E, F, G, J. These mutants bend very poorly in response to unilateral light or in response to gravity and show a poor avoidance response (Figure 5) [74]. Also the light-growth response is defective and characterized by a damped amplitude [20]. For these reasons, they have frequently been referred to as the "stiff" mutants. Class-2 mutants possess normal photodifferentiation including photo-phophorogenesis [79,80], photocarotenogenesis [83] and photosporangiogenesis [81,82]. For this reason their defects had for a long time been attributed to the malfunction of steps close to the output of the transduction chain, e.g. cell wall growth of the sporangiophore. Recently arguments were presented that the defects may, as a matter of fact, reside at very early steps of the transduction chain [78].

Fluence rate-response curves for photogravitropic equilibrium of these mutants are biphasic and not monophasic exponential as those of the wild type (Figure 5) [74,78]. The slope of the first part of the curve is very shallow compared to that of the wild-type curve and the threshold is elevated. The thresholds in the high-intensity region appear to be lower than in the wild type [78,88].

Interestingly, the fluence-rate response curves at 280 nm are only mildly affected in comparison to the drop of sensitivity in the near-UV and blue light which is 4 to 6 orders of magnitude higher (Figure 5) [78]. This finding indicates that the far-UV receptor must be independent from the near-UV/blue-light receptor.

Action spectra for photogravitropic equilibrium have been measured for *madF* and *madJ* (Figure 6) [78]. Above 300 nm the peak height is reduced several orders of magnitude and also the shape of the remaining spectrum is greatly distorted [78]. The action spectrum for the light-growth response of a *madD* mutant is only mildly affected [92]. Unfortunately, no direct comparison with the photogravitropic action spectra of *madF* and *madJ* is possible, because they were measured on the basis of a bichromatic irradiation regime (temporal balance), while the phototropism action spectra were generated on the basis of fluence rate-response curves.

Hypertropic mutants: madH. In these mutants all tropic responses are enhanced. This includes phototropism, gravitropism and the avoidance response [93]. The action spectra for photogravitropic equilibrium resemble those of the wild type. The sensitivity to near-UV is, however, about 10-fold elevated and the peak near 380 nm is thus substantially enhanced [59]. The action spectrum for the light-growth response of strain L85 *madH* is also very similar to that of the wild type [92]. The hypertropic phenotype is preserved even in the double mutant *madCmadH*, in which the photogravitropic threshold is that of the single *madC* mutant while the slope of the corresponding fluence rate-response curve is that of the *madH* mutant [94].

Beside the *madH* mutants there exists a series of other hypertropic mutants which have a similar phenotype. Because these mutants are dominant over the wild type, the classical complementation test cannot be applied and, as a result, they have not yet been characterized genetically [94]. Also the photodifferentiation of hypertropic mutants has not been studied extensively. The dominant mutant, L82 *mad-702* (genetically undefined), has normal photophorogenesis [93].

Uvi mutants. Recently mutants have been isolated which are specifically affected in far-UV, not however, in blue light [95]. This phenotype is thus opposite to that of the class-2 mutants which are almost normal in far-UV but insensitive to near-UV and blue light. The existence of the *uvi* mutants supports thus the concept of an independent far-UV photoreceptor. Unfortunately, however, the *uvi* mutation is lethal so that the effect of the mutation on behavior had to be tested in heterokaryons containing nuclei of wild type and *uvi* mutants [95]. In the *uvi*-wt heterokaryons the kinetics of dark adaptation are altered.

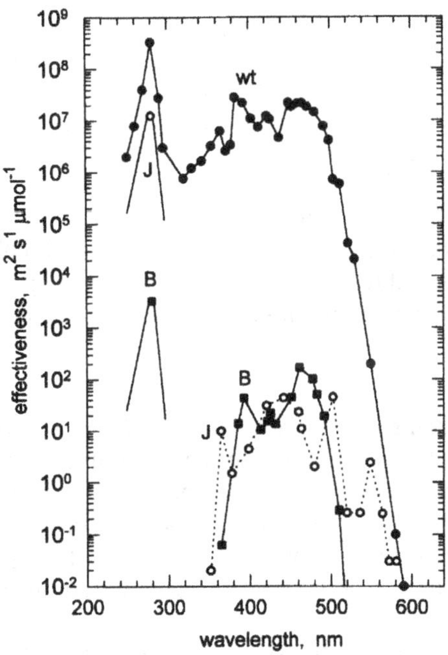

Figure 6. Action spectra of photogravitropic equilibrium (8 h irradiation with unilateral light) for wild type (filled circles, solid line), class-1 mutant C109 *madB* (squares) and class-2 mutant A909 *madJ* (open circles, dotted line). The phototropism below 300 nm is negative. The data for the wild type below 300 nm are adapted from [121]; above 300 nm from [124]. The data for C109 and A909 are from [78] and [59].

Riboflavin-requiring mutants. Mutants with an auxotrophy for riboflavin, *ribA-D*, display normal phototropism in response to blue light when grown in the presence of riboflavin. The thresholds for photogravitropic equilibrium are normal and thus also the shape of the action spectra above 400 nm are normal [90]. Most interestingly, a mutant with a defective *ribC* gene, has an about 100-fold reduced loss of sensitivity in the near-UV, not however, above 400 nm. This defect of the action spectrum correlates with the loss of the compound 6,7-dimethyl-8-ribityllumazine, the direct precursor of riboflavin [90]. The absorption spectrum of 6,7-dimethyl-8-ribityllumazine has a peak near 410 nm while the action spectrum has the defect below 400 nm. To explain this discrepancy it was proposed that a derivative of 6,7-dimethyl-8-ribityllumazine could represent beside the flavin a second chromophore of the photoreceptor. It was further concluded that the near-UV system is independent from the blue-light system.

Piloboloid mutants. The piloboloid mutants (genotype *pil*) have a spherically enlarged growing zone which gives them some resemblance to the related fungus *Pilobolus crystallinus* (Figure 1). They have been isolated on the basis of their morphology rather on that of behavior [96]. The known mutants fall into four complementation groups, *pilA-D* [97]. During the period of increased radial expansion of the growing zone the direction of rotation of the sporangiophore reverses from clockwise to counterclockwise [98].

Piloboloid mutants have a normal light-growth response [62]. The phototropism is nevertheless abnormal in that the sporangiophores display negative instead of positive phototropism [62] (Figure 1). This property has been attributed to the increased optical path length in the growing zone [63]. Negative phototropism is observed for sporangiophores exceeding a critical diameter of 210 μm; for smaller diameters positive bending is observed [62]. The dependence of the negative photogravitropic bending angle on the diameter is complex and follows a biphasic curve. Highest bending angles approaching 90° are obtained for thicker sporangiophores with diameters above 400 μm [99]. The photogravitropic threshold for the negative bending is the same as that of the wild type [62] and there is presently no indication that the *pil*-mutation affects directly the sensory transduction.

Carotene mutants. Mutants which lack carotene have a normal phototropism and display also a normal light-growth response. The phototropic threshold [74,100] and the kinetics for phototropic dark adaptation are unaltered [57]. Even double mutants of the type *carAcarR* or *carBcarR* have normal phototropic sensitivity. These mutants have less than 10^{-5} the amount of β-carotene than the wild type [100]. Because of these findings it can be concluded that the photoreceptor system of *Phycomyces* cannot be a carotenoid or retinal.

In carotene-lacking mutants the slope of the fluence rate-response curves for photogravitropism is steeper than in the wild type [100]. It is assumed that the absence of carotene enhances the lens effect and causes more efficient bending. In carotene-overproducing strains (genotyp *carS*) the screening effect of these pigments counteracts the focus effect of the growing zone and they display, therefore, negative phototropism [63].

22.7 Dark and light adaptation

22.7.1 Sensor and effector adaptation

Phototropism entails the recognition of an internal intensity contrast which is created inside the sporangiophore via the lens effect. The sporangiophore has to be able to evaluate the internal contrast over its entire dynamic range, i.e. between 10^{-9} and 10^2 W m^{-2}. To this end it relies on an efficient adaptation mechanism (also called range adjustment). The adaptation mechanism not only operates for the evaluation of spatial contrast, as in phototropism, but also for the evaluation of temporal contrast, as in the case of the light-growth response, which ensues after a step-up or a step-down stimulus. Even in this case the intensity contrast has to be recognized against a background of changing environmental intensities.

The "adaptation" to a new background intensity does not occur instantaneously but requires some time and also strongly depends on whether an intensity increase or a decrease occurs. The sporangiophore seems to "remember" as it were the light intensity to which it had been adapted. The intensity to which it had been adapted and with which it has come into an equilibrium was defined by Delbrück and Reichardt as the "level of adaptation". It is given the dimension of a fluence rate (W m^{-2}). Because the molecular nature of the "level of adaptation" remains unknown, its actual state needs to be inferred indirectly by physiological experiments.

The time course of the adaptation process is called kinetics of adaptation (Figure 7). The kinetics of adaptation in response to a step-down of intensity (dark adaptation) describes how the sensitivity is restored as a function of time. The kinetics of adaptation in response to a step-up of light (light adaptation) describes how the sensitivity is lost as a function of time. In this context the term "sensitivity" is applied in the universally valid meaning, namely as the reciprocal of the prevailing threshold. Because the mentioned dark- and light-adaptation kinetics refer to sensitivity changes this type of adaptation should be specified as "sensor adaptation" [57,101].

Sensor adaptation should be distinguished from another type of adaptation, i.e. effector adaptation (also called habituation) which is not directly related to sensitivity and threshold [57,101–103]. The time courses for growth rate changes in response to a step-up or a step-down of the fluence rates are to a large extent controlled by effector adaptation. Their kinetics can be very different from the corresponding kinetics of sensor adaptation. For example, after a step-down of the fluence rate by a factor of 10^5 the growth rate resumes its normal rate after about 35 min while the level of adaptation reaches the new fluence rate not before 90 min [57].

22.7.2 Dark-adaptation kinetics

Phototropic latency method. A simple method to measure the dark-adaptation kinetics is the so-called phototropic latency method (Delbrück. In: [6]). The method takes advantage of the fact that sporangiophores which are fully adapted to a given fluence rate have a constant phototropic latency of about 4–5 min. If symmetrically light-adapted sporangiophores are subjected to a step-down of unilateral light, the phototropic latency increases in accordance with the size of the step down. A plot of the fluence rates of the unilateral light against the phototropic latency represents the kinetics of dark adaptation (Figure 7, filled symbols). The highest point of the curve represents the fluence rate, to which the sporangiophore had been adapted. Such kinetics can be described by the empirical relation:

$$A = A_1 \exp(-t/b_1) + A_2 \exp(-t/b_2) \quad (1)$$

where A represents the level of adaptation (expressed as W m^{-2}), A_1 and A_2 represent constants, b_1 and b_2 are the time constants of dark adaptation (dimension = min) and where t is the time (min), i.e. the observed phototropic latency [57]. Phototropic dark

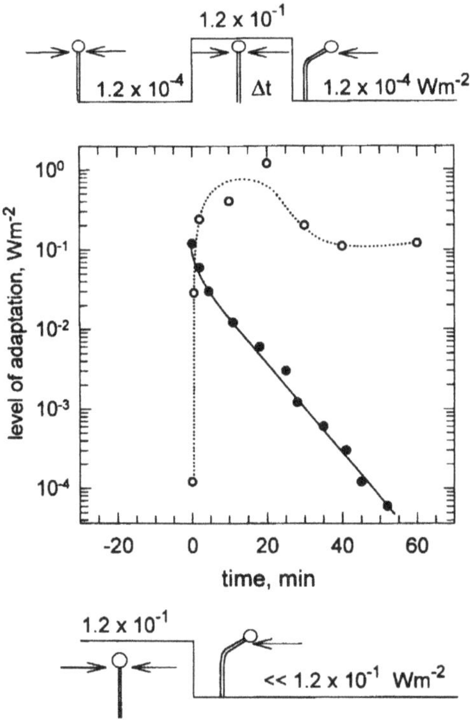

Figure 7. Kinetics of phototropic dark and light adaptation. Dark adaptation (filled symbols): The irradiation protocol is indicated below the abscissa. Sporangiophores were adapted bilaterally to broad-band blue light at a fluence rate of 1.2×10^{-1} W m^{-2}. The fluence rate was stepped down at time 0 and the light was given unilaterally ($<< 1.2 \times 10^{-1}$ W m^{-2}). The ensuing phototropic latency, i.e. the time elapsing between the beginning of the unilateral irradiation and the beginning of bending, is plotted on the abscissa. Light adaptation (open symbols, dotted line): The irradiation protocol is indicated above the graph. Sporangiophores were adapted to broad-band blue light at a fluence rate of 1.2×10^{-4} W m^{-2}. The adapting light was stepped up at time 0 for variable durations Δt (conditioning pulse). After the conditioning pulse the light was stepped down (1.2×10^{-4} W m^{-2}) and given unilaterally. From the ensuing phototropic latency the level of adaptation at the end of the conditioning pulse was calculated from the previously generated dark-adaptation kinetics. For further explanations see text (after [57]).

adaptation can be slightly accelerated and the intrinsic phototropic latency can be slightly shortened by exogenous Ca^{2+} [104].

Mutants with defects in the genes *madA* and *madB* display altered dark-adaptation kinetics with increased time constants of adaptation, while a hypertropic mutant (L82, undefined genotype) has smaller time constants. Since *madB* mutants are probably defective at the photoreceptor level, it was concluded that dark adaptation is at least in part processed at the level of the photoreceptor [57]. The time constants of dark adaptation are biologically meaningful as they are correlated to the photogravitropic thresholds. Mutants with increased thresholds have larger time constants, while mutants with decreased threshold have smaller time constants in comparison to that of the wild type [101]. The meaning of this observation is that a "blind" mutant has presumably a constitutively high level of adaptation.

Surprisingly, the time constants of the dark adaptation kinetics depend strongly on the wavelength. For near-UV and green light for example dark adaptation proceeds much slower than near 400 or 500 nm [105]. In bichromatic experiments it is also crucial, in which sequence the adapting light and the stimulus light are given [105]. For a simple flavin receptor with a single chromophore it had been expected that the adaptation kinetics should be wavelength independent.

The minimum method. To understand this method it is necessary to know that the minimal phototropic latency of about 5 min is obtained only when the fluence rates of the adapting light and the unilateral light are equal. If the fluence rate of the unilateral light is either smaller or higher than the preadapting light, then the phototropic latency is greater than 5 min. This fact can be exploited to measure the level of adaptation under non-steady state conditions, for example after placing the sporangiophore in darkness. If a sporangiophore, which had been transferred to darkness, is exposed to unilateral light of a given fluence rate and if it displays a phototropic latency of 5 min, then the level of adaptation must have been equal to the fluence rate of the unilateral test light.

To reconstruct the kinetics of dark adaptation the sporangiophore is omnilaterally light-adapted and subsequently left for variable periods in darkness, during which the "level of adaptation" decays. The actual state of adaptation is inferred by the phototropic latency in response to a unilateral test light of a constant fluence rate. This way one obtains a "minimum curve" for one particular fluence rate of the test light. In such a curve the minimum indicates the smallest phototropic latency (about 5 min) as a function of the time in darkness [8,106,107]. To reconstruct the kinetics of dark adaptation the procedure is repeated for a series of test lights with different fluence rates. Each curve gives one single minimum point. The actual dark adaptation kinetics is then obtained by plotting the fluence rates of the actinic lights against the time in darkness at which the minima occurred. This method is laborious but has the advantage that the kinetics of adaptation proceed undisturbed from any unilateral measuring light as is the case with the so-called "phototropic latency method".

If this method is employed it becomes apparent, that the dark-adaptation kinetics in blue light can be accelerated if the sporangiophores are irradiated during the adaptation period with long-wavelength light. This puzzling observation implies that orange light which is phototropically inactive is nevertheless capable to accelerate the process of dark adaptation [106,108]. The wavelength dependence for this acceleration of dark adaptation indicates that monochromatic light at 440 and at 600 nm is most effective (Figure 8) [107].

Light-growth response technique. The kinetics of dark adaptation can also be determined through measurement of the light-growth response. The method relies on the fact that the size of the light-growth response depends on the "subjective stimulus", i.e. the ratio of the stimulus intensity to the adapting intensity (Weber-Fechner law). For *Phycomyces* the response size, R, for the light-growth response can be described by the following empirical relation:

$$R = R_0 S/(S + S_0) \tag{2}$$

where R_0 and S_0 are constants and where S is the subjective stimulus:

$$S = \Delta t(I - A)/A \tag{3}$$

In equation (3) I represents the fluence rate of the pulse, A is the level of adaptation, i.e. the fluence rate, to which the specimen had been adapted, and Δt is the stimulus duration.

A prerequisite for the measurement of the dark adaptation is the generation of a series of fluence-response curves for the light-growth response over a broad intensity range

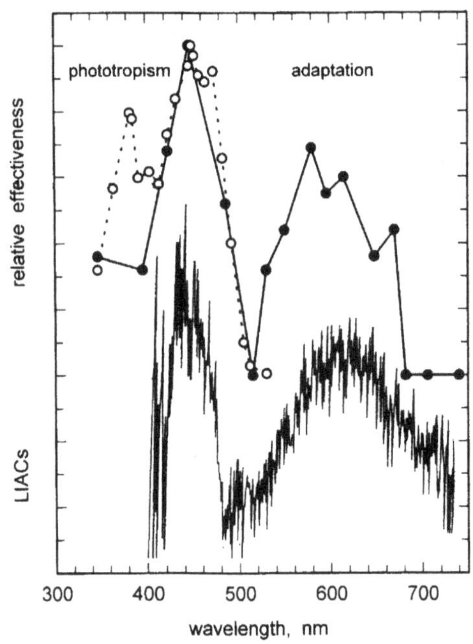

Figure 8. Action spectrum of phototropic balance (open circles, left ordinate) and wavelength dependence of the acceleration of dark adaptation (filled circles, right ordinate). The bottom part shows the in-vivo LIACs (light-induced absorbance changes) which are generated in the growing zone after excitation with white light of moderate fluence rate (10^{-4} W m^{-2}). Data were obtained with the wild-type strain of *Phycomyces* (after [107,120,127]).

that should be identical to the range of intensities for which the dark adaptation kinetics are to be measured. For the generation of the actual dark-adaptation kinetics one proceeds as follows:

Omnilaterally light-adapted sporangiophores are subjected to a step-down of intensity. During this time the level of adaptation decays. The apparent level of adaptation can be inferred from a test pulse of light. From the size of the ensuing response one can infer the actual level of adaptation by using equations (2) and (3). Inversion of equation (2) gives an expression for the subjective stimulus S:

$$S = S_0/(R/R_0 - 1) \tag{4}$$

and inversion of equation (3) gives an expression for the level of adaptation:

$$A = I/(1 + S/\Delta t) \tag{5}$$

The dark adaptation kinetics that were obtained with this method are generally similar to the one obtained with the phototropic latency method [2,109]. Bi-exponential decay kinetics were obtained for the high-intensity range but monophasic decay kinetics were obtained in the middle-intensity range [109]. One can infer that similar, if not identical, adaptation mechanisms are operating for the light-growth response and for phototropism. The method is very labor intensive and has thus not been employed extensively.

22.7.3 Light adaptation

Phototropic latency method. A prerequisite for obtaining the light-adaptation kinetics is the prior generation of the dark-adaptation kinetics. In the actual experiment a sporangiophore which had been omnilaterally adapted to a modest light intensity, is subjected to an omnilateral pulse stimulus of variable duration Δt (conditioning pulse) (Figure 7). The fluence rate of the conditioning pulse may be up to several orders of magnitude greater than the preadapting light. During the conditioning pulse the "level of adaptation" increases rapidly. The new level of adaptation which had been reached at the end of the conditioning pulse can be inferred through a step-down to unilateral test light (fluence rate, I_{test}). The sporangiophore reacts to this new stimulus with a certain phototropic latency. On the basis of the previously measured dark-adaptation kinetics one can infer the level of adaptation which had been reached at the end of the conditioning light stimulus. To calculate this level, A_0, the following formula is employed:

$$A_0 = I_{test} \exp (l_{ph}/b) \tag{6}$$

In this equation I_{test} is the fluence rate of the unilateral test light, l_{ph} is the phototropic latency obtained with that unilateral test light, and b is the dark-adaptation constant which had been determined from the independently measured dark-adaptation kinetics. To reconstruct the actual light adaptation kinetics one needs to repeat the entire procedure for several conditioning pulses of different duration. In the light adaptation-kinetics thus obtained one plots A_0 as a function of time, i.e. the duration of the conditioning pulse (Figure 7).

The prominent features of such light adaptation kinetics are that they are much faster than dark adaptation. For conditioning pulses whose fluence rate exceeds about 10^3-fold the adapting light the level of adaptation, A_0, appears to overshoot the actual intensity level of the adapting light. For moderate conditioning pulses no such overshoot is observed [57]. As was the case with dark adaptation, mutants with defects in genes *madA*, *madB* and *madC* had altered time constants of adaptation. *MadB* mutants show for example the overshoot phenomenon under conditions where the wild type does not display it [57].

Light-growth response method. If the light adaptation kinetics are generated on the basis of light-growth measurements one employs the same principles as described above. Even in this case it is necessary to first generate the dark-adaptation kinetics. One subjects a sporangiophore to a conditioning pulse of constant duration and fluence rate and steps then the fluence rate down. At a given interval one gives a test pulse. From the size of the ensuing growth response one can infer the level of adaptation, A_0, at the end of the conditioning pulse by using the previously generated dark-adaptation kinetics by employing the following equation:

$$A_0 = A_d \exp (t_d/b) \qquad (7)$$

where A_d is the level of adaptation which had been reached at the time of the test pulse, and t_d is the time interval between the conditioning pulse and the test pulse. A_d is obtained as described above with equations (4) and (5).

Even with this method an apparent transient "overshoot" of the level of adaptation was observed [109]. The fact that the overshoot phenomenon is common to the light-growth response and to phototropism argues again in favor of a common adaptation mechanism for both processes.

22.8 The photoreceptor system

22.8.1 *Location of the photoreceptor; photoreceptor dichroism*

On the basis of calculated light-intensity and excitation profiles it was concluded that photoreceptor(s) cannot be located at the tonoplast [110]. These calculations took into account phototropic balance experiments done with sporangiophores immersed in oils of different refractive indices [111].

Photoreceptor dichroism was inferred from experiments with polarized light [112]. In these experiments the light-growth response elicited with horizontally polarized light was more effective than with a vertically oriented one. The transition dipole moment in the UV was different from that in blue light. The relative angle between the transition moments inferred for UV and for blue light were in agreement with a flavin-like photoreceptor [112]. A more thorough mathematical treatment of the problem of photoreceptor orientation and experimental tests with polarized light indicate that the hypothesis of a single receptor can be maintained only under the assumption of a mixture of photoreceptors that are oriented and non-oriented, respectively [66].

22.8.2 *Cryptochrome and other photoreceptors*

Action spectra for *Phycomyces* phototropism [113] and the related light-growth response [114] indicate the presence of a flavin-like blue-light receptor which is also ubiquitously found for diverse protists, fungi and plants (Figures 5 and 7). To explain the dynamics and great variability of blue-light action spectra in the near-UV (see below) it had been proposed that pterins could function in *Phycomyces* and other organisms as secondary chromophores beside the flavins [115–117]. The only blue-light photoreceptor which has been isolated to date and whose gene (*hy4*) has been sequenced is the FAD-containing CRY1-protein of the crucifer *Arabidopsis thaliana*, whose sequence indicates a close relation to prokaryotic DNA-photolyases, which have beside a flavin also a pterin chromophore [118].

The identification of the CRY1-protein as a flavoprotein was in line with decades of action spectroscopy on diverse phototropic organisms including *Phycomyces*. One may, therefore, rightly expect that a single flavin photoreceptor should in principle explain the relevant blue-light physiology of *Phycomyces*. I will show here, that the photophysiology of the *Phycomyces* sporangiophore is far too complex to be explained on the basis of a simple flavin-type photoreceptor which acts through the excited S1-state of an oxidized flavin. A series of observations indicates instead the presence of more or less independent receptor systems (or receptor intermediates) which are specific for far-UV, for near-UV, for blue light and even for red light.

22.8.3 Action spectra for phototropism

The now classical action spectra for phototropism by Curry and Gruen [113] and for the light-growth response by Delbrück and Shropshire [114] have three prominent maxima near 280, 370 and 450 nm and resemble thus very closely the absorption spectra of free flavins or typical flavoproteins (Figures 5, 7). The hypothesis of a flavin photoreceptor was further corroborated by analogue-substitution experiments in which riboflavin was replaced by roseoflavin. The replacement caused a bathochromic shift in the phototropic balance which was in accordance with the bathochromic shift of the absorption spectrum of roseoflavin [119].

In the interpretation of the action spectra little attention had been paid to the fact how these spectra had been generated. In the so-called balance method (Figure 8) sporangiophores are placed between two light sources, a reference light (broad-band blue) and a monochromatic test light. Since no fluence rate-response curves need to be generated with this method it can be employed in any intensity range. It was found much later that the choice of the intensity range is very critical for the shape of the action spectrum. Near the absolute threshold (10^{-8} W m^{-2}) the phototropic balance action spectrum loses its prominent near-UV peak while the blue maximum is relatively enhanced [120]. In the high-intensity region (10^{-1} W m^{-2}) the near-UV peak becomes much stronger. Also the choice of the reference wavelength is of importance. The dynamics of these action spectra are not fully understood. They are not easily explained on the basis of a single photoreceptor pigment [120]. Fukshansky [8] has, however, pointed out that this could nevertheless be the case if the data were analyzed in the framework of a quantitative model of phototropism including optics and photoreceptor dynamics.

When action spectra are measured on the basis of fluence rate-response curves for photogravitropic equilibrium then the shape is not so close to a flavin (Figure 6). Again, there are prominent maxima at 280, 370 and 450 nm and a sharp drop at 500 nm but the fine structure is not as smooth as in the case of the balance action spectra (Figure 8) [59,91,121]. Most of the phototropism mutants have been characterized with this method and their corresponding action spectra are known (Figure 6). Phototropism action spectra which are generated on the basis of fluence-response curves (pulse stimuli) are presently not yet available.

Far-UV. Previously the 280-nm peak in the action spectra for phototropic balance and light-growth response had been attributed to a single flavin-like photoreceptor [113,114]. The analysis of behavioral mutants shows, however, clearly that the peak at 280 nm must be attributed to an independent photoreceptor. As shown in Figure 5 class-2 mutants with defects in genes *madF* and *madJ* are relatively little affected at 280 nm while they are greatly defective for wavelengths above 300 nm [78]. This behavior could not be understood on the basis of a single photoreceptor – whether flavin-like or not. The existence of the *uvi*-mutant with UV-deficiency but normal blue-light sensitivity provides further support for this conclusion [95].

The behavior in far-UV is furthermore complicated by the fact that phototropism is negative [78,121,122] while the light-growth response in response to far-UV stimuli is positive [114]. The relative independence of the far-UV system is also indicated by the fact that the phototropic dark adaptation proceeds much faster than in blue light [95,123].

Near-UV and blue light. Photogravitropic action spectra of mutants with defects in genes *madB* and *madC* and a double mutant *madAmadC* are more affected in the near-UV than above 400 nm [120]. Though this observation does not prove the existence of an independent near-UV receptor it is in line with this assumption. Also the fact that a riboflavin-requiring mutant with a defective *ribC* gene causes a specific loss of near-UV sensitivity argues for somewhat independent receptors for near-UV and blue light [90]. Finally, also the action spectra of *madI* mutants are specifically affected in the near-UV [124].

22.8.4 A blue-light receptor sees red

During the past 4 decades the idea that cryptochromes could possibly mediate red-light responses had appeared rather unlikely if not to say absurd to most blue-light physiologists. Indications that just this could indeed be the case have nevertheless accumulated during the past 15 years. Four lines of evidence indicate the existence of a red-light absorbing pigment, which appears to represent the intermediate of the *Phycomyces* cryptochrome, probably a flavosemiquinone.

The first indication was obtained from phototropism experiments with bichromatic irradiation [125]. When sporangiophores were phototropically stimulated with blue light in the low-intensity region, the ensuing bending could be partially suppressed by high-intensity red light. This effect was taken as an indication for a blue-light receptor with photochromic properties. It was hypothesized that the biologically effector form of the photoreceptor existed in the red-light absorbing form [125].

Further evidence for a red light-absorbing pigment was provided by the behavior of *madC* mutants whose photogravitropic threshold can be lowered about 100-fold by simultaneous irradiation with red light [85]. Also this observation was interpreted in terms of a blue-light receptor with photochromic properties.

Tonic red light can completely eliminate the antagonism which blue light exerts over the action of far-UV light [123,126]. This observation suggests that the antagonism between far-UV and blue light occurs on the level of a red-light absorbing photoreceptor intermediate.

The most convincing analysis on red-light reception was done in the context of analyzing kinetics of dark adaptation, which can be accelerated by long-wavelength light [108]. The wavelength dependence for this effect shows maxima near 450 and 600 nm (Figure 8) [107]. These maxima indicate probably a true photoreceptor intermediate, because the same absorption maxima can be detected spectroscopically in vivo in the growing zone of the sporangiophore after excitation with moderate white or blue light (see below; [127]).

Delbrück and coworkers [128] have generated an action spectrum for the light-growth response which was extended above 600 nm. The authors interpreted a small peak near 610 nm as indicative for the lowest excited triplet state of an oxidized flavin. The purported peak was, however, only apparent after subtracting from the experimental data a calculated spectrum which had been extrapolated from a previously measured action spectrum. The significance of the red-light peak remains thus questionable.

22.8.5 Mycelial photoreceptors

The action spectra for photophorogenesis [129] and for photoaccumulation of β-carotene [130] resemble those for phototropism in that they also have major peaks near 370 and 450 nm and a sharp drop at 500 nm. The photoreceptor system for the mycelial responses must, however, be much more simple than the one for phototropism and the light-growth response, because it does not involve the products of the genes *madD, E, F, G, J*. The fluence-response curves for photophorogenesis and for photocarotenogenesis are both biphasic and the corresponding action spectra for the low- and high-intensity ranges are not exactly identical. This observation has been taken as evidence for different photosystems in the two intensity ranges. The existence for low- and high-intensity photosystems had also been proposed for phototropism [7].

22.9 Light-induced absorbance changes

A major shortcoming in blue-light research had always been the absence of a spectrophotometric test which could have been applied for the isolation of the cryptochrome. A test analogous to the spectrophotometric assay for phytochrome had never been available, because there was no blue-light equivalent to the photoreversibility of phytochrome responses. As a result, several attempts had been devoted to the development of spectrophotometric assays which rested on the measurement of light-induced absorbance changes (LIACs) or fluorescence changes that could be related to the primary photochemistry of cryptochrome. After several unsuccessful attempts (reviewed below) such a spectrophotometric assay has recently become available and has been successfully applied to *Phycomyces*.

The development of a rapid-scan spectrophotometer [131] has allowed the detection of LIACs in sporangiophores in vivo which have all the properties expected for photochemistry associated with cryptochrome [127]. When sporangiophores are irradiated with moderate white or blue light, the absorbance in the growing zone increases near 450 and 610 nm (Figure 8). These maxima coincide with the wavelength dependence for the acceleration of dark adaptation (Figure 8) and thus show that they are biologically relevant. Because these in-vivo LIACs are absent after red-light irradiation and lacking in mutants with defects in the genes *madA-C*, they must be specific for the *Phycomyces* cryptochrome. It was concluded that blue light induces in the cryptochrome the formation of a red light-absorbing flavosemiquinone [127].

In the light of these recent results previous LIAC studies appear to take on mainly historic importance. The first attempt to develop a spectroscopic assay was that of Berns and Vaughn [132]. Irradiation of *Phycomyces* mycelia or sporangiophores with near-UV produced an absorbance changes $\Delta A_{460-345}$ which regenerated in the wild type, not however, in a *madC* mutant. Near-UV also elicited fluorescence changes near 450 and 520 nm, which can be interpreted as the photoreduction of pterins and flavins. The fluorescence changes occur in mycelia as well as in sporangiophores and are reversible. The fact that a *madC* mutant with presumable photoreceptor defect lacked these changes indicates that the LIACs were specific for cryptochrome.

Blue light induces in a number of diverse organisms including protists, fungi and plants or even human cells absorbance changes which indicate the photoreduction of a *b*-type cytochrome [133,134]. Similar absorption changes were obtained also for *Phycomyces* sporangiophores and mycelia. The respective action spectrum for the generation of an absorbance change $\Delta A_{427-457}$ indicates a flavin photoreceptor [135]. No clear conclusion was, however, reached, as to whether or not these absorbance changes were specific for the photoreceptor. Because none of the *mad* mutants displayed gross anomalies with respect to these LIACs, the absorbance changes were not considered to be linked to the primary photochemistry of the blue-light receptor [135].

Other empirical LIACs in the blue region, $\Delta A_{470-445}$, could be elicited with blue light [136,137]. Fluence response curves for these LIACs are biphasic and display a low-intensity component and a high-intensity component. A derived action spectrum for the low-intensity components has maxima in the near-UV, blue region and also beyond 520 nm. The action spectrum for the high-intensity component has a major maximum in the near-UV, a minimum in the blue region and a secondary maximum near 520 nm. The underlying photoreceptor pigments are unknown [136].

References

1. P. Galland (1990). Phototropism of the *Phycomyces* sporangiophore: a comparison with higher plants. *Photochem. Photobiol.*, **52**, 233–248.
2. M. Delbrück, W. Reichardt (1956). System analysis for the light growth reaction in *Phycomyces*. In: D. Rudnick (Ed.), *Cellular Mechanisms in Differentiation and Growth* (pp. 3–44). Princeton University Press, New Jersey.

3. E.D. Lipson (1975). White noise analysis of *Phycomyces* light growth response system.I. Normal intensity range. *Biophys. J.*, **15**, 989–1011.
4. J.K.E. Ortega (1990). Governing equations for plant cell growth. *Physiol. Plant.*, **79**, 116–121.
5. J.K.E. Ortega (1995). Plant and fungal cell growth: governing equations for cell wall extension and water transport. *Biomimetics*, **2**, 215–227.
6. K. Bergman, P.W. Burke, E. Cerdá-Olmedo, C.N. David, M. Delbrück, K.W. Foster, E.W. Goodell, M. Heisenberg, G. Meissner, M. Zalokar, D.S. Dennison, W. Shropshire Jr. (1969). *Phycomyces*. *Bacteriol. Rev.*, **33**, 99–157.
7. P. Galland, E.D. Lipson (1987). Light physiology of *Phycomyces* sporangiophores. In: E. Cerdá-Olmedo, E.D. Lipson (Eds), *Phycomyces* (pp. 49–92). Cold Spring Harbor Laboratory, Cold Spring Harbor, New York.
8. L. Fukshansky (1993). Intracellular processing of a spatially non-uniform stimulus: case study of phototropism in *Phycomyces*. *J. Photochem. Photobiol. B: Biol.*, **19**, 161–186.
9. L. Fukshansky, T. Richter (1990). What do we know about the non-uniform perception of a phototropic stimulus in *Phycomyces*? *Planta*, **82**, 107–112.
10. L. Herrera-Estrella, B. Chavez, J. Ruiz-Herrera (1982). Presence of chitosomes in the cytoplasm of *Phycomyces blakesleeanus* and synthesis of chitin microfibrils. *Exp. Mycol.*, **6**, 385–388.
11. A.J.P. Oort (1931). The spiral growth of *Phycomyces*. *Proc. K. Ned. Akad. Wet.*, **34**, 564–575.
12. R.I. Gamow, B. Böttger (1981). *Phycomyces*: Irregular growth patterns in stage IVB sporangiophores. *J. Gen. Physiol.*, **77**, 65–75.
13. P.A. Ensminger, E.D. Lipson (1992). Growth rate fluctuations in *Phycomyces* sporangiophores. *Plant Physiol.*, **99**, 1376–1380.
14. R.I. Gamow, D.E. Clough (1983). Stochastic growth patterns generated by *Phycomyces* sporangiophores. *Am. Chem. Soc. Symp. S.*, **207**, 403–420.
15. J.-F. Lafay, J. Matricon (1985). Observation on growth rate fluctuations in *Phycomyces blakesleeanus* using a high resolution automated measurement system. *Physiol. Veg.*, **23**, 929–934.
16. R. Cohen, M. Delbrück (1958). Distribution of stretch and twist along the growing zone of the sporangiophore of *Phycomyces* and the distribution of response to a periodic illumination program. *J. Cell. Comp. Physiol.*, **52**, 361–388.
17. J.K.E. Ortega, R.G. Keanini, K.J. Manica (1988). Pressure probe technique to study transpiration in *Phycomyces* sporangiophores. *Plant Physiol.*, **87**, 11–14.
18. J.K.E. Ortega, S.A. Bell, A.J. Erazo (1992). Pressure clamp method to measure transpiration in growing single plant cells. Demonstration with sporangiophores of *Phycomyces*. *Plant Physiol.*, **100**, 1036–1041.
19. D.J. Cosgrove, J.K.E. Ortega, W. Shropshire Jr. (1987). Pressure probe study of the water relations of *Phycomyces blakesleeanus* sporangiophores. *Biophys. J.*, **51**, 413–423.
20. K.W. Foster, E.D. Lipson (1973). The light growth response of *Phycomyces*. *J. Gen. Physiol.*, **62**, 590–617.
21. P. Galland, A. Palit, E.D. Lipson (1985). *Phycomyces*: Phototropism and light-growth response to pulse stimuli. *Planta*, **165**, 538–547.
22. V.E.R. Russo (1980). Sensory transduction in phototropism: Genetic and physiological analysis in *Phycomyces*. In: F. Lenci, G. Colombetti (Eds), *Photoreception and Sensory Transduction in Aneural Organisms* (pp. 373–395). NATO Advanced Study Institutes Series, Plenum Press, New York.
23. J.K.E. Ortega, K.J. Manica, R.G. Keanini (1988). *Phycomyces*: turgor pressure behavior during the light and avoidance growth response. *Photochem. Photobiol.*, **48**, 697–703.

24. L. Herrera-Estrella, J. Ruiz-Herrera (1983). Light response in *Phycomyces blakesleeanus*: evidence for roles of chitin biosynthesis and breakdown. *Exp. Mycol.*, **7**, 362–369.
25. R.C. Poe, E.D. Lipson (1986). System analysis of *Phycomyces* light-growth response with Gaussian white-noise test stimuli. *Biol. Cybernet.*, **55**, 91–98.
26. E.D. Lipson, P. Pratrap (1988). System analysis of *Phycomyces* light-growth response with Gaussian white noise and sum-of-sinusoids test stimuli. *Ann. Biomed. Eng.*, **16**, 95–110.
27. R.C. Poe, P. Pratrap, E.D. Lipson (1986). System analysis of *Phycomyces* light-growth response: double mutants. *Biol. Cybernet.*, **55**, 105–113.
28. V.E.A. Russo, B. Halloran, E. Gallori (1977). Ethylene is involved in the autochemotropism of *Phycomyces*. *Planta*, **134**, 61–67.
29. P.W. Meyer, I.J. Matus, H.C. Berg (1987). Avoidance of *Phycomyces* in a controlled environment. *Biophys. J.*, **51**, 425–437.
30. J.-P. Métraux, H. Kende (1983). Does ethylene mediate the autochemotropic and rheotropic responses in *Phycomyces*? *Planta*, **159**, 286–288.
31. G.P. Parfyonov, R.N. Platanova, M.G. Tairbekov, Y.N. Belenev, V.P. Olkhovenko, A.V. Rostopshina, E.A. Oigenblick (1979). Biological investigations aboard biosatellite Cosmos-782. *Acta Astronaut.*, **6**, 1235–1238.
32. D.S. Dennison, W. Shropshire Jr. (1984). The gravireceptor of *Phycomyces*: its development and following gravity exposure. *J. Gen. Physiol.*, **84**, 845–859.
33. W. Shropshire Jr. (1971). Phototropic bending rate in *Phycomyces* as a function of average growth rate and cell radius. Biophysics of cells and organs. In: E. Broda (Ed.), *Proceedings of 1st European Biophysics Congress* (Vol. 5, p. 111). Wien, Medizinische Akademie, Vienna, Austria.
34. T. Ootaki, K. Ito, M. Abe, G. Lazarova, A. Miyazaki, T. Tsuru (1995). Parameters governing gravitropic response of sporangiophores in *Phycomyces blakesleeanus*. *Mycoscience*, **36**, 263–270.
35. C. Schimek, P. Eibel, T. Horie, P. Galland, T. Ootaki (1999). Protein crystals in *Phycomyces* sporangiophores are involved in graviperception. *Adv. Space Res.*, **24**, 687–696.
36. C. Schimek, P. Eibel, F. Grolig, T. Horie, T. Ootaki, P. Galland (1999). Gravitropism in *Phycomyces*: identification of octahedral protein crystals as statoliths. *Planta*, **210**, 132–142.
37. T. Ootaki, J.J. Wolken (1973). Octahedral crystals in *Phycomyces*. II. *J. Cell Biol.*, **57**, 278–288.
38. J.J. Wolken (1969). Microspectrophotometry and the photoreceptor of Phycomyces. I. *J. Cell Biol.*, **43**, 354–360.
39. G.H. Banbury (1962). Geotropism of lower plants. *Encycl. Plant Physiol.*, **17**, 344–377. Springer-Verlag, Berlin.
40. D.S. Dennison (1961). Tropic responses of *Phycomyces* sporangiophores to gravitational and centrifugal stimuli. *J. Gen. Physiol.*, **45**, 23–38.
41. D.S. Dennison, T.C. Vogelmann (1989). The *Phycomyces* lens: measurement of the sporangiophore intensity profile using a fiber optic microprobe. *Planta*, **179**, 1–10.
42. J. Buder (1918). Die Inversion des Phototropismus bei *Phycomyces*. *Ber. Dtsch. Bot. Ges.*, **36**, 104–105.
43. W. Shropshire Jr. (1962). The lens effect and phototropism of *Phycomyces*. *J. Gen. Physiol.*, **45**, 949–958.
44. T. Popescu, A. Roessler, L. Fukshansky (1989). A novel effect in *Phycomyces* phototropism. *Plant Physiol.*, **91**, 1586–1593.
45. A.R. Steinhardt, W. Shropshire Jr., L. Fukshansky (1987). Invariant properties of absorption profiles in sporangiophores of *Phycomyces blakesleeanus* under balancing illumination. *Photochem. Photobiol.*, **45**, 515–523.

46. T. Tsuru, K. Koga, H. Aoyama, T. Ootaki (1988). Optics in *Phycomyces blakesleeanus* sporangiophores relative to determination of phototropic orientation. *Exp. Mycol.*, **12**, 302–312.
47. A.R. Steinhardt, L. Fukshansky (1985). Diffusion approximation for scattering in a cylinder: optics of phototropism. *J. Opt. Soc. Am.*, **A2**, 1725–1734.
48. D.S. Castle (1931). Phototropic "indifference" and the light sensitive system of *Phycomyces*. *Bot. Gaz.*, **91**, 206–212.
49. G.H. Banbury (1952). Physiological studies in the Mucorales.I. The phototropism of sporangiophores of *Phycomyces blakesleeanus*. *J. Exp. Biol.*, **3**, 77–85.
50. D.S. Dennison (1959). Gallic acid in *Phycomyces* sporangiophores. *Nature*, **184**, 2036.
51. D.S. Castle (1961). Phototropism, adaptation and the light-growth response of *Phycomyces*. *J. Gen. Physiol.*, **45**, 39–46.
52. D.S. Castle (1965). Differential growth and phototropic bending in *Phycomyces*. *J. Gen. Physiol.*, **48**, 409–423.
53. D.S. Dennison (1965). Steady-state phototropism in *Phycomyces*. *J. Gen. Physiol.*, **48**, 93–408
54. D.S. Dennison, K.W. Foster (1977). Intracellular rotation and the phototropic response of *Phycomyces*. *Biophys. J.*, **18**, 103–123.
55. D.S. Castle (1966). A kinetic model for adaptation and the light responses of *Phycomyces*. *J. Gen. Physiol.*, **49**, 925–935.
56. P. Galland, E.D. Lipson (1987). Blue-light reception in *Phycomyces* phototropism: evidence for two photosystems operating in low- and high-intensity ranges. *Proc. Natl. Acad. Sci. USA*, **84**, 104–108
57. P. Galland, V.E.A. Russo (1984). Light and dark adaptation in *Phycomyces* phototropism. *J. Gen. Physiol.*, **84**, 101–118.
58. M. Iino, E. Schäfer (1984). Phototropic response of the *Phycomyces* sporangiophore to a single pulse of light. *Proc. Natl. Acad. Sci. USA*, **81**, 7103–7107.
59. P. Galland, E.D. Lipson (1985). Modified action spectra of photogeotropic equilibrium in *Phycomyces blakesleeanus* with defects in genes *madA*, *madB*, *madC*, and *madH*. *Photochem. Photobiol.*, **41**, 331–335.
60. T. Ootaki, N. Ishikawa, A. Miyazaki, T. Tsuru (1991). The determination of the maximal bending angle in *Phycomyces* sporangiophores. *Exp. Mycol.*, **15**, 336–345.
61. T. Ootaki, K. Koga, H. Ito, T. Tsuru (1991). Diameter and amount of β-carotene determine the maximal phototropic bending angle of *Phycomyces* sporangiophores. *Bot. Mag. Tokyo*, **104**, 323–340.
62. D. Koga, T. Sato, T. Ootaki (1984). Negative phototropism in the piloboloid mutants of *Phycomyces blakesleeanus* Bgff. *Planta*, **162**, 97–103.
63. T. Ootaki, K. Koga, S. Oosawa, R. Okazaki, T. Tsuru (1988). Optical properties of piloboloid and β-carotene-overproducing mutants of *Phycomyces blakesleeanus* responsible for their negative phototropism. *Exp. Mycol.*, **12**, 313–324.
64. P. Galland (1983). Action spectra of photogeotropic equilibrium in *Phycomyces* wild type and three behavioral mutants. *Photochem. Photobiol.*, **37**, 221–228.
65. R.I. Gamow, B. Böttger (1982). *Phycomyces*: Discovery of the aiming error in the avoidance response. *Plant Physiol.*, **69**, 742–744.
66. M. Reinhardt, D. Wenzler, L. Fukshansky (1997). Orientation of the *Phycomyces* blue light receptor – testing the single dipole hypothesis. *J. Photochem. Photobiol. B: Biol.*, **41**, 152–164.
67. W. Reichardt, D. Varjú (1958). Eine Inversionsphase der phototropischen Reaktion (Experimente an dem Pilz *Phycomyces blakesleeanus*). *Z. Phys. Chem., Neue Folge*, **15**, 297–320.

68. D.S. Castle (1961). Phototropic inversion in *Phycomyces*. *Science*, **133**, 1424–1425.
69. P. Galland, V.E.A. Russo (1985). Tropic reversal in *Phycomyces*. *Planta*, **163**, 401–404.
70. S.S. Harris, D.S. Dennison (1979). *Phycomyces*: interference between the light-growth response and the avoidance response. *Science*, **206**, 357–358.
71. D.S. Dennison (1964). The effect of light on the geotropic responses of *Phycomyces* sporangiophores. *J. Gen. Physiol.*, **47**, 651–665.
72. W. Shropshire Jr., J.-F. Lafay (1987). Sporangiophore and mycelial responses to stimuli other than light. In: E. Cerdá-Olmedo, E.D. Lipson (Eds), *Phycomyces* (pp. 127–154). Cold Spring Harbor Laboratory, Cold Spring Harbor, New York.
73. M. Morales, J. Ruiz-Herrera (1990). Light-induced changes in the vesicular apparatus of *Phycomyces* sporangiophores during the phototropic response. *Photochem. Photobiol.*, **52**, 223–227.
74. K. Bergman, A.P. Eslava, E. Cerdá-Olmedo (1973). Mutants of *Phycomyces* with abnormal phototropism. *Mol. Gen. Genet.*, **123**, 1–16.
75. M.I. Alvarez, M.I. Alvarez, E.P. Benito, V. Campuzano, A.P. Eslava (1992). Genetic loci of *Phycomyces blakesleeanus*. In: S.J. O'Brien (Ed.), *Genetic Maps. Locus Maps of Complex Genomes* (6th ed., pp. 3.120–3.126). Cold Spring Harbor Laboratory Press, Cold Spring Harbor, New York.
76. A.P. Eslava (1987). Genetics. In: E. Cerdá-Olmedo, E.D. Lipson (Eds), *Phycomyces* (pp. 27–48). Cold Spring Harbor Laboratory, Cold Spring Harbor, New York.
77. V.E.A. Russo (1977). Ethylene-induced growth in *Phycomyces* mutants abnormal for autochemotropism. *J. Bacteriol.*, **130**, 548–551.
78. V. Campuzano, P. Galland, M.I. Alvarez, A.P. Eslava (1996). Blue-light receptor requirement for gravitropism, autochemotropism and ethylene response in *Phycomyces*. *Photochem. Photobiol.*, **63**, 686–694.
79. P. Galland, V.E.A. Russo (1979). Photoinitiation of sporangiophores in *Phycomyces* mutants deficient in phototropism and in mutants lacking β-carotene. *Photochem. Photobiol.*, **29**, 1009–1014.
80. L.M. Corrochano, E. Cerdá-Olmedo (1991). Photomorphogenesis in *Phycomyces* and in other fungi. *Photochem. Photobiol.*, **54**, 319–327.
81. V.E.A. Russo, P. Galland (1980). Sensory physiology of *Phycomyces blakesleeanus*. *Struct. Bonding*, **41**, 71–110.
82. V.E.A. Russo, P. Galland, M. Toselli, L. Volpi (1980). Blue light induced differentiation in *Phycomyces blakesleeanus*. In: H. Senger (Ed.), *The Blue Light Syndrome* (pp. 563–569). Springer-Verlag, Berlin, Heidelberg, New York.
83. M. Jayaram, L. Leutwiler, M. Delbrück (1980). Light-induced carotene synthesis in mutants of *Phycomyces* with abnormal phototropism. *Photochem. Photobiol.*, **32**, 241–245.
84. N. Hohl, P. Galland, H. Senger (1992). Flavin patterns in photobehavioral mutants of *Phycomyces blakesleeanus*. *Photochem. Photobiol.*, **55**, 247–255.
85. P. Galland, S., Amon, H. Senger, V.E.A. Russo (1995). Blue light reception in *Phycomyces*: red light sensitization in *madC* mutants. *Bot. Acta*, **108**, 344–350.
86. E.D. Lipson, D.T. Terasaka (1981). Photogeotropism in *Phycomyces* double mutants. *Exp. Mycol.*, **5**, 101–111.
87. N. Hohl, P. Galland, H. Senger (1992). Altered pterin patterns in photobehavioral mutants of *Phycomyces blakesleeanus*. *Photochem. Photobiol.*, **55**, 239–246.
88. M.I.Alvarez, A.P. Eslava, E.D. Lipson (1989). Phototropism mutants of *Phycomyces blakesleeanus* isolated at low light intensity. *Exp. Mycol.*, **13**, 38–48.
89. V. Campuzano, J.M. Díaz-Mínguez, A.P. Eslava, M.I. Alvarez (1990). A new gene (*mad I*) involved in the phototropic response of *Phycomyces*. *Mol. Gen. Genet.*, **223**, 148–151.

90. S. Tillmanns, H. Senger, P. Galland (1995). Reduced near-UV sensitivity in *Phycomyces* mutants affected in the biosynthesis of 6,7-dimethyl-8-ribityllumazine. *Photochem. Photobiol.*, **62**, 588–595.
91. P.A. Ensminger, X. Chen, E.D. Lipson (1989). Action spectra for photogravitropism of *Phycomyces* wild type and three behavioral mutants (L150, L152, and L154). *Photochem. Photobiol.*, **51**, 681–687.
92. P.A. Ensminger, E.D. Lipson (1991). Action spectra of the light-growth response in three behavioral mutants of *Phycomyces*. *Planta*, **184**, 506–509.
93. E.D. Lipson, I. López-Díaz, J.A. Pollock (1983). Mutants of *Phycomyces* with enhanced tropism. *Exp. Mycol.*, **7**, 241–252.
94. I. López-Díaz, E.D. Lipson (1983). Genetic analysis of hypertropic mutants of *Phycomyces*. *Mol. Gen. Genet.*, **190**, 318–325.
95. V. Martín-Rojas, H. Greiner, T. Wagner, L. Fukshansky, E. Cerdá-Olmedo (1995). Specific tropism caused by ultraviolet C radiation in *Phycomyces*. *Planta*, **197**, 63–68.
96. D. Koga, T. Ootaki (1983). Growth of the morphological, piloboloid mutants of *Phycomyces blakesleeanus*. *Exp. Mycol.*, **7**, 148–160.
97. D. Koga, T. Ootaki (1983). Complementation analysis among piloboloid mutants of *Phycomyces blakesleeanus*. *Exp. Mycol.*, **7**, 161–169.
98. K. Yoshida, T. Ootaki, J.K.E. Ortega (1980). Spiral growth in the radially-expanding piloboloid mutants of *Phycomyces blakesleeanus*. *Planta*, **149**, 370–375.
99. T. Ootaki, T. Tsuru (1993). Diphasic negative phototropism of sporangiophores of the piloboloid mutant of *Phycomyces blakesleeanus*. *Exp. Mycol.*, **17**, 103–108.
100. D. Presti, W.-J. Hsu, M. Delbrück (1977). Phototropism in *Phycomyces* mutants lacking β-carotene. *Photochem. Photobiol.*, **26**, 403–405.
101. P. Galland, V.E.A. Russo (1984). Threshold and adaptation in *Phycomyces*. Their interrelation and regulation by light. *J. Gen. Physiol.*, **84**, 119–132.
102. P. Galland (1989). Photosensory adaptation in plants. *Bot. Acta*, **102**, 11–20.
103. P. Galland (1991). Photosensory adaptation in aneural organisms. *Photochem. Photobiol.*, **54**, 1119–1134.
104. A.V. Sineshchekov, E.D. Lipson (1992). Effect of calcium on dark adaptation in *Phycomyces* phototropism. *Photochem. Photobiol.*, **56**, 667–676.
105. P. Galland, A. Pandya, E.D. Lipson (1984). Wavelength dependence of dark adaptation in *Phycomyces* phototropism. *J. Gen. Physiol.*, **84**, 739–751.
106. P. Galland, L.M. Corrochano, E.D. Lipson (1989). Subliminal light control of dark-adaptation kinetics in *Phycomyces* phototropism. *Photochem. Photobiol.*, **49**, 485–492.
107. X.Y. Chen, Y.Q. Xiong, E.D. Lipson (1993). Action spectrum for subliminal light control of adaptation in *Phycomyces* phototropism. *Photochem. Photobiol.*, **58**, 425–431.
108. P. Galland, M. Orejas, E.D. Lipson (1989). Light-controlled adaptation kinetics in *Phycomyces*: evidence for a novel yellow-light absorbing pigment. *Photochem. Photobiol.*, **49**, 493–500.
109. E.D. Lipson, S.M. Block (1983). Light and dark adaptation in *Phycomyces* light-growth response. *J. Gen. Physiol.*, **81**, 845–859.
110. A.R. Steinhardt, T. Popescu, L. Fukshansky (1989). Is the dichroic photoreceptor for *Phycomyces* phototropism located at the plasma membrane or at the tonoplast? *Photochem. Photobiol.*, **49**, 79–88.
111. K.L. Zankel, P.V. Burke, M. Delbrück (1967). Absorption and screening in *Phycomyces*. *J. Gen. Physiol.*, **50**, 1893–1906.
112. A.J. Jesaitis (1974). Linear dichroism and orientation of the *Phycomyces* photopigment. *J. Gen. Physiol.*, **63**, 1–21.
113. G.M. Curry, H.E. Gruen (1959). Action spectra for the positive and negative phototropism

of *Phycomyces* sporangiophores. *Proc. Natl. Acad. Sci. USA*, **45**, 797–804.

114. M. Delbrück, W. Shropshire Jr. (1960). Action and transmission spectra of *Phycomyces*. *Plant Physiol.*, **35**, 194–204.
115. P. Galland, H. Senger (1988). The role of pterins in the photoreception and metabolism of plants. *Photochem. Photobiol.*, **48**, 811–820.
116. D. Siefermann-Harms, B. Fritz, H. Ninnemann (1985). Evidence for a pterin-derivative associated with the molybdenum cofactor of *Neurospora crassa* nitrate reductase. *Photochem. Photobiol.*, **42**, 771–778.
117. H. Ninnemann (1995). Some aspects of blue light research during the last decade. *Photochem. Photobiol.*, **61**, 22–31.
118. M. Ahmad, C.A.R. Cashmore (1996). Seeing blue: the discovery of cryptochrome. *Plant Mol. Biol.*, **30**, 851–861.
119. M.K. Otto, M. Jayaram, R.M. Hamilton, M. Delbrück (1981). Replacement of riboflavin by an analogue in the blue-light photoreceptor of *Phycomyces*. *Proc. Natl. Acad. Sci. USA*, **78**, 266–269.
120. P. Galland, E.D. Lipson (1985). Action spectra for phototropic balance in *Phycomyces blakesleeanus*: dependence on reference wavelength and intensity range. *Photochem. Photobiol.*, **41**, 323–329.
121. D. Varjú, L. Edgar, M. Delbrück (1961). Interplay between the reactions to light and to gravity in *Phycomyces*. *J. Gen. Physiol.*, **45**, 47–58.
122. G.M. Curry, H.E. Gruen (1957). Negative phototropism of *Phycomyces* in the ultra-violet. *Nature*, **179**, 1028–1029.
123. P. Galland (1998). Reception of far-ultraviolet light in *Phycomyces*: antagonistic interaction with blue and red light. *Planta*, **205**, 269–276.
124. N. Hohl, P. Galland, H. Senger, A.P. Eslava (1992). Altered pterin patterns in photoreceptor mutants of *Phycomyces blakesleeanus* with defective *madI* gene. *Bot. Acta*, **105**, 441–448.
125. G. Löser, E. Schäfer (1986). Are there several photoreceptors involved in phototropism of *Phycomyces blakesleeanus*? Kinetic studies of dichromatic irradiation. *Photochem. Photobiol.*, **43**, 195–204.
126. P. Galland, A.P. Eslava, M.I. Alvarez (1997). Photoreception and phototropism in *Phycomyces*: antagonistic interactions between far-UV, blue, and red light. *Photochem. Photobiol.*, **66**, 879–884.
127. W. Schmidt, P. Galland (1999). Light-induced absorbance changes in *Phycomyces*: evidence for cryptochrome-associated flavosemiquinones. *Planta*, **208**, 274–282.
128. M. Delbrück, A. Katzir, D. Presti (1976). Responses of *Phycomyces* indicating optical excitation of the lowest triplet state of riboflavin. *Proc. Natl. Acad. Sci. USA*, **73**, 1969–1973.
129. L.M. Corrochano, P. Galland, E.D. Lipson, E. Cerdá-Olmedo (1988). Photomorphogenesis in *Phycomyces*: fluence-response curves and action spectra. *Planta*, **174**, 315–320.
130. E.R. Bejarano, J. Avalos, E.D. Lipson, E. Cerdá-Olmedo (1991). Photoinduced accumulation of carotene in *Phycomyces*. *Planta*, **183**, 1–9.
131. W. Schmidt (1995). Novel single-beam optical spectrophotometer for fast luminescence, absorption, and reflection measurements of turbid materials. *Opt. Engin.*, **34**, 589–595.
132. D.S. Berns, J.R. Vaughn (1970). Studies of the photopigment system in *Phycomyces*. *Biochem. Biophys. Res. Comm.*, **39**, 1094–1103.
133. K.I. Poff, W.L. Butler (1974). Absorbance changes induced by blue light in *Phycomyces blakesleeanus* and *Dictyostelium discoideum*. *Nature*, **248**, 799–801.
134. W. Schmidt, K. Thomson, W.L. Butler (1977). Cytochrome *b* in plasma membrane enriched fractions from several photoresponsive organisms. *Photochem. Photobiol.*, **26**,

407–411.

135. E.D. Lipson, D. Presti (1977). Light-induced absorbance changes in *Phycomyces* photomutants. *Photochem. Photobiol.*, **25**, 203–208.
136. C.H. Trad, E.D. Lipson (1987). Biphasic fluence-response curves and derived action spectra for light-induced absorbance changes in *Phycomyces blakesleeanus*. *J. Photochem. Photobiol. B: Biol.*, **1**, 169–180.
137. C.H. Trad, B.A. Horwitz, E.D. Lipson (1988). Light-induced absorbance changes in extracts of *Phycomyces* sporangiophores: modifications in night-blind mutants. *J. Photochem. Photobiol. B: Biol.*, **1**, 305–313.

Photomovement
D.-P. Häder and M. Lebert, editors.

Chapter 23

Phototropism in higher plants

Moritoshi Iino

Table of contents

Abstract

Phototropism refers to the curvature movements of organs whose directionality is determined by the direction or uneven distribution of impinging light. Optimizing the acquisition of light energy in shade, and directing the growth of shoots away from shade, appear to be the most common roles played by phototropism. Seedling organs such as coleoptiles and hypocotyls are typically phototropic and have served as model systems in laboratory studies. Results with these organs have revealed the following basic aspects of the process of phototropism. The light signal is perceived by the photoreceptor which is most sensitive to blue light. The direction of movement is determined by the transverse light gradient generated within the photosensing zone. The phototropic curvature results from an asymmetry of growth, which is not caused by local light-growth response but by growth redistribution. The plant hormone auxin is laterally translocated in response to the transverse light gradient and mediates the growth redistribution. This chapter summarizes most of the results obtained during the last century, including the recent advances made in molecular genetic studies. The field has experienced controversies over some of its basic aspects. It will be noted that many of the apparently complex and conflicting results represent aspects of the overall phototropic system that is undoubtedly highly sophisticated.

Abbreviations: fPIPP, first pulse-induced positive phototropism; sPIPP, second pulse-induced positive phototropism; PINP, pulse-induced negative phototropism; TDP, time-dependent phototropism

23.1 Introduction

Plants are capable of moving their parts or organs. These movements are induced or controlled by various environmental signals and are thought to be of an adaptive nature. Light is one of such signals and is the most important one. When a light-induced movement shows directionality that is correlated to the direction or uneven distribution of impinging light, it is referred to by the term "phototropism". In the laboratory, one-directional or unilateral light is used to induce phototropism, allowing precise determination of the movement direction. Bending towards or away from the brighter environment, or the light source in a laboratory, is called positive or negative phototropism, respectively. However, phototropic bending does not always occur with such simple directionality. Organs may even twist while bending in association with their dorsiventral properties.

The mechanism of phototropism in higher plants has been most extensively investigated with young seedling shoots. Coleoptiles of Gramineae grasses and hypocotyls (or epicotyls) of dicotyledons are typically phototropic and have served as model systems in laboratory studies. This chapter first deals with the various types of phototropism observed in higher plants. It then discusses various mechanistic aspects of phototropism investigated in the model systems. A large body of literature, from old to new, will be covered. This approach is made primarily to acknowledge the originality of individual studies. In addition, many published results will be evaluated to obtain possible conclusions on the issues which are otherwise complex and controversial. To this effect, pertinent results are examined critically and in detail, providing new interpretations in some cases.

23.2 Diversity and adaptive values

Laboratory and field studies of phototropism have been limited to special cases such as the positive phototropism induced in juvenile shoot organs and the sun-tracking movements observed in leaves and flowers of certain plants. Phototropism is, however, a very common property of plants expressed in stems (both leaf-bearing and flower-bearing) and leaves of diverse taxonomic groups. Table 1 lists examples of phototropism that the author has consistently observed in plants in natural habitats and in cultivated plants growing under natural or greenhouse conditions.

In open habitats, where light is abundant, we do not usually observe plants expressing clear phototropism. However, the same plants growing in shady environments often show marked phototropism, obviously responding to the lateral asymmetry of diffuse daylight. Such environments typically occur at the shaded edge of tree and shrub canopies and can also be found in fenced yards and in the shade of buildings. Many of the examples of phototropism listed in Table 1 have been recorded in such environments.

This section summarizes various forms of phototropism observed in different organ types and considers their adaptive values.

23.2.1 Phototropism by stems

Hypocotyls of many dicotyledonous seedlings are strongly positively phototropic, as has been shown in many laboratory investigations. Generally, elongating stems of seedlings are positively phototropic [1]. This can be demonstrated easily under laboratory conditions. Figure 1 shows examples of such phototropism observed in the hypocotyl of a gymnosperm (*Pinus thunbergii*; A) and dicotyledons (*Licopersicon esculentum*, *Raphanus sativum*, and *Vigna radiata*; B–D) and the epicotyl of a dicotyledon (*Pisum sativum*; E). The seedlings of Gramineae such as oats (*Avena sativa*) and maize (*Zea mays*) initially show a strong phototropic curvature in the coleoptile (see below), but the curvature can extend to the mesocotyl (see Figure 14B). Rothert [1] described some examples of Gramineae seedlings in which the mesocotyl is the major phototropic organ (e.g. *Panicum sanguinale*, *Setaria viridis*, and *Sorghum vulgare*). Although green vegetative shoots have been little used in laboratory investigations, leaf-bearing stems very commonly exhibit positive phototropism under natural light conditions (Table 1). This type of phototropism can be widely observed in dicotyledons (herbaceous species and young plants of woody species). The branches of some shrubs show phototropism (e.g. *Euonymus japonicus*, *Hydrangea macrophylla*, *Nerium oleander*, *Rhododendron* cvs., and *Spiraea thunbergii*). Stem phototropism also occurs in monocotyledons (e.g. *Aloe arborescens*, *Polygonatum odoratum*, *Sasaella ramosa*, *Tradescantia ohiensis*, and *Tricyrtis hirta*) and young plants of gymnosperms (e.g. *Metasequoia glyptostoroboides* and *Podocarpus nagi*).

Table 1. Phototropism of higher plants in natural light environments. This table lists plant species identified by the author to be phototropic under natural light conditions. Identification has been made on plants growing in natural habitats (the temperate region of Japan), those cultivated in fields or gardens, and tropical plants maintained in greenhouses. The parts of plants showing phototropic orientation are indicated by the following abbreviations: **SL**, leaf-bearing stems; **SF**; flower-bearing stems; **L**, leaves (in parentheses; **L**, laminae; **P**, petioles; **M**, pulvini). In the cases of woody species, it is also indicated whether observations were made on juvenile plants (J) or adult plants (A). Phototropism of stems was positive in all cases. The leaf of seed plants that has a petiole (or a petiole and pulvinus) showed in many cases diaphototropic orientation. The monocotyledons that do not have a petiolar structure showed positive phototropism.

Family	Species	Organs
PTERIDOPHYTES		
Blechnaceae	*Blechnum niponicum*	L
Dennstaedtiaceae	*Pteridium aquilinum*	L
Dryopteridaceae	*Cyrtomium fortunei*	L
	Dryopteris erythrosora	L
Gleicheniaceae	*Dicranopteris linearis*	L
Oleandraceae	*Nephrolepis biserrata*†	L
Pteridaceae	*Pteris multifida*	L
GYMNOSPERMS		
Podocarpaceae	*Podocarpus nagi*	J: SL
	Podocarpus macrophyllus	J: SL
Taxodiaceae	*Metasequoia glyptostroboides*	J: SL
DICOTYLEDONS		
Acanthaceae	*Strobilanthes dyerianus*†	SL, L(P)
Aizoaceae	*Lampranthus spectabilis*	SL
Anacardiaceae	*Rhus trichocarpa*	J: SL
	Rhus verniciflua	J: SL
Apocynaceae	*Nerium oleander*	A: SL
Aquifoliaceae	*Ilex integra*	A: L(P), SL
	Ilex latifolia	A: L(P)
	Ilex rotunda	J, A: L(P)
Araliaceae	*Dendropanax trifidus*	J: SL, L(P)
	Fatsia japonica	J: L(P), SL
	Hedera rhombea	L(P)
Asclepiadaceae	*Cynanchum japonicum*	SL
Begoniaceae	*Begonia heracleifolia*†	L(P)
	Begonia manicata†	L(P)
Boraginaceae	*Borago officinalis*	L(P), SL
	Heliotropium arborescens†	SL
	Symphytum officinale	L(P)
Campanulaceae	*Platycodon grandiflorum*	SF
Caprifoliaceae	*Viburnum japonicum*	J: L(P), SL/A: L(P)
	Viburnum odoratissimum	A: L(P)
Caryophyllaceae	*Dianthus superbus*	SL
	Stellaria aquatica	SL, L(P)
Celastraceae	*Euonymus japonicus*	A: L(P), SL

Table 1. Continued.

Family	Species	Organs
DICOTYLEDONS – Continued		
Compositae	*Artemisia princeps*	SL, L(P, L)
	Bellis perennis	SF
	Calendula officinalis	L(L), SL
	Chrysanthemum cvs.	SL, SF
	Cirsium japonicum	SL, SF
	Cosmos bipinnatus	SF
	Erigeron annuus	SL, L(P, L)
	Erigeron philadelphicus	SL
	Euryops pectinatus	SF
	Farfugium japonicum	L(P), SF
	Felicia amelloides	SF
	Galinsoga ciliata	L(P), SF
	Gnaphalium affine	SL
	Gnaphalium pensylvanicum	SL
	Ixeris stolonifera	L(P), SF
	Petasites japonicus	L(P)
	Solidago altissima	SL
	Sonc hus oleraceus	L(P, L)
	Taraxacum japonicum	L(P), SF
	Youngia japonica	SL, SF
Convolvulaceae	*Ipomoea fistulosa*†	L(P)
Cornaceae	*Aucuba japonica*	J: L(P), SL/A: L(P)
Cruciferae	*Brassica nigra*	SF
	Eruca vesicaria	L(P)
	Rorippa indica	SL
Ebenaceae	*Diospyros morrisiana*	J: SL, L(P)
Ericaceae	*Rhododendron macrosepalum*	J: SL
	Rhododendron cvs.	A: L(P), SL
Euphorbiaceae	*Daphniphyllum teijsmannii*	A: L(P)
	Euphorbia characias†	SL
	Euphorbia drupifera†	L(P, L)
	Euphorbia sp.	SF
	Mallotus japonicus	J: SL, L(P)
Fagaceae	*Castanopsis cuspidata*	J: L(P)
	Pasania edulis	J: L(P), SL/A: L(P)
	Quercus glauca	J: L(P), SL/A: L(P)
	Quercus myrsinaefolia	J: L(P)
	Quercus phillyraeoides	A: L(P)
	Quercus salicina	A: L(P)
	Quercus sessilifolia	J: L(P), SL/A: L(P)
Geraniaceae	*Pelargonium* sp.†	SF
Gesneriaceae	*Gloxinia sylvatica*†	SL, SF
Goodeniaceae	*Leschenaultia* sp.	SL
Guttiferae	*Hypericum perfoliatum*†	SL

Table 1. Continued

Family	Species	Organs
DICOTYLEDONS – Continued		
Hamamelidaceae	*Distylium racemosum*	A: L(P)
Hernandiaceae	*Hernandia sonora*†	A: L(P)
Iridaceae	*Sisyrinchium rosulatum*	SL, SF
Labiatae	*Clinopodium gracile*	SL, L(P)
	Clinopodium micranthum	SL
	Coleus blumei	L(P)
	Lamium amplexicaule	SL, L(P)
	Melissa officinalis	SL
	Mentha viridis	SL
	Mentha x *piperita*	SL
	Salvia dorisiana	SL
	Salvia elegans	SF
	Salvia leucantha	SL, SF
Lauraceae	*Cinnamomum camphora*	J: L(P), SL/A: L(P)
	Cinnamomum japonicum	J: L(P), SL
	Litsea coreana	A: L(P)
	Litsea japonica	A: L(P)
	Neolitsea aciculata	J: L(P), SL
	Neolitsea sericea	J: SL/A: L(P)
	Persea japonica	J: L(P)
	Persea thunbergii	J: SL, L(P)
Leguminosae	*Astragalus sinicus*	SL
	Desmodium oxyphyllum	SL
	Erythrina crista-galli	A: L(M)
	Trifolium repens	L(M, P)*, SF*
Lythraceae	*Cuphea micropetala*†	SL, L(P)
Magnoliaceae	*Illicium religiosum*	J: L(P), SL/A: L(P)
	Michelia compressa	A: L(P)
	Michelia figo	A: L(P)
Malvaceae	*Malva sylvestris*	L(M, P)
Melastomataceae	*Melastoma malabathricum*	J: SL, L(P)
Moraceae	*Ficus elastica*†	L(P)
	Ficus erecta	J: SL, L(P)/A: L(P)
Myricaceae	*Myrica rubra*	J: L(P)
Myrsinaceae	*Ardisia crenata*	J: L(P)
	Ardisia japonica	A: SL, L(P)
Nyctaginaceae	*Mirabilis jalapa*	SL, L(P)
Oleaceae	*Ligustrum japonicum*	A: L(P)
Onagraceae	*Oenothera speciosa*	SF*
Oxalidaceae	*Oxalis brasiliensis*	SF
	Oxalis corniculata	L(M, P)
	Oxalis corymbosa	L(M, P), SF
Phytolaccaceae	*Phytolacca esculenta*	SL
Piperaceae	*Peperomia langsdorffii*†	SL, L(P)

Table 1. Continued

Family	Species	Organs
DICOTYLEDONS – Continued		
Pittosporaceae	*Pittosporum tobira*	J: SL/A: L(P)
Plantaginaceae	*Plantago virginica*	SF
Polygonaceae	*Polygonum cuspidatum*	SL, L(P)
Primulaceae	*Primula sieboldii*	SF
Rosaceae	*Duchesnea chrysantha*	L(P), SF
	Photinia glabra	J: L(P), SL/A: L(P)
	Rhaphiolepis umbellata	J: L(P), SL/A: L(P)
	Spiraea thunbergii	A: SL
Rubiaceae	*Galium trifloriforme*	SL
	Gardenia jasminoides	J: L(P)
	Mitchella undulata	L(P)
	Rubia argyi	SL, L(P)
Saururaceae	*Houttuynia cordata*	SL
Saxifragaceae	*Heuchera micrantha*	SF
	Hydrangea macrophylla	A: SL
Scrophulariaceae	*Veronica persica*	SF
Solanaceae	*Solanum melongena*	SL
Theaceae	*Camellia japonica*	J: SL, L(P)/A: L(P)
	Ternstroemia gymnanthera	J, A: L(P)
Urticaceae	*Boehmeria nipononivea*	SL
	Pilea cadierei†	SL, L(P)
	Pilea grandis†	SL, L(P)
Verbenaceae	*Clerodendrum thomsoniae*†	SL
Violaceae	*Viola* cvs.	SF
Vitaceae	*Cayratia japonica*	SL, L(P)
	Ampelopsis brevipedunculata	SL
MONOCOTYLEDONS		
Amaryllidaceae	*Clivia miniata*†	L
	Clivia nobilis†	L
	Cyrtanthus cvs.†	L, S(F)
	Haemanthus coccineus†	L
	Lycoris radiata	L, SF
	Narcissus tazetta	L, SF
Araceae	*Anthurium andraeanum*†	L(P), SF
	Dieffenbachia maculata†	L(P)
	Dieffenbachia picta†	L(P)
	Homalomena sp.†	L(P, L)
	Monstera obliqua†	L(P)
	Philodendron cruentum†	L(P)
	Philodendron pittieri†	L(P)
	Rhaphidophora liukiuensis†	L(P, L)
	Scindapsus aureus†	L(P)
	Spathiphyllum 'Clevelandii'†	L(P)
	Syngonium podophyllum†	L(P)
	Syngonium wendlandii†	L(P)

Table 1. Continued

Family	Species	Organs
MONOCOTYLEDONS – Continued		
Commelinaceae	*Tradescantia ohiensis*	SL
Costaceae	*Costus malortieanus*†	L(P)
Cyperaceae	*Carex lenta*	L
Gramineae	*Isachne globosa*	SL
	Sasaella ramosa	SL
Liliaceae	*Allium fistulosum*	L
	Allium grayi	L
	Aloe arborescens†	SL, L
	Gasteria disticha†	SF
	Gasteria gracilis†	SF
	Gasteria obtusa†	SF
	Haworthia fasciata†	L, SF
	Haworthia margaritifera†	L, SF
	Haworthia radula†	L, SF
	Hosta sieboldiana	SF
	Liriope platyphylla	L, SF
	Ophiopogon japonicus	L
	Ophiopogon ohwii	L
	Polygonatum odoratum	SL
	Tricyrtis hirta	SL
	Tulipa cvs.	SF
Marantaceae	*Calathea insignis*†	L(P)
	Calathea leopardina†	L(P)
	Maranta leuconeura†	L(P)
Orchidaceae	*Bletilla striata*	L
	Calanthe sp.†	SF
	Cymbidium cvs.†	SF
	Ludisia discolor†	SF
	Paphiopedilum insigne†	L, SF
	Paphiopedilum cvs.†	L, SF
	Phalaenopis cvs.†	LF
	Vanilla planifolia†	L
Palmae	*Trachycarpus fortunei*	L(P)
Zingiberaceae	*Globba flavibracteata*†	L

† Tropical plants (mostly observed in the greenhouses of the Botanical Gardens, Faculty of Science, Osaka City University)
* Found to show sun-tracking movement.

Most of the examples in Table 1 showing phototropism by leaf-bearing stems have been recorded in shade environments (see Figure 2, A and B). In many cases, phototropic curvature was much less clear in open habitats, and had a tendency to become more expressed in deeper shade. These observations are probably related to the fact that the phototropism of hypocotyls is typically induced by weak light [2–4]. For

Figure 1. Phototropism of seedling organs. Plants were stimulated with unilateral white light (2.5 μmol m^{-2} s^{-1}) under overhead irradiation with red light (3.5 μmol μmol m^{-2} s^{-1}). Unilateral irradiation was initiated, unless otherwise specified, from the beginning of germination. The white-light source was a fluorescent tube (FL10N-EDL, Mitsubishi/Osram). The overhead red light was obtained by filtering the light from the same type of fluorescent tubes through a layer of red plate acrylic (No. 102, 3-mm thick, Mitsubishi Rayon). The horizontal bar = 10 mm. (**A**) Positive phototropism of the hypocotyl in a gymnosperm, *Pinus thunbergii*. The hypocotyl shows a strong phototropism after germination (bottom left). At a later stage, leaves face the light source while the hypocotyl remains tilted. (**B–D**) Positive phototropism of the hypocotyl in dicotyledons: *Lycopersicon esculentum* (**B**), *Raphanus sativum* (**C**), and *Vigna radiata* (**D**). (**E**) Positive phototropism of the epicotyl in a dicotyledon, *Pisum sativum*. (**F**) Positive phototropism of the coleoptile and negative phototropism of the primary root in a Gramineae, *Oryza sativa* (left panel); the young leaves which emerge from the coleoptile remain tilted towards the light source (right panel). (**G**) Positive phototropism of leaves in a monocotyledon, *Hemerocallis thunbergii*. The first leaf shows a strong phototropism (top left). The second leaf is also phototropic as demonstrated by the two pictures of the same plant: just before the onset of unilateral irradiation (left) and after 2 days of unilateral irradiation (right). The plant was initially raised under overhead white light (4 μmol m^{-2} s^{-1}).

example, the positive phototropism of light-grown cucumber (*Cucumis sativum*) hypocotyls is saturated at about 5 W m^{-2} of white light [2], far lower than daylight fluence rates. Although more thorough laboratory and field investigations are necessary, it is very likely that the system of stem phototropism in most vegetative shoots is optimized to function under light-limiting shade conditions.

It is generally believed that stem phototropism functions to enhance the interception by leaves of photosynthetically active solar radiation. Such enhancement is apparent at least in many dicotyledons that extend leaves laterally from the stem. In view of the conclusion described in the preceding paragraph, this explanation can be refined to state that stem phototropism primarily functions to enhance light interception (i.e. carbon gain) when light is limiting for growth in shade environments. In other words, phototropism is considered to be a physiological system related to shade tolerance. Furthermore, with the ability to express strong stem phototropism in the shade, the plant could orient the growth of its shoot away from the shade of surrounding plants or towards a canopy gap. The far-red-rich light found in vegetation shade induces shade-avoiding accelerated elongation in stems of many herbaceous dicotyledons [5]. This shade-avoiding response is provably made more effective in concert with phototropism. The term **shade-avoiding movement** may represent a major functional role for phototropism. Another useful term is **shade-avoiding growth orientation,** which describes the predictive function of phototropism to orient subsequent shoot growth into a brighter environment.

Positive phototropism by flower-bearing stems (pedicels and peduncles) is commonly observed in entomophilous plants (Table 1). In certain cases, bending accompanies twisting (e.g. *Viola* and *Paphiopedilum*). The result of this phototropism is the orientation of flowers towards the brighter environment. Although such flower orientation may be observed under relatively bright environments, it is also often pronounced in shade conditions. Examples are shown in Figure 2 (C and D). The

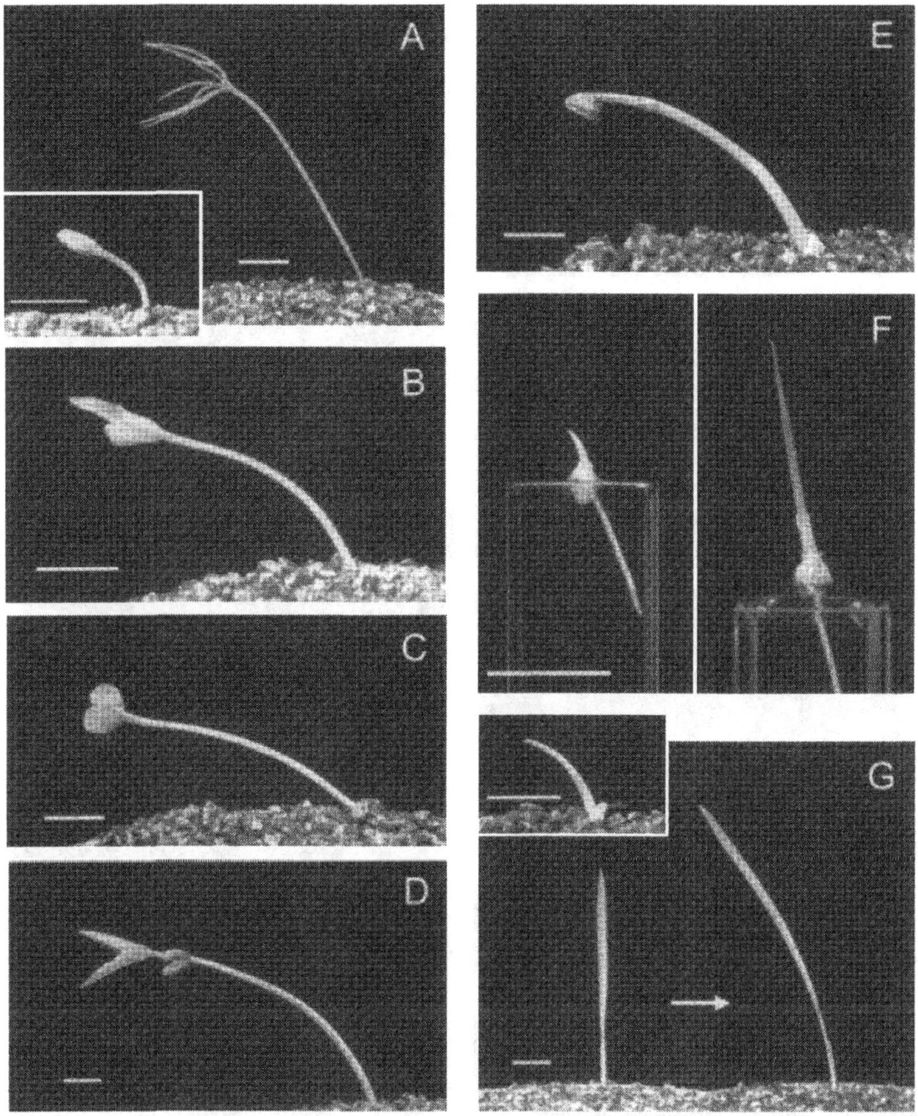

phototropism by flower-bearing stems would function effectively to orient flowers away from the shade of vegetation. Enhancing the rate of visiting by insect pollinators is probably the most important adaptive value of this type of phototropism. In plants that show a marked phototropic curvature by long-elongating pedicels or peduncles (e.g. *Haworthia fasciata* and *Ludisia discolor*), the phototropic flower orientation may also be advantageous for dispersion of seeds into an open space.

There are specialized cases in which stem phototropism functions very effectively under full daylight. Sunflowers and some other *Helianthus* species can track the sun by facing their shoot apex towards the sun [6–10]. This sun-tracking movement or

A
B
C
D
E
F
G
H
I
J

Figure 2. Phototropism observed in natural shade environments. (**A** and **B**) Positive phototropism of leaf-bearing stems: *Boehmeria niponovinea* (**A**) and *Erigeron annuus* (**B**). The entire stem of *Boehmeria* is tilted towards the brighter side of the environment with a curvature at the basal part. The tilt angle probably represents the equilibrium between positive phototropism and negative gravitropism. The apical rapidly elongating part of the stem of *Erigeron* is bent towards the brighter side of the environment. Below this part, the stem is straight and vertical. (**C** and **D**) Positive phototropism of flower bearing stems: *Farfugium japonicum* (**C**) and *Taraxacum japonicum* (**D**). The entire scape of *Farfugium* is tilted towards the brighter side of the environment with a curvature at the basal part. Note also diaphototropic orientation of the leaf laminae. The scape of *Taraxacum* elongates very long in shade environments; if diffuse light impinges more from one direction, the scape bends sharply towards that side (left panel). With this phototropism, flowers can escape from deep vegetation shade (right panel). (**E** and **F**) Diaphototropism of dicotyledonous leaves: *Litsea japonica* (**E**) and *Coleus blumei* (**F**). (**G** and **H**) Positive phototropism of monocotyledonous leaves: *Bletilla striata* (**G**) and *Lycoris radiata* (**H**). Leaves of *Bletilla* are tilted towards the brighter side of the environment. This appearance is maintained by the curvature at the sheath. Leaf laminae of *Bletilla* are curved towards the brighter side of the environment. Under much deeper shade, the curvature is more pronounced. (**I** and **J**) Paraphototropism of leaves in a fern, *Dryopteris erythrosora*; side (**I**) and front (**J**) views of the same plant.

heliotropism[1] is caused by the bending of elongating stems and occurs throughout most of the vegetative growth stage until anthesis begins [7,10]. Since fully opened flowers neither track the sun nor face south, this heliotropism differs from the flower heliotropism described below, and its adaptive value probably rests on enhancing the interception of solar radiation by leaves. The *Helianthus* heliotropism observed under natural conditions may, however, not be a response that is induced directly by the sun's rays. When a sunflower plant is rotated by 180°, the vegetative shoot apex moves from the west to the east during the day and returns to the west during the night, for the next few day/night cycles [8]. The directional light signal appears somehow to be memorized with the help of a circadian clock.

Other examples of stem phototropism adapted to full daylight are found in plants whose flowers track the sun. Such heliotropic flower movement is common in plants that live in arctic and alpine habitats. Examples are *Adonis ramosa* [12], *Anemone patens* [13], *Dryas integrifolia* [14,15], *Oritrophium limnophilum* [13], *Papaver radicatum* [14,15], *Ranunculus adoneus* [16,17], and *Trillium nivale* [13]. These plants direct the interior of the flower to the sun during daytime by positive phototropism of flower-bearing stems. The work on *Ranunculus adoneus* has indicated that this type of phototropism is still based on asymmetric growth [17]. The heliotropic flower

[1] Heliotropism is the classical term used to describe phototropism [11]. Even after its replacement with the new term, heliotropism has often been used when the sun-tracking movement of plant organs observed under natural daylight is referred to. Sometimes the term has also been used when phototropic responses of the organs with a sun-tracking ability are described. This chapter limits its usage to the sun-tracking movement observed under natural daylight or induced under a simulated solar movement. Heliotropism may be defined as a phototropism that can be induced by light as strong as the direct sun's rays and can function as effectively as tracking the sun.

movement causes a substantial increase in temperature within the flower, attracting insect pollinators and facilitating pollen and seed development [12,14–16].

Negative phototropism of stems is rare. The hypocotyl of *Hedera helix* is positively phototropic, but the stem during the juvenile growth stage is negatively phototropic [11]. The stem of *Tropaeolum majus* may be positively phototropic in weak light but negatively phototropic in strong light [11]. The pedicel of *Linaria cymbalaris* is positively phototropic at the flowering stage but negatively phototropic at the fruiting stage [18,19]. The tendril of *Ampelopsis hederacea* is negatively phototropic [20]. The negative phototropism by stems or organs of stem origin observed in these plants is apparently related to their climbing growth habit.

When one cotyledon is shaded from overhead light, the hypocotyl of sunflower seedlings bends towards the side of the non-shaded cotyledon. Lam and Leopold [21] presented this result in support of the hypothesis that the hypocotyl phototropism is caused by the difference in light interception between the two cotyledons [8]. Although the latter hypothesis is now little supported (see Section 23.5.5), the result of Lam and Leopold [21] indicates that sunflower seedlings can detect the quantitative difference in intercepted light between the two cotyledons and bend the hypocotyl to enhance the net light interception. Therefore, the response is a shade-avoiding movement and could also have a role of shade-avoiding growth orientation (see above). Although the direction of curvature is not determined by the direction of light, it is still related to an asymmetry in the light environment. In a broad sense, the response can be classified as phototropism.

23.2.2 Phototropism by dicotyledonous leaves

Many dicotyledons are capable of moving leaves themselves to orient the lamina's upper surface towards the brighter environment. This leaf phototropism, called diaphototropism,[2] is apparently a more effective means of enhancing light interception than the leaf orientation by stem phototropism. The most commonly observed leaf diaphototropism is based on bending of petioles. This type of diaphototropism has been investigated to some extent in *Begonia discolor* [23], *Eranthis hyemalis* [24], *Geranium pratense* [24], *Hedera helix* [11], *Oxalis macra* [25], *Ranunculus ficaria* [22], *Sparmannia africana* [25], and *Tropaeolum majus* [1,22–24,26]. The leaf diaphototropism by petiolar bending can be observed in many herbaceous plants and even in trees (typically broad-leaved evergreen trees) as listed in Table 1. In many of these examples, it was observed that the petiole is also twisted to achieve better light interception by laminae. Leaf diaphototropism can be observed together with positive phototropism of

---

[2] Diaphototropism and paraphototropism are defined as the phototropisms by which the leaf orients itself in such a way that the light interception by the lamina is maximized or minimized, respectively. These terms are the replacements for the original terms, diaheliotropism and paraheliotropism, introduced by Darwin [22]. When irradiated with a one-directional light beam, perfect diaphototropism and paraphototropism bring the upper lamina surface perpendicular or parallel to the light beam, respectively. Even if such an ideal response is not induced, the terms can be used when the light interception is substantially enhanced or reduced by leaf movement.

stems, especially in young plants including tree species. In some cases, leaf diaphototropism is the only apparent phototropism observed in plants. This is especially so in adult trees and in plants with a rosette life form and a climbing growth habit.

The experiments of leaf diaphototropism listed above were apparently conducted using relatively weak light sources. In fact, in the examples shown in Table 1, leaf diaphototropism was typically observed in shade environments (e.g. Figure 2, E and F). Adult trees were found to assume diaphototropic leaf orientation in the shade of other trees or within their own canopy. Perhaps, optimization of carbon gain in vegetation shade is the major role for leaf diaphototropism. Shell et al. [27] provided a rare example of a quantitative field study in which it was shown that cucumber and pepper (*Capsicum annuum*) do show no trend for leaf diaphototropism in full daylight, in contrast to sun-tracking plants.

The petiole shows positive phototropism in response to unilateral light impinging on it, as shown in *Eranthis hyemalis*, *Geranium pratense*, *Oxalis macra*, *Ranunculus ficaria*, *Sparmannia africana*, and *Tropaeolum majus* [1,22–25,28]. Wager [24] carried out many interesting experiments using *Eranthis hyemalis*, *Geranium pratense*, and *Tropaeolum majus*. His results presented in photographs have clearly indicated that the petiole and not the lamina responds to oblique or unilateral light for its positive curvature. He was even able to demonstrate with *Sparmannia africana* and *Tropaeolum majus* that only the top part (within about 10 mm) can respond to unilateral light to cause positive curvature that extends along the long petiole. The petiole of *Oxalis macra* [25] and *Tropaeolum majus* [28] can show positive phototropism after removal of the lamina, also depicting its capability to perceive the directional light signal. (The petiole loses phototropic responsiveness with time after removal of the lamina, but can retain high responsiveness when supplied with auxin [28].)

In contrast to the conclusion by Wager, the earlier work of Haberlandt [23] indicated that irradiation of the lamina is also necessary for complete diaphototropic orientation in *Tropaeolum majus*. He concluded that the lamina also perceives directional light for fine adjustment of its diaphototropic orientation. The experiments by Wager on *Tropaeolum majus* probably do not exclude this possibility. Furthermore, irradiation of the lamina alone could lead to correct diaphototropic orientation of *Begonia discolor* leaves [23]. The plants often used for studies of leaf diaphototropism are those that have a long petiole with a broad lamina not in the plane of the petiole. In such leaves, positive phototropism of the petiole could bring the lamina surface towards the light source. However, dicotyledonous leaves more generally exhibit complex movements to achieve diaphototropic laminar orientation. As mentioned above, phototropic bending of the petiole can accompany twisting or a torsion of the petiole for fine diaphototropic laminar orientation. This twisting is apparently correlated with the dorsiventrality of the leaf [29]. In some cases, bending and twisting extend to the midrib or lamina (e.g. *Plectranthus fruticosus* [30] and *Stachys silvatica* [31]). Furthermore, it is often observed that the petiole is curved to either direction to orient the upper laminar surface towards the brighter environment (e.g. *Aucuba japonica*, *Hedera rhombea*, *Litsea japonica*, and *Mitchella undulata* in Table 1). It seems as if the lamina can detect vectorial light. Leaf diaphototropism probably includes complex mechanisms, although positive petiolar phototropism is clearly the major response component in certain cases.

Some plants can track the sun by orienting the upper surface of leaf laminae to the sun. This sun-tracking movement with diaphototropic leaf orientation is called diaheliotropism.[3] The leaf diaheliotropism is observed in many arid land plants that must grow within a short rainy season [33]. This phototropism has been extensively studied in *Malva neglecta* (Malvaceae), *Lavatera cretica* (Malvaceae), and *Lupinus* species (*L. arizonicus*, *L. palaestinus* and *L. succulentus*; Leguminosae). In *Malva* and *Lavatera*, the bending takes place mainly at the pulvinus-like structure located at the junction between the lamina and the petiole [34,35]. In *Lupinus*, the pulvinus adjacent to each leaflet lamina is the major site of bending [36–38]. The bending at the pulvinus or the pulvinus-like structure is based on turgor-dependent reversible changes in cell volume. For precise diaheliotropic laminar orientation, these plants detect vectorial light and can discriminate light oblique to the tip from that oblique to the base of the lamina [35,39,40]. The vectorial light is not perceived by the pulvinus but by the lamina in *Malva* [41] and *Lavatera* [35] and perhaps also in *Lupinus* [42]. The major vein is the prime photoperception site in *Lavatera* [35]. The leaves of *Malva, Lavatera* and *Lupinus* can assume the east-facing dawn position without any directional light signal from the east at least once (and typically a few times) after showing daily sun-tracking movement [34,36,43]. The sun-tracking movement during the day, however, requires the directional signal of the sun's rays [36,43]. Other examples of Leguminosae plants indicate that the pulvinus itself perceives the light signal for diaheliotropic leaf movement (*Macroptilium atropurpureum* [44], soybean (*Glycine max*) [45], and *Phaseolus vulgaris* [46,47]). As with stem phototropism, the pulvinus detects the brighter side (not the vectorial light) and exhibits positive phototropism [44,47]. This type of photoperception mechanism may allow sun tracking of only a limited extent, as shown in *Phaseolus* trifoliate leaves [48].

Under certain conditions many Leguminosae and Oxalidaceae plants orient their leaf laminae more-or-less parallel to the direction of the sun's rays by curvature of the pulvinus. This leaf orientation response is called paraphototropism,[2] and the term paraheliotropism[3] may be used when plants track the sun with paraphototropic leaf orientation. Leaf paraphototropism is common to plants that show diaheliotropic leaf movement under optimal growth conditions. Examples are alfalfa (*Medicago sativa*) [49], soybean [50], *Phaseolus vulgaris* [51], and *Strophostyles helvola* [52]. These plants exhibit leaf paraphototropism when the water relationship and temperature are not favorable for growth (typically around midday under field conditions) and may follow the sun with this orientation [51,53,54]. The typically diaheliotropic plant *Lupinus arizonicus* tracks the sun with paraphototropic leaf orientation under drought conditions [55]. Leaf paraphototropism can also be induced by sunflecks in shade plants of Oxalidaceae and Leguminosae. Leaves of *Oxalis oregana* are diaphototropic in a forest canopy, but rapidly assume paraphototropic orientation when exposed to a

[3] Diaheliotropism and paraheliotropism are the terms first introduced by Darwin [22] and have been replaced by diaphototropism and paraphototropism.[2] However, the classical terms are in current use, but with inconsistent definitions, when leaf phototropism under natural daylight is described. Here diaheliotropism and paraheliotropism of leaves are defined as the sun-tracking movement of leaves in which laminae maintain diaphototropic or paraphototropic orientation, respectively (see also [32]).

sunfleck [56]. *Amphicarpa bracteata* shows paraphototropic leaf orientation during a long-term sunfleck [52]. With these paraphototropic responses, plants can reduce water loss and/or avoid unfavorably strong solar radiation [57–59]. In addition to the directional light signal, environmental factors such as light fluence rates, air temperature, and water availability are related in a complex manner to the underlying turgor response at the pulvini [60–67]. A part of the orientation response might, however, be nastic[4] rather than tropic.

Using *Sparmannia africana*, Ball [25] showed that when one side of a lamina is shaded from the light impinging from above, the petiole bends slowly to move the lamina away from the shaded part. This response is analogous to the curvature response of sunflower hypocotyls induced by asymmetric light interception by a pair of cotyledons (see above), and is clearly a shade-avoiding movement. The response reported by Ball [25] probably has a significant ecological function, as was discussed by the author. The response is not related to the direction of impinging light. However, the response may still be classified as a phototropism.

The study of leaf phototropism in dicotyledons has been centered on those rapid and reversible responses found in certain plants. However, much more widely observed is the diaphototropism that is based on petiolar movement. As in stem phototropism, this type of phototropism also functions better in shade environments.

23.2.3 Phototropism by monocotyledonous leaves

The elongating coleoptile of Gramineae seedlings is positively phototropic (Figure 1F; see also Figure 14B) and has been the model system most extensively investigated. The coleoptile is a hollow cylindrical organ of leaf origin, lacking any major photosynthetic activity, and is typically phototropic until the enclosed primary leaves break through the tip. The positive phototropism of the coleoptile (often together with that of the mesocotyl) subsequently allows the growth of leaves towards the brighter light environment. These leaves are generally long and extend at a sharp angle from the main shoot axis. It is expected that under a canopy of herbaceous plants the leaves can grow into a canopy gap after coleoptile phototropism, thereby escaping from the shade of neighboring plants. Thus, the major function of coleoptile phototropism could be shade-avoiding growth orientation. This view is supported by the fact that coleoptile phototropism is optimized under weak light conditions [69]. As shown in Figure 1G, the first-appearing leaf of *Hemerocallis thunbergii* (Amaryllidaceae) is positively phototropic. Like the coleoptile, this organ is cylindrical and encloses the next developing leaves. However, it is anatomically distinct from the coleoptile and has chloroplasts. The organ does not intercept more light after expressing positive phototropism. The adaptive function of this phototropism is probably analogous to that suggested for the coleoptile.

---

[4] Nastic movement or curvature refers to the curvature responses whose direction is determined morphologically and genetically (see [68]). Nastic movement is caused by external stimuli, but its direction is not related to the direction of the stimulus.

Araceae and Marantaceae plants can move petioles to undergo diaphototropic leaf orientation (e.g. *Monstera* [23] and species shown in Table 1). However, monocotyledons typically do not have such a petiolar structure, and more commonly observed is the positive phototropism by leaf laminae. Such phototropism is found in plants of the Amaryllidaceae, Cyperaceae, Liliaceae, Orchidacea, and Zingiberaceae (Table 1; see also Figure 1G). Some monocotyledons show positive phototropism at the leaf sheath (e.g. *Bletilla striata*). Again, the phototropism by monocotyledonous leaves is typically observed in shade environments (e.g. Figure 2, G and H). The leaf phototropism may lead to an increase in light interception by leaves (e.g. *Aloe*, *Cilivia*, *Haworthia*, *Paphiopedilum*, and *Vanilla* in Table 1). However, the phototropism expressed in leaves that are initially erect would not necessarily enhance the light interception (e.g. *Allium*, *Carex*, *Hemerocallis*, *Liriope*, *Lycoris*, and *Ophiopogon* in Table 1). In such cases, the adaptive value may rather lie in a shade-avoiding orientation of leaf growth. This is especially apparent in plants that lay down long linear leaves towards the brighter light environment after expressing positive phototropism (e.g. *Allium grayi*, *Carex lenta*, *Lycoris radiata*, and *Ophiopagon ohwii*).

The paraphototropic leaf movement, observed in dicotyledons, is not common to monocotyledons. The positive phototropism of leaves which is widely observed in monocotyledons is phenotypically more similar to the positive phototropism of stems (and also to the simpler form of petiolar positive phototropism).

23.2.4 Phototropism by roots

Roots are phototropic, although they generally grow in dark soil environments. Roots of many seed plants show negative phototropism [70], and young primary roots are typically negatively phototropic (*Arabidopsis thaliana* [71,72], maize [73], *Sinapis alba* [11,74,75], and sunflowers [76,77]). Rice (*Oryza sativa*) is another example showing negative phototropism by young primary roots (Figure 1F). Hubert and Funke [70] observed positive root phototropism in a few species among many investigated. Pilet [78] reported that the young primary root of *Lens culinaris* initially shows positive phototropism, responding to UV-A (to a lesser extent, also to blue light), but becomes negatively phototropic when it has elongated beyond about 20 mm. In rice, the negative phototropism of primary roots and the positive phototropism of coleoptiles show a similar dependence on the fluence rate of blue light (M. Iino, unpublished). In other words, the root does not appear to be more sensitive to phototropically active light than is the aerial part.

At present we can only speculate on the ecological function of root phototropism. When the seed is germinated on the surface of soil or a litter layer, the primary root initially grows in air. Growth towards the soil is primarily guided by positive gravitropism. With the help of negative phototropism, however, the root could grow into shade, thereby reducing its water loss before reaching the soil. Such a role for negative root phototropism might be especially advantageous for seeds that tend to germinate on the soil surface (e.g. the seeds requiring light for germination). Any possible chance that roots could express phototropism at later developmental stages would be when they grow very near to, or beyond the soil surface. Root phototropism might play a role in establishing the root system near the soil surface.

23.2.5 Phototropism in ferns

Under laboratory conditions, the erect petiole of young *Adiantum cuneatum* leaves shows positive phototropism [79]. Phototropism of developing fern leaves can be observed under a forest canopy or a smaller vegetation canopy. In response to diffuse daylight impinging predominantly from one side, many ferns can orient leaves towards that side (e.g. seven species shown in Table 1). This leaf orientation is apparently achieved by positive phototropism of the petiole and midrib. However, the final leaf orientation is often related to the dorsiventral leaf structure. For example, *Dryopteris erythrosora* can assume clear diaphototropic leaf orientation with bending and twisting of the petiole and midrib (Figure 2, I and J).

The protonema and rhizoid of the gametophyte generation show strong phototropism. Generally the protonema is positively phototropic and the rhizoid is negatively phototropic [80]. The protonema of *Schizaea pusilla*, a rare fern that occurs in acidic bogs, shows negative phototropism [81]. These phototropisms are distinct from those described so far, in that they are induced in a single cell. The phototropism of protonemata has been a subject of extensive laboratory investigations (see [82]).

23.2.6 Remarks

As we have seen above, many higher plants can express positive phototropism in stems (seed plants in general), diaphototropism in leaves (typically dicotyledons), and positive phototropism in leaves (typically monocotyledons). The most common roles played by these phototropisms appear to be optimizing carbon gain under a vegetation canopy and escaping from the shade of surrounding plants. Phototropism probably makes a significant contribution to the net productivity of natural vegetation, although this ecological aspect has received little attention.

The way in which plants use phototropism is closely related to their life forms. The stem is an important architectural element used to sustain the shoot against gravitational force, and phototropism of stems may not be advantageous for establishing a stable shoot configuration. The balance between the phototropic curvature and the upright growth guided primarily by gravity-sensing mechanisms is apparently a critical developmental factor. Some herbaceous plants that grow tall by elongating the main stem axis show stem phototropism in the apical elongating part, but resume a vertically straight appearance in the lower elongated parts, even under light-limiting shade conditions (e.g. *Erigeron annuus* and *Phytolacca esculenta* during their early vegetative growth stage; see Figure 2B). Many plants show this tendency, but retain some curvature in the elongated parts, thereby retaining an orientation towards the brighter environment under limiting light conditions (e.g. *Boehmeria nipononivea*, *Desmodium oxyphyllum*, and many young trees; see Figure 2A). Plants with a climbing growth habit or a rosette life form, not dependent on stem architecture, typically show strong leaf phototropism.

Although the phenomenon of phototropism is diverse in many aspects, it includes at least two general features (but with clear exceptions). First, curvature is brought about by the asymmetry of growth. This feature is, in many cases, assumed rather than experimentally verified. However, it is perhaps true because the growth difference between the organ's two sides appears to be the only means by which plants can generally establish a large and stable curvature. The clearest exception to this feature is the curvature made in pulvini or pulvinus-like structures (see Section 23.2.2). Secondly, blue light is most effective, implying participation of blue-light-absorbing photoreceptors (blue-light receptors). Sachs [11] was probably the first to note this feature. He found, using filtered lights, that the phototropism of seedlings was mainly caused by blue light, red light being virtually ineffective. The fact that blue light is much more effective than red or green light has been shown, in addition to stems and coleoptiles (see Sections 23.3.1 and 23.8.1), in leaf diaphototropism by petiolar bending (some dicotyledons [24]), leaf diaphototropism by pulvinar bending (*Lupinus succulentus* [39], *Macroptilium atropurpureum* [44], *Malva nelecta* [34], *Melilotus indicus* [35], *Phaseolus vulgaris* [83], and soybean [45]), leaf paraphototropism by pulvinar bending (soybean [45] and *Oxalis oregana* [56]), flower heliotropism (*Ranunculus adoneus* [84]), and negative phototropism of roots (maize [73], *Sinapis alba* [75], and sunflowers [77]). Exceptions will be described at the end of this chapter (Sections 23.12.1 and 23.12.2).

Our knowledge of the physiological and molecular mechanisms of phototropism is very much limited to special cases. In the subsequent sections, we will focus on the mechanisms of phototropism investigated in seedling organs (coleoptiles, hypocotyls, and epicotyls). Sun-tracking movements of leaves and flowers are reviewed by Koller in this volume. The positive phototropism of coleoptiles may represent the positive phototropism of monocotyledonous leaves. On the other hand, the positive phototropism of hypocotyls and epicotyls probably represents the positive phototropism of stems.

23.3 Fluence-response relationships and response types

The blue-light-sensitive phototropism shows specific and complex relationships with the fluence of light, the product of the fluence rate and the stimulus duration, and additionally with the stimulus duration. These relationships will first be summarized, since they are in many ways related to the mechanisms of phototropism discussed in later sections.

23.3.1 Historical background and experimental systems

Blaauw [85] irradiated dark-adapted oat coleoptiles with unilateral light for a short period and detected phototropic curvature that developed in the dark. With this protocol, he demonstrated that the barely detectable curvature response obeys the Bunsen-Roscoe reciprocity law, which states that the response is constant when the fluence is constant, irrespective of the fluence rate and the stimulation time used. This study laid the experimental and analytical basis for the subsequent studies of the phototropic fluence-response relationship. Blaauw [85] also carried out the earliest systematic study on the wavelength dependence of phototropism. Again using oat coleoptiles, he could demonstrate that wavelengths around the blue region of the light spectrum are most effective in inducing phototropism, while those in the red and yellow regions are ineffective. This spectral sensitivity distribution of oat coleoptiles had been confirmed and well refined by the 1930s [86], and the phototropic fluence-response relationship has since been studied as a property of blue-light-sensitive phototropism.

Within a few decades after the work of Blaauw [85], it became apparent that the relationship between the applied light fluences and the phototropic responses in oats is multiphasic. du Buy and Nuernbergk [87] summarized earlier results and their own and classified the blue-light-sensitive phototropism into several response types in view of the multiphasic fluence-response relationship. Many workers have since investigated the fluence-response relationship in oats and have extended the study to other plant materials.

The following procedures are generally used to obtain the fluence-response data. The entire plant or organ is treated with unilateral light. The light source is polychromatic or monochromatic blue light. (It was white light in some early studies.) The fluence is controlled by changing the duration of irradiation at fixed fluence rates. This duration can be as long as 40 min, by which time a small curvature appears. Another method used to control the fluence is to change the fluence rate at a fixed duration of irradiation. This method is useful when only pulse-induced phototropism is investigated. The phototropic response of the organ is determined at a fixed time (usually in the range between 90 and 120 min) after the onset of the unilateral irradiation. The response is quantified by the angle of curvature. This angle is determined from the line tangent to the long axis at the apical region. Dark-adapted seedlings have often been used. For reasons that will become clear later (Sections 23.4.1–4), however, proper pretreatment with red light is necessary to obtain complete fluence-response curves. To this effect, it is convenient to use the seedlings grown under continuous red light. The measured response is generally plotted against the logarithm of the fluence. With this plot, the fluence-response curve over a wide fluence range can be shown at once, and the evaluation of the sensitivity and responsiveness to light becomes easier. Descriptions of the fluence-response relationship given below are based, unless otherwise specified, on this form of plot.

In addition to the original terms used by du Buy and Nuernbergk [87], later authors introduced other terms to identify the response types [88–90]. This chapter adopts the terms of Iino [90] because they can be used with the least confusion.

23.3.2 In oat coleoptiles

Typical fluence-response curves for phototropism of oat coleoptile are depicted in Figure 3. These curves were obtained using red-light-grown seedlings (i.e. the seedlings maintained under a red light background throughout their growth, including the time of phototropism development). When each fluence is given in a pulse (e.g. 30 s), the fluence-response curve reveals a major peak and a small peak in the positive response range (at about 10^0 and $10^{2.5}$ μmol m^{-2}, respectively, in the case shown in Figure 3A). The trough between the two peaks falls in the negative range. Within the limits of exposure time, reciprocity holds at all fluences. The longer limit is the time at which the time-dependent phototropism begins to be induced (see below). The shorter limit has not been determined. This limit is at least much shorter than 1 s.

The positive responses over the large and small peaks are termed **first pulse-induced positive phototropism** (fPIPP) and **second pulse-induced positive phototropism**

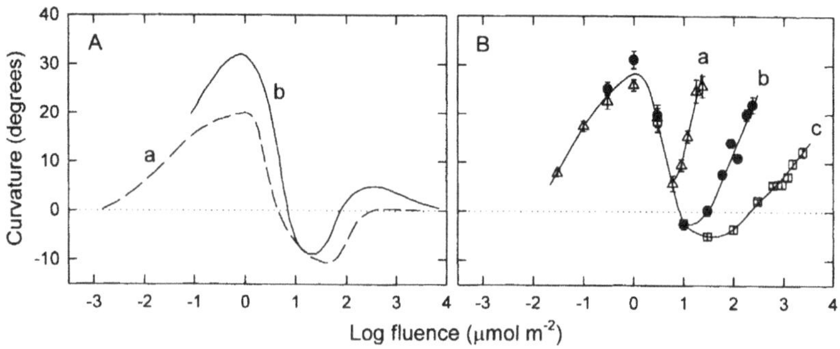

Figure 3. Phototropic fluence-response curves in coleoptiles of red-light-grown oat seedlings. Phototropism was induced with unilateral blue light. (**A**) Fluence-response curves for pulse-induced phototropism. The fitted curves in Iino [90] are redrawn. The fluence was given in a pulse of either 10 or 90 s. Curvature was measured 100 min after the onset of blue-light stimulation. The two lines were obtained with coleoptiles of different length groups: 17–25 mm (solid lines) and 26–36 mm (dashed line). (**B**) Fluence-response curves obtained at fixed fluence rates of 0.01 (**a**), 0.1 (**b**) and 1 (**c**) $\mu mol\ m^{-2}\ s^{-1}$ (Y. Murata and M. Iino, unpublished). In all cases, the exposure time was varied from 3 s to 40 min. Curvature was measured 120 min after the onset of blue-light stimulation. The coleoptile length was between 18 and 22 mm. The means ± SE from 16–32 plants are shown. The light sources and growth conditions were as described in Tarui and Iino [195].

(sPIPP), respectively. The negative response is termed first pulse-induced negative phototropism (fPINP), which can be simplified to **pulse-induced negative phototropism** (PINP) because there is no other pulse-induced negative phototropism. The fPIPP and PINP correspond to the first positive curvature and the first negative curvature, respectively, of du Buy and Nuernbergk [87], and have been characterized by Zimmerman and Briggs [91], Meyer [92], Blaauw and Blaauw-Jansen [89], and Iino [93]. The sPIPP probably corresponds to the second positive curvature of du Buy and Nuernbergk [87], but its clear occurrence has been established much later [89,90,93].

When a given fluence is applied in a period that exceeds a certain limit, the curvature response becomes more positive than that induced by a short pulse, thus invalidating the reciprocity law. This additional response forms an ascending fluence-response curve when the fluence is enhanced by extending the exposure time at a fixed fluence rate (Figure 3B). The curve shifts to higher fluences as a higher fluence rate is used, and can be clearly separated from fPIPP and PINP, and even from sPIPP, when the fluence rate is sufficiently high. Over a wide range of fluence rates (at least two orders of magnitude), the response depends strongly on the exposure time and only weakly on the fluence rate. This response, which increases in an exposure-time-dependent manner, is termed **time-dependent phototropism** (TDP) and corresponds to the third positive curvature of du Buy and Nuernbergk [87] or the "second positive curvature" used frequently by later authors. These properties of TDP have been characterized by Zimmerman and Briggs [88,91], Everett and Thimann [94], Pickard et al. [95], and Blaauw and Blaauw-Jansen [89].

The minimal stimulation time for TDP (i.e. the time at which the response deviates from the pulse-induced response) occurs somewhere between 1 and 10 min in red-light-pretreated coleoptiles [91,94]. Based on a series of fluence-response curves, Blaauw and Blaauw-Jansen [89] estimated the minimal stimulation time to be about 4 min. Stimulation time-response data could be fitted to a straight line [94,95]. On the other hand, Blaauw and Blaauw-Jansen [89] concluded that a better linearity is found when the response is plotted against the logarithm of the stimulation time.

23.3.3 Generality of response types

Of the four response types characterized in oat coleoptiles, fPIPP and TDP are found commonly. The fPIPP has been identified in coleoptiles of other Gramineae cereals [69,96–98], hypocotyls of many dicotyledons [97,99–101], and epicotyls of some dicotyledons [97,102]. Likewise, the ascending response curve corresponding to TDP has been identified in many plants [69,96–99,101,103–106]. The "time dependent" nature of TDP has been demonstrated, with at least two fluence rates, in coleoptiles of maize [96,107], *Hordeum vulgare* [103] and rice [69] and in hypocotyls of *Arabidopsis thaliana* [99,108] and *Fagopyrum esculentum* [106].

An sPIPP is found in maize coleoptiles [93,109]. This sPIPP occurs at fluences similar to the sPIPP of oat coleoptiles and is much more obvious than the latter. Rice coleoptiles also show an sPIPP at comparable fluences [69]. The type of sPIPP found in coleoptiles has not yet been identified in dicotyledons (but, see the next section). The PINP has so far been identified only in oat coleoptiles. Detailed fluence-response investigations have not revealed this type of phototropism in dicotyledonous hypocotyls [99,100] or even in maize coleoptiles [93,110].

Figure 4 shows the phototropic fluence-response curves obtained using red-light-grown maize and rice seedlings. The red-light growth condition and the blue-light source used to induce phototropism are similar to those used to obtain the curves in Figure 3. The data in Figures 3 and 4 indicate that the magnitude of fPIPP and TDP varies among materials. In maize, the magnitude even varies considerably among cultivars [111] (also compare [112] with [109]). However, the relative dependence on the fluence is rather well conserved. For example, the peak of fPIPP occurs between 10^0 and 10^1 μmol m^{-2}. Under similar light conditions, the peak also occurs within this range in the epicotyl of peas (*Pisum sativum*) [102] and the hypocotyl of some dicotyledons [100].

When all the results obtained under similar conditions are compared, it turns out that the fluence-response properties are most exceptional in oat coleoptiles. As mentioned above, PINP has been found only in oats. The fluence causing the peak of fPIPP is lowest in oats; it is definitely lower by one order of magnitude than in maize ([93]; also compare Figure 3 with Figure 4A). The descending arm is much steeper than the ascending arm in oats; cf. maize ([113]; Figure 4A), pea [102], and alfalfa [100]. Furthermore, the quantitative relationship between fPIPP and TDP is quite different from the other materials. The extent of fPIPP, relative to that of TDP, is substantially greater in oats.

Experiments with maize coleoptiles have provided some information that is not known in oat coleoptiles. For dark-adapted maize coleoptiles, the bell-shaped fluence-response curve representing fPIPP could be shown with a blue-light flash of a few ms

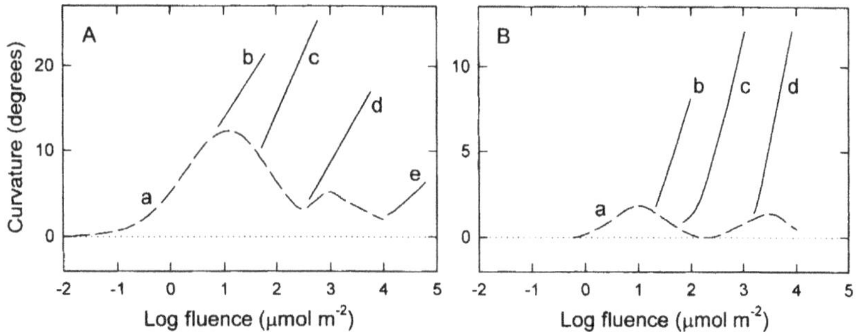

Figure 4. Phototropic fluence-response curves in coleoptiles of red-light-grown maize and rice seedlings. Phototropism was induced with unilateral blue light. (**A**) Maize coleoptiles. The fitted lines in Liu and Iino [107,109] are redrawn. Dashed line (**a**): pulse-induced phototropism (exposure time, 30 or 120 s). Solid lines: time-dependent phototropism induced at 0.022 (**b**), 0.22 (**c**), 2.2 (**d**), and 22 (**e**) μmol m^{-2} s^{-1}. Curvature was determined 120 min after the onset of blue-light stimulation. The time-dependent phototropism became apparent as the exposure time exceeds about 5 min; the longest exposure time was 40 min. (**B**) Rice coleoptiles (submerged in water). The fitted lines in Neumann and Iino [69] are redrawn. Dashed line (**a**): pulse-induced phototropism (exposure time, 10–120 s). Solid lines: time-dependent phototropism induced at 0.025 (**b**), 0.25 (**c**), and 2.5 (**d**) μmol m^{-2} s^{-1}. Curvature was determined 90 min after the onset of blue-light stimulation. The time-dependent phototropism became apparent as the exposure time exceeds about 8 min; the longest exposure time was 50 min.

duration ([90]; see its Figure 3). Therefore, reciprocity appears to hold even in the ms range. Although the TDP response of red-light-grown or red-light-pretreated coleoptiles showed a reasonable fit to a straight line in a plot against the logarithm of the fluence or stimulation time, a unique feature became apparent in the plot against the linear stimulation time [107]. After a minimal stimulation time (3–5 min), a sharp increase in curvature occurred over a short period (5–10 min) of stimulation. Then the slope declined rather abruptly and a more gradual increase followed. It seemed as if the curvature response increased in two phases. Rice coleoptiles showed similar results, except that the minimal stimulation time and the initial phase of curvature development were somewhat longer [69].

23.3.4 More response types

Phototropism of coleoptiles and hypocotyls has been examined over many decades, and yet, as described below, more recent workers have reported phototropic responses of these organs that are induced by blue light, but cannot be related directly to the response types described above. These responses have not yet been confirmed by independent workers.

Ullrich [114] recorded a pulse-induced negative phototropism in maize coleoptiles. This phototropism was induced at a fluence optimal for fPIPP. It was initiated rapidly (within 15 min after stimulation) near the base of the coleoptile and was later overtaken by fPIPP. This negative phototropism is apparently distinct from the PINP of oat coleoptiles; in addition to the difference in the effective fluence, the latter response begins in the tip and develops after a lag time that is longer than that of fPIPP (see Section 23.7.1). The negative curvature of maize coleoptiles has not been recorded by other workers (see above). Time-lapse photographs of red-light-grown maize coleoptiles have never shown any such negative curvature after fPIPP stimulation (M. Iino, unpublished observation). The maize seedlings used by Ullrich were treated daily with red light for a short period and used after dark adaptation of at least 8 h. The occurrence of the response may be related to this growth condition.

Ellis [106] investigated the phototropic fluence-response relationship in the hypocotyls of *Fagopyrum esculentum*. The fluence was controlled by the stimulation time at fixed fluence rates. Either dark-grown or light-grown plants showed a pulse-induced positive phototropism that formed a peak in the fluence response curve. However, this phototropism occurred at fluences much lower than those generally causing fPIPP, and the position of the peak moved depending on the fluence rate used (i.e. reciprocity was not valid).

No negative phototropism has been found to follow continuous unilateral stimulation in coleoptiles, hypocotyls, and epicotyls. Surprisingly, Taylor et al. [115] observed negative phototropism when they continuously exposed the basal region of oat coleoptiles to a unilateral beam (1-mm diameter) of blue light (0.05 $\mu mol\ m^{-2}\ s^{-1}$). Positive phototropism followed when a middle part of the coleoptile was similarly stimulated, but this phototropism was later overcome by a negative phototropism. Irradiation of the tip resulted only in positive phototropism. Interestingly, the negative curvature was initiated first in the tip and later in lower zones following the stimulation of the base. Macleod et al. [116] did not observe any such negative curvature during localized irradiation of coleoptile zones. Taylor et al. [115] used a lower fluence rate and irradiated a narrower zone. Either or both of these conditions may be critical. The negative phototropism is distinct from PINP, because the light signal for PINP is almost exclusively perceived by the coleoptile tip (see Section 23.5.2). It might be related somehow to the pulse-induced negative response of maize coleoptiles described above.

Konjevic et al. [117] obtained phototropic fluence-response curves of dark-adapted *Arabidopsis* hypocotyls by extending the exposure duration at fixed fluence rates. When the fluence rate of blue light was increased from 0.2 to 0.4 $\mu mol\ m^{-2}\ s^{-1}$, an additional peak became apparent in the fluence-response curve, and was retained at a higher fluence rate (0.7 $\mu mol\ m^{-2}\ s^{-1}$). This peak occurred in a narrow fluence range (between 1 and 10 $\mu mol\ m^{-2}$) on the descending arm of fPIPP. The response in the new peak was induced with exposure durations shorter than 30 s, and is an sPIPP, by definition [90]. A similar sPIPP has been suggested to occur in *Vigna radiata* hypocotyls [118]. It is an intriguing possibility that the sPIPP induced in the hypocotyl of *Arabidopsis* (and possibly of other dicotyledons) corresponds, in fact, to the sPIPP of coleoptiles, although the effective fluences are clearly distinct between hypocotyls and coleoptiles and the hypocotyl sPIPP occurs in a narrower fluence range. The kind of fluence-rate dependence found in *Arabidopsis* has not been shown for coleoptile sPIPP. However, it is difficult to investigate the sPIPP of coleoptiles at the low fluence rates with which a lack of sPIPP has been shown in *Arabidopsis*, because the exposure time becomes effective for TDP if the required fluences are to be obtained at such low fluence rates.

23.3.5 Remarks

Plant organs used for phototropism experiments are naturally those which are strongly phototropic. Under continuous stimulation, these organs often bend until their tip establishes a nearly complete orientation towards the light source. This is not always the case, however. For example, the rice coleoptile grown in air establishes a steady curvature angle at any fluence rate of blue light before reaching 30° [69]. The steady-state curvature probably represents the equilibrium between phototropism and gravitropism (see Section 23.7.5). Figure 5 shows the fluence rate-response curve obtained for this curvature [69]. The data provide an estimate of the overall relationship between the fluence rate and the magnitude of phototropic response under continuous stimulation. It is noted that the phototropic response is induced effectively over a wide range of fluence rates, in agreement with the fluence-response relationship resolved for TDP with limited stimulation times. However, as expected, the fluence rate limits the response at low fluence rates. Furthermore, the response decreases as the fluence rate exceeds a certain optimal value (1–3 μmol m^{-2} s^{-1}).

Many of the properties uncovered for TDP probably represent the properties of the phototropism functioning in nature. On the other hand, pulse-induced phototropisms, found under specific laboratory conditions, are unlikely to have an effective role in nature. This conclusion simply follows from the fact that each of the pulse-induced phototropisms occurs in a very specific fluence range and that the extension of fluences leads to substantially reduced or no response. In fact, plants cannot remain in the effective fluence range of either fPIPP or sPIPP for most of the day. It may be expected that pulse-induced phototropisms represent mechanistic aspects of TDP. This view will be extended in later sections. Until we obtain solid conclusions, however, the examined response type should be defined whenever the mechanistic aspects of phototropism are described.

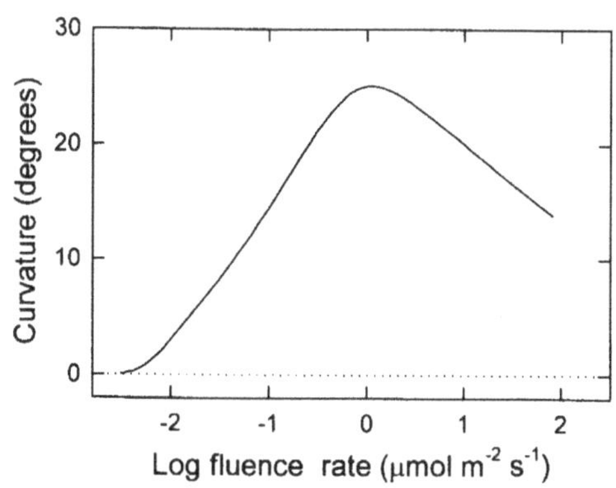

Figure 5. Steady-state curvature established in rice coleoptiles during continuous phototropic stimulation. The fitted line in Neumann and Iino [69] is redrawn. Coleoptiles of red-light-grown rice seedlings, not submerged in water, were continuously stimulated with unilateral blue light. Curvature was determined 5 h after the onset of blue-light stimulation, by which time the curvature reached a steady value.

23.4 Regulatory roles of phytochrome

After the work of Blaauw [85], red light (or orange light) became the standard safelight for phototropism experiments. However, Curry [119] and Blaauw-Jansen [120] found that the fluence-response curve of fPIPP was affected by red-light pretreatment in oat coleoptiles. Since then it has become more and more evident that red light (and also far-red light) affects the phototropic fluence-response relationship at many different levels. It has also turned out that in certain cases the ability of plants to show phototropism depends critically on the preceding light signal. Recent evidence demonstrates clearly that these light effects are mediated by phytochrome.

Information on the phytochrome regulation of phototropism is central to our understanding of how phototropism functions during plant development, and is also necessary for evaluating published results and designing phototropism experiments properly. In contrast to analyzing the process that follows phototropic stimulation, the investigations described here concern how the light pretreatment affects the system of phototropism. Although the experiments include the period of curvature development, the temporal component considered below is primarily the time between the onset of pretreatment and that of phototropic stimulation.

23.4.1 In first pulse-induced positive phototropism (sensitivity regulation)

When dark-adapted oat and maize coleoptiles are pretreated with red light, the position of the entire bell-shaped fluence-response curve of fPIPP is shifted to higher fluences [91,109,111,112,119,121–124]. The maximal shift observed is about one order of magnitude in both oats and maize. The results indicate that the red light somehow makes fPIPP less sensitive to the inductive pulse of blue light, maximally by about 10-fold.

The desensitization of fPIPP progresses over a period of about 2 h after treatment with a short pulse of red light [109,121,122] (Figure 6). The desensitization response is not induced immediately, but after a lag of about 15 min [109,112,121]. A pulse of red light can be as effective as continuous red light in causing the maximal desensitization [121]. When treated with a pulse, however, fPIPP is resensitized (i.e. the fluence-response curve returns to the original position) over a period of 3–5 h after showing the maximal desensitization [109,122]; under continuous red light, fPIPP remains desensitized [113]. The desensitization response is saturated at very low fluences of red light [125] and can also be induced fully by far-red light [123]. The response is probably mediated by phytochrome, with only a small portion of phytochrome being required in the active far-red-light-absorbing form, Pfr, for saturation. This type of response, the so-called "very-low-fluence response", is generally mediated by phytochrome A [126].

The desensitization of fPIPP by red light has not been shown in materials other than oat and maize coleoptiles. Janoudi and Poff [127] concluded that such desensitization does not take place in *Arabidopsis* hypocotyls. However, desensitization of a smaller extent (e.g. 3-fold desensitization) is probably not excluded by the data presented [127,128].

23.4.2 In first pulse-induced positive phototropism (responsiveness regulation)

In addition to desensitizing fPIPP, red-light pretreatment enhances the maximal fPIPP responsiveness in oat and maize coleoptiles. This responsiveness enhancement is

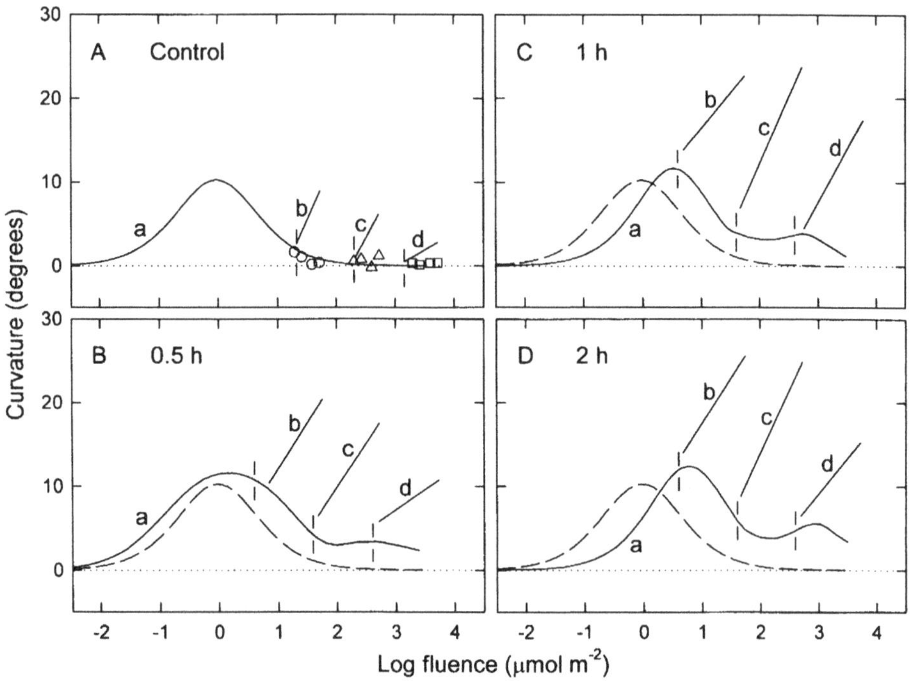

Figure 6. Multiple effects of red-light pretreatment on phototropic fluence-response relationships in maize coleoptiles. The fitted curves in Liu and Iino [107] are redrawn. The seedlings raised initially under red light were dark adapted for 24 h and pretreated with a red-light pulse (3 min at 10 μmol m^{-2} s^{-1}). After a defined period, the coleoptiles were exposed to unilateral blue light to obtain phototropic fluence-response curves. Curvature was determined 120 min after the onset of blue-light stimulation. In addition to the red light for pretreatment purposes, all seedlings received red light from above during blue-light stimulation (except for the data shown by symbols, see below) and for 5 min immediately after blue-light stimulation. These red-light treatments were given to eliminate the phytochrome-dependent phototropism. (**A**) Fluence-response curves obtained without red-light pretreatment. (**B–D**) Fluence-response curves obtained with red-light pretreatment; phototropic stimulation was initiated after the dark interval indicated in each panel. Solid curves in each panel indicate pulse-induced phototropism (**a**) and time-dependent phototropism induced at 0.025 (**b**), 0.25 (**c**), and 2.5 (**d**) μmol m^{-2} s^{-1}. The curve **a** in **A** is reproduced in **B–D** (dashed lines) for comparison. The results indicate that fPIPP is shifted to higher fluences in response to red-light pretreatment, but with non-synchronized shifts of the ascending and descending arms at the dark interval of 15-min (**B**). The peak of fPIPP is slightly enhanced after red-light pretreatment. The sPIPP, absent in dark-adapted coleoptiles, becomes inducible in response to red-light pretreatment. In red-light-pretreated coleoptiles (**B–D**), TDP is induced as the stimulation time exceeds 3 min (indicated by vertical lines). When no pretreatment is given (**A**), TDP is induced only at stimulation times longer than 15 min (indicated by vertical lines). When the red-light treatment during phototropic stimulation but that after phototropic stimulation is omitted, however, no TDP can be induced even if the stimulation time is extended to 40 min. This is demonstrated in **A** by the data sown in symbols, which indicate the mean curvatures induced at fluence rates of 0.025 (circles), 0.25 (triangles), and 2.5 (squares) μmol m^{-2} s^{-1}.

detected as an increase of the peak height of the bell-shaped fluence-response curve [109,112,120,121,123]. Enhancement to a similar extent can be induced by far-red light in oats [120]. Following stimulation with a saturating pulse of red light, the responsiveness enhancement becomes detectable at about 15 min and develops over a period of about 2 h [109,121]. The significance of the enhancement response has been obscure because the observed effect was relatively small (20–50%; see Figure 6) and because not all workers could find this enhancement (see [90]).

Recent studies with *Arabidopsis* hypocotyls have substantially advanced our knowledge of the phytochrome-mediated fPIPP enhancement. Janoudi and Poff [129] observed a 4-fold enhancement by a red-light pulse given 2 h before phototropic stimulation. The action spectrum obtained for this response agrees with the absorption spectrum of the red-light-absorbing form of phytochrome, Pr [129]. The use of *phyA* and *phyB* mutants indicated that phytochrome A is the major phytochrome species responsible for the enhancement response [130,131]. Subsequent work using a *phyAphyB* double mutant has led to the conclusion that phytochrome B can also function when phytochrome A is absent [132]. We obtained similar results using *phyA-201*, *phyB-5*, and *phyA-201/phyB-5* mutants of *Arabidopsis* in the Landsberg *erecta* background (abstract in [133]).

Dark-grown wild-type *Arabidopsis* showed some fPIPP without red-light pretreatment (about 10 degrees curvature at the optimal fluence), while a *phyAphyB* double mutant showed much less curvature (at most, 2 degrees) [132]. We found a small but clear fPIPP in non-pretreated wild-type *Arabidopsis*, but no detectable fPIPP in the *phyAphyB* double mutant [133]. In these studies, no green working light was used. Although the results from *phyAphyB* mutants indicate strict dependence of fPIPP responsiveness on phytochrome, it is somewhat paradoxical that fPIPP was significantly greater without red-light pretreatment in wild-type plants than in *phyAphyB* mutants. In experiments with *Arabidopsis*, imbibed seeds were treated with light to initiate germination and incubated subsequently in the dark for more than 40 h. Whether or not some phytochrome signal could be retained during such a long dark period needs to be investigated.

The pea epicotyl is another material in which the fPIPP responsiveness is strictly controlled by phytochrome. Dark-grown pea epicotyls show a small pulse-induced positive phototropism, but this phototropism is probably mediated by phytochrome rather than a blue-light receptor ([134], see Section 23.12.1). When grown under red light, the epicotyl develops a substantial fPIPP [102]. Strict dependence of fPIPP responsiveness on phytochrome might indeed be common to dicotyledons.

23.4.3 In second pulse-induced positive phototropism and pulse-induced negative phototropism

Blaauw and Blaauw-Jansen [89] have shown that sPIPP cannot be induced in dark-adapted oat coleoptiles, but becomes inducible during 2-h pretreatment with red light. A short pulse of red light can partially bring about the responsiveness. The dependence of sPIPP responsiveness on red light has been demonstrated more clearly in maize. While fPIPP is the only pulse-induced phototropism in dark-adapted maize coleoptiles, sPIPP becomes inducible upon pretreatment with a red-light pulse ([109,112]; Figure 6). The maximal sPIPP responsiveness is established 2 h after the pretreatment and decays gradually afterwards [109]. A pulse of far-red light can bring about the sPIPP responsiveness to a similar extent [107]. Although the fluence requirement appears to differ between oats and maize, these results suggest that the occurrence of sPIPP in these materials depends strictly on phytochrome.

The PINP is inducible in dark-adapted coleoptiles, and neither the maximal extent nor the optimal fluence is obviously affected by red-light pretreatment [121]. It appears that red light does not significantly affect PINP while it desensitizes fPIPP. As described above (Section 23.3.3), the descending arm of fPIPP is much steeper than the ascending arm. However, the two arms are more symmetrically placed in dark-adapted coleoptiles [91,121]. Kinetic data indicate that PINP is distinct from fPIPP with respect to the lag period and that the response representing PINP accompanies fPIPP on the descending arm of the fPIPP fluence-response curve (see Section 23.7.1). Perhaps, the descending arm becomes steeper than the ascending arm in red-light-pretreated oats because the effective fluences of fPIPP overlap more with those of PINP in these plants.

23.4.4 In time-dependent phototropism

In oat coleoptiles, pretreatment with red light results in an enhancement of TDP at optimal fluence rates of blue light [88,89]. This enhancement is detected as an increase in curvature response at a given exposure time, or as an increase in the slope in the plot of curvature response vs. exposure time. When red light is given in a pulse, the optimal dark interval between the red-light pulse and the onset of phototropic stimulation for this enhancement is 1–1.5 h [89]. Another effect identified is the shortening of the minimum stimulation time required for the induction of TDP from about 15 min down to 4 min [89].

In maize coleoptiles, the TDP responsiveness depends almost entirely on phytochrome [107]. Dark-adapted maize coleoptiles can express fPIPP, but no detectable TDP up to at least 40 min of blue-light stimulation (Figure 6A; see the data shown by symbols). Following the treatment with a red-light pulse, TDP rapidly becomes inducible, and the maximal extent of TDP can be obtained when phototropic stimulation is initiated after a dark interval of about 1 h (Figure 6). As the time interval is extended for another few hours, TDP again disappears. The effect of red light can be totally suppressed by a pulse of far-red light given after the red-light pulse, indicating that the response is controlled by phytochrome in a far-red-light reversible fashion. (The lack of TDP in dark-adapted maize was already apparent in the data of Chon and Briggs [122]. In their Figure 1, the fluence-response curve showed a bell-shaped characteristic of fPIPP even though the exposure time was extended up to 30 min. A small response was detected at high fluences beyond the descending arm of fPIPP, but this response did not show a time-dependent increase typical of TDP. The response at high fluences was probably sPIPP that had become inducible by the green working light used.)

Although TDP can be induced in dark-adapted oat coleoptiles, the TDP responsiveness in this material may also depend almost entirely on phytochrome. If a phytochrome-mediated, very-low-fluence response were to contribute to the establishment of TDP responsiveness in oat coleoptiles, then TDP could be made inducible by the Pfr produced during the phototropic blue-light stimulation. This possibility is not unlikely in view of the results described below and is at least supported by the result that the minimal stimulation time for TDP is longer when red-light pretreatment is not given (see above).

Kang and Burg [135] observed that red-light pretreatment of dark-grown pea seedlings results in a greater phototropic response of epicotyls to continuous blue light. Britz and Galston [136] investigated this red-light effect in further detail. When dark-grown pea epicotyls were stimulated with continuous unilateral blue light, phototropic curvature occurred after a lag longer than 2 h. A red-light pulse given 15 h before the onset of phototropic stimulation reduced the lag to a period shorter than 1 h and enhanced the rate of curvature. The phototropic response to 3-h stimulation increased gradually as the dark interval after the red-light pretreatment increased, and the maximal response was obtained at intervals longer than 16 h. A pulse of far-red light was also partially effective. The responsiveness enhancement caused by red light could be reversed by far-red light to the level induced by far-red light alone. Therefore, both very-low-fluence and low-fluence responses appear to participate in the enhancement effect. Although a phototropic response could be induced in non-pretreated seedlings after a long lag time, it is very probable that this response was made inducible by the Pfr generated during phototropic stimulation with blue light. In pea epicotyls, the TDP responsiveness may in fact depend almost entirely on phytochrome. It is to be noted, however, that TDP responsiveness develops much more slowly after red-light treatment in peas than in maize (see above) and *Arabidopsis* (see below).

Recent results indicate that the TDP responsiveness in *Arabidopsis* hypocotyls is also under strict phytochrome control. Janoudi et al. [132] found that the magnitude of TDP is reduced by either *phyA* or *phyB* mutation and is much reduced by *phyAphyB* double mutation. In the *phyAphyB* double mutant, TDP became inducible only when the exposure time was extended beyond 1 h (see also [137]). These mutation effects were observed without red-light pretreatment. We also investigated TDP of *phyA*, *phyB* and *phyAphyB* seedlings [133]. In the wild-type seedling pretreated with a red-light pulse 2 h before the onset of phototropic stimulation, TDP was induced when the stimulation time exceeded about 4 min. On the other hand, the *phyAphyB* double mutant similarly treated with red light showed no detectable TDP for a stimulation time as long as 35 min. It was also found that, in addition to the far-red-light-reversible low-fluence response, a very-low-fluence response and a far-red-light high irradiance response (so called FR-HIR) participate in the establishment of TDP responsiveness, and that the low-fluence response is mediated by phytochrome B and the very-low-fluence response and FR-HIR are mediated by phytochrome A.

Although dark-adapted wild-type *Arabidopsis* can express some TDP without red-light pretreatment ([132]; see above), it is likely that this TDP is made inducible by the Pfr produced during the phototropic blue-light stimulation. In agreement with this view, the dark-grown *Arabidopsis* required a minimal stimulation time of 20 min [132] in contrast to the 4 min found in red-light-pretreated *Arabidopsis* (see above). Although the *phyAphyB* double mutant showed TDP after very long stimulation time [132], phytochrome species other than phytochromes A and B might be responsible in this case. Dark-adapted wild-type *Arabidopsis* probably lacks the TDP responsiveness almost entirely.

Other reported results suggest that the phytochrome control of TDP responsiveness is ubiquitous. In radish (*Raphanus sativus*) hypocotyls, the minimal stimulation time for TDP was shorter in dark-grown plants than in red-light-pretreated plants [104]. The phototropism of cress (*Lepidium sativum*) hypocotyls induced by 2-h stimulation with blue light was much greater in white-light-grown seedlings than in dark-grown seedlings [105]. When white-light-grown sunflower hypocotyls were subjected to continuous blue-light stimulation after 2 h pretreatment with far-red light, phototropic curvature occurred after a lag longer than 30 min; without far-red-light treatment, the lag was much shorter [138]. Also, the lag period is substantially enhanced in white-light-grown sunflower hypocotyls when they are kept in darkness before phototropic stimulation [139]. Dark-grown cucumber hypocotyls showed a phototropic response to continuous blue light after a lag of 4 h, but this lag period was reduced significantly when the hypocotyls were pretreated briefly with red light one day before phototropic stimulation [140]. In hypocotyls of dark-grown mung bean (*Phaseolus aureus*) seedlings, the phototropic response to continuous blue light began after a lag of about 1 h, but no such lag could be observed if the seedlings had been grown under overhead red light or blue light [141]. All of these results support the conclusion that phytochrome acts to enhance TDP responsiveness. Indeed, dark-grown seedlings may generally have little TDP responsiveness (note the long lag time commonly observed when dark-grown seedlings were stimulated; see the discussion in preceding paragraphs).

Woitzik and Mohr [4] investigated the effects of red and far-red light on the phototropic response of sesame (*Sesamum indicum*) hypocotyls to continuous blue light. The results indicated, in line with the results described above, that phototropic responsiveness is enhanced by phytochrome. In addition, they found that higher fluence rates were required to saturate the phototropic response in red-light-grown plants than in dark-grown plants. Therefore, it was concluded that phytochrome reduces the phototropic sensitivity to continuous blue light.

23.4.5 Remarks

The results described above have demonstrated that phytochrome exerts various effects on the overall system of phototropism, but do not clarify whether there is any causal relationship among these effects. For example, the occurrence of sPIPP depends critically on phytochrome in oat and maize coleoptiles, but phytochrome has only modifying roles on fPIPP in these materials. In maize, phytochrome determines the occurrence of TDP in a red/far-red reversible manner, but a non-reversible very-low-fluence response participates significantly in all other effects exerted on pulse-induced phototropisms. At present, it seems that different members of the phytochrome family and different components of the overall phototropic system constitute a complex interacting network.

Although the relationship between phytochrome and phototropism is complex, the prime action of phytochrome appears to be to establish the ability of seedling shoots to show phototropism in prolonged stimulation. It is likely that the plants germinated and growing within the dark soil environment establish full phototropic responsiveness only after the shoot apex or the coleoptile tip emerges from the soil surface (or as it approaches the soil surface). Phytochromes are the major photoreceptors involved in the process of de-etiolation [142]. The establishment of phototropic responsiveness is probably programmed as a part of this de-etiolation process.

After the recognition that red light affects phototropism, green light has been used as a common "safelight". However, the experimental results obtained using a green working light are liable to include uncontrolled phytochrome effects. This conclusion follows the fact that many of the phytochrome-mediated responses in phototropism are induced by very low fluences of red light and that such very-low-fluence responses are effectively induced by green light [143,144]. In this chapter, dark-treated plant materials are specified, primarily to contrast them with white-light-grown plants or de-etiolated green plants, with the words "dark-grown" (when grown in the dark throughout) and "dark-adapted" (when grown in the dark for a day or so after certain light treatment). It should be recognized, however, that green light has often been used to manipulate plants and/or to obtain plant images.

23.5 From photoperception to the curvature response: spatial and temporal bases

In this section, we ask where the directional light is sensed, and how photosensing is correlated spatially and temporally with the curvature response. These aspects have been studied most extensively in oat coleoptiles, but the experimental results have often been contradictory.

The photosensing region is commonly investigated by either stimulating a defined zone with unilateral light or shading a defined zone from unilateral light. On the other hand, the zones delimited by markers are used to find where and when curvature responses occur. The most detailed experiments on marked zones have employed growth measurements on the two sides of individual zones. Although such measurements have been made primarily to analyze underlying growth responses, this section only considers where curvature takes place and when it is initiated. The specific regions of coleoptiles will be described below, in terms of the distance from the tip. Such information must be considered in relation to the overall length of the organs used, although it will not be described in detail here. The lengths of the oat and maize coleoptiles, the major materials used to obtain positional data, have typically ranged between 20 and 25 mm.

23.5.1 Historical experiments on coleoptiles

Charles Darwin, with his son Francis Darwin, conducted a number of experiments on plant movements and published the results in 1880 [22]. Among the movements investigated was the phototropism of *Phalaris canariensis* coleoptiles. For this species, the conclusion was reached that light is perceived exclusively by the tip and, "some influence is transmitted from the upper to the lower, causing the latter to bend." The most basic result was that shading the tip eliminated the phototropic curvature observed at the lower parts. Oat coleoptiles were also used; the result was less clear, but was thought to support the above conclusion. The work of Darwin has had great influence on subsequent workers. As will be described in Section 23.10.1, the discovery of the plant hormone, auxin, and the studies of the role for auxin in phototropism have stemmed from this work.

Rothert [1] repeated the experiments of Darwin and found that shading of the tip did not eliminate the curvature response. Rothert mainly used oat coleoptiles, but similar results could also be obtained in *Phalaris*. Therefore, the tip did not appear to be the sole photoperceptive region. Rothert then provided unilateral stimulation to the tip and to the lower remaining part simultaneously but from opposite directions. The lower part initially showed a positive curvature in response to the light directed to this part, but this curvature was overtaken in time by a curvature to the opposite direction that had begun at the tip. This simple experiment demonstrated a predominant role of the tip and provided support to the notion of the transmittable influence. Furthermore, the observation indicated that the transmission of the influence is expressed as a basipetal movement of curvature response.

All these early experiments were conducted using white light and stimulating coleoptiles continuously. After the work of Rothert, oat coleoptiles became the most popular material for phototropism research. It is ironic, however, that this material showed the clearest photoperceptivity of the lower regions among the several materials Rothert examined.

23.5.2 Photoperceptivity distribution in coleoptiles

Sierp and Seybold [145] and Lange [146] irradiated oat coleoptiles with a narrow beam of light to analyze the photoperceptive zone for fPIPP. The sensitivity to light for each zone was determined from the light fluence needed to induce a barely detectable curvature response (i.e. the method of Blaauw). It was found that the extreme tip had the highest sensitivity and the sensitivity dropped sharply within the apical 1-mm region. For example, the 100-μm zone 1 mm below the tip required a more than 100 times higher fluence than the top 100-μm zone. Meyer [92] obtained fluence-response curves by stimulating the top 350-μm zone and the next 350-μm zone. Stimulation of either zone showed a bell-shaped curve depicting fPIPP, but the stimulation of the lower zone gave a smaller peak of response (about 1/3). The results indicated that what declines basipetally is not the actual sensitivity to light, but the capacity of each stimulated zone to produce curvature response. The earlier results mentioned above would then suggest that the zones below the 1-mm tip could hardly respond to light at any fluence to produce fPIPP. Based on these studies, it has generally been believed that the photoperception for fPIPP occurs almost exclusively within the apical 1-mm zone. The fluence-response data of Meyer have additionally indicated that the photoperceptivity for PINP decreases basipetally in parallel to that for fPIPP.

In agreement with the results of Rothert mentioned above, the TDP induced by either continuous stimulation [147–149] or limited stimulation [94] could not be eliminated by shading the tip (3–5 mm). In fact, the tip shaded coleoptiles showed substantial phototropic response during continuous stimulation. Pickard et al. [95] obtained fluence rate-response curves for TDP (45-min stimulation) with and without shading the 1-mm tip. (Coleoptiles were rotated on a horizontal clinostat after stimulation.) Without shading, a bell-shaped curve was obtained. The portion of response due to non-tip photoperception, evaluated from the response decrease by tip shading, was as high as 60% at the optimal and high fluence rates. At the low fluence rates (on the ascending arm of the curve), the portion of response due to non-tip photoperception was substantially smaller. It was assumed that the response by non-tip perception was small at low fluence rates because the fPIPP, which was thought to depend exclusively on the tip for photoperception, contributed to the overall response.

Altogether the available data indicate that the photosensing tissue for fPIPP is located within the apical 1-mm zone whereas that for TDP extends for a much longer distance. Our unpublished data (Y. Murata and M. Iino) demonstrate, however, that not only TDP but also some fPIPP could be induced in tip-shaded oat coleoptiles. These results will be summarized below, because they provide some insights on how the spatial distribution of photoperceptivity is correlated with the response types.

In our experiments, the coleoptiles (about 20-mm long) of red-light-grown seedlings were shaded from unilateral blue light with transparent red film for different distances from the tip (see [150] for the methods used). When fPIPP was induced at a fluence of $10^{-0.52}$ μmol m^{-2}, which induced a response on the ascending arm (see Figure 3), the curvature response was almost entirely eliminated by shading of the 1-mm tip (Figure 7A). At a fluence of $10^{0.30}$ μmol m^{-2}, which induced a response on the descending arm, the curvature response also dropped sharply when the 1-mm tip was shaded, but a portion (about 20%) of the response remained in the tip-shaded coleoptiles; this remaining response declined only gradually with further shading (Figure 7B). Similar experiments on TDP (0.1 and 1 μmol m^{-2} s^{-1}; 20 min stimulation) revealed a portion of response (about 50%) that drops sharply by 1-mm shading and the remaining response that declines gradually with further shading (Figure 7, C and D). These results suggested that the tip and the lower part show distinct fluence-response features. This difference could be clarified by comparing the fluence-response curves obtained by stimulating only the tip (Figure 7E) with those obtained by shading the tip (i.e. stimulating only the non-tip part) (Figure 7, F and G). The fluence-response curve for tip stimulation showed large fPIPP, relatively small TDP, and clear PINP. On the other hand, the fluence-response curve for non-tip stimulation could be characterized by small fPIPP, relatively large TDP, and lack of PINP. It was also noted that the fPIPP induced by non-tip stimulation was located at higher fluences (i.e. showed lower light sensitivity) than the fPIPP induced by tip stimulation. It is true that the photoperceptivity is more confined to the tip in fPIPP than in TDP, but the above results suggest that the fundamental difference rather resides in the region specificity (i.e. the tip vs. the lower part). In fact, the tip-specific properties can largely account for the unique features of the fluence-response relationships in oats (see Section 23.3.3).

A few notes may be added here in relation to the earlier results described above. The results obtained by stimulating within the apical 1-mm zone [92,145,146] probably reflect the properties of the tip-specific fPIPP. The relatively large contribution of the tip found at low fluences under the TDP stimulus condition [95] could still be related to the tip-specific fPIPP as originally assumed.

In red-light-grown maize coleoptiles, the fluence-response curve that covers the entire fPIPP and the ascending part of sPIPP could be produced by stimulating only the 1-mm

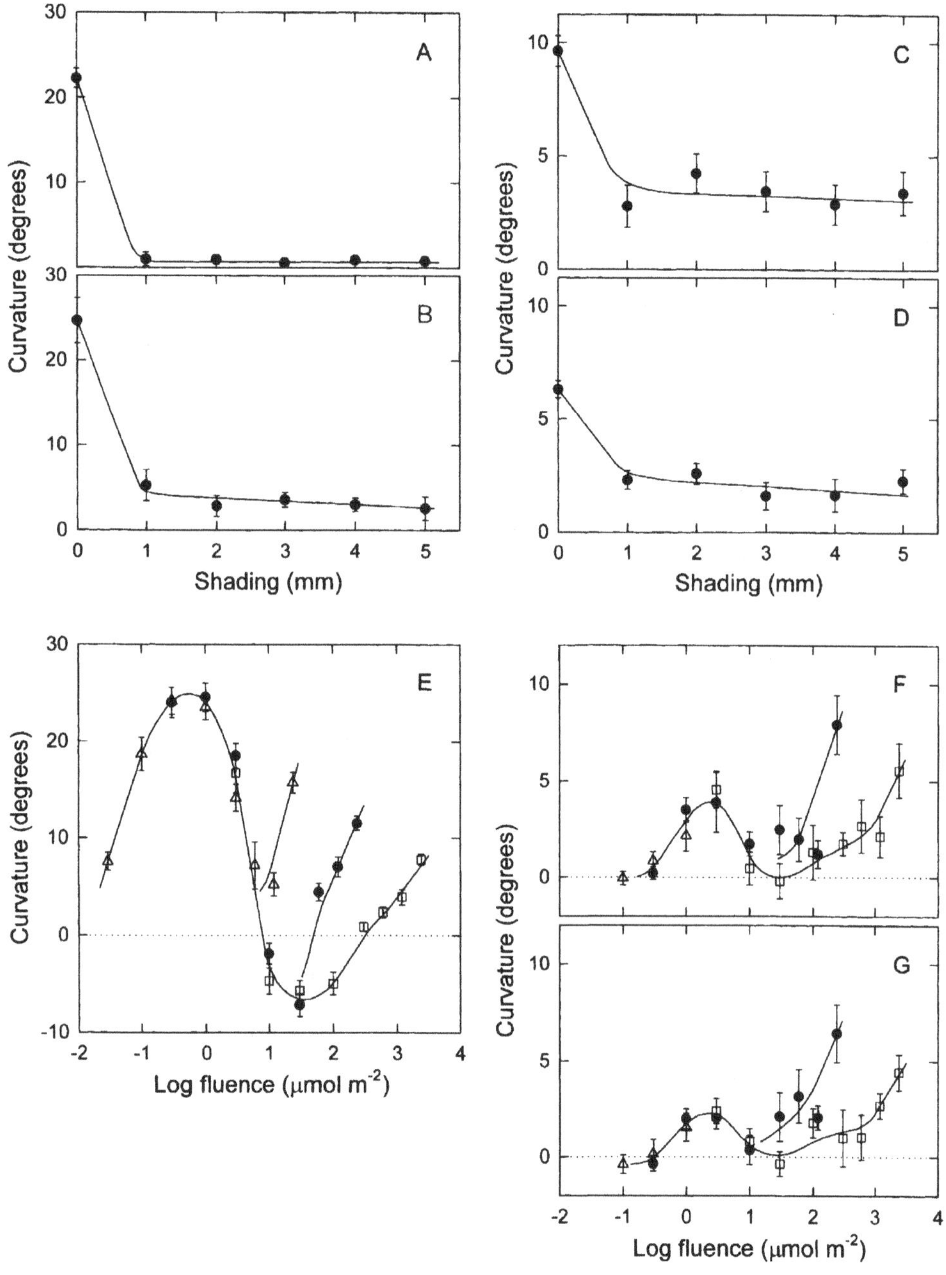

tip, and this curve was similar to that obtained by stimulating the entire coleoptile [151]. The apical 1-mm part appears to be the most photoperceptive zone in both fPIPP and sPIPP of this material. Later work on fPIPP indicated, however, that some photopereceptivity occurs below the apical 1-mm part [152]. Shading the apical 5-mm

Figure 7. Photoperceptivity distribution in oat coleoptiles (Y. Murata and M. Iino, unpublished). Coleoptiles of red-light-grown seedlings were used when 18–22 mm long. Phototropism was induced with unilateral blue light and curvature was measured 120 min after the onset of stimulation. Light sources and growth conditions were as described in Tarui and Iino [195]. The method of shading was as described in Kaldenhoff and Iino [150]. (**A–D**) Effects of tip-shading on the fPIPP induced at a fluence of 0.30 μmol m^{-2} (30-s exposure) (**A**) or 2.1 μmol m^{-2} (20-s exposure) (**B**) and on the TDP induced by 20-min exposure at a fluence rate of 0.1 μmol m^{-2} s^{-1} (**C**) or 1 μmol m^{-2} s^{-1} (**D**). The tip was shaded from unilateral blue light for the indicated distance using a cap made of red plastic. (**E**) Phototropic fluence–response curves obtained by stimulating only the apical 2-mm zone; the lower part was shaded from unilateral blue light with red plastic. The fluence was controlled with exposure times at fixed fluence rates of 0.01 (triangles), 0.1 (circles), and 1 (squares) μmol m^{-2} s^{-1}. (**F** and **G**) Phototropic fluence-response curves obtained by shading the apical 2 mm (**F**) or 4 mm (**G**) zone. The fluence was controlled with exposure times at fixed fluence rates of 0.01 (triangles), 0.1 (circles), and 1 (squares) μmol m^{-2} s^{-1}. The means ± SE from 8–32 plants are shown.

zone from unilateral stimulation reduces the extent of fPIPP by more than 90%, whereas the same shading reduced TDP on average by about 85% [150]. The photoperceptivity for TDP appears to extend more to the lower part than that for fPIPP. In this respect, maize shares a similarity with oats. However, the effect of apical 5-mm shading to reduce TDP is significantly greater in maize than in oats.

23.5.3 Expression of curvature along coleoptiles

When only the tip of oat coleoptiles is exposed to a pulse of unilateral light to induce fPIPP, the curvature response is initiated near the stimulated region and migrates basipetally [153]. Similar results have been obtained for the phototropism induced by continuous stimulation [115]. Dolk [154] was able to observe such basipetal migration of the onset of curvature even when the entire length was stimulated for fPIPP. However, this is not always the case. In red-light-grown oat coleoptiles, basipetal migration was observed when fPIPP was induced at a low fluence on the ascending arm, but such migration was not clear at a higher fluence optimal for fPIPP [155]. The basipetal migration was not clear at the higher fluence probably because the lower zones of red-light-grown coleoptiles could respond to that fluence (see the preceding section). Also, basipetal migration did not follow continuous stimulation of the entire coleoptile [148,149], in agreement with the conclusion that the lower coleoptile zones can perceive the TDP stimulus (see above).

The results summarized above demonstrate that the phototropic signal perceived by the tip is transmitted basipetally and that this signal transmission results in basipetal migration of the curvature response. When the entire coleoptile is stimulated, however, the basipetal migration may be masked by the curvature response induced by non-tip photoperception. All available data indicate that the basipetal migration occurs at a unique velocity in the range between 10 and 30 mm h^{-1} [115,153–155].

The maize coleoptile is another material in which clear basipetal migration of curvature has been demonstrated. When the tip of a red-light-grown maize coleoptile is stimulated for fPIPP, the curvature response begins at the tip and migrates basipetally ([151]; Figure 8). This basipetal migration can be observed even when the entire coleoptile is stimulated [156]. This result is expected because in maize the photoperceptivity is largely confined to the tip at the optimal fluence (see above) or any fluence [151]. The velocity of curvature migration detected in these studies was 20–30 mm h^{-1}.

Macleod et al. [116,157] reported that no substantial migration of phototropic curvature from the stimulated top zone to the lower zones occurred in dark-adapted oat coleoptiles. These workers probably failed to detect a clear basipetal migration because

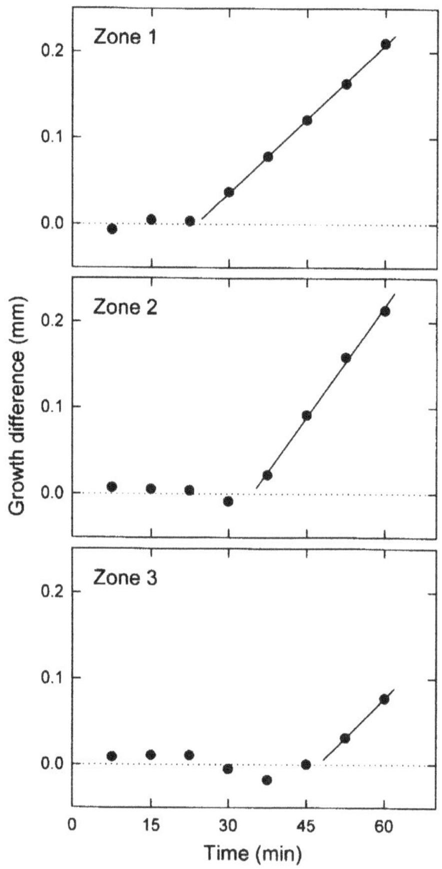

Figure 8. Basipetal migration of phototropic response in maize coleoptiles. The coleoptile of red-light-grown seedlings was marked at 5 mm intervals from the tip (zones, numbered from the tip). The apical 1-mm part was stimulated with a pulse of unilateral blue light (5 μmol m^{-2}) and the growth on either side of each zone was monitored. The data shown are the differences in growth increment between the two sides. The positive sign indicates positive curvature. By the end of the 60-min measurement period, zone 4 had not developed any curvature response. It is noted that the onset of curvature response migrates basipetally. Calculated from the data in Figure 5 of Iino and Briggs [151].

the most photoperceptive part of the tip was located above the marked top zone and was not stimulated. Figure 9 shows the data of Macleod et al. [157] obtained by stimulating each of four zones with continuous blue light. The original growth data are replotted here as the length difference between the two sides of individual zones. Apparently, some curvature response occurred in non-irradiated zones. However, as described by the authors, a large part of the curvature response was confined to the irradiated zone. The results probably indicate that the phototropism of oat coleoptiles characterized by non-tip photoperception does not involve effective longitudinal transmission of the phototropic signal.

23.5.4 Photoperceptivity of decapitated coleoptiles

Oat coleoptiles lose their phototropic responsiveness when decapitated. However, as first noted by Rothert [1], the decapitated coleoptiles again become phototropically responsive some hours after decapitation. This phenomenon has been attributed to the

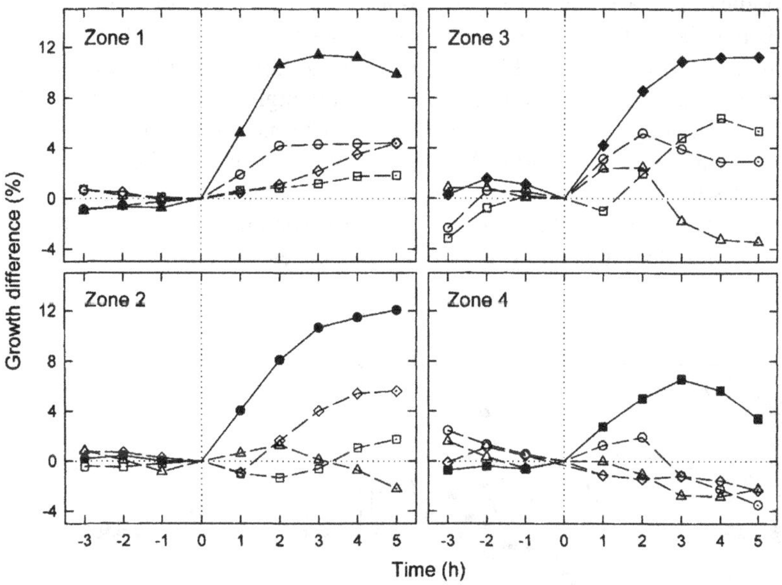

Figure 9. Spatial distribution of phototropic response after localized simulation in oat coleoptiles. The coleoptile of dark-adapted seedlings was divided into zones 1–4 (apical to basal). The apical few mm part was not included in the top zone. The central part of a given zone was stimulated continuously with a unilateral 3-mm beam of blue light and the growth on either side of all zones was monitored. The time zero corresponds the onset of stimulation. The differences in growth increment between the two sides are plotted. The positive sign indicates positive curvature. Each panel shows the data obtained by stimulating the indicated zone with the growth differences determined for zones 1 (triangles), 2 (circles), 3 (diamonds), and 4 (squares). The solid symbols indicate the growth differences in the stimulated zone. Calculated from the data in Macleod et al. [116].

regeneration of auxin-producing activity in the coleoptile stump, or the "regeneration of the physiological tip" [158]. Later studies have substantiated the conclusion that auxin, which is almost solely produced in the tip of intact coleoptiles (see Section 23.10.1), begins to be produced in decapitated coleoptiles [159].

von Guttenberg [160] showed that the phototropic response of decapitated oat coleoptiles to 3-h continuous stimulation became comparable to that of intact coleoptiles when the stimulation was given 10 h after decapitation. In separate experiments, he irradiated decapitated oat coleoptiles with unilateral light for 3 h and decapitated the stumps again. In the subsequent 2-h period in the dark, the stumps developed a small positive curvature (about 6 degrees). When supplied with auxin through the apical cut surface after the second decapitation, the stumps developed a greater curvature (about 25 degrees). These results indicated that the coleoptile stump can perceive and store phototropic stimulus and can express the curvature response once auxin is supplied. Therefore, the restoration of phototropic responsiveness in decapitated coleoptiles was thought to be attributable to the regeneration of auxin-producing activity.

Pohl [161,162] investigated phototropic fluence-response relationships in decapitated oat coleoptiles after a sufficient regeneration time (15 h). The data are reproduced in Figure 10 together with those obtained using intact coleoptiles. Although fPIPP and TDP were clearly induced, no PINP occurred in decapitated coleoptiles. The fluence causing the peak of fPIPP seemed to be a little higher in decapitated coleoptiles. Furthermore, the magnitude of either fPIPP or TDP was only slightly smaller than that in intact coleoptiles. Comparison of these data with those described above (Figure 7, E–G) provides the following suggestions. First, the photoperceptivity of the non-tip part for either fPIPP or TDP is enhanced in response to decapitation. Secondly, the responses of decapitated coleoptiles after photoperceptivity enhancement retain the fluence-

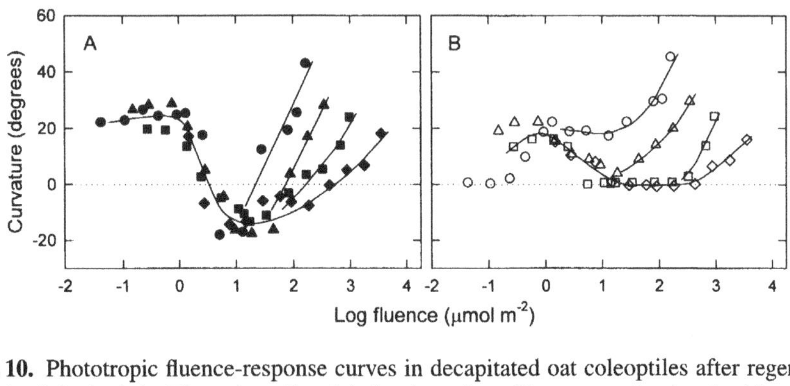

Figure 10. Phototropic fluence-response curves in decapitated oat coleoptiles after regeneration of the physiological tip. The coleoptile of dark-adapted seedlings was stimulated with unilateral blue light. The fluence of blue light was controlled with exposure times at fixed fluence rates of 0.022 (circles), 0.14 (triangles), 0.27 (squares), and 1.4 (diamonds) μmol m^{-2} s^{-1}. (**A**) Fluence-response curves in intact coleoptiles. (**B**) Fluence-response curves in decapitated coleoptiles; the apical 0.2-mm region of dark-adapted coleoptiles was removed and the coleoptiles were subjected to phototropic stimulation 15 h after decapitation. Adapted from Pohl [161].

response features characteristic of the non-tip perception rather than of the tip perception.

Kaldenhoff and Iino [150] re-investigated the effect of decapitation on phototropic responsiveness with red-light-grown maize coleoptiles in which photoperception takes place predominantly in the apical 5-mm zone. It was found that neither fPIPP nor TDP could be induced in the decapitated coleoptile beyond the extent found in tip-shaded intact coleoptiles even if the stump was supplied with auxin. However, the phototropic responsiveness increased sharply from 90 to 120 min after decapitation. These results, together with those described above, indicate that restoration of phototropic responsiveness in decapitated coleoptiles includes an enhancement of photoperceptivity as well as regeneration of an auxin-producing physiological tip. The photoperceptivity enhancement appeared to take place predominantly in the apical part of the coleoptile stump in maize [150]. This is not clear in oats (see above).

23.5.5 Results on hypocotyls and epicotyls

In the phototropism of hypocotyls, the directional light signal is not perceived by the cotyledon but by the hypocotyl itself. This conclusion is based on the results from shading and/or amputation treatments conducted with dark-grown *Sinapis alba* seedlings [163] and de-etiolated seedlings of cucumber [2], sunflowers [2,138], *Lavatera cretica* [43], and cress [105]. Conducting similar experiments with de-etiolated mung bean seedlings, Brennan and Gunckel [164] concluded that the epicotyl perceives the directional light signal for its own phototropism.

No detailed study has been made to resolve the distribution of photoperceptivity within the hypocotyl or epicotyl. In hypocotyls of de-etiolated cress and cucumber seedlings, the curvature response begins rapidly and almost simultaneously in all elongating zones following the onset of continuous stimulation [148,165]. This observation suggests that the photoperception and the curvature response are not spatially separated in the hypocotyl of de-etiolated plants.

It has been noticed that phototropic curvature of hypocotyls is more confined to the upper part in dark-grown seedlings as compared to de-etiolated [105] or red-light-pretreated [4] seedlings. This difference is probably related, at least in part, to the difference in growth rate distribution. The hypocotyl of dark-grown seedlings shows, at least at the age used for phototropism experiments, faster growth in the apical part than in the lower part, whereas growth is more or less evenly distributed in red-light- or white-light-grown hypocotyls [166,167].

Although phototropic curvature of hypocotyls is confined more to the apical zone in dark-grown seedlings, there is some evidence that the phototropic signal is transmitted basipetally in these hypocotyls. Shading of the apex and cotyledons prevented the phototropism of dark-grown cress seedlings induced by continuous stimulation [105]. Since cotyledons cannot perceive the directional light signal (see above), the result suggests that the photoperceptivity is concentrated in the very apical part of the hypocotyl. Woitzik and Mohr [4] observed that the phototropic response of sesame hypocotyls to continuous stimulation was initiated earlier in the more apical part; this tendency was clearer in dark-grown plants than in red-light-grown plants. Orbovic and

Poff [168] observed that the fPIPP curvature of dark-grown *Arabidopsis* hypocotyls was first initiated in the apical part and migrated basipetally. Dark-grown seedlings might perceive the phototropic signal predominantly at the apical part of hypocotyls and transmit the signal to lower parts.

23.5.6 Blue-light-induced enhancement of phototropic responsiveness

As we have already seen (Sections 23.4.2–4), phytochrome causes profound effects in enhancing or establishing the phototropic responsiveness. There is evidence, however, that a blue-light-receptor system also contributes to the establishment of phototropic responsiveness. Franssen and Bruinsma [138] observed that etiolated sunflower hypocotyls showed only a trace amount of phototropism to continuous phototropic stimulation, but that a high phototropic responsiveness is brought about when the hypocotyls were pretreated with overhead white light or blue light. Pretreatment with red light had only a partial enhancement effect. The results suggest that, in addition to the phytochrome-mediated mechanism, a blue-light-sensitive mechanism contributes to the enhancement. Shading experiments indicated that neither the cotyledon nor the shoot apex but rather the hypocotyl itself perceives the light for this pretreatment effect.

Blue-light-sensitive enhancement of phototropic responsiveness may also take place in coleoptiles, especially accounting for the substantial phototropic response induced by continuous stimulation of the lower coleoptile zones (see Sections 23.5.1 and 23.5.2). This view [90] has become very probable in light of the recent finding of Salomon et al. [169] (see Section 23.9.4). The fact that blue light enhances phototropic responsiveness in lower coleoptile zones may be found in the result of Brauner [147,170]. When the subapical 5-mm zone of oat coleoptiles was irradiated continuously with unilateral light, the coleoptile developed a relatively small curvature in the first hour, but the rate of curvature increased and became comparable in the subsequent hour to that induced by stimulation of the apical 5-mm zone. This result was obtained with the coleoptiles filled with India ink, which enhanced the phototropic response attributable to non-tip photoperception (see Section 23.6.1). Although Brauner used white light to induce phototropism, the seedlings were manipulated under orange light and adapted to darkness for 1 h before phototropic stimulation. This is exactly the procedure used to maximize the red-light-pretreatment effect, and the phytochrome-mediated enhancement response was likely to be saturated before phototropic stimulation. The substantial phototropic curvature of decapitated coleoptiles recorded by von Guttenberg after 3-h unilateral irradiation ([160]; see Section 23.5.4) may also include a blue-light-specific enhancement effect.

23.6 Perceiving the directional light signal

Every photobiological system in plants can sense the quality and quantity of light. What is unique about phototropism is that in addition the system can sense the direction of light. We will see how this is achieved.

23.6.1 Significance of internal light gradients

Buder [171] used a kind of light pipe to irradiate one side of an oat coleoptile tip from inside the coleoptile cavity. In this way, he created a light gradient between the two sides with the light direction opposite to the normal situation. He found that the coleoptile bent towards the irradiated side as it did when irradiated from the outside. This result demonstrated that the coleoptile does not obtain the directional light information from the direction of impinging light but from the internal light gradient between the two sides. The significance of the light gradient between the two sides has also been indicated by experiments in which one side of the oat coleoptile tip was irradiated with vertical light from the top [171] or with a unilateral light beam [92].

Recently, Nick and Furuya [172] applied a modern light-piping device to maize coleoptiles to repeat the experiments by Buder. They used dark-adapted coleoptiles and obtained fluence-response curves, with three fluence rates of blue light, for irradiation from outside and inside the coleoptile tip. Irradiation from outside revealed a phototropic fluence-response curve typical of fPIPP (the peak at about 1 μmol m^{-2}) and this curve did not depend on the fluence rates. Irradiation from inside also showed a peak of response at comparable fluences. The curvature occurred away from the light source, thus confirming Buder's result. However, in this case, the peak was smaller when the fluence rate was higher. In addition, a relatively small curvature to the opposite direction (i.e. not to the brighter side but to the direction of light) was induced at high fluences (3–15 μmol m^{-2}). This curvature response formed a peak in the fluence-response curve, but the size and position of the peak changed depending on the fluence rate used. These results could not be explained. The curvature observed at high fluences might indicate that the coleoptile tip can additionally respond to the light gradient occurring within its half side.

Some lines of evidence indicate that the extent of the light gradient between the two sides determines the extent of phototropism. The earliest results were provided by Bünning et al. [173], who could demonstrate that decapitated coleoptiles show greater phototropic response when the primary leaf is not removed and when the cavity is filled with a solution of dyes. Brauner [147,170] carried out many experiments to demonstrate the significance of a light gradient. The phototropic response of oat coleoptiles to continuous stimulation was larger when the coleoptile cavity was filled with India ink rather than with water, and also when the coleoptile (elliptical in cross section) was irradiated from the narrow side rather than from the broad side. Furthermore, when the coleoptile filled with India ink was irradiated from either the narrow or the broad side, the responses in the two cases became comparable. In fact, the data of Brauner [147] indicate that the phototropic response attributable to non-tip photoperception is especially enhanced by the ink (see his Table 2). Using the decapitated coleoptiles which had been irradiated for 3 h with unilateral light and supplied with IAA, von Guttenberg [160] could show that the phototropic response was less when the primary leaf in the coleoptile cavity was removed prior to phototropic stimulation. On the other hand, the response was greater when the primary leaf was replaced by black paper. The light gradient is relatively independent of the wavelength in the blue region of the light spectrum, but decreases in the green region; this decrease in light gradient could be correlated to a reduced maximum of fPIPP [174]. Treatment with the herbicide norflurazon, which inhibits carotenoid biosynthesis, reduced the extent of TDP [175] and fPIPP [90,176]. Although the results obtained using norflurazon may have to be treated with caution [141], the observed effects of norflurazon are probably attributable to a reduction in the internal light gradient that results from a reduction in screening pigments (i.e. carotenoids). In fact, norflurazon was much less effective in inhibiting the TDP induced by UV-A, which is not absorbed by carotenoids [175]. All of these results provide correlative evidence for the significance of the internal light gradient in determining the extent as well as the direction of the phototropic response.

In single-cell organs such as the *Phycomyces* sporangiophores, unilateral light is focused onto the shaded side by the so-called lens effect of the organ, and the shaded side is sensed as the brighter side for the determination of curvature direction (reviewed in [177]; see also [178]). Such a lens effect has once been suggested to play a role in the phototropism of oat coleoptiles ([179]; but see [90] for an interpretation of the results). The actual light-gradient measurements made in coleoptiles and hypocotyls have, however, indicated that light is not focused onto the shaded side [140,180].

It is now generally believed that plants acquire the directional light information for the phototropism of coleoptiles and stems from an internal light gradient. The conclusion concerning the significance of the gradient between the two sides of an organ is almost exclusively based on the results from oat and maize coleoptiles. It is not clear whether the same applies to the phototropism of stems. In this case, it remains possible that the light gradient within a narrow radial region plays a significant role.

23.6.2 Light gradient measurements

Efforts have been made to measure the light distribution within a unilaterally irradiated organ. Because light is scattered in tissue, any point inside the organ receives light from all directions. At the moment, there is no perfect method to allow exact determination of the light fluence rate at a narrow tissue spot.

The most detailed investigation of the light gradient in the organ unilaterally irradiated with blue light was carried out by Vogelmann and Haupt [180]. They used a fiber optic probe to obtain the lateral distribution of fluence rates at different parts of dark-grown maize coleoptiles. At the dome-shaped solid part of the coleoptile tip (a few hundred μm below the top surface), an approximately linear light gradient was found (Figure 11). The ratio of the fluence rate at the irradiated side to that at the shaded side was about 1 : 0.25. (Here, the word "side" means just inside the outer surface.) In the lower part, where the coleoptile is hollow or filled with primary leaves, the gradient profile was more complex. At the basal part filled with the primary leaf, the fluence-rate ratio between the two sides was about 1 : 0.13. This study has also revealed that, owing to the light scattering, the fluence rate at the irradiated side is substantially greater than the incident fluence rate (Figure 11c).

Cosgrove [140] used a fiber optic probe to measure the light gradient in dark-grown cucumber hypocotyls. He estimated the fluence-rate ratio between the two sides to be about 1 : 0.15. Kunzelman et al. [174] used in situ phototransformation of phytochrome to measure the light gradient in dark-adapted maize coleoptiles; vertical slices of the apical 1-cm part of the coleoptile were subjected to a phytochrome assay after unilateral

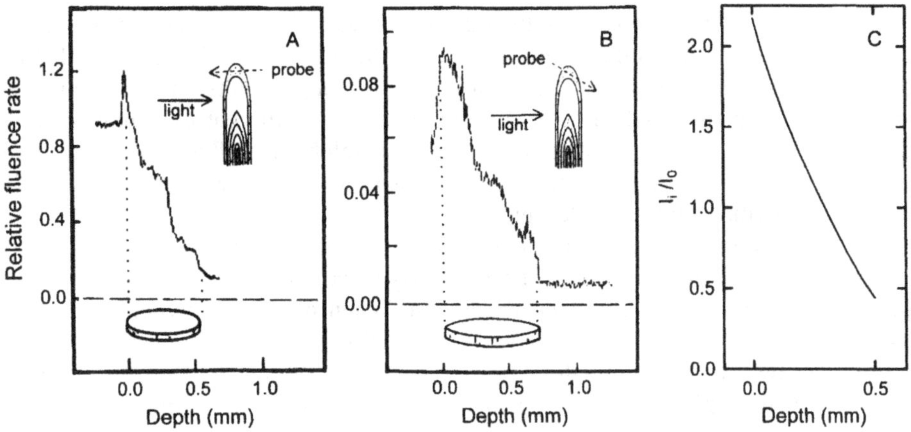

Figure 11. The light gradient in the tip of dark-grown maize coleoptiles unilaterally irradiated with blue light (450 nm). Unilateral blue light was given parallel to the symmetrical plane of the coleoptile (i.e. the plane passing through the two vascular bundles).A fiber optic probe was used to measure the fluence rate distribution along the central plane of the tip; the probe was inserted from the shaded side (**A**) or the irradiated side (**B**) and moved towards the other side as illustrated. Based on the measurements, the internal gradient in net fluence rate was estimated (**C**). Adapted from Vogelmann and Haupt [180].

irradiation of the coleoptile. The fluence-rate ratio between the two sides was found to be 1 : 0.12. Parsons et al. [181] reported much greater gradients, exceeding 1 : 0.02, for all cases examined (dark-adapted oat coleoptiles, and dark-grown and light-grown hypocotyls of sunflowers and *Vicia faba*). They measured the light leaving the cut surface through a narrow slit at different distances from the outer surface on which unilateral light was incident. With this method, back-scattered light is neglected from distribution measurements, but this is unlikely to be the cause of the disagreement. The high estimate of light gradients was probably obtained because only a spot of the organ surface was irradiated with a light beam; in this case, the gradient is expected to be greater than that occurring in the organ widely irradiated with unilateral light. To conclude, the gradient of blue light between the two sides of organs is generally between 1 : 0.12 and 1 : 0.15. In the coleoptile tip, the gradient is somewhat smaller.

The light gradient across either etiolated or de-etiolated hypocotyls was similar for blue light and red light [181]. On the other hand, Kunzelmann and Schäfer [182] found that the gradient is steeper with blue light than with red light in dark-adapted maize coleoptiles (see also [174]). If a steeper light gradient were to occur with blue light, the use of a blue-light receptor would be advantageous for phototropism. This interesting aspect needs to be investigated further.

Internal light gradients can be generated by light scattering within the tissue and by light absorption by screening pigments. As mentioned above, the specific inhibition of the blue-light-induced phototropism by the carotenoid biosynthesis inhibitor has been interpreted on the assumption that screening pigments contribute to the light-gradient formation. Light distribution measurements in green tissues indicate that the light absorption by pigments steepens the light gradient [183]. However, Parsons et al. [181] concluded that scattering is the prime cause of the light gradient. They showed that the light gradient across the hypocotyl of sunflowers and *Vicia faba* was similar for etiolated and de-etiolated plants, and that the light gradient across the oat coleoptile was not significantly affected by the carotenoid-biosynthesis inhibitor norflurazon. At present it seems that scattering is the primary cause and absorption is the secondary cause of the internal light gradient in coleoptiles and hypocotyls.

The light given to cylindrical organs is transmitted axially within the organs [184,185]. Such an optical property of organs may be critical when the photoperceptive region is investigated with localized irradiation. When oat and maize coleoptiles are exposed to unilateral light such that the light direction is parallel to the plane passing through the two vascular bundles (i.e. the standard orientation for phototropism experiments), the light transmitted axially from the dome-shaped tip to lower parts cannot form a gradient between the two sides of the coleoptile [184]. The unilateral light given to hypocotyls is sharply attenuated as it is scattered axially [181]. These data indicate that the phototropic response induced by localized unilateral irradiation largely reflects the photoperceptivity of the zone irradiated. When unilateral light is given continuously without angular adjustment, however, the light-acceptance angle becomes sharper as the organ develops curvature, and the axially transmitted light may form a lateral gradient in non-irradiated zones [184].

23.6.3 Photoproduct-gradient model

We do not yet know any molecular and cellular mechanisms by which plants sense the internal light gradient. However, it would be a reasonable assumption that light-gradient sensing is somehow based on the gradient in the extent of light-induced reactions or in the amount of a photoproduct that occurs across the organ [186]. (The term photoproduct is used here to describe a molecular component or a state of a molecule that is generated in response to light. The photoproduct does not necessarily mean the direct product of a photochemical reaction.)

A model based on the photoproduct gradient has been explored to explain the bell-shaped fluence-response curve of fPIPP [113]. The model considers a photoproduct that limits the subsequent dark reactions at the site of photoperception. It is assumed that the difference in the amount of this photoproduct between the two sides of the photosensing zone determines the phototropic response. Following these assumptions, the bell-shaped fluence-response curve is thought to span from the threshold of the photoproduct formation on the irradiated side to its saturation on the shaded side, the peak of the curve occurring at the fluence causing the largest difference in the photoproduct (Figure 12). The photoproduct-gradient model formulated with second-order kinetics of the photoproduct-formation reaction could be well fitted by the fluence-response data from maize coleoptiles and could predict the fluence-response data obtained with two-sided irradiation [113].

The computed light gradient showed a reasonable agreement with the fluence-rate gradient estimated by Vogelmann and Haupt [180] in the maize coleoptile tip (see

Figure 11C). The photoproduct-gradient model predicts that a reduction in the light gradient results in a reduced extent of fPIPP, especially at the peak and the descending arm of the fluence-response curve [113]. This was, in fact, found when the light gradient was reduced by simultaneous irradiation from the opposite side [174] or by treatment with norflurazon [90].

Hofmann and Schäfer [124] investigated the red-light-induced desensitization of fPIPP by providing red-light pretreatment unilaterally from the same or the opposite direction to the inductive unilateral blue light. The effect of red light differed substantially depending on its direction. The general conclusion is that the effect of red light is local to the tissue site of blue-light perception. Although the authors hypothesized that fPIPP is based on at least two photosystems that are differently modulated by red light, their results can in fact be interpreted in terms of the photoproduct-gradient model. The major results of Hofmann and Schäfer (their Figs. 2, 3 and 6) are predictable if it is assumed that phytochrome locally desensitizes the photoproduct-formation reaction at the irradiated and the shaded side.

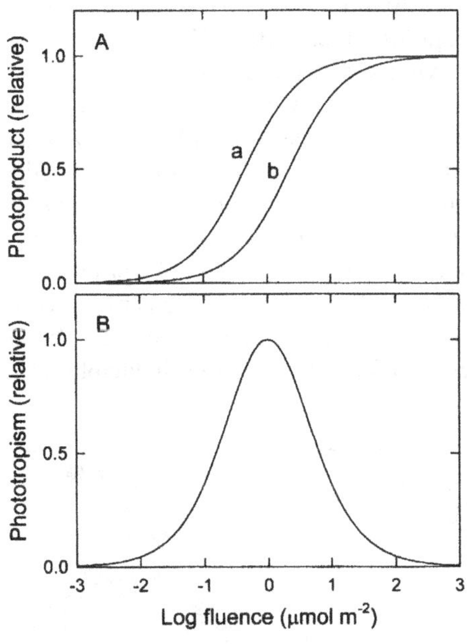

Figure 12. The photoproduct-gradient model describing the bell-shaped fluence-response curve for fPIPP. This model first assumes a photoproduct that limits the subsequent dark reactions at the tissue site of photoperception. When plotted as a function of the fluence of unilateral light, the formation of this hypothetical photoproduct follows the saturation curves diagrammatically shown in **A**, with curve **a** on the irradiated side and curve **b** on the shaded side. It is next assumed that the phototropic response is a function of the difference in photoproduct concentration between the two sides. Ideally this difference is linearly related to the phototropic response. Then curve **b** is subtracted from curve **a** to yield a bell-shaped curve that represents the phototropic fluence-response curve (**B**). The curves in **A** were produced using a hyperbolic function [113] and assuming a light gradient of 1:0.2 (irradiated:shaded side).

23.7 The process of curvature development

This section summarizes some basic information obtained by analyzing the kinetics of curvature development.

23.7.1 Lag period

Phototropic curvature of an organ is initiated after a lag period (also called a latent period) following pulse stimulation or the onset of continuous stimulation. In most cases investigated, the curvature response begins at or near the stimulated zone (see Sections 23.5.3 and 23.5.5). The only exception to this rule is the negative phototropism of oat coleoptiles described by Taylor et al. [115] (see Section 23.3.4). Therefore, in general, the lag period found in an organ represents the minimal time needed before the perceived light signal is transduced to a curvature response without involving longitudinal signal transmission.

Early work on dark-adapted oat coleoptiles has indicated that fPIPP curvature is initiated with a lag of about 40 min [153,154]. However, the lag is as short as 20–25 min in red-light-grown coleoptiles of oats [93,155], maize [151], and rice [69]. In rice coleoptiles, TDP was expressed with a lag similar to that of fPIPP [69]. The development of fPIPP is terminated when the coleoptile is treated with high-fluence bilateral light. In red-light-grown maize coleoptiles, such a bilateral treatment interferes with fPIPP after a lag similar to the one found for the induction of fPIPP (Figure 13). Therefore, the initial inductive signal and the subsequently imposed signal are transduced to the final step of phototropism with a similar lag period.

Pickard et al. [95] noted that the TDP of oat coleoptiles is initiated with a lag period of about 3 min, which is much shorter than the lag period recorded for fPIPP. This short lag means that the coleoptile begins to bend once it has been exposed to light for a minimal time that is needed before TDP becomes inducible. The short lag was estimated

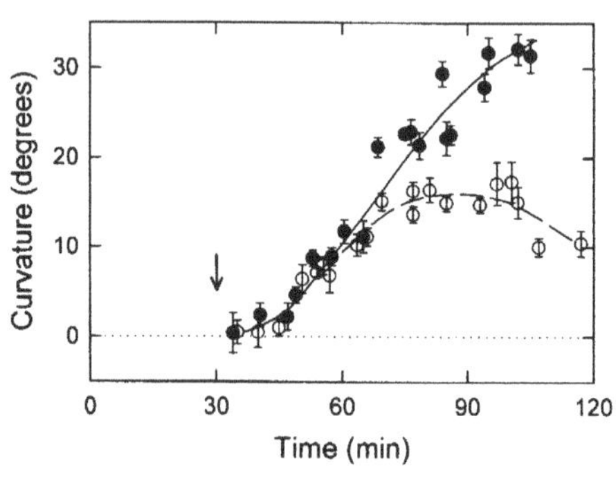

Figure 13. The effect of a bilateral pulse of blue light on fPIPP in maize coleoptiles (M. Iino, unpublished). The coleoptile of red-light-grown seedlings was stimulated at time zero with a 3-s pulse of unilateral blue light (2.6 μmol m^{-2}). Open circles: coleoptiles were again treated at 30 min with a 30-s bilateral pulse of blue light (450 μmol m^{-2}). Closed circles: coleoptiles were not treated with the bilateral pulse. Other experimental details were as described in [113].

after linear transformation of the response at an early parabolic phase of curvature development. No comparable analysis has been made for fPIPP, precluding direct comparison. However, the data of Franssen et al. [149] also indicate that phototropic curvature begins immediately following the onset of continuous stimulation (or at least, with a lag much shorter than 30 min). The lag period of TDP may, in fact, be shorter than that of fPIPP. Pickard et al. [95] examined the lag period under the stimulus condition with which the contribution of non-tip photoperception was substantially. Franssen et al. [149] could record the rapid response at all coleoptile zones. It is an intriguing possibility that the short lag may characterize the TDP induced in the coleoptile by non-tip photoperception (see Section 23.5.2).

The hypocotyl of dark-grown *Arabidopsis* seedlings showed fPIPP with a lag of about 15 min; the lag was somewhat shorter (about 10 min) when seedlings were pretreated with red light [168]. The hypocotyl of dark-grown sesame seedlings showed TDP with a lag of about 30 min; the lag was shorter (about 15 min) in red-light-grown seedlings [4]. The shorter lag period found in red-light-pretreated or red-light-grown seedlings is comparable to, or a little shorter than, the lag period generally found for the fPIPP of red-light-grown coleoptiles.

Hypocotyls of de-etiolated seedlings appear to express TDP with a lag period that is still shorter than that found in red-light-pretreated seedlings. When de-etiolated hypocotyls were subjected to continuous phototropic stimulation, the curvature response was already substantial at the earliest measurement time of 20 min for sunflowers [138] or 30 min for cress [165]. Curvature began as early as 4 min after the onset of stimulation in *Fagopyrum esculentum* [3] and within 15 min in *Sinapis alba* [187]. Feyerabend and Weiler [139] found that the lag period in de-etiolated sunflower hypocotyls was age-dependent; the shortest lag period recorded in 5-day-old seedlings was about 5 min. Therefore, the lag period of TDP can be as short as 4–5 min in de-etiolated hypocotyls. Again, this short lag period is comparable to the minimal stimulation time that is generally needed before TDP becomes inducible.

In the above-mentioned studies with de-etiolated plants, phototropic stimulation was given during the day period of artificial day/night cycles after a brief dark adaptation. Under such conditions, TDP may occur without the minimal stimulation time that is needed when TDP is induced in dark-adapted plants (but with appropriate red-light pretreatment). This is supported by the result of Hart et al. [165] indicating that pretreatment of de-etiolated cress hypocotyls with 4-h darkness leads to a longer lag period (about 30 min). In fact, the minimal stimulation time was much longer in etiolated *Fagopyrum esculentum* seedlings than in de-etiolated ones [106]. Even if the lag period of TDP does not include the minimal stimulation time, it may still be expected that the lag period of TDP is not shorter than that of fPIPP if both types of phototropism share the same signal transduction. It is yet to be determined whether fPIPP can be induced in hypocotyls of de-etiolated plants with a lag as short as 4–5 min.

The PINP of oat coleoptiles is characterized by a long lag. In red-light-grown oat coleoptiles, the lag period of PINP is about 50 min while that of fPIPP is shorter than 30 min [93]. When coleoptiles are stimulated with a fluence in the high fluence range of the descending arm of fPIPP, an initially positive curvature is later overtaken by a curvature to the opposite direction, indicating that fPIPP and PINP can be induced consecutively in the same plant. In fact, it has long been known that at intermediate fluences between fPIPP and PINP, the coleoptile becomes S-shaped, with positive curvature in the lower part and a negative curvature in the upper part [91,188]. It is apparent that S-shaped curvature results because fPIPP and PINP are induced with different lag periods and because the major part of either phototropism begins in the tip and migrates basipetally.

23.7.2 Persistence of curvature response after stimulation

One interesting property of phototropism is that curvature develops over a substantially long period after a brief light stimulation. This is most apparent in pulse-induced phototropisms. When stimulated for fPIPP, plants develop curvature after a lag for a period of 1–1.5 h, as observed in coleoptiles of oats, maize, and rice [69,93,96], hypocotyls of *Arabidopsis* [168], and epicotyls of peas [102]. The other pulse-induced phototropisms of coleoptiles, PINP and sPIPP, also show similarly sustained curvature development [93]. A long sustained curvature response after phototropic stimulation is also a property of TDP. Although TDP is characterized by the requirement for extended stimulation, the stimulation-time-dependent increase in response typically occurs with stimulation shorter than 30 min, and the major part of the curvature response follows the stimulation. For example, curvature develops for 1–1.5 h in coleoptiles of oats [95] and rice [69] after the termination of TDP stimulus.

As an initially erect organ develops positive phototropism, it becomes subjected to gravitropic stimulation. The ensuing negative gravitropism will then counteract the development of positive phototropism. Accordingly, the coleoptiles stimulated for fPIPP or TDP return to the original erect orientation after establishing a maximal curvature [69,95,189,190]. The phototropic response in the absence of gravitropic counteraction has been evaluated using horizontal clinostats, following the notion that the rotation on the clinostat nullifies the gravitropic counteraction. In fact, when rotated on horizontal clinostats after phototropic stimulation, oat coleoptiles can continue to bend for 5–6 h after fPIPP stimulation, [153,188,191] or following the termination of TDP stimulation [192]. A still longer curvature development (about 13 h) has been recorded for the fPIPP of maize coleoptiles [190].

In contrast to the above results, other workers could not find a long sustained curvature on clinostats. Pickard et al. [95] observed that the TDP of oat coleoptiles could develop for no more than 3 h on a clinostat. Steiniz et al. [193] also showed that the fPIPP of oat coleoptiles could continue only for about 3 h. The overall curvature, however, did not diminish after establishing the maximum within the period in which upright seedlings would show a reversal of curvature. Clearly, coleoptiles develop curvature over a longer time and remain curved for longer when rotated on horizontal clinostats, but the results have been contradictory as to how long the curvature response can continue. The reason for the different results is not clear. The speed of clinostat rotation usually ranged between 0.5 and 3 rpm, and there is no clear indication that the different results are related to different rotation speeds. Although red-light pretreatment seemed to somewhat enhance the duration of curvature on a clinostat [193], Pickard et al. [95] recorded the short duration using red-light-pretreated seedlings. The possibility that the duration of curvature development is limited by the maximally attainable curvature also cannot account for the difference (see Figure 1 in [95]).

If the effect of clinostat treatment on phototropism merely represents a removal of gravitropic counteraction, one would predict that the early time-course of phototropic curvature is not very much affected by clinostat rotation. This is because the counteracting gravitropism is induced and begins to affect phototropic curvature after a lag period (typically 20–30 min) and also because the organ receives gravitropic stimulation only gradually as it develops phototropic curvature [194,195]. In agreement with the above prediction, the data of Pickard et al. [95] indicate that clinostat-treated and stationary oat coleoptiles develop TDP curvature similarly for about 40 min. Also, in maize coleoptiles, the fPIPP curvature measured 100 min after the onset of stimulation is not so different between the plants left vertical and those rotated on a horizontal clinostat [110]. However, other results indicate that a substantially greater curvature may already occur at an early stage of clinostat rotation. Shen-Miller and Gordon [191] observed that the fPIPP curvature of oat coleoptiles measured 2 h after the onset of stimulation was about 3 times higher in clinostat-rotated plants than in vertical stationary plants. Heathcote and Bircher [98] observed that either the fPIPP curvature or the TDP curvature of wheat (*Triticum aestivum*) coleoptiles, measured 100 min after the onset of stimulation, was 2–3 times greater in clinostat-rotated plants than in vertically stationary plants.

Härtling [196] has demonstrated that the effect of clinostat rotation on phototropism is not merely a removal of gravitropic counteraction. He found that the phototropic response of sunflower hypocotyls to 1-h phototropic stimulation is substantially enhanced by 10-h pretreatment of the seedlings with a horizontal clinostat. Shen-Miller and Gordon [191] also noted that the fPIPP of oat coleoptiles, allowed to develop on a horizontal clinostat, was somewhat greater when the seedlings were pretreated with the clinostat. These results indicate that phototropic responsiveness is enhanced by clinostat rotation. This effect might, in part, explain the apparently different results described in the preceding paragraph. For example, Pickard et al. [95] rotated oat coleoptiles on the clinostat after TDP stimulation. On the other hand, Heathcote and Bircher [98] adapted wheat seedlings to clinostat rotation for 5 h before phototropic stimulation. In the work mentioned above, Härtling [196] also found that the clinostat treatment enhanced the growth of sunflower hypocotyls. As suggested by the author, this effect might explain the enhancement of phototropic responsiveness. However, Shen-Miller and Gordon [191] found that growth is inhibited, though slightly, by clinostat treatment. Therefore, the effect of clinostat treatment on phototropic responsiveness cannot generally be accounted for by a growth-stimulating effect.

Mere rotation on horizontal clinostats causes a slow and long-sustained curvature in coleoptiles of oats [146], maize [197], and rice [69]. The curvature develops to the predetermined direction along the plane of morphological asymmetry (i.e. the plane perpendicular to the plane of the vascular bundles) and is a nastic response by definition[4]. This nastic curvature could interfere with the measurement of phototropic curvature on clinostats, but this potential problem has received little attention. In the clinostat experiments described above, apart from those by Heathcote and Bircher (see below), phototropism was induced in the plane of morphological symmetry. Therefore, the interference by nastic curvature could be the least. However, Nick and Schäfer [197] showed that the nastic curvature of maize coleoptiles deviates from the asymmetric plane as it develops. This deviation, observed in random directions, is probably caused by twisting of the coleoptile. In phototropically stimulated coleoptiles, this twisting might occur predominantly towards the direction of phototropic curvature, resulting in an overestimation of phototropic curvature. This possibility is supported by the fluence-response data indicating that the curvature in the asymmetric plane decreases as the phototropic curvature in the symmetric plane increases [197]. The nastic curvature that could be sustained in the twisted coleoptile might have contributed to the exceptionally long-sustained phototropic curvature of maize coleoptiles on the clinostat [190].

Heathcote et al. [198] used the microgravity environment in a space laboratory to investigate the phototropism of wheat coleoptiles in the absence of gravitropic stimulation. During the initial growth stage, seedlings were subjected to 1-g centrifugal acceleration. The acceleration was terminated, and phototropic stimulation was initiated after 5-h adaptation to microgravity. The direction of the phototropic stimulus was parallel to the plane of morphological asymmetry. Under microgravity, the coleoptile developed a clear nastic curvature in the same plane [199]. However, the nastic curvature had been completed during the 5-h pre-adaptation period, and phototropism could be induced in the same direction without accompanying the nastic curvature. It was found that both fPIPP and TDP curvatures measured 100 min after the onset of phototropic stimulation were similar to those induced under the ground, 1-g condition [198]. Under the ground condition, the coleoptile more-or-less ceased to develop phototropic curvature 100 min after the onset of phototropic stimulation. Under microgravity, the curvature continued for a longer period, but not very much. These results were surprising because the wheat coleoptiles had developed substantially greater curvature by 100 min after the onset of stimulation when rotated on horizontal clinostats [98].

As in the case of the space experiments, Heathcote and Bircher [98] obtained data on clinostat-treated wheat coleoptiles by pre-adapting the seedlings to clinostat rotation for 5 h and stimulating the coleoptile with unilateral light in the plane perpendicular to the plane of vascular bundles. It was not reported whether or not a nastic curvature had also occurred on the clinostat. It is possible that the results were somewhat complicated by the accompanying nastic curvature, which might have occurred differently between clinostat and microgravity conditions. In the microgravity experiments, the direction of the phototropic stimulation was perpendicular to the direction of the 1-g acceleration used initially to orient coleoptile growth. Therefore, in contrast to the case of ground control experiments, the direction of blue light for phototropic stimulation was not perpendicular to the long axis of coleoptiles under microgravity in which the coleoptile had developed nastic curvature. Although these points need to be evaluated carefully, the results of Heathcote and Bircher probably indicate that wheat coleoptiles do not develop phototropic curvature under microgravity to the extent that they do on horizontal clinostats. This issue will be discussed further in the next section.

23.7.3 Autostraightening

As stated above, the mere removal of gravitropic counteraction could not consistently explain the phototropic curvature investigated on clinostats and in microgravity. In fact, another response that also counteracts phototropic curvature appears to provide a clue to our fuller understanding of the overall process of phototropism. The occurrence of this response has been better substantiated for gravitropism. The results from gravitropism research will be summarized first.

Gravitropically responding coleoptiles and hypocotyls finally assume a straight and vertical appearance, with a curvature retained at the basal part. This is achieved because the apical to middle parts straighten after showing upward curvature. Changes in the strength and the sign of gravitropic stimulus within the responding organ could contribute to the straightening [194]. However, a response that is autonomic in nature rather than gravitropic has been shown to make a greater contribution to the straightening [154,195,200,201]. The occurrence of such an autonomic response was first demonstrated by the result that gravitropically bent oat coleoptiles can straighten on a horizontal clinostat [154]. Maize coleoptiles also underwent straightening on a clinostat [190], although its extent depended somehow on the batches of maize caryopses used [202]. Oat and wheat coleoptiles can straighten before any of their parts reach the vertical position, indicating that the straightening response is entirely autonomic in these organs [195]. It could be demonstrated that the autonomic response actively counteracts the gravitropic curvature [195]. (Here the term autostraightening is used to refer to the autonomic straightening response, which has traditionally been described by the term autotropism [155]. This term has been introduced because the straightening response is not a true tropism in the sense that it is not directly induced by an external stimulus [203].)

In view of the observation that the apical to middle parts of oat coleoptiles straighten during continuous phototropic stimulation, Franssen et al. [148] have suggested that autostraightening also takes place during phototropism (see also [204]). However, counteracting gravitropism could theoretically lead to the observed organ-straightening. If autostraightening were to take place in phototropically bent organs, a straightening response should be observable on horizontal clinostats. No zone of the oat coleoptile straightened on a clinostat after showing an fPIPP curvature within the time span in which clear straightening was found in gravitropically bent coleoptiles [154]. However, earlier studies had clearly indicated that the coleoptiles could straighten when rotated on a clinostat for much longer periods [153,188].

Tarui and Iino [155] reinvestigated the possible occurrence of autostraightening in phototropically bent oat coleoptiles. It was found that when the coleoptile was stimulated with a low fluence that causes an fPIPP curvature on the ascending arm of the fluence-response curve, the most apical 5-mm zone straightened after showing a maximal curvature at about 90 min. On the other hand, no such straightening could be found when the coleoptile was stimulated at a fluence that was optimal for fPIPP. Since it was possible that autostraightening was less apparent because the signal for curvature response remains active for a long period in the case of phototropism, bilateral treatment with a high-fluence blue light, which itself caused no curvature response, was carried out 25 min after phototropic stimulation. In this case, clear straightening could be detected in the apical two 5-mm zones even at the optimal fluence. These results support the idea that autostraightening takes place in phototropically bent coleoptiles. In agreement with the view of Firn and Digby [200] and the model of Chapman et al. [201], it was deduced that the autostraightening of coleoptiles is a response that follows a preceding curvature response without direct relation to the nature of the stimulus causing the curvature.

The spaceflight experiment conducted by Chapman et al. [201] provided interesting results concerning the expression of autostraightening in oat coleoptiles. In this experiment, coleoptiles were subjected to 25-min gravitropic stimulation at 1 *g* and allowed to develop curvature under microgravity. After showing a gravitropic curvature, these coleoptiles underwent autostraightening. The curvature time-course including the straightening phase was very similar to that obtained by the ground experiment in which coleoptiles were subjected to gravitropic stimulation for 25 min and allowed to stand in the original vertical position. However, autostraightening occurred less effectively under ground condition when coleoptiles were subjected to clinostat rotation after the 25-min stimulation. The results suggest that autostraightening, which is clearly the major response accounting for the straightening of gravitropically bent oat coleoptiles (see above), is inhibited by clinostat treatment, and that the inhibition is a clinostat-specific response, not representing the absence of gravity stimulus. This explanation could possibly account for some apparently contradictory results described in the preceding section: The phototropic curvature of wheat coleoptiles was not enhanced very much under the microgravity condition [198] because autostraightening is probably the major component of the organ straightening. On the other hand, substantially greater curvature occurred on the horizontal clinostat [98], possibly because autostraightening, which counteracts the development of phototropic curvature, was inhibited by the clinostat treatment.

Although autostraightening has been little investigated in organs other than coleoptiles, the results by Orbovic and Poff [168] suggest that it occurs in phototropically bent *Arabidopsis* hypocotyls. The hypocotyl of stationary seedlings established the maximal curvature (on average a little less than 15 degrees) 80 min after fPIPP stimulation. In the subsequent period, the hypocotyl began to straighten and the overall curvature decreased gradually. When treated with a clinostat, the hypocotyl bent for a little longer period, but the overall time-course including the phase of curvature decrease was not much affected. The results suggest that the cessation of phototropic bending and the subsequent straightening are caused mainly by autostraightening.

23.7.4 Storage of the phototropic signal

The fact that coleoptiles can develop phototropic curvature for a long period after termination of light stimulation indicates that the perceived light signal is stored in the organ. It would be of interest to know how long the phototropic signal can be stored at the tissue site of photoperception. Such signal storage may be evaluated from the curvature response that takes place in the most apical zone of coleoptiles. In the maize coleoptile simulated for fPIPP, the apical zone continued to bend at the maximal rate for at least 30 min ([151]; Figure 8). When rotated on horizontal clinostats after fPIPP stimulation, the oat coleoptile continued phototropic curvature for 2 h in its apical zone [154,155]. Thus, the phototropic signal perceived by the tip can be stored there for a period as long as 2 h. This period could be the shorter limit because it is possible that phototropic curvature is counteracted by autostraightening. It is suggested that the early signal transduction that takes place at the tissue site of photoperception involves a component that can remain active for at least 2 h after photoperception.

Investigating the long-sustained fPIPP curvature of maize coleoptiles on a horizontal clinostat, Nick and Schäfer [205] have resolved a unique mechanism of signal storage. They stimulated maize coleoptiles with two unilateral pulses of blue light from opposite directions. Each pulse induced the maximal fPIPP if given alone. Curvature was allowed to develop on a horizontal clinostat. When the two opposing stimuli were given with an interval of 30 or 60 min, a sustained curvature occurred towards the second stimulus. When the two pulses were given with an interval of 2 h, the curvature towards the first stimulus was counteracted by the curvature towards the second stimulus; however, this curvature did not last long and the coleoptiles began to show a sustained curvature towards the direction of the first stimulus. The authors concluded that the transverse polarity induced by the first stimulus is somehow stabilized when a certain period elapses. Although the coleoptile of upright stationary plants returns to the vertical orientation after showing phototropic curvature, Nick and Schäfer were able to demonstrate that the stable transverse polarity is induced and sustained similarly in these plants. Therefore, it seems that the curvature response attributable to the stabilized transverse polarity is almost totally compensated for by gravitropism (perhaps, with participation of autostraightening) in upright plants.

Nick and Schäfer [206] extended the experimentation with two opposing pulses of blue light, but providing the first pulse to a narrow zone of either the tip or the base (see their Figure 3). As in the above experiments, the second stimulus was given to the entire length of coleoptiles after different time intervals. The final curvature attained on a horizontal clinostat was determined. The tip or base stimulation alone induced positive curvature; the curvature induced by base stimulation was about a half of that induced by tip stimulation. The response induced by tip stimulation was overtaken by the response to the second opposing stimulus, regardless of the time interval between the two stimuli. The response induced by base stimulation could be reversed by the second opposing stimulus when the interval was short (30 min). However, as the interval was extended, the response to the second stimulus became weaker and, at intervals longer than 2 h, the coleoptiles showed the final curvature as if no second stimulus had been given. The results indicated that the stabilization of transverse polarity did not occur in tip-stimulated coleoptiles but in base-stimulated coleoptiles. It was concluded that the phototropic induction and the induction of stable transverse polarity are not causally linked. The data of Nick and Schäfer, however, provide interesting information with regard to the phototropism attributable to non-tip photoperception. The relatively large curvature attained after base stimulation and clinostat treatment probably corresponds to the fPIPP induced by non-tip photoperception (see Section 23.5.2). It can therefore be suggested that the stabilization of phototropic polarity is a property of the fPIPP (and perhaps also of the TDP) induced by non-tip photoperception. As will be discussed further in Section 23.10.13, the stabilization of transverse polarity could have been explained in terms of the stabilization of the asymmetry of cortical microtubule orientation.

von Guttenberg [160] reported results that are relevant to the concept of signal storage. He was able to induce a large phototropic curvature in decapitated oat coleoptiles by increasing the light gradient and applying auxin (see Sections 23.5.4 and 23.6.1). When the auxin application was delayed for 8 h after 3-h phototropic stimulation of decapitated coleoptiles, a curvature amounting to half of that induced without a delay of auxin application could still be induced. Thus, the phototropic signal perceived by non-tip coleoptile zones was stored beyond 8 h following the termination of stimulation. The results of Diemer [207] obtained using sunflower hypocotyls also indicate that the phototropic signal can be stored for a long period. Two days after decapitation, hypocotyls were stimulated with unilateral light for 16 h. No appreciable curvature could be found at the end of stimulation. However, when the hypocotyls were supplied with IAA through the apical cut surface after stimulation, they developed a substantial curvature. Even when IAA application was delayed for 4.5 h, the hypocotyls could still produce a curvature amounting to half of that induced without a delay. The signal storage indicated by these results might be comparable to that shown to occur in the coleoptile tip, although a contribution of the kind of stable transverse polarity described by Nick and Schäfer (see above) cannot be ruled out.

23.7.5 Photogravitropic equilibrium with contribution of autostraightening

When stimulated with continuous unilateral light, hypocotyls and coleoptiles of many plants show a very similar temporal and spatial pattern of phototropism (two examples are shown in Figure 14). This pattern can be summarized as follows: During the initial phase of rapid curvature development, the organ shows an arc-shaped appearance (b in Figure 14, A and B). This appearance is apparently caused because most parts of the organ can contribute to curvature. While still enhancing the overall curvature (i.e. orienting the tip more to the light source), the shape of the organ changes. The most notable change is the straightening of the apical part. The overall curvature does not decrease by this straightening because the basal part continues to bend. The organ subsequently assumes a more or less steady appearance: the upper half of the organ is straight and the basal part is curved (c in Figure 14, A and B). This appearance is maintained for a long period, although the organ may gradually become less curved in the basal part as it approaches the final growth stage (d in Figure 14B).

The organ straightening observed in the upper part can be induced by either gravitropic counteraction or autostraightening. The straightening might be more apparent in the upper part, which perceives a stronger gravitropic stimulus (i.e. deviates more from the vertical than does the lower part). Alternatively, autostraightening may be expressed more strongly in the apical part than in the lower part. The observation of fPIPP curvature of oat coleoptiles on horizontal clinostats indicates that the upper part has a greater ability to express autostraightening than the basal part [153,155].

Because it is so apparent that the phototropically bending organ is subject to counteracting gravitropic stimulation, there would seem to be no doubt that gravitropism contributes to the observed straightening. On the other hand, the results described in the preceding sections make it very probable that autostraightening also makes a significant contribution in the process of organ straightening. At the moment, it is difficult to evaluate the extent to which autostraightening contributes to the actual straightening observed during continuous phototropic stimulation. The equilibrium between phototropism and gravitropism (photogravitropic equilibrium) alone might be insufficient to establish a precisely straight appearance. One might speculate that the mechanism of autostraightening has been evolved to make fine straightening possible.

The autostraightening can counteract phototropic curvature response, but once the straight appearance is achieved in the organ's upper part, it will no longer provide the counteracting force in this part. Therefore, the photogravitropic equilibrium is probably a major component with which the upper part can maintain a certain angle with respect to the vertical. This view is supported by the following observation made by du Buy and Nuernbergk [192]. During continuous stimulation with unilateral blue light at a high fluence rate, the coleoptile straightened along the upper part after showing positive phototropism, with a curvature being retained at the base. The upper straight part maintained an angle of about 45° from the vertical. Light stimulation was then terminated and the coleoptile was rotated on a clinostat. The upper part began to bend again in the original positive direction. The following observation made with rice seedlings provides further support to the above view (M. Iino, unpublished). Seedlings of one group were stimulated with unilateral light from the upright position. Seedlings of another group were displaced by 90° towards the light source at the start of irradiation. In other words, developing phototropism was subjected to gravitropic counteraction (the first group) or the developing gravitropism was subjected to phototropic counteraction (the second group). The final angle from the vertical made by

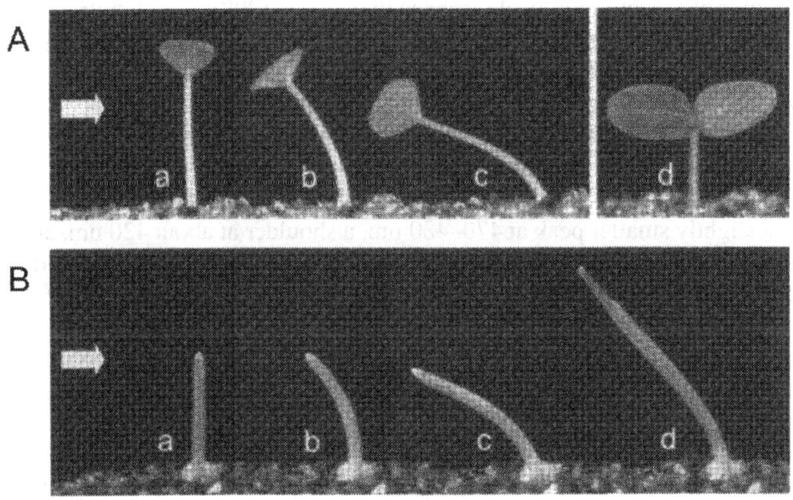

Figure 14. Phototropism of cucumber and maize seedlings during continuous stimulation. Seedlings were raised in vermiculite under overhead irradiation with white light (cucumber; 12 μmol m^{-2} s^{-1}) or red light (maize; 3.5 μmol m^{-2} s^{-1}). Phototropism was induced with unilateral white light (2.5 μmol m^{-2} s^{-1}) under overhead irradiation with red light (3.5 μmol m^{-2} s^{-1}). The light sources were as described in Figure 1. **(A)** Pictures of a cucumber seedling obtained immediately before (**a**) and 2 (**b**) and 8 (**c**, **d**) h after the onset of unilateral irradiation. **(B)** Pictures of a maize seedling obtained immediately before (**a**) and 2 (**b**), 8 (**b**) and 20 (**c**) h after the onset of unilateral irradiation. The arrow indicates the direction of unilateral light. The photograph **d** in **A** is a view from the light source.

the upper part was similar (about 25°). This observation indicates that the final angle represents the photogravitropic equilibrium, no matter whether the initial curvature was induced phototropically or gravitropically.

In conclusion, phototropic curvature is counteracted by gravitropism and the final steady-state angle represents the photogravitropic equilibrium. However, autostraightening probably plays an important role in establishing a precisely straight appearance and could provide a strong counterforce to the development of phototropic curvature at an early stage of curvature development.

23.8 Photoreceptors

Sensory pigments for photobiological responses are often attached to receptor proteins to form photoreceptor molecules. Such a photoreceptor has been implicated for phototropism, but has long remained obscure. Recent molecular genetic approaches are providing promising results on this topic.

23.8.1 Action spectra

Obtaining an action spectrum is a basic analytical step towards elucidating the photoreceptor of a photobiological response in question [208]. For phototropism, the oat coleoptile has again been the most extensively used material. Following the early conclusion that the wavelengths in the blue region of the light spectrum are most effective while those longer than 600 nm are virtually ineffective, the action spectrum of oat coleoptile phototropism has been obtained repeatedly by different methods (summarized in [90]). The most detailed action spectra obtained for fPIPP by Shropshire and Withrow [209] and Thimann and Curry [210] are typified by a major peak at 440 nm, a slightly smaller peak at 470–480 nm, a shoulder at about 420 nm, and another small peak in the UV-A region (around 370 nm). The action spectrum obtained using the balance method for the phototropic response to continuous stimulation [86] and that estimated for TDP [94] were essentially identical to the one for fPIPP. A detailed action spectrum for the fPIPP in hypocotyls of a dicotyledon, alfalfa, was reported much later by Baskin and Iino [100]. This action spectrum (Figure 15A), obtained using red-light-adapted seedlings, closely resembles that for oat coleoptiles. The action spectrum from alfalfa revealed an additional peak in the UV-B, a spectral region not previously investigated. Figure 15B shows an action spectrum obtained additionally by applying the photoproduct-gradient model (see Section 23.6.3) to the spectral fluence-response data of Baskin and Iino [100]. The spectrum is similar to that originally reported, except that the peak at 470 nm is relatively larger and that the UV-B peak is smaller. These results together support the notion that a common type of photoreceptors participates in the phototropism of both coleoptiles and hypocotyls.

Some results require more careful attention, however. The first issue concerns the relative effectiveness of UV-A. In oat coleoptiles, blue light (436 nm) was clearly more effective than UV-A (365 nm) in inducing fPIPP and PINP, but TDP showed a similar sensitivity to the two light sources [186]. This result appears to contradict that of Everett and Thimann [94] mentioned above. However, there is a fundamental difference in the methods used. The former result was obtained by simply providing unilateral light, whereas the latter authors measured the spectral sensitivity of TDP by first providing a high fluence pulse of blue light, which was thought to eliminate the fPIPP component. Secondly, the data of Haig [211,212] indicate that the phototropism of oat coleoptiles induced by stimulating only the tip (tip response) and that induced by shading the tip (non-tip response) may have different spectral sensitivities. Haig investigated the two responses by measuring the lag period before any detectable curvature occurs after phototropic stimulation. These responses were induced by a unilateral pulse (30-s or shorter duration) and apparently corresponded to fPIPP. The tip response showed a

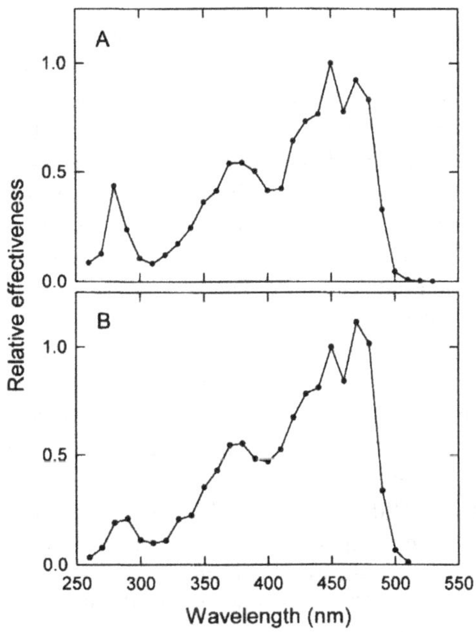

Figure 15. Action spectra for fPIPP in alfalfa hypocotyls. (**A**) An action spectrum reported in Baskin and Iino [100]. (**B**) An action spectrum obtained by applying the photoproduct-gradient model (equation 11 in [113]) to the spectral fluence-response data in [100]. To obtain the spectrum in **B**, one log unit higher than the fluence causing the estimated peak was set as the upper limit of the fluence-response data used for analysis. In this way, the data for wavelengths between 260 and 510 nm could be analyzed with a comparable range of fluence-response curve. When the equation was directly applied, the fitted curves showed some deviation from the data in such a way that the fitted curve overestimates the measured values at and near the bottom of the ascending arm. Therefore, a modified equation was also applied; in this equation, it was assumed that the amount of response per unit of the photoproduct concentration difference increases exponentially with increase in the photoproduct concentration difference. The spectrum shown is the one produced with the modified equation. The spectrum produced with the original equation, however, was very similar to the one shown including the positions and relative heights of the peaks.

major peak at 480 nm and a smaller peak at about 430 nm. The non-tip response showed no clear peak in the blue region. The sensitivity increased from long to short wavelengths, and the highest sensitivity was observed at the shortest wavelength used (about 410 nm). Therefore, it is predicted that the sensitivity to UV-A is relatively greater in the non-tip response than in the tip response. Finally, *nph* mutants of *Arabidopsis* isolated by Liscum and Briggs [213] (see below, Section 23.8.4) showed interesting results with regard to the sensitivity to blue light and UV-A. As compared to the wild type, the *nph-2* mutant showed only a slightly smaller response to continuous UV-A, but a substantially smaller response to continuous blue light. The *nph-4* mutant (see Section 23.10.10.7) showed a smaller response to UV-A, but a much smaller response to blue light. These results on the sensitivity relationship between blue light and UV-A may pose some basic questions concerning the mediating photoreceptors or photosystems.

Atkins [214] used the balance method to obtain action spectra with de-etiolated seedlings of dicotyledons. In addition to the major peak in the blue region, he could resolve a small but distinct peak in the red region. A typical result was obtained for cress hypocotyls. He claimed that similar results were obtained with sunflower hypocotyls. Another action spectrum for *Celosia cristata* hypocotyls also showed a peak in the red region, although smaller than in cress. Furthermore, the occurrence of red-light-sensitive phototropism could be noted, in comparison to the sensitivity to green light, in the coleoptile of de-etiolated maize seedlings and in the stem internode (perhaps, the leaf sheath) of de-etiolated oat seedlings. The peak in the red region was probably not caused by any impurity of the filtered red light, because Atkins did not find any such peak for the coleoptile of etiolated oat seedlings. The peak height (maximal light effectiveness) in the red region amounted at most to 20% of that in the blue region in all cases investigated, but a smaller peak in the action spectrum does not prove that the response is ecologically irrelevant or less significant [90]. Another interesting result relates to the sensitivity to green light. The effectiveness in the blue region dropped sharply from about 480 to 500 nm in de-etiolated cress hypocotyls as in etiolated oat coleoptiles. However, the corresponding drop occurred at wavelengths about 50 nm longer in *Celosia* hypocotyls, indicating that the spectral sensitivity extends more to the green region in this material. (The very long light stimulation practiced to obtain the data for *Celosia* makes this difference less conclusive. To obtain the action spectrum for cress, Atkins provided the test stimulus for several hours after 4-h dark adaptation to obtain the action spectrum for cress. On the other hand, *Celosia* seedlings were subjected to the test stimulus for 80 h continuously from an early stage of germination.)

No action spectroscopic data are available for the sPIPP of coleoptiles. Konjevic et al. [215] presented evidence that the sPIPP of *Arabidopsis* is as sensitive to green light (510 nm) as to blue light (450 nm) and that the single-peaked fluence-response curve produced with green light corresponds mainly to the sPIPP. The spectral fluence-response data obtained by Baskin and Iino [100] for the hypocotyl of red-light-adapted alfalfa, however, revealed no such sPIPP. Certain plants might have a photoreceptor system that is sensitive to both blue light and green light, as also suggested by the data of Atkins mentioned above, and the green-light sensitivity might be related to the occurrence of sPIPP, at least in *Arabidopsis*.

The photoreceptors considered in the following Sections (23.8.2–6) are those participating in the responses specific to blue light and UV-A. Discussion on the issue of red-light sensitivity and an additional issue concerning UV-B sensitivity will be extended later (Sections 23.12.1 and 23.12.2). The issue of green-light sensitivity will be mentioned again in Section 23.8.6.

23.8.2 Receptor pigment: carotenoids vs. flavins

Carotenoids [216] and flavins [217] have long been considered as candidates for the receptor pigment for phototropism. The unique shape of the action spectrum in the spectral region, 400–500 nm, was thought to support the carotenoid hypothesis [209]. However, carotenoids do not usually have an absorption peak in the UV-A region, and the peak of the action spectrum in this region could have been explained better by flavins. Some lines of evidence, although indirect, have favored flavins over carotenoids:

1. carotenoid-deficient albino mutants of sunflowers, maize, and barley show phototropism [218,219];
2. potassium iodide, which quenches the triplet excited state of flavins, inhibits phototropism of oat and maize coleoptiles [92,220];
3. phenylacetic acid, which binds to flavins, inhibits phototropism of maize coleoptiles [220]; and
4. the short lifetime of the excited singlet state of carotenoids is unfavorable, though not exclusively, for a photoreceptor function [221].

More recently, Quiñones and Zeiger [222] provided some evidence that the carotenoid zeaxanthin is a receptor pigment for phototropism of maize coleoptiles. These authors were able to show a close correlation between the zeaxanthin content and the phototropic responsiveness. Dark-adapted coleoptiles, which had no detectable zeaxanthin, showed little phototropic curvature to a pulse of unilateral blue light. The amount of zeaxanthin and the phototropic responsiveness to blue light increased concomitantly over a period of a few hours and decreased in the dark following red-light pretreatment. A parallelism between zeaxanthin and phototropic responsiveness was also found when zeaxanthin formation was inhibited by dithiothreitol. Quiñones et al. [223] subsequently suggested a specialized function of coleoptile plastids as the site of the photoperception by zeaxanthin. To reinvestigate any such possible role for zeaxanthin in phototropism, Palmer et al. [224] used albino mutants of maize in combination with chemical treatments that inhibit carotenoid biosynthesis. These authors found that nearly normal phototropism was induced by a blue-light pulse in the absence of detectable amounts of zeaxanthin and other carotenoids, and concluded that zeaxanthin is not a photoreceptor pigment in maize coleoptiles.

In the experiments of Quiñones and Zeiger [222], phototropism was induced with a 200-s pulse of blue light that provided a fluence of 200 μmol m^{-2}. As pointed out by Horwitz and Berrocal [225], it is most likely that these authors measured sPIPP. This is also supported by the fact that sPIPP is absent in dark-adapted maize coleoptiles and becomes inducible after red-light pretreatment (see Section 23.4.3). On the other hand,

Palmer et al. [224] used a blue-light fluence that is optimal for fPIPP. Therefore, it may be argued that zeaxanthin functions as the receptor pigment for the sPIPP of maize. It should be noted, however, that the maize sPIPP can be made fully inducible by a short pulse of red light, and even of far-red light (see Section 23.4.3). The zeaxanthin accumulation in the coleoptile was detected during continuous irradiation with red light at a high fluence rate. Quiñones and Zeiger investigated the enhancement of phototropic responsiveness only under this red-light condition. Before zeaxanthin is considered as the receptor pigment for sPIPP, it must be demonstrated that it accumulates following a short pulse of red light. It is also required that sPIPP is impaired in carotenoid-less mutants.

Zeaxanthin may function as a receptor pigment for the blue-light-dependent stomatal response [226]. So far, no conclusive evidence has been provided for the hypothesis that zeaxanthin is a receptor pigment for phototropism of coleoptiles and hypocotyls. The action spectrum for the light-induced folding of *Oxalis oregana* leaves showed two peaks in the blue region, but no peak in UV-A ([57]; see also [227]). The action spectrum obtained for stomatal opening in wheat similarly showed two peaks in the blue region and no apparent peak in UV-A [228]. This similarity raises the possibility that a similar type of blue-light receptor is responsible for the two responses, with zeaxanthin being a receptor-pigment candidate. (The leaf-folding response of *Oxalis* shows a paraphototropic feature in the natural habitat, but the response may be a nastic movement induced from the diaphototropic leaf position; see also Section 23.2.2.)

23.8.3 Cryptochrome

Cashmore and his co-workers analyzed the *cry1 Arabidopisis* mutant (previously called the *hy4* mutant) and uncovered cryptochrome 1, the protein encoded by the *CRY1* gene, as the first-identified blue-light receptor acting on physiological processes in plants [229–231]. This photoreceptor has a homology to DNA photolyases and, like the latter, it has flavin adenine dinucleotide (FAD) as the primary chromophore and probably a pterin as the secondary chromophore [232]. The gene for cryptochrome 2 has been subsequently identified as a member of the cryptochrome gene family in *Arabidopsis* [233].

Because the phototropism of *Arabidopsis* hypocotyls was normal in the *cry1* mutant, it was thought that cryptochrome 1 is not the photoreceptor of phototropism [213,234]. A further study by Cashmore's group, however, indicated that the mutant deficient in both cryptochromes 1 and 2 (*cry1cry2* double mutant) shows no detectable fPIPP, although the mutant deficient in only cryptochrome 2 (*cry2* mutant) shows normal fPIPP as the *cry1* mutant does [235]. The double mutant could, however, express some TDP. Briggs and his co-workers reinvestigated fPIPP in *cry1cry2* double mutants [236,237]. They could not reproduce the lack of fPIPP in these mutants, although the curvature response was partially reduced. Another conflicting result was reported with regard to the blue-light-dependent phosphorylation of the membrane-associated 120-kDa protein (see the next section). Ahmad et al. [235] found that the phosphorylation response is absent in the double mutant, whereas Lascève et al. [237] found that the response is not impaired in the double mutant.

The reason for the conflicting results is not clear. At least it seems that cryptochromes 1 and 2 somehow affect the phototropic responsiveness in *Arabidopsis*. Whether or not cryptochromes have any receptor role for phototropism requires further investigations.

23.8.4 Phototropin (NPH1 holoprotein and its homologues) and related proteins

Briggs and his co-workers found that a 120-kDa protein associated with pea epicotyl plasma membranes is rapidly phosphorylated, both in vivo and in vitro, in response to blue light, and characterized the protein in detail [238–241]. Similar proteins are present in many materials investigated [242]; the proteins from dicotyledonous tissues have molecular masses of 120–130 kDa, whereas those from Gramineae coleoptiles have molecular masses near 110 kDa. The proteins from *Arabidopsis* hypocotyls [243], maize coleoptiles [244–247], and oat coleoptiles [169,248–250] have been characterized in some details. The membrane-associated proteins that are phosphorylated in response to blue light are designated here as "blue-light-responsive proteins". On the basis of some correlation with physiological properties of phototropism, it was suggested that the blue-light-responsive proteins mediate phototropism (summarized in [251]). As described in the following paragraphs, subsequent investigations by Briggs' group have led to the conclusion that these proteins have a photoreceptor function for phototropism.

Liscum and Briggs [72,213] screened mutants of *Arabidopsis* that are defective in the hypocotyl phototropism induced by continuous unilateral blue light. The seed populations used for screening were those derived from fast-neutron or T-DNA insertion mutagenesis. They obtained mutants of four loci, designated *nph1* to *nph4* (*nph* for non-phototropic hypocotyl). Of these mutants, *nph1* and *nph3* were most defective in phototropism and showed no detectable curvature response to continuous stimulation with either blue light or UV-A. The other mutants, *nph2* and *nph4*, showed some curvature especially in response to UV-A. Mutants of all loci but one (*nph4*) showed normal gravitropism. The two mutants of *Arabidopsis*, strains JK224 and JK218, previously isolated by Khurana and Poff [252], were found to be alleles of *nph1* and *nph3*, respectively. The blue-light-dependent phosphorylation of a 120-kDa protein occurred normally in all but *nph1* mutants. In fact, all the *nph1* alleles isolated by Liscum and Briggs showed no detectable phosphorylation response and had a greatly reduced amount, if any, of the 120-kDa protein. Based on these results, Liscum and Briggs [72] hypothesized that the product of the *NPH1* locus constitutes the most up-stream component of the sensory transduction pathway for phototropism.

The *NPH1* gene was cloned and sequenced by Huala et al. [253]. It has turned out that the NPH1 protein has a serine-threonine kinase domain and, at the NH_2-terminal region, two repeats of a motif, designated the LOV domain (Figure 16). Sequences homologous to LOV occur as a single copy in proteins of diverse organisms, which include those that are sensitive to light, oxygen, or voltage (on which, the domain name was based). Since two of them are known to bind FAD, it was suggested that the LOV domain might be a flavin-binding site. Indeed, the *nph1* mutant restored hypocotyl phototropism when transformed with a genomic clone containing the *NPH1* gene. Christie et al. [236] transformed insect cells with a recombinant baculovirus containing the NPH1 coding

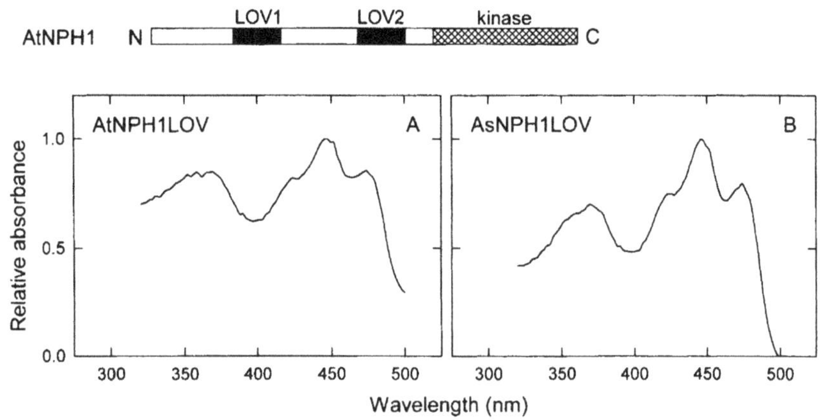

Figure 16. Molecular properties of the NPH1 protein. AtNPH1: the structural features of the NPH1 protein deduced from the *NPH1* gene cloned from *Arabidopsis thaliana* [253]. Absorption spectra of the LOV domains of *Arabidopsis* NPH1 (**A**) and *Avena sativa* NPH1 (**B**) expressed in *E. coli* as fusions with the calmodulin-binding peptide. Adapted from Christie et al. [254].

sequence. The recombinant NPH1 had non-covalently bound flavin, chromatographically identified to be flavin mononucleotide (FMN). The extracted recombinant NPH1 was phosphorylated upon exposure to blue light. At least one allele (*nph1–5*) is protein null [236,253]. Furthermore, the fluorescence excitation spectrum of the recombinant NPH1 showed good agreement with the action spectrum of phototropism [236]. Indeed, the fluorescence excitation spectrum revealed the fine shape in the 400–500 nm region (see above, Section 23.8.2). In a further study, polypeptides of a single LOV domain or those spanning the two LOV domains of NPH1 and a NPH1 homologue of oats (see below) were expressed in *Escherichia coli* as fusion proteins [254]. Each LOV domain was shown to bind FMN stoichiometrically, and the recombinant fusion proteins revealed similar absorption spectra that agree with the action spectrum of phototropism as well as the fluorescence excitation spectrum of the recombinant NPH1 expressed in insect cells (Figure 16, A and B; compare with Figure 15). It is now very probable that the NPH1 holoprotein with FMN chromophores acts as a photoreceptor for phototropism (see also [255] for a review).

Biochemical data have indicated that the blue-light-responsive proteins of various plant species share certain similarities with the blue-light-responsive protein of *Arabidopsis*, now identified to be NPH1 [251]. Full-length cDNAs for NPH1 homologues have been cloned from oats, maize, and rice (see below). The estimated molecular weights of the homologues in oats and maize agree in principle, as NPH1 does, with the apparent molecular mass of the blue-light-responsive proteins identified in these plants. In fact, the molecular weights of the NPH1 homologues in oats, maize, and rice are smaller than that of NPH1, in agreement with the fact that the blue-light-responsive proteins of Gramineae grasses are generally smaller in size than the corresponding proteins of dicotyledons [242]. In addition, the recombinant fusion proteins containing LOV1 or LOV2 (or both) of oats were shown to bind FAM [254]. It is very probable that the blue-light-responsive proteins identified in various plants are the homologues of NPH1. Future studies may indeed reveal that NPH1 homologues are common photoreceptor proteins of higher plant phototropism. The name "phototropin" has now been proposed for the holoproteins of NPH1 and its homologues [254]. In this chapter, this name will be used when discussing general molecular properties of, or general physiological roles for, the NPH1 holoprotein and its homologues.

By now, the following *NPH1* homologues have been cloned: *NPL1* (*NPH1-like*) in *Arabidopsis* [256], *AsNPH1–1* and *AsNPH1–2* of oats [257], *ZmNPH1* of maize (Zacherl et al., unpublished: GenBank Accession No. AF033263), and *OsNPH1a* and *OsNPH1b* of rice [258]. Partial sequences that are similar to *NPH1* have been reported for peas [259,260], *Mesembryanthemum crystallinum* [261], and *Spinacia oleracea* (Baur et al., unpublished: GenBank Accession No. X73298) (see [255]). Figure 17 shows a phylogenetic tree of the corresponding proteins [258]. Most closely related to NPH1 are AsNPH1a, AsNPH1b, and ZmNPH1 of three Gramineae cereals. On the other hand, NPL1 has the least similarity to NPH1, and one of the two rice homologues (OsNPH1b) bears a relatively high similarity to NPL1. The three closely related NPH1

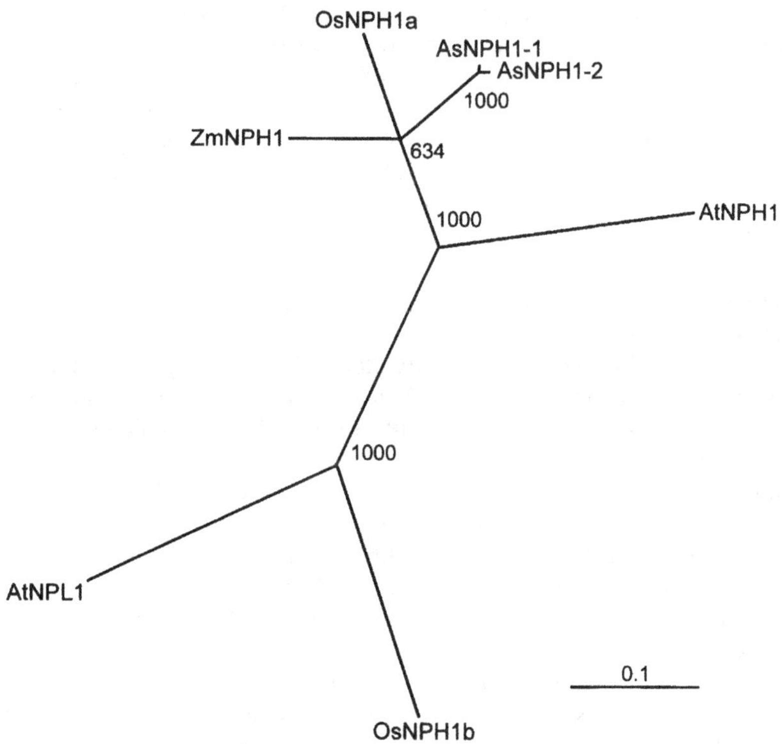

Figure 17. A phenogram of NPH1 and its homologues based on a neighbor-joining method. AtNPH1 and AtNPL1: *Arabidopsis thaliana* NPH1 and NPL1. ZmNPHN1: *Zea mays* NPH1. AsNPH1-1 and AsNPH1-2: *Avena sativa* NPH1-1 and NPH1-2. OsNPH1a and OsNPH1b: *Oryza sativa* NPH1a and NPH1b. Numbers are bootstrap values from 1000 replicates. The scale represents 0.1 substitutions/site. Adapted from Kanegae et al. [258].

homologues in Gramineae cereals are the most probable photoreceptor candidates for phototropism of these plants. It is not excluded, however, that the other homologues (NPL1 and OsNPH1b) also participate to phototropism (see the next section). Even if this is not the case, these homologues that retain the sequences for the two LOV domains are likely to function as photoreceptors in some other photobiological responses. If any such function is uncovered in the future study, then the name phototropin can also be used to indicate the second class of NPH1 homologues, but by discriminating the two classes (e.g. phototropin 1 for NPH1, and phototropin 2 for NPL1).

Phototropin is a highly hydrophilic protein [253]. The way in which phototropin is attached to the membrane appears to be an important part of its function. It is probable that phototropin constitutes, with other proteins, a plasma-membrane-associated **photoreceptor complex**. The biochemical study of Warpeha and Briggs [262] provided the earliest molecular evidence for this idea. When the pea plasma membrane preparation was solubilized with Triton X-100 and subjected to non-denaturing gel electrophoresis, the blue-light-responsive 120-kDa protein migrated as a part of a protein complex of about 335 kDa. The blue-light-responsive protein of oat coleoptile tips becomes more loosely bound to the plasma membrane as the coleoptile ages [248]. This result supports the notion of the plasma-membrane-associated photoreceptor complex and the functional significance of the putative protein(s) needed for the membrane association of phototropin. LOV domains are a subclass of PAS domains, thought not only to be able to bind ligands but to mediate protein/protein interactions [263]. Therefore, the LOV domain of phototropin might also be involved in the formation of the putative photoreceptor complex. Ballario et al. [264] found that the single LOV domain contained in the WC-1 protein of *Neurospora crassa* (the gene product responsible for *white collar-1* mutation) can undergo specific self-dimerization [264]. This result raises yet another possibility, that phototropin exists as a dimer in the photoreceptor complex [255].

Motchoulski and Liscum [265] recently cloned *NPH3*, the gene responsible for the *nph3* mutation (see above). They found that NPH3 is also associated with plasma membranes and obtained evidence that NPH3 can undergo specific binding to NPH1. Close association of NPH3 with NPH1 was also supported by the result that the membrane-associated NPH3 showed a greater mobility on SDS-polyacrylamide gel electrophoresis when the membrane preparation was obtained from the NPH1 null mutant or from the seedlings pretreated with blue light. Thus, NPH3 is very probably a protein component that constitutes the suggested photoreceptor complex. It has turned out that NPH3 is also a hydrophilic protein. Therefore, NPH3 does not seem to be a protein essential for the membrane association of NPH1, in agreement with the result that NPH1 is plasma-membrane-associated in a *nph3* mutant [72]. Because NPH3 is also plasma-membrane-associated in the NPH1 null mutant [265], the two proteins are mutually independent for their association with the plasma membrane. Perhaps, the NPH1/NPH3 complex is attached to the plasma membrane together with other proteins, constituting a larger photoreceptor complex. In support of this view, the NPH3 protein has two motifs that are homologous to the sequences involved in protein-protein interactions (BIB/POZ and coiled-coil domains); of the two domains, the coiled-coil may be involved in the interaction with NPH1 [265].

23.8.5 The relationship between phototropin and phototropism

Liscum and Briggs [72] reported that their *nph1* mutants did not show phototropism to continuous blue-light stimulation. Lascève et al. [237] confirmed this result using the protein-null *nph1–5* mutant. However, Sakai et al. [266] recently found that NPH1 null mutants are not without phototropic responsiveness. They isolated an *nph1* allele (*nph1–101*) that is probably protein null. This mutant showed phototropism at high fluence rates of blue light. They obtained similar results with the *nph1–5* mutant. The fluence rate of blue light used by Liscum and Briggs [72] was 0.1 μmol m^{-2} s^{-1}, and the one used by Lascève et al. [237] was 2 μmol m^{-2} s^{-1}. Sakaki et al. [266] observed no detectable phototropism at 0.01 – 0.1 μmol m^{-2} s^{-1}, confirming the earlier results, but found that clear phototropism is inducible at 10 and 100 μmol m^{-2} s^{-1}. The response at 10 μmol m^{-2} s^{-1} was about half that of the wild type, and the response at 100 μmol m^{-2} s^{-1} was not smaller than that of the wild type. Therefore, NPH1 holoprotein does not appear to be the sole photoreceptor for phototropism of *Arabidopsis* hypocotyls.

Lascève et al. [237] used the *nph1–5* mutant to investigate the phototropic response to five successive pulses of blue light given at 20-min intervals. No detectable response was found in the fluence range from 10^{-1} to 10^{3} μmol m^{-2}. In view of the conclusion that the cumulative response resulting from multiple pulses is based on fPIPP [99], it has been deduced that fPIPP is absent in the *nph1* mutants. The fluence-response relationship obtained in this way probably includes the photosensory adaptational response (see Section 23.9.2). Since Lascève et al. used dark-grown seedlings, it is also probable that the phytochrome-dependent responsiveness enhancement (see Section 23.4.2) developed during the successive pulses. Although these features could complicate the shape of the fluence-response curve, the lack of any detectable curvature response in the wide fluence range most probably indicates that fPIPP is absent in the NPH1 null mutant. The results may also indicate that sPIPP is, too, absent in the mutant. However, because the occurrence of sPIPP in this material seems to depend strictly on the stimulus conditions (see Section 23.3.4), the results are less conclusive for this contention.

The mutant strain JK224 isolated by Khurana and Poff is an allele of *nph1* (see above) and designated *nph1–2* [213]. This mutant shows a greatly reduced or undetectable level of blue-light-dependent phosphorylation in vitro of phototropin [72,243], whereas it can express desensitized fPIPP and apparently normal TDP [252]. The amount of phototropin may be reduced significantly in the mutant, accounting for the reduced level of the phosphorylation response [72]. The *nph1–2* mutant, isolated from EMS-mutagenized seeds, has an NPH1 protein in which one amino acid residue in the kinase domain is replaced by another [253]. It is also possible that the phosphorylation response itself is impaired in the mutant. As will be discussed later (Section 23.9.1), the results obtained using the *nph1–2* mutant provide some significant insights into the relationship between phototropin and phototropic response.

In the coleoptile of dark-grown maize seedlings, the blue-light-responsive protein is most abundant in the tip, but also present in the base; the protein was not detectable in the nodal region and the mesocotyl [267]. In maize coleoptiles treated daily with 1-h red light, the blue-light-responsive protein was located almost entirely in the apical 5-mm zone [245]. The latter result agrees with the tip localization of photoperceptivity found in red-light-grown maize coleoptiles [150] (see Section 23.5.2). In red-light-grown oat seedlings, the blue-light-responsive protein was present along the entire length of the coleoptile [249]. However, the highest concentration was found in the apical few mm, and the concentration declined basipetally. This distribution pattern agrees roughly with the distribution of photoperceptivity in red-light-grown oat coleoptiles (Figure 7). The phototropin distribution seems more or less to represent the photoperceptivity distribution.

Salomon et al. [249,250] were able to show with oat seedlings that the level of blue-light-dependent phosphorylation in vivo of the blue-light-responsive protein is greater in the irradiated half than in the shaded half of the coleoptile tip. This result has provided the first biochemical evidence for the notion that the phototropic response is determined by the gradient in the extent of light-dependent reactions across the organ (see Section 23.9.1 for a further discussion).

The two NPH1 homologues in oats (AsNPH1-1 and AsNPH1-2) are very closely related [257]. Such homologues have not yet been found in other plant species (Figure 17). This result is interesting in view of the unique physiological features uncovered for the phototropism of oats (see Sections 23.3.3, 23.4.3, and 23.7.1).

23.8.6 Remarks

Zeaxanthin and cryptochrome have been proposed as the photoreceptors for phototropism. However, the experimental results are controversial and no clear conclusion has emerged. Meanwhile, evidence has accumulated indicating that phototropin is the prime photoreceptor for phototropism in *Arabidopsis* and probably also in many other plant species.

Although no photoreceptor has been found to be solely responsible for phototropism, the lack of NPH3 protein appears to result in a total loss of phototropism. Liscum and Briggs [213] found no detectable phototropic response in *nph3* mutants after continuous stimulation with either blue light (0.1 μmol m^{-2} s^{-1}) or UV-A (0.2 μmol m^{-2} s^{-1}). Sakai et al. [266] also could not detect any phototropic response in an *nph3* allele in the fluence-rate range from 0.01 to 100 μmol m^{-2} s^{-1}, which includes the high fluence rates with which the phototropic response of the NPH1 null mutants was detected. Two possibilities are apparent. First, the phototropism of *Arabidopsis* hypocotyls is mediated by a single photoreceptor complex that includes a phototropin, NPH3, and at least one additional photoreceptor. The light signal perceived by the photoreceptors is transduced all through NPH3. Secondly, the phototropism is mediated by at least two distinct photoreceptor complexes. Phototropin is a photoreceptor in only one of them, whereas NPH3 is a key component of all. In either possibility, it remains possible that a NPH1 homologue (e.g. NPL1) is a second photoreceptor protein.

Another significant result is that negative phototropism of roots is lost or severely impaired in the *nph1* and *nph3* mutants [72,213]. It can be concluded that phototropin and NPH3 are also involved in root phototropism. This, in turn, suggests that the early signal transduction for phototropism is similar between hypocotyls and roots. We have obtained a mutant of rice in which the phototropism of coleoptiles to continuous blue-light stimulation in the fluence rate range from 0.01 to 100 μmol m^{-2} s^{-1} (Figure 5) is totally lost (abstract in [268]). Root phototropism is also severely impaired in this mutant.

Although green light is capable of inducing substantial phototropism [269], this fact alone does not indicate the occurrence of a separate photoreceptor that is most sensitive to green light or similarly sensitive to blue and green lights. However, Konjevic et al. [117,215] presented evidence that *Arabidopsis* has a blue-light receptor whose sensitivity extends more to green light and is responsible for sPIPP in this material (see Section 23.8.1). The result that the NPH1 null mutant does not show any pulse-induced phototropism suggests that phototropin is also responsible for the sPIPP (see above, Section 23.8.5). Furthermore, the *nph1* and *nph3* mutants of *Arabidopsis* show no phototropic response to continuous green light. Although more thorough investigation is necessary to prove that green light is ineffective even at high fluence rates, the available results do not favor the occurrence of a unique green-light-sensitive photoreceptor. At present it seems more likely that any possible high sensitivity to green light is attributable to the photoreceptor complex in which phototropin and NPH3 play central roles.

23.9 Photosystem

Here the term photosystem represents a molecular system that contributes to early sensory transduction and occurs at the tissue site of photoperception. The concept of the photosystem is not congruent with that of the photoreceptor complex described above. The photoreceptor complex refers specifically to a unique assembly of proteins, whereas the photosystem is more loosely defined and can additionally include molecular components that follow the photoreceptor complex. The multiphasic fluence-response relationship of blue-light-sensitive phototropism could imply either that the plant has multiple photosystems for phototropism or that it has one system with complex regulatory components. Although either explanation may be extreme, it now appears likely that the photosystem for fPIPP is also responsible for a major part of TDP.

23.9.1 Photosystem for first pulse-induced positive phototropism

The fluence-response curve for fPIPP is bell-shaped, being composed of ascending and descending arms. One of the possible explanations of this fluence-response characteristic is that two different photoreceptor systems contribute to fPIPP, one mainly accounting for the ascending arm and the other accounting for the descending arm. This explanation may be favored by the findings that the shifts of the ascending and descending arms are not always synchronized when the bell-shaped fluence-response curve moves to higher fluences in response to red light or far-red light ([109,123,124]; see Figure 6). On the other hand, the idea that a single photosystem is responsible for fPIPP has been favored by the findings that the action spectra for the ascending and descending arms are similar [100] and that the curvature time-courses on the ascending and descending arms are similar [93].

Since the NPH1 null mutant of *Arabidopsis* probably does not show any detectable fPIPP ([237]; see Section 23.8.5), phototropin appears to be the prime photoreceptor of fPIPP. The lack of fPIPP in the NPH1 null mutant may not necessarily disprove the possible contribution of another photoreceptor that only modifies the response amplitude to produce the descending arm. However, the parallel shifts of the two response arms observed in the *nph1–2* mutant ([252], see below) strongly suggest that phototropin is responsible for both the ascending and descending arms. It would be reasonable to adopt the simpler hypothesis based on a single photosystem, at least until more definite evidence is presented for the participation of two systems (e.g. a single mutation that modifies the ascending and descending arms independently). The photoproduct-gradient model (see Section 23.6.3) offers the simplest explanation of the bell-shaped feature of the fPIPP fluence-response curve in terms of a single photosystem.

The *nph1–2* mutant (see Section 23.8.5) has provided an interesting set of results. This mutant showed an fPIPP fluence-response curve shifted to about 20 times higher fluences; the maximal response (i.e. the peak height in the fluence-response curve) was little affected [252]. The result indicates that the fPIPP of this mutant is desensitized by a factor of about 20 without a significant change in the maximal responsiveness. The blue-light-dependent phosphorylation in vivo is severely impaired in this mutant [72,243]. This impairment is probably attributable to a greatly reduced amount of membrane-associated phototropin [72], although the autophosphorylating function of phototropin may also be impaired. Since the maximal response is not reduced in the *nph1–2* mutant, it is apparent that the phototropin concentration in the membrane does not limit the phototropic responsiveness. The concentration of phototropin is apparently correlated with the light sensitivity of fPIPP. According to the photoproduct-gradient model, the fluence-response curve is determined by the difference in a limiting photoproduct between the two sides of an organ. The conclusion above then determines that none of the activated forms of phototropin, including the phosphorylated form, is the limiting photoproduct. The limiting photoproduct must be a component that occurs after phototropin activation.

All the implications given above depend on the conclusion that the concentration of phototropin in the membrane is severely reduced in the *nph1–2* mutant and on the supplemental result that the level of phosphorylation cannot be enhanced at very high fluences to any extent comparable to the level in the wild type. The severely reduced amount of phototropin was deduced from the silver-staining of proteins on the SDS-polyacrylamide gel [72]. Confirmation of the result with a more unequivocal determination method would be necessary before we can reach a definite conclusion on this important issue. Furthermore, examination of the presented data [72,243] cannot exclude the possibility that 3200 μmol m^{-2} is a little above the threshold fluence for the detection of phosphorylation response in vitro and the phosphorylation response can increase significantly at higher fluences. In the wild type, the phosphorylation response became detectable from a fluence of 10 μmol m^{-2} and was saturated at about 3200 μmol m^{-2} [243]. Therefore, the possibility is not entirely eliminated that the phosphorylation response is simply desensitized by a factor of 100–300. If it happens that the *nph1–2* mutant has a normal level of phototropin with only impaired phosphorylation activity, then any activated form of phototropin, including the phosphorylated phototropin, becomes a possible candidate for the limiting photoproduct. As mentioned in Section 23.6.3, the formation of the limiting photoproduct could be explained better by the second-order kinetics [113]. It is tempting to speculate that such kinetics are somehow related to the fact that each phototropin molecule has two chromophore-binding sites (i.e. two LOV domains) or to the possibility that phototropin can self-dimerize (see Section 23.8.4). If, however, any activated form of phototropin cannot be the limiting photoproduct, then this possibility becomes unlikely.

In the oat coleoptile tip, the amount of the blue-light-responsive protein phosphorylated in vivo is greater on the irradiated side than on the shaded side [249,250] (see Section 23.8.5). Although this result supports a condition of the photoproduct-gradient model that the phototropic response depends on local light-induced reactions, the difference in the phosphorylation level between the two sides did not agree with the bell-shaped fluence-response curve of fPIPP; the peak of the difference in phosphorylation level occurred at a fluence much higher than the fluence causing the peak of phototropic response [250]. This result agrees with the above-mentioned conclusion that the limiting photoproduct is not the phosphorylated phototropin.

23.9.2 Blue-light-dependent changes in responsiveness and sensitivity of first pulse-induced positive phototropism

Experiments using two pulses of light have provided results that probably represent the basic properties of the photosystem for fPIPP. Early results were obtained using white light as the light source for coleoptiles of oats [188] and maize [96]. Meyer [270] and Blaauw and Blaauw-Jansen [271] used blue light to extend the two-pulse experiments on oat coleoptiles. These workers did not agree upon one common conclusion, but the results indicated that either the responsiveness or the sensitivity of fPIPP is reduced by a high-fluence pulse and is subsequently restored in the dark. Early results may have to be evaluated carefully, however, because the first pulse of white or blue light is likely to induce phytochrome-mediated changes in fPIPP (desensitization and responsiveness enhancement) when dark-adapted plants are used without red-light pretreatment (see Sections 23.4.1 and 23.4.2). The results from oat coleoptiles will be reviewed after summarizing the more straightforward results from maize.

Briggs [96] stimulated maize coleoptiles with a high-fluence pulse of unilateral light that caused an indifferent response at the bottom of the descending arm and, after different dark intervals, with a low-fluence pulse that is optimal for fPIPP in non-pretreated coleoptiles. It was found that phototropic responsiveness to the second pulse was restored with increasing dark intervals. Nearly full restoration was achieved after an interval of about 20 min. Because Briggs conducted experiments using red working light, the phytochrome effects on phototropism were probably fully expressed before the first inductive pulse. It appeared that the fPIPP responsiveness at the fluence initially optimal for fPIPP is eliminated by the high-fluence pulse and is restored gradually afterwards.

Working with red-light-grown maize coleoptiles, Iino [113] obtained phototropic fluence-response curves at various times after treatment with a high fluence of bilateral blue light (Figure 18A). Unilateral stimulation given immediately after the high-fluence pulse gave no phototropic response. The bell-shaped fluence-response curve became detectable after an interval of several minutes, and the maximal responsiveness (the peak height) increased gradually towards the original level with a further increase of the interval time. Half the original responsiveness was restored at about 10 min. It also became apparent that the recovering fPIPP occurred at higher fluences compared to the original fPIPP. Already at 5 min after the pretreatment, the peak of the fPIPP fluence-response curve was located at about 30 times higher fluences. The position of the peak returned gradually with the interval time. These results have demonstrated, in agreement with the result of Briggs [96], that a high-fluence pulse rapidly eliminates fPIPP responsiveness and that the responsiveness is subsequently restored gradually. In addition, the results have indicated that fPIPP is rapidly desensitized and gradually resensitized following the high-fluence pulse. As we have seen above (Section 23.4.1),

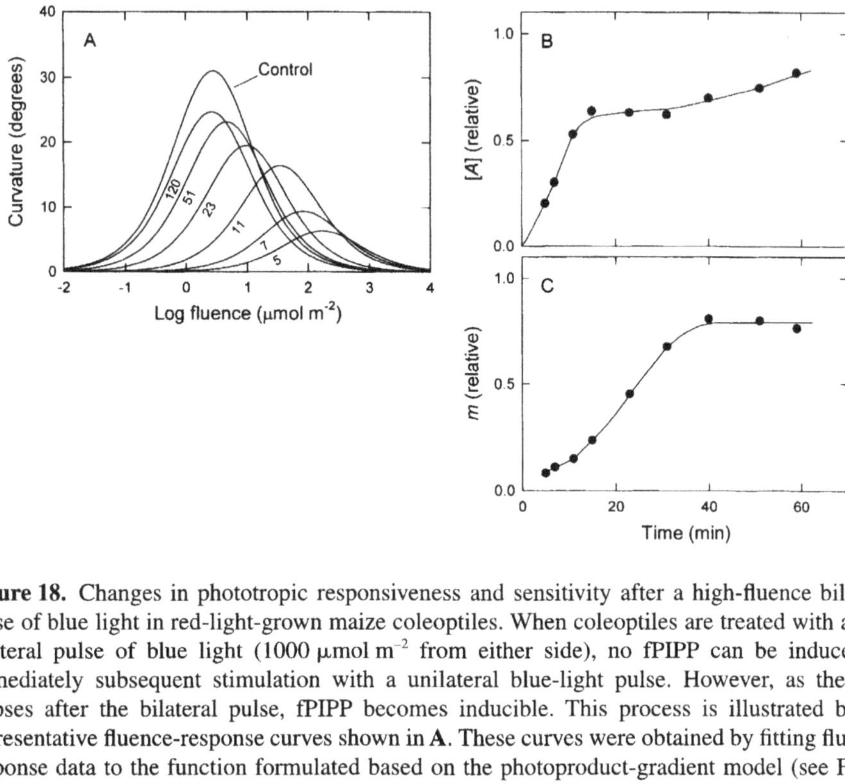

Figure 18. Changes in phototropic responsiveness and sensitivity after a high-fluence bilateral pulse of blue light in red-light-grown maize coleoptiles. When coleoptiles are treated with a 60-s bilateral pulse of blue light (1000 μmol m^{-2} from either side), no fPIPP can be induced by immediately subsequent stimulation with a unilateral blue-light pulse. However, as the time elapses after the bilateral pulse, fPIPP becomes inducible. This process is illustrated by the representative fluence-response curves shown in **A**. These curves were obtained by fitting fluence-response data to the function formulated based on the photoproduct-gradient model (see Figure 12). The number on each curve indicates the time interval between the bilateral and the unilateral test pulse. In addition to the disappearance and subsequent reappearance of phototropic responsiveness, it is noted that the recovering fluence-response curve initially occurs at higher fluences and returns gradually to the original position (compare with the control fluence-response curve obtained without the bilateral pulse). The photoproduct-gradient model was extended to incorporate these kinetic features [113]. In light of the model, changes in the concentration of the hypothetical substance *A*, a precursor of the limiting photoproduct *B*, was computed (**B**). Values of the sensitivity parameter *m* were simultaneously resolved (**C**). Adapted from Iino [113].

the fPIPP of maize coleoptiles is also desensitized by phytochrome. In fact, both phytochrome-mediated and blue-light-dependent desensitization responses can be induced in dark-adapted maize [112]. The two desensitization responses are kinetically distinct and can progress simultaneously and additively.

The two-pulse experiments on oat coleoptiles were somewhat complicated by the occurrence of PINP at high fluences, and the conclusion reached was affected by how the PINP was interpreted [270,271]. The reported results, however, agree in principle with those from maize. The most thorough results obtained by Meyer [270] clearly indicate that responsiveness is eliminated and subsequently restored following a high-fluence pulse of blue light. The data also indicate that desensitization and resensitization follow a high-fluence pulse (see her Figures 4 and 5). (In the experiments of Meyer, however, phytochrome-mediated slow desensitization and responsiveness enhancement, induced by the blue-light pulse or by green working light, probably accompanied the blue-light-dependent changes in responsiveness and sensitivity.) Using red-light-pretreated plants, Blaauw and Blaauw-Jansen [271] were able to demonstrate that fPIPP is desensitized by pretreatment with a high-fluence pulse of blue light. Since the results were obtained with red-light-pretreated plants, the desensitization response was probably not phytochrome-mediated. Blaauw and Blaauw-Jansen [271] failed to observe the changes in responsiveness, apparently because they did not obtain the complete fluence-response curve for the second pulse with sufficiently short intervals.

The data of Blaauw and Blaauw-Jansen ([271], see their Figure 1) indicate that the fPIPP responsiveness is almost fully restored by 20 min after the high-fluence pulse while the fPIPP still remains desensitized. Therefore, the responsiveness seems to be restored more rapidly than sensitivity. This relationship also held in maize coleoptiles [113]. The data by Blaauw and Blaauw-Jansen ([271], their Figure 2) further indicated that the desensitization response requires blue-light fluences much greater than the fluences effective in inducing fPIPP.

Janoudi and Poff [128] conducted two-pulse experiments with hypocotyls of dark-grown *Arabidopsis*. They found that after pretreatment with a high fluence of overhead or bilateral blue light, which eliminates fPIPP to immediately subsequent unilateral blue light, fPIPP responsiveness is restored and enhanced well beyond the dark-control level. Therefore, the responsiveness changes after blue-light pretreatment in *Arabidopsis* hypocotyls as found in maize and oat coleoptiles. The responsiveness was restored to exceed the dark-control level probably because a phytochrome-mediated enhancement, induced by the high-fluence blue light, accompanied the true responsiveness restoration. Because partially and fully restored fPIPP had similar threshold fluences, these authors concluded that the blue-light pretreatment did not affect the sensitivity to blue light. However, any possible sensitivity change induced by the blue-light pretreatment might be masked by the accompanying large phytochrome-mediated increase in responsiveness. In fact, the data by Steiniz and Poff [99] support the idea that fPIPP is desensitized by a preceding high-fluence pulse (see their Figure 7). When a high fluence, which fell in the indifferent region of fPIPP and caused little curvature, was administered five times with 20-min intervals, a significant curvature could be produced. The result indicates that fPIPP is desensitized by a preceding pulse, through a blue-light-sensitive mechanism and/or a phytochrome-mediated mechanism, so that the initially indifferent fluence becomes an effective fluence.

Iino [113] extended the photoproduct-gradient model (see Section 23.6.3) to explain the responsiveness and sensitivity changes found to follow a high-fluence pulse of blue light in maize coleoptiles. The photoproduct-gradient model determines that fPIPP responsiveness is eliminated by the high-fluence pulse because the formation of the limiting photoproduct is saturated in both sides of the coleoptile. In the extended model, it is assumed that the photoproduct precursor A is regenerated from the photoproduct B by a dark reaction, leading to the regeneration of phototropic responsiveness. In addition, it is assumed that the light sensitivity of the photoproduct-formation reaction ($A \rightarrow B$) is rapidly lowered by the preceding blue-light pulse and is subsequently raised to the original level in the dark. These sensitivity changes are thought to underlie the desensitization and resensitization of fPIPP. By fitting the data to the minimal mathematical model, the relative changes in the concentration of the photoproduct precursor A (Figure 18B) and the sensitivity parameter m (Figure 18C) were computed.

The calculated data in Figure 18B indicate that A is regenerated rapidly within 20 min after the high fluence pulse, although the data do not eliminate the possibility that a portion of A is regenerated more slowly. The first-order rate constant of the regeneration reaction, which more or less represents the early regeneration phase, is about 0.001 s^{-1} [113]. In the model, the regeneration of A is inversely related to the decay of the active photoproduct B. Therefore, B is thought to decay within 20 min after the high fluence pulse. It has been suggested that the early signal transduction that takes place in the coleoptile tip contains a component that can store the perceived signal for at least 2 h (see Section 23.7.4). It appears that this component occurs downstream of the hypothetical photoproduct B. A possibility not entirely excluded is that a fraction of B decays slowly and constitutes the long-lived component.

23.9.3 Photosystem for time-dependent phototropism

Why is fPIPP, which does not seem to function effectively in nature (see Section 23.3.5), ubiquitous in higher plants? It may be envisaged that fPIPP simply reflects a mechanistic aspect of TDP or of the phototropism that takes place during continuous stimulation. This view has been extended by some workers [99,113,271]. On the other hand, there are results that preclude a straightforward acceptance of such a simplified view. It has been recognized that the photosensing region for TDP extends more to the basal part as compared to that for fPIPP in coleoptiles of oats [186] and maize [96,272]. As discussed in Section 23.5.2, this difference presumably does not reflect a fundamental separation of TDP from fPIPP. More critical is the issue concerning the lag period (see Section 23.7.1). Although there are still unresolved questions, recent genetic evidence rather supports the view that fPIPP and TDP are not based on fundamentally distinct photosystems (see Sections 23.8.5 and 23.8.6).

Efforts have been made to explain TDP in terms of fPIPP and the properties associated with it. Blaauw and Blaauw-Jansen [271] considered desensitization of fPIPP to be linked to the expression of TDP. Poff and his co-workers provided experimental evidence that the stimulation-time-dependent nature of TDP is somehow related to the ability of the fPIPP system to recover light responsiveness [99,128,193,273]. Iino [113] explained TDP on the bases of both the desensitization of fPIPP and the recovery of fPIPP responsiveness. In light of the extended photoproduct-gradient model (see the preceding section), he hypothesized that the photoproduct *B* and its precursor *A* cycle during unilateral irradiation. It was then thought that, although the blue-light-dependent forward reaction ($A \rightarrow B$) is initially saturated in both sides of the organ, a lateral gradient in *B* is established while *A* and *B* cycle because the reaction is subject to blue-light-dependent desensitization.

In maize coleoptiles, TDP begins to be inducible from about 4 min of irradiation and the extent of TDP increases sharply within the next 10 min or so; with further extension of irradiation time, the increase in response levels off and a slow increase follows [107] (see Section 23.3.3). The model of Iino mentioned above can explain these kinetic features as follows [113]. The maximal extent of desensitization is achieved within several minutes after a high-fluence pulse (see Figure 18C). This desensitization process could account for the minimal stimulation time of about 4 min. The establishment of the gradient in *B* under the desensitized condition is limited by the recovery of *A*, which progresses sharply for 15 min after a high-fluence pulse (see Figure 18B). This recovery process could account for the sharp response increase observed after 4 min of stimulation. This explanation assumes that the phototropic response is more or less determined by the gradient in *B* established at the end of TDP stimulation. This assumption is supported by the conclusion that the phototropic signal is retained for a substantially long period after the end of stimulation (see Section 23.7.4).

Although the dark-adapted maize coleoptile can express fPIPP, pretreatment with red light is necessary before TDP becomes inducible in this coleoptile [107] (see Section 23.4.4). In view of the conclusion that fPIPP and TDP are not based on entirely distinct photosystems, the above fact can be interpreted as indicating that the photosystem that is sufficient to bring about fPIPP needs to be modified, e.g. by the addition of new molecular components, before TDP becomes inducible.

Khurana and Poff [252] found that the *nph1-2* (JK224) mutant, which shows substantially desensitized fPIPP, can express TDP almost normally. However, the results do not exclude the possibility that TDP is similarly desensitized (discussed in [90]). Because TDP is relatively fluence rate-independent over a wide range of fluence rates, any possible difference in the sensitivity of TDP must be evaluated at low fluence rates that limit TDP.

23.9.4 Molecular bases of the responsiveness and sensitivity changes

The plasma membrane preparation does not show the blue-light-dependent phosphorylation of phototropin (here synonymous with the blue-light-responsive protein) when isolated immediately after administration of a saturating blue-light pulse. However, the plasma membrane gradually regains in vivo the capacity for the blue-light response. This recovery process was first detected in pea stems [239], and subsequently found in maize [245,267] and oat [169] coleoptiles. The simplest explanation is that phototropin is dephosphorylated in the dark allowing the recovery of phosphorylation capacity, although to date such a recovery process has not been demonstrated directly.

The level of phosphorylated phototropin in plasma membrane fractions declines in vitro after the protein was phosphorylated in response to a saturating pulse of blue light

[240,243,246]. At 30°C, the phosphorylation level declines with a half time of 5–10 min. Under this in vitro situation, the phosphorylation level could not be enhanced again by a second pulse of blue light [246]. It was noted that the phosphorylation level of phototropin in the plasma membrane preparation declines with similar kinetics even during continuous stimulation with blue light [240]. The changes in the phosphorylation level after pulse stimulation or during continuous stimulation do not represent the equilibrium between phosphorylation and dephosphorylation [240]. In other words, the observed decrease in the phosphorylation level represents only a dephosphorylation process. These results indicate that phosphorylated phototropin can be dephosphorylated in vitro but do not provide evidence for the recovery of phosphorylation capacity.

The in vitro measurements have indicated that phototropin in plasma membrane preparations retains the autophosphorylating activity after a pulse of blue light, the activity decaying with time in the dark [236,246]. This was shown by applying [γ-^{32}ATP] to the membrane preparation at various times after blue-light stimulation. The result demonstrates that phototropin can remain activated in the dark in the non-phosphorylated state. Interestingly, the soluble phototropin isolated from recombinant insect cells can remain activated longer than the phototropin in the *Arabidopsis* plasma membrane fraction [236]. The dephosphorylation and the decay of phosphorylation activity observed in membrane fractions after a blue-light pulse followed very similar kinetics although the two processes are probably not physically correlated; also the dephosphorylation occurred more rapidly than the recovery in vivo of phosphorylation capacity [169,239,245,267]. Perhaps these kinetics resolved by in vitro measurements represent some kind of destructive protein inactivation that results in both dephosphorylation and loss of phosphorylation activity. Phototropin may remain in the active state for a long period in vivo than resolved by in vitro measurements, and the recovery in vivo of the phosphorylation capacity may indeed represent the dephosphorylation process.

It has been suggested that the recovery of the phosphorylation capacity after a saturating blue-light pulse underlies the recovery of fPIPP responsiveness observed after a high-fluence pulse [251]. However, as described above (Section 23.9.1), the available results indicate that the amount of phosphorylated phototropin does not limit the fPIPP responsiveness. At present, it seems that the recovery of the phosphorylation capacity is not directly correlated with the recovery of phototropic responsiveness. The limiting photoproduct that represents the recovery of phototropic responsiveness is probably a component that occurs in the signal transduction downstream of phototropin.

It is an intriguing possibility that the phosphorylation and dephosphorylation of phototropin may rather be related to the desensitization and the subsequent resensitization, respectively, of fPIPP observed after administration of a high-fluence blue-light pulse. Such relationships can occur if the phosphorylated phototropin is less effective in mediating the formation of the limiting photoproduct (i.e. phosphorylated phototropin has a lower quantum yield for the formation of the limiting photoproduct). This idea is supported by some correlations between phosphorylation and desensitization response. The recovery of phosphorylation capacity progresses over a period of 60 min in peas [239], 20–30 min in maize [245,267], and 90 min in oats ([169]; see the data for the coleoptile tip). In essential agreement with such kinetics, resensitization of fPIPP takes place over a period of 40–60 min in maize ([113]; see Figure 18C). The recovery of phosphorylation is slow during the initial several min [239,245], which also agrees with the resensitization kinetics [113]. The fluence-response curve for the blue-light-dependent phosphorylation in vivo spans from $10^{0.5}$ to 10^3 μmol m^{-2} in dark-adapted and red-light-pretreated maize coleoptiles [267] and from $10^{-0.5}$ to 10^3 μmol m^{-2} in red-light-grown oat coleoptiles [250]. These fluences are 10- to 30-fold higher than those effective for the fPIPP in red-light-grown plants (see Figures 3 and 4). This relationship agrees with the requirement of similarly higher fluences for the desensitization response in oats [271]. Reciprocity holds for the pulse-induced phosphorylation response [239], as it does for the desensitization response [271].

As described above, stimulation of the tissue with a saturating pulse of blue light eliminates the membrane-associated phototropin that can be phosphorylated in vitro, but the capacity for phosphorylation is restored with time. Interestingly, Salomon et al. [169] found for oat coleoptiles that the capacity for phosphorylation is restored to exceed the level found initially without any blue-light stimulation. Thus, the level of blue-light-dependent phosphorylation response was actually enhanced following the blue-light stimulation. This enhancement was most apparent in the coleoptile zones below 5 mm from the tip (more than 100% enhancement in 60 min). Little enhancement could be detected in the 2-mm tip. The authors hypothesized that the enhancement in phosphorylation, which probably results because the amount of the protein itself increases in response to the blue-light pulse, is related to the expression of TDP. However, since the coleoptile tip is the region most responsive to TDP stimulus (see Section 23.5.2), it is unlikely that the enhancement response is a component of TDP. The response may rather be related to the blue-light-dependent enhancement of the phototropic responsiveness (see Section 23.5.6).

23.10 The growth mechanism and hormonal mediation

The final step in the process of phototropism is the curvature response. We will now consider the mechanisms of phototropism from this final step backwards. The discussion is centered on the nature of the growth response underlying phototropism and the roles played by auxin and other growth-regulating substances in phototropic growth responses.

23.10.1 Historical background

Blaauw [274] explained the positive phototropic curvature of higher plant organs in terms of an inhibitory effect of light on growth. It was postulated that the growth on the irradiated side of a unilaterally irradiated organ is inhibited to a greater extent than that on the shaded side where light is attenuated. The resulting growth asymmetry was thought to cause the curvature. Blaauw based this hypothesis on his own observation that light inhibits the growth of *Helianthus* hypocotyls and on some correlative relationships found between phototropism and light-dependent growth inhibition. The hypothesis requires that the light-growth response occurs locally at cellular or tissue sites of photoperception. (The term "light-growth response" generally refers to the light-induced response observed in elongation growth.)

A totally different view arose from the studies with oat coleoptiles which were conducted to resolve the nature of the transmittable influence suggested by Darwin ([22]; see Section 23.5.1). Boysen-Jensen [275] was able to demonstrate that the transmittable influence could move from the excised tip to the stump of oat coleoptiles through a layer of gelatin. Thus, the transmittable influence was thought to be of a chemical nature. It was next shown by Paál [276] that decapitated coleoptiles could bend without phototropic stimulus when the excised tip was placed onto one side of the apical cut surface. This curvature occurred away from the side where the coleoptile tip was placed. He put forward the following thoughts:

1. the coleoptile tip internally secretes a growth promoting substance;
2. this substance is transported downwards and distributed symmetrically to sustain symmetric growth of the coleoptile; and
3. phototropic stimulation brings about an asymmetric distribution of this substance, which in turn results in growth asymmetry, i.e. curvature.

The idea that the tip supplies a growth-promoting substance to the lower parts was subsequently supported by Söding [277,278]. He demonstrated that the growth of coleoptiles suppressed by decapitation is restored substantially by replacement of the excised tip.

Went [279] finally succeeded in separating such a growth-promoting substance, later named auxin, from the coleoptile tissue. This substance was shown to diffuse out of the excised tip of oat coleoptiles into a block of agar through the cut surface and to cause curvature of decapitated coleoptiles when applied asymmetrically. By measuring the auxin activity with the *Avena* curvature test he had developed, Went [280] obtained a number of important results confirming and extending earlier conclusions. Most importantly Went was able to demonstrate that the amount of diffusible auxin obtained from the excised coleoptile tip is larger on the shaded half than on the irradiated half of the phototropically stimulated coleoptile. He proposed, with experimental evidence, that auxin is laterally translocated in the tip and that the resulting auxin asymmetry is transmitted to the lower parts, owing to the basipetally polarized transport of auxin, to cause the coleoptile curvature.

Slightly preceding the work of Went described above, Cholodny [281,282] published results on gravitropism of roots and hypocotyls. He explained gravitropism in terms of growth-regulating substances (hormones), as Paál did for phototropism. Cholodny's hypothesis was more specific, however, with regard to the way by which growth asymmetry is induced. He thought that a certain physiological polarity of cells induced by the gravity stimulus disturbs the normal flow of growth-regulating substances, causing a laterally directed flow and an asymmetric distribution of the substances. Cholodny [283] suggested that the phototropism of coleoptiles could be explained similarly. Evidence for this hypothesis was immediately provided by Went, but specifically for auxin (see above). Based on these early works, a generalized hypothesis, known as the *Cholodny-Went theory of tropisms*, was formulated [158]. This hypothesis can be separated into the following three parts:

1. tropisms are caused by an asymmetric distribution of auxin, the idea originally presented by Paál;

2. the asymmetric distribution results from lateral translocation or transport of auxin; and
3. the lateral translocation results from the transverse polarity of cells induced by the tropic stimulus.

The last part does not tell us about any specific mechanism, but implies that stimulus perception and lateral auxin translocation are closely linked within the stimulus-sensing cells. The function of the organ apex as the site of the lateral auxin translocation was specified by Went for the phototropism of oat coleoptiles, but is not included in the generalized hypothesis.

Although indole-3-acetic acid (IAA) was known as early as 1934 to have an auxin activity, it was quite a while before its widespread occurrence in plants was established with unequivocal identification methods [284]. It is now generally believed that IAA is the major, if not the sole, auxin of plants. Later investigations with maize seedlings have confirmed that IAA is produced in the tip of coleoptiles and is transported to the lower parts [159,285,286].

After the work of Went, the role of auxin has been a major topic of phototropism research. By measuring diffusible and extractable auxin with the *Avena* curvature test, early workers could at least confirm that auxin is distributed asymmetrically in oat coleoptiles and other organs in response to phototropic stimulation [158]. The Went diffusion method was later applied to maize coleoptiles, leading to a much clearer demonstration of the lateral auxin redistribution [287,288]. The alternative idea that the auxin asymmetry is caused by a difference in light-dependent auxin breakdown [217] became less likely. Radioisotope-labeled IAA began to be used in the 1950s as a tracer of phototropic auxin distribution. This line of study did not provide conclusive results until Pickard and Thimann [289] could finally demonstrate that ^{14}C-IAA was asymmetrically distributed in maize coleoptile tips following phototropic stimulation. These studies have substantiated the notion of the lateral auxin translocation.

Since the beginning of the 1980s, the Cholodny-Went theory for phototropism has been subjected to various criticisms. Firn and Digby [290] listed a number of basic issues that had not been clarified. Trewavas [291] also listed a number of criticisms, one of which was that the extent of auxin asymmetry reported would be insufficient to account for the curvature response observed. A more serious challenge was that of Hasegawa and his co-workers, who could not find any asymmetric distribution of IAA in the phototropically stimulated organs including oat and maize coleoptiles (summarized in [292]). As an alternative, Blaauw's hypothesis has been explored [139,149]. In essential support of this hypothesis, Hasegawa and co-workers have presented evidence that growth inhibitors accumulate on the irradiated side of hypocotyls and coleoptiles (see [292]). Meanwhile, evidence also accumulated in support of the Cholodny-Went theory for phototropism. Generally speaking, it is difficult to prove that one mechanism is universal or, alternatively, that it does not occur at all. Nevertheless, as will be discussed below, many experimental results now indicate that Blaauw's hypothesis cannot explain any major part of the blue-light-sensitive phototropism.

The topic of this section is closely related to the growth mechanism extensively investigated for gravitropism, although this section will make no detailed reference to the corresponding results on gravitropism. These results from gravitropism research have been reviewed in detail by Pickard [293]. More recent reports include those by Clifford et al. [294], Harrison and Pickard [295], Parker and Briggs [296], Iino [297], Li et al. [298,299], and Bennett et al. [300]. These reports have provided strong support for the Cholodny-Went theory for gravitropism. Various controversies that have continued until recently concerning the Cholodny-Went theory of tropisms can been found in the written debate edited by Trewavas [301].

23.10.2 Evaluation of Blaauw's hypothesis

Substantial phototropism, either fPIPP or TDP, follows stimulation of the tip alone in oat and maize coleoptiles (see Section 23.5.3). Blaauw's hypothesis, which is based on a local light-growth response, fails to explain this type of phototropism. This argument does not apply to the phototropism induced by non-tip photoperception. However, the early work of Beyer [302] already indicated that phototropic response to continuous white light is not exactly correlated with the light-growth response in oat coleoptiles. Beyer [303] also found that decapitated coleoptiles of oats and barley could show a light-growth response without phototropic curvature. Furthermore, Cholodny [304] indicated that a pulse of unilateral white light could induce phototropism (fPIPP) without causing any light-growth response. Later work has in fact revealed that the blue-light-dependent growth inhibition found in coleoptiles does not match kinetically with the phototropic response. In oat coleoptiles, growth is rapidly inhibited following the onset of blue-light irradiation, and the inhibition disappears within 30 min after the termination of irradiation [305,306]. Phototropism develops more slowly and persists for a much longer period (see Section 23.7.2). In red-light-grown maize coleoptiles, a pulse of blue light induces a transient growth inhibition; even for continuous irradiation, a large part of blue-light-dependent growth inhibition occurs transiently following the onset of irradiation [307]. These transient growth responses complete within a period in which any major phototropic curvature takes place in unilaterally irradiated coleoptiles.

In dark-adapted oat coleoptiles, the most apparent growth response that followed continuous irradiation with unilateral blue light was inhibition on the irradiated side [148,149,192], and this inhibition occurred in locally irradiated zones [116]. Such data seemed to favor Blaauw's hypothesis. However, the subsequent investigation by Macleod et al. [306] has led to the conclusion that Blaauw's hypothesis cannot explain phototropism. To induce phototropism of dark-adapted oat coleoptiles, they used two fluence rates of blue light differing by a factor of 10 (Figure 19). At either fluence rate, phototropism could be described by strong growth inhibition on the irradiated side; growth on the shaded side was either slightly stimulated or little affected. They next conducted bilateral irradiation using the lower fluence rate in one direction and the higher fluence rate in the other direction. The coleoptile bent towards the higher fluence-rate side, with growth changes similar to those found in the unilaterally irradiated coleoptiles; that is, the growth on the side receiving the weaker light was not at all inhibited, although unilateral irradiation with this light caused a substantial inhibition on the irradiated side. Clearly, the decrease in growth rate observed on the irradiated side does not result from any direct growth-inhibiting effect of light.

Evidence has also accumulated indicating that phototropism of dicotyledonous hypocotyls and epicotyls cannot be explained by growth-inhibiting effects of blue light. It was noted that a low fluence-rate blue light, that was too weak to cause any detectable growth inhibition, could induce phototropism in mustard seedlings [308]. Unilateral blue light inhibited the growth of dark-grown cucumber hypocotyls rapidly and maximally well before phototropic curvature occurred [140]. In *Arabidopsis* hypocotyls, CRY1 mediates a large part of blue-light-dependent growth inhibition. The mutant of *Arabidopsis* that turned out later to be a mutant lacking CRY1-mediated growth inhibition showed normal fPIPP and TDP [234].

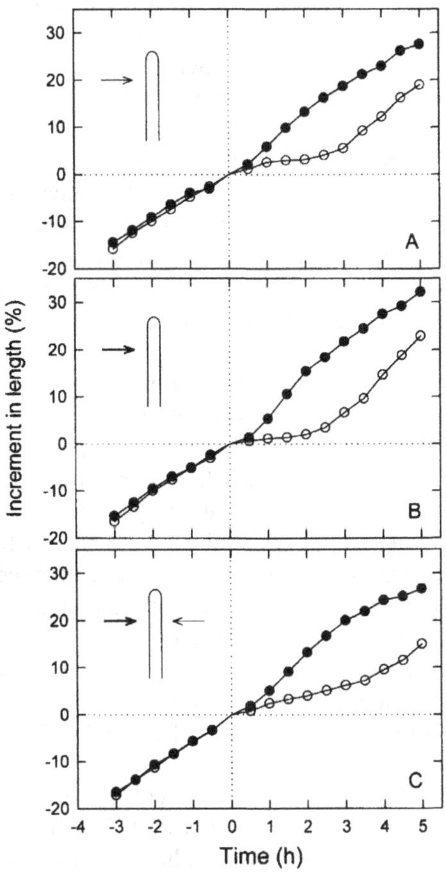

Figure 19. The experiments demonstrating that the growth inhibition on the irradiated side of oat coleoptiles does not result from a local growth-inhibiting effect of blue light. Dark-adapted coleoptiles were stimulated continuously with unilateral or bilateral blue light. The growth on the two sides was monitored. (**A**) Unilateral stimulation with a low fluence rate (0.03 μmol m^{-2} s^{-1}). (**B**) Unilateral stimulation with a high fluence rate (0.3 μmol m^{-2} s^{-1}). (**C**) Bilateral stimulation with low and high fluence rates. Solid circles: the shaded side or the low-fluence-rate side. Open circles: the irradiated side or the high-fluence-rate side. Adapted from Macleod et al. [306]; the original data shown for zones 2 and 3 are averaged.

Our knowledge that phototropism is a growth response has originated from the work of Blaauw. However, the results described above and those described in the next section indicate that his hypothesis based on a local light-growth response cannot explain any major part of blue-light-sensitive phototropism.

23.10.3 Growth redistribution as the primary growth response

When dark-adapted plants are stimulated with unilateral blue light to induce phototropism, light-growth responses will also be induced. If any such response were not causally related to phototropic curvature, then it would mean that the light-growth response simply accompanies the phototropic growth response. This possibility calls for careful treatment of the growth data obtained to analyze the nature of the growth response underlying phototropism. The most extensive study conducted with oat coleoptiles has provided complex results. Before discussing these results, more straightforward results obtained from maize coleoptiles will be introduced.

In the first place, it is necessary to consider that phytochrome-mediated growth responses, which are clearly distinct from the blue-light-sensitive phototropism, could accompany phototropic growth responses, even if pure blue light is used to induce phototropism. This follows the fact that blue light can effectively trigger the very-low-fluence, phytochrome-mediated growth responses [110,143]. Iino and Briggs [151] used red-light-grown maize coleoptiles to analyze the phototropic growth response. The red-light condition was thought to eliminate any possible induction of phytochrome-mediated responses by blue light. This study has established that fPIPP of this material results from redistribution of growth, i.e. inhibition on the irradiated side and compensating stimulation on the shaded side. Figure 20 depicts such a redistribution pattern observed during the fPIPP induced at an optimal fluence. The results shown in this figure also indicate that nearly identical growth changes follow irradiation of the entire coleoptile and irradiation of only the 1-mm tip. Fluence-response analysis has indicated that the overall growth was stimulated at high fluences on the descending arm of fPIPP. However, this growth stimulation could be separated from the phototropic growth redistribution when only the 1-mm tip was exposed to unilateral light. Furthermore, the growth responses on the two sides occurred about simultaneously in the three successive 5-mm zones, but with a basipetal delay of the onset of growth responses, a feature described in Section 23.5.3.

The growth response during fPIPP of oat coleoptiles has been investigated by Blaauw-Jansen [120], Curry [186], Macleod et al. [157], Iino [90], and Tarui and Iino [155]. Growth inhibition on the irradiated side has been observed consistently. Growth on the shaded side was either little affected or stimulated depending on the conditions. Blaauw-Jansen [120] found a clear redistribution pattern in red-light-pretreated coleoptiles. Iino ([90]; see his Figure 2) could also find a redistribution pattern using red-light-grown seedlings, but only at a sub-optimal fluence inducing fPIPP on the ascending arm of the fluence-response curve. At a higher fluence, optimal for fPIPP, growth on the shaded side was stimulated only slightly, while that on the irradiated side was inhibited substantially. The fluence of blue light (0.08 μmol m^{-2}) used by Blaauw-Jansen was clearly sub-optimal for fPIPP, in agreement with the result of Iino. These

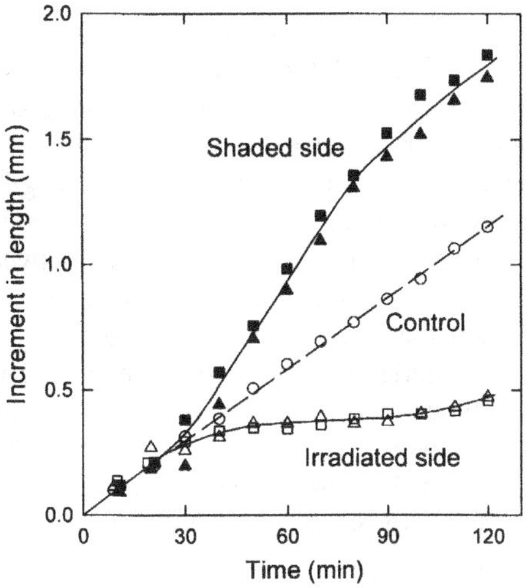

Figure 20. Redistribution of growth during fPIPP of red-light-grown maize coleoptiles. Coleoptiles were stimulated with a unilateral pulse of blue light (5 μmol m^{-2} s^{-1}) and the growth on irradiated and shaded sides of the apical 15-mm part was monitored. The entire coleoptile (squares) or the apical 1-mm part (triangles) was subjected to phototropic stimulation. Control plants (circles) received no stimulation. Adapted from Iino and Briggs [151].

results indicate that fPIPP can be caused by a simple growth redistribution, but it remains to be answered why the net growth was inhibited at the optimal fluence.

In the work mentioned above, Iino [90] found that the decrease in net growth is less at a high fluence causing PINP than at the fluence optimal for fPIPP. Blaauw-Jansen [120] also noted a similar situation in dark-adapted coleoptiles not pretreated with red light. The fact that the decrease in net growth occurs specifically under fPIPP stimulus conditions has been confirmed by Tarui and Iino [155]. In this study, red-light-grown oat coleoptiles were subjected to fPIPP stimulation and the growth on the two sides of three successive 5-mm zones was monitored. At an optimal fluence, the net growth was inhibited in all zones. A 30-times higher fluence given bilaterally did not induce any significant inhibition in all zones. Such results may be explained by assuming that two light-growth responses, one inhibiting growth at low fluences and another stimulating growth at high fluences, accompany phototropic growth redistribution. However, the time-course data did not reveal any complex changes in net growth at all fluences investigated, i.e. no kinetic separation between net growth change and the development of growth asymmetry [90]. Also, after the high-fluence pulse, no subtle change in net growth rate was observed in any investigated zones [155]. It appears that the observed decrease in net growth is not based on a growth-inhibiting effect of blue light but is somehow related to the mechanism of fPIPP itself. (The work of Tarui and Iino [155] has also indicated that the irradiated side of the most apical zone shrinks when the net growth is inhibited during fPIPP. The implication of this result will be discussed in Section 23.11.2.)

The phototropism of continuously stimulated oat coleoptiles has also provided various results. When dark-adapted coleoptiles were exposed continuously to unilateral blue light, growth on the irradiated side was markedly inhibited while that on the shaded side was little affected [148,149,192,306]. Therefore, net growth decreased during phototropism. In the most apical zone, however, a clear growth stimulation on the shaded side could be noted [306]. Curry [186] used red-light-grown coleoptiles and measured growth on the two sides of nearly the entire length of coleoptiles during continuous unilateral stimulation with blue light. A decrease in net growth occurred initially, but the overall curvature response that developed between 30 min and 120 min of stimulation could be characterized by a clear redistribution pattern. Macleod et al. [157] used red-light-adapted oat coleoptiles to measure growth of their zones; the coleoptiles were maintained under red light for 10 h before the onset of blue-light irradiation and also during blue-light irradiation. In contrast to the case of dark-adapted coleoptiles, growth on the shaded side was not stimulated in the most apical zone and the net growth was inhibited in this zone. However, a clear growth stimulation on the shaded side was found in the most basal zone. Hasegawa and Sakoda [309] observed a redistribution pattern in all investigated zones. In this work, dark-adapted coleoptiles were stimulated with unilateral white light; the coleoptiles were briefly pretreated with red light 1 h before the onset of unilateral light.

As mentioned in the preceding section, the growth of dark-adapted oat coleoptiles is inhibited by continuous blue-light irradiation. This blue-light response, which is evidently not the cause of phototropism, would explain the net-growth decrease generally observed during the phototropic response of dark-adapted coleoptiles to continuous stimulation. The data of Macleod et al. [306] indicate that the blue-light-induced growth inhibition is not expressed significantly in the most apical zone and becomes greater basipetally (see their Figure 4). This explains why a clear redistribution pattern was found in the apical zone, while the net growth on the lower zones decreased during continuous stimulation [157]. On the other hand, the decrease in net growth found in the most apical zone of red-adapted coleoptiles [157] agrees with the result obtained for fPIPP of red-light-grown coleoptiles (see above). It may be suggested that the marked decrease in net growth thought to be associated with the mechanism of fPIPP also takes place during phototropic response to continuous irradiation, the occurrence of the net-growth decrease being related to the red-light condition. The net-growth decrease in the apical zone does not occur in the coleoptiles pretreated briefly with red light [309] or handled under green working light in advance of phototropic stimulation [157,306]. These results suggest that the net-growth decrease possibly associated with the mechanism of phototropism becomes apparent only when coleoptiles are exposed to red light for a relatively long period. All these arguments lead to a plausible conclusion that redistribution is the primary growth response underlying the phototropic response of oat coleoptiles to continuous stimulation, although the mechanism of phototropism may allow a decrease in net growth under certain conditions.

The growth of dicotyledonous hypocotyls and epicotyls is inhibited markedly by light, being mediated by both phytochromes and blue-light receptors. As mentioned in the preceding section, the blue-light-sensitive phototropism of hypocotyls can be separated kinetically and genetically from the major growth-inhibiting effect of blue light. If dark-adapted dicotyledonous seedlings were exposed to unilateral blue light, the phototropic response would be accompanied by such non-phototropic blue-light response, and possibly also by phytochrome-mediated growth response. Rich et al. [310,311] adapted mustard seedlings to a high fluence-rate orange light from low-pressure sodium lamps, before and during continuous phototropic stimulation with blue light. A high and constant Pfr level was expected to be maintained throughout the experiment. These workers found that phototropism is induced by growth inhibition on the irradiated side and growth stimulation on the shaded side. Using red-light-grown pea seedlings, Baskin obtained similar results for the fPIPP [102] and TDP (cited in [312]) of epicotyls. These studies have not revealed any substantial inhibition of net growth which might be caused by a non-phototropic blue-light system. Perhaps, the blue light used to induce phototropism was not optimal for the induction of blue-light-dependent growth inhibition. It is also possible that blue-light-dependent growth inhibition was induced only transiently under the phytochrome-saturated condition, as was shown in maize coleoptiles [307]. In view of the above results, it can be concluded that the phototropism of dicotyledonous stems is caused, in principle, by a redistribution of growth.

Clearly, local light-growth response cannot explain blue-light-sensitive phototropisms of coleoptiles and hypocotyls, and growth redistribution is the primary growth response underlying these phototropisms. It can be stated that the process of phototropism begins with a "gradient" in light-driven reactions and ends up with a "redistribution" of growth. This conclusion agrees with the concept of the transverse polarity that mediates phototropic response, and the Cholodny-Went theory is the only one available hypothesis that can adequately explain the underlying process. As far as the redistribution matter is concerned, however, any future hypothesis that bases the growth asymmetry on a lateral translocation of a substance would equally be reconciled with the above conclusion.

It appears to be a rational strategy of plants to use directional light information for phototropism independently of the light information for photomorphogenetic responses that control the overall growth pattern [90]. As described in the first part of this chapter, shade-avoiding growth orientation is apparently an important function of phototropism. In this connection, it is especially reasonable that plants can express phototropism independently of the blue-light-dependent growth inhibition, which would counteract the shade-avoiding growth acceleration.

23.10.4 Asymmetric distribution of endogenous auxin

To begin our discussion on the relationship between auxin and phototropism, we will review the evidence that endogenous auxin is distributed asymmetrically following or during phototropic stimulation (the most basic condition for the first part of the Cholodny-Went theory; see Section 23.10.1).

Evidence for the asymmetric auxin distribution was first provided by Went [280] with the following experimental procedure:

1. excising the tip from the coleoptile which has been subjected to an fPIPP stimulus;
2. setting the tip on a set of agar blocks separated by a razor blade, in such a way that the irradiated and shaded halves rest separately on blocks;

3. allowing auxin to diffuse into the blocks for a defined period; and
4. assaying the auxin activity in the blocks with the *Avena* curvature test.

With a diffusion time of 2.5 h, an auxin gradient of about 1 : 2 (irradiated:shaded half) was detected. This result could be reproduced by subsequent workers [189,313,314].

Went's diffusion method and the *Avena* curvature test were applied to other response types and different materials. Asana [314] found an opposite auxin gradient (a greater amount on the irradiated side) with the oat coleoptile tip stimulated for PINP. Using the excised shoot of light-grown radish seedlings from which the cotyledons but not the petioles were removed, van Overbeek [313] found that diffusible auxin obtained from the basal cut end of hypocotyls is asymmetrically distributed (with a greater amount on the shaded half). In this study, the explants were stimulated continuously with unilateral white light during the diffusion period. Later, Briggs et al. [287] showed that auxin is asymmetrically distributed in continuously stimulated maize coleoptile tips (4 mm). Extending the work on maize coleoptiles, Briggs [288] could show further that an asymmetric distribution is induced by an fPIPP stimulus as well as by a TDP stimulus (20-min irradiation). The auxin gradient detected was about 1 : 2 in either fPIPP or TDP. In these experiments, the phototropic stimulation was given just after the placement of the excised tips on agar blocks and auxin was allowed to diffuse for 3–3.5 h. Briggs [288] also applied a high fluence in a 100-s pulse. This fluence caused a small positive curvature, which probably corresponded to sPIPP, and a small auxin gradient in favor of the curvature response.

The work of Wilden [315] will also be mentioned. She excised the phototropically stimulated tip of oat coleoptiles and placed it directly and asymmetrically on the decapitated coleoptile of the curvature test (i.e. an extension of Paál's method). Either the irradiated or the shaded side of the tip was kept in contact with the test plant. With the coleoptile tips stimulated for fPIPP and PINP, a greater curvature of the test plant was induced when the shaded side and the irradiated side, respectively, were in contact with the test plants. Thus, this simple procedure provided results that agree with those obtained with Went's diffusion method. Wilden also examined a fluence that was higher than the fluence used to induce PINP. This high fluence caused a positive curvature and, in agreement with the curvature response, the stimulated tip induced a greater curvature of the test plant when the shaded side was in contact with the test plant. The high fluence was given in 50 s. Thus, the phototropism induced was not TDP but sPIPP.

The possible occurrence of auxin asymmetry within the tissue was investigated by measuring the auxin activity of tissue extracts with the *Avena* curvature test. The first demonstration was achieved by Boysen-Jensen [316]. He stimulated the epicotyl of either dark-grown or light-grown *Phaseolus multiflorus* seedlings continuously with unilateral light and measured the auxin activity of the chloroform extracts. A gradient of about 1 : 2 was detected between the irradiated and the shaded half of the epicotyl segments. Successful detection of an auxin gradient in coleoptile tissue was achieved by Oppenoorth [189]. He measured the auxin activity of the ether extracts obtained from the irradiated and shaded halves of oat coleoptile tips (3 mm). Auxin asymmetry was found to follow the light stimulation causing fPIPP, PINP, and TDP, although the gradients detected were smaller than those found for diffusible auxin. The direction of the gradient agreed with the direction of each curvature response.

It is essential to confirm the bioassay data with direct physicochemical determinations of IAA. Until now, this confirmation has been made only for the fPIPP of red-light-grown maize coleoptiles. To determine IAA in a small amount of plant materials, Iino [297] used an improved indolo-α-pyrone fluorescence method. The accurate determination of endogenous IAA does not depend only on the final assay method [317]. He used extraction and purification methods that were improved to alleviate the problems related to non-enzymatic decomposition of IAA and the conversion of indole-3-pyruvic acid (IPyA) to IAA during extraction and purification of IAA. The reliability of the entire procedure had been checked in comparison with the determination by a GC-MS method [318]. A gradient of about 1 : 2 was detected in the diffusible IAA obtained from the two half-sides of the coleoptile tip (3 mm). A similar gradient was found for the IAA extracted from the subapical coleoptile zone, 2–7 mm from the tip (Figure 21).

Li et al. [298] used tobacco plants transformed with the soybean small auxin up-regulated RNA (SAUR) promoter fused to a GUS reporter to investigate the relationship between endogenous auxin and phototropism. Seven to 10 days old plants were exposed continuously to unilateral light after 24-h dark adaptation. A greater GUS expression could be detected in the shaded half than in the irradiated half of the phototropically responding zone of the stem. The gradient was a little smaller than 1 : 2. The result could be attributed to an asymmetric distribution of endogenous auxin in the stem (see [299] for an extended study).

In summary, the earlier results indicate that the auxin detected by the *Avena* curvature test is asymmetrically distributed in phototropically stimulated coleoptiles, hypocotyls, and epicotyls. In coleoptiles, all response types could have been correlated with auxin asymmetry. In one case (fPIPP of maize coleoptiles), the asymmetry was directly shown for IAA. Transgenic tobacco was used successfully to demonstrate auxin asymmetry in stems.

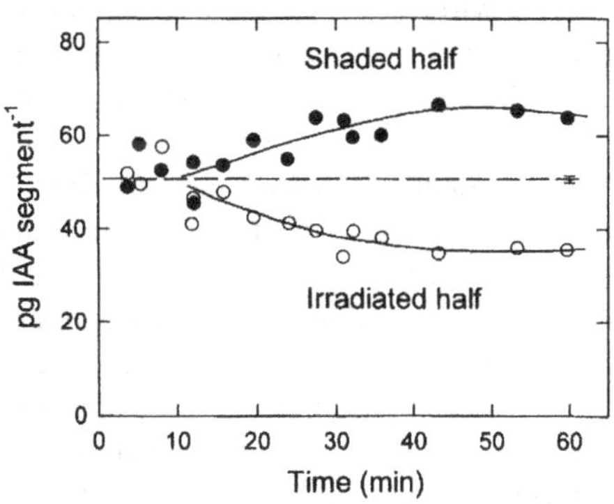

Figure 21. Redistribution of IAA during fPIPP of red-light-grown maize coleoptiles. Coleoptiles were stimulated with a unilateral pulse of blue light (2.6 μmol m^{-2}). At the indicated time after stimulation, the subapical zone, 2–7 mm from the tip, was excised and divided into irradiated and shaded halves. The amount of extractable free IAA was determined. The horizontal dashed line indicates the mean IAA amount per half segment determined without phototropic stimulation; the vertical line at the right end of the dashed line indicates ± SE (n = 6). Adapted from Iino [297].

23.10.5 Lateral translocation of auxin

We will now review the evidence for the second, most original part of the Cholodny-Went theory, that the asymmetric auxin distribution is caused by lateral translocation of auxin. In the experiment measuring diffusible auxin from the tip of oat coleoptiles, Went [280] found that the irradiated side of the tip diffused less auxin and the shaded side more auxin as compared to the non-irradiated control. Furthermore, the auxin gradient increased during the two successive 75-min diffusion periods without a change in the net amount. Based on these results, he concluded that auxin was laterally translocated within the tip. The original data of Went were somewhat ambiguous for the conclusion, however, because the amount of auxin obtained from the whole tip was reduced by phototropic stimulation, and the claimed increase on the shaded side was relatively small. This ambiguity invited some criticism.

The later measurements of diffusible auxin in maize coleoptile tips have provided clear evidence that auxin is asymmetrically distributed without a significant change in the net amount during continuous stimulation [287] and after the limited stimulation causing fPIPP and TDP [288]. Furthermore, no auxin asymmetry occurred when the tip was totally bisected in the plane perpendicular to the direction of unilateral light [287]. The result that diffusible auxin is redistributed following the fPIPP stimulus was confirmed for IAA [297]. It was also shown that the tissue content of extractable IAA was similarly redistributed in the subapical zone [297]. The asymmetry of extractable IAA detected in the 2-mm tip was much smaller than that in the subapical zone. This probably arose because the coleoptile tip contains a pool of IAA that has been produced but has not yet been translocated laterally. These results from maize coleoptiles have provided straightforward evidence for the idea that endogenous auxin is laterally translocated in the tip following phototropic stimulation. The results on endogenous auxin, however, do not entirely disprove the alternative possibility that the auxin asymmetry is caused by an asymmetry of auxin biosynthesis activity.

The work of van Overbeek with radish hypocotyl segments [313] provided the first evidence that exogenously applied auxin can be laterally translocated in response to phototropic stimulation. He placed an agar block containing auxin on the whole area or the irradiated half of the apical cut end of hypocotyl segments, which were obtained from the seedling shoots decapitated 4 h before the experiment. The amounts of auxin transported to the receiver blocks placed at the two half-sides of the base were measured by the *Avena* curvature test. (Van Overbeek used the auxin isolated from urine, which was very probably IAA.) It was shown that auxin could in fact be translocated laterally while segments are continuously irradiated with unilateral light. This type of experiments is impractical with coleoptiles because coleoptile segments begin to produce auxin and it is difficult to eliminate the contribution of endogenous auxin [159]. Such an initiation of auxin production would not generally occur in stems [319,320], so auxin could well be depleted from the hypocotyl after removal of the cotyledons and apex, the major sources of auxin in the seedling [313].

The fact that plant organs have an ability to translocate auxin laterally in response to unilateral light has been shown more clearly by tracer experiments with isotope-labeled IAA. The first success was achieved by Pickard and Thimann [289]. They applied 1-^{14}C-IAA to the top surface of the coleoptile tip (6.5 mm) excised from red-light-pretreated maize coleoptiles, and measured the radioactivity transported to the agar blocks placed at the irradiated and shaded halves of the basal cut end. (As shown by Goldsmith and Thimann [321] and supported by many later workers, the radioactivity obtained in the agar block can almost entirely be attributed to non-degraded tracer IAA.) Stimulation for fPIPP and TDP of the tip resulted in gradients of 1 : 2 and 1 : 3, respectively, after a 3-h diffusion period. The gradients detected were similar to those of endogenous auxin shown by Briggs [288] under comparable conditions (see above).

Gardner et al. [322] used a micropipette to apply a solution of ^{3}H-IAA asymmetrically to the tip of maize and oat coleoptiles, and measured lateral and longitudinal distributions of radioactivity in the coleoptile. The radioactivity collected from the tissue was identified to be due mainly to non-degraded ^{3}H-IAA. The study allowed a more direct demonstration that lateral auxin translocation is induced during both fPIPP and TDP. These workers used dark-adapted coleoptiles, with and without red-light pretreatment. They also used both intact and excised coleoptiles. Although a statistically significant lateral translocation could not be demonstrated in some cases, it is possible that the conditions used were not optimal in these cases. Phototropic translocation of ^{3}H-IAA in the coleoptile might have been counteracted by gravitropic translocation (note the difference from the experiments in which only the excised tip was used). In fact, in many of the cases in which statistically significant lateral translocation was not shown, the distribution determination was made at 100–120 min, when phototropic curvature should have slowed down (see also [90] for a discussion). In addition, asymmetric application of ^{3}H-IAA had to be made within a very narrow region of the dome-shaped tip. Although this requirement agrees with the conclusion that lateral translocation occurs in the extreme tip [288], it in turn makes the approach with one-sided application less optimal.

The pea epicotyl is another material in which isotope-labeled IAA has been used. In all the studies described below, pea seedlings were used when the third internode was the most apical elongating internode. Kang and Burg [135] found that ^{14}C-IAA applied from the apical cut end of decapitated third internodes is distributed asymmetrically in the bending region of the internode after 4-h unilateral irradiation with blue light. The gradient detected was 1 : 1.8 (irradiated:shaded half). Using intact seedlings pretreated with red light and applying ^{14}C-IAA to the apical bud, Kuhn and Galston [323] obtained complex results on the longitudinal and lateral distributions of radioactivity in the phototropically responding third internode. Because unilateral blue light was given parallel to the plane of the hook, the results were complicated by the asymmetric auxin distribution related to the morphological asymmetry.

Our unpublished data obtained using red-light-grown intact pea seedlings (K. Haga and M. Iino) demonstrated the occurrence of lateral translocation in the third internode. We found that ^{3}H-IAA applied to the apical bud was distributed asymmetrically in this internode (about 15 mm) in response to the optimal fPIPP stimulus. A gradient of 1 : 1.4 (irradiated:shaded half) occurred in an apical 5-mm zone (about 3–8 mm below the hook) when curvature of the zone had just begun (25 min after stimulation). The net amount of ^{3}H-IAA in the zone was not lower than in the non-stimulated control. Therefore, lateral translocation is the most likely cause of the asymmetric distribution. Because it was possible that a greater asymmetry occurs in the peripheral tissues, we also measured ^{3}H-IAA in the epidermal peels obtained from the irradiated and shaded halves of the same zone. An asymmetry of ^{3}H-IAA was again detected, but its extent was not significantly greater than that measured for bisected segments. In this work, the ^{3}H-IAA mixed with lanolin (0.6 μg g^{-1}) was applied to the top surface (about 20 mm^2) of the apical bud 4 h before phototropic stimulation. The unilateral blue light was provided perpendicular to the plane of the hook. The radioactivity due to ^{3}H-IAA was measured after extraction of ^{3}H-IAA with aqueous methanol and its purification by water/ether partitioning and TLC. The values were corrected for losses during purification using the recovery of the cold IAA added in large excess at the start of extraction. The distribution could be measured with ^{3}H-IAA concentrations that were lower than one hundredth the concentration of endogenous IAA reported for the third internode of peas [324].

The relationship between the distribution of photoperceptivity and the region for lateral auxin translocation in coleoptiles would merit further discussion. Briggs [288] showed that the apical 0.5-mm zone of maize coleoptiles is the major region for lateral auxin translocation during either fPIPP or TDP. On the other hand, the decapitated maize coleoptiles could respond to unilateral light to show some fPIPP curvature when an appropriate amount of IAA is applied through the apical cut surface; about 35% and 15% of the response in intact coleoptile could be found in coleoptiles decapitated at 1 and 2 mm, respectively [152]. If phototropism is mediated by lateral auxin translocation, the latter result must indicate that the region below the apical 1-mm zone has some activity for lateral auxin translocation. Together with the information on the photoperceptivity distribution described in Section 23.5.2, the following picture probably represents the most likely spatial relationship in maize coleoptiles. The apical zone less than 1 mm long, which encompasses the solid part of the tip, is the most active region for both photoperception and lateral auxin translocation (especially in terms of the activity per unit length). The region below the apical 1-mm zone is, however, photoperceptive and can induce some lateral auxin translocation. In oat coleoptiles, high photoperceptivity is clearly confined within the apical 1-mm zone (see Section 23.5.2). As in the case of maize, lateral auxin translocation may take place most actively within this zone. Although this has not been investigated, the finding that ^{14}C-IAA must be applied to the extreme tip for successful demonstration of its lateral translocation [289] is in agreement with this prediction.

In the study mentioned above, Pickard and Thimann [289] also investigated whether auxin could be laterally translocated in non-tip coleoptile zones. They applied ^{14}C-IAA to the apical cut end of maize and oat coleoptile segments from which the tip (1.5 and 2 mm, respectively) had been removed, and then collected radioactivity from the irradiated and shaded half of the basal end using donor and receiver blocks of agar. A small asymmetry (1 : 1.17; greater radioactivity in the shaded half) was detected in maize coleoptile segments after a 3-h diffusion period, during which the segments were continuously stimulated with unilateral white light. A similar result was obtained for oats. dela Fuente and Leopold [325] applied ^{14}C-IAA to one side of decapitated maize coleoptile segments and measured the radioactivity which had migrated to the opposite side of the segment. The segments were irradiated continuously with unilateral blue light for 4 h from either the ^{14}C-IAA-applied side or the opposite side. The data demonstrated that auxin was translocated from the irradiated to the shaded side. In fact, comparison with the non-stimulated control indicated that the lateral translocation occurred in such a way that the migration of ^{14}C-IAA to the irradiated side was reduced and the migration to the shaded side was enhanced. The extent of lateral translocation detected by dela Fuente and Leopold [325] was apparently much greater than that detected by Pickard and Thimann [289]. The former workers stimulated coleoptile segments for a longer period. Also, while the latter workers collected all diffusible ^{14}C-IAA during stimulation, the former workers measured radioactivity distribution at the end of stimulation. The greater translocation might be related to the possibility that the phototropic responsiveness in the lower parts of coleoptiles is enhanced during continuous stimulation (see Sections 23.5.6 and 23.8.4). Another possibility is that the lateral translocation of IAA results in a relatively large asymmetry between the two halves, while the asymmetry of basipetally transportable IAA is not induced to that extent (see Section 23.10.12 for a further discussion).

23.10.6 Difficulties in demonstrating the redistribution of endogenous auxin

If unilateral light irradiation causes a light response that is not directly involved in phototropism but affects the auxin status, then demonstration of phototropic auxin redistribution would become difficult. There are at least two such responses that can obscure the demonstration of auxin redistribution. This problem is analogous to that confronted when analyzing the phototropic growth redistribution (see Section 23.10.3).

The amount of diffusible auxin obtainable from the excised coleoptile tip is reduced in response to red light, as shown in oat [120,326–328], maize [111], and rice [329]. The response is probably mediated by phytochrome. Detailed study with maize coleoptiles has indicated that red light also reduces the tissue content of IAA throughout the coleoptile [286], and that the decrease in diffusible and extractable IAA results from the inhibition of the IAA biosynthesis from tryptophan in the tip [286,330]. The response in endogenous IAA progressed over a few hours following a pulse of red light and included a very-low-fluence and a low-fluence component, the former being the major one [286]. Therefore, exposure of dark-adapted coleoptiles to unilateral light would cause a decrease in the net amount of either diffusible or extractable auxin even if pure blue light were used, and the decrease would complicate the demonstration of phototropic auxin redistribution.

Huisinga [328] presented bioassay data suggesting that red light reduces diffusible auxin in oat coleoptiles by inhibiting the basipetal transport of auxin. However, in view of the results obtained for maize, it is probable that phytochrome-mediated inhibition of auxin biosynthesis occurs similarly in oat coleoptiles. Stewart and Went [331] observed that the auxin extracted from totally dark-grown oat coleoptiles decreases after a brief exposure to white light. Perhaps, this response largely represents the phytochrome-mediated inhibition of auxin biosynthesis. Whatever is the cause of the decrease in diffusible auxin in oat coleoptiles, this response could complicate the demonstration of phototropic auxin redistribution in this material.

Another response likely to accompany the phototropic auxin redistribution is the blue-light-dependent inhibition of basipetal auxin transport, which has been shown in oat and maize coleoptiles. As will be discussed below (Section 23.10.11), this blue-light response is probably not causally related to phototropism. Inhibition of auxin transport would result in a decrease in the amount of diffusible auxin. On the other hand, the content of extractable IAA in a region of an organ would be either reduced or enhanced, depending on where auxin transport is inhibited.

In the experiments by Briggs et al. [287] and Briggs [288] demonstrating redistribution of diffusible auxin, dark-adapted seedlings were exposed to red working light one hour or more prior to the experiments. Under this experimental condition, the phytochrome-mediated changes in diffusible auxin would have been saturated and were probably expressed similarly in phototropically stimulated and non-stimulated coleoptile tips. In the study of Iino [297], red-light-grown seedlings were used. Therefore, the results were free from any major phytochrome-mediated changes.

Went's original data indicated that the net amount of diffusible IAA obtained from the oat coleoptile tip decreased somewhat after fPIPP stimulation (see Section 23.10.5). The experiments were conducted under orange working light, and the phytochrome response was probably not included in the difference between the phototropically-stimulated and the control plants. It is possible that, in this case, the blue-light-dependent, non-phototropic inhibition of auxin transport occurred in the tip. This possibility is supported by the conclusion of Gardner et al. [322] that inhibition of basipetal transport by blue light is detectable in oat coleoptiles but not in maize coleoptiles. Although IAA was redistributed without a significant change in the net level in the apical and subapical zones of red-light-grown maize coleoptiles, a small and transient decrease in IAA was observed in a basal zone [297]. Such a response might represent some contribution of the blue-light-dependent inhibition of basipetal auxin transport in maize. In the study of Oppenoorth [189] demonstrating asymmetric distribution of extractable auxin in the tip of dark-adapted oat coleoptiles, the net auxin content was reduced by the fluence causing fPIPP and was enhanced by the higher fluences causing PINP and TDP. The reduction occurred rapidly, whereas the enhancement occurred gradually. Although the accuracy of the auxin determination based on the *Avena* curvature test might be questioned, it is possible that the observed changes in the net amount were caused by multiple effects of light that are distinct from phototropic auxin redistribution. Although van Overbeek [313] could demonstrate lateral translocation of auxin by using isolated segments of radish hypocotyls (see Section 23.10.5), his experiments conducted with tip-intact segments showed that asymmetric distribution of endogenous diffusible auxin involved a substantial decrease in the net amount. A non-phototropic inhibition of auxin biosynthesis or basipetal auxin transport might be responsible for the decrease.

23.10.7 Controversy regarding the occurrence of auxin asymmetry

In spite of the evidence that lateral auxin translocation and the ensuing auxin asymmetry occur in phototropically stimulated coleoptiles, hypocotyls, and epicotyls, some workers have reported that auxin asymmetry does not occur in these organs, questioning its involvement in phototropism. At the moment, this controversy is the most serious issue in this field.

Bruinsma and his co-workers could not detect IAA asymmetry between the irradiated and the shaded half of the hypocotyl of de-etiolated sunflower seedlings [138,332,333]. They failed to observe any asymmetry of either extractable or diffusible IAA. Feyerabend and Weiler [139] conducted a more detailed time-course study. They were also unable to find any IAA asymmetry in continuously stimulated sunflower hypocotyls. So far there is no evidence that auxin is asymmetrically distributed during phototropism of sunflower hypocotyls.

Hasegawa and his co-workers published a series of papers reporting that auxin asymmetry does not occur in phototropically stimulated organs including oat coleoptiles [309], maize coleoptiles [334], radish hypocotyls [335], and pea epicotyls [336]. Hasegawa et al. [337] also reported that the result of Went demonstrating the asymmetric distribution of diffusible auxin could not be reproduced when measurements were made specifically for IAA (see also Section 23.10.9). These results of Hasegawa's group conflict with those described above (especially the result of Iino [297] and the results from tracer experiments).

In the work mentioned above, different physicochemical methods were used to assay IAA or its derivative: an indolo-α-pyrone fluorescence method [138,332,333], an immunological method [139], gas-chromatography (electron-capture detection) [309,337], and HPLC (fluorescence detection) [334,336]. Also, different internal standards were used: 1-^{14}C-IAA [139,333], 2-^{14}C-IAA [138,309,332,337] and indole-3-propionic acid [334,336]. Accurate determination of endogenous IAA depends not only on the final assay method, but also on the extraction and purification procedures [317,318]. It is difficult to make a full assessment of the methods used from published papers.

The above controversy must be settled by further studies. Any further study should consider the following problems. There is no doubt that IAA must be sufficiently purified for the assay method chosen. In addition, any endogenous substances must not be converted to IAA during extraction and purification. In particular, IPyA is highly labile and is converted to IAA spontaneously. Such conversion should not occur until IAA is separated from IPyA during purification. Furthermore, IAA itself is a rather unstable substance and is liable to undergo non-enzymatic decomposition. If IAA is decomposed substantially during extraction and purification, serious errors could occur, especially in relation to the nature of the internal standard used.

23.10.8 Participation of other plant hormones

Asymmetric application of the gibberellic acid GA_3 was found to induce no curvature response in decapitated oat coleoptiles [338] and first-internode segments of sunflowers [339] and *Phaseolus vulgaris* [340]. Furthermore, ^{3}H-GA_3 applied to the apical cut end of pea internode segments was not asymmetrically distributed after 4-h stimulation with unilateral blue light, although ^{3}H-IAA was asymmetrically distributed under the same condition [135]. The phototropic curvature of pea internode segments could be stimulated by applied GA_3; the stimulation is probably attributable to the effect of GA_3 in enhancing the overall growth rate [135]. This interpretation is supported by the result that the applied GA_3 stimulates the curvature of decapitated oat coleoptiles induced by asymmetric IAA application [338].

In spite of these results, Phillips [341] reported that diffusible gibberellin-like substances measured by a bioassay (lettuce hypocotyl elongation test) were asymmetrically distributed in the phototropically stimulated shoot apex of sunflowers. He excised the 10-mm shoot apex from 14-day-old, light-grown sunflowers and placed it onto two agar blocks as practiced to demonstrate the asymmetric distribution of diffusible IAA. After a 20-h diffusion period during which the shoot apex was unilaterally irradiated with white light, the diffusates collected in the agar blocks were extracted with aqueous methanol and subjected to paper chromatography. The gibberellin activity of each chromatogram zone was determined. He found that the total activity from the shaded side is greater than that from the irradiated side at a ratio of 1:7. Phillips [342] found similar results for the gravitropically stimulated shoot apex. He suggested that the two tropisms of sunflower stems are mediated by asymmetric distribution of gibberellins rather than of auxin. Furthermore, Railton and Phillips [343] found that diffusible gibberellin-like substances were asymmetrically distributed in gravitropically stimulated maize coleoptile tips. It has not been shown in any of the cases, however, that the endogenous gibberellin-like substances found to be distributed asymmetrically can cause curvature when applied asymmetrically.

Unfortunately the results of Phillips [341] have not been reflected in the later development of gibberellin research. It is of particular interest to know whether the endogenous gibberellin determined to be directly active in the growth of sunflowers [344] or any given material are distributed asymmetrically following phototropic stimulation. Asymmetric distribution of any precursor of the active gibberellin could also result in asymmetric distribution of the active gibberellin. It also remains to be determined whether or not any such gibberellins or precursors can cause curvature when applied asymmetrically.

Scott and Most [345] found that the longitudinal movement of applied $^{3}H\text{-}GA_1$ in internode segments of sugar cane is slow and non-polarized, but becomes substantial with an establishment of basipetal polarity when applied together with IAA. GA_1 is now believed to be the major gibberellin directly active in many higher plant species (see [346]). The above finding, already discussed by Pickard [347] in relation to tropisms, can be extended to provide the following view. If the polar movement of GA_1 from the shoot apex to lower zones is dependent on auxin, then lateral redistribution of auxin could result in an asymmetric distribution of GA_1 in lower responding zones (i.e. a greater GA_1 concentration in the side with a higher auxin concentration). A synergistic effect of GA on IAA-dependent growth could then lead to an amplification of the growth asymmetry caused by auxin asymmetry.

Ethylene was distributed asymmetrically in the hypocotyl of *Phaseolus vulgaris* seedlings irradiated unilaterally with white light for 5 h [348]. A higher ethylene level was found in the shaded half than in the irradiated half, at a ratio of 1.3:1. This asymmetry is probably attributable to the auxin asymmetry and the auxin-dependent ethylene biosynthesis. Since ethylene is generally inhibitory to growth, the observed ethylene asymmetry may result in an inhibition of the curvature response. Brennan and Gunckel [164] found that the phototropism of mung bean is partially inhibited by hypobaric treatment that reduces the level of endogenous ethylene, and suggested that ethylene is involved in the process of phototropism. On the other hand, Kang and Burg [135] could not observe such an effect of hypobaric treatment in pea stems. They also found that the asymmetry of applied ^{3}H-IAA induced by phototropic stimulation is not affected by ethylene application. Applied ethylene inhibits phototropism of pea stems [135] and oat coleoptiles [349]. Ethylene is stimulatory on the growth of rice coleoptiles, and, in this material, ethylene is also stimulatory on phototropism [349].

Perhaps, ethylene affects phototropism indirectly by modifying the rate of elongation growth. To date there is no evidence that ethylene participates in the induction of phototropic curvature, although its contribution to autostraightening is not excluded.

23.10.9. Growth inhibitor hypothesis

Franssen and Bruinsma [138] reported that the growth inhibitor xanthoxin, partially purified by TLC and assayed by the cress-seed germination test, was asymmetrically distributed in the phototropically responding hypocotyl of de-etiolated sunflowers. A gradient of 2 : 1 was detected between the irradiated and the shaded half of rapidly bending hypocotyls. Shen-Miller et al. [333] found a still greater asymmetry, close to 3 : 1, in the biologically active form of xanthoxin (*cis*-xanthoxin) by conducting more extensive purification and determining it with gas chromatography. Because no asymmetry could be detected for IAA (see Section 23.10.7), these authors concluded that the asymmetry of xanthoxin is responsible for the phototropism of sunflower hypocotyls. The suggested involvement of xanthoxin in light-grown sunflower hypocotyls was questioned by Feyerabend and Weiler [139]. They were not able to detect any asymmetry in immunologically detectable xanthoxin over the entire phase of curvature development. No clear asymmetry could be found even after separation of *cis*-xanthoxin from *trans*-xanthoxin.

Hasegawa et al. [350] isolated three neutral growth-inhibiting substances from light-grown radish seedlings and identified them as *cis*- and *trans*-raphanusanins and 6-methoxy-2,3,4,5-tetrahydro-1,3-oxazepin-2-one (designated raphanusamide). When applied as a lanolin mixture along one side of dark-grown radish hypocotyls, these substances inhibited growth more on the applied side than on the opposite side, causing a curvature toward the applied side [351]. The levels of these inhibitors were enhanced in dark-grown hypocotyls by white-light treatment; when the hypocotyls were unilaterally irradiated, the enhancement was greater in the irradiated half than in the shaded half [351,352]. The asymmetry of inhibitors occurred at the onset of curvature response [351,352]. The inhibitor asymmetry could also be detected during fPIPP [101]. In these studies, white light was used as the light source. Sakoda et al. [353] subsequently used blue light, and showed that *cis*- and *trans*-raphanusanins were distributed asymmetrically during fPIPP and TDP, although the asymmetry of raphanusamide could be detected only during TDP.

Preceding the work summarized above, Hasegawa and co-workers isolated and identified two growth-inhibiting neutral substances, designated raphanusol A and B, from light-grown radish seedlings. These related substances, but distinct from the inhibitors mentioned above, were found to be distributed asymmetrically during the fPIPP and TDP of radish hypocotyls [354]. Thus, Hasegawa's group has shown that five neutral growth inhibitors are asymmetrically distributed in phototropically stimulated radish hypocotyls.

All of the data obtained by Hasegawa's group have indicated that the level of growth inhibitors is enhanced in response to light treatment and that the asymmetric distribution after unilateral irradiation results from the difference in the extent of this enhancement. These workers showed, in agreement with these results, that the phototropism of radish hypocotyls is caused by greater inhibition of growth on the irradiated side than on the shaded side. This conclusion, however, does not agree with the results from hypocotyls of other plant species (see Sections 23.10.2 and 23.10.3). It is possible that, although the level of growth inhibitors is enhanced by light treatment, the asymmetric distribution of the inhibitors involves a light perception mechanism that is distinct from that responsible for inhibitor accumulation. Asymmetric distribution of raphanusanins occurred in radish hypocotyls after 30-s unilateral irradiation (0.1 W m^{-2}) even though the plants were maintained under overhead irradiation (the same fluence rate) for 1 h before the unilateral irradiation [101]. Such results would be difficult to explain only in terms of the local light action on inhibitor accumulation.

Hasegawa and co-workers extended the inhibitor study to investigate the possible contribution of inhibitors in coleoptile phototropism. Bioassay data indicated that acidic growth inhibitor(s) was asymmetrically distributed in unilaterally irradiated oat coleoptiles [309]. The inhibitor was at least distinct from ABA. Hasegawa et al. [337] subsequently found that the diffusates collected into agar blocks from excised tips of oat coleoptiles contained at least two substances that inhibit the curvature response in the *Avena* curvature test when applied together with IAA, and that the inhibitors were asymmetrically distributed in unilaterally irradiated coleoptiles. In the same papers, Hasegawa et al. reported that extractable as well as diffusible IAA, determined physicochemically, was not distributed asymmetrically, although an asymmetry of auxin activity could be found if the agar receiver blocks were subjected directly to the *Avena* curvature test. Based on these results, they concluded that Went's result demonstrating asymmetric distribution of diffusible auxin was in fact caused by asymmetric distribution of growth inhibitors. The results showing no IAA asymmetry in these studies are, however, in directly conflict with the results of other workers, as has already been discussed (Section 23.10.7).

The results from radish hypocotyls have provided the strongest evidence for the participation of neutral growth inhibitors in phototropism. To date, the involvement of neutral growth inhibitors has been shown only for radish hypocotyls, and no asymmetric distribution of neutral growth inhibitors could have been detected in coleoptiles [309]. Further clarification of the issue whether acidic growth inhibitors participate in the phototropism of coleoptiles [337] would require the identification of the inhibitors and detailed studies in relation to the various physiological properties uncovered for coleoptile phototropism.

We are left with interesting but puzzling questions about the underlying mechanisms. How can the asymmetric distribution of many growth inhibitors, that are chemically distinct, be induced specifically for phototropism? How can different plants use chemically distinct inhibitors? One might speculate that the growth inhibitors are distributed asymmetrically in response to auxin asymmetry (i.e. translocation of inhibitors towards the side of low auxin concentration), and that for this mechanism, plants use various growth inhibitors. If such inhibitor translocation were induced, then an amplification of the auxin-mediated growth asymmetry would also occur.

23.10.10 Further evidence for auxin mediation and some unresolved problems

The occurrence of auxin asymmetry in phototropically stimulated organs is most fundamental for the Cholodny-Went theory. However, closer examination is required to establish that the auxin asymmetry is causally related to phototropic curvature. This causality is indeed supported by a number of results, although several critical issues still remain to be examined.

23.10.10.1 Implications of the tip-splitting effect

Boysen-Jensen [355] demonstrated that an oat coleoptile whose tip is split vertically and separated by a transparent barrier (cover slip) is less responsive to unilateral light given perpendicular to the barrier than that given parallel to the barrier. The difference in curvature response was, on average, 1 : 0.18 (parallel : perpendicular). In these experiments, the tip was split to the depth of 3–4 mm and only the split part was irradiated continuously for 3 h. The result has led to a general conclusion that physical continuity is required between the irradiated and the shaded half of the tip for the full expression of phototropism. In view of the Cholodny-Went theory, the result suggests that physical continuity is required for the lateral translocation of auxin.

Brauner [147] repeated the experiment of Boysen-Jensen. Curvature was measured every 30 min for a period of 4 h. When the light direction was parallel to the barrier, the coleoptiles developed curvature linearly from 0.5 h to 4 h after the onset of stimulation. When the light direction was perpendicular to the barrier, the coleoptiles showed no curvature during the first 1 h and only a slight curvature at 1.5 h. Afterwards the coleoptile developed curvature linearly at a rate about 60% of that measured when the light was parallel to the barrier. Briggs [96] conducted tip-splitting experiments with maize coleoptiles. In his experiments, nearly the entire length of the coleoptile was irradiated with unilateral light. The coleoptile split to the depth of about 2 mm showed no significant curvature when an fPIPP stimulus was given perpendicular to the barrier, while the same stimulus caused a clear curvature when given parallel to the barrier. On the other hand, a TDP stimulus given perpendicular to the barrier yielded a curvature only partially smaller than when the stimulus was given parallel to the barrier.

Together, the results indicate the following points. First, fPIPP is very severely inhibited when the tip is split perpendicular to the direction of light. This point, clarified for maize coleoptiles, supports the conclusion that the tip is nearly the sole photoperceptive site (see Section 23.5.2). Second, TDP is not inhibited to any comparable extent. The partial inhibition of TDP observed by Briggs could represent the participation of non-tip photoperception (see Section 23.5.2). However, Boysen-Jensen and Brauner observed clear TDP by stimulating only the split part. It is possible that auxin is laterally translocated within the half side, causing some lateral auxin asymmetry in the lower part [355]. The delayed curvature response observed by Brauner might then suggest that the half part gains (or recovers) the ability to translocate auxin about 1 h after the splitting treatment. The ability of the half tip to respond to unilateral light recalls the observation of Wager [24] that the bisected petiole of *Geranium pratense* could show a positive phototropism in response to unilateral light given perpendicular to the cut surface.

23.10.10.2 Implications of the basipetal migration of growth asymmetry

In coleoptiles, the phototropic signal perceived by the tip moves downwards, as depicted by the basipetal movement of the onset of phototropic curvature (see Section 23.5.3). This basipetal movement of the phototropic signal provides strong support for the involvement of auxin asymmetry in coleoptile phototropism. The basipetal migration of curvature response (10–30 mm h^{-1}; see Section 23.5.3) is in general agreement with the mean velocity of basipetal auxin transport measured in oat and maize coleoptiles (10–20 mm h^{-1}; see e.g. [296,297,321,356]).

Using red-light-grown maize coleoptiles, Baskin et al. [156] demonstrated that the phototropic growth responses (inhibition on the irradiated side and stimulation on the shaded side) move basipetally at a velocity comparable to that of the basipetal migration of the growth stimulation induced by apical application of IAA. Although the velocity determined (about 30 mm h^{-1}) was greater than the mean velocity of basipetal IAA transport measured in maize coleoptiles, the agreement found under comparable conditions and using the same analytical methods supports the conclusion that the basipetal movement of phototropic growth changes represent the basipetal transport of auxin. Earlier work of Newman [357] provides supplemental support for the close link between the basipetal movement of the phototropic signal and the basipetal auxin transport. He found that a wave of electrical surface potential moves basipetally along the shaded side of oat coleoptiles at a velocity of about 15 mm h^{-1} when only the tip was stimulated continuously with unilateral light. Application of IAA to the apical cut end of decapitated coleoptiles caused a similar electric wave moving basipetally at nearly the same velocity.

In an extended study, Baskin et al. [358] were able to show that a spot of IAA applied unilaterally to the tip of intact maize coleoptiles stimulates growth on the applied side while little affecting the growth on the opposite side. The results indicate that the auxin asymmetry established in the tip can migrate basipetally to cause an effective growth asymmetry in the lower zones. The conclusion that the auxin asymmetry induced in the tip migrates basipetally via basipetally polar transport of auxin has been substantiated by measurements of endogenous IAA [297]. The asymmetric distribution of IAA was initiated at about 10 min in the subapical zone (2–7 mm from the tip) and at about 35 min in the basal zone (12–17 mm) following fPIPP stimulation.

23.10.10.3 Kinetic relationship between auxin and growth asymmetries

If the auxin asymmetry were the cause of the curvature response, the former should precede the later. This was found by Iino [297]; in either the subapical or the basal zone, the onset of IAA asymmetry preceded by about 10 min the onset of curvature response. As described above (Section 23.7.4), the coleoptile tip retains the signal for curvature response for 2 h or more. If this fact is to be explained in terms of auxin asymmetry, the lateral auxin translocation in the tip must continue similarly. This is supported by the result that the asymmetry of diffusible auxin is detectable after a diffusion period as long as 3 h (see Section 23.10.4). The asymmetry of endogenous IAA is retained in the subapical zone for at least 1 h after fPIPP stimulation [297]. The results of Pickard and Thimann [289] provide still stronger evidence for the sustained lateral translocation. In the experiments determining the asymmetric distribution of ^{14}C-IAA diffusing out of the excised maize coleoptile tips, they also measured the amounts of ^{14}C-IAA remaining in the irradiated and shaded halves after the diffusion period of 3 h. An asymmetry that is only slightly smaller than that measured for the diffusible ^{14}C-IAA was found under both fPIPP and TDP stimulus conditions.

All these results obtained with coleoptiles characterize the phototropism that is associated with the lateral auxin translocation induced in the tip. In the phototropism of the hypocotyl of de-etiolated seedlings (and perhaps also in the phototropism of oat coleoptiles induced by non-tip photoperception), the curvature response can develop with a lag as short as 5 min following the onset of irradiation (see Section 23.7.1). The rate of curvature initially increases gradually, and the short lag has been determined by resolving this initial phase. Isolated tissue segments generally respond to applied auxin with a lag of about of 10 min [359]. It has been reported that the lag period for the auxin-dependent growth stimulation approaches zero when the initial phase is analyzed with high-resolution methods [360,361]. The rapid response observed by these workers might be caused by the acidic pH of the applied auxin solution [359]. However, there is also evidence that a lag can be very short when an increase in auxin level is moderate [362]. At the moment, the short lag is a critical challenge to the Cholodny-Went theory. This problem will be reflected in further discussion (Section 23.11.5).

23.10.10.4 Quantitative relationships between auxin and growth asymmetries

The growth responses underlying phototropic curvature generally occur as predicted by the Cholodny-Went theory (see Section 23.10.3). Directly comparable growth data and auxin distribution data are available for the fPIPP of red-light-grown maize coleoptiles [151,297]. These data show that, at least in the upper part of the coleoptile, both the IAA level and the growth rate decrease on the irradiated side and increase on the irradiated side (Figures 20 and 21). Although such data provide strong evidence for the Cholodny-Went theory, the quantitative relationship between auxin and growth asymmetries is still to be investigated.

The growth data were obtained by measuring elongation along the two opposite flanks of organs, and it has been commonly observed that the growth on the irradiated side decreases substantially or even ceases. On the other hand, measurements of either endogenous auxin or isotope-labeled exogenous IAA in the organ's two halves typically yielded a ratio of around 1 : 2 (see Sections 23.10.4 and 23.10.5). A substantial auxin difference between the two sides might occur in the organ's peripheral region thought to be the primary tissue target for auxin in the control of growth ([363–365]; cf. [366]). To date, however, no evidence has been provided in support of this contention (see the result obtained with pea internodes, Section 23.10.5). Although oat coleoptiles show a substantial curvature in response to continuous light perceived by non-tip zones (see Sections 23.5.1 and 23.5.2), only a small IAA asymmetry could be detected in decapitated coleoptile segments (see Section 23.10.5).

More detailed study is clearly needed to clarify whether or not the measured auxin asymmetry can adequately explain the phototropic growth asymmetry. Future study may have to consider the possibility that the auxin-dependent asymmetry of growth is amplified beyond the extent expected from the measured auxin asymmetry. This **amplification hypothesis** is not an unlikely one because there are possible mechanisms through which the suggested amplification could occur. These mechanisms include

1. induction by auxin asymmetry of asymmetric distribution of other growth-regulating substances (see Sections 23.10.8 and 23.10.9),
2. a greater asymmetry in protoplasmic concentration of auxin than the measured asymmetry in tissue content of auxin (see below, Section 23.10.10.5),

3. curvature-dependent microtubule orientation response (see Section 23.10.13), and
4. auxin-dependent response that is induced specifically in response to auxin asymmetry (e.g. turgor-driven cell-volume changes; see Section 23.11.2).

23.10.10.5 Implications of the cellular localization of auxin receptors

Evidence has accumulated indicating that the auxin-binding protein 1 (ABP1), a putative auxin receptor with a probable function for growth control, occurs on the outer surface of plasma membranes and interacts with auxin on the apoplastic side of the membranes (see [367]). This conclusion, however, does not demonstrate that the ABP1 located on the outer surface of plasma membranes is the sole or the major receptor responsible for growth regulation. In fact, there is evidence that the major auxin receptor responsible for growth control is located in the protoplasm, or has its auxin-interacting site on the protoplasmic side of plasma membrane [368,369]. At the moment, it is controversial as to whether ABP1 has any major receptor role in growth response [370,371].

Auxin uptake by cells depends on the activity of the plasma membrane H^+-pump, and the H^+-pump activity itself is a function of auxin concentration (see Section 23.11.1). Therefore, the proportion of auxin partitioned into the protoplasm would be greater in a tissue site having a greater auxin concentration. It is predicted that the asymmetry in protoplasmic auxin concentration is greater than the auxin asymmetry at the tissue level. This could be a mechanism by which the growth asymmetry is more effectively induced by auxin asymmetry, provided that auxin interacts with the growth-limiting receptor on the protoplasmic side of cells. If the critical auxin-receptor interaction occurs on the apoplastic side, however, the auxin asymmetry found between the two sides would be less effective in causing growth asymmetry. Therefore, the Cholodny-Went theory favors the idea that the growth-limiting auxin receptor is located on the protoplasmic side of cells.

23.10.10.6 Effects of the growth-saturating dose of auxin

If phototropism results from auxin asymmetry, it should be eliminated by application of growth-saturating doses of IAA. The early results of Ball [372], obtained with oat coleoptiles, appeared to contradict this prediction. However, as previously discussed [90], these results are probably not sufficient for the conclusion that auxin asymmetry does not play a role in phototropism. In maize coleoptiles, fPIPP could be inhibited by a ring of IAA/lanolin applied to the position 5 mm below the tip just after or before the phototropic stimulation [272]. The curvature was eliminated at the growth-saturating concentrations of IAA. This result is expected if the lateral auxin translocation shown to occur in the tip of maize coleoptiles (Section 23.10.5) accounts almost solely for the fPIPP of this material. The phototropic response to 2-h continuous stimulation could also be inhibited by the same treatment, but a portion of the response (about one fifth) remained even at the saturating IAA concentrations. This result agrees with the idea that the lateral auxin translocation in the tip mainly accounts for the TDP of maize coleoptiles, but again indicates the participation of non-tip photoperception in this phototropism (see Section 23.10.10.1). The result does not seriously contradict the view that the phototropism due to non-tip photoperception is mediated by lateral auxin translocation, because a curvature could be induced before growth saturating auxin reaches the lower responding zones. Rice coleoptiles grow normally and show phototropism while submerged in water [69]. The coleoptiles cannot develop phototropism during 2.5-h continuous stimulation with unilateral blue light (0.3 μmol m^{-2} s^{-1}) when IAA (0.1 mM) is added to the surrounding water before phototropic stimulation (R. Neumann and M. Iino, unpublished).

23.10.10.7 Implications of long-term auxin-growth relationships

For effective regulation of growth by changes in the level of auxin, the level of endogenous auxin must be limiting for growth in intact plants. This point has long remained obscure, because it was often noted that externally applied auxin could not clearly stimulate organ growth in intact plants. By now, however, substantial results have accumulated indicating that applied IAA can stimulate growth in intact plants (*Citrullus lanatus* hypocotyls [373], maize coleoptiles [152,358,374], oat coleoptiles [372], and pea epicotyls [375–378]).

The growth in intact plants is stimulated, however, only for a short period (typically a few hours) following continuous IAA application (red-light-grown maize coleoptiles [320, 374], and red-light- and white-light-grown pea epicotyls [320,378]). This transient growth stimulation was found in zones of the organs when IAA (mixed with lanolin) was applied as a ring at a position located above the zones. In maize coleoptiles, the growth of a zone remained stimulated for at least 7 h when IAA was applied directly to the zone [374]. On the other hand, growth stimulation was transient in pea epicotyls even after such direct application [320,378]. The transient growth stimulation occurs in such a way that the elevated growth rate returns to the pre-stimulation rate. It was also noted that the pre-stimulation rate, or the rate in non-treated intact plants, was approximately half the rate maximally achieved by IAA application [320,374,378]. The transient nature of growth stimulation appears to represent an adaptive change in growth rate that follows an enhanced IAA level. The molecular and cellular mechanisms enabling such adaptive growth responses are not clear. It would be of interest to resolve how these responses are related to tropisms.

The *nph4* mutant of *Arabidopsis* impaired in hypocotyl phototropism and gravitropism [213] has provided interesting results. This mutant was found to be impaired in the expression of several auxin-responsive genes [379]. The growth of mutant hypocotyls is also less sensitive to applied auxin [379,380]. These results are in agreement with the Cholodny-Went theory, or at least indicate that auxin is a key element of tropisms. Surprisingly, however, intact seedlings of this mutant show normal hypocotyl growth. The normal growth with impaired auxin-dependent processes may be related somehow to the ability of plants to undergo adaptive growth responses. It is suggested that the net growth is adjusted adaptively to a pre-determined rate irrespective of the impairment caused by *nph4* mutation, although the auxin-dependent processes impaired by *nph4* mutation are essential for the establishment of growth asymmetry in tropisms.

23.10.11 Mechanism of lateral auxin translocation

According to the original Cholodny-Went theory, auxin is translocated laterally across the tissue in response to the transverse polarity generated in individual cells (see Section

23.10.1). In gravitropism, the gravity sensed by each cell could theoretically lead to its transverse polarity. In phototropism, however, the lateral light gradient occurring across each cell may be too small to provide the information for the polarity establishment. Many experimental results from coleoptiles indicate the significance of the light gradient between the two sides of the organ (see Section 23.6.1). In fact, the results obtained with half-sided irradiation, either from above or the side, and irradiation from inside the coleoptile cavity argue strongly against the requirement of a light gradient across individual cells. It can be stated that the transverse polarity induced across the organ, rather than the individual cells, is responsible for the lateral auxin translocation.

The observation that blue light inhibits basipetal transport of auxin in oat and maize coleoptiles has led to a hypothesis that the lateral auxin translocation results from differential inhibition of basipetal auxin transport [381–387]. The hypothesis, here referred to as the "transport-inhibition hypothesis", can be generalized as follows. In unilaterally irradiated organs, basipetal transport of auxin is inhibited more on the irradiated side than on the shaded side, and this differential inhibition of transport results in a flow of auxin to the shaded side. The lateral flow occurs in principle by concentration-dependent diffusion, whether mediated by specific or non-specific transport channels.

Although the differential inhibition of basipetal auxin transport could result in a lateral flow of auxin, it would not necessarily lead to a greater level of auxin on the shaded side (see also [347]). When inhibition occurs in a narrow zone of an organ, the concentration of auxin on the irradiated side is enhanced above the zone (which causes a negative curvature) and reduced below the zone (which causes a positive curvature). When inhibition occurs along the length of an organ, the pattern of auxin distribution would vary, depending on how the extent of inhibition is distributed. In all cases, there should be a region where the auxin concentration is enhanced on the irradiated side. Investigation of the transport-inhibition hypothesis must consider these problems.

In the case of coleoptiles, a specific inhibition of basipetal transport in the tip could result, in the lower parts, in a reduced level of auxin on the irradiated side and an enhanced level on the shaded side. The amount of diffusible auxin obtained from the coleoptile tip could also be redistributed without a significant change in the net amount as demonstrated. (Although inhibition of basipetal auxin transport in the tip, whether induced symmetrically or asymmetrically, would result in a decrease in the net amount of auxin transported in a given time, this decrease would be transient even if the inhibitory effect is sustained when the production of auxin in the tip remains unchanged.) The concentration of auxin is expected, however, to increase on the irradiated side of the tip.

Shen-Miller and Gordon [381] could show that the amount of IAA diffusing out of the basal cut end of the maize coleoptile tip (5 mm), which had been subjected to fPIPP simulation, was greater from the shaded half than from the irradiated half. In contrast, they found that the tissue content of extractable IAA was greater in the irradiated half. In this study, the IAA partially purified by paper chromatography was assayed by the *Avena* curvature test. The results appeared to provide strong evidence for the transport-inhibition hypothesis. However, Pickard and Thimann [289] found that the tissue content of applied ^{14}C-IAA is still greater on the shaded half of the coleoptile tip. Furthermore, Iino [297] showed that the concentration of endogenous IAA in the irradiated half of the coleoptile tip (2 mm) was not greater than that in the shaded half or the non-irradiated control. These results do not reproduce the results of Shen-Miller and Gordon, and do not favor the transport-inhibition hypothesis. If the hypothesis were to be supported, it would be necessary to demonstrate that an enhancement of the auxin level on the irradiated side occurs in the extreme tip.

Shen-Miller and co-workers reported that the fluence-response relationship and the action spectrum agreed between the light-induced inhibition of basipetal auxin transport and phototropism [386,387]. However, the inhibition of basipetal auxin transport by blue light is generally more transient than phototropism [305]. Blue-light-induced inhibition of basipetal transport has been typically demonstrated with non-tip segments of coleoptiles [305,384], and there is no evidence that the inhibition is pronounced in the tip. Thus, even the correlative evidence for the transport-inhibition hypothesis is far from complete. Hager and Schmidt [384,385] suggested that 3-methyleneoxindole, an oxidative degradation product of IAA, is generated more on the irradiated side and causes a greater inhibition of basipetal auxin transport on that side. This line of study, however, has not yielded any clear conclusion [388]. Naqvi and Engvild [389] found that abscisic acid (ABA) inhibits basipetal transport of IAA in maize coleoptiles, and suggested that a greater concentration of ABA on the irradiated side could cause a larger inhibition of basipetal auxin transport on that side. However, there is no evidence that ABA is asymmetrically distributed during phototropism [139,333]. See also Pickard [347] for critical treatments of the results reported in favor of the transport-inhibition hypothesis.

The following logical problems require further attention. In coleoptiles, the asymmetric auxin distribution established in the tip is maintained against any substantial lateral diffusion while auxin is transported basipetally [297,358]. The results of Naqvi [390], although interpreted differently, do in fact indicate that the basipetal transport predominates over the lateral diffusion in subapical coleoptile segments. The transport-inhibition hypothesis would require that auxin easily diffuses laterally in the tip in response to a concentration gradient while protected from lateral diffusion in the lower parts. Also, the hypothesis cannot easily explain the cases in which photoperceptivity is not limited to an apical narrow zone. If the hypothesis were to be generalized to incorporate such cases, more subtle assumptions would be required. So far there is no strict evidence for the transport-inhibition hypothesis. The transport inhibition which has been observed in coleoptiles and is not specifically associated with the tip tissue is at least unlikely to be the cause of the lateral translocation.

It seems more likely that plants have a specific cellular system for the phototropic lateral translocation of auxin. Pickard [347] has provided some theoretical account of such systems on the basis of a plasma-membrane-located, putative efflux carrier functioning for lateral transport. The simplest model is that the density or activity of the carrier becomes asymmetric within individual cells. This idea faces a difficulty in view of the conclusion that the light gradient is not perceived by individual cells (see above). It would have to be assumed that the carrier asymmetry in each cell is induced indirectly in response to the transverse polarity occurring between the two sides of the organ. Another feasible model is that the density or activity of the carrier is enhanced differently in the cells of the irradiated and shaded sides. A greater net activity of the carrier on the irradiated side could lead to an accumulation of auxin to the shaded side.

Any possible models based on the lateral-transport carrier have to consider the relationship between this carrier and the carrier for basipetally polar transport. If the same cells contribute to both basipetal transport and phototropic lateral translocation, many of the models face the problem that the basipetal transport is also significantly affected by the changes occurring in the lateral-transport carrier. This problem becomes less critical if the tissue contributing to phototropic lateral translocation is separated from that for basipetal transport. In this case, the efflux carrier responsible for basipetal transport (see [391]) could still be allocated as the key carrier for lateral translocation.

23.10.12 Unified hypothetical views on lateral auxin translocation

The tip of coleoptiles is the most active site of phototropic lateral auxin translocation and is also a site of auxin biosynthesis (see [286] for the latter fact). It is of interest to know how the cells contributing to the two processes are spatially correlated within the tip tissue. In maize coleoptiles, the most active lateral auxin translocation takes place in an apical zone as narrow as 0.5 mm under both fPIPP and TDP stimulus conditions ([288]; see Section 23.10.5). This result would suggest that the cells contributing to lateral auxin translocation are not clearly separated from the auxin-producing cells. Another important point is that the concentration of auxin is substantially higher in the tip tissue than in the rest of the coleoptile tissue. Early bioassay data revealed that the auxin level increases acropetally in oat coleoptiles [392]. Later measurements of IAA in maize coleoptiles, in fact, indicated that IAA is highly concentrated in the tip [297,318]. In red-light-grown coleoptiles, for example, a subapical region (2–7 mm from the tip) and a basal region (12–17 mm) contain IAA at concentrations of 15 ng and 10 ng g^{-1} fresh weight, respectively, whereas the tip (2 mm) contains 59 ng IAA g^{-1} fresh weight [297]. The high auxin concentration in the tip suggests that the auxin produced in the tip cannot immediately enter the basipetal transport system. If the two lines of information are put together, it would appear that the lateral translocation of auxin is carried out in the tip by the cells contributing to auxin production and have little or no basipetally polarized transport activity. This also explains why an effective lateral translocation can occur within a very narrow tip zone (note that if auxin is effectively transported downwards, then the tip zone has to have an extremely effective lateral translocation system).

The above conclusion would lead to interesting cell models. The following is an example: The auxin transport carriers are not oriented in favor of basipetally polarized transport in the auxin-producing cells (and perhaps also in some neighboring cells). Auxin moves by non-polarized fluxes through these cells before reaching the cells in which the carriers are oriented for basipetally polarized transport. Upon phototropic stimulation, the transport of auxin through the auxin-producing cells is laterally polarized in favor of the movement to the shaded side. This may be made by reorientation of the transport carrier in individual cells, or by differential activation of the carrier between the cells of the two sides (see the preceding section).

Although the tip is most photoperceptive in oat and maize coleoptiles, the lower coleoptile zones can also perceive phototropic stimulus. This is typical for TDP and, in particular, for phototropism induced by continuous stimulation (see Section 23.5.2). As the tracer experiments of Pickard and Thimann [289] indicate, the non-tip parts of coleoptiles can cause only slight translocation of diffusible auxin from the irradiated to the shaded half during continuous stimulation. It seems that the diffusible auxin transported basipetally is only slightly differentiated between the two halves, although the curvature response induced could be substantial. Another interesting feature of the phototropism induced by non-tip perception is that a large part of the curvature response is confined to the site of photoperception; i.e. the curvature response does not effectively migrate from the stimulated to the lower zone [116] (see Section 23.5.3). Thus the phototropism induced by non-tip photoperception can be characterized by a relatively small asymmetry of basipetally transportable auxin and by a curvature response more-or-less confined to the stimulated zone. These features might be shared with the phototropism of other organs (e.g. hypocotyls of de-etiolated seedlings) in which the photoperceptivity is not confined to the apical part (see Section 23.5.5).

The features described above are explicable if it is assumed that the lateral auxin translocation occurs along the radial tissue that does not much contribute to basipetal auxin transport. Hertel and Leopold [356] have presented evidence that basipetal transport of auxin in maize coleoptiles takes place in a wide cross-sectional area that is mostly occupied by parenchyma tissue. The above hypothesis, however, requires that coleoptile tissues are somewhat radially differentiated for the basipetal transport and phototropic lateral transport of auxin. In stems, the basipetally polarized transport occurs mainly in the tissues associated with the vascular bundle (e.g. [319,393,394]; cf. [395]). The diagram in Figure 22 represents the hypothesis applied to dicotyledonous stems. Here, it is additionally assumed that the radial auxin movement occurs more along the peripheral region. The movement along the peripheral region could allow effective and rapid regulation of the auxin level in the growth-limiting tissues (see Section 23.10.10.4).

If lateral translocation of auxin results from the asymmetry of carrier activity between the organ's two sides, it is unlikely that a lateral auxin asymmetry occurs in apoplastic concentration in favor of the curvature response. An expected asymmetry would, however, occur in protoplasmic concentration. Therefore, the explanation of lateral

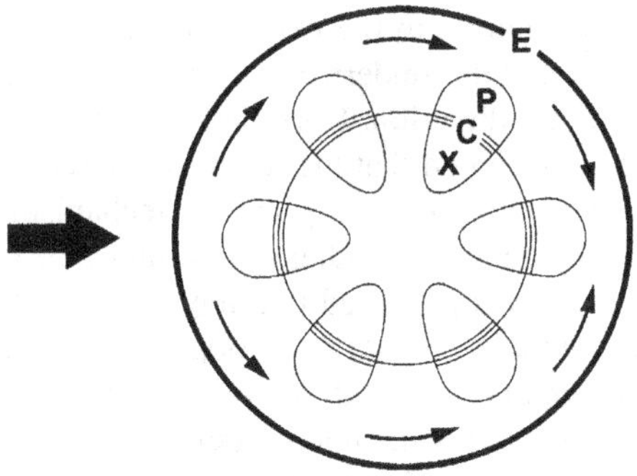

Figure 22. A model for lateral auxin translocation. The diagram shows a cross sectional view of a stem. Unilateral light is given from the left. The arrows in the stem indicate the direction of auxin movement. E, epidermis; P, phloem; C, cambium; X, xylem.

auxin translocation based on the difference in carrier activity between the two sides can be appreciated only when the mediating auxin receptor is located in the protoplasmic side of the cell (see also Section 23.10.10.5).

23.10.13 Participation of microtubules

Nick et al. [396,397] found that cortical microtubules are reoriented during the phototropism of maize coleoptiles and sunflower hypocotyls. The clearest response identified was the change from the transverse to the longitudinal position on the irradiated side. In maize coleoptiles, the microtubule orientation response showed a bell-shaped fluence-response curve that matched that of fPIPP [397]. The longitudinal microtubule orientation would lead to corresponding microfibril orientation, which in turn could result in a reduced rate of elongation growth. Therefore, it could be hypothesized that phototropic curvature results from the difference in microtubule orientation established between the organ's two sides. The auxin applied to maize coleoptile segments induced microtubule reorientation with a positive correlation with auxin-dependent growth [398], so it also seemed a reasonable hypothesis that the asymmetry of microtubule reorientation results from auxin asymmetry. However, subsequent work by Nick et al. [399] led them to conclude that the microtubule orientation response is not the prime cause of phototropic response (see the chapter by Nick in this volume). This conclusion was based on the observation that phototropic curvature could occur against the microtubule orientation (i.e. away from the side having a greater proportion of longitudinal microtubules). This was noted in maize coleoptiles during the curvature response to the second opposing fPIPP stimulus given 2 h after the first stimulus (i.e. after stabilization of the transverse polarity; see Section 23.7.4). Also, the organ straightening that followed an fPIPP curvature occurred against the microtubule orientation.

Further progress on this topic has been achieved by the work of Fischer and Schopfer [400]. They found that when phototropic curvature was prevented by a mechanical counterforce, the microtubules on either side of the coleoptile did not undergo the reorientation response that developed when no counterforce was present. Furthermore, when the counterforce was applied to the extent that the coleoptile was curved to the opposite direction, the microtubules underwent reorientation as if a phototropic stimulus had been given from the other direction. Essentially identical results were obtained for gravitropism. The authors concluded that the microtubule reorientation during tropisms is not caused by auxin asymmetry, or any other mechanism that is needed for the induction of the curvature response, but rather as a result of the curvature response itself. Following this conclusion, it was proposed that the microtubule-reorientation response constitutes a positive feedback regulatory loop. This idea provides a possible mechanism by which the auxin-dependent growth asymmetry is amplified. The data of Nick and Schäfer [206] indicate that the microtubule reorientation on the irradiated side is initiated within 30 min and is established about 60 min after fPIPP stimulation. If this result is to be explained by the conclusion of Fischer and Schopfer, then it must be assumed that the curvature-dependent microtubule reorientation is induced very effectively in response to the developing curvature.

Nick et al. [399] and Nick and Schäfer [206] obtained evidence that the stable transverse polarity induced in the maize coleoptiles during fPIPP (see Section 23.7.4) is attributable to the asymmetry of microtubule orientation that is stabilized and becomes non-reversible. The microtubule asymmetry induced by fPIPP simulation could also be stabilized against the orientation response to applied auxin when 2 h was allowed between fPIPP stimulation and auxin application. (The response to auxin was tested in isolated segments.) In an extended experiment, the removal of a half side of the coleoptile tip was found to result in a curvature of the coleoptile towards the tip-removed side and in an asymmetry of microtubule orientation that can be explained in terms of the auxin asymmetry caused by half-tip removal. In this case, the transverse asymmetry was not stabilized against the orientation response to applied auxin. Therefore, it was concluded that the microtubule-stabilization response does not simply follow the induction of microtubule asymmetry. The result also demonstrates that, although the microtubule asymmetry observed in half-tip-removed coleoptiles might be caused as a consequence of curvature response as described in the preceding paragraph, the stabilization response itself does not follow the curvature response.

Nick and Schäfer [206] provided further useful information. They were able to demonstrate that stimulation with a pulse of overhead blue light induces longitudinal microtubule orientation in the two sides of the maize coleoptile and that the induced microtubule orientation is stabilized when 2 h has elapsed after blue-light stimulation. Therefore, both the microtubule reorientation and the stabilization response were found to follow non-phototropic blue-light stimulation. The relationship between the microtubule orientation response after overhead blue-light stimulation and that after phototropic unilateral stimulation is not clear. However, the fact that microtubule orientation can be stabilized after overhead irradiation suggests that the stabilization of microtubule orientation found in phototropically stimulated maize coleoptiles is a blue-light response that is not directly related to the induction of phototropism. Furthermore, Nick and Schäfer found that the stabilization of transverse polarity, investigated by stimulating the coleoptile with two-opposing blue-light pulses and measuring the curvature response on a horizontal clinostat, was induced when the base was stimulated, but not when the tip was stimulated. The phototropic response induced by base stimulation, which was smaller in magnitude than that found after tip stimulation, is apparently related to the phototropism by non-tip photoperception. Although it has not been investigated whether base stimulation can cause an asymmetry of microtubule orientation, it is very likely that the stabilization of asymmetric microtubule orientation is induced by base stimulation but not by tip stimulation. A feasible explanation of all these results would be that the unilateral blue light perceived locally at the basal part (or non-tip parts) induces asymmetric microtubule orientation (which might be a consequence of the curvature response itself) and microtubule stabilization, independently. Because the microtubule-stabilization response occurs in the period between 1 and 2 h after pulse stimulation, it would be suggested that the blue-light signal perceived locally for stabilization response is stored or processed slowly during the period of 60 min without stabilizing the microtubule orientation.

The information described here generally indicates that microtubules do not play any major role in the process of phototropism induction. However, the results do not entirely rule out the possibility that the asymmetry of microtubule orientation is involved in the induction of the phototropism attributable to non-tip photoperception. It also remains to be investigated whether the microtubule-stabilization response participates in this type of phototropism (see also Section 23.7.4).

23.10.14 Remarks

Phototropic curvature results, in principle, from redistribution of growth. So far, the Cholodny-Went theory provides the most successful explanation of how this growth redistribution is achieved. Although it is not excluded that growth-regulating substances other than auxin are also asymmetrically distributed and participate in curvature response, it is arguable that the asymmetric distribution of non-auxin substances functions in close association with the auxin asymmetry. In some cases, however, the growth asymmetry may be established without involving any detectable auxin asymmetry, as shown most convincingly in sunflower hypocotyls. This problem will be reflected in a further discussion (Section 23.11.6).

The relationship between cell growth and cell division will be briefly mentioned. It is generally assumed that phototropic growth asymmetry results from the asymmetry of cell elongation growth, without involving an asymmetry of cell division. In fact, coleoptiles and hypocotyls have been used in experiments at relatively late developmental stages, when cell division does not occur to any significant extent. The cell division per se cannot increase the net volume of dividing cells, although it contributes to growth by increasing the number of cells that can grow. Therefore, it is unlikely that any major part of the growth asymmetry induced within a period of a few to several hours is accounted for by the asymmetry of cell division. There would be no doubt that the growth asymmetry measured in laboratory studies represents the asymmetry of cell elongation growth. This conclusion, however, does not mean that phototropism is generally independent of cell division. A possibility that has not yet been investigated is that an asymmetry of cell division, being induced at an early developmental stage of juvenile organs or in the apical part of stems, contributes to the curvature often sustained in fully elongated organs.

23.11 Participation of ions

Inorganic ions have been shown to play important roles in plant growth and its regulation. These roles are expected to be closely related to the mechanism by which phototropic growth asymmetry is established. Furthermore, it is becoming increasingly evident that Ca^{2+} and H^{+} make specific contributions to the transduction of phototropic signals. Below the relationships between ions and growth are first described in overview, and then the roles of ions in phototropism will be discussed. Results on gravitropism are included because they provide a useful comparison. Insights will be provided on the issues that could not have been explained easily in terms of the Cholodny-Went theory of tropisms.

23.11.1 Cell growth and ions: some background views

The plant cell grows by irreversibly expanding cell walls and supplying new cell wall materials. The irreversible expansion of the cell wall is achieved by the process known as the cell wall loosening [401], and any immediate change in cell growth is achieved by this process. In addition, the cell must maintain adequate turgor, or a turgor pressure in excess of the wall yield threshold pressure, to undergo growth [402]. The plant cell generates turgor by maintaining a higher solute concentration (and thus a lower water potential) in the protoplasm, relative to the concentration in the apoplast. Inorganic ions are the major solutes used to sustain and control turgor. The plant cell takes up K^+, the major cation component, against the concentration gradient through inwardly rectifying cation channels that reside in the plasma membrane and have a high selectivity for K^+ [403]. The uptake is driven by the inside negative membrane potential, which is maintained by the activity of the H^+ pump (H^+-ATPase) that also resides in the plasma membrane. The plant cell also has to take up anions to maintain a charge balance and to sustain the electrical function of the H^+ pump. For example, the plant cell takes up Cl^-, a major anion component, probably through Cl^-/H^+ symporters against both the concentration gradient and the membrane potential (see e.g. [404]). The H^+ gradient necessary for this process is again generated by the H^+ pump.

The hypothesis known as the "acid-growth theory" provides an explanation for the process of auxin-dependent cell wall loosening. This hypothesis, based on the independent suggestions by Hager et al. [405] and Cleland [406], states that auxin acidifies cell walls by enhancing the plasma membrane H^+-pump activity and that this acidification causes the cell wall loosening. The acid-growth theory has been supported by a number of results [401]. However, some difficulties of the hypothesis were reported by Schopfer and his co-workers [407,408] (see [409] for counter-arguments and [359] for a critical discussion).

More recent studies have provided direct evidence for the notion that the plasma membrane H^+-pump activity is enhanced by auxin. Hargar et al. [410] found with maize coleoptile segments that the application of IAA substantially enhances the level of the antibody-detectable H^+-ATPase of the plasma membrane. It was subsequently shown, in agreement with this finding, that the expression of the gene for a major plasma membrane H^+-ATPase is enhanced [411]. The increase in H^+-ATPase level began within several minutes and was saturated by about 40 min after IAA application. The response may be too slow to account for the initial step of auxin-induced growth. Rück et al. [412] applied a patch-clamp technique (whole-cell configuration) to protoplasts of maize coleoptiles to demonstrate that IAA enhances an outwardly directed current of positive charge, which most probably represents the H^+-pump activity. The current increased with a lag as short as 30 s and reached a steady state in the next 2 min or so. This result most probably indicates that IAA can activate the H^+ pump already present in the plasma membrane. This explanation, though not the authors' first choice, is supplemented by the earlier results that the ATPase activity of plasma membrane fractions can be stimulated by auxin [413–415]. It appears that auxin can enhance the activity of the H^+ pump already present in the plasma membrane and, in addition, the density of the pump. These results substantiate a condition of the acid-growth theory that auxin stimulates the H^+-pump activity. More recently, Philippar et al. [416] found that the expression of the gene for an inwardly rectifying plasma membrane K^+ channel, that functions for K^+ uptake, is enhanced by the application of a synthetic auxin NAA in maize coleoptile segments. The level of the K^+-channel mRNA began to increase within 15 min and reached the maximum about 75 min after auxin application. The same study also demonstrated that the density of the inwardly rectifying K^+ channels is actually enhanced by auxin and that the properties of the channel itself are not affected by auxin. These results on K^+ channels, together with those on the H^+ pump, indicate that auxin induces a sequence of cellular responses that activate the net ion uptake.

Although an enhanced activity for ion uptake would contribute to auxin-induced growth by providing solutes necessary to maintain turgor, it might also have a more direct effect on cell growth. One such possibility is that plants control cell wall loosening by sensing the change in turgor caused by an altered ion uptake activity. Investigations using a pressure probe have not revealed turgor increasing after auxin application [417], and it has often been noted that coleoptile segments contain less solutes when growing rapidly in response to auxin [418–420]. The most accepted current view is that the auxin-dependent cell wall loosening is not mediated by turgor pressure [401,402]. On the other hand, it has been noted that the process of cell wall loosening stops when turgor pressure is reduced to the extent that no further growth takes place [421]. Also, investigations using protoplasts suggested that blue-light-induced growth inhibition might involve a drop in turgor [307,404]. Even if the primary auxin-dependent growth control is not mediated by turgor pressure, it is possible that plants have a mechanism to exert growth control through turgor sensing. Claussen et al. [422] found that auxin-induced growth of maize coleoptile segments depends strictly on the presence of external K^+. This finding at least indicates that K^+ uptake is an important step required for auxin-dependent cell wall loosening.

One of the earliest cellular events to follow auxin application is a rise in cytosolic Ca^{2+} concentration. This was first shown by Felle [423] with a microelectrode in the epidermal cells of maize coleoptiles. Gehring et al. [424] subsequently used a fluorescent Ca^{2+} indicator and a scanning laser confocal microscope to demonstrate that the level of cytosolic Ca^{2+} is elevated following the application of 2,4-D, a synthetic auxin, in maize coleoptiles and parsley (*Petroselinum hortense*) hypocotyls. These studies indicated that the rise in Ca^{2+} begins immediately after auxin application and progresses over a period of a few to several minutes. Therefore, Ca^{2+}, shown to have a second-messenger role in various cellular signal-transduction pathways, appears to play an important role in the transduction of the auxin signal for growth control. Furthermore, the above-mentioned studies [423,424] have also uncovered the fact that the cytosolic pH drops following auxin application. The response was as rapid as that in cytosolic Ca^{2+}. Therefore, a rise in cytosolic H^+ concentration also appears to represent an early step of the auxin signal transduction.

23.11.2 Participation as osmotic solutes

Goswami and Audus [425] measured the distribution of ^{45}Ca, ^{42}K, and ^{32}P in phototropically stimulated maize coleoptiles. Salts of these isotopes (^{32}P, in the form of phosphate) were dissolved in the culture medium and supplied from roots in advance of phototropic stimulation. It was found that the shaded half contained more ^{42}K and ^{32}P than the irradiated half. (The results on the ^{45}Ca distribution are described in the next section.) The concentration difference had already occurred at an early stage of curvature development. The same workers carried out more extensive experiments with sunflower hypocotyls, although these were restricted to gravitropism and the curvature response induced by asymmetric IAA application. In these cases, too, greater amounts of ^{42}K and ^{32}P were recovered from the convex side than from the concave side. It is not surprising to find more ^{42}K and ^{32}P in the convex side where the greater growth is taking place. What is interesting is that the concentration of these isotopes on either a fresh-weight or dry-weight basis was higher in the convex side than the concave side. The higher concentration in the convex side could also be found in the plants whose roots were removed before the experimental treatments, indicating that the asymmetry was induced with the isotopes already transported to the investigated organ. The authors concluded that K^+ and phosphate are actively translocated from the concave- to the convex-side. Because the ion asymmetry could be generated by asymmetric application of IAA, the authors reached another conclusion that the ion asymmetry detected during tropisms is mediated by auxin asymmetry. In further support of this conclusion, they showed that the ion asymmetry was inhibited by application of the auxin-transport inhibitor, *N*-1-naphthylphthalamic acid (NPA), in all cases including the phototropism of maize coleoptiles.

The ratio of the K^+ concentrations between the two halves of coleoptiles or hypocotyls recorded by Goswami and Audus [425] was around 45 : 55. Because a large proportion of K^+ in plant tissues occurs in the protoplasm [426], the ions must be actively excreted from the cells on the irradiated side and/or actively taken up by the cells on the shaded side, to achieve the measured concentration gradient. This explanation is indeed supported by the recent results on H^+ pump and K^+ channels which indicate that auxin enhances ion uptake (see Section 23.11.1). There is evidence that the H^+-pump activity becomes asymmetric during phototropism and gravitropism (see Section 23.11.5). Moreover, Philippar et al. [416] demonstrated that the K^+-channel mRNA is asymmetrically expressed during the gravitropism of red-light-grown maize coleoptiles. This asymmetry occurred in such a way that the expression is slightly enhanced on the convex side and is substantially reduced on the concave side. This distribution pattern is in essential agreement with the pattern of the IAA distribution measured during gravitropism of red-light-grown maize coleoptiles [297].

The occurrence of ion asymmetry during tropisms suggests that the turgor pressure decreases on the concave side and/or increases on the convex side. Rich and Tomos [187] used a pressure probe to measure turgor in the cells located on the irradiated and shaded sides of *Sinapis alba* hypocotyls. They could obtain no evidence that the turgor pressure changes during phototropism. The turgor pressure was very stable, within the limit of resolution, for a period of 60 min after the onset of stimulation, by which time the maximal rate of phototropic curvature had been established.

However, Tarui and Iino [155] found that the irradiated side of the top 5-mm zone of oat coleoptiles shrinks during fPIPP, indicating that the turgor pressure may change during phototropism (Figure 23B). Shrinkage could also be observed during gravitropism on the concave side of the top 5-mm zone. Although not claimed by the authors, some shrinkage on the irradiated side of the apical oat coleoptile zone is found in the data of Macleod et al. [306] and Taylor et al. [115]. A slight shrinkage of the concave side was also recorded during gravitropism of cucumber and cress hypocotyls [427,428]. Such small shrinkage was thought to be due to the mechanical compression caused by organ-bending [428]. However, the shrinkage observed by Tarui and Iino was substantial. The shrinkage had already occurred at a very early stage of curvature development and, while the irradiated side was shrinking, the shaded side did not elongate more than in the non-stimulated control (Figure 23, A and B). In other words, the curvature of the top zone occurred mainly by shrinkage of the irradiated side. These results indicated that the shrinkage was not caused by mechanical compression, but by a drop in turgor pressure on the irradiated side. When the coleoptile stimulated for fPIPP was treated bilaterally after 25 min with a high-fluence blue light, the shrinkage was prevented and the zone resumed nearly normal elongation from about 1 h after the high fluence pulse, as if it had not received any light (Figure 23D). The high-fluence pulse alone induced neither shrinkage nor any significant inhibition of elongation (Figure 23C; compare with Figure 23A). It is most probable that the shrinkage was induced as

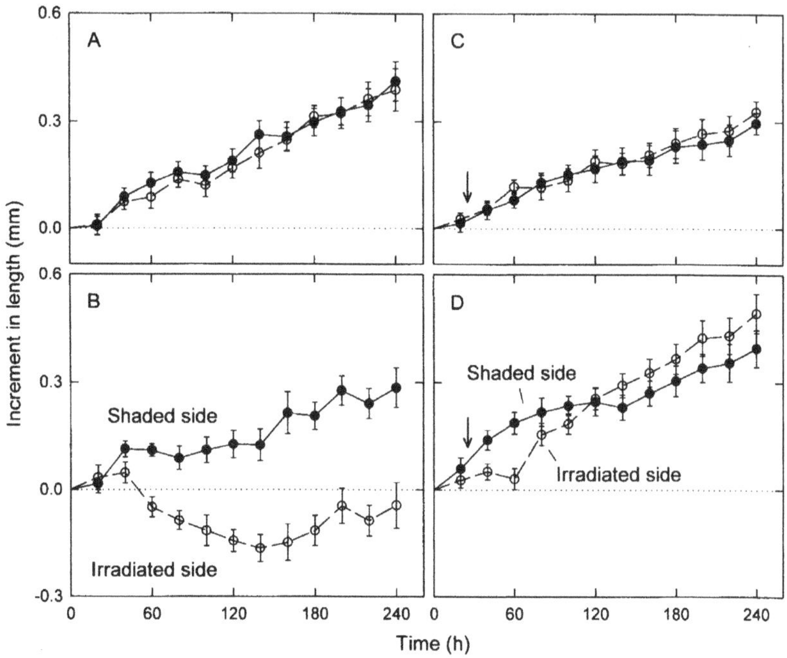

Figure 23. Shrinkage in the apical zone of oat coleoptiles during fPIPP. The length on the two sides of the apical 5-mm zone of red-light-grown coleoptiles was monitored. The seedling was rotated on a horizontal clinostat after time zero except the time of blue-light treatment. (**A**) Coleoptiles received no blue light. (**B**) Coleoptiles were treated at time zero with a pulse of unilateral blue light (1 μmol m^{-2}). (**C**) Coleoptiles were treated at 25 min (indicated by an arrow) with a bilateral pulse of blue light (15 μmol m^{-2} from either side). (**D**) Coleoptiles were treated with a unilateral pulse of blue light as in **B** and also with a bilateral pulse of blue light as in **C**. The minus value indicates that the length became shorter than the initial length. Adapted from Tarui and Iino [155].

a part of the phototropic response (see also Section 23.10.3). The shrinkage was only evident in the top zone. In non-stimulated coleoptiles, the top zone elongated at a rate less than half of the rate in the lower two zones, although the extent of curvature was not smaller than in the other zones. It was therefore suggested that the shrinkage on the irradiated side is also a component of the curvature in the other zone, but is not apparent because it is masked by greater net growth [155]. An additional possibility is that the shrinkage is apparent only in the top zone because the cells constituting this zone have cell walls that are uniquely elastic.

These results of Tarui and Iino are in agreement with the conclusion that K^+ moves from the concave- to the convex-side during tropisms. Tarui and Iino did not observe that the shaded side of the top coleoptile zone expands more than in the non-stimulated control (Figure 23). It might be that phototropic stimulation enhances K^+ efflux differentially between the two sides (more enhancement on the irradiated side). In this case, the net movement of K^+ to the shaded side could occur with a minor change in turgor on the shaded side. The oat coleoptile appears to be exceptional in its growth response during phototropism (see Section 23.10.3). One might speculate that K^+ uptake is enhanced on the shaded side in other materials (or in oat coleoptiles, under certain conditions) that show a growth enhancement on the shaded side.

Clearly it is an important future task to resolve whether or not a control of the turgor pressure is involved in the tropisms of coleoptiles and stems. This issue is closely related to the question whether or not auxin can control growth through a turgor-sensing mechanism (see Section 23.11.1). To summarize all the available information, the following view may be presented: Auxin induces cellular responses that enhance the net ion uptake activity. These responses are not absolutely required for cell wall loosening, though they may partially mediate the latter. In the straight growth system, an auxin-induced increase in ion uptake activity is not easily reflected in an increase in turgor because the actual ion uptake is limited by the availability of ions. On the other hand, the asymmetry of ion uptake activity induced by auxin asymmetry results more effectively in a turgor asymmetry, because ions can be redistributed between the two sides without requiring an enhanced net supply of ions. Depending on the materials and conditions, the turgor asymmetry accompanies the true growth asymmetry to cause a curvature greater than that established only by the latter asymmetry. This seems to be the case, at least in red-light-grown oat coleoptiles.

23.11.3 Apoplastic Ca^{2+}

Arslan-Cerim [429] supplied $^{45}Ca^{2+}$ to intact sunflower seedlings through their roots, and measured the radioactivity distribution in the two halves of gravitropically stimulated hypocotyls. A gradient of 53:47 (concave:convex, on a dry-weight basis) was detected 1 h after the onset of stimulation. Using a similar method, Goswami and Audus [425] repeated the experiment on sunflower hypocotyls and extended it to maize coleoptiles. In either material, gradients ranging from 65:45 to 60:40 (concave:convex, on either a fresh-weight or dry-weight basis) could be detected 1 h after the onset of stimulation. In these studies, it was also noted that the gradient increased only slightly during the subsequent 2-h period. Therefore, the calcium asymmetry seemed to be established at an early phase of curvature development. Goswami and Audus [425] could also show that ^{45}Ca is distributed asymmetrically in maize coleoptiles during phototropic stimulation. A gradient of 55 : 45 (irradiated : shaded half) was measured. These results suggest that Ca^{2+} is translocated form the convex- to the concave-side during either gravitropism or phototropism. Because a substantial portion of Ca^{2+} is generally located in the apoplastic space, the measured asymmetry of Ca^{2+} would largely represent the asymmetry of apoplastic Ca^{2+}.

The occurrence of lateral Ca^{2+} translocation during gravitropism could be demonstrated more clearly by Migliaccio and Galston [430]. They applied $^{45}Ca^{2+}$ directly to pea epicotyls through the lightly abraded surface and measured the radioactivity distribution in the epidermal peels. A gradient of 55 : 45 (concave : convex, on a fresh-weight basis) could be found 90 min after the onset of gravitropic stimulation. Slocum and Roux [431] used the antimonate precipitation method to visualize the distribution of calcium in light and electron micrographs. They found that calcium accumulates on the concave side of oat coleoptiles. This accumulation occurred in peripheral cell layers, and the cell wall could be identified as the site of accumulation. Electron micrographs provided evidence that Ca^{2+} moves from the convex- to the concave-side. Moreover, it could be observed that the calcium in the vacuole decreases in the convex side. This observation suggested that the Ca^{2+} translocated to the concave side is not only the apoplastic Ca^{2+} but also the Ca^{2+} stored in the vacuole.

Bagshaw and Cleland [432] reinvestigated the distribution of Ca^{2+} during gravitropism of sunflower hypocotyls, pea epicotyls, and maize coleoptiles, determining calcium with atomic absorption spectrometry. In all cases, they could not find any significant asymmetry of calcium. These authors concluded that the asymmetry found in the previous studies must be attributed to a very small fraction of apoplastic Ca^{2+}. It is in fact possible that membrane-bound Ca^{2+} is not easily equilibrated with newly supplied Ca^{2+} and that the results obtained using $^{45}Ca^{2+}$ represent the movement of non-bound Ca^{2+}. This idea might explain why Goswami and Audus [425] found a smaller $^{45}Ca^{2+}$ gradient when de-rooted plants were subjected to gravistimulation after a longer pre-incubation with $^{45}Ca^{2+}$ (6 h vs. 3 h). The termination of $^{45}Ca^{2+}$ supply from the root and the longer pre-incubation could increase the proportion of cell-wall-bound $^{45}Ca^{2+}$. The results of Slocum and Roux [431] cannot be explained, however, by this principle. As pointed out by Bagshaw and Cleland [432], the antimonite-precipitation method used for the histochemical analyses might have failed to detect a large portion of cell-wall-bound Ca^{2+}. Whether or not this explanation can entirely account for the large calcium asymmetry detected by Slocum and Roux [431] in oat coleoptiles needs to be clarified. The results of Bagshaw and Cleland [432] do not disprove the occurrence of lateral Ca^{2+} movement during tropisms. Given the strong evidence for such a movement, it may be concluded that a mobile fraction of apoplastic Ca^{2+} is distributed asymmetrically with a movement of Ca^{2+} from the convex- to the concave-side.

Goswami and Audus [425] showed that asymmetric application of IAA causes an asymmetric distribution of $^{45}Ca^{2+}$ in sunflower hypocotyls. Furthermore, application of the auxin transport inhibitor prevented the development of $^{45}Ca^{2+}$ asymmetry that occurred during gravitropism of sunflower hypocotyls and phototropism of maize coleoptiles. Migliaccio and Galston [430] also demonstrated that the $^{45}Ca^{2+}$ asymmetry induced in pea epicotyls by gravitropic stimulation is prevented by auxin transport inhibitors. These results lead to the conclusion that lateral Ca^{2+} translocation is caused by auxin asymmetry. This conclusion is substantiated by the finding that application of IAA to one side of maize coleoptile segments can induce Ca^{2+} movement away from the IAA-applied side [433]. Migliaccio and Galston [430] obtained evidence that the Ca^{2+} mobility in the tissue is enhanced by low pH and treatments with IAA and fusicoccin, both of which induce apoplastic acidification. Based on such results, they concluded that the Ca^{2+} bound to cell walls is displaced to a greater extent by H^+ on the convex side, where a greater H^+-pump activity occurs, and that the released Ca^{2+} diffuses passively to the concave side.

In the above-mentioned study, Goswami and Audus [425] found that asymmetric application of mersalyl induces an asymmetric distribution of $^{45}Ca^{2+}$ but no detectable curvature in sunflower hypocotyls. They concluded that the Ca^{2+} asymmetry is not the cause of the curvature response. However, this toxic chemical might have interfered with other processes that are essential for curvature response. Slocum and Roux [431] hypothesized that the accumulation of Ca^{2+} observed on the concave side of oat coleoptiles contributes to the curvature response by inhibiting auxin-dependent growth on that side. This idea was based on the results that the application of Ca^{2+} inhibited the auxin-dependent growth of oat coleoptile [434] and pea internode [435] segments. On the other hand, Migliaccio and Galston [430] expressed the opinion that the asymmetry of apoplastic Ca^{2+} cannot be the prime cause of curvature response, based on the contention that displacement of cell-wall-bound Ca^{2+} by H^+ is unlikely to be a major step in auxin-induced growth stimulation. Bagshaw and Cleland [436] subsequently showed with sunflower hypocotyls that apoplastic Ca^{2+} at physiological concentrations is not inhibitory on growth. The growth and gravitropism of the hypocotyl were stimulated when seedlings were incubated in 1 mM and 10 mM $CaCl_2$, although a high concentration (50 mM) was inhibitory. The stimulatory effect is probably attributable to an electrolyte effect that is not specific to Ca^{2+} [437]. At the moment it appears unlikely that asymmetric distribution of apoplastic Ca^{2+} contributes to curvature by causing asymmetric growth inhibition. The results, however, do not exclude the possibility that the Ca^{2+} asymmetry plays some other specific role in tropisms.

23.11.4 Cytosolic Ca^{2+}

Measurements of cytosolic Ca^{2+} have become feasible with new techniques. Gehring et al. [438] used a fluorescent Ca^{2+} indicator to find that the level of cytosolic Ca^{2+} increases on the shaded side in maize coleoptiles. In their experiments, the excised apical portion (1–2 cm) of a coleoptile was loaded with the fluorescent Ca^{2+} indicator, fluo-3, and the cytosolic distribution of Ca^{2+} was recorded with a scanning laser confocal microscope during unilateral irradiation of the tip with white light. The microscopic viewing was made from the basal cut end. Upon unilateral irradiation of the tip, cytosolic Ca^{2+} in the cells of epidermal and peripheral cortical tissues increased. The rise in cytosolic Ca^{2+} became detectable within 5 min of irradiation and reached a stable level within 15 min. No clear fluorescence change could be detected on the irradiated side. However, when the coleoptile was turned by 180° (i.e. the unilateral light was directed to the previously shaded side) after the fluorescence change resulting from the initial unilateral irradiation was stabilized, a rapid decrease in the Ca^{2+} level on the irradiated side could be detected. The Ca^{2+} level did not decrease beyond the initial pre-stimulation level, but the result indicates that the change in stimulus direction is rapidly reflected in the Ca^{2+} status. The response to gravitropic stimulation was also investigated. The peripheral cells were visualized from the lower side of the horizontally oriented coleoptile segments. At a position 5 mm from the tip, the cytosolic Ca^{2+} level in the peripheral cells increased rapidly and stabilized within 3 min after the onset of stimulation.

In the above experiments on phototropism, Gehring et al. [438] directed the beam of unilateral light to the tip zone less than 3 mm in length and detected the rise in cytosolic Ca^{2+} at the transverse cell layers located between 5 and 10 mm from the tip (C.A. Gehring, personal communication). The response in cytosolic Ca^{2+} appears to occur at a distance from the stimulated tip, although it is not excluded that some scattered light reached the lower coleoptile zone to induce the response. Decapitated coleoptile segments did not show any detectable change in cytosolic Ca^{2+} (C.A. Gehring, personal communication). This result at least indicates that the change in cytosolic Ca^{2+} requires the presence of the tip.

After finding that auxin can also induce an increase in cytosolic Ca^{2+}, Gehring et al. [438] hypothesized that the increase observed on the shaded side during phototropic stimulation is caused by an increase in auxin concentration on that side. The tip dependence of the Ca^{2+} response (see above) also appeared to support this hypothesis. However, the hypothesis faces a serious problem if the time course for the development of IAA asymmetry is taken into consideration. In the subapical zone of maize coleoptiles (2–7 mm below the tip), which is not more basal than the region used by Gehring et al. [438] for the Ca^{2+} measurement, the IAA asymmetry began to be detectable 10 min after fPIPP stimulation, and the asymmetry increased gradually during the next 30 min or so ([297]; see Figure 21). The auxin asymmetry also developed during gravitropism, with similar kinetics [297]. Therefore, the response in cytosolic Ca^{2+}, which already saturates within 15 min (phototropism) or 3 min (gravitropism), is too fast to follow the IAA asymmetry. One may argue that a substantial change in IAA level occurs in a peripheral region before the asymmetry between the two halves develops. However, the phototropic and gravitropic curvatures followed the measured IAA asymmetry with a delay of about 10 min [297]. It seems very probable that the response in cytosolic Ca^{2+} precedes the auxin asymmetry.

In working on the phototropism of sunflower hypocotyls, Ma and Sun [439] found that EGTA enhanced the curvature response when applied on the irradiated side and reduced it when applied on the shaded side. This result does not agree with the idea that the higher apoplastic Ca^{2+} on the irradiated side results in a curvature, but is in accord with the idea that the higher cytosolic Ca^{2+} on the shaded side is responsible for the curvature response. The Ca^{2+}-channel blocker, verapamil, was also found to enhance the phototropic curvature when applied on the irradiated side, in agreement with the result from the EGTA application, although it little affected the curvature response when applied on the shaded side. Surprisingly, one-sided application of the calcium ionophore A23187 caused a substantial curvature away from the applied side. These results of Ma and Sun [439] can be largely explained by the idea that a greater concentration of cytosolic Ca^{2+} results in a greater rate of cell growth. If phototropic stimulation were found to cause an asymmetry of cytosolic Ca^{2+} in sunflower hypocotyls, then the causality between Ca^{2+} asymmetry and phototropic curvature would be substantiated.

Recently, Baum et al. [440] used transgenic *Arabidopsis* and tobacco (*Nicotiana plumbaginifolia*) plants that express cytosolic apoaequorin to investigate the effect of blue light on cytosolic Ca^{2+}. When de-etiolated seedlings of these plants were treated with a 10-s pulse of blue light after in vivo reconstitution of aequorin in the dark, chemiluminescence representing an increase in cytosolic Ca^{2+} could be detected. The Ca^{2+} level increased from about 50 nM to 300 nM in 20–30 s after the onset of the blue-light pulse, and returned to the dark level during the next 50 s. Baum et al. produced apoaequorin-producing transgenic *Arabidopsis* plants that carry *cry1*, *cry2*, and *nph1* mutations. On either the *cry1* or *cry2* background, the Ca^{2+} response was normal. However, on the *nph1* background, the response was reduced by half. This strongly suggests that the response is related to phototropism. Since they used the NPH1 null, *nph1–5* mutant, half the response appears to be mediated by the NPH1 holoprotein or phototropin (see Section 23.8.4) and the other half by another blue-light receptor that also differs from cryptochromes 1 and 2 (see Section 23.8.3).

In the experiments of Baum et al. [440], seedlings were unilaterally irradiated with a pulse of blue light (A.J. Trewavas, personal communication). Therefore, the response recorded by Baum et al. might be explained by the one observed by Gehring et al. [438], i.e. an increase in cytosolic Ca^{2+} that occurs only on the shaded side. However, the former response was much faster than the latter. Furthermore, since Baum et al. [440] applied a pulse of high-fluence blue light, the photochemical reactions on both sides were probably saturated. Given the available information, it may be concluded that the two responses are distinct in nature and the response recorded by Baum et al. occurs closer to photoperception.

Some of the photobiological data presented by Baum et al. [440] merit further discussion. They found that pretreatment of seedlings with a red-light pulse reduced the Ca^{2+} response substantially in tobacco, although no comparable red-light effect was found in *Arabidopsis*. The authors attempted to relate this effect to the red-light effects known in the literature. However, the response observed is probably unrelated to those reported previously. In their experiments, the red-light pulse was given 5 min before the blue-light pulse. The effect of red-light pretreatment on phototropic responsiveness or sensitivity becomes detectable with a lag of about 15 min and progresses gradually over a period of 1–2 h (see Sections 23.4.1–4). The pretreatment effect detected in de-etiolated tobacco is probably a new kind of red-light effect that has not been identified before. Baum et al. also found that responsiveness to a second blue-light pulse was restored gradually in the dark over a period of 3 h or more. The kinetics were much slower than those resolved for the restoration of phototropic responsiveness and for resensitization response (Figure 18; see Section 23.9.2). The data, however, suggest that the responsiveness restoration involves two kinetic components (see their Figure 5). During the 2-h period after the first blue-light pulse, the restored response peaked at about 30 s after the onset of the test blue light. On the other hand, when the test blue light was given after 3 h, the peak of Ca^{2+} response was found near 20 s. The control measurement obtained with only one pulse showed a peak at about 20 s. It seems that the blue-light-induced rise in cytosolic Ca^{2+} is composed of fast and slow responses and that the fast response is restored more slowly than the slow response. Although exact comparison is not possible, the restoration kinetics of the slow response is roughly comparable to the kinetics of either phototropic responsiveness restoration or resensitization response. Therefore, the slow response might be associated with the limiting photoproduct deduced from the analysis of fPIPP (see Section 23.9.4). It is also tempting to speculate that the Ca^{2+} response observed in the phototropin null *nph1–5* mutant is related to the phototropic responsiveness remaining in this mutant (see Section 23.8.5). In fact, the data of Baum et al. (see their Figure 2C) indicate that the Ca^{2+} response in the *nph1–5* mutant peaks earlier than in the wild type, and therefore suggest that the fast response characterized by slow restoration kinetics is related to the phototropic responsiveness remaining in the phototropin null mutant.

23.11.5 Apoplastic and cytosolic H^+

According to the acid-growth theory (see above, Section 23.11.1), tropisms can be ascribed to the asymmetry of apoplastic H^+ caused by auxin asymmetry. In fact, it has been shown that asymmetric treatment with an acidic buffer causes curvature away from the applied side [441,442]. Using dark-adapted maize coleoptiles and de-etiolated sunflower hypocotyls, Mulkey et al. [443] demonstrated that the pH on the organ's abraded surface, detected with a pH indicator, becomes asymmetric following phototropic and gravitropic stimulation. Although exactly quantitative measurements were not possible, they could resolve that the convex side became more acidic than the concave side. In the maize coleoptiles stimulated for either phototropism or gravitropism, the pH asymmetry occurred in such a way that the concave side was alkalinized and the convex side was acidified. In similarly stimulated sunflower hypocotyls, acidification developed on the two sides, but was greater on the convex side. The results for maize coleoptiles agree with the auxin asymmetry detected during phototropism of maize coleoptiles (Figure 21) and gravitropism [297,444]. However, any auxin asymmetry has not been shown to occur in sunflower hypocotyls during phototropism (see Section 23.10.7) and gravitropism [444]. Therefore, the link between the auxin asymmetry and the pH asymmetry is not clear in this material.

In the study investigating the effects of phototropic and gravitropic stimuli on the cytosolic Ca^{2+} of maize coleoptiles, Gehring et al. [438] also measured cytosolic pH using a pH indicator probe. Following gravitropic stimulation, a drop in pH could be recorded on the shaded side. The pH continued to drop during the measurement period of 9 min. The authors stated that a drop in pH could also be observed during phototropic stimulation on the shaded side. Because the drop in cytoplasmic pH also followed IAA treatment ([424]; see also Section 23.11.1), it was suggested that the pH drop found to follow phototropic and gravitropic stimulation is mediated by an increase in auxin. However, as in the case of the response in cytosolic Ca^{2+}, the pH response is still too rapid to be correlated to the auxin asymmetry induced during tropisms.

Measurements of the surface electrical potential have provided some related information. Backus and Schrank [445] investigated in oat coleoptiles the electrical changes that follow phototropic stimulation. They found that the convex side becomes more positive than the concave side. This result was obtained for both fPIPP and PINP (the positive side was opposite with respect to the light direction between fPIPP and

PINP). Some of the reported individual time-course measurements suggest that the asymmetry of electric potential developed very rapidly following light stimulation. Unfortunately, these measurements showed uncontrolled fluctuations, and no solid statement can be made on this point. More extensive studies have been conducted to characterize the electrical changes induced by gravitropic stimulation. Most results have indicated that the lower side becomes more positive relative to the upper side in horizontally displaced organs (reviewed in [446]). The electrical changes recorded in maize coleoptiles could be attributed to the changes in H^+-pump activity caused by asymmetric auxin distribution [447]. However, rapid changes that cannot be correlated with auxin asymmetry have also been detected in soybean hypocotyls [448] and *Phaseolus angularis* epicotyls [449,450]. Shigematsu et al. [450] could distinguish the rapid and transient response and the slow and long lasting response in *Phaseolus angularis* epicotyls.

23.11.6 Sequence of events: hypothetical views

As we have discussed above, both phototropic and gravitropic stimuli have been found to induce the following responses:

1. Ca^{2+} is translocated from the convex- to the concave-side, causing a higher apoplastic concentration of Ca^{2+} ($[Ca^{2+}]_{ap}$) on the concave side (Section 23.11.3);
2. the concentration of cytosolic Ca^{2+} ($[Ca^{2+}]_{cy}$) increases on the convex side (Section 23.11.4);
3. the concentration of apoplastic H^+ ($[H^+]_{ap}$) becomes higher on the convex side than on the concave side (Section 23.11.5); and
4. the concentration of cytosolic H^+ ($[H^+]_{cy}$) increases on the convex side (Section 23.11.5).

In view of the results that both tropisms include very similar responses with regard to Ca^{2+} and H^+, it may be suggested that the two tropisms share similar mechanisms for auxin translocation and other cellular responses involved in the establishment of growth asymmetry. However, studies with oat and maize coleoptiles have indicated some fundamental differences between the two tropisms. The phototropism of coleoptiles is generally characterized by high photoperceptivity of the tip (see Section 23.5.2), but such a specialized function of the tip is not a property of coleoptile gravitropism. This is noted from the result that basipetal migration of the onset of curvature does not occur in gravitropism [154,297]. Furthermore, gravitropic stimulation can effectively induce lateral auxin translocation in tip-removed segments, as was originally demonstrated by Dolk [154] (see [293] for later results). On the other hand, phototropic stimulation induces lateral translocation only to a limited extent in such segments (see Section 23.10.5). These differences between the two tropisms may be unified as follows: Plant organs have the system for lateral auxin translocation and other cellular systems for the induction of growth asymmetry along their entire length (or elongating part). The photosystem for phototropism and the gravisensing system for gravitropism share these mechanisms to transduce the signals. The regional difference in signal perceptivity is determined by the density of the signal-sensing system and/or the effectiveness of the system to transduce the signal to the next steps.

According to the starch-statolith theory, the gravity is perceived by statocytes, the putative gravity-sensing cells containing sedimentable amyloplasts [451]. This theory implies, but not exclusively, that the transverse polarity occurs in individual statocytes. Then, the simplest model would be that the transverse polarity established in statocytes causes lateral auxin translocation within these cells, eventually leading to the translocation across the organ [452]. In contrast, the view presented in the preceding paragraph determines that the transverse polarity is established at the organ level. This view, however, does not critically contradict the starch-statolith theory, because the transverse polarity at the organ level can still be based on the transverse polarity in statocytes. This explanation is supported by the result that $[Ca^{2+}]_{cy}$ and $[H^+]_{ap}$ are elevated very rapidly in the lower-side peripheral cells of horizontally displaced organs, apparently not involving polarization within each of the responding cells [438].

We will first examine **Model 1**, shown schematically in Figure 24. This model has been constructed by considering the original claims that the responses listed above are due to auxin asymmetry. The first step of the process after signal perception is the induction of lateral translocation of IAA (*a*), which results in [IAA] asymmetry (*b*). In response to this asymmetry, $[Ca^{2+}]_{cy}$ and $[H^+]_{cy}$ increase on the convex side (*c*). Then an increase in the plasma membrane H^+-pump activity follows on the same side. The H^+-pump activation can be a direct consequence of the increase in $[H^+]_{cy}$ (i.e. an increase in the substrate concentration). The H^+-pump activation causes an increase in $[H^+]_{ap}$ on the convex side (*d*). In response to this increase, cell-wall-bound Ca^{2+} is displaced, and the freed Ca^{2+} moves to the concave side by diffusion (*e*), and an asymmetry of $[Ca^{2+}]_{ap}$ is generated (*f*). The increase in $[Ca^{2+}]_{cy}$ at step *c* might be attributable, at least in part, to the release of Ca^{2+} from the vacuole, especially in view of the observation that vacuolar Ca^{2+} decreases on the convex side during gravitropism of oat coleoptiles [431]. It is assumed here that cytosolic Ca^{2+} is pumped out into the cell wall in response to the release of vacuolar Ca^{2+} into cytoplasm; the vacuolar Ca^{2+} serves as a source of apoplastic Ca^{2+} translocated towards the concave side (the dashed line connecting *c* and *e*). If the $[Ca^{2+}]_{ap}$ asymmetry leads to lateral IAA translocation (a process discussed below for an alternative model), then the step *f* is linked to the step *a* and a loop is formed for self-acceleration. Growth asymmetry (Δ growth) is caused by $[H^+]_{ap}$ asymmetry (Δ growth from the step *d*) and also via other IAA-dependent processes (Δ growth from the step *b*). The asymmetry of $[Ca^{2+}]_{ap}$ might also contribute to growth asymmetry via Ca^{2+}-dependent inhibition of cell wall loosening [453]. However, this growth mechanism is now little supported ([401]; see Section 23.11.3).

The model contradicts the results that the increases in $[Ca^{2+}]_{cy}$ and $[H^+]_{cy}$ occur too rapidly to be described by auxin asymmetry (see Sections 23.11.4 and 23.11.5). The model does not explain why $[Ca^{2+}]_{cy}$ and $[H^+]_{cy}$ on the concave side do not decrease although [IAA] decreases. Because the first problem is very critical, the step *c* must be removed from the sequence of reactions or we must look for alternative models.

Model 2 (Figure 24) retains all the response components, but their sequence is altered. The first step after signal perception is the increase in $[Ca^{2+}]_{cy}$ and $[H^+]_{cy}$ that occurs more extensively on the convex side than on the concave side (*a*). This step is followed by an increase in H^+-pump activity on the convex side. The resulting rise in $[H^+]_{ap}$ (*b*) causes displacement of cell-wall-bound Ca^{2+}; the released Ca^{2+} moves to the concave side (*c*), leading to the asymmetry of $[Ca^{2+}]_{ap}$ (*d*). As described for Model 1,

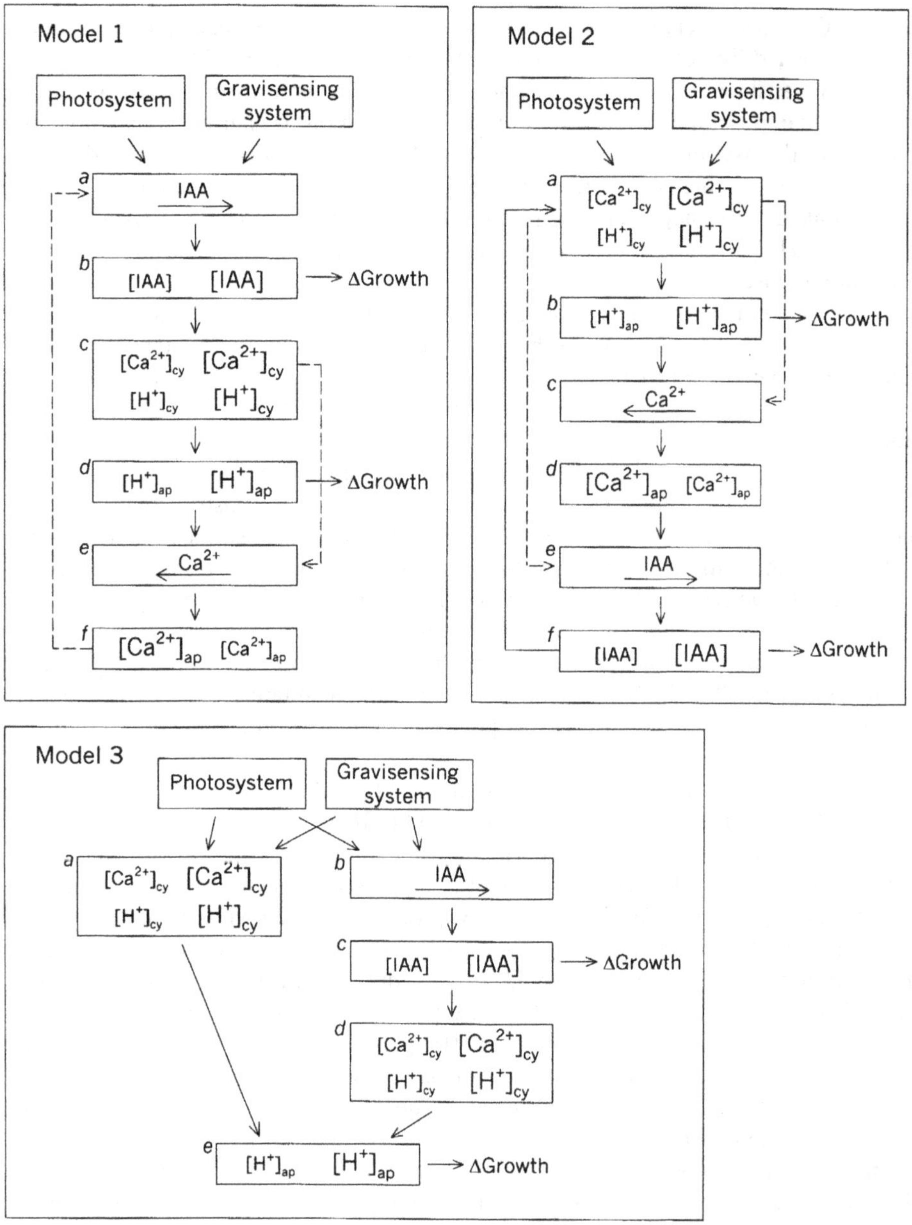

Figure 24. Models describing the causal relationships among IAA, Ca^{2+}, and H^+ in phototropism and gravitropism. [IAA], tissue concentration of IAA; $[Ca^{2+}]_{cy}$, cytosolic concentration of Ca^{2+}; $[Ca^{2+}]_{ap}$, apoplastic concentration of Ca^{2+}; $[H^+]_{cy}$, cytosolic concentration of H^+; $[H^+]_{ap}$, apoplastic concentration of H^+. The left and right sides of each small box represent the two sides of a stimulated organ; the left side corresponds to the irradiated side (phototropism) or the upper side (gravitropism). The lateral arrow inside the box indicates the direction of movement. The difference in letter size represents the relative difference in concentration between the two sides. The arrows connecting the boxes indicate the sequence of events.

vacuolar Ca^{2+} could serve as a source of the apoplastic Ca^{2+} translocated to the concave side (the dashed line connecting *a* and *c*). In response to the $[Ca^{2+}]_{ap}$ asymmetry, IAA is laterally translocated (*e*), and the asymmetry of [IAA] is formed (*f*). This model also considers the possibility that lateral IAA translocation is caused more directly in response to the asymmetry of $[Ca^{2+}]_{cy}$ and/or $[H^+]_{cy}$ (the dashed line connecting *a* and *e*). The [IAA] asymmetry then leads to the asymmetry of $[Ca^{2+}]_{cy}$ and $[H^+]_{cy}$. Thus, the step *f* is linked to the step *a*, and forms a self-accelerating loop. The asymmetry of either $[Ca^{2+}]_{cy}$ or $[H^+]_{cy}$ occurs in such a way that the concentration decreases on the concave side and increases on the convex side (i.e. the responses in $[Ca^{2+}]_{cy}$ and $[H^+]_{cy}$ depend on IAA concentration). The asymmetry of $[H^+]_{ap}$ occurs in a similar manner. Therefore, the steps *a* and *b* that result from the step *f* are not identical to the steps that directly follow signal perception. During continuous stimulation, the two types of reactions in steps *a* and *b* continue to operate. The steps linked to Δ growth are as described for Model 1.

In Model 2, the asymmetry of $[Ca^{2+}]_{ap}$ (step *d*) is considered to be a prime cause of the lateral IAA translocation. Gross and Sauter [433] have demonstrated that one-sided, lateral application of IAA to the segment of maize coleoptiles can lead to a movement of Ca^{2+} towards the IAA-applied side, providing experimental evidence for this sequence. The activity of IAA efflux through the plasma membrane might depend on $[Ca^{2+}]_{ap}$. If so, the asymmetry of $[Ca^{2+}]_{ap}$ could induce the asymmetry of IAA mobility, causing a net movement of IAA from the high-$[Ca^{2+}]_{ap}$ side to the low-$[Ca^{2+}]_{ap}$ side (see Section 23.10.11). The IAA efflux through the plasma membrane efflux carrier might be associated with Ca^{2+} influx [454]. This possibility could provide an explanation for the $[Ca^{2+}]_{ap}$-dependent IAA efflux activity.

Model 2 can explain the rapid increase in $[Ca^{2+}]_{cy}$ and $[H^+]_{cy}$ observed following the onset of phototropic and gravitropic stimulation [438] and the fact that a similar increase follows auxin application [423,424]. The movement of apoplastic Ca^{2+} has been shown to occur in oat coleoptiles within 10 min after the onset of gravitropic stimulation [431]. It is possible that the $[Ca^{2+}]_{ap}$ asymmetry precedes the [IAA] asymmetry [297], as determined by the model. The loop of the model is in agreement with the result that the lateral asymmetry of either Ca^{2+} or IAA can induce lateral translocation of the other [433]. The rapid change in electrical surface potential difference between the two sides (see Section 23.11.5) can be ascribed to the $[H^+]_{ap}$ asymmetry (step *b*) that occurs before a significant [IAA] asymmetry is established. It is necessary to evaluate whether the induced $[Ca^{2+}]_{ap}$ asymmetry is sufficient to describe the lateral IAA translocation, especially in view of the results of Bagshaw and Cleland ([432]; see Section 23.11.3). As included in the model, the asymmetry of $[Ca^{2+}]_{cy}$ and $[H^+]_{cy}$ might be linked more directly to the lateral IAA translocation, although so far there is no experimental evidence to support this linkage.

The model provides reasonable explanations for the paradoxical results on sunflower hypocotyls that no auxin asymmetry could be detected in this material during phototropism (see Section 23.10.7) and gravitropism [444]. It is possible that the reaction sequence more-or-less stops at step *d*. Sunflowers may have a very effective mechanism by which curvature response is induced in response to step *b*. This explanation is supported by the observation of Mulkey et al. [443] on sunflower hypocotyls: i.e. the asymmetry of apoplastic pH occurs with different extents of acidification between the two sides during their phototropism and gravitropism (see Section 23.11.5). The effects of EGTA observed by Ma and Sun ([439]; see Section 23.11.4) cannot be explained if the steps *c* to *f* mainly account for the curvature response of this plant species, but are explicable if the steps *a* to *b* directly coupled to the signal-sensing system are the prime route for the curvature response. Furthermore, the very short lag period found for the phototropism of this plant [139] can be explained adequately (see Section 23.7.1). In fact, the unique ability of sunflower plants to express very effective phototropism during most of their vegetative growth stage (see Section 23.2.2) might indeed be associated with their high ability to generate $[H^+]_{ap}$ asymmetry through the direct route.

The proportion of the $[H^+]_{ap}$ asymmetry attributable to the direct route to that generated via the auxin-mediated route might vary depending on the plant species and growth conditions. The hypocotyls of de-etiolated plants do not show clear spatial separation between photoperception and curvature response (see Section 23.5.5) and can express phototropism with a very short lag (see Section 23.7.1). In the phototropism of these plants, the direct route might generally make a greater contribution than the auxin-mediated route. The phototropism of oat coleoptiles induced by non-tip photoperception also does not show clear spatial separation between photoperception and curvature response (see Section 23.5.3) and is perhaps also characterized by a very short lag time (see Section 23.7.1). In this case, too, the direct route might supplement the curvature response mediated by lateral auxin translocation. This possibility provides an additional explanation for the question why the curvature response by non-tip photoperception can be substantial, although the measured auxin asymmetry is relatively very small (see Section 23.10.12).

Model 3 (Figure 24) is added as another alternative. In this model, the asymmetry of $[H^+]_{ap}$ is generated via two separate routes. Each step of the model operates as already explained. This model accommodates most of the critical and paradoxical results that can be explained by Model 2. However, the model does not indicate how the IAA asymmetry is generated and does not have an amplification loop. The lateral translocation of apoplastic Ca^{2+} is omitted from the model, although it could occur as a side-effect of the $[H^+]_{ap}$ asymmetry. This model may be considered when the mechanism of lateral auxin translocation cannot be explained by $[Ca^{2+}]_{ap}$ asymmetry or by any other preceding reactions in Model 2.

The phototropic curvature of pulvini is made by the asymmetry of turgor-dependent cell volume, and ions (mainly K^+ and Cl^-) move across the pulvinus to account for the turgor asymmetry [455]. It has been generally regarded that this type of phototropism is fundamentally different from the phototropism of coleoptiles and stems, which is based on growth asymmetry. However, as discussed in Section 23.11.2, coleoptile and stem phototropism probably involves active translocation of osmotic ions from the concave- to the convex-side, and this translocation could possibly cause an asymmetry of turgor. There are a few other lines of evidence indicating phenomenological and mechanistic similarities between the two types of phototropism. The first of these is the fact that the pulvinus of *Phaseolus vulgaris* can bend in response to one-sided application of IAA [456,457]. The second line of evidence comes from the experiments with isolated protoplasts. Keller and Von Volkenburgh [458] reported that the IAA applied to the bathing medium could induce swelling of the protoplasts isolated from oat coleoptiles. Similarly, the protoplasts of *Phaseolus vulgaris* pulvini showed swelling in response to applied IAA; at 1–10 μM IAA, the protoplast volume increased at a rate of about 4% h^{-1} (X. Wang and M. Iino, unpublished). These results do not prove that auxin mediates the phototropic curvature of pulvini, but do suggest that the cells of pulvini share some basic properties related to phototropism with those of coleoptiles and stems. The third line of evidence is provided by the results of Koller and Ritter [47]. They showed that the phototropic curvature of *Phaseolus vulgaris* pulvinus involves shrinking on the irradiated side and swelling on the shaded side. Thus, the pulvinar curvature cannot easily be explained by a local light-induced response; the redistribution principle considered to underlie the phototropism of coleoptiles and stems (Section 23.10.3) provides a simpler explanation.

It is tempting to speculate that the steps *a* to *b* in Model 2 (or the steps *a* to *e* in Model 3) also operate in pulvinar phototropism, the growth difference originating in the step *b* being replaced by turgor-driven cell volume difference. The steps *b* to *f* and the loop in Model 2 (or the steps *b* to *e* in Model 3) may also operate. Young leaves of Leguminosae plants often show phototropic curvature in both petioles and pulvini. It is perhaps not unlikely that the phototropic system of the pulvinus, a specialized structure of the petiole, has originated from that of petioles, which in turn could be similar to the phototropic system of stems.

23.12 Phototropism sensitive to red light and UV-B

Although there is no doubt that the phototropism of seed plants is generally mediated by photosystems that have a high sensitivity to blue light (see Sections 23.2.6 and 23.8.1), there have been reports on phototropic responses that show totally distinct spectral sensitivities. These results will be summarized below.

23.12.1 Red-light-sensitive phototropism

Red light is often regarded as phototropically ineffective in seed plants. This view originated primarily from the detailed action-spectroscopic studies conducted with oat coleoptiles (see Section 23.8.1) and has been supplemented by the result of Shropshire and Mohr [459] on etiolated seedlings of *Fagopyrum esculentum* and *Sinapis alba*. The latter authors could not observe any phototropic curvature in hypocotyls after exposure to unilateral red light or far-red light that induced a lateral gradient in the phytochrome-dependent anthocyanin accumulation. However, some workers reported that red light could induce phototropism of de-etiolated plants [2,214,460,461]. The purity of the red-light source used might be questioned. However, the action-spectroscopic data of Atkins [214] most convincingly indicate that red-light-sensitive phototropism occurs at least in de-etiolated plants (see Section 23.8.1).

In fact, Iino et al. [110,462] were able to demonstrate that phytochrome can mediate phototropism. They found that a positive phototropism is induced by a pulse of unilateral red light in dark-adapted maize seedlings. The curvature occurred in the mesocotyl and the basal coleoptile zone in response to the light signal perceived in these regions. When irradiated unilaterally with far-red light after overhead irradiation with red light, the seedlings developed a negative phototropism. These results could be explained in terms of the lateral gradient of Pfr generated across the responding zones. Both very-low-fluence and low-fluence responses contributed to the curvature response. Blue light also induced the phytochrome-mediated phototropism in addition to the blue-light-specific phototropism, and the phototropic fluence-response curves obtained using blue light for the mesocotyl and the coleoptile were very complex. In fact, when seedlings were pretreated with overhead red light just before or after phototropic stimulation with blue light, the fluence-response curve revealed only the blue-light-specific fPIPP of coleoptiles. Because the light-sensitive region and the fluence-response characteristics of the phytochrome-mediated phototropism agreed with the phytochrome-mediated inhibition of elongation growth, it was deduced that the phototropic curvature is based on the lateral asymmetry in the extent of growth inhibition. Thus, Blaauw's hypothesis (see Section 23.10.2) was found to be applicable to this phytochrome-mediated phototropism. Kunzelmann and Schäfer [182] measured Pfr gradients across unilaterally irradiated maize mesocotyls. It was found that the gradient was greater in blue light than in red light, in agreement with the greater phytochrome-mediated curvature response to blue light [110]. Parker et al. [134] obtained evidence that phytochrome can also mediate a phototropism in the epicotyl of dark-grown pea seedlings. In this phototropism, too, the phytochrome-mediated growth inhibition observable in this material could possibly account for the curvature response.

In the studies mentioned above, the phototropic response was induced by a pulse of red light. Phototropism may not be induced with continuous irradiation because the phototransformation of phytochrome would be saturated on both sides of the organ. In fact, the curvature response disappears when the fluence of red light is increased to the extent that Pfr formation is saturated on the two sides. However, when maize seedlings were continuously exposed to weak unilateral red light, the nodal region and the short mesocotyl maintained a positive curvature, while the coleoptile was bent to the other direction at the base and assumed an erect appearance in the upper part (M. Iino, unpublished observation). Such a specific curvature at the node and mesocotyl does not occur during continuous blue-light stimulation; the coleoptile remains tilted towards the light source with curvature at the coleoptile base and the mesocotyl. Oat seedlings showed a similar curvature response to continuous unilateral red light. It appears that a lateral gradient in phytochrome action is somehow maintained during continuous irradiation and can cause a sustained curvature in the mesocotyl region.

Ballaré et al. [463] subsequently demonstrated that phytochrome can mediate a phototropism of hypocotyls in de-etiolated cucumber seedlings. When the seedlings, maintained under white light (160 μmol m^{-2} s^{-1}), were exposed to unilateral far-red light, the hypocotyl developed a negative phototropism. A slight curvature was detected at a fluence rate of 17 μmol m^{-2} s^{-1}, and the response increased with fluence rate. At the highest fluence rate tested (120 μmol m^{-2} s^{-1}), a steady curvature of about 15° was achieved. The *lh* mutant that lacks phytochrome B did not show any such curvature. The curvature response is probably based on phytochrome-dependent growth inhibition (i.e. less Pfr in the far-red-light irradiated side, and thus negative curvature).

The phototropism induced by a pulse of red light in maize seedlings is relatively small as compared to the blue-light-sensitive phototropism induced in the coleoptile [110].

The corresponding phototropism in pea epicotyls is also very small [134] compared to the blue-light-sensitive phototropism induced in red-light-adapted plants [102]. However, plants may use a fluence-rate-dependent phytochrome action to cause a significant curvature response during continuous stimulation (see above). Furthermore, green stems of de-etiolated plants might include an additional mechanism: It is expected that the lateral light gradient in unilaterally irradiated green stems is greater in red light than in far-red light. Thus, a greater red/far-red light ratio, and hence a greater steady-sate level of Pfr, could occur on the irradiated side during continuous stimulation, leading to a phytochrome-mediated positive phototropism (a view presented in [90]).

Although the de-etiolated cucumber hypocotyl can express a curvature in response to a lateral Pfr gradient, field experiments conducted by Ballaré et al. [464] did not reveal any clear participation of phytochrome-mediated phototropism in the phototropism induced by the direct rays of the sun. The light might have been too strong for an appreciable induction of the phytochrome-mediated phototropism. Although it is not yet clear whether phytochrome-mediated phototropism occurs in nature in response to unilateral light or uneven distribution of light fluence rates, the work of Ballaré et al. [464] has uncovered another way in which plants use phytochrome to cause an ecologically relevant curvature response. Vegetation shade provides a unique light environment in which the proportion of far-red light is greater than that of red light (see [5]). If a plant is fenced by other plants on one side, it receives a far-red-light-enriched light from the direction of the neighboring plants. Ballaré et al. [463] provided convincing evidence that cucumber seedlings detect this light environment using a phytochrome system and can induce a shade-avoiding movement of the hypocotyl (see Section 23.2.1). The uneven spectral quality of incident light is the directional information used in this new type of phototropism.

In ferns, phytochrome is clearly a major photoreceptor for phototropism. The protonema of ferns can respond to both red light and blue light to exhibit positive phototropism [465]. Detailed investigations by Wada and his co-workers on the phototropism and the related polarotropism in *Adiantum capillus-veneris* have indicated that phytochrome is a major photoreceptor [466,467], although a blue-light receptor also participates in this material [468]. The protonema probably detects the light fluence rate gradient between the two sides for the induction of phytochrome-mediated phototropism [469,470]. Young leaves of *Adiantum* are also phototropic. Wada and Sei [79] obtained evidence that a blue-light receptor and phytochrome function equally in the positive phototropism observed in initially erect petioles. A phytochrome gene (cDNA) of *Adiantum* was cloned by Okamoto et al. [471]. Nozue et al. [472] subsequently cloned another gene having a homology to the *Adiantum* phytochrome. Surprisingly, the deduced protein has been revealed to be a fusion protein of phytochrome and phototropin, which is probably expressed in the plant. At present, the causal relationship between these gene products and phototropism is not clear.

The possibility that phytochrome can function as a photoreceptor for phototropism in seed plants should be reflected in experimental planning and data analyses when the blue-light-sensitive phototropism is investigated. A simple approach for overcoming this problem is to use red-light-grown plants. In summary, the use of red-light-grown plants has the following advantages in the study of the process of blue-light-sensitive phototropism:

1. the phototropic responsiveness that depends on phytochrome is fully established (Sections 23.4.2–4);
2. the accompanying phytochrome-dependent responses in growth and auxin status are minimized (Sections 23.10.3 and 23.10.6);
3. the accompanying phytochrome-mediated phototropism is minimized (this section); and
4. the plants can be handled easily for experiments.

23.12.2 UV-B-sensitive phototropism

It has been noted that UV-B (radiation in the spectral region around 300 nm) can effectively induce positive phototropism in oat seedlings. The occurrence of such UV-sensitive phototropism was first noted by Mills and Schrank [473]. When they irradiated the seedlings grown under a weak red light with UV (spectral lines between 265 and 435 nm from a low-pressure mercury arc lamp), the coleoptile developed a positive curvature in its basal zone. The response was nearly saturated at a fluence of 300 J m^{-2} (100-s irradiation), and the curvature detected was about 15°. Using filter combinations, it could be shown that the radiation between 255 and 350 nm is mainly responsible for this curvature. A little later, Curry et al. [474] found that 280-nm radiation (isolated from a mercury lamp) could induce positive phototropism in dark-adapted coleoptiles. The curvature occurred at the base of the shoot (the coleoptile base and the short mesocotyl). A substantial curvature (about 30°) could be induced at a fluence of 0.6 J m^{-2} (1.4 μmol m^{-2}). The response detected by these workers was apparently much more sensitive to UV than that measured by Mills and Schrank. Interestingly, localized irradiation of either the tip (about 5 mm) or the base (nodal region) caused a partial response, but irradiation of the middle zone did not cause any appreciable response. The action spectrum obtained by shading the tip during test irradiation revealed a peak at 280–300 nm.

The action spectrum obtained by Baskin and Iino [100] for the fPIPP of red-light-adapted alfalfa hypocotyls showed a peak at about 280 nm in addition to the peaks in the blue and UV-A regions (Figure 15). The fact that this peak is attributable to fPIPP was shown by the continuity of the bell-shaped fluence-response curve from the longer wavelengths. It was noted, however, that another fluence-dependent positive phototropism takes place at wavelengths shorter than 300 nm and at fluences higher than those causing fPIPP. The threshold of this additional response was near 1000 μmol m^{-2}, and the curvature response increased sharply with fluence. Because the response observed at such high fluences of UV-B is unlikely to be of physiological consequence, the authors suggested that the curvature resulted from non-specific cellular damage that occurs more strongly on the irradiated side of the hypocotyl [100].

Iino [90] re-evaluated the data of Curry et al. [474] described above, in view of the action spectroscopic data from alfalfa hypocotyls. It turned out that the 280-nm curvature response induced by tip irradiation is explicable in terms of fPIPP. He also presented the view that the 280-nm curvature response induced by irradiation of the nodal region can be phytochrome-mediated. This interpretation was based on the agreement in effective fluences between the phytochrome phototransformation by UV-B and the phytochrome-dependent phototropism in maize coleoptiles (see the preceding section). The base sensitivity of the phytochrome-mediated phototropism also supports this view. The UV-sensitive response observed by Mills and Schrank [473] is, however, unlikely to be phytochrome-mediated, because these workers grew plants and carried out experiments under red light. As mentioned above, the fluences causing this curvature response are far greater than those found to be effective in the experiments of Curry et al., but are lower than the threshold fluence for the UV-B-specific response in alfalfa. The results are insufficient to conclude that the response observed by Mills and Schrank represents a UV-B-specific phototropism. At present, there is no clear evidence that seed plants use a specific UV-B-absorbing photoreceptor for phototropism. Perhaps, plants are not prepared to bend actively towards harmful UV-B radiation.

23.13 Concluding remarks

Phototropism was made a subject of modern plant physiology by Julius von Sachs and Charles Darwin in the late 19th century and has since been one of the major areas of plant physiology. The study of phototropism contributed greatly in the first half of the 20th century in establishing two major and broader fields: plant growth physiology (especially in relation to auxin) and plant photobiology (in particular the blue-light responses). Our knowledge of phototropism itself, however, has not advanced much in the subsequent decades. In fact, it seemed that the physiology of phototropism becomes more and more complex with each new report, instead of being resolved. This situation is highlighted by the fact that the field has experienced, and is still experiencing, repeated controversies on the basic issues such as whether auxin asymmetry is involved in phototropism, how auxin asymmetry is established, and whether carotenoids can also function as photoreceptor pigments.

It is apparent that this confusing situation has arisen largely because the system of phototropism itself is highly sophisticated. For example, it has been made clear that the expression of phototropism is controlled in a very complex manner by a photoreceptor family of phytochromes. It has also become clear that the overall photosystem of phototropism can adjust its light sensitivity and responsiveness, being probably based on complex molecular systems. Furthermore, phototropism is closely related to growth, which itself is based on complex molecular and cellular mechanisms. Apparently the confusing situation is a step forwards to resolving all such complexities. It should also be mentioned, however, that experimental results and their interpretation have often been unnecessarily complicated by not considering and reflecting on available information. The uncontrolled effect of green "safelights" is a typical source of complications. Phytochrome activation in dark-grown or dark-adapted plants by the phototropic blue-light stimulus is another source. Although we have had sufficient information on these potential problems since the early 1980s, recent work still appears to be inviting such problems.

In the last decade, our understanding of phototropism has been enhanced greatly by the introduction of molecular genetic approaches and modern cell biological techniques. In particular, the photoreceptor phototropin has been identified and closely related molecular components are being resolved. In the next decade, the photoreceptor complexes, which probably reflect many of the detailed aspects of phototropic fluence-response relationships, will be clarified. A unique part of the overall phototropic system is the way in which plants detect the side receiving more light. This question may be answered in the continuation of the molecular study of photoreceptor complexes. Other unique parts of the overall phototropic system are how auxin is laterally translocated and how this lateral translocation is correlated with growth asymmetry. These questions remain largely unanswered. Perhaps, the isolation of mutants impaired in these processes, and their analyses, will provide clues on these very biological parts of phototropism. Finally, as discussed at the beginning of this chapter, phototropism probably has significant ecological functions. Close examination of these possible functions will certainly deepen our understanding of phototropism.

In this chapter, I have attempted to summarize most of the results of phototropism research carried out until the end of the 20th century. To sum up a large body of results, however, it was necessary to review individual reports in detail, and to join the results under hypothetical views. The hypothetical views, which are partially or entirely original to this chapter, include the following:

1. The shade-avoiding movement is the major role for phototropism (Sections 23.2.1–3 and 23.12.1) and is the means by which growth is oriented to a brighter environment (shade-avoiding growth orientation) (Sections 23.2.1 and 23.2.3).
2. The seedlings growing in a dark soil environment generally have little ability to express phototropism (Section 23.4.5); the phototropic responsiveness is established in these seedlings by light-dependent reactions mediated by phytochrome (Section 23.4.4) and a blue-light receptor (Section 23.5.6).
3. The phototropic signal can be stored in the tissue site of photoperception for at least 2 h (Section 23.7.4).
4. During continuous phototropic stimulation, phototropism equilibrates with gravitropism with a contribution of autostraightening (Section 23.7.5).
5. The phototropism of coleoptiles is based on a specific mechanism that is associated with the tip but, in addition, on the photoperception in non-tip parts, which leads to curvature response without an effective longitudinal signal transmission (Sections 23.5.2, 23.5.3, 23.10.12).
6. The lateral auxin translocation accounting for the coleoptile phototropism induced by non-tip photoperception and the phototropism of stems occurs along the radial direction without significantly interfering with basipetally polarized auxin transport (Section 23.10.12).
7. Auxin asymmetry is amplified for growth asymmetry (Section 23.10.10.4).
8. Lateral asymmetries in apoplastic Ca^{2+} and H^+, cytosolic Ca^{2+} and H^+, and auxin are intimately correlated for the establishment of growth asymmetry in both phototropism and gravitropism, and the growth asymmetry can occur without depending on the auxin asymmetry (Section 23.11.6).
9. The pulvinar phototropism, based on turgor-driven cell-volume changes, is not fundamentally different from the phototropisms of coleoptiles and stems (Section 23.11.6).

Although many of these views are speculative at the present stage, it is nevertheless hoped that they will stimulate further progress in phototropism research.

Acknowledgements

Many of the early papers were written in German. It would have been impossible for me to prepare this chapter without the help of Rainer Hertel, Ralf Kaldenhoff, Winfried S. Peters, and Christine Schimek, who kindly read some of these papers and provided me with necessary information. I am indebted to Winslow R. Briggs and Rainer Hertel for valuable comments and suggestions on the manuscript and to Ken Haga for help in preparation of the Figures. Thanks are also due to all those who kindly responded to my specific questions regarding their publications. The preparation of this chapter has been supported by a Grant-in-Aid (No. 10440242) from the Ministry of Education, Science and Culture in Japan.

References

1. W. Rothert (1896). Ueber heliotropismus. *Beiträge zur Biologie der Pflanzen*, **7**, 1–212.
2. J.E. Shuttleworth, M. Black (1977). The role of cotyledons in phototropism of de-etiolated seedlings. *Planta*, **135**, 51–55.
3. R.J. Ellis (1984). Kinetics and fluence–response relationships of phototropism in the dicot *Fagopyrum esculentum* Moench. (buckwheat). *Plant Cell Physiol.*, **25**, 1513–1520.
4. F. Woitzik, H. Mohr (1988). Control of hypocotyl phototropism by phytochrome in a dicotyledonous seedling (*Sesamum indicum* L.). *Plant Cell Environ.*, **11**, 653–661.
5. H. Smith (1994). Sensing the light environment: the functions of the phytochrome family. In: R.E. Kendrick, G.H.M. Kronenberg (Eds.), *Photomorphogenesis in Plants* (2nd ed., pp. 491–535). Kluwer Academic, Netherlands.
6. J.H. Schaffner (1898). Observations on the nutation of *Helianthus annuus*. *Bot. Gaz.*, **25**, 395–403.
7. J.H. Schaffner (1900). The nutation of *Helianthus*. *Bot. Gaz.*, **29**, 197–200.
8. H. Shibaoka, T. Yamaki (1959). Studies on the growth movement of sunflower plant. *Scientific Papers of the College of General Education University of Tokyo*, **9**, 105–126.
9. Y. Leshem (1977). Sunflower: a misnomer? *Nature*, **269**, 102.
10. A.R.G. Lang, J.E. Begg (1979). Movements of *Helianthus annuus* leaves and heads. *J. Appl. Ecol.*, **16**, 299–305.
11. J.V. Sachs (1887). *Lectures on the physiology of plants* (translated by H.M. Ward). Clarendon, Oxford.
12. G. Kudo (1995). Ecological significance of flower heliotropism in the spring ephemeral *Adonis ramosa* (Ranunculaceae). *OIKOS*, **72**, 14–20.
13. R.M. Knutson (1981). Flowers that make heat while the sun shines. *Nat. Hist.*, **90**, 75–80.
14. B. Hockings, C.D. Sharplin (1965). Flower basking by arctic insects. *Nature*, **206**, 215.
15. P.G. Kevan (1975). Sun-tracking solar furnaces in high arctic flowers: significance for pollination and insects. *Science*, **189**, 723–726.
16. M.L. Stanton, C. Galen (1989). Consequences of flower heliotropism for reproduction in an alpine buttercup (*Ranunculus adoneus*). *Ecologia*, **78**, 477–485.
17. R.A. Sherry, C. Galen (1998). The mechanism of floral heliotropism in the snow buttercup *Ranunculus adoneus*. *Plant Cell Environ.*, **21**, 983–993.
18. E.M. Schmitt (1922). Beziehungen zwischen der Befruchtung und den postfloralen Blüten- bzw. Fruchtstielbewegungen bei *Digitalis purpurea*, *Digitalis ambigua*, *Althaea rosea* und *Linaria cymbalaria*. *Z. Bot.*, **14**, 625–675.
19. H.D. Zinsmeister (1960). Das phototropische Verhalten der Blütensteile von *Cyclamen*

persicum hybr. *Planta*, **55**, 647–668.
20. C. Darwin (1875). *The movements and habits of climbing plants*. John Murray, London.
21. S.-L. Lam, A.C. Leopold (1966). Role of leaves in phototropism. *Plant Physiol.*, **41**, 847–851.
22. C. Darwin (1880). *The power of movement in plants*. John Murray, London.
23. G. Haberlandt (1904). Die Perzeption des Lichtreizes durch das Laubblatt. *Ber. der Deutsch. Bot. Ges.*, **22**, 105–104.
24. H. Wager (1915). Behavior of plants in response to light. *Nature*, **96**, 468.
25. N.G. Ball (1923). Phototropic movements of leaves. The functions of the lamina and the petiole with regard to the perception of the stimulus. *Scientific Proceedings, Royal Dublin Society*, **17**, 281–286.
26. W.N. Jones (1938). Observations on the response of leaves of *Limnanthemum* and *Tropaeolum* to light and gravity. *Ann. Bot.*, **2**, 819–825.
27. G.S.G. Shell, A.R.G. Lang, P.J.M. Sale (1974). Quantitative measures of leaf orientation and heliotropic response in sunflower, bean, pepper and cucumber. *Agr. Meteorol.*, **13**, 25–37.
28. L. Brauner, Y. Vardar (1950). Über die Funktion der Lamina bei der phototropischen und geotropischen Reaktion des *Tropaeolum*-Blattes. *Rev. Fac. Sci. Univ. Istanbul Ser.*, **B15**, 269–299.
29. R. Snow (1962). Geostrophism. In: W. Ruhland (Ed.), *Encyclopedia of Plant Physiology* (Vol. 17, part 2, pp. 378–389). Springer, Berlin.
30. G. Raydt (1925). Über die Bewegungen euphotometrischer Blätter. *Jb. Wiss. Bot.*, **64**, 731–769.
31. R.A. Snow (1947). A test of Sachs's theory of the plagiotropism of laminae. *New Phytol.*, **46**, 258–261.
32. J.W. Hart (1990). *Plant Tropisms and other growth movements*. Unwin Hyman, London.
33. J. Ehleringer, I. Forseth (1980). Solar tracking by plants. *Science*, **210**, 1094–1098.
34. H.C. Yin (1938). Diaphototropic movement of the leaves of *Malva neglecta*. *Am. J. Bot.*, **25**, 1–6.
35. A. Schwartz, D. Koller (1978). Phototropic response to vectorial light in leaves of *Lavatera cretica*. *Plant Physiol.*, **61**, 924–928.
36. C.M. Wainwright (1977). Sun-tracking and related leaf movements in a desert lupine (*Lupinus arizonicus*). *Am. J. Bot.*, **64**, 1032–1041.
37. T.C. Vogelmann (1984). Site of light perception and motor cells in a sun-tracking lupine (*Lupinus succulentus*). *Physiol. Plant.*, **62**, 335–340.
38. E. Werker, T. Shak, D. Koller (1991). Photobiological and structural studies of light-driven movements in the solar-tracking leaf of *Lupinus palaestinus* Bioss. (Fabaceae). *Bot. Acta*, **104**, 144–156.
39. T.C. Vogelmann, L.O. Björn (1983). Response to directional light by leaves of a sun-tracking lupine (*Lupinus succulentus*). *Physiol. Plant.*, **59**, 533–538.
40. D. Koller, I. Levitan, W.R. Briggs (1985). The vectorial photo-excitation in solar-tracking leaves of *Lavatera cretica* (Malvaceae). *Photochem. Photobiol.*, **42**, 717–723.
41. H. Vöchting (1888). Ueber die Lichtstellung der Laubblätter. *Bot. Zeit.*, **46**, 501–516.
42. D. Koller, T. Shak (1990). Light-driven movements in the solar tracking leaf of *Lupinus palaestinus* Boiss. (Fabaceae). *Photochem. Photobiol.*, **52**, 187–195.
43. A. Schwartz, D. Koller (1986). Diurnal phototropism in solar tracking leaves of *Lavatera cretica*. *Plant Physiol.*, **80**, 778–781.
44. D.W. Sheriff, M.M. Ludlow (1985). Diaheliotropic responses of leaves of *Macroptilium atropurpureum* cv. Siratro. *Asut. J. Plant Physiol.*, **12**, 151–171.

45. R. Donahue, V.S. Berg (1990). Leaf orientation of soybean seedlings: II. Receptor sites and light stimuli. *Crop Sci.*, **30**, 638–643.
46. H.C. Wien, D.H. Wallace (1973). Light-induced orientation in *Phaseolus vulgaris* L. *Crop Sci.*, **13**, 721–724.
47. D. Koller, S. Ritter (1994). Phototropic responses of the pulvinules and associated laminar reorientation in the trifoliate leaf of bean *Phaseolus vulgaris. J. Plant Physiol.*, **143**, 52–63.
48. S. Ritter, D. Koller (1994). Movements of the trifoliate leaf of bean (*Phaseolus vulgaris* L.) during simulated day, and their consequences for solar-tracking fidelity and interception of solar radiation. *J. Plant Physiol.*, **143**, 64–71.
49. R.L. Travis, R. Reed (1983). The solar tracking pattern in a closed alfalfa canopy. *Crop Sci.*, **23**, 664–668.
50. D.M. Oosterhuis, S. Walker, J. Eastham (1985). Soybean leaflet movements as an indicator of crop water stress. *Crop Sci.*, **25**, 1101–1106.
51. V.S. Berg, T.C. Hsiao (1986). Solar tracking: light avoidance induced by water stress in leaves of kidney bean seedlings in the field. *Crop Sci.*, **26**, 980–986.
52. J.M. Prichard, I.N. Forseth (1988). Photosynthetic responses of two heliotropic legumes from contrasting habitats. *Plant Cell Environ.*, **11**, 591–601.
53. K.A. Shackel, A.E. Hall (1979). Reversible leaflet movements in relation to drought adaptation of cowpeas, *Vigna unguiculata* (L.) Walp. *Aust. J. Plant Physiol.*, **6**, 265–276.
54. G. Rajendrudu, V.S. Rama Das (1981). Solar tracking and light interception by leaves of some dicot species. *Current Sci.*, **50**, 618–620.
55. I.N. Forseth, J.R. Ehleringer (1980). Solar tracking response to drought in a desert annual. *Oecologia*, **44**, 159–163.
56. O. Björkman, S.B. Powles (1981). Leaf movement in the shade species *Oxalis oregana*.I. Response to light level and light quality. *Carnegie Institution of Washington Year Book*, **80**, 59–62.
57. S.B. Powles, O. Björkman (1981). Leaf movement in the shade species *Oxalis oregana*. II. Role in protection against injury by intense light. *Carnegie Institution of Washington Year Book*, **80**, 63–66.
58. I.N. Forseth (1990). Function of leaf movements. In: R.L. Satter, H.L. Gorton, T.C. Vogelmann (Eds.), *The Pulvinus: Motor Organ for Leaf Movement* (pp. 238–261). The American Society of Plant Physiologists, Rockville.
59. F. Yu, V.S. Berg (1994). Control of paraheliotropism in two *Phaseolus* species. *Plant Physiol.*, **106**, 1567–1573.
60. M. Ludlow, M.O. Björkman (1984). Paraheliotropic leaf movement in Siratro as a protective mechanism against drought-induced damage to primary photosynthetic reactions: damage by excessive light and heat. *Planta*, **161**, 505–518.
61. R. Reed, R.L. Travis (1987). Paraheliotropic leaf movements in mature alfalfa canopies. *Crop Sci.*, **27**, 301–304.
62. Q.A. Fu, J.R. Ehleringer (1989). Heliotropic leaf movements in common beans controlled by air temperature. *Plant Physiol.*, **91**, 1162–1167.
63. V.S. Berg, S. Heuchlin (1990). Leaf orientation of soybean seedlings.I. Effect of water potential and photosynthetic photon flux density on paraheliotropism. *Crop. Sci.*, **30**, 631–638.
64. V.S. Berg, T.C. Hsiao (1990). Leaf orientation of soybean seedlings.I. Effect of water potential and photosynthetic photon flux density on paraheliotropism. *Crop Sci.*, **30**, 631–638.
65. L.M. Ross, L.R. Dillenberg, I.N. Forseth (1991). Responses of soybean leaf angle, photosynthesis and stomatal conductance to leaf and soil water potential. *Ann. Bot.*, **67**,

51–58.
66. W.-Y. Kao, I.N. Forseth (1992). Responses of gas exchange and phototropic leaf orientation in soybean to soil water availability, leaf water potential, air temperature, and photosynthetic photon flux. *Environ. Exp. Bot.*, **32**, 153–161.
67. W.-Y. Kao, T.-T. Tsai (1998). Tropic leaf movements, photosynthetic gas exchange, leaf $\delta^{13}C$ and chlorophyll *a* fluorescence of three soybean species in response to water availability. *Plant Cell Environ.*, **21**, 1055–1062.
68. G.P. Findlay (1984). Nastic movements. In: M.B. Wilkins (Ed.), *Advanced Plant Physiology* (pp. 186–200). Pitman, London.
69. R. Neumann, M. Iino (1997). Phototropism of rice (*Oryza sativa* L.) coleoptiles: fluence–response relationships, kinetics and photogravitropic equilibrium. *Planta*, **201**, 288–292.
70. B. Hubert, G.L. Funke (1937). The phototropism of terrestrial roots. *Biol. Jaarboek*, **4**, 286–315.
71. K. Okada, Y. Shimura (1992). Mutational analysis of root gravitropism and phototropism of *Arabidopsis thaliana* seedlings. *Aust. J. Plant Physiol.*, **89**, 493–497.
72. E. Liscum, W.R. Briggs (1995). Mutations in the *NPH1* locus of *Arabidopsis* disrupt the perception of phototropic stimuli. *Plant Cell*, **7**, 473–485.
73. H.J. Schneider (1965). Kritische Versuche zum Problem des Phototropismus bei Wurzeln. *Zeit. Bot.*, **52**, 451–499.
74. A. Blaauw (1918). Licht und Wachstum. *III. Meded Landbouwhhoogeschool*, **15**, 87–205.
75. R. Kohlbecker (1957). Die Abhängigkeit des Langenwachstums und der phototropischen Krümmungen von der Lichtqualität bei Keimwurzeln von *Sinapis alba*. *Z. Bot.*, **45**, 507–524.
76. A. Sperlich (1912). Über Krümmungsursachen bei Keimstengeln und beim Monokotylenkeimblatte nebst Bemerkungen über den Phototropismus der positiv geotropischen Zonen des Hypokotyls und über das Stemmorgan bei Cucurbitaceen. *Jb. Bot.*, **50**, 502–653.
77. G. Naundorf (1940). Untersuchungen über den Phototropismus der Keimwurzeln von *Helianthus annuus*. *Planta*, **30**, 639–663.
78. P.-E. Pilet (1952). Le phototropisme des raciness de *Lens culinaris* Medikus. *Bull. Soc. Vaud. Sci. Naur.*, **65**, 197–211.
79. M. Wada, H. Sei (1994). Phytochrome-mediated phototropism in *Adiantum cuneatum* young leaves. *J. Plant Res.*, **107**, 181–186.
80. J.H. Miller (1968). Fern gametophytes as experimental material. *Bot. Rev.*, **34**, 361–440.
81. J.Z. Kiss (1994). Negative phototropism in young gametophytes of the fern *Schizaea pusilla*. *Plant Cell Environ.*, **17**, 1339–1343.
82. M. Wada, M. Sugai (1994). Photobiology of ferns. In: R.E. Kendrick, G.H.M. Kronenberg (Eds.), *Photomorphogenesis in Plants* (2nd ed., pp. 783–802). Kluwer Academic, Netherlands.
83. D. Koller, S. Ritter, D. Fork (1996). Light-driven movement of the trifoliate leaves of bean (*Phaseolus vulgaris* L.). Spectral and functional analysis. *J. Plant Physiol.*, **149**, 384–392.
84. M.L. Stanton, C. Galen (1993). Blue light controls solar tracking by flowers of an alpine plant. *Plant Cell Environ.*, **16**, 983–989.
85. A.H. Blaauw (1909). Die Perzeption des Lichtes. *Rec. Trav. Bot. Neerl.*, **5**, 209–372.
86. E.S. Johnston (1934). Phototropic sensitivity in relation to wave length. *Smithsonian Miscellaneous Collections*, **92**, 1–17.
87. H.G. du Buy, E. Nuernbergk (1934). Phototropismus und Wachstum der Pflanzen. *Ergebnisse der Biologie*, **10**, 207–322.
88. B.K. Zimmerman, W.R. Briggs (1963). A kinetic model for phototropic responses of oat coleoptiles. *Plant Physiol.*, **38**, 254–261.

89. O.H. Blaauw, G. Blaauw-Jansen (1970). The phototropic responses of *Avena* coleoptiles. *Acta Bot. Neerl.*, **19**, 755–763.
90. M. Iino (1990). Phototropism: mechanisms and ecological implications. *Plant Cell Environ.*, **13**, 633–650.
91. B.K. Zimmerman, W.R. Briggs (1963). Phototropic dosage-response curves for oat coleoptiles. *Plant Physiol.*, **38**, 248–253.
92. A.M. Meyer (1969). Versuche zur 1. positiven und zur negativen phototropischen Krümmung der *Avena*koleoptile. *Lichtperception und Absorptionsgradient. Z. Pflanzenphysiol.*, **60**, 418–433.
93. M. Iino (1988). Pulse-induced phototropisms in oat and maize coleoptiles. *Plant Physiol.*, **88**, 823–828.
94. M. Everett, K.V. Thimann (1968). Second positive phototropism in the *Avena* coleoptile. *Plant Physiol.*, **43**, 1786–1792.
95. B.G. Pickard, K. Dutson, V. Harrison, E. Donegan (1969). Second positive phototropic response patterns of the oat coleoptile. *Planta*, **88**, 1–33.
96. W.R. Briggs (1960). Light dosage and phototropic responses of corn and oat coleoptiles. *Plant Physiol.*, **35**, 951–962.
97. B. Steyer (1967). Die Dosis-Wirkungsrelationen bei geotroper und phototroper Reizung: Vergleich von Mono- mit Dicotyledonen. *Planta*, **77**, 277–286.
98. D.G. Heathcote, B.W. Bircher (1987). Enhancement of phototropic response to a range of light doses in *Triticum aestivum* coleoptiles in clinostat-simulated microgravity. *Planta*, **170**, 249–256.
99. B. Steiniz, K.L. Poff (1986). A single positive phototropic response induced with pulsed light in hypocotyls of *Arabidopsis thaliana* seedlings. *Planta*, **168**, 305–315.
100. T.I. Baskin, M. Iino (1987). An action spectrum in the blue and ultraviolet for phototropism in alfalfa. *Photochem. Photobiol.*, **46**, 127–136.
101. K. Hasegawa, H. Noguchi, C. Tanoue, S. Sando, M. Takada, M. Sakoda, T. Hashimoto (1987). Phototropism in hypocotyls of radish. IV. Flank growth and lateral distribution of *cis*- and *trans*-raphanusanins in the first positive phototropic curvature. *Plant Physiol.*, **85**, 379–382.
102. T.I. Baskin (1986). Redistribution of growth during phototropism and nutation in the pea epicotyl. *Planta*, **169**, 406–414.
103. E.J.A. Asomaning, A.W. Galston (1961). Comparative study of phototropic response and pigment content in oat and barley coleoptiles. *Plant Physiol.*, **36**, 453–464.
104. M. Everett (1974). Dose-response curves for radish seedling phototropism. *Plant Physiol.*, **54**, 222–225.
105. J.W. Hart, I.R. MacDonald (1981). Phototropic responses of hypocotyls of etiolated and green seedlings. *Plant Sci. Lett.*, **21**, 151–158.
106. R.J. Ellis (1987). Comparison of fluence–response relationships of phototropism in light- and dark-grown buckwheat. *Plant Physiol.*, **85**, 689–692.
107. Y.J. Liu, M. Iino (1996). Phytochrome is required for the occurrence of time-dependent phototropism in maize coleoptiles. *Plant Cell Environ.*, **19**, 1379–1388.
108. J.P. Khurana, Z. Ren, B. Steinitz, B. Parks, T.R. Best, K.L. Poff (1989). Mutants of *Arabidopsis thaliana* with decreased amplitude in their phototropic response. *Plant Physiol.*, **91**, 685–689.
109. Y.J. Liu, M. Iino (1996). Effects of red light on the fluence–response relationship for pulse-induced phototropism of maize coleoptiles. *Plant Cell Environ.*, **19**, 609–614.
110. M. Iino, W.R. Briggs, E. Schäfer (1984). Phytochrome-mediated phototropism in maize seedling shoots. *Planta*, **160**, 41–51.
111. W.R. Briggs (1963). Red light, auxin relationships, and the phototropic responses of corn

and oat coleoptiles. *Am. J. Bot.*, **50**, 196–207.
112. M. Iino (1988). Desensitization by red and blue light of phototropism in maize coleoptiles. *Planta*, **176**, 183–188.
113. M. Iino (1987). Kinetic modelling of phototropism in maize coleoptiles. *Planta*, **171**, 110–126.
114. C.-H. Ullrich (1976). *Kontinuierlich registrierende Messung der Krümmungsbewegung von Mais-Koleoptilen: Schnelle Effekte von Auxinapplikation, Schwerkraft und Blaulicht.* Dissertation, University of Freiburg.
115. A. Taylor, J.W. Hart, A.R. Berg (1988). Different forms of phototropic response induced by microbeam irradiation of discrete regions of an organ. *Plant Cell Environ.*, **11**, 645–651.
116. K. Macleod, F. Brewer, J. Digby, R.D. Firn (1984). The phototropic response of *Avena* coleoptiles following localized continuous unilateral illumination. *J. Exp. Bot.*, **35**, 1380–1389.
117. R. Konjevic, B. Steinitz, K.L. Poff (1989). Dependence of the phototropic response of *Arabidopsis thaliana* on fluence rate and wavelength. *Proc. Natl. Acad. Sci. USA*, **86**, 9876–9880.
118. R. Konjevic, D. Grubisic, M. Neskovic (1989). Growth retardant-induced changes in phototropic reaction of *Vigna radiata* seedlings. *Plant Physiol.*, **89**, 1085–1087.
119. G.M. Curry (1957). *Studies on the spectral of phototropism.* Ph.D. Thesis, Harvard University.
120. G. Blaauw-Jansen (1959). The influence of red and far red light on growth and phototropism of the *Avena* seedling. *Acta Bot. Neerl.*, **8**, 1–39.
121. O.H. Blaauw, G. Blaauw-Jansen (1964). The influence of red light on the phototropism *Avena* coleoptiles. *Acta Bot. Neerl.*, **13**, 541–552.
122. H.P. Chon, W.R. Briggs (1966). Effect of red light on the phototropic sensitivity of corn coleoptiles. *Plant Physiol.*, **41**, 1751–1724.
123. E. Schäfer, M. Iino, W.R. Briggs (1984). Red-light-induced shift of the fluence–response curve for first positive phototropic curvature of maize coleoptiles. In: H. Senger (Ed.), *Blue Light Effects in Biological Systems* (pp. 476–479). Springer, Berlin.
124. E. Hofmann, E. Schäfer (1987). Red light-induced shift of the fluence–response curve for first positive curvature of maize coleoptiles. *Plant Cell Physiol.*, **28**, 37–45.
125. W.R. Briggs, H.P. Chon (1966). The physiological versus the spectrophotometric status of phytochrome in corn coleoptiles. *Plant Physiol.*, **41**, 1159–1166.
126. T. Shinomura, A. Nagatani, H. Hanzawa, M. Kubota, M. Waranabe, M. Furuya (1996). Action spectra for phytochrome A- and B-specific photoinduction of seed germination in *Arabidopsis thaliana*. *Proc. Natl. Acad. Sci. USA*, **93**, 8129–8133.
127. A.-k. Janoudi, K.L. Poff (1991). Characterization of adaptation in phototropism of *Arabidopsis thaliana*. *Plant Physiol.*, **95**, 517–521.
128. A.-k. Janoudi, K.L. Poff (1990). A common fluence threshold for first positive and second positive phototropism in *Arabidopsis thaliana*. *Plant Physiol.*, **94**, 1605–1608.
129. A.-k. Janoudi, K.L. Poff (1992). Action spectrum for enhancement of phototropism by *Arabidopsis thaliana* seedlings. *Photochem. Photobiol.*, **56**, 655–659.
130. B.M. Parks, P.H. Quail, R.P. Hangarter (1996). Phytochrome A regulates red-light induction of phototropic enhancement in *Arabidopsis*. *Plant Physiol.*, **110**, 155–162.
131. A.-k. Janoudi, W.R. Gordon, D. Wagner, P. Quail, K.L. Poff (1997). Multiple phytochromes are involved in red-light-induced enhancement of first-positive phototropism in *Arabidopsis thaliana*. *Plant Physiol.*, **113**, 975–979.
132. A.-k. Janoudi, R. Konjevic, G. Whitelam, W. Gordon, K.L. Poff (1997). Both phytochrome A and phytochrome B are required for the normal expression of phototropism in *Arabidopsis thaliana* seedlings. *Physiol. Plant.*, **101**, 278–282.

133. Y.J. Liu, M. Iino (1997). Dependence on phytochromes A and B of phototropic responsiveness in *Arabidopsis* hypocotyls (abstract no. 1462). *Plant Physiol.*, **114**, S-281.
134. E.P. Parker, T.I. Baskin, W.R. Briggs (1989). Evidence for a phytochrome-mediated phototropism in etiolated pea seedlings. *Plant Physiol.*, **89**, 493–497.
135. B.G. Kang, S.P. Burg (1974). Red light enhancement of the phototropic response of etiolated pea stems. *Plant Physiol.*, **53**, 445–448.
136. S.J. Britz, A.W. Galston (1983). Physiology of movements in the stems of seedlings of *Pisum sativum*. *Plant Physiol.*, **71**, 313–318.
137. R.P. Hangarter (1997). Gravity, light and plant form. *Plant Cell Environ.*, **20**, 796–800.
138. J.M. Franssen, J. Bruinsma (1981). Relationships between xanthoxin, phototropism, and elongation growth in the sunflower seedling *Helianthus annuus* L. *Planta*, **151**, 365–370.
139. M. Feyerabend, E.W. Weiler (1988). Immunological estimation of growth regulator distribution in phototropically reacting sunflower seedlings. *Physiol. Plant.*, **74**, 185–193.
140. D.J. Cosgrove (1985). Kinetic separation of phototropism from blue-light inhibition of stem elongation. *Photochem. Photobiol.*, **42**, 745–751.
141. R. Konjevic, D. Grubisic, M. Neskovic (1987). The effect of norflurazon on phototropic reaction in *Phaseolus aureus* Roxb. seedlings. *Photochem. Photobiol.*, **45**, 821–824.
142. H. Smith, G.C. Whitelam (1997). The shade avoidance syndrome: multiple responses mediated by multiple phytochromes. *Plant Cell Environ.*, **20**, 840–844.
143. D.F. Mandoli, W.R. Briggs (1981). Phytochrome control of two low-irradiance responses in etiolated oat coleoptiles. *Plant Physiol.*, **67**, 733–739.
144. M. Iino, D.J. Carr (1981). Safelight for photomorphogenetic studies: infrared radiation and infrared-scope. *Plant Sci. Lett.*, *23*, 263–268.
145. H. Sierp, A. Seybold (1926). Untersuchungen über die Lichtempfindlichkeit der Spitze und des Stumpfes in der Koleoptile von *Avena sativa*. *Jahrb. Wiss. Bot.*, **65**, 592–610.
146. S. Lange (1927). Über autonome Krümmungen der Koleoptile von *Avena* auf dem Klinostaten. *Ber. Dtsch. Bot. Ges.*, **43**, 438–443.
147. L. Brauner (1955). Über die Function der Spitzenzone beim Phototropismus der *Avena*-Koleoptile. *Z. Bot.*, **43**, 467–498.
148. J.M. Franssen, S.A. Cooke, J. Digby, R.D. Firn (1981). Measurements of differential growth causing phototropic curvature of coleoptiles and hypocotyls. *Z. Pflanzenphysiol.*, **103**, 207–216.
149. J.M. Franssen, R.D. Firn, J. Digby (1982). The role of the apex in the phototropic curvature of *Avena* coleoptiles: positive curvature under conditions of continuous illumination. *Planta*, **155**, 281–286.
150. R. Kaldenhoff, M. Iino (1997). Restoration of phototropic responsiveness in decapitated maize coleoptiles. *Plant Physiol.*, **114**, 1267–1272.
151. M. Iino, W.R. Briggs (1984). Growth distribution during first positive phototropic curvature of maize coleoptiles. *Plant Cell Environ.*, **7**, 97–104.
152. M. Iino (1995). Gravitropism and phototropism of maize coleoptiles: evaluation of the Cholodny-Went theory through effects of auxin application and decapitation. *Plant Cell Physiol.*, **36**, 361–367.
153. H.G. du Buy, E. Nuernbergk (1929). Weitere Untersuchungen über den Einfluss des Lichtes auf das Wachstum von Koleoptile und Mesokotyl bei *Avana sativa*. II. *Proc. Kon. Akad. Wetensch. Amsterdam.* **32**, 808–817.
154. H.E. Dolk (1936). Geotropism and the growth substance. *Rec.Trav. Bot. Neerl.*, **33**, 509–585.
155. Y. Tarui, M. Iino (1999). Gravitropism and phototropism of oat coleoptiles: post-tropic autostraightening and tissue shrinkage during tropism. *Adv. Space Res.*, **24**, 743–753.
156. T.I. Baskin, M. Iino, P.B. Green, W.R. Briggs (1985). High-resolution measurement of

growth during first positive phototropism in maze. *Plant Cell Environ.*, **8**, 595–603.
157. K. Macleod, R.D. Firn, J. Digby (1986). The phototropic responses of *Avena* coleoptiles. *J. Exp. Bot.*, **37**, 542–548.
158. F.W. Went, K.V. Thimann (1937). *Phytohormones*. Macmillan, New York.
159. M. Iino, D.J. Carr (1982). Sources of free IAA in the mesocotyl of etiolated maize seedlings. *Plant Physiol.*, **69**, 1109–1112.
160. H. von Guttenberg (1959). Über die Perzeption des phototropen Reizes. *Planta*, **53**, 412–433.
161. R. Pohl (1960). Beiträge zum Phototropismus der *Avena*-Koleoptile. I. *Phyton*, **15**, 145–157.
162. R. Pohl (1960). Beiträge zum Phototropismus der *Avena*-Koleoptile. II. *Ber. Deut. Bot. Ges.*, **73**, 481–486.
163. H. Mohr, E. Peters (1960). Der Phototropismus und das lichtabhängige Längenwachstum der Hypokotyle von *Sinapis alba* L. *Planta*, **55**, 637–646.
164. T. Brennan, J.E. Gunckel (1976). Stem sensitivity and ethylene involvement in phototropism of mung bean. *Plant Physiol.*, **57**, 286–289.
165. J.W. Hart, D.C. Gordon, I.R. MacDonald (1982). Analysis of growth during phototropic curvature of cress hypocotyls. *Plant Cell Environ.*, **5**, 361–366.
166. J.R. Shinkle, S.K. Sooudi, R.L. Jones (1992). Adaptation to dim-red light leads to a nongradient pattern of stem elongation in *Cucumis* seedlings. *Plant Physiol.*, **99**, 808–811.
167. E. Gendreau, J. Traas, T. Desnos, O. Grandjean, M. Caboche, H. Höfte (1997). Cellular basis of hypocotyl growth in *Arabidopsis thaliana. Plant Physiol.*, **114**, 295–305.
168. V. Orbovic, K.L. Poff (1991). Kinetics for phototropic curvature by etiolated seedlings of *Arabidopsis thaliana. Plant Physiol.*, **97**, 1470–1475.
169. M. Salomon, M. Zacherl, L. Luff, W. Rüdiger (1997). Exposure of oat seedlings to blue light results in amplified phosphorylation of the putative photoreceptor for phototropism and in high sensitivity of the plants to phototropic stimulation. *Plant Physiol.*, **115**, 493–500.
170. L. Brauner (1957). *The perception of the phototropic stimulus in the oat coleoptile. Soc. Exp. Biol. Symposium, No. 11* (pp. 86–94). Cambridge University Press, Cambridge.
171. J. Buder (1920). Neue phototropische Fundamentalversuche. *Ber. Deut. Bot. Ges.*, **38**, 10–19.
172. P. Nick, M. Furuya (1996). Buder revised: cell and organ polarity during phototropism. *Plant Cell Environ.*, **19**, 1179–1187.
173. E. Bünning, I. Dorn, G. Schneiderhöhn, I. Thorning (1953). Zur Funktion von Lactoflavin und Carotin beim Phototropismus und bei lichtbedingten Wachstumsbeeinflussungen. *Ber. Deut. Bot., Ges.*, **66**, 333–340.
174. P. Kunzelmann, M. Iino, E. Schäfer (1988). Phototropism of maize coleoptiles. *Influences of light gradients. Planta*, **176**, 212–220.
175. R.D. Viestra, K.L. Poff (1981). Role of carotenoids in the phototropic response of corn seedlings. *Plant Physiol.*, **68**, 798–801.
176. C.J. Piening, K.L. Poff (1988). Mechanism of detecting light direction in first positive phototropism in *Zea mays* L. *Plant Cell Environ.*, **11**, 143–146.
177. D.S. Dennison (1979). Phototropism. In: W. Haupt, M.E. Feinleib (Eds.), *Encyclopedia of Plant Physiology* (New Series, Vol. 7, pp. 506–556). Springer, Berlin.
178. M. Iino, E. Schäfer (1984). Phototropic response of the stage I *Phycomyces* sporangiophore to a pulse of blue light. *Proc. Natl. Acad. Sci. USA*, **81**, 7103–7107.
179. W. Shropshire Jr. (1974). Phototropism. In: G.O. Schenk (Ed.), *Proceedings of the VI International Congress on Photobiology* (pp. 1–6). Springer, Berlin.

180. T.C. Vogelmann, W. Haupt (1985). The blue light gradient in unilaterally irradiated maize coleoptiles: measurement with a fiber optic probe. *Photochem. Photobiol.*, **41**, 569–576.
181. A. Parsons, K. Macleod, R.D. Firn, J. Digby (1984). Light gradients in shoots subjected to unilateral illumination – implications for phototropism. *Plant Cell Environ.*, **7**, 325–332.
182. P. Kunzelmann, E. Schäfer (1985). Phytochrome-mediated phototropism in maize mesocotyls. Relation between light and Pfr gradients, light growth response and phototropism. *Planta*, **165**, 424–429.
183. T.C. Vogelmann (1994). Light within the plant. In: R.E. Kendrick, G.H.M. Kronenberg (Eds.), *Photomorphogenesis in Plants* (2nd ed., pp. 491–535). Kluwer Academic, Netherlands.
184. D.F. Mandoli, W.R. Briggs (1982). Optical properties of etiolated plant tissues. *Proc. Natl. Acad. Sci. USA*, **79**, 2902–2906.
185. B.M. Park, K. Poff (1985). Phytochrome conversion as an *in situ* assay for effective light gradients in etiolated seedlings of *Zea mays*. *Photochem. Photobiol.*, **41**, 317–322.
186. G.M. Curry (1969). Phototropism. In: M.B. Wilkins (Ed.), *The Physiology of Plant Growth and Development* (pp. 241–273). McGraw-Hill, New York.
187. T.C.G. Rich, A.D. Tomos (1988). Turgor pressure and phototropism in *Sinapis alba* L. seedlings. *J. Exp. Bot.*, **39**, 291–299.
188. W.H. Arisz (1915). Untersuchungen über den Phototropismus. *Rec. Trav. Bot. Neerl.*, **12**, 44–216.
189. W.F.F. Oppenoorth (1942). On the role of auxin in phototropism and light–growth reactions of *Avena*-coleoptiles. *Rec. Trav. Botan. Neerl.*, **38**, 287–372.
190. P. Nick, E. Schäfer (1988). Interaction of gravi- and phototropic stimulation in the response of maize (*Zea mays* L.) coleoptiles. *Planta*, **173**, 213–220.
191. J. Shen-Miller, S.A. Gordon (1967). Gravitational compensation and the phototropic response of oat coleoptiles. *Plant Physiol.*, **42**, 352–360.
192. H.G. du Buy, E. Nuernbergk (1930). Über das Wachstum der Koleoptile und des Mesokotyls von *Avana sativa* unter verschiedenen Bedingungen. *Proc. Kon. Akad. Wetensch. Amsterdam*, **33**, 542–556.
193. B. Steiniz, T. Best, K.L. Poff (1988). Phototropic fluence–response relations for *Avena* coleoptiles on a clinostat. *Planta*, **176**, 189–195.
194. M. Iino, Y. Tarui, C. Uematsu (1996). Gravitropism of maize and rice coleoptiles: dependence on the stimulation angle. *Plant Cell Environ.*, **19**, 1160–1168.
195. Y. Tarui, M. Iino (1997). Gravitropism of oat and wheat coleoptiles: dependence on the stimulation angle and involvement of autotropic straightening. *Plant Cell Physiol.*, **38**, 1346–1353.
196. C. Härtling (1964). Untersuchungen über den Einfluss der Rotation am Klinostaten auf das tropistiche Reaktionsvermögen von *Helianthus*-Keimlingen. *Planta*, **63**, 43–64.
197. P. Nick, E. Schäfer (1989). Nastic response of maize (*Zea mays* L.) coleoptiles during clinostat rotation. *Planta*, **179**, 123–131.
198. D.G. Heathcote, A.H. Brown, D.K. Chapman (1995). The phototropic response of *Triticum aestivum* coleoptiles under conditions of low gravity. *Plant Cell Environ.*, **18**, 53–60.
199. D.G. Heathcote, D.K. Chapman, A.H. Brown (1995). Nastic curvatures of wheat coleoptiles that develop in true microgravity. *Plant Cell Environ.*, **18**, 818–822.
200. R.D. Firn, J. Digby (1979). A study of the autotropic straightening reaction of a shoot previously curved during geotropism. *Plant Cell Environ.*, **2**, 149–154.
201. D.K. Chapman, A. Johnson, C. Karlsson, A. Brown, D. Heathcote (1994). Gravitropically-stimulated seedlings show autotropism in weightlessness. *Physiol. Plant.*, **90**, 157–162.
202. H. Sailer, P. Nick, E. Schäfer (1990). Inversion of gravitropism by symmetric blue light on the clinostat. *Planta*, **180**, 378–382.

203. B. Stankovic, D. Volkmann, F.D. Sack (1998). Autotropism, automorphogenesis and gravity. *Physiol. Plant.*, **102**, 328–335.
204. R.D. Firn (1994). Phototropism. In: R.E. Kendrick, G.H.M. Kronenberg (Eds.), *Photomorphogenesis in Plants* (2nd ed., pp. 659–681). Kluwer Academic, Netherlands.
205. P. Nick, E. Schäfer (1988). Spatial memory during the tropism of maize (*Zea mays* L.) coleoptiles. *Planta*, **175**, 380–388.
206. P. Nick, E. Schäfer (1994). Polarity induction versus phototropism of maize: auxin cannot replace blue light. *Planta*, **195**, 63–69.
207. R. Diemer (1961). Untersuchungen des phototropischen Induktionsvorganges an Helianthus-keimlingen. *Planta*, **57**, 111–137.
208. E. Schäfer, L. Fukushansky, W. Schropshire Jr. (1983). Action spectroscopy of photoreversible pigment systems. In: W. Schropshire Jr., H. Mohr (Eds.), *Encyclopedia of Plant Physiology* (New Series, Vol. 16, pp. 39–68). Springer, Berlin.
209. W. Shropshire Jr., R.B. Withrow (1958). Action spectrum of phototropic tip-curvature of *Avena. Plant Physiol.*, **33**, 360–365.
210. K.V. Thimann, G.M. Curry (1960). Phototropism and phototaxis. In: M. Florkin, H.S. Mason (Eds.), *Comparative Biochemistry* (Vol. 1, pp. 243–309). Academic Press, New York.
211. C. Haig (1934). The spectral sensitivity of *Avena. Proc. Natl. Acad. Sci. USA*, **20**, 476–479.
212. C. Haig (1935). The phototropic responses of *Avana* in relation to intensity and wavelength. *Biol. Bull.*, **69**, 305–324.
213. E. Liscum, W.R. Briggs (1996). Mutations of *Arabidopsis* in potential transduction and response components of the phototropic signaling pathway. *Plant Physiol.*, **112**, 291–296.
214. G.A. Atkins (1936). The effect of pigment on phototropic response: A comparative study of reactions to monochromatic light. *Ann. Bot.*, **50**, 197–218.
215. R. Konjevic, J.P. Khurana, K.L. Poff (1992). Analysis of multiple photoreceptor pigments for phototropism in a mutant of *Arabidopsis thaliana. Photochem. Photobiol.*, **55**, 789–792.
216. G. Wald, H.G. du Buy (1936). Pigments of the oat coleoptile. *Science*, **84**, 247.
217. A.W. Galston, R.S. Baker (1949). Studies on the physiology of light action. II. The photodynamic action of riboflavin. *Am. J. Bot.*, **36**, 773–780.
218. R.H. Wallace, A.E. Schwarting (1954). A study of chlorophyll in a white mutant strain of *Helianthus annuus. Plant Physiol.*, **29**, 431–436.
219. A.W. Galston (1959). Phototropism of stems, roots and coleoptiles. In: W. Ruhland (Ed.), *Encyclopedia of Plant Physiology* (pp. 492–529). Springer, Berlin.
220. E. Schmidt, J. Hart, P. Filner, K.L. Poff (1977). Specific inhibition of phototropism in corn seedlings. *Plant Physiol.*, **60**, 736–738.
221. P.S. Song (1980). Spectroscopic and photochemical characterization of flavoproteins and carotenoproteins as blue light photoreceptors. In: H. Senger (Ed.), *The Blue Light Syndrome* (pp. 157–171). Springer, Berlin.
222. M.A. Quiñones, E. Zeiger (1994). A putative role of the xanthophyll, zeaxanthin, in blue light photoreception of corn coleoptiles. *Science*, **264**, 558–561.
223. M.A. Quiñones, Z. Lu, E. Zeiger (1996). Close correspondence between the action spectra for the blue light responses of the guard cell and coleoptile chloroplasts, and the spectra for blue light-dependent stomatal opening and coleoptile phototropism. *Proc. Natl. Acad. Sci. USA*, **93**, 2224–2228.
224. J.M. Palmer, K.M.F. Warpeha, W.R. Briggs (1996). Evidence that zeaxanthin is not the photoreceptor for phototropism in maize coleoptiles. *Plant Physiol.*, **110**, 1323–1328.

225. B.A. Horwitz, T.G.M. Berrocal (1997). A spectroscopic view of some recent advances in the study of blue light photoreception. *Bot. Acta*, **110**, 360–368.
226. E. Zeiger, J. Zhu (1998). Role of zeaxanthin in blue-light photoreception and the modulation of light-CO_2 interactions in guard cells. *J. Exp. Bot.*, **49**, 433–442.
227. W.R. Briggs, M. Iino (1983). Blue-light-absorbing photoreceptors in plants. *Phil. Trans. R. Soc. Lond. B*, **303**, 347–359.
228. P.E. Karlsson (1986). Blue light regulation of stomata in wheat seedlings. II. Action spectrum and search for action dichroism. *Physiol. Plant.*, **66**, 207–210.
229. M. Ahmad, A.R Cashmore (1993). *HY4* gene of *A. thaliana* encodes a protein with characteristics of a blue-light photoreceptor. *Nature*, **366**, 162–166.
230. C. Lin, M. Ahmad, D. Gordon, A.R. Cashmore (1995). Expression of an *Arabidopsis* cryptochrome gene in transgenic taboco results in hypersensitivity to blue, UV-A, and green light. *Proc. Natl. Acad. Sci. USA*, **92**, 8423–8427.
231. C. Lin, D.E. Robertson, M. Ahmad, A.A. Raibekas, M.D. Jones, P.L. Dutton, A.R. Cashmore (1995). Association of flavin adenine dinucleotide with the *Arabidopsis* blue light receptor CRY1. *Science*, **269**, 968–970.
232. K. Malhotra, S.-T. Kim, A. Batchauer, L. Dawut, A. Sancar (1995). Putative blue-light photoreceptors from *Arabidopsis thaliana* and *Sinapis alba* with a high degree of sequence homology to DNA photolyase contain the two photolyase cofactors but lack DNA photolyase activity. *Biochemistry*, **34**, 6892–6899.
233. C. Lin, M. Ahmad, J. Chan, A.R. Cashmore (1996). *CRY2*, a second member of the *Arabidopsis* cryptochrome gene family. *Plant Physiol.*, **110**, 1047.
234. E. Liscum, J.C. Young, K. Poff, R.P. Hangarter (1992). Genetic separation of phototropism and blue light inhibition of stem elongation. *Plant Physiol.*, **100**, 267–271.
235. M. Ahmad, J.A. Jarillo, O. Smirnova, A.R. Cashmore (1998). Cryptochrome blue-light photoreceptors of *Arabidopsis* implicated in phototropism. *Nature*, **392**, 720–723.
236. J.M. Christie, P. Reymond, G.K. Powell, P. Bernasconi, A.A. Raibekas, E. Liscum, W.R. Briggs (1998). *Arabidopsis* NPH1: a flavoptrotein with the properties of a photoreceptor for phototropism. *Science*, **282**, 1698–1701.
237. G. Lascève, J. Leymarie, M.A. Olney, E. Liscum, J.M. Christie, A. Vavasseur, W.R. Briggs (1999). *Arabidopsis* contains at least four independent blue light-activated signal transduction pathways. *Plant Physiol.*, **120**, 605–614.
238. S. Gallagher, T.S. Short, P.M. Ray, L.H. Pratt, W.R. Briggs (1988). Light-mediated changes in two proteins found associated with plasma membrane fractions from pea stem sections. *Proc. Natl. Acad. Sci. USA*, **85**, 8003–8007.
239. T.W. Short, W.R. Briggs (1990). Characterization of a rapid, blue light-mediated change in detectable phosphorylation of a plasma membrane protein from etiolated pea (*Pisum sativum* L.) seedlings. *Plant Physiol.*, **92**, 179–185.
240. T.W. Short, M. Porst, W.R. Briggs (1992). A photoreceptor system regulating in vivo and in vitro phosphorylation of a pea plasma membrane protein. *Photochem. Photobiol.*, **55**, 773–782.
241. T.W. Short, P. Reymond, W.R. Briggs (1993). A pea plasma membrane protein exhibiting blue light-induced phosphorylation retains photosensitivity following Triton solubilization. *Plant Physiol.*, **101**, 647–655.
242. P. Reymond, T.W. Short, W.R. Briggs (1992). Blue light activates a specific protein kinase in higher plants. *Plant Physiol.*, **100**, 655–661.
243. P. Reymond, T.W. Short, W.R. Briggs, K.L. Poff (1992). Light-induced phosphorylation of a membrane protein plays an early role in signal transduction for phototropism in *Arabidopsis thaliana*. *Proc. Natl. Acad. Sci. USA*, **89**, 4718–4721.
244. A. Hager M. Brich, I. Bazlen (1993). Redox dependence of the blue-light-induced

phosphorylation of a 100-kDa protein on isolated plasma membranes from tips of coleoptiles. *Planta*, **190**, 120–126.
245. A. Hager, M. Brich (1993). Blue-light-induced phosphorylation of a plasma-membrane protein from phototropically sensitive tips of maize coleoptiles. *Planta*, **189**, 567–576.
246. J.M. Palmer, T.W. Short, S. Gallagher, W.R. Briggs (1993). Blue light-induced phosphorylation of a plasma membrane-associated protein in *Zea mays* L. *Plant Physiol.*, **102**, 1211–1218.
247. A. Hager (1996). Properties of a blue-light-absorbing photoreceptor kinase localized in the plasma membrane of the coleoptile tip region. *Planta*, **198**, 294–299.
248. M. Salomon, M. Zacherl, W. Rüdiger (1996). Changes in blue-light-dependent protein phosphorylation during the early development of etiolated oat seedlings. *Planta*, **199**, 336–342.
249. M. Salomon, M. Zacherl, W. Rüdiger (1997). Phototropism and protein phosphorylation in higher plants: unilateral blue light irradiation generates a directional gradient of protein phosphorylation across the oat coleoptile. *Bot. Acta*, **110**, 214–216.
250. M. Salomon, M. Zacherl, W. Rüdiger (1997). Asymmetric, blue light-dependent phosphorylation of a 116-kilodalton plasma membrane protein can be correlated with the first- and second-positive phototropic curvature of oat coleoptiles. *Plant Physiol.*, **115**, 485–491.
251. T.W. Short, W.R. Briggs (1994). The transduction of blue light signals in higher plants. *Ann. Rev. Plant Physiol. Plant Mol. Biol.*, **45**, 143–171.
252. J.P. Khurana, K.L. Poff (1989). Mutants of *Arabidopsis thaliana* with altered phototropism. *Planta*, **178**, 400–406.
253. E. Huala, P.W. Oeller, E. Liscum, I.-S. Han, E. Larsen, W.R. Briggs (1997). *Arabidopsis* NPH1: A protein kinase with putative redox sensing domain. *Science*, **278**, 2120–2123.
254. J.M. Christie, M. Salomon, K. Nozue, M. Wada, W.R. Briggs (1999). LOV (light, oxygen, or voltage) domains of the blue-light photoreceptor phototropin (nph1): binding sites for the chromophore flavin mononucleotide. *Proc. Natl. Acad. Sci. USA*, **96**, 8779–8783.
255. W.R. Briggs, E. Huala (1999). Blue light photoreceptors in higher plants. *Ann. Rev. Cell. Dev. Biol.*, **15**, 33–61.
256. J.A. Jarillo, M. Ahmad, A.R. Cashmore (1998). NPH2 (Accession No. AF053941): a second member of the NPH serine/threonine kinase family of *Arabidopsis*. *Plant Physiol.*, **117**, 719.
257. M. Zacherl, E. Huala, W. Rüdiger, W.R. Briggs, M. Salmon (1998). Isolation and characterization of cDNAs from oat encoding a serine/threonine kinase: an early component in signal transduction for phototropism (PGR98–028). *Plant Physiol.*, **116**, 869.
258. H. Kanegae, M. Tahir, F. Savazzini, K. Yamamoto, M. Yano, T. Sakaki, T. Kanegae, M. Wada, M. Takano (2000). Rice *NPH1* homologues, *OsNPH1a* and *OsNPH1b*, are differently photoregulated. *Plant Cell Physiol.*, **41**, 415–423.
259. X. Lin, J.C. Watson (1992). cDNA sequence of PsPK5, a protein kinase homolog from *Pisum sativum* L. *Plant Physiol.*, **100**, 1072–1074.
260. R. Khanna, X. Lin, J.C. Watson (1999). Photoregulated expression of the *PsPK3* and *PsPK5* genes in pea seedlings. *Plant Mol. Biol.*, **39**, 231–242.
261. B. Baur, K. Fischer, K. Winter, K.-J. Diets (1994). cDNA sequence of a protein kinase from the inducible crassulacean acid metabolism plant *Mesembryanthenum crystallinum* L. encoding a SNF-1 homolog. *Plant Physiol.*, **106**, 1225–1226.
262. K.M.F. Warpeha, W.R. Briggs (1993). Blue light-induced phosphorylation of a plasma membrane protein in pea: a step in the signal transduction chain for phototropism. *Aust. J. Plant Physiol.*, **20**, 393–403.

263. B.L. Taylor, I.B. Zhulin (1999). PAS domains: internal sensors of oxygen, redox potential, and light. *Microbiol. Mol. Biol. Rev.*, **63**, 479–506.
264. P. Ballario, C. Talora, D. Galli, H. Linden, G. Macino (1998). Roles in dimerization and blue light photoresponse of the PAS and LOV domains of *Neurospora crassa* white collar proteins. *Mol. Microbiol.*, **29**, 719–729.
265. A. Motchoulski, E. Liscum (1999). *Arabidopsis* NPH3: a NPH1 photoreceptor-interacting protein essential for phototropism. *Science*, **286**, 961–964.
266. T. Sakai, T. Wada, S. Ishiguro, K. Okada (2000). RPT2: a signal transducer of the phototropic response in *Arabidopsis*. *Plant Cell*, **12**, 225–236.
267. J.M. Palmer, T.W. Short, W.R. Briggs (1993). Correlation of blue light-induced phosphorylation to phototropism in *Zea mays* L. *Plant Physiol.*, **102**, 1219–1225.
268. M. Iino, R. Neumann (2000). Phototropism of rice seedlings: characterization and mutant isolation (abstract no. 121). *Plant Cell Physiol.*, **41**, s56.
269. B. Steiniz, S. Ren, K.L. Poff (1985). Blue and green light-induced phototropism in *Arabidopsis thaliana* and *Lactuca sativa* L. seedlings. *Plant Physiol.*, **77**, 248–251.
270. A.M. Meyer (1969). Versuche zur Trennung von 1. positiver und negativer Krümmung der *Avena*koleoptile. *Z. Pflanzenphysiol.*, **60**, 135–146.
271. O.H. Blaauw, G. Blaauw-Jansen (1970). Third positive (C-type) phototropism in the *Avena* coleoptile. *Acta Bot. Neerl.*, **19**, 764–776.
272. M. Iino (1992). Auxin and phototropism in maize coleoptiles. In: Y. Masuda (Ed.), *Plant Cell Walls as Biopolymers with Physiological Functions* (pp. 211–217). Yamada Science Foundation, Osaka.
273. A.-k. Janoudi, K.L. Poff (1993). Desensitization and recovery of phototropic responsiveness in *Arabidopsis thaliana*. *Plant Physiol.*, **101**, 1175–1180.
274. A.H. Blaauw (1915). Licht und Wachstum. *II. Z. Botanik*, **7**, 465–532.
275. P. Boysen-Jensen (1913). Über die Leitung des phototropischen Reizes in der Avenakoleotile. *Ber. D. Bot. Ges.*, **31**, 59–566.
276. A. Paál (1919). Über phototropische Reizleitung. *Jahrb. Wiss. Biot.*, **58**, 406–458.
277. H. Söding (1923). Werden von der Spitze der Haferkoleoptile Wuchschormone gebildet? *Der. D. Bot. Ges.*, **41**, 396–400.
278. H. Söding (1925). Zur Kenntnis der Wuchshormone in der Haferkoleoptile. *Jahrb. Wiss. Bot.*, **64**, 587–603.
279. F.W. Went (1926). On growth-accelerating substances in the coleoptile of *Avena sativa*. *Proc. Kon. Akad. Wetensch. Amsterdam*, **30**, 10–19.
280. F.W. Went (1928). Wuchsstoff und Wachstum. *Rec. Trav. Bot. Neerl.*, **25**, 1–116.
281. N. Cholodny (1924). Über die hormonale Wirkung der Organspitze bei der geotropischen Krümmung. *Ber. D. Bot. Ges.*, **42**, 356–362.
282. N. Cholodny (1926). Beiträge zur Analyse der geotropischen Reaction. *Jahrb. Wiss. Bot.*, **65**, 447–459.
283. N. Cholodny (1927). Wuchshormone und Tropismen bei den Pflanzen. *Biol. Zentralbl.*, **47**, 605–626.
284. J.R. Bearder (1980). Plant hormones and other growth substances – their background, structures and occurrence. In: J. MacMillan (Ed.), *Encyclopedia of Plant Physiology* (New Series, Vol. 10, pp. 9–112.). Springer, Berlin.
285. M. Iino (1982). Inhibitory action of red light on the growth of the maize mesocotyl: evaluation of the auxin hypothesis. *Planta*, **156**, 388–395.
286. M. Iino (1982). Action of red light on indole-3-acetic acid status and growth in coleoptiles of etiolated maize seedlings. *Planta*, **156**, 21–32.
287. W.R. Briggs, R.D. Tocher, J.F. Wilson (1957). Phototropic auxin redistribution in corn coleoptiles. *Science*, **126**, 210–212.

288. W.R. Briggs (1963). Mediation of phototropic responses of corn coleoptiles by lateral transport of auxin. *Plant Physiol.*, **38**, 237–247.
289. B.G. Pickard, K.V. Thimann (1964). Transport and distribution of auxin during tropistic response. II. The lateral migration of auxin in phototropism of coleoptiles. *Plant Physiol.*, **39**, 341–350.
290. R.D. Firn, J. Digby (1980). The establishment of tropic curvatures in plants. *Ann. Rev. Plant Physiol.*, **31**, 131–148.
291. A. Trewavas (1981). How do plant growth substances act? *Plant Cell Environ.*, **4**, 203–228.
292. J. Bruinsma, K. Hasegawa (1989). Phototropism involves a lateral gradient of growth inhibitors, not of auxin. A review. *Environ. Exp. Bot.*, **29**, 25–36.
293. B.G. Pickard (1985). Roles of hormones, proteins and calcium in gravitropism. In: R.P. Pharis, D.M. Reid (Eds.), *Encyclopedia of Plant Physiology* (New Series, Vol. 11, pp. 193–281). Springer, Berlin.
294. P.E. Clifford, D.M.A. Mousdale, S.J. Lynd, E.L. Oxlade (1985). Differences in auxin level detected across geostimulated dandelion peduncles: evidence supporting a role for auxin in geotropism. *Ann. Bot.*, **55**, 293–296.
295. M.A. Harrison, B.G. Pickard (1989). Auxin asymmetry during gravitropism by tomato hypocotyls. *Plant Physiol.*, **89**, 652–657.
296. E.P. Parker, W.R. Briggs (1990). Transport of indole-3-acetic acid during gravitropism in intact maize coleoptiles. *Plant Physiol.*, **94**, 1763–1769.
297. M. Iino (1991). Mediation of tropisms by lateral translocation of endogenous indole-3-acetic acid in maize coleoptiles. *Plant Cell Environ.*, **14**, 279–286.
298. Y. Li, G. Hagen, T.J. Guilfoyle (1991). An auxin-responsive promoter is differentially induced by auxin gradients during tropisms. *Plant Cell*, **3**, 1167–1175.
299. Y. Li, Y.H. Wu, G. Hagen, T. Guilfoyle (1999). Expression of the auxin-inducible GH3 promoter/GUS fusion gene as a useful molecular marker for auxin physiology. *Plant Cell Physiol.*, **40**, 675–682.
300. M.J. Bennett, A. Marchant, H.G. Green, S.T. May, S.P. Ward, P.A. Millner, A.R. Walker, B. Schulz, K.A. Feldmann (1996). *Arabidopsis AUX1* gene: a permease-like regulator of root gravitropism. *Science*, **273**, 948–950.
301. A. Trewavas et al. (1992). What remains of the Cholodny-Went theory? *Plant Cell Environ.*, **15**, 761–794.
302. A. Beyer (1927). Experimentelle Studien zur Blaauwschen Theorie.I. Mitteilung: Die Wachstumsverhältnisse bei der phototropischen Krümmung vorbelichteter *Avena*keimlinge. *Planta*, **4**, 411–436.
303. A. Beyer (1928). Experimentelle Studien zur Blaauwschen Theorie. II. Mitteilung: Neue Vergleichende Untersuchungen über Lichtkrümmung und Lichtwachstum bei normalen und dekapitierten Haferkeimlingen. *Planta*, **5**, 478–519.
304. N. Cholodny (1933). Beiträge zur Kritik der Blaauwschen Teorie des Phototropismus. *Planta*, **20**, 549–576.
305. R.M. Thornton, K.V. Thimann (1967). Transient effects of light on auxin transport in the *Avena* coleoptile. *Plant Physiol.*, **42**, 247–257.
306. K. Macleod, J. Digby, R.D. Firn (1985). Evidence inconsistent with the Blaauw model of phototropism. *J. Exp. Bot.*, **36**, 312–319.
307. X. Wang, M. Iino (1997). Blue light-induced shrinking of protoplasts from maize coleoptiles and its relationship to coleoptile growth. *Plant Physiol.*, **114**, 1009–1020.
308. H. Drumm-Herrel, H. Mohr (1985). Relative importance of blue light and light absorbed by phytochrome in growth of mustard (*Sinapis alba* L.) seedlings. *Photochem. Photobiol.*, **42**, 735–739.

309. K. Hasegawa, M. Sakoda (1988). Distribution of endogenous indole-3-acetic acid and growth inhibitor(s) in phototropically responding oat coleoptiles. *Plant Cell Physiol.*, **29**, 1159–1164.
310. T.C.G. Rich, G.C. Whitelam, H. Smith (1985). Phototropism and axis extension in light-grown mustard (*Sinapis alba* L.) seedlings. *Photochem. Photobiol.*, **42**, 789–792.
311. T.C.G. Rich, G.C. Whitelam, H. Smith (1987). Analysis of growth rates during phototropism: modifications by separate light-growth responses. *Plant Cell Environ.*, **10**, 303–311.
312. W.R. Briggs, T.I. Baskin (1988). Phototropism in higher plants – controversies and caveats. *Bot. Acta*, **101**, 133–139.
313. J. van Overbeek (1932). An analysis of phototropism in dicotyledons. *Roc. Kon. Akad. Wetensch. Amsterdam*, **35**, 1325–1335.
314. R.D. Asana (1938). On the relation between the distribution of auxin in the tip of the *Avena* coleoptile and the first negative phototropic curvature. *Ann. Bot.*, **2**, 955–957.
315. M. Wilden (1939). Zur Analyse der positiven und negativen phototropischen Krümmungen. *Planta*, **30**, 286–288.
316. P. Boysen-Jensen (1936). Über die Verteilung des Wuchsstoffes in Keimstengeln und Wurzeln während der phototropischen und geotropischen Krümmung. Kgl. Danske Videnskab. *Selskab. Biol. Med.*, **13**, 1–31.
317. M. Iino, R. Yu, D.J. Carr (1980). Improved procedure for the estimation of nanogram quantities of indole-3-acetic acid in plants extracts using the indolo-α-pyrone fluorescence method. *Plant Physiol.*, **66**, 1099–1105.
318. M. Iino, D.J. Carr (1982). Estimation of free, conjugated, and diffusible indole-3-acetic acid in etiolated maize shoots by the indolo-α-pyrone fluorescence method. *Plant Physiol.*, **69**, 950–956.
319. J. Sánchez-Bravo, A. Ortuño, J.M. Botia, M. Acosta, F. Sabater (1991). Lateral diffusion of polarly transported indoleacetic acid and its role in the growth of *Lupinus albus* L. hypocotyls. *Planta*, **185**, 391–396.
320. K. Haga, M. Iino (1998). Auxin-growth relationships in maize coleoptiles and pea internodes and control by auxin of the tissue sensitivity to auxin. *Plant Physiol.*, **117**, 1473–1486.
321. M.H. Goldsmith, K.V. Thimann (1962). Some characteristics of movement of indoleacetic acid in coleoptiles of *Avena*. I. Uptake, destruction, immobilization ,and distribution of IAA during basipetal translocation. *Plant Physiol.*, **37**, 492–505.
322. G. Gardner, S. Shaw, M.B. Wilkins (1974). IAA transport during the phototropic responses of intact *Zea* and *Avena* coleoptiles. *Planta*, **121**, 237–251.
323. H. Kuhn, A.W. Galston (1992). Physiological asymmetry in etiolated pea epicotyls: relation to patterns of auxin distribution and phototropic behavior. *Photochem. Photobiol.*, **55**, 313–318.
324. F.J. Behringer, P.J. Davies (1992). Indole-3-acetic acid levels after phytochrome-mediated changes in the stem elongation rate of dark- and light-grown *Pisum* seedlings. *Planta*, **188**, 85–92.
325. R.K. dela Fuente, A.C. Leopold (1968). Lateral movement of auxin in phototropism. *Plant Physiol.*, **43**, 1031–1036.
326. J. van Overbeek (1936). Growth hormone and mesocotyl growth. *Rec. Trav. Bot. Neerl.*, **33**, 333–340.
327. R. Muir, C. Chang (1974). Effect of red light on coleoptile growth. *Plant Physiol.*, **54**, 286–288.
328. B. Huisinga (1976). The export of auxin from tips and from sections of *Avena* coleoptiles as influenced by red light. *Acta Bot. Neerl.*, **25**, 313–320.

329. M. Furuya, C.-J. Pjon, T. Fujii, M. Ito (1969). Phytochrome action in *Oryza sativa* L. III. The separation of photoperceptive site and growing zone in coleoptiles, and auxin transport as effector system. *Dev. Growth Different.*, **11**, 62–76.
330. T. Koshiba, Y. Kamiya, M. Iino (1995). Biosynthesis of indole-3-acetic acid from L-tryptophan in coleoptile tips of maize (*Zea mays* L.). *Plant Cell Physiol.*, **36**, 1503–1510.
331. W.S. Stewart, F.W. Went (1940). Light stability of auxin in *Avena* coleoptiles. *Bot. Gaz.*, **101**, 706–714.
332. J. Bruinsma, C.M. Karssen, M. Benschop, J.B. van Dort (1975). Hormonal regulation of phototropism in the light-grown sunflower seedling, *Helianthus annuus* L.: immobility of endogenous indoleacetic acid and inhibition of hypocotyl growth by illuminated cotyledons. *J. Exp. Bot.*, **26**, 411–418.
333. J. Shen-Miller, E. Knegt, E. Vermeer, J. Bruinsma (1982). Purification and lability of *cis*-xanthoxin, and its occurrence in phototropically stimulated hypocotyls of *Helianthus annuus* L. *Z. Pflanzenphysiol.*, **108**, 289–294.
334. S. Togo, K. Hasegawa (1991). Phototropic stimulation does not induce unequal distribution of indole-3-acetic acid in maize coleoptiles. *Physiol. Plant.*, **81**, 555–557.
335. M. Sakoda, K. Hasegawa (1989). Phototropism in hypocotyls of radish. VI. No exchange of endogenous indole-3-acetic acid between peripheral and central cell layers during first and second positive phototropic curvatures. *Physiol. Plant.*, **76**, 240–242.
336. K. Hasegawa, K. Yamada (1992). Even distribution of endogenous indole-3-acetic acid in phototropism of pea epicotyls. *J. Plant Physiol.*, **139**, 455–459.
337. K. Hasegawa, M. Sakoda, J. Bruinsma (1989). Revision of he theory of phototropism in plants: a new interpretation of a classical experiment. *Planta*, **178**, 540–544.
338. E. Libbert, I. Gerdes (1964). Können Gibberelline an tropistischen Krümmungen beteiligt sein? *Planta*, **61**, 245–258.
339. I.D.J. Phillips, W. Hartung (1976). Longitudinal and lateral transport of [3,4-^{3}H] gibberellin A_1 and 3-indolyl (acetic acid-2-^{14}C) in upright and geotropically responding green internode segments from *Helianthus annuus*. *New Phytol.*, **76**, 1–9.
340. W.J. Meudt, H.W. Bennett (1978). Rapid bioassay for auxin. *Physiol. Plant.*, **44**, 422–428.
341. I.D.J. Phillips (1972). Diffusible gibberellins and phototropism in *Helianthus annuus*. *Planta*, **106**, 363–367.
342. I.D.J. Phillips (1972). Endogenous gibberellin transport and biosynthesis in relation to geotropic induction of excised sunflower shoot-tips. *Planta*, **105**, 234–244.
343. I.D. Railton, I.D.J. Phillips (1973). Gibberellins and geotropism in *Zea mays* coleoptiles. *Planta*, **109**, 121–126.
344. D.W. Pearce, D.M. Reid, R.P. Pharis (1991). Ethylene-mediated regulation of gibberellin content and growth in *Helianthus annuus* L. *Plant Physiol.*, **95**, 1197–1201.
345. T.K. Scott, B.H. Most (1972). The movement of growth hormones in sugar cane. In: Y. Vadar (Eds.), *Hormonal Regulation in Plant Growth and Development* (pp. 57–67). Verlag Chemie, Weinheim.
346. P. Hedden, Y. Kamiya (1997). Gibberellin biosynthesis: enzymes, genes and their regulation. *Ann. Rev. Plant Physiol. Plant Mol. Biol.*, **48**, 431–460.
347. B.G. Pickard (1985). Roles of hormones in phototropism. In: R.P. Pharis, D.M. Reid (Eds.), *Encyclopedia of Plant Physiology* (New Series, Vol. 11, pp. 365–417). Springer, Berlin.
348. F.B. Abeles, B. Rubinstein (1964). Regulation of ethylene evolution and leaf abscission by auxin. *Plant Physiol.*, **39**, 963–969.
349. V.R. Humphry (1980). Effect of ethylene on the phototropic response of *Avena* and *Oryza* coleoptiles. *Naturwissenschaften*, **67**, 198.

350. K. Hasegawa, H. Noguchi, T. Iwagawa, T. Hase (1986). Phototropism in hypocotyls of radish. I. Isolation and identification of growth inhibitors, *cis-* and *trans-*raphanusanins and raphanusamide, involved in phototropism of radish hypocotyls. *Plant Physiol.*, **81**, 976–979.
351. H. Noguchi, K. Nishitani, J. Bruinsma, K. Hasegawa (1986). Phototropism in hypocotyls of radish. II. Role of *cis-* and *trans-*raphanusanins, and raphanusamide in phototropism of radish hypocotyls. *Plant Physiol.*, **81**, 980–983.
352. H. Noguchi, K. Hasegawa (1987). Phototropism in hypocotyls of radish. III. Influence of unilateral or bilateral illumination of various light intensities on phototropism and distribution of *cis-* and *trans-*raphanusanins and raphanusamide. *Plant Physiol.*, **83**, 672–675.
353. M. Sakoda, T. Matsuoka, S. Sando, K. Hasegawa (1988). Phototropism in hypocotyls of radish. V. Lateral distribution of *cis-* and *trans-*raphanusanins and raphanusamide in various phototropisms induced by unilateral broad blue light. *J. Plant Physiol.*, **133**, 110–112.
354. K. Hasegawa, S. Togo (1989). Phototropism in hypocotyls of radish. VII. Involvement of the growth inhibitors, raphnusol A and B in phototropism of radish hypocotyls. *J. Plant Physiol.*, **135**, 110–113.
355. P. Boysen-Jensen (1928). Die phototropische Induckion in der Spitze der Avenacoleoptile. *Planta*, **5**, 464–477.
356. R. Hertel, A.C. Leopold (1963). Versuche zur Analyse des Auxintransports in der Koleoptile von *Zea mays* L. *Planta*, **59**, 535–562.
357. I.A. Newman (1963). Electric potentials and auxin translocation in *Avena*. *Aust. J. Biol. Sci.*, **16**, 629–646.
358. T.I. Baskin, W.R. Briggs, M. Iino (1986). Can lateral redistribution of auxin account for phototropism of maize coleoptiles. *Plant Physiol.*, **81**, 306–309.
359. M.L. Evans (1985). The action of auxin on plant cell elongation. *Critical Rev. Plant Sci.*, **2**. 317–365.
360. D. Nissl, M.H. Zenk (1969). Evidence against induction of protein synthesis during auxin-induced initial elongation of *Avena* coleoptiles. *Planta*, **89**, 323–341.
361. R.K. dela Fuente, A.C. Leopold (1970). Time course of auxin stimulations of growth. *Plant Physiol.*, **46**, 186–189.
362. C.-H. Ullrich (1978). Continuous measurement of initial curvature of maize coleoptiles induced by lateral auxin application. *Planta*, **140**, 201–211.
363. Y. Masuda, R. Yamamoto (1972). Control of auxin-induced stem elongation by the epidermis. *Physiol. Plant.*, **27**, 109–115.
364. D. Brummell, J.L. Hall (1980). The role of the epidermis in auxin-induced and fusicoccin-induced elongation of *Pisum sativum* stem segments. *Planta*, **150**, 371–379.
365. U. Kutschera, W.R. Briggs (1987). Differential effect of auxin on in vivo extensibility of cortical cylinder and epidermis in pea internodes. *Plant Physiol.*, **84**, 1361–1366.
366. R.E. Cleland (1991). The outer epidermis of *Avena* and maize coleoptiles is not a unique target for auxin in elongation growth. *Planta*, **186**, 75–80.
367. R.M. Napier (1997). Trafficking of the auxin-binding protein. *Trends Plant Sci.*, **2**, 251–255.
368. M.J. Vesper, C.L. Kuss (1990). Physiological evidence that the primary site of auxin action in maize coleoptiles is an intracellular site. *Planta*, **182**, 486–491.
369. M. Claussen, H. Lüthen, M. Böttger (1996). Inside or outside? Localization of the receptor relevant to auxin-induced growth. *Physiol. Plant.*, **98**, 861–867.
370. R. Hertel (1995). Auxin binding protein 1 is a red herring. *J. Exp. Bot.*, **46**, 461–462.
371. M.A. Venis (1995). Auxin binding protein 1 is a red herring? Oh no it isn't! *J. Exp. Bot.*,

46, 463–465.

372. N.G. Ball (1962). The effects of externally applied 3-indolylacetic acid on phototropic induction and response in the coleoptile of *Avena*. *J. Exp. Bot.*, **13**, 45–60.
373. C.M. Carrington, J. Esnard (1988). The elongation response of watermelon hypocotyls to indole-3-acetic acid: a comparative study of excised segments and intact plants. *J. Exp. Bot.*, **39**, 441–450.
374. M. Iino (1996). Short-term stimulation of growth induced by the apical application of IAA to intact maize coleoptiles. *Plant Cell Physiol.*, **37**, 27–33.
375. J.L. Hall, D.A. Brummell, J. Gillespie (1985). Does auxin stimulate the elongation of intact plant stems? *New Phytol.*, **100**, 341–345.
376. T. Yang, D.M. Law, P.L. Davies (1993). Magnitude and kinetics of stem elongation induced by endogenous indole-3-acetic acid in intact light-grown pea seedlings. *Plant Physiol.*, **102**, 717–727.
377. T. Yang, P.J. Davies, J.B. Reid (1996). Genetic dissection of the relative roles of auxin and gibberellin in the regulation of stem elongation in intact light-grown peas. *Plant Physiol.*, **110**, 1029–1034.
378. K. Haga, M. Iino (1997). The short-term growth stimulation induced by external supply of IAA to internodes of intact pea seedlings. *Aust. J. Plant Physiol.*, **24**, 215–226.
379. E.L. Stowe-Evans, R.M. Harper, A.V. Motchoulski, E. Liscum (1998). NPH4, a conditional modulator of auxin-dependent differential growth responses in *Arabidopsis*. *Plant Physiol.*, **118**, 1265–1275.
380. M.K. Watahiki, K.T. Yamamoto (1997). The *massugu1* mutation of *Arabidopsis* identified with failure of auxin-induced growth curvature of hypocotyl confers auxin insensitivity to hypocotyl and leaf. *Plant Physiol.*, **115**, 419–426.
381. J. Shen-Miller, S.A. Gordon (1966). Hormonal relations in the phototropic response. III. The movement of C^{14}-labeled and endogenous indoleacetic acid in phototropically stimulated *Zea* coleoptiles. *Plant Physiol.*, **41**, 59–65.
382. J. Shen-Miller, S.A. Gordon (1966). Hormonal relations in the phototropic response. IV. Light-induced changes of endogenous auxins in the coleoptile. *Plant Physiol.*, **41**, 831–841.
383. S.M. Naqvi, S.A. Gordon (1967). Auxin transport in *Zea mays* coleoptiles. II. Influence of light on the transport of indoleacetic acid-2-^{14}C. *Plant Physiol.*, **42**, 138–143.
384. A. Hager, R. Schmidt (1968). Auxintransport und Phototropismus.I. Die lichtbedingte Bildung eines Hemmstoffes für den Transport von Wuchsstoffen in Koleoptilen. *Planta*, **83**, 347–371.
385. A. Hager, R. Schmidt (1968). Auxintransport und Phototropismus. II. Der Hemmechanismus des aus IES gebildeten Photooxidationsproduktes 3-Methylenoxindol beim Transport von Wuchsstoffen. *Planta*, **83**, 372–386.
386. J. Shen-Miller, P. Cooper, S.A. Gordon (1969). Phototropism and photoinhibition of basipetal transport of auxin in oat coleoptiles. *Plant Physiol.*, **44**, 491–496.
387. W.M. Elliott, J. Shen-Miller (1976). Similarity in dose responses, action spectra and red light responses between phototropism and photoinhibition of growth. *Photochem. Photobiol.*, **23**, 195–199.
388. R. Menschick, V. Hild, A. Hager (1977). Decarboxylierung von Indolylessigsäure im Zusammenhang mit dem Phototropismus in *Avena*-Koleoptilen. *Planta*, **133**, 223–228.
389. S.M. Naqvi, K.C. Engvild (1974). Action of abscisic acid on auxin transport and its relation to phototropism. *Physiol. Plant.*, **30**, 283–287.
390. S.M. Naqvi (1972). Lateral transport of indoleacetic acid-2-^{14}C in *Zea mays* L. coleoptile segments. *Physiol. Vég.*, **10**, 495–501.

391. T.L. Lomax, G.K. Muday, P.H. Rubery (1995). Auxin transport. In: P.J. Davies (Ed.), *Plant Hormones* (2nd ed., pp. 509–530). Kluwer Academic, Dordrecht.
392. K.V. Thimann (1934). Studies on the growth hormone of plants. VI. The distribution of the growth substance in plant tissues. *J. Gen. Physiol.*, **18**, 23–34.
393. A.R. Sheldrake (1973). Auxin transport in secondary tissues. *J. Exp. Bot.*, **24**, 87–96.
394. M. Jacobs, S. Gilbert (1983). Basal localization of the presumptive auxin transport carrier in pea stem cells. *Science*, **220**, 1297–1300.
395. A.M. Jones (1990). Location of transported auxin in etiolated maize shoots using 5-azidoindole-3-acetic acid. *Plant Physiol.*, **93**, 1154–1161.
396. P. Nick, R. Bergfeld, E. Schäfer, P. Schopfer (1990). Unilateral reorientation of microtubules at the outer epidermal wall during photo- and gravitropic curvature of maize coleoptiles and sunflower hypocotyls. *Planta*, **181**, 162–168.
397. P. Nick, E. Schäfer, M. Furuya (1992). Auxin redistribution during first positive phototropism in corn coleoptiles. Microtubule reorientation and the Cholodny-Went theory. *Plant Physiol.*, **99**, 1302–1308.
398. R. Bergfeld, V. Speth, P. Schopfer (1988). Reorientation of microfibrils and microtubules at the outer epidermal wall of maize coleoptiles during auxin-mediated growth. *Bot. Acta*, **101**, 57–67.
399. P. Nick, Furuya M., E. Schäfer (1991). Do microtubules control growth in tropism? Experiments with maize coleoptiles. *Plant Cell Physiol.*, **32**, 999–1006.
400. K. Fischer, P. Schopfer (1998). Physical strain-mediated microtubule reorientation in the epidermis of gravitropically or phototropically stimulated maize coleoptiles. *Plant J.*, **15**, 119–123.
401. R.E. Cleland (1995). Auxin and cell elongation. In: P.J. Davies (Ed.), *Plant Hormones* (pp. 214–227). Kluwer Academic, Netherlands.
402. D.J. Cosgrove (1987). Linkage of wall extension with water and solute uptake. In: D.J. Cosgrove, D.P. Knievel (Eds.), *Physiology of Cell Expansion During Plant Growth* (pp. 88–100). The American Society of Plant Physiologists.
403. F.J.M. Maathuis, A.M. Ichida, D. Sanders, J.I. Schroeder (1997). Roles of higher plant K^+ channels. *Plant Physiol.*, **114**, 1141–1149.
404. X. Wang, M. Iino (1998). Interaction of cryptochrome 1, phytochrome, and ion fluxes in blue-light-induced shrinking of *Arabidopsis* hypocotyl protoplasts. *Plant Physiol.*, **117**, 1265–1279.
405. A. Hager, H. Menzel, A. Krauss (1971). Versuche und Hypothese zur Primärwirkung des Auxins beim Streckungswachstum. *Planta*, **100**, 47–75.
406. R.E. Cleland (1971). Cell wall extension. *Ann. Rev. Plant Physiol.*, **22**, 197–222.
407. U. Kutschera, P. Schopfer (1985). Evidence against the acid-growth theory of auxin action. *Planta*, **163**, 483–493.
408. E. Edelmann, P. Schopfer (1989). Role of protein and RNA synthesis in the initiation of auxin-mediated growth in coleoptiles of *Zea mays* L. *Planta*, **179**, 475–485.
409. D.L. Rayle, R.E. Cleland (1992). The acid growth theory of auxin-induced cell elongation is alive and well. *Plant Physiol.*, **99**, 1271–1274.
410. A. Hager, G. Debus, H.-G. Edel, H. Stransky, R. Serrano (1991). Auxin induces exocytosis and the rapid synthesis of a high-turnover pool of plasma-membrane H^+-ATPase. *Planta*, **185**, 527–537.
411. I. Frías, M.T. Caldeira, J.R. Pérex-Castiñeira, J.P. Navarro-Aviñó, F.A. Culiañez-Maciá, O. Kuppinger, H. Stransky, M. Pagés, A. Hager, R. Serrano (1996). A major isoform of the maize plasma membrane H^+-ATPase: characterization and induction by auxin in coleoptiles. *Plant Cell*, **8**, 1533–1544.
412. A. Rück, K. Palme, M.A. Venis, R.M. Napier, H.H. Felle (1993). Patch-clamp analysis

establishes a role for an auxin binding protein in the auxin stimulation of plasma membrane current in *Zea mays* protoplasts. *Plant J.*, **4**, 41–46.

413. K. Kasamo, T. Yamaki (1974). Effect of auxin on Mg^{2+}-activated and -inhibited ATPase from mung bean hypocotyls. *Plant Cell Physiol.*, **15**, 965–970.
414. G.F.E. Scherer (1981). Auxin-stimulated ATPase in membrane fractions from pumpkin hypocotyls (*Cucurbita maxima* L.). *Planta*, **151**, 434–438.
415. V. Santoni, G. Vansuyt, M. Rossignol (1990). Differential auxin sensitivity of proton translocation by plasma membrane H^+-ATPase from tobacco leaves. *Plant Sci.*, **68**, 33–38.
416. K. Philippar, I. Fuchs, H. Lüthen, S. Hoth, C.S. Bauer, K. Haga, G. Thiel, K. Ljung, G. Sandberg, M. Böttger, D. Becker, R. Hedrich (1999). Auxin-induced K^+ channel expression represents an essential step in coleoptile growth and gravitropism. *Proc. Natl. Acad. Sci. USA*, **96**, 12186–12191.
417. D.J. Cosgrove, R.E. Cleland (1983). Osmotic properties of pea stem internodes in relation to growth and auxin action. *Plant Physiol.*, **72**, 332–338.
418. L. Ordin, T.H. Applewhite, J. Bonner (1956). Auxin-induced water uptake by *Avena* coleoptile sections. *Plant Physiol.*, **31**, 44–53.
419. T.T. Stevenson, R. Cleland (1981). Osmoregulation in the *Avena* coleoptile in relation to auxin and growth. *Plant Physiol.*, **67**, 749–753.
420. U. Kutschera, P. Schopfer (1986). Effects of auxin and abscisic acid on cell wall extensibility in maize coleoptiles. *Planta*, **167**, 527–535.
421. R.E. Cleland, D.L. Rayle (1972). Lack of auxin-induced stored growth in *Avena* coleoptiles and its implication concerning the mechanism of wall extension. *Planta*, **106**, 61–71.
422. M. Claussen, H. Lüthen, M. Blatt, M. Böttger (1997). Auxin-induced growth and its linkage to potassium channels. *Planta*, **201**, 227–234.
423. H. Felle (1988). Auxin causes oscillations of cytosolic free calcium and pH in *Zea mays* coleoptiles. *Planta*, **174**, 495–499.
424. C.A. Gehring, H.R. Irving, R.W. Parish (1990). Effects of auxin and abscisic acid on cytosolic calcium and pH in plant cells. *Proc. Natl. Acad. Sci. USA*, **87**, 9645–9649.
425. K.K.A. Goswami, L.J. Audus (1976). Distribution of calcium, potassium and phosphorus in *Helianthus annuus* hypocotyls and *Zea mays* coleoptiles in relation to tropic stimuli and curvatures. *Ann. Bot.*, **40**, 49–64.
426. R.G. Wyn Jones, C.J. Brady, J. Speirs (1979). Ionic and osmotic relationships in plant cells. In: D.L. Laidman, R.G. Wyn Jones (Eds.), *Recent Advances in the Biochemistry of Cereals* (pp. 63–103). Academic Press, New York.
427. I.R. MacDonald, J.W. Hart, D.C. Gordon (1983). Analysis of growth during geotropic curvature in seedling hypocotyls. *Plant Cell Environ.*, **6**, 401–406.
428. D.J. Cosgrove (1990). Rapid, bilateral changes in growth rate and curvature during gravitropism of cucumber hypocotyls: Implications for mechanism of growth control. *Plant Cell Environ.*, **13**, 227–234.
429. N. Arslan-Cerim (1966). The redistribution of radioactivity in geotropically stimulated hypocotyls of *Helianthus annuus* pretreated with radioactive calcium. *J. Exp. Bot.*, **17**, 236–240.
430. F. Migliaccio, A.W. Galston (1987). On the nature and origin of the calcium asymmetry arising during gravitropic response in etiolated pea epicotyls. *Plant Physiol.*, **85**, 542–547.
431. R.D. Slocum, S.J. Roux (1983). Cellular and subcellular localization of calcium in gravistimulated oat coleoptiles and its possible significance in the establishment of tropic curvature. *Planta*, **157**, 481–492.

432. S.L. Bagshaw, R.E. Cleland (1993). Is wall-bound calcium redistributed during the gravireaction of stems and coleoptiles? *Plant Cell Environ.*, **16**, 1081–1089.
433. J. Gross, M. Sauter (1987). Are auxin and calcium movements in corn coleoptiles linked processes? *Plant Sci.*, **49**, 189–198.
434. R E. Cleland (1977). Reevaluation of the effect of calcium ions on auxin-induced elongation. *Plant Physiol.*, **60**, 709–712.
435. J.S. Coartney, D.J. Morré (1980). Studies on the role of wall extensibility in the control of cell expansion. *Bot. Gaz.*, **141**, 56–62.
436. S.L. Bagshaw, R.E. Cleland (1993). The effects of enhanced levels of calcium on the gravireaction of sunflower hypocotyls. *Plant Cell Environ.*, **16**, 1091–1097.
437. E.P. Sparding, D.J. Cosgrove (1993). Influence of electrolytes on growth, phototropism, nutation and surface potential in etiolated cucumber seedlings. *Plant Cell Environ.*, **16**, 445–451.
438. C.A. Gehring, D.A. Williams, S.H. Cody, R.W. Parish (1990). Phototropism and geotropism in maize coleoptiles are spatially correlated with increases in cytosolic free calcium. *Nature*, **345**, 528–530.
439. L.-G. Ma, D.-Y. Sun (1997). The involvement of calcium in light signal transduction chain for phototropism in sunflower seedlings. *Biol. Plant.*, **39**, 569–574.
440. G. Baum, J.C. Long, G.I. Jenkins, A.J. Trewavas (1999). Stimulation of the blue light phototropic receptor NPH1 causes a transient increase in cytosolic Ca^{2+}. *Proc. Natl. Acad. Sci. USA*, **96**, 13554–13559.
441. S. Iwami, Y. Masuda (1973). Hydrogen ion-induced curvature in cucumber hypocotyls. *Plant Cell Physiol.*, **14**, 757–762.
442. L. Wright, D.L. Rayle (1983). Evidence for a relationship between H^+ excretion and auxin in shoot gravitropism. *Plant Physiol.*, **72**, 99–104.
443. T.J. Mulkey, K.M. Kuzmanoff, M.L. Evans (1981). Correlations between proton-efflux patterns and growth patterns during geotropism and phototropism in maize and sunflower. *Planta*, **151**, 239–241.
444. R. Mertens, E.W. Weiler (1983). Kinetics on the redistribution of endogenous growth regulators in gravireacting plant organs. *Planta*, **158**, 339–348.
445. G.E. Backus, A.R. Schrank (1952). Electrical and curvature response of the *Avena* coleoptile to unilateral illumination. *Plant Physiol.*, **27**, 251–262.
446. M.H. Weisenseel, A.J. Meyer (1997). Bioelectricity, gravity and plants. *Planta*, **203**, 98–104.
447. L. Grahm, C.H. Hertz (1962). Measurements of the geoelectric effect in coleoptiles by a new technique. *Physiol. Plant.*, **15**, 96–114.
448. T. Tanada, C. Vinten-Johansen (1980). Gravity induces fast electrical field change in soybean hypocotyls. *Plant Cell Environ.*, **3**, 127–130.
449. K. Imagawa, K. Toko, S. Ezaki, K. Hayashi, K. Yamafuji (1991). Electrical potentials during gravitropism in bean epicotyls. *Plant Physiol.*, **97**, 193–196.
450. H. Shigematsu, K. Toko, T. Matsuno, K. Yamafuji (1994). Early gravi-electrical responses in bean epicotyls. *Plant Physiol.*, **105**, 875–880.
451. F.D. Sack (1997). Plastids and gravitropic sensing. *Planta*, **203**, s63–s68.
452. R. Godbolé, W. Michalke, P. Nick, R. Hertel (2000). Cytoskeletal drugs and gravity-induced lateral auxin transport in rice coleoptiles. *Plant Biol.*, **2**, 1–6.
453. S.S. Virk, R.E. Cleland (1988). Calcium and the mechanical properties of soybean hypocotyl cell walls: possible role of calcium and protons in cell-wall loosening. *Planta*, **176**, 60–67.
454. R. Hertel (1983). The mechanism of auxin transport as a model for auxin action. *Z. Pflanzenphysiol.*, **112**, 53–67.

455. M.S. Irving, S. Ritter, A.D. Tomos, D. Koller (1997). Phototropic response of the bean pulvinus: movement of water and ions. *Bot. Acta*, **110**, 118–126.
456. L. Brauner, N. Arslan (1951). Experiments on the auxin reactions of the pulvinus of *Phaseolus multiflorus. Rev. Fac. Sci. Univ. Istanbul Ser. B*, **16**, 257–300.
457. K.G. Krieger (1978). Early time course and specificity of auxin effects on turgor movement of the bean pulvinus. *Planta*, **140**, 107–109.
458. C.P. Keller, E. Van Volkenburgh (1996). Osmoregulation by oat coleoptile protoplasts. Effect of auxin. *Plant Physiol.*, **110**, 1007–1016.
459. W. Shropshire Jr., H. Mohr (1970). Gradient formation of anthocyanin in seedlings of *Fagopyrum* and *Sinapis* unilaterally exposed to red and far-red light. *Photochem. Photobiol.*, **12**, 145–149.
460. J.B. Dandeno (1903). Phototropism under light-rays of different wave-lengths. *Science*, **18**, 604–606.
461. E. Bünning (1938). Phototropismus und Carotinoide. *III. Weitere Untersuchungen an Pilzen und höheren Pflanzen. Planta*, **27**, 583–610.
462. M. Iino, E. Schäfer, W.R. Briggs (1984). Photoperception sites for phytochrome-mediated phototropism of maize coleoptiles. *Planta*, **162**, 477–479.
463. C.L. Ballaré, A.L. Scopel, S.R. Radosevich, R.E. Kendrick (1992). Phytochrome-mediated phototropism in de-etiolated seedlings. *Plant Physiol.*, **100**, 170–177.
464. C.L. Ballaré, J.J. Casal, R.E. Kendrick (1991). Responses of wild-type and *lh*-mutant seedlings of cucumber to natural and simulated shadelight. *Photochem. Photobiol.*, **54**, 819–826.
465. H. Mohr (1956). Die Abhängigkeit des Protonemawachstums und der Protonemapolarität bei Farnen vom Licht. *Planta*, **47**, 127–158.
466. A. Kadota, M. Wada, M. Furuya (1982) Phytochrome-mediated phototropism and different dichroic orientation of Pr and Pfr in protonemata of the fern *Adiantum* capillus-veneris L. *Photochem. Photobiol.*, **35**, 533–536.
467. A. Kadota, M. Koyama, M. Wada, M. Furuya (1984). Action spectra for polarotropism and phototropism in protonemata of the fern *Adiantum capillus-veneris*. *Physiol. Plant.*, **61**, 327–330.
468. J. Hayami, A. Kadota, M. Wada (1986). Blue light-induced phototropic response and the intracellular photoreceptive site in *Adiantum* protonemata. *Plant Cell Physiol.*, **27**, 1571–1577.
469. M. Iino, A. Kadota, K. Shitanishi, M. Wada (1990). Phytochrome-mediated phototropism in *Adiantum* protonemata.I. Phototropism as a function of lateral Pfr gradient. *Photochem. Photobiol.*, **65**, 1032–1038.
470. M. Iino, K. Shitanishi, M. Wada (1997). Phytochrome-mediated phototropism in *Adiantum* protonemata. II. Participation of phytochrome dark reversion. *Photochem. Photobiol.*, **65**, 1032–1038.
471. H. Okamoto, Y. Hirano, H. Abe, K. Tomizawa, M. Furuya, M. Wada (1993). The deduced amino acid sequence of phytochrome from *Adiantum* includes consensus motifs present in phytochrome B from seed plants. *Plant Cell Physiol.*, **34**, 1329–1334.
472. K. Nozue, T. Kanegae, T. Imaizumi, S. Fukuda, H. Okamoto, K.-C. Yeh, C. Lagarias, M. Wada (1998). A phytochrome from the fern *Adiantum* with features of the putative photoreceptor NPH1. *Proc. Natl. Acad. Sci. USA*, **95**, 15826–15830.
473. K.S. Mills, A.R. Schrank (1954). Electrical and curvature responses of the *Avena* coleoptile to unilateral ultraviolet irradiation. *J. Cell. Comp. Physiol.*, **43**, 39–55.
474. G.M. Curry, K.V. Thimann, P.M. Ray (1956). The base curvature response of *Avena* seedlings to the ultraviolet. *Physiol Plant.*, **9**, 429–440.

Photomovement
D.-P. Häder and M. Lebert, editors.

Chapter 24

Role of the microtubular cytoskeleton in coleoptile phototropism

Peter Nick

Table of contents

24.1 Introduction

Phototropism is based on the ability to regulate cell elongation in response to asymmetric illumination. The classical system to study phototropism has been the Graminean coleoptile, and studies on the phototropism of grass coleoptiles date back to the time of Darwin [1]. So far, all work concerning the role of the cytoskeleton during the phototropic response of higher plants has been performed in coleoptiles, and this review will therefore focus on coleoptile phototropism. The coleoptile, a leaf-like organ, is formed late during embryogenesis from the base of the scutellum and sheaths the apical meristem. Physiologically it behaves like a leaf although it is not correct to treat it as a homologue of the cotyledons. The coleoptile has the biological function to protect and guide the young leaves during the growth through the soil. As soon as the coleoptile reaches the surface it ceases to grow almost immediately and the primary leaves pierce through the coleoptile tip. Later, crown roots emerge at the node separating the coleoptile from the mesocotyl, and these crown roots replace the primary root that dies soon afterwards.

From the biological function of the coleoptile, the advantage of this organ for plant physiology can be easily derived:

1. Coleoptiles grow fast – the leaves must reach the light, before the resources of the seed become exhausted.
2. Coleoptiles respond sensitively to light and gravity – these are the stimuli that guide the seedlings towards the surface.
3. Coleoptiles grow exclusively by cell expansion – this is the most economic way of growth.
4. Coleoptiles possess a clear apico-basal polarity – the tip must direct the growth of the whole organ.

The first experiments by Darwin and Darwin [1] demonstrated already that the perception of the phototropic stimulus is situated in the very tip of the coleoptile, whereas the bending response moves towards the base of the organ. This implies the transport of signals from the tip to the base of the coleoptiles. These signals transmit two types of information:

1. the fact that the coleoptile tip has perceived light, and
2. the direction of the light stimulus.

Independently, Cholodny [2] and Went [3] discovered that a growth promoting substance is redistributed during tropistic bending across the coleoptile cross section. The activity of this substance could be measured by a famous bioassay and the substance was eventually identified as the first plant hormone – auxin [4].

There have been attempts by the school of Blaauw [5] to explain phototropism independently of signal transport – if the light causes a localized growth inhibition, then a gradient of light should lead to a gradient of cell growth that does not require the exchange of intercellular signals. The debate between the disciples of the Cholodny-Went theory [2,3] and those of the Blaauw hypothesis [5] has continued ever since. However, the evidence for a displacement of growth [6] and a displacement of auxin [7] towards the shaded flank of the coleoptile are overwhelming, and it is certainly justified to discuss phototropism together with the auxin response.

The growth response to auxin is situated in the epidermis that limits and guides the expansion of the coleoptile [8]. Auxin causes epidermal wall loosening and thus releases the constraint put by the epidermis upon the expansion of the inner tissues. In the framework of the Cholodny-Went theory, it is the control of epidermal elongation by auxin that is the central element of phototropic bending. Cortical microtubules have been considered in this context, because they define the direction of cellulose deposition in the cell wall and thus the axis of cell expansion [9]. This review will therefore consider the evidence supporting a role of microtubules for phototropism and contrast this with findings where microtubule orientation and growth are discongruent. In a final part, these contradicting observations will be reconciled by introducing the concept of a stable phototropic polarity that can be separated from phototropic bending. It is this phototropic polarity that is intimately linked to a blue-light induced reorientation and fixation of cortical microtubules.

24.2 Do microtubules respond to phototropic stimulation? Yes!

The driving force for cell growth is a gradient of water potential with a more negative water potential in cytoplasm and vacuole as compared to the apoplast [8]. The resulting pressure by itself is not directional, and the cell is therefore expected to grow isotropically. In fact, if the cell wall is removed, the cell will assume an isodiametric shape. In intact cells, the direction of growth is actively controlled by the cell wall. From biophysical considerations it is expected that epidermal cells that are approximately cylindrical in shape, should grow preferentially in lateral direction. In other words, they must be endowed with some kind of **reinforcement mechanism** to maintain their original axiality [10]. This reinforcement mechanism seems to reside in the cell wall and was first uncovered for the long internodal cells of the alga *Nitella* [11]. In these elongate cells, the cellulose microfibrils were demonstrated by electron microscopy to be arranged in transverse rings, especially in the newly deposited inner layers of the wall. It is evident that the transverse arrangement of microfibrils can account for the reinforcement mechanism that maintains the longitudinal growth axis in cylindrical cells. In the meanwhile, such a correlation between transverse microfibrils and cell elongation has been demonstrated in numerous cases [reviewed in 9,12,13].

Moreover, reorientation of the growth axis is often accompanied by a loss or a reorientation in the anisotropy of cellulose deposition [14–16]. Therefore, the correlation between guided cellulose deposition and cell growth seems to be very tight. The so-called terminal complexes responsible for cellulose synthesis are usually organized in rosette-like hexagonal arrays [13]. It is generally believed that these rosettes slide within the membrane leaving behind them bundles of crystallizing cellulose fibers, the microfibrils (Figure 1A). Cortical microtubules seem to be responsible for the guided movement of the terminal complexes and thus for the axiality of cell expansion. The evidence for this statement can be summarized as follows:

1. Cortical microtubules are closely associated with the plasma membrane, and in plasmolyzing cells a direct contact between cortical microtubules and newly formed cellulose microfibrils has been detected by electron microscopy [12].

2. The prospective sites where secondary wall thickenings will form are already marked by parallel thick bundles of cortical microtubules [17,18].
3. In those cases, where changes of the preferential axis of cellulose deposition (and, concomitantly, the axis of cell growth) occur, this reorientation is heralded by a reorientation of cortical microtubules ([14] for the ethylene response, [19] for the auxin response, [20] for the gibberellin response, [21] for wood formation).
4. Elimination of cortical microtubules by antimicrotubular drugs results in a gradual loss of growth anisotropy and a block of cell elongation leading, in extreme cases, to lateral swelling (reviewed in [12]).

The exact mechanism by which microtubules drive and guide cellulose deposition has been under debate since the discovery of cortical microtubules by Ledbetter and Porter [22], and a manifold of different hypotheses have been proposed (reviewed in [12,13]). The principal debate can be summarized into two alternative models:

1. According to the original model by Heath [23], cortical microtubules are physically linked to the terminal complexes and the linking molecule(s) can be pulled along the microtubules by dynein-like motor proteins. Thus the whole complex will be moved in a direction parallel to the adjacent microtubules (Figure 1B). It has been observed

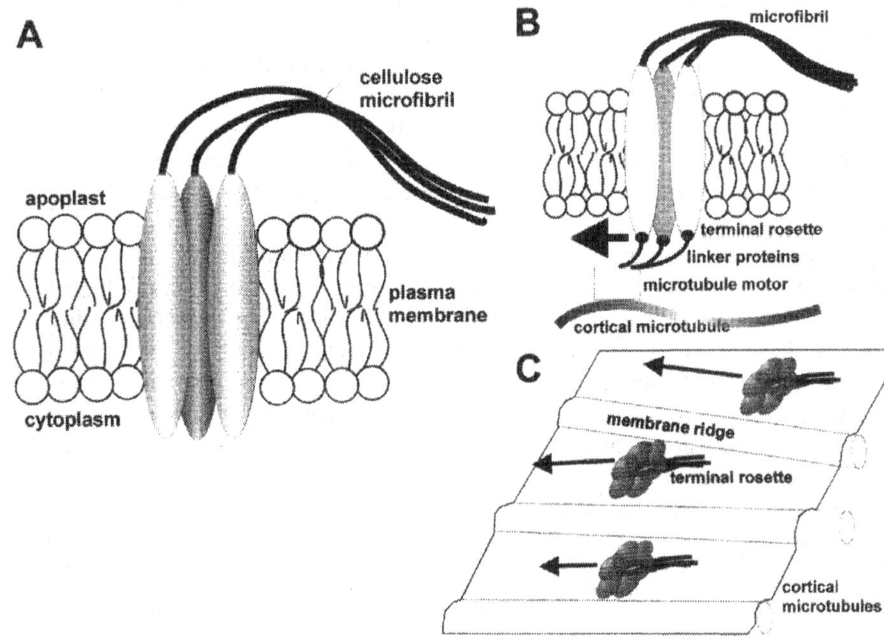

Figure 1. Role of microtubules in directional cellulose synthesis. A Transmembrane localization of a cellulose-synthesizing complex (terminal complex). B Original model by Heath [23], where the terminal complexes are connected to microtubule motors and actively moved along cortical microtubules. C Model by Giddings and Staehelin [13], where cortical microtubules induce membrane ridges that confine the movement of the terminal complexes. According to this model, the driving force for the movement originates from the crystallization of cellulose.

in several cases that removal of the cell wall during formation of protoplasts causes a dramatic restructuring of cortical microtubules [24] and makes them susceptible to cold [25]. These observations demonstrate a stabilization of cortical microtubules by the cell wall and indicate a physical link between microtubules and microfibrils across the plasma membrane.
2. The alternative model is based on the observation that, in some cases, the terminal complexes have been observed in the interspaces outlined by the microtubules rather than being directly attached to them [13]. The guiding of rosette movement, according to this model, is not caused by a physical link of the terminal complexes to microtubule motors. Microtubules are rather supposed to induce membrane channels that impede lateral deviations of rosette movement (Figure 1C). The driving force for the movement would be cellulose crystallization itself, propelling the terminal rosette through the microtubule-dependent membrane channels.

At the present stage, it is difficult to decide between the two models. Moreover, none of them seems to be complete and able to accommodate all observations. It is necessary to understand, on the molecular level, the interaction between microtubules and the plasma membrane, and the potential role of motor proteins for guided cellulose deposition. In this context microtubule-binding proteins are interesting. Such proteins might mediate the association of microtubules with the plasma membrane and could interact through the membrane with the terminal complexes. The exact mechanism by which cortical microtubules guide the deposition of cellulose and thus define the axis of cellular growth remains to be elucidated. Nevertheless, the close relation between microtubule orientation and the direction of growth suggests that the main function of cortical microtubules has to be sought in the control of cell shape by external and internal signals. This view is supported by the microtubular response to auxin and phototropic stimulation:

In coleoptiles that undergo rapid elongation, cortical microtubules are found to be transverse in both, the cells of the inner tissue and in the epidermis (maize [19,26], rice [20]). Consistently, cellulose microfibrils are deposited in transverse direction reinforcing the elongation of the cell (maize [19], rice [20]). Upon excision of the coleoptile tip (the major source of auxin) and incubation of the coleoptile segments in water microtubules change their orientation from transverse to longitudinal and the cellulose-microfibrils are deposited in longitudinal direction [19]. This results in a loss of growth reinforcement and, consequently, in a block of coleoptile elongation. This process can be reversed by addition of exogenous indole-acetic acid [19, 26,27]. Microtubules respond within 10 to 15 min after addition of indole-acetic acid [26,27] and they complete their reorientation within one hour.

Phototropic stimulation of intact coleoptiles causes a reorientation of cortical microtubules in the lighted, auxin-depleted flank, whereas the microtubules in the shaded, auxin-enriched, flank reinforce their transverse orientation [26,27]. This gradient of microtubule orientation is correlated with a gradient of growth (inhibition of growth in the flank, where microtubules are longitudinal, stimulation of growth in the flank, where microtubules are transverse) resulting in tropistic bending. Microtubule reorientation becomes detectable within 10 min after stimulation and is complete within 1 h. Phototropic curvature, in contrast, becomes detectable within 20 to 30 min after stimulation and reaches a maximum at 2 h after stimulation [26]. Thus, for phototropic stimulation, the microtubular response clearly precedes the growth response. This is not trivial, because there exist blue-light responses such as the light-inhibition of stem elongation in peas, where microtubule reorientation occurs at a slower rate as compared to the growth response [28].

The analysis of microtubule reorientation in response to auxin and phototropic stimulation lead to a model (Figure 2), where phototropic stimulation caused a displacement of auxin across the coleoptile (Cholodny-Went theory). The resulting auxin gradient will then produce a gradient of microtubule orientation with longitudinal microtubules in the auxin-depleted flank, and transverse microtubules in the auxin-enriched flank of the coleoptile. The gradient of microtubule orientation will then be translated into a corresponding gradient of cellulose microfibril deposition and thus a gradient of cell wall extensibility between the lighted and the shaded coleoptile flank. This will eventually culminate in an inhibition of growth in the lighted flank and a stimulation of growth in the shaded flank causing phototropic bending towards the light. This model is supported by the observation that other stimuli that produce lateral auxin

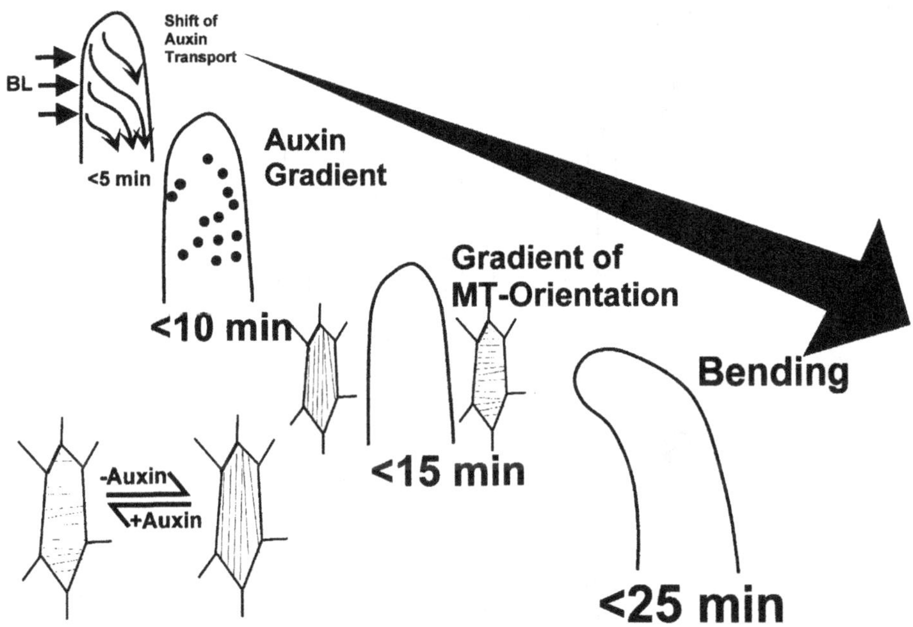

Figure 2. Original model about the role of microtubules in the phototropic response according to [19]. The phototropic stimulus causes a lateral displacement of auxin transport and a gradient of auxin with elevated levels of the hormone in the shaded side, whereas the lighted side is depleted from auxin. Microtubule orientation responds to the local level of auxin with transverse microtubules in auxin-enriched areas and longitudinal microtubules in auxin-depleted areas. The gradient of microtubule orientation is translated into a corresponding gradient in the direction of cellulose microfibrils. The longitudinal extensibility of the epidermal wall will decrease in the lighted side causing a reduction of growth. The growth gradient will eventually drive phototropic bending.

transport, such as gravitropism, produce a similar gradient in the orientation of microtubules [26].

24.3 Do microtubules control tropistic curvature? No!

This model for the role of the microtubular cytoskeleton mediating between auxin displacement and asymmetric growth seems to be oversimplified, though. Microtubule reorientation and bending can be separated experimentally [29].

1. Following phototropic stimulation by a pulse of blue light, curvature reaches a maximum at two hours after stimulation. Curved coleoptiles experience a gravitropic counterstimulation and therefore they straighten again within a few hours. Microtubules are found to be longitudinal in the lighted side, but remain transverse in the shaded side. Interestingly, this gradient of orientation develops within one hour after the light pulse and is maintained throughout the period of gravitropic straightening (Figure 3A). The original hypothesis (Figure 2) would have predicted an inversion of the orientation gradient prior to the onset of gravitropic straightening, if the two tropistic responses produce asymmetric growth by the same mechanism.

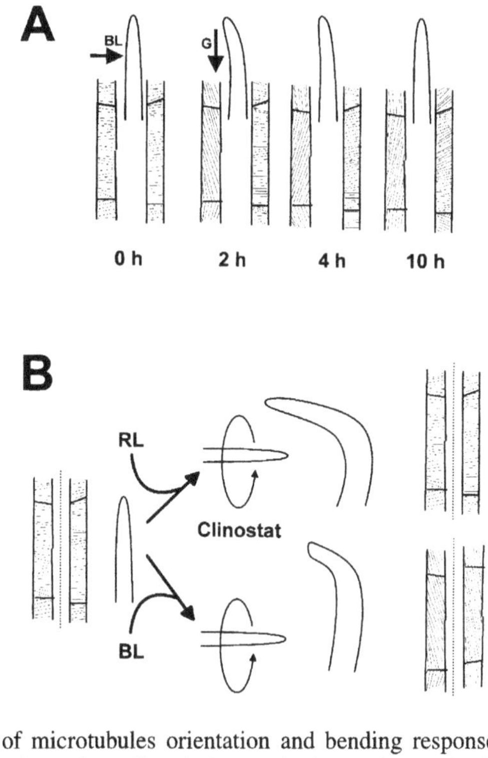

Figure 3. Separation of microtubules orientation and bending response. A During gravitropic straightening, the direction of bending is inverted, the gradient of microtubule orientation is maintained. B During nastic curvature, microtubule orientation is identical on both side of the bending organ.

2. Upon rotation of maize coleoptiles on a horizontal clinostat in the absence of tropistic stimulation, a nastic curvature develops in the dorsiventral axis of the coleoptile [30]. This curvature is strong, when the coleoptiles are symmetrically irradiated by red light, and it becomes weaker, when the plants are irradiated by blue light. Under red light, the microtubules are transverse on both flanks of the bending organ, and they are longitudinal under blue light (Figure 3B). However, not any gradient of microtubule orientation across the organ could be detected [29] in contradiction to the original model (Figure 2).
3. When, two hours after a phototropic stimulus, the coleoptiles are subjected to a phototropic counterstimulation of equal strength, but opposing direction, and the coleoptiles are rotated on a clinostat (to exclude gravitropic counterstimulation), they interrupt their bending towards the first stimulus and start bending towards the counterstimulus [31]. This dominance of the counterstimulus is transient, though: Already one hour later, the coleoptiles „remember" the direction of the first stimulus and return to their original mode of curving, that is subsequently maintained for many hours. The orientation of microtubules is longitudinal in the side that had been hit by the first stimulus. In contrast to bending itself, microtubules do not respond at all to the second, opposing light [29,32]. The original hypothesis (Figure 2) would have predicted an inversion of the orientation gradient in response to the counterstimulus and a second inversion back to the original gradient.

This analysis leads to the conclusion that a gradient of microtubule orientation is neither necessary nor sufficient for asymmetric growth. In other words: microtubule orientation and phototropic curvature appear not to be causally linked but seem to develop as parallel phenomena.

24.4 If microtubules do not control phototropism, why do they respond to phototropic stimulation?

To give an answer to this question, it is important to introduce the concept of phototropic polarity. In contrast to the hypothesis proposed by Blaauw [5], the light gradient is not immediately translated into a gradient of growth. Consistent with the model proposed by Cholodny-Went [2,3], there exists an intermediate step: The gradient of blue light results in a transverse polarization of the tissue. This polarity controls asymmetric growth. It is possible to separate this polarity from tropistic bending and to make it manifest in the form of a so-called directional memory [31]:

The bending response of maize coleoptiles to blue light is transient, reaching a maximum at two hours after induction and disappearing subsequently, a phenomenon that can be understood in terms of gravitropic straightening [31]. However, when the gravitropic counterstimulation experienced by the curving coleoptile is eliminated by rotation on a horizontal clinostat, a stable bending towards the inducing pulse is observed. When the coleoptiles are transferred to the clinostat after the phototropic bending has already vanished due to gravitropic straightening, nevertheless a stable curvature in direction of the first pulse develops (Figure 4). This demonstrates the existence of a directional memory that had been induced by the stimulus and that

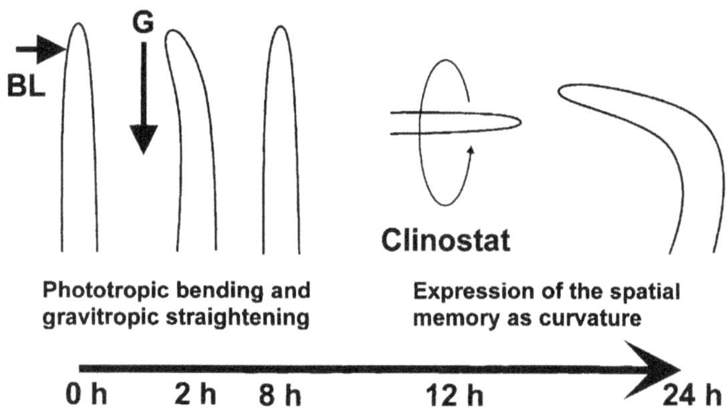

Figure 4. Demonstration of phototropic polarity. The phototropic bending response remains transient, due to gravitropic straightening. When gravitropic counterstimulation is removed (by rotation on a horizontal clinostat), the stable phototropic polarity can be expressed as stable curvature. Despite being suppressed by gravity for many hours, the phototropic polarity has remained stable.

persisted even during the time of gravitropic straightening. This memory can thus be separated from bending itself.

A similar directional memory has been repeatedly described for both gravitropism [33,34] and phototropism [35]. In order to detect a directional memory following tropistic stimulation, the expression of the tropistic response was suppressed by cold or auxin depletion [33,34] or by specific ion-channel blockers [35]. If the suppressive treatment was eventually removed, the response to the stimulus developed.

Such experiments indicate that the stimulus can induce a tropistic polarity that can be separated from its expression as curvature and can nevertheless persist over a long time even if it is prevented from becoming manifest. The suppressive condition in the case of the phototropic memory of maize coleoptiles is the gravitropic counterstimulation experienced by curved coleoptiles – it is removed by clinostat rotation allowing for expression of the memory as stable bending [31].

When does the spatial memory become stable? This question can be addressed by challenging the memory induced by a stimulus with a counterdirected stimulus after variable time intervals [31]. If the opposing pulse is administered early after the first pulse, it can reverse the memory completely and a strong stable bending response towards the second pulse is observed (Figure 5A). If the opposing pulse is administered later, it fails to reverse the memory. It first does reverse, however, the bending response (Figure 5B). This reversal in the sign of bending remains transient though, and subsequently the original response (directed towards the first pulse) is restored and maintained over a long time [31]. A detailed fluence-response study for the first and the second stimulation [36] demonstrated that, independently of fluence, the spatial memory becomes irreversibly fixed at two hours after the first, inducing pulse. This fixation time is independent of the interaction between the two stimuli and the direction of the spatial memory.

The relevance of microtubules for phototropism seems to be linked to the establishment and fixation of phototropic polarity rather than to phototropic bending per se. This conclusion is supported by the following evidence:

1. The gradient of microtubule orientation [29,32] as well as the spatial memory [31] both persist during gravitropic straightening.
2. The gradient of microtubule orientation [29,32] as well as the spatial memory [31,36] do not respond – in contrast to curvature – to a phototropic counterstimulus that is administered late (two hours) after the inducing pulse.
3. The gradient of microtubule orientation [32] as well as the spatial memory [31,36] can be reversed by a counterstimulus that is administered early (one hour) after the first pulse.
4. The orientation of microtubules is irreversibly fixed at the same time, when the spatial memory becomes irreversibly fixed [32,36].

These observations suggest that the reorientation of microtubules is the cellular expression of phototropic polarity, and that the fixation of microtubule orientation is the cellular correlate to the irreversible fixation of this memory.

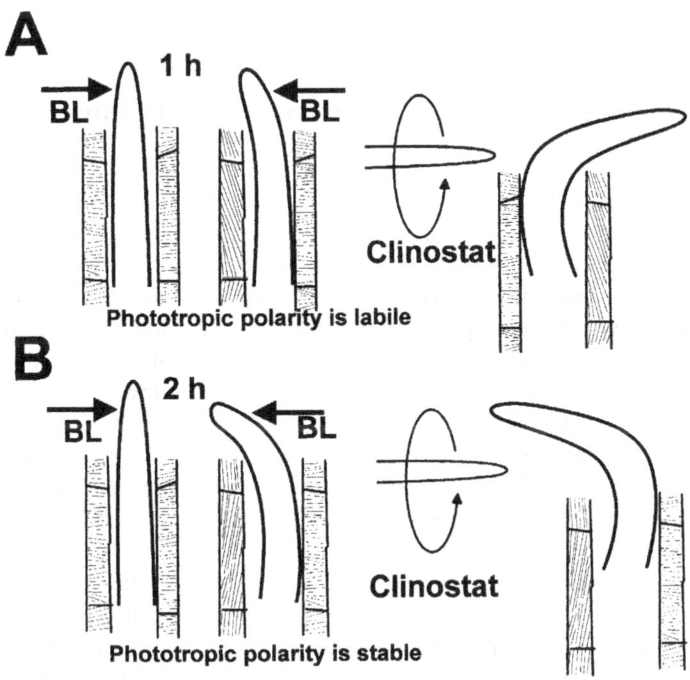

Figure 5. Assay for fixation of the phototropic polarity. The phototropic polarity induced by a stimulus is questioned by a counterstimulus of equal strength administered at various time intervals after the first irradiation. A If the time interval is short, the phototropic polarity (and the gradient of microtubule orientation) can still be inverted. B If the time interval is long, the phototropic polarity (and the gradient of microtubule orientation) is irreversibly fixed and resists a counterstimulation.

A further characterization indicated that the blue-light effect on microtubule reorientation is mediated by auxin [32], whereas the blue-light induced fixation of microtubule orientation is neither mediated by a local depletion of auxin nor by a gradient of auxin. The blue-light induced fixation of microtubules does not require a light gradient and resides in the coleoptile base as shown by partial irradiation using a light pipe. The perception for the blue-light induced reorientation of microtubules, in contrast, resides in the coleoptile tip.

The results of this analysis can be summarized as follows: Tropistic stimulation triggers three chains of events:

1. Perception in the coleoptile tip causing a displacement of auxin towards the shaded side, resulting in an inhibition of cell elongation in the lighted side and a stimulation of growth in the shaded side. The asymmetric growth causes then the bending towards the light.
2. Reorientation of microtubules, caused by a displacement of auxin.
3. Perception of blue light in the coleoptile base triggering the production of a factor (that is independent of auxin) that can fix microtubules (irrespective of their actual orientation). This fixed microtubule orientation seems to be the cellular base of a stable transverse polarity that can be rendered manifest by removal of gravitropic counterstimulation.

24.5 How is the phototropic polarity established? Intercellular cross-talk versus autistic cells

Phototropic polarity might be a function of tissue polarity, i.e. it could arise from mutual interactions between the individual cells across the tissue. Alternatively, the phototropic polarity could arise as a cell polarity, i.e. as an autonomous response of the individual cell that does not require intercellular communication. The findings of Cholodny [2] and Went [3] that a growth promoting substance (auxin) is redistributed across the coleoptile suggests that the individual cells interact during the establishment of phototropic polarity. However, the Cholodny-Went theory does not allow any conclusion to be drawn about the way in which the plant is able to sense the direction of the stimulus. It is still possible that each cell is able to recognize the direction of light individually by producing a radial cell polarity. This cell polarity would subsequently determine the direction of auxin efflux. The tissue response described by the Cholodny-Went theory would then arise only at that stage by mere summation of individual cell responses. Alternatively, the gradient of light might be recognized by the cell population as a whole, and a true tissue polarity might emerge from intercellular signaling, for instance by locally self-amplifying activation in concert with far-ranging mutual inhibition [37]. Such a tissue polarity would then induce a parallel cell polarity leading to transverse auxin transport.

A debate between Heilbronn [38] and Buder [39] at the beginning of this century contributes to this problem. Heilbronn claimed that the plant perceives the direction of the light. Buder, in contrast, insisted on the gradient of light as the signal to be perceived by the coleoptile. This dispute stimulated an ingenious experiment by Buder [39], in which the gradient of light and light direction were opposed to each other. To achieve this, the coleoptile was irradiated from inside out using a prototype of a light-piping device. Buder was able to demonstrate bending of the coleoptiles towards the lighted flank, although the direction of the incident light should have induced bending into the opposite direction.

The debate between Blaauw [5] and Cholodny-Went [2,3] and the debate between Heilbronn [38] and Buder [39] both touch the question, by what mechanism the phototropic polarity is established. This process can be divided into two subsequent steps:

1. sensing of the light direction and
2. induction of a parallel phototropic polarity.

The sensing could be based upon the gradient of the light (Buder) or it could be brought about by actually sensing the direction of the light (Heilbronn). The phototropic polarity could evolve as a cell polarity or it could emerge as a tissue polarity. From these considerations one can design four alternative possibilities, how a phototropic stimulus might result in a phototropic polarity in a multicellular organ such as a Graminean coleoptile (Figure 6). In order to distinguish between these four alternatives, it is necessary to find a marker for cell polarity. Tropistic curvature cannot be the appropriate marker, because it reflects merely the integrated response of the individual cells over the entire coleoptile. In contrast, the orientation of cortical microtubules permits the analysis of individual cells and it has been, so far, the closest marker for phototropic polarity [32].

For this reason, the Buder experiment (irradiation from inside out by means of a light pipe) was combined with immunofluorescence analysis of microtubule orientation in the outer epidermis of maize coleoptiles [40]. By measuring fluence-response curves for phototropic curvature the original findings of Buder could be confirmed, i.e. the coleoptiles curved according to the gradient of the light, although this gradient was opposed to the direction of the incident light. Surprisingly, this type of irradiation caused a reversal of cell polarity in most of the epidermal cells: In irradiated cells that were stimulated conventionally (from outside in), only those microtubules that were adjacent to the outer epidermal wall showed a reorientation from transverse to longitudinal (Figure 7A). Upon irradiation from inside out, reorientation was more frequent at the inner face of the epidermal cells (Figure 7B). In a few cases, transitions between both responses could be observed. These findings might be interpreted such that microtubules can sense the direction of the light. However, the site, where the light is perceived, and the site, where the microtubular response takes place, are separated by several millimeters as shown by microirradiation of individual cells [40]. This means that the information about the light gradient must be transmitted from the tip downwards in form of unknown signals. To detect such signals, the growth response of individual cells was followed over time in response to microirradiation of the coleoptile tip. In fact, a growth-regulating signal could be detected that travels in basal direction with a speed of around 60 mm h^{-1}, i.e. five times faster than auxin. It is possible to imagine that such a signal can embody the information about the tissue polarity – there might be less of this signal in one flank of the coleoptile and more in the other. It is harder to imagine how this signal could convey the information about intracellular light gradients (light direction). But this is exactly implied by the reversal of cell polarity expressed in the inverted microtubular response. To resolve this apparent contradiction it is important to consider endogenous polarities – cells that are organized in a tissue are never symmetric, but usually exhibit distinct endogenous polarization. In case of the epidermis, this endogenous polarity becomes manifest by the different microtubular response to auxin – the reorientation from transverse to longitudinal in response to auxin depletion is

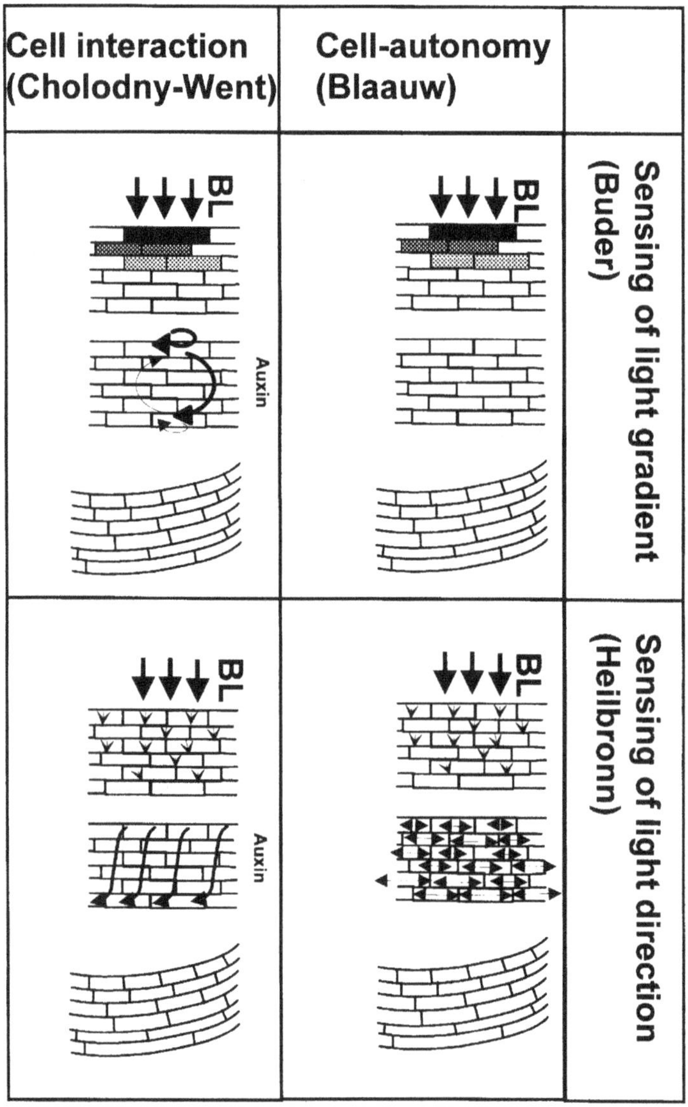

Figure 6. Different models of the establishment of phototropic polarity. The four models can be understood as combinations from two parameters: (1) Do the cells act autonomously or do they communicate? (2) Does the plant sense the gradient of light over the tissue or the direction of the light (i.e. the intracellular light gradient)?

confined to those microtubules that are adjacent to the outer wall [19,26]. The reversal of this endogenous polarity upon irradiation from inside out might be brought about in response to superoptimal local levels of the fast growth-regulating signal that is released

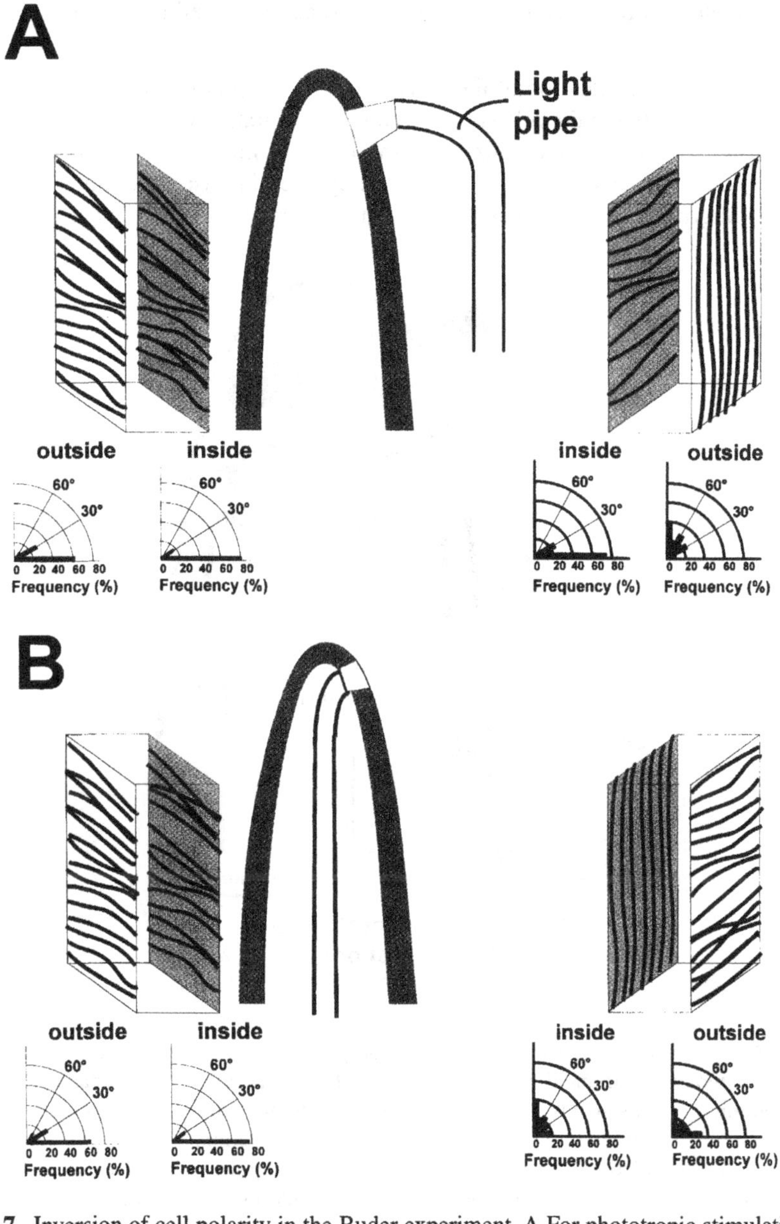

Figure 7. Inversion of cell polarity in the Buder experiment. A For phototropic stimulation from outside in, only the microtubules at the outer cell wall in the lighted coleoptile flank respond by reorientation. B For phototropic stimulation from inside out the microtubules reorient at the inner cell wall of the lighted coleoptile flank (inversion of epidermal cell polarity).

from the coleoptile tip after phototropic stimulation. Such a mechanism has been invoked to explain the blue-light induced inversion of the gravitropic response during clinostat rotation [41].

How is phototropic polarity established? The combination of the Buder experiment with the analysis of microtubule orientation as indicators of cell polarity can detect the following steps (Figure 8):

1. Upon phototropic irradiation of the coleoptile tip, the light gradient is sensed.
2. This gradient results in the release of signals that migrate towards the coleoptile base much faster (60 mm h^{-1}) than auxin transport (10–12 mm h^{-1}).
3. In parallel, a gradient of auxin is established by lateral transport of auxin. This auxin gradient migrates downwards as well, but at a lower speed (10–12 mm h^{-1}).

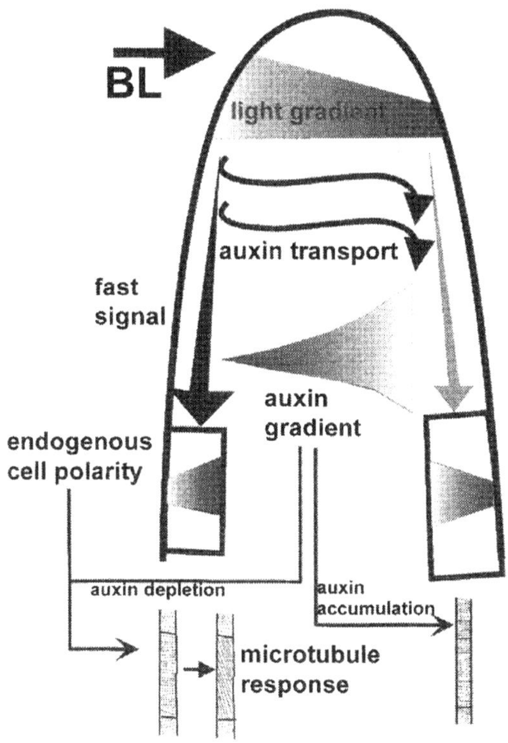

Figure 8. Model of the role of microtubules in the phototropic response. The gradient of blue light induces two events in parallel: (1) Release of a fast signal that migrates basipetally with a speed of 60 mm h^{-1}. This signal interacts with the endogenous radial polarity of epidermal cells and defines their microtubular response to auxin or auxin depletion. (2) Lateral auxin transport towards the shaded coleoptile flank. The auxin gradient migrates basipetally with a speed of 10–12 mm h^{-1}. The local auxin depletion in the lighted side induces a reorientation of cortical microtubules from transverse to longitudinal. The microtubule orientation is fixed after two hours, in response to a blue-light induced signal that is not auxin. The orientation of microtubules defines a phototropic polarity that remains stable over days and participates in developmental imprinting of the node by the coleoptile that is relevant for crown-root formation.

4. The fast blue-light induced signal interacts with the endogenous cell polarity of epidermal cells and determines the cellular (microtubular) response to auxin.
5. The depletion of auxin reaches the cells in the lighted flank of the coleoptile base and causes a reorientation of microtubules. This reorientation response to auxin depletion depends on the interaction with the fast signal: there can be reorientation at the outer cell wall, there can be reorientation at the inner wall, or there can be no reorientation response (if the phototropic polarity has become fixed). This microtubular response is the cellular base for phototropic polarity.
6. Phototropic curvature develops in parallel to phototropic polarity by a direct response of cell elongation to the local concentration of auxin. Therefore the phototropic curvature is initiated at the tip and migrates downwards with about the same speed as auxin transport itself (i.e. around 10–12 mm h^{-1}).
7. Phototropic curvature vanishes within a few hours after the end of stimulation, phototropic polarity instead persists over several days and can induce a stable **photomorphosis** in the coleoptilar node (see below).

24.6 Why does the coleoptile need phototropic polarity? Is phototropic curvature not enough?

The coleoptile is usually regarded as an ephemeral organ that has merely the function to guide the primary leaves through the soil towards the surface. It is soon afterwards torn open by the emerging primary leaves. This behavior poses the question, what the physiological significance of a stable phototropic polarity might be. At a closer look the coleoptile is not as ephemeral as usually written in the text books – it remains alive for several weeks and can even grow actively. However, the axis of cell growth is tilted from elongation to stem thickening. In addition, it seems to influence the emergence of the nodal crown roots that will later form the major root system [42]. The dorsiventrality of the coleoptile (it is homologous to a leaf) appears to be imprinted upon the coleoptilar node: the crown roots emerge later at the ventral flank of the coleoptilar node (adjacent to the caryopsis). Interestingly, this gradient can be shifted by phototropic stimulation [42]. The emergence of crown roots is delayed in the lighted flank and promoted in the shaded flank. This effect of phototropic stimulation occurs several days after a stimulation with a light pulse lasting for a few seconds. The phototropic curvature, induced by this pulse, is transient though and disappears within a few hours. The light effect on crown-root formation is not correlated with phototropic curvature and cannot be mimicked by gravitropic stimulation that induces a similar degree of curving [42]. On the other hand, this effect can be reverted by counterstimulation within one hour, it becomes irreversibly fixed at two hours, i.e. the time, when the microtubule-dependent phototropic polarity becomes stable [42]. If microtubules in the basal part of the node are depolymerized by antimicrotubular herbicides, this leads to a suppression of the phototropic shift of crown-root emergence.

These observations indicate that the coleoptile can imprint a polarity upon the node and that this polarity is then expressed as a gradient of crown-root emergence at a time, when the coleoptile itself has already ceased to elongate. The spatial pattern of crown-root emergence is thus defined by events that take place in the coleoptile several days earlier. These early events are polarized by phototropic stimulation, they are dependent on intact microtubules, they can be separated from phototropic bending as such, and they become irreversibly fixed in space two hours after phototropic induction.

What is the physiological significance of microtubule-based stable phototropic polarity? It is not required for the fast, transient bending that is induced by phototropic pulse stimulation. It is intimately linked to a long-lasting transverse polarization of the coleoptile that is induced by phototropic stimulation and that can persist for several days. In the experiment, this transverse polarization can be rendered manifest by removing gravitropic counterstimulation. Under natural conditions, this microtubule-dependent polarity becomes imprinted upon the coleoptilar node and guides the emergence of nodal crown roots several days after the phototropic stimulus. The shift of crown-root emergence towards the shaded side of the coleoptilar node is certainly of adaptive value and helps the seedling to align its axis with respect to direction of the light. The polarizing mechanism seems to be flexible within a temporal window of about two hours after phototropic induction before it becomes fixed irreversibly. This flexibility is ecologically meaningful, because it allows the seedling to distinguish between stochastic changes of light quantity (caused, for instance, by passing clouds) and more stable light gradients (caused, for instance, by neighboring plants).

24.7 Summary and outlook

The reorientation of cortical microtubules following phototropic induction was initially interpreted in terms of a role for microtubules in the mediation of the phototropic response [26]. This model did not bear closer scrutiny but lead to the discovery of long-lasting effects of phototropic stimulation (stable phototropic polarity). These stable effects develop in parallel to phototropic curvature, but they are not involved in the phototropic response per se. The response of cortical microtubules to phototropic stimulation is linked to the establishment and stabilization of this stable phototropic polarity rather than to phototropism itself. The functional significance of the stable phototropic polarity seems to be related to a developmental imprinting of the coleoptilar node by the coleoptile guiding the emergence of crown roots and thus shaping the architecture of the root system. It appears not appropriate and even misleading to discuss these stable effects in the context of phototropism – it would be more correct to designate them as **photomorphosis**. The developmental impact of photomorphosis should not be underestimated – it helps the seedling to adjust axis and architecture with respect to the light distribution in the canopy and thus supports and stabilizes the adaptive role of phototropism itself.

To understand photomorphosis in molecular terms, it is necessary to understand the mechanism of blue-light induced reorientation of microtubules. The focus will be on those proteins that control the assembly and disassembly of microtubules. The isolation of a microtubule-associated protein from maize that is expressed in response to phytochrome and associates with bundles of cortical microtubules [43] stimulates the search for similar proteins that are formed in response to blue light. The recent finding that blue light causes an increased acetylation of tubulin in maize coleoptiles indicates that the dynamics of assembly and disassembly becomes reduced in response to blue light, what might be the molecular base for the fixation of microtubule orientation and the fixation of the blue-light induced polarity. The isolation and cloning of the factor that is responsible for this reduced dynamics should allow a molecular approach to the problem of photomorphosis.

References

1. C. Darwin, F. Darwin (1881). *The Power of Movements in Plants*. Murray, London.
2. N. Cholodny (1927). Wuchshormone und Tropismen bei Pflanzen. *Biol. Zentralbl.*, **47**, 604–626.
3. F.W. Went (1928). Wuchsstoff und Wachstum. *Recueil des Travaux Botaniques Néerlandais*, **25**, 1–116.
4. F.W. Went, K.V. Thimann (1937). *Phytohormones*. MacMillan, New York.
5. O.H. Blaauw (1918). Licht und Wachstum. *Meddelingen from de Landbouwhogeschool*, **15**, 89–204.
6. M. Iino, W.R. Briggs (1984). Growth distribution during first positive phototropic curvature of maize coleoptiles. *Plant Cell Environm.*, **7**, 97–104.
7. M. Iino (1991). Phototropism: mechanisms and ecological implications. *Plant Cell Environm.*, **13**, 633–650.
8. U. Kutschera, R. Bergfeld, P. Schopfer (1987). Cooperation of epidermal and inner tissues in auxin-mediated growth of maize coleoptiles. *Planta*, **170**, 168–180.
9. R.E. Williamson (1991). Orientation of cortical microtubules in interphase plant cells. *Internat. Rev. Cytol.*, **129**, 135–206.
10. P.B. Green (1980). Organogenesis – a biophysical view. *Annu. Rev. Plant Physiol.*, **31**, 51–82.
11. P.B. Green, A. King (1966). A mechanism for the origin of specifically oriented textures with special reference to *Nitella* wall texture. *Austr. J. Biol. Sci.*, **19**, 421–437.
12. D.G. Robinson, H. Quader (1982). The microtubule-microfibril syndrome. In: C.W. Lloyd (Ed.), *The Cytoskeleton in Plant Growth and Development* (pp. 109–126). Academic Press, London.
13. T.H. Giddings, A. Staehelin (1991). Microtubule-mediated control of microfibril deposition: a re-examination of the hypothesis. In: C.W. Lloyd (Ed.), *The Cytoskeletal Basis of Plant Growth and Form* (pp. 85–99). Academic Press, London.
14. J.M. Lang, W.R. Eisinger, P.B. Green (1982). Effects of ethylene on the orientation of microtubules and cellulose microfibrils of pea epicotyls with polylamellate cell walls. *Protoplasma*, **110**, 5–14.
15. J.M. Hush, C.R. Hawes, R.L. Overall (1990). Interphase microtubule re-orientation predicts a new cell polarity in wounded pea roots. *J. Cell Sci.*, **96**, 47–61.
16. P.B. Green, J.M. Lang (1981). Towards a biophysical theory of organogenesis: birefringence observations on regenerating leaves in the succulent *Graptopetalum paraguayense*. *Planta*, **151**, 413–426.
17. H. Fukuda, H. Kobayashi (1989). Dynamic organization of the cytoskeleton during tracheary-element differentiation, Development. *Growth Different.*, **31**, 9–16.
18. G. Jung, W. Wernicke (1990). Cell shaping and microtubules in developing mesophyll of wheat (*Triticum aestivum* L.). *Protoplasma*, **153**, 141–148.
19. R. Bergfeld, V. Speth, P. Schopfer (1988). Reorientation of microfibrils and microtubules at the outer epidermal wall of maize coleoptiles during auxin-mediated growth. *Bot. Acta*, **101**, 57–67.
20. T. Toyomasu, H., Yamane, N. Murofushi, P. Nick (1994). Phytochrome inhibits the effectiveness of gibberellins to induce cell elongation in rice. *Planta*, **194**, 256–263.

21. H. Abe, R., Funada, H., Imaizumi, J. Ohtani, K. Fukazawa (1995). Dynamic changes in the arrangement of cortical microtubules in conifer tracheids during differentiation. *Planta*, **197**, 418–421.
22. M.C. Ledbetter, K.R. Porter (1963). A "microtubule" in plant cell fine structure. *J. Cell Biol.*, **12**, 239–250.
23. I.B. Heath (1974). A unified hypothesis for the role of membrane-bound enzyme-complexes and microtubules in plant cell wall synthesis. *J. Theor. Biol.*, **48**, 445–449.
24. G. Jung, A. Hellmann, W. Wernicke (1993). Changes in the density of microtubular networks in mesophyll cell derived protoplasts of *Nicotiana* and *Triticum* during leaf development. *Planta*, **190**, 10–16.
25. T. Akashi, S. Kawasaki, H. Shibaoka (1990). Stabilization of cortical microtubules by the cell wall in cultured tobacco cells. Effects of extensin on the cold-stability of cortical microtubules. *Planta*, **182**, 363–369.
26. P. Nick, R., Bergfeld, E. Schäfer, P. Schopfer (1990). Unilateral reorientation of microtubules at the outer epidermal wall during photo- and gravitropic curvature of maize coleoptiles and sunflower hypocotyls. *Planta*, **181**, 162–168.
27. P. Nick, E. Schäfer, M. Furuya (1992). Auxin redistribution during first positive phototropism in corn coleoptiles – microtubule reorientation and the Cholodny-Went theory. *Plant Physiol.*, **99**, 1302–1308.
28. M. Laskowski (1990). Microtubule orientation in pea stem cells: a change in orientation follows the initiation of growth rate decline. *Planta*, **181**, 44–52.
29. P. Nick, M. Furuya, E. Schäfer (1991). Do microtubules control growth in tropism? Experiments with maize coleoptiles. *Plant Cell Physiol.*, **32**, 873–880.
30. P. Nick, E. Schäfer (1989). Nastic response of maize (*Zea mays* L.) coleoptiles during clinostat rotation. *Planta*, **179**, 123–131.
31. P. Nick, E. Schäfer (1988). Spatial memory during the tropism of maize (*Zea mays* L.) coleoptiles. *Planta*, **175**, 380–388.
32. P. Nick, E. Schäfer (1994). Polarity induction versus phototropism in maize: Auxin cannot replace blue light. *Planta*, **195**, 63–69.
33. F. Czapek (1895). Untersuchungen über Geotropismus. *Jb. wiss. Bot.*, **27**, 243.
34. L. Brauner, A. Hager (1957). Über die geotropische "Mneme". *Naturwiss.*, **44**, 429–430.
35. E. Hartmann (1984). Influence of light on phototropic bending of moss protonemata of *Ceratodon purpureus* (Hedw.). *Brid. J. Hattori Bot. Lab.*, **55**, 87–98.
36. P. Nick, E. Schäfer (1991). Induction of transverse polarity by blue light: An all-or-none response. *Planta*, **185**, 415–424.
37. A. Gierer (1981). Generation of biological pattern and form: Some physical, mathematical, and logical aspects. *Progress in Biophysics and Molecular Biology*, **37**, 1–47.
38. A. Heilbronn (1917). Lichtabfall oder Lichtrichtung als Ursache der heliotropischen Reizung? *Ber. Dtsche. Bot. Ges.*, **35**, 641–642.
39. J. Buder (1920). Neue phototropische Fundamentalversuche. *Ber. Dtsche. Bot. Ges.*, **28**, 10–19.
40. P. Nick, M. Furuya (1996). Buder revisited: cell and organ polarity during phototropism. *Plant, Cell Environm.*, **19**, 1179–1187.
41. H. Sailer, P. Nick, E. Schäfer (1990). Inversion of gravitropism by symmetric blue light on the clinostat. *Planta*, **180**, 378–382.
42. P. Nick (1997). Phototropic stimulation can shift the gradient of crown root emergence in maize. *Bot. Acta*, **110**, 291–297.
43. P. Nick, A.M. Lambert, M. Vantard (1995). A microtubule-associated protein in maize is expressed during phytochrome-induced cell elongation. *Plant J.*, **8**, 835–844.

Photomovement
D.-P. Häder and M. Lebert, editors.

Chapter 25

Solar navigation by plants

Dov Koller

Table of contents

Abstract

Plants are capable of reorientating in space their developing buds, flowers, inflorescences, fruits, leaves, leaflets, or even entire shoots, using biophysical motors that subtend them. These motors operate by creating differential changes in volume in their opposite sides, either by means of irreversible cell growth, or by fully reversible hydraulic changes in cell volume. Movement of mature leaves is accomplished by means of a special motor organ – the pulvinus – situated in strategic junctions within the leaf, principally at the base of the leaf(let) lamina. Photomovements are driven by specific light signals. Photonastic movements take place in a predetermined direction and are independent of the direction of the light signals. In phototropic movements, the direction of movement is tightly coupled to the direction of the light signal. In some leaves, the light signal is perceived in the pulvinus itself and is transduced to photonastic and/or phototropic pulvinar responses that result in corresponding laminar movements. In other leaves, the direction of light is perceived in the lamina with precision, and the resulting signal is transmitted to the subtending pulvinus, where its response moves the lamina to face the light. In nature, such leaves track the changes in the solar azimuth and elevation throughout most of each clear day.

Abbreviations: AI – angle of light incidence on laminar surface (0° = normal; positive or negative = tip and base oriented, respectively); BAP – blue light absorbing pigment systems; Ca^{2+}_{cyt} – cytosolic Ca^{2+}; DFS, WFS – Donnan and water free space; EOD – end of day phototransformation of phytochrome to its active form P_{fr}; HIR – high irradiance response of phytochrome; IP_3 – inositol 1,4,5-trisphosphate; K^+_{in} – hyperpolarization-dependent, inward-rectifying K^+ channels; K^+_{out} – depolarization-dependent, outward-rectifying K^+ channels; L, D – light, dark; LE – angle of laminar elevation (0° = horizontal); OP, TP – osmotic and turgor pressure (potential); PAR – photosynthetically active radiation (400–700 nm); PFD – photon fluence rate; PI – phosphatidylinositol; Ψ – water potential.

25.1 Introduction

Most terrestrial plants use the diurnal L ↔ D transitions as signals by which to regulate their exchange of CO_2 and water vapor through their stomata. Some use these transitions as signals by which to unfold their leaves to increase their harvesting of sunlight and to fold them into a compact, protective configuration in the absence of light. Some use these transitions as signals by which to open their flowers, or inflorescences, and close them into a compact, protective configuration in concert with the diurnal activity of their biological pollinators.

In their incessant quest for PAR, most terrestrial plants use the directional signals of sunlight to navigate their light-harvesting organs so as to optimize their utilization of its energy. Some also use these signals to navigate their reproductive organs to maximize their attraction to their biological pollinating vectors. These light-driven responses use sensory mechanisms to perceive the specific signals that are transduced into specific movements.

The flux of solar radiation intercepted by most terrestrial plants may exceed the photosynthetic capacity of their chloroplasts and may lead to photoinhibition, as well as to over-heating, when it exceeds the capacity of the leaves to lose thermal energy through re-radiation, convection and evaporative cooling. These hazards are reduced primarily by mechanisms within the chloroplast for harmless dissipation of the excess light energy it absorbs, as well as by leaf and chloroplast movements that reduce the interception of solar radiation. Certain plants are not deterred by such hazards and reorient the laminae of their mature leaves to maximize their interception of direct solar radiation throughout each day. PAR fluence may also limit growth, particularly of young seedlings, whose light-harvesting surface is limited. Such plants usually maximize the interception of PAR by reorienting their apical bud and its developing leaves to face the direction of the predominant flux [1].

This chapter is concerned with movements by which higher plants in the terrestrial environment navigate their shoot organs in response to directional light signals generated, directly or indirectly, by the sun. These organs may also move in response to L$\leftrightarrow$D transitions. Movements in response to such non-directional signals are included, to place the subject in its proper perspective. A brief discussion of relevant terminology follows, to provide a common basis.

25.2 Terminology

In *physical* terms, plants move as their cells grow. In *physiological terms*, plants move only by changing their spatial orientation, or configuration. Such movement may be *growth-mediated*, or *turgor-mediated*, by expansion/contraction that involve changes in turgor pressure TP, without growth. Higher plants can move stomatal guard cells, entire leaves, the leaf lamina, its leaflets, or its opposite parts, apical buds and their subtending cluster of developing leaves, flowers, or inflorescences, fruits, or entire shoots. Such movements invariably take place by *anisotropic changes in cell volume* (in a preferred direction). Differential, anisotropic changes in cell volume in tissues situated in parallel, opposite sides result in movement by curvature.

Growth-mediated movements take place by *differential changes in rates of growth* of cells in tissues located in different (usually opposite) tissues, resulting in curvature. Cell growth is a direct, irreversible consequence of increased extensibility of the cell wall, resulting in irreversible expansion of cell walls by osmotic uptake of water, balanced by osmotic adjustment of solutes. Therefore, the capacity for growth-mediated movements ends at maturity. However, as long as the "motor" tissue is capable of growth, growth-mediated movements may be reversed by compensatory differential growth in its opposite side. Growth-mediated movement may also be induced in non-growing parts that maintain a potential for growth (for instance at and around the base of the grass internodes, in leaf petioles and in leaves of certain insectivorous plants).

Turgor-mediated movements take place by *differential changes in volume* (expansion, or contraction) of fully mature cells as a specific result of transport of osmotically active solutes, followed by water, from, or into their vacuole. Walls of mature cells can only expand, or contract *elastically*. Therefore, volume changes in such cells depend on the *elastic modulus of their wall* and are associated with corresponding changes in their TP.

Expanding/contracting cells may also exhibit deformation that enables large changes in cell volume in response to relatively smaller changes in turgor pressure.

Autonomous movements are driven by signals that are generated endogenously and take place in circadian cycles (~24-h) under control of the universal biological oscillator ("clock"). The biological clock is genetically determined. It is therefore a specific *chronometer* that controls the free-running, circadian activity of each individual. The direction of autonomous movements remains constant and is predetermined endogenously by the bilateral structure and the fixed, opposite location of the motor tissue that responds to these signals. All other plant movements are controlled by exogenous (environmental) signals.

Nastic movements are driven by exogenous signals. The direction of nastic movements remains constant, predetermined by their fixed, bilateral motor, as in autonomous movements. Nastic movements are therefore totally independent of the direction of any external signal that controls them. That signal may also be *non directional*. *Light-mediated movements* are controlled by the interaction between the biological clock and L↔D transitions, brief exposures to light, or changes in PFD. *Skotonastic* movements take place in response to a L→D transition, and the reverse *photonastic* movements take place in response to the opposite, D→L transition. In nature, these transitions take place diurnally and repeatedly rephase the circadian oscillations of leaves [2–5] and stomatal guard cells [6] to 24-h diurnal cycles. Leaves, most flowers and inflorescences fold around nightfall, which led to the term "*sleep*", or *nyctinastic* movements.

Tropic movements take place in response to *directional*, or *unilateral* signals. The direction of these movements is tightly linked to the direction of the signal (stimulus), because specific sensors are assigned to specific targets (motor tissues). *Positive and negative tropisms* describe movements towards and away from the source of the signal, respectively. *Phototropic movements* are driven by differential interception of unilateral light (*unilateral excitation*), or by interception of directional (more or less collimated) light at an oblique angle (*vectorial excitation*). The movement is *diaphototropic* when curvature reorients the coleoptile tip, shoot apex and its cluster of young leaves, leaf(let) lamina, flower, or inflorescence to face the light source, and *paraphototropic* when it results in a more parallel orientation to the incident light (reported in laminae of certain leaves). Phototropic movements of leaves, flowers, or inflorescences, whose direction changes throughout the day with the changing position of the sun are *heliotropic* [7], "*sun-tracking*" [8], or "*solar-tracking*" [9].

The distinction between "the action of light in modifying the periodic movements of leaves, and in causing them to bend towards its source" is attributed to Julius von Sachs. "... heliotropic movements are determined by the direction of light, whilst periodic movements are affected by changes in its intensity and not by its direction. The periodicity of the circumnutating movements often continues for some time in darkness ... whilst heliotropic bending ceases very quickly when the light fails." [7].

25.3 Motors

An object can be moved only by applying a force. Plants move their entire body, or parts of their body by means of forces generated in their *biological motors*. Growth-mediated movements are reversible only by compensatory growth on the opposite side, which is possible only as long as the zone of curvature is capable of growth. Their motor is characterized by inherent obsolescence. Turgor-mediated movements do not involve growth and are fully reversible and repeatable throughout maturity. The turgor-mediated motor has a lifetime warranty. Turgor-mediated movements are much less costly than growth-mediated ones in resources and metabolic energy. Both types of movements invest metabolic energy in the transport of osmotically active solutes across cell membranes for expansion (in both) and contraction (in the former), but only the latter invest additional metabolic energy, as well as resources in the biosynthesis of new cell walls and cytoplasmic components.

25.3.1 Motors for growth-mediated movements

Movements of immature leaves and other immature plant organs are all mediated by differential growth, resulting from acceleration, or inhibition of elongation in one flank, that are usually, but not necessarily accompanied by opposite changes in the opposite flank [10]. The resulting growth-mediated curvature changes the spatial orientation of more distal part(s) of the plant, without deforming them. The apical bud, and its cluster of developing leaves/cotyledons may reorient by curvature of their subtending young stem (or hypo-/epicotyl), the leaf lamina may reorient by curvature of its petiole, flowers and inflorescences may reorient by curvature of the stalks bearing them. However, a leaf lamina, or floral organ(s), may also move by deformation, changing their own spatial conformation by differential growth in their own opposite tissues. Growth-mediated movements may be controlled by endogenous signals from the biological oscillator, and by exogenous signals, principally gravity, L$\leftrightarrow$D transitions, directional or unilateral light and temperature alternation, as well as by mechanical stimulation.

Cell growth involves auxin-mediated increase in extensibility of the cell walls, as well as uptake of osmotically active solutes, followed by water [11]. Growth may be inhibited by endogenous substances [12]. None of these processes can be eliminated as the prime target for the signals that drive growth-mediated movements. Darwin and Darwin ([7], Chapter VI) noted that as "... growth is preceded by ... increased turgescence ... it does not appear to be advisable to separate them into two distinct classes". Von Sachs [13] argued that as growth itself is also mediated by turgor, the only difference between growth- and turgor-mediated movements lies in the extensibility versus the elasticity of the cell walls of the motor tissue.

25.3.2 The hydro-electric motor of turgor-mediated leaf movement

The most prevalent and vital turgor-mediated motor is that in stomatal guard-cells. However, mature, as well as younger, expanding leaves of many plants belonging to several unrelated taxonomic groups (principally Leguminosae, Oxalidaceae and Malvaceae) exhibit turgor-mediated movements. Of all the plant organs, only the leaf has evolved specialized tissues that are structurally adapted to facilitate its rapid, turgor-mediated reorientation in space. These tissues are organized in a distinct, well delimited *pulvinus*.

25.3.2.1 Structural features

The pulvinus is strategically located for moving the part of the leaf which it subtends. It forms a flexible joint at the junction between the leaf(let) lamina and its axial support (petiole, rachis, rachilla), but may also be found at the leaf base, or intermediate junctions of pinnate leaves. It acts as *crane*, consisting of a flexible "*pivot*" that connects a rigid stationary "*post*" (rachis, or petiole) to a movable, rigid "*boom*" (midrib of the lamina) and equipped with a *hydraulic motor*, able to develop sufficient *torque* to displace the laminar "*boom*" over considerable angles.

The pulvinus differs distinctly in structure from its neighboring parts on either side, and is clearly delimited from them. It is a short cylinder, consisting of a multi-layered sheath ("*cortex*") of thin-walled, anatomically undifferentiated cells, that surrounds a central vascular core and is enclosed by a single layer of epidermal cells. Each of these tissues is uniquely adapted to the function of the pulvinus in leaf movements, providing the means for moving the leaf(let) lamina, the entire leaf, or its parts [14,15]. The pulvinus is capable of unlimited, fully reversible and repeatable curvature throughout the life of the mature leaf. Curvature takes place by opposite changes in volume of its cortical tissue in opposite sides of its vascular core. The contracting sector becomes concave, its opposite, expanding sector convex. The parenchymatous "*motor*" cells of the pulvinar cortex are structurally adapted to undergo reversible, turgor-mediated changes in volume, by elastically stretching or relaxing their walls, as well as by changing their shape. The low bulk modulus of elasticity of the pulvinar motor tissue (*Phaseolus*) enables the cells to undergo extensive changes in volume in response to small changes in mechanical stress (TP) and is thus conducive for efficient conversion of osmotic work into cell expansion [16,17]. Similar properties have been described in stomatal guard cells [18,19]. Expansion, or contraction of pulvinar motor cells are structurally constrained along the pulvinar axis, as a result of the orientation of the cellulose microfibrils in their walls transverse to the pulvinar axis, but also because they are ellipsoid, stacked in parallel along that axis [16]. In addition, the epidermal sleeve presents mechanical constraints to radial expansion. Transverse corrugation of the epidermis along the pulvinar axis [20,21] contributes to reduce mechanical resistance to axial expansion/contraction of the subtending motor tissue. The central vascular core is formed by coalescence of the vascular tissues of the veins, and separates into a number of peripheral bundles at the transition to the subtending petiole (Figure 1), or rachis. It is flexible, but non-extensible, allowing the pulvinus to curve, without changing its length [22]. Motor cells (of *Phaseolus*) contain well-developed functional chloroplasts [23] and mitochondria, large prominent nuclei, extensive endoplasmic reticulum (rough and smooth), and abundant ribosomes, all indicative of high metabolic activity. The pulvinar epidermis is entirely devoid of stomata, but its cells contain small chloroplasts (Koller and Zamski, unpublished observations). Contracted motor cells frequently exhibit transverse folds in the walls parallel to the pulvinar axis [15,23,24], as well as fragmentation of the large central vacuole into numerous small vacuoles, or vesicles [20,25,26], as in stomatal guard cells [27]. The multivacuolate state is reversed by fusion during expansion. Vacuolar fragmentation may conserve the tonoplast membrane during

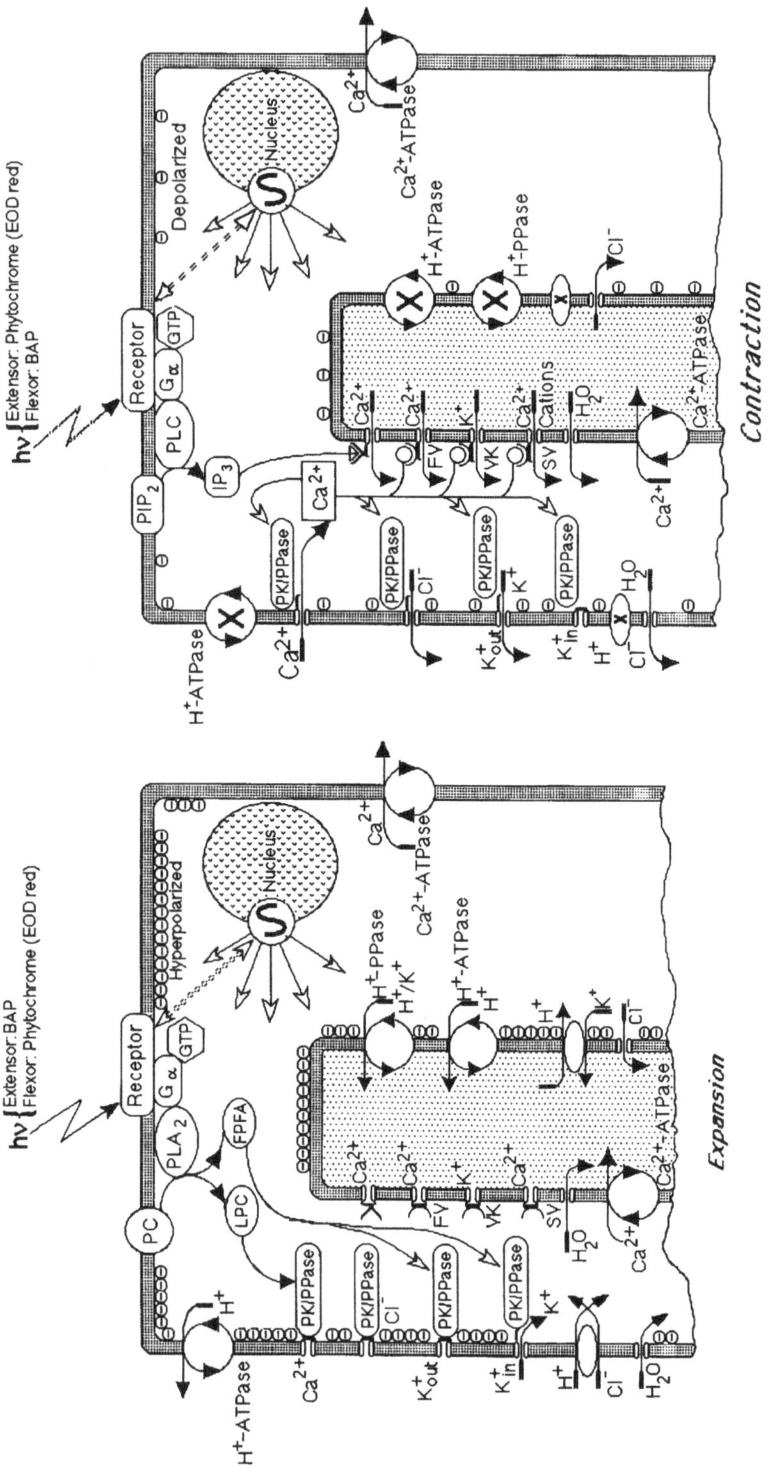

Figure 1. Summary partial model of probable components of the controls of transport processes involved in contraction (right panel) and expansion (left panel) of pulvinar motor cells and/or stomatal guard cells. Details in text. Biological timekeeper shown associated with nucleus. Model was developed in consultation with Dr. Nava Moran and is published with her approval.

the extensive, rapid and reversible changes in cell volume [28]. There is no information on how the plasma membrane copes with such changes.

25.3.2.2 Functional features

The leguminous *pulvinus is an entirely self-contained operational unit*, equipped with its own circadian clock, osmotically-active solutes and the transport systems that move them between its contracting and expanding sectors during curvature [4]. Pulvinar movements take place even when the lamina has been excised [29,30]. Excised pulvini maintain autonomous diurnal movements when transferred to constant environmental conditions, even in darkness. Under such conditions the amplitude of the movements gradually decays, except when the pulvini are supplied with sugar and exposed to red light pulses [5]. Furthermore, although opposite sectors of the pulvinus undergo their (opposite) volume changes with perfect coordination, they do so independently. Excision of either tissue from the pulvinus (*Samanea*) does not prevent the opposite motor tissue in the remaining, intact part of the pulvinus, from continuing rhythmical changes in volume [31]. Partial excision of either flexor or extensor of the pulvinus of the primary leaf of bean does not change the period or the phase of the circadian leaf movements. The excised part of the pulvinus starts to regenerate within 36 h and regenerates completely after 12 days [32]. It remains to be seen how the export and import of solutes and water are managed in the absence of the opposite sector of the pulvinus. Pulvinar motor cells are also self-contained operational units [33]. Protoplasts isolated from pulvinar motor tissue also exhibit circadian oscillations in their volume [34].

A unique feature of pulvinar operation is the *precise physiological coordination* exhibited in the operation of the pulvinar motor tissue. In the terminal pulvinus of the trifoliate leaf of bean, opposite volume changes take place simultaneously in opposite sectors of the pulvinus during its curvature: contraction along one sector matches expansion along the opposite sector. This coupling is supported by the flexibility of the non-extensible vascular core. Export of solutes and water from the contracting sectors and their import into the expanding one take place simultaneously. The volume of water and amounts of osmotically active ions lost from the contracting sector are gained by the opposite, expanding sector. The directions are reversed when the expanding sector contracts and vice versa. [17,22]. The same solutes and water are shuttled back and forth, but different cellular processes are involved in expansion and contraction. The *trans-pulvinar* transport of ions, and water between the contracting sector and its opposite, expanding sector, takes place predominantly through the apoplast [35]. This process starts by bioelectric *trans-membrane* transport of solutes between the protoplasts of the motor cells and their apoplast. Solute influx into the vacuole from the WFS of the apoplast makes Ψ more negative in the former and less negative in the latter. The resulting, ingoing $\Delta\Psi$ gradient leads to uptake of water into the vacuole and expansion of the cell. Solute efflux results in an opposite change in the direction of $\Delta\Psi$ and leads to cell contraction. Most of the water and solutes ($\sim 95\%$) are stored in the vacuole and must therefore be transported across the vacuolar membrane (tonoplast), as well as the cell membrane (plasmalemma). Both membranes are basically impermeable to these solutes, but their transport is carried out by means of specialized trans-membrane proteins.

Autonomous, as well as all types of nastic leaf movements that are turgor-mediated are repeated with remarkable precision in each cycle. The direction of autonomous and nastic leaf movements is predetermined endogenously, by functional bi-lateral symmetry of the pulvinus, and not by the direction from which the (exogenous) signal is intercepted. Pulvinar motor tissue is permanently organized in two opposite sectors that undergo the opposite and reversible volume changes: The extensor expands, while the flexor contracts along the pulvinar axis as the pulvinus curves to unfold the leaf and vice versa [36]. The direction in which the pulvinus curves is species specific, which accounts for the great variety of ways in which leaves of different species fold [7], but may differ in different leaflet of the same compound leaf [22]. In contrast, phototropic movements that are mediated by the pulvinus take place by positive phototropic curvature of the pulvinus, in a direction dictated uniquely either by the direction from which oblique light is intercepted by its lamina (diaheliotropic leaves; see section 25.8.1), or by the direction of unilateral light incident on the pulvinus itself (trifoliate leguminous leaves; see section 25.8.2).

Changes in volume of pulvinar motor cells involve differential changes in components of their water relations. OP of the expanded extensor of *Samanea* is greater than in its contracted flexor and this difference is associated with a more negative Ψ in the former. These parameters exhibit opposite relationships in the contracted extensor and expanded flexor. The (calculated) TP is similar in the two sectors when the leaf is unfolded, but is higher in the flexor than in the extensor when the leaf is folded [37]. However, expressed sap was used for measuring OP, which therefore represents fluid from the vacuole and apoplast. In contrast, OP (measured microscopically at incipient plasmolysis) remains remarkably stable throughout the volume changes that take place during diaheliotropic curvature of the pulvinus of *Malva neglecta* [38]. Similarly, changes in volume of the (abaxial) extensor and (adaxial) flexor of the primary leaf of bean during its circadian movements take place at a relatively stable OP of their cell sap, and are positively correlated only with [K^+]/protein (i.e. per cell) [39]. A linear relationship exists between changes in stomatal opening and in volume of the guard cells, as well as with their K^+-content [18]. These results support earlier findings that "... the concentration of the cell sap remains constant on both ... sides of the joint [pulvinus of *Phaseolus*] during movement" ([40], p. 392). Assays of individual motor cells in the terminal pulvinus of the trifoliate leaf of bean in the course of its phototropic response, by means of a cell pressure probe, showed that although changes in volume throughout pulvinar movement are associated with changes in TP, OP remains stable [17].

The pulvinar motor is controlled by a variety of signals that are transduced initially into electrochemical energy, then into osmotic work, and finally into mechanical work. In absence of environmental signals, opposite volume changes take place in the flexor and extensor motor cells in circadian rhythmicity, controlled uniquely by the ubiquitous circadian oscillator (the biological clock). Light is the most prevalent exogenous signal that controls the pulvinar responses, and the resulting reorientation of the lamina changes its interception of radiant energy. Temperature and water-stress may modify these responses. The circadian oscillator controls the changes in responsiveness of the motor cells to the light signals that cause their expansion/contraction. Clearly, mechanisms controlling trans-membrane transport are the ultimate target of the transduction path initiated by these signals.

25.3.2.3 Operational features

Generation of ion fluxes. In general, ions are transported across the impermeable cell membranes by means of specific, trans-membrane proteins. Some of those act as *carriers* and use either metabolic energy (ATP) to move the ions against their electrochemical gradient by active transport or the electrochemical energy gradient of other ions or organic molecules moving simultaneously, in the same or the opposite direction (*symport* and *antiport*, respectively). Other trans-membrane proteins act as *channels*, through which ions move according to their electrochemical gradient, by *passive transport.*

Proton pumps (electrogenic H^+-ATPases and H^+-PPases) in the cell membranes of pulvinar motor cells and stomatal guard cells are a key component of the mechanism of their reversible volume changes, by contributing to the transport of ions across the membrane [41,42]. Using energy released by ATP hydrolysis they pump protons out of the cytosol and into the apoplast, or vacuole, thereby changing the electrical charge of their membranes. Activation of the proton pumps *hyperpolarizes* the membrane and increases the trans-membrane ΔpH, their deactivation allows the membrane to become *depolarized.* Changes in the membrane potential determine the activity of voltage-gated ion channels in the cell membranes. K^+ and Cl^- are the most abundant osmotically active solutes in motor cells. Their transport through these channels makes the greatest contribution to the osmotic changes and movement of water that result in differential volume changes across the pulvinus and in its deformation [5,17,21,33].

The DFS of walls of pulvinar motor cells, particularly their middle lamella, may serve as a temporary reservoir for cations that are exchanged between the symplast; and apoplast. Activation of the proton pumps in the cell membrane of motor cells acidifies the apoplast, providing the protons to replace these cations and releasing them to be taken up into the symplast. Deactivation of the proton pump enables the reversal of this process. The abundance of fixed negative charges of the carboxylic acid residues of the pectin matrix offers a large capacity DFS for exchange of protons and cations (mainly K^+) with the symplast [35,43–46,17]. In contrast, studies of changes in the elemental composition of pulvinar motor cells of *Robinia pseudoacacia* during leaflet movement showed that K^+ and Cl^- are simultaneously depleted in the apoplast and symplast [47].

Ion channels are much more efficient for transporting electrical charges across the membrane than proton pumps, because ions move through them much more rapidly, by three to four orders of magnitude. Ion channels in the plasma membrane of motor cells and stomata are "*gated*" and can open or close. Changes in the electric charge of the membrane, normally more negative on the cytoplasmic face [21], as well as specific conditions of light, specific hormones, or ligands, particularly Ca^{2+} and IP_3 may change the frequency and/or duration of the open state of the channels by several orders of magnitude.

J.I. Schroeder and co-workers have contributed much to define the roles played by mechanisms that control ion transport processes in volume changes of stomatal guard cells. They have identified voltage-gated K^+_{in} and K^+_{out} channels as major pathways for K^+ movement during contraction and expansion (reviewed by [48], and [49]). They have shown that contraction involves release of anions (mainly Cl^- and malate) through "slow" anion channels, that are activated by increase in $[Ca^{2+}]_{cyt}$ and are controlled by their state of phosphorylation. Furthermore, they have shown that $[Ca^{2+}]_{cyt}$ is controlled by the activity of voltage-dependent Ca^{2+} channels in the plasma membrane, as well as by mobilization of vacuolar Ca^{2+} through the ubiquitous "slow" vacuolar channel. Blatt and Grabov [50] identify $[Ca^{2+}]_{cyt}$, trans-membrane ΔpH and channel-protein phosphorylation as the signaling pathways that control K^+ and anion channel activities during stomatal movement, and suggest that they may integrate stomatal responses to different signals by redundancy. The role of ion channels in signal transduction in guard cells is reviewed by MacRobbie [51].

What follows is an attempt to construct an integrated picture of the transport mechanisms involved in the control of reversible volume changes in pulvinar motor cells and stomatal guard cells. A number of elements are based on circumstantial or indirect evidence, others are hypothetical, but some have a solid factual foundation. Nava Moran contributed much to the concepts included in this picture. Figure 1 integrates these concepts in a model.

Transport processes differ in expansion and contraction, but are basically similar in pulvinar motor cells [52] and in stomatal guard cells [53]. However, all stomatal guard cells in a single leaf exhibit the same response to a light signal, whereas in *skoto-/photonastic* leaves extensor and flexor cells of the same pulvinus exhibit simultaneous, opposite volume changes in response to the same light signals. Extensor cells contract in darkness, probably by similar mechanisms as guard cells, but this does not explain the concomitant expansion of flexor cells, except perhaps as a passive outcome of contraction of the extensor. In *phototropic* pulvini, solutes and water transported out of the contracting sector in response to its exposure to (blue) light are driven through the WFS to the opposite, shaded sector where their arrival apparently triggers uptake and expansion [17].

Cell expansion results from influx of K^+ and Cl^-, accompanied by water. The process starts with activation of the electrogenic H^+-ATPase in the plasmalemma, creating a proton-motive-force (pmf) and increasing the gradient in electric potential (ΔpH) across the plasma membrane. These changes energize the influx of ions, as follows. The pmf hyperpolarizes the negatively-charged membrane, which causes its K^+_{in} channels to open and provide a major pathway for K^+ influx [54], while the ΔpH provides the electrochemical energy for uptake of cations, enabling influx of K^+ through these channels [55]. Apoplastic and cytoplasmic pH affects activity of K^+_{in} and K^+_{out} channels in the plasma membrane of cultured cells of *Arabidopsis* [56]. The influx of K^+ depolarizes the membrane and also reduces the imbalance in electric charge caused by efflux of protons. However, increase in ΔpH also provides the energy for uptake of anions, predominantly Cl^-, into the cell, which acts to hyperpolarize the membrane. Cl^- is transported either through Cl^- channels or by means of an H^+/Cl^- symporter. In the absence of Cl^-, synthesis of malic acid may provide the necessary anions [57]. The increased $[ion]_{in}$ produces an inward-directed ΔΨ, which results in an influx of water from the apoplast.

Cell contraction involves deactivation of the proton pump, depolarization of the plasma membrane and control by $[Ca^{2+}]_{cyt}$. Depolarization of the membrane activates selective

K^+_{out} channels [58,59] when its potential becomes more positive than their activation voltage. Efflux of K^+ alone clamps the membrane potential at the K^+ equilibrium potential, which prevents further, long-term efflux of K^+. The additional, long-term depolarization is provided by activation of Cl^- channels [49], and in particular the S-type (slow) [54], enabling efflux of Cl^- down its chemical gradient, and also stops influx of K^+ by closing K^+_{in} channels [55]. Thus, Cl^- channels play a dual role in contraction, by contributing directly to efflux of Cl^- (and malate) and indirectly to efflux of K^+ through activated K^+_{out} channels [54]. The decreased $[ion]_{in}$ produces an outward-directed $\Delta\Psi$ efflux of water and contraction.

In addition to being activated by depolarization, Cl^- channels and K^+_{out} channels open, and K^+_{in} channels close by phosphorylation. Activation of K^+_{out} channels by depolarization of the plasma membrane may depend on their phosphorylation by a tightly associated kinase [60]. The appropriate state of phosphorylation is achieved by activity of a protein kinase (PK), balanced by simultaneous activity of a protein phosphatase (PPase). These enzymes are activated by an increase in $[Ca^{2+}]_{cyt}$, resulting from import of exogenous Ca^{2+}, or release from endogenous stores (such as the vacuole). Exogenous Ca^{2+} enters via voltage-gated Ca^{2+} channels in the plasma membrane. These channels open transiently upon depolarization of the membrane. The membrane contains a large number of Ca^{2+} channels, but the majority of these channels are quiescent and are not activated by "normal" depolarization. These channels are activated by large pre-polarizing pulses, positive to 0 mV, which also induce recovery of the transient activity of the other channels. This "recruitment" increases with intensity and duration of pre-depolarization. Such modulation might play a role in regulating transport processes that are dependent on $[Ca^{2+}]_{cyt}$ [61,62].

The voltage-mediated, electro-osmotic efflux of Cl^- from its higher intracellular concentration, via anion channels, is a key reaction in contraction, because it is accompanied by efflux of water. The concentration of other ions in the contracting vacuole increases, and as it exceeds the steady state value, their efflux is enhanced, particularly K^+ through opened K^+_{out} channels [63]. Guard cell anion channel 1 (GCAC 1) mediates large, rapid anion efflux and is also an essential element in depolarization of the plasma membrane. Malate, $[Ca^{2+}]_{cyt}$ and nucleotides control the number and/or probability of opening, transport capacity, position of the voltage-sensor, and consequently the voltage threshold of activation of these channels. The voltage range for activity of this channel overlaps that of the K^+_{out} channel [64].

Control of ion fluxes. Under constant environmental conditions (light, dark and temperature) the opposite changes in volume that take place in the extensor and flexor sectors of the pulvinus of *Samanea* are reversed with a circadian periodicity, that is 180° out of phase in the extensor and flexor. This periodicity is expressed in the behavior of K^+_{in} channels in their protoplasts. In uninterrupted darkness, protoplasts isolated from *Samanea* flexor have open channels, while those from the extensor have closed channels, during their circadian "dark" period (leaf folded), and this situation is reversed during the circadian "light" period (leaf unfolded). In both cases, closure of these channels is accompanied by increases in $[IP_3]$. These channels are voltage-gated, under control of the activity of the proton pump in their plasma membrane. This suggests that the rhythmic changes in the state of these channels result from the control by the biological clock over the activity of the proton pumps [65,66]. In the pulvinus of *Phaseolus*, the circadian rhythmical changes in volume of motor cell protoplasts from the extensor and flexor (under constant conditions) are nearly in phase with each other. Volume changes in extensor protoplasts can be synchronized with diurnal light/dark cycles [34]. There is no information on the endogenous signals from the biological clock in its two phases, on the cellular receptors for these signals in the flexor and extensor, or on the transduction steps between them and the rhythmical changes in volume of motor cells that contain them.

Stomata open in response to blue light. Their guard cells expand in blue light, which activates the H^+ pump in their membranes, thereby driving uptake of ions and water. Calmodulin and calmodulin-dependent myosin light-chain kinase are involved in H^+ pumping by guard-cell protoplasts from *Vicia faba* in response to blue light [67]. Light-induced acidification by epidermal leaf cells of *Pisum sativum* is inhibited differentially in red and blue light by substances affecting Ca^{2+}-calmodulin signaling, but a different process activated by blue light is independent of Ca^{2+} [68]. In skoto-/photonastic leaves, flexor cells expand, and extensor cells contract in response to a L→D transition, and this is facilitated by EOD (Section 5.3). Extensor cells expand, and flexor cells contract at the end of nocturnal phase of the endogenous cycle, and this can be advanced by exposure of the pulvinus to blue light. The response of extensor cells to blue light is similar to that of guard cells. In the latter, activation of the enzyme phospholipase 2 (PLA_2) is probably a component of the transduction of the blue light signal to opening K^+_{in} channels and closing K^+_{out} channels, as part of the expansion process. PLA_2 activity produces free, polyunsaturated fatty acids (FPFA) and lysolipids, such as lysophosphatidyl choline (LPC), by hydrolysis of phosphatidyl choline (PC). The LPC produced may activate the proton pump in the plasma membrane, while the FPFA may open of K^+_{in} channels and close K^+_{out} channels. These effects of the products of PLA_2 activity may be direct or by activation of a protein kinase. G-proteins attached to the photoreceptors may become activated by the intercepted light signals, and then activate PLA_2 [69]. The role played by G-protein coupled receptors in signal transduction in plant cells is not clear, but is suggested by presence of G-proteins and of classical downstream signaling elements, such as PLA_2, in plant cells [70].

$[Ca^{2+}]_{cyt}$ may act as a second messenger for a reaction in the leaf movements of *Cassia fasciculata* that involves calmodulin, or other Ca^{2+}-binding enzymes [71]. Stomatal closure in response to a variety of signals is invariably associated with increase in $[Ca^{2+}]_{cyt}$, suggesting that Ca^{2+} acts as a second messenger in the control of ion efflux from stomatal guard cells, by inactivation of the K^+_{in} channel and activation of the slow anion channel in the plasma membrane. Control of the K^+_{out} channel is independent of $[Ca^{2+}]_{cyt}$. Membrane depolarization can activate Ca^{2+} channels, as part of signal transduction [72].

Ca^{2+} plays a major role in the control of ion efflux from pulvinar motor cells by light and other signals. Influx of Ca^{2+} into intact cells is controlled by membrane voltage. It is strongly stimulated by depolarization. The number of Ca^{2+} channels greatly exceeds the requirement for nutrition [73]. Closure of K^+_{in} channels is associated with increase in $[Ca^{2+}]_{cyt}$, which may result from increase in cytoplasmic, soluble $[IP_3]$ [74]. Light stimulates turnover of inositol phospholipids in the *Samanea* pulvinus, resulting in increased $[IP_3]_{cyt}$. IP_3 is apparently produced at the plasma membrane by hydrolysis of

phosphoinositides, such as phosphatidylinositol 4,5-bisphosphate (PIP), catalyzed by phospholipase C (PLC) and possibly activated by a trimeric G protein. Increase in $[IP_3]_{cyt}$ may mediate opening of K^+_{out} channels, as well as in closure of K^+_{in} channels [65,66,75–77]. In stomatal guard cells, increase in $[IP_3]_{cyt}$ activates Ca^{2+} channels in the tonoplast. The resulting elevation of $[Ca^{2+}]_{cyt}$ reversibly inactivates K^+_{in} channels and the H^+-ATPase and activates voltage-dependent depolarizing conductance with a permeability to anions in the plasma membrane. The resulting efflux of Cl^- and K^+, accompanied by water, leads to contraction [78–82]. IP_3 also binds to a specific Ca^{2+} channel (presumably in the tonoplast), that enables increase in $[Ca^{2+}]_{cyt}$ by mobilization from vacuolar (or ER) storage [83]. IP_3 also induces activation of a Ca^{2+}-dependent PPase that participates in the control of phosphorylation of various proteins, such as ion channels [60,84]. On the other hand, exogenous diacylglycerol (DAG) induces stomatal opening. DAG, the other product of phosphoinositide hydrolysis, may act through protein phosphorylation by protein kinase C (PKC), thereby providing the signals that mediate light-induced H^+-ATPase activation and stomatal opening [85]. (DAG-pyrophosphate is a metabolic product of phosphatidic acid during G-protein activation in plant cells [86]). A calcium-dependent protein kinase present in guard cells phosphorylates the KAT1 K^+ channel [87]. The cytoskeleton, by means of its actin filaments, modulates stomatal opening and the associated activity of K^+_{in} channels [88] (see Section 25.6.2).

Where does the tonoplast fit in? The vacuole is the largest compartment of mature cells and provides the major store for osmotically active ions, primarily K^+ and Cl^-, as well as for mobile Ca^{2+}, available for release for signal transduction. The ion pool in the cytosol remains virtually unchanged. Therefore, changes in cell volume must start with transport of solutes across the tonoplast (reviewed by [89]). However, it is not clear whether the control of vacuolar transport processes is direct, or by means of transport processes at the cell membrane (plasmalemma).

Considerable information is available on vacuolar transport processes involved in volume changes of stomatal guard cells. Activation of vacuolar H^+-ATPase (by Cl^-) and H^+-PPase (by K^+) lowers the vacuolar pH and hyperpolarizes the tonoplast, making the membrane potential more negative on its cytosolic face. The resulting proton motive force supplies energy for secondary active transport of anions and organic acids into the vacuole, through ion channels. K^+ is probably imported into the vacuole by H^+/K^+ symport by means of the vacuolar H^+-PPase, but also accumulates via a H^+/K^+ antiport [89–91]. Vacuolar H^+-ATPase activity is greatly enhanced by activation of its closely associated H^+-PPase [92]. Uptake of Ca^{2+} into the vacuole is predominantly by ATP-dependent transport and its release from the vacuole is mediated by mM $[IP_3]$ [93]. Trans-tonoplast voltage is regulated by the transport of K^+ and/or Cl^-. Increasing vacuolar $[Cl^-]$ may activate anion-selective channels, facilitating Cl^- influx [91]. Uptake of Cl^- and malate into expanding guard cell vacuoles take place through a tonoplast channel activated by a calcium- (and ATP-) dependent protein kinase (CDPK) [94]. Studies on the *control by light* of ion transport across the tonoplast of pulvinar motor cells, or of stomatal guard cells, have not been reported.

$[Ca^{2+}]_{cyt}$ plays a crucial role in the release of ions from the vacuole and is maintained at 100–200 nM by ATP-dependent Ca^{2+} pumps, or H^+ gradient driven Ca^{2+}/H^+ antiport

at the plasma membrane or intracellular membranes. Vacuolar ion channels are reviewed by Allen and Sanders [95]. Several types of ion channels have been characterized in the tonoplast: *VK channels* are K^+-selective, voltage independent and activated by increase in $[Ca^{2+}]_{cyt}$ in the physiological range. Mass efflux of K^+ from the vacuole takes place through open VK channels. This shifts the vacuolar membrane potential to less negative values on the cytosolic side, thereby activating voltage-gated ion channels, such as the *SV* (slow vacuolar) *channels*. These channels are highly cation-selective and impermeable to Cl^-. They are gated by $[Ca^{2+}]_{cyt}$, mediated via calmodulin and play an important role in Ca^{2+}-induced release of Ca^{2+} during contraction of guard cells, since their activation allows efflux of Ca^{2+} from the vacuole (permeability ratio for Ca^{2+}:K^+ ~3:1) [96]. IP_3 releases Ca^{2+} from isolated vacuoles of red beet root by voltage-dependent activation of Ca^{2+} channels in the tonoplast [97]. Activation of the tonoplast H^+-ATPase energizes the efflux of K^+, and activates Ca^{2+} channels in the tonoplast. The *FV* (fast vacuolar) *channels* open at low $[Ca^{2+}]_{cyt}$ and are active at physiological tonoplast voltages. These channels are also used for efflux of K^+ from the vacuole during Ca^{2+}-independent guard-cell contraction [49,72]. FV channels are activated by $[Ca^{2+}]_{cyt}$ up to 100 nM, VK channels are activated as $[Ca^{2+}]_{cyt}$ increases beyond 100 nM, while the SV channels are activated when $[Ca^{2+}]_{cyt}$ exceeds 600 nM [98]. Lysophosphatidylcholine and similar phospholipids stimulate proton transport and phosphorylation of tonoplast-specific polypeptides, one of which may be part of the tonoplast ATPase [99]. Ligand-gated Ca^{2+} channels provide a possible mechanism for linking signal perception to intracellular Ca^{2+} mobilization. Three types of vacuolar Ca^{2+} channels that are insensitive to $[Ca^{2+}]_{cyt}$ have been identified in guard cells: one is gated by hyperpolarization of the tonoplast, another by IP_3 and a third by cyclic ADP-ribose (cADPR), a metabolite of NAD^+. The latter releases Ca^{2+} by activating a "ryanoside receptor". None of these are likely to mediate amplification of the $[Ca^{2+}]_{cyt}$ signal, or its long-term duration. However, the first and last of these trigger release of Ca^{2+} through the SV channel, thus indirectly amplifying this signal (77,83,98).

25.3.3 Rapid transport of water across the cell membranes

The large-scale changes in volume that take place in opposite sectors of the pulvinar motor tissue, to cause its movement, result from massive and rapid efflux of water from vacuoles in the contracting motor cells and equally rapid influx of water into vacuoles in expanding motor cells. The bulk of water in cell is in the vacuole, enclosed by the tonoplast. Transport of water during contraction, or expansion of the cell takes place across the tonoplast and plasmalemma, that act as resistances in series to this transport. Thus, flow of water is controlled by the greater resistance. The resistance of these membranes is mainly by their lipid bilayer. However, this resistance may be greatly modified by intrinsic proteins (MIP in the plasmalemma, TIP in the tonoplast) that act as trans-membrane channels, specifically for the rapid transport of water. These *aquaporins* (reviewed by [100], and [101])) have an estimated diameter of 0.3–0.4 nm. Polar groups lining these channels appear to be similar to those of the bulk solution (water). As a result, the osmotic water permeability, or filtration coefficient (P_f) of the membrane exceeds that of its phospholipid bilayer (P_d), and the transport of water takes place at a lower activation energy (E_a) which is similar to that of self-diffusion of water, or of its viscous transport. Aquaporins are generally much more abundant in the tonoplast. P_f of the plasmalemma and tonoplast, measured by osmotic contraction kinetics of microsomal fractions (purified by free flow electrophoresis) from tobacco suspension cells is higher by two orders of magnitude in the tonoplast (690 $\mu m\ s^{-1}$) than in the plasmalemma (6.1 $\mu m\ s^{-1}$). E_a of the tonoplast is about one fifth that in the plasmalemma (2.5 vs. 13.7), suggesting a dominant role for a lipid solubility mechanism for water transport through the plasmalemma. Water transport through the tonoplast is inhibited by mercuric chloride, a specific inhibitor for many aquaporins, suggesting a dominant role for aquaporins [102]. The high P_f of the tonoplast suggests a novel osmoregulatory role for water channels in the vacuole in buffering osmotic fluctuations in the cytoplasm, in case of sudden changes in OP of the apoplast [100,103]. Transport of water in pulvinar motor cells and stomatal guard cells is much more extensive than in tobacco suspension cells. Water permeability of their plasmalemma must therefore be quite considerable. Aquaporins (PIP1) were identified in the plasmalemma of *Arabidopsis thaliana* mesophyll, where they are concentrated in invaginations of the plasmalemma deep into the vacuolar lumen (plasmalemmasomes), suggesting a role in facilitating water transport between the vacuole and apoplast [104]. In mature tissue, both types of aquaporins are expressed in and around vascular tissue and endodermis [101,105]. Aquaporins may be activated by phosphorylation. A major intrinsic protein of the plasma membrane (PM28A) exhibits water-channel activity upon phosphorylation of a serine residue (S-247), that is dependent in vivo on increase in apoplastic Ψ and in vitro upon sub-μM [Ca^{2+}] [106]. There is as yet no information on the regulation of aquaporin activity in pulvinar motor cells, or in stomatal guard cells.

25.4 Nastic, growth-mediated movements synchronized by solar timekeeping

These are the most familiar diurnal plant movements and were already observed by Pliny and by Linnaeus [7]. *Flowers* of water lilies and species of Cactaceae, Oeotheraceae and Oxalidaceae exhibit diurnal cycles of closing by inward curvature of the perianth leaves at their base and opening by unfolding them. *Inflorescences* of Compositae with ligulate ray florets open and close their floral disc diurnally by similar movements of their surrounding bracts. Most flowers and inflorescences are open during daytime and closed at night, but some open at night and close in daytime. Such diurnal movements are growth-mediated, taking place only during (part of) development of the flower/inflorescence (some consist of a single cycle, terminated by senescence), and are generally controlled by the diurnal L$\leftrightarrow$D transitions (skoto-/photonastic), and/or by the diurnal temperature alternation.

In leaves that lack discrete pulvini, diurnal photonastic movements are generally growth-mediated [13,107]. The diurnal, growth-mediated movements of flowers are very spectacular, but the diurnal photonastic turgor-mediated folding and unfolding of leaves are much more widespread and prevalent. In the latter, it is invariably the L$\rightarrow$D transition that causes folding. However, flowers of some plants (such as *Oenothera* spp.) open at night. It must be assumed that the latter close either in response to the D$\rightarrow$L transition, or in a delayed response to the preceding L$\rightarrow$D transition, that took place several hours before. The mechanism of these light-driven diurnal movements, their relationship with the biological clock, their spectral dependence or the location of their photoreceptors are unknown.

25.5 Nastic, turgor-mediated leaf movements synchronized by solar time keeping

25.5.1 Functional features

Pulvinated leaves (particularly of plants belonging to the Leguminosae, or Oxalidaceae) may exhibit diurnal skoto-/photonastic folding/unfolding repeatedly throughout their mature life [7,13,108–110]. These movements continue rhythmically for several cycles in the absence of any exogenous signal, under strictly controlled, constant environmental conditions (even in total darkness, when sucrose is provided). Under such conditions, the cycle length may change gradually to a "free-running", circadian length [5]. Movements of pulvinated leaves in general are therefore controlled by the endogenous biological clock (attributed by Darwin to the "circumnutation" of leaves), in a circadian mode, which is reset daily by the L↔D transitions [111]. The movements resulting from these opposing transitions reverse each other precisely, but the transitions themselves differ operationally in their interaction with the phases of the clock. Skotonastic folding can be induced arbitrarily during daytime, by transfer to darkness, with progressively reduced effectiveness, while photonastic unfolding by the reverse transfer to light cannot take place for several hours after normal folding [112]. The rhythm of stomatal opening of *Phaseolus vulgaris* in continuous darkness is phased primarily by the preceding D→L transition, while that of stomatal closure is phased by the L→D transition [113].

The nocturnal folded configuration and unfolded one during the day are apparently the universal manifestation of the skoto-/photonastic responses in leaves. However, in *Oxalis oregana*, adapted to deep shade, the trifoliate leaves are unfolded during daytime and folded down skotonastically at night, but they also fold down very rapidly in response to an abrupt increase in irradiance (sun-fleck) [114].

The pulvinus contains the complete mechanism for the perception of non-directional light signals and for their transduction to the nastic response [4]. To obtain skoto-/photonastic responses it is necessary, as well as sufficient to expose the pulvinus itself to the corresponding L↔D transition. In a compound leaf, each leaflet responds independently [115–117]. Isolated protoplasts from pulvinar motor tissue to exhibit volume changes in response to light [65,118].

The site of perception of non-directional light within the pulvinus for its photonastic responses has not yet been identified. By virtue of the pulvinar response to non-directional light, the flexor and extensor must independently perceive blue, as well as EOD red light signals and exhibit opposite responses to them. As these light signals need not be directional, excitation of either the flexor, or the extensor must be capable of dictating indirectly the response of its partner. At the same time, in pulvini that also respond photo*tropically*, every pulvinar sector, including its extensor and flexor, must exhibit the same (qualitative) response to blue light. The distinction between photo*nastic* and photo*tropic* responses of the extensor and flexor to blue light is generally overlooked.

25.5.2 *Operational features*

The opposite responses of the extensor and flexor to the biological clock and to L↔D transitions are expressed in protoplasts isolated from them. Rhythmic, as well as light-driven processes in pulvinar motor cells are controlled by activity of K^+_{in} and K^+_{out} channels. Interaction between the biological clock and L↔D transitions controls opening and closing of K_{in} channels in flexor and extensor protoplasts of *Samanea*. Skotonastic folding is associated with hyperpolarization of the (expanding) flexor, and photonastic unfolding with hyperpolarization of the (expanding) extensor. Hyperpolarization takes place by activation of a proton pump in the cell membranes, which enables uptake of K^+ by the cells (and their resulting expansion) by opening K_{in} channels [52].

K^+_{in} channels in flexor protoplasts are open and during the circadian "dark" phase and close upon exposure to blue light, while those in the extensor are closed in the dark and open in light [65]. Influx of K^+ into expanding flexor cells (through K^+_{in} channels) and a concomitant efflux of K^+ from contracting extensor cells (through K^+_{out} channels) take place in response to EOD, which enhances skotonastic folding [119]. K^+_{in} channels in extensor protoplasts that are open in light, closed upon premature transfer to dark, with or without EOD. In contrast, flexor protoplasts that are closed in light remain closed after premature transfer to darkness, but open in response to EOD. Phototransformation of phytochrome enables uptake of K^+ by flexor protoplasts, because it controls the opening of their K^+_{in} channels. This suggests that the enhancement of the nyctinastic folding by EOD is a result of the response of K^+_{in} channels in the flexor cells [66]. Photonastic unfolding of *Samanea* in blue light appears to be associated with the same effects on membrane polarization in extensor and flexor motor cells, but opposite effects on the state of their K^+_{in} channels, expressed by opposite changes in their volume. In darkness, a brief exposure to blue (or white, but not red) light transiently hyperpolarizes membranes in protoplasts from both extensor and flexor. Following hyperpolarization by blue light, protoplasts from extensor, but not flexor, are rapidly depolarized by addition of K^+, indicating that closed K^+_{in} channels in the extensor had opened, while open ones in the flexor had closed. In contrast, addition of K^+ in darkness, or following exposure to red light, results in rapid depolarization of motor cell protoplasts from flexor, not the extensor, indicating that K^+_{in} channels in the flexor remain open in darkness, and are unaffected by EOD. Hyperpolarization by blue light is inhibited by vanadate, suggesting that it results from activation of H^+-ATPase, but the subsequent K^+-mediated depolarization is not similarly inhibited, suggesting that activation of the proton pump is not the sole factor controlling opening of K^+_{in} channels [65]. Studies on ion transport processes involved in autonomous, or photonastic pulvinar movements have focused on K^+. Transport of Cl^- is at least as important.

$[Ca^{2+}]_{cyt}$ may act as a phytochrome messenger in the photocontrol of the ion fluxes that drive turgor changes in pulvinar motor cells. Phytochrome regulates trans-membrane fluxes of Ca^{2+} by controlling the activity of Ca^{2+} channels [120]. Red light regulates Ca^{2+}-activated K^+ channels in the filamentous alga *Mougeotia* [121]. The

phytochrome-regulated swelling of etiolated wheat protoplast (probably responsible for unrolling of the leaf) is Ca^{2+}-dependent, but independent of K^+-uptake, and involves a G-protein [122,123]. Increase of $[Ca^{2+}]_{cyt}$ mimics the effects of EOD on folding in leaves of *Albizzia lophanta* and counteracts its reversion by far-red light. It also mimics the effects of EOD on phase shift of the circadian movement of *Robinia* leaflets. Action of an intracellular Ca^{2+} channel antagonist depends on phototransformation of phytochrome to P_{fr} [124,125]. Ca^{2+} chelators inhibit the nyctinastic folding, as well as the photonastic unfolding responses in the leaf of *Cassia fasciculata*, while a Ca^{2+} ionophore increases their rate (only marginally for the unfolding response). However, Ca^{2+} channel blockers inhibit the EOD-mediated nyctinastic folding, but not their blue light-mediated unfolding, suggesting that activation of Ca^{2+} channels enhances the former, but has no effect on the latter. Ca^{2+} may be mobilized in different ways for these opposite movements, possibly from external sources for the phytochrome response, internal sources for the blue light response [71]. However, studies using Ca^{2+} channel antagonists should be regarded with caution, because these substances may also block K^+_{out} channels [126].

Dark-adapted extensor protoplasts expand upon transfer to light. Increase in $[Ca^{2+}]_{cyt}$ from exogenous sources is required for expansion and takes place via Ca^{2+} channels in the plasma membrane that open in response to the light. Light-induced expansion in these cells is prevented when uptake of extracellular Ca^{2+} is inhibited (by verapamil or La^{3+}), whereas inhibition of intracellular transport (by TMB-8) has no effect. Extensor protoplasts can be induced to expand in the dark, in the presence of Ca^{2+} ionophores (A 23187) or Ca^{2+} agonists (Bay K-8644) or inhibitors of Ca^{2+}-ATPase at endomembranes (thapsigargin). These results suggest that D→L transitions induce opening of Ca^{2+} channels in the plasma membrane. The resulting increase in $[Ca^{2+}]_{cyt}$ from extracellular sources acts as a signal in the transduction chain that ends in light-mediated expansion, because it activates H^+-ATPase and opens K^+ channels, leading to ion uptake. In contrast, contraction of light-adapted extensor protoplasts upon transfer to darkness occurs in absence of extracellular Ca^{2+}, but is inhibited by TMB-8, not by verapamil. The latter is reversed by A23187, or BAY-8644, which by themselves have no effect. Contraction of light-adapted extensor protoplasts in response to an L→D transition is also inhibited by inhibitors of the phosphoinositide pathway for trans-membrane signaling (neomycin, Li^+). The response to such transition probably takes place by inducing PI hydrolysis. The increase in $[IP_3]_{cyt}$ that is generated as a result may mobilize Ca^{2+} from intracellular stores, leading to closing of K^+_{in} channels and activation of outwardly rectifying, voltage-dependent Cl^- channels [65,77,118]. Increase in soluble $[IP_3]$ may activate the proton pump that controls the activity of K^+_{in} channels, by stimulating a protein kinase [84]. The rate of turnover of PI is also controlled by the phases of the circadian clock [74]. Contraction of stomatal guard cells in the dark is associated with closing of K^+_{in} channels and opening of outgoing Cl^- channels, both of which may take place as intracellular Ca^{2+} is mobilized by IP_3 [78,81].

25.5.3 Photoreceptors

Phytochrome and BAP are involved in perception of non-directional light signals that control turgor-mediated, nastic leaf movements. The skotonastic folding response takes place in response to a L→D transition and therefore is not mediated by photoreceptors. However, it is enhanced by EOD, which increases the rate, as well the extent of folding [4,36,127–129]. Enhancement of skotonastic folding by EOD is saturated by brief exposure of the pulvinus to red light at low fluence rates, is reversible by subsequent far-red light and is characteristic of other photomorphogenic responses to phytochrome [129,130]. Photonastic unfolding is probably controlled by BAP and takes place in response to prolonged exposure to blue light (λ_{max} at 470 and >720 nm) at a high irradiance. Blue light is also required to maintain the unfolded configuration [117,128,131–133]. Responsiveness to each of these photosystems depends on the diurnal phase of the circadian oscillator [125,134]. Skotonastic folding of *Albizzia* leaflets is delayed by blue (430–470 nm) and far-red (710 nm) light. Red (660 nm) light and longer wavelength far-red (>730 nm) light are each ineffective by itself, but in combination they delay folding. Green (550 nm) light is also ineffective by itself, but reverses completely the delay in folding caused by 710 nm far-red. These result suggest activity of a photoreceptor with λ_{max} 710 nm, plus broad band activity at $\lambda > 660$ nm [135,136].

The identity of the BAP is unknown. The action spectrum for photonastic unfolding in leaves of *Albizzia julibrissin* and *Vicia faba* (in presence or absence of 660 nm background light) exhibits two major peaks in the blue (440 and 480 nm), falling off sharply beyond 500 nm. 720 nm light is also effective, but after a considerable lag phase. On the basis of its action spectrum, photonastic unfolding was attributed to a BAP, as well as to the HIR of phytochrome [130]. The photonastic unfolding of *Oxalis oregana* leaves following a D→L transition, as well as their rapid downfolding following a sharp increase in irradiance, exhibit identical action spectra (λ_{max} at 450 and 485, a sharp cutoff at $\lambda > 500$ nm and no response at λ 700–2,400 nm). [114]. Clearly, the same photoreceptors must be involved in the opposite responses to light at low and high irradiance. The low irradiance probably excites only the adaxial (flexor) photoreceptors, while the higher irradiance is able to traverse the pulvinus and excite the abaxial (extensor) ones as well. The trans-pulvinar differential in osmotic relations, which maintains the unfolded configuration at low irradiance, collapses at the higher irradiance, resulting in downfolding.

25.6 Growth-mediated movements in search of light

Etiolated and light-grown dicotyledonous seedlings and older plants reorient their apical bud and its complement of young leaves *diaphototropically* toward the predominant direction of light by growth-mediated, positive phototropic curvature of their subtending growing stem. However, certain climbing plants exhibit negative phototropism, by which they locate and become appressed to their support. The young shoot of ivy (*Hedera helix*) exhibits negative diaphototropism, because it curves away from (blue) light to become appressed to vertical supports (walls, trees) [137]. Seedlings of the tropical vine *Monstera gigantea* detect and grow in the direction of the trunk of their host tree from a distance exceeding 100 cm. This response has been attributed to *skototropism*, not to negative phototropism, because growth is the direction of the darkest sector of the horizon, rather than away from the brightest sector, and the response increases with the diameter of the target tree. However, the shoot reverts to positive phototropism when light is too low, and permanently some time after it had started climbing its host, coinciding with the change in the morphology of its new leaves [138]. In contrast, the prostrate shoot of the fern *Selaginella* grows along the ground by differential positive phototropism, greater on its "ventral" underside. Growth of the "dorsal" side of the shoot is also greater in the dark, irrespective of the direction of gravitropic stimulus. The response does not take place in the absence of the small dorsal leaves [139,140]. Growth-mediated, positive phototropic curvature of the stem/hypocotyl, can also take place by long-distance transmission of a *directional light signal* perceived by the foliar organs (vegetative, or reproductive) that it subtends (see Section 25.6.2).

25.6.1 Functional features

Positive phototropic curvature of the stem/hypocotyl, as well as rapid suppression of its elongation, take place by its direct exposure to blue light, *unilateral* in the case of phototropism, *omni- or bilateral* in the case as suppression of elongation. Both responses are *direct results of inhibition of growth of the cells that are exposed to blue light*. Positive phototropic curvature of the stem (hypocotyl) takes place by differential growth, which may result from direct growth inhibition constrained to the cells that intercept light [141]. Phototropic (and gravitropic) signals acting on the stem (hypocotyl) from different directions are integrated in the response [142,143]. Inhibition of growth by blue light has been invoked to account for phototropic curvature in unilateral blue light, which imposes a transverse light-gradient across the stem [144]. However, *the phototropic response is kinetically distinguishable from the general suppression of elongation*: The former exhibits a 4.5 h lag phase, while the latter takes place within 30 s. The 5- to 6-fold difference in fluence rate of (unilateral) blue light measured across the hypocotyl should have caused (calculated) curvature within 30 to 60 min [145]. Support for such separation comes from experiments with seedlings of mustard (*Sinapis alba*) grown under low-pressure sodium lamps, to eliminate growth responses to phytochrome. Exposure to bilateral blue light does not inhibit hypocotyl elongation, but exposure to unilateral blue light results in positive phototropic curvature [141]. While the primary direct response of the exposed cells may be similar, if not identical in blue light phototropism and growth suppression, the differences between them are probably in downstream parts of the transduction pathway [146]. The suppression of stem elongation by blue light is rapid, probably because cell wall extensibility is affected. In contrast, phototropic effects of blue light exhibit a considerable time-lag, probably because they involve translocation of the components of growth (auxin, ions, water), from cells in the exposed flank to those in the opposite flank, and establishment of a differential TP across the stem.

Different mutants of *Arabidopsis thaliana* have been identified each of which exhibiting only one of these responses, respectively, but not the other [147,148]. Therefore, effects of blue light on suppression of stem elongation and phototropic curvature may be controlled by different photoreceptors, and this may also be the case for the stomatal response. The different photoreceptors may nevertheless have the same chromophore (see Section 25.6.3). Detailed discussion of the transduction of unilateral light signals for growth-mediated phototropism is outside the scope of this chapter.

25.6.2 Operational features

Growth of cells that intercept blue light is inhibited. This inhibition does not involve corresponding changes in TP of these cells (measured by cell pressure probe), but is characterized by reduced extensibility of their wall, with little change in its yield threshold. Hydraulic conductivity is too large to limit growth [11,149]. Growth inhibition of cucumber hypocotyls by exposure to blue (but not green or red) light is preceded by a large, albeit transient depolarization of the exposed cells. Such depolarization initially involves inactivation of the plasma membrane H^+-ATPase, with subsequent activation of Ca^{2+} channels, and/or Cl^- channels, allowing these ions to move down their electrochemical gradient [146,150]. An anion channel in the plasma membrane of *Arabidopsis* hypocotyls is activated by blue light [151] and may be a step in signal transduction leading to growth inhibition. Such activation probably takes place via a pathway that is indirectly dependent on $[Ca^{2+}]_{cyt}$, by means of intermediates, such as Ca^{2+}-dependent kinases and/or phosphatases [152,153]. Phototropic curvature in sunflower seedlings is associated with reduction in $[Ca^{2+}]_{cyt}$ and in the activity of calmodulin and of protein kinases in the hypocotyl flank that is exposed to light [154]. Exposure to unilateral blue light generates a directional gradient of protein phosphorylation across the oat coleoptile, reducing it to 32% and 50% of the dark controls, on the exposed and its opposite side, respectively [155].

Growth inhibition in response to blue light may be associated with reorganization of the cytoskeleton. Redistribution of growth during curvature of sunflower hypocotyls (and maize coleoptiles) in response to unilateral light (or gravity) is accompanied by reorientation of microtubules at the outer epidermal wall: transverse to longitudinal along the concave (growth-inhibited) flank, increased transverse along the convex (growth-stimulated) flank [156]. Growth is inhibited and microtubules are reoriented from transverse to longitudinal, or oblique, in individual cells of dark-grown gametophytes of the fern *Ceratopteris richardii* exposed to blue light [157]. The cytoskeleton, and in particular myosin, may take part in sensory functions and in light signal transduction [158,159] (cf. Section 25.3). Furthermore, organization of the cytoskeleton may change dramatically in blue light in processes that are not directly concerned with growth inhibition, but in ways that might suggest such a role. The helical chloroplasts of the filamentous alga *Spirogyra* become tightly coiled where they intercept a microbeam of blue light (λ_{max} at 430, 476 and 500 nm, coinciding with those of phototactic migration of chloroplasts in *Vaucheria*, *Selaginella*, *Lemna*, and *Funaria*), and their response to centrifugation is modified [160]. Microbeam exposure of the filamentous alga *Vaucheria sessilis* to blue light induces localized reticulation of the longitudinal cortical fibers of the cytoskeleton, by forming cross-linkages, resulting in chloroplast aggregation [161]. This response is associated with a light-dependent electric current [162]. Organization of the cytoskeleton may be affected by $[Ca^{2+}]_{cyt}$.

Growth-mediated phototropism may take place by long-distance transmission. This phenomenon is well known in etiolated grass coleoptiles, but is also exhibited by green plants. The shade-evading leaf movements involved in the formation of leaf mosaics in *Sparmannia africana* were attributed to a greater production of auxin in the shaded part of the lamina in the course of leaf development, which results in greater elongation in the subtending flank of the petiole [163]. Directional light perceived by the lamina of cotyledons and/or young leaves may cause positive phototropic curvature of their subtending petiole and/or hypocotyl/stem. Transmission of the signal from the leaves to the stem has been attributed to differential interception of the directional light by leaves/cotyledons growing in opposite inclination on opposite sides of the stem, resulting in differential supply of growth-regulating substances (promotive and/or inhibitory) to their subtending flank. The resulting differential growth on opposite flanks of the stem/hypocotyl is expressed in positive phototropic curvature [12,164,165]. This approach is supported by a number of studies. Hypocotyls of *Helianthus* grown under vertical light, with one of the cotyledons shaded, curve towards the exposed cotyledon. Diffusates from the hypocotyl on the side with the shaded cotyledon exhibit greater growth-promoting activity. In the absence of one cotyledon, the hypocotyl curves away from the remaining cotyledon, but to a lesser extent in light than in darkness [166]. Curvature of the hypocotyl away from the shaded cotyledon has also been attributed to reduced transpiration by the shaded cotyledon, resulting in higher Ψ in the vascular bundles and higher water content of the peripheral tissues on that side [167]. (A similar attempt to attribute phototropic curvature of the stem in unilateral light to enhanced transpiration along its exposed flank [168] has been challenged [169]). Hypocotyls of de-etiolated seedlings of *Cucumis sativus* and *Helianthus* exhibit positive phototropic curvature in response to direct unilateral exposure to blue, but not to red light. The hypocotyls also curve away from the shaded cotyledon when the other is vertically exposed to red light, but curve toward the shaded cotyledon when the other is vertically exposed to blue light. These results suggested direct control by a BAP in the hypocotyl and an independent control by phytochrome in the cotyledons [170,171]. The cotyledons and apical bud contribute to the phototropic response of the hypocotyl when the entire seedling is continuously exposed to blue light, to a greater extent in green than in etiolated seedlings (cress, lettuce, mustard, radish). Shading the cotyledons and apical bud inhibits curvature in etiolated seedlings, but only delays and reduces curvature in green seedlings. In green seedlings curvature corresponds to the level of PFD [172].

Young leaves may also exhibit growth-mediated diaphototropic responses in the course of their expansion. Young, growing leaves of *Tropaeolum* spp. reorient their laminae normal to an oblique light beam by curvature of their subtending petioles toward the light. Positive phototropic curvature of the hypocotyl may also be an indirect result of *vectorial excitation* of its cotyledons by oblique light (see Section 25.8.1.3). Exposure of the petiole itself to unilateral light leads to its positive phototropic curvature, by which its (shaded) lamina reorients to face the light. However, exposure of the lamina to directional light also leads to positive phototropic curvature of its subtending, shaded petiole. Clearly, the petiole exhibits a long-distance phototropic response to vectorial excitation by directional light signals that are perceived in the lamina [173,174]. This conclusion is supported by results showing that detached leaves of *Tropaeolum* and *Limnanthemum* exhibit continuous positive phototropic curvature of the shaded petiole when the lamina is floating on water and continuously exposed to excitation by oblique light [175]. Thus, positive phototropic curvature of the petiole may take place by its direct exposure to unilateral light, as well as indirectly by means of a signal transmitted from the lamina as a result of its exposure to oblique light (*vectorial excitation*). The differential growth of the petiole caused by its direct unilateral exposure is greatly reduced in the absence of the lamina and this is reversed by replacing the excised lamina with auxin [176]. The well-known "compass plants" (such as *Lactuca serriola* and *Silphium laciniatum*) received their name from the predominantly North-South azimuth and rotation of the lamina of their mature leaves, resulting in their vertical reorientation, to face permanently the rising or setting sun. Newly emerged leaves are vertical, with their adaxial face vertically appressed to the stem. Their azimuth orientation is phyllotactic. Laminar reorientation takes place during leaf expansion. The expanding lamina responds to several consecutive cycles of solar transit across the sky by progressively changing its azimuth and/or rotating around its midvein, to face sunrise or sunset ([177,178] and unpublished observations). Presumably, the young lamina responds diaphototropically and rotates axially to face the sun only when the incident light is oblique, transverse to its midvein (sunrise and sunset). Studies on the mechanisms involved in this growth-mediated movement have not been published, but laminar phototropism may be involved (see Section 25.8.1).

25.6.3 Photoreceptors

Flavins are leading candidates for the chromophore of the blue light photoreceptor for phototropism. KI, NaN_3, phenyl acetate specifically inhibit phototropism in response to unilateral blue light (e.g. [139,140]), probably by interacting with the excited state of flavins. Simultaneous irradiation with phototropically inert light also inhibits this response, probably by depopulating the first triplet state of flavins [179]. A soluble protein (cryptochrome 1) has been identified as a flavin-type blue light photoreceptor that mediates blue light-dependent regulation of plant growth and development, specifically hypocotyl elongation in *Arabidopsis* [180]. Responses to blue light may take place by cooperation between a flavin enzyme (such as NADH-dependent oxidoreductase) and a b-type cytochrome in the plasma membrane [181]. Quiñones and Zeiger [182] have provided evidence suggesting a role for the xanthophyll zeaxanthin in phototropism of corn coleoptiles. This suggestion has been refuted by experiments showing that blue light-dependent phototropism, as well as phosphorylation responses to blue light, are the same in seedlings containing normal levels of carotenoids, and in those that are deficient in carotenoids, either through a genetic lesion, or by chemical blocking of carotene biosynthesis [183]. Non-phototropic-hypocotyl mutants have led to identification of the NPH1 locus, that may encode the apoprotein for a dual- or multi-chromophoric holoprotein photoreceptor capable of absorbing UV-A, UV-B and green light, and regulating all phototropic responses. This gene is genetically and biochemically distinct from the HY4 gene that encodes the photoreceptor for blue light-mediated inhibition of hypocotyl elongation. Loci NPH2 and NPH3 appear to act as downstream signal carriers for the phototropism-specific pathway. NPH4 acts in gravitropism as well, and may function directly in the controlling differential growth [184,185]. Blue light activates a kinase that phosphorylates the protein encoded by NPH1 [153]. *Arabidopsis* NPH1 contains a kinase domain, as well as two additional repeat domains (LOV1 and LOV2) that may be flavin-binding and regulate kinase activity in response to blue light-induced redox changes. Similar structural features of proteins with LOV domains exist in totally unrelated organisms, some of which exhibit responses to light, voltage or oxygen [186]. Properties and transduction chains of blue light photoreceptors have recently been reviewed [187]. Light in the blue spectral region is invariably active in growth-mediated phototropism, but $\lambda > 600$ nm are also sometimes active.

Involvement of phytochrome. Vertically growing, young leaves (crozier stage) of the fern *Adiantum cuneatum* exhibit positive phototropic curvature of their midrib in response to unilateral red light, reversible by far red light, as well as in response to blue light, that is not similarly reversible [188]. De-etiolated seedlings of cucumber (*Cucumis vulgaris*) exhibit positive phototropic curvature in continuous exposure to unilateral blue light and negative curvature in continuous exposure to unilateral far-red light. De-etiolated seedlings of the *lh* mutant, deficient in phytochrome B, exhibit the former, but not the negative phototropism mediated by far-red. The magnitude of the negative phototropic response to far-red depends on PFD, and is apparently mediated by HIR of phytochrome B [1]. The negative phototropic response may be an expression of the enhancement of stem elongation by ambient light with a low fluence-rate ratio between red and far-red [189], but kinetic studies [145] may suggest otherwise. Etiolated maize coleoptiles exhibit (time-dependent) 2nd positive curvature in response to unilateral blue light only after exposure to red light, which is reversible by subsequent far-red light [190]. Phytochromes A and B are both required for the phototropic response of *Arabidopsis thaliana* seedlings to blue light [191].

25.7 Solar-tracking by growth-mediated movements

25.7.1 Functional aspects

The capacity for sustained, growth-mediated heliotropic movements is superimposed on the diaphototropic response. The most familiar and conspicuous manifestation of heliotropic movements is that of sunflowers (*Helianthus annuus*). The apical bud and its cluster of young leaves, and eventually its dish-shaped developing inflorescence, move by positive phototropic curvature of the young, growing part of the subtending stem to face the sun with high fidelity during the course of each day, throughout reproductive development. At maturity the flowering heads remain facing in the general direction of sunrise. These movements are unequivocally growth-mediated. The developing leaves contribute to the diaheliotropic response of the stem, since their excision results in partial loss of the response [192]. Positive phototropic curvature of the stem may be accounted for by an inhibitor of stem elongation and of auxin transport that is produced by the young, growing leaves and is translocated more rapidly on the side exposed to light [164]. However, differential interception of directional light by leaves on opposite sides of the shoot, or vectorial excitation, cannot be ruled out (see Section 25.8.1). Diaheliotropic responses have also been observed in the apical buds and their subtending cluster of developing leaves of *Crozophora tinctoria* (Plate 1) and *Xanthium strumarium* (unpublished observations see Section 25.7.2) and in single, open flowers

Plate 1. Growth-mediated (±turgor-mediated?) diaheliotropism of the flowering shoot apices and individual leaves of *Crozophora tinctoria* (Euphorbiaceae) in nature. Post-sunrise (~8 AM, top panel) and pre-sunset (~5 PM, bottom panel) orientation of leaves and apical buds in tagged plant, photographed from a fixed position. Note curvature of stems and petioles.

[193–196], mediated by blue light [197]. The site, or mechanism for perception of directional signals from the sun by plants exhibiting diaheliotropic growth-mediated movements have not been studied.

Nocturnal reorientation. Some time after sunset the apical bud, or the developing inflorescence of sunflower starts reorienting in the opposite direction, and ends up facing in the anticipated direction of sunrise. This nocturnal reorientation is an outstanding feature of these diaheliotropic responses (see Section 25.8.1.3). It is more rapid (~26° h^{-1}) than tracking the sun (at ~15° h^{-1}) during the day, and is completed several hours before sunrise, but the eastward inclination increases at sunrise. In mature plants, the direction of the nocturnal reorientation is maintained even when the preceding day was overcast, and for 3–4 days after the (potted) plant had been rotated 180° around its axis, suggesting endogenous control [198]. In navigational terms, the plant uses the position of the sun to guide the orientation of its apical bud and developing inflorescence by day and an automatic pilot during the night. Endogenous oscillations do not take place in the absence of mature leaves.

25.7.2 Growth-mediated heliotropic responses of leaves

Young leaves of sunflower exhibit diaheliotropic movements even before flower initiation [199]. The (normal) laminar orientation of the leaves lags by ~12° behind that of the sun, but maximum easterly and westerly orientation of leaves precede sunrise and sunset, respectively, by several minutes [200]. The amplitude of the diurnal oscillations (to the West) decreases progressively with flower development through flower opening, and they finally cease at anthesis, leaving the heads fully inclined East. Leaves growing facing East and West reorient diaphototropically by curvature of their petioles and midribs. Those facing North and South do so by axial rotation (torsion) [201]. Expanding leaves of *Stachys sylvatica* [202], *Crozophora tinctoria* (Plate 1) *Xanthium strumarium* and *Hyoscyamus* spp. (unpublished observations) reorient their laminae diaheliotropically by curvature and/or torsion of their petiole and/or of the lamina.

25.8 Solar tracking by turgor-mediated leaf movements

The capacity of mature leaves to continuously reorient their laminae during the day in response to directional signals from the sun (Plate 2) and maintain a constant angle of incidence with its radiation, has been reported in species belonging to diverse taxonomic groups [203–206]. Solar-tracking movements of pulvinated leaves are mediated by turgor. Heliotropic leaves maximize light-use efficiency, because they intercept a nearly constant flux of solar radiation (PAR, as well as thermal) throughout the day, maintaining a constant photosynthetic rate and efficiency of photosystem II (expressed by their chlorophyll fluorescence ratio, F_v/F_m, and photosynthetic oxygen evolution). Leaves with fixed spatial orientation exhibit midday depression in these activities [206]. The mechanism of heliotropic movement of pulvinated leaves has been studied in species of Malvaceae and Leguminosae, whose perception of directional light signals and response to them differ fundamentally.

25.8.1 *Laminar heliotropism*

25.8.1.1 *Perception of directional light*

Laminar heliotropism operates by perception of directional light signals by the lamina. Mature leaves of *Malva* spp., *Lavatera* spp. (Malvaceae) are diaheliotropic. They reorient their laminae throughout every clear day to remain virtually normal to the sun with remarkable accuracy [38,207,208] by curvature of their pulvinus, which sometimes extends to their petioles. Selective shading of the periphery, or center of the lamina (which includes the pulvinus) does not interfere with the diaheliotropic response of the leaf, a result that led to the conclusion that directional light is perceived over the entire lamina, not by the pulvinus [38]. The directional signal must therefore be transmitted from the lamina, probably in a transduced form, to the site of response in the subtending pulvinus. The resulting pulvinar curvature is positively phototropic and reorients the lamina diaphototropically to face the light. However, in certain leguminous plants, pulvinar phototropism may co-exist with laminar phototropism and modify the otherwise normal (~ 0°) angle of light incidence (see Section 25.8.2.5).

Laminar heliotropism does not operate by differential interception of light. Phototropism was discovered and most extensively studied in coleoptiles of grass

Plate 2. Turgor-mediated diaheliotropism of leaves on tagged plant of *Lupinus pilosus* in nature. Post-sunrise (~ 9:30 AM, left panel) and pre-sunset (~ 4:00 PM, right panel), photographed from a fixed position.

seedlings and subsequently also in hypocotyls and epicotyls of dicot seedlings, exposed to unilateral light. In all these phenomena, light direction is perceived as differential interception by the exposed and its opposite, shaded sectors of quasi-cylindrical organs. Perception of directional light by the quasi co-planar lamina of mature phototropic leaves presented a challenge to this concept. Several studies have been addressed to other mechanisms by which the leaf lamina may perceive directional light signals, without invoking differential interception. Specialized cells in the upper leaf epidermis of certain leaves act as optical lens [174,209] and it has been suggested that they contribute to perception of oblique light by focusing it on specific receptive areas in the cytoplasm. To test this hypothesis, it was assumed that the refractive index of water was similar to that of these epidermal cells and that covering them with a flat layer of water may eliminate, or weaken their lens effect. One half-lamina of *Tropaeolum* was covered with water under a thin sheet of mica, and the opposite half left uncovered and dry. When the opposite halves of the lamina were exposed to equivalent, but opposite oblique beams, the lamina reoriented towards the oblique beam incident on its dry half, supporting this hypothesis [174,210]. However, previous results by Kniep [211] showed that the lamina reorients towards the light even when the lens effect was similarly eliminated over its entire surface by means of paraffin oil, suggesting that focusing of directional light by the lens-shaped epidermal cells could not explain the perception of oblique light in the diaphototropic response [209]. Moreover, lens-shaped cells are not a common feature in the upper epidermis of phototropic leaves. Attempts were also made to account for laminar perception of directional light by invoking differential interception of oblique light by leaves/cotyledons inclined in opposite, or otherwise divergent azimuth angles [163,166]. Similar reasoning was the basis for invoking non-planar topography of the laminar surface, resulting in local differences in angle of light incidence and a differential pattern of interception of light. These differences become progressively accentuated as the angle of incident light is more oblique and change with its azimuth angle. On this basis, perception of directional light signals in the lamina of *Lavatera* was attributed to increasingly differential interception of PAR by opposite surfaces on either side of the vein as the azimuth angle of the light beam diverges more from that of the vein. The resulting pattern of assimilate partitioning is therefore differential and these differences are somehow transmitted to different sectors of the pulvinar motor tissue, causing the pulvinus to bend [212]. This hypothesis was eventually retracted because the vascular connections did not conform to the prediction [213]. Nevertheless, CO_2 [214] and a certain level of photosynthetic activity [215] may be required for expression of the diaheliotropic response.

Plate 3. Top panel: Leaf lamina of *Lavatera cretica* (palmate venation). Bottom panel: Final laminar orientation in leaves of *Lavatera cretica* following continuous diaphototropic response to a constant level of vectorial photo-excitation (light beam maintained tip-oriented at +30° for the older leaf, base-oriented at –30° for the younger, opposite leaf). Young distal part of petiole of older leaf responds phototropically to its exposure to the tip-oriented beam. Same part of younger leaf is not exposed to light and shows no response. [216].

25.8.1.2 The concept of vectorial excitation

The lamina of diaheliotropic species of Malvaceae forms an incomplete, nearly circular disk, with the pulvinus at its center, from which the major veins diverge palmately in azimuthal directions that differ from each other by ~50° (Plate 3). When one half the lamina of *Lavatera cretica* is shaded and its opposite half is exposed to light, the lamina reorients when the light is oblique to the surface, not when the light is normal to it, a result that is incompatible with the concept of differential interception of light. These results suggested that directional light, incident on the lamina at an oblique angle, is perceived by the lamina as *vectorial excitation* [208]. To validate this assumption, the lamina was exposed to an oblique light beam that was continuously displaced along the median plane of the lamina to maintain a constant angle of incidence with the lamina, as it moved in response to that light. Under such conditions, laminar reorientation proceeds at a constant angular velocity ($>15°\ h^{-1}$ required to track the solar transit see Figure 6) [216]. The diaheliotropic laminar reorientation continues with high fidelity throughout the day in nature [217], as well as under simulated conditions, by means of a "solar simulator" (see Figure 5). The normal to the lamina trails the moving oblique beam of "sunlight" by a minimal threshold angle [218]. [A similar lag characterizes growth-mediated, diaheliotropic leaf movements in sunflower (see Section 25.7.2).]

Selective shading of either the central, or the two opposite lateral sectors of the lamina showed that the diaheliotropic response takes place only if the directional signal of the oblique beam is oriented along the exposed sector(s) and not at all when oriented transversely to it (Figure 2). The only obvious structurally directional tissues in the lamina are those associated with the veins. Selective shading of laminar strips along the major veins (of cotyledons, for technical reasons), or of the areas between them, showed that perception of the directional signal is associated with photoreceptors that are probably located in as yet unidentified cell files along the major veins [208]. Leaves from which the lamina was excised, with the exception of the mid-sector, responded diaheliotropically to an oblique beam along the plane of symmetry of the midrib, supporting this conclusion. These results led to the hypothesis that the photoreceptors

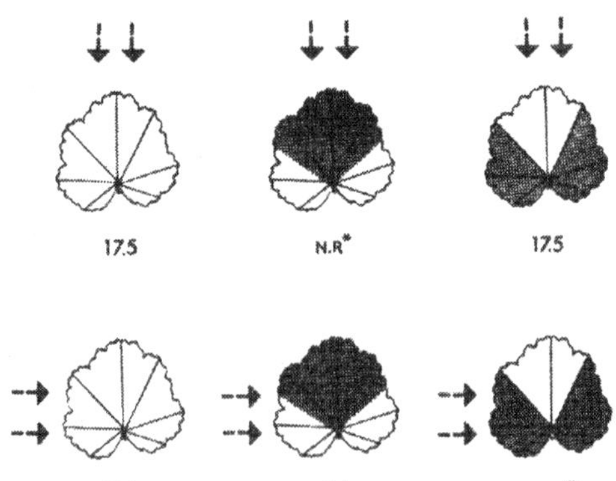

Figure 2. Initial identification of vectorial excitation. Reorientation of lamina of *Lavatera cretica* in response to base-oriented (top row) and lateral (bottom row) light beam initially incident on the lamina at −45°. Figures show initial rates (deg h^{-1}). N.R.* no response (lamina remained within 3° of initial) [208].

are oriented anisotropically, with their transition moments aligned preferentially parallel to the median plane of the vein. The array of photoreceptors associated with the vein oriented along the oblique beam determines the azimuth direction of the response, presumably because they are the ones that are maximally excited. The lamina tracks solar azimuth, as well as elevation. Accordingly, the diaheliotropic response to light at a constant fluence rate should also be a function of angle of light incidence. To validate this assumption, the response (steady-state angular velocity of reorientation) of the lamina was measured under continuous vectorial excitation over a range of constant PFD and constant angle of light incidence. Under such conditions, the response depends on PFD, as well as on the angle of interception of light by the lamina, confirming the vectorial nature of the excitation. The response increases sharply with increase in the angle of incidence, from zero when the light was normal to the lamina (angle of incidence = 0°), peaking between 40° and 50°, but decreases progressively at greater angles. As the intercepted fluence rate is proportional to the cosine of angle of the incident light, this angular dependence suggests that the photoreceptors are not located in a paradermal cytoplasmic layer, but in the layer adjacent to the transverse walls of the cell files along the vein. Analysis of the angular dependence suggests that *the transition moments of the photoreceptors are inclined at some preferential angle to the transverse cell walls, in addition to being preferentially parallel to the plane of symmetry of the vein.* Yin's results with partial shading suggest that the photoreceptors are present along the entire length of the vein.

Vectorial excitation along the median plane of the midrib at equivalent, but opposite angles, results in equivalent, but diametrically opposite responses: increase, or decrease in LE when the beam is directed towards the tip of the vein, or its base ("tip", or "base oriented"), respectively (Plate 3), at equivalent angular velocities when vectorial excitation is maintained constant (see Figure 6). These results suggest that the array of photoreceptors along each vein is equally capable of perceiving tip-oriented and base-oriented vectorial photo-excitation. According to the model that has evolved from these studies, the opposite responses to opposite vectorial excitations could be accounted for by assuming that *the photoreceptors that are immobilized at or in the cell membrane, adjacent to transverse walls of each cell along the file, have an opposite orientation of their transition moments at the proximal and distal poles* (Figure 3). Directional light signals are therefore perceived as differential excitation of the photoreceptors at opposite poles of each cell along the file. This differential may create a potential gradient between the two poles of each cell, as well as across the junction between neighboring cells, which may be expressed as a signal current. Since the pulvinar response is opposite when the vectorial excitation is in opposite directions (base-, or tip-oriented), *the direction of the (presumptive) potential gradient is determined by the direction of the vectorial light signal.* This signal is transmitted along the cell file to the target in the pulvinus. The vascular bundles of the major veins coalesce to form the vascular core of the pulvinus, but each maintains its identity and continuity. This enables the signal generated by excitation of the photoreceptors along a vein to be transmitted selectively to its subtending (target) sector of pulvinar motor tissue, where it is transduced into osmotic activity by controlling trans-membrane transport of ions and water. The signal has opposite consequences for its target sector when the excitation is tip-oriented (expansion), or base-oriented (contraction) [216,219].

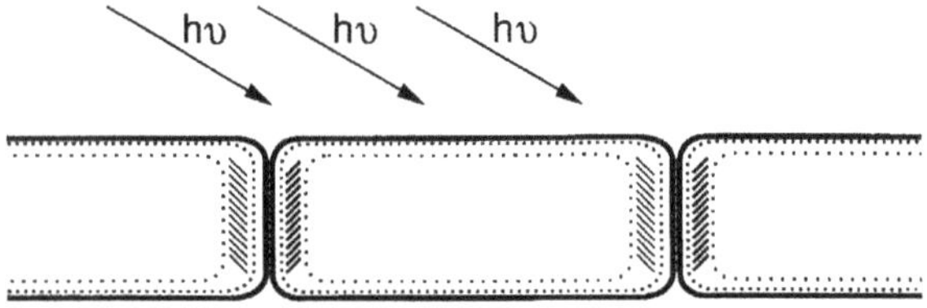

Figure 3. Hypothetical organization of photoreceptors for vectorial photo-excitation in the leaf lamina of *Lavatera*. Perception of light azimuth is by anisotropic arrays of photoreceptors located in a cell file(s) along each of the major veins that radiate in the lamina in different directions (Plate 3). Photoreceptors are immobilized in cytoplasm (dotted lines), adjacent to transverse walls, with their transition moment ~ along the plane of symmetry of the vein, at a ~ similar angle with the wall, opposite at the opposite poles. This arrangement is compatible with perception of angle of light elevation, as well as with differentiation between tip- and base-oriented light. Thickness of photoreceptors indicates level of excitation [220].

Laminar diaheliotropism exhibits action dichroism. The concept that diaheliotropic responses in leaves of *Lavatera*, resulting from vectorial excitation, depend on anisotropic arrangement of immobilized photoreceptors in the lamina is supported by action dichroism exhibited by using plane-polarized light for vectorial excitation. Preliminary kinetic analysis showed that the integrated net response (angular velocity of laminar reorientation) of leaves of *Lavatera* to two simultaneous, diametrically opposite vectorial excitations (transverse to the median plane of the midrib) is proportional to their PFD ratio, and that the lamina is capable of discriminating between opposite directional light that differs by as little as 10 percent in PFD. This relationship provided the basis for studying the dependence of the responsiveness to vectorial excitation with polarized light on the plane of its polarization. When the lamina is similarly exposed to two diametrically opposite vectorial excitations by polarized light, with identical PFD, one with its plane of polarization parallel to the plane of symmetry of the midvein ($\parallel$) and the other transverse to that plane ($\perp$), it invariably reorients towards the ($\parallel$) beam. When the fluence rate of the more effective ($\parallel$) beam is reduced, the rate and direction of laminar reorientation are linearly related to its ratio with respect to the opposite beam ($\perp$), and the two beams are balanced at a $\parallel/\perp$ ratio of 0.62 (Figure 4). The action dichroism could not be accounted for by dichroic optical properties of the lamina [220]. Similar results were obtained with the diaheliotropic leaf of *Lupinus palaestinus* [221].

25.8.1.3 Functional features

Nocturnal reorientation. An outstanding feature of the diaheliotropic responses of malvaceous leaves is their nocturnal reorientation (remarkably similar to that of the apical bud of sunflower see Section 25.7.1). After sunset, leaves of *Malva neglecta* start to reorient their sunset-facing lamina, to end facing the direction of the anticipated sunrise several hours before it occurs. Nocturnal reorientation is the dark part of diaheliotropism, because its direction is predetermined by that of the preceding sunrise. Plants that are rotated by 180° (or 90°) at sunset persist in their original direction of nocturnal reorientation to face the preceding sunrise [38]. Facing directly opposite to the new sunrise, their adaxial face is not exposed to the sun for several hours, during which they are unable to start their daytime heliotropic movement. As a result, the leaves resume normal nocturnal reorientation to face the "new" sunrise only after the plants had adapted to their new position during several cycles. These results were confirmed

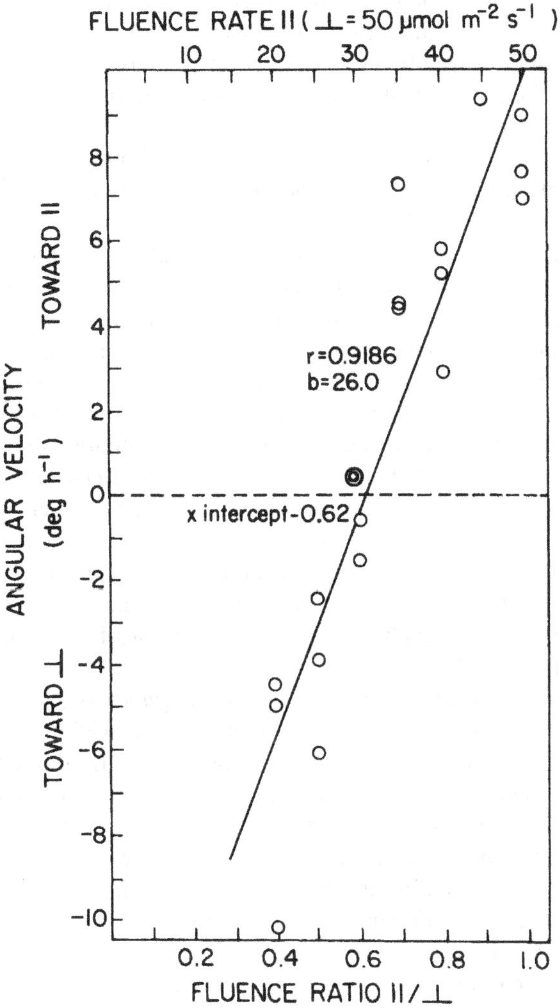

Figure 4. Action dichroism in the diaphototropic response of the leaf of *Lavatera cretica*. Lamina was simultaneously exposed to vectorial excitation (AI = +30°) from opposite sides, transverse to the midvein, using light polarized in the vertical plane of the beam (II) and transverse to it (⊥), respectively. Angular velocity of laminar reorientation was measured over a range of PFD of the II beam, when PFD of the ⊥ beam was maintained constant (50 μmol m^{-2} s^{-1}) [220].

and expanded in studies with the related *Lavatera cretica*. Nocturnal reorientation of the lamina consist of three phases:

1. pulvinar relaxation from the strained, sunset-facing configuration (duration of this phase depends on the extent of laminar displacement required),
2. pulvinar equilibrium (time measuring) and
3. reorientation to face sunrise.

Cotyledons acquired the capacity for nocturnal reorientation after the seedlings had performed 3 to 4 cycles of diaheliotropic movements under simulated conditions [217]. Nocturnal reorientation was observed in time-course studies of several consecutive cycles of diurnal diaheliotropism under simulated conditions, by means of a "solar simulator" [218] (Figure 5).

After-effects of vectorial excitation. Sustained diaheliotropic response of *Lavatera* leaves requires continuous vectorial excitation, but the lamina continues to move in the same direction for some time after transfer to darkness. Eventually, laminae previously exposed to base-oriented excitation continue to decline downward, whereas those previously exposed to tip-oriented excitation reverse their upward inclination, to decline downward as well, to a low steady state. The immediate laminar movement following the L→D transition results from inertia of the trans-pulvinar mass transport of water and solutes, rather than from residual excitation of the photoreceptors, but the subsequent decrease in LE appears to be an inevitable, dominant feature of this transition [222]. Laminar reorientation reverses its direction after the direction of vectorial excitation is reversed, but exhibits the inertial after-effects of the preceding excitation. The rate (angular velocity) of the reversed reorientation is enhanced considerably (by ~40% and ~70% for a base- and tip-oriented excitation, respectively). The enhancement appears to be a result of change(s) at the site of perception. It depends on the preceding opposite vectorial excitation (Figure 6), but is not a result of adaptation to the preceding diaphototropic movement, or of the preceding mechanical stress of the pulvinus [223].

Vectorial excitation can be transmitted beyond the pulvinus (cf. Sections 25.6.2 and 25.7.2). The hypocotyl of seedlings of *Lavatera cretica* exhibits positive phototropic curvature in response to its direct exposure to unilateral light, reorienting the (shaded) apical bud and cotyledons passively to face the light. Exposure of the cotyledons to vectorial excitation results in a diaphototropic response, reorienting their laminae normal to the light by curvature of their pulvini, while the (shaded) hypocotyl remains upright. When the entire seedling is exposed to directional light, positive curvature of the hypocotyl satisfies the diaphototropic requirement of the cotyledons, without pulvinar intervention. In explants of seedlings (hypocotyl and a single cotyledon), exposure of the lamina alone to continuous vectorial excitation results in curvature of the shaded petiole, as well as of the (shaded) hypocotyl. The hypocotyl curves in the direction determined by the directional light: toward the cotyledon or away from it when the excitation is base- or tip-oriented, respectively [224]. These results suggest that the signal generated by the vectorial excitation of the lamina may cause differential growth of the hypocotyl by the same changes that drive the turgor-mediated pulvinar curvature, namely by driving fluxes of solutes and water from its concave to its convex flank (see

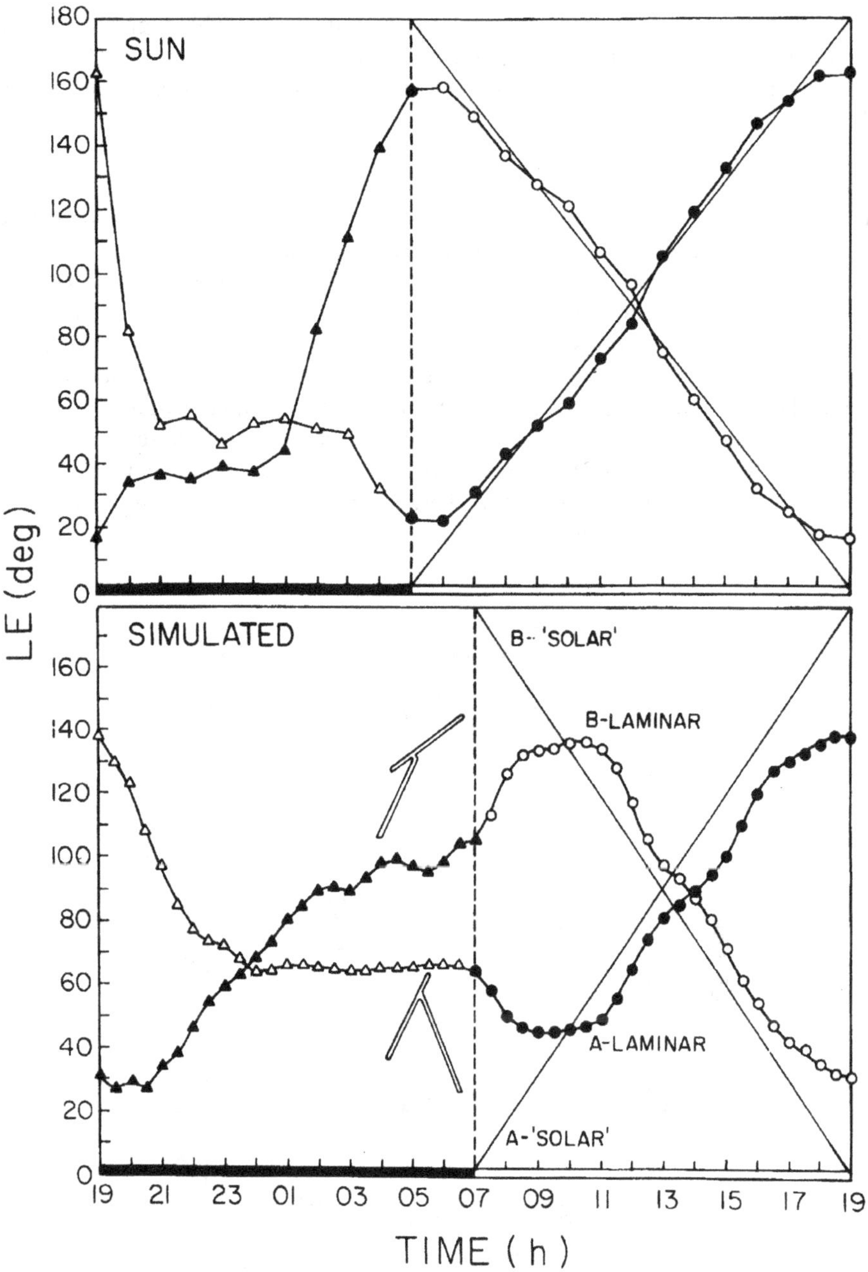

Figure 5. Diurnal laminar reorientation of the diaheliotropic leaf of *Lavatera cretica* in nature (upper panel daytime 5:00–19:00 h) and in response to simulated "sun" directed on lamina and rotated in a vertical arc at an angular velocity of 15° h^{-1} (lower panel "daytime" 7:00–19:00 h). Time-course of laminar reorientation (LE): Group A leaf azimuth facing sunrise, daytime (△) and nocturnal (▲) LE. Group B leaf azimuth facing sunset; daytime (○) and nocturnal (●) LE [218].

Section 25.3.2.2). This concept is supported by instances where growth-mediated curvature in response to unilateral stimulation by light [141] or by gravity [225] results from elongation along the convex flank, coupled with *contraction* along the concave flank.

25.8.1.4 Structural and operational features

Structural studies of the pulvinar response have been reported in *Lavatera* [15] and *Lupinus* [24]. The transduction pathway between vectorial excitation of the lamina and the pulvinar response has not been studied. Differential volume changes in the motor tissue of the contracting and its opposite, expanding sector of the pulvinus apparently result from simultaneous trans-pulvinar transport of solutes and water between them. OP of the cell sap in the contracting and expanding motor tissue of *Malva* remains unchanged during pulvinar curvature [38]. Trans-membrane and trans-pulvinar transport of ions and water probably operate by similar mechanisms to those in leguminous pulvini (see Section 25.3.2.2). However, the transduction of the light signals must include additional downstream elements, between the laminar site of signal perception and that of the pulvinar response.

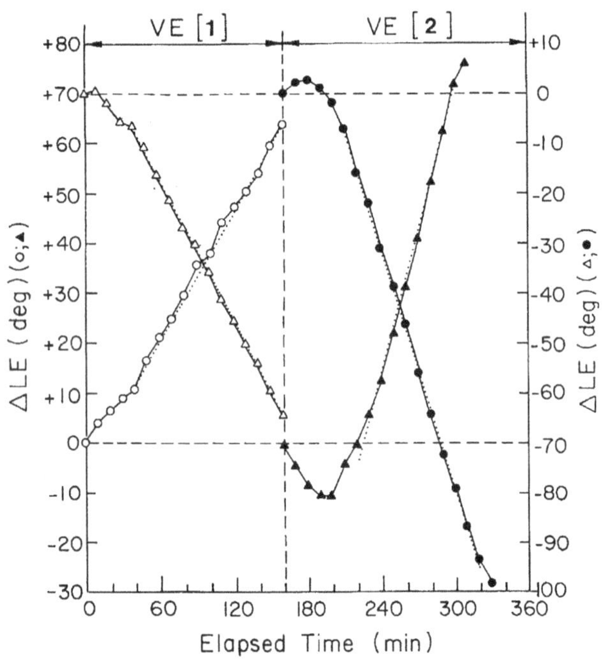

Figure 6. Laminar diaphototropism in *Lavatera cretica* under a constant vectorial excitation in the vertical plane of the mid-vein (AI +30° ○, △; AI –30° ●, ▲). VE[I], VE[II] initial, and subsequent excitation in the opposite direction (empty and full symbols, respectively). Dotted lines linear regressions (r ≥ 0.997). Angular velocity (deg h^{-1}) ratio VE[II]/VE[I]: for AI +30° 55.7/31.4 = 1.78; for AI –30° 42.0/30.1 = 1.40. Note inertial reorientation following reversal of excitation. [223].

25.8.1.5 Spectral dependence

Detailed action spectra for laminar heliotropism are not available. Continuous vectorial excitation with blue light is required to drive laminar heliotropism [38,208,226]. Exposure of the diaphototropic leaf of *Lavatera* to directional red light is equivalent to absence of vectorial excitation. An osmotic balance is established across the pulvinus, which reorients the lamina normal to its petiole [216]. Leaves of *Malva neglecta* respond diaphototropically to blue light, even at relatively low PFD, but the response requires additional PAR in excess of the light compensation point. The rate of response increases progressively with PFD of total PAR. A role for photosynthesis-mediated translocation of the directional signal from the leaf to the pulvinus has been suggested [215]. Normal levels of CO_2 are required to sustain the response to vectorial excitation [214], supporting this suggestion.

25.8.2 Pulvinar phototropism

Pulvinated leaves of many leguminous plants growing in the open exhibit movements that are phototropic because they are clearly affected by the direction of incident light during the daily transit of the sun. Such movements are superimposed on their nycti-/photonastic movements of solar time-keeping.

In phototropic leguminous leaves, perception of the directional light signals, as well as the response to them, are localized in the pulvinus at the base of the leaf(let) lamina, not in the lamina itself. This conclusion is based on selective shading of the pulvinus or the lamina, from directional light, as well as on the pulvinar responses of debladed leaves to directional light [227–230]. Pioneering studies with the primary leaf of *Phaseolus multiflorus* and the pinnate leaf of *Robinia pseudacacia*, showed that unilateral exposure of the upper (adaxial) or lower (abaxial) surface of the pulvinus to light results in an increase or decrease of LE, respectively, while lateral exposure results in rotation of the lamina toward the light [29,30]. Similar responses to adaxial and lateral exposure to unilateral light, and their pulvinar consequences for the movement of the attached lamina, were observed in leaflets of several trifoliate leguminous species and may therefore combine changes in *azimuth* and *axial rotation* of the lamina with changes in its *elevation* [227,229,231–233]. Clearly, *laminar reorientation in directional light is a passive result of the (modified) phototropic response of the subtending laminar pulvinus* to differential interception of (unilateral) light. Complex laminar reorientation result when phototropic responses of the pulvinus combine curvature with torsion. There are as yet no adequate explanations for the mechanism of torsional rotation [234].

Leaves exhibiting pulvinar phototropism are neither diaphototropic nor paraphototropic. Trifoliate legumes exhibit leaf movements that generally appear diaheliotropic in the morning and late afternoon, and paraheliotropic, to varying extents, around midday [235–237]. It is widely accepted that such movements are primarily diaheliotropic, and are modified to paraheliotropic when the leaf experiences water stress or stress by high irradiance or high temperature [199,235,238–242].

With few exceptions, reports of dia-, or paraheliotropic movements in leaves [203,204] have only been supported by field observations. The leaf of *Dolichos lablab* has been designated paraheliotropic, but in contrast with all diaheliotropic species, its azimuth remains stationary throughout the changes in the solar azimuth [206]. The trifoliate leaf of *Erythrina herbacea* is downfolded at night. After sunrise, the leaflet laminae incline upwards well beyond the horizontal and also rotate axially. This orientation remains virtually unchanged until nightfall, exhibiting no dependence on the changes in light angle throughout the day [243]. The leaf exhibits an apparent diaphototropism in the morning and afternoon, and apparent paraphototropism around midday, despite the fact that its lamina remains stationary. As we shall see, laminar reorientation in *Melilotus alba* does not correspond to changes in angle of light incidence [237], suggesting that pulvinar phototropism is incapable of producing genuine dia- or paraheliotropic laminar reorientation.

25.8.2.1 Functional features

The phototropic response of the pulvinus to unilateral light from any direction (Plate 4) suggests that the *pulvinar photoreceptors for the phototropic response are located in every sector.* Exposure of the adaxial, or abaxial sectors to equivalent light treatments may result in responses of different magnitude [22,30,229,237], suggesting that the distribution or effectiveness of these photoreceptors or their capacity for interception of light, may differ in different pulvinar sectors. The pulvinus responds phototropically to *differential interception* of light in its opposite sectors. Selective shading of the adaxial, abaxial or lateral surfaces of the light-adapted pulvinus (*Glycine max*) results in laminar reorientation or rotation towards the opposite side [230]. Conversely, phototropic curvature of the pulvinus (*Phaseolus*) by exposure of one of its faces to unilateral light is reversed when its opposite face is simultaneously exposed to similar unilateral light. These observations led to the conclusion that *the pulvinus integrates the phototropic responses of different sectors*, so that its curvature represents their net effect [22] as in hypocotyls (see Section 25.6.1).

Phototropic responses of the pulvinus to directional light may be extensively modified by the pulvinar topology. The photonastically light-adapted leaf(let) lamina of *Phaseolus* is horizontal, but is supported by a pulvinus that is inherently curved upwards. As a result, the concave (adaxial) face of the pulvinus may be shaded by the lamina from base-oriented directional light, in which case the lamina will not reorient upward and may remain horizontal until the incident light changes its direction sufficiently to be intercepted by the pulvinus [227]. This conclusion is supported by

→

Plate 4. Pulvinar phototropism in *Phaseolus*. **Terminal leaflet**. A–D. Profile view before (A, C) and after continuous exposure of the pulvinus to adaxial light (from above) (B, D). A, B pulvinus (rachis on left); C, D leaflet. E same pulvinus following simultaneous exposure to abaxial light (from below). F–J top view of pulvinus (F, G rachis on top) and front view of the leaflet (H, J) before (F, H) and after (G, J) continuous lateral exposure of the pulvinus to a horizontal beam from right. **Lateral leaflets**. Front view of leaflets before (K) and after (L) continuous exposure to adaxial light (terminal leaflet excised; rachis stump facing observer). M–O abaxial view of pulvini before (M) and after their continuous exposure of their front (N) or rear (O) to light. P–R lateral view of single leaflet before (P) and after continuous exposure of its pulvinus from the front (Q) or rear (R) to light (other leaflets and parts of the lamina excised to access light [22]).

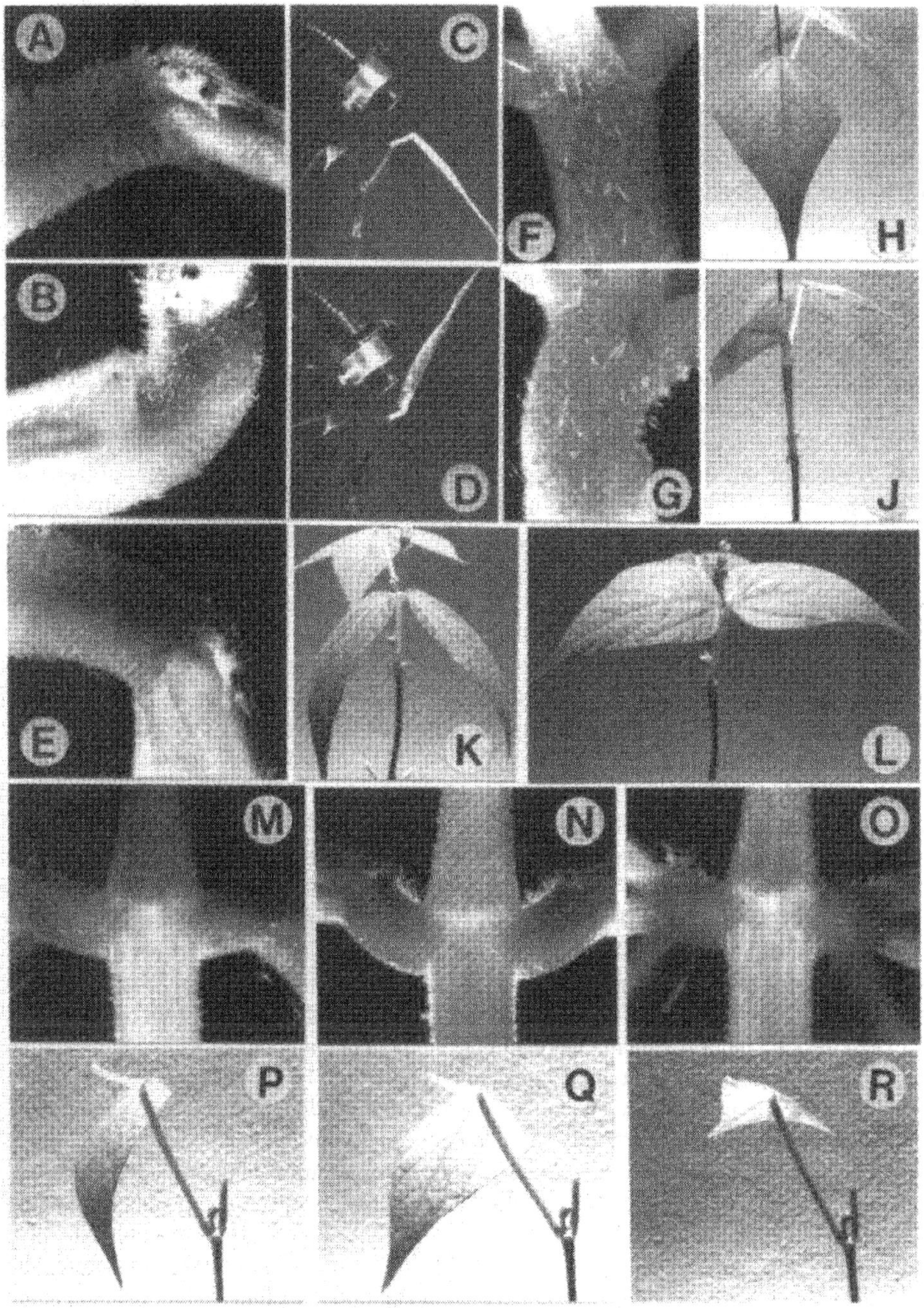
A
C
F
H
B
D
G
J
E
K
L
M
N
O
P
Q
R

studies with the light-adapted, horizontal terminal leaflet of *Melilotus alba*, exposed to directional light at various angles in a vertical arc along its midrib. LE does not exceed the horizontal unless the angle of light incidence enables interception of light by the upcurved pulvinus [237]. Moreover, the pulvinus undergoes topological changes in the course of its phototropic curvature. As we shall see, such changes may affect its interception of directional light, resulting in feed-back control of laminar orientation. Studies with an optical microprobe showed that penetration of (blue) light into the adaxial motor tissue in the pulvinus of *Glycine max* is affected by the angle of light incidence, as well as by the topology of the pulvinus [244].

Quantitative kinetic analysis of the phototropic movements of the individual leaflets of the trifoliate leaf of *Phaseolus vulgaris* in response to adaxial, abaxial or lateral exposure of their pulvinus to light showed that the rate of laminar reorientation depends on angle of the incident light, as well as on changes in pulvinar topology in the course of the pulvinar response to light. The dependence on pulvinar topology provides feed-back control of the eventual steady-state of the response. Adaxial, or abaxial exposure of the *terminal* pulvinus results in upward, or downward curvature, respectively. LE increases or decreases, in both cases reorienting it less normal to the light. Lateral exposure to light results in positive curvature (laminar azimuth changes), combined with torsional axial rotation. The change in azimuth reorients the lamina less normal to the light, while its concomitant rotation towards the light reorients it more normal to the light. Pulvini of the *lateral* leaflets are inherently curved, as well as rotated to the front (away from the leaf base) and this superimposes additional torsional rotation over their positive phototropic response to exposure from all directions. The resulting laminar reorientations reflect these asymmetries (Plate 4 [22]). Individual leaflets of a compound leaf move independently. In trifoliate leaves the pulvinar axes of each leaflet are at right angles to each other, and each pulvinus intercepts the same directional light differently. Moreover, since the topology of pulvini of the lateral leaflets also differs from that of the terminal leaflet, when the entire trifoliate leaf is exposed to directional light from the front, rear or side, the different leaflets exhibit independent, different phototropic responses. These aspects are usually overlooked in studies of heliotropic movements of leguminous leaves and do not support the concept of their *dia-* or *para*phototropic movements. The consequences of the differences in topology and azimuth of the pulvini for their heliotropic movements were studied in the trifoliate leaf of *Phaseolus vulgaris*, exposed to a transit of a simulated "sun" in a vertical arc (12 h at 15° h^{-1}), transverse or along its major axis. Independent movements of the individual leaflets take place during different intervals of the "solar" transit and their fidelity of "solar" tracking is considerably less than in malvaceous leaves exhibiting laminar heliotropism [245].

The phototropic responses of the pulvinus of *Phaseolus* to unilateral blue light from any direction (see Section 25.8.2.3) suggest that BAP photoreceptors are present in (or under) its entire surface and are not confined to its extensor and flexor. In nature, the overhead flux of light almost invariably exceeds the opposite flux, even in the absence of direct sunlight, and the lateral fluxes will be very similar to each other. The effect of the blue light intercepted by the more exposed sector of the pulvinus (the flexor, expanded in darkness) are balanced by the opposite effect in its opposite side (the extensor, contracted in darkness). Under these conditions, the pulvinus integrates the light signals and the leaf unfolds photonastically. The pulvinus starts to respond heliotropically when any of its sectors intercepts (blue) light at PFD that disrupt this balance.

25.8.2.2 Operational features

Phototropic curvature of the pulvinus is a result of transport of ions, primarily K^+ and Cl^-, out of the contracting sector and into the opposite, expanding one [17]. Therefore, mechanisms of ion transport that cause the contraction and expansion of pulvinar motor cells of skoto-/photonastic leaves probably also operate in those of phototropic pulvini. However, light-driven nastic and tropic responses of the pulvinus differ fundamentally.

1. Light-driven nastic responses may be controlled by brief exposure to light of the appropriate wavelength at relatively low PFD [65], whereas the phototropic responses are driven by continuous exposure at high irradiance, and exhibit dependence on PFD.
2. Nastic responses involve opposite changes in *specific flexor and extensor sectors* of the pulvinus, whereas tropic responses invariably result from contraction of *any pulvinar sector* that is exposed to light and in concomitant expansion of its opposite sector [22].
3. The phototropic response of the intact pulvinus is not reflected in the response to light of protoplasts isolated from it. Protoplasts isolated from the abaxial (extensor), or adaxial (flexor) sector of the pulvinus of *Phaseolus* exhibit opposite responses to increase in irradiance: expansion of the former and contraction of the latter [246], but in the intact pulvinus, exposure to light of either of these sectors results in its contraction, with concomitant expansion of the opposite, shaded sector [22].

Information on the mechanism by which continuous excitation of the BAP causes and maintains the differential activities across the pulvinus is fragmentary. The following sequence of events probably takes place in the course of phototropic curvature of the pulvinus in *Phaseolus*. Continuous exposure of motor cells to blue light of sufficient irradiance affects critical components of their membrane transport, which causes efflux of ions from them into their apoplast WFS. Water follows these ions into the apoplast (TP decreases, but OP remains unchanged) and creates a trans-pulvinar gradient in hydrostatic pressure across the apoplast WFS, to the opposite sector. The resulting bulk flow of solution changes the ionic environment as well as Ψ of the apoplast WFS in the opposite sector. These changes in the apoplast of motor cells in the opposite sector affect their membrane transport, enhancing ion uptake, followed by water into them (TP increases, but OP remains unchanged), and result in their expansion [17]. The latter suggestion is supported by expansion of protoplasts isolated from extensor and flexor sectors of *Phaseolus* in response to increase in $[K^+]_{out}$ [247]. In stomatal guard cells, K^+_{in} channels may function as a K^+-sensing valves, that open whenever the K^+ gradient is inward, allowing K^+ uptake only [49]. The activity of K^+ channels depends on the voltage of the membrane, which in turn is affected by activity or inactivity of its H^+-ATPase. Phototropic responses of the pulvinus of *Phaseolus* are inhibited by activation of H^+-ATPase, as well as by its inactivation, and by TEA, a K^+ channel blocker [248], suggesting that inactivation of the proton pump in the contracting sector, as well as its continued activity in the opposite, expanding one, take part in establishing the turgor differential for the phototropic response of the pulvinus. Furthermore, contraction of the extensor or flexor of the pulvinus of *Phaseolus* when exposed unilaterally to blue light is associated with depolarization of their motor cell membranes, supporting the concept that the phototropic pulvinar response to blue light results by inhibition of H^+-ATPase activity, causing loss of the electrogenic component of membrane potentials. As a result, H^+/K^+ fluxes are suppressed and TP falls. Even a brief exposure to blue light causes transient depolarization and transient alkalization of the extracellular pH (greater in the extensor) as well as a transient change in LE, and these changes increase with irradiance [249,250]. These responses differ from those reported above for *Samanea* [65]. However, whereas pulvini of *Samanea* and *Phaseolus* are both photonastic, only the latter are phototropic as well. This contradiction may therefore be explained by a photo*tropic* response of the latter. Continuous adaxial exposure of the primary leaf pulvinus of *Phaseolus* to blue light changes the electric potential difference between the motor tissue in the exposed and shaded sectors from –40 to +20 mV, in parallel with the change in LE. This change in potential difference between the contracting and expanding tissue may be important for the movement of ions between them [251]. (Light induces a PFD-dependent electric polarity across the leaf lamina of *Elodea canadensis*: acidification of the abaxial side by means of active proton efflux and concomitant alkalization of the adaxial side by passive OH^- efflux. The electric potential difference results in net cation flux from the abaxial to the adaxial side [252].) Exposure of the *Phaseolus* pulvinus to red light causes membrane hyperpolarization, that is inhibited by DCMU, while the ionophore CCCP depolarizes the cells and prevents further light-induced changes. Motor cells depolarize under conditions of anoxia and do not respond to blue light under such conditions. Exposure of the anoxic pulvinus to red light leads to progressive recovery (hyperpolarization) of the membrane potential and of its capacity to depolarize transiently in response to a unilateral blue light pulse. Under anoxia, unilateral exposure to red light results in negative phototropic curvature, presumably by enhancing hyperpolarization in the exposed side. These results support the conclusion that the primary effect of blue light is depolarization of the hyperpolarized membrane, by inactivating a H^+-ATPase, the activity of which depends on a supply of ATP from respiration. DCMU completely inhibits the hyperpolarization by red light, but has no effect on the transient depolarization by blue light pulses. The proton ionophore CCCP depolarizes the membrane and progressively reduces the magnitude of the transient depolarization by blue light pulses, suggesting that the source of ATP may also be photophosphorylation, probably by pulvinar chloroplasts ([250] cf. [23]). Nishizaki [250] concludes that unilateral blue light induces a decreasing gradient in depolarization of motor cells across the pulvinus, resulting in greater contraction of the exposed sector than in its opposite. However, only the exposed pulvinar sector contracts in the course of phototropic curvature, while its opposite, shaded sector expands [22]. Blue light activates a $[Ca^{2+}]_{cyt}$ and ATP-dependent anion channel in the plasma membrane of mesophyll cells of pea leaves. This may be an early step in the transduction pathway. Efflux of Cl^- through this channel depolarizes the membrane, which promotes leaf growth by activating a proton pump, H^+ efflux, hyperpolarization and uptake of K^+ [253,254]. If the Cl^- channel also exists in the plasma membrane of pulvinar motor cells, it may be involved in the phototropic pulvinar response (as well as the photonastic response of its flexor), by enabling the massive efflux of Cl^- from motor cells upon exposure to blue light.

25.8.2.3 *Modification of pulvinar phototropism by stress*

Pulvinar phototropism of leguminous leaves may be strongly modified when the leaf experiences a variety of stresses. The most frequently cited modification is by *high irradiance intercepted by the pulvinus*, by which the apparent diaheliototropic movement is transformed into an apparent paraheliotropic one. The magnitude of the response corresponds to the excess irradiance [233,255,256]. However, LE is a function of the (adaxial) PFD intercepted by its pulvinus. Therefore the lamina reorients normal to the light over some critical, intermediate range of irradiance, but progressively away from the normal as irradiance is *either greater or smaller* ([22]; cf. [229]). The observed change from dia- to paraheliotropic orientation in response to high-irradiance stress is only apparent. Heliotropic leaf movement in *Phaseolus* are progressively modified by increase in *ambient air temperature*, under constant atmospheric humidity and [CO_2] or mesophyll [CO_2]. Selective exposure of the pulvinus to radiant heat showed that the effects of air temperature take place only when the pulvinus is exposed to light and depend uniquely on its tissue temperature, but are not mediated by effects of temperature on leaf Ψ, transpiration, or leaf conductance [257]. LE of primary leaves of bean under overhead light is greater at higher PFD and increases progressively with leaf water stress, expressed by increasingly negative leaf Ψ [233,255]. Phototropic curvature of excised pulvini of bean increases progressively with temperature [246].

Temperature and water stresses modify leaf movements, but only in light [257]. It may therefore be assumed that these stresses enhance the processes by which ions are transported from the exposed to the shaded sector of the pulvinus. This enhancement may involve stress-mediated increases in levels of abscisic acid (ABA) in the exposed sector. Abscisic acid inhibits circadian leaf movements of *Oxalis regnelii*, suggesting effects on membrane permeability [258]. ABA inhibits the blue light-dependent proton pumping and depolarizes the plasma membrane in protoplasts from stomatal guard cells of *Vicia*. This is a first step of contraction in daytime in response to elevated ABA, as an indirect result of inactivation of the plasma membrane H^+-ATPase, and/or inhibition of the blue light signaling pathway [259]. Increase in ABA promotes stomatal closure in light by increasing the efflux of ions from pulvinar motor cells, possibly by increasing levels of IP_3, acting as a messenger in cascades leading to opening of K^+_{out} and anion channels (primarily Cl^-) and possibly closing K^+_{in} channels in the cell membrane of guard cells [53,260].

25.8.2.4 *Modification of turgor-mediated laminar heliotropism*

The palmately compound lamina of *Lupinus arizonicus* exhibits nyctinastic down-folding and true diaheliotropic movement during daytime (Plate 2). In the absence of water stress, laminae of the leaflets remain co-planar and normal to the sun throughout most of the day. In water-stressed leaves, the entire compound lamina maintains its diaheliotropic movement normal to the sun, but during several hours around midday the individual leaflets increase their LE to the same extent. In this "cupped" configuration, the leaflet laminae intercept sunlight at the same acute angle and therefore at the same reduced level [8]. The similar leaf of the related *L. succulentus* responds to directional light only if its center (from which the individual leaflet pulvini diverge) is not shaded, leading to the conclusion that perception of directional light signals was localized in the pulvinus of the individual leaflets [261]. However, the leaf of the related *L. palaestinus* exhibits sustained diaheliotropic reorientation when exposed to constant vectorial excitation, even when a circle of the lamina centered on and including the pulvinar crown is shaded from the oblique beam. The magnitude of the response (angular velocity of laminar reorientation) is reduced as the diameter of the shaded circle extends to cover a greater laminar segment (Figure 6). Shading the leaflet pulvini at the center of the lamina does not interfere with the sustained diaheliotropic response, but causes the characteristic skotonastic downfolding of the leaflets. Conversely, the diaheliotropic response takes place when vectorial excitation is confined within such a circle, but is reduced as the diameter of the exposed circle becomes smaller (Figure 7). The diaphototropic response also takes place when vectorial excitation is confined to a narrow ring at the base of the leaflets, leaving in shade the pulvinar crown, as well as the more distal parts of the leaflet laminae. Therefore, at least in *L. palaestinus*, vectorial excitation is perceived in the basal part of the lamina of each leaflet, not in the pulvinus and must be transmitted to the pulvinar site of response [262]. A similar role for the laminar base is also suggested by results with selective shading in leaflets of *Phaseolus* [228] and *Macroptilium* [229]. Individual leaflets of *Lupinus palaestinus* exhibit diaphototropic responses when their pulvinus is exposed, by removing their neighbors. In the intact leaf of *Lupinus*, the leaflet pulvini are arranged side-by-side in a tight "crown" over the juncture with the petiole, so that their lateral flanks are shaded by their neighbors. As the entire lamina moves to remain normal to the sun, only the adaxial surfaces of the pulvini, facing the center of the crown, remain exposed to direct sunlight (Figure 8 [24]). When strong light is piped obliquely into the center of the pulvinar crown by means of a shielded optical fiber and the entire lamina is exposed to *vectorial* excitation, the entire lamina responds to the latter, but the leaflets exhibit "cupping", by simultaneous and equivalent increase in LE [262], probably by a simultaneous phototropic response to increase in irradiance.

25.8.3 Photoreceptors

The spectral dependence of pulvinar phototropism suggests control by blue as well as red light. Pulvinar phototropism in leguminous leaves is driven by continuous unilateral excitation by blue light [30,226,228,229,235], λ_{max} 420 > 470–490 nm [230]. The action spectrum for the light-pulse induced membrane depolarization of pulvinar motor cells in *Phaseolus* exhibits a major peak at 460 nm and two lesser peaks at 380 and 420 nm. No activity was observed at wavelengths < 360 and > 520 nm [263]. Prolonged exposure of the pulvinus of *Phaseolus* to blue (450 nm) or far-red (722 nm) light results in proton efflux, increase of osmotically active solutes, a more negative water potential and expansion of the extensor. Similar changes take place in the flexor in response to continuous exposure to red (671 nm) light. The conclusion that phytochrome is present in both sectors of the pulvinus, but the BAP is only in the extensor [264,265], is incompatible with the positive phototropic response exhibited by the pulvinus to adaxial, abaxial as well as lateral exposure to blue light [266]. However, the responses to blue light at low and high fluence rate may be mediated by different photosensory systems. The fluence-response relationships for the phototropic response of light-grown seedlings of *Fagopyrum esculentum* to blue light exhibit reciprocity at fluence rates $< 0.2\ \mu mol\ m^{-} s^{-1}$ and dependence on total fluence at higher levels [267].

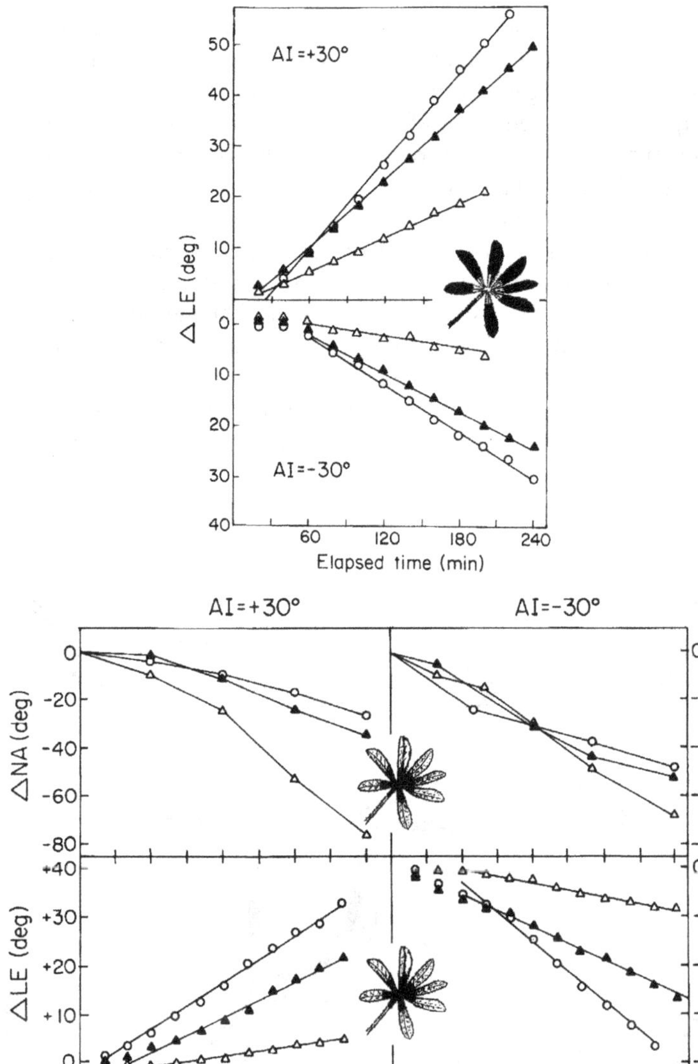

A

B

Figure 7. Perception of vectorial excitation in the diaheliotropic leaf of *Lupinus palaestinus* is localized in the basal part of individual leaflets. Time-course of changes in overall LE (ΔLE) and of change in nyctinastic downfolding ("cupping") of individual leaflets (ΔNA) under a constant level of vectorial photo-excitation in the vertical plane of the mid-vein (AI +30°). A Beam confined to a circle, centered on leaflet junction; no downfolding (ΔNA = 0°); leaflets remained co-planar. B Beam excluded from circle, centered on leaflet junction. Diameter of circle 11 (○), 6 (▲), or 3 mm (Δ). [262].

The rate (initial angular velocity of increase in LE), as well as the eventual steady-state (LE) of the terminal leaflet of *Phaseolus* in response to continuous pulvinar exposure to overhead blue light depend on PFD. However, the effectiveness of unfiltered white light equals or exceeds the effectiveness of blue light at equivalent PFD (200–800 μmol m^{-2} s^{-1}). The response to blue light is enhanced progressively by supplementary red light at increasing PFD. Adding red light to blue light is more effective in accelerating the initial response than adding blue light at equivalent PFD, whereas adding blue light is more effective in increasing the steady-state LE. The pulvinus also responds to overhead exposure to red or far-red light, at high PFD (red – 500 μmol m^{-2} s^{-1} far-red – 880 μmol m^{-2} s^{-1} between 700 and 800 nm, 1,760 between 700 and 2400 nm), but at substantially lower rates than in blue light (at 50 μmol m^{-2} s^{-1}). The kinetics of the response to red light differ qualitatively from those to far-red light. However, the kinetics of the response to red light, alone or during enhancement of the response to blue light, remain unchanged in the presence of far-red

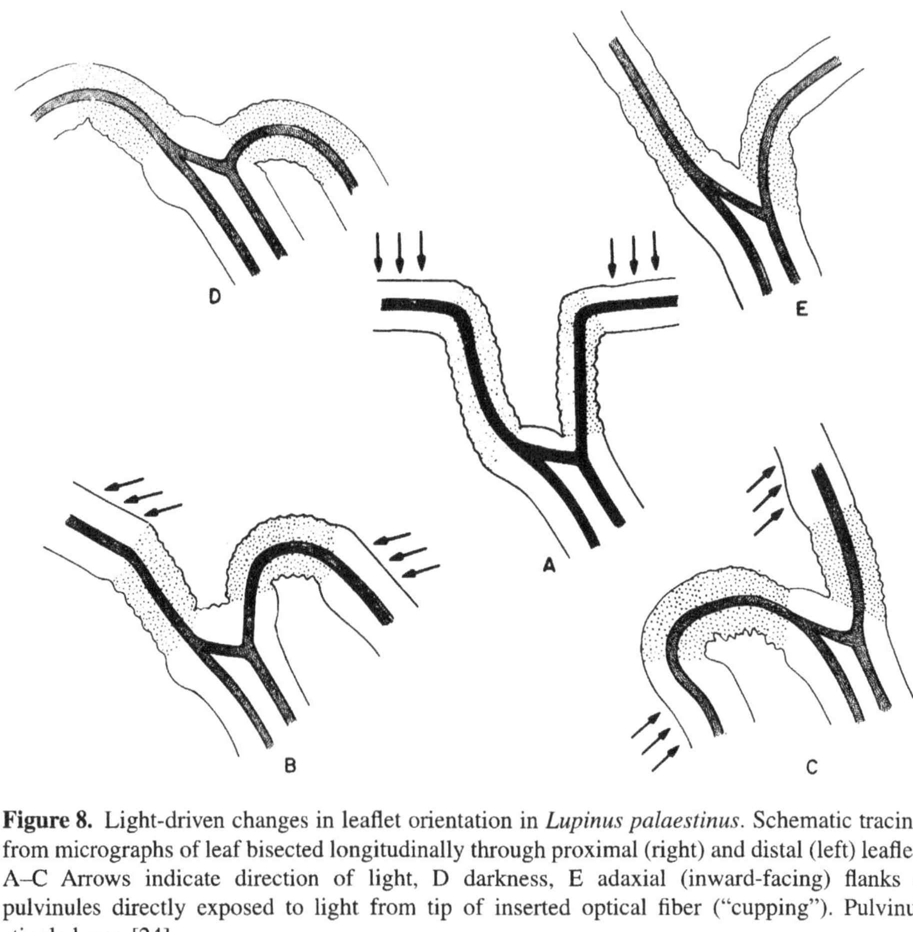

Figure 8. Light-driven changes in leaflet orientation in *Lupinus palaestinus*. Schematic tracings from micrographs of leaf bisected longitudinally through proximal (right) and distal (left) leaflets. A–C Arrows indicate direction of light, D darkness, E adaxial (inward-facing) flanks of pulvinules directly exposed to light from tip of inserted optical fiber ("cupping"). Pulvinule stippled area [24].

light. In contrast, the response to blue alone, or enhanced by mixture with red, was partially inhibited by simultaneous exposure to far-red. The kinetics of the of the *inhibition* of the response to blue plus red by simultaneous far-red light are similar to those of the *positive response* to far-red alone. The results suggest that the response to blue resulted mostly from a BAP, but may involve some absorption by phytochrome, while responses to red or far-red, with and without blue, may be mediated by HIR of phytochromes A and B ([266,268]; cf. [269,270]). HIR involvement in leaf movements and in flowering of long-day plants has been attributed to interaction between phytochrome and an unidentified "*heliochrome*" [271]. Co-action between phytochrome and BAP is widespread [272,273]. It is exhibited in inhibition of hypocotyl elongation of de-etiolated castor bean [274]. Opening of the apical hook in etiolated seedlings of *Arabidopsis* is stimulated by red as well as by low-fluence blue light, apparently acting by excitation of phytochrome, because it is inductive, reversible by far-red and exhibits reciprocity. Hook opening is also stimulated by prolonged exposure to far-red, apparently acting by means of a HIR and by high-fluence blue light, apparently acting by means of a BAP. HIR is apparently not involved in the stimulation by red light, presumably because phytochrome B is absent in etiolated tissue [275]. Excitation of phytochrome by red light and of a BAP by blue light controls cell expansion in leaves of *Phaseolus*, and H^+ efflux, hyperpolarization and cation uptake in epidermal strips from pea leaves. In the latter, the effects of near-saturating red and blue light are additive. Expansion is enhanced in continuous irradiation, with peak activity in blue (460 nm) and red (660 nm) light, whether photosynthesis is active or inactivated. Activity of red, but not blue light is reduced in presence of far-red (730 nm) light [276,277]. Although excitation of either photoreceptor by light enhances leaf growth, their transduction follows different paths. Growth in continuous exposure to blue light is greater and exhibits a significantly shorter lag than in red light. However, the effects of red light on acidification of the leaf surface (normally associated with growth) greatly exceed those of blue light. Inhibition of cation uptake by DCMU in red and blue light suggests a contribution by photosynthesis to growth [278].

Functional differences between red and blue light become apparent in the pulvinar responses to abaxial or lateral red and blue light, separately and in combination. Abaxial, adaxial and lateral blue light, cause positive phototropic responses. LE increases in adaxial exposure and decreases in abaxial exposure, while lateral exposure causes laminar rotation and azimuth change toward the light, without affecting LE. In contrast, abaxial and lateral red light only increase LE, as overhead red, but have virtually no effect on laminar rotation, or azimuth. Whereas the response to adaxial blue light is enhanced in the presence of red light, the opposite response to abaxial blue light is reversed in the presence of red light. In contrast, lateral red, which by itself has no effect on laminar azimuth and rotation, enhances the azimuth change, but not the rotation caused by lateral blue light. Lateral blue, which by itself has little or no effect on LE, enhances the increase in elevation caused by red light. This analysis suggests that red controls the photonastic unfolding of the pulvinus, whereas blue controls its phototropic responses. These responses co-exist in the same tissue, but are separate and additive. Exposure to red light has opposite effects on motor cells in the extensor (expansion) and flexor (contraction) sectors of the pulvinus, while exposure to blue light has the same effect (contraction) on either [268].

Pulvinar chloroplasts contribute to its response to red light. Their pigment composition and activity of the xanthophyll cycle are similar to those of mesophyll chloroplasts. They perform photosynthetic electron transport and quench fluorescence non-photochemically, showing that they build up a considerable transthylakoid proton gradient in the light. Results of pretreatment of the pulvinus with DCMU on its response to red (500 μmol $m^{-2} s^{-1}$) and blue light (50 μmol $m^{-2} s^{-1}$) and with the ionophore (uncoupler) CCCP on its response to red light suggest that the pulvinar response to red, but not to blue light, requires non-cyclic electron transport and the resulting generation of ATP [23]. Red light acts in stomatal opening by stimulating electrogenic proton pumping at the plasmalemma, using ATP supplied by photophosphorylation in the guard cell chloroplasts, but other products of photosynthesis may also be involved [279]. Adaxial blue or red light has different effects on membrane polarization of the motor cells in the exposed (adaxial) sector of the pulvinus. DCMU inhibits hyperpolarization of pulvinar motor cells in the primary leaves of *Phaseolus* by red light, but has no effect on their depolarization by blue light. CCCP depolarized the membranes and prevented the changes in polarization by red or blue light. These results support the concept that blue light inhibits activity of a proton-ATPase, causing loss of the electrogenic component of membrane potential. As a result, H^+/K^+ fluxes are suppressed, turgor falls and the exposed sector contracts [249,280].

References

1. C.L. Ballaré, A.L. Scopel, S.R. Radosevich, R.E. Kendrick (1992). Phytochrome-mediated phototropism in de-etiolated seedlings. Occurrence and ecological significance. *Plant Physiol.*, **100**, 170–177.
2. E. Bünning (1959). Tagesperiodische Bewegungen. In: W. Ruhland (Ed.), *Encyclopedia of Plant Physiology, XVII/1* (pp. 579–656). Springer-Verlag, Berlin, Heidelberg, New York.
3. E. Bünning (1973). *The physiological clock* (3rd. ed.). The English Universities Press, London.
4. R.L. Satter (1979). Leaf movements and tendril curling. In: W. Haupt, M.E. Feinleib (Eds), *Encyclopedia of Plant Physiology, 7* (pp. 442–484). Physiology of Movements, Springer-Verlag, Berlin, Heidelberg, New York.
5. R.L. Satter, M.J. Morse (1990). Light-modulated circadian rhythmic leaf movements in nyctinastic legumes. In: R.L. Satter, H.L. Gorton, T.C. Vogelmann (Eds), *The Pulvinus: Motor Organ for Leaf Movement, American Society of Plant Physiologists* (pp. 10–24). Rockville, MD.
6. H.L. Gorton (1990). Stomates and pulvini: A comparison of two rhythmic, turgor-mediated movement systems. In: R.L. Satter, H.L. Gorton, T.C. Vogelmann (Eds), *The Pulvinus: Motor Organ for Leaf Movement* (pp. 223–237). American Society of Plant Physiologists, Rockville, MD.
7. C. Darwin, F. Darwin (1881). *The Power of Movement in Plants*. D. Appleton & Co, (Reprinted 1966: De Capo Press), New York.
8. C.M. Wainwright (1977). Suntracking and related leaf movements in a desert lupin (*Lupinus arizonicus*). *Amer.J. Bot.*, **64**, 1032–1041.
9. H.A. Mooney, J. Ehleringer (1978). The carbon gain benefits of solar tracking in a desert annual. *Plant Cell Env.*, **1**, 301–311.
10. R.D. Firn (1994). Phototropism. In: R.E. Kendrick, R.H.M. Kronenberg (Eds), *Photomorphogenesis in Plants* (2nd ed., pp. 659–681). Kluwer, Dordrecht, Boston, London.

11. D.J. Cosgrove (1983). Photocontrol of extension growth. *Proc. Roy. Soc. London*, **303**, 453–465.
12. J. Bruinsma, K. Hasegawa (1990). A new theory of phototropism – Its regulation by a light-induced gradient of auxin-inhibiting substances. *Physiol. Plant.*, **79**, 700–704.
13. J. von Sachs (1887). *Lectures on the Physiology of Plants*. The Clarendon Press, Oxford, Translated by H.M. Ward from Vorlesungen der Pflanzenphysiologie, Würzburg.
14. J.M. Morse, R.L. Satter (1979). Relationships between motor cell ultrastructure and leaf movements in *Samanea saman*. *Physiol. Plant.*, **46**, 338–346.
15. E. Werker, D. Koller (1987). Structural specialization of the site of response to vectorial photo-excitation in the solar-tracking leaf of *Lavatera cretica*. *Amer. J. Bot.*, **74**, 1339–1349.
16. W.-E. Mayer, D. Flach, M.V.S. Raju, N. Starrach, E. Wiech (1985). Mechanics of circadian pulvini movements in *Phaseolus coccineus* L. Shape and arrangement of motor cells, micellation of and bulk moduli of extensibility. *Planta*, **163**, 381–390.
17. M.S. Irving, S. Ritter, A.D. Tomos, D. Koller (1997). Phototropic response of the bean pulvinus: movement of water and ions. *Bot. Acta*, **110**, 118–126.
18. K. Raschke (1975). Stomatal action. *Annu. Rev. Plant Physiol.*, **26**, 309–340.
19. P.J.H. Sharpe, H. Wu, R.D. Spence (1987). Stomatal mechanics. In: E. Zeiger, G.D. Farquhar, I.R. Cowan (Eds), *Stomatal Function* (pp. 91–114). Stanford University Press, Stanford, CA.
20. R.L. Satter, D.D. Sabnis, A.W. Galston (1970). Phytochrome controlled nyctinasty in *Albizzia julibrissin*.I. Anatomy and fine structure of the pulvinule. *Amer. J. Bot.*, **57**, 374–381.
21. R.L. Satter, N. Moran (1988). Ionic channels in plant cell membranes. *Physiol. Plant.*, **72**, 816–820.
22. D.Koller, S. Ritter (1994). Phototropic responses of the pulvinule and associated laminar reorientation in the trifoliate leaf of bean *Phaseolus vulgaris* (Fabaceae). *J. Plant Physiol.*, **143**, 52–63.
23. D. Koller, S. Ritter, O. Björkman (1995). Role of pulvinar chloroplasts in light-driven leaf movements of the trifoliate leaf of bean (*Phaseolus vulgaris* L.). *J. Exp. Bot.*, **290**, 1215–1222.
24. E. Werker, T. Shak, D. Koller (1991). Photobiological and structural studies of light-driven movements in the solar-tracking leaf of *Lupinus palaestinus* Boiss. (Fabaceae). *Bot. Acta*, **104**, 144–156.
25. M. Weintraub (1951). Leaf movement in *Mimosa pudica*. *New Phytol.*, **50**, 357–382.
26. N. Campbell, W.W. Thomson (1977). Multivacuolate motor-cells in *Mimosa pudica* L. *Ann. Bot.*, **41**, 1361–1362.
27. W.W. Thomson, R.D. Journett (1970). Studies on the ultrastructure of guard cells of *Opuntia*. *Amer. J. Bot.*, **57**, 309–316.
28. N.A. Campbell, R.C. Garber (1980). Vacuolar reorganization in the motor cells of *Albizzia* during leaf movement. *Planta*, **148**, 251–255.
29. M. Brauner (1932). Untersuchungen über die Lichtturgorreaktionen des Primärblattgelenkes von *Phaseolus multiflorus*. *Planta*, **18**, 288–337.
30. L. Brauner, M. Brauner (1947). Untersuchungen über der Mechanismus der phototropischen Reaktion der Blattfiedern von *Robinia pseudacaia*. *Rev. Fac. Sci. Univ. Istanbul*, **12B**, 35–79.
31. J.H. Palmer, G.F. Asprey (1958). Studies in the nyctinastic movement of leaf pinnae of *Samanea saman* (Jacq.) Merrill. II. The behavior of upper and lower half pulvini. *Planta*, **51**, 770–785.
32. B. Millet, L. Coillot, R.D. Agosti (1989). The rhythmic leaf movements after regeneration

of the partially excised pulvinus in *Phaseolus vulgaris* L. *Plant Cell Physiol.*, **30**, 643–648.
33. W.-E. Mayer, R. Hampp (1995). Movement of pulvinated leaves. *Progress in Botany*, **56**, 236–162.
34. W.-E. Mayer, C. Fischer (1994). Protoplasts from *Phaseolus coccineus* L. pulvinar motor cells show circadian volume oscillations. *Chronobiol. Int.*, **11**, 156–164.
35. N.A. Campbell, R.L. Satter, R.G. Garber (1981). Apoplastic transport of ions in the motor organ of *Samanea*. *Proc. Natl. Acad. Sci. USA*, **78**, 2981–2984.
36. R.L. Satter, G.T. Geballe, P.B. Applewhite, A.W. Galston (1974). Potassium flux and leaf movement in *Samanea saman*. I. Rhythmic movement. *J. Gen. Physiol.*, **64**, 413–430.
37. H.L. Gorton (1987). Water relations in pulvini from *Samanea saman*. I. Intact pulvini. *Plant Physiol.*, **83**, 945–950.
38. H.C. Yin (1938). Diaphototropic movement of the leaves of *Malva neglecta*. *Amer. J. Bot.*, **25**, 1–6.
39. K. Kiyosawa, H. Tanaka (1976). Changes in potassium distribution in a *Phaseolus* pulvinus during circadian movement of the leaf. *Plant Cell Physiol.*, **17**, 289–298.
40. R. Harder, W. Schumacher, F. Firbas, D.v. Denffer (1965). *Strassburger's Textbook of Botany* New English Edition (translated from the 28th German edition, Gustav Fischer Verlag, Berlin), Longmans Green, Berlin.
41. A. Iglesias, R.L. Satter (1983). H^+ fluxes in excised *Samanea* motor tissue. I. Promotion by light. *Plant Physiol.*, **72**, 564–569.
42. A. Iglesias, R.L. Satter (1983). H^+ fluxes in excised *Samanea* motor tissue. II. Rhythmic properties. *Plant Physiol.*, **72**, 570–572.
43. N. Starrach, D. Flach, W.-E. Mayer (1985). Activity of fixed negative charges of isolated extensor cell walls of the laminar pulvinus of the primary leaves of *Phaseolus*. *J. Plant Physiol.*, **120**, 441–445.
44. C. Freudling, N. Starrach, D. Flach, D. Gradmann, W.-E. Meyer (1988). Cell walls as reservoirs of potassium ions for reversible changes in pulvinar motor cells during rhythmic leaf movements. *Planta*, **175**, 193–203.
45. N. Starrach, W.-E. Mayer (1989). Changes in the apoplastic pH and K^+ concentration in the *Phaseolus* pulvinus *in situ* in relation to rhythmic leaf movements. *J. Exp. Bot.*, **40**, 856–873.
46. W.-E. Mayer (1990). Walls as potassium reservoirs in *Phaseolus* pulvini. In: R.L. Satter, H.L. Gorton, T.C. Vogelmann (Eds), *The Pulvinus: Motor Organ for Leaf Movements* (pp. 160–174). American Society of Plant Physiologists, Rockville, MD.
47. L. Moysset, L. Solé-sugrañes, E. Simon (1991). Changes in morphometry and elemental composition of *Robinia pseudoacacia* pulvinar motor cells during leaflet movement. *J. Exp. Bot.*, **42**, 1315–1323.
48. G. Thiel, A.H. Wolf (1997). Operation of K^+-channels in stomatal movement. *Trends Pl. Sci.*, **2**, 339–345.
49. F.J.M. Maathuis, A.M. Ichida, D. Sanders, J.I. Schroeder (1997). Roles of higher plant K^+ channels. *Plant Physiol.*, **114**, 1141–1149.
50. M.R. Blatt, A. Grabov (1997). Signal redundancy, gates and integration in the control of ion channels for stomatal movement. *J. Exp. Bot.*, **48**, 529–537.
51. E.A.C. MacRobbie (1997). Signaling in guard cells and regulation of ion channel activity. *J. Exp. Bot.*, **48**, 515–528.
52. Y. Lee (1990). Ion movements that control pulvinar curvature in nyctinastic legumes. In: R.L. Satter, H.L. Gorton, T.C. Vogelmann (Eds), *The Pulvinus: Motor Organ for Leaf Movement* (pp. 130–141). American Society of Plant Physiologists, Rockville, MD.
53. S.M. Assman (1993). Signal transduction in guard cells. *Annu. Rev. Cell Biol.*, **9**,

345–375.

54. J.M. Ward, Z.-M. Pei, J.I. Schroeder (1995). Roles of ion channels in initiation of signal transduction in higher plants. *Plant Cell*, **7**, 833–844.
55. N. Moran (1990). The role of ion channels in osmotic volume changes in *Samanea* motor cells analyzed by patch-clamp methods. In: R.L. Satter, H.L. Gorton, T.C. Vogelmann (Eds), *The Pulvinus: Motor Organ for Leaf Movement* (pp. 142–159). American Society of Plant Physiologists, Rockville, MD.
56. L. Giromini, G. Camattari, R. Cerana, R. Colombo (1997). Effect of apoplastic and cytoplasmic pH on inward and outward rectifying plasma membrane K^+ channels of *Arabidopsis thaliana* cultured cells. *J. Plant Physiol.*, **151**, 682–688.
57. J. Bialczyk, Z. Lechowski (1989). Malic acid synthesis in relation to K^+ and Cl^- availability in *Phaseolus coccineus* L. pulvini. *Biochem. Physiol. Pflanz.*, **184**, 79–86.
58. N. Moran, G. Ehrenstein, K. Iwasa, C. Mischke, C. Bare, R.L. Satter (1988). Potassium channels in motor cells of *Samanea saman*. A patch-clamp study. *Plant Physiol.*, **88**, 643–648.
59. N. Moran, D. Fox, R.L. Satter (1990). Interaction of the depolarization-activated K^+ channel of *Samanea saman* with inorganic ions. A patch-clamp study. *Plant Physiol.*, **94**, 424–431.
60. N. Moran (1996). Membrane-delimited phosphorylation enables activation of outward-rectifying K^+ channels in motor cell protoplasts of *Samanea saman*. *Plant Physiol.*, **111**, 1281–1292.
61. P. Thuleau, M. Moreau, J.I. Schroeder, R. Ranjeva (1994). Recruitment of plasma membrane voltage-dependent calcium permeable channels in carrot cells. *EMBO J.*, **13**, 5843–5847.
62. P. Thuleau, M.J. Ward, R. Ranjeva, J.I. Schroeder (1994). Voltage-dependent calcium-permeable channels in the plasma membrane of a higher plant cell. *EMBO J.*, **13**, 2970–2975.
63. Freudling, W.-E. Meyer, D. Gradmann (1980). Electrical membrane properties and circadan rhythm in extensor cells of laminar pulvini of *Phaseolus coccineus* L. *Plant Physiol.*, **65**, 966–968.
64. R. Hedrich, D. Becker (1994). Green circuits – The potential of plant-specific ion channels. *Plant Mol. Biol.*, **26**, 1637–1650.
65. H.Y. Kim, G.G. Coté, R.C. Crain (1992). Effects of light on the membrane potential of protoplasts from *Samanea saman* pulvini. *Plant Physiol.*, **99**, 1532–1539.
66. H.Y. Kim, G.G. Coté, R.C. Crain (1993). Potassium channels in *Samanea saman* protoplasts controlled by phytochrome and the biological clock. *Science*, **260**, 960–962.
67. K. Shimazaki, T. Kinoshita, N. Nishimura (1992). Involvement of calmodulin and calmodulin-dependent myosin light-chain kinase in blue light-dependent H^+ pumping by guard cell protoplasts from *Vicia faba*. *Plant Physiol.*, **99**, 1416–1421.
68. J.T.M. Elzenga, M. Staal, H.B.A. Prins (1997). Calcium-calmodulin signaling is involved in light-induced acidification by epidermal leaf cells of pea, *Pisum sativum* L. *J. Exp. Bot.*, **48**, 2055–2060.
69. Y. Lee, S. Suh, N. Moran, R.C. Crain (1996). Phospholipid metabolism and light regulation of stomatal opening and leaf movement. In: W.R. Briggs, R.L. Heath, E. Tobin (Eds), *Regulation of Plant Growth and Development by Light* (pp. 89–97). American Society of Plant Physiologists, Rockville, MD.
70. P.A. Millner, B.E. Causier (1996). G-protein coupled receptors in plant cells. *J. Exp. Bot.*, **47**, 983–992.
71. G. Roblin, P. Fleurat-Lessard, J. Bonmort (1989). Effects of compounds affecting calcium channels on phytochrome- and blue pigment-mediated pulvinar movements of *Cassia*

fasciculata. Plant Physiol., **90**, 697–701.
72. M.R. McAinsh, C. Brownlee, A.M. Hetherington (1997). Calcium ions as second messengers in guard cell signal transduction. *Physiol. Plant.*, **100**, 16–29.
73. R.J. Reid, M.A. Tester, F.A. Smith (1997). Voltage control of calcium influx in intact cells. *Aust. J. Plant Physiol.*, **24**, 805–810.
74. R.L. Satter, Y. Lee, M.J. Morse, R.C. Crain, G.G. Coté, N. Moran (1988). Light- and clock-controlled leaflet movements in *Samanea saman*: A physiological. biophysical and biochemical analysis. *Bot. Acta*, **101**, 205–213.
75. J.M. Morse, R.C. Crain, G.G. Coté, R.L. Satter (1989). Light-stimulated inositol phospholipid turnover in *Samanea saman* leaf pulvini: increased levels of diacylglycerol. *Plant Physiol.*, **89**, 724–727.
76. J.M. Morse, R.L. Satter, R.C. Crain, G.G. Coté (1989). Signal transduction and phosphatidylinositol turnover in plants. *Physiol. Plant.*, **76**, 118–121.
77. H.Y. Kim, G.G. Coté, R.C. Crain (1996). Inositol 1,4,5-trisphosphate may mediate closure of K^+ channels by light and darkness in *Samanea saman* motor cells. *Planta*, **198**, 279–287.
78. J.I. Schroeder, S. Hagiwara (1989). Cytosolic calcium regulates ion channels in plasma membrane of *Vicia faba* guard cells. *Nature (Lond.)*, **338**, 427–430.
79. M.R. Blatt, G. Thiel, D.R. Trentham (1990). Reversible inactivation of K^+ channels of *Vicia* stomatal guard cells following the photolysis of caged inositol 1,4,5-trisphosphate. *Nature (Lond.)*, **346**, 766–769.
80. S. Gilroy, N.D. Read, A.J. Trewavas (1990). Elevation of cytoplasmic calcium by caged calcium or caged inositol trisphosphate initiates stomatal closure. *Nature (Lond.)*, **346**, 769–771.
81. R. Hedrich, H. Bush, K. Raschke (1990). Ca^{2+} and nucleotide dependent regulation of voltage dependent anion channels in plasma membrane of guard cells. *EMBO J.*, **9**, 3889–3992.
82. T. Kinoshita, M. Nishimura, K. Shimazaki (1996). Cytosolic concentration of Ca^{2+} regulates the plasma membrane H^+-ATPase in guard cells of the fava bean. *Plant Cell*, **7**, 1333–1342.
83. S.R. Muir, M.A. Bewell, D. Sanders, G.J. Allen (1997). Ligand-gated Ca^{2+} channels and Ca^{2+} signaling in higher plants. *J. Exp. Bot.*, **48**, 589–597.
84. G.G. Coté, Y.G. Yueh, R.C. Crain (1996). Phosphoinositide turnover and its role in plant signal transduction. In: B.B. Biswas, R. Biswas (Eds), *Subcellular Biochemistry, 26* (pp. 317–343). Plenum Press, New York.
85. Y. Lee, S. Assman (1991). Diacylglycerols induce both ion pumping in patch-clamped guard-cell protoplasts and opening of intact stomata. *Proc. Natl. Acad. Sci. (USA)*, **88**, 2127–2131.
86. T. Munnik, T. de Vrije, R.F. Irvine, A. Musgrave (1996). Identification of diacylglycerol pyrophosphate as a novel metabolic product of phosphatidic acid during G-protein activation in plants. *J. Biol. Chem.*, **271**, 15708–15715.
87. J. Li, Y.-R.J. Lee, S.M. Assman (1998). Guard cells possess a calcium-dependent protein kinase that phosphorylates the KAT1 potassium channel. *Plant Physiol.*, **116**, 785–795.
88. J.U. Hwang, S. Suh, J. Kim, Y. Lee (1997). Actin filaments modulate both stomatal opening and inward K^+ channel activities in guard cells of *Vicia faba* L. *Plant Physiol.*, **115**, 335–342.
89. Martinoia (1992). Transport processes in vacuoles of higher plants. *Bot. Acta*, **105**, 232–245.
90. M. Maeshima, Y. Nakanishi, C. Matsuura-Endo, Y. Tanaka (1996). Proton pumps of the vacuolar membrane in growing plant cells. *J. Plant Res.*, **109**, 119–125.

91. J.M. Davies (1997). Vacuolar energization: pumps, shunts and stress. *J. Exp. Bot.*, **48**, 633–641.
92. E. Fischer-Schliebs, J.-B. Mariaux, U. Lüttge (1997). Stimulation of H^+ transport activity of vacuolar H^+-ATPase by activity of H^+-PPase in *Kalanchoë blossfeldiana*. *Biol. Plant.*, **39**, 169–177.
93. C. Lommel, H.H. Felle (1997). Transport of Ca^{2+} across the tonoplast of intact vacuoles from *Chenopodium album* L. suspension cells: ATP-dependent import and inositol-1,4,5-trisphosphate-induced release. *Planta*, **201**, 477–486.
94. Z.-M. Pei, J.M.Ward, J.F. Harper, J.I. Schroeder (1996). A novel chloride channel in *Vicia faba* guard cell vacuoles activated by the serine/threonine kinase, CDPK. *EMBO J.*, **15**, 6564–6574.
95. G.J. Allen, D. Sanders (1997). Vacuolar ion channels of higher plants. *Adv. Bot. Res.*, **25**, 217–252.
96. J.M. Ward, J.I. Schroeder (1994). Calcium-activated K^+ channels and calcium-induced calcium release by slow vacuolar ion channels in guard cell vacuoles implicated in the control of stomatal closure. *Plant Cell*, **6**, 669–683.
97. J. Alexandre, J.P. Lassalles, R.T. Kado (1990). Opening of Ca^{2+} channels in isolated red beet vacuole membrane by inositol 1,4,5-trisphosphate. *Nature*, **343**, 567–570.
98. G. Allen, S. Muir, D. Sanders (1996). Control of Ca^{2+} channels at the vacuolar membrane of plant cells. *J. Exp. Bot.*, **47**, 65.
99. G. Martiny-Baron, M.F. Manolson, R.J. Poole, D. Hecker, G.F.E. Scherer (1992). Proton transport and phosphorylation of tonoplast polypeptides from zucchini are stimulated by the phospholipid platelet-activating factor. *Plant Physiol.*, **99**, 1635–1641.
100. C. Maurel (1997). Aquaporins and water permeability of plant membranes. *Annu. Rev. Plant Physiol. Plant Mol. Biol.*, **48**, 399–429.
101. A.R. Schäfner (1998). Aquaporin function, structure, and expression: are there more surprises to urface in water-relations? *Planta*, **204**, 131–139.
102. C. Maurel, F. Tacnet, J. Güclü, J. Guern, P. Ripoche (1997). Purified vesicles of tobacco cell vacuolar and plasma membranes exhibit dramatically different water permeability and water channel activity. *Proc. Natl. Acad. Sci. (USA)*, **94**, 7103–7108.
103. M.J. Chrispeels, C. Maurel (1994). Aquaporins: The molecular basis for facilitated water movement through living plant cells. *Plant Physiol.*, **105**, 9–15.
104. D.G. Robinson, H. Sieber, W. Kammerlohrer, A.R. Schäffner (1996). PIP1 aquaporins are concentrated in plasmalemmasomes of *Arabidopsis thaliana* mesophyll. *Plant Physiol.*, **111**, 645–649.
105. S. Yamada, D.E. Nelson, E. Let, S. Marquez, J.H. Bohnert (1997). The expression of an aquaporin promoter from *Mesembryanthemum crystallinum* in tobacco. *Plant Cell Physiol.*, **38**, 1326–1332.
106. I. Johansson, M. Karlsson, V.K. Shukla, M.J. Chrispeels, C. Larsson, P. Kjellbom (1998). Water transport activity of the plasma membrane aquaporin PM28A is regulated by phosphorylation. *Plant Cell*, **10**, 451–460.
107. D.F. Wetherell (1990). Leaf movements in plants without pulvini. In: R.L. Satter, H.L. Gorton and T.C. Vogelmann (Eds), *The Pulvinus, Motor Organ For Leaf Movement, American Society of Plant Physiologists* (pp. 72–78). Rockville.
108. J. von Sachs (1875). Über das Bewegungsorgan und die periodischen Bewegungen der Blätter von *Phaseolus* und *Oxalis*. *Bot. Zeitung*, **15**, 793–802; 809–815.
109. W. Pfeffer (1875). *Die periodischen Bewegungen der Blattorgane*. Engelmann, Leipzig.
110. W. Pfeffer (1881). *The Physiology of Plants* (Vol. 3) (translated into English and edited by A.J. Ewart). Clarendon, Oxford.
111. T. Roennenberg, R.G. Foster (1997). Twilight times: light and the circadian rhythm.

Photochem. Photobiol., **66**, 549–561.
112. P.K. Burkholder, R. Pratt (1936). Leaf movements of *Mimosa pudica* in relation to intensity and wavelength of the incident radiation. *Amer. J. Bot.*, **23**, 212–220.
113. M.G. Holmes, W.H. Klein (1986). Photocontrol of dark circadian rhythms in stomata of *Phaseolus vulgaris* l. *Plant Physiol.*, **82**, 28–33.
114. O. Björkman, S.B. Powles (1981). Leaf movement in the shade species *Oxalis oregana*. I. Response to light level and light quality. *Carnegie Inst. Wash. Yearb.*, **80**, 59–62.
115. W.L. Koukkari, W.S. Hillman (1968). Pulvini as the photoreceptors in the phytochrome effect on nyctinasty in *Albizzia julibrisin*. *Plant Physiol.*, **43**, 698–704.
116. S. Watanabe, T. Sibaoka (1973). Site of photo-reception to opening response in *Mimosa* leaflets. *Plant Cell Physiol.*, **14**, 1221–1224.
117. R.L. Satter, S.E. Guggino, T.A. Lonergan, A.W. Galston (1981). The effects of blue and far red light on rhythmic leaf movements in *Samanea* and *Albizzia*. *Plant Physiol.*, **67**, 965–968.
118. W.-E. Mayer, C. Hohloch, A. Kalkuhl (1997). Extensor protoplasts of the *Phaseolus* pulvinus: light-induced swelling may require extracellular Ca^{2+} influx, dark-induced shrinking inositol 1,4,5-trisphosphate-induced Ca^{2+} mobilization. *J. Exp. Bot.*, **48**, 219–228.
119. C.Z. Lowen, R.L. Satter (1989). Light-promoted changes in apoplastic K^+ activity in the *Samanea saman* pulvinus, monitored with liquid membrane electrodes. *Planta*, **179**, 421–427.
120. Tretyn, R.E. Kendrick, G. Wagner (1991). The role(s) of calcium ions in phytochrome action. *Photochem. Photobiol.*, **54**, 1135–1155.
121. R.R. Lew, B.S. Serlin, C.L. Schauf, M.E. Stockton (1990). Red light regulates calcium-activated potassium channels in *Mougeotia* plasma membrane. *Plant Physiol.*, **92**, 822–830.
122. M.E. Bossen, H.H.A. Dassen, R.E. Kendrick, W.J. Vredenberg (1988). The role of calcium ions in phytochrome-controlled swelling of etiolated wheat (*Triticum aestivum* L.) protoplasts. *Planta*, **174**, 94–100.
123. M.E. Bossen, R.E. Kendrick, W.J. Vredenberg (1990). The involvement of a G-protein in phytochrome-regulated, Ca^{2+}-dependent swelling of etiolated wheat protoplasts. *Physiol. Plant.*, **80**, 55–62.
124. L. Moysset, E. Simon (1989). Role of calcium in phytochrome-controlled nyctinastic movements of *Albizzia lophanta* leaflets. *Plant Physiol.*, **90**, 1108–1114.
125. L.A. Gomez, E. Simon (1995). Circadian rhythm of *Robinia pseudoacacia* leaflet movements: role of calcium and phytochrome. *Photochem. Photobiol.*, **61**, 210–215.
126. S. Thomine, S. Zimmermann, B. Vanduijn, H. Barbierbrygoo, J. Guern (1994). Calcium channel antagonists induce indirect inhibition of the outward rectifying potassium channel in tobacco protoplasts. *FEBS Lett.*, **340**, 1–2.
127. J.C. Fondeville, H.A. Borthwick, S.B. Hendricks (1966). Leaflet movements in *Mimosa pudica* L. indicative of phytochrome action. *Planta*, **69**, 357–364.
128. J.C. Fondeville, M.J. Schneider, H.A. Borthwick, S.B. Hendricks (1967). Photocontrol of *Mimosa pudica* leaflet movement. *Planta*, **75**, 228–238.
129. M.J. Jaffe, A.W. Galston (1967). Phytochrome control of rapid nyctinastic movements and membrane permeability in *Albizzia julibrissin*. *Planta*, **77**, 135–141.
130. L.T. Evans, W.G. Allaway (1972). Action spectrum for the opening of *Albizzia julibrissin* pinnules, and the role of phytochrome in the closing movements of pinnules and stomata of *Vicia faba*. *Aust. J. Biol. Sci.*, **25**, 885–893.
131. P.K. Burkholder, R. Pratt (1936). Leaf movements of *Mimosa pudica* in relation to light. *Amer. J. Bot.*, **23**, 46–52.

132. C.W. Williams, V. Raghavan (1966). Effects of light and growth substances on the diurnal movements of the leaflets of *Mimosa pudica*. *J. Exp. Bot.*, **17**, 742–749.
133. S. Watanabe, T. Sibaoka (1983). Light- and auxin-induced leaflet opening in detached pinnae of *Mimosa pudica*. *Plant Cell Physiol.*, **24**, 641–647.
134. W.S. Hillman, W.L. Koukkari (1967). Phytochrome effects in nyctinastic leaf movements of *Albizzia julibrissin* and some other legumes. *Plant Physiol.*, **42**, 1413–1418.
135. T. Tanada (1982). Effect of far-red and green irradiation on the nyctinastic closure of *Albizzia* leaflets. *Plant Physiol.*, **70**, 901–904.
136. T. Tanada (1984). Interactions of green or red light with blue light on the dark closure of *Albizzia* pinnules. *Physiol. Plant.*, **61**, 35–37.
137. M. Negbi, E. Zamski, O. Ze'evi (1982). Photo- and thigmomorphogenetic control of the attachment of the ivy (*Hedera helix*) to its support. *Z. Pflanzenphysiol.*, **108**, 9–19.
138. D.R. Strong, T.S. Ray (1975). Host tree location behavior of a tropical vine (*Monstera gigantea*) by skototropism. *Science*, **190**, 804–807.
139. D.E. Bilderback (1984). Phototropism in *Selaginella*: the differential response to light. *Amer. J. Bot.*, **71**, 1323–1329.
140. D.E. Bilderback (1984). Phototropism in *Selaginella*: the role of the small dorsal leaves and auxin. *Amer. J. Bot.*, **71**, 1330–1337.
141. T.C.G. Rich, G.C. Whitelam, H. Smith (1985). Phototropism and axis extension in light-grown mustard (*Sinapis alba*) seedlings. *Photochem. Photobiol.*, **42**, 789–792.
142. R.D. Firn (1990). Phototropism – The need for a sense of direction. *Photochem. Photobiol.*, **52**, 225–260.
143. D. Gleed, R.D. Firn, J. Digby (1994). How is phototropic stimulus perceived by hypocotyls? *J. Exp. Bot.*, **45**, 409–412.
144. M. Iino (1990). Phototropism: mechanics and ecological implications. *Plant Cell Env.*, **13**, 633–650.
145. D.J. Cosgrove (1985). Kinetic separation of phototropism from blue light inhibition of stem elongation. *Photochem. Photobiol.*, **42**, 745–751.
146. E.P. Spalding, D.J. Cosgrove (1989). Large plasma membrane depolarization precedes rapid blue light-induced growth inhibition in cucumber. *Planta*, **178**, 407–410.
147. J.P. Khurana, K.L. Poff (1989). Mutants of *Arabidopsis thaliana* with altered phototropism. *Planta*, **178**, 400–406.
148. E. Liscum, R.P. Hangarter (1991). *Arabidopsis* mutants lacking blue light-dependent inhibition of hypocotyl elongation. *Plant Cell*, **3**, 685–694.
149. D.J. Cosgrove (1988). Mechanism of rapid suppression of cell expansion in cucumber hypocotyls after blue light irradiation. *Planta*, **176**, 109–116.
150. E.P. Spalding, D.J. Cosgrove (1992). Mechanism of blue light-induced plasma-membrane depolarization in etiolated cucumber seedlings. *Planta*, **188**, 199–205.
151. M.H. Cho, E.P. Spalding (1996). An anion channel in *Arabidopsis* activated by blue light. *Proc. Natl. Acad. Sci. (USA)*, **93**, 8134–8138.
152. B.D. Lewis, G. Karun-Neumann, G. Davis, E.P. Spalding (1997). Ca^{2+}-activated anion channels and membrane depolarization induced by blue light and cold in *Arabidopsis* seedlings. *Plant Physiol.*, **114**, 1327–1334.
153. P. Reymond, T.W. Short, W.R. Briggs (1992). Blue light activates a specific protein kinase in higher plants. *Plant Physiol.*, **100**, 655–671.
154. L.-G. Ma, D.-Y. Sun (1997). The involvement of calcium in light signal transduction chain for phototropism in sunflower seedlings. *Biol. Plant.*, **39**, 569–574.
155. M. Salomon, M. Zacherl, W. Rüdiger (1997). Phototropism and protein phosphorylation in higher plants: Unilateral blue light irradiation generates a directional gradient of protein phosphorylation across the oat coleoptile. *Bot. Acta*, **110**, 214–216.

156. P. Nick, R. Bergfeld, E. Schäfer, P. Schopfer (1990). Unilateral reorientation of microtubules at the outer epidermal wall during photo- and gravitropic curvature of maize coleoptiles and sunflower hypocotyls. *Planta*, **181**, 162–168.
157. T. Murata, A. Kadota, M. Wada (1997). Effects of blue light on cell elongation and microtubule orientation in dark-grown gametophytes of *Ceraopteris richardii*. *Plant Cell Physiol.*, **38**, 201–209.
158. D. Miller, W. Hable, J. Gottwald (1997). Connections: The hard wiring of the plant cell for perception, signaling, and response. *Plant Cell*, **9**, 2105–2116.
159. V. Mermall, P.L. Post, M.S. Mooseker (1998). Unconventional myosins in cell movement, membrane traffic, and signal transduction. *Science*, **279**, 527–533.
160. T. Ohiwa (1977). Response of *Spirogyra* chloroplast to local illumination. *Planta*, **136**, 7–11.
161. M. Blatt, W.R. Briggs (1980). Blue light induced cortical fiber reticulation concomitant with chloroplast aggregation in the alga *Vaucheria sessilis*. *Planta*, **147**, 355–362.
162. M. Blatt, M.H. Weisenseel, W. Haupt (1981). A light-dependent current associated with chloroplast aggregation in the alga *Vaucheria sessilis*. *Planta*, **152**, 513–526.
163. N.G. Ball (1923). Phototropic movements of leaves. The functions of the lamina and the petiole with regard to the perception of the stimulus. *Proc. Roy. Dublin Soc. N.S.*, **17**, 281–286.
164. H. Shibaoka (1961). Studies on the mechanism of growth-inhibiting effect of light. *Plant Cell Physiol.*, **2**, 175–197.
165. J. Bruinsma, C.M. Karssen, M. Benschop, J.B.V. Dort (1976). Hormonal regulation of phototropism in the light-grown sunflower seedling, *Helianthus annuus* L.: Immobility of endogenous indoleacetic acid and inhibition of hypocotyl growth by illuminated cotyledons. *J. Exp. Bot.*, **26**, 411–418.
166. S.-L. Lam, A.C. Leopold (1966). Role of leaves in phototropism. *Plant Physiol.*, **41**, 847–851.
167. G.I. McIntyre, K.P. Browne (1996). Effect of darkening the cotyledons on the growth and curvature of the sunflower hypocotyl, evidence of hydraulic signaling. *J. Exp. Bot.*, **47**, 1561–1566.
168. G.I. McIntyre (1980). The role of water distribution in plant tropisms. *Aust. J. Plant Physiol.*, **7**, 401–413.
169. J.M. Franssen, J. Digby, R.D. Firn (1982). An investigation of the importance of transpiration gradients in the control of phototropic curvature in coleoptiles of *Avena sativa* cultivar Victory. *Aust. J. Plant Physiol.*, **9**, 583–586.
170. M. Black, J.E. Shuttleworth (1974). The role of the cotyledons in the photocontrol of hypocotyl extension in *Cucumis sativus* L. *Planta*, **117**, 57–66.
171. J.E. Shuttleworth, M. Black (1977). The role of the cotyledons in phototropism of de-etiolated seedlings. *Planta*, **136**, 51–55.
172. J.W. Hart, I.R. MacDonald (1981). Phototropic responses of hypocotyls of etiolated and green seedlings. *Plant Sci. Lett.*, **21**, 151–158.
173. G. Haberlandt (1905). *Die Lichtsinnesorgane der Laubblätter*. Leipzig.
174. G. Haberlandt (1914). The motor system. Optical sense organs. In: Physiological Plant Anatomy, 1965 reprint edition (Translated from the 4th German edition by M. Drummond), Today and Tomorrow's Printers & Publishers, India, pp. 613–631.
175. W.N. Jones (1938). Observations on the response of leaves of *Limnanthemum* and *Tropaeolum* to light and gravity. *Ann. Bot. N.S.*, **2**, 819–825.
176. L. Brauner, Y. Vardar (1950). Über die Funktion der Lamina bei der phototropischen und geotropischen Reaktion des *Tropaeolum*-Blattes. *Rev. Fac. Sci. Univ. Istambul*, **15B**, 269–299.

177. K.S. Werk, J. Ehleringer (1984). Non-random leaf orientation in *Lactuca serriola* L. *Plant Cell Env.*, **7**, 81–87.
178. H. Zhang, J.M. Pleasants, T.W. Jurik (1991). Development of leaf orientation in the prairie compass plant, *Silphium laciniatum* L. *Bulletin of the Torrey Botanical Club*, **118**, 33–42.
179. W. Schmidt, J. Hart, P. Filner, K.L. Poff (1977). Specific inhibition of phototropism in corn seedlings. *Plant Physiol.*, **60**, 736–738.
180. C. Lin, M. Ahmad, A.R. Cashmore (1996). *Arabidopsis* cryptochrome 1 is a soluble protein mediating blue light-dependent regulation of plant growth and development. *Plant Cell*, **10**, 893–902.
181. H. Asard, N. Horemans, W.R. Briggs, R.J. Caubergs (1995). Blue light perception by endogenous redox components of the plant plasma membrane. *Photochem. Photobiol.*, **61**, 518–522.
182. M.A. Quiñones, E. Zeiger (1994). A putative role of the xanthophyll, zeaxanthin, in blue light photoperception of corn coleoptiles. *Science*, **264**, 558–561.
183. J.M. Palmer, K.M.F. Warpeha, W.R. Briggs (1996). Evidence that zeaxanthin is not the photoreceptor for phototropism in maize coleoptiles. *Plant Physiol.*, **110**, 1323–1328.
184. E. Liscum, W.R. Briggs (1995). Mutations in the NPH1 locus of *Arabidopsis* disrupt the perception of phototropic stimuli. *Plant Cell*, **7**, 473–485.
185. E. Liscum, W.R. Briggs (1996). Mutations of *Arabidopsis* in potential transduction and response components of the phototropic signaling pathway. *Plant Physiol.*, **112**, 291–296.
186. E. Huala, P.W. Oeller, E. Liscum, I.-S. Han, E. Larsen, W.R. Briggs (1998). *Arabidopsis* NPH1: A protein kinase with a putative redox-sensing domain. *Science*, **278**, 2120–2123.
187. B.A. Horwitz (1994). Properties and transduction chains of the UV and blue light photoreceptors. In: R.E. Kendrick, G.H.M. Kronenberg (Eds), *Photomorphogenesis in Plants* (2nd ed., pp. 327–350). Kluwer, Dordrecht, Boston, London.
188. M. Wada, H. Sei (1994). Phytochrome-mediated phototropism in *Adiantum cuneatum* young leaves. *J. Plant Res.*, **107**, 181–186.
189. H. Smith (1994). Sensing the light environment. In: R.E. Kendrick, G.H.M. Kronenberg (Eds), *Photomorphogenesis in Plants* (2nd ed., pp. 377–416). Kluwer, Dordrecht, Boston, London.
190. Y.J. Liu, M. Iino (1996). Phytochrome is required for the occurrence of time-dependent phototropism in maize coleoptiles. *Plant Cell Env.*, **19**, 1379–1388.
191. A.-K. Janoudi, R. Konjevic, G. Whitelam (1997). Both phytochrome A and phytochrome B are required for the normal expression of phototropism in *Arabidopsis thaliana* seedlings. *Physiol. Plant.*, **101**, 278–282.
192. H. Shibaoka, T. Yamaki (1959). Studies on the growth movement of sunflower plant. *Scientific Papers College of General Education, Tokyo 9*, 105–126.
193. P.G. Kevan (1972). Heliotropism in some Arctic flowers. *Canadian Field Naturalist*, **86**, 41–44.
194. P.G. Kevan (1975). Sun-tracking solar furnaces in high Arctic flowers: significance for pollination and insects. *Science*, **189**, 723–726.
195. A.P. Smith (1975). Insect pollination and heliotropism in *Oritrophium limnophilum* (Sch. Bip.) Cuatr. (Compositae) of the Andean paramos. *Biotropica*, **7**, 284–286.
196. B. Kjellberg, S. Karlsson, I. Kestensson (1982). Effects of heliotropic movements of flowers of *Dryas octopetalla* on gynoecium temperature and seed development. *Oecologia*, **54**, 10–13.
197. M.L. Stanton, C. Galen (1993). Blue light controls solar tracking by flowers of an alpine plant. *Plant Cell Env.*, **16**, 983–989.
198. Y.Y. Leshem (1977). Sunflower: a misnomer? *Nature*, **270**, 102.
199. J.E. Begg, B.W.R. Torssell (1974). Diaphotonastic and parahelionastic leaf movements in

Stylosanthes humilis HBK (Townsville Stylo). *Roy. Soc. N.Z. Bull.*, **12**, 277–283.
200. G.S.G. Shell, A.R.G. Lang (1976). Movements of sunflower leaves over a 24-h period. *Agric. Meteorol.*, **16**, 161–170.
201. A.P.G. Lang, J.E. Begg (1979). Movements of *Helianthus annuus* leaves and heads. *J. Appl. Ecol.*, **16**, 299–305.
202. R.A. Snow (1947). A test for Sachs's theory of the plagiotropism of laminae. *New Phytol.*, **46**, 258–261.
203. J. Ehleringer, I. Forseth (1980). Solar tracking by plants. *Science*, **210**, 1094–1098.
204. G. Rajendrudu, V.S. Rama Das (1981). Solar tracking and light interception by leaves of some dicot species. *Current Science*, **50**, 618–620.
205. D. Koller (1990). Light-driven leaf movements. *Plant Cell Env.*, **13**, 615–632.
206. M.V. Sailaja, V.S. Rama Das (1996). Leaf solar tracking response exhibits diurnal constancy in photosystem II efficiency. *Env. Exp. Bot.*, **36**, 431–438.
207. H. Vöchting (1888). Über der Lichtstellung der Laubblätter. *Bot. Zeitung*, **46**, 501–560.
208. A. Schwartz, D. Koller (1978). The phototropic response to vectorial light in leaves of *Lavatera cretica* L. *Plant Physiol.*, **61**, 924–928.
209. T.C. Vogelmann, J.F. Bornman, D.J. Yates (1996). Focusing of light by leaf epidermal cells. *Physiol. Plant.*, **98**, 43–56.
210. H. Smith (1984). Plants that track the sun. *Nature (Lond.)*, **308**, 774.
211. H. Kniep (1907). Über die Lichtperzeption der Laubblätter. *Biol. Zbl.*, **27**, 97–106; 129–142.
212. F.J.F. Fisher, P.M. Fisher (1983). Photosynthetic patterning: a mechanism for sun-tracking. *Can. J. Bot.*, **61**, 2632–2640.
213. F.J.F. Fisher, D.L. Ehert, J. Hollingdale (1987). The pattern of vascular deployment near the pulvinus of the solar-tracking leaf of *Lavatera cretica* (Malvaceae). *Can. J. Bot.*, **65**, 2109–2117.
214. F.J.F. Fisher, M. Wright (1984). The dependence of sun-tracking in *Lavatera cretica* L. upon carbon dioxide availability. *New Phytol.*, **98**, 241–248.
215. F.J.F. Fisher, D.L. Ehert, G.R. Lister, J. Hollingdale (1989). Light quality and sun tracking in *Malva neglecta*. *Can. J. Bot.*, **67**, 515–520.
216. D. Koller, I. Levitan, W.R. Briggs (1985). Components of vectorial photo-excitation in solar-tracking leaves of *Lavatera cretica* (Malvaceae). *Physiol. Vég.*, **23**, 913–920.
217. A. Schwartz, D. Koller (1986). Diurnal phototropism in the solar-racking leaves of *Lavatera cretica*. *Plant Physiol.*, **80**, 778–781.
218. D. Koller, I. Levitan (1989). Diurnal phototropism in leaves of *Lavatera cretica* L. under conditions of simulated solar-tracking. *J. Exp. Bot.*, **218**, 1059–1064.
219. D. Koller, I. Levitan, W.R. Briggs (1985). The vectorial photo-excitation in solar-tracking leaves of *Lavatera cretica* (Malvaceae). *Photochem. Photobiol.*, **42**, 717–723.
220. D. Koller, S. Ritter, W.R. Briggs, E. Schäfer (1990). Action dichroism in perception of vectorial photo-excitation in the solar-tracking leaf of *Lavatera cretica*. *Planta*, **181**, 184–190.
221. D. Koller, S. Ritter (1991). Diaphototropic responses to polarized light in the solar-tracking leaf of *Lupinus palaestinus* Boiss. (Fabaceae). *J. Plant Physiol.*, **138**, 322–325.
222. D. Koller, M. Cohen (1989). After-effects of vectorial photo-excitation in solar-tracking leaves of *Lavatera cretica* L. A kinetic analysis. *J. Plant Physiol.*, **133**, 702–709.
223. D. Koller, T. Shak, W.R. Briggs (1989). Enhanced diaphototropic response to vectorial excitation in solar-tracking leaves of *Lavatera cretica* by an immediately preceding opposite vectorial excitation. *J. Plant Physiol.*, **135**, 601–607.
224. A. Schwartz, D. Koller (1980). Role of the cotyledons in the phototropic response of *Lavatera cretica* seedlings. *Plant Physiol.*, **66**, 82–87.

225. J. Kohji, R. Yamamoto, Y. Masuda (1995). Gravitropic response of *Eichhornia crassipes* (water hyacinth) I. Process of gravitropic bending in the peduncle. *J. Plant Res.*, **108**, 387–393.
226. T.C. Vogelmann, L.O. Björn (1983). Response to directional light by leaves of a sun-tracking lupine (*Lupinus succulentus*). *Physiol. Plant.*, **59**, 533–538.
227. H.C. Wien, D.H. Wallace (1973). Light induced leaflet orientation in *Phaseolus vulgaris* L. *Crop Sci.*, **13**, 721–724.
228. H. Sato, K. Gotoh (1983). Studies on leaf orientation movements in kidney beans (*Phaseolus vulgaris* L.) I. The response to light intensity and location of the photoreceptor. Jap. J. Crop Sci., **52**, 515–520.
229. D.W. Sheriff, M.M. Ludlow (1985). Diaheliotropic responses of leaves of *Macroptilium atropurpureum* cv. Siratro. *Aust. J. Plant Physiol.*, **12**, 151–171.
230. R. Donahue, V.S. Berg (1989). Leaf orientation of soybean (*Glycine max*) seedlings. II. Receptor sites and light stimuli. *Crop Sci.*, **30**, 638–643.
231. W.S. Meyer, S. Walker (1981). Leaflet orientation in water-stressed soybean. *Agron. J.*, **73**, 1071–1074.
232. R. Reed, R.L. Travis (1984). Phototropic leaflet orientation in alfalfa in response to vectorial light. *Crop Sci.*, **24**, 593–597.
233. S.V. Berg, T.C. Hsiao (1986). Solar tracking: Light avoidance influenced by water stress in leaves of kidney bean seedlings in the field. *Crop Sci.*, **26**, 980–986.
234. R. Snow (1959). Some tests of theories of plant torsions. *Proc. Roy. Soc. Lond.*, **151B**, 26–38.
235. W.P. Wilson, J.M. Greenman (1892–1897). Preliminary observations on he movement of the leaves of *Melilotus alba* (L.) and other plants. *Contributions from the Penn State University Botanical Laboratories*, **1**, 66–73.
236. R. Kawashima (1969). Studies on the leaf orientation-adjusting movement in soybean plants. I. The leaf orientation-adjusting movement and light intensity on leaf surface. *Crop Sci. Soc. Japan Proc.*, **38**, 718–729.
237. A. Schwartz, S. Gilboa, D. Koller (1987). Photonastic control of leaflet orientation in *Melilotus indicus* (Fabaceae). *Plant Physiol.*, **84**, 318–323.
238. S. Dubetz (1969). An unusual photonastism induced by drought in *Phaseolus vulgaris*. *Can. J. Bot.*, **47**, 1640–1641.
239. K.A. Shackel, A.E. Hall (1979). Reversible leaflet movements in relation to drought adaptation of cowpeas, *Vigna unguiculata* (L.) Walp. *Aust. J. Plant Physiol.*, **6**, 265–276.
240. I. Forseth, J. Ehleringer (1980). Solar tracking response to drought in a desert annual. *Oecologia*, **44**, 159–163.
241. R.L. Travis, R. Reed (1983). The solar tracking pattern in a closed alfalfa canopy. *Crop Sci.*, **23**, 664–668.
242. M.M. Ludlow, O. Björkman (1984). Paraheliotropic leaf movement in Siratro as a protective mechanism against drought-induced damage to primary photosynthetic reactions damage by excessive light and heat. *Planta*, **161**, 505–518.
243. T.J. Herbert (1984). Axial rotation of *Erythrina herbacea* leaflets. *Amer.J. Bot.*, **71**, 76–79.
244. R.A. Donahue, V.S. Berg, T.C. Vogelmann (1990). Assessment of the potential of blue light gradient in the soybean pulvini as a leaf orientation signal. *Physiol. Plant.*, **79**, 593–598.
245. S. Ritter, D. Koller (1994). Light-driven movements of the trifoliate leaf of bean (*Phaseolus vulgaris* L.) Activity of blue light and red light. *J. Exp. Bot.*, **45**, 335–341.
246. F. Yu, V.S. Berg (1994). Control of paraheliotropism in two *Phaseolus* species. *Plant Physiol.*, **106**, 1567–1573.
247. F. Erath, W.A. Ruge, W.-E. Mayer, R. Hampp (1988). Isolation of functional extensor and

flexor protoplasts from *Phaseolus coccineus* L. pulvini: Potassium induced swelling. *Planta*, **173**, 447–452.

248. S.L. Cronlund, I.N. Forseth (1995). Heliotropic leaf movement response to H^+/ATPase activation, H^+/ATPase inhibition, and K^+ channel inhibition *in vivo*. *Amer. J. Bot.*, **82**, 1507–1513.
249. Y. Nishizaki (1992). Effects of inhibitors on the light-induced changes in electric potential in pulvinar motor cells of *Phaseolus vulgaris* L. *Plant Cell Physiol.*, **33**, 1073–1078.
250. Y. Nishizaki (1996). Effects of blue light on electrical potential and turgor in pulvinar motor cells of *Phaseolus*. *J. Plant Res.*, **109**, 93–97.
251. Y. Nishizaki (1986). Rhythmic and blue light-induced turgor movements and electric potential in laminar pulvinus of *Phaseolus vulgaris* L. *Plant Cell Physiol.*, **27**, 155–162.
252. J.T.M. Elzenga, H.B.A. Prins (1989). Light induced polar pH changes in leaves of *Elodea canadensis*.I. Effects of carbon concentration and light intensity. *Plant Physiol.*, **91**, 62–67.
253. J.T.M. Elzenga, E. Van Volkenburgh (1997). Characterization of a light-controlled anion channel in the plasma membrane of mesophyll cells of pea. *Plant Physiology*, **113**, 1419–1426.
254. J.T.M. Elzenga, E. Van Volkenburgh (1997). Kinetics of Ca^{2+}- and ATP-dependent, voltage-controlled anion conductance in the plasma membrane of mesophyll cells of *Pisum sativum*. *Planta*, **201**, 415–423.
255. V.S.Berg, S. Heuchelin (1990). Leaf orientation of soybean seedlings. I. Effects of water potential and photosynthetic photon flux density. *Crop Sci.*, **30**, 631–638.
256. Q.A. Fu, J.R. Ehleringer (1991). Modification of paraheliotropic leaf movements in *Phaseolus vulgaris* by photon flux density. *Plant Cell Env.*, **14**, 339–344.
257. Q.A. Fu, J.R. Ehleringer (1989). Heliotropic leaf movements in common beans controlled by air temperature. *Plant Physiol.*, **91**, 1162–1167.
258. D. Skrove, T. Rinnar, A. Johnsson (1982). Effects of abscissic acid on the circadian leaf movements of *Oxalis regnelli*. *Physiol. Plant.*, **55**, 221–225.
259. C.-H. Goh, T. Kinoshita, T. Oku, K. Shimazaki (1996). Inhibition of blue light-dependent H^+ pumping by abscisic acid in *Vicia* guard cell protoplasts. *Plant Physiol.*, **111**, 433–440.
260. G.G. Coté, R.C. Crain (1994). Why do plants have phosphoinositides? *BioEssay*, **16**, 39–46.
261. T.C. Vogelmann (1984). Site of light perception and motor cells in a sun-tracking lupine (*Lupinus succulentus*). *Physiol. Plant.*, **62**, 335–340.
262. D. Koller, T. Shak (1990). Light-driven movements in the solar-tracking leaf of *Lupinus palaestinus* Boiss. (Fabaceae). *Photochem. Photobiol.*, **52**, 187–195.
263. Y. Nishizaki, M. Kubota, K. Yamamiya, M. Watanabe (1997). Action spectrum of light-pulse induced membrane depolarization in pulvinar motor cells of *Phaseolus*. *Plant Cell Physiol.*, **35**, 526–529.
264. J. Bialczyk, Z. Lechowski (1990). Effects of light quality on malic acid synthesis in *Phaseolus coccineus* pulvini. *Plant Physiol. Biochem.*, **28**, 315–320.
265. J. Bialczyk, Z. Lechowski (1992). Influence of different spectral regions of light and Ca^{2+} channel blockers on Ca^{2+} and K^+ levels in *Phaseolus coccineus* L. pulvini. *Acta Soc. Bot. Pol.*, **61**, 241–252.
266. S. Ritter, D. Koller (1994). Movements of the trifoliate leaf of bean (*Phaseolus vulgaris* L.) during a simulated day, and their consequences for solar-tracking fidelity and interception of solar radiation. *J. Plant Physiol.*, **143**, 64–71.
267. R.J. Ellis (1984). Kinetics and fluence response relationships of phototropism in the dicot *Fagopyrum esculentum* Moench (Buckwheat). *Plant Cell Physiol.*, **25**, 1513–1520.

268. D. Koller, S. Ritter, D.C. Fork (1996). Light-driven movements of the trifoliate leaf of bean (*Phaseolus vulgaris* L.). Spectral and functional analysis. *J. Plant Physiol.*, **149**, 384–392.
269. J.W. Reed, P. Nagpal, D.S. Poole, M. Furuya, J. Chory (1993). Mutations in the gene for red/far-red light receptor phytochrome B alter cell elongation and physiological responses throughout *Arabidopsis* development. *Plant Cell*, **5**, 147–157.
270. B.M. Parks, P.H. Quail, R.P. Hangarter (1996). Phytochrome A regulated red-light induction of phototropic enhancement in *Arabidopsis*. *Plant Physiol.*, **110**, 155–162.
271. T. Tanada (1997). The photoreceptors in the high irradiance response of plants. *Physiol. Plant.*, **101**, 451–454.
272. A.L. Mancinelli (1989). Interaction between cryptochrome and phytochrome in higher plant photomorphogenesis. *Amer. J. Bot.*, **76**, 143–154.
273. H. Mohr (1994). Coaction between pigment systems. In: R.E. Kendrick, G.H.M. Kronenberg (Eds), *Photomorphogenesis in Plants* (2nd ed., pp. 353–373). Kluwer, Dordrecht, Boston, London.
274. J. Kigel, A. Schwartz (1990). Co-operative effects of blue and red light in the inhibition of hypocotyl elongation of de-etiolated castor bean. *Plant Sci. Lett.*, **21**, 83–88.
275. E. Liscum, R.P. Hangarter (1993). Light-stimulated apical hook opening in wild type *Arabidopsis thaliana* seedlings. *Plant Physiol.*, **101**, 567–572.
276. E. Van Volkenburgh, R.E. Cleland, M. Watanabe (1990). Light-stimulated cell expansion in bean (*Phaseolus vulgaris* L.) leaves II. Quantity and quality of light required. *Planta*, **182**, 77–80.
277. M. Staal, J.T.M. Elzenga, A.G. van Elk, H.B.A. Prins, E. Van Volkenburgh (1994). Red and blue light-stimulated proton efflux by epidermal leaf cells of the Argenteum mutant of *Pisum sativum*. *J. Exp. Bot.*, **45**, 1213–1218.
278. D.E. Blum, T.M. Elzenga, P.A. Linnemeyer, E. Van Volkenburgh (1990). Stimulation of growth and ion uptake in bean leaves by red and blue light. *Plant Physiol.*, **100**, 1968–1975.
279. E.E. Serrano, E. Zeiger, S. Hagiwara (1988). Red light stimulates an electrogenic proton pump on *Vicia* guard cells. *Proc. Natl. Acad. Sci. (USA)*, **85**, 436–440.
280. Y. Nishizaki (1990). Effects of anoxia and red light on changes induced by blue light in the membrane potential of pulvinar motor cells and leaf movement in *Phaseolus vulgaris* L. *Plant Cell Physiol.*, **31**, 591–596.

Photomovement
D.-P. Häder and M. Lebert, editors.

Chapter 26

Light-controlled chloroplast movement

Masamitsu Wada and Takatoshi Kagawa

Table of contents

26.1 Introduction

Light-directed chloroplast movement is a wide-spread phenomenon in plant cells which need to perform highly efficient photosynthesis both under low and high fluence rate conditions. Chloroplasts in a mesophyll cell of a seed plant, for example, spread over the surface of the cell faced toward the light source under low and moderate light conditions (called low fluence rate response, LFR), but under very high fluence rate, they move to the cell sides, i.e. anticlinal walls, to avoid photodamage of the chloroplasts (high fluence rate response, HFR). In this chapter we call this phenomenon "chloroplast relocation" (Figure 1). In the case of some algal cells, such as *Mougeotia* and *Mesotaenium,* which have one large ribbon-shaped chloroplast in the midst of the cell, between two large vacuoles, the chloroplast turns in the cell to face the light source at low fluence rates but does not move toward the light source. We call this phenomenon "photoorientation" (Figure 1). These phenomena have been known and analyzed for more than 100 years [1], but up to now the photoreceptors which are involved in the chloroplast relocation in seed plants and in cryptogams as well are not known. The objects used for the analytical studies on this phenomenon are very restricted, such as green algae (*Mougeotia, Mesotaenium*), ferns (*Adiantum, Pteris*) and seed plants (*Vallisneria, Lemna*), which are all water plants except fern and have simple organization. Microscopic observation of intact cells in multi-layered tissues of land plants is not easy, but had been unavoidable before a new method was introduced using

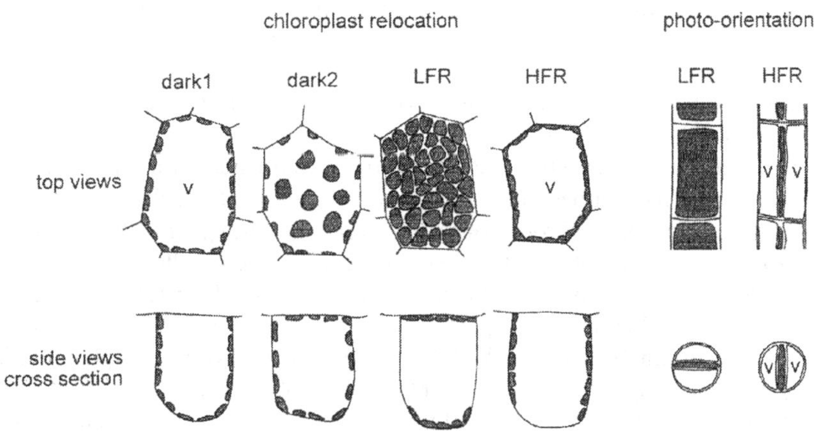

Figure 1. Schematic illustration of chloroplast relocation in multi-chloroplast cells such as of seed plants, ferns and mosses (left) and photoorientation in a cell with a ribbon-shaped chloroplast such as *Mougeotia* (right). Chloroplasts change their location according to light condition, i.e. dark condition (dark 1 and dark 2), low fluence rate (LFR) and high fluence rate (HFR) conditions. Patterns of chloroplast distribution in a cell are shown as top and side views. There are two patterns of chloroplast relocation under dark conditions, with or without chloroplasts at a periclinal wall. In LFR chloroplasts distribute to the periclinal walls and the bottom of the cell. In HFR they locate near anticlinal walls. In a photoorientation, a chloroplast shows a face position under LFR and edge position in HFR. V indicates a vacuole. Shaded circles or rectangles are chloroplasts.

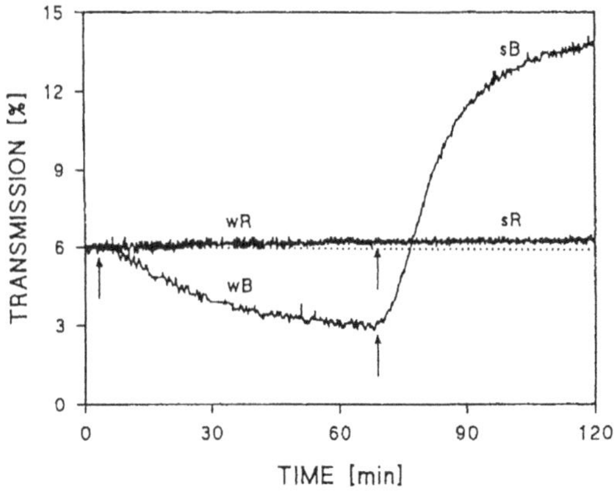

Figure 2. Spectroscopic detection of chloroplast relocation. Transmission of a measuring beam of red light changes according to chloroplast relocation, high in HFR and low in LFR. Chloroplasts under dark conditions in an *Arabidopsis thaliana* leaf cell respond to continuous blue, weak and then strong light irradiated from the points indicated by the arrows. A red light control is also shown as weak and strong light irradiated at the same time as blue light. The plant was dark-adapted for 12 h. wB, weak blue (0.1 W m^{-2}); sB, strong blue (30 W m^{-2}); wR, weak red (0.07 W m^{-2}); sR, strong red (20 W m^{-2}) (from [4]).

photometric monitoring of chloroplast photomovement ([2–4], Figure 2). However, the photometric detection based on light transmittance or light absorption may include artificial signals other than chloroplast relocation itself. But the photometric method is more sensitive than microscopic observation and applicable to multicellular systems, such as leaves of land seed plants [2,4].

The mechanisms of chloroplast relocation or photoorientation are not yet resolved. The photoreceptor has not been identified in a strict sense, although it is clear that blue light receptor(s) and phytochrome (in a few cases, i.e. *Mougeotia, Mesotaenium* and *Adiantum*) are involved, but not chlorophyll (at least in the low fluence rate response). In relation to blue light responses, two cryptochrome genes (CRY1, CRY2) were found in *Arabidopsis* as blue light receptors [5] and 5 (CRY1-CRY5) in *Adiantum* [6]. Recently, phototropin (formerly called NPH1), which is involved in phototropism of *Arabidopsis,* was suggested to be a blue light photoreceptor [7,8]. At least 3 phytochrome genes were cloned and all three were found to be expressed in *Adiantum* (AdiPHY1-AdiPHY3, [9,10]), one in *Mougeotia* [11] and six in *Mesotaenium* (*mesphy,* [12]).

Since the light-induced chloroplast movement is a very fundamental and important phenomenon in plant life, many reviews on this subject have been published within the last 10 years [13–18]; especially, the recent review by Haupt [13] summarizes much old, but basic information before Senn [1], so that it is very useful to know the historical background of this field. In contrast, not many original papers were published during this time. In this article, we overview the historical background of chloroplast movement briefly and then discuss mainly the data obtained in the last ten years. The old information can be read in detail in the above-mentioned reviews. Light-induced chloroplast movement is not a simple phenomenon but is quite different in different plant groups. It is still possible that even the photoreceptor and/or the mechanism of chloroplast photomovement vary from group to group (see Figure 3). Therefore we will describe the phenomenon of each plant group in separate sections.

26.2 Ferns

26.2.1 Fern gametophytes as experimental systems

Fern gametophytes are a good model system for studying light-induced chloroplast relocation on the cellular level [19] because protonemata are composed of linearly arranged cells (Figure 4) and gametophytes of two-dimensional but single cell layers except in the central mat area of aged gametophytes. Since fern gametophytes are autonomous organisms and not surrounded by any other tissue, it is easy to treat them with light and observe them under a microscope. Blue as well as red light are effective

directional movement	small numerous chloroplasts cell	HFR	B		most of seed plants[1,2,3], Adiantum[4,5], Funaria[6]
			R		Adiantum [4]
		LFR	B		most of seed plants[1], Adiantum[4], Dryopteris[7], Funaria[6], Ceratodon[8]
			R		Adiantum[4], Dryopteris[7]
	single large chloroplast cell	HFR	B	directional movement	Mougeotia[9]
				elevating sensitive	Mesotaenium[10]
		LFR	B		Mougeotia[11]
			R		Mougeotia[12], Mesotaenium[13]
non-directional movement		HFR	B		Vaucheria[14]
		LFR	B		Vallisneria[14], Vaucheria[15]
			R		Vallisneria[15]

HFR, high fluence rate response; LFR, low fluence rate respose; B, blue light; R, red light
[1]Senn (1908); [2] Zurzycki and Lelatko (1969); [3]Zurzycki et al. (1983); [4]Yatsuhashi et al. (1985); [5]Kagawa and Wada (1998); [6]Zurzycki (1980); [7]Yatsuhashi and Kobayashi (1993); [8]Kagawa et al. (1997); [9]Gabrys et al. (1985); [10]Haupt and Häder (1994); [11]Gabrys (1984); [12]Haupt (1972); [13]Haupt and Thiele (1961); [14]Schönbohm (1980); [15]Izutani et al. (1990)

Figure 3. Classification of chloroplast movement.

to induce chloroplast relocation [20]. The red light effect is mediated by phytochrome and blue light effect by an unknown blue light receptor. On chloroplast relocation, *Adiantum capillus-veneris* is the best-analyzed species in the last decade, not only in ferns but in all plants. In *Pteris vittata*, which is another well-studied fern species on photomorphogenesis [19,21], chloroplast relocation is only induced by blue light [22], which indicates that the results described below are not always true for all fern species.

Before describing the experimental results on chloroplast relocation in *Adiantum*, the cell system will be explained briefly. When imbibed spores were cultured under continuous red light for a week, long single-celled protonemata were obtained, in which chloroplasts were dispersed evenly except in the apical part of the cell where about 200 chloroplasts were counted [23]. When the cells were irradiated with white or blue light, cell division occurred at the apical part of the cell (Figure 4a). In the long filamentous part of a basal cell of the two-celled protonema chloroplast relocation is more pronounced than in the same part before cell division, although we do not know the reason [24].

When protonemal cells are cultured under white light for several weeks, a two dimensional heart-shaped prothallus develops (Figure 4b). The prothallial cells were

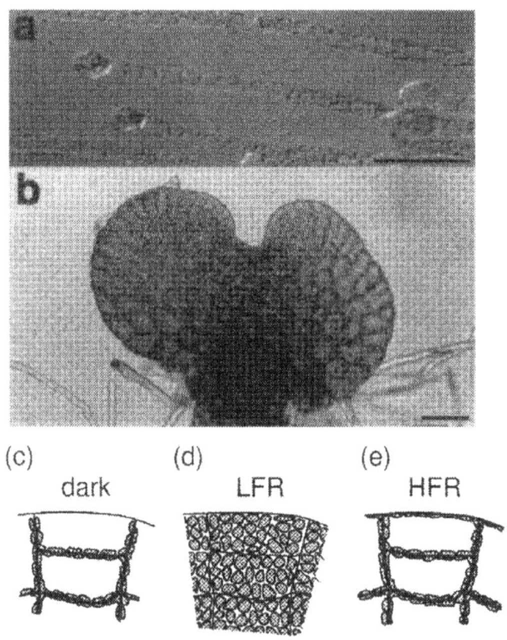

Figure 4. Fern *Adiantum* gametophytes, two-celled protonemata (a), a two-dimensional prothallus (b) and chloroplast relocation in prothallial cells (c-e). In LFR (d) chloroplasts accumulate near periclinal walls but in HFR (e) and dark condition (c) they accumulate near anticlinal walls. The only difference between HFR and dark conditions is whether chloroplasts accumulate at the cell wall of the prothallus margin in the former and not in the latter (bar 100 μm).

found to be good experimental material to study chloroplast relocation [25]. Under weak light condition of red, blue or white light, chloroplasts spread over the cell surface (Figure 4d). Under very strong light, on the contrary, chloroplasts moved to the anticlinal walls between adjacent cells (Figure 4e), probably to avoid photodamage of the chloroplasts. Under continuous darkness for 2–3 days, chloroplasts accumulated at the anticlinal walls (dark position of chloroplasts, Figure 4c) like under very high fluence rates. There is one difference between the dark-position and the position under high fluence rate light, namely the existence of chloroplasts at the mergin of prothallia in the latter but not in the former case (compare Figure 4c and 4e). Dark adaptation is thought to cause high sensitivity to light in both protonemata and prothalli.

26.2.2 Chloroplast relocation in protonemal cells

Chloroplast relocation in protonemal cells could be induced effectively by a partial irradiation of the cells with a blue or red microbeam or by whole cell irradiation with polarized red or blue light [20]. A microbeam of low fluence rate light (<1 W m^{-2} of blue light and <100 W m^{-2} of red light) induces chloroplast accumulation in the irradiated area and that of the high fluence rate light (>10 W m^{-2} of blue and >230 W m^{-2} of red) induces a chloroplast avoidance response, that is, the chloroplasts move out of the beam. Polarized red and blue light coming from the tip or bottom of the cell (Figure 5c,d) or coming from cell side and vibrating perpendicular to the cell axis (Figure 5b) induces chloroplast accumulation in the area where the plasma membrane and vibration plane are parallel to each other in the former case (Figure 5c, d), or in the area where the plasma membrane and the incident light are perpendicular in the latter case (Figure 5b). Polarized light from the side but vibrating parallel to the cell axis (Figure 5a), however, does not induce cell accumulation (compare to Figure 5b). These patterns of chloroplast accumulation could be explained by a Pfr gradient in the cell periphery as shown in Figure 5e-h.

The photoreceptor for red light is phytochrome and that for blue light is a blue light receptor but not the phytochrome blue peak, because a reversible effect between red and far-red light could be observed in the former, but not between blue and far-red light in the latter [20]. Chloroplasts move towards the high Pfr area, not like in *Mougeotia* where the edge of the chloroplast avoids the high Pfr and moves towards the low Pfr area [26]. Mutants which do not show phytochrome-mediated phototropism as well as chloroplast movement in protonemata were screened in *Adiantum* [27], indicating that both physiological responses share the same phytochrome species or at least its early signal transduction pathway. The specific phytochrome species has not yet been identified, but is not the same one as for spore germination, because phytochrome-dependent spore germination is still functional in these mutants [27].

The dual effects of phytochrome on chloroplast relocation was studied in protonemata of *Adiantum* [28] and *Dryopteris sparsa* [29]. When dark-adapted two-celled protonemata of *Adiantum* were irradiated with polarized red light of about 0.1–10 W m^{-2} for 3 or 10 min and then transferred to darkness, directional movement of chloroplasts was detected as early as 10 min after the irradiation, and continued for over 30–60 min in the subsequent dark period, then the chloroplasts gradually dispersed

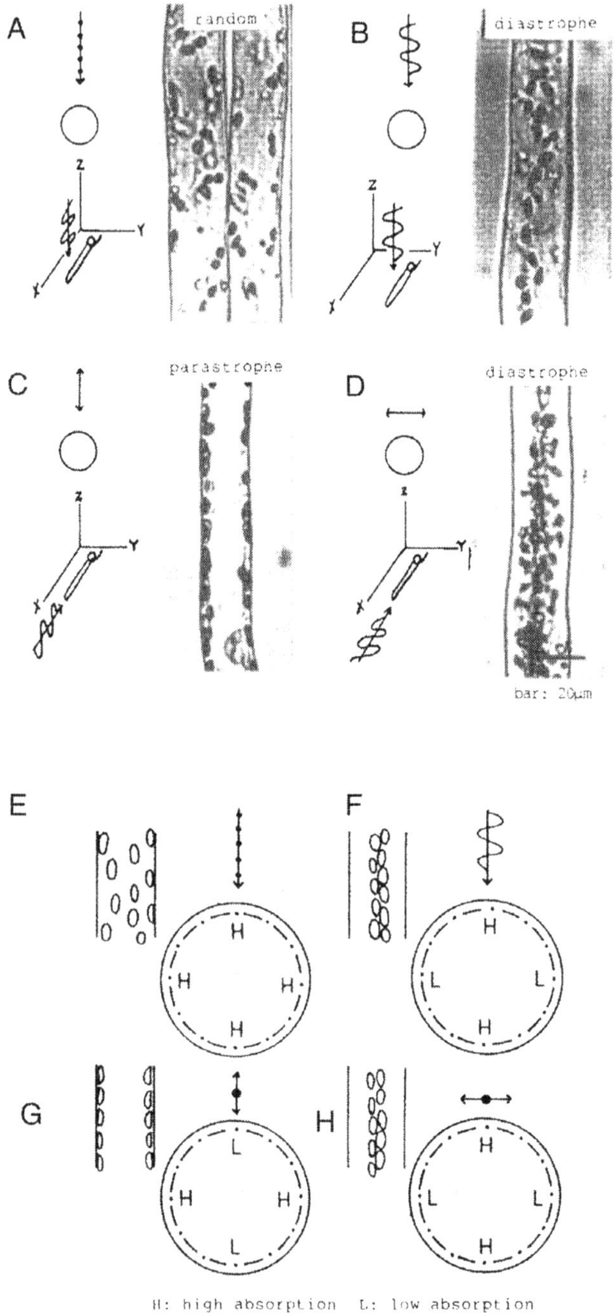
A
random
B
diastrophe
C
parastrophe
D
diastrophe
bar: 20μm
E
F
G
H
H: high absorption L: low absorption

Figure 5. Chloroplast relocation in *Adiantum* protonemal cells induced by polarized light with different vibration planes from different directions (a–d) and its implication (e – h) by means of tetrapolar distribution of high and low densities of light absorbed by the photoreceptor, phytochrome or blue light absorbing pigment. Light direction and a vibration plane of polarized light and the direction of a protonemal cell are shown three dimensionally (x, y, z) in the left bottom and its cross sectional view (in yz plane) in left top of a–d. Photographs on the right show the results (bar 20 μm). In e–h, chloroplast relocation is shown schematically in the left, and in the right highest (H) and lowest (L) absorption of light by photoreceptors in a cortical region illustrated as bars parallel to the cell membrane and dots are shown. All except e show tetrapolar distribution, that is chloroplasts accumulate at the highest light absorption, at the top and bottom or at both flanks. In e on the contrary, all dots absorbed light but not bars, so that no absorption difference is induced, and no accumulation of chloroplasts is observed.

again. When dark-adapted protonemata were irradiated with far-red light some time after the red light, the magnitude of the relocation response was smaller than in the control without far-red light and the chloroplasts dispersed more quickly than those exposed to red light alone (Figure 6). A similar phenomenon was observed in dark-adapted *Drypteris* protonemata [29]. Chloroplast relocation induced by continuous irradiation with polarized red light was maintained for approximately 3 h in the following darkness, but if far-red light was given immediately after the red light, the

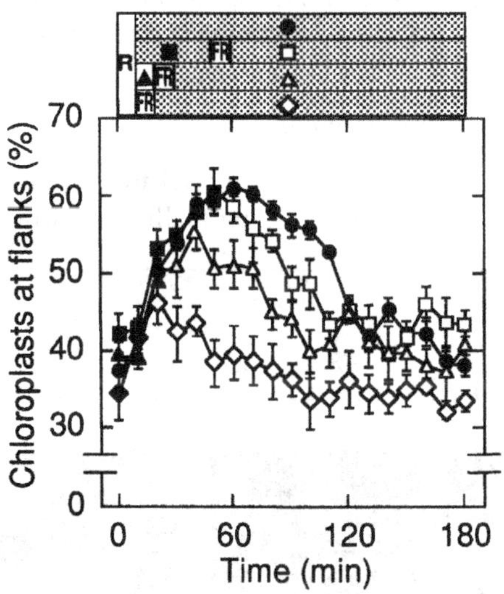

Figure 6. Dual effects of Pfr in chloroplast relocation in *Adiantum* dark-adapted protonemal cells. When vertically vibrating polarized red light ($3\ W\ m^{-2}$) was irradiated for 10 min chloroplasts started to accumulate at the flanks. But if non polarized far-red light ($5\ W\ m^{-2}$, 10 min) was irradiated at various times during accumulation (see white triangle and diamond), chloroplasts moved back from the points. When far-red was irradiated after accumulation, the chloroplasts moved back faster than without far-red light (compare dark circle and white square). Thus Pfr promotes chloroplast accumulation and anchorage. Closed symbols represent the response after R irradiation, open symbols after far-red light (after [28]).

chloroplasts began to disperse after 1 h [28]. These results suggest that phytochrome is effective not only to determine the direction of the movement but also to hold the chloroplast at the sites where they have relocated. It is not known whether these dual effects are mediated by one molecular species of phytochrome or by different species, because phytochrome is encoded by a gene family in *Adiantum* and all three phytochromes are expressed in protonemal cells [10]. It is possible that there exists a population of phytochrome molecule(s) of a long-lasting form, as proposed for *Mougeotia* [30].

In connection with the above results, chloroplast anchorage at the relocated area was detected by a video-tracking system in *Adiantum* (Figure 7a, [31]) and was confirmed by the observation of a circular structure of microfilaments formed between the plasma membrane and the periphery of convex-shaped chloroplasts (Figure 7b, [32]). The circular structure was formed when the chloroplasts reached the appropriate area and disappeared before chloroplast dispersion occurred [33].

In dark-adapted *Adiantum* protonemata, irradiation with polarized blue light of about 0.1–1.0 W m^{-2} for 10 min was not effective to induce chloroplast relocation in the following darkness, while irradiation with the blue light for a longer time at the same fluence rate clearly induced the relocation. It is likely that long-term irradiation is necessary for the blue light-induced response in *Adiantum* protonemata [28], although in dark-adapted prothallial cells a short pulse of blue light is sufficient (see below). This is the same in *Dryopteris* protonemata, for which long-term irradiation is needed. Moreover, in *Dryopteris* protonemata blue light-induced relocation decayed as soon as the blue light was switched off and no anchorage effect could be observed [29].

26.2.3 Chloroplast relocation in prothalli

Chloroplast relocation could be induced by a brief irradiation with red or blue light in dark-adapted prothalli [25]. When whole prothalli were irradiated with horizontally

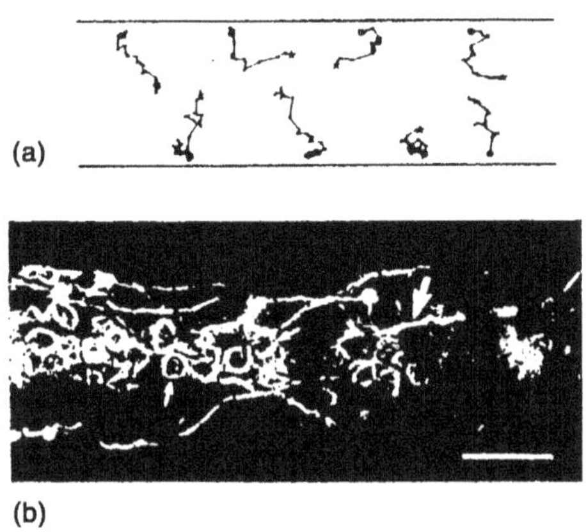

Figure 7. Individual tracks (traced for 0.5 h every 72 s) during chloroplast relocation induced by continuous polarized red light (0.27 W m^{-2}) in an *Adiantum* protonemal cell (a) and ring-like structures observed between chloroplast margin and cell membrane after relocation (b) (a from [31], b from [32]).

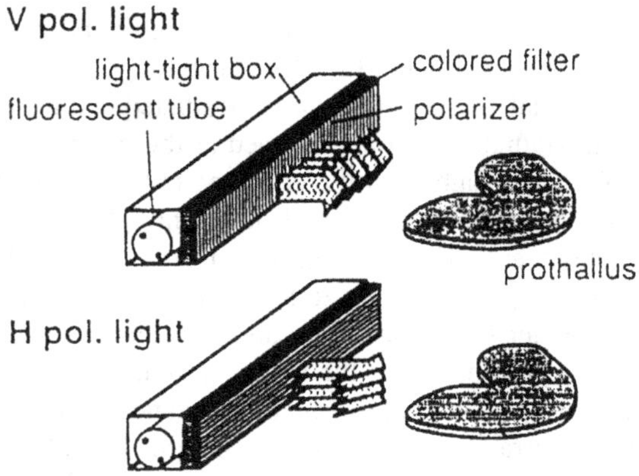

Figure 8. Schematic illustration of a method of irradiation with polarized light of different vibrating planes of the electric vector. Prothallial cells were irradiated horizontally from their lobes (after [25]).

vibrating polarized red or blue light from a horizontal direction for 3 or 10 min (Figure 8), chloroplasts moved from the anticlinal wall to the cell surface within 1–2 h in the following darkness and returned to the anticlinal walls within 10 h in the former (red light) and 4 h in the latter (blue light), indicating that the lifetime of red and blue light signals are different. The red light effect could be cancelled by the subsequent far-red light but the blue light effect could not, similar to the responses in protonemata. This result indicates that the red light was absorbed by phytochrome but blue light was not by phytochrome, but a blue light absorbing pigment. Vertically vibrating polarized light did not have such an effect.

Partial cell irradiation at a mid part of a dark-adapted prothallial cell with a microbeam ($10 \times 10\ \mu m^2$) of low fluence rate red or blue light induced a chloroplast relocation movement towards the beam (Figure 9a). However, in the neighboring cells, no chloroplast movement could be induced either by blue or red light, indicating that the signals raised by blue and red light could not be transferred across cell walls (Figure 9b,c, [25]). Using this experimental system, further studies were carried out with various fluence rates [34,35]. The higher the fluence rate of the light, the greater the distance from which chloroplasts could be attracted in both blue and red microbeam. Although the blue microbeam is less effective than the red one, chloroplasts began to move much faster by blue (within 10 min) than by red microbeam (more than 20 min for the farthest ones). Although it is clear that both light effects are mediated by different photoreceptors, i.e. phytochrome and blue light-absorbing pigment(s), when red and blue microbeams of sub-thresholds intensities for chloroplast relocation were given at the same area, they worked additively, and chloroplast relocation was induced (Figure 10), indicating that some parts of the signal transduction pathways of both photoreceptors might be shared before chloroplast movement occurs [34]. Accordingly, it is quite reasonable that the velocity of chloroplast movement was constant, about 0.3 μm/min, under both red and blue light with different fluence rates tested.

When a very strong blue microbeam was irradiated at a mid part of a dark-adapted prothallial cell, chloroplasts which had been at the anticlinal walls began to move toward the irradiated area, but they stopped at the edge of the beam and could not enter the beam-irradiated area (Figure 11a,b, [35]). If the same light treatment was performed in a light-adapted prothallial cells, chloroplasts irradiated with the strong blue light moved out of the beam irradiated area but stopped at the periphery of the beam when they escaped from it (Figure 12a,b). On the contrary, chloroplasts which had been far from the irradiated area, began to move towards the light field until they were forced to stop by other chloroplasts escaped from the beam area (Figure 12a). These results suggest that the blue light signal effective to accumulate chloroplasts (the low fluence response) can be raised under strong blue light and transferred a long distance from the irradiated area to the cell periphery, but the signal for chloroplast avoidance (the high fluence response) can not be transferred over long distance. The latter signal could be functional only at the area irradiated with the strong light. Thus, the signal transduction pathways of high and low fluence rate responses might be different. More support for this assumption was obtained by different experiments [35]. In the low fluence rate response, red and blue light are additively functional (Figure 10, [34]), but in the high fluence rate response simultaneous irradiation of a blue microbeam and red background illumination or sequential irradiation with blue and red microbeams did not show any additive effect. The lifetime of a strong blue light signal is less than about 10 min [35]. In contrast, the signal of weak blue light lasts about 30–40 min [34]. In summary, the

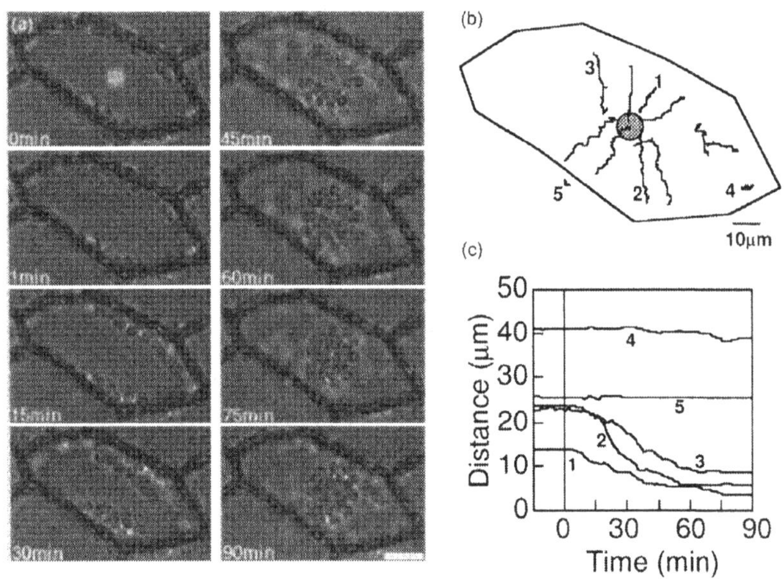

Figure 9. Chloroplast relocation induced by a microbeam irradiation in a dark-adapted prothallial cell of *Adiantum*. (a) When one minute of red (30 W m^{-2}) microbeam was given, chloroplasts moved towards this position after the beam was switched off over 60 min and then dispersed (bar 20 μm). Individual tracks (traced for 120 min every 1 min) during chloroplast accumulation induced by red microbeam (30 W m^{-2}, 1 min) are shown in (b) and their speeds, i.e. the distances between each chloroplast and beam center in terms of time are plotted in (c). The same numbers in both panels mean the same chloroplasts.

high fluence rate response and low fluence rate response of chloroplast relocation might be controlled by different signal transduction pathways (probably including photoreceptors), although both responses have been thought to be controlled by a single photoreceptor, because similar shapes of action spectra for both low and high fluence rate responses were obtained in *Lemna* and *Selaginella* [36]. The assumption of

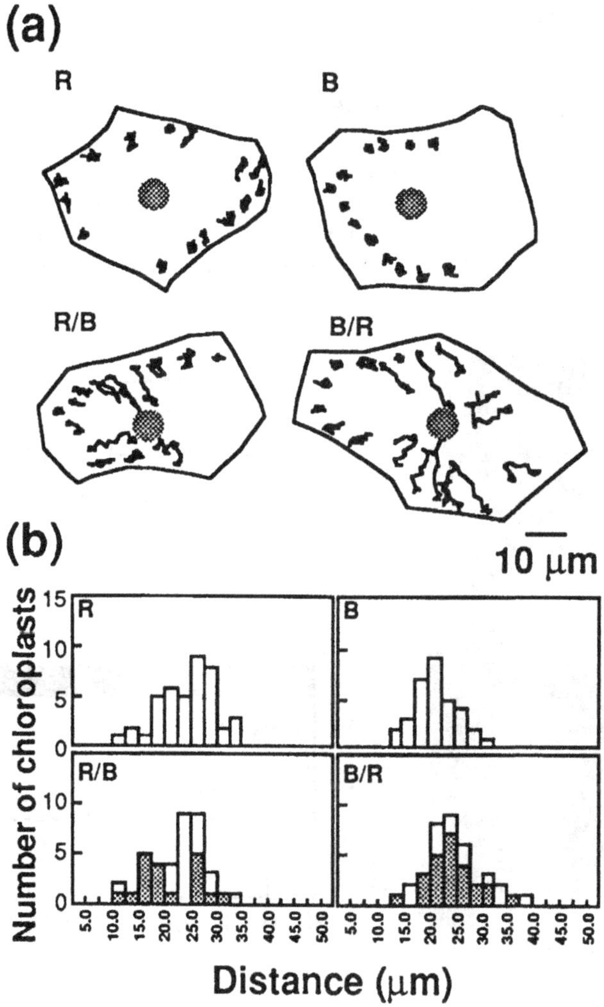

Figure 10. Additive effects of red and blue microbeams on the chloroplast accumulation response. A dark-adapted *Adiantum* prothallial cell was irradiated with a microbeam of under threshold of blue (30 W m^{-2} for 2 s) and red light (5 W m^{-2}, 2 s) at the same part of the cell and the tracks of individual chloroplast were traced (a). Histograms of accumulated (shaded columns) or non accumulated (open columns) chloroplasts were obtained from the data of a (b). Note that single irradiation with red or blue microbeam has no effect on accumulation, but double beams with red and blue show an additive effect (after [34]).

different photoreceptors in low and high fluence rate responses is easy to understand, because *Adiantum* has five cryptochrome genes [6] and all of them are expressed in gametophytes [37]. Furthermore, mutants which are deficient of the low fluence rate response but normal in the high fluence rate response and vice versa have been screened in *Adiantum* (Wada unpublished data). In *Adiantum*, three phytochrome genes have been sequenced [9,10] and one fragment of phytochrome-related gene is cloned. Among five cryptochromes and three (or four) phytochromes, it is not yet known which CRY and PHY genes are involved in the above chloroplast responses.

26.2.4 Chloroplast relocation in sporophyte leaves

Chloroplast relocation is also observed in the cells of young leaves (Kadota personal communication). Since phototropism could be induced by red light as well as blue light not only in protonemata but also in young leaves in *Adiantum capillus-veneris* and *A. cuneatum* [38], low fluence rate response of chloroplasts relocation under red as well as

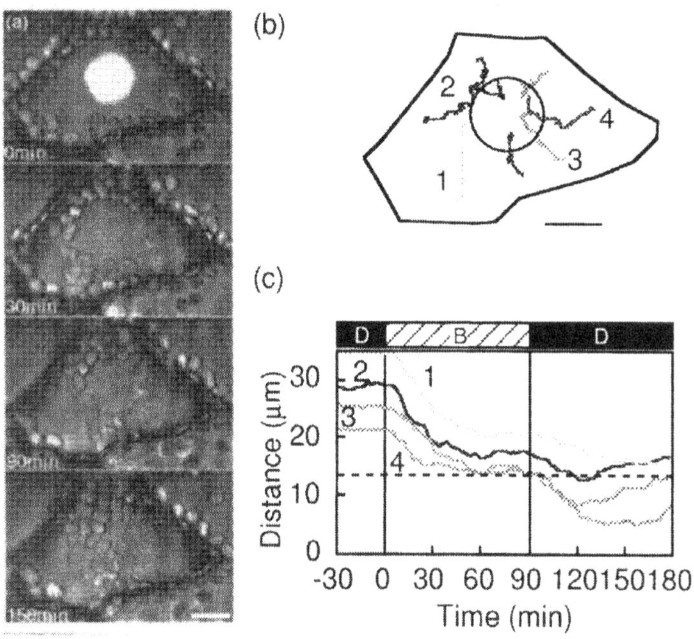

Figure 11. Chloroplast behavior at the edge of strong blue light beam in a dark-adapted *Adiantum* prothallial cell. (a) A strong blue (30 W m^{-2}) microbeam (27 µm in diameter) was applied continuously from 0 to 90 min at the center of a dark-adapted cell, and then the cell was kept in the dark (bar 20 µm). The tracks of individual chloroplast were traced (b) and the distances between each chloroplast and beam center were plotted against time (c). The same numbers in both panels mean the same chloroplasts. Note that the chloroplasts moved towards the beam close to its edge but could not enter the beam area. Once blue light was switched off, they moved into the formerly irradiated area.

blue light was tested in young leaves of *A. capillus-veneris*. In the epidermal cells along the petiole chloroplast relocation could be induced by partial cell irradiation with a microbeam as well as whole cell irradiation with polarized light, similar to prothallial cells. The high fluence rate response under blue light is also observed. If the chloroplast behavior is different between gametophytes and sporophytes, i.e. haplophase and diplophase, respectively, it is worth to continue the experiments, but so far no difference has been observed in these systems. Recently, Augusynowicz and Gabrys [39] tested chloroplast relocation photometrically in leaves of 4 different species of ferns, *A. capillus-veneris*, *A. caudatum*, *A. diaphanum* and *Pteris cretica* and reported that the species showed low and high fluence response under blue light. Red light is effective only in *A. capillus-veneris* for low fluence rate response. Ferns living in varied light conditions show the dynamic chloroplast movement, but species living under rather constant light condition show very small movement [39].

26.3 *Mougeotia*

Mougeotia has long been the best model system for studying chloroplast photoorientation movement. Haupt and his coworkers studied this organism intensely in Germany

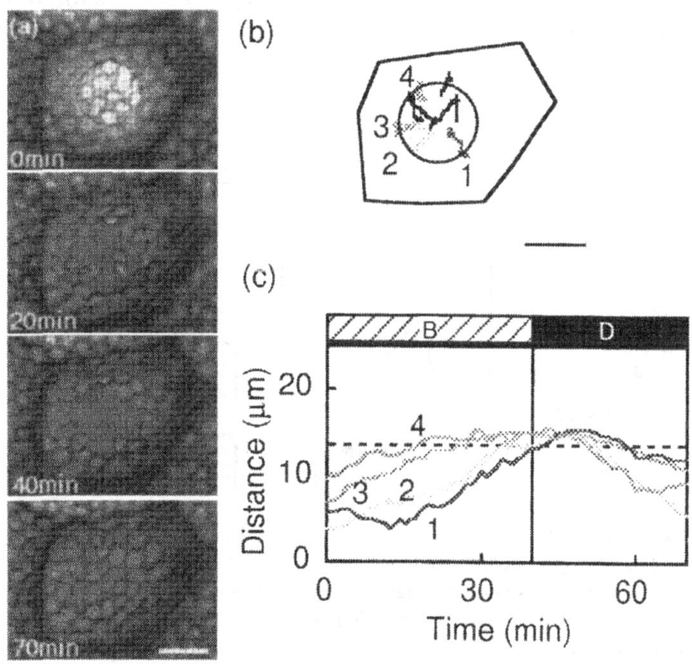

Figure 12. Chloroplast behavior at the border of a strong blue microbeam in a light-adapted *Adiantum* prothallial cell. A strong blue microbeam was applied continuously from 0 to 40 min, and then the cell was kept in the dark (a). The tracks of individual chloroplasts were traced (b) and the distance between chloroplast and beam were plotted against time (c). Other details as in Figure 11. Note that the chloroplasts in the beam moved out of the beam during blue light irradiation and came back into the beam area after the light was switched off.

for many years. Since most of the results are included in many text books and various review articles [16–18,40], there is no need to repeat the fundamental results here in detail. Further, not much new information was obtained on the mechanism of chloroplast photoorientation in *Mougeotia* during the last decade. Here we summarize briefly the historical background of the photoorientation movement of *Mougeotia* chloroplasts and then describe recent results.

Mougeotia is a filamentous green alga consisted of many identical cells chained end to end (Figure 13a). Each cell contains a large, single chloroplast which is flat rectangular and sandwiched by two large vacuoles. A nucleus sits on the center of the "ribbon-shaped" chloroplast. Under weak white light, the chloroplast turns in a cell facing the light source of the incident light ("face position"). But at high fluence rates of white light, the chloroplast changes the direction to the "edge position", being parallel to the incident light, to reduce light absorption. The former is called "low fluence rate response" and the latter "high fluence rate response".

The low fluence rate response is induced by red light or weak white light and cancelled by subsequent far-red light, indicating that phytochrome is the photoreceptor [41]. However, the photoorientation response is not so simple. Under polarized red light, chloroplast orientation is dependent on the direction of vibration plane of the polarized light, but not simply on the direction of incident light (Figure 13a). To explain this complicated phenomenon, Haupt studied the intracellular localization of phytochrome

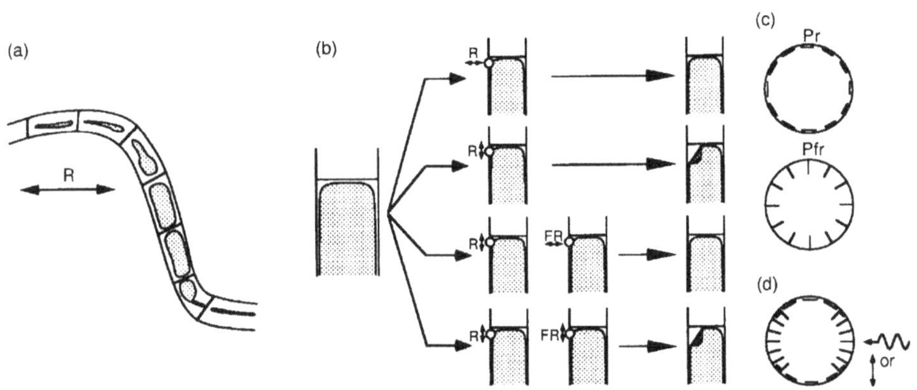

Figure 13. Schematic illustration of *Mougeotia* chloroplast photoorientation under polarized red light vibrating horizontally indicated by the double headed arrow (a). Note that the chloroplasts in cells parallel to the cell axis showed edge position and those in cells perpendicular to the cell axis showed face position (arranged from [83]). Phytochrome distribution in cell cortex was studied by Haupt with a skilful technique of microbeam irradiation with polarized red and far-red light on a cell edge (b). Note that polarized red light vibrating parallel to the cell axis is effective, but that of perpendicular is not. On the contrary, polarized far-red light vertical to the cell axis is effective, but not the parallel one to reverse the red-light induced effect. From these results the transition moment of Pr was thought to be parallel to the cell surface and that of Pfr was perpendicular (c) (arranged from [43]). Tetrapolar distribution of the Pr and Pfr level induced by polarized red light irradiation in a, causes chloroplast photo-orientation. Chloroplast edges move from high Pfr to low Pfr regions.

molecules and the orientation of their transition moment in *Mougeotia* with a skilful technique of partial cell irradiation with a microbeam of polarized red and/or far-red light (Figure 13b, [42,43]). The answer is as follows: phytochrome is localized in or near the plasma membrane and the transition moment of the red absorbing form of phytochrome (Pr) is parallel to the plasma membrane, and the far-red light absorbing form (Pfr) is perpendicular to it (Figure 13c). Phytochrome is not on a chloroplast itself. Because of the dichroic orientation of phytochrome molecules, a tetrapolar distribution of Pfr gradient is established under polarized red light (Figure 13d). Both chloroplast edges, which attach to the cell periphery, move from high Pfr to low Pfr. His model of the phytochrome distribution and its dichroic orientation in the cell explains all chloroplast photoorientation movement of *Mougeotia* not only under polarized light but also under non polarized light conditions [16,40]. The dichroic orientation of phytochrome on or close to the plasma membrane is also shown in fern protonemata [44,45].

The blue light effect on chloroplast photoorientation in *Mougeotia* is also complicated. Low fluence rate blue light is not so effective comparing to red light, so that the responses under low fluence rate blue light are not well understood. The effect of high fluence rate blue light, on the contrary, is well studied compared to the red light effect [16,18]. When high fluence rate blue light is given with red light, chloroplast edges move from the area of low Pfr to high Pfr, just opposite to red light alone, indicating that the blue light changes the direction of response to the red light. In this case the direction of blue light is not important. When a cell is irradiated with a strong blue light alone (without red light), the chloroplast moves into the edge position as "high fluence rate response", because, besides the effect of the high fluence rate blue light, phytochrome molecules in the front and rear end of the cell absorb more blue light at the second absorption peak of phytochrome than those in the flanks and increase the Pfr gradient. The interaction between phytochrome and the blue light photoreceptor is not a pigment-pigment interaction (see below).

26.3.1 Photoreceptors

Cloning and sequencing of *Mougeotia* phytochrome cDNA [11] were performed recently. Only one phytochrome gene was found in *Mougeotia*, which indicates that this is the phytochrome involved in chloroplast photoorientation. The deduced amino acid sequence of the phytochrome has a microtubule binding site, which is not common in phytochromes of other species [10]. Although the effect of microtubules on chloroplast photoorientation is not clear in this species [46], further analysis of this phytochrome may shed light on the signal transduction pathway of chloroplast photoorientation in *Mougeotia*.

The involvement of the third photoreceptor in *Mougeotia* chloroplast photoorientation was postulated by Lechowski and Bialczyk [47,48] based on experiments in which chloroplast movement was induced by two different light beams, one at 730 nm and another varying from 500 – 680 nm. Walczak et al. [49] repeated the experiment with high fluence rate far-red light, but could not find the 620 nm peak for photoorientation, although they confirmed that the peak appeared when far-red fluence rate was weak. The reason why 620 nm is more effective than 660 or 680 nm under weak far-red light still remains unknown.

26.3.2 Signal transduction pathways

Calcium has been thought to be a second messenger of the chloroplast photoorientation in *Mougeotia* and was studied for many years. However, as positive (e.g. [50–53]) and negative [54] data have been accumulated, it is still an open question whether calcium is a real messenger in this phenomenon.

The involvement of Ca^{2+} on chloroplast photoorientation has been suggested [15,55]. Schönbohm et al. [54,56] tested the hypothesis by Ca^{2+} deficiency, or with Ca^{2+}-entry blockers (diltiazem, nifedepine, ruthenium-red, La^{3+} and Co^{2+}) and calmodulin antagonists (trifluoperazine and W-7). However, no evidence supporting the Ca^{2+} hypothesis could be obtained in their experiments. Calcium ions must be important not only to accomplish chloroplast photoorientation but also to maintain basic conditions for cellular activities; therefore it may be difficult to prove that Ca^{2+} is a specific messenger for chloroplast photoorientation.

If *Mougeotia* is irradiated with blue light of high fluence rates simultaneously with red light, a chloroplast which had faced the red light source turn into the profile position irrespective of the blue light direction (face-to-profile movement) [3,57]. To explain this response, a "signal carrier" which might interact with Pfr was postulated to be formed in the blue light-mediated reaction and to have a lifetime of about 2 min at 17°C [57]. Gabrys et al. [58] studied the nature of the blue signal carrier recently with light flashes of 6 ms at low temperature. Cells with chloroplasts in face position were cooled to 2°C and inducing blue and polarized red light was applied. At 2°C, the interaction between the hypothetical blue signal carrier and Pfr proceeded effectively and the lifetime of the signal carrier increased, suggesting that the decay of the carrier is a chemical process.

26.3.3 Motor system

The actomyosin system is believed to be the motile system responsible for chloroplast movement. Actin bundles connecting chloroplast and cell periphery were observed by electron microscopy in *Mougeotia* [51]. However, this result had not been confirmed thereafter, probably because plant actin is rather unstable and reliable methods for actin fixation have not been developed. Since a pre-treatment with *m*-maleimidobenzoyl *N*-hydroxysuccinimide ester (MBS) was recently found to stabilize actin filaments [59], it became possible to observe actin filaments associated with a "moving chloroplast", fixed when the chloroplast was rotating, by fluorescence microscopy [60]. A filamentous structure appeared at the front edge of the chloroplast when it began to rotate under irradiation with high-fluence rate white light and disappeared after completion of the orientation movement. No such structure could be observed before chloroplast movement occurred in cells that had been irradiated with low-fluence rate red light. In *Mougeotia*, red and blue light are effective for chloroplast photoorientation, and moreover high and low fluence rate responses are involved. In this very complicated system, the structure observed under high fluence rate white light should be confirmed during the movement under all light conditions where chloroplast photoorientation occurs.

To estimate the driving force acting on the edge of a *Mougeotia* chloroplast, the angular velocity of a rotating chloroplast was measured using a NIR laser diffractometer [61]. The chloroplast rotated 90° in 20 min at the mean angular velocity of 1 mrad/s. The velocity was constant immediately (shorter than 30 s) after red light irradiation, indicating that the net torque acting on the chloroplast is almost zero. Since the net torque acting on the chloroplast is the sum of the torque generated by the driving force and the viscous force, the driving force on the chloroplast is balanced by the viscous force. The authors estimated the maximum driving force acting on the chloroplast to be 1–10 pN for one cell, which is nearly equal to that generated by a single actomyosin system. However, the authors did not discuss whether this driving force is reasonable on the bases of actin filaments observed as bundles at the edge of a chloroplast [51,60]. Further analyses are needed.

26.4 *Mesotaenium*

A single cell alga, *Mesotaenium*, is another well-studied organism for chloroplast photoorientation. Chloroplast structure and its behavior under red light are similar to those of *Mougeotia*. Phytochrome involvement and its dichroic effect on chloroplast photoorientation is also the same as in *Mougeotia* although the response is much slower and needs continuous irradiation with red light [62]. The cell is a little bit smaller than that of *Mougeotia*, but the single cell organization of *Mesotaenium* makes it easier to handle than the filamentous, multicellular *Mougeotia*, because of cell uniformity and rapid proliferation in shaking culture. Moreover, protoplast preparation of this alga is rather easy [63]. The alga is, therefore, good for biochemical studies, which have been performed in Lagarias' laboratory [63–65]. As a step to investigate the possibility of calcium involvement of chloroplast photoorientation Berkelman and Lagarias [63] prepared membrane fractions from algal protoplasts and showed the activity of ATP-dependent calcium transport through the membrane. It is still unknown, however, whether the result indicates a part of the signal transduction pathway of the chloroplast photoorientation movement. *Mesotaenium* is a potent material for studying chloroplast photoorientation in vitro.

26.4.1 Photoreceptor

Phytochrome was extracted from *Mesotaenium* and its spectral characteristics were studied [65]. A main difference from the angiosperm phytochrome is the absorbance maxima, which are blue-sifted both for Pr and Pfr, at 650 and 722 nm, respectively. The phytochrome level is regulated by light [64]. In darkness, the level increases 4 fold during dark adaptation for 6 to 8 days and decreases again when transferred to light.

Recently, phytochrome genes of *Mesotaenium* were cloned and sequenced [12]. A gene family of 6 genes is detected and two of them (MesPHY1a and MesPHY1b) are highly related in sequence. The phytochrome has a region related to the photoactive yellow protein of purple bacterium, *Ectothiorhodospira*, which regulates negative phototaxis of this organism. It is not yet revealed, however, whether MesPHY1 is the real photoreceptor of the photoorientation movement in this alga.

26.4.2 Interaction between red and blue light

Under continuous red light, chloroplasts of this alga move from edge position to face position taking at least half an hour, but no response could be observed by a single pulse of red light. When blue light was irradiated continuously following a red light pulse for 1 min, however, fast chloroplast photoorientation occurred [66]. Red light was absorbed by phytochrome and blue light by a blue light receptor other than phytochrome. Hermann and Kraml [67] analyzed the interaction of phytochrome and the blue light receptor. A single red flash of about 4 ms was sufficient to generate a signal raised by Pfr, although 1 min irradiation was necessary to saturate its level. The lifetime of the signal is rather short. The competence of the signal to interact with the activated blue light receptor decreased within 5 min. Neither of the characteristics of the signal nor the type of interaction with a blue light receptor is known.

26.5 Seed plants

26.5.1 Direct observation of chloroplast relocation

Chloroplast relocation in seed plants has not been studied in detail, mostly because multicellular leaf organization is complicated and direct observation of chloroplast relocation under a microscope is not easy. Light transmission through chloroplast rich mesophyll cells is very much disturbed, so that the observation of chloroplasts in a mesophyll cell through epidermal cells is difficult. Therefore, the response was recently monitored photometrically as absorption or transmission of light through leaves [2,68]. The photometric detection of chloroplast relocation is quite sensitive so that even a few percents difference could be detected, although various artifacts cannot be excluded.

Direct observation of chloroplast relocation in seed plant leaves has been done in some plants, such as *Lemna, Sambucus, Sedum, Asparagus* etc. [69]. In these plants an action dichroism was detected under polarized blue light. An action spectrum was also obtained in *Lemna* [70]. However, the chloroplast relocation movement itself has not been analyzed by microscopy in detail.

Recently, Trojan and Gabrys [68] photometrically detected the chloroplast relocation of both avoidance and accumulation responses in *Arabidopsis*. The low fluence rate response occurred at less than 1 W m^{-2} of blue light and the high fluence rate response occurred at >5 W m^{-2}, although the detected response was less than 10% to the total transmission (see Figure 2). To prove that the photometrical data correspond to chloroplast relocation, they showed the patterns of chloroplast distribution in cells during low and high fluence rate responses in cross sections of fixed leaves. Chloroplasts distributed throughout leaves at walls parallel to the leaf surface in the former case and at the side wall in the latter case.

Chloroplast behavior during relocation induced by microbeam irradiation on a part of *Arabidopsis* mesophyll cell surface was observed microscopically (Figure 14a,b, [71]). Under continuous irradiation with a microbeam of high fluence rate of blue light (30 W m^{-2}), chloroplasts moved away from the irradiated area and stayed outside the beam spot. The avoidance response could be observed within a few minutes. When part of a cell was irradiated continuously with a microbeam of weak blue light (1 W m^{-2}), the chloroplasts gathered toward the beam. The movement could be detected within 10 min. The data obtained photometrically by Trojan and Gabrys [4] was confirmed as real chloroplast movement by continuous and direct observation of chloroplast relocation [71].

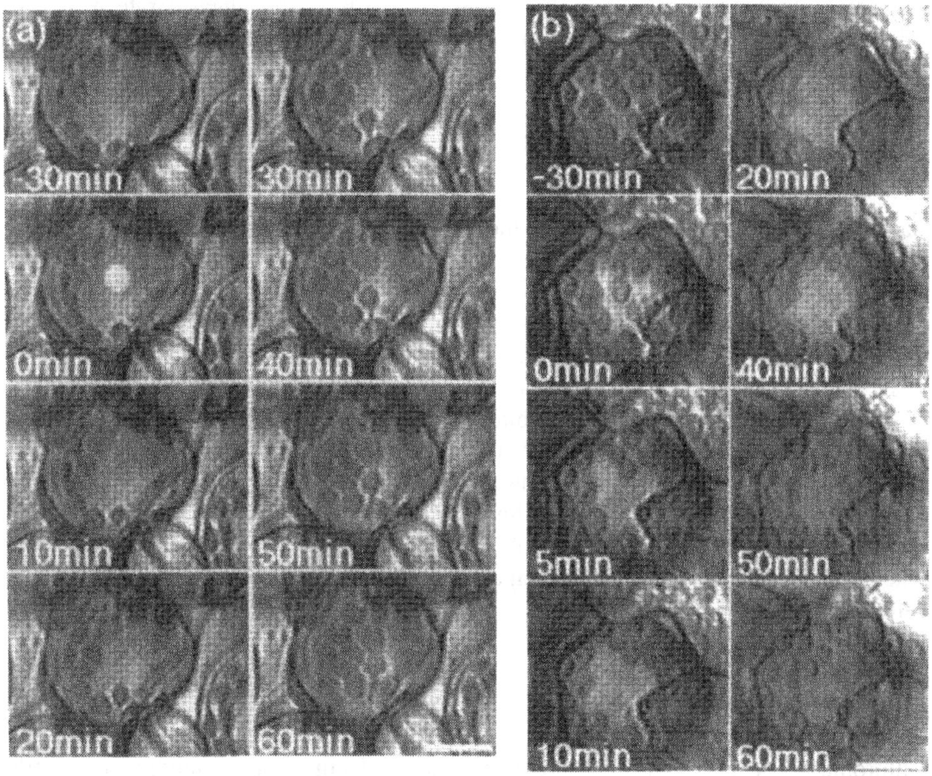

Figure 14. Chloroplast relocation induced by a blue microbeam in *Arabidopsis* mesophyll cells. (a) A center of a dark-adapted cell was continuously irradiated from 0 min with a blue microbeam (10 μm in diameter, 30 W m^{-2}). Note that chloroplasts moved towards the irradiated area after the beam was switched on. (b) A strong blue microbeam (20 μm in diameter, 30 W m^{-2}) was applied to the center of a light-adapted cell from 0 to 40 min. Note that chloroplasts move out of the beam but come back after the beam was switched off (bar 20 μm).

26.5.2 Mutant analysis of chloroplast relocation

The reason to use *Arabidopsis* as an experimental material, although the detection of chloroplast relocation is not easy in this organism, is to study the phenomenon from the genetic as well as molecular aspects. Blue light receptors have been studied in *Arabidopsis* using mutants which are deficient in blue light responses. Cryptochrome which is a flavoprotein similar to photolyases but with a C-terminus extension was first screened using the hy4 mutant [72]. Cryptochrome is a small gene family and now CRY1 and CRY2 [73,74] are known. Gabrys and her coworkers are trying to select mutants of chloroplast relocation in *Arabidopsis* [68]. They tested photometrically the chloroplast relocation in hypocotyl elongation mutants (*hy*) and phototropism mutants (*bru*), but these mutants showed normal chloroplast behavior [4]. Kagawa and Wada [71] also observed chloroplast relocation under a microscope in blue light receptor mutants, *cry1 and cry2* (kind gift from Dr. C.T. Lin) and their double mutant (kind gift from Dr. M. Ahmad, [75]) and *nph1–5* (kind gift from Dr. W.R. Briggs), and found that these mutants were normal in both the high as well as low fluence rate responses, indicating that the blue light receptor(s) for chloroplast relocation in *Arabidopsis* is different from those already known.

26.5.3 Signal transduction

Gabrys' group used *Lemna* as another model plant for chloroplast relocation. Tlalka and Gabrys [76] showed that calcium is essential for blue light induced chloroplast movement and suggested that calcium influx is not essential but the internal calcium concentration is important. Because only a long period of incubation (12 h) in ethylene glycol-bis(2-aminoethylether)-*N,N,N′,N,*-tetraacetic acid (EGTA) or La^{3+} is effective in this response. Malec et al. [77] showed that a contractile actomyosin system is involved in chloroplast movement from inhibitor works using *m*-maleimidobenzoic acid *N*-hydroxysuccinimide ester (MBS), *N*-ethylmaleimide (NEM) and cytochalasin B. In their semi-in vitro experiment using a glycerinated cell model, aggregation of chloroplasts could be induced by adding Mg-ATP and inhibited by adding NEM and cytochalasin B. Judging from their photographs (in [77]), however, it is hard to tell whether the results obtained in their paper reflect the real chloroplast movement in intact cells.

26.6 Miscellaneous

Chloroplast relocation by light has been studied from various standpoints in other plants, such as *Vaucheria* (algae, [78]), *Funaria hygrometrica* (moss, [36]), and *Vallisneria* (aquatic seed plant, [79,80]). These are rather old works and further analytical studies have not been carried out. As a part of history of chloroplast relocation studies, however, some points of these works are shortly described here.

In *Vallisneria* epidermal cells, chloroplasts accumulate at the periclinal walls under weak red light (i.e. 0.2 W m^{-2}), but move to anticlinal walls under strong blue light (ca.

3 W m^{-2}). The responses are quite similar to common chloroplast relocation except that red light, but not blue light, is effective for weak light response. According to time lapse video tracking analysis of individual chloroplasts by Izutani et al. [80] in *Vallisneria gigantea*, the direction of chloroplast movement during accumulation is random under both weak red and strong blue light, but movement rates are different between both light conditions, slower in the former and faster in the latter, respectively. Even if the movement is non-directional, if they move rapidly in one area and slowly in other area, chloroplast accumulate in the area where chloroplasts move slowly. This phenomenon is called photodinesis, and is not a true directional movement when observed each chloroplast precisely. The chloroplast accumulation in *Vallisneria* epidermal cell is probably this case. Izutani et al. [80] also propose an anchoring effect of phytochrome near periclinal walls, although no reliable data were shown. In contrast, Seitz [79] showed that centrifugability of cytoplasm in *V. spiralis* epidermal cells was enhanced when blue light was irradiated, indicating that the decrease of cytoplasm viscosity enables a faster organelle movement and consequently induce the photodinesis phenomenon.

In the siphonacious alga *Vaucheria*, organelles including chloroplasts and nuclei move directionally along the long tubular cell without a septum. Chloroplast photo-accumulation could be induced by local blue light irradiation [78], where reticulation of the microfilament system was observed [81]. When blue light is given at one part of a cell, the microfilament system, which has a role in the transfer of organelles, is reticulated, so that organelles could not be transported any more and accumulated at the site. It is obviously not the directional movement, but the change of moving speed. This phenomenon also induces photodinesis.

Funaria was used to study the characteristics of the blue light photoreceptor from the data of action dichroism of chloroplast accumulation by Zurzycki [36]. He measured an action spectrum of chloroplast accumulation using both parallel or oblique vibrating polarized light and found that action dichroism occurs in polarized light at wavelengths longer than 400 nm but not at shorter ones. From these data he concluded that the blue light receptor is a flavin, because the transition moment of the flavin molecule changes at this point. It is still an open question whether his hypothesis is correct or not, because we do not know yet what is the photoreceptor of chloroplast movement, not only in *Funaria* but also in any other plant.

26.7 Concluding remarks

Research on chloroplast relocation and photoorientation in the 90s was not as progressive as in the 70s and 80s. One of the reasons is a decrease in the number of researchers who study this phenomenon, probably because the application of molecular biology to this field was not easy, especially in *Mougeotia*, the main plant material for the study, so that we could not attract younger people to this field. Furthermore the key researchers in this field have retired. Another reason is the difficulty of screening mutants in chloroplast movement. Only one mutant is available at this moment in *Adiantum*, a red-light aphototropic mutant, which is deficient of red light-induced chloroplast relocation [27]. To make this field active again, molecular biological techniques should be applied. Screening of mutants of chloroplast relocation in *Arabidopsis* is urgent, although mutant screening is very tough work, if not difficult, because we have to look at the chloroplast movement itself under a microscope one by one, or measure a small change in the transmission of the measuring beam in spectroscopy in every mutagenized plant leaf.

Chloroplast movement is an excellent model system to study signal transduction pathways in photobiology, because the response is rather simple and occurs in the cell periphery without involvement of gene expression [82]. Not many components might be involved. Moreover, the first step is photoperception by a photoreceptor and the last step is movement by the acto-myosin system, that means the first and last components could clearly be focused, although the molecular species are not yet identified. We do not know what the signal transferred from the photoreceptor to the acto-myosin system is. Is it calcium? Many people have searched for a long time the signal connecting the first and the last steps. We are afraid that it must be very difficult to clarify it, even if we use modern technique, such as caged probes. We rather think that to identify signal(s) it must be better to first identify the molecular species of the photoreceptor, myosin and actin and then characterize these molecules. What we have to do now is to screen *Arabidopsis* mutants as many as possible.

References

1. G. Senn (1908). *Die Gestalts- und Lageveränderung der Pflanzen-Chromatophoren*. Engelmann, Leipzig.
2. Y. Inoue, K. Shibata (1973). Light-induced chloroplast rearrangements and their action spectra as measured by absorption spectrophotometry. *Planta*, **114**, 341–358.
3. H. Gabrys, T. Walczak, W. Haupt (1985). Interaction between phytochrome and the blue light photoreceptor system in *Mougeotia. Photochem. Photobiol.*, **42**, 731–734.
4. A. Trojan, H. Gabrys (1996). Chloroplast distribution in *Arabidopsis thaliana* (L.) depends on light conditions during growth. *Plant Physiol.*, **111**, 419–425.
5. A.R. Cashmore (1998). The cryptochrome family of blue/UV-A photoreceptors. *J. Plant Res.*, **111**, 267–270.
6. T. Kanegae, M. Wada (1998). Isolation and characterization of homologues of plant blue-light photoreceptor (cryptochrome) genes from the fern *Adiantum capillus-veneris. Mol. Gene Genet.*, **259**, 345–353.
7. E. Huala, P.W. Oeller, E. Liscum, I.-S. Han, E. Larsen, W.R. Briggs (1997). *Arabidopsis* NPH1: A protein kinase with a putative redox-sencing domain. *Science*, **278**, 2120–2123.
8. J.N. Christie, P. Reymond, G.K. Powell, P. Bernasconi, A.A. Raibekas, E. Liscum, W.R. Brigss (1998). Arabidopsis NPH1: Glavoprotein with the properties of a photoreceptor for phototropism. *Science*, **282**, 1698–1701.
9. M. Wada, T, Kanegae, K. Nozue, S. Fukuda (1997). Cryptogam phytochrome. *Plant Cell Environ.*, **20**, 685–690.
10. K. Nozue, T. Kanegae, T. Imaizumi, S. Fukuda, H. Okamoto, K.-C. Yeh, J.C. Lagarias, M. Wada (1998). A phytochrome from the fern *Adiantum* with features of the putative photoreceptor NPH1. *Proc. Natl. Acad. Sci. USA*, **95**, 15826–15830.
11. A. Winands, G. Wagner (1996). Phytochrome of the green alga *Mougeotia*: cDNA sequence, autoregulation and phylogenetic position. *Plant Mol. Biol.*, **32**, 589–597.

12. D.M. Lagarias, S.-H. Wu, J.C. Lagarias (1995). Atypical phytochrome gene structure in the green alga *Mesotaenium caldariorum*. *Plant Mol. Biol.*, **29**, 1127–1142.
13. W. Haupt (1999). Chloroplast movement: from phynomenology to molecular biology. In: K. Esser et al. (Eds), *Progress in Botany*, **60**, 3–36. Springer-Verlag, Berlin, Heidelberg.
14. H. Yatsuhashi (1996). Photoregulation systems for light-oriented chloroplast movement. *J. Plant Res.*, **109**, 139–146.
15. G. Wagner(1996). Intracellular movement. *Prog. Bot.*, **57**, 68–80.
16. W. Haupt, R. Scheuerlein (1990). Chloroplast movement. *Plant Cell Environ.*, **13**, 595–614.
17. M. Wada, F. Grolig, W. Haupt (1993). Light-oriented chloroplast positioning. Contribution to progress in photobiology. *J. Photochem. Photobiol. B: Biol.*, **17**, 3–25.
18. E. Schönbohm (1987). Movement of *Mougeotia* chloroplasts under continuous weak and strong light. *Acta Physiol. Plant.*, **9**, 109–135.
19. M. Wada, M. Sugai (1994). Photobiology in ferns. In: R.E. Kendrick, G.H.M. Kronenberg (Eds), *Photomorphogenesis in Plants* (2nd ed., pp. 783–802). Kluwer Academic Publishers, Dordrecht.
20. H. Yatsuhashi, A. Kadota, M. Wada (1985). Blue- and red-light action in photoorientation of chloroplasts in *Adiantum* protonemata. *Planta*, **165**, 43–50.
21. M. Wada, A. Kadota (1989). Photomorphogenesis in lower green plants. *Annu. Rev. Plant Physiol. Plant Mol. Biol.*, **40**, 169–191.
22. A. Kadota, I. Kohyama, M. Wada (1989). Polarotropism and photomovement of chloroplasts in the fern *Pteris* and *Adiantum* protonemata: Evidence for the possible lack of dichroic phytochrome in *Pteris*. *Plant Cell Physiol.*, **30**, 523–531.
23. M. Wada (1988). Chloroplast proliferation based on their distribution and arrangement in *Adiantum* protonemata growing under red light. *Bot. Mag. Tokyo*, **101**, 519–528.
24. H. Yatsuhashi, M. Wada, T. Hashimoto (1987). Dichroic orientation of phytochrome and blue-light photoreceptor in *Adiantum* protonemata as determined by chloroplast movement. *Acta Physiol. Plant.*, **9**, 163–173.
25. T. Kagawa, M. Wada (1994). Brief irradiation with red or blue light induces orientational movement of chloroplasts in dark-adapted prothallial cells of the fern *Adiantum*. *J. Plant Research*, **107**, 389–398.
26. W. Haupt, G. Mörtel, I. Winkelnkemper (1969). Demonstration of different dichroic orientaion of phytochrome Pr and Pfr. *Planta*, **88**, 183–186.
27. A. Kadota, M. Wada (1999). Red light-aphototropic (rap) mutants lack red light-induced chloroplast relocation movement in the fern *Adiantum capillus-veneris*. *Plant Cell Physiol.*, **40**, 238–247.
28. T. Kagawa, A. Kadota, M. Wada (1994). Phytochrome-mediated photoorientation of chloroplasts in protonemal cells of the fern *Adiantum* can be induced by a brief irradiation with red light. *Plant Cell Physiol.*, **35**, 371–377.
29. H. Yatsuhashi, H. Kobayashi (1993). Dual involvement of phytochrome in light-oriented chloroplast movement in *Dryopteris sparsa* protonemata. *J. Photochrom. Photobiol. B: Biol.*, **19**, 25–31.
30. W. Haupt (1966). Die Inversion der Schwachlichtbewegung des *Mougeotia*-Chloroplasten: Versuche zur Kinetik der Phytochrom-Umwandlung. *Z. Pflanzenphysiol.*, **54**, 151–160.
31. A. Kadota, M. Wada (1992). Photoorientation of chloroplasts in the fern *Adiantum* protonemal cells as analyzed by video-tracking system. *Bot. Mag. Tokyo*, **105**, 265–279.
32. A. Kadota, M. Wada (1989). Photoinduction of circular F-actin on chloroplast in a fern protonemal cell. *Protoplasma*, **151**, 171–174.
33. A. Kadota, M. Wada (1992). Photoinduction of formation of circular structures by microfilament on chloroplasts during intracellular orientation in protonemal cells of the fern *Adiantum capillus-veneris*. *Protoplasma*, **167**, 97–107.

34. T. Kagawa, M. Wada (1996). Phytochrome- and blue light-absorbing pigment-mediated directional movement of chloroplasts in dark-adapted prothallial cells of fern *Adiantum* as analyzed by microbeam irradiation. *Planta*, **198**, 488–493.
35. T. Kagawa, M. Wada (1999). Chloroplast avoidance response induced by blue light of high fluence rate in prothallial cells of the fern *Adiantum* as analyzed by microbeam irradiation. *Plant Physiology*, **119**, 917–923.
36. J. Zurzycki (1980). Blue light-induced intracellular movement. In: H. Senger (Ed.), *Blue light syndrome* (pp. 50–68). Springer-Verlag, Berlin.
37. T. Imaizumi, K. Kanegae, M. Wada (2000). Cryptochrome nucleocytoplasmic distribution and gene expression are regulated by light quality in the fern *Adiantum capillus-veneris*. *Plant Cell*, **12**, 81–96.
38. M. Wada, H. Sei (1994). Phytochrome-mediated phototropism in *Adiantum cuneatum* young leaves. *J. Plant Res.*, **107**, 181–186.
39. J. Augustynowicz, H. Gabrys (1999). Cloroplast movements in fern leaves: correlation of movement dynamics and environmental flexibility of the species. *Plant Cell Environ.*, **22**, 1239–1248.
40. W. Haupt (1982). Light-mediated movement in chloroplasts. *Ann. Rev. Plant Physiol.*, **33**, 205–233.
41. W. Haupt (1959). Die Chloroplastendrehung bei *Mougeotia*: I. Mitteilung. Über den quantitativen und qualitativen Linchtbedarf der Schwachlichtbewegung. *Planta*, **53**, 484–501.
42. G. Bock, W. Haupt (1961). Die Chloroplastendrehung bei *Mougeotia* III. Die Frage der Lokalisierung des Hellrot-Dunkelrot-Pigmentsystems in der Zelle. *Planta*, **57**, 518–530.
43. W. Haupt (1970). Über den Dichroismus von Phytochrom_{660} und Phytochrom_{730} bei *Mougeotia*. *Z. Pflanzenphysiol.*, **62**, 287–298.
44. H. Etozld (1965). Der Polarotropismus und Phototropismus der Chloronemen von *Dryopteris filix-max* (L.) Schott. *Planta*, **88**, 183–186.
45. M. Wada, A. Kadota, M. Furuya (1983). Intracellular localization and dichroic orientation of phytochrome in plasma membrane and/or ectoplasm of a centrifugated protenema of fern *Adiantum capillus-veneris* L. *Plant Cell Physiol.*, **24**, 1441–1447.
46. B.S. Serlin, S. Ferrell (1989). The involvement of microtubules in chloroplast rotation in the alga *Mougeotia*. *Plant Sci.*, **60**, 1–8.
47. Z. Lechowski, J. Bialczyk (1987). Interaction between green and far-red light on the low fluence rate chloroplast orientation in *Mougeotia*. *Plant Physiol.*, **85**, 581–584.
48. Z. Lechowski, J. Bialczyk (1988). Action spectrum for interaction between visible and far-red light on face chloroplast orientation in *Mougeotia*. *Plant Physiol.*, **88**, 189–193.
49. T. Walzak, H. Gabrys, K.J. Appenroth (1990). Is there a third photoreceptor involved in the control of chloroplast movements in *Mougeotia*? *Plant Physiol.*, **94**, 221–226.
50. E.M. Dreyer, M.H. Weisenseel (1979). Phytochrome mediated uptake of calcium in *Mougeotia* cells. *Planta*, **146**, 31–39.
51. G. Wagner, K. Klein (1978). Differential effect of calcium on chloroplast movement in *Mougeotia*. *Photochem. Photobiol.*, **27**, 137–140.
52. B.S. Serlin, S.J. Roux (1984). Modulation of chloroplast movement in the green alga *Mougeotia* by the Ca^{2+} ionophore A23187 and by calmodulin antagonists. *Proc. Natl. Acad. Sci. USA*, **81**, 6368–6372.
53. U. Russ, F. Grolig, G. Wagner (1991). Changes of cytoplasmic free calcium in the green alga, *Mougeoria scalaris*: monitored with indo-1, and effect on the velocity of the chloroplast movements. *Planta*, **184**, 105–112.
54. E. Schönbohm, J. Meyer-Wegener, E. Schönbohm (1990). No evidence for Ca^{2+} influx as an essential link in the signal transduction chains of either light-oriented chloroplast movements

or Pfr-mediated chloroplast anchorage in *Mougeotia*. *J. Photochem. Photobiol, B: Biology*, **5**, 331–341.
55. G. Wagner, U. Russ, H. Quader (1992). Calcium, a regulator of cytoskeletal activity and cellular competence. In: D. Menzel (Ed.), *The Cytoskeleton of the Algae* (pp. 411–433). CRC Press, Boca Raton.
56. E. Schönbohm, E. Schönbohm, J. Meyer-Wegener (1990). On the signal-transduction chains of two Pfr-mediated short-term processes: Increase of anchorage and movement of *Mougeotia* chloroplasts. *Photochem. Photobiol.*, **52**, 203–209.
57. H. Gabrys, T. Walczak, R. Scheuerlein (1988). Time characteristics of the interaction between phytochrome and the blue light photoreceptor system in *Mougeotia*. *J. Photochem. Photobiol. B: Biol.*, **2**, 464–474.
58. H. Gabrys, T. Walczak, P. Malec (1997). Interaction between phytochrome and the blue light photoreceptor system in *Mougeotia*: temperature dependence. *J. Photochem. Photobiol. B: Biology*, **38**, 35–39.
59. S. Sonobe, H. Shibaoka (1989). Cortical fine actin filaments in higher plant cells visualized by rhodamine-phalloidin after pretreatment with *m*-maleimidobenzoyl-*N*-hydroxysuccinimide ester. *Protoplasma*, **148**, 80–86.
60. Y. Mineyuki, H. Kataoka, Y. Masuda, R. Nagai (1995). Dynamic changes in the actin cytoskeleton during the high-fluence rate response of the *Mougeotia* chloroplast. *Protoplasma*, **185**, 222–229.
61. N. Ytow, T. Yamada, S. Ishizaka (1992). Mechanics of the chloroplast rotation in *Mougeotia*: Measurement of angular velocity by laser diffractometry. *Cell Motil. Cytoskel.*, **23**, 102–110.
62. M. Kraml, G. Büttner, W. Haupt, H. Herrmann (1988). Chloroplast orientation in *Mesotaenium*: the phytochrome effect is strongly potentiated by interaction with blue light. *Protoplasma*, **S11**, 72–179.
63. T. Berkelman, J.C. Lagarias (1990). Calcium transport in the green alga *Mesotaenium caldariorum*. Preliminary characterization and subcellular distribution. *Plant Physiol.*, **93**, 748–757.
64. L.Z. Morand, D.G. Kidd, J.C. Lagarias (1993). Phytochrome levels in the green alga *Mesotaenium caldariorum* are light regulated. *Plant Physiol.*, **101**, 97–103.
65. D.G. Kidd, J.C. Lagarias (1990). Phytochrome from the green alga *Mesotaenium caldariorum*. Purification and preliminary characterization. *J. Biol. Chemstry*, **265**, 7029–7035.
66. M. Kraml, H. Hermann (1991). Red-blue-interaction in *Mesotaenium* chloroplast movement: Blue seems to stablize the transient memory of the phytochrome signal. *Photochem. Photobiol.*, **53**, 255–259.
67. H. Hermann, M. Kraml (1997). Time-dependent formation of Pfr-mediated signals for the interaction with blue light in *Mesotaenium* chloroplast orientaion. *J. Photochem. Photobiol. B: Biol.*, **37**, 60–65.
68. A. Trojan, H. Gabrys (1993). Light induced chloroplast movement in *Arabidopsis thaliana* screening and selection of movement mutants, Proceedings of the Fifth Congress of The European Society for Photobiology, Marburg, Germany.
69. J. Zurzycki, Z. Lelatko (1969). Action dichroism in the chloroplasts rearrangements in various plant species. *Acta Soc. Bot. Polon.*, **38**, 495–506.
70. J. Zurzycki (1962). The action spectrum for the light depended movements of chloroplasts in *Lemna trisulca*. *Acta Soc. Bot. Polon.*, **31**, 489–538.
71. T. Kagawa, M. Wada (2000). Blue light-induced chloroplast relocation in Arabidopsis thaliana as analyzed by microbeam irradiation. *Plant Cell Physiol.*, **41**, 84–93.

72. M. Ahmad, A.R. Cashmore (1993). *HY4* gene of *A. thaliana* encodes a protein with characteristics of a blue-light photoreceptor. *Nature*, **366**, 162–166.
73. P.D. Hoffman, A. Batschauer, J.B. Hays (1996). *PHH1*, a novel gene from *Arabidopsis thaliana* that encodes a protein similar to plant blue-light photoreceptors and microbial photolyases. *Mol. Gen. Genet.*, **253**, 259–265.
74. C. Lin, M. Ahmad, J. Chan, A.R. Cashmore (1996). *CRY2*, a second member of the *Arabidopsis* cryptochrome gene family. *Plant Physiol.*, **110**, 1047.
75. M., Ahmad, J.A., Jarillo, O. Sirnova, A. Cashomore (1998). Cryptochrome blue-light photoreceptors of *Arabidpsis* implicated in phototropism. *Nature*, **392**, 720–723.
76. M. Tlalka, H. Gabrys (1993). Influece of calcium on blue-light-induced chloroplast movement in *Lemna trisulca* L. *Planta*, **189**, 491–498.
77. P. Malec, R.A. Rinaldi, H. Gabrys (1996). Light-induced chloroplast movements in *Lemna trisulca*. Idetification of the motile system. *Plant Sci.*, **120**, 127–137.
78. M.R. Blatt, W.R. Briggs (1980). Blue-light-induced cortical fiber reticulation concomitant with chloroplast aggregation in the alga *Vaucheria sessilis*. *Planta*, **147**, 355–362.
79. K. Seitz (1967). Wirkungsspektren für die Starklichtbewegung der Chloroplasten, die Photodinese und die lichtabhängige Viskositätsänderung bei *Vallisneria spiralis* ssp. *torta*. *Z. Plantzenphisiol.*, **65**, 246–261.
80. Y. Izutani, S. Takagi, R. Nagai (1990). Orientation movements of chloroplasts in *Vallisneria* epidermal cells: Different effect of light at low- and high-fluence rate. *Photochem. Photobiol.*, **51**, 105–111.
81. M.R. Blatt, N.K. Wessels, W.R. Briggs (1980). Actin and cortical fiber reticulation concomitan with chloroplast aggregation in the alga *Vaucheria sessilis*. *Planta*, **147**, 363–375.
82. M. Wada (1988). Chloroplast photoorientation in enucleated fern protonemata. *Plant Cell Physiol.*, **29**, 1227–1232.
83. W. Haupt (1972). Localization of phytochome within the cell. In: Phytochrome, K. Mitrakos, K. Shropshire (Eds), (pp. 553–569). Academic Press, London, 1972, .

Index

Edwards Brothers Malloy
Ann Arbor MI. USA
December 3, 2013